AF660742

ESE 2024

UPSC ENGINEERING SERVICES EXAMINATION

Main Examination

Civil Engineering

Topicwise *Conventional* Solved Papers

Paper-I

25 Years SOLVED PAPERS

- ✓ Thoroughly revised and updated
- ✓ Fully solved with explanations
- ✓ Also useful for State Engineering Services Exams

MADE EASY PUBLICATIONS

www.madeeasypublications.org

MADE EASY
PUBLICATIONS

MADE EASY Publications Pvt. Ltd

Corporate Office: 44-A/4, Kalu Sarai (Near Hauz Khas Metro Station), New Delhi-110016
E-mail: infomep@madeeasy.in
Contact: 0-9021300500
Visit us at: www.madeeasypublications.org

ESE-2024 : Main Examination
Civil Engineering : Paper-I | Conventional Solved Questions : (1999-2023)

1st Edition : 2010
2nd Edition : 2011
3rd Edition : 2012
4th Edition : 2013
5th Edition : 2014
6th Edition : 2015
7th Edition : 2016
8th Edition : 2017
9th Edition : 2018
10th Edition : 2019
11th Edition : 2020
12th Edition : 2021
13th Edition: 2022 (Feb)
14th Edition: 2022 (Aug)

15th Edition : 2023

B. Singh (Ex. IES)

Director's Message

During the last few decades of engineering academics, India has witnessed geometric growth in engineering graduates. It is noticeable that the level of engineering knowledge has degraded gradually, while on the other hand competition has increased in each competitive examination including GATE and UPSC examinations. Under such scenario higher level efforts are required to take an edge over other competitors.

The objective of **MADE EASY books** is to introduce a simplified approach to the overall concepts of related stream in a single book with specific presentation. The topic-wise presentation will help the readers to study & practice the concepts and questions simultaneously.

The efforts have been made to provide close and illustrative solutions in lucid style to facilitate understanding and quick tricks are introduced to save time.

Following tips during the study may increase efficiency and may help in order to achieve success.

- Thorough coverage of syllabus of all subjects
- Adopting right source of knowledge, i.e. standard reading text materials
- Develop speed and accuracy in solving questions
- Balanced preparation of Paper-I and Paper-II subjects with focus on key subjects
- Practice online and offline modes of tests
- Appear on self assessment tests
- Good examination management
- Maintain self motivation
- Avoid jumbo and vague approach, which is time consuming in solving the questions
- Good planning and time management of daily routine
- Group study and discussions on a regular basis
- Extra emphasis on solving the questions
- Self introspection to find your weaknesses and strengths
- Analyze the exam pattern to understand the level of questions
- Apply shortcuts and learn standard results and formulae to save time

B. Singh (Ex. IES)
CMD, MADE EASY Group

ESE 2024 Main Examination

Civil Engineering
Conventional Solved Questions

Paper-I

Contents

1. **Building Materials 1-83**
   1. Cement 1
   2. Mortar 19
   3. Concrete 22
   4. Stones, Bricks and Bricks Masonry 47
   5. Timber 64
   6. Miscellaneous 80

2. **Solid Mechanics 84-257**
   1. Simple Stress-Strain & Elastic Constants 84
   2. Shear Force and Bending Moment 112
   3. Principal Stress-Strain & Theories of Failure 159
   4. Deflection of Beams 197
   5. Bending & Shear Stresses in Beams 220
   6. Thick & Thin Cylinders and Spheres 240
   7. Torsion in Shafts & Springs 242
   8. Theory of Columns 252
   9. Shear Centre, Moment of Inertia & Principal Axes 254

3. **Structural Analysis 258-440**
   1. Determinacy and Indeterminacy 258
   2. Influence Line Diagram and Rolling Loads 260
   3. Arches and Suspended Cables 274
   4. Methods of Structural Analysis 289
   5. Trusses 373
   6. Matrix Method of Structural Analysis 424
   7. Structural Dynamics 436

4. **Design of Steel Structures 441-547**
   1. Structural Fasteners 441
   2. Tension Member 468
   3. Compression Member 475
   4. Beams 493
   5. Plate Girders and Industrial Roofs 504
   6. Plastic Analysis 525

5. **Design of Concrete and Masonry Structures 548-695**
   1. Fundamentals of RCC 548
   2. Beams and Slabs 553
   3. Shear, Bond, Torsion and Development Length 613
   4. Staircase, Wall, Columns and Footings 616
   5. Prestress Concrete 653
   6. Miscellaneous 677

6. **Construction Practice, Planning and Management 696-779**
   1. Project Management and Network Techniques 696
   2. PERT and CPM 703
   3. Crashing, Resource Allocation, Updating and Engineering Economy 741
   4. Construction Equipments 752
   5. Miscellaneous 769

■■■■

1 Building Materials

Revised Syllabus of ESE : *Stone, Lime, Glass, Plastics, Steel, FRP, Ceramics, Aluminum, Fly Ash, Basic Admixtures, Timber,* ***Bricks and Aggregates:*** *Classification, properties and selection criteria;*
Cement: *Types, Composition, Properties, Uses, Specifications and various Tests; Lime & Cement Mortars and* ***Concrete:*** *Properties and various Tests;* ***Design of Concrete Mixes****: Proportioning of aggregates and methods of mix design.*

---

1. Cement

1.1 Describe the hydration of portland cement and outline the ways in which the Vicat apparatus and the Le-Chatelier apparatus can be used to assess the properties of fresh and hardened pastes.

[15 marks : 1999]

Solution:

The chemical reactions that takes place between cement and water is known as **hydration of cement**. On account of hydration certain products are formed. These products are important because they have cementing or adhesive value. The quality, quantity, continuity, stability and the rate of formation of the hydration products are important.

Anhydrous cement compounds when mixed with water, react with each other to form hydrated compounds of very low solubility. The hydration of cement can be visualized in two ways. The first is "through solution" mechanism. In this the cement compounds dissolve to produce a super saturated solution from which different hydrated products get precipitated. The second possibility is that water attacks cement compounds in the "solid state" converting the compounds into hydrated products starting from the surface and proceeding to the interior of the compounds with time. It is probable that both "through solution" and "solid state" types of mechanism may occur during the course of reactions between cement and water. The former mechanisms may predominate in the early stages of hydration in view of large quantities of water being available and the latter mechanism may operate during the later stages of hydration.

The reaction of cement with water is exothermic. The reaction liberates a considerable quantity of heat. This liberation of heat is called heat of hydration. The hydration process is not an instantaneous one. The reaction is faster in the early period and continues indefinitely at a decreasing rate. Complete hydration can not be obtained under a period of one year or more unless the cement is very finely ground and reground with excess of water to expose fresh surfaces at intervals. During the course of reaction of C_3S and C_2S with water, calcium silicate hydrate (C–S–H) and calcium hydroxide $Ca(OH)_2$ are formed. Calcium silicate hydrate is the essence that determines the properties of concrete. It makes up 50-60 per cent of the volume of solids in a completely hydrated cement paste. On the other hand, calcium hydroxide is a compound which is responsible for the lack of durability. The calcium hydroxide also reacts with sulphates presents in soils or water to form calcium sulphate which reacts further with C_3A and cause deterioration of concrete which is known as sulphate attack. The only advantage of $Ca(OH)_2$ is that, being alkaline in nature, it maintain pH value around 13 in the concrete which resist the corrosion of reinforcements.

The hydration of aluminates (C_3A) results in a calcium aluminate system CaO-Al_2O_3-H_2O. This compound do not contribute anything to the strength of concrete. On the other hand their presence is harmful to the durability of the concrete particularly where the concrete is likely to be attacked by sulphates. As it hydrates fast it may contribute a little to early strength.

On hydration, C_4AF is believed to form a system of the form CaO-Fe_2O_3-H_2O. The hydrates of C_4AF also do not contribute anything to the strength but they show a comparatively higher resistance to the attack of sulphates than the hydrates of calcium aluminate.

Vicat apparatus is used for determining the **normal consistency and setting time for cement**. A known weight of cement is taken and a paste is prepared with a weighed quantity of water (24% by weight of cement) for the first trial. The Paste is then filled in a mould and the plunger of the apparatus is brought down to touch the surface of the paste in test block and quickly released allowing it to sink into the paste by its own weight. Similarly, trials are conducted with higher and higher water cement ratios till such time the plunger penetrates for a depth of 33-35 mm from the top. That particular percentage of water is known as the percentage of water required to produce a cement paste of standard consistency.

For setting times the plunger is replaced by a needle (for initial setting time) or a circular attachment (for final setting time).

Le Chatelier apparatus can be used to determine **soundness in cement**. Unsoundness in cement is due to excess of lime, magnesia or sulphates. Cement is gauged with 0.78 times the water required for standard consistency in a standard manner and filled into the mould kept on a glass plate. The mould is covered on the top with another glass plate. The whole assembly is immersed in water at a temperature of 27°C - 32°C and kept there for 24 hours. Now the distance is measured between the indicator points. The mould is again submerged in water and water is heated to brought to boiling point in about 25-30 minutes and it is kept boiling for three hours. The mould is now removed from water and allowed to cool. The distance between indicator points is measured again. The difference between these two measurements represent the expansion of cement. This must not exceed 10 mm. The Le Chatelier test detects the unsoundness due to free lime only.

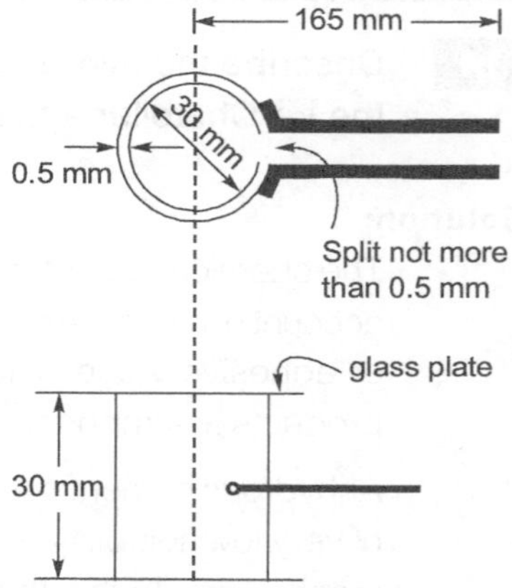

1.2 **Explain how sulphate resisting cement and rapid hardening portland cement differ from OPC and specific circumstance in which these cements would be used.**

[15 marks : 1999]

Solution:

Sulphate Resisting Cement: Ordinary portland cement is vulnerable to sulphate attack. Sulphate attack is greatly accelerated if accompanied by alternate wetting and drying which normally takes place in marine structures in the zone of tidal variations.

To prevent the sulphate attack, the use of cement with low C_3A content is found to be effective. Such a cement with low C_3A content and comparatively low C_4AF content is known as sulphate resisting cement. In other words, this cement has a higher silicate content than OPC.

It is not often possible (feasible) to reduce the Al_2O_3 content of the raw materials. So Fe_2O_3 may be added to the mix so that C_4AF content increase at the expense of C_3A. Many of the physical properties of sulphate resisting cement are similar to ordinary portland cement.

The use of sulphate resistant cement is recommended under the following conditions:

(i) Concrete to be used in marine conditions.

(ii) Concrete to be used in foundation and basement, where soil is infested with sulphates.

(iii) Concrete used for fabrication of pipes which are likely to be marshy region or sulphate bearing soils.

(iv) Concrete to be used in the construction of sewage treatment works.

Rapid Hardening Cement: This cement is similar to OPC but with higher C_3S content and finer grinding. It gains strength more quickly than OPC, though the final strength is only slightly higher. The one day strength of this cement is equal to three day strength of OPC with the same water-cement ratio. This cement is used where a rapid strength development is required. The rapid gain of strength is accompanied by a higher rate of heat development during the hydration of cement. It is about 10 per cent costlier than OPC.

The use of rapid hardening cement is recommended in the following situations:

(i) In prefabricated concrete construction.

(ii) Where formwork is required to be removed early for reuse elsewhere.

(iii) Road repair works.

(iv) In cold weather concrete where the rapid rate of development of strength reduces the vulnerability of concrete to the frost damage.

1.3 **Name the principal compounds in portland cement, their relative rates of reaction with water and their approximate proportions.**

[10 marks : 1999]

Or

List of the products of hydration and their influence on the properties of cement.

[10 marks : 2001]

Or

Which are the four important compounds formed during the setting action of cement (four principal minerals in ordinary portland cement)? Mention their relative proportions expressed as percentages and also functions of these compounds.

[10 marks : 2007]

Solution:

The principal compounds in portland cement are known as Bogue's compounds. They are as follows:

(i) Tricalcium silicate (C_3S) or Alite

(ii) Dicalcium silicate (C_2S) or Belite

(iii) Tricalcium aluminate (C_3A) or Celite

(iv) Tetracalcium aluminoferrite (C_4AF) or Felite

The reaction of water with C_3A is very fast and in the process **flash setting** i.e. stiffening without strength development can occur because the C-A-H phase prevents the hydration of C_3S and C_2S. To prevent this flash set, gypsum is added at the time of grinding the cement clinker. The hydrated aluminates do not contribute anything to the strength of concrete. On the other hand, their presence is harmful to the durability of concrete particularly, where the concrete is likely to be attacked by sulphates. As it hydrates fast, it may contribute a little to the early strength.

$$C_3A + H_2O \rightarrow C_3AH_6$$

On hydration, C_4AF is believed to form a system of the form C-F-H. A hydrated calcium ferrite of this form is comparatively more stable. This hydrated product also does not contribute anything to the strength. However, the hydrates of C_4AF show a comparatively higher resistance to the attack of sulphates than the hydrates of calcium aluminate.

$$C_3AF + H_2O \rightarrow C_3AH_6 + CFH$$

When C_3S and C_2S reacts with water, calcium silicate hydrate (C-S-H) and calcium hydroxide are formed. Calcium silicate hydrates are the most important products. It is the essence that determines the good properties of concrete.

$$C_3S + H_2O \rightarrow C_3S_2H_3 + 3Ca(OH)_2$$

$$C_2S + H_2O \rightarrow C_3S_2H_3 + Ca(OH)_2$$

It can be seen that C_3S produces a comparatively lesser quantity of calcium silicate hydrates and more quantity of $Ca(OH)_2$ than that formed in the hydration of C_2S. C_2S hydrates rather slowly than C_3S.

The relative rates of hydration of Bogue's compounds can be shown with the help of a graph as shown.

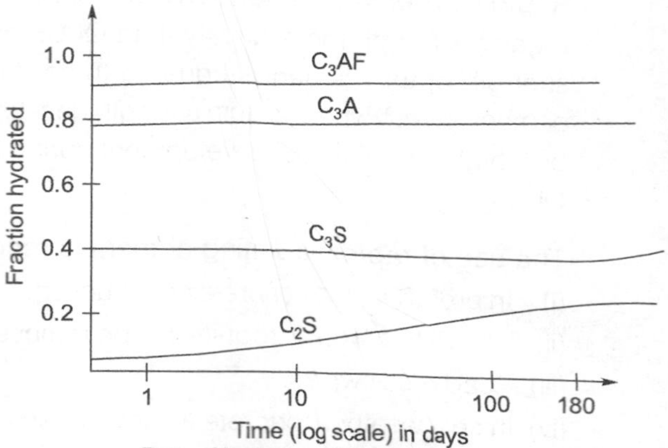

Rate of hydration of Bogue's Compounds

The approximate proportions of Bogue's compound are:

Compound	Percentage
C_3S	30-50
C_2S	20-45
C_3A	8-12
C_4AF	6-10

1.4 **Explain how do the portland pozzolana cement and super sulphate cement differ from OPC. Under what specific circumstances these cements would be used?**

[10 marks : 2000]

Solution:

Portland Pozzolana Cement: It is manufactured either by grinding together portland cement clinker and pozzolana with the addition of gypsum or calcium sulphate or by intimately and uniformly blending portland cement and fine pozzolana. PPC produces less heat of hydration and offers greater resistance to the attack of impurities in water than normal portland cement. However, it is important to appreciate that the addition of pozzolana does not contribute to strength at early ages. Strengths similar to those of OPC can be expected in general only at later ages. In PPC costlier clinker is replaced by cheaper pozzolanic material making it more economical than OPC. PPC consumes calcium hydroxide and does not produce calcium hydroxide as much as that of OPC. It generates lower heat of hydration than OPC and that too at a lower rate. As the fly ash is finer and of lower density, the bulk volume of 50 kg bag is slightly more than OPC. The long term strength of PPC beyond a couple of months is higher than OPC if enough moisture is available for continued pozzolanic action.

The PPC is particularly suitable under the following conditions:

(i) For hydraulic structures.

(ii) For mass concrete structure like dam, bridge piers and thick foundations.

(iii) For marine structures.

(iv) For sewage disposal works, sewers etc.

Super Sulphate Cement: It is manufactured by grinding together a mixture of 80-85 per cent granulated slag, 10-15 per cent hard gypsum and about 5 per cent portland cement clinker. It is rather more sensitive to deterioration during storage than portland cement. Super sulphated cement has a low heat of hydration than OPC. This cement has high sulphate resistance. This cement like high alumina cement, combines with more water on hydration than OPC.

This cement is recommended for the following:

(i) Due to high sulphate resistance, it is particularly used in foundations, where chemically aggressive conditions exist.

(ii) It can be used in marine works.

(iii) It is used in fabrication of reinforced concrete pipes which are likely to be buried in sulphate bearing soils.

1.5 Explain the difference between various grades of OPC.

[10 marks : 2002]

Solution:

The commonly used portland cement in India is branded as **33-grade (IS:269-1989)**, **43-grade (IS:8112-1989)** and **53 grade (IS:12269-1987)** having 28 days mean compressive strengths exceeding 33 MPa, 43 MPa and 53 MPa respectively. All the three grades are produced from same materials. The higher strengths are achieved by increasing C_3S content and also by finely grinding the clinker. The fineness of 53-grade OPC obtained by Blaine's air permeability test is specified to be of the order of 350000 mm^2/g. The initial and final setting times are same for all the three grades. The 33-grade cement has virtually disappeared and has been displaced by high strength 43-grade cement. The minimum compressive strengths of the 43-grade cement are 23 MPa and 33 MPa at the end of 3 and 7 days respectively. At higher water cement ratios, the concrete produced with high strength cement has about 80% higher strength and at lower water cement ratios, it has 40% higher strength than that of concrete using 33-grade cement. Greater fineness of 43 and 53 grade cements increase workability due to reduction of friction between aggregates. Moreover, due to shorter setting time and faster development of strength, the stripping time is shorter. Although cements of grade 43 and 53 are desirable for economical design of high grade concretes but they can also be used for lower grade concretes.

1.6 Explain pozzolanic action.

[10 marks : 2003]

Solution:

A pozzolana is a finely ground siliceous material which as such does not possess cementitious property in itself but reacts in the presence of water with calcium hydroxide at normal temperature to form compounds of low solubility having cementitious properties. The action is known as pozzolanic action.

The reaction can be shown as

$$\text{Pozzolana} + Ca(OH)_2 + H_2O \rightarrow \text{C-S-H (gel)}$$

This reaction is called pozzolanic reaction. The characteristic feature of pozzolanic reaction is initially slow, with the result that heat of hydration and strength development will accordingly be slow. The reaction involves the consumption of $Ca(OH)_2$ and not production of $Ca(OH)_2$. It may be noted that on hydration of C_3S and C_2S present in cement, $Ca(OH)_2$ is formed as one of the products of hydration. This compound has no cementitious value and it is soluble in water and may be leached out by the percolating water. It is pointed out that $Ca(OH)_2$, otherwise, a water soluble material is converted into insoluble cementitious material by reaction of pozzolanic materials.

The reduction of $Ca(OH)_2$ also improves the durability of cement paste by making the paste dense and impervious. Pozzolanic materials can be natural or artificial. Clay and shales, opaline cherts, diatomaceous earth and volcanic tuffs are natural pozzolanic materials. Fly ash, blast furnace slag, silica fume, rice husk ash are artificial pozzolanic materials.

The pozzolanic action also reduce the expansion caused by the alkali-aggregate reaction in concrete. Excessive expansion causes pattern cracking of concrete. This expansion can usually be controlled by using of pozzolana ranging from 2 to 35% by mass of cement depending upon the type of aggregate and alkali content of cement.

1.7 **List the various laboratory tests for assessing the quality of cement and their importance.**
[10 marks : 2008]

Solution:

The following tests are usually conducted in laboratory to assess the quality of cement:

(i) **Fineness test:** The fineness of cement has an important bearing on the rate of hydration and hence on the rate of gain of strength and also on the rate of evolution of heat. Fineness of cement can be tested in two ways viz. by sieving and by determination of specific surface using air permeability apparatus.

(ii) **Setting time test:** In actual construction dealing with cement paste, mortar or concrete, certain time is required for mixing, transporting, placing, compacting and finishing. During this time cement paste, mortar or concrete should be in plastic condition. This time is known as initial setting time. Once the concrete is placed in the final position, compacted and finished, it should lose its plasticity in the earliest possible time so that it is least vulnerable to damages from external destructive agencies. This time is known as final setting time. Setting time test is carried out with the help of Vicat apparatus.

(iii) **Compressive strength test:** The compressive strength of hardened cement is the most important of all the properties. Therefore, it is not surprising that the cement is always tested for its strength in laboratory before the cement is used in important works.

(iv) **Soundness test:** It is very important that the cement after setting shall not undergo any appreciable change of volume. The testing of soundness of cement, to ensure that the cement does not show any appreciable subsequent expansion is of prime importance. Unsoundness in cement is due to excess of lime, excess of magnesia or excessive proportions of sulphates. Unsoundness due to lime can be tested using Le Chatelier apparatus. Unsoundness due to magnesia can be tested using Autoclave test. Unsoundness due to calcium sulphate can be tested using chemical analysis.

(v) **Heat of hydration test:** The reaction of cement with water is exothermic. It is estimated that about 120 calories of heat is generated in the hydration of 1 gm of cement. The total quantum of heat produced in a conservative system such as the interior of a mass concrete dam, a temperature rise of about 50°C has been observed. This unduly high temperature developed at the interior of a concrete dam causes serious expansion of the body of dam and with the subsequent cooling considerable shrinkage takes place resulting in serious cracking of concrete. Heat of hydration test can be easily carried out over a few days by vacuum flask methods, or over a longer period in an adiabatic calorimeter.

(vi) **Chemical composition test:** The raw materials used for the manufacture of cement consist mainly of lime, silica, alumina and iron oxide. The relative proportions of these oxide compositions are responsible for influencing the various properties of cement, in addition to rate of cooling and fineness of grinding. Thus the chemical composition test is carried out in laboratory.

1.8 **What are the different types of Portland cement as per Indian code of practice? Discuss any two.**
[10 marks : 2011]

Solution:

The commonly used portland cement in India is branded as, 33 grade (IS 269 - 1989), 43 grade (IS 8112 - 1989) and 53 grade (IS 122 - 1989) having 28 days mean compressive strength exceeding, 33 MPa,

43 MPa and 53 MPa respectively.

All the three grades are produced from same materials.

- High strength are achieved by increasing C_3S content and also by finely grinding the clinker.
- The initial and final setting time are same for all the three grades.
- The 33 grade cement has virtually disappeared and has been replaced by high strength 43 grade cement.

43 grade OPC cement: The 43 grade OPC is the most popular general purpose cement in the country today. The production of 43 grade OPC is nearly 50% of total production of cement in country.

43 grade OPC can be used for the following applications:

- General Civil Engineering construction work.
- RCC work (preferably where grade of concrete is upto M30).
- Precast item such as blocks, tiles, pipes etc.
- The minimum compressive strength of 43 grade cement are 23 MPa and 33 MPa, at the and of 3 and 7 days respectively.
- Asbestos products such as sheets and pipes
- Non structural works such as plastering flooring etc.

53 Grade OPC cement: 53 grade OPC is higher strength cement to meet the needs of the consumer for higher strength concrete. As per BIS requirements the minimum 28 days compressive strength of 53 grade OPC should not be less than 53 MPa. For certain specialized works such as prestressed concrete and certain items of precast concrete requiring consistently high strength concrete, 53 grade OPC is found very useful. 53 grade OPC produce higher grade concrete at very economical cement content. In concrete mix design for concrete M20 and above grades a saving of 8-10% of cement may be achieved with the use of 53 grade OPC. 53 grade OPC can be used for the following applications.

- RCC works (preferably where grade of concrete is M25 and above)
- Precast concrete items such as paving blocks, tiles, building blocks etc.
- Prestress concrete component.
- Railways, Concrete road, bridges etc.

  Greater fineness of 43 grade and 53 grade cement increase workability due to reduction of friction between aggregates, moreover due to shorter setting time and faster development of strength, the stripping time is shorter.

1.9 **Discuss how consistency of cement is determined.**

[5 marks : 2011]

Solution:

The purpose of consistency test is to determine the percentage of water required for preparing cement pastes for other test. **Following procedure is adopted for the test:**

(i) Take 400 gm of cement and add 27% by weight or 108 gm of water to it.

(ii) Mix water and cement on a non-porous surface. The mixing should be done thoroughly.

(iii) Fill the mould of vicat apparatus. The interval between the addition of water to the commencement of filling the mould is known as the time of gauging and it should be 3 min. to 5 min.

(iv) Vicat apparatus consists of a frame to which is attached a movable rod weighing 300 gm and having 10 mm dia and 50 mm length respectively. An indicator is attached to the movable rod. This indicator moves on a vertical scale and it gives the penetration value. The vicat mould is in the form of a cylinder and it can be split into two halves. The vicat would is placed on a non porous plate. These are three attachments square needle, plunger and needle with annular collar. The square needle is used for initial setting time test, the plunger is used for consistency test and the needle with annular collar is used for final setting time test.

(v) The plunger is attached to the movable rod of vicat apparatus. The plunger is gently lowered on the paste in the mould.

(vi) The settlement of plunger is noted. If the penetration is between 5 mm to 7 mm from the bottom of would, the water added is correct. If penetration is not proper the process is repeated with different percentages of water till the desired penetration is obtained.

1.10 What do you understand by the initial and final setting times of cement? What are the typical initial and final setting times of 43 grade OPC cement and Portland Pozzolana Cement (PPC) as per IS code?

[5 marks : 2014]

Solution:

When water is added to cement, the resulting paste starts gaining strength upon hydration. A stage is reached when paste loose its consistency and become stiff. The time to reach this stage is termed as setting time.

The setting time is divided into two parts.

(i) Initial setting time

(ii) Final setting time

Initial and final setting times are determined with the help of vicat's apparatus. To measure these times a cement paste of 0.85 times standard consistency is filled into the vicat's mould in specified manner and time of specified penetration is recorded.

(i) **Initial setting time:** Initial setting time is the time which is measured from the instant water is added to the cement upto the instant when it looses its plasticity. Initial setting time is equal to the time required by the needle to penetrate upto the depth of 33-35 mm from top.

(ii) **Final setting time:** Final setting time is the time which is measured from the instant water is added to the time mix in vicats apparatus achieve sufficient stiffness to resist pressure. Final setting time is equal to the time after which the paste become so hard that annular attachment attached to the needle under standard weight fails to leave any mark on hardened cement paste and needle is able to make a mask.

S.N	Cement	Initial setting time (minutes)	Final setting time (minutes)
1.	43 grade OPC cement	30 minutes	600 minutes
2.	Portland Pozzolana cement (PPC)	30 minutes	600 minutes

1.11 List out eight chemical ingredients of Portland cement and briefly explain their functions.

[12 marks : 2018]

Solution:

Chemical ingredients of portland cement are:

(i) Lime (CaO) : 62-67%

- This is an important ingredient of cement and its proportion is to be carefully maintained.
- Lime in excess makes the cement unsound and causes the cement to expand and disintegrate.
- Lime in deficiency reduces the strength of cement and causes it to set quickly.

(ii) Silica (SiO_2) : 17-25%

- Imparts strength to the cement due to formation of dicalcium and tricalcium silicates.
- If it is present in excess, strength of cement increases but setting time gets prolonged.

(iii) Alumina (Al_2O_3) : 3-8%

- Imparts quick setting property.
- Acts as a flux and lowers clinkering temperature. So, suitable cement type is not formed.
- In excess amount, its presence weakens the cement.

(iv) Calcium sulphate ($CaSO_4$) : 3-4%

- Added in the form of gypsum.
- Increases initial setting time of cement.

(v) Iron Oxide (Fe_2O_3) : 3-4%

- Imparts colour, hardness and strength to cement.

(vi) Magnesia (MgO) : 0.1-3%

- Imparts hardness and colour if present in small amount.
- High content causes unsoundness.

(vii)Sulphur (S) : 1-3%

- Very small amount makes the cement sound.
- If it is in excess, it causes unsoundness in cement.

(vii)Alkalies (Soda and Potash; $Na_2O + K_2O$) – (0.5-1.3%)

- Causes alkali-aggregate reaction, efflorescence and staining.

1.12 What are Bogue's compounds? Briefly mention their functions.

[8 marks : 2018]

Solution:

- The raw materials used for the manufacture of cement consist mainly of lime, silica, alumina and iron oxide.
- These oxides when subjected to high clinkering temperature combine with each other to form complex compounds.
- The identification of major complex compounds is based on R.H. Bogue's work and hence these are called Bogue's compound.

There are 4 Bogue's compounds:

1. Tricalcium Silicate (C_3S)

   Chemical formula : 3 $CaOSiO_2$

   Percentage : 30 - 50%

   - It undergoes hydration within a week and is responsible for development of early strength in cement.
   - It is the best cementing material and is well burnt.
   - Increases resistance to freezing and thawing.
   - Renders the clinker easier to grind.
   - Its proportion can be increased where early gain of strength is required.

   Example : Emergency repair work, cold weather concreting, prefabricated construction, etc.

2. **Dicalcium silicate (C_2S)**
   Chemical formula : $2CaOSiO_2$
   Percentage : 20-45%
   - It hydrates and hardens slowly and takes long time to add to strength (1 year or more).
   - Imparts resistance to chemical attack.
   - Proportion is increased when early strength is not required and higher heat of hydration should not be there.

   Example : dam or bridge construction.
3. **Tricalcium Aluminate (C_3A)**
   Chemical formula : $3CaOAl_2O_3$
   Percentage : 8-12%
   - Responsible for flash set of cement as it undergoes hydration within 24 hours after water addition.
   - Highest heat of hydration and tendency to volume changes causing cracking.
   - If present in higher amount, resistance to sulphate attack decreases.
4. **Tetracalcium Alumino Ferrite (C_4AF)**
   Chemical formula : $4CaOAl_2O_3Fe_2O_3$
   Percentage : 6-10%
   - Responsible for flash set but generates less heat.
   - Poorest cementing value.
   - Raising its content reduces the strength slightly.

1.13 **Explain the products of hydration of C_3S and C_2S (Bogues compounds) giving the relevant equations involving the reactions.**

[4 marks : 2019]

Solution:

- The chemical reactions that take place between cement and water is referred to as hydration of cement.
- The hydration of cement can be visualized in two ways viz. "through solution" and "solid state" type of mechanisms
- The reaction of cement with water is exothermic i.e. it liberates a considerable quantity of heat and this liberated heat is called as **heat of hydration**.
- The hydration process is not an instantaneous one. The reaction is faster in the early periods and continues indefinitely at a decreasing rate.
- During hydration, C_3S and C_2S react with water and calcium silicate hydrate (C-S-H) is formed along with calcium hydroxide [$Ca(OH)_2$].
- Calcium silicate hydrate is the most important product of hydration and it determines the good properties of concrete.
- $2(3CaO \cdot SiO_2) + 6H_2O \rightarrow 3CaO \cdot 2SiO_2 \cdot 2H_2O + 3Ca(OH)_2$
  or it can be written as:

$$2C_3S + 6H \rightarrow C_2S_2H_3 + 3Ca(OH)_2$$
$$100 + 24 \rightarrow 75 + 49$$

The corresponding weights involved are

Similarly,

$$2(2CaO \cdot SiO_2) + 4H_2O \rightarrow 3CaO \cdot 2SiO_2 \cdot 3H_2O + Ca(OH)_2$$

or it can be written as:

$$2C_3S + 6H \rightarrow C_3S_2H_3 + Ca(OH)_2$$

$$100 + 21 \rightarrow 99 + 22$$

The corresponding weights involved are:

$$2C_3S + 6H \longrightarrow C_3S_2H_3 + 3Ca(OH)_2$$

$$2C_2S + 4H \longrightarrow C_3S_2H_3 + Ca(OH)_2$$

1.14 **Describe the various tests performed to assess the suitability of Lime as a cementing material.**

[8 marks : 2019]

Solution:

Testing of Lime

Visual inspection: (a) A sample of lime is examined for its colour and lumps. (b) Lumps of lime indicate quick lime of unburnt lime

Field Tests: The field tests usually performed on lime at site are as below:

- **Test for physical properties:** Hydraulic limes are brush grey, brown or dark coloured. Hydraulic limestones have a clayey teste and give out earthy smell. White colour of lime is an indication of pure variety of limestone. Shining particles on the surface of limestones indicate the presence of free salt.
- **Workability test:** A handful of mortar (lime sand mortar in the ratio 1 : 3) is thrown on the surface on which it is to be used. The area covered by the mortar and its quantity is recorded. These data indicate the workability of the lime mortar. It is a very crude field test performed with the actual mortar.
- **Impurity test:** A known weight of lime is mixed with water in a breaker. The solution is decanted. The residue is dried for 8 hours in the natural heat of sun and is then weighted. If the residue is less than 10 per cent of the weight of the lime taken initially it is considered to be of good quality. In case, if it is 10-20 percent, the lime is fat and if more than 20 per cent the lime is classed as poor.
- **Heat test:** Limestone is heated for four hours on open fire; CO_2 escapes and it loses weight. From this the carbonate percentage in the limestone can be worked out. Lumpy form indicate quick lime or unburnt limestone. A porous structure may indicate quick lime.
- **Hydraulic acid test:** The test is carried to know the classification and the carbonate content of lime.

  When a teaspoon full of lime is put in a test tube containing 10 ml of 50 per cent hydrochloric acid, effervescence takes place. Too much of effervescence indicates high percentage of calcium carbonate in limestone. The residue at the bottom of the tube indicates percentage of inert materials present in the limestone. Abundant of liberation of CO_2 indicates unburnt lime.

  In case of class-A lime, a good get is formed above the layer of inert material. A thick gel indicate class-B lime and absence of gel means class-C lime.

  For eminently hydraulic lime, the get formed is thick and does not flow. Absence of gel indicates non-hydraulic of fat lime. If get flows, it indicates feebly hydraulic lime.

- **Ball test:** Balls of stiff lime paste are made and left for six hours. They are placed in a basin of water. If expansion and disintegration of balls is observed, the lime is of type C. Little expansion and numerous cracks indicate it to be class-B lime. Class-A lime will have no adverse effect.

Physical Tests

- Sampling of testing lime should be done as quickly as possible so that the material does not deteriorate. From each lot, three test samples are taken for quick lime as well as for hydrated lime.

Fineness Test [IS : 6932 (Part IV)]

- The sieves are arranged one above the other with the coarser sieves at the top and the finer sieves at the bottom. Sieving is done with a gental wrist motion. 100 g of the hydrated lime is placed on the top sieve and is washed through the sieves with a moderate jet of water for not more than 30 minutes. The residue on each sieve is dried at 100 ± 10°C to constant mass and weighted. The result is expressed as a percentage of mass of hydrated lime taken.

Determination of Residue on Slaking of Quick Lime [IS : 6932 (Part III)]

- Sample of quick lime is sieved through 2.36 mm IS sieve and the residue, if any, is broken and sieved again until the whole quantity passes through the sieve. The quantity of water required for slaking is usually 4 times the mass of quick lime, however, it may be as high as 8 times for certain high calcium limes.

Workability Test [IS : 6932 (Part VIII)]

- The test is conducted on a standard flow table and a truncated conical mould. For testing hydrated lime, the lime putty is prepared by thoroughly mixing 500 g of hydrated lime with an equal mass of water at a temperature of 27 ± 2°C and kept for 24 hours. The soaked material is then thoroughly mixed and knocked up to produce a plastic putty, by passing the material twice through the mixer.

Setting Time Test

- The initial and final setting times of hydrated lime are determined using Vicat's apparatus in the same way as that for Portland cement. Here in this test lime putty is used instead of cement mortar.

Soundness Test [IS : 632 (Part IX)]

- The test is done to find the quality, i.e., the unsoundness or disintegration property of lime using the Le-chatelier apparatus.

Popping and Pitting Test [IS : 6932 (Part X)]

- To determine the soundness of fat lime, pats are prepared by mixing hydrated lime, Plaster of Paris and water. The pats are subjected to steam and then tested for disintegration, popping and pitting. If any of these occurs the lime is considered to be unsound.

1.15 Write briefly about the following:
(A) Air Entraining admixtures
(B) Role of Fly ash as a part replacement of cement

[10 marks : 2019]

Solution:

(A) **Air Entraining Admixture:** This cement is manufactured by mixing small quantity of air-entraining agent like alkali salts of natural wood resins, synthetic detergents of alkylaryl sulphate type, sodium salts of sulphonates, calcium lignosulphate, salts of fatty acids, etc., with ordinary Portland cement or Portland blast furnace slag cements. These agents in powder or in liquid forms are added to the extent of 0.025 to 0.100 per cent by weight of OPC cement clinker at the time of grinding. At the time of mixing concrete ingredients, these cements produce tiny, discrete noncoalescing air bubbles in the concrete mass which enhances workability and reduces tendency to segregation and bleeding. The air entrainment increases the frost and sulphate water resistance of concrete.

(B) **Role of Fly ash as a part replacement of cement:** The fly ash or Pulverised Fuel Ash (PFA) is the residue from the combustion of coal collected by the mechanical dust collectors or electrostatic preceptors or separators from the fuel gases of thermal power plants. Like Portland cement, fly ash contains oxides of calcium, aluminium and silicon, but the amount of calcium oxide is considerably less. The properties of fly ash depend on the type of coal burnt. In general, silicious fly ash is pozzolanic, while calcareous fly ash has latent hydraulic properties.

The pozzolanic activity is due to the presence of finely divided glassy silica and lime which produce calcium silicate hydrate as is produced in hydration of Portland cement. The carbon content is fly ash should be as low as possible, whereas the silica content should be as high as possible.

The fly ash obtained from Electrostatic Precipitators (ESP) is finer than the Portland cement. The most important mineral aspect of fly ash is the presence of 60-90 per cent glassy spherical silicon dioxide (SiO_2). The fly ash is generally used in the following three ways:

1. **As a part replacement of cement:** This simple replacement of Portland cement up to 60 per cent by mass reduces the strength at ages up to 3 months.
2. **As a simultaneous replacement of cement and fine aggregate:** This replacement enables the strength as a specified age to be equalled depending on the water content.
3. As a part of cement in the form of blended cement.

1.16 **Describe how the compounds of clinker affect the properties of cement.**

[12 marks : 2020]

Solution:

The raw materials used for the manufacture of cement consist mainly of lime, silica, alumina and iron oxide.

These oxides when subjected to high clinkering temperature combine with each other to form complex compounds.

The identification of major complex compounds is based on R.H. Bogue's work and hence these are called Bogue's compound.

There are 4 Bogue's compounds:

1. **Tricalcium Silicate (C_3S)**

   Chemical formula : 3 $CaOSiO_2$

   Percentage : 30 - 50%

   - It undergoes hydration within a week and is responsible for development of early strength in cement.
   - It is the best cementing material and is well burnt.
   - Increases resistance to freezing and thawing.
   - Renders the clinker easier to grind.
   - Its proportion can be increased where early gain of strength is required.

     Example : Emergency repair work, cold weather concreting, prefabricated construction, etc.

2. **Dicalcium silicate (C_2S)**

   Chemical formula : $2CaOSiO_2$

   Percentage : 20-45%

   - It hydrates and hardens slowly and takes long time to add to strength (1 year or more).
   - Imparts resistance to chemical attack.

- Proportion is increased when early strength is not required and higher heat of hydration should not be there.

Example : Dam or bridge construction.

3. **Tricalcium Aluminate (C_3A)**

   Chemical formula : $3CaOAl_2O_3$

   Percentage : 8-12%

- Responsible for flash set of cement as it undergoes hydration within 24 hours after water addition.
- Highest heat of hydration and tendency to volume changes causing cracking.
- If present in higher amount, resistance to sulphate attack decreases.

4. **Tetracalcium Alumino Ferrite (C_4AF)**

   Chemical formula : $4CaOAl_2O_3Fe_2O_3$

   Percentage : 6-10%

- Responsible for flash set but generates less heat.
- Poorest cementing value.
- Raising its content reduces the strength slightly.

1.17 **What do you mean by normal consistency of cement? What is its significance? How is it tested?**

[8 marks : 2020]

Solution

Normal consistency of cement:

- The normal (standard) consistency of a cement paste is defined as that consistency which will permit a Vicat plunger having 10 mm diameter and 50 mm length to penetrate a depth of 33 to 35 mm from the top (or 5 to 7 mm from the bottom) of the mould.

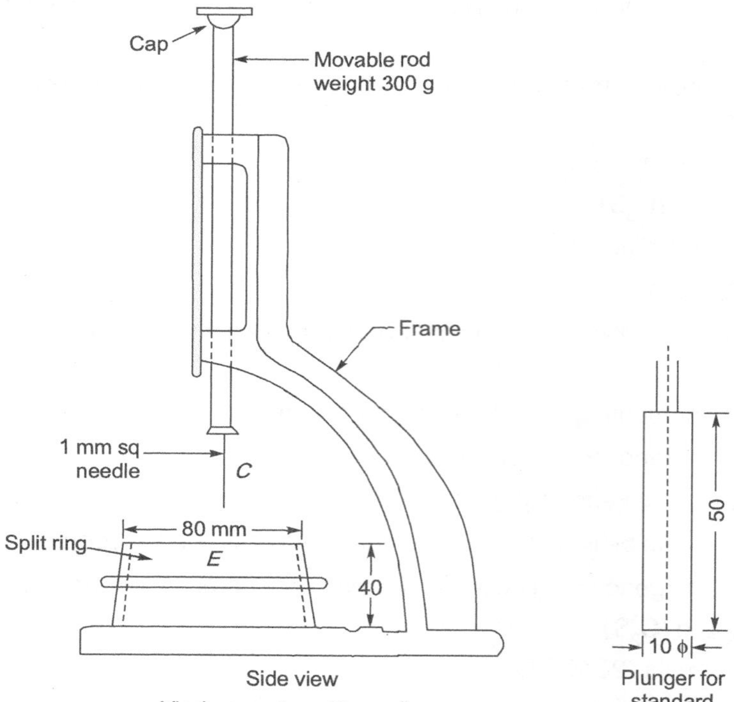

Side view

Vical apparatus with needle for initial section time test

Plunger for standard consistency test

Significance:

- It is necessary to determine consistency because the amount of water affects the setting time of the cement. It is resistance to shear deformation. Thus, it plays vital role in the determination of compressive strength or workability of concrete.

Vicat Apparatus:

- Vicat apparatus assembly consists of a plunger 300 gm in weight with a length of 50 mm and diameter of 10 mm and a mould which is 40 mm deep and 80 mm in diameter.

Test Procedure:

- To prepare the paste, take weighed quantity (300 g) of cement and place it in a crucible.
- Mix a weighed quantity of water (approximately 24% by weight of cement) for the first trial.
- The time of mixing or gauging should not be less than 3 minutes nor more than 5 minutes and gauging time should be counted from the time of adding water to the dry cement until commencing to fill the mould.
- The Vicat mould is filled with the paste, which is levelled off at its top.
- The mould is placed under the Vicat plunger.
- The vicat plunger is brought down to touch the surface of paste in the mould and quickly released allowing it to sink into the paste by its own weight.
- Take the reading by noting the depth of penetration of the plunger.
- Similarly conduct the trials with increasingly water cement ratio till such time the plunger penetrates for a depth of 33 to 35 mm from the top (or 5 to 7 mm from the bottom).
- That particular percentage of water which allows the plunger to penetrate only to a depth of 33 to 35 from the top (or 5 to 7 mm from the bottom) is known as the percentage of water required to produce a cement paste of normal (standard) consistency.
- This percentage is generally denoted by *P*.
- This test should be conducted at a constant temperature of 27° ± 2°C and a constant humidity of 90%.

1.18 **It is quite common to replace/partially substitute, cement with flyash. However, all the concretes made by cement substituted by flyash may not be used for all the applications. Specify some applications where using flyash concrete is useful and the applications where we should avoid using flyash products and provide reasons.**

[12 marks : 2022]

Solution

- Flyash is pozzolanic material obtained by burning the pulverized coal.
- Replacement of cement with flyash modifies property of cement and concrete and entire cementaneous material is combination of cement clinker and pozzolanic material like flyash.
- Addition of flyash by reduction in amount of basic cement clinker results in reduction in rate and amount of heat of hydration.
- Lesser rate of hydration maintains workability of concrete for a longer time, i.e., concrete will remain with it's internal energy to achieve a higher degree of compaction.

  So, best use of replacement of cement with flyash is found for self compacted concrete, where water powder ratio is kept to be 0.3 with high powder content.
- Flyash does not leach out portlandite while formation of C-S-H gel, so finished structure will not become porous. Moreover, less amount and rate of heat evolution do not cause shrinkage. Hence, overall durability of structure is improved e.g. Mass concreting.
- Due to less amount of C_3A clinker, sulphate resistance can also be improved. Hence, cement with flyash can be used against sulphur attack as well.

- Replacement with flyash reduces the cement demand, which helps in reducing emission of CO_2 (green house gases) in atmosphere. Hence, concrete prepared by flyash is a green concrete as well.
- But as amount of C_3A clinker is less, the rate of hydration is less. Hence, it can not be useed against quick setting like for underwater concreting, grounting, shortcrete etc.
- Slow rate of hydration is not desirable for strength gaining construction like precast construction, pavement repair, cold weather construction etc.
- Still replacement of flyash is very common as it gives better workability to mix, high durability to the finished structure, economical and environment friendly.

1.19 **Determine the proportion of aggregates required to give a suitable grading based on their fineness moduli. Assume the following gradings of aggregates:**

IS sieve designation (mm)	Aggregate size associated with percentage passing IS sieve (mm)		
	40	20	10
40	90	—	—
20	10	90	—
10	0.2	10	90
4.75	—	0.3	20

How, fineness modulus affect the characteristics of concrete? What is the optimum value of FM for fine aggregates?

[20 marks : 2022]

Solution

IS sieve designation (mm)	Aggregate size associated with percentage passing IS sieve (mm)		
	40	20	10
40	90	—	—
20	10	90	—
10	0.2	10	90
4.75	—	0.3	20

The data provided for the percentage passing from 20 mm size sieve is inconsistent.
In accordance with IS 383-2016 of 40 mm maximum size aggregate:

IS sieve designation (mm)	Desired grading for 40 mm maximum size aggregate as IS 383 : 2016
40	90 to 100
20	30 to 70
10	10 to 35
4.75	0 to 5

The fractions of coarse aggregates are combined to get a combined grading.
Start with trial fraction of 4:87:9, check whether the combined grading is obtained.

IS sieve designation (mm)	Fraction I 40 mm 4%	Fraction II 20 mm 87%	Fraction III 10 mm 9%	Combined grading
40	90	100	—	90.60
20	10	90	—	78.70
10	2	10	90	16.88
4.75	—	3	20	4.41

From the table it is observed that the with the of 4:87:9, the desired combined grading as per IS 838:2016 is obtained.

(i) The significance of fineness modulus [FM] is in specifying the proportions of fine and coarse aggregates when designing the concrete mixes.

(ii) Higher value of FM indicates that the aggregates are coarser.

(iii) Lower value of FM results in more paste which helps in making the concrete easier to finish.

(iv) It helps in proper gradation of aggregates.

1.20 **(i) Draw a neat sketch of complete operational sequence of dry process in making cement. Also indicate the place/equipment/part of plant where the process is taking place.**

(ii) Indicate the advantages of Dry process over the Wet process.

[10 + 10 = 20 marks : 2022]

Solution

(i)

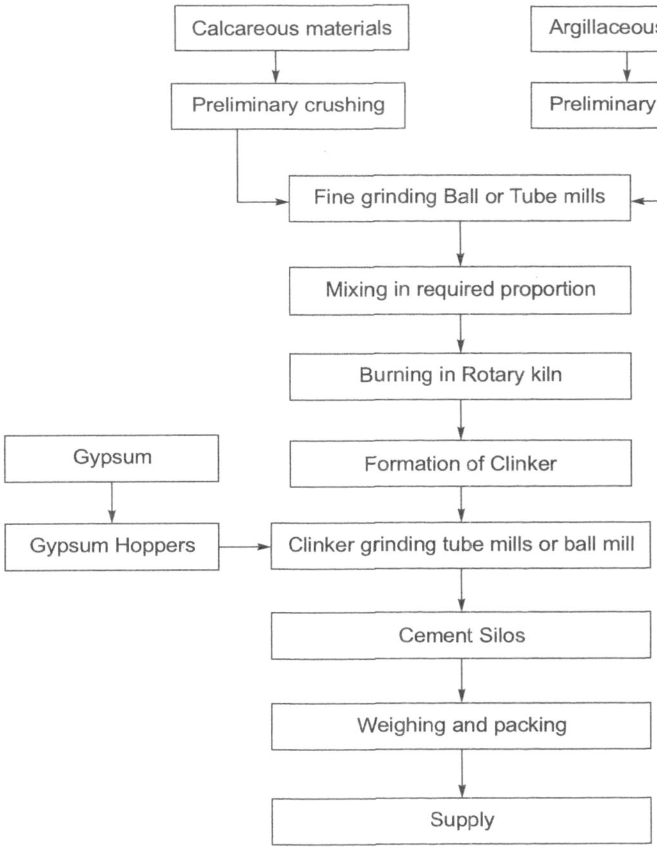

A rotary kiln is formed of steel tubes. Its diameter varies from 2.50 m to 3 m. Its length varies from 90 m to 120 m. It is laid at a gradient of about 1 in 25 to 1 in 30. The kiln is supported at intervals by columns of masonry or concrete. The kiln rotates at about one to three revolutions per minute about its longitudinal axis. The corrected slurry is injected at the upper end kiln. The hot gases or flames are forced through the lower end of kiln.

Exhaust gases from kiln in dry process can be used for preheating and hence helps to make the manufacturing economical.

At the portion of the kiln near its upper end, the water of slurry is evaporated. As the slurry gradually descends, there is rise in temperature and in the next section of kiln, the carbon dioxide from slurry is evaporated. The small lumps, known as the nodules, are formed at this stage. These nodules then gradually roll down passing through zones of rising temperature and ultimately reach to the burning zone, where temperature is about 1400°C to 1500°C. In burning zone, the calcined product is formed and nodules are converted into small hard dark greenish blue balls which are known as the clinkers.

The size of clinkers varies from 3 mm to 20 mm and they are very hot when they come out of burning zone of kiln. The clinker temperature at the outlet of kiln is nearly 1000°C. The ball mills are used to have preliminary grinding and the tube mills are used to carry out final grinding. The ball mill is in the form of steel cylinder of diameter about 2 m to 2.50 m and length about 1.80 m to 2 m.

The cylinder is placed in horizontal position and it rotates around a steel shaft. On the inside of cylinder, the perforated curved plates are fixed. The ends of these plates overlap each other. The cylinder is filled partly with steel balls of size varying from 50 to 120 mm.

When the mill is rotated about its horizontal axis, the steel balls strike against the perforated curved plates and in doing so, they crush the material. This crushed material passes through an inner sieve plate and then through an outer sieve plate. It is collected from an outlet at the bottom of outer casing of mill. Tube mill is in the form of a long horizontal steel cylinder of diameter about 1.50 m and of length about 7 m to 10 m. The cylinder is filled partly with steel balls of size varying from 20 mm to 25 mm.

The action of tube mill is similar to that of ball mill. But fine grinding is achieved due to steel balls of smaller size. To combine preliminary and final grinding, the compartment mill or multiple chamber mill may be adopted. Such a mill has different chambers or sections in which steel balls of different sizes are placed. The material to be ground is allowed to pass through chambers in succession. The chambers with steel balls of bigger size are placed first and they are followed by chamber having steel balls of smaller size.

(ii) Advantages of dry process over wet process

- In dry process of cement manufacture, raw materials are fed into the kiln in dry powder form and hence this saves cost of energy consumption needed for evaporation of moisture from the raw materials.
- Cost of production of cement with dry process comes out to be low as compared to wet process.
- Size of kiln required for cement manufacture is small as compared to the size of kiln required in the wet process.
- Dry process has a great adaption of materials which runs the device steadily. Cement clinker calcined in the kiln has better quality.
- Dry process can take full advantage of the waste heat thereby reducing the cost of production of cement.

1.21 **How is PPC different from OPC on the basis of their ingredients? Describe the advantages of using PPC in comparison to OPC.**

[6 marks : 2023]

Solution

- Portland Pozzolona cement is prepared by inter grinding of OPC clinker with 15-35% pozzolonic additive like fly ash, granulated blast furnance slag, surkhi, silica fume and rice husk etc.
- Pozzolonic materials are artificial mineral admixture obtained as different industrial waste and used as replacement of certain amount of clinker. Hence, production of clinker can be reached which makesPPC becomes environment friendly cement.
- As clinker is replaced by pozzolonic materials, hence PPC is found to be more economical.
- Pozzolonic material doesn't have binding property but when it reacts with portlandite $Ca(OH)_2$ which was supposed to be during hydration of OPC, it forms C-S-H gel. Hence, PPC results into more durability and impermeability.
- As PPC is finer than OPC. So, mix becomes more cohesive due to which resistance against segregation is more.
- PPC reduces requirement of water due to which chances of bleeding are reduced.
- PPC results in more volume of mortar than OPC due to its lower density.
- PPC results in improvement over strength.
- PPC results in resistance against external attack like chloride attack, carbonation and also provide electrolyte resistance.
- PPC also reduces chances of alkali-aggregate reaction due to replacement of clinker.

2. Mortar

2.1 **What are the properties of a good mortar? Enumerate four main uses of mortar.**

[10 marks : 2004]

Solution:

The properties of a good mortar are:

(i) The most important properties of a good mortar are mobility, placeability and water retention.
(ii) It should be capable of developing good adhesion with the building units such as bricks, stones, etc.
(iii) It should be capable of developing the designed stresses.
(iv) It should be cheap.
(v) It should be durable.
(vi) It should be easily workable.
(vii) It should not affect the durability of materials with which it comes into contact.
(viii) It should set quickly so that speed in construction may be achieved.
(ix) The joints formed by mortar should not develop cracks they should be able to maintain their appearance for a sufficiently long period.

Four main uses of mortar are:

(i) Cement mortar is used for damp-proof course below the ground level, exposed work such as exterior free standing walls and parapets.
(ii) Cement mortar is also used for pointing the joints of masonry, for plastering the surface of masonry to protect it from weather and to provide a pleasing and smooth finish.
(iii) Lime mortar is used in masonry to bind stones, bricks or concrete blocks together. It is suitable for masonry and plastering in cheap and light load bearing wall construction.

(iv) Gauged mortar is used for bedding and for thick brick walls.

2.2 **Explain the compressive strength test of cement mortar and state its importance.**

[10 marks : 2005]

Solution:

The compressive strength test of cement mortar is carried out as per the guidelines specified by **IS:4031 (Part 6)-1988**. The material for each cube shall be mixed separately and the quantity of cement, standard sand and water shall be as follows:

(i) Cement — 200 gm

(ii) Standard sand — 600 gm

(iii) Water — $\left(\frac{P}{4}+3.0\right)$ per cent of combined mass of cement and sand, where P is the percentage of water required to produce a paste of standard consistency.

The mixture of cement and standard sand is placed on a non-porous plate. This mixture is mixed dry with a trowel for one minute and then with water until the mixture is of uniform colour. The time of mixing shall in any event be not less than 3 minute and should the time taken to obtain a uniform colour exceed 4 min, the mixture shall be rejected and the operation repeated with a fresh quantity of cement, sand and water. Immediately after mixing the mortar, it is filled into a cube mould of size 7.06 cm. The area of the face of the cube should be 50 cm^2. The mortar shall be prodded 20 times in about 8 seconds to ensure elimination of entrained air and honey combing. Place the remaining quantity of mortar in the hopper of cube mould and prod again as specified for the first layer and then compact the mortar by vibration The period of vibration shall be two minutes at the specified speed of 12000 ± 400 vibration per minute. Keep the compacted cube in the mould at a temperature of 27°C ± 2°C and at least 90 per cent relative humidity for 24 hours. After 24 hours the cubes are removed from the mould and immersed in clean fresh water until taken out for testing. Test three cubes for compressive strength for each period of curing. The compressive strength shall be the average of the strengths of the three cubes for each period respectively.

The compressive strength of hardened cement is the most important of all the properties. Therefore it is not surprising that the cement is always tested for its strength at the laboratory before the cement is used in important works.

2.3 **Discuss the types of mortar which can be used for the following types of masonry work with suggested proportions:**

(i) Masonry in foundation and plinth

(ii) Masonry in superstructure

(iii) Plastering work

(iv) Pointing

[5 marks : 2011]

Solution:

(i) **Masonry in foundation and plinth:** A mortar in the ratio of 1 : 6 that means 1 part of cement and 6 parts of sand is considered of medium quality and strength and is considered suitable for heavy walls and foundation of heavy or multi-storey building.

(ii) **Masonary in super structure:** For lightly loaded above ground parts of building lime mortar can be used. It has high plasticity possesses good cohesiveness with other surfaces and shrinks very little.

For construction of multi-storey buildings, building pillars, heavy or load effective walls, boundary wall and other walls that require high strength, cement sand mortar in the ratio of 1 : 4 is used. For temporary structures, light walls and single storey construction etc. cement sand mortar in the proportion of 1 : 7 may be used.

(iii) The mortar used for **plastering work** are:

1. **Lime mortar:** Lime used for plastering may be either fat lime or hydraulic lime. The mix proportion (i.e., lime : sand) varies from 1 : 3 to 1 : 4 fcr fat lime and 1 : 2 for hydraulic or kankar lime.
2. **Cement mortar:** Cement mortar is the best mortar for external plastering work since it is practically non-absorbant. The mix proportion (i.e., cement : sand) may vary from 1 : 4 to 1 : 6. Sand used for plastering should be clean, coarse and angular.
3. **Lime-cement mortar:** Lime-cement mortar contains properties of both the lime mortar as well as cement mortar. Mix proportions generally used are 1 : 1 : 6 (cement : lime : sand), 1 : 1 : 8 or 1 : 2 : 8. Generally, fat lime is used.

(iv) **Pointing:** Pointing is done with the following mortar mixes:

1. Lime mortar 1 : 2 mix (1 lime : 2 sand or surkhi).
2. Cement mortar 1 : 3 mix (1 cement : 3 sand).

2.4 **List the factors that influence the basic permissible compressive stresses for masonry of different mortars. Also elaborate the modification due to each factor.**

[10 marks : 2016]

Solution:

Due to use of different mortars, following factors will influence the basic permissible compressive stresses of masonry:

(i) **Strength of bond between the mortar and the masonry unit:** A strong bond will make masonry stronger by holding individual bricks or blocks together and it will function as a single unit. A weaker bond leads to cracking in masonry construction.

(ii) **Moisture holding capacity of the mortar:** Due to moisture absorption, we get the ability to retain water and thus preventing its escape into masonry units.

(iii) **Compressive strength of mortar:** It is always desirable to have weak mortar as compared to compressive strength of masonry unit so that any crack that develops must remain within the joint where it can be easily repaired.

(iv) **Strength of masonry unit:** It should gain enough strength and harden in a reasonable time so that further courses of masonry can be laid without excessive racking movements of courses below.

(v) **Presence of voids at mortar-masonry unit interface:** A mortar in which high voids are present offer very little resistance to water penetration and hence aids in disintegration of masonry.

(vi) **Conditions at the time of curing like temperature, humidity, wind etc.**

2.5 What are the impurities in lime and how do they affect the cementing properties?

[8 marks : 2022]

Solution

Impurities in lime:

(a) **Magnesium carbonate:** Lime contains Magnesium Carbonate in varying proportions. Presence of this constituent allows the lime to slake and set slowly. The excess of $MgCO_3$ imparts hydraulicity even in the absence of clay.

The makes the magnesium oxide "hard burned" and therefore slow slaking and less active. This will ultimately affect the binding properties of lime.

(b) **Clay:** It is responsible for hydraulic properties of lime. It also makes lime insoluble in water. Clay in excess of 10% to 30%, arrests slaking. If it is in small quantity, the slaking is retarded and do not display any hydraulic properties which ultimately does not set and harden under water.

(c) **Silica:** In its free from (sand), it has considerable effect on the cementitious properties of lime. Lime having high silica content shows poor cementing and hydraulic properties.

(d) **Iron compounds:** Iron oxides, carbonates or sulphides at lower temperature of calcination get converted in Fe_2O_3. But at higher temperature, iron combines with lime and silicates and forms complex silicate compounds. Pyrite or iron sulphide are regarded as highly undesirable.

(e) **Carbonaceous matter:** Its presence in the lime is an indication of poor quality of lime.

(f) **Sulphates:** They slow down the slaking action and increase the setting rate of lime.

(g) **Alkalis:** When pure lime is required, the alkalis are undesirable.

3. Concrete

3.1 **What is non-destructive testing of concrete? What are its relative merits? Name methods of non-destructive testing and explain briefly anyone method.**

[10 marks : 1998]

Or

Discuss the following:

(a) Non-destructive tests on concrete

(b) Fiber reinforced concrete

[10 marks : 2001]

Solution:

Non-destructive testing of concrete is an indirect method in which specimen are not loaded to failure and as such the strength inferred or estimated cannot be expected to yield absolute values of strength. These methods therefore attempt to measure some other properties of concrete from which an estimate of its strength, durability and elastic parameters are obtained. Some such properties of concrete are hardness, resistance to penetration of projectiles, rebound number, resonant frequency and ability to allow ultrasonic pulse velocity to propagate through it.

Though non-destructive testing methods are relatively simple to perform, the analysis and interpretation of test results are not so easy. Therefore special knowledge is required to analyse the properties of hardened concrete.

Some of the non-destructive test methods are:

(i) Surface hardness tests
(ii) Rebound tests
(iii) Penetration and pull out techniques
(iv) Dynamic or vibration tests
(v) Combined methods
(vi) Radioactive and nuclear methods
(vii) Magnetic and electrical methods
(viii) Acoustics emission techniques

Dynamic or Vibration Method: This is an important method used in testing concrete strength and other properties. The fundamental principle on which the dynamic or vibration methods are based is velocity of sound through a material. A mathematical relationship could be established between the velocity of sound through specimen and its resonant frequency and the relationship of these two to the modulus of elasticity of the same material. The relationships which are derived for solid mediums considered to be homogeneous, isotropic and perfectly elastic, but they may be applied to heterogeneous material like concrete.

Fiber Reinforced Concrete: Plain concrete possesses a very low tensile strength, limited ductility and little resistance to cracking. Internal microcracks are inherently present in the concrete and its poor tensile strength is due to the propagation of such microcracks, eventually leading to brittle fracture of the concrete.

It has been recognised that the admission of small, closely spaced and uniformly dispersed fibres to concrete would act as crack arrester and would substantially improve its static and dynamic properties. This type of concrete is known as fiber reinforced concrete.

Fiber reinforced concrete can be defined as a composite material consisting of mixtures of cement, mortar or concrete and discontinuous, discrete, uniformly dispersed suitable fibres. Continuous meshes, woven fabrics and long wires or rods are not considered to be discrete fibres.

The fibers can be imagined as an aggregate with an extreme deviation in shape from the rounded smooth aggregate. The fibers interlock and entangle around aggregate particles and considerably reduce the workability, while the mix becomes more cohesive and less prone to segregation. The fibers suitable for reinforcing the concrete have been produced from steel, glass and organic polymers.

The major factors affecting characteristics of fiber reinforced concrete are water-cement ratio, Percentage (volume fraction) of fibers, diameter and length of fibers. Fiber reinforced concrete has special application in hydraulic structures, air field and highway pavements, bridge decks, heavy duty floors and tunnel linings.

3.2 **Describe the workability of fresh concrete and its measurement in the light of statement that workability is a composite property and each test measures only a particular aspect of it.**

[10 marks : 2000]

Solution:

As per **IS:6461 (Part VII)-1973** workability can be defined as that property of freshly mixed concrete or mortar which determines the ease and homogeneity with which it can be mixed, placed, compacted and finished."

The diverse requirements of mixibility, stability, transportability, placeability, mobility, compactability and finishability of fresh concrete are collectively referred to as workability. The workability of fresh concrete is thus a composite property. The optimum workability of fresh concrete varies from situation to situation e.g., the concrete which can be termed as workable for pouring into large sections with minimum reinforcement may not be equally workable for pouring into heavily reinforced thin sections. A concrete may not be workable when compacted by hand but may be satisfactory when vibration is used.

Sometimes the terms consistency and plasticity are used to denote the workability of concrete mix. The consistency of the mix really means the wetness of the mix, and a wetter mix need not have all the above desired properties. On the other hand an extremely wet mix may cause segregation and may be difficult to place in moulds. Plasticity is the cohesiveness of the mix to hold the individual grains together by the cement matrix.

The tests which are widely used to measure workability in fact measures only one aspect of workability at a time. The **slump test** indicates the behaviour of a compacted concrete cone under the action of gravitational forces. The slump test is essentially a measure of consistency or the wetness of the mix. The compacting factor test gives the behaviour of fresh concrete under the action of external forces. It measures the compactibility of concrete which is an important aspect of workability, by measuring the amount of compaction achieved for a given amount of work. The **compacting factor test** has been held to be more accurate than slump test, especially for concrete mixes of medium and low workabilities.

The **Vee-Bee consistometer** test is suitable for stiff concrete mixes having low and very low workability. This is a good laboratory test to measure indirectly the workability of concrete using vibration technique.

Flow test is also a laboratory test which gives an indication of the quality of concrete with respect to consistency, cohesiveness and the proneness to segregation. In this test a standard mass of concrete is subjected to jolting. The spread or the flow of the concrete is measured and this flow is related to workability.

3.3 **Using a mix design procedure, mix proportions for the desired grade of concrete have been obtained as 1:2.1:3.5 (by mass) with water cement ratio of 0.5 and air content of 3%. Calculate the weights of individual ingredients required to make 0.25 m³ concrete. The specific gravity of cement, sand and aggregate were 3.15, 2.65 and 2.7 respectively.**

[10 marks : 2000]

Solution:

Given Data:

$w_c : w_s : w_a = 1 : 2.1 : 3.5$, Water cement ratio = 0.5, Air content = 3%, Total volume of concrete = 0.25 m³

Specific gravity of cement, $G_c = 3.15$; Specific gravity of sand, $G_s = 2.65$; Specific gravity of aggregate, $G_a = 2.7$; Let the weight of cement = w_c kg

Let the weight of sand = w_s kg

Let the weight of aggregate = w_a kg

Now, Net volume of concrete = Total volume of concrete – Volume of air in concrete

$$= 0.25 - \left(\frac{3}{100} \times 0.25\right) = 0.25 - 0.0075$$

∴ Net volume of concrete = 0.2425 m³

We have $\frac{w_c}{w_s} = \frac{1}{2.1}$ and $\frac{w_c}{w_a} = \frac{1}{3.5}$

But we also know that

Net volume of concrete = Volume of water + Volume of solids

$$\Rightarrow \quad 0.2425 = \frac{\text{Weight of water}}{\text{Density of water}} + \frac{\text{Weight of solids}}{\text{Density of solids}}$$

$$\Rightarrow \quad 0.2425 = \frac{0.5 \times w_c}{1000} + \left[\frac{w_c}{G_c \times 1000} + \frac{w_s}{G_s \times 1000} + \frac{w_a}{G_a \times 1000}\right]$$

$$\Rightarrow \quad 0.2425 \times 1000 = 0.5\,w_c + \left[\frac{w_c}{G_c} + \frac{2.1w_c}{G_s} + \frac{3.5\,w_c}{G_a}\right]$$

$$\Rightarrow \quad 242.5 = w_c\left[0.5 + \frac{1}{3.15} + \frac{2.1}{2.65} + \frac{3.5}{2.7}\right]$$

$$\Rightarrow \quad w_c = \frac{242.5}{2.9062} = 83.44 \text{ kg}$$

$$\therefore \quad w_s = 2.1 \times w_c = 2.1 \times 83.44 = 175.22 \text{ kg}$$

$$w_a = 3.5\,w_c = 3.5 \times 83.44 = 292.04 \text{ kg}$$

3.4 **What is meant by "workability of concrete"? List the methods used for measurement of workability of concrete. Clearly indicate the aspect of workability measured by each method.**

[10 marks : 2002]

Or

Give a brief description of slump test for measurement of workability of concrete and its merits and demerits. What are the slump values recommended for different types of works?

[10 marks : 2008]

Solution:

Hundred per cent compaction is required for contributing to the **maximum strength**. To enable the concrete to be fully compacted with given efforts, normally a higher water/cement ratio than that calculated by theoretical considerations may be required. That is to say the function of water is also to lubricate the concrete so that the concrete can be compacted with specified effort forthcoming at the site of work. The lubrication required for handling concrete without segregation, for placing without loss of homogeneity, for compacting with the amount of efforts forthcoming and to finish it sufficiently easily, the presence of a certain quantity of water is of vital importance. The quality of concrete satisfying above requirements is termed as workable concrete. It is very difficult to define precisely all the aspects of the workability in a single definition. **IS : 6461 (Part VII)-1973** defines workability as that property of freshly mixed concrete or mortar which determines the ease and homogeneity with which it can be mixed, placed, compacted and finished. The optimum workability of fresh concrete varies from situation to situation e.g., the concrete which can be termed as workable for pouring into large sections with minimum reinforcement may not be equally workable for pouring into heavily reinforced thin sections.

Sometimes the terms "consistency" and "plasticity" are often used loosely for workability. Consistency is a general term to indicate **degree of fluidity or the degree of mobility**. A concrete which has high consistency and which is more mobile, need not to be of right workability for a particular job. On the other hand plasticity is the cohesiveness of the mix to hold the individual grains together by cement matrix.

The following tests are commonly employed to measure workability:

(i) Slump test
(ii) Compacting factor test
(iii) Flow test
(iv) Kelly ball test
(v) Vee Bee consistometer test

The insitu determination of workability can be carried out by **slump test**. The test is carried out with a mould called the slump cone. The slump cone is placed on a horizontal and non-absorbent surface and filled in three equal layers of fresh concrete, each layer being tamped 25 times with a standard tamping rod. The top layer is struck off level and the mould lifted vertically without disturbing the concrete cone. The subsidence of concrete in millimeters is termed as the slump. The concrete after the test when slumps evenly all round is called true slump. In the case of very lean concrete, one half of the cone may slide down the other which is called a shear slump, or it may collapse in case of very wet concretes. The slump test is essentially a measure of consistency or the wetness of the mix. The test is suitable only for concretes of medium to high workabilities (i.e. having slump values of 25 mm to 125 mm). For very stiff mixes having zero slump, the slump test does not indicate any difference in concrete of different workabilities.

A concrete with slump value between 25 mm and 75 mm can be used in mass concrete foundations without vibration or lightly reinforced sections with vibration. For a slump value of 50 mm-100 mm concrete can be used in normal reinforced sections manually compacted and heavily reinforced sections with vibrations. For a slump value of 100 mm-150 mm, concrete can be used for sections with congested reinforcement. It can be used for pumping and tremie placing.

3.5 **Calculate the quantities of cement, sand and coarse aggregate required to produce one cubic meter of concrete for mix proportions of 1:1:40:2.80 (by volume) with water cement ratio of 0.48 (by mass). Bulk densities of cement, sand and coarse aggregates are 14.7, 16.66 and 15.68 kN/m^3 respectively. Percentage of entrained air is 2.0. Specific gravity of cement, sand and coarse aggregate are 3.15, 2.6 and 2.5 respectively.**

[10 marks : 2003]

Solution:

Let volume of cement be V_c m^3, volume of sand be V_s m^3, volume of aggregate be V_a m^3, weight of cement be w_c kN, weight of sand be w_s kN, weight of aggregate be w_a kN, weight of water be w_w kN

Given that: $V_c : V_s : V_a = 1 : 1.40 : 2.80$

$$\Rightarrow \quad \frac{V_c}{V_s} = \frac{1}{1.40}$$

$$\Rightarrow \quad V_s = 1.4\, V_c$$

and $$\frac{V_c}{V_a} = \frac{1}{2.80}$$

$$\Rightarrow \quad V_a = 2.8\, V_c$$

also $$\frac{w_w}{w_c} = 0.48 \text{ and entrained air} = 2\%$$

Now we know that

Net volume of concrete = Total volume – entrained air volume

$$\text{Net volume of concrete} = 1-\left(\frac{2}{100}\times 1\right) = 0.98\text{ m}^3$$

But net volume of concrete = volume of water + volume of solids

$$\Rightarrow \quad 0.98 = \frac{\text{Weight of water}}{\text{Density of water}} + \frac{\text{Weight of solids}}{\text{Density of solids}}$$

$$\Rightarrow \quad 0.98 = \frac{0.48\times w_c}{9.81} + \left[\frac{w_c}{G_c\times 9.81} + \frac{w_s}{G_s\times 9.81} + \frac{w_a}{G_a\times 9.81}\right]$$

$$\Rightarrow \quad 0.98\times 9.81 = 0.48\, w_c + \frac{w_c}{G_c} + \frac{w_s}{G_s} + \frac{w_a}{G_a}$$

$$\Rightarrow \quad 0.98\times 9.81 = 0.48\times 14.7\times V_c + \frac{14.7\, V_c}{3.15} + \frac{16.66\times 1.4\, V_c}{2.6} + \frac{15.68\times 2.8\, V_c}{2.5}$$

$$\Rightarrow \quad V_c = 0.2513\text{ m}^3$$

$$\therefore \quad V_s = 1.4\, V_c = 1.4\times 0.2513 = 0.35182\text{ m}^3$$

$$V_a = 2.8\, V_c = 2.8\times 0.2544 = 0.70364\text{ m}^3$$

$$\text{Thus weight of cement in kg} = \frac{0.2513\times 14.7\times 10^3}{9.81} = 376.57\text{ kg}$$

$$\text{weight of sand in kg} = \frac{0.35182\times 16.66\times 10^3}{9.81} = 597.48\text{ kg}$$

$$\text{weight of coarse aggregate in kg} = \frac{0.70364\times 15.68\times 10^3}{9.81} = 1124.68\text{ kg}$$

3.6 **Describe in brief the rebound hammer method on nondestructive testing of concrete.**

[10 marks : 2004]

Or

Explain in detail the non-destructive testing of concrete using Rebound hammer.

[10 marks : 2010]

Solution:

Rebound hammer test is one of the commonly adopted equipments for measuring the surface hardness.

This method was developed by a Swiss Engineer, Ernst Schmidt as a practical rebound test. The Schmidt test hammer weight less than 2 kg and has an impact energy of 2.2 N-m. The spring controlled hammer mass slides on a plunger within a tubular casing.

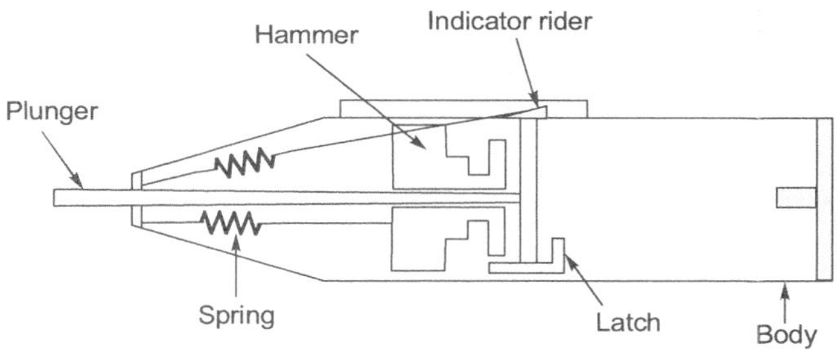

The plunger retracts against a spring when pressed against the concrete surface and this spring is automatically released when fully tensioned causing the hammer mass to impact against the concrete through the plunger. When the spring controlled mass rebounds, it takes with it a rider which slides along a graduated scale and is visible through a small window in the side of the casing. The rider can be held in position on the scale by depressing the locking button. The equipment can be operated horizontally or vertically. The plunger is pressed strongly and steadily against the concrete surface to be tested at right angles, until the spring loaded mass is triggered from its locked position. The measurement taken is an arbitrary quantity referred to as rebound number. A calibration curve relating the compressive strength of the concrete with the rebound number is plotted. The test is suitable for concrete having strength in the range of 20-60 MPa. The concrete surface must be smooth, clean and dry. Loose material should be ground off. This test gives a measure of relative hardness of the test zone. It provides useful information about a surface layer of concrete upto 30 mm deep.

3.7 Define workability of concrete. What are the factors affecting workability of concrete?

[10 marks : 2005]

Solution:

As per **IS:6461 (Part VII)-1973** workability can be defined as that property of freshly mixed concrete or mortar which determines the ease and homogeneity with which it can be mixed, placed, compacted and finished.

The factors affecting the workability of concrete are:

(i) **Water Content:** Water content in a given volume of concrete will have significant influences on the workability. The higher the water content per cubic meter of concrete, the higher will be the fluidity of concrete, which is one of the important factors affecting workability. It should be noted that from the desirability point of view, increase of water content is the last recourse to be taken for improving the workability even in the case of uncontrolled concrete.

(ii) **Mix Proportions:** Aggregate cement ratio is an important factor influencing workability. The higher the aggregate cement ratio, the leaner is the concrete. In lean concrete less quantity of paste is available for providing lubrication per unit surface area of aggregate and hence the mobility of aggregate is restrained to given bad workability. In case of rich concrete the phenomenon is reversed.

(iii) **Size of Aggregates:** The bigger the size of aggregate, the less is the surface area and hence less amount of water is required for wetting the surface and less matrix of paste is required for lubricating the surface to reduce internal friction. For a given quantity of water and paste, bigger size of aggregate will give higher workability.

(iv) **Shape of Aggregates:** The shape of aggregates influences workability in good measure. Angular, elongated or flaky aggregate makes the concrete very harsh when compared to rounded aggregates or cubical shaped aggregates. Contribution to better workability of rounded aggregate will come from the fact that for the given volume or weight it will have less surface area and less voids than angular or flaky aggregate. Not only that, being round in shape the frictional resistance is greatly reduced.

(v) **Surface Texture:** The influence of surface texture on workability is again due to the fact that the total surface area of rough textured aggregate is more than the surface area of smooth rounded aggregate of same volume. Thus the rough textured aggregate will show poor workability and smooth or glassy textured aggregate will give better workability.

(vi) **Grading of Aggregates:** This is one of the factors which will have maximum influence on workability. A well graded aggregate is the one which has least amount of voids in a given volume. Other factors being constant, when the total voids are less, excess paste is available to give better lubricating effect giving high workability, also well graded aggregates provide domino effect which provides better workability.

(vii) **Use of Admixtures:** This is the most important factor which affects the workability. Admixtures reduce the internal friction between the particles resulting in better workability.

3.8 **Estimate the quantities of cement, fine aggregate and coarse aggregate per cubic meter of concrete if the void ratio in cement is 62%, fine aggregate is 41% and coarse aggregate is 45%. The material properties are as follows:**
Mix is 1:2:4 with a w/c of 0.55, one bag of cement contains 50 kg of cement and its density is 1440 kg/m^3. The density of fine aggregate is 1700 kg/m^3 and coarse aggregate is 1600 kg/m^3 respectively. One bag of cement is equal to 34.7 litres.

[10 marks : 2006]

Solution:

Let the weight of cement in 1 m^3 of concrete be x kg.

$\therefore$ Weight of fine aggregate = $2x$ kg; Weight of coarse aggregate = $4x$ kg

We know that,

$$\text{Void ratio} = \frac{\text{Volume of voids}}{\text{Volume of solids}}$$

Volume of concrete = volume of voids + volume of solids + volume of water

$$\therefore \quad \text{Volume of voids} = \frac{x}{1440}\times\frac{62}{100}+\frac{2x}{1700}\times\frac{41}{100}+\frac{4x}{1600}\times\frac{45}{100}$$

$$= \frac{x}{1440}\times 0.62+\frac{2x}{1700}\times 0.41+\frac{4x}{1600}\times 0.45$$

$$\text{Volume of solids} = \frac{x}{1440}+\frac{2x}{1700}+\frac{4x}{1600}$$

$$\text{Volume of water} = \frac{0.55x}{1000}$$

But volume of concrete = 1 m^3

$$\therefore \quad \frac{x}{1440}\times 1.62+\frac{2x}{1700}\times 1.41+\frac{4x}{1600}\times 1.45+\frac{0.55x}{1000} = 1$$

$\Rightarrow$ x = 143.70 kg

Weight of cement = 143.70 kg

Weight of fine aggregate = 143.70 × 2 = 287.40 kg

Weight of coarse aggregate = 143.70 × 4 = 574.80 kg

Weight of water = 0.55 × 143.70 = 79.035 kg

3.9 **Write about various modulii of elasticity of plain cement concrete. Which values are used in design? What are the factors affecting modulus of elasticity of concrete?**

[10 marks : 2007]

Solution:

Static Modulus of Elasticity: The term Young's modulus of elasticity can strictly be applied only to the straight part of the stress-strain curve. In case of concrete, since no part of the graph is a straight line, the modulus of elasticity is found out with reference to the tangent drawn to the curve at origin. The modulus found from this tangent is called **initial tangent modulus.** This gives satisfactory results only at low stress value.

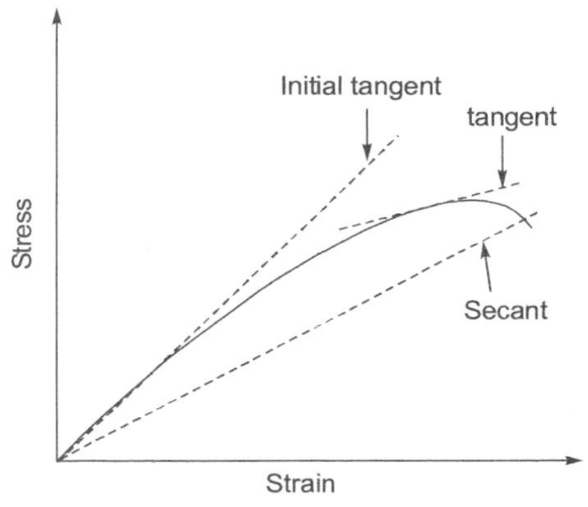

Tangent can also be drawn at any other point on the stress strain curve. The modulus of elasticity calculated with reference to this tangent is then called **tangent modulus.** The tangent modulus also does not give a realistic value of modulus of elasticity for the stress level much above or much below the point at which the tangent is drawn.

A line can be drawn connecting a specified point on the stress strain curve to the origin of the curve. If the modulus of elasticity is calculated with reference to the slope of this line, the modulus of elasticity is called as **secant modulus.**

If the modulus of elasticity found out with reference to the chord drawn between two specified points on the stress-strain curve then such value of the modulus of elasticity is known as **chord modulus.**

The static modulus of elasticity is used in design. Secant modulus is the most commonly used modulus of elasticity. Since the value of secant modulus decreases with increase in stress, the stress at which the secant modulus has been found out should always be stated.

Dynamic Modulus of Electricity: The modulus of elasticity can also be determined by subjecting the concrete member to longitudinal vibration at their natural frequency. This modulus of elasticity is known as dynamic modulus of elasticity. The value of dynamic modulus is somewhat higher than that of static or secant modulus. This is because dynamic modulus is unaffected by creep.

Factors affecting modulus of elasticity:

(i) The most important factor affecting the modulus of elasticity of concrete is the strength of concrete.

(ii) The modulus of elasticity also depends upon the state of wetness of concrete when other conditions being the same. Wet concrete will show higher modulus of elasticity than dry concrete.

(iii) The quality and quantity of aggregate will have a significant effect on the modulus of elasticity.

(iv) The age of concrete is also a factor which affect modulus of elasticity. Older the concrete, stronger it will be giving high modulus of elasticity.

3.10 **Briefly describe the admixtures generally used in concrete and properties they impart to the concrete.**

[10 marks : 2009]

Solution:

Admixtures are the chemical compounds in concrete other than hydraulic cement (OPC), water and aggregates and mineral additives that are added to the concrete mix immediately before or during mixing to modify one of more of the specific properties of concrete in the fresh or hardened state.

The commonly used admixtures are as follows:

(i) **Accelerating admixtures or accelerators:** An admixture used to speed up the initial setting of concrete is called an accelerator. These are added to concrete either (a) to increase the rate of hydration of hydraulic cement and hence to increase the rate of development of strength, or (b) to shorten the setting time. Accelerators are also used when the concrete is to be placed at low temperatures. The most widely used accelerator is calcium chloride ($CaCl_2$). The other less commonly used accelerators consist of NaCl, Na_2SO_4, NaOH, Na_2CO_3, K_2SO_4 and KOH.

(ii) **Retarding admixtures or Retarders:** The retarding admixtures slow down the initial rate of hydration of cement or prolong the setting of the cement paste in concrete. They are used primarily to offset the accelerating and damaging effect of high temperature and to keep concrete workable during the entire placing period which should be sufficiently long so that the succeeding lifts can be placed without the development of cold joints or discontinuities in a structural unit. Thus a retarding admixture holds back the hydration process, leaving more water for workability and allowing concrete to be finished and protected before drying out. The most commonly known retarder is calcium sulphate. Other retarders are sugar, lignosulphonic acids and their salts, hydroxylated carboxylic acids and their salts etc.

(iii) **Air-entraining Admixtures:** These admixtures help to incorporate a controlled amount of air, in the form of millions of minute non-coalescing bubbles distributed throughout the body of concrete during mixing, without significantly altering the setting or the rate of hardening of concrete. It is generally recognized that a proper amount of entrained air results in improved properties of plastic concrete like workability, easier placing and finishing, increased durability, better resistance to frost action and reduction in bleeding and segregation. The compounds used for air-entrainment are a number of natural wood resins containing abietic and pimeric acid salts, various sulphonated compounds and some animal and vegetable fats and oils such as tallow, olive oil and their fatty acids such as stearic and oleic acids.

(iv) **Water reducing admixtures or Plasticizers:** To achieve the desired workability and hardened physical properties, it is often necessary to add a water reducer to disperse the ingredients and reduce the amount of water to be added during mixing. Plasticizers enable a given fresh concrete mix to have higher flowability (workability) without increasing the water content. Benefits of water reduction in hardened state of concrete are increased strength, density, durability, volume stability, abrasion resistance, reduced permeability and cracking. The most commonly used plasticizer is lignosulphonic acid in the form of either its calcium or sodium salt.

3.11 **(i) Give a detailed account of the cylinder splitting test of concrete.**

(ii) What are the limitations of the cylinder splitting test in evaluating the real tensile strength of concrete?

[8 + 2 = 10 marks : 2009]

Solution:

(i) The cylinder splitting test is a well known indirect test used for determining the tensile strength of concrete, sometimes referred as the splitting tensile strength of concrete. The test is carried out by placing a cylindrical specimen (300 mm long and 150 mm in diameter) horizontally between the loading surfaces of a compression testing machine and the load is applied until failure of the cylinder, along the vertical diameter. When the load is applied a fairly uniform tensile stress is induced over two-third of the loaded diameter as obtained from an elastic analysis.

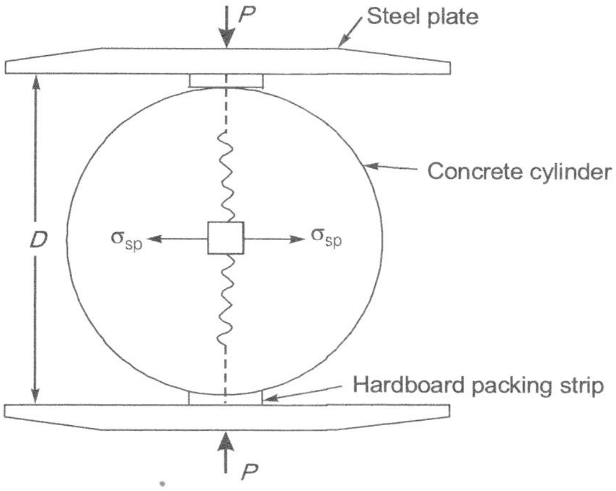

The magnitude of this tensile stress is given by

$$\sigma_{sp} = \frac{2P}{\pi DL}$$

where P is the applied load, D is the diameter and L is the length of cylinder.

(ii) The load is applied without shock and increased continuously until failure at a nominal rate within the range 1.2 MPa/min to 2.4 MPa/min. In order to reduce the magnitude of the high compression stresses near the points of application of the load narrow packing strips of hardboard are placed between the cylinder and loading platens of the testing machine. The packing strips should be soft enough to allow distribution of load over a reasonable area, yet narrow and thin enough to prevent large contact area. The main advantage of this test is that the same type of specimen and the same testing machine used for the compression test can be employed for this test. The test is simple to perform and gives more uniform results than other tension tests. The strength determined is closer to the actual tensile strength of the concrete than that given by the modulus of rupture test.

3.12 Draw a sketch showing the typical creep strain-time curve under uniaxial compression for concrete.

[4 marks : 2009]

Solution:

The typical creep strain-time curve is drawn below:

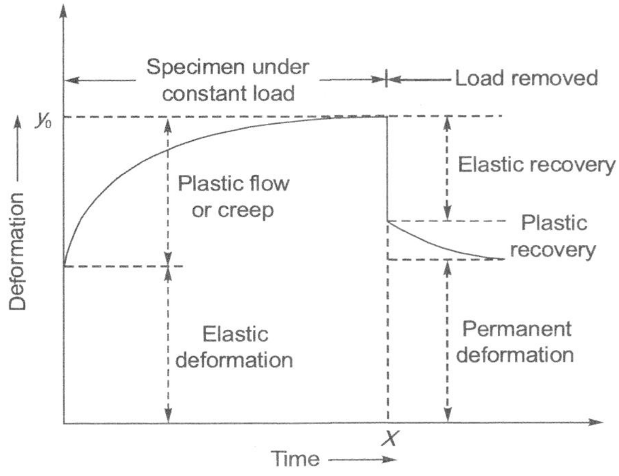

Creep strain-time curve

3.13 What is ferrocement? List the properties of ferrocement.

[10 marks : 2010]

Solution:

Ferrocement is a type of thin reinforced concrete construction where cement mortar matrix is reinforced with many layers of continuous and relatively small diameter wire meshes. While the mortar provides the mass, the wire mesh imparts tensile strength and ductility to the material. In terms of structural behaviour ferrocement exhibits very high tensile strength to weight ratio and superior anti cracking performance. The distribution of a small diameter wire mesh reinforcement over the entire surface, and sometimes over the entire volume of the matrix, provides a very high resistance against cracking. Moreover many other engineering properties such as toughness, fatigue resistance, impermeability, etc. are also considerably improved, superior resistance to cracking.

3.14 Explain in detail the ultrasonic pulse velocity method of non-destructive testing of concrete.

[5 marks : 2011]

Solution:

Ultrasonic Pulse Velocity Method: At present the ultrasonic pulse velocity method is the only one of this type that show potential for testing concrete strength in situ. It measures the time of travel of an ultrasonic pulse passing through the concrete. The fundamental design feature of all commercially available units are similar, consisting of a pulse generator and a pulse receiver. Pulses are generated by shock exciting piezoelectric crystals, with similar crystal used in the receiver. The time taken for the pulse to pass through the concrete is measured by electronics measuring circuits.

Based on the principle that velocity of sound (V) in a solid material, is a function of square root of Modulus of Elasticity (E) to its density (ρ)

$$V = f\left[\frac{gE}{\rho}\right]^{1/2}$$

Pulse velocity tests can be carried out on both laboratory sized specimens and complete concrete structures.

The factors that affect the measurements are:

- There must be smooth contact with the surface under test; a coupling medium such as a thin film of oil is mandatory.
- It is desirable for path lengths to be at least 12 inch in order to avoid any errors introduced by heterogeneity.
- The presence of reinforced steel in concrete has an appreciable effect on pulse velocity. It is therefore desirable and often mandatory to choose pulse paths that avoid the influence of reinforcing steel or to make correction if steel is in the path of pulse.

Applications and Limitations: The pulse velocity method is an ideal tool for estabilishing whether concrete is uniform. It can be used for both existing structures and those under construction. Usually if large difference in pulse velocities are found within a structure for no apparent reason, these is a strong season to presume that defective or deteriorated concrete is present. High pulse velocity reading are generally indicative of good quality concrete. A general relation between concrete quality and pulse velocity is given table.

S.No.	Pulse Velocity in km/sec	Concrete Quality Grading
1.	Above 45	Excellent
2.	3.5 to 4.5	Good
3.	3.0 to 3.5	Medium
4.	Below 3.0	Doubtful

3.15 **List the principal constituents of fly ash. Explain its pozzolanic action when used in concrete.**

[10 marks : 2012]

Solution:

The principal constituents of fly ash are **silica, alumina, iron oxide, and calcium** with varying amounts of **carbon** as measured by loss on ignition. Lignite and sub bituminous coal fly ash are characterized by higher concentrations of calcium and magnesium oxide and reduced percentages of silica and iron oxide, as well as a lower carbon content compared with bituminous coal fly ash.

The given table compares the normal range of chemical constituents of bituminous coal ash with those of lignite coal ash and sub bituminous coal fly ash.

From the table we can see that lignite and sub bituminous coal fly ashes have a higher calcium oxide content and lower loss on ignition than fly ashes from bituminous coals.

A pozzolana is a finely ground silicious glass material which as such does not possess cementitious property in itself, but reacts in the presence of water with lime at normal temperature to form compounds of low solubility having cementitious properties. This action is called pozzolanic action.

Component	Bituminous	Sub-bituminous	Lignite
SiO_2	20 - 60	40 - 60	15 - 45
Al_2O_3	5 - 35	20 - 30	10 - 25
Fe_2O_3	10 - 40	4 - 10	4 - 15
CaO	1- 12	5 - 30	15 - 40
MgO	0 - 5	1- 6	3 - 10
SO_3	0 - 4	0 - 2	0 - 10
Na_2O	0 - 4	0 - 2	0 - 6
K_2O	0 - 3	0 - 4	0 - 4
LOI	0 - 15	0 - 3	0 - 5

The incorporation of classified fly ash in cement paste results in lower pore size compared with the use of original fly ash. This is due to the gradual filling of large pores from factors such as hydration reaction, nucleation effect and pozzolanic action of fly ash particles. The pozzolanic action results from the reaction of $Ca(OH)_2$ with SiO_2 and Al_2O_3 from fly ash. The smaller spherical particle of fly ash could be easily dispersed into the blended cement paste and make the cement paste matrix more homogeneous.

3.16 **For a mix design of proportion 1 : 2 : 3.6 (by mass) with w/c ratio of 0.45 and air content 3% of the concrete volume, calculate the weights of water, cement, fine aggregate and coarse aggregate to make 1 m³ of concrete. The specific gravities of cement, F.A. and C.A. are 3.15, 2.65 and 2.6 respectively.**

[10 marks : 2012]

Solution:

Let weight of cement $= w_c$ kg,

Weight of fine aggregate, F.A. $= w_f$ kg,

Weight of coarse aggregate, C.A. $= w_a$ kg

Given, $w_c : w_f : w_a = 1 : 2 : 3.6$l; Water cement ratio, $w_w/w_c = 0.45$; Air content = 3 %; Total volume of concrete = 1 m³; Specific gravity of cement $G_c = 3.15$; Specific gravity of F.A. $G_f = 2.65$; Specific gravity of C.A. $G_a = 2.6$

Now, Net volume of concrete = Total volume of concrete – volume of air in concrete

$$= 1 - \frac{3}{100} \times 1 = 0.97\,\text{m}^3$$

∴ Net volume of concrete $= 0.97\,\text{m}^3$

We have
$$\frac{w_c}{w_f} = \frac{1}{2} \text{ and } \frac{w_c}{w_a} = \frac{1}{3.6}$$

We know that

$$\text{Net volume of concrete} = \text{volume of water} + \text{volume of solids}$$

$$\Rightarrow \quad 0.97 = \frac{\text{weight of water}}{\text{density of water}} + \frac{\text{weight of solids}}{\text{density of solids}}$$

$$\Rightarrow \quad 0.97 = \frac{0.45 \times w_c}{1000} + \left[\frac{w_c}{G_c \times 1000} + \frac{w_f}{G_f \times 1000} + \frac{w_a}{G_a \times 1000}\right]$$

$$\Rightarrow \quad 0.97 \times 1000 = 0.45 w_c + \left[\frac{w_c}{G_c} + \frac{w_f}{G_f} + \frac{w_a}{G_a}\right]$$

$$\Rightarrow \quad 970 = 0.45 w_c + \left[\frac{w_c}{G_c} + \frac{2w_c}{G_f} + \frac{3.6w_c}{G_a}\right]$$

$$\Rightarrow \quad 970 = w_c\left[0.45 + \frac{1}{3.15} + \frac{2}{2.65} + \frac{3.6}{2.6}\right]$$

$$\Rightarrow \quad w_c = \frac{970}{2.9068}$$

$$\therefore \quad w_c = 333.70 \text{ kg}$$

$$w_w = 0.45 \times 333.70 = 150.165 \text{ kg}$$

and $$w_f = 2 \times w_c = 2 \times 333.70 = 667.40 \text{ kg}$$

$$\Rightarrow \quad w_a = 3.6 \times w_c = 3.6 \times 333.70 = 1201.32 \text{ kg}$$

3.17 **Write a short answer to the following:**

(i) When testing cements as per Indian Standards, it may not be possible to decide on their quality only on the basis of the compressive strength results. Why?

(ii) What are some of the steps that can be taken to prevent the formation of 'thermal cracks' is mass concrete?

(iii) Explain the concept of 'maturity' of concrete.

[15 marks : 2013]

Solution:

(i) Quality of cement cannot be totally relied upon only on the basis of compressive strength. Various other factors also affect the quality of cement apart from compressive strength.

For instance, "fineness of cement" determines the rate of gain of strength. Presence of lime and magnesia in cement determines the soundness of cement. Variation in temperature decides the setting time of cement.

Desired quality of cement depends upon:

(a) Chemical composition of cement

(b) Fineness of cement

(c) Temperature of hydration

(d) Soundness of cement

(e) Heat of hydration of cement

(ii) **Thermal Cracks:** During concrete mixing, heat of hydration is generated causing a rise in concrete temperature. The temperature in the middle region of concrete is often higher than the temperature of surface.

Due to temperature gradient across the concrete thickness, the surface is in tension and concrete in middle region is in compression. If overall tensile strain which is maximum at the surface, exceeds the ultimate tensile stress capacity of hardening concrete then cracks are likely to occur.

Such thermal cracking can be prevented by using low percent of C_3A, C_3S, using Pozzolanic admixtures, mixing ice with concrete to control the mixing temperature etc. None delaying the process of cooling, insulating the concrete also helps in reducing thermal cracks.

(iii) **Maturity of concrete:** Strength of concrete depends on period of curing and the temperature during the curing process.

Thus concrete strength = f(curing period, curing temperature)

The quantity "curing period × curing temp" is called as maturity of concrete.

It is measured in "°C hours" or °C days".

Strength of concrete is found to increase with increase in maturity.

For OPC, maturity factor recommended is 4200°C hours.

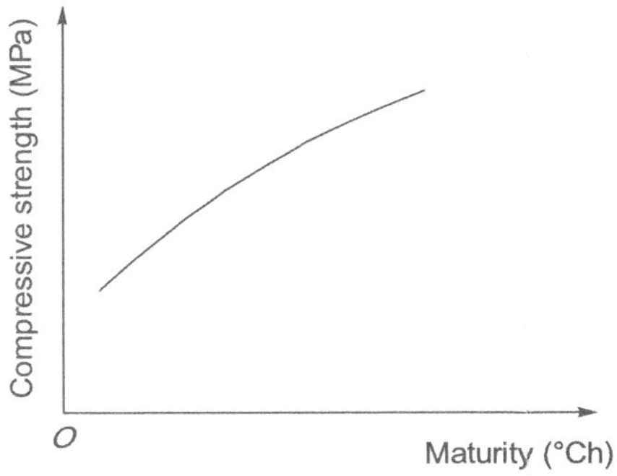

3.18 **A site is using a concrete where the unit content of water, cement, sand and coarse aggregate are 180 kg/m³, 360 kg/m³, 700 kg/m³ and 1210 kg/m³, respectively. For a portion of the work, the Engineer permits volume batching and rectangular boxes measuring 35 cm by 45 cm have to be fabricated to measure coarse aggregate.**

Assume the following:

(i) The mixer available will mix concrete with one bag of cement (of 50 kg) at one time.

(ii) 2 (two) boxes of coarse aggregate will be used in a batch (as defined above).

(iii) When filled in a normal manner, the void content in the box is 40%.

(iv) Specific gravity of the coarse aggregate is 2.75.

Find the height of the box.

[10 marks : 2013]

Solution:

For 1 m³ cement used = 360 kg

When 50 kg cement is used. Mix volume $= \dfrac{50}{360} \times 1 = 0.1388 \text{m}^3$

Now, for 1 m³ coarse aggregate used = 1210 kg

So, for mix volume of 0.1388 m³,

Weight of coarse aggregate used, $W_{CA} = 1210 \times \dfrac{0.1388}{1} = 167.948 \text{kg}$

Specifc gravity of coarse aggregate, $G_{CA} = 2.75$

$\therefore$ Volume of $CA = \dfrac{W_{CA}}{G_{CA} \times 1000} = \dfrac{167.948}{2.75 \times 1000} = 0.061072 \text{m}^3$ ...(i)

Given that 2 boxes of coarse aggregate is used.

Let h be the height of the box.

$\therefore$ Volume of box $= 0.35 \times 0.45 \times h = 0.1575(h)$ m³

Now, given that, void content in the box = 40%.

$\therefore$ Solid content in the box = 60%

Two boxes are used.

$\therefore$ Total volume of coarse aggregate used $= (0.6 \times 0.1575(h)) \times 2 = 0.189\,(h)$ m³ ...(ii)

Now, equating (i) and (ii), we get

$$0.189\,h = 0.061072$$

$$\therefore \quad h = 0.32313 \text{ m} = 32.313 \text{ cm} \approx 35 \text{ cm}$$

3.19 **List some of the important factors that are known to affect the readings taken while carrying out non-destructive testing of concrete using Schmidt rebound hammer and explain the effect briefly.**

[5 marks : 2014]

Solution:

The rebound hammer test is indirect method of determining the strength of concrete, i.e., compressive strength of concrete is determined from the calibration relationship between parameter of non-destructive testing and strength of concrete.

The basic factors affecting the hammer test readings are:

(a) **Concrete composition:** Like cement type, cement content and coarse aggregate type, affects hammer rebound value. It also affects the usage of basic calibration relationship for concrete strength determination.

(b) **Shape, size and smoothness of specimen:** It also affect the hammer rebound value i.e., greater the smoothness greater will be the reading.

(c) **Concrete humidity:** As in case of wet concrete cubes, the hardness of concrete surface is lower than dry surface. A wet surface test may lead to an under estimation of strength of upto 20%.

(d) **Hammer position:** It is normally reflected in the tables for determining the concrete strength (for the same rebound value depending on the location of hammer indicates various compressive strength of concrete)

(e) **Age of concrete:** After a certain time, the hardness of the surface layer decreasing with increasing age due to physico-chemical process in the hardened concrete mass and it also affect hardness strength relationship of concrete. Most often it is known as process of carbonation. Therefore, for testing it is desirable to remove loose mortar and obtain heavily textured of concrete by grounding.

(f) **Position of hammer** with respect to gravity shows higher value if return of hammer is supported by gravity. Thus, for a given concrete rebound number is smaller for a floor than a soffit.

3.20 **Explain different types of vibrators and the application of each in construction works.**

[5 marks : 2015]

Solution:

1. **Needle Vibrator:** It is also called as immersion vibrator and is the most commonly used type of vibrator. The diameter of the needle varies from 40 mm to 100 mm depending on the spacing required between the reinforcing bars. The needle is inserted into the concrete at intervals of not more than 600 mm.

2. **External or Shutter Vibrator:** It is used at locations where concrete is required to be vibrated within the form work. These vibrators are clamped to the form work at predetermined locations so as to vibrate the concrete and the form work. Power requirement for these vibrators is high as compared to needle vibrators.

3. **Surface Vibrator:** This is suitable for vibrating shallow depth of concrete like slab etc. These vibrators are placed directly on the concrete mass. Usually the depth of concrete should not exceed 250 mm.

4. **Vibrating Table:** It is a rigidly built steel table mounted on springs. This vibrating table is driven by an electric motor. This type of vibrator is most suitable for vibrating very stiff concrete mixes for manufacture of precast concrete elements and also for testing of samples in the laboratory.

3.21 What is an admixture? List four different mineral and chemical admixtures.

[10 marks : 2016]

Solution:

Admixture: A material other than water, aggregates, or cement that is used as an ingredient of concrete or mortar; to control setting and early hardening, workability, or to provide additional cementing properties is called an *Admixture*. Over decades, attempts have been made to obtain concrete with certain desired characteristics such as high compressive strength, high workability, and high performance and durability parameters to meet the requirement of complexity of modern structures. The properties commonly modified are the heat of hydration, accelerate or retard setting time, workability, water reduction, dispersion and air-entrainment, impermeability and durability factors.

Chemical admixtures: Accelerators, retarders, water-reducing agents and super plasticizers.

Mineral admixtures: Fly-ash, blast-furnace slag, silica fume and rice husk ash.

3.22 The data of material for making a cement concrete mix may be assumed as follows:
(i) Water-cement ratio by mass = 0.46
(ii) Entrained air = 2%

S. No.	Material	Specific gravity	Bulk density (kg/m^3)	Proportion in mix by dry volume
1	Cement	3.15	1500	1 part
2	Fine aggregate	2.61	1620	1.359 part
3	20 mm nominal size crushed coarse aggregate	2.70	1530	2.79 part

Determine the following for cement concrete:

1. Absolute volume of fully compacted fresh concrete (ignoring air content) produced by one bag of cement of 50 kg.
2. Cement content per cubic metre of concrete (ignoring air content) produced by one bag of cement of 50 kg.
3. Quantity of materials to make one cubic metre of concrete.

[20 marks : 2017]

Solution:

1. Let volume of cement be V_C m^3, volume of sand be V_s m^3, volume of aggregates be V_a m^3, weight of cement be W_c kN, weight of sand be W_s kN, weight of aggregate be W_a kN, weight of water be W_w kN.

Given $$V_c : V_s : V_a = 1 : 1.359 : 2.79$$

$$\Rightarrow \quad \frac{V_c}{V_s} = \frac{1}{1.359} \Rightarrow V_s = 1.359\,V_c$$

Also, $$\frac{V_c}{V_a} = \frac{1}{2.79}$$

$$\Rightarrow \quad V_a = 2.79\,V_c$$

Also, $$\frac{W_w}{W_c} = 0.46 \text{ and entrained air} = 2\%$$

Now, we know that

$$\text{Net volume of concrete} = \text{Total volume} - \text{entrained air volume}$$
$$= 1 - 0.02 = 0.98 \text{ m}^3$$

But, Net volume of concrete = volume of water + volume of solids

$$0.98 = \frac{\text{Weight of water}}{\text{Density of water}} + \frac{\text{Weight of solids}}{\text{Density of solids}}$$

$$\Rightarrow \quad 0.98 = \frac{0.46W_c}{9.81} + \left[\frac{W_c}{G_c \times 9.81} + \frac{W_s}{G_s \times 9.81} + \frac{W_a}{G_a \times 9.81}\right]$$

2. $$50 \text{ kg of cement} = \frac{50}{1500} = 0.333 \text{ m}^3 \text{ of cement}$$

Which produces 0.1196 m^3 of concrete (ignoring air content)

$$\therefore \quad \text{Cement content} = \left(\frac{0.0333}{0.1196}\right) = 0.279 \text{ m}^3\text{/cubic metre of concrete}$$

$$\Rightarrow \quad 0.98 \times 9.81 = \frac{0.46 \times 1500 \times 9.81 \times V_c}{1000} + \left[\frac{1500 \times 9.81 V_c}{G_c \times 1000} + \frac{1620 \times 9.81 \times V_s}{G_s \times 1000} + \frac{1530 \times 9.81 \times V_a}{G_a \times 1000}\right]$$

$$\Rightarrow \quad 0.98 \times 10^3 = 0.46 \times 1500 V_c + \left[\frac{1500V_c}{3.15} + \frac{1620V_s}{2.61} + \frac{1530 \times V_a}{2.70}\right]$$

$$\Rightarrow \quad 0.98 \times 10^3 = 690\, V_c + 476.19\, V_c + 620.68 \times 1.359\, V_c + 566.67 \times 2.79\, V_c$$

or, $V_c = 0.273 \text{ m}^3$

$\Rightarrow$ $V_s = 0.371 \text{ m}^3$

$\Rightarrow$ $V_a = 0.762 \text{ m}^3$

Weight of cement in kg = 0.273 × 1500 = 409.5 kg

Weight of sand in kg = 0.371 × 1620 = 601.02 kg

Weight of coarse aggregate in kg = 0.762 × 1530 = 1165.86 kg

3. 409.5 kg cement produces = 0.98 m^3 concrete (ignoring air content)

$$50 \text{ kg cement will produce} = \frac{50}{409.5} \times 0.98 = 0.1196 \text{ m}^3 \text{ (full compacted concrete)}$$

3.23 **(i) Name the tests to measure workability of fresh concrete.**
(ii) What is a super plasticizing admixture?
(iii) What are the three basic qualities of high performance concrete? Discuss the contradictions.
(iv) What is creep of concrete?

[20 marks : 2017]

Solution:

(i) Tests to measure workability of fresh concrete: (IS : 1199)

1. Slump test
2. Compacting Factor Test
3. Vee-Bee consistometer method
4. Flow test
5. Kelly Ball test

(ii) **Super plasticizing admixture:** Superplasticizers are hydrodynamic lubricators which impart high workability by reducing friction between the grains or by reducing the amount of water to be added. They are improved versions of plasticizers interact both physically and chemically with cement particles. The mechanism of action of superplasticizers is same as that of plasticizer. Superplasticizers are anionic in nature and impart a negative charge to the cement particles, causing them to repel each other.

Effects of superplasticizers:

1. The workability of the superplasticised concrete decreases more with time than that of ordinary concrete.
2. The strength, water cement ratio and creep of concrete with the additive are the same as that of concrete before the addition of plasticizer while shrinkage and surface absorption are reduced slightly and then resistance to thawing and freezing is improved.
3. By addition of superplasticizers it is possible to obtain same strength with a reduced cement content.
4. The water-cement ratio of the superplasticised concrete may be reduced upto 30 percent to obtain the initial equality of workability.

Examples: Sulphonated melamine formaldehyde condensates, napthalene sulphonate formaldehyde condensates, modified lignosulphonate etc.

(iii) **Basic qualities of high performance concrete:**

1. **Workability :** In early stages of its development, the HSC has a tendency to be sticky and stiff due to large amount of fines (high cement content and pozzolana, a low water cement ratio and a normal water reducing admixture.
2. **Strength :** The most important point about HSC is their capacity to develop strength at a rapid rate without steam curing. Concrete can develop 20 to 27 MPa on normal curing within 24 hours and the ultra HSC can develop 42 MPa in 12 hours and 64 MPa in 24 hours.
3. **Microstructure, stress-strain relation, creep and fracture:** As a result of reduction in the size and number of micro-cracks in HSCs, its stress-strain relation, creep and fracture behaviour is different from the normal concrete.
4. **Durability :** It has been found that primarily due to low permeability HSCs exhibit excellent durability to various physical and chemical agents that are normally responsible for concrete deterioration. Due to high cement content, thermal cracking can lead to durability problem in structures using HSC.

(iv) **Creep of concrete:** The continued deformation under applied loads is called creep. It increases strain in concrete with time under sustained stress. This is also known as plastic flow or time yield. The rate of creep decrease with time and the creep strains at five years are taken as ultimate values.

Creep increases rapidly with stress.

The deformation of hardened concrete is shown.

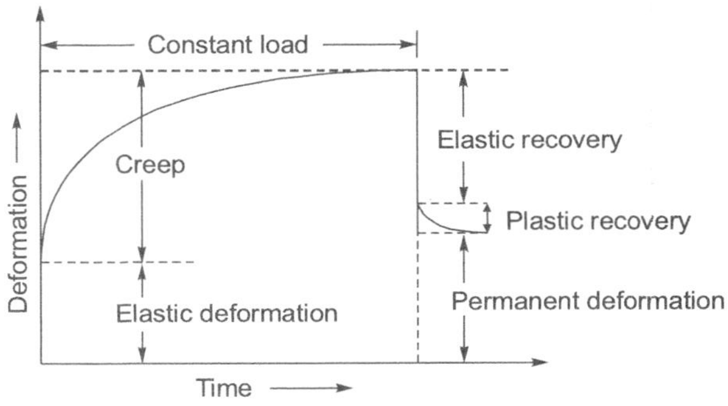

3.24 The strength of a sample of fully matured concrete is found to be 50 MPa. Find the strength of identical concrete at the age of 7 days when cured at an average temperature of 25°C during day time and 15°C during the night time. Take constants A and B as 32 and 54 respectively. These are the Plowman's Coefficients for Maturity Equation.

[12 marks : 2019]

Solution:

Maturity, $$M = \sum_{0}^{t}(T - T_0)\Delta t$$

where, T = Average concrete temperature

T_0 = Datum temperature (–11°C)

Δt = Time interval

$$M = 7 \times 12\ [25 - (-11)] + 7 \times 12\ [15 - (-11)]$$

$$M = 5208°\text{C-hours}$$

Strength of maturity, $$f = A + B\log_{10}\left(m \times 10^{-3}\right)$$

Where A and B are Plowman's coefficient

A = 32

and B = 54

$$f = 32 + 54\log_{10}\left[\frac{5208}{10^3}\right] = 70.7\%$$

Strength of fully matured concrete = 50 MPa

So strength of concrete after 7 day = 50 × 0.707 = 35.35 MPa

3.25 Calculate the quantities of ingredients required to produce one cubic metre of structural concrete. The mix is to be used in proportions of 1 part of cement to 1.42 parts of sand to 2.94 parts of 20 mm nominal size crushed coarse aggregate by dry volumes with a w/c ratio of 0.49 (by mass). Assume the bulk densities of cement, sand and coarse aggregate to be 1500, 1700 and 1600 kg/m^3 respectively. The percentage of entrained air is 2.0. Take specific gravity of cement, sand and coarse aggregate as 3.15, 2.6 and 2.6 respectively.

[10 marks : 2019]

Solution:

Volume of concrete = Vol. of air + Vol. of water + Vol. of solids of [Cement + Sand + Coarse aggregate]

$$1 = 0.02 + V_W + [V_{CS} + V_{SS} + V_{CAS}]$$

Assume volume of cement = V_C (dry volume)

So, Weight of cement = $V_C \times 1500$

$$\text{Volume of cement solids} = \frac{V_C \times 1500}{3.15 \times 1000}$$

Using same concept for sand of coarse aggregate

$$1 = 0.02 + \frac{V_C \times 1500 + 0.49}{1000} + \frac{V_C \times 1500}{3.15 \times 100} + \frac{1.42 \times V_C \times 1700}{2.6 \times 100} + \frac{2.94 \times V_C \times 1600}{2.6 \times 1000}$$

$$V_C = 0.248\ \text{m}^3$$

$$V_{Sand} = 1.42 \times 0.248 = 0.352\ \text{m}^3$$

$$V_{CA} = 2.94 \times 0.248 = 0.729 \text{ m}^3$$

$$\text{Mass of cement} = 0.248 \times 1500 = 372 \text{ kg}$$

$$\text{Mass of sand} = 598.67 \text{ kg}$$

$$\text{Mass of coarse aggregate} = 1166.59 \text{ kg}$$

$$\text{Mass of water} = 372 \times 0.49 = 182.28 \text{ kg}$$

3.26 **How is workability of concrete defined as per Indian Standard Specification IS 1199 - 1959? Briefly explain the method of measurement of workability through compacting factor test.**

[6 marks : 2020]

Solution

As per IS 1199-1959 workability defined as the property of concrete which determines the amount of useful internal work necessary to produce complete compaction.

Compaction factor test:

- In the compaction factor test, the degree of workability of concrete is measured in terms of internal energy required to compact the concrete thoroughly.
- A compaction factor of 0.95 represents flowing concrete having high workability; 0.92 plastic concrete having medium workability; 0.85 stiff plastic concrete having low workability and a compaction factor of 0.75 represents stiff concrete having very low workability.
- The compacting factor test is designed primarily for use in the laboratory but it can also be used in the field.
- It is more precise and sensitive than the slump test and is particularly useful for concrete mixes of very low workability and is normally used when concrete is to be compacted by vibration like, dry concrete which is insensitive to slump test.
- The sample of concrete to be tested is placed in the upper hopper up to the brim.
- The trap-door is opened so that the concrete falls into the lower hopper.
- Then the trap-door of the lower hopper is opened and the concrete is allowed to fall into the cylinder.
- In the case of a dry mix, it is likely that the concrete may not fall on opening the trap-door. In such a case, a slight packing by a rod may be required to set the concrete in motion.
- The excess concrete remaining above the top level of the cylinder is then cut off with the help of plane blades supplied with the apparatus.
- The outside of the cylinder is wiped clean.
- The concrete is filled up exactly up to the top level of the cylinder.
- It is weighed to the nearest 10 gm.
- This weight is known weight of partially compacted concrete.
- The cylinder is emptied and then **refilled with the concrete from the same sample** in layers approximately 5 cm deep.
- The layers are heavily rammed or preferably vibrated so as to obtain full compaction.
- This weight is known as weight of fully compacted concrete

$$\text{C.F.} = \frac{\text{Weight of partially compacted concrete}}{\text{Weight of fully compacted concrete}}$$

3.27 (i) How do super plasticizers increase the mobility and make the cement concrete flow? What is the advantage of using fly ash in concrete for massive dam construction work?

(ii) What are the disadvantages of destructive methods of testing concrete and advantages of nondestructive testing of concrete?

[10 + 10 = 20 Marks : 2021]

Solution

(i) Superplasticisers are hydrodynamic lubricants which impart high workability by reducing friction between the grains or by reducing the amount of water to be added. Superplasticizers are principally surface reactive agents (surfactants). They confer negative charge on individual cement particles (and also its hydrated particles) such that they are kept in a dispersed or suspended state due to inter-particle repulsion. Thus they confer high mobility to the particles.

Following are the advantages of using fly ash in concrete for massive dam construction work.

- Improved workability with lesser amount of water.
- Lower heat of hydration and thermal shrinkage.
- Improved resistance to attack from salts and sulfates from soils and sea water.
- Reduced susceptibility to dissolution and leaching of calcium hydroxide.
- Reduce permeability
- Lower costs

(ii) **Disadvantages of destructive test:**

1. A large number of specimens are required which could be tested to destruction, at various ages.
2. Test results are not very reliable as all specimens are not identical in quality with the entire mass of concrete.
3. There is much delay in obtaining the results.
4. It is impossible to obtain requisite information on in-situ concrete without damaging the concrete.
5. Crushed samples used for destructive testing creates high amount of debris which in-turn degrades the environment.

Advantages of destructive test:

1. The measurement can be done on concrete in-situ and thus representative samples are not required.
2. Non-destructive testing makes its possible to study the variation in quality of concrete with time and external influences.
3. In N.D.T method the concrete is not loaded to destruction. Its quality is judged by measuring certain of its physical properties, which are related to its quality.
4. In N.D.T there is no wastage of material as in destructive methods of testing.
5. Non-destructive methods are quick and can be performed both in laboratory and in-situ with convenience.
6. Non-destructive test (NDT) can be performed on fresh (green) as well as hardened concrete with equal ease and are described as follows.

3.28 Withe the help of neat sketch of a typical grading curve, describe the term 'gap-graded aggregate' and the adverse effects of using such types of aggregates in concrete.

[6 marks : 2023]

Solution

- A mix of aggregate which doesn't carry a particular range of grain size or in a aggregate mix if a particular range of grain size is absent, then it is known as gap graded aggregate.
- It contains higher proportion of coarse aggregate and low percentage of fine aggregate which results in lesser surface area. Therefore, it requires lesser cement and lower w/c ratio.
  As it contains higher proportion of coarse aggregate hence chances of segregation will be more and may create problem of blockage if it is transported through reinforcements. Hence it is recommended to make low workable mix with gap graded aggregate and then compact it by vibration.

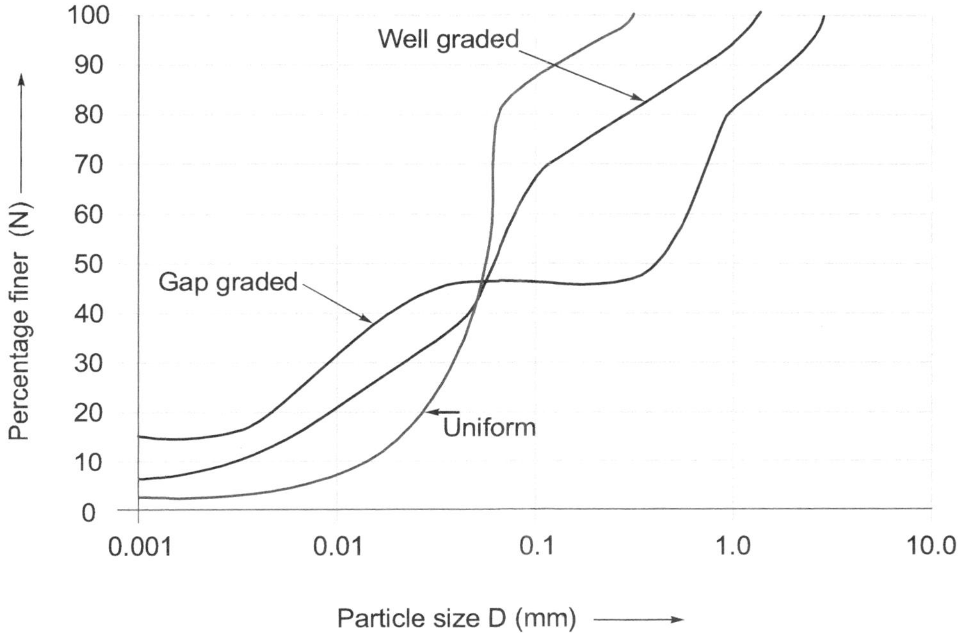

3.29 Explain the mechanism of alkali-aggregate reaction in concrete. How can it be controlled?

[20 marks : 2023]

Solution

Alkali aggregate reaction is reaction between Silica in aggregate and alkali in cement that results into extensive expansion and may lead to complete disruption and disintegration of concrete, also known as cancer of concrete. A reactive aggregate in finely grounded state will inhibit the reaction which exits in range of 150 mm to 300 mm especially.

Reaction between cement and aggregate can be of two types: reaction between the alkalis of the cement and either silica or carbonates in the aggregate. The former appears to be more common. The alkali-aggregate reaction takes place only in the presence of water or water vapour. The water forms strong caustic solute with alkalis of cement. This caustic liquid attacks reactive silica to form alkali-silica-gel of unlimited swelling type. When the conditions congenial, progressive manifestation of concrete, particularly in thin concrete sections, results cracks and eventual failure of concrete structures. Conspicuous effect may not be notified in mass concrete sections. The formation of pattern cracks result in subsequent loss in strength and elasticity because of stresses induced by the growth of silica gel. Many destructive forces become operative on the concrete disrupted by alkali-aggregate reaction which further hasten the total disintegration of concrete

The carbonate rocks which are reactive range from pure limestone to pure dolomites and those rocks which have excess of Mg or Ca ions over the ideal proportion are likely to give trouble. Apparently there are several different types of alkali-carbonate reactions not all of which are expansive and harmful. The reaction zones are up to 2 mm wide around the aggregate particles and are visible on polished surface in thin petrographic sections.

Control of Alkali-aggregate Reaction

- By Selecting Non-reactive Aggregate:
- **By Using Low Alkali Cement:** Cements with alkali less than 0.6 per cent should be used.
- **By Controlling Moisture :** Old concrete should not be allowed to come in contact with water. The best way is to apply mortar with water proofing agents on concrete surface.
- **By Puzzolanas:** The aggregates are found to be reactive when they contain silica in a particular proportion and fineness. When fly-ash or surkhi or crushed stone dust is added, this optimum condition of silica being in particular proportion and fineness is disturbed and aggregate turns out to be innocuous.
- **By Air Entraining Agents:** The alkali-silica-gel imparts osmotic pressure over the set cement gel and this is mainly responsible for formation of cracks. When air entraining agents are added they absorb the osmotic pressure and control the expansion.

3.30 **(i) What is gel-space ratio? How is it estimated? Discuss its effect on strength of concrete.**

(ii) What are the factors affecting durability of concrete?

[10 + 10 = 20 marks : 2023]

Solution:

(i)

Gel-space ratio is defined as the ratio of volume of hydrated cement paste to the sum of the volumes of the hydrated cement and that of the capillary pores. Power and Brownyard established a relationship between the strength and gel-space ratio as given below.

$$S = 240x^3$$

where S = strength of concrete x = gel-space ratio

240 = intrinsic strength of gel in N/mm^2 for the tyep of cement and specimen used

NOTE: Strength calculated on the basis of gel-space ratio is independent of the age, whereas of that calculated on the basis of w/c ratio depends upon the age.

A typical curve relating gel-space ratio to the compressive strength is shown in Figure.

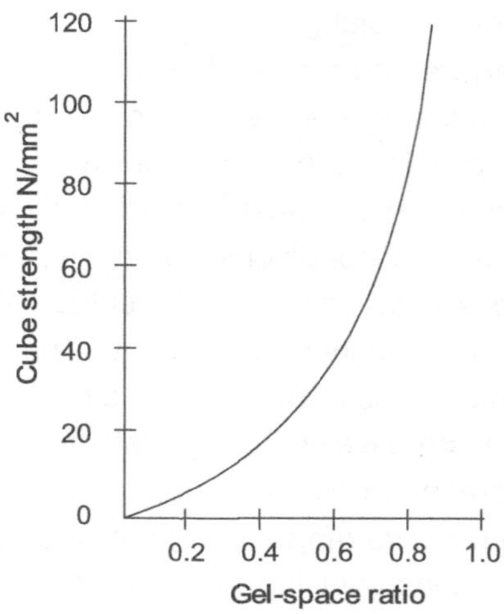

Calculation of Gel Space Ratio:

For Complete Hydration

Let C = weight of cement in g, V_c = specific volume of cement ml/g, and W_0 = volume of mixing water in ml.

Assuming 1 ml of cement on hydration produces 2.06 ml of gel.

$$\text{Volume of gel} = C \times 0.319 \times 2.06 = 0.657\ C$$

$$\text{Space available} = C \times 0.319 + W_0$$

$$\text{Gel-space ratio} = \frac{0.6571\,C}{0.319\,C + W_0}$$

For Partial Hydration

Let α = fraction of cement that has hydrated.

$$\text{Volume of gel} = C \times \alpha \times 0.319 \times 2.06 = 0.657\ C\alpha$$

$$\text{Space available} = C \times 0.319 \times \alpha + W_0$$

$$\text{Gel-space ratio} = \frac{0.6571\,C\alpha}{0.319\,C\alpha + W_0}$$

(ii)

Major factors affecting durability are as follows:

(a) **Permeable surface:** Due to deficiency of compaction of concrete the entrapped voids are left over finished concrete surface, which are non-uniformly distributed and interconnected. Permeable surface can also be caused due to leaching out Portlandite during hydration.
- Permeable surface allows all form of deterioration in concrete like all external chemical can penetrate into concrete in liquid form.
- Through permeable surface, effect of frost action can accelerate.

(b) **Excess water-cement Ratio:** Excess water-cement Ratio results into high porosity of mix which results in high volume changes which causes heavy shrinkage cracks.

(c) **Carbonation:** Concrete is alkaline in nature and has an initial pH value of 12-13. So long as the steel reinforcement is in alkaline environment, its corrosion cannot start. However, carbon dioxide present in the atmosphere reacts with concrete in presence of water (the moist conditions). Infact, in the presence of moisture, carbon dioxide changes to carbonic acid and attacks the concrete. $Ca(OH)_2$ present in concrete is converted to $CaCO_3$ by CO_2 causing reduction in pH of the concrete. The concrete thus turns acidic. Once pH value in the concrete-cover drops below 10, the corrosion of steel reinforcement begins due to the reduction in the alkaline environment. Because of the corrosion of steel reinforcement, its volume increases and consequently the concrete cracks and spalling takes place. However, if concrete is dense and well compacted, carbonation will reduce porosity, specific surface of cement paste and water permeability well which consequently increases sulphate and aggressive ion penetration.

Other than above, frost action, sulphur attack, attack of mineral oil, attack of organic acid, attack of sugar, mineral oil, and temperature changes will affect durability of concrete.

Note: If coefficient of thermal expansion differ by more than 5.4×10^{-6}/°C between two material of concrete, durability will be affected by freezing and thawing.

3.31 (i) **What are the factors affecting rheological properties of concrete?**

[10 marks : 2023]

(ii) **How are the properties of concrete affected by seawater and industrial wastewater, if they are used for making the concrete?"**

[10 marks : 2023]

Solution:

(i)

- Consistency and homogeneity of concrete mixture signifies Rheological behaviour of concrete.
- Fluid behaviours of concrete is similar to Bingham plastic fluid.

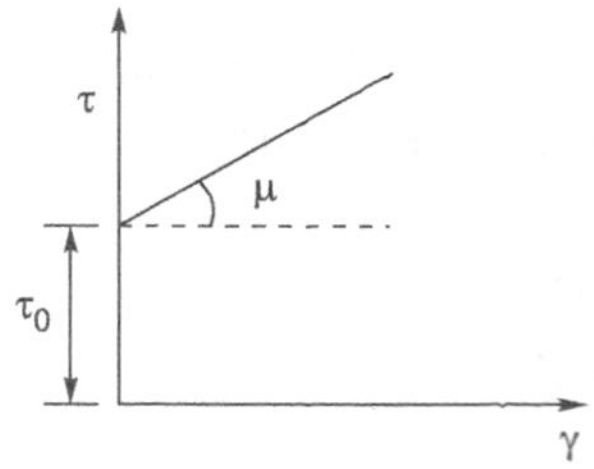

$\tau = \tau_0 + \mu\gamma$
τ = shear stress
γ = rate of shear strain

- Consistency of mix controls over viscosity of mix. Homogeneity signifies internal resistance against segregation, bleeding, honey combing etc. Homogeneity gives significance to shear stress development.

Rheology of concrete deals with flow and deformation of material which is concerned with relationship between stress (shear), shear stain, rate of shear strain and time.

Rheology indirectly signify the stability of concrete mix, flow ability (viscosity, cohesion, internal friction).

Factors affecting Rheological Properties

1. **Mix Proportions:** Excess amount of coarse aggregate in a mix lacks in sufficient mortar required to fill the voids; there will be a loss of cohesion and mobility of the whole mix. If amount of fine aggregate is in excess or air is entrained, the cohesion increases and affects the mobility of the mix. Moreover, more the fine aggregate, more will be the cement requirement (because of the greater surface area of the mix) to obtain the desired mobility; the drying shrinkage may increase resulting in cracks.
2. **Consistency:** More than required water content (slump) to produce a workable mix results in greater fluidity. Water content more than that needed will not improve the rheological properties of concrete. Segregation and bleeding in concrete have higher chance to occur with increase in fluidity. A very low water content reduces the compactability, resulting in difficulties in placement and compaction.
3. **Hardening and Stiffening:** Use of rapid hardening cement, cement deficient in gypsum and use of accelerating admixtures, increase the rate of hardening, and consequently the mobility of concrete reduces.
4. **Aggregate Shape and Texture:** The rough aggregate particles have higher percentage of voids. Consequently, more mortar, higher fine aggregate content and correspondingly higher water content is required to fill the voids. Similarly angular fine aggregate increases internal fraction in the concrete mix and require higher water contents than well rounded aggregates.
5. **Aggregate Grading:** Well graded aggregates results in good workability; this helps in filling the voids systematically. Gap graded aggregate affects void system and workability. The increase in the fine aggregate requires more water and result in sticky concrete mix.
6. **Maximum Size of Aggregate:** For same workability, an increase in the maximum size of aggregate reduces the fine aggregate requirement so as to maintain a given workability and thereby reduces the surface area to be wetted.
7. **Admixtures:** A variety of admixtures are added in concrete mixes for specific purposes. Rheology of concrete is significantly influenced by the use of plasticizers and super-plasticizers, air-entraining agents, and accelerators and retarders.

(ii)

Sea Water : Sea water should be avoided to make the concrete due to following reasons:

- Salt in sea water cause efflorescence and unsoundness (due to Alkali-aggregate reaction).
- The salt of chloride increase the chances of corrosion.
- Due to presence of salt of chloride and sulpahte both, ultimate strength reduces however 3 day strength may be more due to accelerator salts.
- Sea water can be suitable for PCC.

Industrial Waste Water

- Most waters carrying industrial waste have less than 3,000 ppm of total solids. When such water is used as mixing water in concrete, the reduction in compressive strength is generally less than about 10 per cent. Waste waters from paint factories, coke plants, chemical and galvanizing plants may contain harmful impurities. It is advisable to test any waste water that contains even few hundred parts per million-of unusual solids before using it for mining concrete.
- One way of using sewage containing large organic matters (say 400 ppm) is to dilute it in a good disposal system to reduce the concentration to about 30 ppm or less, an amount too low to have any significant effect on concrete strength.

3.32 Explain roller-compacted concrete. What are the advantages of roller compacted concrete?

[6 marks : 2023]

Solution

Roller-compacted concrete (RcC) presents a relatively recent development in construction technology of dams and locks. It is based on the concept that a no-slump concrete mix transported, placed, and compacted with the same construction equipment that is used for earth and rockfill dams can meet the design specifications for conventional mass concrete. It is a mixture of aggregates, cement (with or without puzzolanas), water and sometimes water reducing admixtures proportioned to support external compaction equipment. The stiff (dry or lean), zero slump concrete mixture has consistency of damp gravel. The air entraining, water reducing and set controlling admixtures are effective in reducing the vibration time required for full consolidation. This lowers the entrapped air-void content, increases strength and lowers the permeability of concrete.

RcC differs from conventional concrete principally in its consistency requirement. For effective consolidation, concrete must be dry to prevent sinking of the vibratory roller equipment but wet enough to permit adequate distribution of the binder mortar throughout the material during the mixing and vibratory compaction operations. From the stand point of workability, fly ash is commonly included in RcC mixtures.

RcC mixes contain less cement and cementitious material paste (250-350 kg/m^3) and significantly high fly-ash proportion (25-70%) by weight of cementitious material. The 28 day strength is comparable with OPC.

Since consistency of RcC is low, the bonding of fresh concrete to hardened concrete is not adequate. The creep and thermal properties of RcC are within the range of those of conventional normal concrete.

Advantages

1. Cement consumption is lower because much leaner concrete can be used.
2. Form work costs are lower because of the layer placement method.
3. Cost of transporting concrete is lower because concrete can be hauled by end dump trucks.
4. The construction period can be shortened considerably.

4. Stones, Bricks and Bricks Masonry

4.1 Give the engineering properties of bricks. What are the characteristics of a good brick? Enumerate various factors that govern the thickness of brick wall.

[10 marks : 2000]

Solution:

The thickness of a brickwall depends upon the following factors:

(i) Super imposed load per unit length of the wall.

(ii) Overall height of the wall.

(iii) Height of the wall between floors.

(iv) Length of the wall between piers, buttresses, crosswalls

(v) Strength of brick masonry which depends upon the quality of bricks, quality of mortar and method of bonding.

The characteristics of good bricks are:

(i) The bricks should be table moulded, well burnt in kilns, copper coloured, free from cracks and with sharp and square edges. The colour should be uniform and bright.

(ii) The bricks should be uniform in shape and should be of standard size.

(iii) The bricks should give a clear metallic ringing sound when struck with each other.

(iv) The brick when broken or fractured should show a bright, homogenous and uniform compact structure free from voids.

(v) The brick should not absorb water more than 20 per cent by weight for first class bricks and 22 per cent by weight for second class bricks, when soaked in cold water for a period of 24 hours.

(vi) The bricks should be sufficiently hard. No impression should be left on brick surface, when it is scratched with finger nail.

(vii) The bricks should not break into pieces when dropped flat on hard ground from a height of a about one metre.

(viii) The bricks should have low thermal conductivity and they should be sound proof.

(ix) The bricks when soaked in water for 24 hours, should not show deposits of white salts when allowed to dry in shade.

(x) No brick should have the crushing strength below 5.5 MPa.

A brick is generally subjected to the following tests to find out its suitability for the construction work:

(i) Absorption

(ii) Crushing strength

(iii) Hardness

(iv) Presence of soluble salts

(v) Shape and size

(vi) Soundness

(vii) Structure

Absorption: No bricks should absorb water more than 20% of its weight when kept immersed in it for 24 hours.

Crushing strength: As per **IS: 1077-1992**, the minimum crushing strength or compressive strength of bricks is 35 kg/cm^2.

Hardness: A scratch is made on the brick with the help of finger nail. If no impression is left on the surface, the brick is treated to be sufficiently hard.

Soundness: Two bricks are taken and struck with each other. The bricks should not break and a clear ringing sound should be produced.

Structure: A brick is broken and its structure is examined. It should be homogeneous, compact and free from any defects such as holes, lumps etc.

4.2 Write short notes on the following bricks:

(a) Table moulded (b) Refractory

(c) Pressed (d) Machine moulded

[10 marks : 2001]

Solution:

(a) **Table moulded bricks:** These bricks are moulded on a table of size 2 m × 1 m. The clay, mould, water pots, stock board, strikes and pallet boards are placed on this table. The bricks are moulded on this table and sent for the further process of drying. However the efficiency of moulding decreases gradually because of standing at the same place for long duration. The cost of brick moulding also increase when table moulding is adopted.

(b) **Refractory bricks:** A refractory material is able to stand very high temperatures without losing its shape. The bricks manufactured with the help of refractory materials are known as refractory bricks. The most common refractory material is fire clay hence the name fire bricks. These bricks are usually white or yellowish write in colour. The weight of a fire brick is about 30 to 35 N. The fire bricks can resist high temperature without softening or melting. Hence they are used for linings of interior surfaces of furnaces, chimneys, kilns, ovens, fireplaces etc. The compressive strength of these bricks varies from 200 to 220 MPa. The percentage of absorption for these bricks varies from 5 to 10. The three types of fire bricks are acidic bricks, basic bricks and neutral bricks.

(c) **Pressed bricks:** These are made by subjecting moist powdered clay to a great pressure of about 40 kg/cm^2. These bricks do not require drying. They are burnt directly

(d) **Machine moulded bricks:** These types of bricks are moulded with machines. Machine moulded bricks are economical when bricks in huge quantity are to be manufactured at the same in a short time. It is also helpful for moulding hard and strong clay. The machines are broadly classified into two categories viz. plastic clay machines and dry clay machines.

4.3 Explain the following:

(a) Grading zones of sand

(b) Grading of coarse aggregate

[10 marks : 2002]

Solution:

Grading zones of sand: The particle size distribution of an aggregate as determined by sieve analysis is termed as grading of aggregates. The proper grading of an aggregate produces dense concrete and needs less quantity of fine aggregate and cement paste.

The grading of coarse aggregate may be varied through wider limits than that for fine aggregate since it does not affect much the workability, uniformity and finishing qualities. It is difficult to control the grading of fine aggregate. If it is done or large scale, it can be effected by combining two or more different kinds of sand from different sources.

The sands are generally divided into different zones according to the percentage passing the IS : 600 μ sieve **IS:383-1970** classifies the sand into **four zones, I, II, III and IV** so that the range of percentage passing the 600μ sieve in each zone does not overlap. From grading zone I to IV, the fine aggregate becomes progressive finer and the ratio of fine to coarse aggregate should be progressively reduced. An aggregate is considered as belonging to the zone in which its percentage passing the 600μ sieve falls and it is allowed to fall outside the limits fixed for other sieves by not more than a total of 5%. For crushed stone sands the permissible limit on IS:150 micron sieve is increased by 20%. However this does not affect the 5% allowance permitted above as applied to other sieve sizes.

Grading of coarse aggregates: The grading of an aggregate is expressed in terms of percentage by weight retained on or passing through a series of sieves taken in order **80 mm, 40 mm, 20 mm, 10 mm, 4.75 mm for coarse aggregate.** These sieves are arranged in such an order that the square openings are half for each succeeding smaller size. The curve showing the cumulative percentages of the material passing the sieves represented on the ordinate with the sieve openings to the logarithmic scale represented on the abscissa is termed as the grading curve. The grading curve indicates whether the grading of a given sample conforms to that specified, or is too coarse or too fine, or deficient in a particular size.

The grading of the aggregate affects the workability which in turn controls the water and cement requirements, segregation and influences the placing and finishing of concrete. The main factors governing the desired aggregate grading are the surface area of aggregate, the relative volume occupied by the aggregate, the workability of the mix and the tendency to segregate. Surface area of the aggregate is an important parameter which varies with the shape but is inversely proportional to the particle size. The smaller the size of aggregate, the greater is the surface area per unit mass or unit volume. The aim must, therefore, be to have as large a maximum aggregate size as possible and to grade it down in such a way that the voids in the coarse aggregates are filled with the minimum amount of fine aggregate.

4.4 **Explain absorption and saturation factors with regard to bricks.**

[10 marks : 2002]

Solution:

Absorption of bricks depends upon their porosity which is due to voids of various sizes present in the bricks. Absorption can be broken into two distinct categories viz. absorption and initial rate of absorption (IRA).

Absorption of a brick is expressed as a percentage and is defined as the ratio of weight of water that is taken up in its body divided by the dry weight of the unit.

It is measured in two ways:

(i) **Cold water test:** The specimen is dried in a ventilated oven at a temperature of 105 to 115°C till it attains substantially constant mass (M_1). It is then kept immersed in clean water at a temperature of 27 ± 2°C for 24 hours. The specimen is removed and weighed (M_2) to determine the weight of water absorbed and water absorption is given by

$$\text{24 hour water absorption by weight } (W_{24}) = \frac{M_2 - M_1}{M_1} \times 100$$

(ii) **Boiling water test:** The specimen is dried in a ventilated oven at a temperature of 105 to 115°C till it attains substantially constant mass (M_1). The dried specimen is immersed in a tank such that the water can circulate freely on all sides of the specimen. Water is heated to boiling in one hour and boiled continuously for five hours. The water is allowed to cool to 27°C by natural loss of heat for 16 to 19 hours. The specimen is again weighed (M_3) and the water absorption percentage is given by

$$\text{5 hour water absorption by weight } (W_5) = \frac{M_3 - M_1}{M_1} \times 100$$

These tests are carried out to calculate the saturation factor, which is given by

$$\text{Saturation factor} = \frac{W_{24}}{W_5}$$

The saturation factor is helpful in predicting the durability of brick. The average water absorption (W_{24}) of common building bricks should not be more than 20% by weight upto class 12.5 MPa and 15% by weight for higher classes.

The initial rate of absorption (IRA) or suction is the rate of how much water a brick draws in during first minute after contact with water. The suction has a direct bearing on the bond between brick and mortar.

4.5 List four important tests conducted on bricks. Describe briefly the method used to determine compressive strength of bricks.

[10 marks : 2003]

Solution:

According to IS:3495-1992, bricks can be tested for:

(i) Compressive strength (ii) Water absorption

(iii) Efflorescence (iv) Warpage

Compressive strength test: The specimen brick is immersed in water for 24 hours. The frog of the brick is filled flush with 1: 3 mortar and the brick is stored under damp jute bags for 24 hours followed by immersion in clean water for three days. The specimen is then placed between the plates of the compression testing machine. Load is applied axially at a uniform rate of 14 MPa (140 kg/cm^2) and the maximum load at which the specimen fails is noted for determination of compressive strength of brick given by

$$\text{Compressive strength} = \frac{\text{Maximum load at failure}}{\text{Loaded area of brick}}$$

The average of five results should be reported.

4.6 Describe in brief the classification of ordinary bricks according to their qualities into four categories.

[10 marks : 2004]

Solution:

The ordinary bricks according to their qualities can be classified as:

(i) **First class bricks** :These bricks are table moulded and of standard shape and they are burnt in kilns. The surface and edges of these bricks are in shape, square, smooth and straight. They comply with all the quantities of good bricks i.e. water absorption, compressive strength, hardness, efflorescence etc. These bricks are used for superior work of permanent nature.

(ii) **Second class bricks:** The bricks are ground moulded and they are burnt in kilns. The surface of these bricks is somewhat rough and shape is also slightly irregular. These bricks may have hair cracks and their edges may not be sharp and uniform. These bricks are commonly used at places where brickwork is to be provided with a coat of plaster.

(iii) **Third class bricks:** These bricks are ground moulded and they are burnt in clamps. These bricks are not hard and they have rough surfaces with irregular and distorted edges. These bricks give dull sound when struck together. They are used for unimportant and temporary structures and at places where rainfall is not heavy.

(iv) **Fourth class bricks:** These are overburnt bricks with irregular shape and dark colour. These bricks are used as aggregates for concrete in foundations, floors, roads, etc. because of the fact that the overburnt bricks have a compact structure and hence they are sometimes found to be even stronger than the first class bricks.

4.7 What are the properties governing the quality of bricks? Discuss the important of water absorption and strength under compression:

[10 marks : 2005]

Solution:

As per **IS:1077-1992** code bricks are classified on the basis of **minimum compressive strength** which has been kept at 3.5 MPa. Each class of bricks which have been already divided based on compressive strength shall further be divided into subclasses *A*, *B*, etc.

The properties which are associated with this classification are:

(i) **General quality of bricks:** Bricks having smooth rectangular faces and sharp corners and emitting a clear ringing sound when struck against each other are classified under subclass A. Bricks having a slight distortion and round edges, provided no difficulty arise on this account in laying of uniform courses have been classified under subclass *B*.

(ii) **Dimensions and tolerance:** According to **IS:1077-1992**, the standard size of common building bricks is as follows:

19 cm × 9 cm × 9 cm

19 cm × 9 cm × 4 cm

The dimension of bricks when tested by stacking 20 bricks should be within the limits specified by **IS:1077-1992**.

(iii) **Water absorption of bricks:** When tested in accordance with the procedure laid down in **IS:3495-1992**, after immersion in cold water for 24 hours the average water absorption of common building bricks should not be more than 20% by weight upto class 125 and 15% by weight for higher classes.

(iv) **Efflorescence:** According to **IS:3495-1992** the rating efflorescence should not be more than moderate upto class 12.5 MPa and slight for higher classes.

(v) **Strength of bricks:** As per **IS:1077-1992** common building bricks should have a minimum strength of 3.5 MPa. Also, the compressive strength of any individual brick should not fall below the average compressive strength specified for the corresponding class of brick by more than 20%.

4.8 **Explain how bulking of fine aggregates take place and how it is taken care of in the field. Also explain the method of preparation of bulking chart in the laboratory.**

[10 marks : 2006]

Solution:

The presence of free moisture in sand increases the volume of sand which is known as bulking of sand. Free moisture forms a film around each particle. This film of moisture exerts what is known as surface tension which keeps the neighboring particles away from it. Similarly, the force exerted by surface tension keeps every particle away from each other. Therefore, no point contact is possible between the particles. This causes bulking of the volume.

The extent of surface tension and consequently how far the adjacent particles are kept away will depend upon the percentage of moisture content and the particle size of the fine aggregate. It is interesting to note that the bulking increases with the increase in moisture content upto a certain limit and beyond that with further increase in the moisture content, it results in the decrease in volume and at a moisture content representing saturation point, the fine aggregate shows no bulking. It can be seen from the figure below that fine sand bulks more and coarse sand bulks less. From this it follows that the coarse aggregate also bulks but the bulking is so little that it is always neglected.

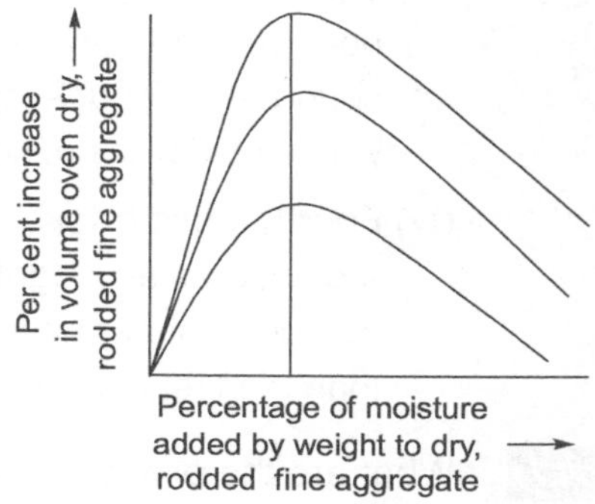

Due to the bulking, fine aggregates show completely unrealistic volume. Thus, the extent of bulking can be estimated by a simple field test. A sample of moist fine aggregate is filled into a measuring cylinder in the normal manner. The level is noted down say h_1. Pour water in the measuring cylinder and completely inundate the sand and shake it. Since the volume of the saturated sand is same as that of dry sand, the inundated sand completely offsets the bulking effects. The level of the sand is again noted down say h_2. Then $h_1 - h_2$ shows the bulking of the sample of sand under test.

$$\% \text{ of bulking} = \left(\frac{h_1 - h_2}{h_2}\right) \times 100$$

4.9 Explain various defects in bricks.

[10 marks : 2006]

Solution:

The various defects in bricks are as follows:

(i) **Over-burning of bricks:** Bricks should be burned at temperatures at which incipient, complete and viscous vitrification occur. However, if the bricks are overburnt, a soft molten mass is produced and the bricks loose their shape. Such bricks are not used for construction purposes.

(ii) **Under burning of bricks:** When bricks are not burnt to cause complete vitrification, the clay is not softened because of insufficient heat and the pores are not closed. This results in higher degree of water absorption and less compressive strength. Such bricks are not recommended for construction works.

(iii) **Bloating:** This defect is observed as spongy swollen mass over the surface of burned bricks and is caused due to the presence of excess carbonaceous matter and sulphur in brick clay.

(iv) **Black core:** When brick clay contains bituminous matter or carbon and they are not completely removed by oxidation, the brick results in black core mainly because of improper burning.

(v) **Efflorescence:** This defect is caused due to presence of alkalies in bricks. When bricks come in contact with moisture, water is absorbed and the alkalies crystallize. On drying, grey or white powder patches appear on the brick surface. This can be minimized by selecting proper clay materials for brick manufacturing, preventing moisture to come in contact with masonry by providing water-proof coping, using water repellant materials in mortar and by providing damp proof course.

4.10 Write about the following tests on clay bricks and mention the desired test limits as per Indian standards:

(i) Water absorption
(ii) Crushing strength
(iii) hardness
(iv) Soundness
(v) Presence of soluble salts

[10 marks : 2007]

Solution:

The various test which are performed on clay bricks are:

(i) **Water absorption test:** The clay bricks are tested in accordance with the procedure laid down in **IS:3495-1992**. The code recommends that this test should be performed in both cold water and boiling water. This is because water absorption in bricks occur in the pores and many times pores are completely sealed and hence inaccessible to water under ordinary conditions. As per the code the average water absorption of common building bricks should not be more than 20% by weight upto class 12.5 MPa and 15% by weight for higher classes.

(ii) **Crushing strength test:** Bricks often have to withstand great compressive stresses. The durability of the masonry depends upon the strength of the bricks. The crushing or compressive strength of a brick is found out by placing it in a compression testing machine. It is pressed till it breaks. As per **IS:1077-1992**, the minimum crushing strength of bricks is 3.5 MPa.

(iii) **Hardness test:** In this test a scratch is made on brick surface with the help of a finger nail. If no impression is left on the surface, the brick is treated to be sufficiently hard.

(iv) **Soundness test:** In this test the two bricks are taken and they are struck with each other. The bricks should not break and a clear ringing sound be produced. The bricks will fail soundness test if there is excess of lime present in it because excess lime causes the brick to melt and disintegrate. As a result the shape of the bricks will be lost. However, a small quantity of lime not exceeding 5% is desirable in good bricks.

(v) **Presence of soluble salts test:** The soluble salts, if present in bricks will cause efflorescence on the surface of bricks. For finding out the presence of soluble salts in a bricks it is immersed in water for 24 hours. It is then taken out and allowed to dry in shade. The absence of grey or white deposits on its surface indicates absence of soluble salts. The liability of efflorescence is reported as 'nil', 'slight', 'moderate', 'heavy' and 'serious' according to the limits specified by **IS 3495 : 1992.**

(a) **Nil :** When there is no perceptible deposit of efflorescence.

(b) **Slight :** When no more than 10% of the area of the brick is covered with a thin deposit of salts.

(c) **Moderate :** When deposition is more than 10% and covering upto 50% of the exposed area of the brick surface but unaccompanied by powdering or flaking of the surface.

(d) **Heavy :** When there is a heavy deposit of salts covering 50% or more of the exposed area of the brick surface but unaccompanied by powdering or flaking of the surface.

(e) **Serious :** When there is a heavy deposit of salts accompanied by powdering and/or flaking of the exposed surfaces.

4.11 **What are the principal constituents of brick earth and how do they influence the quality of brick?**
[10 marks : 2008]

Solution:

The principal constituents of brick earth are as follows:

(i) **Silica:** It exists in clay either as free or combined. A good brick earth should contain about 50 to 60% of silica. It enables the brick to retain its shape and imparts durability, prevents shrinkage and warping. The durability of bricks depends on the proper proportion of silica in brick earth. The excess of silica destroys the cohesion between particles and the bricks become brittle and weak on burning. A large percentage of sand or uncombined silica in clay is undesirable. However, it is added to decrease shrinkage in burning and to increase the refractoriness of low alumina clays.

(ii) **Alumina:** A good brick earth should contain about 20 to 30% of alumina. It absorbs water and imparts plasticity to the earth so that it can be moulded. If alumina is present in excess of the specified quantity, it produces cracks in bricks on drying and the bricks become too hard when burnt. Clays having exceedingly high alumina content are likely to be very refractory.

(iii) **Lime:** It normally constitutes less than 10% of clay. The lime prevents shrinkage of raw bricks on drying. The sand alone is infusible, but it slightly fuses at kiln temperature in presence of lime. In carbonated form, lime lowers the fusion point. Excess of lime causes the brick to melt and the brick looses its shape.

(iv) **Oxide of iron:** A small quantity of iron oxide to the extent of about 5 to 6% is desirable in good brick earth. Like lime, it also helps to fuse sand. It also imparts red colour to the bricks. However the excess of oxide of iron makes the bricks dark blue or blackish. If on the other hand, the quantity of iron oxide is comparatively less, the bricks will be yellowish in colour. It improves impermeability and durability of bricks. Its also imparts strength and hardness to bricks.

(v) **Magnesia:** It rarely exceeds 1 per cent, affects the colour and makes the brick yellow in colour. On burning, it causes the clay to soften at slower rate than in most cases as lime and reduces warping. Excess of magnesia leads to decay of bricks.

4.12 **Write briefly on the composition and properties of refractory bricks.**

[6 marks : 2009]

Solution:

Refractory bricks or fire clay bricks as they are sometimes called are made from fire clay. Fire clay is a term loosely applied to include those sedimentary or residual clays which vitrify at a very high temperature and which, when so, burnt, possess great resistance to heat.

Fire clays are pure hydrated silicates of alumina and contain a large proportion of silica (55-75%), alumina (20-35%), iron oxide (2-5%) with about 1 per cent of lime, magnesia and alkalis. The greater the percentage of alumina, the more refractory the clay will be. Fire clays are capable of resisting very high temperatures upto 1700°C without melting or softening. Iron oxide or other alkalis reduce refractory qualities of fire clay. The process of manufacture of refractory bricks is same as that of ordinary bricks.

The properties of refractory bricks are as follows:

(i) These bricks are whitish yellow or light brown in colour.

(ii) The weight of refractory bricks is about 30 to 35 N.

(iii) The refractory bricks can resist high temperature without softening or melting. Hence, they are used for lining blast furnace, ovens, kilns, boilers and chimneys.

(iv) The water absorption of refractory bricks various from 4 to 10%.

(v) The compressive strength of refractory bricks varies from 150 to 220 N/mm^2.

4.13 **Write short notes on:**

(i) Harmful ingredients in brick earth.

(ii) Tests on brick to assess its suitability–list the names.

[3 + 3 = 6 marks : 2010]

Solution:

(i) Harmful ingredients in brick earth:

(a) **Pebbles of stone and gravel:** These don't allow the clay to be mixed uniformly and thoroughly and result in weak and porous brick. Bricks containing grits are likely to crack and cannot be readily cut or worked.

(b) **Alkaline salts:** Alkaline salts if present, act as hygroscopic substances. They absorb moisture from the atmosphere in due course of time and create damp conditions. The moisture on drying leaves behind a greyish white deposit known as efflorescence.

(c) **Limestone and Kankar:** Presence of large quantity of lime and limestone in lumps is detrimental to brick earth, as lumps of limestone if burnt in a brick, slake afterwards and split the brick. Thus limestone should be present in very finely divided state.

(d) **Vegetation and organic matter:** Organic matter if present in brick earth will produce porous bricks. This is due to the evolution of gas during the burning of the carbonaceous matter, resulting in the formation of small pores.

(ii) A brick is generally subjected to the following tests to find out its suitability:

(a) Absorption
(b) Crushing strength
(c) Hardness
(d) Presence of soluble salts
(e) Shape and size
(f) Soundness
(g) Structure

4.14 Briefly describe the various defects in bricks.

[5 marks : 2011]

Solution:

The defects in bricks are caused due to certain harmful ingredients which affects the strength of a brick.

The ingredients and their effect are enlisted below:

(i) **Lime:** The excess of lime cause the brick to melt and hence its shape is lost. The lumps of lime are converted into quick lime after burning and this quick lime slakes and expands in presence of moisture. Such an action results in splitting of bricks into pieces.

(ii) **Iron pyrites:** If iron pyrites are present in brick earth, the bricks are crystallized and disintegrated during burning because of the oxidation of the iron pyrites.

(iii) **Alkalies:** These are mainly in the form of soda and potash. The alkalies act as a flux in the kiln during burning and they cause bricks to fuse, twist and warp. As a result the bricks are melted and they loose their shape. Further, the alkalies remaining in bricks will absorb moisture from the atmosphere, when bricks are used in masonry. Such moisture when evaporated, leaves behind grey or white deposits on the wall surface. The appearance of the building as a whole is then seriously spoiled.

(iv) **Pebbles:** The presence of pebbles or grits of any kind is undesirable in brick earth because it will not allow the clay to be mixed uniformly and thoroughly which will result in weak and porous bricks. Also the brick containing pebbles will not break regularly as desired.

(v) **Vegetation and organic matter:** The presence of vegetation and organic matter in brick earth assists in burning but if such matter is not completely burnt the bricks becomes porous. This is due to the fact that the gases will be evolved during the burning of the carbonaceous matter and it will result in the formation of small pores. Hence it is necessary to see that all these gases are removed during the process of burning for getting bricks of good quality.

4.15 Briefly explain the processes involved in the manufacture of bricks in order.

[5 marks : 2011]

Solution:

The processes involved in the manufacture of bricks are as follows:

(i) **Preparation of clay:** The clay is dug out from the ground after the top layer of soil is removed. The clay obtained is cleaned of pebbles, stones and vegetable matter etc and is exposed to atmosphere for softening. Then water is added into clay. Any ingredient to be added to clay is mixed and proper blending with water is done before tempering the clay upto a proper degree of hardness.

(ii) **Moulding:** Moulding is done using hand moulding or machine moulding. Moulding is the process of filling the prepared clay into moulds of definite size and pattern. This gives an even surface and shape to the brick.

(iii) **Drying:** The moulded bricks are dried to remove moisture from the clay. The bricks are stacked by providing proper spacing so as to allow tree circulation of air. The bricks should be allowed to dry till they become leather hard or bone-dry with moisture content of about 2 percent.

(iv) **Burning:** Burning imparts hardness and strength to the bricks and makes them dense and durable. Burning is done either in clamps or kilns. The clamps are temporary structures and they are adopted to manufacture bricks on a small scale. The kilns are permanent structure and are adopted to manufacture bricks on a large scale.

4.16 **Discuss the properties imparted to brick-earth by its constituents alumina and silica.**
[10 marks : 2012]

Solution:

(i) **Silica:** It exists in clay either as free or combined. A good brick earth should contain about 50 to 60% of silica. It enables the brick to retain its shape and imparts durability, prevents cracking, shrinkage and warping. The durability of brick depends on the proper proportion of silica in brick earth. The excess of silica destroys cohesion between particle and bricks become brittle and weak on burning. A large percentage of sand or uncombined silica in clay is undesirable. However it is added to decrease shrinkage in burning and to increase the refractoriness of low alumina clay.

(ii) **Alumina:** A good brick earth should contain 20 to 30% of alumina. It absorbs water and imparts plasticity to the earth so that it can be moulded. If alumina is present in excess of specified quantity, it produce cracks in bricks due to shrinkage on drying and the bricks become too hard when burnt. Clays having exceedingly high alumina content are likely to be very refractory. High quantity of alumina also causes warping.

4.17 **What do you understand by bulking of sand? How does it affect quantity of sand by volume batching?**
[5 marks : 2014]

Solution:

The increase in the volume of given mass of sand caused by the presence of water is known as bulking. Free moisture forms a fine film around each particle. This film of moisture exert surface tension which keep particles away from each other resulting into bulking of the volume of sand.

Bulking of sand depend upon the percentage moisture content and the particle size of sand. It is noted that the bulking increase with the increase in moisture content upto a certain limit and beyond that with further increase in the moisture content, it results in the decrease in volume and moisture content corresponding to maximum volume is known as saturation point. For most of the sand it is 4% to 6%.

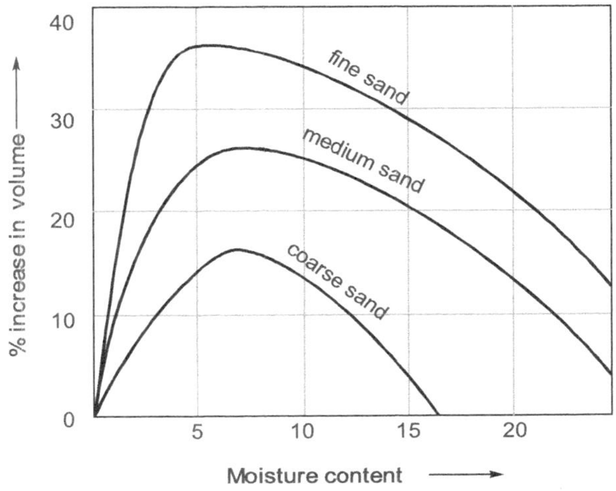

In volume batching, if the sand is measured by volume and no allowance is made for bulking, then the actual quantity of sand in each batch will be less than the recommended volume and consequently the mix will be rich in cement i.e., the proportions of cement and sand to be provided will be different.

This also increase the chances of **segregation and honey combing of concrete.**

It also increase water to cement ratio in mix as sand is having certain extra percentage of water above surface dry condition.

4.18 **Give a detailed procedure for determining the compressive strength of bricks as per I.S. code. Also write about water absorption test. Mention the usual limit.**
[5 marks : 2015]

Solution:

Compressive strength of bricks is tested as per **IS 3495 (Part-1): 1992**. As per this, all dimensions of the bricks must be measured nearest to 1 mm. The compression testing machine to be used shall have a ball seating. The following procedure is adopted:

(i) Place the specimen with its flat surfaces horizontal and mortar filled face facing upwards between two, 3 plywood sheets each of 3 mm thickness and adequately centered between the plates of the testing machine.

(ii) Apply load uniformly at a uniform rate of 14 N/mm^2 per minute till failure occurs and note down the ultimate load at failure.

The compressive strength is given by,

$$\text{Compressive strength (N/mm}^2) = \frac{\text{Maximum load at failure (N)}}{\text{Mean area of bed faces (mm}^2)}$$

Water absorption test on bricks is done as per **IS 3495 (Part-2): 1992**. This test is also referred to as 24 hour Immersion Cold Water Test. The weighing balance should be sensitive to within 0.1%. The following procedure is adopted.

(i) Dry the specimen in a ventilated oven at a temperature of 105 to 115°C till the specimen attains a constant mass.

(ii) Allow the specimen to cool at room temperature and measure its weight (M_1). The specimen used should not be warm to touch.

(iii) Immerse completely dried specimen in clean water at 27 ± 2 °C for 24 hours.

(iv) Remove the specimen after 24 hours and wipe the excess water. Weigh the specimen (W_2) for 3 minutes.

(v) **Water absorption as percentage by mass is given by,**

$$\text{Water absorption (in\%)} = \frac{M_2 - M_1}{M_1} \times 100$$

Water absorption should not exceed 20% by weight of dry brick.

4.19 **List the tests for bricks and the corresponding IS code.**

[4 marks : 2017]

Solution:

Various tests for bricks:

1. Dimension Test (IS : 1077)
2. Compressive Strength Test (IS : 3495 – Part I)
3. Water Absorption test (IS : 3495 – Part II)
4. Efflorescence Test (IS : 3495 – Part III)
5. Warpage Test (IS : 3495 – Part IV)

4.20 **List out at least eight tests required to determine the suitability of stone for engineering use.**

[8 marks : 2018]

Solution:

Following are different tests on stones for engineering use:

1. **Acid Test :** This test is done by the help of a hydrophobic acid to detect the presence of calcium carbonate in the stone, which is responsible for poor weathering quality of stones. This test is usually carried out on sandstones.

2. **Attrition Test :** This test is carried out to find out the rate of wear of stones, which are used in road construction. The results of this test indicate the resisting power of stones against the grinding action under traffic.
3. **Crushing Test :** This test is carried out to find out the compressive strength of the stones. Test is undertaken on $50 \times 50 \times 50$ mm^3 cube with rate of loading of 140 kg/cm^2 per minute.

$$\text{Crushing strength} = \frac{\text{Max load at which sample crushes}}{\text{Area of bearing face}}$$

4. **Crystalline Test :** This test is carried out to find out the durability of stones by calculating the degree of decay of stone due to 30 repeated cycles of placing the stone in 14% sodium sulphate solution for 16-18 hours and then in air for 4 hours.
5. **Freezing and Thawing Test :** Stone specimen is kept immersed in water for 24 hours. It is then placed in freezing machine at –12°C for 24 hours. Then it is thawed or warmed at atmospheric temperature. The procedure is repeated several times and behaviour of stone is carefully observed.
6. **Hardness Test :** To determine the hardness of stone test is carried out by the help of Darcy's Abrasion Machine.

$$\text{Coefficient of hardness} = \left[\frac{20 - \text{Loss of weight in gram}}{3}\right]$$

7. **Impact Test :** To determine the toughness of the sample impact test is carried out. Blow at which the specimen break is noted. If it is the nth blow, 'n' represents the toughness index.

   Specimen size : 25 mm diameter, 25 mm height

   Steel hammer : 2 kg

   Height of first flow = 1 cm, that of second = 2 cm, that of third = 3 cm and soon.
8. **Smith's Test :** Thin test is undertaken to find out the amount of soluble matter in a sample of stone. Few pieces of stones are taken in a glass tube along with water and vigorously shaken. Presence of earthy matter will convert the clear water into dirty.
9. **Microscopic Test :** The sample of the test is subjected to microscopic examination. The sections of stones are taken and placed under the microscope to study the various properties such as :
   (i) Average grain size
   (ii) Existence of pores, fissure, veins and shakes.
   (iii) Mineral constituents.
   (iv) Nature of cementing material
   (v) Presence of any harmful substance
   (vi) Texture of stones etc.
10. **Water Absorption Test :** Water absorption test is carried out with aim to get percentage absorption by wt after 24 hours in order to know how much water the stone absorbs.

4.21 **Briefly explain the purpose and the procedure for Attrition Test on stone.**

[4 marks : 2018]

Solution:

Purpose :

- To determine the rate of wear of stones.
- To determine the resistance power of stones against the grinding action.
- It is mainly conducted to find out the suitability of particular stone for road construction.

Procedure:

- Sample of stone is broken into pieces about 60 mm size.
- Such pieces, weighing 5 kg are put in both the cylinder of Deval's attrition test machine. Diameter and length of cylinder are respectively 20 cm and 34 cm.
- Cylinders are closed. Their axes make an angle of 30° with the horizontal.
- Cylinders are rotated about the horizontal axis for 5 hours at the rate of 30 rpm.
- After this period, the contents are taken out from the cylinders and they are passed through a sieve of 1.5 mm mesh.
- Quantity of material which is retained on the sieve is weighed.
- % wear worked out as follows:

$$\% \text{ wear} = \frac{\text{Loss in weight}}{\text{Initial weight}} \times 100$$

4.22 **Explain briefly the various tests conducted on Bricks mentioning the relevant codal provisions.**

[8 marks : 2019]

Solution:

1. Absorption
   - A brick is taken and it is weighed dry. It is then immersed in water for a period of 16 hours.
   - Then weigh again and the difference in weights should not, in any case, exceed
     (a)20 per cent of weight of dry brick for first class bricks.
     (b)22.5 per cent for second class bricks.
     (c)25 per cent for third class bricks.
2. Crushing Strength
   - Minimum crushing strength for first class bricks $\not<$ 10 N/mm^2 and for second class bricks $\not<$ 7.5 N/mm^2.
3. Hardness
   - In this test, a scratch is made on brick surface with the help of a finger or nail. If no impression is left on the surface, brick is treated to be sufficiently hard.
4. Presence of Soluble Salts
   - Soluble salts, if present in bricks, will cause efflorescence on the surface of bricks.
   - It is immersed in water for 24 hours. It is then taken out and allowed to dry in shade. Absence of grey or white deposits on its surface indicates absence of soluble salts.
   - If the white deposits cover about 10% surface, the efflorescence is said to be slight.
   - When white deposit cover about 50% of surface then it is said to be moderate.
   - If grey or white deposits are found on more than 50% of surface, the efflorescence becomes heavy and it is treated as serious.
5. Shape and Size
   - Its shape should be truly rectangular with sharp edges.
   - 20 bricks are randomly selected of standard size (19 × 9 × 9 cm). For good quality bricks, the results should be within the following permissible limits:

Length	–	368 cm to 392 cm
Width	–	174 cm to 186 cm
Height	–	174 to 186 cm

6. Soundness
   - In this test, two bricks are taken and they are struck with each other.
   - Bricks should not break and a clear ringing sound should be produced.

7. Structure
   - It should be homogeneous, compact and free from any defects such as holes, lumps, etc.
   - High duty fireclays can resist temperature range of 1482°C to 1648°C; medium duty fireclays can resist temperature range of 1315°C to 1482°C and low duty fireclays can resist temperature up to 870°C only.

4.23 What are the various factors that promote the Alkali Aggregate Reaction? How can this be controlled? [8 marks : 2019]

Solution:

Alkali aggregate reaction (AAR) is a chemical reaction of alkali in concrete and certain alkaline reactive minerals i.e., aggregate producing a hygroscopic gel which, when moisture present, absorbs water and expand. This gel expansion causes cracking in the concrete.

Factors promoting alkali-aggregate reaction are:

(i) **Reactive type aggregate:** The petrographic examination of thin rock sections may also immensely help to asses the potential reactivity of the aggregate. This test often requires to be supplemented by other tests.

(ii) **High alkali content in cement:** The high alkali content in cement is one of the most important factors contributing to the alkali-aggregate reaction.

(iii) **Availability of moisture:** Progress of chemical reactions involving alkali-aggregate reaction in concrete requires the presence of water. It has been seen in the field and laboratory that lack of water greatly reduces this kind of deterioration.

(iv) **Optimum temperature conditions:** The ideal temperature for the promotion of alkali-aggregate reaction is in the range of 10 to 38°C. If the temperature condition is more than or less than the above, it may not provide an ideal situation for the alkali-aggregate reaction.

4.24 How are aggregates classified based on particle size? What is bulking of sand? [6 marks : 2020]

Solution:

(a) Coarse aggregate
   - Size = 4.75 mm - 80 mm
   - This composes the main matrix of concrete.
   - These are generally obtained from natural disintegration or artificial crushing of rocks.

(b) Fine aggregate
   - Size = 45 micron - 4.75 mm
   - This composes the filler matrix in between the coarse aggregate
   - These are generally obtained from natural sand deposited by river or crushed stones.

Bulking of sand:

The presence of free moisture in sand increases the volume of sand which is known as bulking of sand. Free moisture forms a film around each particle. This film of moisture exerts what is known as surface tension which keeps the neighbouring particles away from it. Similarly, the force exerted by surface tension keeps every particle away from each other. Therefore, no contact point is possible between the particles. This causes bulking of the volume.

The extent of surface tension and consequently how far the adjacent particles are kept away will depend upon the percentage of moisture content and the particle size of the fine aggregate. It is interesting to note that the bulking increases with the increase in moisture content upto a certain limit and beyond that with further increase in the moisture content, it results in the decrease in volume and at a moisture content representing saturation point, the fine aggregate shows no bulking. It can be seen from the figure below that fine sand bulks more and coarse sand bulks less. From this it follows that the coarse aggregate also bulks but the bulking is so little that it is always neglected.

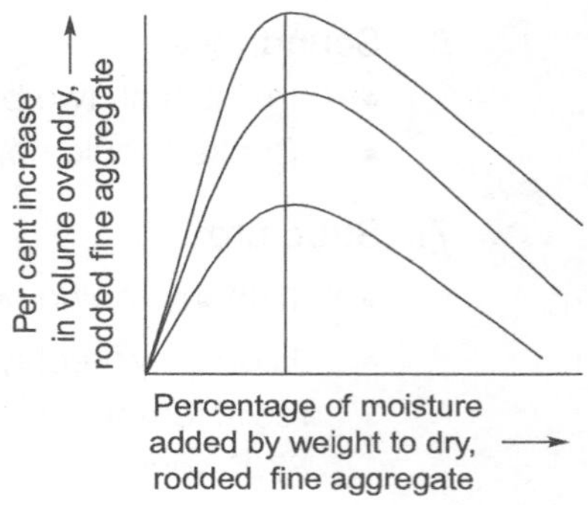

Due to the bulking, fine aggregates show completely unrealistic volume. Thus, the extent of bulking can be estimated by a simple field test. A sample of moist fine aggregate is filled into a measuring cylinder in the normal manner. The level is noted down say h_1. Pour water in the measuring cylinder and completely inundate the sand and shake it. Since the volume of the saturated sand is same as that of dry sand, the inundated sand completely offsets the bulking effects. The level of the sand is again noted down say h_2. Then $h_1 - h_2$ shows the bulking of the sample of sand under test.

$$\% \text{ of bulking} = \left(\frac{h_1 - h_2}{h_2}\right) \times 100$$

4.25 **(i) What are the functions of cement mortar in brick masonry work?**
(ii) What are the purposes of making frog in bricks during moulding?
Explain with the help of neat sketches.

[6 + 6 = 12 Marks : 2021]

Solution:

(i) Uses of cement mortar:

- To carry out pointing and plaster work on exposed surfaces of masonry.
- To form an even and soft bedding layer for building units.
- To hide the open joints of brickwork.
- To improve the general appearance of structure.
- To fill up the cracks detected in the structure during maintenance process etc.

(ii) **Frog or kick:** A frog is an indentation in the face of a brick to form a key for holding the mortar. It serves two purposes:

1. It indicates the trade name of the manufacturer.
2. In brickwork, the bricks are laid with frog uppermost. It thus affords a key for mortar when the next brick is placed over it.

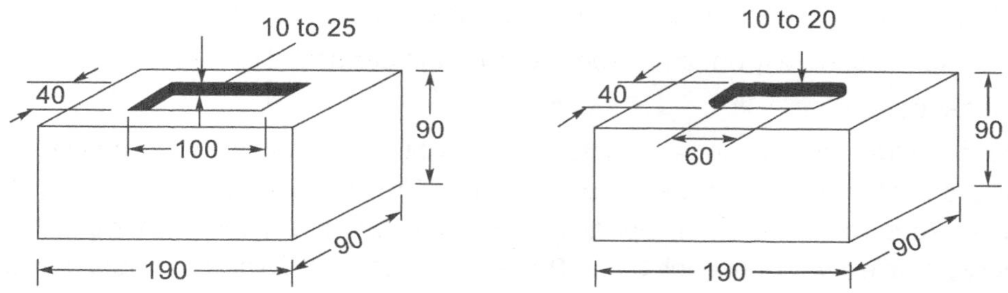

4.26 Name any six tools for cutting and dressing stones.

[4 marks : 2022]

Solution

Stone cutting and dressing tools: The dressing/cutting tools for stones are (i) wedge, (ii) pitching tool (iii) boaster (iv) scabbing hammer (v) mash hammer (vi) separate pick (vii) punch (viii) scabbing pick (ix) crow bar (x) axe punch (xii) dressing knife and (xii) splitting chisel.

4.27 (i) Classify the bricks based on the physical and mechanical properties. Indicate the properties of first class bricks. Which are the applications where first class bricks are highly recommended and other classes of bricks should not be used?

(ii) Discuss the scale of sampling for physical characteristics of bricks. Indicate its importance.

[10 + 10 = 20 marks : 2022]

Solution

(i) Clay bricks are classified as first class, second class, third class and fourth class bricks based on their physical and mechanical properties:

First class bricks:

1. These are thoroughly burnt and are of deep red, cherry or copper colour.
2. The surface should be smooth and rectangular , with parallel, sharp and straight edges and square corners.
3. These should be free from flaws, cracks and stones.
4. These should have uniform texture.
5. No impression should be left on the brick when a scratch is made by a finger nail.
6. The fractured surface of the brick should not show lumps of lime.
7. A metallic or ringing should come when two bricks are struck against each other.
8. Water absorption should not be more than 15% by its dry weight and compressive strength of brick is ≥ 12.5 N/m^2.

Use of first class bricks is recommended for pointing, exposed facework in masonry structure, flooring and reinforced brick work.

(ii) About fifty piece of bricks are taken at random from different parts of the stack to perform various tests. For the purpose of sampling , a lot should contain maximum of 50,000 bricks. The scale of sampling for physical characteristics as per IS : 5454 is

Scale of sampling and permissible number of defectives for visual and dimensional characteristics

No. of bricks in the lot	For characteristics specified for individual brick		For dimensional characteristics specified for group of 20 bricks-No. of bricks to be selected
	No. of bricks to be selected	Permissible No. of defectives in the sample	
2001 to 10000	20	1	40
10001 to 35000	32	2	60
35001 to 50000	50	3	80

Scale of sampling for physical characteristics

Lot size	Sampling size for compressive strength, breaking load, transverse strength, bulk density, water absorption and efflorescence	Permissible No. of defectives for eflorescence	Warpage	
			Sample size	Permissible No. of defectives
2001 to 10000	5	0	10	0
10001 to 35000	10	0	20	1
35001 to 50000	15	1	30	2

- If lot contains 2000 or less bricks, the sampling shall be subjected to agreement between purchaser and supplier.

4.28 What are the causes and remedies of efflorescence in bricks?

[6 marks : 2023]

Solution

Efflorescence is a kind of defect that is caused due to alkalis present in bricks. When bricks come in contact with moisture, water is absorbed by them. This absorbed water dries out by evaporation from the exposed faces, and as it does so, the soluble salts it contains crystallize out on the surface. On drying, grey or white powder patches appear on the brick surface. This process depends on the quantity of salts present in the bricks and their solubility.

Efflorescence can be minimized by selecting proper clay materials for brick manufacturing, preventing moisture to come in contact with the masonry. This can be achieved by providing waterproof coping and by using water repellent materials in mortars and by providing damp proof course.

Surface affected	Efflorescence degree
0%	nil
0 - 10%	Slight
10 - 50%	Moderate
> 50%	Heavy

Efflorescence will be serious when deposits are heavy and powder or flake gets detatched from the brick surface.

5. Timber

5.1 Explain the following:

(i) Plywood
(ii) Laminated board
(iii) Batten board
(iv) Fiber board

[10 marks : 2001]

Solution:

(i) **Plywood:** The meaning of term ply is thin layer. The plywoods are boards which are prepared from thin layers of woods or veneers. The number of veneers in plywood is always kept odd for a balanced construction and to prevent shrinkage. This gives rise to 3 ply, 5 ply, 7 ply etc. Plywoods vary in thickness depending upon the number of veneers and the thickness of veneer used. The outside layers of veneer are called faces, the central layer is called core and those between the face and the core are called cross bands. The veneers are held in position by application of suitable adhesives. The placing of veneers normal to each other increases the longitudinal and transverse strengths of plywoods. The plywoods are available in different forms such as batten board, laminated board, three ply, veneered plywood etc.

(ii) **Laminated board:** It is a board having a core of strips, each not exceeding 7 mm in thickness, glued together face to face to form a slab which in turn is glued between two or more outer veneers, with the direction of the grain of the core strip running at right angles to that of the adjacent outer veneers.

(iii) **Batten board:** It is a board having a core made up of strip of wood usually 8 cm wide, each laid separately or glued or otherwise joined to form a slab which is glued between two or more outer veneers with the direction of the grain of the core running at right angles to that of the adjacent outer veneers.

(iv) **Fiber board:** It is a rigid board and also known as pressed wood or reconstructed wood. It is made up of wood, cane or other vegetable fibres and chippings. Depending upon their form and composition, the fiber boards are classified as insulating boards, medium hard boards, hard boards, superhard boards and laminated boards. They are also available under various trade names such as Euraka Indiamite, Insulite etc. The fiber boards form an ideal base for practically all types of decorative finishes such as distemper, oil paint, etc.

5.2 **What is seasoning of timber and why is it done? Discuss in brief the different methods of seasoning timber.**

[10 marks : 2004]

Or

Write briefly about five methods of artificial seasoning of timber. **[10 marks : 2007]**

Or

Why is seasoning of timber required? List out the methods of seasoning. **[4 marks : 2009]**

Or

Write a short note on electrical seasoning of wood.

[4 marks : 2010]

Solution:

Seasoning of Timber: It is a process by which the moisture content of timber is reduced to a suitable level depending upon the use. Seasoned timber is definitely superior to unseasoned timber. While seasoning does not make timber safe from all destructive agencies, this is a prime process before timber can be put to use. By nature timber is hygroscopic and takes in moisture from air depending upon many factors. Timber should be seasoned as early as possible after felling because felled timber is nothing but dead vegetation that will rot and decay due to many environmental agencies.

The reasons or objectives for which seasoning is done are:

(i) To reduce the tendency to split, warp and shrink.

(ii) To make it immune from attack by insects.

(iii) To increase the strength, durability, workability and resilience.

(iv) To make the timber receptive to finish like preservative, paints and varnish.

(v) To reduce the weight and minimise cost of transportation.

(vi) To make the timber burn readily, if used as a fuel.

Seasoning methods can be basically classified into two groups namely, natural seasoning and artificial seasoning:

Natural Seasoning: In this method, the seasoning of timber is carried out by natural air and hence it is also sometimes referred to as air seasoning. The timber has to be stacked properly depending upon the species of timber and the environmental conditions. The timber log is cut and sawn into suitable section of planks or scantlings. Then they are stacked either horizontally or vertically. The stack is to be protected from fast blowing wind, rain and extreme heat of sun. Hence the stack should preferably be covered by a roof material. The advantage of natural seasoning is that it does not require skilled supervision. This method of seasoning timber is cheap and simple. The disadvantage of natural seasoning is that, as the process depends on the natural air, it some times become difficult to control it as well as the drying of different surface may not be even and uniform.

Artificial Seasoning: It is a quicker method of seasoning timber to the desired moisture content under controlled conditions. The drying conditions required for different species of timber are different and artificial seasoning makes it possible to provide conditions cited to each species. The advantages of artificial seasoning are speed, adaptability and precision.

The various methods of artificial seasoning are boiling, chemical seasoning, electrical seasoning and water seasoning which are as follows:

(i) **Boiling** : In this method of artificial seasoning, the timber is immersed in water and water is then boiled. This is a very quick method. The timber is thus boiled with water for about three to four hours. It is then dried very slowly under a shed. The periods of seasoning and shrinkage are reduced by this method, but it affects the elasticity and strength of wood. In place of boiling water, the timber may be exposed to the action of hot steam. This method of seasoning proves to be costly.

(ii) **Chemical Seasoning** : This is also known as salt seasoning. In this method, the timber is immersed in a solution of suitable salt. It is then taken out and seasoned in the ordinary way. The interior surface of timber dries in advance of exterior one and chances of formation of external cracks are reduced.

(iii) **Electrical seasoning** : In this method, the use is made of high frequency alternating currents. The timber, when it is green offers less resistance to the flow of electric current. The resistance increases as the wood dries internally which also results in the production of heat. This is the most rapid method of seasoning. But the initial and maintenance costs are so high that it becomes uneconomical to season timber on commercial basis by this method.

(iv) **Kiln Seasoning:** In this method, the drying of timber is carried out inside an airtight chamber or oven. Depending upon the mode of construction and operation, the kilns are of two types namely stationary kilns and progressive kilns. Stationary kiln is adopted for seasoning timber which requires a close control of humidity and temperature. It gives better results. Progressive kiln is used for seasoning timber on a large scale. If not properly attended the drying in this kiln may prove to be unsatisfactory. The kiln seasoning though costly gives well seasoned timber as it controls three important conditions namely circulating air, relative humidity and temperature.

(v) **Water seasoning:** The water seasoning is a quick method and it renders timber which is less liable to shrink or warp. It also removes organic materials contained in sap of timber. It however weakens the timber and make it brittle.

5.3 **Draw the related stress-strain curves for a timber member under tension, compression and bending parallel to grain. Discuss the importance of compression test results.**

[10 marks : 2005]

Solution:

The relative stress-strain curves for direct tension, direct compression and bending stress intensities parallel to the grains are given in figure:

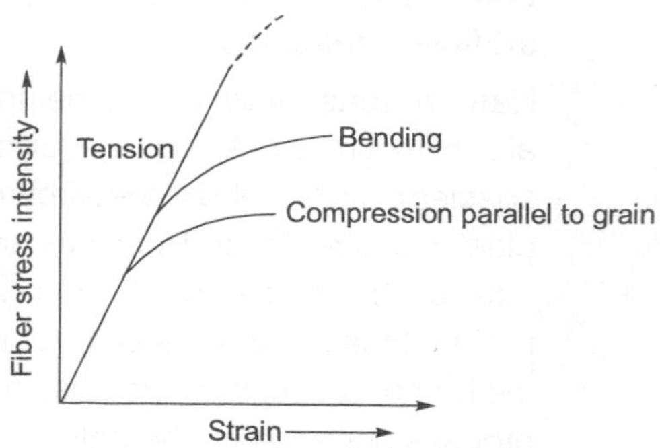

These curves show that in both direct compression and bending, the proportional limit is in the vicinity of 65 to 75 per cent of the ultimate strength. For all practical purposes, there is no limit in direct tension. Modulus of elasticity of the grain is practically the same in direct tension, direct compression and bending, if shear deformation in bending is eliminated.

Compressive strength parallel to fibres at 15 per cent moisture content varies from 30.0 to 77.5 MPa. When subjected to compressive force acting parallel to the axis of growth, timber is found to be one of the strongest structural material. However compressive strength perpendicular to fibers of wood is much lower than that parallel to fibres of wood. Furthermore, a knowledge of the compressive strength is of value in estimating strength in bending, since experiments have demonstrated that the yield point of wooden beams is determined by compressive strength of the wood.

5.4 Explain methods of improving fire resistance of timber.

[10 marks : 2006]

Solution:

Wood is highly combustible but not inflammable. That is, while in favourable circumstances any wood may be burnt to ash, timber is not readily ignited expect in small thickness. With respect to the fire resistance, the timber is classified as refractory timber and non-refractory timber. The refractory timber is non-resinous and it does not catch fire easily. The examples of refractory timbers are sal, teak, etc. The non-refractory timber is resinous and it catches fire easily. The examples of non-refractory timbers are chir, deodar, fir etc.

To make timber more fire-resistant, the following methods are adopted:

(i) **Application of special chemicals:** The timber surface is coated with the solution of certain chemicals. It is found that two coats of solution of borax or sodium arsenate with strength of 2% are quite effective in rendering the timber fire resistant.

These special chemicals are known as the fire protection compounds or antipyrines and they are more reliable. When the temperature rises, they either melt or give off gases which hinder or forbid combustion. When the wood is treated with antipyrine, it does not inflame even at high temperature, but it merely smoulders i.e. burns slowly without flame. The antipyrines containing salts of ammonium or boric acid and phosphoric acids are considered to be the best in making the timber fire-resistant.

(ii) **Sir Abel's Process:** In this process, the timber surface is cleaned and it is coated with a dilute solution of sodium silicate. A cream-like paste of slaked fat lime is then applied and finally, a concentrated solution of silicate of soda is applied on the timber surface. This process is quite satisfactory in making the timber fire-resistant.

5.5 Give a short description of preservation of wood using various wood preservatives.

[10 marks : 2008]

Solution:

Preservation of Timber: No timber is immune to deterioration and ultimate disintegration if exposed for a sufficiently long period to ordinary atmospheric conditions. The principal causes of deterioration of wood in service are fungal infection, termite and other insects attack, mechanical failure and fire. The resistance of wood to these agents of destruction may be increased by the application of a suitable chemical to the wood. This application of chemicals to woods in order to enhance their resistance against decay, fire etc. is known as preservation of timber.

Preservation can basically be classified into the following four types:

(i) **Oil type:** Coal tar creosote with or without admixture of petroleum, coal tar, fuel oil or other suitable oils having high boiling range. This group includes creosote, a complex substance derived from coal or wood distillation. The advantages of creosote as a wood preservative are that it is an indigenous product with high toxicity, has relatively high performance and is non-corrosive. The disadvantages of creosote are that it has an unpleasant odour, not clean to handle and the treated timber is not very receptive to other finishes. This limits its use to timber for external use.

(ii) **Organic solvent type:** These consist of toxic chemical compounds dissolved in a non-aqueous solvent like volatile oils, spirits, naphtha etc.

Some of the compounds are:

(i) Copper and zinc naphthanates
(ii) Pentachlorophenol
(iii) Benzene hexachloride
(iv) DDT
(v) Dieldrin

These are clean to handle and are permanent. The timber after treatment can be painted, waxed or polished in most cases. The only disadvantage is that some of these are inflammable and hence require great care in handling and application.

(iii) **Water soluble (leachable) type:** These preservatives are inorganic or organic salts soluble in water. Some of these are zinc chloride, boric acid and borax, sodium fluoride, sodium pentachlorophenol and benzene hexa-chloride. These preservatives are non-creeping, non-staining and non-inflammable. Timber treated with such preservatives can be painted or varnished when dry. But these are subject to leaching, hence not satisfactory for external use.

(iv) **Water soluble (fixed) type:** These consist of the various salts described in water soluble (leachable) type preservatives with the addition of a fixative salt usually sodium or potassium dichromates. The effect of chromium is to fix the toxic elements such as arsenic, copper, zinc, etc. to the wood so that these are not leached out by water. These should be applied cold as they are liable to get precipitated when heated.

Some of the preservative compositions are:

(i) Copper chrome arsenic composition
(ii) Acid cupric chromate composition
(iii) Chromated zinc chloride
(iv) Copper chrome borax composition
(v) Zinc meta arsenite
(vi) Zinc chrome boric composition

Timber treated with such chemicals can be used for outside locations and can be painted over also.

Characteristics of a good preservative are:

(i) It should be strongly toxic to insects, fungi, etc.
(ii) It should penetrate readily into the timber.
(iii) It should cover a large surface area with a comparatively small quantity.
(iv) It should have no destructive influence on the strength of the timber.
(v) It should be highly resistant to water and dampness.

Some of the methods of treatment of timber are:

(i) Surface application by brushing, spraying or dipping
(ii) Soaking
(iii) Hot and cold process
(iv) Boucherie process
(v) Diffusion process
(vi) Pressure or pneumatic process
(vii) Full cell or Bethel process
(viii) Empty cell process which consists of Lawry process or Rupeing process.

5.6 **(i) Draw a neat sketch of macrostructure of exogenous tree.**
(ii) Draw neat sketches showing various types of shakes.

[3 + 3 = 6 marks : 2009]

Solution:

(i) Sketch of macrostructure of exogenous tree is given below:

(a) Pith is the innermost central portion or core of the tree.

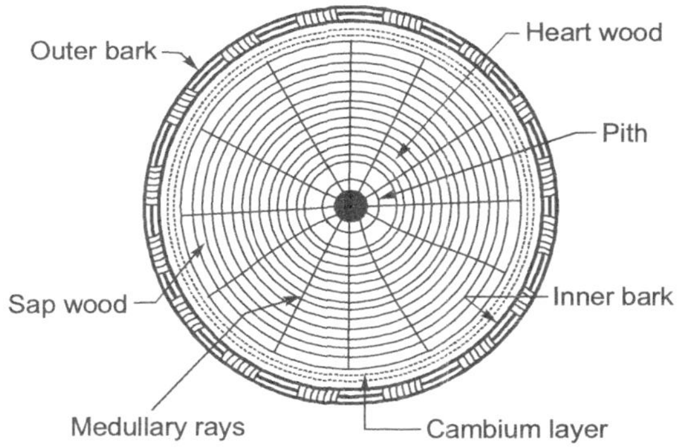

Cross-section of an exogenous tree

(b) **Heartwood** are the inner annual rings surrounding the pith.

(c) **Sap wood** are the outer annual rings between heartwood and cambium layer.

(d) **Cambium layer** is a thin layer of sap between sap wood and inner bark.

(e) **Inner bark** is the inner skin or layer covering the cambium layer.

(f) **Outer bark** is the outer skin or outer cover of the tree.

(g) **Medullary rays** are the thin radial fibres extending from pith to cambium layer.

(ii) **Shakes are cracks which partly or completely separate the fibers of wood.**

(a) Cup shakes are caused by the rupture of tissue in a circular direction. It is a curved crack and it separates one annual ring from the other.

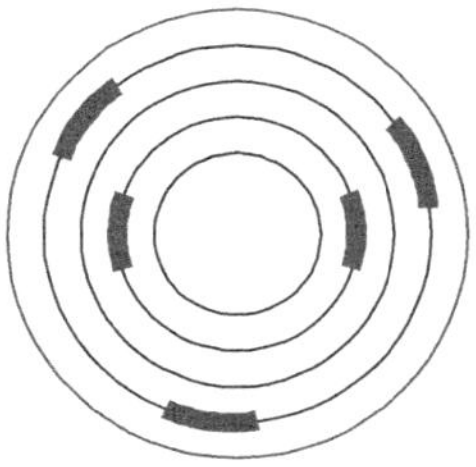
Cup Shakes

(b) Heart shakes occur in the centre of cross-section of tree and they extend from pith to sap wood in the direction of medullary rays.

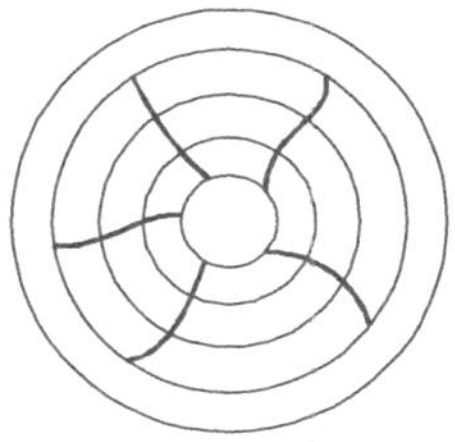
Heart shakes

(c) Ring shakes occur when cup shakes cover the entire ring.

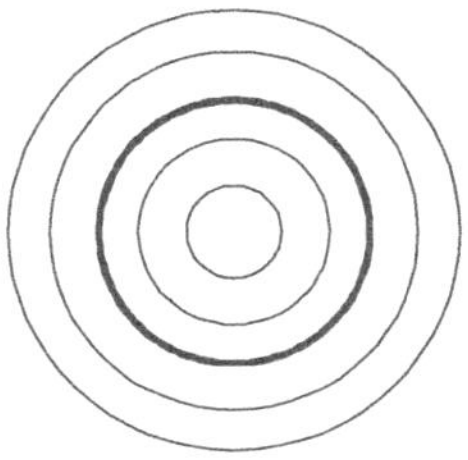
Ring Shakes

(d) Star shakes are cracks which extend from bark towards the sap wood.

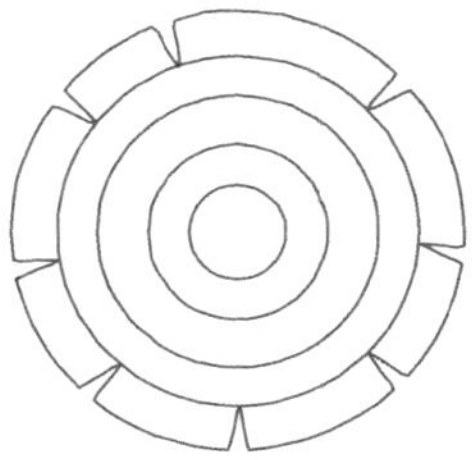
Star Shakes

5.7 Discuss the factors affecting the strength of timber.

[5 marks : 2011]

Solution:

The factors which affect the strength of timber are:

1. **Knots:** These are the sections of the branches of the tree which will be present on the surface of wood in the form of hard dark pieces. Knots are source of weakness in timber when used to carry compression.
2. **Shakes:** These are the cracks and splits in the felled log due to many causes. They can be cup shake, heart shake or circumferential shrinkage
3. **Twisted grain or fibre:** This defect is caused in the tree itself due to (action of) wind.
4. **Upset or rupture:** This is caused due to an injury during the growth of the tree due to strong winds or bad felling of trees resulting in discontinuity of fibres.
5. **Wane:** It is the part of the original outside rounded surface of a tree that remains in the timber during conversion. It is important when the timber is used for important uses, but is not considered important when timber is used for works like shoring, piling, etc.
6. **Presence of Sapwood:** Sapwood is less durable than heartwood and it should not be present in wood to be used at important places.
7. **Sloping grains:** In living trees, the cells do not always grow perfectly vertical or straight and parallel to the length of the trunk. The taper from bottom to the top causes sloping of grain in timber sawn parallel to the pitch. Such sloping grains are considered as a defect.
8. **Cracks, fissures, resin pockets:** Cracks and fissures are fibre disruptions which appear in timber due to many causes. These disruptions affect the strength of timber. Resin pockets and fissures containing resins are defects which affect strength and suitability for decoration of the wood.
9. **Seasoning:** A well seasoned timber has more strength and is more durable than an unseasoned timber.
10. **Conversion:** Conversion of timber tangentially or parallel to medullary says also effects strength of timber.

5.8 What are the varieties of industrial timber? Indicate the procedure followed for making fibre boards.

[10 marks : 2012]

Solution:

The Varieties of industrial timber are:

1. Plywood
2. Particle board or chipboard
3. Hardboard
4. Fibreboard
5. Blackboard
6. Flushdoor shutters

The fibre boards are manufactured from 100 per cent raw agro-wastes such as bagasse, cotton stalks, rise husk, jute, corn etc. The agro based raw material are chipped to graded sizes, converted into fibres which are dried, mixed with synthetic resin binder to form mats which are pressed to form fibre board with uniformly distributed fibres. The binders generally used are phenol formaldehyde and paraffin wax.

Fibre board is manufactured in varying densities, soft, medium and hard. The unit weight of fibre board lies in the range of 2.4 to 10 kN/m^3. The boards are claimed to possess good mechanical, moisture resistant and termite resistant properties. These boards can be cut into any size like natural wood.

The soft board can be used for thermal insulation and as a resilient material for floors; medium density board is suitable for panelling, wall lining, ceilings, doors and windows. Hard boards can be used for doors and windows, as they are claimed to be water and abrasion resistant.

Fibre boards, ranging from soft boards having good thermal insulation to hardboards having properties similar to plywood, are principally made of wood chips or other vegetable fibres. Fibres are reduced to pulp by mechanical or pressure heating methods which interlock fibres mechanically requiring no adhesive such as the lignin in the fibres to act as the bonding agent. In wet process, the pulp is mixed with water and other additives such as water repellants, insecticides, fungicides and formed into sheets on a flat surface. The sheets are hot pressed card boards or simply dried without pressing (soft board). Board types are differentiated by manufacturing process whether produced by the wet or a dry process and their density.

5.9 **Describe four common defects in timber.**

[10 marks : 2003]

Or

Describe the defects in timber with the help of neat sketches.

[5 marks : 2014]

Solution:

The defects occurring in the timber are grouped into the following division:

1. **Defects due to conversion:** During the process of converting timber into commercial form, the following defects may occur.
   (a) **Chip mark:** The defect is indicated by marks or signs placed by chips on the finished surface of timber
   (b) **Diagonal grain:** This defect is formed due to improper sawing of timber.
   (c) **Torn grain:** This defect is caused when a small depression is formed on the finished surface of timber by falling of a tool.
   (d) **Wane:** This defect is denoted by the presence of original rounded surface on the manufactured piece of timber. Fungi defect takes place when moisture > 20% and air is present in surrounding.

2. **Defects due to fungi:** The following defects are caused in timber by the fungi.
   (a) **Blue stain:** The sap of wood is stained to bluish colour by action of certain kind of fungi.
   (b) **Dry rot:** The fungi of certain type feed on wood and during feeding, they attack on wood and convert it into dry powder form. This is known as the dry rot. It occurs where there is no free circulation of air, absence of sunlight, presence of sap i.e., softwood.
   (c) **Brown rot:** The fungi of certain type remove cellulose compounds from wood and hence the wood assumes the brown colour. This is known as brown rot.
   (d) **Wet rot:** Some varieties of fungi causing chemical decomposition of wood of the timber and in doing so, they convert timber into a greyish brown powder alternate drying and wetting condition.
   (e) **White rot:** Fungi attack lignin of wood and wood assumes the appearance of a white mass consisting of cellulose compounds.

3. Defects due to Natural forces

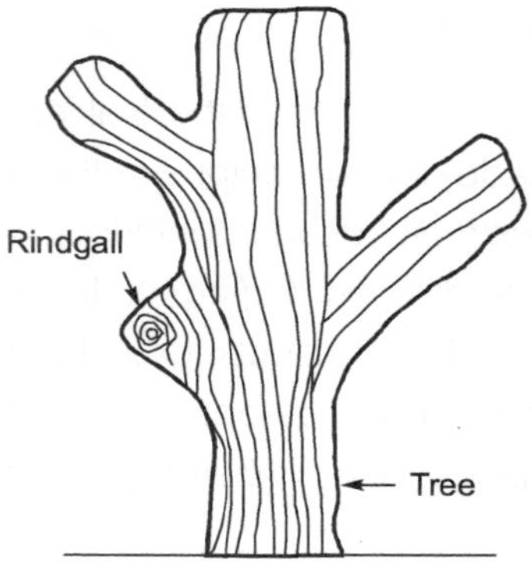

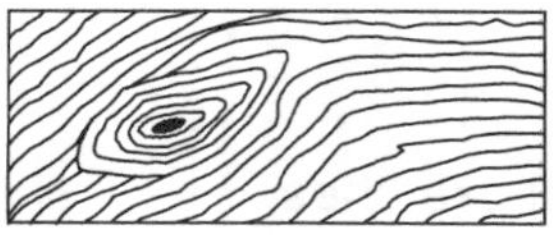

(a) Knot

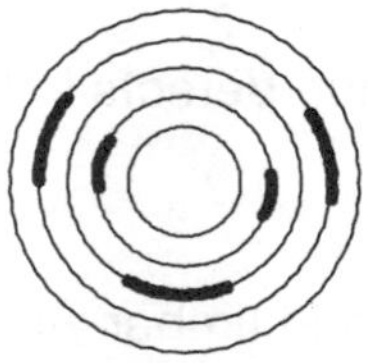

(b) Cup shake

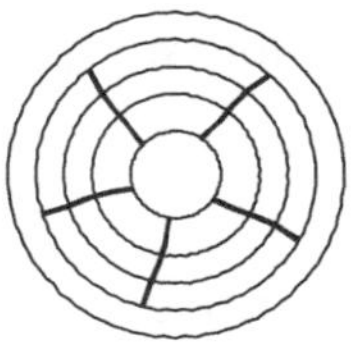

(c) Heart shake

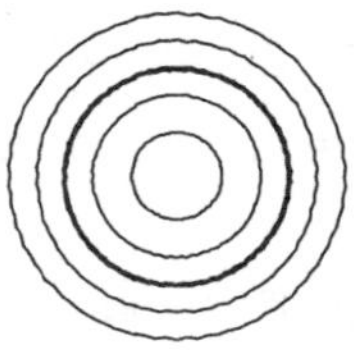

(d) Ring shake

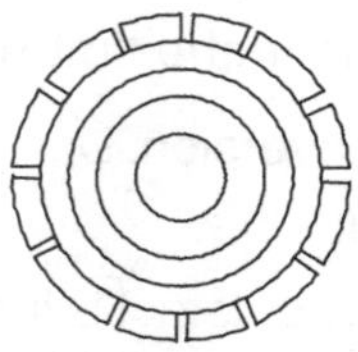

(e) Star shake

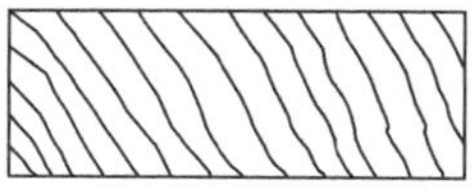

(f) Twisted fibres

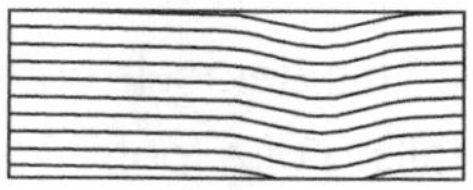

(g) Upset

(a) **Shakes:** Shakes are longitudinal separations in the wood between rings. These lengthwise separations reduce the allowable shear strength.
   - **Heart shake:** Occurs due to shrinkage of heart wood, when tree is overmatured. Cracks starts from pitch and run towards sap wood.
   - **Cup shake:** Appears as curved split which partly or wholly separates annual rings from one another.
   - **Star shake:** These are radial splits or cracks wide at circumference and diminishing towards the centre of the tree.

(b) **Rindgall:** Rindgall is characterized by swelling caused by the growth of layers of sap wood over wounds after the branch has been cut off in an irregular manner.

(c) **Knots:** Knots are bases of twigs or branches buried by cambial activity of the mother branch.

(d) **Twisted fibres:** Twisted fibres are caused by high speed wind which constantly turns the trunk of young tree in one direction.

4. **Defects due to seasoning:** There may be following defects due to seasoning of timber.
   (a) **Bow:** It is a curvature of the timber in the direction of its length.
   (b) **Cup:** A curvature in the transverse direction of the timber.

(c) **Twist:** It is a spiral distortion along the length of the timber.

(d) **Spring:** It is a curvature of the timber in its own plane due to unequal shrinkage in radial and tangential direction.

(e) **Warp:** When a piece of timber has twisted out of shape, it is said to have warped.

5. **Defects due to insects:**

(a) **Beetle:** They form pin holes to attack the sap wood and convert timber into flour like powder. They do not attack shell or cover of timber, hence it looks sound from outside but weak from inside.

(b) **Marine Borers:** Found in salty water make holes of bore tunnels in wood for taking shelters.

(c) **Termites:** Known as white ants found abundantly in tropical and subtropical climate eat core of wood very fast.

(a) Bow

(b) Cup

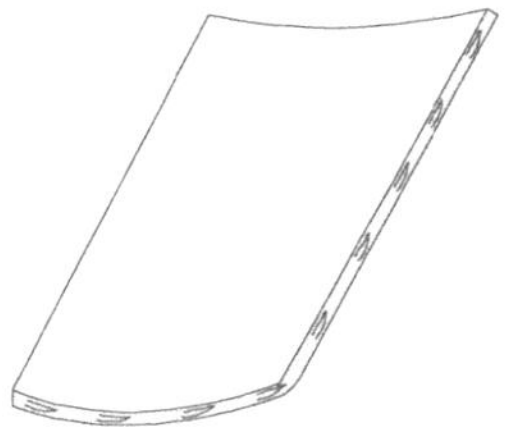

(c) Twist

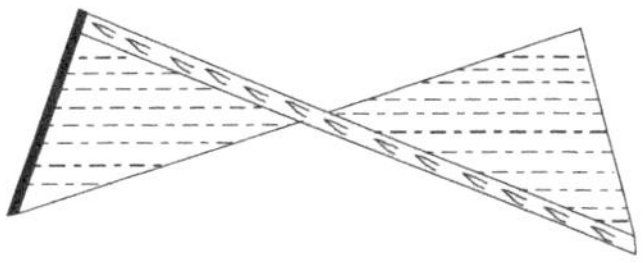

(d) Warp

5.10 Elaborate the terms: Grade of timber, location and factor of safety with reference to structural timber.

[10 marks : 2016]

Solution:

(a) **Grade of timber:** Timber used in construction are graded as per strength and stiffness. Grading into different classes allows the user to select the best timber based on his/her requirements. Grading of timber also ensures that the timber which is unsuitable due to various reasons like low strength, twist, warp etc. must not be used structurally. There are three methods of timber grading viz. machine stress grading, proof grading and the visual stress grading.

(b) **Location of structural timber:**

- **Seasoned softwood:** Seasoned softwood is widely used in residential frame construction (wall frames: studs, plates, headers; floor and roof truss components) and all other internal fit-out elements.
- **Unseasoned softwood:** It is generally used as desk joints, beam or posts.
- **Seasoned hardwood:** It is greatly valued in structural applications for its high durability and strength-to-weight ratio as well as its inherent beauty. It is used in high strength floor and roof beams and joists, as well as support lintels over doors and window openings.
- **Unseasoned hardwood:** Historically, unseasoned or 'green' hardwood was extensively used as both framing and large structural members in residential housing; it has now largely been replaced by seasoned softwood and hardwood. However, in rural areas, where availability is greater, it remains a viable option for joist and beam construction.

(c) Factor of Safety

The belief that a timber with a "factor of safety" of 3 or 4 will carry three or four times the load for which it is designed is erroneous and has been the cause of failures due to overloading of structures. Only a small part of the usual "factor of safety" for wood is available for taking care of overloading; most of it is required to adjust for the known variability in the strength of clear wood.

A general explanation of how this "factor of safety" is taken up largely in adjusting laboratory test data to service conditions is given below:

- **Variations in the strength of clear wood:** The strength of clear wood varies a great deal within a species. It is not uncommon to find one piece of wood twice as strong as another piece of the same species, although both pieces may be clear, straight-grained, and sound. It is evident, therefore, that part of the "factor of safety" as measured by comparison of working stresses with average strength values for a species would be used up in making the working stresses safe for the weaker timbers.
- **Effect of defects:** Defects have about the same effect on strength in all species; that is, a given defect at a given location in a given timber reduces the strength of the timber from the strength of clear wood by about the same percentage, whatever the species.
- **Moisture conditions of service:** Dry wood fibres are stronger then wet wood fibres. For large timbers in damp locations the working stresses must be decreased to make allowance for some deterioration which in such timbers is not offset by any gain in strength due to the dryness of the fibres.
- **Duration of load:** Part of the "factor of safety" is necessary to make timbers safe for loads that may be left on for a long time. Laboratory tests indicate that if a certain load will cause failure in a structure if it is left on for a given time, nine-tenths of that load would cause failure if left on ten times as long.
- **Overloading:** Part of the "factor of safety" plays the role of a true factor of safety; that is, it makes allowance for small accidental overloads that may be left on a structure for a short time.

5.11 **(i) What are various defects in timber? Explain through neat sketches.**

(ii) What is CFRP? Indicate its ultimate tensile strength and modulus of elasticity.

[8 marks : 2017]

Solution:

(i) **Defects in timber:** Following are some of the important defects commonly found in wood due to abnormal growth or rupture of tissues due to natural forces.

(a) **Defects due to Abnormal agrowth:**

- **Checks:** It is a longitudinal crack which is usually normal to the annual rings. These adversely affect the durability of timber because they readily admit moisture and air.
- **Shakes:** These are longitudinal separations in the wood between the annual rings. These lengthwise separations reduce the allowable shear strength without having much effect on compressive and tensile strength values.
  1. Heart shake
  2. Cup shake
  3. Star shake
- **Rind gall:** It is characterised by swelling caused by the growth of layers of sapwood over wounds after the branch has been cutoff in the irregular manner. The newly developed layers do not write properly with old rot, thereby leaving cavities, from where decay starts.

- **Knots:** They are bases of twigs or branches buried by cambial activity of the mother branch. The root of the branch is embedded in the stem with formation of annual rings at right angles to those of the stem.
- **End splits:** End splits are caused by greater exoploration of sap at the end grains of log and can be reduced by painting the exposed.
- **Twisted fibres:** They are caused by wind constantly turning the trunk of young tree in one direction.
- **Upsets:** Upsets are caused by crushing of fibres running transversely during the growth of tree due to strong winds and unskilled felling consequently resulting in discontinuity of fibres.
- **Foxiness:** It is sign or decay appearing in the form of yellow or red tinge or discolouration of overmatured trees.
- **Rupture:** It is caused due to injury or impact.

Star shake

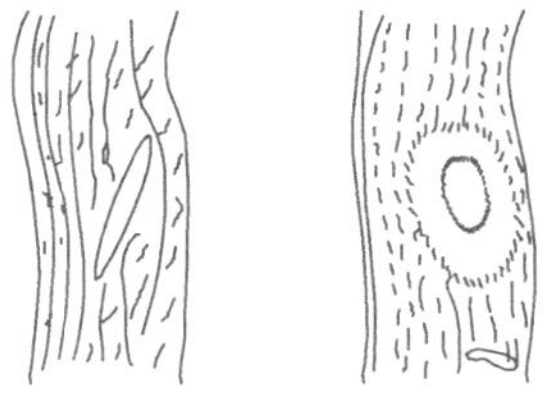

Knots

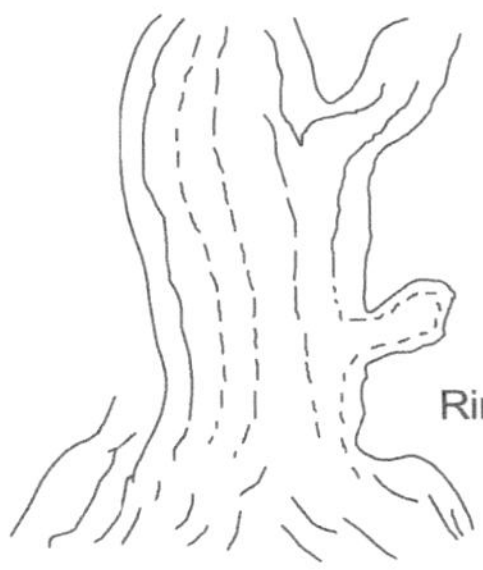

Rindgall

Twisted fibres

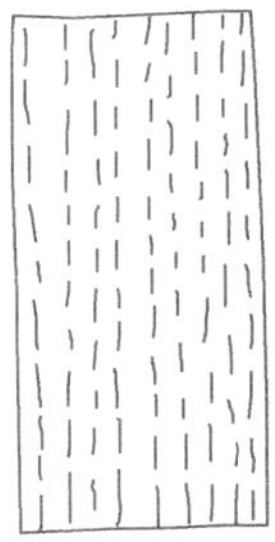

Upsets

(ii) **CFRP (Carbon Fiber Reinforced Polymer):** CFRP is an extremely strong and light fibre reinforced plastic which contains carbon fibres.

In this case, composite materials consists of two parts: a matrix and reinforcement. In CFRP, the reinforcement is carbon fibre which provides the strength. The matrix is usually a polymer resin such as epoxy, to bind the reinforcements together.

The reinforcement will give the CFRP its strength and rigidity measured by stress and modulus respectively. Unlike isotropic materials like steel and aluminium, CFRP has directional strength properties. Despite its high initial strength to weight ratio, a design limitation of CFRP is its lack of a definable endurance limit.

Tensile strength: High tensile strength which may reach upto 700 MPa.

Modulus of Elasticity: They may have modulus of elasticity exceeding 450 GPa.

CFRP are characterized by following properties:

- Light weight
- High Fatigue strength
- Very low coefficient of thermal expansion
- High electric conductivity
- High strength to weight ratio.
- Good Corrosion resistance
- Low impact resistance
- High cost

5.12 Explain the following defects in timber with neat sketches:
(A) Shakes (B) Knots

[4 marks : 2019]

Solution:

A. **Shakes:** These are cracks which partly or completely separate the fibres of wood. Following are the different varieties of shakes:

(a) **Cup Shakes:** These are caused by the rupture of tissue in a circular direction as shown in **Fig.** It is a curved crack and it separates partly one annual ring from the other. It develops due to non-uniform growth. It may not prove to be harmful, if it covers only a portion of ring.

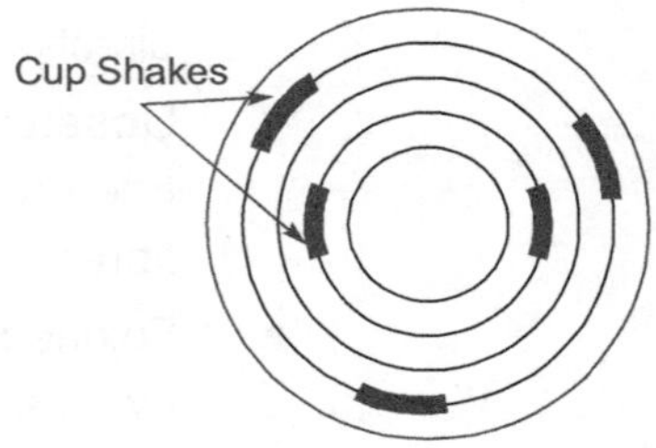

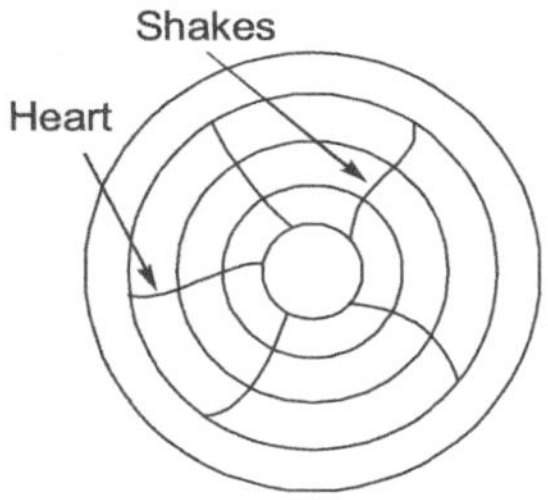

(b) **Heart Shakes:** These cracks occur in the centre of cross-section of tree and they extend from pith to sap wood in the direction of medullary rays as shown in **Fig.** These cracks occur due to shrinkage of interior part of tree which is approaching maturity. Heart shakes divide the tree cross-section into two to four parts.

(c) **Ring Shakes:** When cup shakes cover the entire ring, they are known as ring shakes as shown in **Fig.**

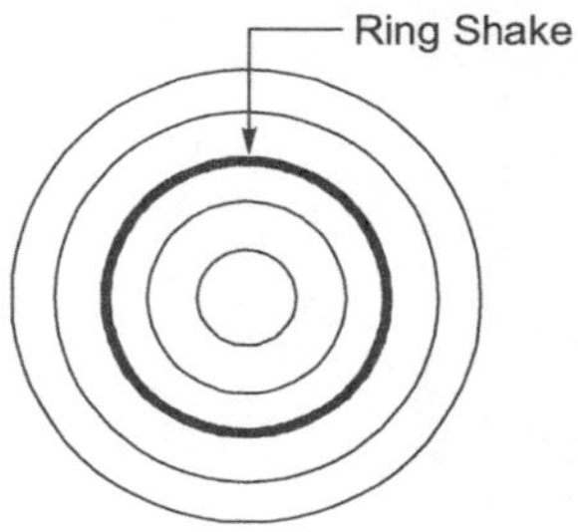

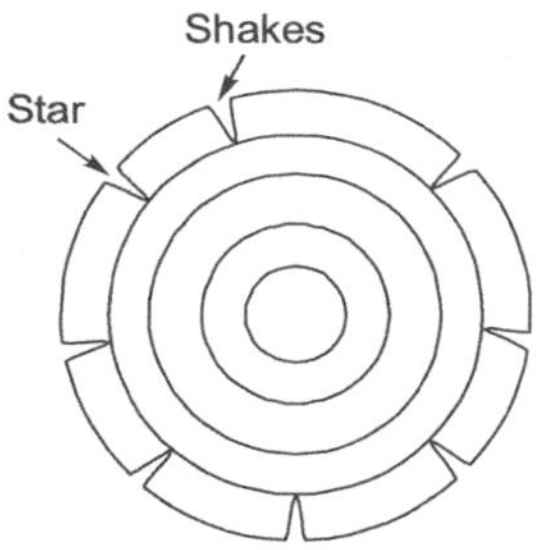

(d) **Star Shakes:** These are cracks which extend from bark towards the sap wood. They are usually confined up to the plane of sap wood. They are usually formed due to extreme heat or frost.

(e) **Radial Shakes:** These are similar to star shakes. But they are fine, irregular and numerous. They usually occur when tree is exposed to sun for seasoning after being felled down.

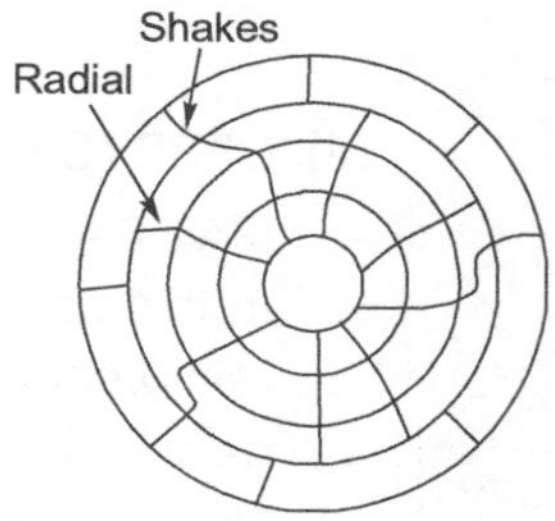

B. **Knots:** These are base of branches or limbs which are broken or cut off from the tree.

- The portion from which the branch is removed receives nourishment from the stem for a pretty long time and it ultimately results in the formation of dark hard rings which are known as the knots.

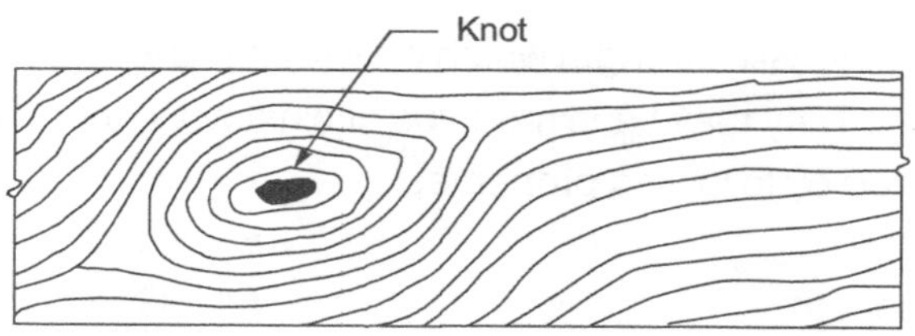

- These are caused by wood limb encased by wood of tree trunk.
- Caused when branch base embedded in timber by natural growth.

5.13 Discuss in brief the methods of preserving timbers by water soluble preservatives.

[12 marks : 2020]

Solution:

Preservative treatment

Preservatives: Timber in exposed application should be treated to minimize fungal decay and attack by insects. Preservative fall into three main groups:

1. Creosote oil preservative
2. Water borne preservative
3. Organic solvent preservative

Water soluble perspectives are odourless organic or inorganic salts and are adopted for inside locations only. If applied over outside surfaces, the salts can be leached by rainwater . Examples of leachable type of preservatives are zinc chloride, boric acid (borex), etc. Zinc chloride, sodium fluoride and sodium-penta-chloro-phenate are toxic to fungi. These are expensive and odourless (except for sodium-penta-chloro-phenate). Benzene-hexa-chloride is used as spray against borers. Boric acid is used against Lyctus borers and to protect plywood in tea chests.

Some of the other water soluble preservatives are fixed type and are as follows:

Copper-chromate-arsenic composition is made of three chemicals:

Arsenic-pentraoxide	$As_2O_5\,2H_2O$	1 part
Copper sulphate	$CuSO_4\,5H_2O$	3 parts
Sodium or potassium dichromate	$(Na\text{ or }K)_2\,Cr_2CO_7$	4 part

The preservatives is in the form of a powder and is used with water. Six parts of this powder is mixed with 100 parts by weight of water. The solution is applied in two coats. The timber is then allowed to dry for six weeks. This treatment renders the timber immune to the attacks of white ants and is known as AsCu treatment.

Acid-cupric-chromate composition:

Chromic acid	1.7 parts
Copper sulphate	50 parts
Sodium dichromate	48.3 parts

Chromate-zinc chloride composition:

Zinc chloride	1 part
Sodium or potassium dichromate	1 part

Copper-chrome-boric composition:

Boric acid	1.5 parts
Copper sulphate	3 parts
Sodium or potassium dichromate	4 parts

Zinc-meta-arsenic composition:

Arsenious trioxide	3 parts
Zinc oxide	2 parts
Acetic acid	Just to keep the above in solution under operating conditions

Zinc-chrome-boric compassion:

Boric acid	1 part
Zinc chloride	3 parts
Sodium dichromate	4 parts
Water	100 parts

5.14 (i) Briefly explain the following with the help of neat sketches:

I. Structure of an exogenous tree

II. Heart shakes and star shakes defects in timber

(ii) Determine the proportion of aggregates A (with Fineness Modulus FM = 7.83) and B (FM = 6.81) required to suitably combine to provide the following grading of such a combination:

IS sieve designation (mm)	80 mm	40 mm	20 mm	10 mm	4.75 mm	2.36 mm	1.18 mm	600 μm	300 μm	150 μm
Cumulative Percentage retained on each sieve	0	6	45	72	95	100	100	100	100	100

[10 + 10 = 20 Marks : 2021]

Solution:

(i) The structure of wood visible to the naked eye or at a small magnification is called the macrostructure.

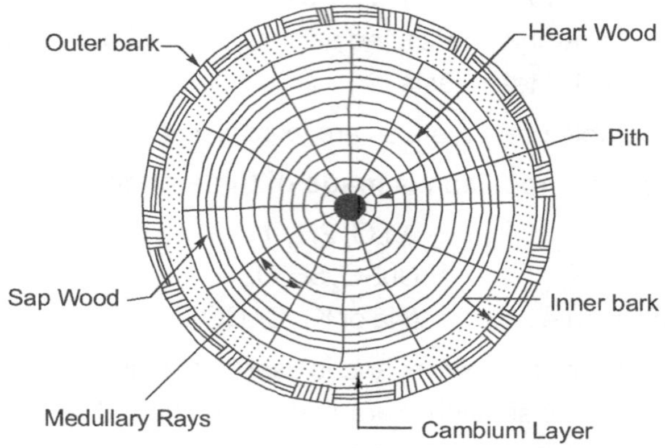

1. **Pith:** The innermost central portion or core of the tree is called the pith or medulla.
2. **Heart Wood:** The inner annual rings surrounding the pith is known as heart wood. It is usually dark in colour.

   It does not take active part in the growth of tree. But it imparts rigidity to tree and hence, it provides strong and durable timber for various engineering purposes.
3. **Sap Wood:** The outer annual rings between heart wood and cambium layer is known as sap wood. It is usually light in colour and weight. It indicates recent growth and it contains sap.

   It takes active part in the growth of tree and sap moves in an upward direction through it. Sap wood is also known as laburnum.
4. **Cambium Layer:** The thin layer of sap between sap wood and inner bark is known as cambium layer. It indicates sap which has yet not been converted into sap wood.
5. **Inner Bark:** It gives protection of cambium layer from any injury.
6. **Outer Bark:** It consists of cells of wood fibre and is also known as cortex.
7. **Medullary Rays:** The thin radial fibres extending from pith to cambium layer are known as *medullary rays*. The function of these rays is to hold together the annual rings of heart wood and sap wood.

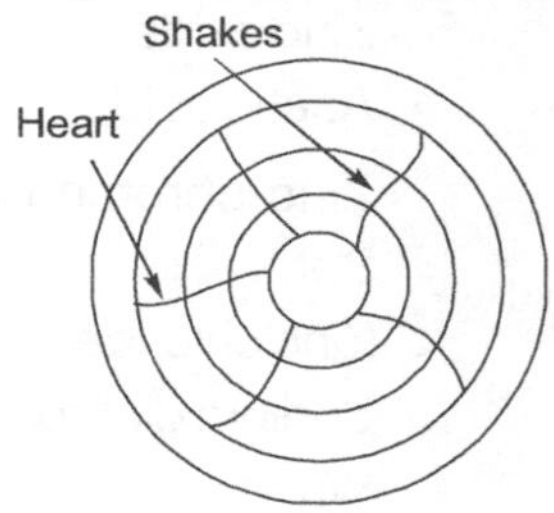

(ii) **Heart Shakes:** These cracks occur in the centre of cross-section of tree and they extend from pith to sap wood in the direction of medullary rays as shown in **Fig.** These cracks occur due to shrinkage of interior part of tree which is approaching maturity. Heart shakes divide the tree cross-section into two to four parts.

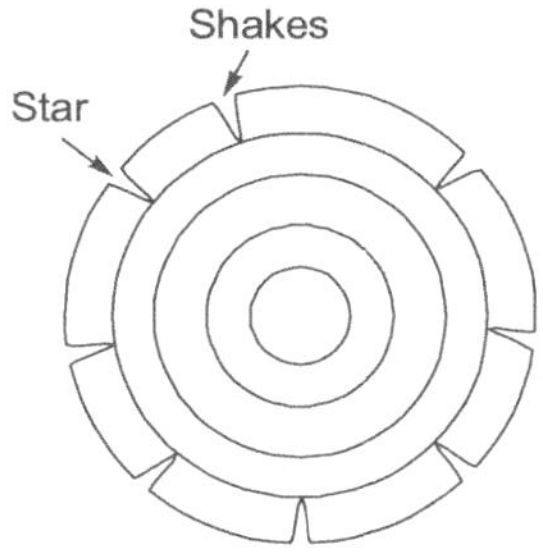

Star Shakes: These are cracks which extend from bark towards the sap wood. They are usually confined up to the plane of sap wood. They are usually formed due to extreme heat or frost.

(ii) Fineness modulus of A $= 7.83$

Fineness modulus of B $= 6.81$

Fineness of modulus of combination of both A and B to be calculating as per given table to get the proportion.

Sieve	Cumulative retention on each sieve
80 mm	0
40 mm	6
20 mm	45
10 mm	72
4.75 mm	95
2.36 mm	100
1.18 mm	100
600 μm	100
300 μm	100
150 μm	100
	Sum of cumulative retained = 718

Fineness modulus of combination,

$$= \frac{\text{Sum of \% cumulative wt. retained}}{100}$$

$$= \frac{718}{100} = 7.18$$

Let the proportion of aggregate, A is P

Then, proportion of aggregate B in mix is

$$= 1 - p$$

Now, $p \times 7.83 + (1 - p) \times 6.81 = 7.18$

$$1.02\,p = 0.37$$

$$p = 0.3627$$

Now, Proportion of A in mix $= 36.27\%$

Proportion of B in mix $= 63.72\%$

Proportion of $A : B = 36.27 : 63.72 = 1 : 1.756$

6. Miscellaneous

6.1 Briefly explain Thermoplastic and Thermosetting materials. List out six differences between them.

[4 + 8 marks : 2018]

Solution:

Thermoplastic materials:

- These are the polymers which can be easily softened repeatedly when heated and hard when cooled with little change in their properties.
- When heated, they melt and form a fluid which can be changed to any desirable shape and then cooled to get desired product.
  Examples : Polythene, polystyrene, PVC, teflon etc.

Thermosetting materials:

- These are the polymers which undergo permanent change on heating. They become hard and infusible on heating.
- They are generally prepared from low molecular mass semifluid substances.

  Examples : Bakelite, melamine, formaldehyde etc.

Thermoplastic	Thermosetting
1. Hardness is a tempering property subjected to rise or fall of temperature.	1. Chemical reaction in this process cannot be reversed. Once solidified they cannot be softened.
2. Formed by addition polymerisation.	2. Formed by condensation polymerisation.
3. Have long chain molecular structure.	3. Aquire three dimensional cross-linked structure.
4. Aesthetically-superior finishes.	4. More difficult to surface finish.
5. More expensive	5. Cost effective
6. Can melt if heated	6. More resistant to high temperature

6.2 Briefly explain manufacture of Aluminium and state at least six physical and mechanical properties of Aluminium.

[8 marks : 2018]

Solution:

Aluminium manufacturing process:

The aluminium manufacturing is accomplished in two phases:

A. The Bayer process of refining the Bauxite ore to obtain aluminium oxide.

B. And the Hall Heroult process of Smelting the aluminium oxide to release pure aluminium.

Step-1: The Bauxite ore is mechanically crushed. Then, the crushed ore is mixed with caustic soda and processed in a grinding mill to produce a slurry containing very fine particles of ore.

Step 2: The slurry is pumped into a digester, a tank that works like a pressure cooker. The slurry is heated to 110-270°C under a pressure of 340 kPa and these conditions are maintained for a time ranging from half an hour to several hours. Additional caustic soda may be added to ensure that all aluminium containing compounds are dissolved.

Step 3: The hot slurry, which is sodium aluminate solution, passes through a series of flash tanks that reduce the pressure and recover heat that can be reused in the refining process.

Step 4: The slurry is pumped into setting tank and the impurities settle at the bottom of the tank.

Step 5: After the impurities have settled out, the remaining liquid is pumped through a series of cloth filters.

Step 6: The filtered liquid is pumped through a series of six-storey tall precipitation tanks. Seed crystals of Alumina hydrate an added through the top of each tank. The seed crystals grow as they settle through the liquid and dissolved alumina attaches to them.

Step 7: The crystals precipitate and are removed. After washing, they are transferred to a kiln for calcining (heating to release the water molecules that are chemically bonded to alumina molecules).

B. The Hall-Heroult Process

Smelting alumina into metallic aluminium takes place in a steel vat called a reduction pot. The bottom of the pot is lined with carbon, which acts as one electrode of the system. The opposite electrode consists of set of carbon rods suspended above the pot.

Step 8: Within the reduction pot, alumina crystals are dissolved in molten cryolite at a temperature of 960°-970°C to form an electrolyte solution, that will conduct electricity from carbon rods to carbon lined bed. A direct current is passed through the solution and freed aluminium atoms settle to the bottom as molten metal.

$$2Al_2O_3 + 3C \rightarrow 4Al + 3CO_2$$

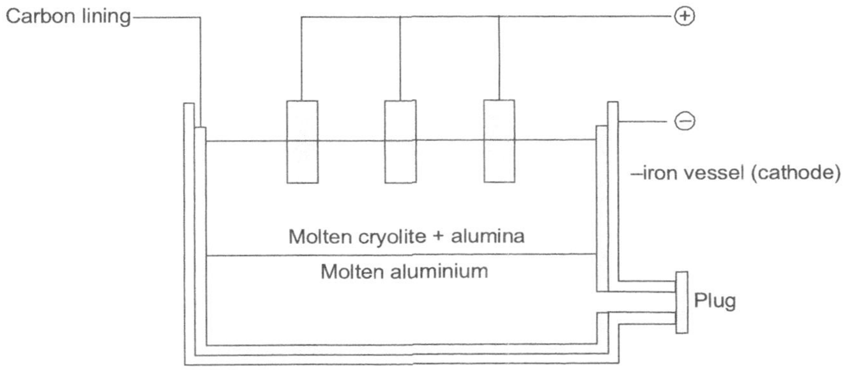

Hall-Heroult Process

Physical and mechanical properties of aluminium are:

- Aluminium is silvery gray metal with symbol Al and atomic no. 13.
- It is soft, non magnetic and ductile metal.
- It is IIIrd most abundant element after oxygen and silicon, and most abundant metal in the crust.
- It has low density of 2700 kg/m^3.
- It has high corrosion resistance due to oxide layer formation over it.
- Though having low density, high mechanical strength may be achieved by suitable alloying and heat treatments.
- Other valuable properties include - high thermal and electrical conductance, its reflectivity, high ductility, magnetic neutrality, high scrap value and non poisonous and colourless nature of its corrosion products.

6.3 Describe the thermal and electrical properties of ceramics.

[8 marks : 2019]

Solution:

Thermal properties of ceramic: Thermal capacity, conductivity and resistance to shocks need to be considered while using a ceramic.

The specific heat for refractories to be used in regenerator chambers should be more. The heat for refractories to be used in regenerator chamber should be more. The heat in ceramics is conducted by photon conductivity and by the interaction of lattice vibration. The ceramic materials do not have enough free electrons to bring out electronic thermal conductivity. At high temperatures, conduction takes place by transfer of radian energy. The thermal conductivity of refractories should be minimum for lining and maximum for crucibles and retorts.

Electrical Properties of ceramic: Ceramic materials have no free electrons so they have low electrical conductivity. However, at high temperatures the ionic diffusion is accelerated.

Clay displays a very high dielectric constant – a property of material related to its behaviour when located within an electric field between two electrodes – under static conditions. However, for alternating current, the dielectric constant in clay arises from ion and electron movement.

6.4 **How is the presence of surface oxide film responsible for excellent corrosion resistance of Aluminium?**

[4 marks : 2019]

Solution:

Resistance to Corrosion: Aluminium is inherently corrosion-resistant. Like stainless steel, aluminium alloys can also be rendered corrosion-resistant by the formation of a protective oxide film. However, in the case of aluminium it is the oxide of the base metal itself that has this characteristic. A transparent layer of aluminium oxide forms on the surface of aluminium almost immediately upon exposure to the atmosphere. Colour can be introduced to this oxide film by the anodising process, which can also be used to develop a thicker protective layer than one that would occur naturally.

Carbon steel, on the other hand, has a tendency to self-destruct over time by virtue of the continual conversion of the base metal to iron oxide, commonly known as rust. Thus, carbon steels require coatings or painting which also have to be maintained and periodically replaced.

Furthermore, aluminium is often used without any finish coating or painting. The cost of the initial painting alone may result in steel being more expensive than aluminium, depending on the quality of coating that is specified. In addition to the direct cost of painting, increasing environmental and worker safety concerns are associated with painting and paint preparation practices. The costs of maintaining steel, then, give aluminium a further advantage in life-cycle cost.

6.5 **Differentiate in brief between Thermoplastic and Thermosetting plastic.**

[8 marks : 2020]

Solution:

Thermo-plastics

- Thermo-plastics variety softens by heat and hardens when cooled down. It can be used by remoulding as many times as required.
- The thermo-plastic or heat non-convertible group is the general term applied to the plastics which becomes soft when heated and hard when cooled.
- Thermoplastic materials can be cooled and heated several times.
- When thermoplastics are heated, they melt to a liquid. They also freeze to a glassy state when cooled enough.
- Thermoplastic can be moulded into any shape.

Thermo-setting plastics

- **Thermo-setting plastics** cannot be reused. This variety requires a great pressure and momentary heat during moulding which hardens on cooling.

- The thermo-setting or heat convertible group is the general term applied to the plastics which become rigid when moulded at suitable pressure and temperature.
- This type of plastic passes originally through thermo-plastic stage. When they are heated in temperature range of 127°C to 177°C, they set permanently and further application of heat does not alter their form or soften them.
- But at the temperature of about 343°C, the charring occurs. This charring is a peculiar characteristic of the organic substances.

6.6 **Explain the types of glazing used for the clay products.**

[6 marks : 2023]

Solution:

Glazing the clay products seals porosity of product, make the product water proof, protects surface from external attacks and also decorates the product.

Types of glazing can be as follows:

1. **Transparent glazing:** It is also known as Salt glazing. It is a process in which NaCl is thrown in kiln when burning is in its peak. Salts at that high temperature reacts with silica of clay and makes sodium silicate which makes product impermeable.
2. **Lead glazing:** It is related with dipping clay product in solution of lead oxide and tin oxide. After this product is again returned to potter's kiln where these adhered particle of lead and tin melts and form a thin transparent layer on outer surface. This metal of glazing suits for product which can't withstand against high temperature as in transparent glazing.
3. **Opaque glazing:** It is also known as Enamelling. It is a process in which borax, chalk and other colouring matter is fired with feldspar fint and lead oxide. The resulting molten glass is poured into water and grounded with remaining material in order to make product having consistency like cream known as slip. After that fully burn earthen wave is dipped into slip and a thin layer glaze is formed over the surface which can be burnt after drying to fuse the glaze over surface.

6.7 **Describe the mechanical properties of ceramics.**

[6 marks : 2023]

Solution:

Mechanical Properties of Ceramics:

- **Tensile Strength:** Theoretically, the tensile strength of ceramics is very high but in practice it is quite low. Tensile failures of ceramics are attributed to the stress concentrations at the pores and micro-cracks at grain corners. Even minute cracks and scratches may drastically reduce the tensile stresses. The modulus of elasticity ranges from 7×10^4 to 42×10^4 N/mm^2. Glass fibres have tensile strength of the order of 700 N/mm^2.
- **Compressive Strength:** The compressive strength is high. So, it is usual to use ceramics like clay, cement and glass products in compression.
- **Shear Strength:** Ceramics have very high shear strength with resistance to failing in a brittle manner.
- **Transverse Strength** is difficult to ascertain and ceramics are not used in places where such strength is the criteria. However, slight flexure may not be a problem.

■■■■

2 Solid Mechanics

Revised Syllabus of ESE : *Elastic constants, Stress, plane stress, Strains, plane strain, Mohr's circle of stress and strain, Elastic theories of failure, Principal Stresses, Bending, Shear and Torsion.*

1. Simple Stress-Strain & Elastic Constants

1.1 (A) he figure below shows three metal cubes *A*, *B*, *C* of side 100 mm in direct contact, resting on a rigid base and confined in x-coordinate direction between two rigid end plate. If the upper face of the center cube *B* is subjected to uniform compressive stress of 0.5 kN/mm^2, compute for cube *B*, the following:

(*i*) The direct stress in x-direction (σ_x)

(*ii*) The direct strains in the three coordinates directions x, y and z

(*iii*) The volumetric strain

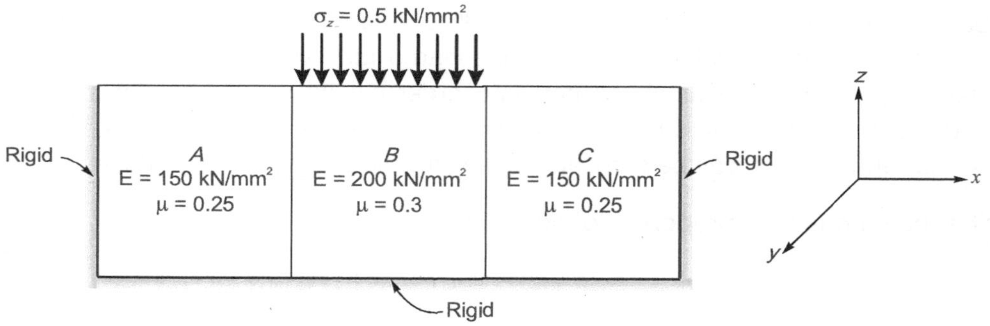

(B) State all assumptions made.

[30 + 10 marks : 1999]

Solution:

(A) (i) Since the cubes are confined in x-direction, hence

$$\Delta_{xA} + \Delta_{xB} + \Delta_{xC} = 0$$

where, Δ_{xA} is change in length of cube *A* in x-direction

Δ_{xB} is change in length of cube *B* in x-direction

Δ_{xC} is change in length of cube *C* in x-direction

$$\frac{\Delta_{xA}}{L} + \frac{\Delta_{xB}}{L} + \frac{\Delta_{xC}}{L} = 0$$

$$\in_{xA} + \in_{xB} + \in_{xC} = 0 \qquad ...(i)$$

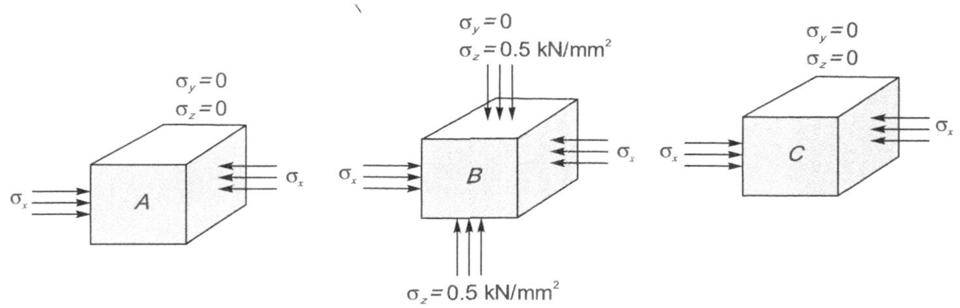

From above FBD, we have,

$$\epsilon_{xA} = -\frac{\sigma_x}{E_A}; \quad \epsilon_{xB} = -\frac{\sigma_x}{E_B} + \mu_B\frac{\sigma_z}{E_B}; \quad \epsilon_{xC} = -\frac{\sigma_x}{E_C}$$

Putting these values in (*i*), we get

$$\left[\frac{-\sigma_x}{E_A}\right] + \left[\frac{-\sigma_x}{E_B} + \mu_B\frac{\sigma_z}{E_B}\right] + \left[\frac{-\sigma_x}{E_C}\right] = 0 \qquad [\because E_A = E_C]$$

$$-\frac{\sigma_x}{E_A} - \frac{\sigma_x}{E_B} + \mu_B\frac{\sigma_z}{E_B} - \frac{\sigma_x}{E_c} = 0$$

$$-\frac{2\sigma_x}{E_A} - \frac{\sigma_x}{E_B} + \mu_B\frac{\sigma_z}{E_B} = 0$$

$$\sigma_x\left[\frac{2}{E_A} + \frac{1}{E_B}\right] = \frac{\mu_B\sigma_z}{E_B}$$

Direct stress in x-direction,

$$\sigma_x = \frac{\mu_B\sigma_z}{2\cdot\frac{E_B}{E_A}+1} = \frac{0.3\times 0.5}{\left(2\times\frac{200}{150}\right)+1} = 0.041 \text{ kN/mm}^2$$

(ii) Direct strain in x,

$$\epsilon_{xB} = -\frac{\sigma_x}{E_B} + \frac{\mu_B\sigma_z}{E_B} = \frac{-0.041}{E_B} + \frac{0.3\times 0.5}{E_B} = 5.45\times 10^{-4}$$

Direct strain in y,

$$\epsilon_{yB} = \frac{+\mu_B\sigma_x}{E_B} + \frac{\mu_B\sigma_z}{E_B}$$

$$= \frac{0.3\times 0.041}{E_B} + \frac{0.3\times 0.5}{E_B} = 8.11\times 10^{-4}$$

Direct strain in z,

$$\epsilon_{zB} = -\frac{\sigma_z}{E_B} + \frac{\mu_B\sigma_x}{B} = \frac{-0.5}{E_B} + \frac{0.3\times 0.041}{E_B} = -2.44\times 10^{-3}$$

(iii) Volumetric strain,

$$\epsilon_V = \epsilon_{xB} + \epsilon_{yB} + \epsilon_{zB}$$

$$= 5.45\times 10^{-4} + 8.11\times 10^{-4} - 2.44\times 10^{-3} = -1.084\times 10^{-3}$$

(B) The assumptions made in the above analysis are:

1. The material is linearly elastic which means it obeys Hooke's law.
2. The material is homogeneous i.e. properties of the material are same at any point of cross-section.
3. Material is isotropic i.e. properties are equal in all directions.
4. The cubes are placed on a perfectly rigid base i.e. base doesn't undergo any deformations.
5. The cubes do not have frictional force acting on their surfaces and also at the ends.

1.2 A steel specimen of 12 mm diameter extends by 6.31×10^{-2} mm over a gauge length of 150 mm when subjected to an axial load of 10 kN. The same specimen undergoes a twist of 0.5° an a length of 150 mm over a twisting moment of 10 N-m. Using the above data, determine the elastic constants E, μ, G and K.

[12 marks : 2001]

Solution:

Tensile Test: $P = 10$ kN

Length of specimen, $L = 150$ mm

Cross-sectional area, $A = \frac{\pi}{4} \times 12^2 = 113.09 \text{ mm}^2$

Change in length of specimen, $\Delta = 6.31 \times 10^{-2}$ mm

Let E N/mm² is modulus of elasticity of material.

We know, axial deformation due to axial load is given by

$$\Delta = \frac{PL}{AE}$$

$$\therefore \quad E = \frac{PL}{A\Delta} = \frac{10 \times 1000 \times 150}{113.09 \times 6.31 \times 10^{-2}} = 2.10 \times 10^5 \text{ N/mm}^2$$

Torsion test:

We know, $$\frac{T}{I_P} = \frac{G\theta}{L} \quad \text{...(i)}$$

$\therefore$ Modulus of rigidity, $$G = \frac{TL}{I_P \theta}$$

$$I_P = \frac{\pi}{32} D^4 = \frac{\pi}{32} \times (12)^4 = 2035.75 \text{ mm}^4$$

Angle of twist, $$\theta = \frac{0.5 \times \pi}{180} \text{ radian} = 8.73 \times 10^{-3} \text{ radian}$$

From eq. (i), we get

$$G = \frac{10 \times 10^3 \times 150}{2035.75 \times 8.73 \times 10^{-3}} = 8.44 \times 10^4 \text{ N/mm}^2$$

We know, $$E = 2G(1 + \mu)$$

$$\frac{E}{2G} = 1 + \mu$$

$$\therefore \quad \mu = \frac{E}{2G} - 1 = \frac{2.10 \times 10^5}{2 \times 8.44 \times 10^4} - 1 = 1.24 - 1 = 0.24$$

Also $$E = 3k(1 - 2\mu)$$

$$k = \frac{E}{3(1 - 2\mu)} = \frac{2.10 \times 10^5}{3(1 - 2 \times 0.24)} = 1.35 \times 10^5 \text{ N/mm}^2$$

1.3 A prismatic bar is fastened between two rigid walls at A and B and subjected to loads as shown in figure below. Determine the reactions at supports.

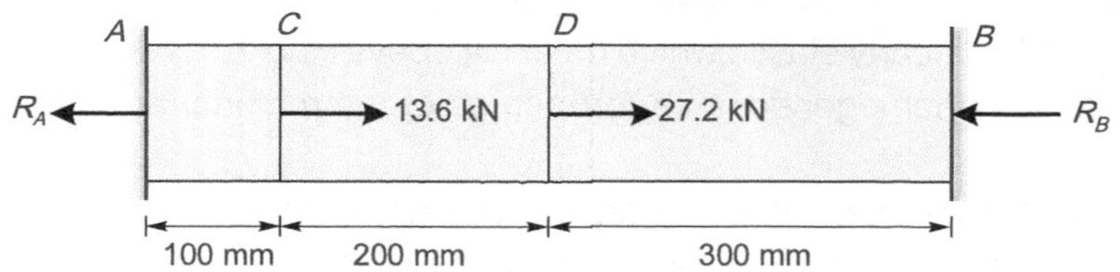

[10 marks : 2001]

Solution:

Consider FBDs of portion AC, CD and DB individually

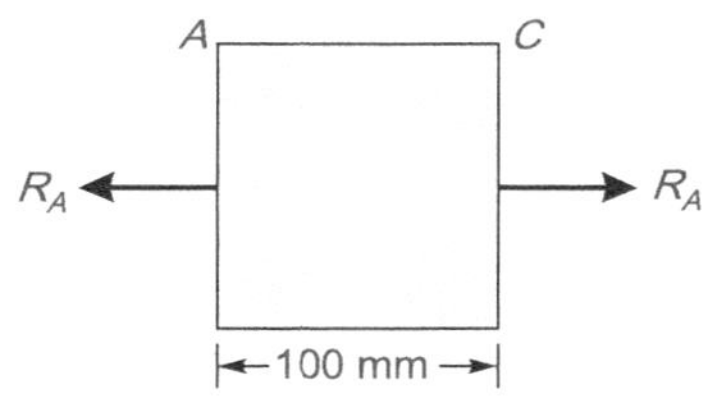

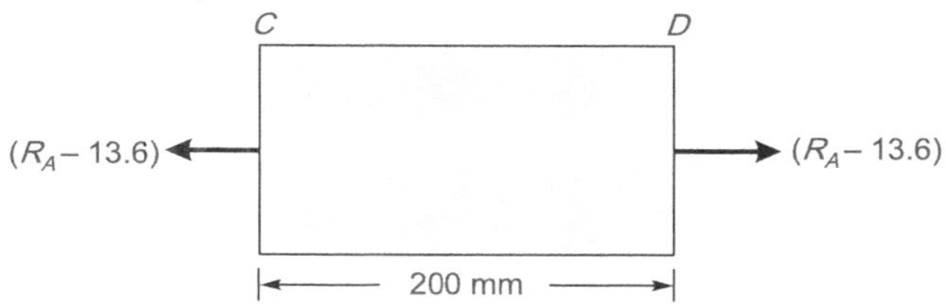

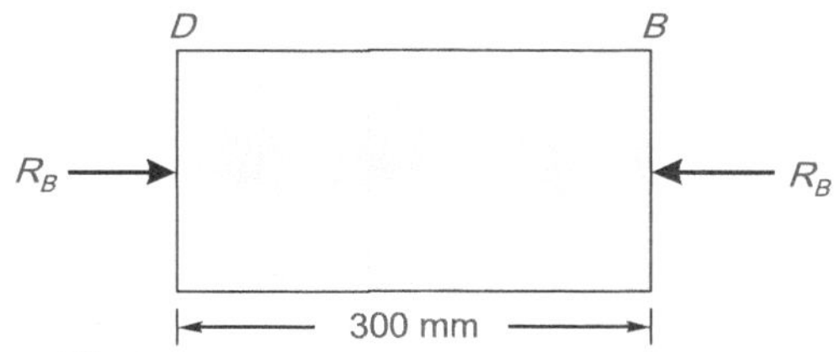

From the condition of static equilibrium,

$$R_A + R_B = 13.6 + 27.2$$

$$R_A + R_B = 40.8 \text{ kN} \quad \text{...(i)}$$

Ends A and B are fixed, then

$$\Delta_{AC} + \Delta_{CD} + \Delta_{DB} = 0$$

$$\frac{R_A \times 100}{AE} + \frac{(R_A - 13.6) \times 200}{AE} - \frac{R_B \times 300}{AE} = 0$$

$$R_A + 2(R_A - 13.6) - 3R_B = 0$$

$$R_A + 2(R_A - 13.6) - 3(40.8 - R_A) = 0$$

$$R_A + 2\,R_A - 27.2 - 122.4 + 3\,R_A = 0$$

$$6\,R_A - 149.6 = 0$$

$$R_A = 24.93 \text{ kN}$$

$\therefore$ From (i), we get $\quad R_B = 15.87$ kN

1.4 A floating column AB of length L, is modelled with top end fixed and bottom end yielding under column load. The yielding support can be visualised as an axial spring of stiffness k. If an axial force P is applied at a distance 'a' from the top, determine reactions at the ends A and B and displacement at the end B.

[10 marks : 2003]

Solution:

The given arrangement can be shown as given figure:

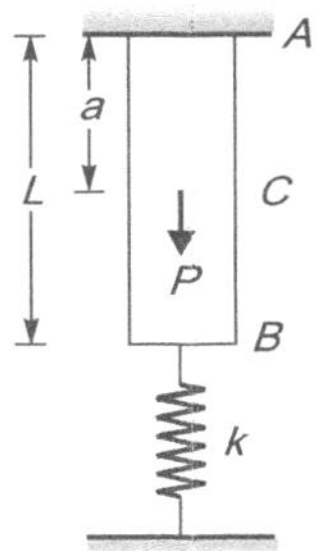

Let reactions at A and B be R_A and R_B respectively.

From FBD of figure (ii), we get

$$P - R_A = R_B$$

$\Rightarrow \quad R_A + R_B = P$

The deflection in the spring,

$$\Delta_S = \frac{R_B}{k} = \frac{P - R_A}{k}$$

Net elongation in column is given by

$$\Delta = \Delta_{AC} + \Delta_{CB} = \frac{R_A a}{AE} - \frac{(P-R_A)(L-a)}{AE}$$

Net elongation in bar = Net compression in spring

i.e. $\Delta = \Delta_s$

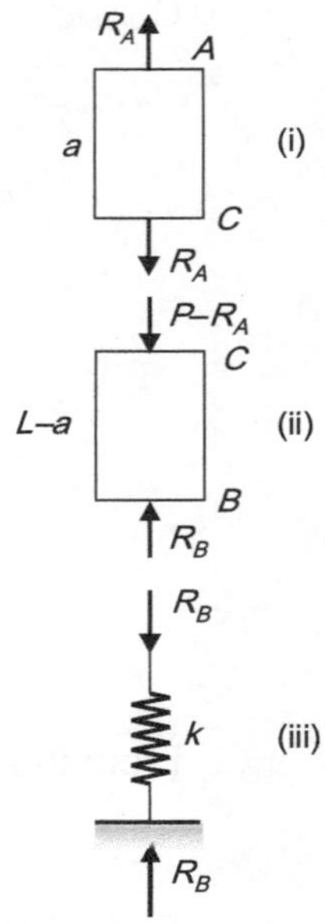

$$\Rightarrow \quad \frac{R_A a}{AE} - \frac{(P-R_A)(L-a)}{AE} = \frac{P-R_A}{k}$$

$$\Rightarrow \quad \frac{R_A a}{AE} - \frac{P(L-a)}{AE} + \frac{R_A(L-a)}{AE} = \frac{P}{k} - \frac{R_A}{k}$$

$$\Rightarrow \quad \frac{R_A a}{AE} + \frac{R_A(L-a)}{AE} + \frac{R_A}{k} = \frac{P}{k} + \frac{P(L-a)}{AE}$$

$$\Rightarrow \quad R_A\left[\frac{a}{AE} + \frac{L-a}{AE} + \frac{1}{k}\right] = P\left[\frac{1}{k} + \frac{L-a}{AE}\right]$$

$$\Rightarrow \quad R_A\left[\frac{L}{AE} + \frac{1}{k}\right] = P\left[\frac{1}{k} + \frac{L}{AE} - \frac{a}{AE}\right]$$

$$\Rightarrow \quad R_A\left(\frac{L}{AE} + \frac{1}{k}\right) = P\left(\frac{1}{k} + \frac{L}{AE}\right) - \frac{Pa}{AE}$$

$$\Rightarrow \quad R_A = P - \frac{Pa}{AE\left(\frac{L}{AE} + \frac{1}{k}\right)}$$

$$\Rightarrow \quad R_A = P - \frac{Pa}{L + \frac{AE}{k}}$$

Also,

$$R_B = P - R_A = P - P + \frac{Pa}{L + \frac{AE}{k}} = \frac{Pa}{L + \frac{AE}{k}}$$

$$\text{Deflection at } B = \Delta_s = \frac{R_B}{k} = \frac{Pa}{k\left(L + \frac{AE}{k}\right)} = \frac{Pa}{kL + AE}$$

1.5 In a tensile test, a test piece of 25 mm diameter is tested over a gauge length of 125 mm. The elongation over this length is 0.0875 mm under a pull of 68,725 N. In a torsion test, a test piece was made of the same material and of same diameter and it twisted 0.025 rad. Over a length of 250 mm at a torque of 0.3068 kNm. Find Poisson's ratio and three elastic modulii of the test piece material.

[15 marks : 2007]

Solution:

Tension test:

Given Data: Diameter of specimen, D = 25 mm; Gauge length, L = 125 mm; Elongation of the specimen, Δ = 0.0875 mm; Tensile force, P = 68725 N

We know, axial deformation due to axial load is given by

$$\Delta = \frac{PL}{AE}$$

$$\therefore \quad E = \frac{PL}{A\Delta} = \frac{68725 \times 125}{\frac{\pi \times 25^2}{4} \times 0.0875}$$

$$E = 2.00 \times 10^5 \text{ N/mm}^2$$

Torsion test:

$$L = 250 \text{ mm}$$
$$T = 0.3068 \text{ kNm}$$

and $\theta = 0.025$ rad

We know, $$\frac{T}{I_P} = \frac{G\theta}{L}$$

$$\therefore \quad G = \frac{TL}{I_P\theta} = \frac{0.3068 \times 10^6 \times 250}{\frac{\pi \times (25)^4}{32} \times 0.025}$$

$$\therefore \quad G = 8 \times 10^4 \text{ N/mm}^2$$

Let k and μ be the Bulk modulus and Poisson's ratio of material respectively,

We know that

$$E = 2G(1 + \mu)$$
$$2.00 \times 10^5 = 2 \times 8 \times 10^4 (1 + \mu)$$
$$1 + \mu = 1.25$$
$$\therefore \quad \mu = 0.25$$

Also, we know that,

$$E = 3K(1 - 2\mu)$$

$$k = \frac{E}{3(1-2\mu)} = \frac{2.00 \times 10^5}{3(1-2 \times 0.25)}$$

$$k = 1.33 \times 10^5 \text{ N/mm}^2$$

1.6 A homogenous isotropic material of size 20 cm × 20 cm × 20 cm is placed inside a rigid box of internal dimension 20 × 20 × 20 cm. The top of the box is open and the material inside is subjected to a uniform compressive stress of 10 N/mm² from the top. Assuming there is no friction between the surface of contact, find the stresses on the side faces of the material. Also find maximum shear stress, the normal stresses on the plane of maximum shear stress and the change of volume of the material. Young's modulus E is 1×10^5 N/mm² and Poisson's ratio, $m = 0.3$ for the material.

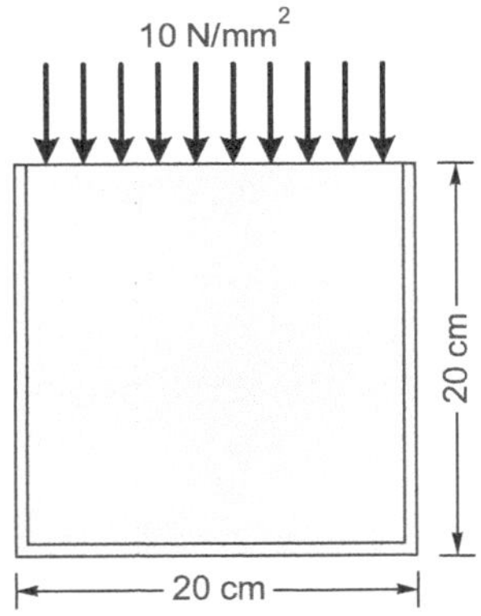

[15 Marks : 2008]

Solution:

Given Data: $\sigma_z = 10$ N/mm²; $E = 1 \times 10^5$ N/mm² and $\mu = 0.3$

The free body diagram of the material is given by

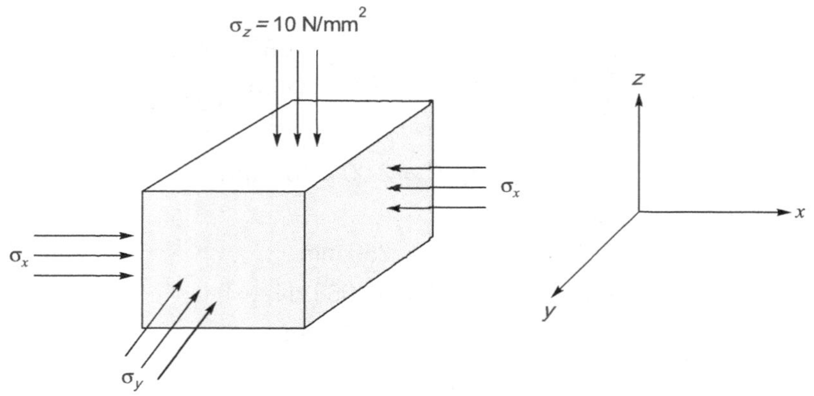

The stresses are induced in x and y directions, since the material is confined in x and y directions. Thus, change in length of material in x-direction

$$\Delta x = 0$$

and change in length of material in y-direction

$$\Delta y = 0$$

Since the material is homogenous and isotropic, therefore the stresses in x and y directions will be equal

i.e. $\sigma_x = \sigma_y$

But, $\Delta x = 0$

or $\dfrac{\Delta x}{L} = 0$

so, $\epsilon_x = 0$

also, $$-\frac{\sigma_x}{E} + \mu\frac{\sigma_y}{E} + \mu\frac{\sigma_z}{E} = 0 \qquad [\because \sigma_x = \sigma_y]$$

$$-\sigma_x + \mu\sigma_x = -\mu\sigma_z$$

$$\sigma_x(1-\mu) = \mu\sigma_z$$

$$\sigma_x = \frac{\mu\sigma_z}{1-\mu} = \frac{0.3\times 10}{1-0.3} = 4.28 \text{ N/mm}^2 \text{ (Compressive)}$$

$\therefore$ $$\sigma_y = \sigma_x = 4.28 \text{ N/mm}^2 \text{ (Compressive)}$$

Since there is no shear stress in plane of x, y and z

So, σ_x, σ_y and σ_z are principal stresses

$$\sigma_1 = -\sigma_z = -10 \text{ N/mm}^2$$

$$\sigma_2 = -\sigma_x = -4.28 \text{ N/mm}^2$$

We know maximum shear stress is given by

$$\tau_{max} = -\left(\frac{\sigma_1 - \sigma_2}{2}\right)$$

Here, $$\tau_{max} = -\left[\frac{-10-(-4.28)}{2}\right] = 2.86 \text{ N/mm}^2$$

The normal stress on the plane of maximum shear stress is given by

$$\sigma_n = \frac{\sigma_1 + \sigma_2}{2} = \frac{-10+(-4.28)}{2} = -7.14 \text{ N/mm}^2$$

Volumetric strain, $$\frac{\Delta V}{V} = \frac{\sigma_x + \sigma_y + \sigma_z}{E}(1-2\mu)$$

$$\frac{\Delta V}{V} = \frac{-4.28 + (-4.28) + (-10)}{1 \times 10^5} \times (1 - 2 \times 0.3)$$

$$\frac{\Delta V}{20 \times 20 \times 20} = -7.424 \times 10^{-5}$$

$$\Delta V = -0.5939 \text{ cm}^3$$

1.7 A stepped vertical steel bar *ABCD* is fixed at the top end *A*. Each segment of the bar *AB*, *BC* and CD is 1 m long and has cross-sections 20 mm × 20 mm, 10 mm × 10 mm and 20 mm × 20 mm respectively. A 5 kN load is applied directly at *D* and 6 kN loads are applied on the levers attached to the stepped bar at *B* and *C* as shown in figure. Find the vertical displacement of *D* and the change in volume of the bar.

$E = 2 \times 10^5$ MPa and Poisson's ratio $\mu = 0.25$. Connections between the levers and bar at *B* and *C* are hinged.

[15 marks : 2009]

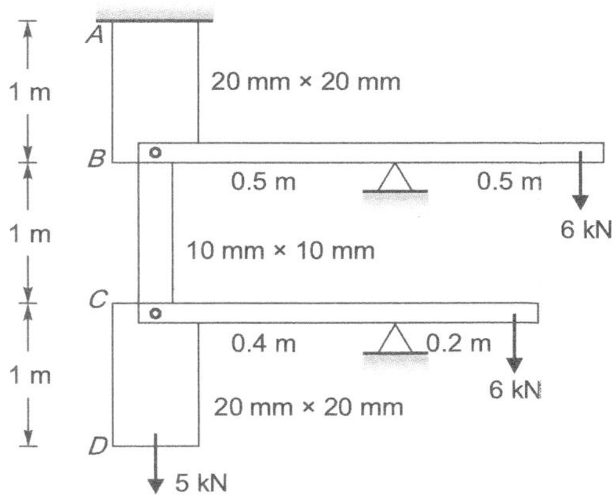

Solution:

Let the reactions at *B* and *C* be $V_B(\downarrow)$ and $V_C(\downarrow)$ respectively.

Since *B* and *C* are hinged, hence there will be no moments induced at *B* and *C*.

Considering the lever attached at *B*, the bending moment at the support should be equal when calculated from either of the ends i.e.

$$V_B \times 0.5 = 6 \times 0.5$$

$$V_B = 6 \text{ kN} (\downarrow)$$

Similarly,

$$V_C \times 0.4 = 6 \times 0.2$$

$$V_C = 3 \text{ kN} (\downarrow)$$

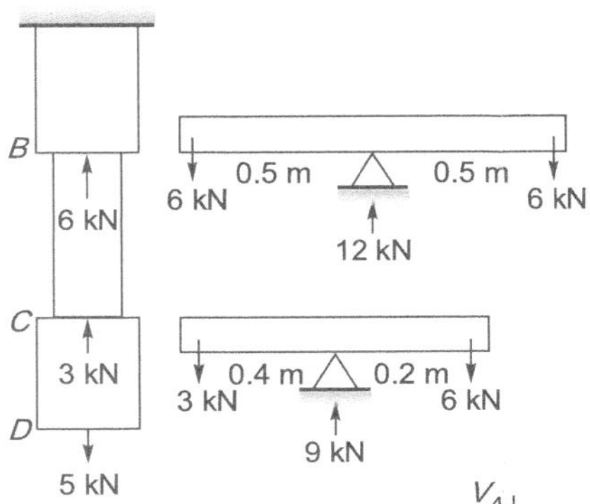

Free body diagram of the stepped bar and levers.

Further free body diagram of the stepped bar.

There will be tension in portion *BC* i.e.

$$6 - V_A = 5 - 3$$

$$\Rightarrow \quad 6 - 2 = V_A$$

$$\Rightarrow \quad V_A = 4 \text{ kN}$$

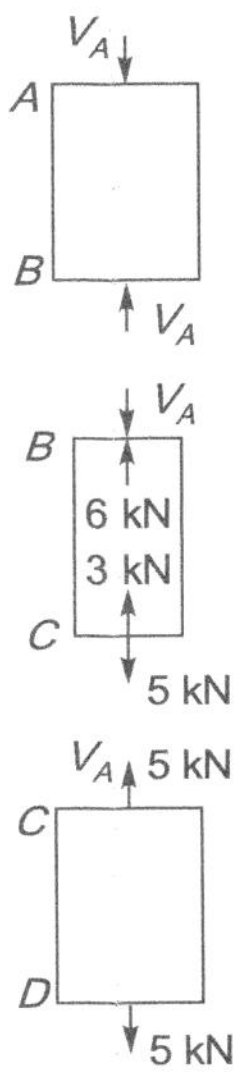

Thus *AB*, will expreience a compressive force of 4 kN, *BC* and *CD* will experience a tensile force of 2 kN and 5 kN respectively.

$$\text{Vertical deflection of } D = -\Delta_{AB} + \Delta_{BC} + \Delta_{CD}$$

$$= -\frac{4 \times 10^3 \times 1 \times 10^3}{20 \times 20 \times 2 \times 10^5} + \frac{2 \times 10^3 \times 1 \times 10^3}{10 \times 10 \times 2 \times 10^5} + \frac{5 \times 10^3 \times 1 \times 10^3}{20 \times 20 \times 2 \times 10^5}$$

$$= -0.05 + 0.1 + 0.0625 = 0.1125 \text{ mm}$$

$$\text{Volumetric strain for } AB, \quad \frac{\Delta V_{AB}}{V_{AB}} = -\frac{\sigma}{E} + \mu \times \frac{\sigma}{E} + \mu \times \frac{\sigma}{E}$$

$$\Rightarrow \quad \Delta V_{AB} = \frac{\sigma}{E} \times (2\mu - 1) \times V_{AB}$$

$$\Rightarrow \quad \Delta V_{AB} = \frac{4\times10^3}{20\times20}\times\frac{1}{2\times10^5}\times(2\times0.25-1)\times20\times20\times1000$$

$$\Rightarrow \quad \Delta V_{AB} = -10 \text{ mm}^3 \text{ (Decrease)}$$

Volumetric strain for BC, $\dfrac{\Delta V_{BC}}{V_{BC}} = \dfrac{\sigma}{E} - \mu\times\dfrac{\sigma}{E} - \mu\times\dfrac{\sigma}{E}$

$$\Rightarrow \quad \Delta V_{BC} = \frac{\sigma}{E}\times(1-2\mu)\times V_{BC}$$

$$\Delta V_{BC} = \frac{2\times10^3}{10\times10\times2\times10^5}\times(1-2\times0.25)\times10\times10\times1000$$

$$\Rightarrow \quad \Delta V_{BC} = 5 \text{ mm}^3 \text{ (increase)}$$

Volumetric strain for CD, $\dfrac{\Delta V_{CD}}{V_{CD}} = \dfrac{\sigma}{E} - \mu\times\dfrac{\sigma}{E} - \mu\times\dfrac{\sigma}{E}$

$$\Rightarrow \quad \Delta V_{CD} = \frac{\sigma}{E}\times(1-2\mu)\times V_{CD}$$

$$\Rightarrow \quad \Delta V_{CD} = \frac{5\times10^3}{20\times20}\times\frac{1}{2\times10^5}\times(1-2\times0.25)\times20\times20\times1000$$

$$\Rightarrow \quad \Delta V_{CD} = 12.5 \text{ mm}^3 \text{ (increase)}$$

Total change in volume of stepped bar, $ABCD = \Delta V_{AB} + \Delta V_{BC} + \Delta V_{CD}$

$$= -10 + 5 + 12.5 = 7.5 \text{ mm}^3 \text{ (increase)}$$

1.8 **A stepped bar with circular cross-section and supported at top, hangs vertically under its own weight. Dimensions of the bar are as shown in the figure below. Calculate the elongation of the bar under its own weight. What is the change in diameter of the bar at section *A-A* shown in the figure?**

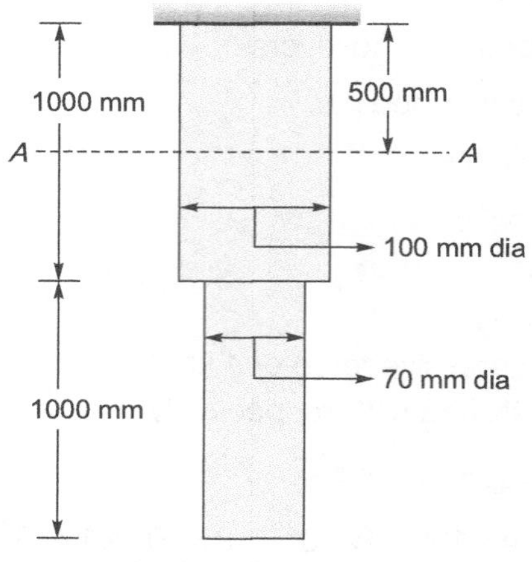

$E = 2\times10^5$ N/mm², density $\gamma = 8\times10^{-5}$ N/mm³ and Poisson's ratio $\mu = 0.2$.

[20 marks : 2010]

Solution:

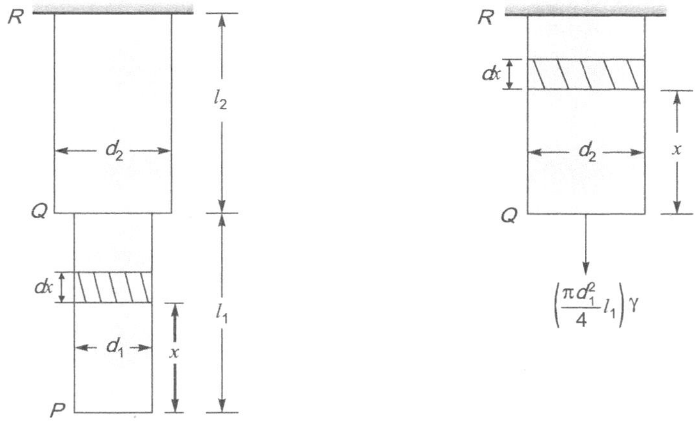

$$\text{Elongation of bar } PQ = \int_0^{l_1} \frac{\left(\frac{\pi d_1^2}{4} \times \gamma\right) dx}{\frac{\pi d_1^2}{4} \times E} = \int_0^{l_1} \frac{\gamma x\, dx}{E} = \frac{\gamma}{E} \frac{x^2}{2}\Bigg|_0^{l_1} = \frac{\gamma l_1^2}{2E}$$

$$\text{Elongation of bar } QR = \int_0^{l_2} \frac{\left(\frac{\pi d_1^2}{4}\gamma l_1 + \frac{\pi d_2^2}{4} x\gamma\right) dx}{\frac{\pi d_2^2}{4} E} = \frac{1}{E}\int_0^{l_2}\left(\frac{\gamma d_1^2 l_1}{d_2^2} + \gamma x\right) dx = \frac{\gamma d_1^2 l_1 l_2}{d_2^2 E} + \frac{\gamma l_2^2}{2E}$$

$$\text{Total Elongation of stepped bar} = \frac{\gamma l_1^2}{2E} + \frac{\gamma l_2^2}{2E} + \frac{\gamma d_1^2 l_1 l_2}{d_2^2 E}$$

$$= \frac{8.0 \times 10^{-5}}{2 \times 10^5 \times 2}\left[(1000)^2 + (1000)^2 + 2 \times (1000)(1000) \times \left(\frac{70}{100}\right)^2\right]$$

$$\text{Total elongation} = 5.96 \times 10^{-4} \text{ mm}$$

Change in diameter at *A-A*

$$\text{Lateral strain} = \frac{-\mu\sigma}{E} = \frac{\Delta d}{d_2}$$

Stress at *A-A*

$$\sigma = \frac{\gamma \frac{\pi}{4}\left(d_1^2 l_1 + d_2^2 \frac{l_2}{2}\right)}{\frac{\pi d_2^2}{4}} = \gamma\left[\frac{d_1^2}{d_2^2} l_1 + \frac{l_2}{2}\right]$$

$$\sigma = 8 \times 10^{-5}\left[\left(\frac{70}{100}\right)^2 \times 1000 + 500\right] = 0.0792 \text{ N/mm}^2$$

$$\Delta d = \frac{-\mu\sigma}{E} d_2 = -\frac{0.2 \times 0.0792}{2 \times 10^5} \times 100 = -7.92 \times 10^{-6} \text{ mm}$$

Negative sign indicates that diameter at *AA* decreases.

1.9 **A compound bar consists of a circular rod of steel of diameter 25 mm, rigidily fitted into a copper tube of internal diameter 25 mm and thickness 2.5 mm. If the bar is subjected to an axial load of 100 kN, find the stresses developed in the two materials.**
Given $E_s = 2 \times 10^5$ N/mm and $E_c = 1.2 \times 10^5$ N/mm².

[10 marks : 2011]

Solution:

Given: Steel bar dia = 25 mm; Internal dia of copper tube = 25 mm; Outer dia of copper tube = 25 + 2.5 + 2.5 = 30 mm; Bar is subjected to an axial load P = 100 kN; $E_s = 2 \times 10^5$ N/mm², $E_c = 1.2 \times 10^5$ N/mm²

Let, A_1 = cross-sectional area of steel bar = $\frac{\pi}{4} \times 25^2 = 490.87$ mm²

A_2 = cross-sectional area of C_u tube $\frac{\pi}{4} \times (30^2 - 25^2) = 215.98$ mm²

Let P_1 and P_2 be the forces acting in steel bar and C_u tube respectively. Also let σ_1 and σ_2 be the stresses developed in steel bar and copper tube.

By the condition of static equilibrium,

the external force will be resisted by steel (P_1) and copper tube (P_2),

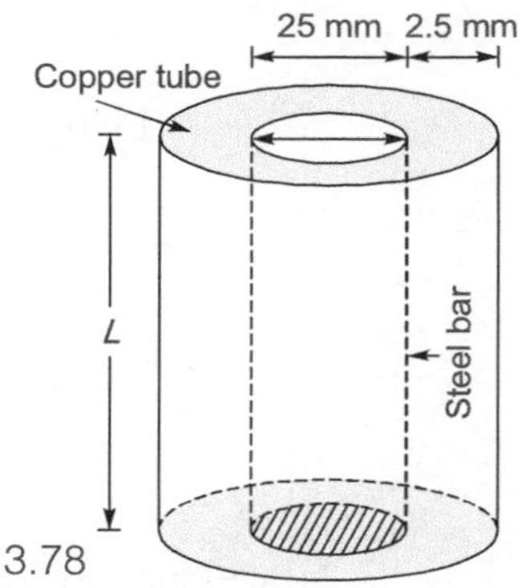

$$\therefore \quad P = P_1 + P_2 \quad \ldots(i)$$

Also, by compatibility condition,

$$\Delta_{steel} = \Delta_{cu}$$

or
$$\frac{P_1 L}{A_1 E_s} = \frac{P_2 L}{A_2 E_c}$$

or
$$\frac{P_1}{P_2} = \frac{A_1 E_s}{A_2 E_c} = \frac{\frac{\pi}{4} \times 25^2 \times 2 \times 10^5}{\frac{\pi}{4} \times (30^2 - 25^2) \times 1.2 \times 10^5} = 3.78$$

$$\therefore \quad P_1 = 3.78 P_2$$

From equation (i),
$$P = 3.78 P_2 + P_2 = 4.78 P_2$$

$$P_2 = \frac{100}{4.78} = 20.92 \text{ kN}$$

and
$$P_1 = P - P_2 = 100 - 20.90 = 79.08 \text{ kN}$$

so, **stresses in steel bar,**
$$\sigma_1 = \frac{P_1}{A_1} = \frac{79.08}{\frac{\pi}{4} \times 25^2} = 0.161 \text{ kN/mm}^2$$

and **stresses in Cu tube,**
$$\sigma_2 = \frac{P_2}{A_2} = \frac{20.92}{\frac{\pi}{4}(30^2 - 25^2)} = 0.097 \text{ kN/mm}^2$$

1.10 **Compute the total elongation caused by an axial load of 100 kN applied to a flat bar 20 mm thick, tapering from a width of 120 mm to 40 mm in a length of 10 m as shown in figure. Assume E = 200 GPa.**

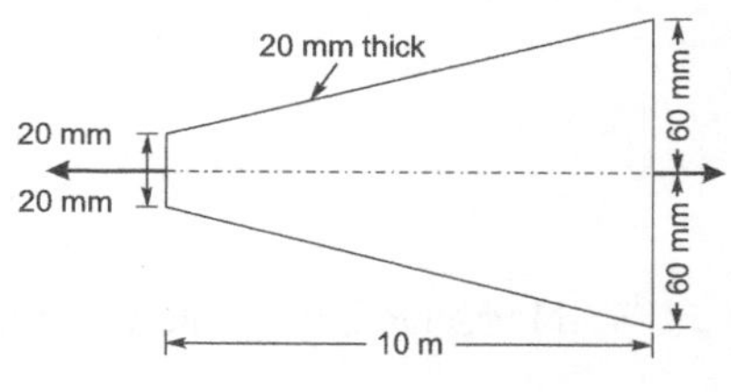

[20 marks : 2013]

Solution:

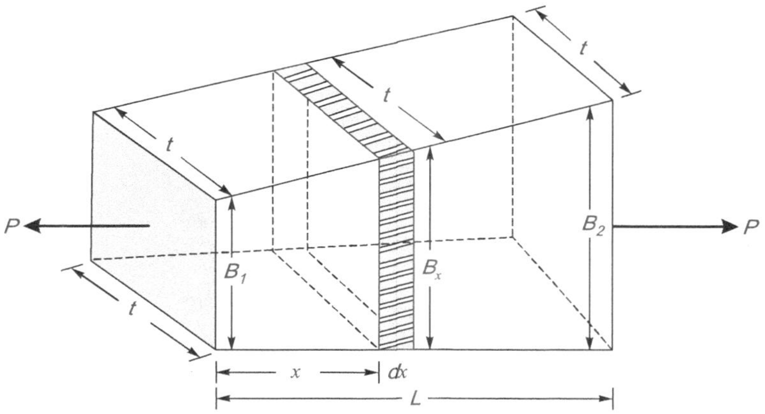

Let us consider a uniform tapering bar of rectangular cross-section, length L and thickness t. The width of the bar at one end is B_1 and the width at the other end is B_2, where $B_2 > B_1$ and bar is subjected to an axial force P.

Consider an element of length dx, width B_x and thickness t,

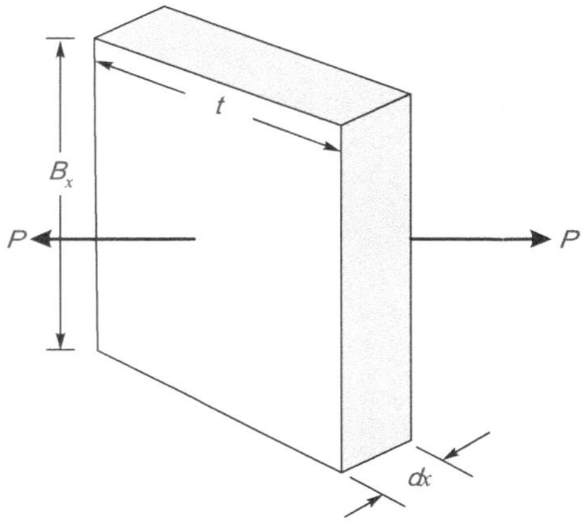

Width, $$B_x = B_1 + \frac{B_2 - B_1}{L}\cdot x$$

Cross-sectional area of element
$$A_x = B_x \cdot t$$

Now this element can be consider as prismatic bar of length dx and area cross-section A_x

$$d\Delta = \frac{P\cdot dx}{A_x E}$$

Hence the deflection of whole length L

$$\int d\Delta = \Delta = \int_0^L \frac{P\,dx}{A_x E}$$

$$\Delta = \int_0^L \frac{P\,dx}{\left(B_1 + \frac{B_2 - B_1}{L}\cdot x\right)\cdot tE} = \frac{P}{tE}\int_0^L \frac{dx}{\left(B_1 + \frac{B_2 - B_1}{L}\cdot x\right)}$$

$$\Delta = \frac{PL}{tE}\left[\frac{\log\left(B_1 + \frac{B_2 - B_1}{L}\cdot x\right)}{(B_2 - B_1)}\right]_0^L = \frac{PL\log_e\left(\frac{B_2}{B_1}\right)}{(B_2 - B_1)\,tE}$$

Given: $P = 100$ kN ; $L = 10$ m, $B_2 = 120$ mm, $B_1 = 40$ m, t = 20 mm

$$= \frac{(100\times 10^3\text{N})(10\text{m})}{(0.08\text{m})\times(0.02)\times\left(2\times 10^{11}\ \text{N/m}^2\right)}\ln\left(\frac{120}{40}\right) = 3.433 \text{ mm}$$

1.11 A non-prismatic rod is attached between two supports as shown in figure. It is subjected to axial loads of 80 kN and 100 kN. Determine the support reactions and elongations. Take E = 200 GPa.

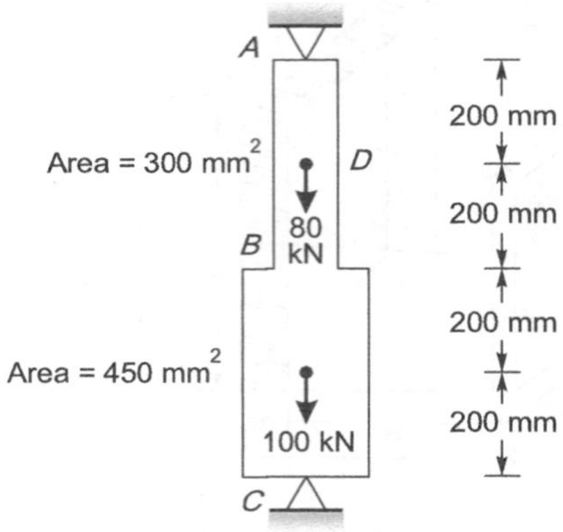

[25 marks : 2016]

Solution:

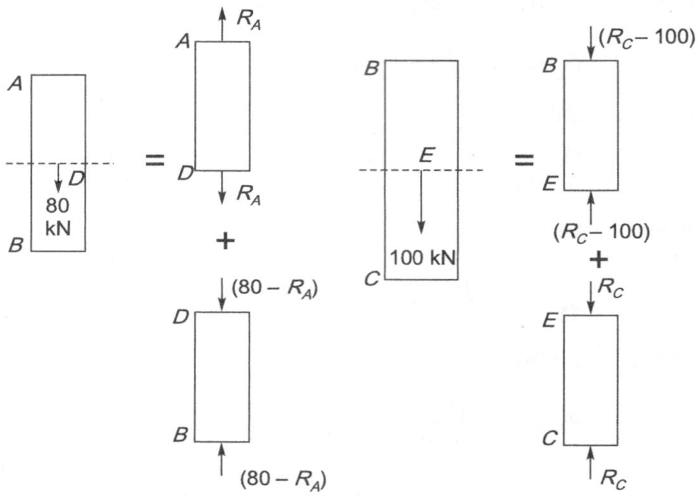

$\therefore$ At interface B, $\quad 80 - R_A = R_C - 100$

$$\Rightarrow \quad R_A + R_C = 180 \text{ kN} \quad \text{...(i)}$$

Now supports A and C are unyielding.

$\therefore$ Change in length of composite bar $AC = 0$

$$\Rightarrow \quad \Delta_{AD} + \Delta_{DB} + \Delta_{BE} + \Delta_{EC} = 0$$

$$\Rightarrow \quad \frac{R_A(200)}{300E} + \left[\frac{-(80 - R_A)200}{300E}\right] + \left[\frac{-(R_C - 100)200}{450E}\right] + \left[\frac{-R_C 200}{450E}\right] = 0$$

$$\Rightarrow \quad \frac{R_A}{300} - \frac{(80 - R_A)}{300} - \frac{(R_C - 100)}{450} - \frac{R_C}{450} = 0$$

$$\Rightarrow \quad 3R_A - 3(80 - R_A) - 2(R_C - 100) - 2R_C = 0$$

$$\Rightarrow \quad 6R_A - 4R_C = 40$$

$$\Rightarrow \qquad 3R_A - 2R_C = 20$$

Solving (i) and (ii),

$$R_C = 104 \text{ kN}$$
$$R_A = 76 \text{ kN}$$

$\therefore$ Elongation of bar portion *AD*,

$$(\Delta_{AD}) = \frac{R_A(200)}{300 \times 2 \times 10^5} = \frac{76000 \times 200}{300 \times 2 \times 10^5} = 0.2533 \text{ mm}$$

Compression of bar portion *DB*,

$$(\Delta_{DB}) = \frac{(80 - R_A)200}{300 \times 2 \times 10^5} = \frac{4 \times 10^3 \times 200}{300 \times 2 \times 10^5} = 0.0133 \text{ mm}$$

Compression of portion *BE*,

$$(\Delta_{BE}) = \frac{(R_C - 100)200}{450 \times 2 \times 10^5} = \frac{4 \times 10^3 \times 200}{450 \times 2 \times 10^5} = 0.00889 \text{ mm}$$

Compression of portion *EC*,

$$(\Delta_{EC}) = \frac{R_C(200)}{450 \times 2 \times 10^5} = \frac{104 \times 1000 \times 200}{450 \times 2 \times 10^5} = 0.231 \text{ mm}$$

Check: $\Delta_{DB} + \Delta_{BE} + \Delta_{EC}$ = Total compression

$= 0.25319$ mm

$\simeq \Delta_{AD}$ = Total elongation

1.12 A rectangular bar of 30 mm × 60 mm cross-section and 200 mm length is restrained from expansion along its 30 mm × 200 mm sides by surrounding material. Find the change in dimensions and volume when a compressive force of 180 kN acts in axial direction. $E = 2 \times 10^5$ MPa; $\nu = 0.3$

[12 marks : 2017]

Solution:

Let us take coordinate directions as shown below. The bar is restrained from expanding in *y*-direction. Let the stresses that develop to restrain be σ_y (compressive).

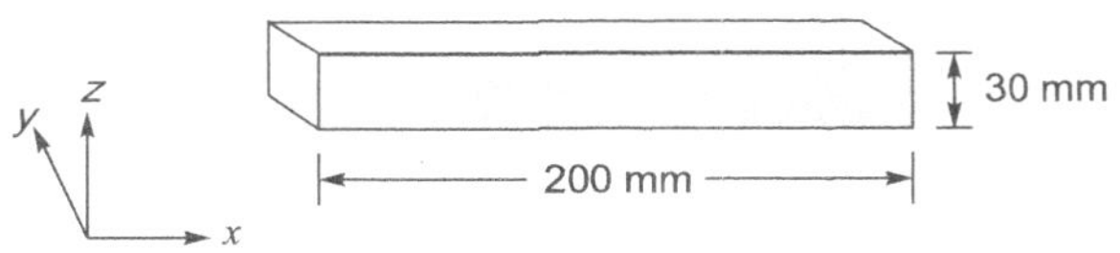

Now, $\sigma_z = 0$ and $\sigma_x = \dfrac{180 \times 10^3}{30 \times 60} = 100 \text{ N/mm}^2$ (compressive)

Taking compressive strain as positive.,

$$\varepsilon_x = \frac{\sigma_x}{E} - \mu\frac{\sigma_y}{E} = \frac{100 - 0.3\sigma_y}{2 \times 10^5} \qquad \text{...(i)}$$

($\sigma_z = 0$ since there is no restriction on 200 mm × 60 mm face i.e., in z-direction)

$$\varepsilon_y = \frac{\sigma_y}{E} - \mu\frac{\sigma_x}{E} = \frac{\sigma_y - 0.3\times 100}{2\times 10^5} \quad ...(ii)$$

$$\varepsilon_z = -\mu\frac{\sigma_y}{E} - \mu\frac{\sigma_x}{E} = \frac{-0.3}{2\times 10^5}(100+\sigma_y) \quad ...(iii)$$

When it is fully restrained.

Given, $\varepsilon_y = 0$

$\therefore$ $\sigma_y = 30\ \text{N/mm}^2$

$\therefore$ $$\varepsilon_x = \frac{100 - 0.3\times 30}{2\times 10^5} = \frac{91}{2\times 10^5}$$

and $$\varepsilon_z = \frac{-0.3\times(100+30)}{2\times 10^5} = \frac{-39}{2\times 10^5}$$

Now, Change in length $= \varepsilon_x \times 200 = \dfrac{91\times 200}{2\times 10^5} = 0.091\text{mm}$

Change in width $= \varepsilon_y \times 60 = 0$

Change in depth $= 30\times\varepsilon_z = \dfrac{30(-130\times 0.30)}{2\times 10^5} = -0.00585\ \text{mm}$ (extension)

$\therefore$ Change in volume $= (\varepsilon_x + \varepsilon_y + \varepsilon_z)\times V$

$$= \left(\frac{91 - 130\times 0.3}{2\times 10^5}\right)\times 30\times 60\times 200$$

$\Rightarrow$ $\delta V = 93.60\ \text{mm}^3$

1.13 **A very long steel drill pipe got stuck in hard clay at an unknown depth. The drill pipe was applied a large upward force and observed that the drill pipe came out elastically by 500 mm. It was also observed that there was elongation of 0.04 mm in a gauge length of 200 mm. Estimate the depth of hard clay bed. Following consideration may be taken into account:**
Resistance offered by all material/elements may be taken as zero.

[12 marks : 2018]

Solution:

Very long drill pipe got stuck at an unknown depth.
Let the depth be L.
Elongation in gauge length of 200 mm = 0.04 mm
Elongation (ΔL) in overall length of L = 500 mm

$\therefore$ Strain $= \dfrac{0.04}{200} = 2\times 10^{-4}$

Since resistance offered by all materials is to be taken as zero.
Strain will be same for the long drill pipe

$\therefore$ $$\frac{500\ \text{mm}}{L} = 2\times 10^{-4}$$

$\therefore$ $L = 2.5\times 10^6$ mm

or Length = 2500 m

1.14 A square bar (50 mm × 50 mm cross-section) of 100 mm length is subjected to an axial compressive load of 10 kN. Calculate the change in volume of the bar, if all lateral strains of the bar are prevented by a uniform pressure on its four lateral faces. Calculate this pressure value and the change in volume. Also calculate the value of bulk modulus K and shear modulus G for the material of bar. Following parameters may be used if required.

1. $E = 2 \times 10^5$ N/mm^2
2. $\mu = 0.25$

[20 marks : 2018]

Solution:

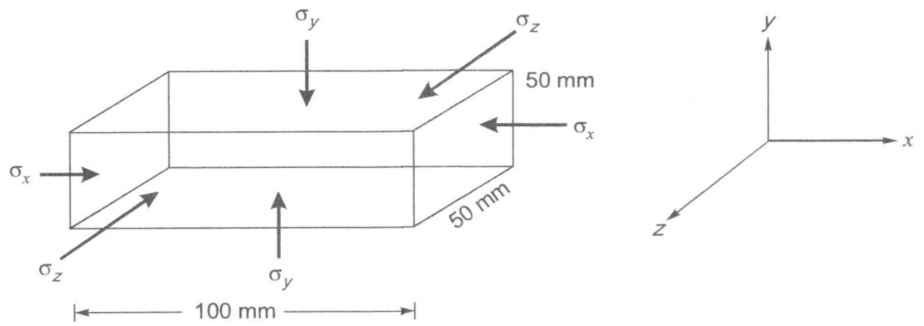

Given : 1. $E = 2 \times 10^5$ N/mm^2; 2. $\mu = 0.25$

(i)

$$\sigma_x = \frac{\text{Force}}{\text{Area}} = \frac{10 \times 1000}{50 \times 50} = 4 \text{ N/mm}^2 \text{(compressive)}$$

As lateral strain is prevented,

So, $\varepsilon_y = \varepsilon_z = 0,$

Let, $\sigma_y = \sigma_z = P = $ Applied pressure

$$\Rightarrow \quad \epsilon_\psi = 0 \Rightarrow \quad \frac{P}{E} - \frac{\mu P}{E} - \frac{\mu \sigma_x}{E} = 0$$

$$\Rightarrow \quad P(1-\mu) = \mu \times \sigma_x$$

$$\Rightarrow \quad P = \frac{0.25 \times 4}{(1-0.25)}$$

$$\Rightarrow \quad P = 1.33 \text{ N/mm}^2 \quad \text{(Compressive)}$$

(ii)

$$\varepsilon_v = \varepsilon_x + \varepsilon_y + \varepsilon_z$$

$$\varepsilon_x = \frac{\sigma_x}{E} - \frac{\mu\sigma_y}{E} - \frac{\mu\sigma_z}{E}$$

$$\varepsilon_y = \frac{\sigma_y}{E} - \frac{\mu\sigma_z}{E} - \frac{\mu\sigma_x}{E}$$

$$\varepsilon_z = \frac{\sigma_z}{E} - \mu\frac{\sigma_x}{E} - \mu\frac{\sigma_y}{E}$$

Adding above 3 equations,

$$\varepsilon_v = \varepsilon_x + \varepsilon_y + \varepsilon_z$$

$$\Rightarrow \quad \varepsilon_v = \frac{\sigma_x}{E}(1-2\mu) + \frac{\sigma_y}{E}(1-2\mu) + \frac{\sigma_z}{E}(1-2\mu)$$

$$\Rightarrow \quad \varepsilon_v = \frac{(\sigma_x + \sigma_y + \sigma_z)}{E}(1-2\mu)$$

$$\Rightarrow \quad \frac{\Delta V}{V} = \frac{(-4-1.33-1.33)(1-2\times 0.25)}{2\times 10^5}$$

$$\Rightarrow \quad \Delta V = -1.667 \times 10^{-5} \times 50^2 \times 100$$

$$\Rightarrow \quad \Delta V = -4.167 \text{ mm}^3$$

(iii) Bulk modulus

$$E = 3k(1-2\mu)$$

$$\Rightarrow \quad k = \frac{2\times 10^5}{3(1-2\times 0.25)}$$

$$\Rightarrow \quad k = 1.33 \times 10^5 \text{ N/mm}^2$$

(iv) Shear modulus

$$E = 2G(1+\mu)$$

$$\Rightarrow \quad G = \frac{2\times 10^5}{2\times(1+0.25)}$$

$$\Rightarrow \quad G = 8 \times 10^4 \text{ N/mm}^2$$

1.15 A compound tube consists of a steel tube 150 mm internal diameter and 170 mm external diameter and a brass tube of 170 mm internal diameter and 190 mm external diameter. The two tubes are of the same length. The compound tube carries an axial load of 1000 kN. Find the stresses and the load carried by each tube and the amount it shortens. Length of each tube is 140 mm. Take $\bar{E}$ for steel as 2×10^5 N/mm^2 and for brass as 1×10^5 N/mm^2.

[8 marks : 2019]

Solution:

For **steel tube**

Given, Internal diameter, $D_{is} = 150$ mm

External diameter, $D_{os} = 170$ mm

For **brass tube**

Internal diameter, $D_{iB} = 170$ mm

External diameter, $D_{oB} = 190$ mm

Also, $L_B \Rightarrow L_S = 140$ mm

Axial load, $P = 1000$ kN

$E_{steel} = 2 \times 10^5$ N/mm^2

and $E_{brass} = 1 \times 10^5$ N/mm^2

Let us assume

The load carried by steel tube = P_s

And the load carried by brass tube = P_b

Using compatibility equation,

$$\Delta_{steel} = \Delta_{brass}$$

$$\frac{P_s L_s}{A_s E_s} = \frac{P_b L_b}{A_b E_b} \qquad (\because L_S = L_B)$$

$$\frac{P_s \times 1000}{\frac{\pi}{4}(170^2 - 150^2)\times 2\times 10^5} = \frac{P_b \times 1000}{\frac{\pi}{4}(190^2 - 170^2)\times 10\times 10^5}$$

$$\frac{P_b}{(190^2 - 170^2)} = \frac{P_s}{(170^2 - 150^2)\times 2}$$

$$P_s = 1.778\ P_b$$

Also,
$$P_s + P_b = 1000 \text{ kN}$$
$$1.778\ P_b + P_b = 1000$$
$$P_b = 360 \text{ kN}$$
$$P_s = 640 \text{ kN}$$

So,
$$\sigma_s = \frac{640\times 1000}{\frac{\pi}{4}(170^2 - 150^2)} = 127.3239 \simeq 127.32 \text{ N/mm}^2$$

$$\sigma_b = \frac{360\times 1000}{\frac{\pi}{4}(190^2 - 170^2)} = 63.66 \text{ N/mm}^2$$

Shortening, i.e.,
$$\Delta_b = \Delta_s = \frac{P_s L_s}{A_s E_s} = \frac{640\times 1000\times 140}{\frac{\pi}{4}(170^2 - 150^2)\times 2\times 10^5}$$
$$= 0.089 \text{ mm}$$

1.16 On a steel bar specimen of 15 mm diameter and 150 mm gauge length, when tested as a tensile test specimen, a force of 15 kN produces an extension of 0.063 mm. When the specimen of same diameter and same length is tested under torsion, a twisting moment of 6.94 Nm produces an angular twist of 0.15°. Determine the Poisson's ratio of the material of the bar.

[4 marks : 2019]

Solution:

Diameter of bar, $d = 15$ mm

Gauge length, $l = 150$ mm

Let E be the Young's modulus of elasticity.

In Tensile test,

$$\text{Elongation} = \frac{PL}{AE}$$

$$0.063 = \frac{15\times 1000\times 150}{\frac{\pi}{4}\times 15^2\times E}$$

$$E = 202101.515 \text{ MPa}$$

In Torsion test,

Let G be the modulus of rigidity

So,
$$\theta = \frac{TL}{GJ}$$

$$\frac{0.15°}{360} \times 2\pi = \frac{6.94 \times 1000 \times 150}{G \times \frac{\pi}{32} \times 15^4}$$

$$G = 80005 \text{ MPa}$$

Now as we know, if μ is Poisson's ratio then

$$E = 2G(1 + \mu)$$

$$202101.515 = 2 \times 80005 (1 + \mu)$$

$$\mu = 0.263$$

1.17 Three vertical rods carry a tensile load of 100 kN. Area of cross-section of each rod is 500 mm^2. Their temperature is raised by 60°C and the load is now so adjusted that they extend equally. Determine the load shared by each. The outer two rods are of steel and the middle one is of brass. $E_s = 2\ E_b = 210$ GPa. $\alpha_s = 11 \times 10^{-6}$/°C; $\alpha_\beta = 18 \times 10^{-6}$/°C

[12 marks : 2019]

Solution:

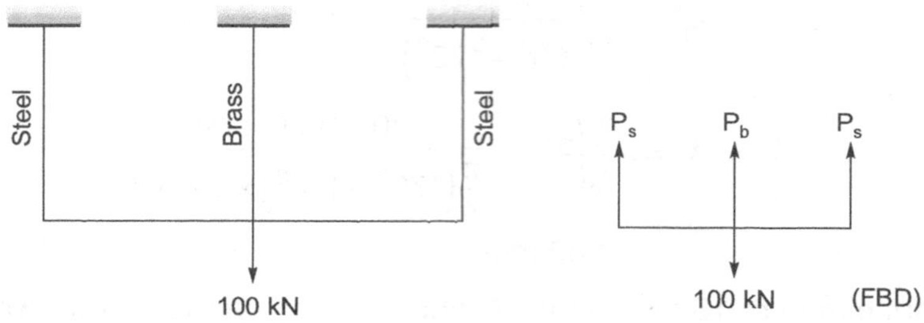

Given, the temperature rise = 60°C

The area of bars, $A_s = A_b = 500$ mm^2

And $E_s = 2E_b = 210$ GPa

Assuming the length of the bars to be equal.

Also, $2P_s + P_b = 100$ kN ...(i)

Using compatibility equation,

$$\Delta_s = \Delta_b$$

$$\frac{P_s L_s}{A_s E_s} + L_s \alpha_s \Delta T = \frac{P_b L_b}{A_b E_b} + L_b \alpha_b \Delta T$$

$$\frac{P_s \times 1000}{500 \times 2.1 \times 10^5} - 11 \times 10^{-6} \times 60 = \frac{P_b \times 1000}{500 \times 1.05 \times 10^5} + 18 \times 10^{-6} \times 60 \qquad [\because L_s = L_b]$$

After solving,

$$9.5238 \times 10^6\ P_s - 1.09047 \times 10^{-5}\ P_b = -0.42 \times 10^{-3}$$

$$9.5238\ P_s - 10.9047\ P_b = -420 \qquad \text{...(ii)}$$

Solving equation (i) and (ii)

We get, $P_s = 48.82$ kN (Tension)

$P_b = -2.36$ kN (Tension)

1.18 Calculate the modulus of rigidity and bulk modulus of a cylindrical bar of diameter 30 mm and of length 2.0 m if the longitudinal strain in a bar during a tensile stress is six times the lateral strain. Find the change in the volume, when the bar is subjected to a hydrostatic pressure of 120 N/mm^2. Take $E = 1 \times 10^5$ N/mm^2.

[12 marks : 2019]

Solution:

For a cylindrical bar, $\epsilon_1 = \text{Longitudinal strain} = \frac{\sigma}{E}$

$$\epsilon_2 = \text{Lateral strain} = -\frac{\mu\sigma}{E}$$

But, $|\epsilon_1| = 6|\epsilon_2|$

$$\frac{\sigma}{E} = 6\frac{\mu\sigma}{E}$$

$$\mu = \frac{1}{6}$$

Now using relation, $E = 3K(1 - 2\mu)$

$$1 \times 10^5 = 3K\left(1 - \frac{2}{6}\right)$$

Bulk modulus, $K = 0.5 \times 10^5$ MPa

Modulus of rigidity, $G = \frac{E}{2(1+\mu)}$

$$G = \frac{1\times 10^5}{2\left(1+\frac{1}{6}\right)} = \frac{3}{7}\times 10^5 \text{ MPa}$$

Now change in volume = ΔV.

We know, $K = \frac{\sigma}{\epsilon_v}$

$$\epsilon_v = \frac{\sigma}{K} = \frac{120}{0.5\times 10^5}$$

$$\Delta V = \frac{120}{0.5\times 10^5}\times\frac{\pi}{4}\times 30^2 \times 2000$$

Reduction in volume, $\Delta V = 3392.92$ mm^3

1.19 A semi-composite steel bar as shown in figure is loaded at free end with an axial load of 50 kN. Determine the axial stiffness of the system and extension of the free end. Diameter of steel bar is 40 mm. Outer diameter of the concrete portion is 200 mm. Modulus of elasticity of steel = 200 GPa, modulus of elasticity of concrete = 20 GPa. Central portion of the bar is embedded with concrete.

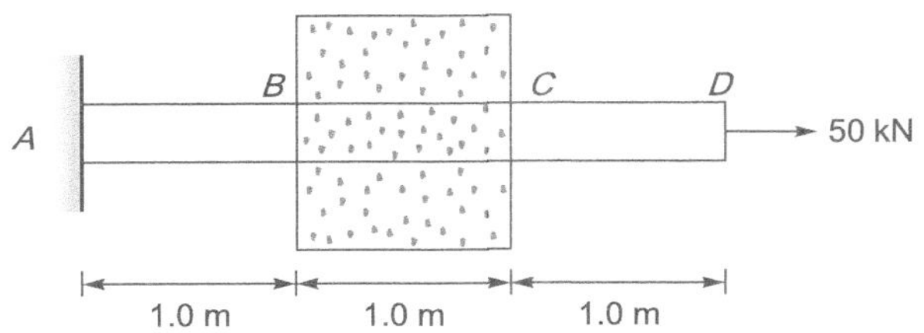

[12 marks : 2020]

Solution:

The bar could be divided into three parts *AB*, *BC* and *CD* and each part will have axial loads as shown below:

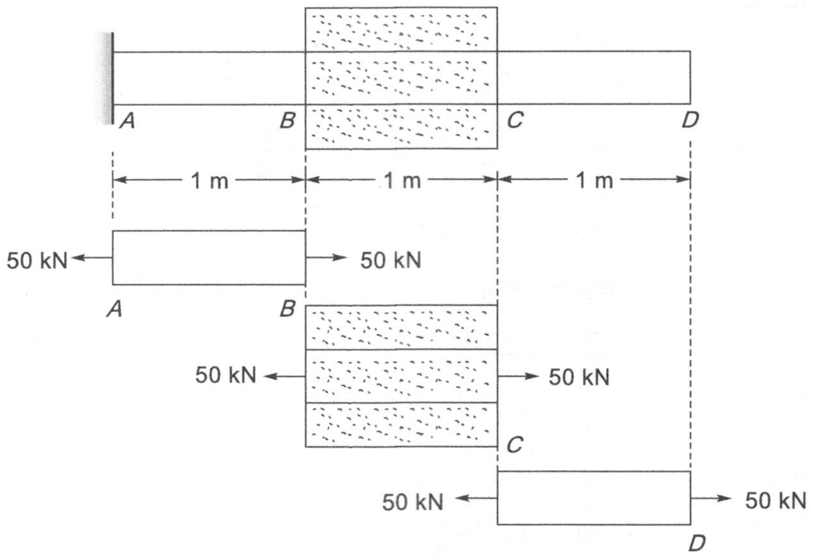

Now deformation due to axial load in *CD*,

$$\Delta_{CD} = \left(\frac{PL}{AE}\right)_{CD} = \frac{50\times10^3\times1000}{\frac{\pi}{4}\times40^2\times200\times10^3} = 0.1989 \text{ mm}$$

Similarly,

$$\Delta_{AB} = \left(\frac{PL}{AE}\right)_{AB} = \frac{50\times10^3\times1000}{\frac{\pi}{4}\times40^2\times200\times10^3} = 0.1989 \text{ mm}$$

Now, for BC, compatibility condition,

$$\Delta_{BC} = \Delta_{conc.} = \Delta_{bar}$$

$$\Rightarrow \quad \left(\frac{PL}{AE}\right)_{conc} = \left(\frac{PL}{AE}\right)_{bar}$$

$$\Rightarrow \quad \frac{P_{conc}\times1000}{\frac{\pi}{4}\left(200^2-40^2\right)\times20\times10^3} = \frac{P_{bar}\times1000}{\frac{\pi}{4}\times40^2\times200\times10^3}$$

$$P_{conc} = 2.4\,P_{bar}$$

As we know,

$$P_{conc} + P_{bar} = 50 \text{ kN}$$

$$(2.4+1)P_{bar} = 50 \text{ kN}$$

$$P_{bar} = 14.706 \text{ kN}$$

Thus,

$$\Delta_{BC} = \frac{14.706\times10^3\times1000}{\frac{\pi}{4}\times40^2\times200\times10^3} = 0.0585 \text{ mm}$$

Total elongation of system,

$$\Delta_{total} = \Delta_{AB} + \Delta_{BC} + \Delta_{CD}$$

$$= 0.1989 + 0.0585 + 0.1989$$

$$\Delta_{total} = 0.4563 \text{ mm}$$

As we know, stiffness of system

$$k_{\text{system}} = \frac{\text{Force}}{\text{Deformation}} = \frac{50\times10^3}{0.4563\times10^{-3}} = 109.577 \times 10^6 \text{ N/m}$$

1.20 A steel bar AB of length 3.5 m and diameter 25 mm is connected by four inextensible cables of length 2.5 m each, forming a rhombus with AB as diagonal. A 100 kN force acts at the points C and D. Determine the decrease in length of the member AB and increase in the length between points C and D. Take modulus of steel as 201 GPa.

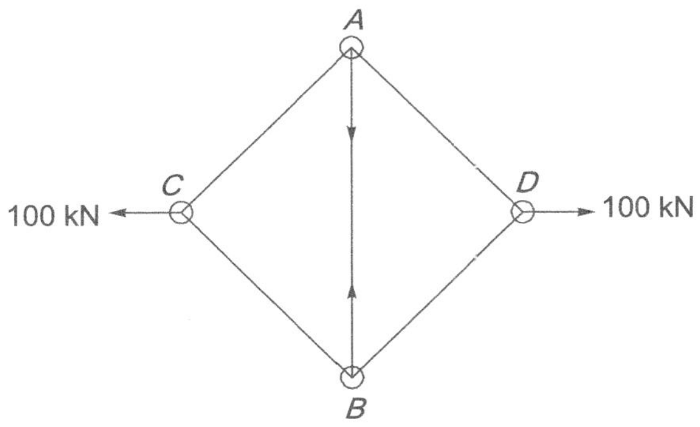

[20 Marks : 2021]

Solution:

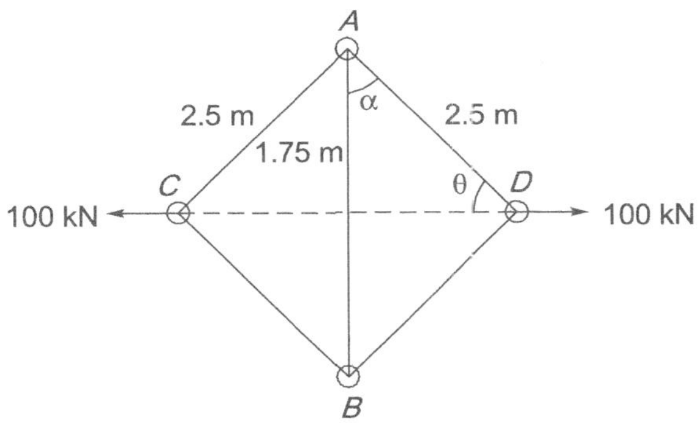

$$\sin\theta = \frac{1.75}{2.5} = 44.43°$$

$$\cos\alpha = \frac{1.75}{2.5} = 45.57°$$

At joint D:

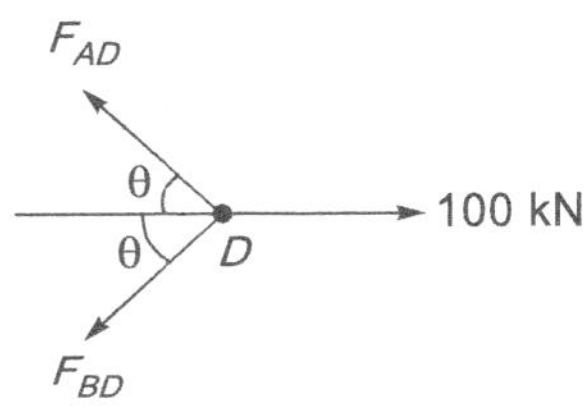

$$\Sigma F_x = 0$$

$$F_{AD}\cos\theta + F_{BD}\cos\theta = 100$$

$$\Sigma F_y = 0$$

$$F_{AD}\sin\theta = F_{BD}\sin\theta$$

$$F_{AD} = F_{BD}$$

$$2F_{AD}\cos\theta = 100$$

$$F_{AD} = \frac{50}{\cos 44.43} = 70.02 \text{ kN}$$

At joint A:

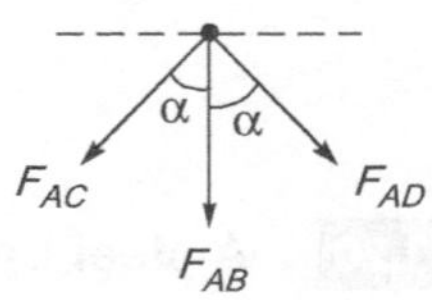

$$\Sigma F_x = 0$$
$$F_{AC} = F_{AD}$$
$$\Sigma F_y = 0$$
$$F_{AB} + 2F_{AD}\cos\alpha = 0$$
$$F_{AB} = -2 \times 70.02 \times \cos 45.57° = -98.03 \text{ kN}$$
$$F_{AB} = 98.03 \text{ (compression)}$$

Decrease in the length of member (Δ_{AB}), $AB = \dfrac{PL}{AE}$

$$A = \frac{\pi}{4}(0.025)^2 = 4.91\times10^{-4} \text{ m}^2$$
$$E = 201 \times 10^6 \text{ kN/m}^2$$
$$= \frac{98.03\times3.5}{4.91\times10^{-4}\times201\times10^6}$$
$$= 0.347 \times 10^{-2} \text{ m} = 3.47 \text{ mm}$$

As cables are inextensible, total work done by the external force will be equal to the strain energy stored in the bar *AB*.

Strain energy stored in bar $AB = \dfrac{1}{2}F_{AB}\,\Delta_{AB}$

$$= \left[\frac{1}{2}\times98.02\times3.47\right] \quad \text{...(i)}$$

$$\text{Work done} = \text{Strain energy stored}$$
$$\frac{1}{2}\times100\times\Delta_{CD} = \frac{1}{2}\times98.02\times3.47$$
$$\Delta_{CD} = 3.401 \text{ mm} \simeq 3.4 \text{ mm}$$

Alternative solution:

Unit load method

$$\Delta_{CD} = \Sigma K_i \Delta_i$$

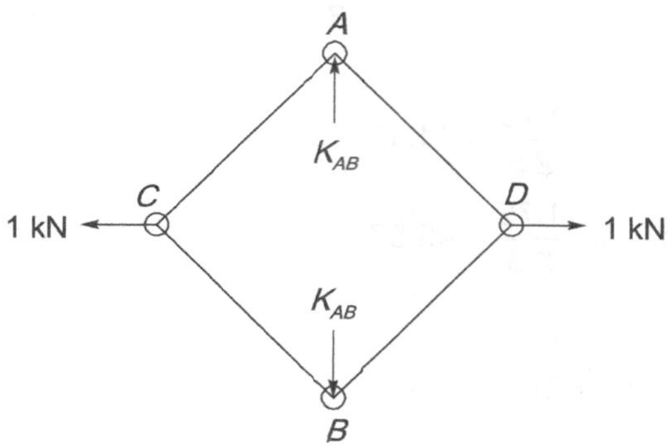

$$\Delta_{AC} = \Delta_{CB} = \Delta_{BD} = \Delta_{AD} = 0$$
$$\Delta_{AB} = -3.47 \text{ mm}$$
$$K_{AB} = \frac{F_{AB}}{100} = \left(-\frac{98.03}{100}\right) = -0.9803$$
$$\delta_{CD} = K_{AB}\Delta_{AB}$$
$$= (-0.9803)(-3.47) = 3.401 \text{ mm}$$

1.21 Two loads W_1 (equal to 1.5 kN) and W_2, resting on two inclined rough planes OA and OB are connected by a horizontal link *PQ* as shown in the figure. Find the maximum and minimum values of W_2 for which the equilibrium can exist. Take angle of friction for both the planes as 20°.

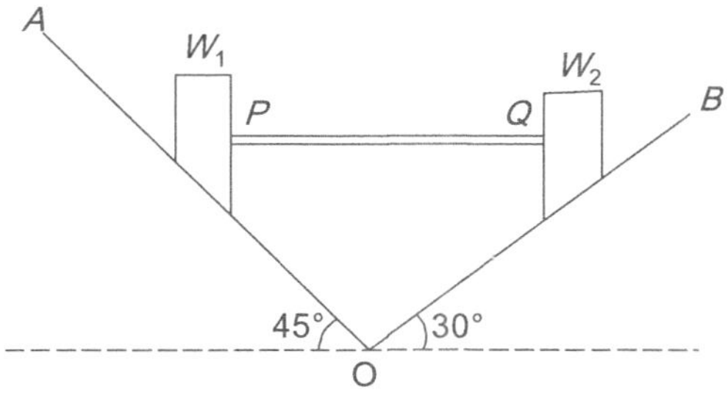

[20 Marks : 2021]

Solution:

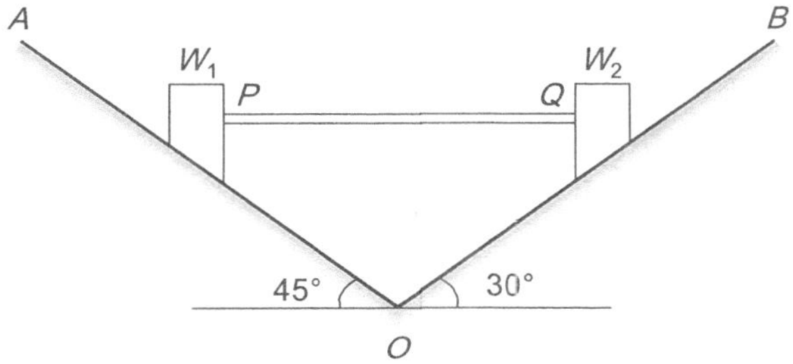

Given, $W_1 = 1.5$ kN

Angle of friction, $\phi = 20°$

$\mu = \tan\phi = \tan 20° = 0.364$

FBD of W_1 (For minimum value of W_2)

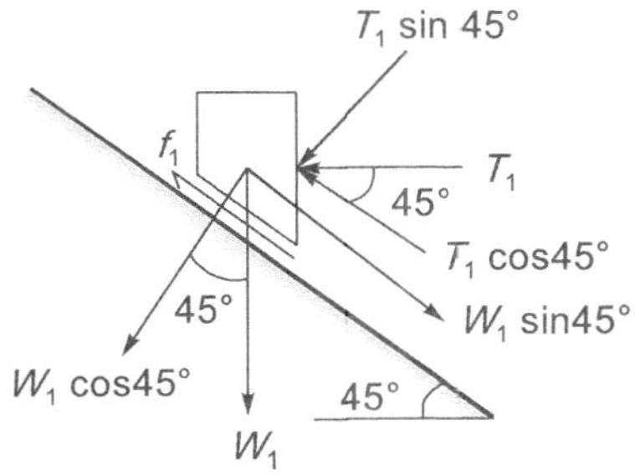

Normal reaction, $N_1 = T_1 \sin 45° + W_1 \cos 45°$

$N_1 = 1.06 + 0.707\,T_1$ ..(i)

For equilibrium of load,

$$f_1 + T_1 \cos 45° = W_1 \sin 45°$$

$$\mu N_1 + T_1 \cos 45° = 1.06$$

$$(0.364)(1.06 + 0.707\,T_1) + 0.707T_1 = 1.06$$

$$0.385 + 0.257T_1 + 0.707T_1 = 1.06$$

$$T_1 = 0.7 \text{ kN}$$

FBD of W_2, (for minimum value of W_2)

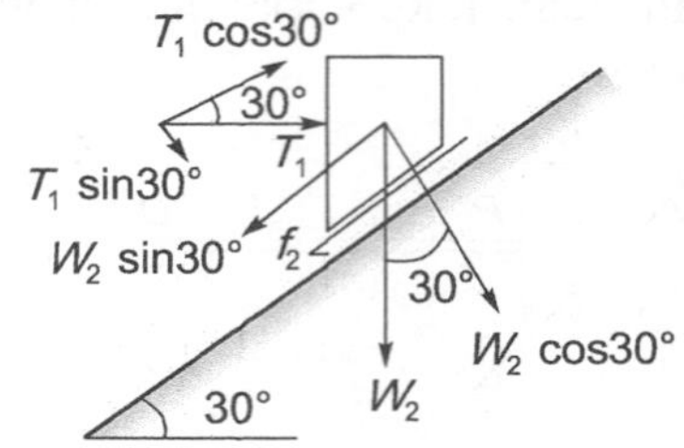

Normal reaction,

$$N_2 = T_1 \sin 30° + W_2 \cos 30°$$
$$= 0.7 \sin 30° + W_2 \cos 30°$$
$$= 0.35 + 0.866 W_2$$

For equilibrium of W_2 load,

$$f_2 + W_2 \sin 30° = T_1 \cos 30°$$
$$\mu N_2 + W_2 \sin 30° = 0.7 \cos 30°$$
$$0.364(0.35 + 0.866 W_2) + W_2 \times 0.5 = 0.606$$
$$W_2 = \frac{0.7786}{0.8152} = 0.587 \text{ kN}$$

FBD of W_1 (for maximum value of W_2)

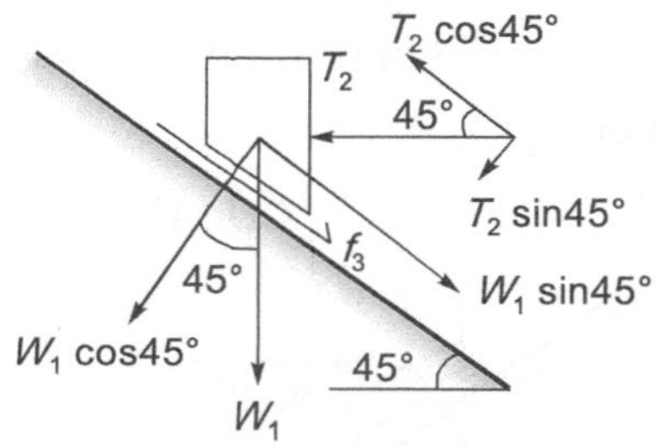

Normal reaction,

$$N_3 = W_1 \cos 45° + T_2 \sin 45°$$
$$N_3 = 1.06 + 0.707 T_2$$

For equilibrium of load W_1,

$$f_3 + W_1 \sin 45° = T_2 \cos 45°$$
$$\mu N_3 + (1.5) \sin 45° = T_2 \cos 45°$$
$$0.364(1.06 + 0.707 T_2) + 1.06 = T_2 \times 0.707$$
$$1.445 = 0.4496 \, T_2$$
$$T_2 = 3.21 \text{ kN}$$

FBD of W_2, (for maximum value of W_2)

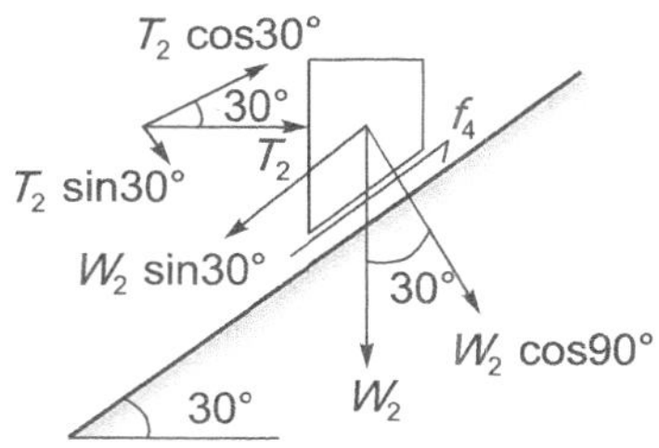

Normal reaction,
$$N_4 = T_2 \sin 30° + W_2 \cos 30°$$
$$N_4 = 3.213 \sin 30° + W_2 \cos 30°$$
$$N_4 = 1.606 + 0.866 W_2 \quad \text{...(ii)}$$

For equilibrium of W_2 load,
$$f_4 + T_2 \cos 30° = W_2 \sin 30°$$
$$\mu N_4 + 3.213 \cos 30° = W_2 \sin 30°$$
$$0.364(1.606 + 0.866 W_2) + 2.782 = 0.5 W_2$$
$$0.5845 + 0.315 W_2 + 2.782 = 0.5 W_2$$
$$0.184 W_2 = 3.366$$
$$W_2 = 18.293 \text{ kN}$$

1.22 A rectangular plate in biaxial stress is subjected to normal stress σ_x = 65 MPa and σ_y = –20 MPa. The plate ($b \times h \times t$) has dimensions 200 × 300 × 15 and is made of aluminium with E = 75 GPa an v = 0.33. All dimensions are in mm. Determine
(i) The maximum in-plane shear strain (γ_{max}) in the plate,
(ii) The change in the thickness of the plate (Δt), and
(iii) The change in volume of the plate (Δv).

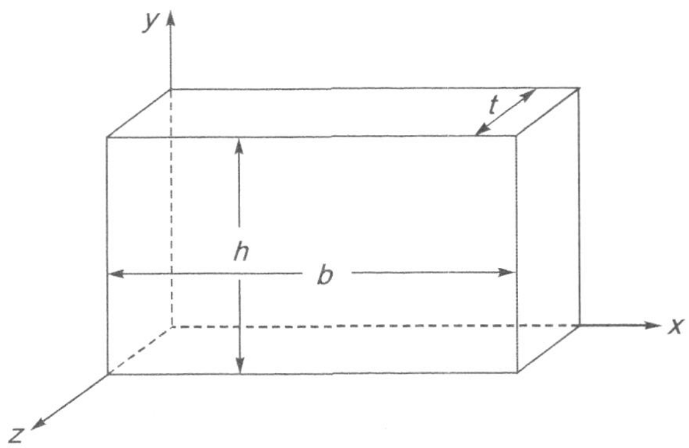

[12 Marks : 2021]

Solution:

Given: σ_x = 65 MPa, σ_y = –20 MPa, E = 75 GPa = 75 × 10^3 MPa, ν = 0.33, l = 200 mm, h = 300 mm, t = 15 mm

Strain in x-direction,
$$\epsilon_x = \frac{\sigma_x}{E} - \nu\frac{\sigma_y}{E} = \frac{1}{75 \times 10^3}[(65) - (0.33)(-20)]$$

$$\in_x = 9.54 \times 10^{-4}$$

Strain in y-direction, $\in_y = \dfrac{\sigma_y}{E} - \nu\dfrac{\sigma_x}{E} = \dfrac{1}{75\times10^3}[(-20)-(0.33)(65)]$

$$\in_y = -5.52 \times 10^{-4}$$

Strain in z-direction, $\in_z = \dfrac{1}{E}[0 - \nu(65) - \nu(-20)]$

$$\in_z = -1.98 \times 10^{-4}$$

(i) Maximum in-plane shear strain

$$\gamma_{max} = \in_{major} - \in_{minor} = \in_x - \in_y$$
$$= (9.54 \times 10^{-4}) - (-5.52 \times 10^{-4})$$
$$\gamma_{max} = 15.06 \times 10^{-4}$$

(ii) Change in thickness of plate

$$\Delta t = \in_z(t)$$
$$= (-1.98 \times 10^{-4})(15)$$
$$\Delta t = -2.97 \times 10^{-3} \text{ mm}$$
$$= 2.97 \times 10^{-3} \text{ mm} \quad \text{(Reduction in thickness)}$$

(iii) Change in volume of plate

$$\in_v = \frac{\Delta V}{\text{Volume}} = \in_x + \in_y + \in_z$$
$$\Delta V = [(9.54 - 5.52 - 1.98)\times 10^{-4}][200 \times 300 \times 15]$$
$$\Delta V = 183.6 \text{ mm}^3$$

or, $\in_v = \dfrac{\sigma_x + \sigma_y + \sigma_z}{E}(1-2\mu) = \dfrac{65-20}{75\times10^3}(1-2\times0.33)$

$$= 2.04 \times 10^{-4}$$
$$\Delta V = 2.04 \times 10^{-4} \times 200 \times 300 \times 15$$
$$= 183.6 \text{ mm}^3$$

1.23 **(i) An axial pull of 45 kN is applied on a steel bar of diameter 13 mm and length of 2500 mm. Determine the change in length, diameter and volume of the steel bar if the Poisson's ratio is 0.25. Use $E = 200 \times 10^3$ N/mm^2.**

(ii) Determine the value of Poisson's ratio and Modulus of Elasticity, if the modulus of rigidity of a given material is 50×10^3 N/mm^2 and the bulk modulus is 80×10^3 N/mm^2.

[6 + 6 = 12 marks : 2022]

Solution:

(i) Axial pull, $P = 45$ kN

Diameter, $d = 13$ mm

Length, $L = 2500$ mm

Poisson's ratio, $\mu = 0.25$

(a) Change in length $\Delta l = \dfrac{PL}{AE} = \dfrac{45\times10^3\times2500}{\frac{\pi}{4}\times13^2\times2\times10^5}$

$\Rightarrow$ $\Delta l = 4.24$ mm

(b) $$\mu = \frac{-\Delta d/d}{\Delta l/L}$$

$$\Rightarrow \quad -\Delta d = \left[\mu\left(\frac{\Delta l}{L}\right)\right]d$$

$$\Rightarrow \quad -\Delta d = \left[0.25 \times \frac{4.24}{2500}\right] \times 13$$

$$\Rightarrow \quad \Delta d = -0.0055 \text{ mm}$$

(c) Change in volume,

$$\epsilon_v = \epsilon_l + 2\epsilon_d$$

$$\Delta V = \epsilon_v V$$

$$\Rightarrow \quad \Delta V = \left[\frac{4.24}{2500} + 2\left(\frac{-0.0055}{13}\right)\right]\left[\frac{\pi}{4} \times 13^2 \times 2500\right]$$

$$\Rightarrow \quad \Delta V = 282.005 \text{ mm}^3$$

(ii) Modulus of rigidity, $G = 50 \times 10^3$ N/mm^2

Bulk modulus, $K = 80 \times 10^3$ N/mm^2

Modulus of elasticity 'E'

$$E = \frac{9KG}{3K + G}$$

$$\Rightarrow \quad E = \frac{9 \times 80 \times 10^3 \times 50 \times 10^3}{3 \times 80 \times 10^3 + 50 \times 10^3} = 124137.93 \text{ N/mm}^2$$

But $$E = 3K(1 - 2\mu)$$

$$\Rightarrow \quad 124137.93 = 3 \times 80 \times 10^3[1 - 2\mu]$$

$$\Rightarrow \quad \mu = 0.24$$

1.24 A bar specimen of 38 mm diameter was subjected to a pull of 98 kN during a tensile test. The extension on a gauge length of 200 mm was measured to be 0.092 mm and the change in diameter of 0.0048 mm. Determine the Poisson's ratio, modulus of elasticity, modulus of rigidity and bulk modulus of the material of bar specimen.

[12 marks : 2023]

Solution:

Given:

Length of bar, $L = 200$ mm

Diameter of bar, $d = 38$ mm

Axial force, $P = 98$ kN

Axial deflection, $\Delta L = 0.092$ mm

Now,

$$\Delta L = \frac{PL}{AE}$$

where E is Young's modulus of elasticity and A is cross-sectional area of bar.

Putting values, we get

$$0.092 = \frac{98 \times 10^3 \times 200}{\frac{\pi}{4} \times 38^2 \times E} \qquad \left[\because A = \frac{\pi}{4}d^2\right]$$

$\Rightarrow$ $E = 1.88 \times 10^5$ N/mm^2

Also, Change in diameter, $\Delta d = -0.0048$ mm

Now, Poisson's ratio, $\mu = \dfrac{\text{Lateral strain}}{\text{Longitudinal strain}}$

$$= \frac{-\left(\dfrac{\Delta d}{d}\right)}{\dfrac{\Delta L}{L}}$$

$$\Rightarrow \quad \mu = \frac{-\left(\dfrac{-0.0048}{38}\right)}{\dfrac{0.092}{200}} = 0.275$$

Now, $E = 2G(1 + \mu)$ where G is modulus of rigidity

$$\Rightarrow \quad G = \frac{E}{2(1+\mu)} = \frac{1.88\times 10^5}{2\times(1+0.275)}$$

$= 0.74 \times 10^5$ N/mm^2

Also, $E = 3K(1 - 2\mu)$ where K is bulk modulus of elasticity

$$\Rightarrow \quad K = \frac{E}{3(1-2\mu)} = \frac{1.88\times 10^5}{3\times(1-2\times 0.275)}$$

$= 1.39 \times 10^5$ N/mm^2

So, Young's modulus of elasticity, $E = 1.88 \times 10^5$ N/mm^2

Modulus of rigidity, $G = 0.74 \times 10^5$ N/mm^2

Bulk modulus of elasticity, $K = 1.39 \times 10^5$ N/mm^2

Poisson's ratio, $\mu = 0.275$

2. Shear Force and Bending Moment

2.1 Draw the shear force, load, deflected shape diagrams corresponding to the BMD given in the figure below. Specify the values at all change of load positions and at all points of maximum moments.

[25 marks : 2000]

Solution:

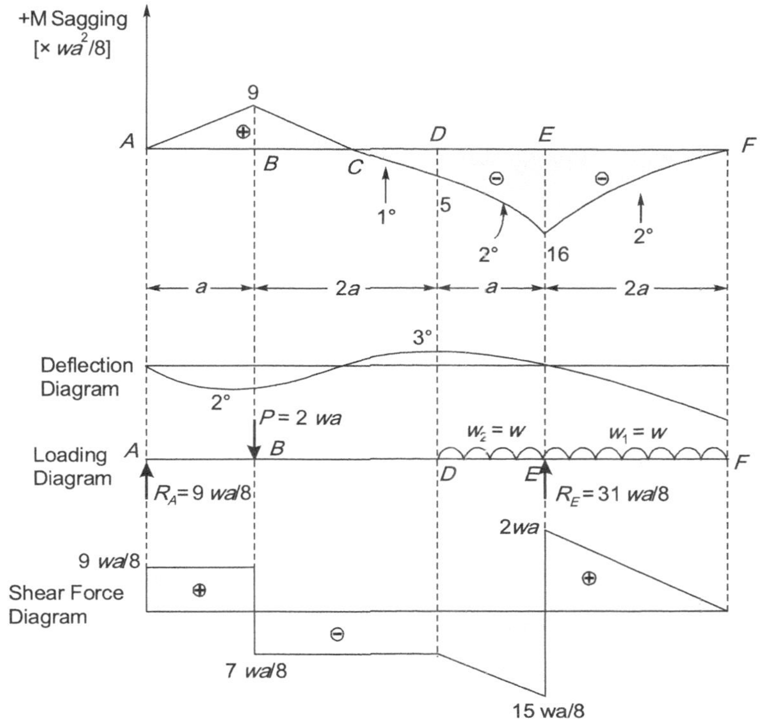

Loading diagram:

Portion *AB*: BMD is incline line between *A* and *B* and bending moment is positive (sagging). It is possible when a upward point load acts at *A*. Generally upward point loads are reactions provided by support. Hence there is a support at *A* and let R_A is the support reaction at *A*.

Also, at point *B*
$$M_B = +\frac{9wa^2}{8}$$
$$R_A \times a = \frac{9wa^2}{8}$$
$$\therefore \quad R_A = \frac{9wa}{8}\text{kN}$$

it is also noted BMD varies uniformly.
Hence there will be no load between *A* and *B*.

Portion *BD*:

Between *B* and *D*, BM varies linearly and decreases from $\frac{9wa^2}{8}$ to $-\frac{5wa^2}{8}$ it mean at point *B* there should be a downward point load. Say P_1

Also, at
$$M_D = -\frac{5wa^2}{8}$$
$$R_A \times 3a - P_1 \times 2a = M_D$$
$$\frac{9wa}{8} \times 3a - P_1 \times 2a = -\frac{5wa^2}{8}$$

$$P_1 \times 2a = \frac{27wa^2}{8} + \frac{5wa^2}{8}$$

$$P_1 = 2\,wa \text{ kN}$$

Portion *EF*:

In portion *EF* BMD is 2° curve it mean this portion is loaded by UDL.

Let intensity of loading is w_1 kN/m downwards

Also at point *E*,

$$M_E = -\frac{16wa^2}{8}$$

$$M_E(\text{from right side}) = -\frac{w_1 \times (2a)^2}{2} = -\frac{16wa^2}{8}$$

$$w_1 = w \text{ kN/m}$$

Portion *DE*:

In portion *DE* BMD is 2° curve it mean this portion is loaded by UDL.

Let intensity of loading is w_2 kN/m downwards

Also at point *E*,

$$M_E = -\frac{16wa^2}{8}$$

$$M_E(\text{from left side}) = R_A \times 4a - P_1 \times 3a - \frac{w_2 a^2}{2} = -\frac{16wa^2}{8}$$

$$\frac{9wa}{8} \times 4a - 2wa \times 3a - \frac{w_2 a^2}{2} = -2wa^2$$

$$-\frac{w_2 a^2}{2} = -2wa^2 - 4.5wa^2 + 6wa^2$$

$$-\frac{w_2}{2} = -0.5w$$

$$w_2 = w \text{ kN/m}$$

At *E*, *BM* is same from both sides, for the stability of the beam there should be a support there. Let reaction at the support at R_E

$$M_D(\text{from right}) = +R_E \times a - w \times (3a) \times 1.5a = -\frac{5wa^2}{8}$$

$$R_E a = -\frac{5}{8}wa^2 + 4.5wa^2$$

$$R_E = \frac{31wa}{8} \text{ kN}$$

Alternatively:

Since BM varies parabolically in portion *CE* and continuously decreasing but after point *E* bending moment starts increasing and become zero. It mean there should be a upward reaction at *E* which provide sagging BM. Let reaction at *E* is R_E

From $\Sigma f_y = 0$

$$R_A + R_E = P_1 + w_1 \times 2a + w_2 \times a$$

$$R_E = 2wa + 3wa - R_A$$

$$R_E = 5wa - \frac{9wa}{8}$$

$$R_E = \frac{31wa}{8} \text{ kN}$$

For SFD:

Portion AB:

$$S_x(x \text{ from } A) = +R_A \qquad [0 \le x < a]$$

$$S_x = +\frac{9wa}{8} \text{ kN} \qquad \text{(Constant)}$$

Portion BD:

$$S_x(x \text{ from } A) = +R_A - 2wa \qquad [a < x < 3a]$$

$$S_x = \frac{9wa}{8} - 2wa \Rightarrow -\frac{7wa}{8}$$

$$S_x = -\frac{7wa}{8} \text{ kN} \qquad \text{(Constant)}$$

Portion DE:

$$S_x(x \text{ from } D) = -\frac{9wa}{8} - 2wa - wx \qquad [0 \le x < a]$$

$$S_x = -\frac{7wa}{8} - wx \qquad \text{(Linear)}$$

at $x = 0$, $$S_D(\text{just right of } D) = -\frac{7wa}{8} \text{ kN}$$

at $x = a$, $$S_E(\text{just left of } E) = -\frac{7wa}{8} - wa \Rightarrow -\frac{15wa}{8}$$

Portion EF:

$$S_x(x \text{ from } F) = wx \qquad [0 \le x < 2a]$$

at $x = 0$, $S_F = 0$

at $x = 2a$, $S_E(\text{just right of } E) = w \times 2a$

$= 2\ wa$ kN

Deflected Shape:

Portion ABC: BM in this portion is sagging, hence the deflected shape for ABC will be concave upwards as support A is simply supported.

Porton CDE: As BM changes sign at C thus point C will act as point of contraflexure (for BMD) and point of inflection (for deflected shape) also for portion CDE BM is hogging, hence deflected shape will be convex upwards.

Portion EF: As E is a support the beam will neither deflect nor break (i.e., slope of deflected shape at left of E = slope at right of E) also BM is hogging, hence deflected shape will be convex upwards.

2.2 For the SFD of a loaded beam shown in the figure below, draw the corresponding loading diagram and BMD.

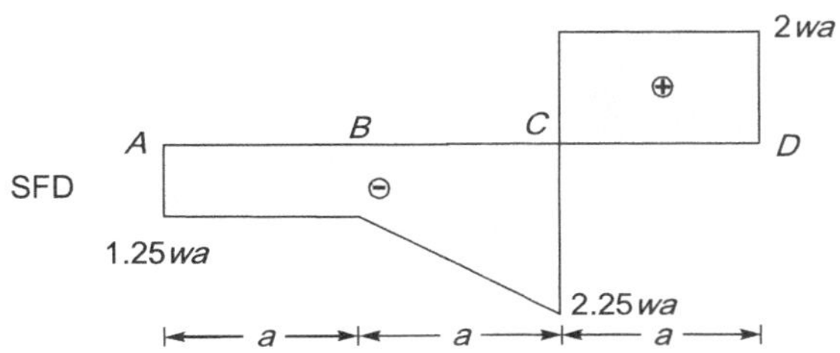

[20 marks : 2002]

Solution:

For loading diagram:

Point A:

At point A, SF is negative. Hence at A there will be a downward point load.

$$P_A = 1.25\text{ wa } (\downarrow)$$

Portion AB:

In portion AB, SF is constant. Hence there will be no loading in portion AB.

Portion BC:

Portion BC will be subjected to UDL of loading rate $= \dfrac{-2.25wa + 1.25wa}{a}$

$$w_1 = w \text{ unit/length}$$

Point C:

At point C, SF drastically change from –2.25 wa to 2 wa.

Hence there will be a support at C and the reaction at the support C will be

$$R_C = 2.25\ wa + 2\ wa = 4.25\ wa$$

Portion CD:

In portion CD, SF is constant. Hence there will be no loading is portion AB.

Point D:

At point D, SF drastically changes from + 2 wa to 0.

Hence there will be a downward point load of magnitude 2 wa.

Support:

There are two supports at A and C.

For support A

$$\Sigma M_A = -2wa \times 3a + 4.25\ wa \times 2a - wa\,(a + 0.5a)$$
$$= +wa^2$$

Hence support A will be fixed.

Therefore a clockwise reacting moment of magnitude wa^2 is acting at A.

For BMD:

Portion AB:

$$M_x\ (x \text{ from } A) = +M_A - R_A x \qquad [0 \le x \le a]$$
$$M_x = wa^2 - 1.25\ wax \qquad \text{(Linear)}$$

at $x = 0$, $\quad M_A = wa^2$

at $x = a$, $\quad M_B = wa^2 - 1.25\ wa^2 \Rightarrow -0.25\ wa^2$

Bending moment will be zero if $M_x = 0$

Then $\quad wa^2 - 1.25\ wax = 0$

$$x = \frac{wa^2}{1.25wa} = 0.8a$$

Portion BC: $\quad M_x\ (x \text{ from } B) = M_A - R_A(x+a) - \dfrac{wx^2}{2}$ [Parabolic] $\quad [0 \le x \le a]$

at $x = 0$, $\quad M_B = -0.25\text{ wa}^2$

at $x = \text{a}$, $\quad M_C = wa^2 - 1.25wa \times 2a - \dfrac{wa^2}{2} \Rightarrow -2wa^2$

Portion DC: $M_x(x \text{ from } D) = -2wax$ (Linear) $[0 \le x \le a]$

at $x = 0$, $M_D = 0$

at $x = a$, $M_C = -2wa^2$

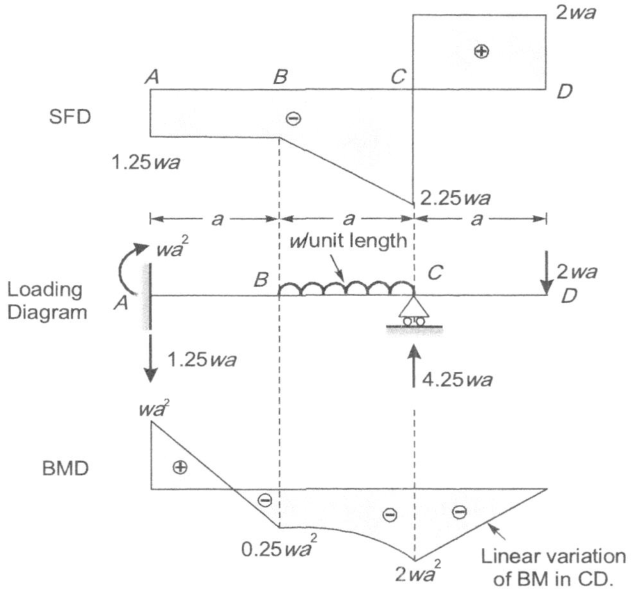

2.3 A horizontal beam of length '6*a*' has built-in-ends. The beam has two horizontal normally projecting cantilevers of length '*a*' at one third points which carry vertical loads of magnitudes *P* and 2*P*, one at each end. Draw bending moment, shear force and torsion diagrams.

[20 marks : 2002]

Solution:

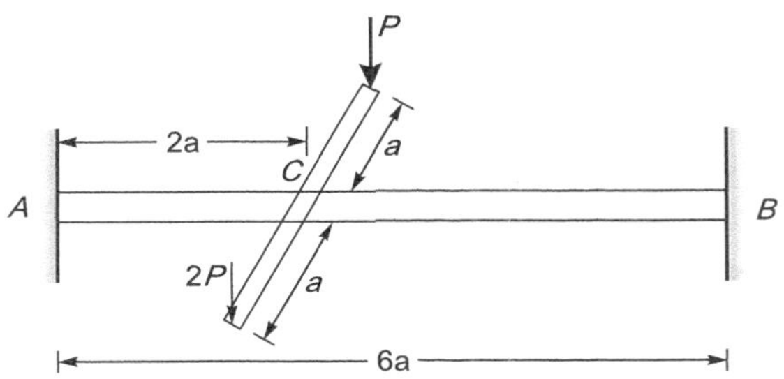

Net torsional moment at C, $T = 2Pa(\to\to) - Pa(\leftarrow\leftarrow)$

$\therefore$ $T = Pa(\to\to)$

Net vertical point load at C,

$= 2P + P = 3P$

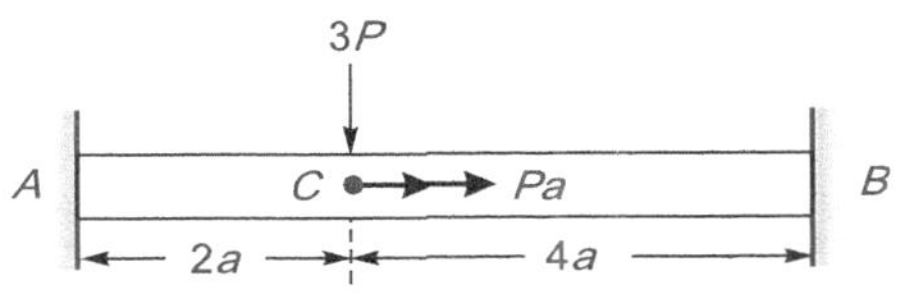

Fixed end moment: Let M_A and M_B be the fixed end moments at supports A and B respectively.

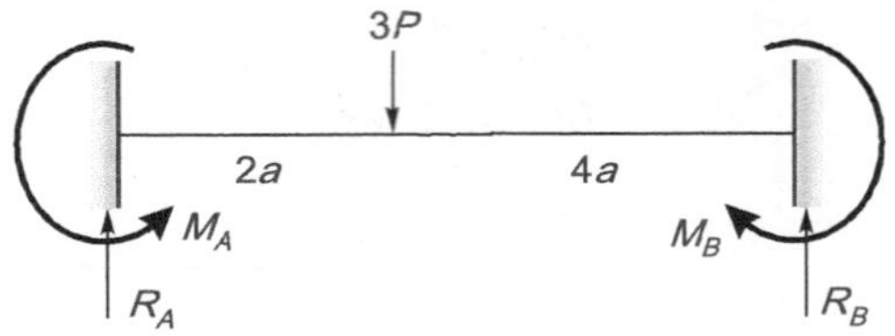

Fixed end moment at A, $$M_A = \frac{-3P\times(2a)\times(4a)^2}{(6a)^2} = -\frac{8Pa}{3}$$

Fixed end moment at B, $$M_B = \frac{3P\times(2a)^2\times(4a)}{(6a)^2} = +\frac{4Pa}{3}$$

Reactions:

$$\Sigma F_y = 0,\ R_A + R_B = 3P \quad \text{...(i)}$$

Also, $$\Sigma M_A = 0$$

$$-(R_B\times 6a) - M_A + M_B + (3P\times 2a) = 0$$

$$-6aR_B - \frac{8Pa}{3} + \frac{4Pa}{3} + 6Pa = 0$$

$$6aR_B = \frac{14Pa}{3}$$

$\therefore$ $$R_B = \frac{7P}{9}$$

From eq. (i), we get

$$R_A = \frac{20P}{9}$$

Torsional Reactions: Let T_A and T_B be the torsional reactions at supports A and B.

Since beam in equilibrium

$\therefore$ $$\Sigma T = 0$$

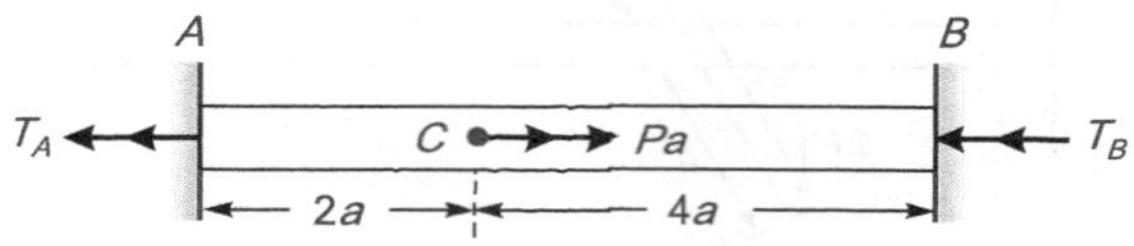

$$T_A + T_B = Pa \quad \text{...(ii)}$$

Since ends A and B are fixed, hence total change of angle from A to B will be zero.

i.e., $$\theta_{AB} = 0$$

$\therefore$ $$\theta_{AC} + \theta_{CB} = 0 \quad \text{...(iii)}$$

Drawing the *FBD* of portions,

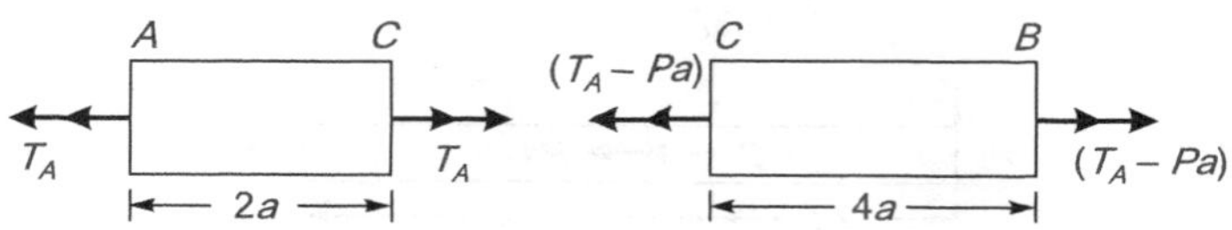

$$T_{AC} = -T_A$$

and
$$T_{CB} = -(T_A - Pa) = Pa - T_A$$

From eq. (iii)

$$\frac{T_{AC}L_{AC}}{GI_P} + \frac{T_{CB}L_{CB}}{GI_P} = 0$$

$$-\frac{T_A 2a}{GI_P} + \frac{(Pa - T_A).4a}{GI_P} = 0$$

$$-2aT_A + 4Pa^2 - 4aT_A = 0$$

$$-6aT_A + 4Pa^2 = 0$$

$\therefore$
$$T_A = \frac{2Pa}{3}$$

From eq. (ii), we get

$$T_B = \frac{Pa}{3}$$

(i) **Shear force diagram (SFD)**

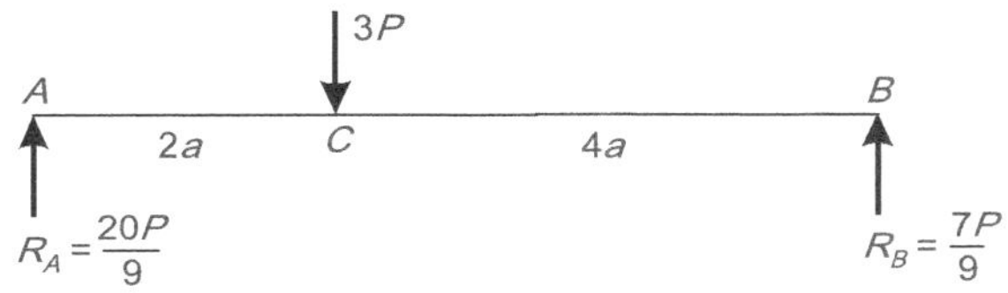

Portion *AC*:
$$S_x \text{ (x from A)} = +\frac{20P}{9} \qquad [0 \le x < 2a]$$

$$S_x = +\frac{20P}{9} \qquad \text{(constant)}$$

Portion *CB*:
$$S_x \text{ (x from A)} = +\frac{20P}{9} - 3P$$

$$S_x = -\frac{7P}{9} \qquad \text{(constant)}$$

(ii) **Bending moment diagram (BMD)**

Portion *AC*:
$$M_x \text{ (x from A)} = +\frac{20}{9}Px - M_A \qquad [0 \le x \le 2a]$$

$$M_x = +\frac{20}{9}Px - \frac{8Pa}{3} \qquad \text{(linear)}$$

at $x = 0$,
$$M_A = -\frac{8Pa}{3}$$

at $x = 2a$,
$$M_C = +\frac{16Pa}{9}$$

Portion *CB*:
$$M_x \text{ (x from B)} = +\frac{7P}{9}x - M_B \qquad [2a < x \le 6a]$$

$$M_x = +\frac{7P}{9}x - \frac{4Pa}{3} \qquad \text{(linear)}$$

at $x = 0,$ $M_B = -\frac{4Pa}{3}$

at $x = 4a,$ $M_C = +\frac{16Pa}{9}$

(iii) Torsional moment diagram (TMD)

Portion AC: $T_x (x \text{ from } A) = -T_A$ $[0 \le x < 2a]$

$T_x = -\frac{2Pa}{3}$ (constant)

Portion CB: $T_x (x \text{ from } A) = -T_A + Pa$ $[2a < x \le 6a]$

$T_x = -2\frac{Pa}{3} + Pa = \frac{Pa}{3} = +\frac{Pa}{3}$ (constant)

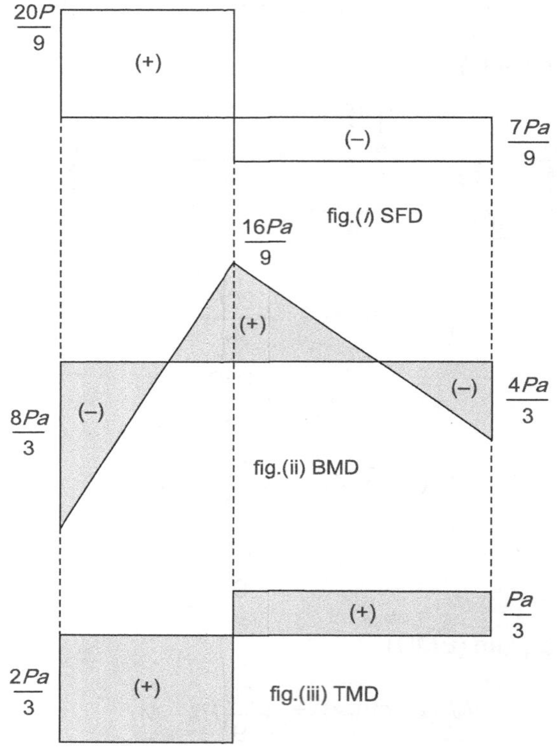

2.4 A continuous beam $ABCD$ of total length '5a' is fixed at A and has a simple support at point C distant '4a' from A. The portion CD of length 'a' is an overhang. The beam carries uniformly distributed load 'w' per unit length throughout it length and a concentrated load of magnitude 2.5 wa at the free end D. Draw the shear force and bending moment diagrams, if there is a hinge at the mid-span point B of the span AC. Clearly mark the position of points of contraflexure.

[10 marks : 2003]

Solution:

Let R_A and R_C are the vertical reactions at supports A and C respectively. Since there is no horizontal load so horizontal reaction at fixed support A will be zero.

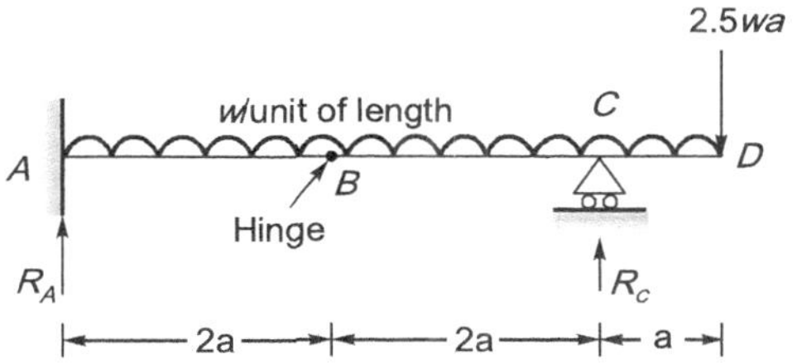

Reactions:

$$\Sigma F_y = 0$$
$$R_A + R_C = 5wa + 2.5wa$$
$$R_A + R_C = 7.5wa \quad \text{...(i)}$$

Since there is an internal hinge at B, therefore moment at B either from right or left is zero. Taking moment about B from right, we get

$$\Sigma M_B = 0$$
$$R_C \times 2a - (2.5wa \times 3a) - (w \times 3a \times 1.5a) = 0, \; R_C = 6\,wa$$

Putting value of R_C in (i), we get

$$R_A = 7.5wa - 6wa$$
$$R_A = 1.5wa$$

For SFD:

Portion DC:

$$S_x(x \text{ from } D) = 2.5wa + wx \quad [0 \le x < a]$$

at $x = 0$, $S_D = 2.5wa$

at $x = a$, $S_C(\text{just right of } C) = 2.5wa + wa = 3.5wa$

Portion CA:

$$S_x(x \text{ from } D) = 2.5wa + wx - 6wa \quad [a < x < 5a]$$

at $x = a$, $S_C(\text{just left of } C) = 2.5wa + wa - 6wa = -2.5wa$

at $x = 5a$, $S_A(\text{at support } A) = 2.5wa + 5wa - 6wa = +1.5wa$

If $S_x = 0$, then we have (in CA)

$$2.5wa + wx - 6wa = 0$$
$$3.5wa = wx$$
$$x = 3.5a \text{ from } D \text{ or } 1.5a \text{ from } A$$

For BMD:

Portion DC:

$$M_x(x \text{ from } D) = -2.5wax - \frac{wx^2}{2} \quad [0 \le x \le a]$$

$$M_x = -2.5wax - \frac{wx^2}{2} \quad \text{(Parabola)}$$

at $x = 0$, $M_D = 0$

at $x = a$, $M_C = -2.5wa^2 - \frac{wa^2}{2} = -3wa^2$

Portion CA:

$$M_x(x \text{ from } D) = -2.5wax - \frac{wx^2}{2} + R_C(x - a) \quad [a \le x \le 5a]$$

$$M_x = -2.5wax - \frac{wx^2}{2} + 6wa(x-a) \quad \text{(Parabola)}$$

at $x = a$,
$$M_C = -2.5wa^2 - \frac{wa^2}{2} = -3wa^2$$

at $x = 5a$.
$$M_A = -2.5wa(5a) - \frac{w}{2}(5a)^2 + 6wa(5a-a)$$
$$= -12.5wa^2 - 12.5wa^2 + 24wa^2$$
$$= -wa^2$$

For M_{max},

$$\frac{dM_x}{dx} = 0$$
$$-2.5\ wa - wx + 6wa = 0$$
$$x = 3.5a$$
$$M_{max}(\text{ at } x = 3.5a) = -2.5wa(3.5a) - \frac{w}{2}(3.5a)^2 + 6wa(3.5a-a)$$
$$= (-8.75 - 6.125 + 15)\ wa^2$$
$$= 0.125wa^2$$

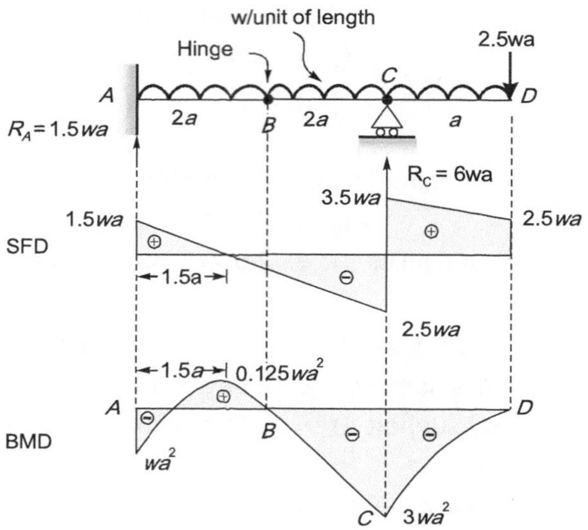

There are two points of contraflexure and they are D_1 and D_2 at a distance 'a' and '$2a$' respectively from fixed end.

2.5 A beam *ABCD* is hinged at *A* and simply supported at *D*. A vertical strut *BE*, 1 m long is fixed at *B*, 1 m from *A* and a frictionless pulley is attached to the beam at *C*, 2 m from *A*. A flexible string carries a load and passes over the pulley and is attached to the strut at *E* as shown in the figure. Draw the *BM*, *SF* and axial force diagrams of the beam.

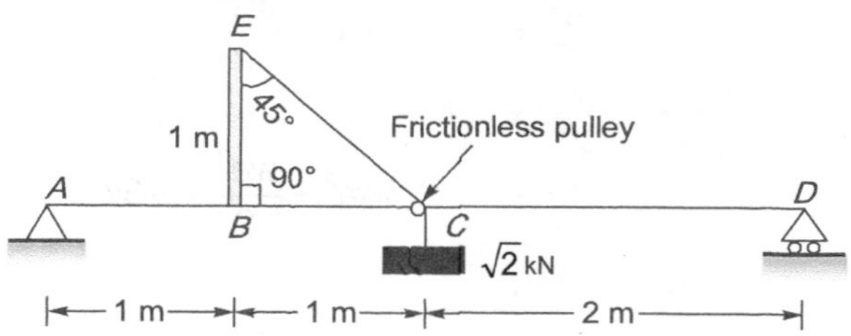

[10 marks : 2004]

Solution:

Cut string between E and C and transmit forces at E and C.

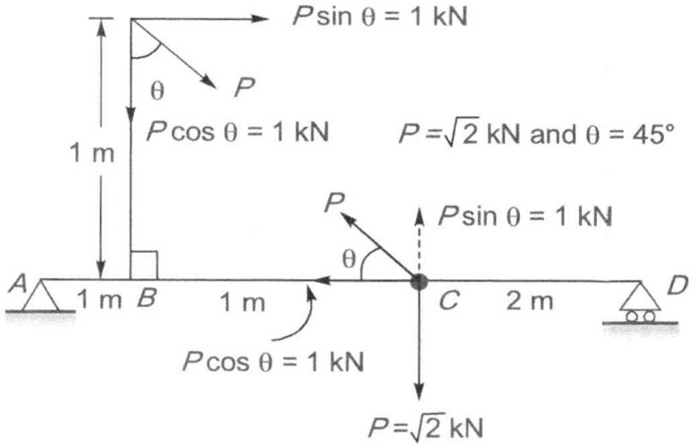

$$\theta = 45,\ P = \sqrt{2}\ \text{kN}$$

$$\therefore \quad P\cos\theta = \sqrt{2}\cos 45° = 1\ \text{kN}$$

$$P\sin\theta = \sqrt{2}\sin 45 = 1\ \text{kN}$$

Now above beam can be considered as the following loading.

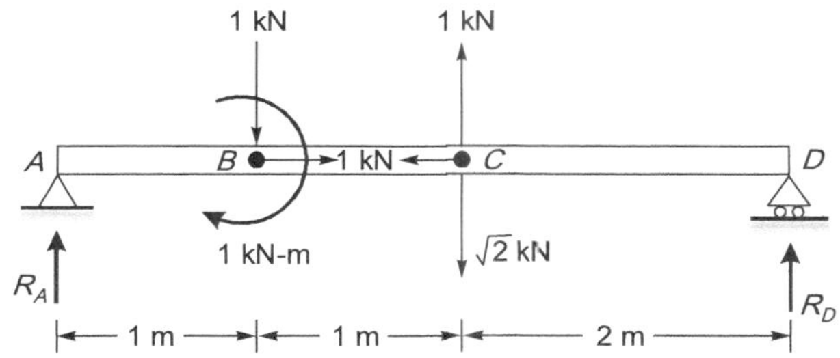

Reactions:

$\Sigma f_y = 0,$
$$R_A + R_D = \sqrt{2} - 1 + 1 = \sqrt{2} \quad \text{...(i)}$$

Also, $\Sigma M_A = 0,$
$$R_D \times 4 = \left[\left(\sqrt{2} - 1\right) \times 2\right] + (1 \times 1) + 1$$

$$4R_D = 2\sqrt{2} - 2 + 1 + 1$$

$$R_D = \frac{2\sqrt{2}}{4}$$

$$R_D = \frac{1}{\sqrt{2}} = 0.707\ \text{kN}$$

$$R_A = \sqrt{2} - R_D \Rightarrow \sqrt{2} - 0.707 \Rightarrow 0.707\ \text{kN}$$

For SFD:

Portion AB:

$$S_x\ (x \text{ from } A) = +R_A \quad [0 \le x < 1]$$

$$S_x = 0.707\ \text{kN} \quad \text{(Constant)}$$

Hence SF will be constant between A and B

Portion BC:

$$S_x\ (x \text{ from } A) = +R_A - 1\ \text{kN} \quad [1 < x < 2]$$

$$S_x = 0.707 - 1$$

$$S_x = -0.293\ \text{kN} \quad \text{(Constant)}$$

Hence SF will be constant between right of B to left of C.

Portion CD:

$$S_x\ (x \text{ from } A) = +R_A - 1 - \sqrt{2} + 1 \quad [2 < x \le 4]$$

$$S_x = 0.707 - \sqrt{2}$$

$$S_x = -0.707\ \text{kN} \quad \text{(Constant)}$$

Hence SF will be constant between just right of C to D.

For BMD:

Portion *AB*:

$$M_x \text{ (x from A)} = +R_A x \quad [0 \le x < 1]$$
$$M_x = 0.707x \quad \text{(Linear)}$$

at $x = 0$, $M_A = 0$

at $x = 1$, M_B (just left of B) $= 0.707 \times 1 \Rightarrow 0.0707$ kNm

Portion *BC*:

$$M_x \text{ (x from A)} = +R_A x + 1 - 1(x-1) \quad [1 < x \le 2]$$
$$M_x = 0.707x + 1 - (x-1) \quad \text{(Linear)}$$

at $x = 1$, M_B (just right of B) $= 0.707 \times 1 + 1 - 0 = 1.707$ kNm

at $x = 2$, $M_C = 0.707 \times 2 + 1 - (2-1) = 1.414$ kNm

Portion *CD*:

$$M_x \text{ (x from A)} = +R_A x + 1 - 1(x-1) - (\sqrt{2}-1)(x-2) \quad [2 \le x \le 4]$$
$$M_x = 0.707x + 1 - (x-1) - (\sqrt{2}-1)(x-2) \quad \text{(Linear)}$$

at $x = 2$, $M_C = 1.414$ kNm

at $x = 4$, $M_D = 0$

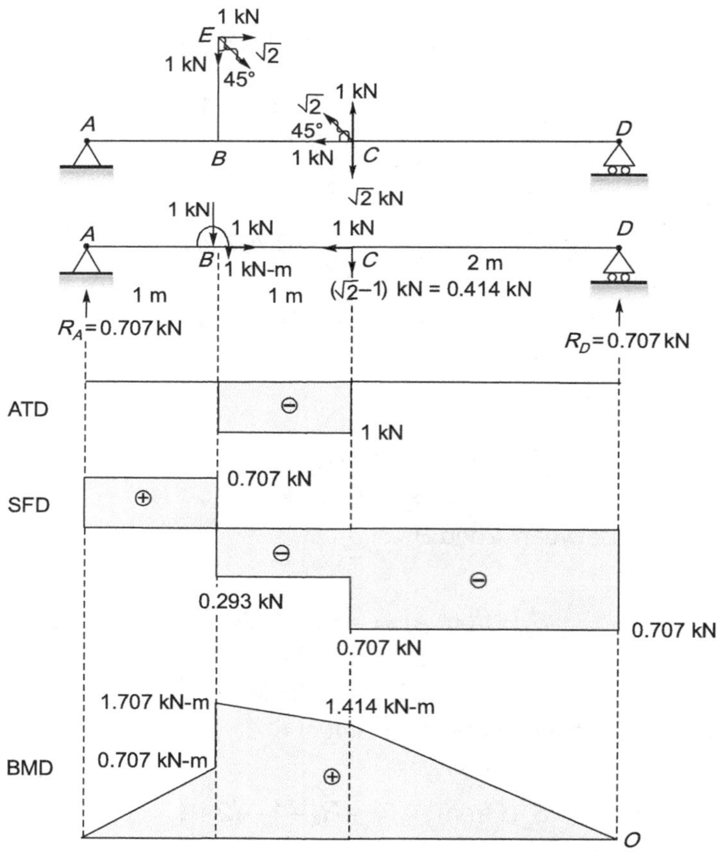

2.6 Determine all the reactions components at supports *A* (fixed) and *B* (hinged). Joint *C* and *D* are hinged. Effect of axial forces in various beams parts, *AC*, *CD*, *DB* need not to be considered. Also draw SFD and BMD for beam *ACDB*.

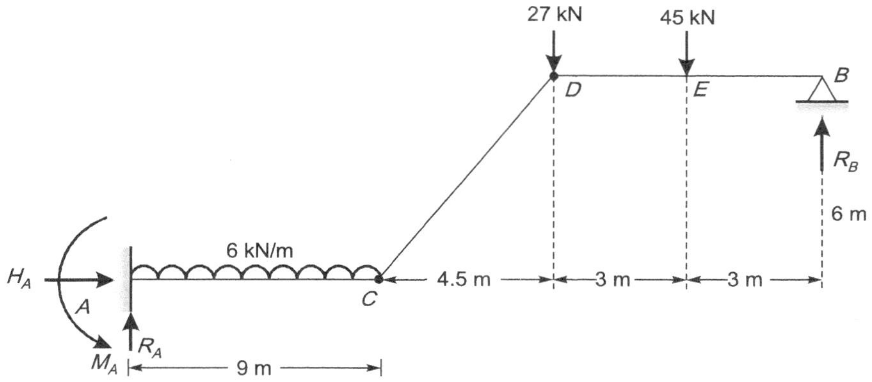

[20 marks : 2005]

Solution:

$\Sigma f_x = 0,\quad H_A - H_B = 0$

$H_A = H_B \quad \text{...(i)}$

Also, $\Sigma f_y = 0,$

$R_A + R_B - 6 \times 9 - 27 - 45 = 0$

$R_A + R_B = 126 \quad \text{...(ii)}$

Since *D* is hinge, hence $M_D = 0$

Taking moment about *D* from right

$R_B \times 6 - 45 \times 3 = 0$

$R_B = 22.5 \text{ kN} \quad \text{...(iii)}$

Also, taking moment about *C* from right

$\because \quad M_C = 0 \quad \text{(hinged)}$

$R_B \times 10.5 + H_B \times 6 - 45 \times 7.5 - 27 \times 4.5 = 0$

$22.5 \times 10.5 + H_B \times 6 - 45 \times 7.5 - 27 \times 4.5 = 0$

$6H_B = 222.75$

$H_B = 37.125 \text{ kN}$

Hence from (i) $\quad H_A = 37.125 \text{ kN}$

Also from (ii) and (iii) $\quad R_A = 126 - R_B$

$R_A = 126 - 22.5 = 103.5 \text{ kN}$

Taking moment about *C* from left

$R_A \times 9 - M_A - 6 \times 9 \times 4.5 = 0$

$103.5 \times 9 - M_A - 6 \times 9 \times 4.5 = 0$

$M_A = 688.5 \text{ kNm}$

Reactions:

$R_A = 103.5 \text{ kN} (\uparrow)$

$R_B = 22.5 \text{ kN} (\uparrow)$

$H_A = 37.125 \text{ kN} (\rightarrow)$

$H_B = 37.125 \text{ kN} (\leftarrow)$

$M_A = 688.5 \text{ kNm}$

For SFD:

Consider top face as reference face.

Portion *AC*:

$$S_x\ (x \text{ from } A) = +R_A - 6x \qquad [0 \le x < 9]$$

$$S_x = 103.5 - 6x \qquad \text{(Linear)}$$

at $x = 0$, $S_A = 103.5$ kN

at $x = 9$, S_C (just left of C) $= 103.5 - 6 \times 9 = 49.5$ kN

Portion *CD*:

$$S_x\ (x \text{ from } C) = +R_A \cos\theta - (6 \times 9)\cos\theta - H_A \sin\theta \qquad [0 < x < 4.5]$$

$$S_x = (103.5 - 54)\cos\theta - 37.125 \sin\theta$$

$$S_x = \frac{49.5 \times 4.5}{\sqrt{4.5^2 + 6^2}} - 37.125 \times \frac{6}{\sqrt{4.5^2 + 6^2}} = 0$$

Hence SF will be zero in member *CD*.

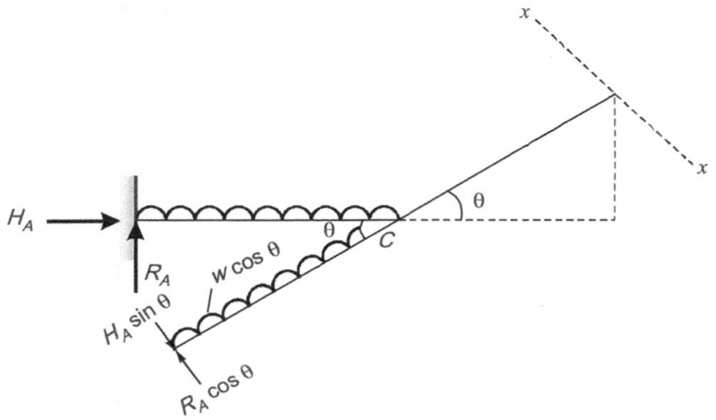

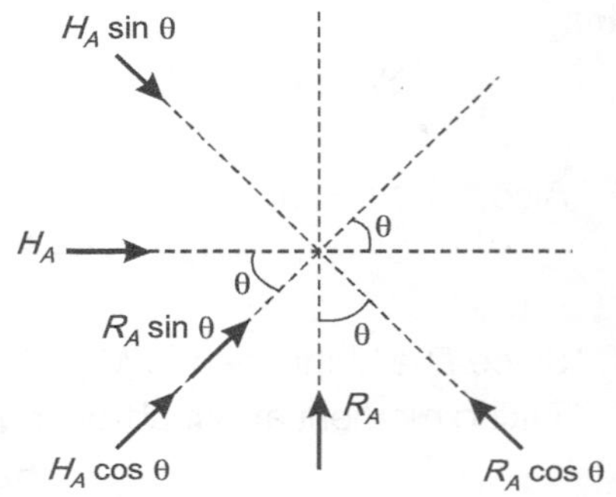

For SF in CD

Portion *BE*:

$$S_x\ (x \text{ from } B) = -R_B \qquad [0 \le x < 3]$$

$$S_x = -22.5 \text{ kN}$$

Hence SF will be constant from *B* to just right of *E*.

Portion *ED*:

$$S_x\ (x \text{ from B}) = -R_B + 45 \qquad [3 < x < 6]$$

$$S_x = -22.5 + 45$$

$$S_x = +22.5 \text{ kN} \qquad \text{(Constant)}$$

Hence *SF* will be constant between *ED*.

For BMD:

Consider outer (top) face as reference face.

Portion *AC*:

$$M_x\ (x \text{ from } A) = R_A x - M_A - \frac{6x^2}{2} \qquad [0 \le x \le 9]$$

$$M_x = 103.5x - 688.5 - 3x^2 \qquad \text{(Parabolic)}$$

at $x = 0$, $M_A = -688.5$ kNm

at $x = 9$, $M_C = 103.5 \times 9 - 688.5 - 3 \times 9^2 = 0$

Portion *CD*:

$$M_{CD} = 0$$

If there is member which is hinged at both ends and there is no load applied between these two hinges. Then BM between these hinges at any section will be zero.

Portion *BE*:

$$M_x \text{ (x from } B) = 22.5x \qquad \text{(Linear) } [0 \le x \le 3]$$

at $x = 0$, $M_B = 0$

at $x = 3$, $M_E = 22.5 \times 3 \Rightarrow 67.5$ kNm

Portion *ED*:

$$M_x \text{ (x from } D) = R_B x - 45(x-3) \qquad [3 \le x \le 6]$$

$$M_E = 22.5x - 45(x-3) \qquad \text{(Linear)}$$

at $x = 3$, $M_E = 22.5 \times 3 - 0 \Rightarrow 67.5$ kNm

at $x = 6$, $M_D = 22.5 \times 6 - 45(6-3)$

$$M_D = 0$$

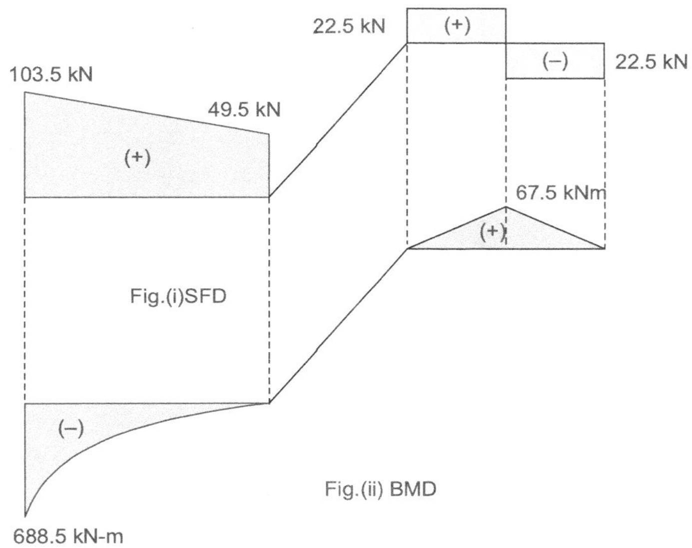

2.7 A portal frame *ABCD* as shown in the figure has hinged end at *A* and fixed end at *D*. Vertical leg *AB* = 4 m, horizontal *BC* = 7 m. Joint *C* and mid point *E* of inclined leg *CD* may be acting as hinges. *A* and *D* are at same level.

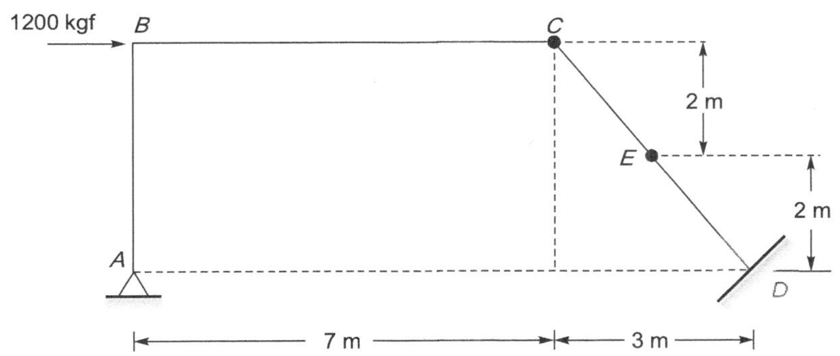

Calculate the components of reactions at *A* and *D* when a horizontal force of 1200 kgf is applied at joint *B*.

[20 marks : 2007]

Solution:

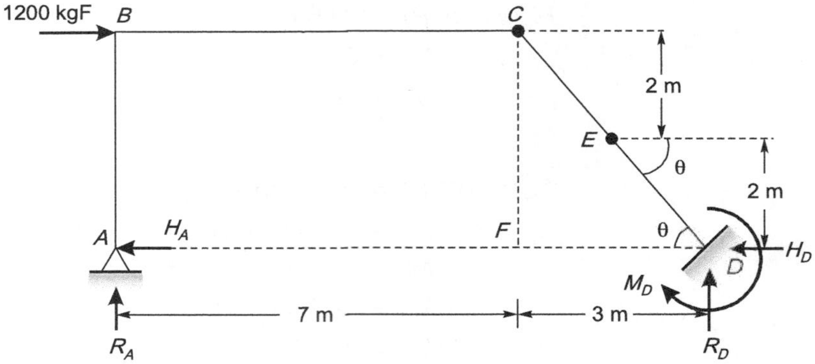

Given that horizontal load at *B* is 1200 kgF = 12 kN

Let vertical reactions at *A* and *D* be $R_A(\uparrow)$ and $R_D(\uparrow)$ respectively. Also let horizontal reactions at *A* and *D* be $H_A(\leftarrow)$ and $H_D(\leftarrow)$ respectively. Also, let the reactive moment at *D* be M_D (↷).

Let $\angle CDA = \theta$

$\therefore$ $\tan\theta = \dfrac{4}{3}$

$\Sigma F_x = 0$ $H_A + H_D = 12$ ...(i)

and $\Sigma F_y = 0$ $R_A + R_D = 0$ ...(ii)

Now taking moments about *A*, we have

$\Sigma M_A = 0$ $(R_D \times 10) - (12 \times 4) - M_D = 0$

$$10R_D - M_D = 48 \quad \text{...(iii)}$$

Since *C* is a hinge, therefore moments about *C* either from right or left will be zero.

M_C from left = 0 $R_A \times 7 + H_A \times 4 = 0$

$$4H_A + 7R_A = 0 \quad \text{...(iv)}$$

Taking moments about *E* from right, we have

M_E from right = 0

$$R_D \times \frac{2}{\tan\theta} - H_D \times 2 - M_D = 0$$

$$R_D \times \frac{2}{4/3} - 2H_D - M_D = 0$$

$$1.5\,R_D - 2H_D - M_D = 0 \quad \text{...(v)}$$

Substituting value of R_D, H_D and M_D from (i), (ii) and (ii) respectively in (v) we get

$$1.5\,(-R_A) - 2\,(12 - H_A) - (10R_D - 48) = 0$$

$$-1.5\,R_A - 24 + 2H_A - 10\,R_D + 48 = 0$$

$$-1.5\,R_A - 10\,(-R_A) + 2\,H_A + 24 = 0$$

$$8.5\,R_A + 2\,H_A = -24 \quad \text{...(vi)}$$

Solving (iv) and (vi), we get $H_A = 8.4$ kN

$R_A = -4.8$ kN

$\therefore$ $R_D = -R_A = -(-4.8) = 4.8$ kN

$H_D = 12 - H_A = 12 - 8.4 = 3.6$ kN

$M_D = 10\,R_D - 48 = 10 \times 4.8 - 48 = 0$ kN-m

Thus the reactions will be

$H_A = 8.4$ kN ($\leftarrow$)

$R_A = 4.8$ kN ($\downarrow$)

$H_D = 3.6$ kN ($\leftarrow$)

$R_D = 4.8$ kN ($\uparrow$)

$M_D = 0$ kN-m

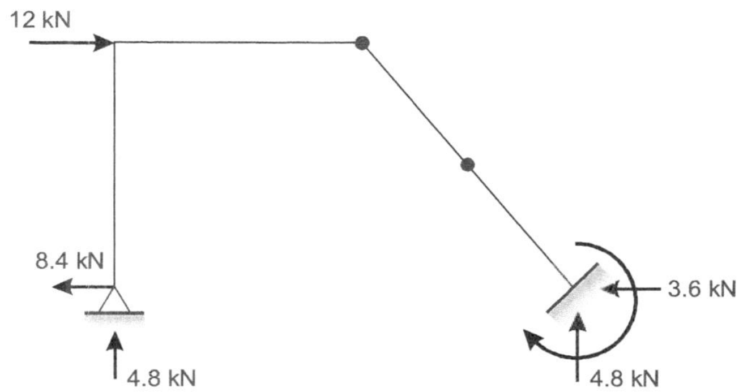

2.8 A portal frame *ABCDEF* with inclined legs has hinges at *B*, *D* and *E* as shown in figure below. Joint at *C* is monolithic. Supports at *A* and *F* are fixed. Calculate all components of reactions.

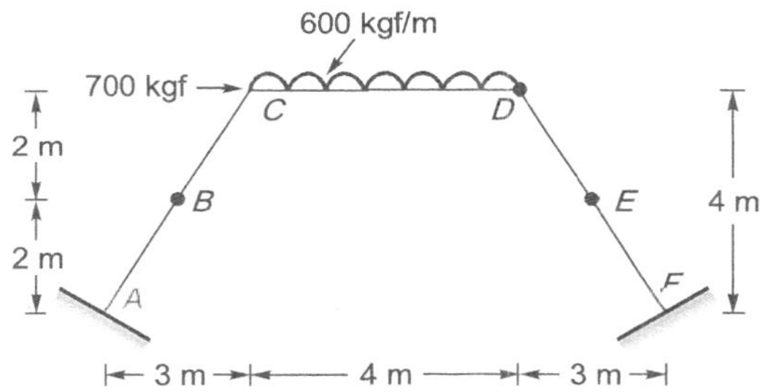

[20 marks : 2009]

Solution:

Let the reactions at *A* and *F* be $R_A(\uparrow)$, $H_A(\leftarrow)$, M_A (anticlockwise), $R_F(\uparrow)$, $H_F(\leftarrow)$ and M_F (anticlockwise). Taking 100 kgf = 1 kN

Free body diagram of the portal frame

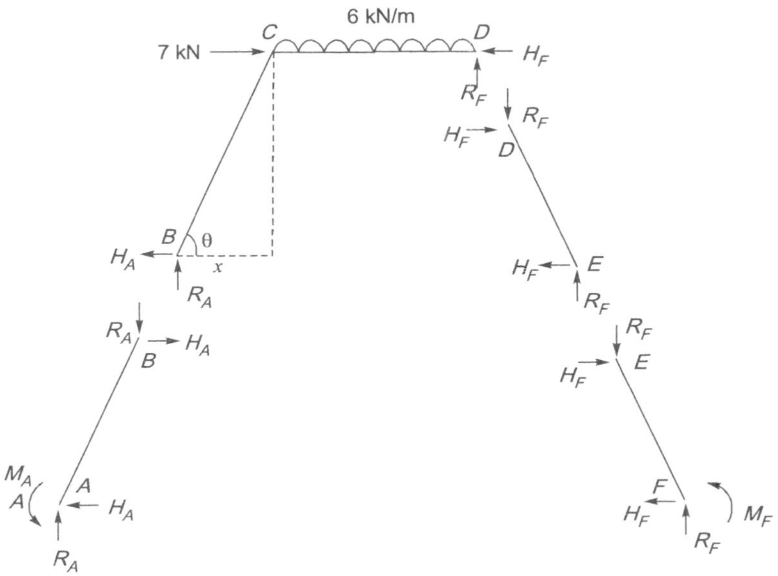

From the given portal frame, we know that

$$\tan\theta = \frac{4}{3}$$

Now, from portion BCD of the free body diagram, we have

$$\tan\theta = \frac{2}{x}$$

$$\Rightarrow \quad x = \frac{2}{4/3}$$

$$\Rightarrow \quad x = 1.5 \text{ m}$$

Taking portion BCD of the free body diagram, we get

$\Sigma F_y = 0$ $\quad R_A + R_F - 6 \times 4 = 0$

$$\Rightarrow \quad R_A + R_F = 24 \quad \text{...(i)}$$

Also, $\Sigma F_x = 0$ $\quad H_A + H_F - 7 = 0$

$$\Rightarrow \quad H_A + H_F = 7 \quad \text{...(ii)}$$

Taking moments about *D*, we get

$$R_A \times (1.5 + 4) + H_A \times 2 - 6 \times 4 \times 2 = 0$$

$$\Rightarrow \quad 5.5\,R_A + 2\,H_A = 48 \quad \text{...(iii)}$$

Taking portion *DE* of the free body diagram and calculating moments about *D*, we get

$$R_F \times 1.5 - H_F \times 2 = 0$$

$$\Rightarrow \quad H_F = 0.75\,R_F \quad \text{...(iv)}$$

Substituting values of H_A and H_F from (iii) and (iv) in (ii), we get

$$H_A + H_F = 7$$

$$\Rightarrow \quad \frac{48 - 5.5\,R_A}{2} + 0.75\,R_F = 7$$

$$\Rightarrow \quad 48 - 5.5\,R_A + 1.5\,R_F = 14$$

$$\Rightarrow \quad 5.5\,R_A - 1.5\,R_F = 34 \quad \text{...(v)}$$

Substituting value of R_F from (i) in (v), we get

$$5.5\,R_A - 1.5(24 - R_A) = 34$$

$$\Rightarrow \quad 7R_A = 34 + 36$$

$$\Rightarrow \quad R_A = 10 \text{ kN } (\uparrow)$$

$$\therefore \quad R_F = 24 - R_A = 24 - 10 = 14 \text{ kN } (\uparrow)$$

$$H_F = 0.75\,R_F = 0.75 \times 14 = 10.5 \text{ kN } (\leftarrow)$$

$$H_A = 7 - H_F = 7 - 10.5 = -3.5 \text{ kN} = 3.5 \text{ kN } (\rightarrow)$$

Considering portion *AB* of free body diagram and taking moments about *B*, we get

$$R_A \times 1.5 - M_A + H_A \times 2 = 0$$

$$\Rightarrow \quad 10 \times 1.5 - M_A - 3.5 \times 2 = 0$$

$$\Rightarrow \quad 15 - M_A - 7 = 0$$

$$\Rightarrow \quad M_A = 8 \text{ kN-m (anticlockwise)}$$

Considering portion *EF* of free body diagram and taking moments about *E*, we get

$$R_F \times 1.5 + M_F - H_F \times 2 = 0$$

$$\Rightarrow \quad 14 \times 1.5 + M_F - 10.5 \times 2 = 0$$

$$\Rightarrow \quad M_F = 0$$

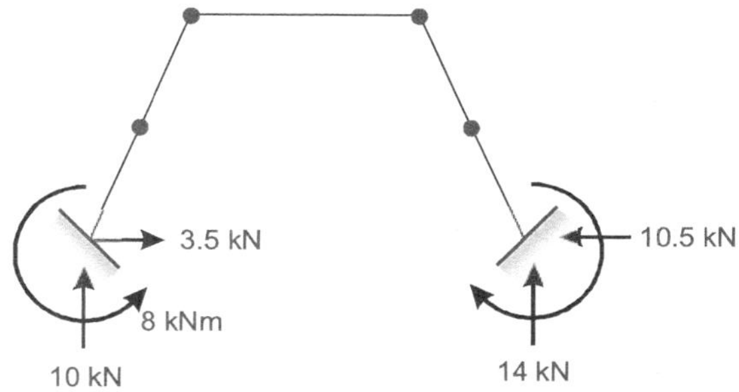

The final reactions are:

$$R_A = 10\text{ kN}(\uparrow),\ R_F = 14\text{ kN }(\uparrow)$$

$$H_A = 3.5\text{ kN }(\rightarrow),\ H_F = 10.5\text{ kN }(\leftarrow)$$

$$M_A = 8\text{ kN-m (anticlockwise)},\ M_F = 0$$

2.9 Bending moment diagram of a balanced cantilever beam is shown below. Draw the beam and find the loads acting on the beam.

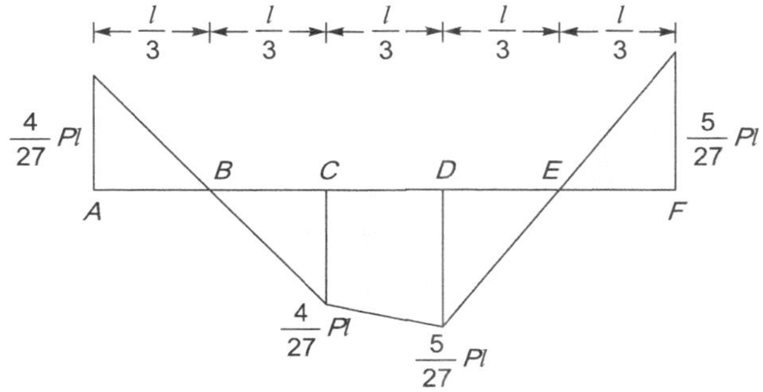

[10 marks : 2010]

Solution:

At point *A*:

Bending moment at A is non-zero, so a clockwise moment exist at point *A*.

Between point *A* and *C*:

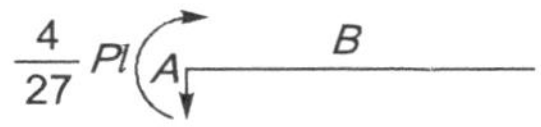

BMD is a straight line

$$\therefore \qquad \frac{dM}{dx} = \frac{\left[-\frac{4}{27} - \frac{4}{27}\right]Pl}{\frac{2}{3}l} = -\frac{4}{9}P \text{ (constant)}$$

We know that,
$$V = \frac{dM}{dx}$$

$$\therefore \qquad V = -\frac{4}{9}P$$

$\Rightarrow$ SF is constant between *A* and *C* and at point *A*.

Negative shear force implies that there is downward force of 4P/9 at *A* and no load intensity between *A* and *C*.

At point *C*:

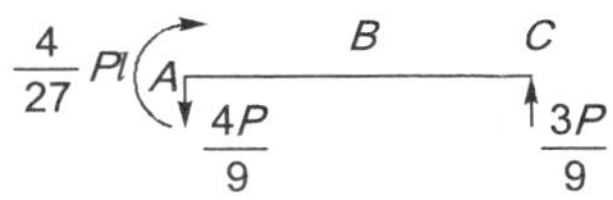

$$\left.\frac{dM}{dx}\right|_{\text{L.H.S. of}C} = -\frac{4}{9}P$$

$$\left.\frac{dM}{dx}\right|_{\text{R.H.S. of}C} = \frac{M_D - M_C}{\frac{l}{3}} = \left(\frac{\frac{-5}{27} - \left(-\frac{4}{27}\right)}{\frac{l}{3}}\right)Pl = -\left(\frac{P}{9}\right)$$

$$\Rightarrow \qquad \text{Shear force from left} = \frac{-4P}{9} \text{ and shear force from right} = \frac{-P}{9}$$

Hence, at C shear force changes from $\frac{-4P}{9}$ to $\frac{-P}{9}$

$$\therefore \quad \text{Upward force at } C = \left(\frac{-P}{9}+\frac{4P}{9}\right) = \frac{3P}{9}$$

Hence at point C an upward force exist of magnitude = $\frac{3P}{9}$

Between point C and D:

The slope is constant between C and D,

$$V = \frac{dM}{dx} = \text{constant}$$

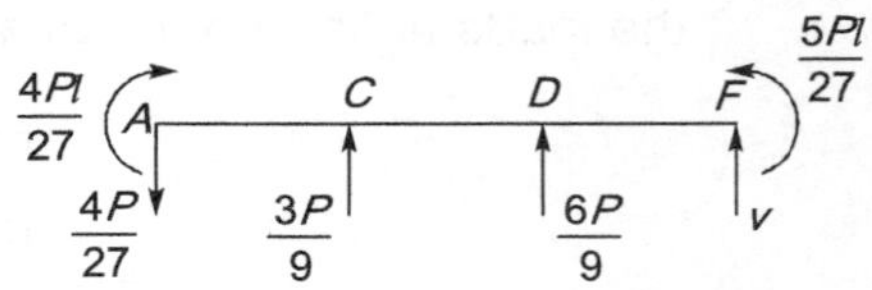

$\therefore$ No loading exist between C and D.

At point D:

$$\left.\frac{dM}{dx}\right|_{\text{L.H.S. of } D} = \frac{-P}{9}$$

$$\left.\frac{dM}{dx}\right|_{\text{R.H.S. of } D} = \frac{\left(\frac{5}{27}-\left(-\frac{5}{27}\right)\right)Pl}{\frac{2l}{3}} = \frac{5}{9}P$$

$\Rightarrow$ Shear force changes from $\frac{-P}{9}$ to $\frac{5P}{9}$

$$\therefore \quad \text{Upward point load at } D = \frac{5P}{9}-\left(-\frac{P}{9}\right) = \frac{6P}{9}$$

Between point D and F:

Since slope is constant for BMD between D and F,

Hence no loading will exist in span DF.

At point F:

Let the vertical force at E is V

$$\Sigma F_y = 0, \quad V+\frac{6P}{9}+\frac{3P}{9}-\frac{4P}{9} = 0$$

$$\therefore \quad V = -\frac{5P}{9}$$

There exists *a* positive bending moment at F equal to $\frac{5Pl}{27}$

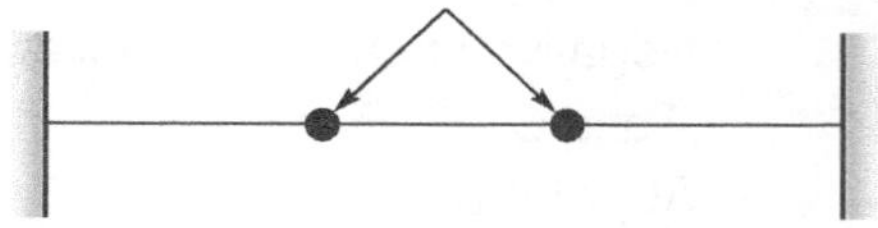

Balanced cantilever beam

But it is given that the beam is a balanced cantilever beam.

As BM is zero at B and E, hence the final beam with loading will be

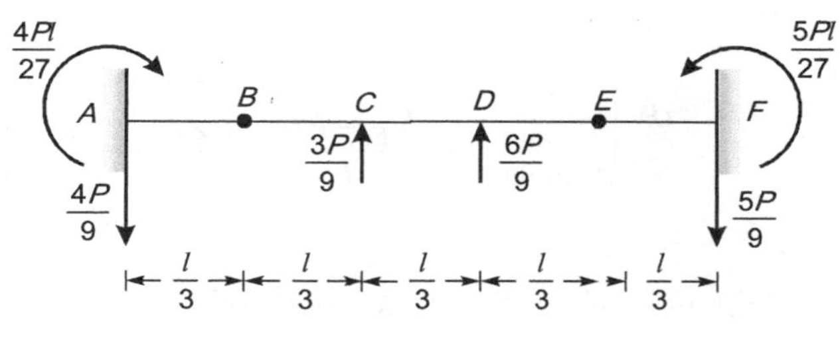

2.10 Find the reactions at supports A and D of the structure shown below. Draw BM diagram indicating BM values at salient points. Supports A and D are at the same level.

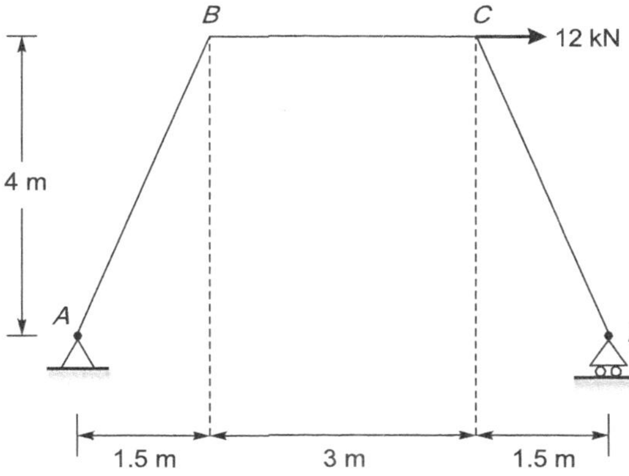

[10 marks : 2010]

Solution:

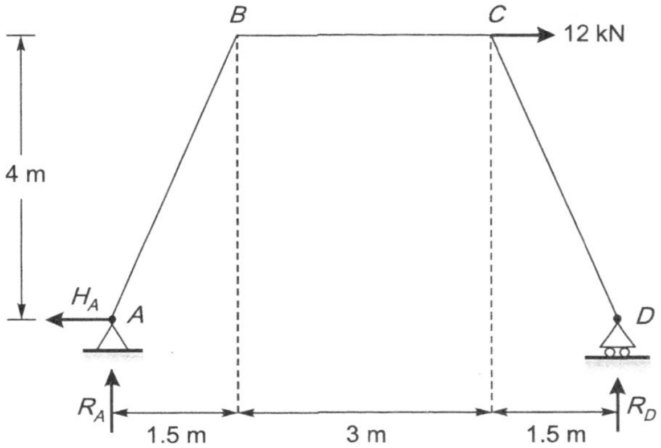

$\Sigma F_x = 0,$ $\quad H_A = 12 \text{ kN} (\leftarrow)$

$\Sigma M_D = 0,$ $\quad R_A = -\dfrac{12 \times 4}{6} \text{kN} = -8 \text{ kN}$

$\Sigma M_y = 0,$ $\quad R_D = -R_A = 8 \text{ kN}$

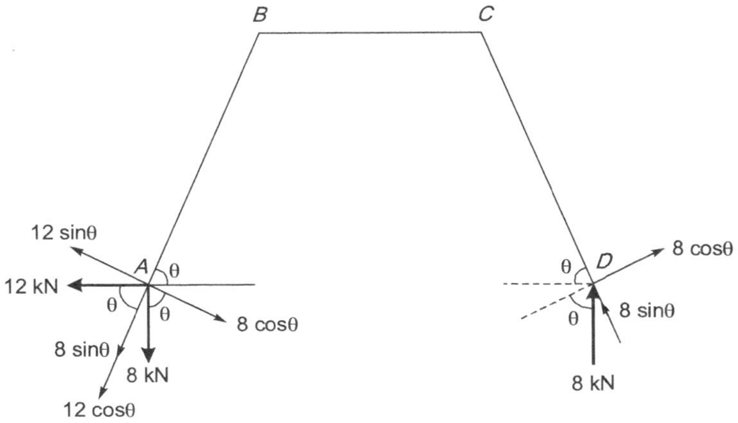

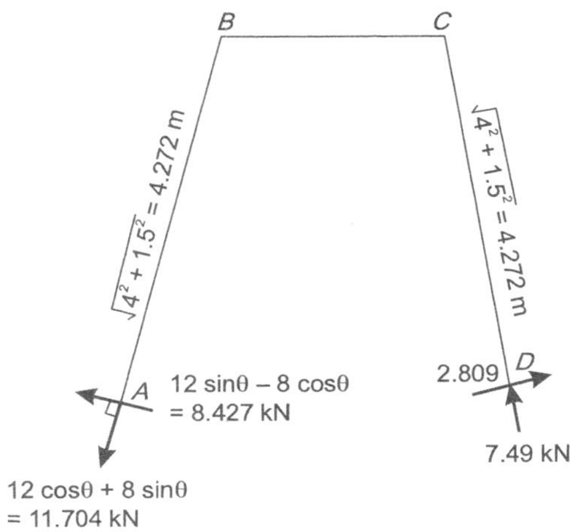

$$\tan\theta = \frac{4}{1.5} = \frac{8}{3}, \ \sin\theta = \frac{8}{\sqrt{73}}, \ \cos\theta = \frac{3}{\sqrt{73}}$$

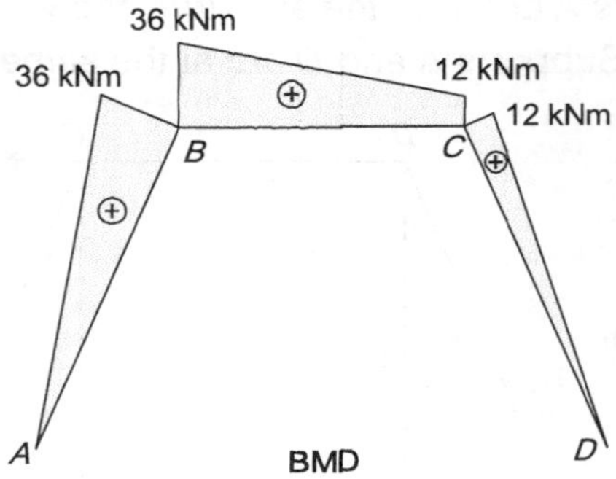

2.11 Draw the bending moment and shear force diagrams for overhanging beam shown in the figure below. Indicate the significant values including the point of contraflexure.

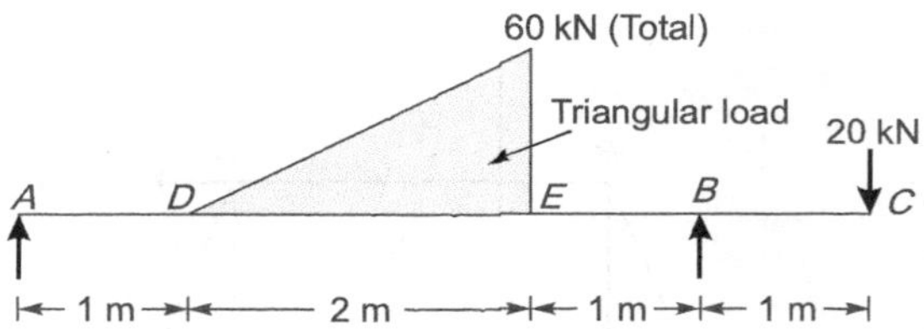

[10 marks : 2011]

Solution:

Reactions: Let R_A and R_B be the support reaction at A and B respectively

Given total load of UVL = 60 kN

Let w is rate of loading for UVL, then

$$\frac{1}{2} \times 2 \times w = 60$$

$\therefore$ $$w = 60 \text{ kN/m}$$

$\Sigma F_y = 0$ $$R_A + R_B = 60 + 20$$

$$R_A + R_B = 80 \quad \text{...(i)}$$

$\Sigma M_A = 0$

$$20 \times 5 - R_B \times 4 + \frac{1}{2} \times 2 \times 60 \times \left(1 + 2 \times \frac{2}{3}\right) = 0$$

$$4 R_B = 100 + 60\left(\frac{7}{3}\right)$$

$\therefore$ $$R_B = 60 \text{ kN}$$

$$R_A = 80 - 60 = 20 \text{ kN}$$

For SFD:

Portion *AD*:

Consider a section x - x at a distance x from support A between A and D

$$S_x (x \text{ from } A) = +R_A \quad [0 \le x < 1]$$

$$S_x = +20 \text{ kN} \quad \text{(Constant)}$$

at $x = 0$, $$S_A = 20 \text{ kN}$$

at $x = 1$, $$S_D \text{ (just left of } D) = 20 \text{ kN}$$

Portion DE:

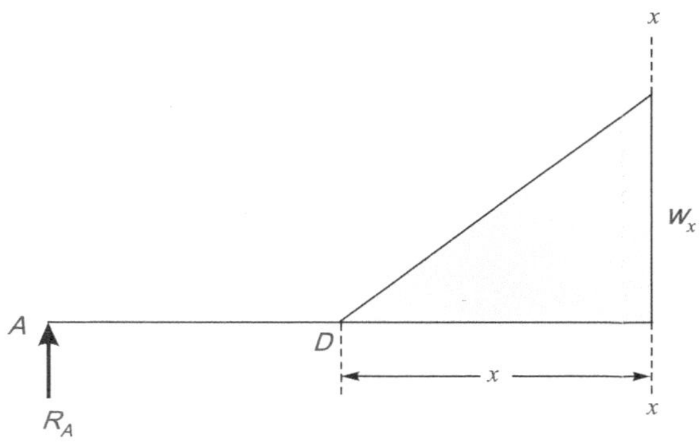

Again consider section x-x at a distance x from D between D and E

From the property of triangle,

$$\frac{w}{L} = \frac{W_x}{x}$$

$$\Rightarrow \quad \frac{60}{2} = \frac{W_x}{x}$$

$$\therefore \quad W_x = 30x \text{ kNm}$$

Take D as origin,

$$S_x (x \text{ from } D) = +R_A - \text{Total load of UVL between } D \text{ and section}$$

$$= +20 - \frac{1}{2} \times 30x \times x$$

$$= 20 - 15x^2 \qquad \text{(Parabolic) } [0 < x < 2]$$

at $x = 0$, $\quad S_D$ (just right of D) $= +20$ kN

at $x = 2$, $\quad S_E$ (just left of E) $= +20 - 15 \times 2^2 = -40$ kN

Portion EB:

$$S_x (x \text{ from } A) = +R_A - \text{Total UVL load} \qquad [3 < x < 4]$$

$$= +20 - \frac{1}{2} \times 2 \times 60 = -40 \text{ kN} \qquad \text{(Constant)}$$

Hence SF will be constant between E and B.

Portion CB:

Now consider section x-x from C

$$S_x (x \text{ from } C) = +20 \text{ (Constant)} \qquad [0 \le x < 1]$$

Hence SF will constant between C and D.

For BMD:

Portion AD:

$$M_x (x \text{ from } A) = R_A x \qquad [0 \le x \le 1]$$

$$= +20x \qquad \text{(Linear)}$$

at $x = 0$, $\quad M_A = 0$

at $x = 1$, $\quad M_D = 20 \times 1 = 20$ kN-m

Portion DE:

Consider section x-x at a distance x from D between D and E

at section loading rate will be $w_x = 30x$

$$M_x (x \text{ from } D) = 20(1+x) - \left(\frac{1}{2} x \times (30x) \times \frac{x}{3}\right) \qquad [0 \le x \le 2]$$

$$M_x = 20 + 20x - 5x^3 \qquad \text{(Cubic)}$$

at $x = 0$, $\quad M_D = 20$ kNm

at $x = 2$, $\quad M_E = 20 + 20 \times 2 - 5 \times 2^3 = 20$ kNm

Since BM at two extreme points of section D and E is equal. Hence there must be a point of zero slope. So within DE, BM may be maximum or minimum.

$$\therefore \quad \frac{dM_x}{dx} = 0$$

$$\frac{d}{dx}(20+20x-5x^3) = 20-15x^2 = 0$$

$$\Rightarrow \quad x = 1.15 \text{ m}$$

Here, $\frac{d^2M_x}{dx^2} < 0$ (negative) mean concavity downward.

Hence the BM will be max at $x = 1.15$ m from D.

Portion EB:

$$M_x \text{ (x from A)} = R_A x - 60\times\left[x-\left(1+2\times\frac{2}{3}\right)\right]$$

$$= 20x - 60\left(x-\frac{7}{3}\right) \text{[Linear } 3\le x\le 4]$$

at $x = 3$, $$M_E = 20\times3-60\left(3-\frac{7}{3}\right) = 20 \text{ kNm}$$

at $x = 4$, $$M_B = 20\times4-60\left(4-\frac{7}{3}\right) = -20 \text{ kNm}$$

Portion CB:

$$M_x(x \text{ from } C) = -20x \quad [0\le x\le 1]$$

$$M_x = -20x \quad \text{(Linear)}$$

at $x = 0$, $M_C = 0$

at $x = 1$, $M_B = -20\times1 = -20$ kNm

Point of contraflexure:

In portion BE bending moment changes sign. Hence there will be a point of zero BM between BE

$$M_x = 0 \quad \text{(for portion } BE)$$

$$20x-60\left(x-\frac{7}{3}\right) = 0$$

$$x = 3.5 \text{ m from support } A$$

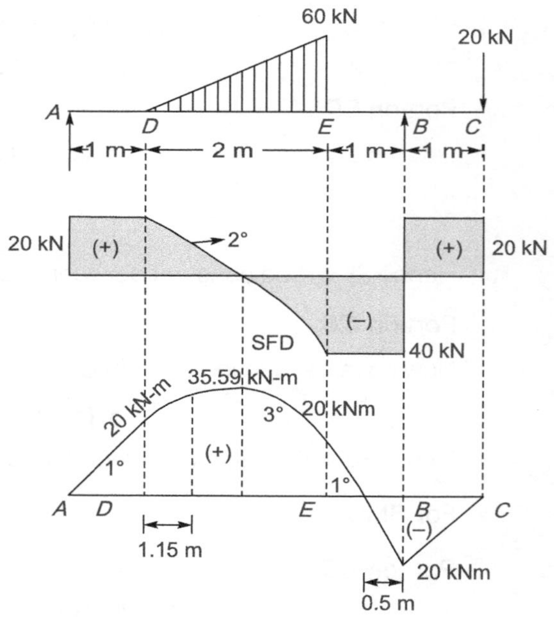

BMD

2.12 **Find all the support reactions and draw BM diagram for the frame shown below. Frame has hinged supports at A and B and internal hinge at C.**

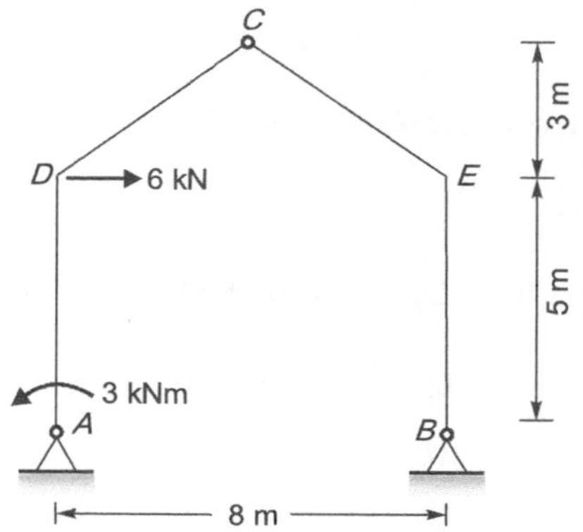

[15 marks : 2012]

Solution:

The given frame has been made determinate by providing a internal hinge at C

$\therefore \quad D_S = 0$

Let vertical reactions at A and B be $V_A(\uparrow)$ and $V_B(\uparrow)$ respectively.

Also let horizontal reactions at A and B be $H_A(\leftarrow)$ and $H_B(\leftarrow)$ respectively.

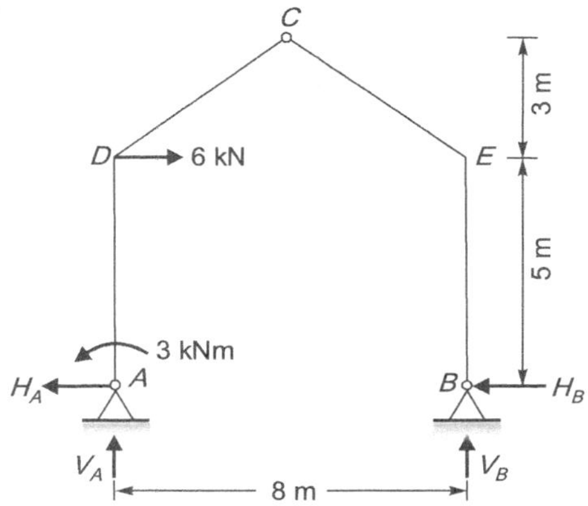

Calculation of reactions:

Equilibrium conditions:

$$\Sigma F_y = 0;\ V_A + V_B = 0$$

$$\Sigma F_x = 0;\ H_A + H_B = 6$$

$$\Sigma M_B = 0$$

$$V_A \times 8 + 6 \times 5 - 3 = 0$$

$$\Rightarrow \quad V_A = -3.375 \text{ kN}$$

$$\therefore \quad V_B = 3.375 \text{ kN}$$

Since C is a hinge, therefore moments about C either from right or left will be zero.

Taking M_C from left = 0

$$V_A \times 4 + H_A \times 8 - 3 - 6 \times 3 = 0$$

$$\Rightarrow \quad (-3.375 \times 4) + 8H_A - 3 - 18 = 0$$

$$\Rightarrow \quad 8H_A = 34.5$$

$$\Rightarrow \quad H_A = +4.3125 \text{ kN}$$

$$\therefore \quad H_B = 6 - H_A$$

$$= 1.6875 \text{ kN}$$

Thus the reactions will be, $\quad V_A = 3.375 \text{ kN } (\downarrow)$

$$V_B = 3.375 \text{ kN } (\uparrow)$$

$$H_A = 4.3125 \text{ kN } (\leftarrow)$$

$$H_B = 1.6875 \text{ kN } (\leftarrow)$$

Bending Moment Diagram

Portion AD

$$M_y(y \text{ from } A) = H_A \times y - 3 \qquad [0 \le y \le 5]$$

$$M_y = 4.3125\,y - 3$$

$$M_A = -3 \text{ kN-m}$$

$$M_D = 4.3125 \times 5 - 3$$

$$= 18.5625 \text{ kN-m}$$

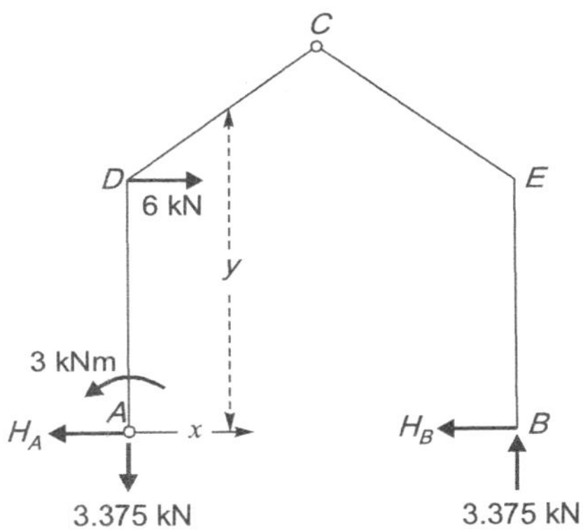

For $\quad M_y = 0$

We have, $\quad 4.3125\,y - 3 = 0$

$$\Rightarrow \quad y = 0.696 \text{ m}$$

Portion DC

$$\Rightarrow \quad M(x \text{ and } y \text{ from } A) = V_A x + H_A y - 3 - 6(y - 5)$$

$$\Rightarrow \quad = -3.375x + 4.3125y - 3 - 6y + 30$$

$$= -3.375x - 1.6875y + 27$$

$$M_C\,(x = 4,\ y = 8) = -3.375 \times 4 - 1.6875 \times 8 + 27$$

$$= 0$$

Portion *BE*

$$M_y(y \text{ from } B) = -H_B \times y$$

$$= -1.6875y$$

$$M_E = -8.4375 \text{ kN-m}$$

Portion *EC*

$$M(x \text{ and } y \text{ from } B) = V_B x - H_B \times y$$

$$= 3.375x - 1.6875y$$

$$\therefore \quad M_C(x = 4,\ y = 8) = 3.375 \times 4 - 1.6875 \times 8$$

$$= 0$$

Taking outer face as reference face

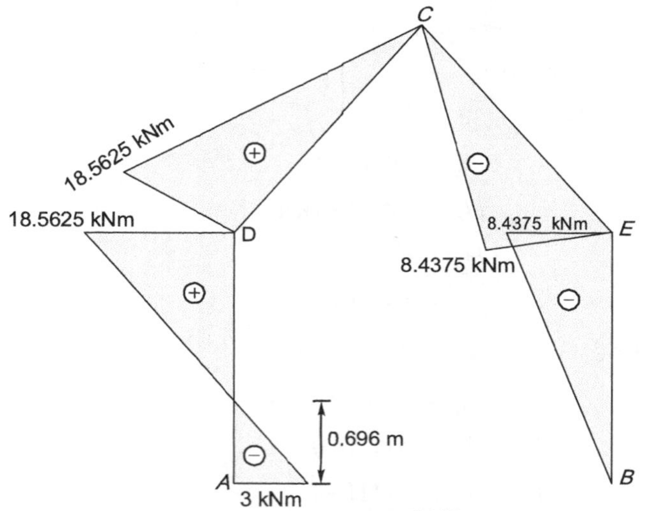

BMD

2.13 A rigid bent *ABC* is of uniform cross-section and is in a horizontal plane. It is fixed at *A* and free at *C* as shown in figure. It carries two vertical loads 10 kN each at *B* and *C*. Draw Bending Moment, shear force and torsional moment in the bent.

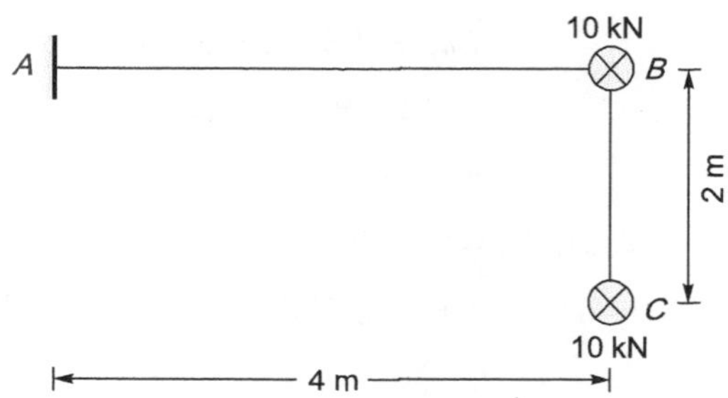

[10 marks : 2013]

Solution:

Bending moment diagram

Portion *BC*:

Fig. (i) Plan

Fig. (ii) Elevation

$$M_x(x \text{ from } C) = -10x \text{ (linear) } [0 \le x < 2]$$

$$M_x = -10x$$

at $x = 0$, $M_C = 0$

at $x = 2$, M_B (just right of B) $= -20$ kNm

Portion *AB*:

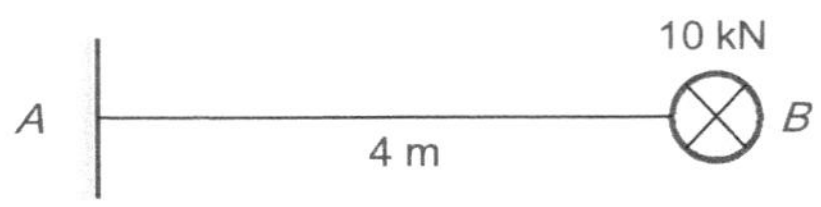

Fig. (i) Plan

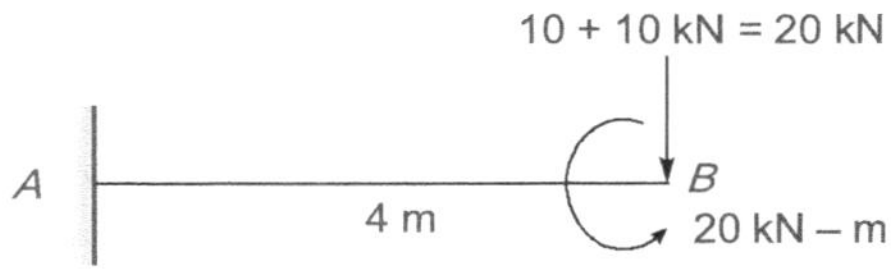

Fig. (ii) Elevation

Due to portion *BC* a concentrated load of 10 kN and concentrated torsional moment acts at point *B*

$\therefore$ M_x (x from B) $= -20x$ $[0 < x \le 4]$

$M_x = -20x$ (linear)

at $x = 0$, M_B(just left of B) $= 0$

at $x = 4$, $M_A = -80$ kNm

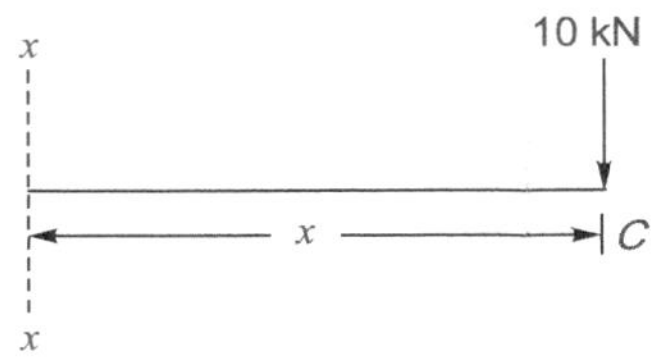

Shear force diagram

Portion *BC*:

S_x(x from C) $= +10$ kN (constant) $[0 \le x < 2]$

at $x = 0$, $S_C = 10$ kN

at $x = 2$, S_B (just right of B) $= 10$ kN

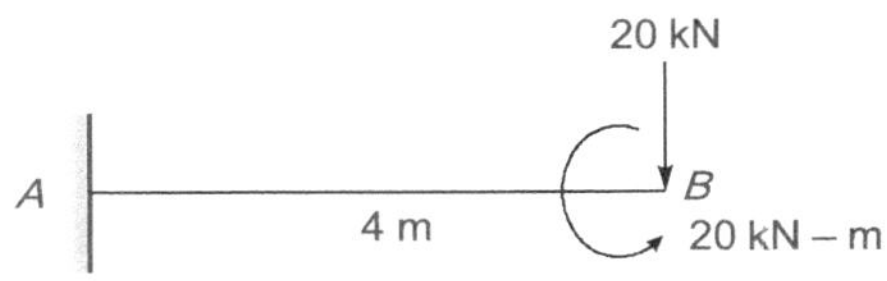

Portion *BA*:

S_x (x from B) $= +20$ kN (constant) $[0 < x \le 4]$

at $x = 0$, S_B (just left of B) $= +20$ kN

at $x = 4$, $S_A = +20$ kN

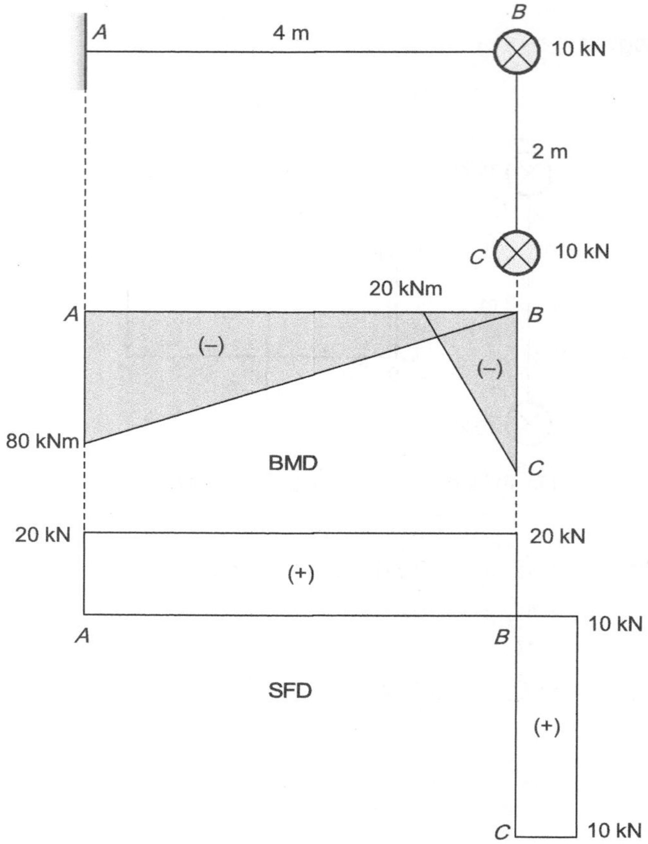

Torsional Moment Diagram (TMD)

Portion *BC*:

$$T_x \text{ (} x \text{ from } C) = 0 \qquad [0 \leq x < 2]$$

$$T_x = 0$$

Portion *BA*:

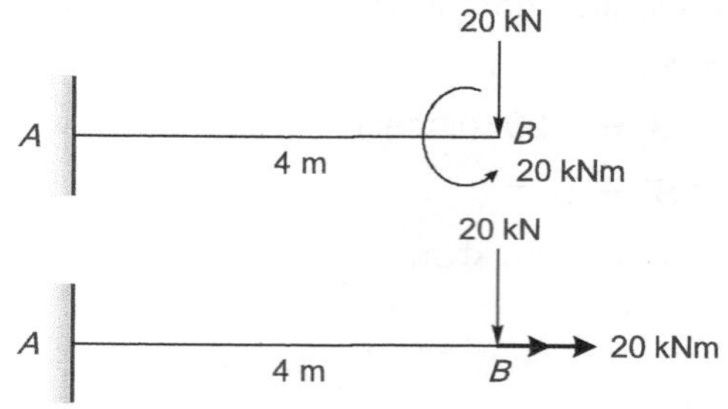

$$T_x \text{ (} \boldsymbol{x} \text{ from } B) = -20 \text{ kNm} \qquad [0 < \boldsymbol{x} \leq 4]$$

$$T_x = -20 \text{ kNm (constant)}$$

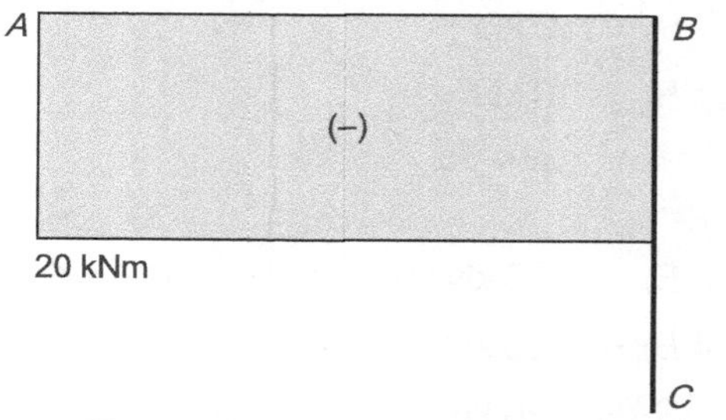

Torsional Moment Diagram

2.14 Draw the bending moment diagram of a beam from the shear force diagram of the beam.

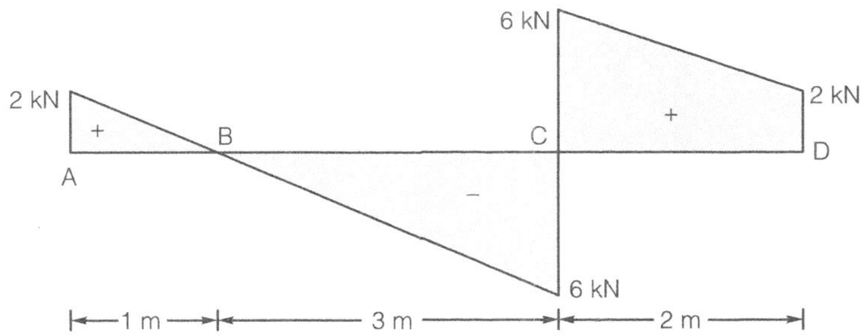

[4 marks : 2013]

Solution:

For loading diagram:

Point *A*: At point *A* SF is +2 kN .Hence a upward reaction should be present at *A*. So at point *A* there will be a support.

$$\therefore \qquad R_A = +2 \text{ kN}$$

Portion *AB*: SFD consists incline line in portion *AB*. So portion *AB* will be subjected to UDL and loading rate is given by the slope of SFD

$$w_1 = \frac{+0-2}{1} = -2 \text{ kN/m}$$

Negative sign indicate downward loading rate.

Portion *BC*: SFD consists incline line is portion *BC*. So portion *BC* will be subjected to UDL and loading rate is given by the slope of SFD

$$w_2 = \frac{-6-0}{3} = -2 \text{ kN/m}$$

Negative sign indicate downward loading rate.

Point *C*: *A* point *C SF* drastically changes from –6 kN to +6 kN. Hence there will be support at *C* and reaction at support *C* will be

$$R_C = 6 + 6 = 12 \text{ kN}$$

Portion *CD*: SFD consists of inclined line in portion *CD*.
Hence portion *CD* will be subjected to UDL of loading rate

$$w_3 = \frac{2-6}{2} = -2 \text{ kN/m}$$

Point *D*: At point *D*, *SF* drastically changes from +6 kN to 0.
Hence there will be a downward point load of 2 kN at *D*.

Supports: There are two supports at *A* and *C*

at support *A*, $\Sigma M_A = 2 \times 6 \times 3 - 12 \times 4 + 2 \times 6 = 0$

Hence support A will be hinge or roller.

For BMD:

Portion *AC*:

$$M_x \text{ (x from A)} = +R_A x - 2\frac{x^2}{2} \qquad [0 \le x \le 4]$$

$$M_x = 2x - x^2 \qquad \text{(Parabolic)}$$

at $x = 0$, $M_A = 0$

at $x = 4$, $M_C = 2 \times 4 - 4^2 \Rightarrow -8 \text{ kNm}$

Bending moment will be zero if $M_x = 0$

$$2x - x^2 = 0$$
$$x(2 - x) = 0$$
$$x = 0, 2$$

For bending moment to be maximum

$$\frac{dM}{dx} = 0$$
$$2 - 2x = 0$$
$$x = 1 \text{ m}$$
$$M_{max} = 2 \times 1 \Rightarrow 1 \text{ kNm}$$

Portion *DC*:

$$M_x \,(x \text{ from } D) = -2x - \frac{2x^2}{2} \qquad [0 \le x \le 2]$$
$$M_x = -2x - x^2 \qquad \text{(Parabolic)}$$

at $x = 0$, $M_D = 0$

at $x = 2$, $M_C = -2 \times 2 - 2^2 = -8$ kNm

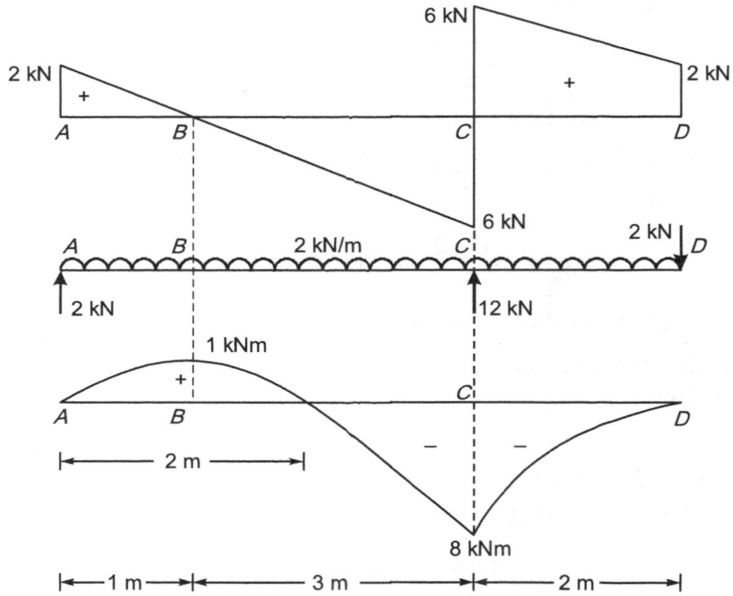

Bending Moment Diagram

2.15 Draw the bending moment and torsional moment diagram for the beam as shown. The load is perpendicular to the beam in plane.

[5 marks : 2015]

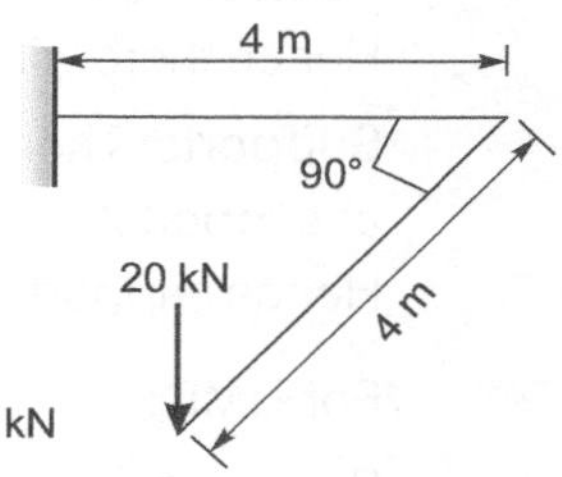

Solution:

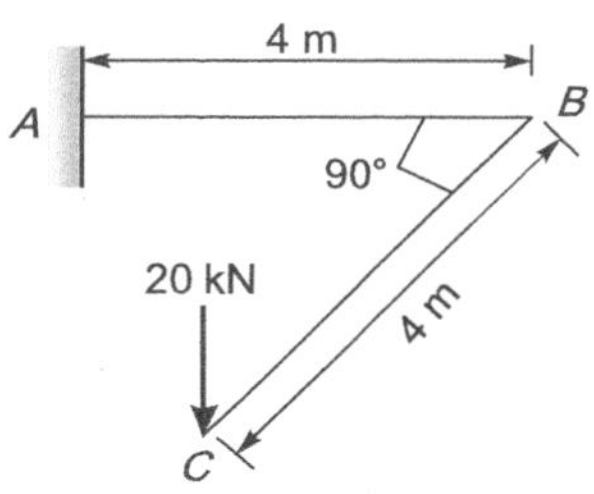

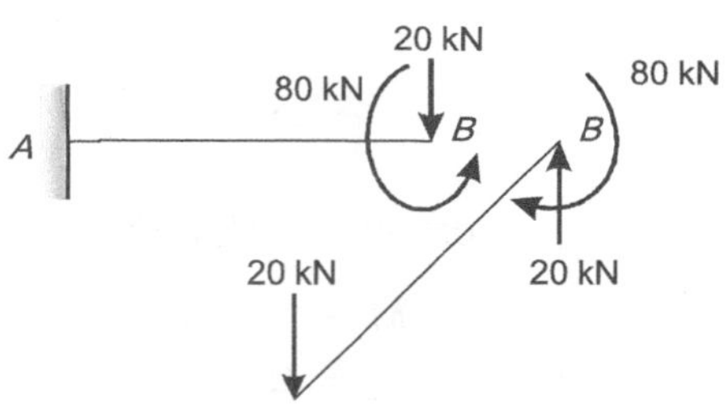

FBD

There will be no torsion in beam BC

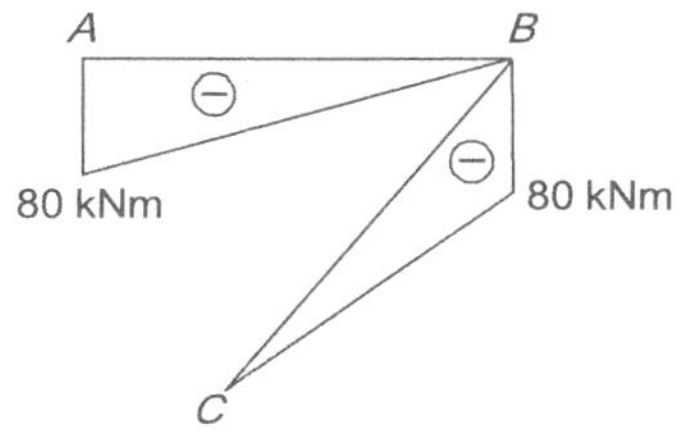

Bending moment diagram (BMD)

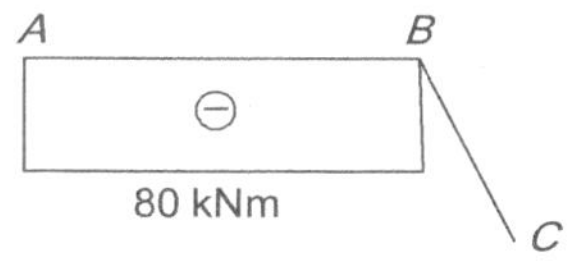

Torsion moment diagram (TMD)

2.16 Compute and draw the vertical shear, bending moment and torque diagrams of the cantilever circular bow girder shown in figure in the shape of a quadrant of a circle of radius 3 m and carrying a uniformly distributed load of 3 kN/m (normal load). The girder lies in the horizontal plane.

[20 marks : 2017]

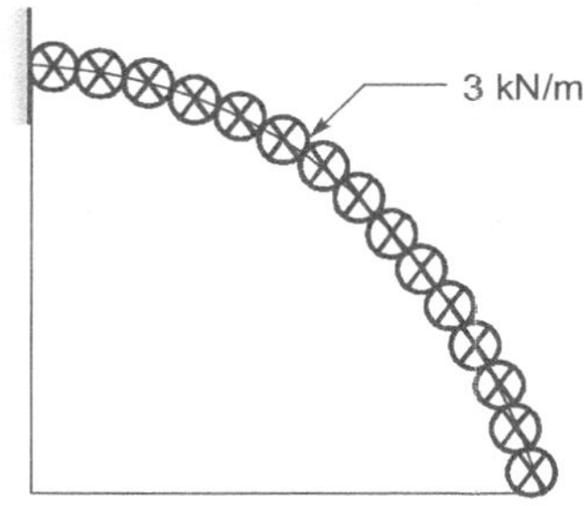

Solution:

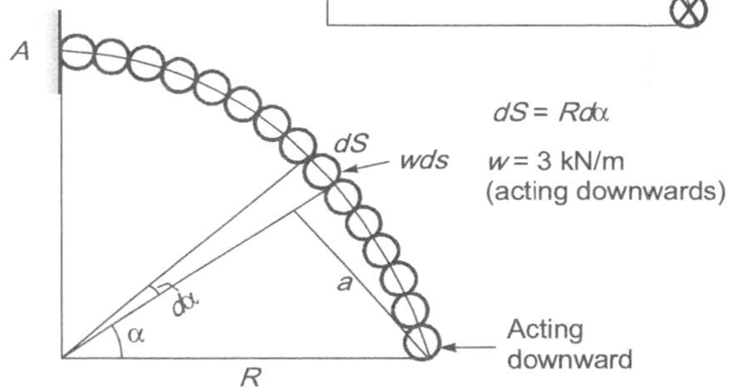

Vertical shear diagram: Let ds be small element.

$$dF = wds \text{ (at any angle } \alpha \text{ with horizontal)}$$

$$\Rightarrow \quad dF = wRd\alpha$$

Integrating from 0 to α.

$$F_S = \int_0^\alpha wRd\alpha = wR(\alpha - 0) = wR\alpha$$

At $\alpha = 0, F = 0$

and at $\alpha = \dfrac{\pi}{2} \Rightarrow F = \dfrac{\pi wR}{2}$

Given, $R = 3$ m and $w = 3$ kN/m

$\therefore$ At A, $F = \dfrac{\pi \times 3 \times 3}{2} = +14.14$ kN

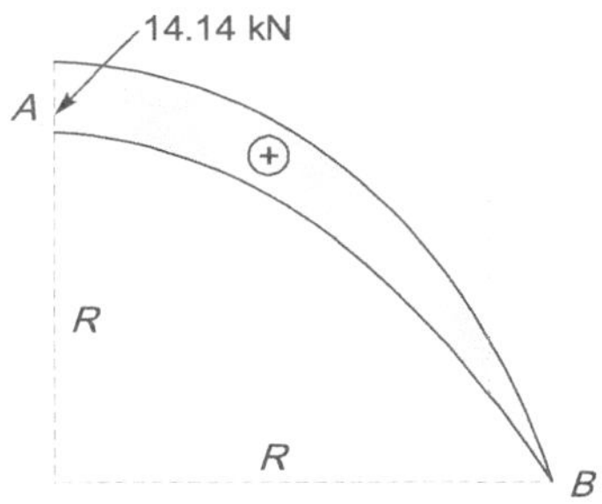

Vertical shear diagram

Bending moment diagram:

$$dM = wds \times a = wRd\alpha \times R\sin\alpha$$

$$\Rightarrow \quad dM = wR^2 \sin\alpha \, d\alpha$$

Integrating w.r.t. α, we get

$$\int dM = \int_0^\alpha wR^2 \sin\alpha d\alpha = wR^2[-\cos\alpha]_0^\alpha$$

$$\Rightarrow \quad M = wR^2[-\cos\alpha + 1] = wR^2[1 - \cos\alpha]$$

At $\alpha = 0, M_s = 0$

and $\alpha = \dfrac{\pi}{2}, M_s = wR^2 = 3 \times (3)^2 = 27$ kNm

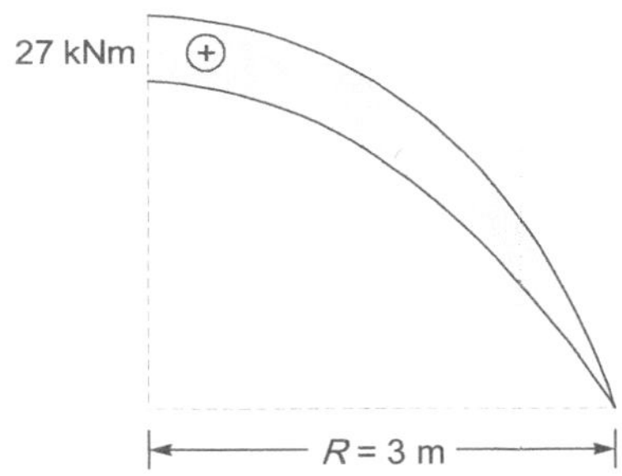

Bending moment diagram

Torque diagram:

$$dT = wds \times b = wRd\alpha \times (R - R\cos\alpha)$$

$$\Rightarrow \quad dT = wR^2(1 - \cos\alpha)\, d\alpha$$

Integrating w.r.t α, we get,

$$\int dT = \int_0^{\alpha} wR^2(1-\cos\alpha)d\alpha = wR^2[\alpha - \sin\alpha]_0^{\alpha}$$

$\Rightarrow$ $T = wR^2[\alpha - \sin\alpha]$

At $\alpha = 0,\ T = 0$

and at $\alpha = \pi/2$

$$\Rightarrow \quad T = wR^2\left[\frac{\pi}{2} - 1\right]$$

$$= 3 \times (3)^2 \times \left[\frac{\pi}{2} - 1\right] = 15.412 \text{ kNm}$$

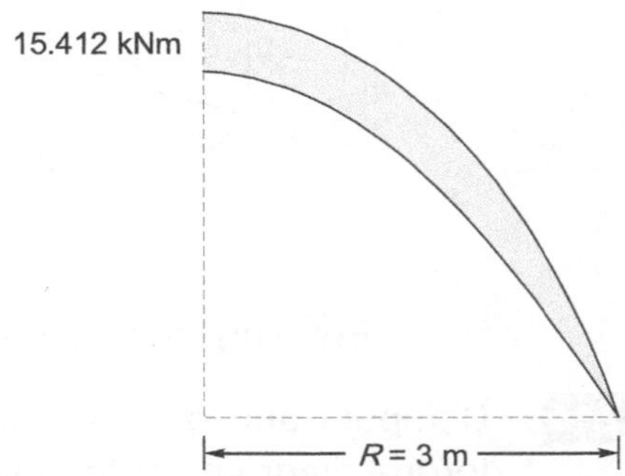

Torsion diagram

2.17 **A girder and floor beam construction subjected to concentrated loads is shown in figure. Draw the free body diagrams and determine support reactions.**

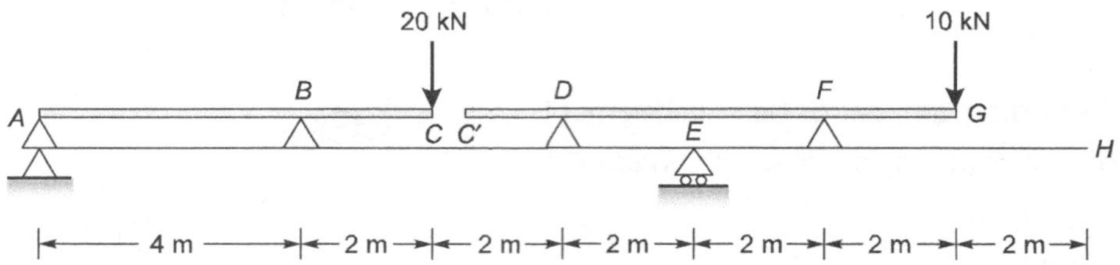

[12 marks : 2017]

Solution:

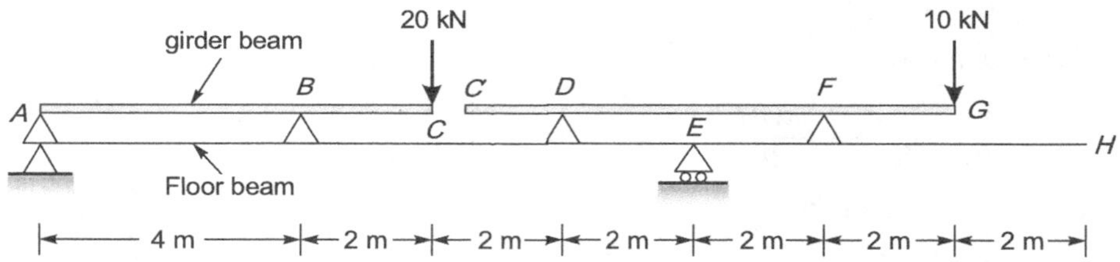

Free body diagram

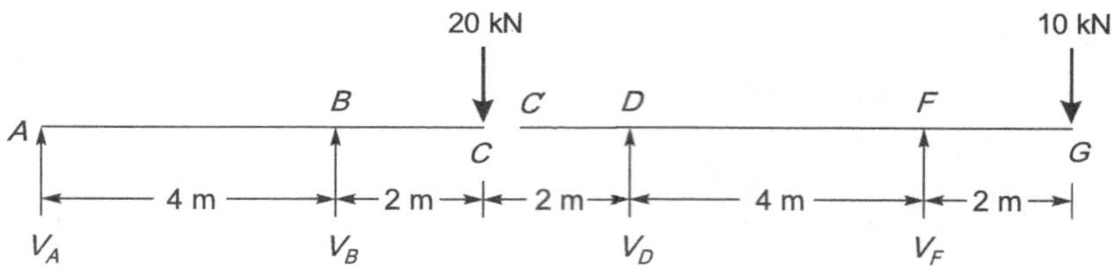

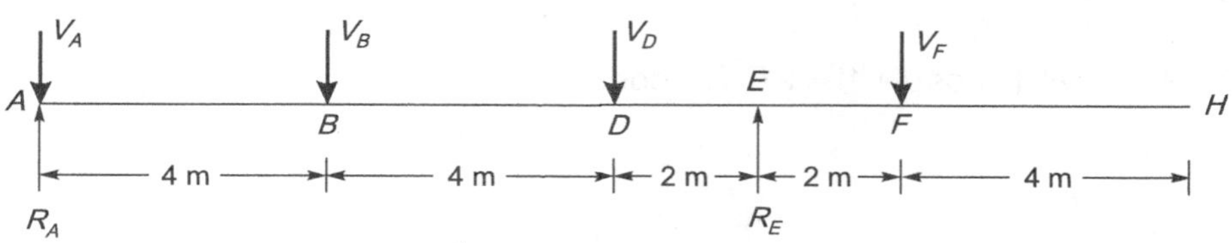

The support reactions developed in girder and floor beam construction are shown above.

For girder *ABC*;

$$\Sigma M_A = 0$$

and $$\Sigma F_y = 0$$

$\Rightarrow \quad V_B \times 4 = 20 \times 6$

$\Rightarrow \quad V_A + V_B = 20 \text{ kN}$

$\Rightarrow \quad V_B = 30 \text{ kN}(\uparrow)$

$\Rightarrow \quad V_A = -10 \text{ kN}(\downarrow)$

For girder $C'DFG$; $\quad \Sigma M_D = 0$

and $\quad \Sigma F_y = 0$

$\Rightarrow \quad V_F \times 4 = 10 \times 6$

$\Rightarrow \quad V_F + V_D = 10 \text{ kN}$

$\Rightarrow \quad V_F = 15 \text{ kN}(\uparrow)$

$\Rightarrow \quad V_D = -5 \text{ kN}(\downarrow)$

Now for floor beam. Let R_A and R_E be the support reactions.

$$\Sigma M_A = 0$$

$\Rightarrow \quad R_E \times 10 = V_B \times 4 + V_D \times 8 + V_F \times 12$

$\Rightarrow \quad R_E \times 10 = 30 \times 4 + (-5) \times 8 + 15 \times 12$

$$\Rightarrow \quad R_E = \frac{120 - 40 + 180}{10} = \frac{260}{10} = 26 \text{ kN}(\uparrow)$$

Now, $\quad \Sigma F_y = 0$

$\Rightarrow \quad R_A + R_E = V_A + V_B - V_D + V_F = -10 + 30 - 5 + 15$

$\Rightarrow \quad R_A + R_E = 30 \text{ kN}$

$\Rightarrow \quad R_A = (30 - 26) = 4 \text{ kN}(\uparrow)$

2.18 The compound beam is subjected to a uniform dead load of 1.5 kN/m and a single live load of 10 kN at B. Determine

(i) The maximum negative moment created by these load at 'A', and

(ii) Maximum positive shear at 'B'.

Assume A is a fixed support, B is a pin and C is a roller.

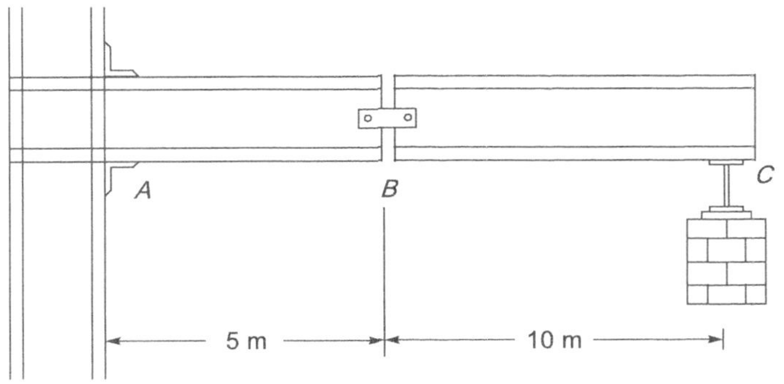

[12 Marks : 2021]

Solution:

(i) Maximum -ve bending moment at A:

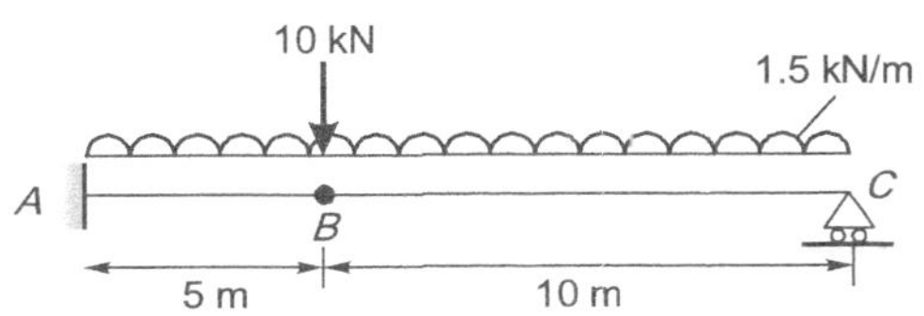

$$\text{Maximum} - \text{ve BM at A} = -1.5\left(\frac{1}{2}\times 15\times 5\right) - 10\times 5$$

$$= -56.25 - 50 = -106.25 \text{ kN-m}$$

(ii) Maximum +ve SF at B,

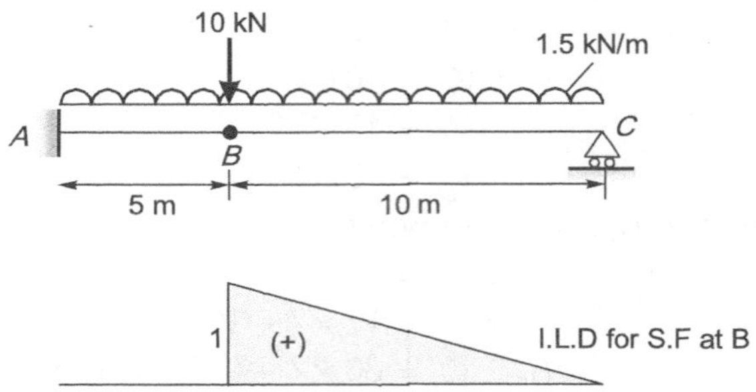

$$\text{Maximum SF at } B = 1.5\left(\frac{1}{2}\times 10\times 1\right) + 10\times 1$$

$$= 7.5 + 10 = 17.5 \text{ kN}$$

2.19 A T-section beam is constructed by ruling two pieces of wood together as shown in the figure. The maximum stress in the glue joints is to be limited to 2 MPa in tension and the maximum shear stress is to be limited to 1.7 MPa.

(i) Determine the stress components on element at pint 'P'. Point 'P' is located at glued joint.

(ii) Determine principal stresses at point 'P'.

(iii) Show these stresses on properly oriented 2-D element.

(iv) Determine the maximum value for load w.

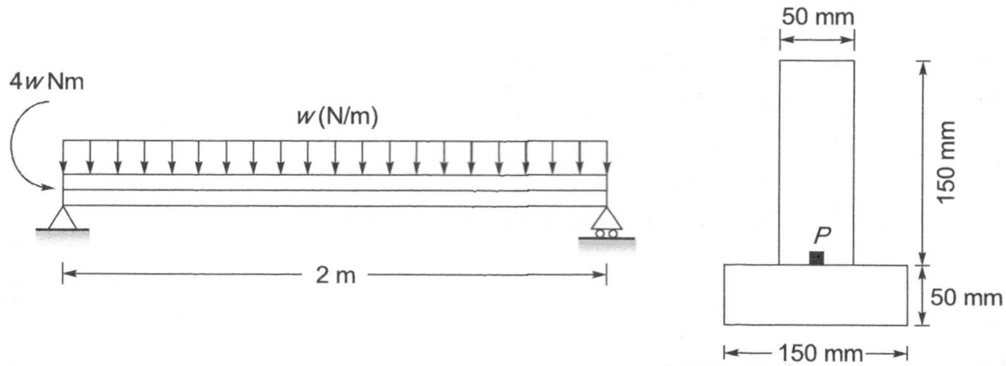

Sign convention for shear force and bending moment are as follows:

[20 Marks : 2021]

Solution:

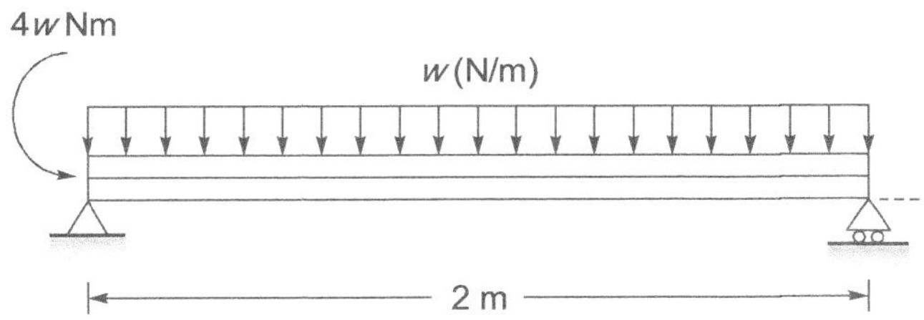

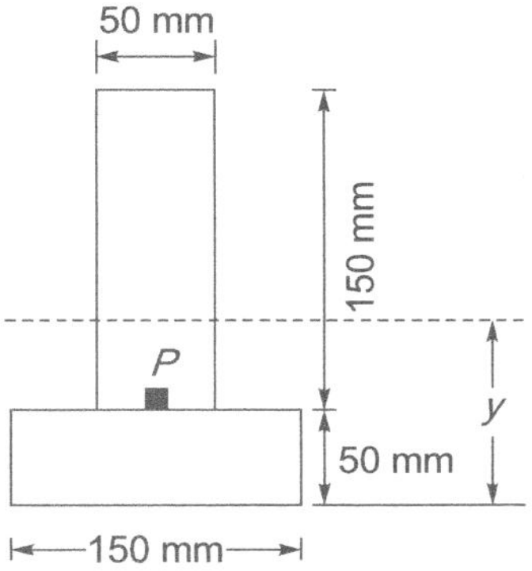

$$\bar{y} = \frac{A_1 y_1 + A_2 y_2}{A_1 + A_2} = \frac{(150\times 50\times 25)+(150\times 50\times 125)}{(150\times 50)+(150\times 50)} = 75 \text{ mm}$$

$$I_{NA} = I_1 + I_2 = \left[\frac{150(50)^3}{12}+\left(150\times 50\times 50^2\right)\right]+\left[\frac{50(150)^3}{12}+\left(50\times 150\times 50^2\right)\right]$$

$$I_{NA} = \left(20.31\times 10^6\right)+\left(32.81\times 10^6\right) = 53.12\times 10^6 \text{ mm}^4$$

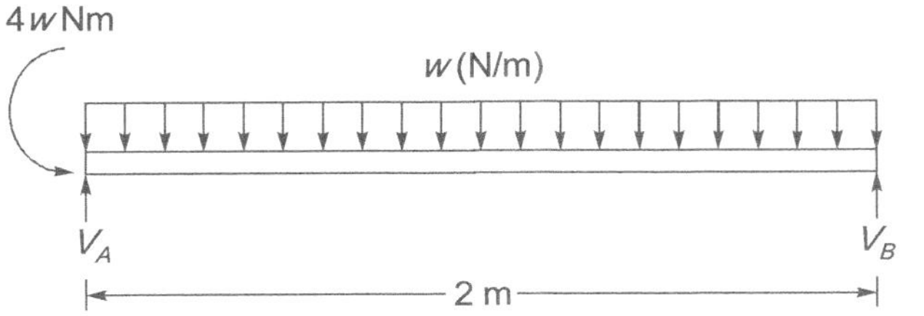

$$V_A + V_B = 2w \qquad \text{...(i)}$$

$$\Sigma M_A = 0$$

$$(-V_B\times 2)+(w\times 2\times 1)+(-4w) = 0$$

$$V_B = -w \text{ N}$$

$$V_A = 3w \text{ N}$$

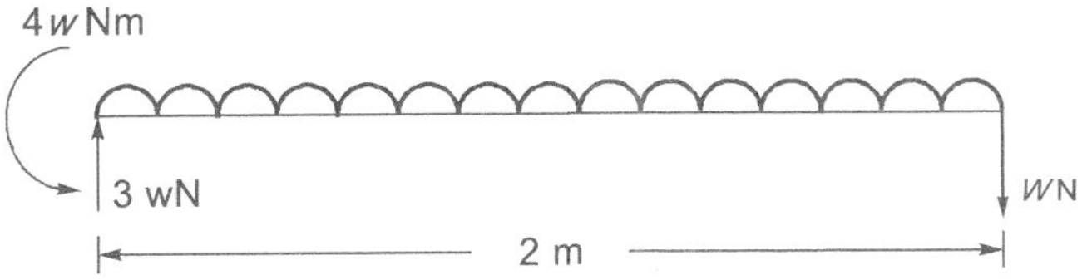

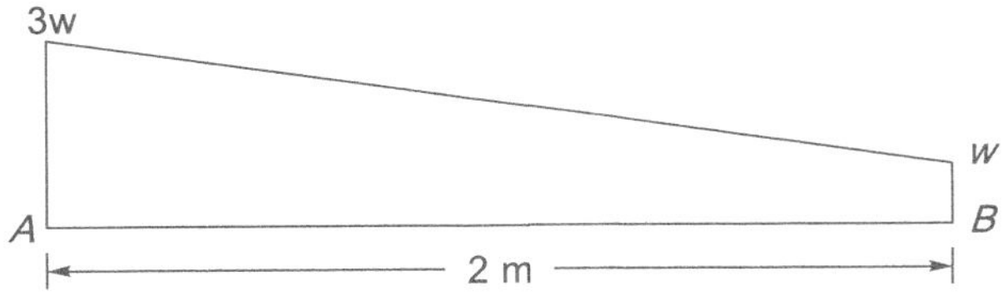

SFD

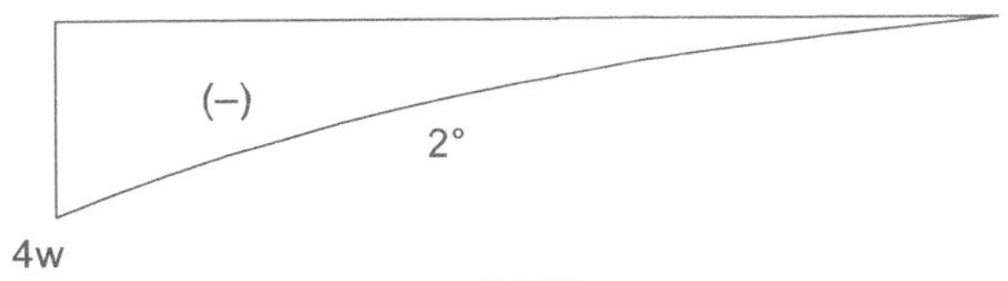

BMD

Bending stress at 'P'

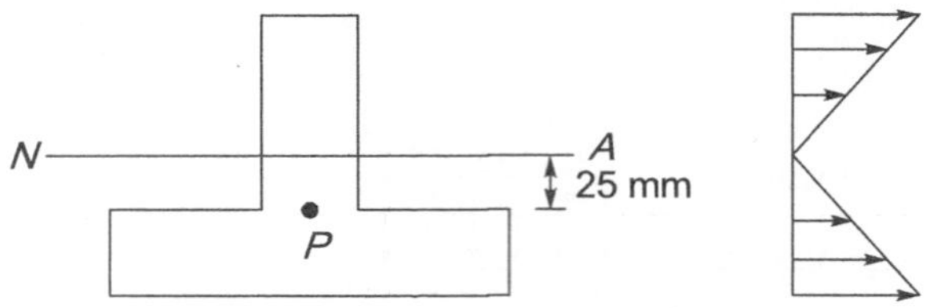

Bending stress distribution

$$\sigma = \frac{My}{I} = \frac{(4w)(25)}{53.12 \times 10^6} = 1.88 \times 10^{-6} w$$

$$\sigma = 18.82 \times 10^{-4} w$$

Direct shear stress at 'P'

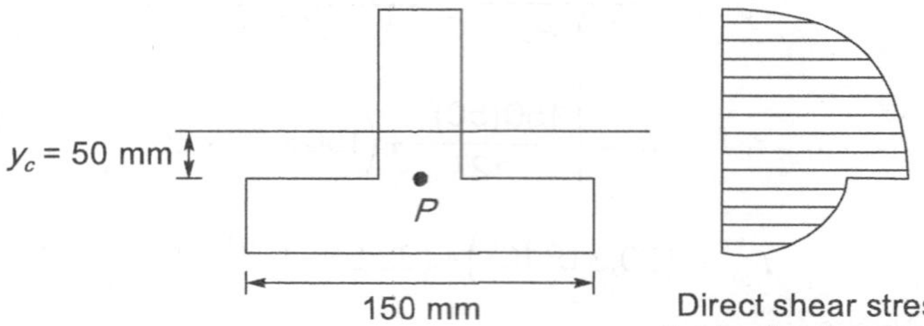

Direct shear stress distribution (parabolic)

$$\tau = \frac{VQ}{It} = \frac{(3w) \times (150 \times 50)(50)}{(53.12 \times 10^6)(50)} = 4.23 \times 10^{-4} w$$

$$(4.23 \times 10^{-4})\, w \leq 1.7$$

$$w \leq 4.014 \text{ kN/m}$$

$$\sigma \leq 2 \text{ MPa}$$

$$18.82 \times 10^{-4} w = 2$$

$$w = 1062.7 \text{ N/m} \simeq 1.062 \text{ kN/m}$$

(It seems error in problem because in hogging moment below NA at point 'P' compressive bending stress will develop)

So, $w \leq 1.062$ kN/m

(i) Stress element as 'P'

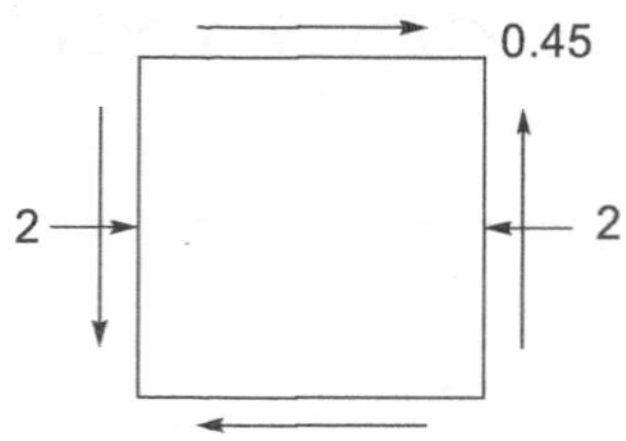

$$\tau_p = \frac{(3w)(150 \times 50 \times 50)}{(53.125 \times 10^6)(50)} = 0.449 \text{MPa} \simeq 0.45 \text{ MPa}$$

$$\sigma_x = -2 \text{ MPa}$$

$$\sigma_y = 0$$

$$\tau_{xy} = 0.45 \text{ MPa}$$

(ii)

$$\sigma_{p_1} / \sigma_{p_2} = \frac{\sigma_x + \sigma_y}{2} \pm \frac{1}{2}\sqrt{(\sigma_x - \sigma_y)^2 + 4\tau_{xy}^2}$$

$$= \frac{(-2+0)}{2} \pm \frac{1}{2}\sqrt{(0+2)^2 + 4(0.45)^2}$$

$$\sigma_{p_1} = -2.096 \text{ MPa}$$

$$\sigma_{p_2} = +0.096 \text{ MPa}$$

(iii)

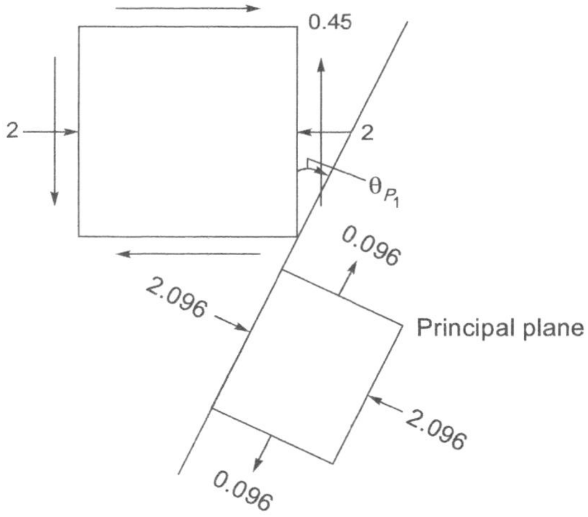

$$\tan 2\theta_{P_1} = \frac{2\tau_{xy}}{\sigma_x - \sigma_y} = \frac{2(0.45)}{(-2)-(0)}$$

$$\theta_{P_1} = -12.11°$$

$$\theta_{P_2} = 77.86°$$

Crosscheck:

$$\theta = -12.11°$$

$$\sigma_x' = (-2)\cos^2(-12.11) + 2(0.45)\cos(-12.11)\sin(-12.11)$$

$$= -2.096 \text{ MPa}$$

2.20 Analyse the propped cantilever beam shown in the figure. During loading the prop B sinks by 6 mm. E = 2 × 10^5 MPa, I = 86.04 × 10^6 mm^4. Draw shear force and bending moment diagrams. Find the maximum BM value and its location from either of the supports.

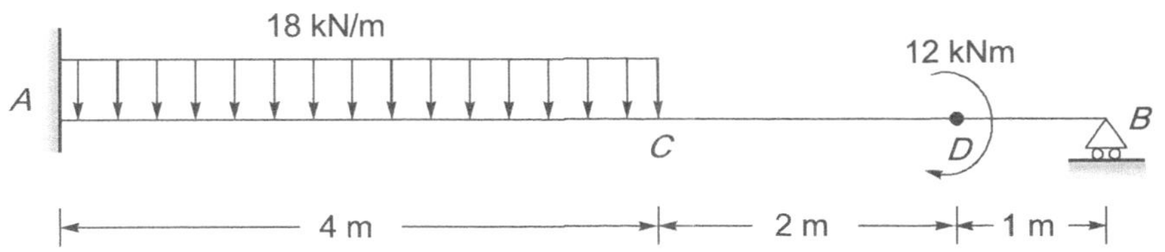

[12 Marks : 2021]

Solution:

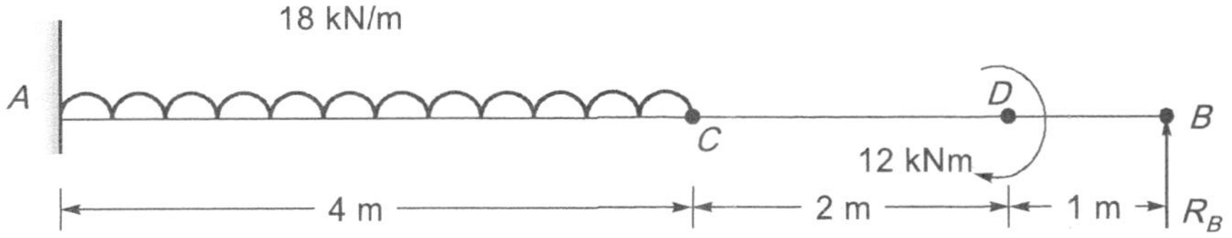

Case -1: Deflection at support B due to UDL

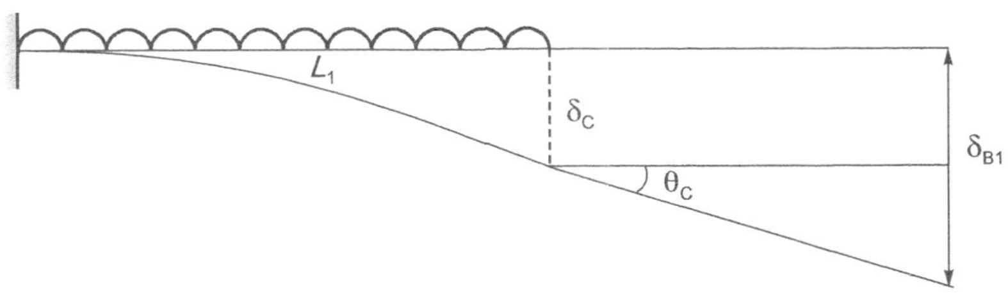

$$\delta_{B_1} = \frac{wL_1^4}{8EI} + \frac{wL_1^3}{6EI} \times 3(\downarrow) \qquad [L_1 = 4 \text{ m}]$$

Deflection at support B due to concentrated moment

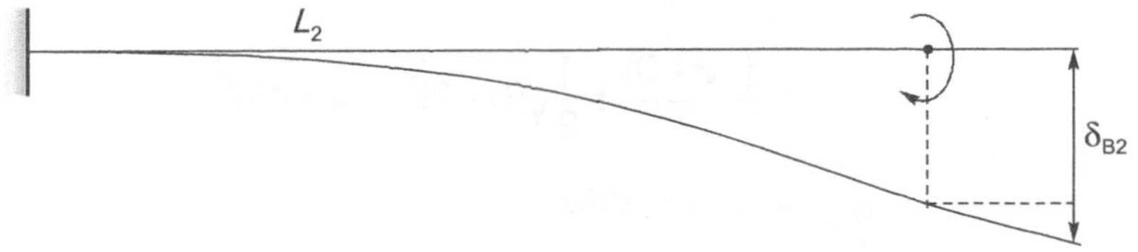

$$\delta_{B2} = \frac{ML_2^2}{2EI} + \frac{ML_2}{EI} \times 1(\downarrow) \quad (L_2 = 6\text{m})$$

Deflection at support B due to R_B

$$\delta_{B3} = \frac{R_B L^3}{3EI}(\uparrow)$$

Net downward deflection at B = 6 mm

$$\Rightarrow \quad \delta_{B_1} + \delta_{B_2} - \delta_{B_3} = 6 \text{ mm}$$

$$\text{or,} \quad \delta_{B_1} + \delta_{B_2} - 6\times10^{-3} = \delta_{B_3}$$

$$\text{or,} \quad \frac{wL_1^4}{8EI} + \frac{wL_1^3}{3EI}\times3 + \frac{ML_2^2}{2EI} + \frac{ML_2}{EI}\times1 - 6\times10^{-3} = \frac{R_B 7^3}{3EI}$$

$$= \frac{(18\times10^3)\times4^4}{8EI} + \frac{(18\times10^3)\times4^3}{6EI}\times3 + \frac{(12\times10^3)\times6^2}{2EI} + \frac{12\times6}{EI}\times1 - 6\times10^{-3}$$

$$= \frac{R_B\times7^3}{3EI}$$

$$\Rightarrow \quad \frac{576\times10^3}{EI} + \frac{576\times10^3}{EI} + \frac{216\times10^3}{EI} + \frac{72\times10^3}{EI} - 6\times10^{-3} = \frac{R_B\times114.33}{EI}$$

$$\Rightarrow \quad 1440\times10^3 - 6\times10^{-3}\times EI = R_B\times114.33$$

$$\text{EI} = \left[2\times10^5\times10^6\times86.04\times10^6\times10^{-12}\right] = [172.08\times10^5]\ \text{Nm}^2$$

$$1440000 - 103248 = R_B \times 114.33,$$

$$R_B = 11692.05 \text{ N} = 11.69 \text{ kN} \simeq 11.7 \text{ kN}$$

$$\Sigma F_y = 0$$

$$R_A + R_B = 18\times4$$

$$R_A = 72 - 11.7 = 60.3 \text{ kN}$$

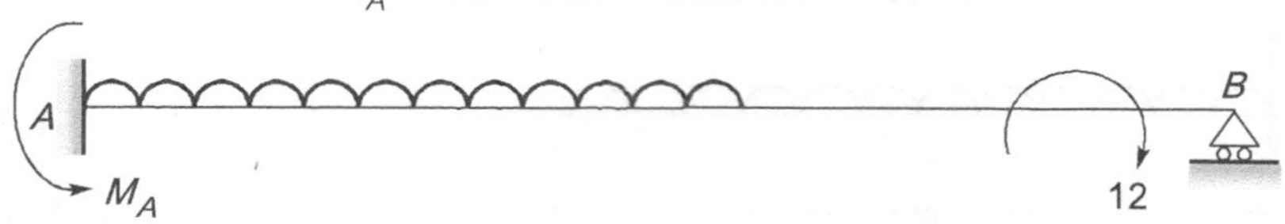

$$\Sigma M_B = 0$$

$$-M_A + R_A\times7 - w\times4\times5 + 12 = 0$$

$$-M_A + 60.3\times7 - 18\times4\times5 + 12 = 0$$

$$M_A = 74.1 \text{ kNm}$$

BMD

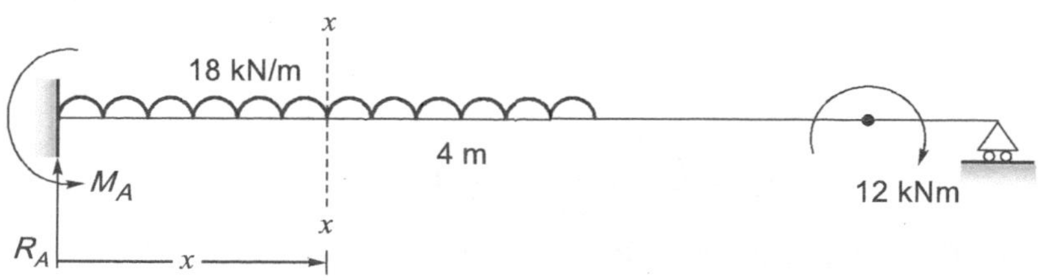

$$M_x = -M_A - R_A x - \frac{wx^2}{2} = -74.1 + 60.3x - \frac{18x^2}{2}$$

For maximum bending moment:

$$\frac{dM_x}{dx} = 60.3 - 18x = 0$$

$$x = 3.35 \text{ m}$$

$$BM_{\text{at support A }(x=0)} = -M_A = -74.1 \text{kNm}$$

$$\text{BM at C } (x = 4 \text{ m}) = -74.1 + 60.3 \times 4 - 23.1 \text{ kNm}$$

$$BM_{\text{at left of D}} = -74.1 + 60.3 \times 6 - (18 \times 4) \times 4 = -0.3 \text{ kNm}$$

$$BM_{\text{at right of D}} = -0.3 + 12 = 11.7 \text{ kNm}$$

$$BM^{max} = -74.1 + 60.3 \times 3.35 - \frac{18 \times (3.35)^2}{2} = 26.9 \text{ kNm}$$

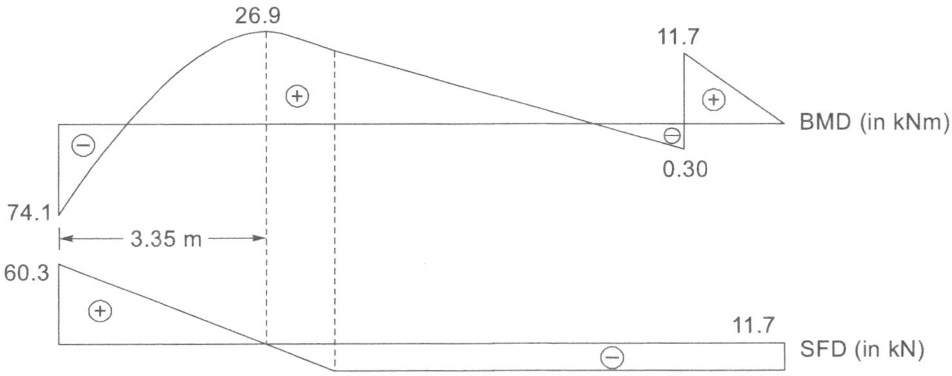

2.21 (i) Define determinate and indeterminate structures with suitable examples.

(ii) Draw the shear force diagram of the beam shown in figure. Supports (1) and (2) are simply supported.

(iii) Draw the bending moment diagram for the beam shown in figure.

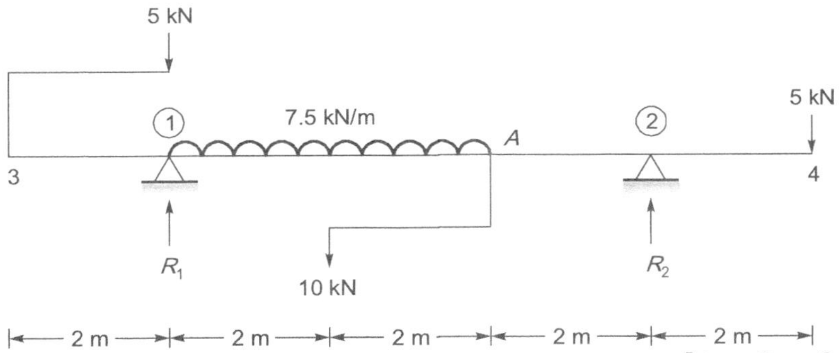

[4 + 8 + 8 = 20 marks : 2022]

Solution:

(i) Determinate structures:

A structure is said to be determinate if conditions of static equilibrium are sufficient to analyse the structure.

(a) In determinate structures, bending moment and shear force are independent of properties of material and cross-sectional area of members.

(b) No stresses are induced due to temperature changes.

(c) No stresses are induced due to lack of fit and support settlement.

Examples:

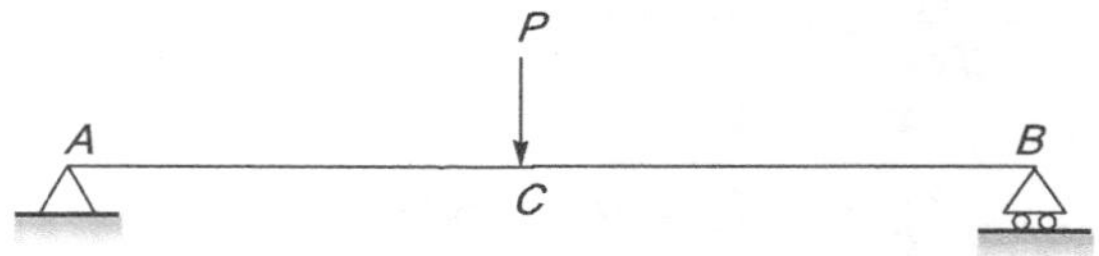

Here a simply supported beam is shown. The total number of unknown reactions will be 3 (2 at *A* and 1 at *B)* and number of equilibrium equations is also 3.

Therefore, condition of static equilibrium can be used to calculate value of unknowns.

Indeterminate structures:

A structure is said to be indeterminate if conditions of static equilibrium are not sufficient to analyse the structure. To analyse these structures, compatibility conditions are required.

(a) In these structure, bending moment and shear force depend on properties of material and cross-sectional area of members.

(b) Stresses are induced due to temperature variation.

(c) Stresses are induced due to lack of fit and support settlement.

Example:

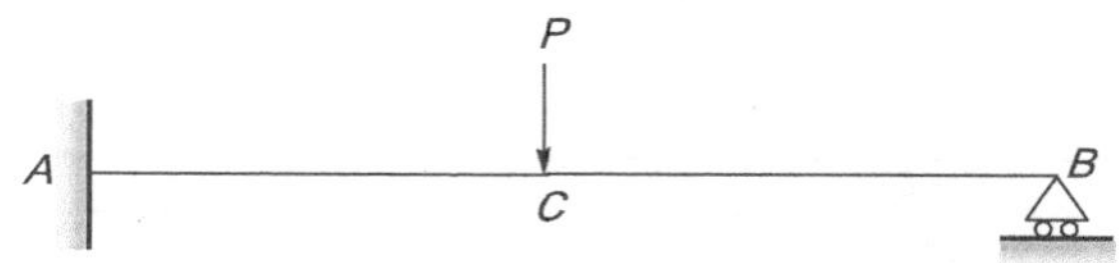

Here, a propped cantilever beam is shown. The total number of unknowns will be 3 at A and 1 at *B*. So it can't be analysed using conditions of static equilibrium alone. A compatibility equation is required to calculate values of unknowns.

(ii)

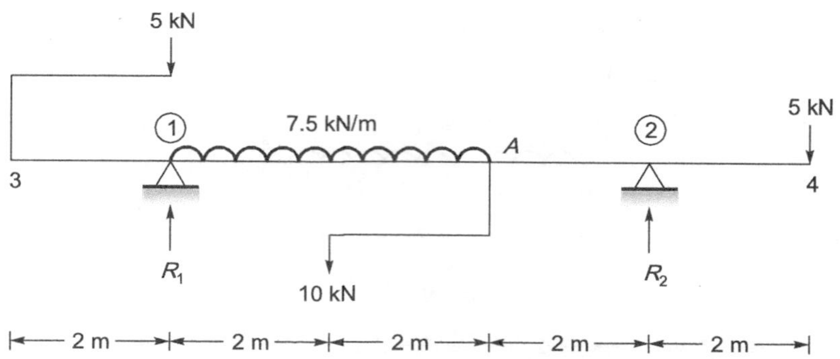

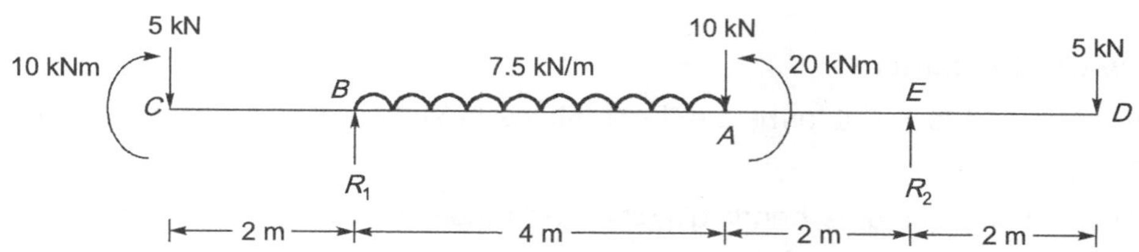

Reactions: $\Sigma F_y = 0$

$R_1 + R_2 = 50$ ...(i)

and $\Sigma M_B = 0$

$\Rightarrow \quad 10 - (5 \times 2) + (7.5 \times 4 \times 2) + (10 \times 4) - 20 - (R_2 \times 6) + 5 \times 8 = 0$

$\Rightarrow \quad R_2 = 20 \text{ kN}$

From eq. (i),

$R_1 = 50 - 20 = 30 \text{ kN}$

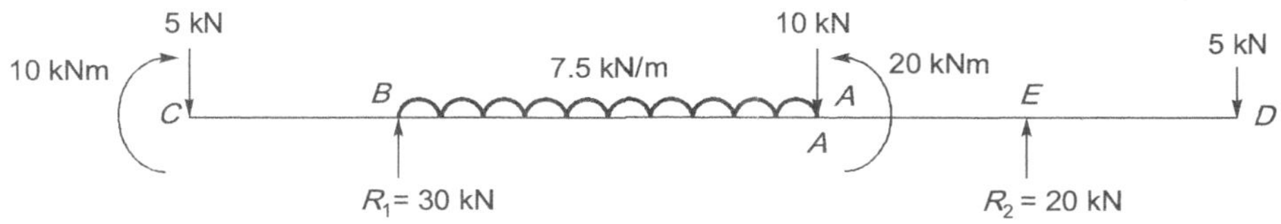

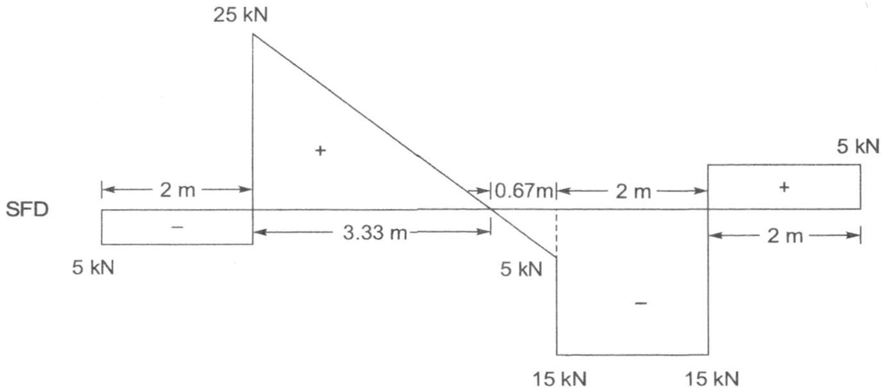

Portion *CB*:

S_x (x from C) = –5 kN $[0 \le x < 2]$

At $x = 0$, $S_C = 5$ kN (Constant)

At $x = 2$ m,

S_B (just left of B) = – 5 kN

S_B (just right of B) = –5 + 30 = 25 kN

Portion *BA*: (linear variation)

S_x (x from B) = $25 - 7.5x$ $[0 \le x < 4]$

$S_x = 0 = 25 - 7.5x$

$$x = \frac{25}{7.5} = 3.33 \text{ m}$$

At $x = 0$,

S_B (just right of B) = 25 kN

At $x = 4$ m, $S_A = 25 - 7.5 \times 4$

S_A = –5 kN (just left of A)

Portion *AE*:

$S_x = -5 + (-10)$ $(0 \le x < 2)$

$S_x = -15$ kN

At $x = 0$, S_A = –15 kN (just right of A)

At $x = 2$ m, S_E = –15 kN (just left at E)

Portion *ED*:

$$S_x = -15 + 20 = +5 \text{ kN} \qquad [0 \le x < 2]$$

At $x = 0$, $S_E = 5$ kN (just right of *E*)

At $x = 2$ m, $S_D = 5$ kN (just left of *D*)

$S_D = 5 - 5 = 0$ (just right of *D*)

(iii) Bending moment diagram

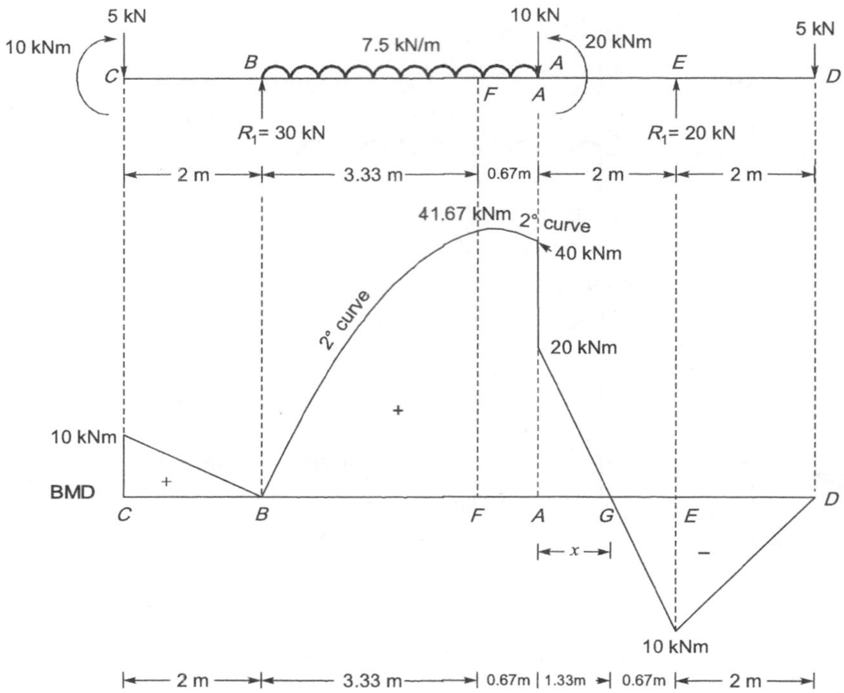

Portion *CB*:

M_x (*x* from *C*) = $(10 - 5x)$ $\qquad (0 \le x < 2)$

At $x = 0$, $M_C = 10$ kNm

At $x = 2$ m, $M_B = 0$

Portion *DE*:

M_x (*x* from *D*) = $-5x$ $\qquad (0 \le x < 2)$

At $x = 0$, $M_D = 0$ kNm

At $x = 2$ m, $M_E = -10$ kNm

Portion *EA*:

M_x (*x* from *E*) = $-5(x + 2) + 20x$ $\qquad (0 \le x < 2)$

At $x = 0$, $M_E = -10$ kNm

At $x = 2$ m, $M_A = 20$ kNm (Just right of *A*)

Portion *AF*:

$$M_x \text{ (x from A)} = -5(4+x)+20(x+2)+20-10x-\left(\frac{7.5\times x^2}{2}\right)$$

$$(0 \le x < 0.67)$$

At $x = 0$, $M_A = 40$ kNm (just left of A)

At $x = 0.67$ m, $M_F = 41.67$ kNm

Portion FB:

$$M_x \,(x \text{ from } F) = -5(4.67+x)+20(2.67+x)+20-10(0.67+x)-\frac{\left[7.5(x+0.67)^2\right]}{2}$$

$(0 \le x < 3.33)$

At $x = 0$, $M_x = 41.67$ kNm

At $x = 3.33$ m, $M_x = 0$

In portion AE, location of zero BM (point of contraflexure)

$$\frac{20}{x} = \frac{10}{2-x}$$

$\Rightarrow$ $x = 1.33$ m

2.22 Draw the bending moment and shearing force diagrams for the overhanging beam loaded as shown in the figure below. Determine the positions of maximum bending moment, maximum shearing force and locate the locations of zero bending moment:

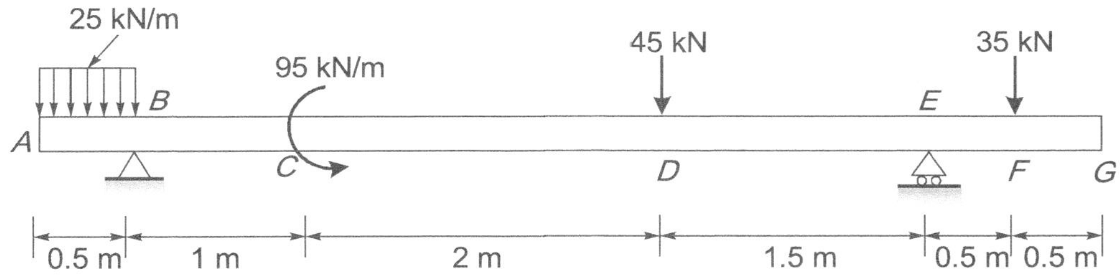

[20 marks : 2023]

Solution:

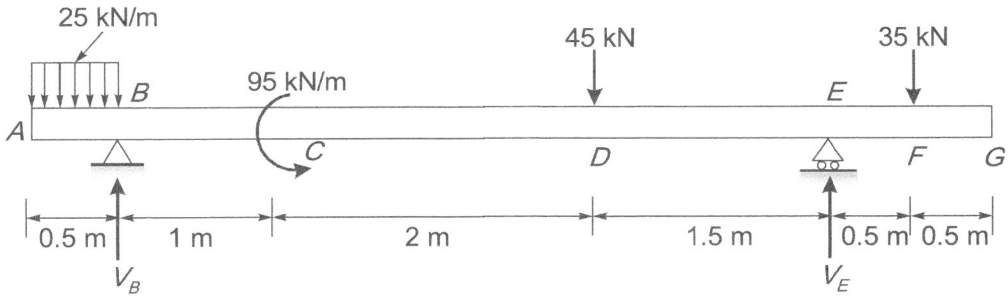

Let, V_B and V_E be the vertical reactions developed at B and E respectively.

Now, $\Sigma F_y = 0$

$\Rightarrow -25 \times 0.5 + V_B - 45 + V_E - 35 = 0$

$\Rightarrow$ $V_B + V_E = 92.5$ ..(i)

Also, $\Sigma M_E = 0$

$$\Rightarrow -25\times 0.5\times\left(4.5+\frac{0.5}{2}\right)+V_B\times 4.5-95-45\times 1.5+35\times 0.5 = 0$$

$\Rightarrow -59.375 + V_B \times 4.5 - 95 - 67.5 + 17.5 = 0$

$\Rightarrow \quad V_B = 45.42$ kN

Putting V_B in (i), we get

$$V_E = 92.5 - V_B$$
$$= 92.5 - 45.42 = 47.08 \text{ kN}$$

Shear force diagram

Take a section x-x at 'x' distance from A

In portion *AB*

$S_x(x \text{ from A}) = -25x \qquad [0 \le x < 0.5]$

At $x = 0$, $\quad S_A = -25 \times 0 = 0$

At $x = 0.5$ m, S_B(just left of B) $= -25 \times 0.5$

$= -12.5$ kN

In portion *BD*

$S_x = -25 \times 0.5 + V_B \qquad [0.5 \le x < 3.5]$

At $x = 0.5$ m, S_B(just right of B) $= -12.5 + 45.42$

$= 32.92$ kN

At $x = 3.5$ m, S_D(just left of D) $= 32.92$ kN

NOTE: In portion *BD*, shear force will be constant.

In portion *DE*

$S_x = -25 \times 0.5 + V_B - 45 \qquad [3.5 \le x < 5]$

At $x = 3.5$ m, S_D(just right of D) $= -12.5 + 45.42 - 45$

$= -12.08$ kN

At $x = 5$ m, $\quad S_E$(just left of E) $= -12.08$ kN

NOTE: In portion *DE*, shear force will be constant.

In portion *EF*

$S_x = -25 \times 0.5 + V_B - 45 + V_E \qquad [5 \le x < 5.5]$

At $x = 5$ m, S_E(just right of E) $= -12.5 + 45.42 - 45 + 47.08$

$= 35$ kN

At $x = 5.5$ m, S_F(just left of F) $= 35$ kN

NOTE: In portion *EF*, shear force will be constant.

In portion *FG*

$S_x = -25 \times 0.5 + V_B - 45 + V_E - 35 \qquad [5.5 \le x < 6]$

At $x = 5.5$ m, S_F(just right of F) $= -12.5 + 45.42 - 45 + 47.08 - 35$

$= 0$

NOTE: In portion *FG*, shear force will be constant.

Bending moment diagram:

Take a section x-x at 'x' distance from A

In portion *AB*

$$M_x(x \text{ from } A) = -25 \times x \times \frac{x}{2} \qquad [0 \le x < 0.5]$$

$$= \frac{-25x^2}{2}$$

At $x = 0$, $\quad M_A = -25 \times 0 = 0$

At $x = 0.5$, M_B(just left of B) $= -25\times\frac{0.5^2}{2} = -3.125$ kN-m

In portion *BC*

$$M_x = -25\times0.5\times\left(x-\frac{0.5}{2}\right)+V_B\times(x-0.5) \qquad [0.5\le x<1.5]$$

At $x = 0.5$ m, M_B(just right of B) $= -25\times0.5\times\left(0.5-\frac{0.5}{2}\right)+45.4\times(0.5-0.5)$

$= -3.125$ kN-m

$x = 1.5$ m, M_C(just left of C) $= -25\times0.5\times\left(1.5-\frac{0.5}{2}\right)+45.42\times(1.5-0.5)$

$= 29.795$ kN-m

In portion *CD*

$$M_x = -25\times0.5\times\left(x-\frac{0.5}{2}\right)+V_B\times(x-0.5)-95 \qquad [1.5\le x<3.5]$$

At $x = 1.5$ m, M_C(just right of C) $= -25\times0.5\times\left(1.5-\frac{0.5}{2}\right)+45.42\times(1.5-0.5)-95$

$= -25\times0.5\times\left(1.5-\frac{0.5}{2}\right)+45.42\times(1.5-0.5)-95$

$=-65.205$ kN-m

At $x = 3.5$ m, M_D(just left of D)$= -25\times0.5\times\left(3.5-\frac{0.5}{2}\right)+45.42\times(3.5-0.5)-95$

$= 0.63$ kN-m

In portion *DE*

$$M_x = -25\times0.5\times\left(x-\frac{0.5}{2}\right)+V_B\times(x-0.5)-95-45(x-3.5)$$

$[3.5\le x<5]$

So, at $x = 3.5$ m,

M_D(just right of D) $= -25\times0.5\times\left(3.5-\frac{0.5}{2}\right)+45.42\times(3.5-0.5)-95-45(3.5-3.5)$

$= 0.63$ kN-m

So, at $x = 5$ m,

M_E(just left of E) $= -25\times0.5\times\left(5-\frac{0.5}{2}\right)+45.42\times(5-0.5)-95-45(5-3.5)$

$= -17.5$ kN-m

In portion *EF*

$$M_x = -25\times0.5\times\left(x-\frac{0.5}{2}\right)+V_B\times(x-0.5)-95-45\times(x-3.5)+V_E(x-5)$$

$[5\le x<5.5]$

So, at $x = 5$ m,

$$M_E(\text{just right of } E) = -25\times0.5\times\left(5-\frac{0.5}{2}\right)+45.42\times(5-0.5)-95-45\times(5-3.5)+47.08(5-5)$$
$$= -17.5 \text{ kN-m}$$

So, at $x = 5.5$ m,

$$M_F(\text{just left of } F) = -25\times0.5\times\left(5.5-\frac{0.5}{2}\right)+45.42\times(5.5-0.5)$$
$$-95-45(5.5-3.5)+47.08\times(5.5-5)$$
$$= 0$$

NOTE: In portion *FG*, there will be no bending moment.

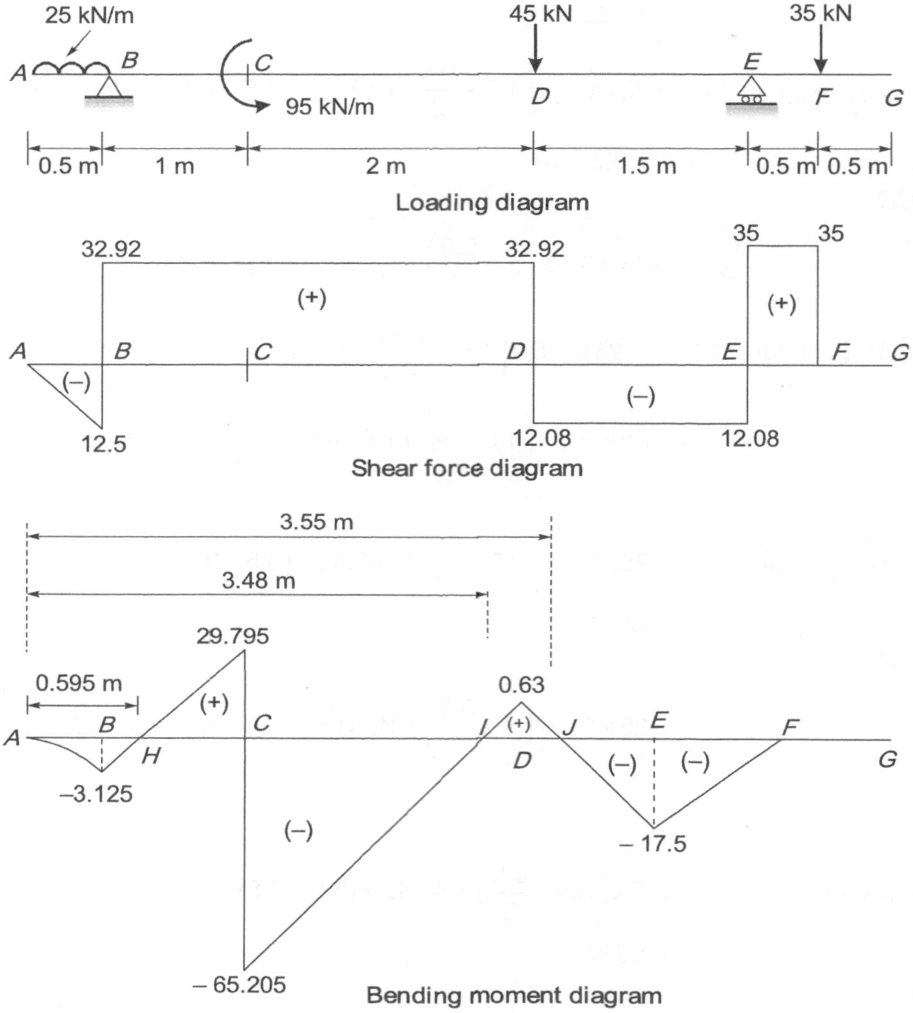

Loading diagram

Shear force diagram

Bending moment diagram

Maximum shear force,

Shear force will be maximum in portion EF and its value will be 35 kN.

Maximum bending moment.

Maximum bending moment will be at just right of C i.e. at 1.5 m from A and its value will be 65.205 kN-m (hogging)

Locations of zero bending moment.

(i) Bending moment is zero in portion BC i.e., at H.

Now, In portion *BC*, $M_x = -25\times0.5\times\left(x-\frac{0.5}{2}\right)+45.42\times(x-0.5)$

$\Rightarrow$ $0 = -12.5\times(x-0.25)+45.42\times(x-0.5)$

$\Rightarrow$ $x = 0.595$ m

(ii) Bending moment is zero in portion *CD* i.e. at *I*.

Now, In portion *CD*, $M_x = -25\times0.5\times\left(x-\frac{0.5}{2}\right)+45.42\times(x-0.5)-95$

$\Rightarrow$ $0 = -12.5\times(x-0.25)+45.42\times(x-0.5)-95$

$\Rightarrow$ $x = 3.48$ m

(iii) Bending moment is zero in portion *DE* i.e. at *I*.

Now, In portion *DE*, $M_x = -25\times0.5\times\left(x-\frac{0.5}{2}\right)+45.42\times(x-0.5)-95-45(x-3.5)$

$\Rightarrow$ $0 = -12.5\times(x-0.25)+45.42\times(x-0.5)-95-45(x-3.5)$

$\Rightarrow$ $x = 3.55$ m

3. Principal Stress-Strain & Theories of Failure

3.1 A simply supported beam of rectangular cross-section of size 200 × 300 mm (deep) supports a uniformly distributed load of 6.0 kN/m over an effective span of 4 m. Calculate the magnitude and direction of principal stresses at a point located 500 mm from left support and 50 mm above the neutral axis.

[15 marks : 2000]

Solution:

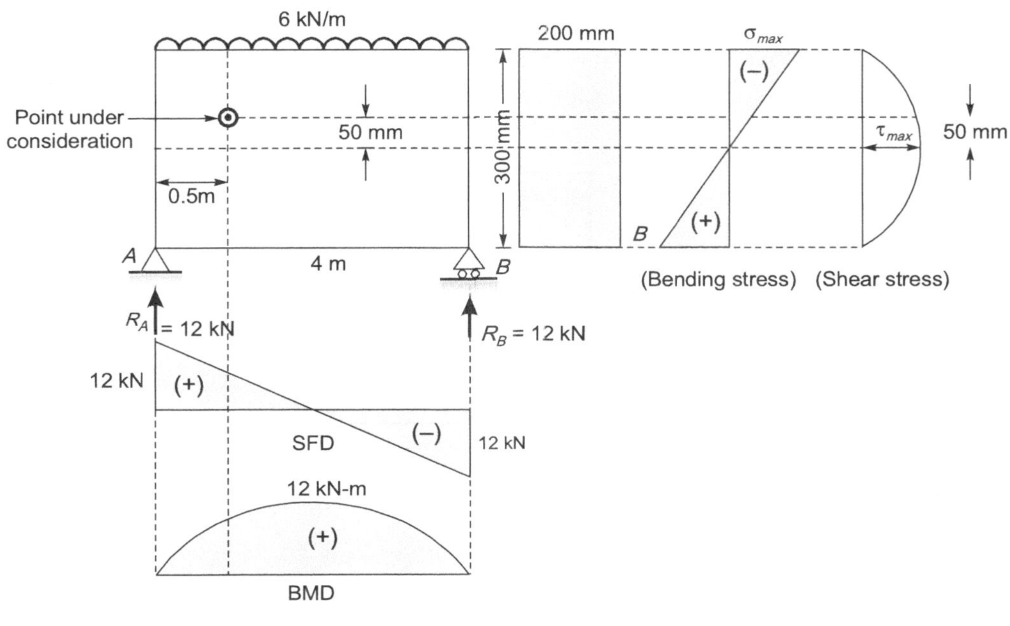

Shear force in beam ***AB*** is given by

$$S_x\ (x \text{ form } A) = R_A - wx \qquad [0 \le x \le 4]$$

$$S_x = 12 - 6x$$

$$\text{Shear force at } (x = 0.50 \text{ m}) = 12 - 6 \times 0.50 = 9 \text{ kN}$$

Bending moment in beam ***AB*** is given by

$$M_x\ (x \text{ from A}) = R_A x - \frac{wx^2}{2} \qquad [0 \le x \le 4]$$

$$M_x = 12x - \frac{6x^2}{2}$$

$$M_x = 12x - 3x^2$$

M_x at $x = 0.50$ m,

$$M_x(x = 0.50 \text{ m}) = 12 \times 0.5 - 3 \times 0.5^2$$

$$M = 5.25 \text{ kNm}$$

and MOI about neutral axis is given by $I = \frac{bd^3}{12} = \frac{200 \times 300^3}{12}$

$$I = 4.5 \times 10^8 \text{ mm}^4$$

Now we have $M = 5.25$ kNm, $S = 9$ kN, $I = 4.5 \times 10^8$ and normal bending stresses and shear stresses at $y = 50$ mm above NA can be calculated as

Bending stresses may be given as

$$\sigma = \frac{M}{I}y = \frac{5.25 \times 10^6 \times 50}{4.5 \times 10^8} \text{ N/mm}^2$$

$$= 0.5833 \text{ N/mm}^2 \text{ (compressive)}$$

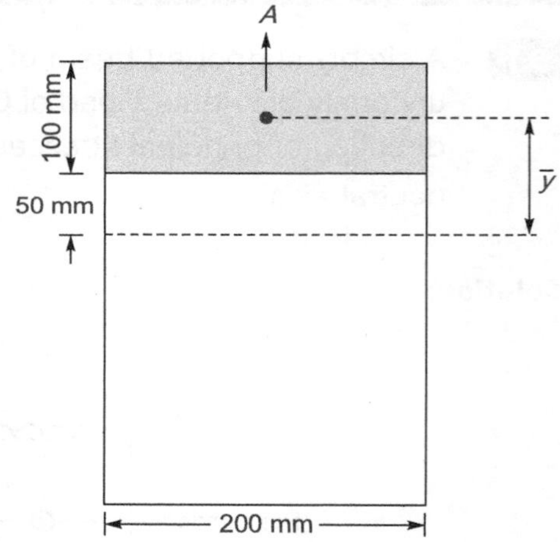

and **shear stress** may be given as

$$\tau = \frac{SA\bar{y}}{Ib} = \frac{9 \times 10^3 \times 100 \times 200 \times 100}{4.5 \times 10^8 \times 200}$$

$$= 0.2 \text{ N/mm}^2$$

Now point can be considered as stress element shown below:

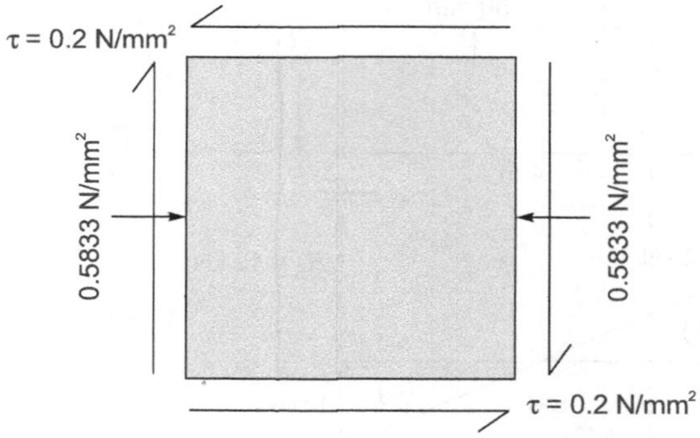

Here $\sigma_x = -0.5833$ N/mm², $\sigma_y = 0$ and $\tau_{xy} = -0.2$ N/mm²

(τ_{xy} = –ve, because positive shear force will produce –ve shear stress element)

Principal stresses may be calculated as

$$\sigma_1/\sigma_2 = \frac{\sigma_x+\sigma_y}{2} \pm \sqrt{\left(\frac{\sigma_x-\sigma_y}{2}\right)^2 + \tau_{xy}^2}$$

$$= \frac{-0.5833+0}{2} \pm \sqrt{\left(\frac{-0.5883-0}{2}\right)^2 + (-0.2)^2}$$

$$= -0.2916 \pm 0.3536$$

$\therefore$ $\sigma_1 = 0.062$ N/mm² and $\sigma_2 = -0.645$ N/mm²

For **direction of principal stresses,** we have

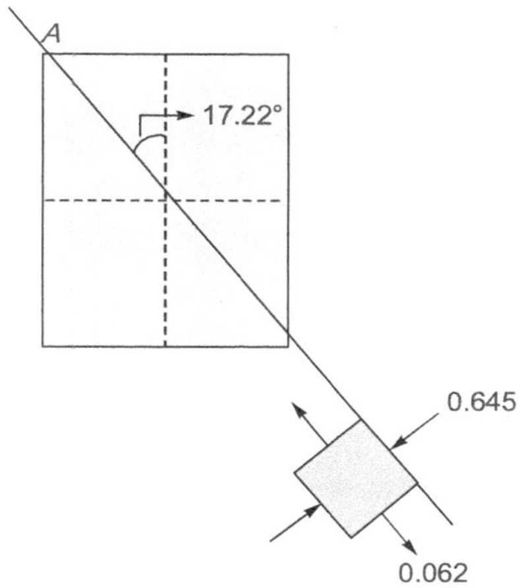

$$\tan 2\theta_P = \frac{2\tau_{xy}}{\sigma_x-\sigma_y} = \frac{2\times(-0.2)}{-0.5833-0} = \frac{0.4}{0.5833}$$

$\theta_{P_1} = 17.22°$ (anti-clockwise)

and $\theta_{P_2} = 90° + \theta_{P_1}$

$\theta_{P_2} = 107.22°$ (anti-clockwise)

Putting $\theta_{P_1} = 17.22°$

$$\sigma_x' = \frac{\sigma_x+\sigma_y}{2} + \left(\frac{\sigma_x-\sigma_y}{2}\right)\cos 2\theta_{p_1} + \tau_{xy}\sin 2\theta_{p_1}$$

$$= -0.645 \text{ N/mm}^2 = \sigma_2$$

Hence, angle θ_{p_1}(= 17.22°) is corresponding to major principal plane (= –0.645 N/mm²)
and angle θ_{p_2}(= 107.22°) is corresponding to minor principal plane (= 0.062 N/mm²)

3.2 **In a stressed member two strain gauges are fixed such that they are inclined at 30° to the known directions of the principal stresses. The strains measured in these two gauges are +445 × 10⁻⁶ and –32 × 10⁻⁶ respectively. E = 2.1 × 10⁵ N/mm² and Poisson's ratio (μ) = 0.3. Determine the magnitude of the principal stress.**

[18 marks : 2001]

Solution:

Let normal principal strain in x and y direction are ϵ_1 and ϵ_2 respectively.

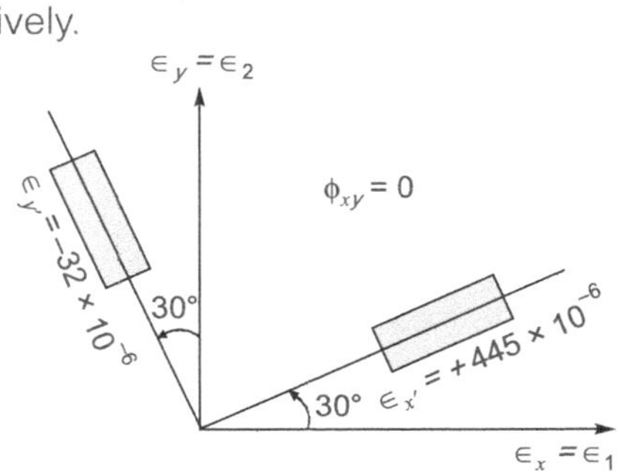

$$\epsilon_x' = \frac{\epsilon_x+\epsilon_y}{2} + \frac{\epsilon_x-\epsilon_y}{2}\cos 2\theta_1 + \frac{\phi_{xy}}{2}\sin 2\theta_1$$

Here, $\theta_1 = 30°$ and $\epsilon_x = \epsilon_1$, $\epsilon_y = \epsilon_2$

$$445\times10^{-6} = \frac{\epsilon_1+\epsilon_2}{2} + \frac{\epsilon_1-\epsilon_2}{2}\cos(2\times30°) + 0 \quad [\because \phi_{xy} = 0]$$

$$3\epsilon_1 + \epsilon_2 = 1780\times10^{-6} \quad \text{...(i)}$$

We know that

$$\epsilon_1 + \epsilon_2 = \epsilon_x' + \epsilon_y'$$

$$\epsilon_1 + \epsilon_2 = +445\times10^{-6} - 32\times10^{-6}$$

$$\epsilon_1 + \epsilon_2 = +413\times10^{-6} \quad \text{...(ii)}$$

From eq. (i) and (ii)

$$\epsilon_1 = +683.5\times10^{-6}, \quad \epsilon_2 = -270.5\times10^{-6}$$

Principal stresses may be calculated as

$$\sigma_1 = \frac{E}{1-\mu^2}(\epsilon_1 + \mu \epsilon_2) = \frac{2.1\times10^5}{1-(0.3)^2}\left[683.5\times10^{-6} + 0.3\times(-270.5\times10^{-6})\right] = 139.00 \text{ N/mm}^2$$

and $$\sigma_2 = \frac{E}{1-\mu^2}(\epsilon_2 + \mu \epsilon_1) = \frac{2.1\times10^5}{1-(0.3)^2}\left[-270.5\times10^{-6} + 0.3\times683.5\times10^{-6}\right] = -15.10 \text{ N/mm}^2$$

3.3 A prismatic beam *ABC* is simply supported at *A* and *B*. *AB* = 2.0 m, *BC* = 1 m, *C* is free end. The entire beam is uniformly loaded with 10 kN/m. The cross section of the beam is (*I*) with the following particulars:

Flange width = 150 mm Flange thickness = 10 mm

Web thickness = 10 mm Overall depth of the beam = 300 mm

Determine the maximum value of principal stresses occuring anywhere in the beam. Specify the location.

[20 marks : 2002]

Solution:

$$\Sigma f_y = 0$$

$$R_A + R_B = 30 \qquad \text{...(i)}$$

Taking moment about *A*, $\Sigma M_A = 0$

$$R_B \times 2 - 10\times3\times\frac{3}{2} = 0$$

$\therefore$ $R_B = 22.5$ kN

and $R_A = 30 - 22.5 = 7.5$ kN

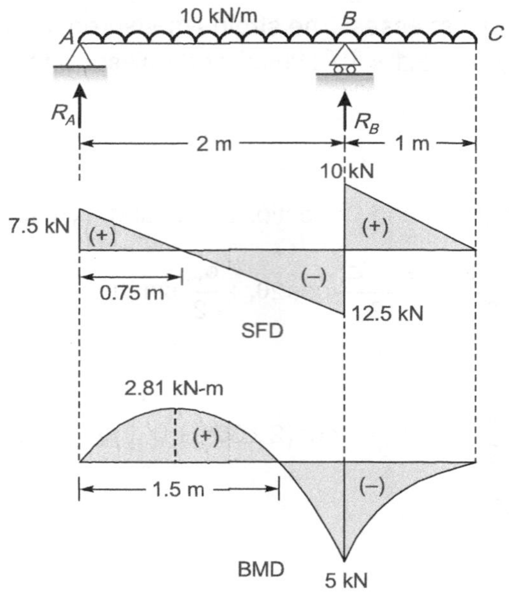

Shear Force Diagram (SFD)

Portion *AB*:

$$S_x(x \text{ from } A) = R_A - 10x \quad [0 \le x \le 2]$$

$$S_x = 7.5 - 10x \quad \text{(Linear)}$$

at $x = 0$, $S_A = 7.5$ kN

at $x = 2$, S_B (just left of B) $= 7.5 - 10 \times 2 = -12.5$ kN

and point of zero shear will be given by

$$S_x = 7.5 - 10x = 0$$

$$\therefore \quad x = \frac{7.5}{10} = 0.75 \text{ m from } A$$

Portion *CB*:

$$S_x \text{ (} x \text{ from } C) = 10x \quad \text{(Linear)} \qquad [0 \le x \le 1]$$

at $x = 0$, $S_C = 0$

at $x = 1$, S_B (just right of B) $= 10 \times 1 = 10$ kN

Bending Moment Diagram:

Taking top face as reference face.

Portion *AB*:

$$M_x \text{ (} x \text{ from } A) = R_A x - \frac{10x^2}{2} \qquad [0 \le x \le 2]$$

$$M_x = 7.5x - 5x^2 \quad \text{(Parabolic)}$$

at $x = 0$, $M_A = 0$

at $x = 2$, $M_B = 7.5 \times 2 - 5 \times 4 = -5$ kNm

For maximum bending moment M_{max},

$$\frac{dM_x}{dx} = 0$$

$$7.5 - 10x = 0$$

$$x = 0.75 \text{ m (also the point of zero shear)}$$

$$\therefore \quad M_{max} (x = 0.75 \text{ m}) = 7.5 \times 0.75 - 5 \times 0.75^2 = 2.81 \text{ kNm}$$

Location of zero bending moment is given by

$$M_x = 0$$

$$7.5x - 5x^2 = 0$$

$$x(7.5 - 5x) = 0$$

$$x = 0, 1.5 \text{ m}$$

So, bending moment is zero at $x = 0$ and $x = 1.5$ m from A

Portion *CB*:

$$M_x \text{ (} x \text{ from } C) = \frac{-10x^2}{2} = -5x^2 \quad \text{(Parabolic)} \qquad [0 \le x \le 1]$$

at $x = 0$, $M_C = 0$

at $x = 1$, $M_B = -5$ kNm

Now, there can be two locations where maximum principal stresses can occur. First is at a distance 0.75 m from A and second is at the support B. We will calculate and compare them whichever is maximum will be our result.

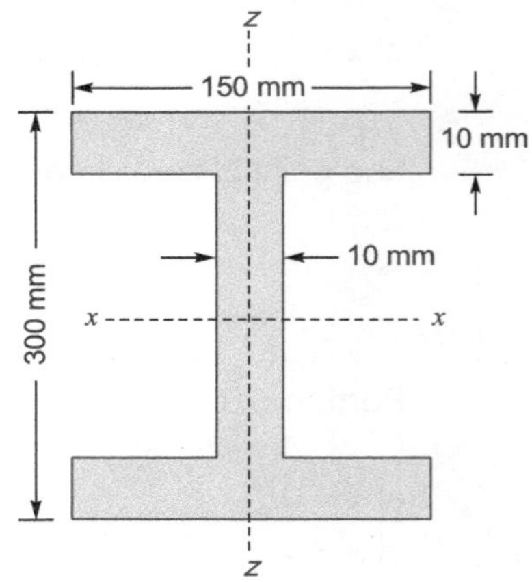

$$I_{xx} = 2\left[\frac{150\times10^3}{12} + 150\times10\times145^2 + \left(\frac{10\times140^3}{12} + 10\times140\times70^2\right)\right]$$

$$I_{xx} = 8.14\times10^7 \text{ mm}^4$$

Case-I:

At x = 0.75 m, M = 2.81 kNm, S = 0 and y = 150 mm

Bending stress, $\sigma = \frac{M}{I}y = \frac{2.81\times10^6\times150}{8.14\times10^7}$ [y = 150 mm]

$\therefore$ σ = 5.18 N/mm²

and shear stress, $\tau = \frac{SA\bar{y}}{Ib}$ But $S = 0$

$\therefore$ $\tau = 0$

Here we have

σ_x = 5.18 N/mm²

$\sigma_y = 0$

$\tau = 0$

Principal stresses may be given as

$$\sigma_1/\sigma_2 = \frac{\sigma_x+\sigma_y}{2} \pm \sqrt{\left(\frac{\sigma_x-\sigma_y}{2}\right)^2 + \tau_{xy}^2} = \frac{5.18+0}{2} \pm \sqrt{\left(\frac{5.18-0}{2}\right)^2 + 0^2}$$

$\therefore$ σ_1 = 5.18 N/mm²

and $\sigma_2 = 0$

Case II:

At x = 2 m, M = 5 kNm and S = 12.5 kN

One more point is to be considered, i.e., the location of a point in the beam where shear stress as well as normal stress are optimum and such a location is the junction between web and flange of the beam i.e., y = 140 mm.

$$\therefore \quad \sigma = \frac{M}{I}y = \frac{5\times10^6\times140}{8.14\times10^7} = 8.6 \text{ N/mm}^2$$

and shear stress, $\tau = \frac{SA\bar{y}}{Ib} = \frac{12.5\times10^3\times150\times10\times145}{8.14\times10^7\times10}$ = 3.34 N/mm²

Now we have, σ_x = 8.6 N/mm²

$\sigma_y = 0$

τ = 3.34 N/mm²

Principal stresses may be given as

$$\sigma_1/\sigma_2 = \frac{\sigma_x+\sigma_y}{2} \pm \sqrt{\left(\frac{\sigma_x-\sigma_y}{2}\right)^2+\tau_{xy}^2} = \frac{8.6+0}{2} \pm \sqrt{\left(\frac{8.6-0}{2}\right)^2+(3.34)^2}$$

$$= 4.3 \pm 5.44$$

$\therefore$ $\sigma_1 = 9.74\text{ N/mm}^2$ and $\sigma_2 = -1.14\text{ N/mm}^2$

Hence maximum values of principal stresses are (maximum of above two cases)

$$\sigma_1 = 9.74\text{ N/mm}^2 \quad\text{and}\quad \sigma_2 = -1.14\text{ N/mm}^2$$

3.4 A simply supported beam is 4 m long and carries a UDL of 12 kN/m over its entire length. The cross-section of the beam is 300 × 400 mm (deep). Find the principal stresses and their directions at a point 100 mm below the neutral axis and at 1/4 of the span.

[20 marks : 2003]

Solution:

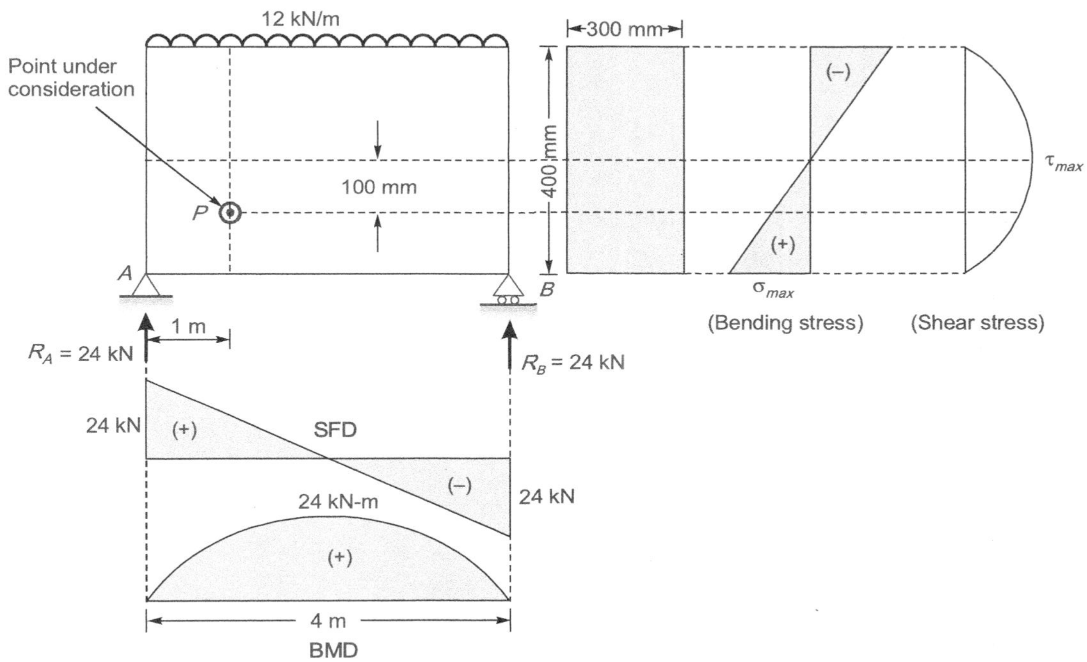

Shear force in beam AB is given by

$$S_x(x \text{ from } A) = R_A - wx = 24 - 12x$$

$$\text{Shear force at } (x = 1\text{ m}) = 24 - 12 \times 1 = 12\text{ kN}$$

Bending moment in beam AB is given by

$$M_x(x \text{ from } A) = R_A x - \frac{wx^2}{2} = 24x - \frac{12x^2}{2} = 24x - 6x^2$$

M_x at $(x = 1.0\text{ m})$

$$M = 24 \times 1 - 6 \times 1^2 = 18\text{ kNm}$$

and MOI about neutral axis is given by

$$I = \frac{bd^3}{12} = \frac{300 \times 400^3}{12} = 1.6 \times 10^9\text{ mm}^4$$

Now we have $M = 18$ kN, $S = 12$ kN and $I = 1.6 \times 10^9$ mm^4. Hence normal and shear stresses at $y = 100$ below NA can be calculated as

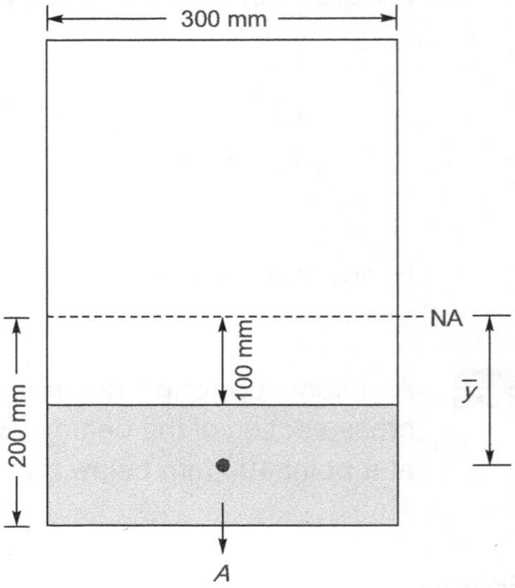

Bending stresses may be given as

$$\sigma = \frac{M}{I}y = \frac{18\times10^6\times100}{1.6\times10^9}$$

$$= 1.125 \text{ N/mm}^2 \text{ (tensile)}$$

and **shear stress** τ at a point P ($y = 100$ mm) below NA may be given as

$$\tau = \frac{SA\bar{y}}{Ib} = \frac{12\times10^3\times100\times300\times150}{1.6\times10^9\times300}$$

$$= -0.1125 \text{ N/mm}^2$$

Note: Positive shear will produce negative shear stress element

Now point can be considered as stress element shown below.

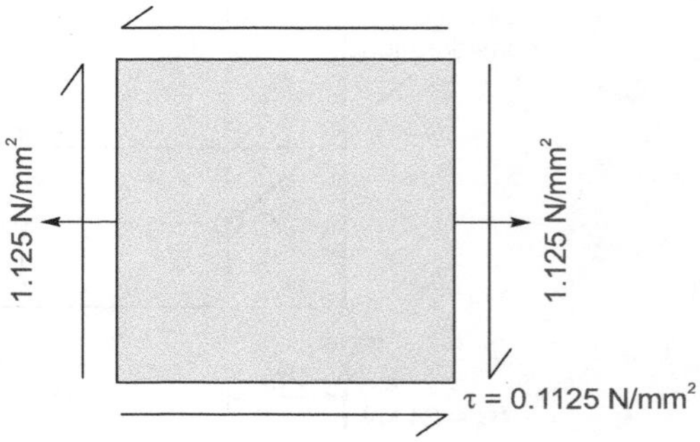

Here we have

$$\sigma_x = 1.125 \text{ N/mm}^2$$

$$\sigma_y = 0$$

$$\tau = -0.1125 \text{ N/mm}^2$$

The **principal stresses** may be given as

$$\sigma_1/\sigma_2 = \frac{\sigma_x+\sigma_y}{2} \pm \sqrt{\left(\frac{\sigma_x-\sigma_y}{2}\right)^2 + \tau_{xy}^2}$$

$$= \frac{1.125+0}{2} \pm \sqrt{\left(\frac{1.125-0}{2}\right)^2 + (-0.1125)^2}$$

$$= 0.5625 \pm 0.5736$$

$\therefore$ $$\sigma_1 = 1.136 \text{ N/mm}^2$$

and $$\sigma_2 = -0.011 \text{ N/mm}^2$$

For **direction of principal stresses**, we have

$$\tan 2\theta_P = \frac{2\tau_{xy}}{\sigma_x-\sigma_y} = \frac{-2\times0.1125}{1.125-0} = -0.2$$

$$\theta_{P_1} = -5.65° \text{ (clockwise)}$$

and $$\theta_{P_2} = 90° + \theta_{P_1}$$

$$= -90° - 5.65° = -95.65° \text{ (clockwise)}$$

Check: Principal stress normal to $\theta_{P_1} = -5.65°$

$$\sigma_x' = \frac{\sigma_x+\sigma_y}{2} + \frac{\sigma_x-\sigma_y}{2}\cos2\theta_{P_1} + \tau_{xy}\sin2\theta_{P_1}$$

$$= \frac{1.125+0}{2}+\frac{1.125-0}{2}\cos(2\times -5.65^\circ)$$

$$-0.1125\sin(2\times -5.65^\circ)$$

$$= 0.5625 + 0.5516 + 0.0220$$

$$\sigma_x' = 1.136\ \text{N/mm}^2 = \sigma_1$$

and $\sigma_2 = -0.011\ \text{N/mm}^2$

Hence major principal plane is 5.65° clockwise and minor principal plane is 95.65° clockwise from the vertical

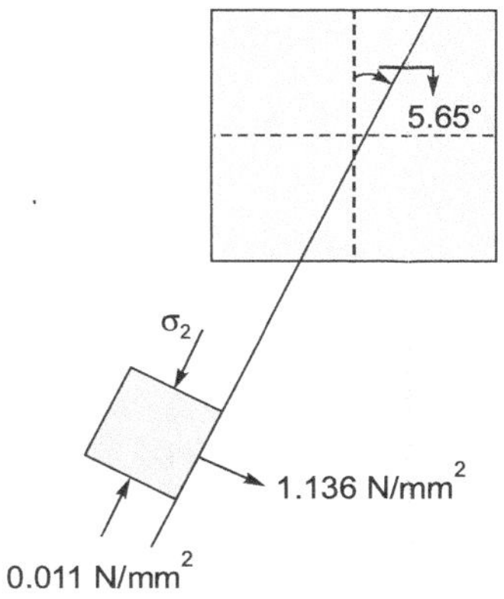

3.5 The principal tensile stresses at a point are: σ_x = 80 MPa and σ_y = 40 MPa. Find the magnitudes of normal, tangential and the resultant stress on a plane at 20° with the major principal plane. What is the angle of obliquity of the resultant stress with the major principal plane? Show by means of a sketch how normal stresses (σ_n) and tangential stress (σ_t) act on the triangular element.

[10 marks : 2005]

Solution:

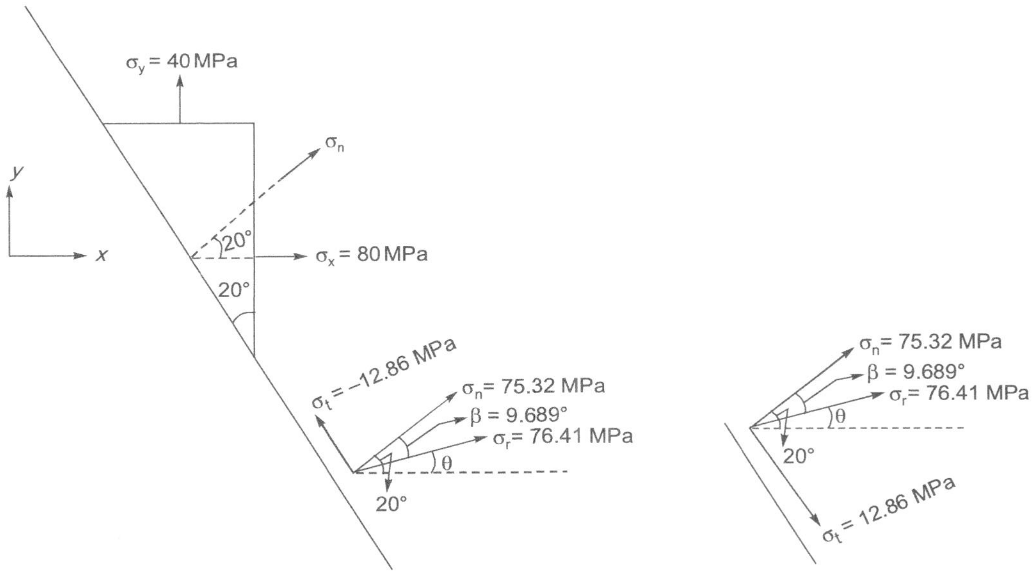

We know that

Normal stress, $$\sigma_n = \frac{\sigma_x+\sigma_y}{2}+\frac{\sigma_x-\sigma_y}{2}\cos 2\theta = \frac{80+40}{2}+\left(\frac{80-40}{2}\right)\cos 40^\circ = 60 + 15.32$$

$$\sigma_n = 75.32\ \text{N/mm}^2$$

Also **Tangential stress,** $$\sigma_t = \frac{-(\sigma_x-\sigma_y)}{2}\sin 2\theta = \frac{-(80-40)}{2}\sin 40^\circ = -12.86\ \text{N/mm}^2$$

[(–)ve sign indicates that σ_t is acting in a direction opposite to the direction assumed.]

Resultant stress, $$\sigma_R = \sqrt{(\sigma_n)^2+(\sigma_t)^2} = \sqrt{(75.32)^2+(-12.86)^2} = 76.41\ \text{N/mm}^2$$

Let the resultant makes an angle β with the normal stress plane as shown in the figure,

$$\tan\beta = \frac{\sigma_t}{\sigma_n}$$

$$\Rightarrow \quad \beta = \tan^{-1}\left(\frac{-12.86}{75.32}\right)$$

$$\Rightarrow \quad \beta = -9.689^\circ$$

(–) sign indicates that the obliquity obtained is opposite to the actual obliquity i.e. the resultant makes an angle of 9.689° with the normal stress plane in clockwise sense as shown in figure.

Let θ be the angle of obliquity of the resultant stress with major principal plane.

$$\therefore \quad \theta = 20 - \beta = 10.311^\circ$$

3.6 The stresses in a flat steel plate in a condition of plane stress are:
σ_x = 10000 N/mm²; σ_y = – 6000 N/mm² and τ_{xy} = 8000 N/mm²
Find the magnitude and orientation of the principal stresses in the plane of the plate.

[15 marks : 2006]

Solution:

We know that
$$\sigma_{x'} = \frac{\sigma_x+\sigma_y}{2}+\frac{\sigma_x-\sigma_y}{2}\cos 2\theta+\tau_{xy}\sin 2\theta$$

where, $\sigma_{x'}$ is the stress at any plane at an angle θ from the vertical plane.

$\sigma_{y'}$ can be found by substituting θ + 90° in place of θ and shear stress on this plane will be given by

$$\tau_{x'y'} = -\frac{(\sigma_x-\sigma_y)}{2}\sin 2\theta+\tau_{xy}\cos 2\theta$$

Now for $\sigma_{x'}$ and $\sigma_{y'}$ to be principal stress, shear stress on their plane i.e. $\tau_{x'y'}$ should be equal to zero.

$$\therefore \quad \tau_{x'y'} = -\left(\frac{\sigma_x-\sigma_y}{2}\right)\sin 2\theta+\tau_{xy}\cos 2\theta = 0$$

$$\Rightarrow \quad \frac{\sin 2\theta}{\cos 2\theta} = \frac{2\tau_{xy}}{\sigma_x-\sigma_y}$$

$$\Rightarrow \quad \tan 2\theta = \frac{2\tau_{xy}}{\sigma_x-\sigma_y}$$

The above equation can be used for finding the orientation of principal planes. Now if this value of θ is substituted in the equation which gives the value of $\sigma_{x'}$ and $\sigma_{y'}$, then they will change into equation which will give values of major principal stress and minor principal stress denoted by σ_1 and σ_2 respectively.

$$\therefore \quad \sigma_1/\sigma_2 = \frac{\sigma_x+\sigma_y}{2}\pm\sqrt{\left(\frac{\sigma_x-\sigma_y}{2}\right)^2+\left(\tau_{xy}\right)^2}$$

$$\Rightarrow \quad \sigma_1/\sigma_2 = \frac{10000+(-6000)}{2}\pm\sqrt{\left(\frac{10000+6000}{2}\right)^2+(8000)^2} = 2000 \pm 11313.71$$

$$\therefore \quad \sigma_1 = 13{,}313.71 \text{ N/mm}^2 \quad \text{and} \quad \sigma_2 = -9313.71 \text{ N/mm}^2$$

For orientation of principal plane, we have

$$\tan 2\theta_P = \frac{2\tau_{xy}}{\sigma_x - \sigma_y}$$

$$\Rightarrow \quad \tan 2\theta_P = \frac{2\times 8000}{10000+6000}$$

$$\Rightarrow \quad \tan 2\theta_P = 1$$

$$\Rightarrow \quad 2\theta_P = 45°$$

$$\Rightarrow \quad \theta_{P_1} = 22.5° \text{ (ACW)}$$

and $$\theta_{P_2} = 22.5 + 90 = 112.5°$$

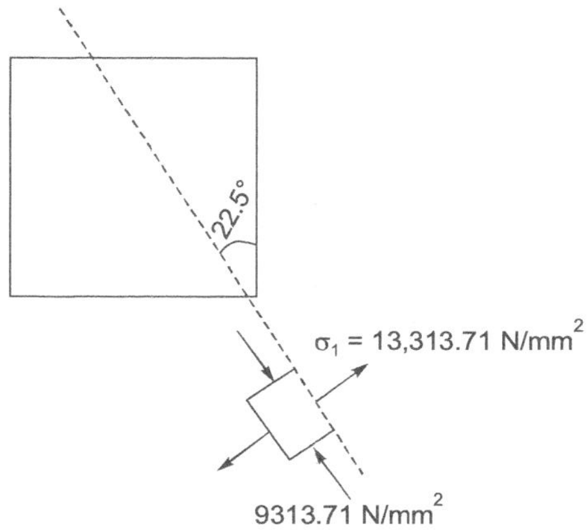

Check: $$\sigma_{x'} = \frac{\sigma_x+\sigma_y}{2} + \frac{\sigma_x-\sigma_y}{2}\cos 2\theta + \tau_{xy}\sin 2\theta$$

$$= \frac{10000-6000}{2} + \frac{10000+6000}{2}\cos 45° + 8000\sin 45°$$

$$= 2000 + 11313.71 = 13{,}313.71 \text{ N/mm}^2$$

i.e. plane which is at an angle of 22.5° to the vertical, will be the plane of major principal stress σ_1 and the other plane will be the minor principal plane.

3.7 **A circle of 100 mm diameter is inscribed on a steel plate before it is stressed and then the plate is loaded as shown in figure. Then the circle is deformed into an ellipse. Determine the major and minor axes of the ellipse and their directions.**

$E = 2.1 \times 10^6$ kg/cm²; $\frac{1}{m} = \mu = 0.28$

[10 Marks : 2006]

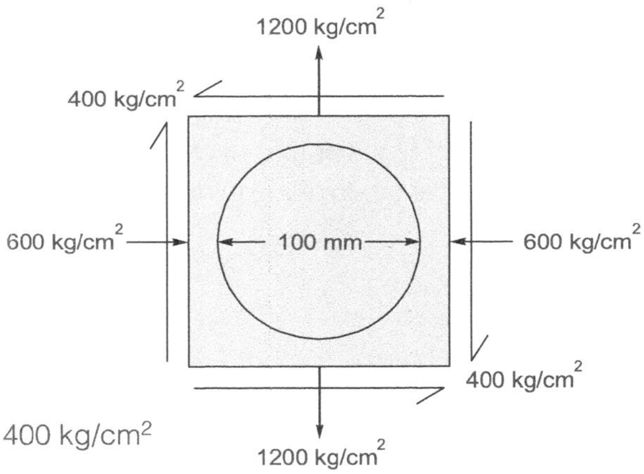

Solution:

Given: $\sigma_x = -600$ kg/cm²; $\sigma_y = +1200$ kg/cm²; $\tau_{xy} = -400$ kg/cm²

The principal stresses may be given as

$$\sigma_1/\sigma_2 = \frac{\sigma_x+\sigma_y}{2} \pm \sqrt{\left(\frac{\sigma_x-\sigma_y}{2}\right)^2 + (\tau_{xy})^2}$$

$$= \frac{-600+1200}{2} \pm \sqrt{\left(\frac{-600-1200}{2}\right)^2 + (-400)^2} = 300 \pm 984.9$$

$$\therefore \quad \sigma_1 = 1284.9 \text{ kg/cm}^2 \text{ and } \sigma_2 = -684.9 \text{ kg/cm}^2$$

and $$\tan 2\theta_P = \frac{2\tau_{xy}}{\sigma_x-\sigma_y} = \frac{2\times(-400)}{-600-1200} = \frac{800}{1800}$$

$$\theta_{P_1} = 11.98° \text{ (anti-clockwise)}$$

and $$\theta_{P_2} = 11.98° + 90° = 101.98° \text{ (anti-clockwise)}$$

Check the principal stresses along θ_{P_1} and θ_{P2}

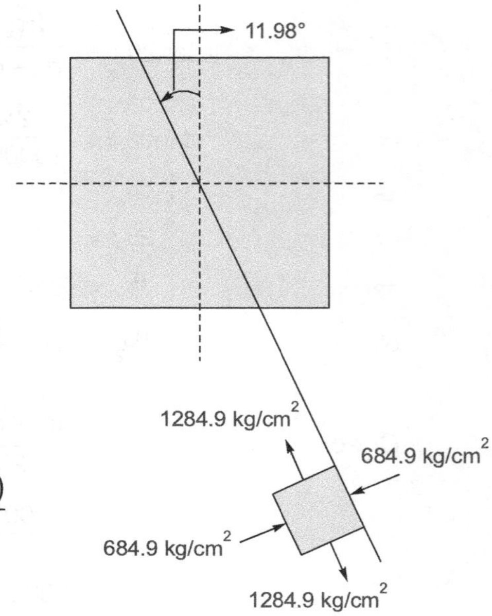

$$\sigma_x' = \frac{\sigma_x + \sigma_y}{2} \pm \frac{\sigma_x - \sigma_y}{2}\cos 2\theta + \tau_{xy}\sin 2\theta$$

$$= \frac{-600+1200}{2} + \frac{-600-1200}{2}$$

$$\times \cos(2\times 11.98) - 400\sin(2\times 11.98)$$

$$\sigma_x' = 300 - 822.45 - 162.44$$

$$= -684.9 \text{ kg/cm}^2$$

So, σ_2 = –684.9 kg/cm² will be normal to the principal plane incline 11.98° with vertical and σ_1 = 1284.9 kg/cm² will be normal to the plane inclined 101.98° from vertical

Linear strain along major principal plane

$$\in_{major} = \frac{\sigma_1}{E} - \mu\frac{\sigma_2}{E} = \frac{1284.9}{E} - \frac{\mu(-684.9)}{E}$$

$$= \frac{1284.9 + 0.28\times 684.9}{2.1\times 10^6}$$

$$\in_{major} = 7.03 \times 10^{-4}$$

$$\frac{\delta d_1}{d} = 7.03 \times 10^{-4}$$

$$\delta d_1 = 7.03 \times 10^{-4} \times d \approx 0.007 \text{ cm} = 0.07 \text{ mm (elongation)}$$

∴ Length of major axis of ellipse = 100 + 0.07 = 100.07 mm

Also linear strain along minor principal plane

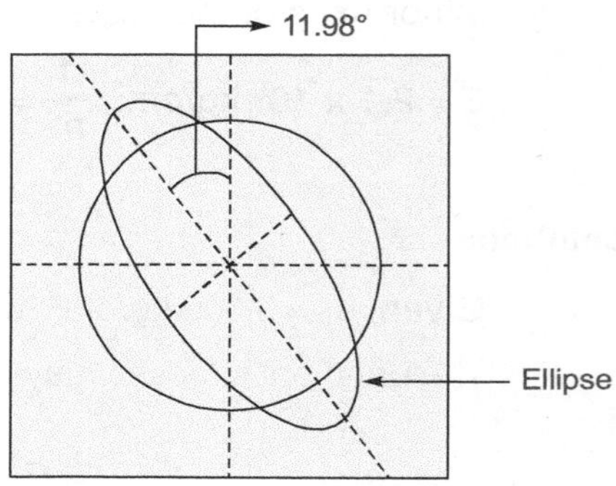

$$\in_{minor} = \frac{\sigma_2}{E} - \mu\frac{\sigma_1}{\mu}$$

$$= -\frac{684.9}{2.1\times 10^6} - \frac{0.28\times 1284.9}{2.1\times 10^6}$$

$$= -4.97 \times 10^{-4}$$

$$\frac{\delta d_2}{d} = -4.97 \times 10^{-4}$$

$$\delta d_2 = 4.97 \times 10^{-4} \times d$$

$$\delta d_2 = -0.0497 \text{ mm (contraction)}$$

∴ Length of minor axis of ellipse

$$= 100 - 0.0497 = 99.9503 \text{ mm}$$

3.8 **Given the state of stress shown in the figure below, find**

(i) Principal stress and

(ii) Maximum shearing stresses and the associated normal stresses.

Calculate the principal planes and maximum shear stress plane.

[10 marks : 2007]

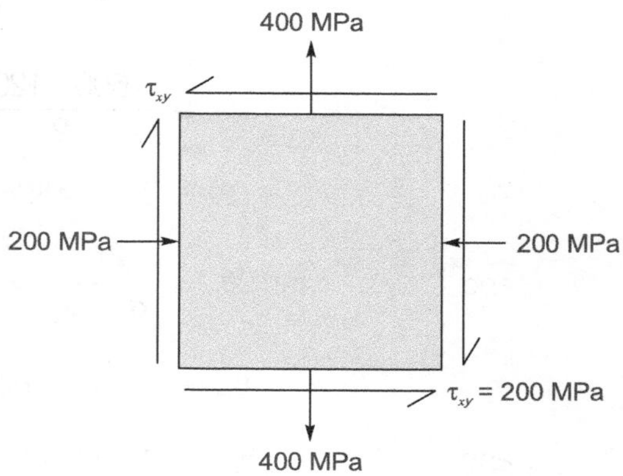

Solution:

(i) Given, $\sigma_x = -200$ MPa; $\sigma_y = 400$ MPa; $\tau_{xy} = -200$ MPa

Principal stresses can be calculated by

$$\sigma_1/\sigma_2 = \frac{\sigma_x+\sigma_y}{2} \pm \sqrt{\left(\frac{\sigma_x-\sigma_y}{2}\right)^2 + \tau_{xy}^2}$$

$$= \frac{-200+400}{2} \pm \sqrt{\left(\frac{-200-400}{2}\right)^2 + (-200)^2}$$

$$= 100 \pm 360.55$$

So,
$$\sigma_1 = 460.55 \text{ MPa}$$
$$\sigma_2 = -260.55 \text{ MPa}$$

Inclination of the principal plane can be found by

$$\tan 2\theta_P = \frac{2\tau_{xy}}{\sigma_x - \sigma_y}$$

$$\tan 2\theta_P = \frac{2\times(-200)}{-200-400}$$

$$2\theta_P = \tan^{-1}\left(\frac{2}{3}\right)$$

$$\theta_{P_1} = 16.85° \text{ (anti-clockwise)}$$

and principal plane perpendicular to θ_{P_1}

$$\theta_{P_2} = 90° + \theta_{P_1} = 106.85° \text{ (anti-clockwise)}$$

(ii) **Maximum shear stress** can be calculated as

$$\tau_{max} = \left|-\frac{(\sigma_1-\sigma_2)}{2}\right| = \frac{(460.55+260.55)}{2} = 360.55 \text{ MPa}$$

Maximum shear stress plane lies at an angle of 45° with the principal plane

Normal stress on plane of τ_{max} can be given as

$$\sigma_n = \frac{\sigma_1+\sigma_2}{2} = \frac{460.55-260.55}{2} = 100 \text{ MPa}$$

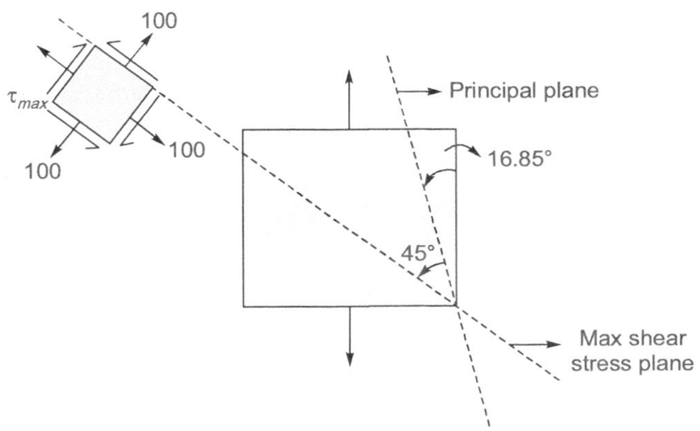

3.9 Describe any "five" Theories of Failure (Elastic) giving necessary formulae and comparing with the values obtained from direct tension test.

[15 marks : 2007]

Solution:

The five theories of failure are as follows:

1. **Maximum principal stress theory or Rankine's Theory:** According to this theory for no failure, maximum principal stress developed in a strained body must not exceed the yield stress in uniaxial loading.

   $\therefore$ For no failure, $|\sigma_1| \leq \sigma_y|$

   For design a suitable factor of safety is applied.

   $\therefore$ For design, $\sigma_1 \leq \dfrac{\sigma_y}{F.S}$

   This theory has certain limitations such as:

   (i) It is not suitable for ductile metals.
   (ii) It is not suitable for hydrostatic loading.
   (iii) It is not suitable for the case of pure shear.

2. **Maximum principal strain theory or St. Venant's Theory:** According to this theory, a ductile material begins to yield when the maximum principal strain reaches the strain at which yielding occurs in simple tension, or when the minimum principal strain (i.e.the compressive strain) equals the yield point strain in simple compression.

   $$\epsilon_1 \leq \frac{\sigma_y}{E} \text{...... for no failure}$$

   3-dimensional case $\quad \epsilon_1 \leq \dfrac{\sigma_1}{E} - \dfrac{\mu\sigma_2}{E} - \dfrac{\mu\sigma_3}{E}$

   2- dimensional case $\quad \epsilon_1 \leq \dfrac{\sigma_1}{E} - \dfrac{\mu\sigma_2}{E}$

   Some limitations of this theory are:

   (i) It is not suitable for hydrostatic loading.
   (ii) It cannot be applied for pure shear.
   (iii) It overestimates the elastic strength of ductile materials.

3. **Maximum shear stress theory or Guest's theory:** According to this theory for no failure, maximum shear stress developed in a strained body should be less than or equal to maximum shear stress in uniaxial loading.

   $$\frac{\sigma_1 - \sigma_2}{2} \leq \frac{\sigma_y}{2}$$

   $$\Rightarrow \quad \sigma_1 - \sigma_2 \leq \sigma_y$$

   Limitations of this theory are:

   (i) It is not applicable when principle stresses are nearly equal and alike.
   (ii) In case of pure shear though it gives safe results but is still inaccurate.
   (iii) It is not suitable for hydrostatic loading also.

4. **Maximum strain energy theory or Haigh's Theory:** According to this theory inelastic action at any point in a body due to any state of stress begins only when the energy per unit volume absorbed at the point is equal to the energy absorbed per unit volume of the material when subjected to the elastic limit under a uniaxial state of stress, as occurs in a simple tensile test.

$$U \le \frac{\sigma_y^2}{2E}$$

3- dimensional case $$U = \frac{1}{2E}\left[\sigma_1^2 + \sigma_2^2 + \sigma_3^2 - 2\mu\left(\sigma_1\sigma_2 + \sigma_2\sigma_3 + \sigma_3\sigma_1\right)\right]$$

2- dimensional case $$U = \frac{1}{2E}\left[\sigma_1^2 + \sigma_2^2 - 2\mu\sigma_1\sigma_2\right]$$

Limitations of this theory are:

(i) It is not suitable for brittle materials.

(ii) It is not applicable for the case of pure shear. However, it gives better result than maximum principal strain theory and maximum principal stress theory.

5. **Maximum shear strain energy theory or Mises-Henky Theory:** The maximum shear strain energy theory, also known as distortion energy theory states that inelastic action at any point in a body under any combination of stresses begins when the strain energy of distortion per unit volume absorbed at the point is equal to the strain energy of distortion absorbed per unit volume at any point in a bar stressed to the elastic limit under the state of uniaxial stress as occurs in a simple tension (or compression) test.

$$U_s \le \frac{\sigma_y^2}{6G}$$

3- dimensional case $$U_s \le \frac{1}{12G}\left[(\sigma_1 - \sigma_2)^2 + (\sigma_2 - \sigma_3)^2 + (\sigma_3 - \sigma_1)^2\right]$$

2- dimensional case $$U_s = \frac{1}{6G}\left[\sigma_1^2 + \sigma_2^2 - \sigma_1\sigma_2\right]$$

This theory is well justified for case of pure shear in ductile materials.

Limitations of this theory are:

(i) It is not applicable to brittle materials.

(ii) It cannot be applied for materials under hydrostatic pressure.

3.10 **A solid circular shaft of diameter 50 mm is subjected to pure bending of 3.5 kNm. Find the maximum twisting moment that can be applied on this shaft such that the material of the shaft does not yield. Use Tresca's theory (maximum shear stress theory) of failure. The yield stress of the material in uniaxial tension is 400 N/mm².**

[15 marks : 2009]

Solution:

Diameter, $D = 50$ mm, $M = 3.5$ kNm $= 3.5 \times 10^6$ Nmm

Let maximum twisting moment is T kNm

Bending stress due to bending moment at top or bottom of shaft is given by

$$\sigma = \frac{M}{Z} = \frac{32M}{\pi D^3} = \frac{32 \times 3.5 \times 10^6}{\pi \times 50^3} = 285.21 \text{ N/mm}^2$$

Shear stress due to torque T,

$$\tau = \frac{T}{Z_P} = \frac{16T}{\pi D^3} = \frac{16 \times 10^6 \times T}{\pi \times 50^3} = 40.74T \text{ N/mm}^2$$

Due to the combined bending and torsion, principal stresses will occur at the surface of shaft,

$$\sigma_1/\sigma_2 = \frac{\sigma_x + \sigma_y}{2} \pm \sqrt{\left(\frac{\sigma_x - \sigma_y}{2}\right)^2 + \tau_{xy}^2}$$

$$\sigma_1/\sigma_2 = \left(\frac{285.21}{2}\right) \pm \sqrt{\left(\frac{285.21}{2}\right)^2 + (40.74T)^2}$$

$$= 142.605 \pm \sqrt{(142.605)^2 + 1660.04T^2}$$

$\therefore$
$$\sigma_1 = 142.605 + \sqrt{20336.19 + 1660.04T^2}$$

and
$$\sigma_2 = 142.605 - \sqrt{20336.19 + 1660.04T^2}$$

As per Tresca's theory,

$$\frac{\sigma_1 - \sigma_2}{2} \le \frac{\sigma_y}{2}$$

$$\therefore \quad 142.605 + \sqrt{20336.19 + 1660.04T^2} - 142.605 + \sqrt{20336.19 + 1660.4T^2} \le 400$$

$$2 \times \sqrt{20336.19 + 1660.04T^2} \le 400$$

$$T^2 \le 19663.81$$

$$T^2 \le 11.8453$$

$$\therefore \quad T \le 3.44 \text{ kNm}$$

Hence maximum twisting moment T is 3.44 kNm.

3.11 **A square plate of side 500 mm is subjected to pure shear of intensity 100 MPa as shown in figure below. Young's modulus of the material is 2×10^5 MPa and Poisson's ratio is 0.2. Find the principal stresses, their directions and the change in length of the diagonals of the plate.**

[10 marks : 2012]

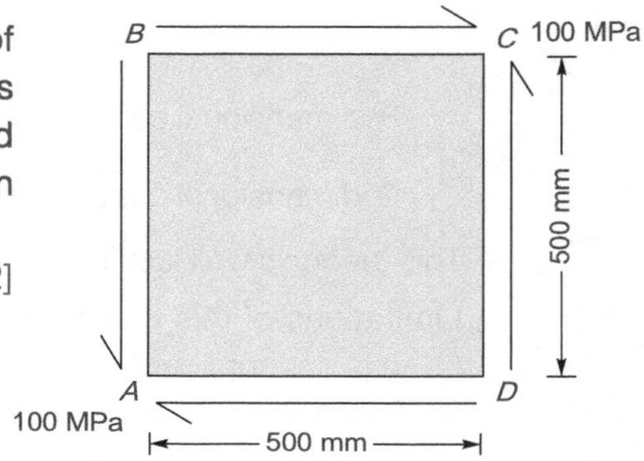

Solution:

We know that, for pure shear, normal stresses at any plane at an angle θ from the vertical plane

$$\sigma_x' = \tau_{xy} \sin 2\theta$$

and normal stress on plane perpendicular to above plane

$$\sigma_y' = -\tau_{xy} \sin 2\theta$$

Shear stress on this plane will be

$$\tau_{x'y} = \tau \cos 2\theta$$

For principal stress $\tau_{x'y} = 0$ and $\theta = 45°$ or $135°$.

So
$$\sigma_1 = +\tau_{xy} = +100 \text{ MPa}$$

and
$$\sigma_2 = -\tau_{xy} = -100 \text{ MPa}$$

direction of principal planes

$$\theta_{P_1} = 45° \text{ (anticlockwise)}$$

and
$$\theta_{P_2} = 135° \text{ (anticlockwise)}$$

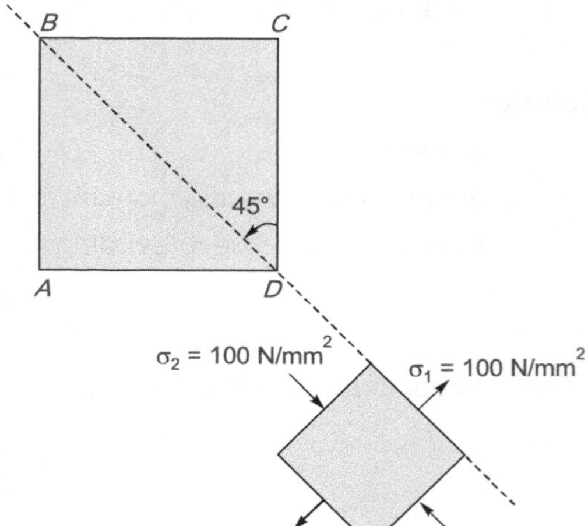

Since diagonal coincide with the principal plane so $\sigma_1 = 100$ MPa will be normal to principal plane along BD and $\sigma_2 = -100$ MPa will be normal to the plane perpendicular to plane BD

Now, linear strain along the diagonals is given as

$$\epsilon_1 = \frac{\sigma_1}{E} - \frac{\mu\sigma_2}{E} = \frac{100}{2\times10^5} - \frac{0.2\times(-100)}{2\times10^5} = \frac{100}{2\times10^5} + \frac{0.2\times100}{2\times10^5} = 6 \times 10^{-4}$$

and $\epsilon_2 = \frac{\sigma_2}{E} - \frac{\mu\sigma_1}{E} = \frac{-100}{2\times10^5} - \frac{0.2\times100}{2\times10^5} = -6 \times 10^{-4}$

Length of diagonals of square, $l = a\sqrt{2} = 500\sqrt{2}$

Hence $\epsilon_1 = \frac{\delta l_1}{l} = 6\times10^{-4}$

$$\delta l_1 = 6\times10^{-4}\times500\sqrt{2} = 0.424 \text{ mm (along AC)}$$

and $\epsilon_2 = \frac{\delta l_2}{l} = -6\times10^{-4}$

$$\delta l_2 = -6\times10^{-4}\times500\sqrt{2} = -0.424 \text{ mm (along BD)}$$

3.12 **At a certain cross-section, a circular shaft 90 mm in diameter is subjected to a BM of 3 kNm and twisting moment of 6 kNm. Find the principal stresses induced in the section using maximum normal stress theory.**

[15 marks : 2014]

Solution:

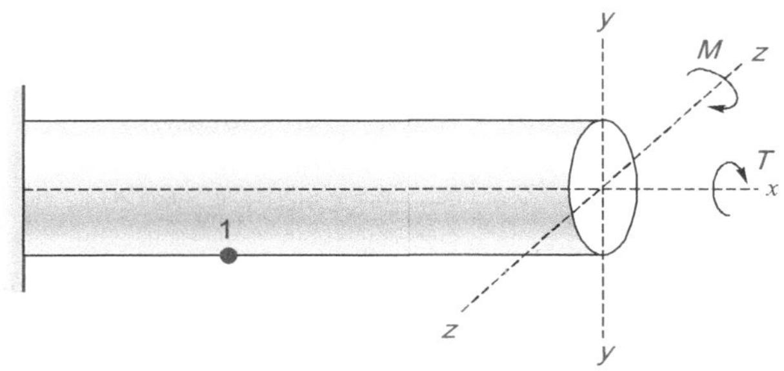

Given, $D = 90$ mm, $M = 3$ kN-m or 3×10^6 N-mm, $T = 6$ kN-m or 6×10^6 N-mm

Maximum bending stresses will occur either at top or bottom of shaft. Consider point 1 at bottom of shaft.

$$\sigma = \frac{M}{Z} = \frac{M}{\frac{I}{y_{max}}} = \frac{M}{I}\times y_{max} = \frac{M}{\frac{\pi D^4}{64}}\times\frac{D}{2} = \frac{32M}{\pi D^3}$$

$$= \frac{32\times3\times10^6}{\pi\times(90)^3} = 41.92 \text{ N/mm}^2$$

Shear stresses at surface of shaft (point 1) is given by

$$\tau = \frac{T}{Z_P}$$

where, Z_P = polar sectional modulus $= \frac{I_P}{y_{max}}$

$$\therefore \quad \tau = \frac{T}{I_P}\times y_{max} = \frac{T}{\frac{\pi D^4}{32}}\times\frac{D}{2}$$

$$= \frac{16T}{\pi D^3} = \frac{16\times6\times10^6}{\pi\times(90)^3} = 41.92 \text{ N/mm}^2$$

Now point 1 can be considered as stress element shown below:

Here, $\sigma_x = \sigma = 41.92$ N/mm^2

$\sigma_y = 0$

$\tau_{xy} = \tau = 41.92$ N/mm^2

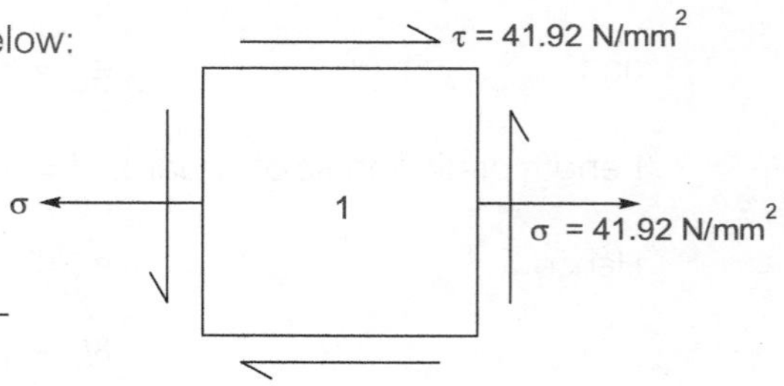

Hence **principal stresses** may be given as

$$\sigma_1/\sigma_2 = \frac{\sigma_x + \sigma_y}{2} \pm \sqrt{\left(\frac{\sigma_x - \sigma_y}{2}\right)^2 + \tau_{xy}^2}$$

$$= \frac{41.92+0}{2} \pm \sqrt{\left(\frac{41.92}{2}\right)^2 + (41.92)^2} = 20.96 \pm 46.87$$

$\therefore \quad \sigma_1 = 67.83$ N/mm^2 and $\sigma_2 = -25.91$ N/mm^2

Orientation of principal planes is given by,

$$\tan 2\theta_P = \frac{2\tau_{xy}}{\sigma_x - \sigma_y} = \frac{2 \times 41.92}{41.92 - 0}$$

$\Rightarrow \quad \tan 2\theta_P = 2$

$\Rightarrow \quad 2\theta_P = \tan^{-1}(2) = 63.43°$

$\theta_{P_1} = 31.72°$ (anticlockwise)

and $\quad \theta_{P_2} = \theta_{P_1} + 90° = 121.72°$ (anticlockwise)

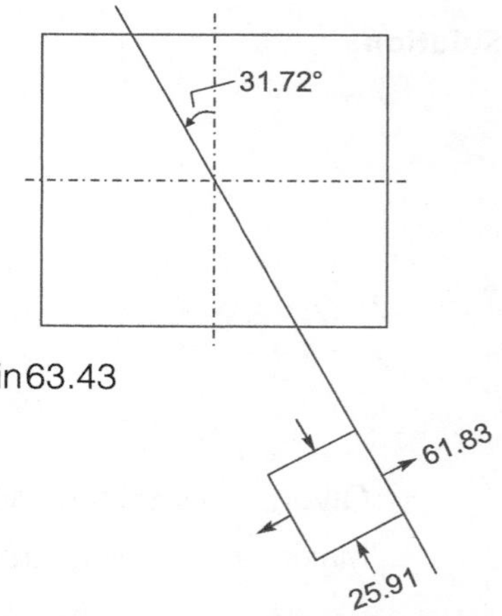

Check: $$\sigma_x' = \frac{\sigma_x + \sigma_y}{2} + \frac{\sigma_x - \sigma_y}{2}\cos 2\theta + \tau_{xy}\sin 2\theta$$

$$\sigma_x' = \frac{41.92+0}{2} + \frac{41.92-0}{2}\cos 63.43 + 41.92 \sin 63.43$$

$$\sigma_x' = 67.83 \text{ N/mm}^2$$

i.e., plane which is at an angle of 31.72° to the vertical will be the plane of major principal stress σ_1 and the other plane will be the minor principal plane.

3.13 **In a 2-*D* body, normal stress $\sigma_x = -10$ MPa, $\sigma_y = -10$ MPa, and shear stress $\tau_{xy} = 8$ MPa. Draw the Mohr circle and determine the principal stresses, principal planes, maximum shear stresses and their planes.**

[10 marks : 2014]

Solution:

Given, $\sigma_x = -10$ MPa, $\sigma_y = -10$ MPa, $\tau_{xy} = 8$ MPa

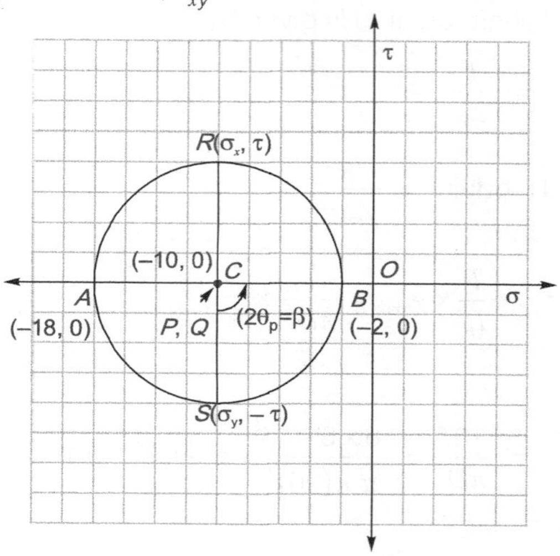

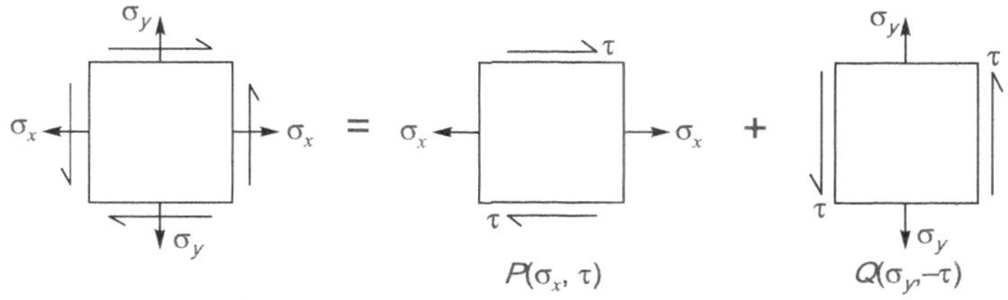

Procedure:

1. Select origin O, the axis passing through O represents σ and τ axis.
2. Assume scale, 1 cm = 2 MPa
3. $OP = OQ = -\dfrac{10}{2} = -5$ cm
4. Draw perpendicular from P and Q such that PR and QS represents τ and $-\tau$ respectively

   $\therefore\ PR = QS = \dfrac{8}{2} = 4$ cm
5. Now point R represents (σ_x, τ) and point S represents $(\sigma_y, -\tau)$
6. Joint point R and S by straight line, which cut x-axis at C
7. Draw a circle of radius $CR = CS$ from C, and resulting circle will represent Mohr's circle of stress.

Check:

Centre of Mohr's circle, $C = \left(\dfrac{\sigma_x + \sigma_y}{2}, 0\right)$

$\therefore$ Centre are (–10, 0) (OK)

Radius of Mohr's circle, $r = \sqrt{\left(\dfrac{\sigma_x - \sigma_y}{2}\right)^2 + \tau_{xy}^2} = \sqrt{\left(\dfrac{-10+10}{2}\right)^2 + 8^2}$

$= \sqrt{0+8^2} = 8$ unit (OK)

For Mohr's circle,
Principal stresses

$\sigma_1 = OA \times \text{scale} = -9 \text{ cm} \times 2 \text{ MPa} = -18 \text{ MPa}$

and $\sigma_2 = OB \times \text{scale} = -1 \text{ cm} \times 2 \text{ MPa} = -2 \text{ MPa}$

Principal planes: $\angle PCR = 2\theta_{P_1} = 90$

$\therefore\ \theta_{P_1} = 45$

and $\theta_{P_2} = \theta_{P_1} + 90° = 135°$

Maximum shear stress: τ_{max} = Radius of Mohr's circle × Scale

= 4 cm × 2 MPa = 8 MPa

Maximum shear stress planes: $\theta_1 = 0°$ and $\theta_2 = 90°$

3.14 **A plane element of a body is subjected to stresses as shown in figure.**

Find the factor of safety as per the following theories, if the yield stress given for the material is 200 N/mm² and Poisson's ratio = 0.3:

(i) Maximum principal stress theory
(ii) Maximum principal strain theory
(iii) Maximum shear stress theory
(iv) Maximum strain energy theory.

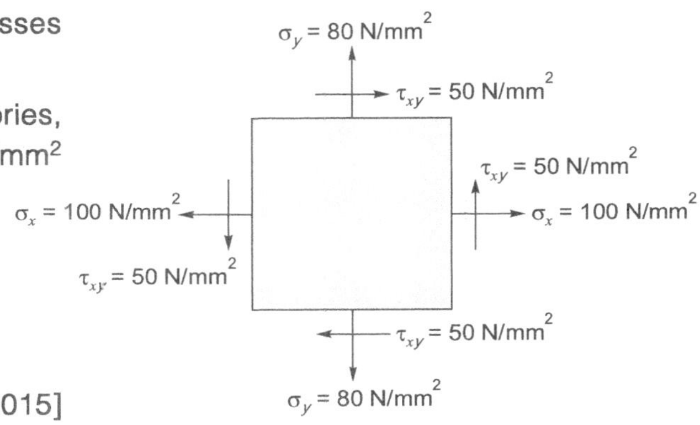

[15 marks : 2015]

Solution:

Given

Yield stress of material (f_y) = 200 N/mm^2

Poisson's ratio (μ) = 0.3

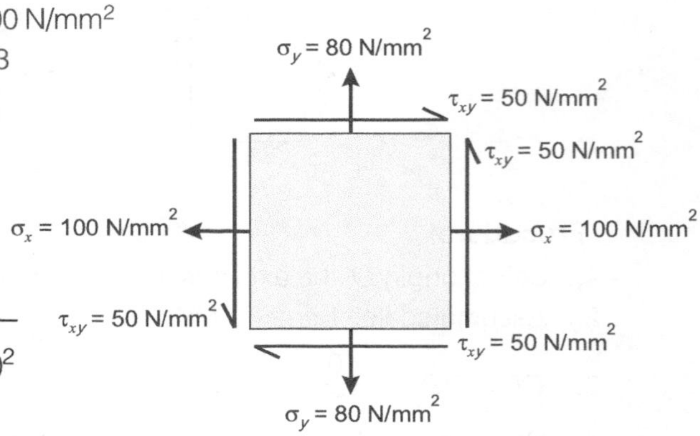

(i) Maximum principal stress theory

Maximum principal stress (σ_1)

$$= \frac{\sigma_x + \sigma_y}{2} + \sqrt{\left(\frac{\sigma_x - \sigma_y}{2}\right)^2 + \tau_{xy}^2}$$

$$= \left(\frac{100+80}{2}\right) + \sqrt{\left(\frac{100-80}{2}\right)^2 + 50^2}$$

$$= 141 \text{ N/mm}^2$$

$$\therefore \quad \text{FOS} = \frac{f_y}{\sigma_1} = \frac{200}{141} = 1.418$$

(ii) Maximum principal strain theory

$$\text{Minimum principal stress } (\sigma_2) = \frac{\sigma_x + \sigma_y}{2} - \sqrt{\left(\frac{\sigma_x - \sigma_y}{2}\right)^2 + \tau_{xy}^2}$$

$$= \left(\frac{100+80}{2}\right) - \sqrt{\left(\frac{100-80}{2}\right)^2 + 50^2} = 39 \text{ N/mm}^2$$

$$\therefore \quad \text{Maximum strain } (\epsilon_1) = \frac{\sigma_1}{E} - \frac{\mu\sigma_2}{E} = \frac{141}{E} - \frac{0.3(39)}{E} = \frac{129.3}{E}$$

$$\text{Maximum strain in simple tension } (\epsilon_y) = \frac{f_y}{E} = \frac{200}{E}$$

$$\therefore \quad \text{FOS} = \frac{\epsilon_y}{\epsilon_1} = \frac{200/E}{129.3/E} = 1.546$$

(iii) Maximum shear stress theory

$$\text{Maximum shear stress,} \quad \tau_{max} \le \frac{(f_y/FOS)}{2}$$

$$\tau_{max} = \max\left[\left|\frac{\sigma_1 - \sigma_2}{2}\right|, \left|\frac{\sigma_2 - \sigma_3}{2}\right|, \left|\frac{\sigma_3 - \sigma_1}{2}\right|\right]$$

$$= \max\left[\left|\frac{141-39}{2}\right|, \left|\frac{39-0}{2}\right|, \left|\frac{0-141}{2}\right|\right]$$

$$= \max\left[|51|, |19.5|, |-70.5|\right]$$

$$= \max[51, 19.5, 70.5] = 70.5$$

$$\text{Maximum shear in simple tension} = \frac{f_y}{2} = \frac{200}{2} = 100 \text{N/mm}^2$$

$$\therefore \quad \text{FOS} = \frac{f_y/2}{\tau_{max}} = \frac{100}{70.5} = 1.4184$$

(iv) Maximum strain energy theory

$$\text{Strain energy per unit volume } (U) = \frac{1}{2E}\left(\sigma_1^2 + \sigma_2^2 - 2\mu\sigma_1\sigma_2\right)$$

$$= \frac{1}{2E}\left[141^2 + (39)^2 - 2(0.3)(141)(39)\right] = \frac{18102.6}{2E}$$

Strain energy in simple tension (U_y) $= \frac{(f_y/\text{FOS})^2}{2E} = \frac{18102.6}{2E} = \frac{(f_y/\text{FOS})^2}{2E}$

$$18102.6 = \left(\frac{200}{\text{FOS}}\right)^2$$

$$\text{FOS} = \frac{200}{134.546} = 1.486$$

3.15 **A cylindrical shaft made of steel for which the tensile yield strength σ_y = 550 MPa is subjected to static loads consisting of a bending moment of 700 kNmm and torsion of 400 kNmm. Determine the diameter of shaft for a factor of safety equal to 1.5 if E = 200 GPa and Poisson's ratio ν = 0.25; Adopt maximum shear stress failure theory.**

[15 marks : 2016]

Solution:

Given: Tensile yield strength (σ_y) = 550 MPa; Bending moment (M) = 700 kNmm; Torsional moment (T) = 400 kNmm; Modulus of elasticity (E) = 200 GPa = 2×10^5 N/mm^2; Poisson's ratio (ν) = 0.25; FOS = 1.5

Theoretical concept

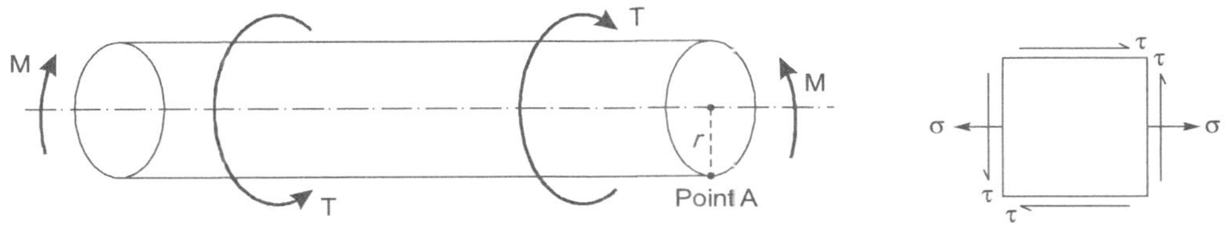

Loading on cylindrical shaft

State of stress at point A on cylindrical shaft

Bending stress, $\sigma = \frac{My}{J} = \frac{M.\frac{d}{2}}{\frac{\pi}{64}d^4} = \frac{32M}{\pi d^3}$

Shear stress, $\tau = \frac{T\nu}{J} = \frac{T\frac{d}{2}}{\frac{\pi}{32}d^4} = \frac{16T}{\pi d^3}$

The maximum shear stress, $\tau_{max} = \frac{1}{2}\sqrt{(\sigma - 0)^2 + 4\tau^2}$

$$\tau_{max} = \frac{1}{2}\sqrt{\left(\frac{32M}{\pi d^3}\right)^2 + 4\left(\frac{16T}{\pi d^3}\right)^2} = \frac{16}{\pi d^3}\sqrt{M^2 + T^2} \quad \text{...(i)}$$

Putting values of M&T in eq. (i)

$$\tau_{max} = \frac{16}{\pi d^3}\sqrt{\left(700 \times 10^3\right)^2 + \left(400 \times 10^3\right)^2} = \frac{4106.074 \times 10^3}{d^3}$$

As per maximum shear stress theory, $\tau_{max} \le \frac{1}{2}\left(\frac{64}{\text{FOS}}\right)$

$$\frac{4106.074 \times 10^3}{d^3} \leq \frac{1}{2}\left(\frac{55.0}{1.5}\right)$$

$$\Rightarrow \quad d^3 \leq 22396.71 \text{ mm}^3$$

$$d \leq 28.19 \text{ mm} \simeq 28.5 \text{ mm (say)}$$

3.16 **Explain with appropriate figures the failure criteria based on maximum shear stress theory. Also draw its Mohr circle representation.**

[15 marks : 2016]

Solution:

The **maximum shear stress theory** assumes that yielding occurs when the absolute maximum shear stress equals the maximum shear stress at yielding in the uniaxial loading.

$$\tau_{max} = \tau_{cr} = \left|\pm\frac{\sigma_1}{2}\right| = \frac{\sigma_{yp}}{2} \qquad \text{...(i) } (\sigma_1 \neq 0)$$

In applying this criterion to a biaxial plane stress problem, two different cases arise.

Case-1: The signs of the principal stresses σ_1 and σ_2 are same.

Let both σ_1 and σ_2 are tensile in nature and $\sigma_3 = 0$, the resulting Mohr's stress circles are shown below:

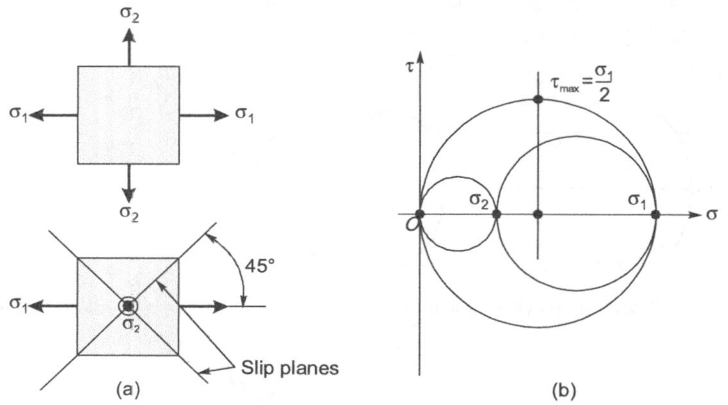

(a) (b)

Here the maximum shear stress is of the same magnitude as would occur in a simple uniaxial stress.

$\therefore$ If $|\sigma_1| > |\sigma_2|$, then according to eq. (i), $|\sigma_1|$ must not exceed σ_{yp}. Similarly, if $|\sigma_2| > |\sigma_1|$, $|\sigma_2|$ must not be greater than σ_{yp}.

$$\therefore \quad |\sigma_1| \leq \sigma_{yp}$$

$$\text{and} \quad |\sigma_2| \leq \sigma_{yp}$$

Case-2: The signs of σ_1 and σ_2 are opposite and $\sigma_3 = 0$

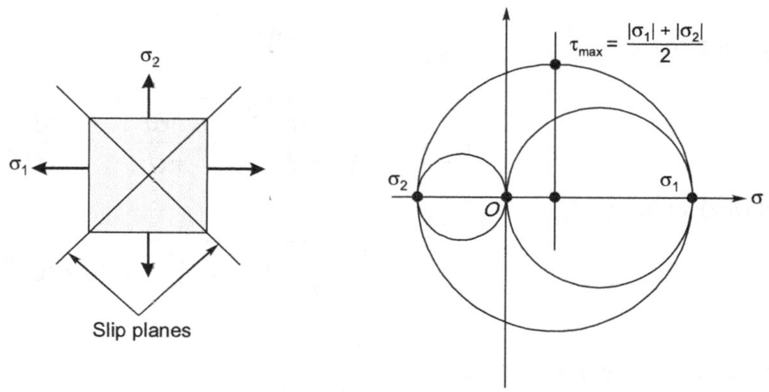

(c) (d)

The largest Mohr's circle passes through σ_1 and σ_2 and the maximum shear stress

$$z_{max} = \frac{|\sigma_1| + |\sigma_2|}{2}$$

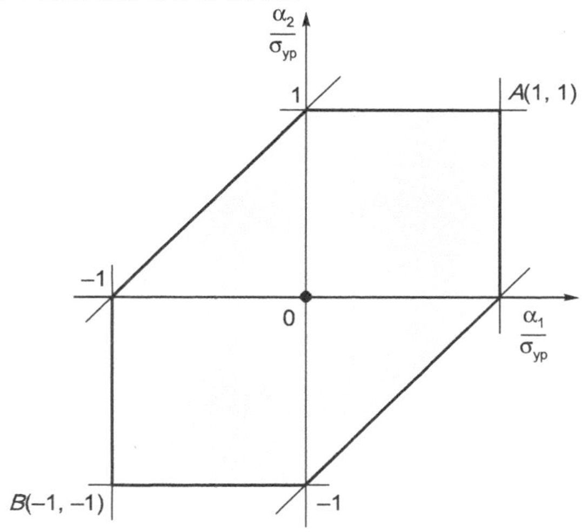

This maximum shear stress cannot exceed the shear yield criterion in simple tension.

$$\therefore \quad \left|\pm\frac{\sigma_1 - \sigma_2}{2}\right| \le \frac{\sigma_{yp}}{2}$$

or for impending yield

$$\frac{\sigma_1}{\sigma_{yp}} - \frac{\sigma_2}{\sigma_{yp}} = \pm 1$$

If a point defined by $\frac{\sigma_1}{\sigma_{yp}}$ and $\frac{\sigma_2}{\sigma_{yp}}$ falls on the hexagon shown above then a material begins and continues to yield. The stress points falling within the hexagon indicate that a material behaves elastically.

3.17 **A cylindrical shaft of 75 mm diameter is subjected to a maximum bending moment of 2500 Nm and a twisting moment of 4200 Nm. Find the maximum principal stresses developed in the shaft. If the yield stress of the shaft material is 360 N/sq. mm, determine the factor of safety of the shaft using maximum shear stress theory.**

[20 marks : 2017]

Solution:

This is the case of combined bending and torsion, hence shear stress will be developed due to torsion and bending stress due to bending respectively.

1. **Shear stress due to torsion:**

We know,
$$\frac{\tau}{r} = \frac{T}{I_P}$$

$$\Rightarrow \quad \frac{\tau_{max}}{R} = \frac{T}{I_P}$$

$$\Rightarrow \quad \tau_{max} = \frac{T}{I_P} \times R = \frac{T}{I_P} \times \frac{D}{2} \quad \text{where D = diameter of shaft}$$

$$I_P = \text{Polar moment of inertia} = \frac{\pi D^4}{32}$$

$$\therefore \quad \tau_{max} = \frac{T}{\frac{\pi D^4}{32}} \times \frac{D}{2} = \frac{16T}{\pi D^3}$$

2. **Bending stress due to bending moment:**

We know that,
$$\frac{f}{y} = \frac{M}{I}$$

$$\Rightarrow \quad f = \frac{M}{I} \times y$$

$$\Rightarrow \quad f_{max} = \frac{M}{I} \times y_{max}$$

$$\Rightarrow \quad f_{max} = \frac{M}{\frac{\pi}{64} D^4} \times \frac{D}{2} = \frac{32M}{\pi D^3}$$

The maximum principal stress under the combined effect of bending and torsion will occur either at top or at bottom surface.

We know,
$$\sigma_1, \sigma_2 = \frac{\sigma_x + \sigma_y}{2} \pm \sqrt{\left(\frac{\sigma_x - \sigma_y}{2}\right)^2 + \tau_{xy}^2}$$

Here,
$$\sigma_x = f_{max} = \frac{32M}{\pi D^3}$$
$$\sigma_y = 0$$
$$\tau_{xy} = \tau_{max} = \frac{16T}{\pi D^3}$$

$\therefore$
$$\sigma_1, \sigma_2 = \frac{16}{\pi D^3}\left[M \pm \sqrt{M^2 + T^2}\right]$$

$\Rightarrow$
$$\sigma_1 - \sigma_2 = \frac{16}{\pi D^3}\sqrt{M^2 + T^2} + \frac{16}{\pi D^3}\sqrt{M^2 + T^2} = \frac{32}{\pi D^3}\sqrt{M^2 + T^2}$$

Given, Shaft diameter, $D = 75$ mm
Maximum bending moment, $M = 2500$ Nm
Maximum Twisting moment, $T = 4200$ Nm

$\therefore$ Maximum principal stress,
$$\sigma_1 = \frac{16}{\pi D^3}\left[M + \sqrt{M^2 + T^2}\right]$$
$$= \frac{16}{\pi (75)^3}\left[2500 \times 10^3 + \sqrt{(2500 \times 10^3)^2 + (4200 \times 10^3)^2}\right]$$
$$= 89.186 \text{ N/mm}^2 \simeq 89.19 \text{ N/mm}^2$$

According to maximum shear stress theory,
$$\sigma_1 - \sigma_2 = \frac{\sigma_y}{FOS} \qquad \text{[Given } \sigma_y = 360 \text{ N/mm}^2\text{]}$$

$$\Rightarrow \frac{32}{\pi (75)^3} \times 10^3 \times \sqrt{(2500)^2 + (4200)^2} = \frac{360}{FOS}$$

$$\Rightarrow 118.01 = \frac{360}{FOS}$$

$$\Rightarrow FOS = 3.05 \simeq 3$$

3.18 **A plane element in a body is subjected to stresses as shown in figure. Determine the principal stresses and maximum shear stresses and the planes on which they act. Indicate them in separate sketches.**

[12 marks : 2017]

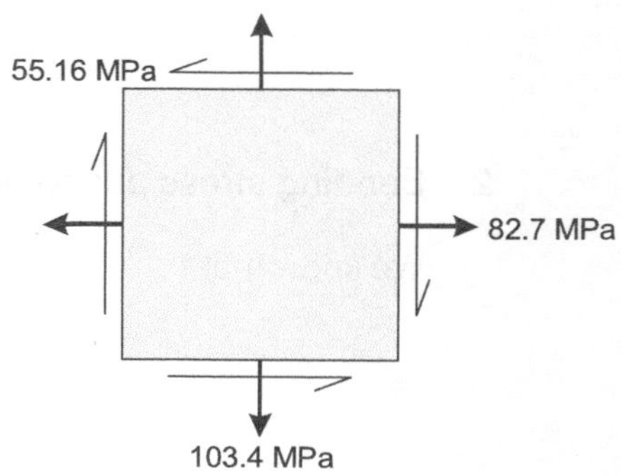

Solution:

Given, $\sigma_x = 82.7$ MPa, $\sigma_y = 103.4$ MPa, $\tau_{xy} = -55.16$ MPa

Principal stresses can be calculated by

$$\sigma_1, \sigma_2 = \frac{\sigma_x + \sigma_y}{2} \pm \sqrt{\left(\frac{\sigma_x - \sigma_y}{2}\right)^2 + \tau_{xy}^2}$$

$$= \left(\frac{82.70 + 103.40}{2}\right) \pm \sqrt{\left(\frac{82.7 - 103.4}{2}\right)^2 + (55.16)^2}$$

$$= 93.05 \pm \sqrt{107.1225 + 3042.626}$$

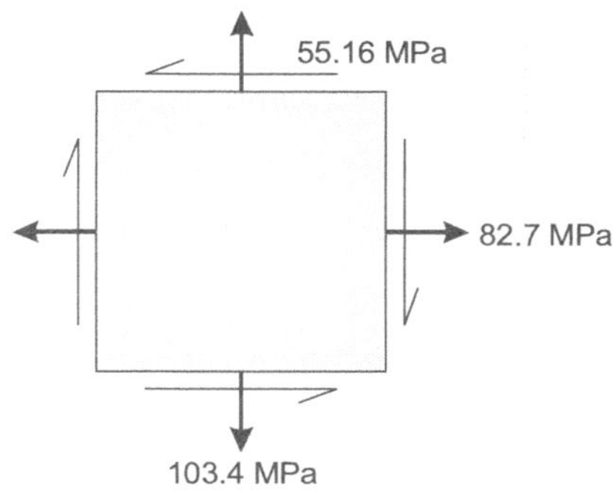

or $= 93.05 \pm 56.123$

$\therefore$ $\sigma_1 = 149.173$ MPa

and $\sigma_2 = 36.927$ MPa

Inclination of the principal plane is given by

$$\tan 2\theta_p = \frac{2\tau_{xy}}{(\sigma_x - \sigma_y)} = \frac{2 \times (-55.16)}{(82.7 - 103.4)}$$

or, $2\theta_p = \tan^{-1}(+5.329)$

$\Rightarrow$ $\theta_p = 39.41'11''$ or $39.69°$

Maximum shear stress is given by,

$$\tau_{max} = -\left(\frac{\sigma_1 - \sigma_2}{2}\right) = -\frac{(149.173 - 36.927)}{2} = -56.123 \text{ N/mm}^2$$

Maximum shear stress plane lies at an angle of 45° with the principle plane i.e., at $\theta_p + 45° = 84.69°$

Normal stresses on maximum shear stress plane can be given as;

$$\sigma'_x = \frac{\sigma_1 + \sigma_2}{2} + \frac{\sigma_1 - \sigma_2}{2}\cos 2\theta$$

But $\theta = 45°$

$$\therefore \quad \sigma'_x = \frac{\sigma_1 + \sigma_2}{2} = \left(\frac{149.173 + 36.927}{2}\right) = 93.05 \text{ MPa}$$

Also $\sigma'_x + \sigma'_y = \sigma_x + \sigma_y$

$\Rightarrow$ $93.05 + \sigma'_y = 82.7 + 103.40$

$\Rightarrow$ $\sigma'_y = 93.05$ MPa

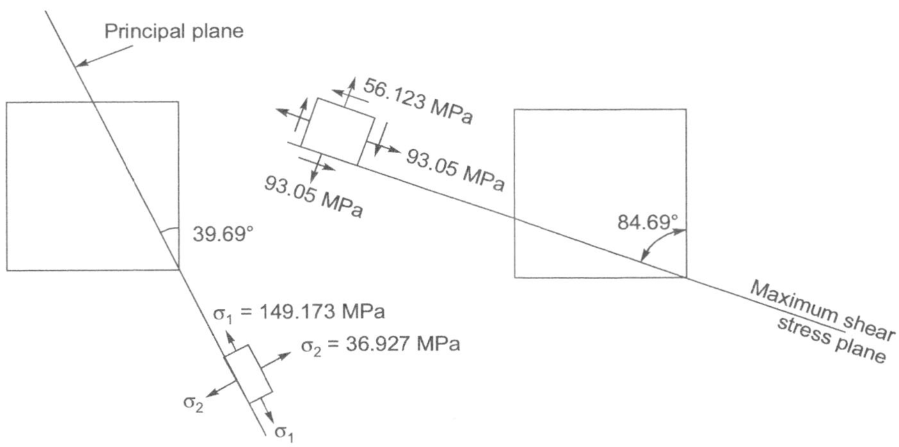

3.19 In a steel flat plate a state of plate stress available. Calculate σ_y and principal stresses if $\sigma_x = 140$ N/mm², $\mu = 0.25$, $\tau_{xy} = 40$ N/mm², $\epsilon_z = -3.6 \times 10^{-4}$ and $E = 2 \times 10^5$ N/mm².

[20 marks : 2018]

Solution:

Given data: Plane stress condition

$\sigma_x = 140$ N/mm², $\mu = 0.25$, $\tau_{xy} = 40$ N/mm², $\varepsilon_z = -3.6 \times 10^{-4}$, $E = 2 \times 10^5$ N/mm²

In plane stress condition

$$\therefore \quad \sigma_z = \tau_{zx} = \tau_{zy} = 0$$

(If z-axis is considered to be perpendicular to the face of the stress element on which no stress is acting)

Hence the remaining stress components are σ_y, σ_x and τ_{xy}

In the given data, ε_z has be provided,

So from Hooke's law,
$$\varepsilon_z = \frac{\sigma_z}{E} - \mu\frac{\sigma_x}{E} - \mu\frac{\sigma_y}{E} \qquad [\because \sigma_z = 0]$$

$$\varepsilon_z = -\frac{\mu}{E}(\sigma_x + \sigma_y)$$

$$-3.6 \times 10^{-4} = -\frac{0.25}{2 \times 10^5}(140 - \sigma_y)$$

$$\Rightarrow \quad \sigma_y = 148 \text{ N/mm}^2$$

Calculation for Principal stress

$$\sigma_{1,2} = \frac{\sigma_x + \sigma_y}{2} \pm \frac{1}{2}\sqrt{(\sigma_x - \sigma_y)^2 + 4\tau_{xy}^2}$$

$$\sigma_{1,2} = \frac{140 + 148}{2} \pm \frac{1}{2}\sqrt{(140 - 148)^2 + 4 \times 40^2}$$

$$\sigma_{1,2} = (144 \pm 40.2) \text{ N/mm}^2$$

$$\sigma_1 = 184.2 \text{ N/mm}^2$$

$$\sigma_2 = 103.8 \text{ N/mm}^2$$

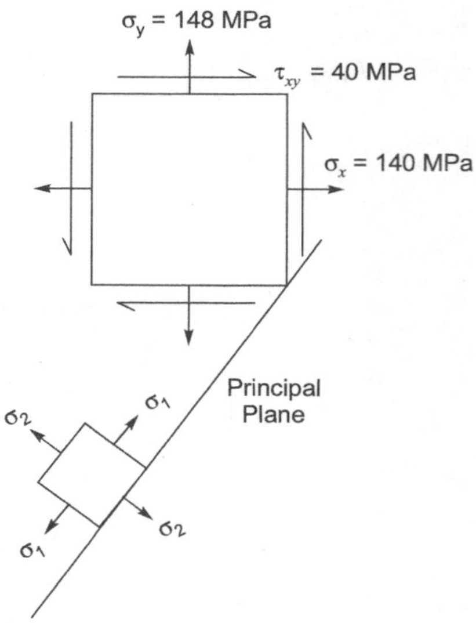

Direction of principal plane,

$$\tan 2\theta_{p_1} = \frac{2\tau_{xy}}{\sigma_x - \sigma_y}$$

$$\tan 2\theta_{p_1} = \frac{2 \times 40}{140 - 148}$$

$$\theta_{p_1} = -42.14°$$

Check: $\sigma_{x'} = \sigma_x \cos^2\theta + \sigma_y \sin^2\theta + 2\tau_{xy} \cos\theta \sin\theta \; [\because \theta = \theta_{p1}]$

$\sigma_{x'} = 103.8 \text{ N/mm}^2 = \sigma_2$

3.20 Direct stresses of 120 MN/m² in tension and 90 MN/m² in compression are applied to an elastic material at a certain point on planes at right angles to each other. If the maximum principal stress is not to exceed 150 MN/m² in tension, to what shearing stress can the material be subjected? What is then the maximum resulting shearing stress in the material? Also find the magnitude of the other principal stress and its inclination to 120 MN/m² stress.

[8 marks : 2019]

Solution:

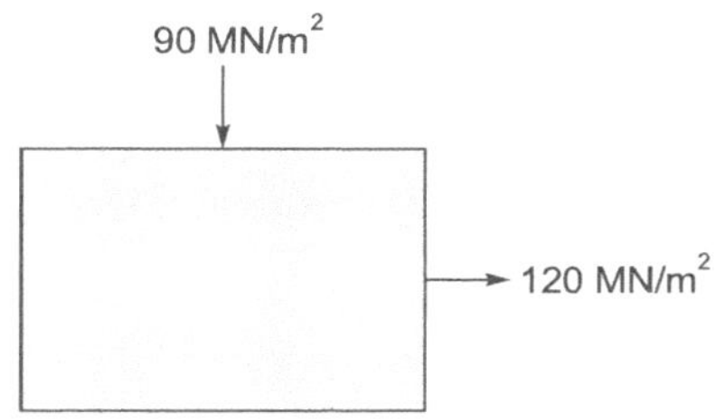

$$\sigma_1 = 120 \text{ MN/m}^2$$
$$\sigma_2 = -90 \text{ MN/m}^2$$

If $\sigma_{P_1} \not> 150 \text{ MN/m}^2$

Then, $\tau_{xy} = ?$

$$\sigma_{P_1} = \frac{\sigma_1 + \sigma_2}{2} + \sqrt{\left(\frac{\sigma_1 - \sigma_2}{2}\right)^2 + \tau_{xy}^2}$$

$$150 = \frac{120 - 90}{2} + \sqrt{\left(\frac{120 + 90}{2}\right)^2 + \tau_{xy}^2}$$

$$150 = 15 + \sqrt{105^2 + \tau_{xy}^2}$$

$$135^2 = 105^2 + \tau_{xy}^2$$

$$\tau_{xy} = 84.85 \text{ MN/m}^2$$

Other principle stress, $$\sigma_{P_2} = \frac{\sigma_1 + \sigma_2}{2} - \sqrt{\left(\frac{\sigma_1 - \sigma_2}{2}\right)^2 + \tau_{xy}^2}$$

$$\sigma_{P_2} = \frac{120 - 90}{2} - \sqrt{\left(\frac{120 + 90}{2}\right)^2 + 84.85^2}$$

$$\sigma_{P_2} = 15 - \sqrt{105^2 + 84.85^2}$$
$$= -119.998 \text{ MN/m}^2 \simeq -120 \text{ MN/m}^2$$

Maximum shear stress $$= \frac{\sigma_{P_1} - \sigma_2}{2}$$

$$\tau_{max} = \frac{150 - (-120)}{2} = 135 \text{ MN/m}^2$$

Orientation with plane,

$$\tan 2\theta_{P_1} = \frac{2\tau_{xy}}{\sigma_1 - \sigma_2} = \frac{2 \times 84.85}{120 - (-90)}$$

$$\Rightarrow \quad 2\theta_{P_1} = 38.94°$$

$$\theta_{P_1} = 19.47°$$

3.21 What combination of Principal stresses will give the same factor of safety for failure by yielding according to the maximum shear stress theory and distortion energy theory. Consider only a two dimensional case.

[10 marks : 2019]

Solution:

According to maximum shear stress theory

$$\frac{\sigma_{P_1}}{2} \text{ or } \frac{\sigma_{P_2}}{2} \text{ or } \left(\frac{\sigma_{P_1} - \sigma_{P_2}}{2}\right) = \left(\frac{\sigma_y}{2 \text{ FOS}}\right) \quad \text{...(i)}$$

According to Distortion Energy theory

$$\frac{1}{2}\left[\left(\sigma_{P_1}-\sigma_{P_2}\right)^2+\left(\sigma_{P_2}-\sigma_{P3}\right)^2+\left(\sigma_{P_3}-\sigma_{P_1}\right)^2\right] = \left(\frac{\sigma_y}{FOS}\right)^2$$

For 2-dimensional system, $\sigma_{P_3} = 0$

$$\sigma_{P_1}{}^2+\sigma_{P_2}{}^2-\sigma_{P_1P_2} = \left(\frac{\sigma_y}{FOS}\right)^2 \quad ...(ii)$$

Now, put value of $\left(\frac{\sigma_y}{FOS}\right)$ from eq. (i)

$$\sigma_{P_1}{}^2+\sigma_{P_2}{}^2-\sigma_{P_1P_2} = \left(\sigma_{P_1}-\sigma_{P_2}\right)^2$$

$$\sigma_{P_1}\sigma_{P_2} = 0 \quad \text{(which is not possible)}$$

So taking another case in max shear stress theory

$$\left(\frac{\sigma_{P_1}}{2}\right) = \left(\frac{\sigma_y}{2FOS}\right) \quad ...(iii)$$

From eq. (i) and (ii)

$$\sigma_{P_1}{}^2+\sigma_{P_2}{}^2-\sigma_{P_1}\sigma_{P_2} = \sigma_{P_1}{}^2$$

$$\sigma_{P_2}\left[\sigma_{P_2}-\sigma_{P_2}\right] = 0$$

$$\sigma_{P_2} \neq 0$$

So, $\sigma_{P_2} = \sigma_{P_1}$

So both principle stresses must be equal.

3.22 **A 200 mm × 150 mm × 10 mm aluminium plate is subjected to uniform bi-axial stresses σ_x and σ_y. Two strain gauges A and B are attached to the surface of the plate as shown in the figure. If readings in strain gauges are $\varepsilon_A = 200 \times 10^{-6}$ and $\varepsilon_B = 285 \times 10^{-6}$, what are the value of σ_x and σ_y? What is the reduction in thickness of the plate as a result of stresses?**
Take Young's modulus E = 75 GPa and Poisson's ratio ν = 0.33.

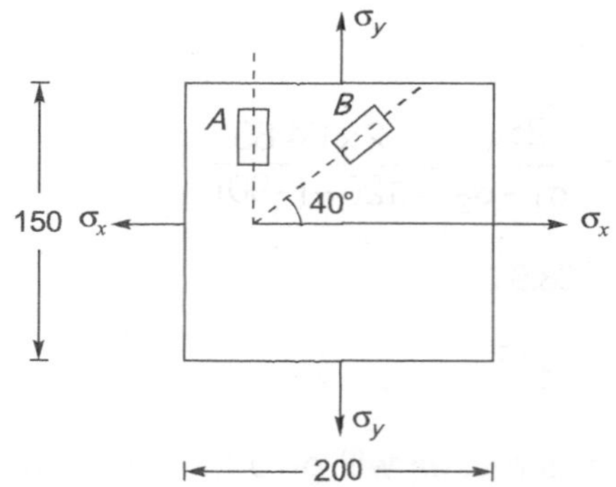

[12 marks : 2020]

Solution:

$\varepsilon_A = 200 \times 10^{-6}$, $\varepsilon_B = 285 \times 10^{-6}$, $E = 75$ GPa $= 75 \times 10^3$ MPa, $\nu = 0.33$, $t = 10$ mm

Strain gauge *A* is along *y*

Thus $\varepsilon_A = \varepsilon_y = 200 \times 10^{-6}$

From strain transformation,

$$\varepsilon_B = \frac{\varepsilon_x+\varepsilon_y}{2}+\frac{\varepsilon_x-\varepsilon_y}{2}\cos 2\theta+\frac{\gamma_{xy}}{2}\sin 2\theta$$

$$\varepsilon_B = \frac{\varepsilon_x + \varepsilon_y}{2} + \frac{\varepsilon_x - \varepsilon_y}{2}\cos(2\times 40) + \frac{\gamma_{xy}}{2}\sin(2\times 40)$$

Since no shear is acting on plane normal to x and y thus $\gamma_{xy} = 0$

So, $$285 \times 10^{-6} = \frac{\varepsilon_x + 200\times 10^{-6}}{2} + \frac{\varepsilon_x - 200\times 10^{-6}}{2}\cos 80°$$

$$570 \times 10^{-6} = \varepsilon_x(1+\cos 80°) + 200\times 10^{-6} \times (1-\cos 80°)$$

Thus, $$\varepsilon_x = 344.847 \times 10^{-6}$$

Now, $$\sigma_x = \frac{E}{1-\nu^2}(\varepsilon_x + \nu\varepsilon_y) = \frac{75\times 10^3}{1-0.33^2}(344.847 + 0.33\times 200)\times 10^{-6}$$

$$\sigma_x = 34.579 \text{ MPa}$$

Similarly, $$\sigma_y = \frac{E}{1-\nu^2}(\varepsilon_y + \nu\varepsilon_x) = 26.411 \text{ MPa}$$

For change in thickness, $$\varepsilon_z = \frac{\sigma_z}{E} - \nu\frac{\sigma_y}{E} - \nu\frac{\sigma_x}{E}$$

Since $\sigma_z = 0$, $$\varepsilon_z = \frac{-0.33(26.411 + 34.579)}{75\times 10^3}$$

$$\varepsilon_z = \frac{\Delta t}{t} = -2.684\times 10^{-4}$$

$$\Delta t = -10 \times 2.674 \times 10^{-4} \text{ mm}$$

Thus, reduction in thickness, $$\Delta t = 2.684 \times 10^{-3} \text{ mm}$$

3.23 In a strained body the normal stresses on three planes inclined as shown in figure are 60 MPa (Tensile), 80 MPa (Tension) and 100 MPa. Determine the shear stresses acting on these planes. Also find the principal stresses.

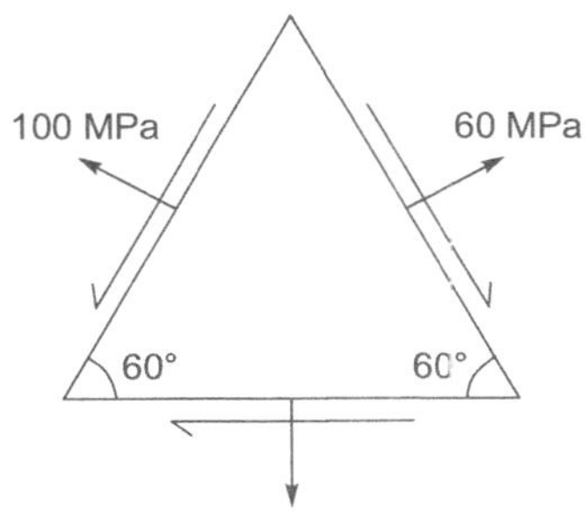

[20 marks : 2020]

Solution:

Let us assume the plane arrangements as shown in figure.

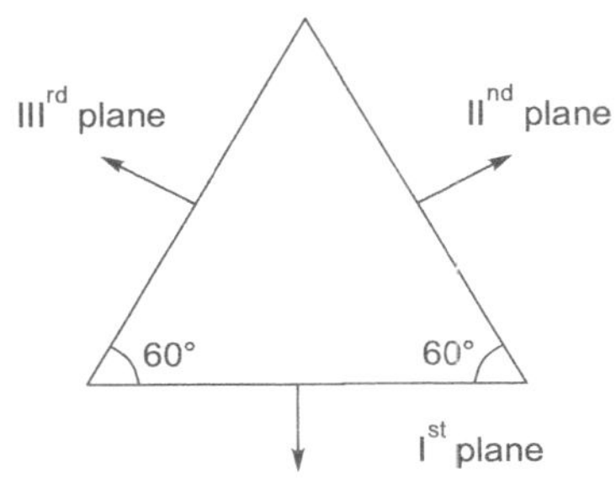

For IInd plane

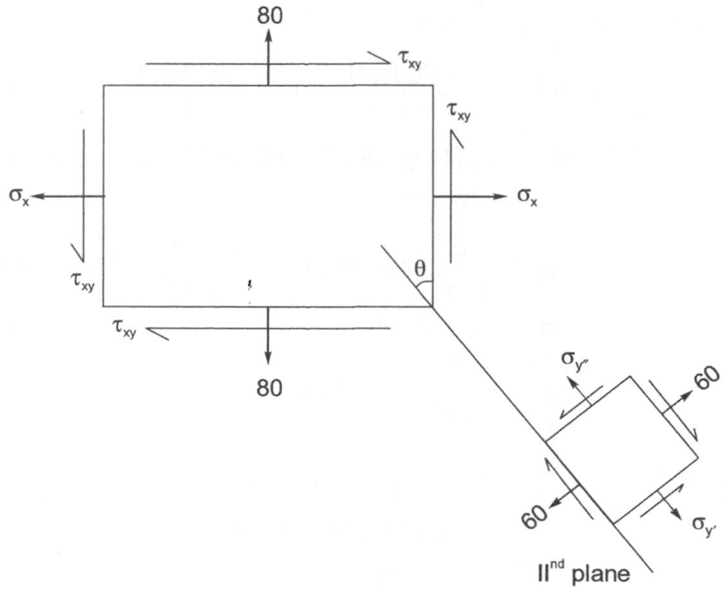

$$\sigma_y = 80 \text{ MPa}$$
$$\theta = 90° - 60° = 30°$$
$$\sigma_x' = 60 \text{ MPa}$$

As we know,
$$\sigma_x' = \sigma_x \cos^2\theta + \sigma_y \sin^2\theta + 2\tau_{xy} \sin\theta \cos\theta$$

So,
$$60 = \sigma_x \cos^2 30° + 80 \sin^2 30° + 2\tau_{xy} \cos 30° \sin 30°$$

$$60 = \frac{3}{4}\sigma_x + \frac{80}{4} + \frac{2\sqrt{3}\,\tau_{xy}}{4}$$

$$160 = 3\sigma_x + 2\sqrt{3}\,\tau_{xy} \quad \text{...(i)}$$

For IIIrd plane

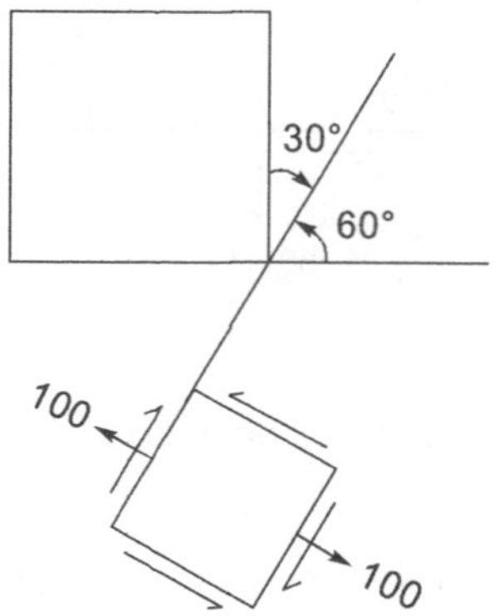

$$\sigma_x'' = 100 \text{ MPa}$$
$$\theta'' = -30° \quad \text{[Since clockwise]}$$
$$\sigma_x'' = \sigma_x \cos^2\theta'' + \sigma_y \sin^2\theta'' + 2\tau_{xy} \sin\theta'' \cos\theta''$$
$$100 = \frac{3\sigma_x}{4} - \frac{2\sqrt{3}}{4}\tau_{xy} + 20$$

$$320 = 3\sigma_x - 2\sqrt{3}\,\tau_{xy} \quad \text{...(ii)}$$

From eq. (i) and (ii)

$$\sigma_x = 80 \text{ MPa}$$
$$\tau_{xy} = -23.09 \text{ MPa}$$

Similarly for plane II,

$$\tau_{x'y'} = (\sigma_y - \sigma_x)\cos\theta\sin\theta + \tau_{xy}(\cos^2\theta - \sin^2\theta)$$
$$= 0 - 23.09 \times (\cos^2 30° - \sin^2 30°)$$
$$= -11.55 \text{ MPa}$$

Similarly for plane III,

$$\tau_{x''y''} = -11.55 \text{ MPa}$$

So final arrangement is

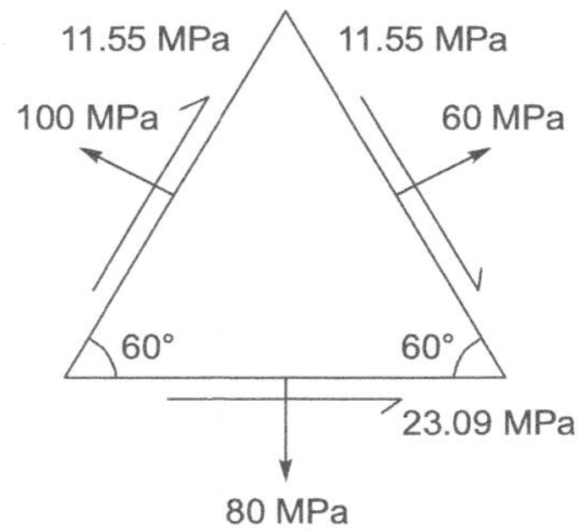

Since,

$$\sigma_x = 80 \text{ MPa}$$
$$\sigma_y = 80 \text{ MPa}$$
$$\tau_{xy} = -23.09 \text{ MPa}$$

$$\sigma_{major}/\sigma_{minor} = \frac{\sigma_x + \sigma_y}{2} \pm \sqrt{\left(\frac{\sigma_x - \sigma_y}{2}\right)^2 + \tau_{xy}^2}$$

$$= \frac{80+80}{2} \pm \sqrt{\left(\frac{80-80}{2}\right) + (-23.09)^2}$$

$$= 80 \pm 23.09$$

$$\sigma_{major} = 80 + 23.09 = 103.09 \text{ MPa}$$
$$\sigma_{minor} = 80 - 23.09 = 56.91 \text{ MPa}$$

3.24 A bar of length 1.2 m, diameter 40 mm is subjected to an axial tensile load of 130 kN and a twisting moment of 600 N-m. If the same material yielded at an axial stress of 200 N/mm^2, determine the safety factor associated with the bar, considering.

(i) Principal stress failure theory
(ii) Maximum shear stress theory
(iii) Distortional strain energy theory

Take E = 200 GPa and μ = 0.25.

[20 marks : 2020]

Solution:

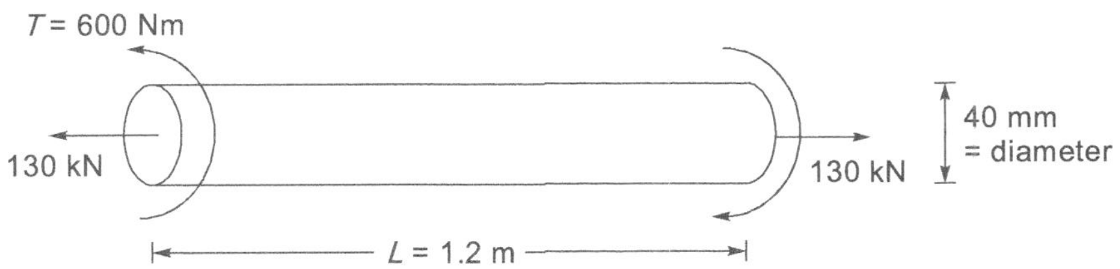

σ_y = 200 MPa, E = 200 GPa, μ = 0.25, $E = 2G(1 + \mu)$

$$G = \frac{200}{2(1.25)} = 80 \text{ GPa}$$

$$\sigma = \frac{P}{A} = \frac{130 \times 10^3}{\frac{\pi}{4} \times 40^2} = 103.451 \text{ MPa}$$

By torsion formula,

$$\frac{T}{J} = \frac{\tau}{r} = \frac{G\theta}{l}, \qquad \tau = \frac{Tr}{J} = \frac{600 \times 10^3 \times \left(\frac{40}{2}\right)}{\frac{\pi}{32}(40)^4}$$

$$= \frac{16 \times 600 \times 10^3}{\pi \times (40)^3} = 47.746 \text{ MPa}$$

Principal stresses,

$$\sigma_1/\sigma_2 = \frac{\sigma_x + \sigma_y}{2} \pm \sqrt{\left(\frac{\sigma_x - \sigma_y}{2}\right)^2 + \tau_{xy}^2}$$

$$= \frac{103.451}{2} \pm \sqrt{\left(\frac{103.451}{2}\right)^2 + (47.746)^2}$$

$$\sigma_1 = 122.12 \text{ MPa (Tensile)}$$

$$\sigma_2 = -18.667 \text{ MPa (Compressive)}$$

Calculating FOS considering,

(i) Principal stress theory,

$$\sigma_1 \le \sigma_y \text{ (Tensile)}$$

$$122.12 \times \text{FOS} = 200$$

$$\text{FOS} = 1.637$$

(ii) Maximum shear stress theory

$$\tau_{max} \le \left(\frac{\sigma_y}{2}\right)$$

$$\tau_{max} = \frac{\sigma_{major} - \sigma_{minor}}{2}$$

$$= 70.393 \text{ MPa}$$

$$70.393 \times \text{FOS} = 100$$

$$\text{FOS} = 1.420$$

(iii) Distortional strain energy theory,

$$\frac{1}{12G}\left[(\sigma_1 - \sigma_2)^2 + (\sigma_2 - \sigma_3)^2 + (\sigma_3 - \sigma_1)^2\right] \le \frac{1}{6G}\left(\frac{\sigma_y}{\text{FOS}}\right)^2$$

$$\frac{1}{2}\left[(122.12)^2 + (18.667)^2 + (122.12 + 18.667)^2\right] \le \left(\frac{\sigma_y}{\text{FOS}}\right)^2$$

$$17541.365 = \left(\frac{\sigma_y}{\text{FOS}}\right)^2$$

$$\text{FOS} = 1.51$$

3.25 A point on a thin plate is subjected to the two successive states of stress as shown in the figure below. Using Mohr's circle method, determine

(i) The resultant state of stress represented on the element oriented as shown on the right.

(ii) The principal stresses and principal planes caused by the superposition of these two stress states.

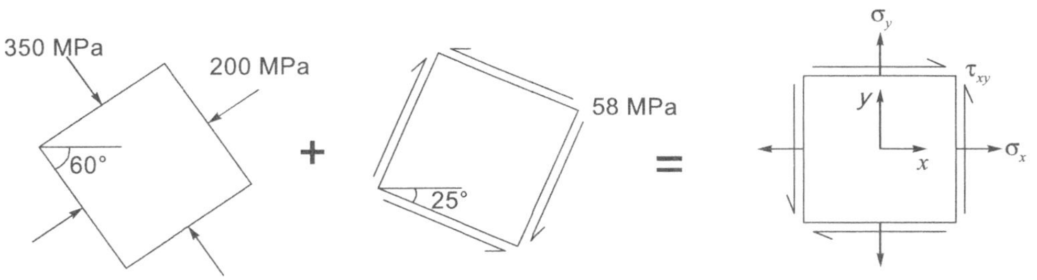

[20 Marks : 2021]

Solution:

Case-1

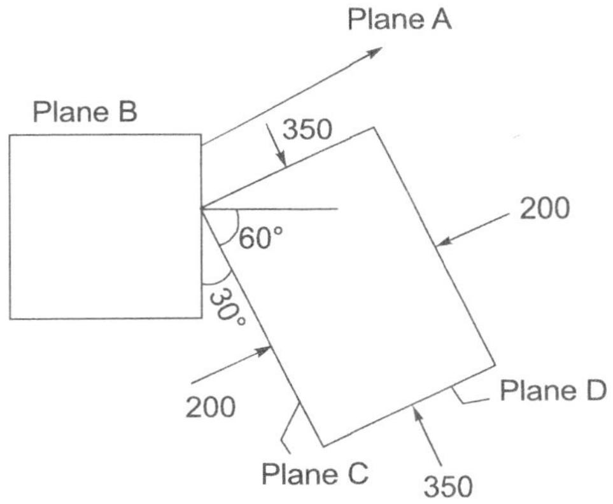

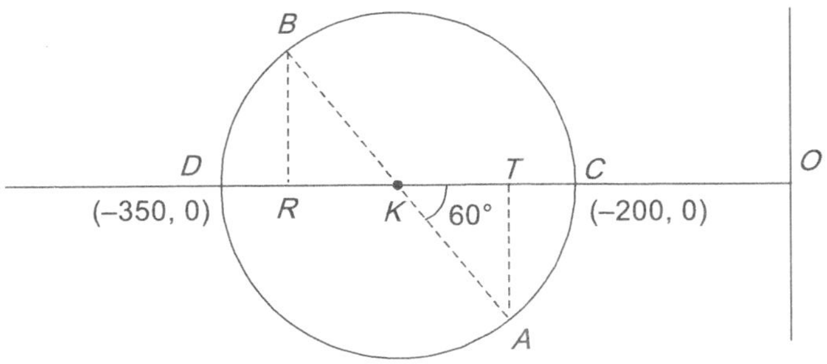

$$KC = \frac{350 - 200}{2} = 75$$

$$\sigma_{x_1} = -275 + 75 \cos 60° = -237.5 \text{ MPa}$$

$$\sigma_{x_2} = -275 - 75 \cos 60° = -312.5 \text{ MPa}$$

$$\tau_{x_1y_1} = 75 \sin 60° = 64.95 \simeq 65 \text{ MPa}$$

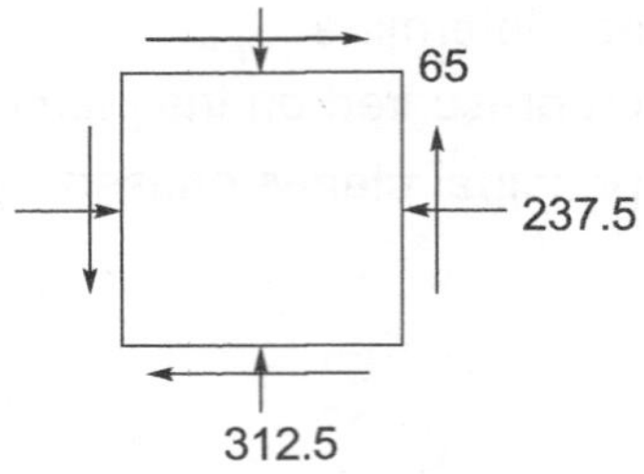

Case-2

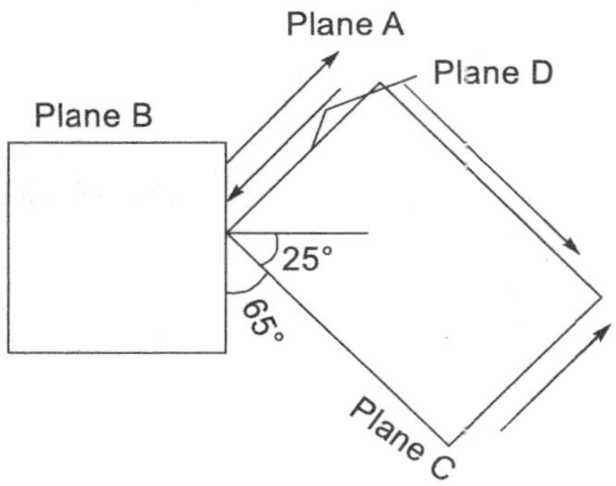

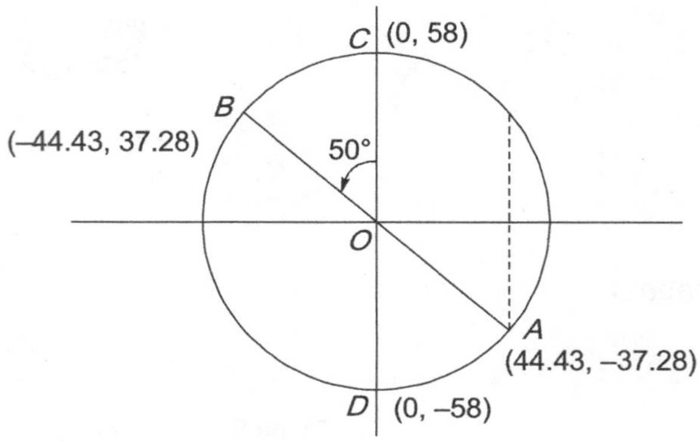

$$\sigma_{x_2} = 58 \sin 50° = 44.43 \text{ MPa}$$
$$\sigma_{y_2} = -44.43 \text{ MPa}$$
$$\tau_{x_2y_2} = 58 \cos 50° = 37.28 \text{ MPa}$$

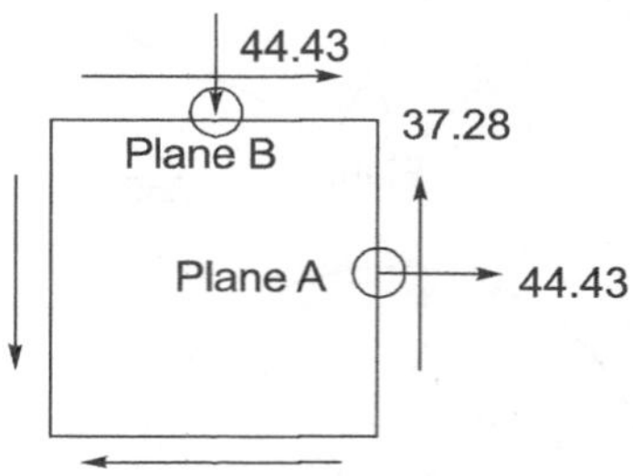

Total stress

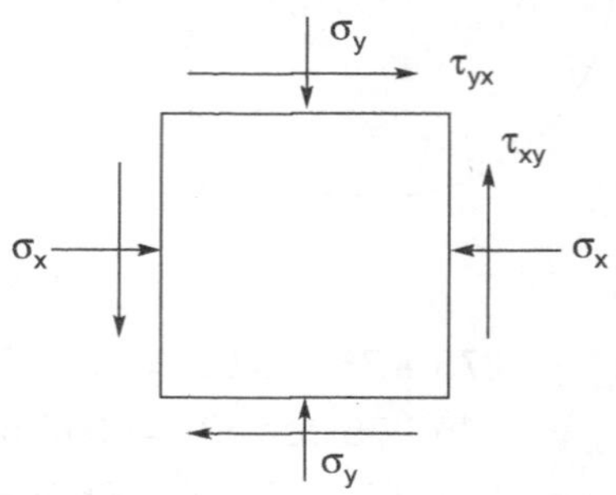

$$\sigma_x = (-237.5) + (+44.43) = -193.07 \text{ MPa}$$
$$\sigma_y = (-312.5) + (-44.43) = -356.93 \text{ MPa}$$
$$\tau_{xy} = 65 + 37.28 = 102.28 \text{ MPa}$$

Principal stress

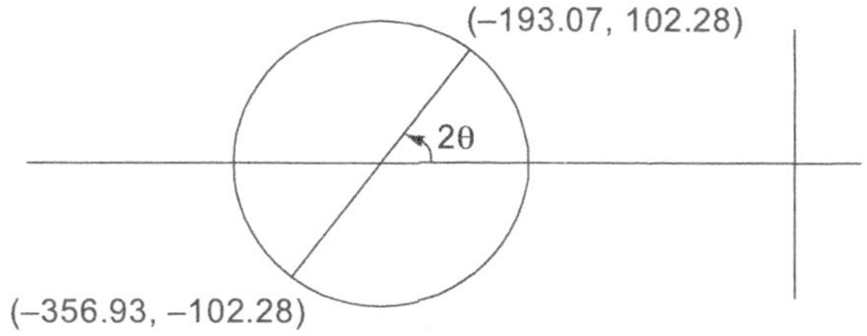

$$\text{Centre} = \left(\frac{\sigma_x + \sigma_y}{2}, 0\right) = \left(\frac{-193.07 - 356.93}{2}, 0\right) = (-275, 0)$$

$$R = \sqrt{\left(\frac{-356.93 + 193.07}{2}\right)^2 + 102.28^2} = 131.05$$

$$\sigma_1 = -275 - R = -275 - 131.05 = -406.5$$

$$\sigma_2 = -275 + R = 275 + 131.05 = -143.95$$

$$\sin 2\theta = \frac{\tau}{R} = \frac{102.28}{131.05} = 0.78$$

$$2\theta = 51.26$$

$$\theta = 25.63 \text{ (Anticlockwise from } x\text{-axis)}$$

IInd Method

By Graphical Mohr's circle method:

Scale: 1 cm = 25 MPa

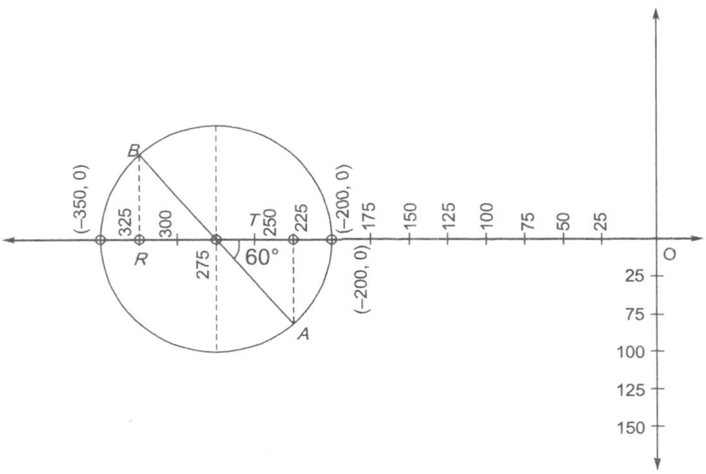

From scale

Distance,	OR = 12.5 cm
So stress,	OR = 12.5 × 25 = 312.5 MPa (C)
Distance,	OT = 9.5 cm
So, stress,	OT = 9.5 × 25 = 237.5 MPa (C)
Distance,	TA = 2.6 cm
Stress,	TA = 2.6 × 25 = 65 MPa

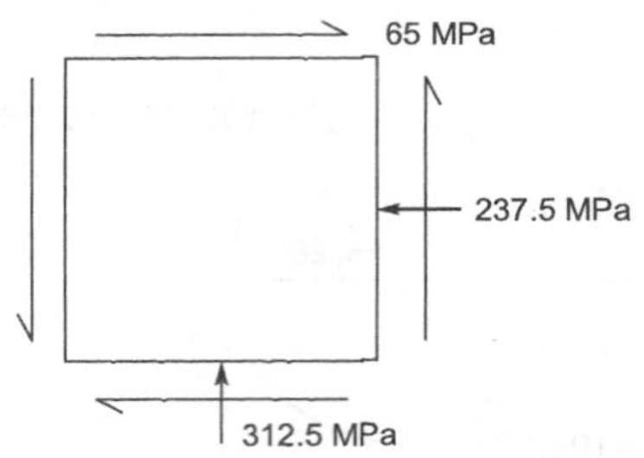

Case-2

Scale, 1 cm = 10 MPa

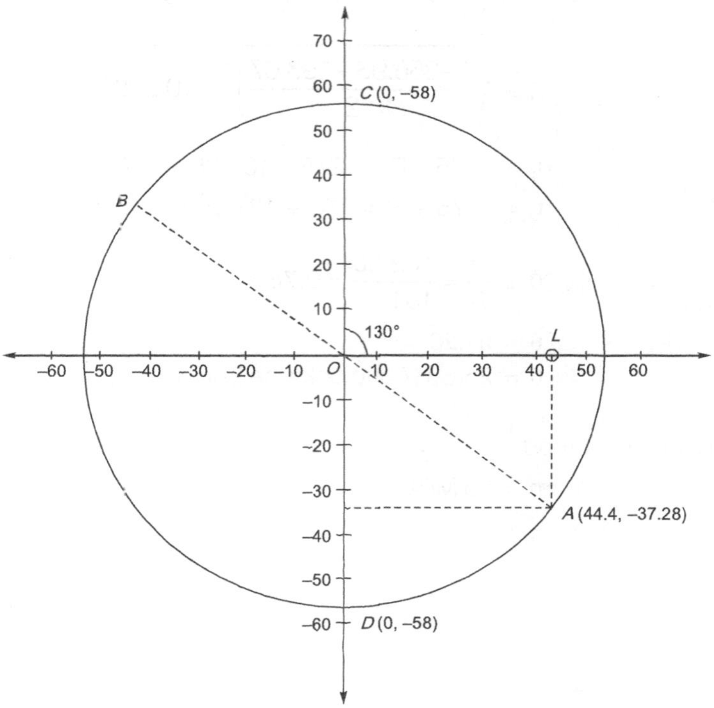

Distance, OL = 4.44 cm

Stress, OL = 4.44 × 10 = 44.4 MPa

Distance, LA = 3.728 cm

Stress, LA = 37.28 MPa

Total stress, 1 cm = 25 cm

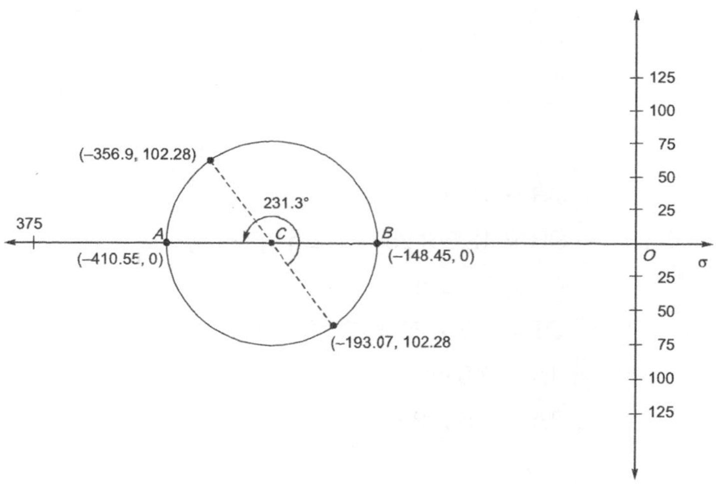

$$\sigma_x = -237.5 + (44.4) = -193.1 \text{ MPa} = 7.724 \text{ cm}$$
$$\sigma_y = -312.5 - 44.4 = -356.9 \text{ MPa} = 14.276 \text{ cm}$$
$$\tau_{xy} = 65 + 37.28 = 102.28 \text{ MPa} = 4.09 \text{ cm}$$

From graph:

Major principal stress, $\sigma_1 = OA = 16.242 \text{ cm} \times 25 = -406.05 \text{ MPa}$

Minor principal stress, $\sigma_2 = OB = 5.758 \text{ cm} \times 25 = -143.95 \text{ MPa}$

Principal angles: From Graph

$$2\theta_{p2} = 51.31^\circ$$
$$\Rightarrow \quad \theta_{p2} = 25.65^\circ$$
$$2\theta_{p1} = 231.31^\circ$$
$$\Rightarrow \quad \theta_{p1} = 115.65^\circ$$

3.26 Determine the maximum principal stresses and maximum shearing stresses at a point A on the surface of a 25 mm diameter shaft. The shaft is subjected to a torque of 175.93 Nm and axial tension of 31.415 kN. Show the orientation of principal planes and planes of maximum shearing stress with respect to an axis parallel to the axis of the shaft.

[12 marks : 2022]

Solution:

Consider the figure shown below.

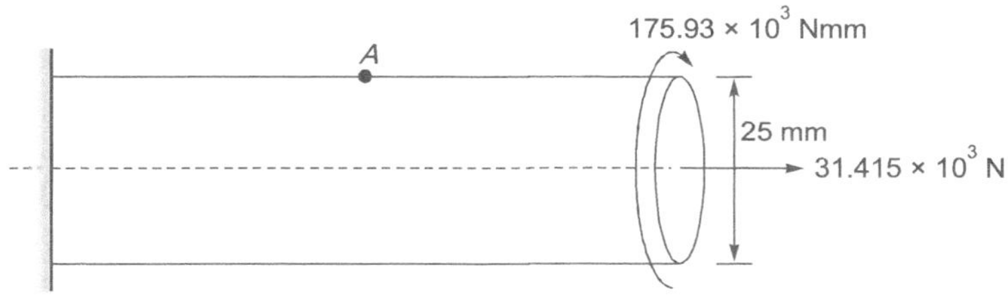

Stress element for point *A* on surface of shaft:

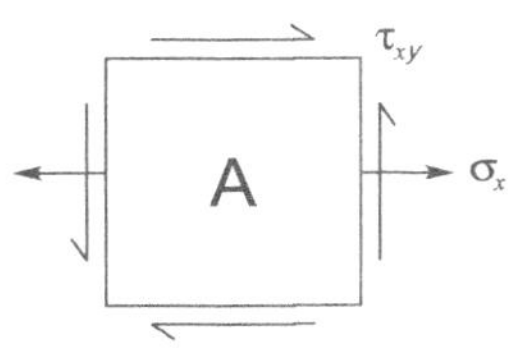

$$\sigma_x = \frac{\text{Axial force}}{\text{Cross-sectional area}} = \frac{31.415 \times 10^3}{\frac{\pi}{4} \times 25^2} = 64 \text{N/mm}^2$$

$$\tau_{xy} = \frac{16T}{\pi d^3} = \frac{16 \times 175.93 \times 10^3}{\pi \times 25^3} = 57.34 \text{ N/mm}^2$$

Principal stresses:

$$\sigma_{p1}/\sigma_{p2} = \left(\frac{\sigma_x + \sigma_y}{2}\right) \pm \sqrt{\left(\frac{\sigma_x - \sigma_y}{2}\right)^2 + \tau_{xy}^2}$$

$\sigma_y = 0$

So,
$$\sigma_{p1}/\sigma_{p2} = \frac{64}{2} \pm \sqrt{\left(\frac{64}{2}\right)^2 + (57.34)^2}$$

$\therefore$
$$\sigma_{P1} = 97.66 \text{ N/mm}$$

$$\sigma_{P2} = -33.665 \text{ N/mm}^2 \simeq -33.67 \text{ N/mm}^2$$

Maximum shear stress,

$$\tau_{xy} = \left|\frac{\sigma_{P1} - \sigma_{P2}}{2}\right|$$

$$= \frac{97.66 - (-33.67)}{2} = 65.665 \text{ N/mm}^2$$

Orientation of planes w.r.t. vertical plane,

$$\tan 2\theta_p = \frac{2\tau_{xy}}{\sigma_x - \sigma_y} = \frac{2 \times 57.34}{64 - 0}$$

$$\therefore \quad 2\theta_p = 60.84°$$

$$\theta_p = 30.42°$$

Check: σ_x' at $\theta = 30.42°$

$$\sigma_x' = \left(\frac{\sigma_x + \sigma_y}{2}\right) + \left(\frac{\sigma_x - \sigma_y}{2}\right)\cos 2\theta + \tau_{xy} \sin 2\theta$$

$$= \left(\frac{64}{2}\right) + \left(\frac{64}{2}\right)\cos 60.84° + 57.34 \sin 60.84°$$

$$= 97.66 \text{ N/mm}^2$$

Hence 97.56 N/mm² stress will be at 30.42° from vertical plane and will be at 59.58° w.r.t. an axis parallel to the axis of shaft.

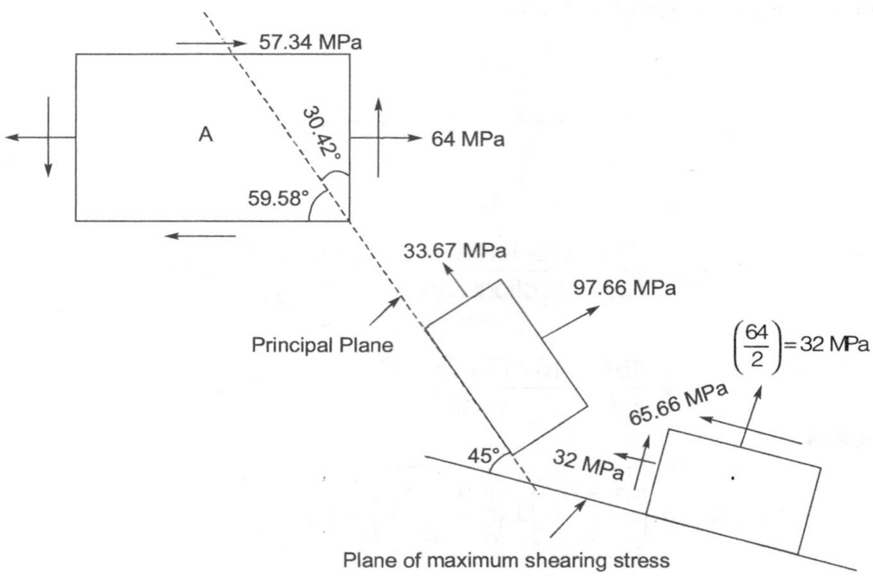

4. Deflection of Beams

4.1 In the structure shown in the figure below, calculate the distance by which the points A move away under the action of forces P.

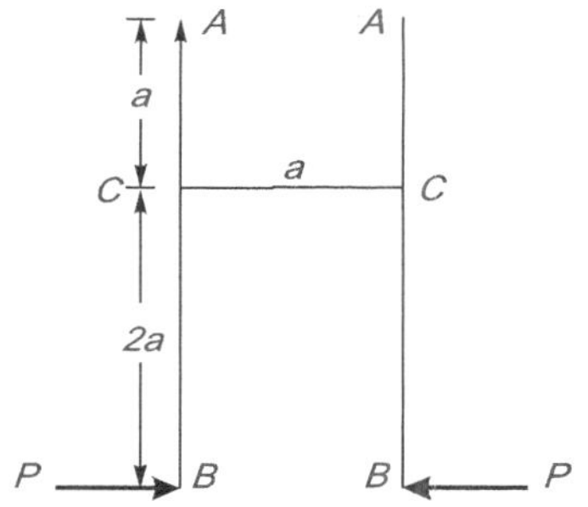

[10 marks : 2000]

Solution:

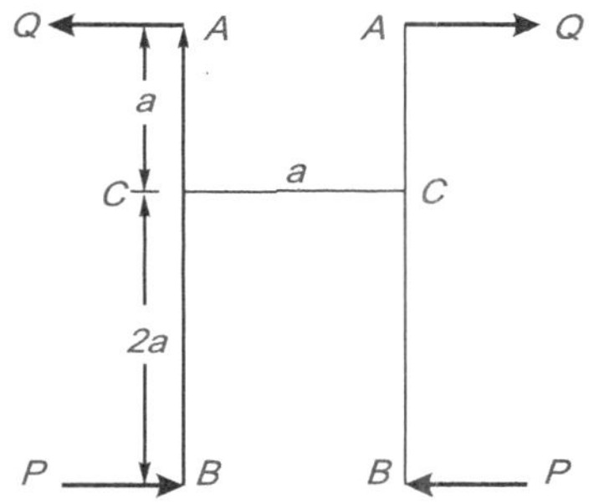

Using strain energy method obtain find distance between AA

Since no load is acting at AA. Hence apply a Pseudo load at ends A.

Taking outer face as reference.

The free body diagram,

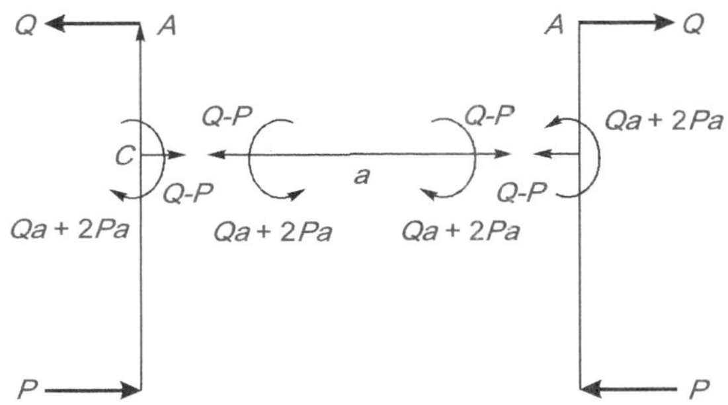

In the question, it is not mentioned that the beam is axially rigid, hence we cannot ignore the contribution of axial strain energy.

$$U = 2U_{AC} + 2U_{BC} + U_{CC}$$

$$U = 2\int_0^a \frac{(Qy)^2\,dy}{2EI} + 2\int_0^{2a} \frac{(-Py)^2\,dy}{2EI} + \int_0^a \left[\frac{(Qa+2Pa)^2\,dx}{2EI} + \frac{(Q-P)^2\,a}{2AE}\right]$$

$$U = \left[\frac{Q^2y^3}{6EI}\right]_0^a + 2\left[\frac{P^2y^3}{6EI}\right]_0^{2a} + \int_0^a \left[\frac{Q^2a^2 + P^2a^2 + 4PQa^2}{2EI}dx + \frac{(Q-P)^2\,a}{2AE}\right]$$

$$U = \frac{2\times Q^2a^3}{6EI} + 2\times\frac{4P^2y^3}{3EI} + \int_0^a\left[\frac{Q^2q^2+4P^2q^2+4PQa^2}{2EI}dx + \frac{(Q-P)^{2a}}{2AE}\right]$$

$$\Delta_{AA} = \left.\frac{\partial U}{\partial Q}\right|_{Q=0} = 0+0+\frac{4Pa^3}{2EI}-\frac{Pa}{AE}$$

$$\Delta_{AA} = \frac{2Pa^3}{2EI}-\frac{Pa}{AE}$$

4.2 The equation for the deflected shape of a beam carrying a UDL and simply supported at the ends is given below:

$$y = \frac{1}{EI}\left(-2x^3+\frac{x^4}{6}+36x\right)$$

Determine the load carrying capacity of the beam and draw the BMD and SFD. The unit of EI is kNm^2.

[14 marks : 2001]

Solution:

Given that,
$$y = \frac{1}{EI}\left(-2x^3+\frac{x^4}{6}+36x\right) \quad ...(i)$$

Differentiating (*i*), we get

$$\frac{dy}{dx} = \frac{1}{EI}\left(-6x^2+\frac{4x^3}{6}+36\right) \quad ...(ii)$$

Again differentiating (*ii*), we get

$$\frac{d^2y}{dx^2} = \frac{1}{EI}\left[-12x+\frac{4\times 3x^2}{6}\right] = \frac{1}{EI}(2x^2-12x)$$

Also,
$$EI\frac{d^2y}{dx^2} = 2x^2-12x \quad ...(iii)$$

we know that,
$$EI\frac{d^2y}{dx^2} = M$$

on comparing with (*iii*), we get

$$M = 2x^2-12x \quad ...(iv)$$

we know that,
$$\frac{dM}{dx} = 4x-12 = S_x$$

$\therefore$ $S_x = 4x-12$

we also know, at $x = \frac{L}{2}, S_x = 0$

$\therefore$ $S_x = 0$ or $4x-12 = 0$

$x = 3$ m

$\therefore$ $L = 2x = 6$ m

We also know that,
$$\frac{dS_x}{dx} = -\text{Loading rate } (w)$$

$\therefore$ $\frac{d}{dx}(4x-12) = w$

$\therefore$ $w = 4$ kN-m

Alternate **Solution:**

From equation (*ii*), $$\frac{dy}{dx} = \frac{1}{EI}\left[-6x^2 + \frac{4x^3}{6} + 36\right]$$

and from equation (*i*), $$y = \frac{1}{EI}\left[-2x^3 + \frac{x^4}{6} + 36x\right]$$

we know, for simply supported beam carrying UDL

$$\theta_A = \theta_B = \frac{wL^3}{24EI}$$

and $$\delta_C = \frac{5}{384} \times \frac{wL^4}{EI}$$

Hence, $$\frac{wL^3}{24EI} = \frac{1}{EI}[0 + 0 + 36] \qquad [\because x = 0]$$

$$wL^3 = 36 \times 24 = 864 \qquad \text{...(A)}$$

and $$\frac{5}{384} \times \frac{wL^4}{EI} = \frac{1}{EI}\left[-2\frac{L^3}{8} + \frac{L^4}{6 \times 16} + \frac{36 \times L}{2}\right] \qquad \left[\because x = \frac{L}{2}\right]$$

$$5\,wL^4 = \frac{384[-24L^3 + L^4 + 1728L]}{96}$$

$$5\,wL^4 = 4[-24\,L^3 + L^4 + 1728L]$$

$$5\,wL^3 = -96\,L^2 + 4\,L^3 + 6912 \qquad \text{...(B)}$$

From equation (*A*), $$wL^3 = 864$$

$\therefore$ $$5 \times 864 = -96\,L^2 + 4\,L^3 + 6912$$

$$4320 = 4\,L^3 - 96\,L^2 + 6912$$

$$4\,L^3 - 96\,L^2 + 2592 = 0$$

$\therefore$ $$L = 6\text{ m}$$

From equation (*A*), $$w \times (6)^3 = 864$$

$\therefore$ $$w = 4\text{ kN-m}$$

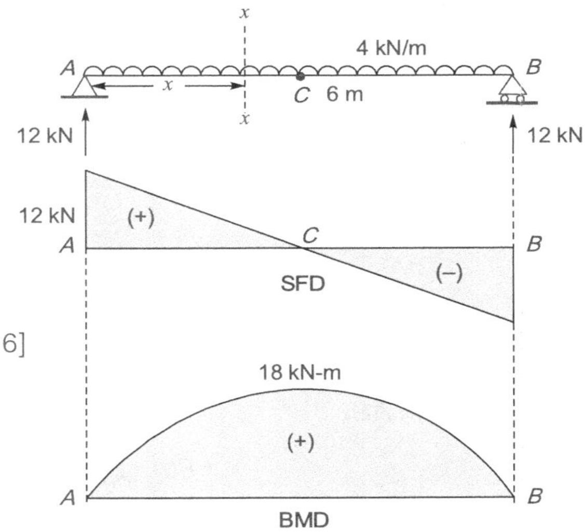

SFD and BMD:

Reactions:

By symmetry, $$R_A = R_B = \frac{wl}{2}$$

$\therefore$ $$R_A = R_B = \frac{4 \times 6}{2} = 12\text{ kN}$$

SFD: $$S_x\,(x\text{ From }A) = +R_A - wx \quad [0 \le x \le 6]$$

$$S_x = +12 - 4x \quad \text{(Linear)}$$

at $x = 0$, $S_A = +12$ kN

at $x = 3$ m $S_C = 0$

at $x = 6$ m $S_B = -12$ kN

BMD: $$M_x\,(x\text{ From }A) = 12x - \frac{4x^2}{2} \qquad [0 \le x \le 6]$$

$$M_x = 12x - 2x^2 \qquad \text{(Parabolic)}$$

at $x = 0$, $M_A = 0$

at $x = 6$ m $M_B = 0$

For maximum moment, $$\frac{dM_x}{dx} = 0$$

$$12 - 4x = 0$$

$$x = 3 \text{ m}$$

$$M_{max} \text{ (at } x = 3 \text{ m)} = 12 \times 3 - 2 \times 3^2 = 18 \text{ kN-m}$$

4.3 Find the horizontal displacement of the end *D* of the bent rod of circular cross-section of 20 mm diameter. A horizontal force of 1 kN acts at *D* as shown in figure. Consider only bending energy $E = 2 \times 10^5$ MPa.

[10 Marks : 2004]

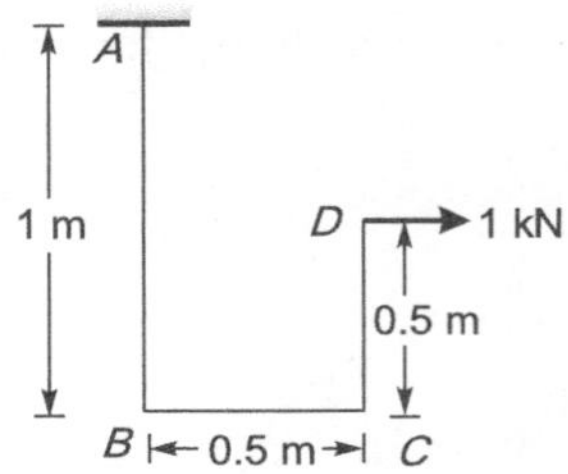

Solution:

Given: $P = 1$ kN, $L = 1$ m, $E = 2 \times 10^5$ MPa, $d = 20$ mm

Total strain energy stored in bent rod will be given by

$$U = U_{AB} + U_{BC} + U_{CD}$$

Taking outer face as reference

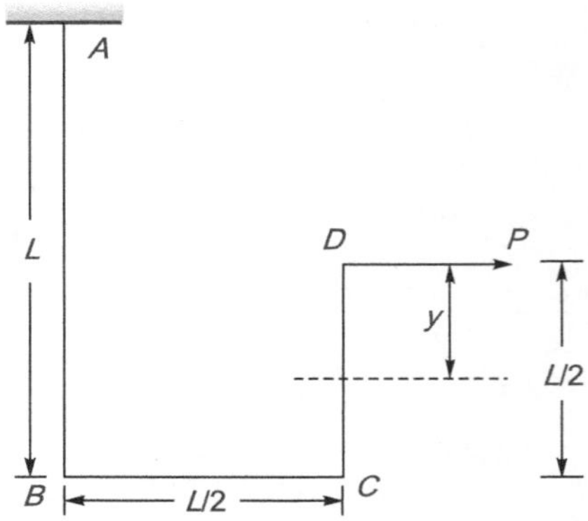

Portion *CD*:

$$U_{CD} = \int_0^{L/2} \frac{(Py)^2 dy}{2EI}$$

$$= \int_0^{L/2} \frac{P^2 y^2 dy}{2EI} = \frac{P^2}{2EI}\left[\frac{y^3}{3}\right]_0^{L/2} = \frac{P^2 L^3}{6EI \times 8}$$

$$\therefore \quad U_{CD} = \frac{P^2 L^3}{48EI}$$

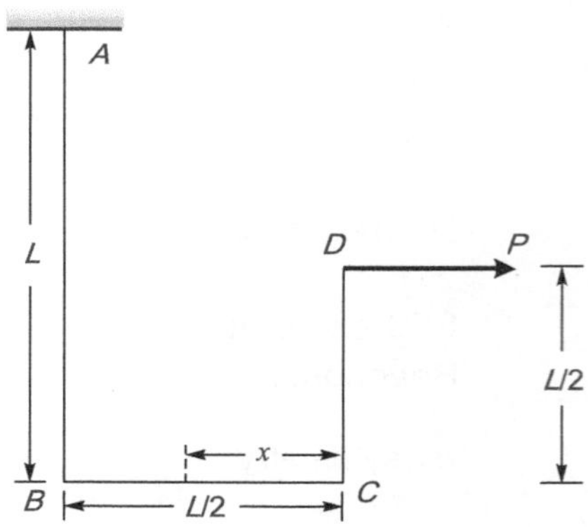

Portion *BC*:

$$M_x = +\frac{P \times L}{2}$$

$$\therefore \quad U_{BC} = \int_0^{L/2} \frac{\left(\frac{PL}{2}\right)^2 dx}{2EI} = \frac{P^2 L^2}{8EI}\int_0^{L/2} dx$$

$$= \frac{P^2 L^2}{8EI}\left[\frac{L}{2}\right]$$

$$\therefore \quad U_{BC} = \frac{P^2 L^2}{16EI}$$

Portion *AB*:

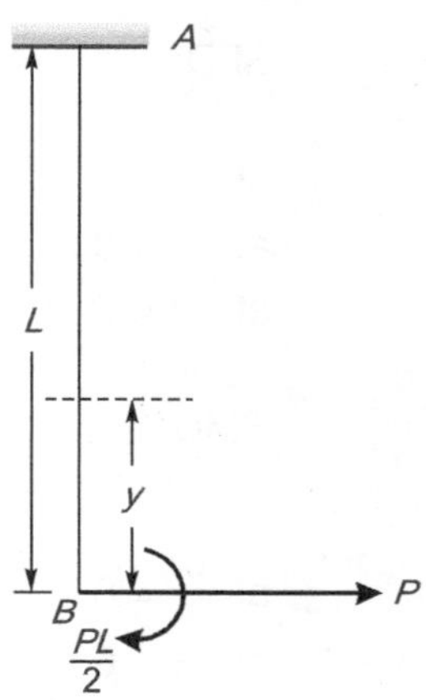

$$U_{AB} = \int_0^{L} \frac{\left(\frac{PL}{2} - Py\right)^2 dy}{2EI}$$

$$U_{AB} = \int_0^{L} \frac{\left(\frac{P^2 L^2}{4} + P^2 y^2 - P^2 Ly\right) dy}{2EI}$$

$$= \int_0^L \frac{(P^2L^2 + 4P^2y^2 - 4P^2Ly)dy}{8EI}$$

$$= \frac{1}{8EI}\left[P^2L^2(y) + 4P^2\left(\frac{y^3}{3}\right) - 4P^2L\left(\frac{y^2}{2}\right)\right]_0^L$$

$$= \frac{1}{8EI}\left[P^2L^3 + \frac{4P^2L^3}{3} - \frac{4P^2L^3}{2}\right] = \frac{1}{8EI}\left[\frac{6P^2L^3 + 8P^2L^3 - 12P^2L^3}{6}\right]$$

$\therefore$ $$U_{AB} = \frac{P^2L^3}{24EI}$$

Hence, $$U = U_{AB} + U_{BC} + U_{CD} = \frac{P^2L^3}{48EI} + \frac{P^2L^3}{16EI} + \frac{P^2L^3}{24EI}$$

$\therefore$ $$U = \frac{P^2L^3}{8EI}$$

For horizontal deflection, by Castigliano's theorem

$$\delta_H = \frac{\partial U}{\partial P} = \frac{\partial}{\partial P}\left[\frac{P^2L^3}{8EI}\right] = \frac{PL^3}{4EI}$$

Hence, $$\delta_H = \frac{1000 \times (1000)^3}{4 \times 2 \times 10^5 \times 7853.98} \qquad \left[\because \frac{\pi d^4}{64} = 7853.98 \text{ mm}^4\right]$$

$$\delta = 159.15 \text{ mm}$$

4.4 The free end B of a cantilever AB '4a' long is supported from below on the free end C of the cantilever CD '2a' long through a roller as shown in figure. The two beams have uniform and equal flexural rigidities and carry a UDL of intensity w/unit length. Draw the BMD for AB and CD.

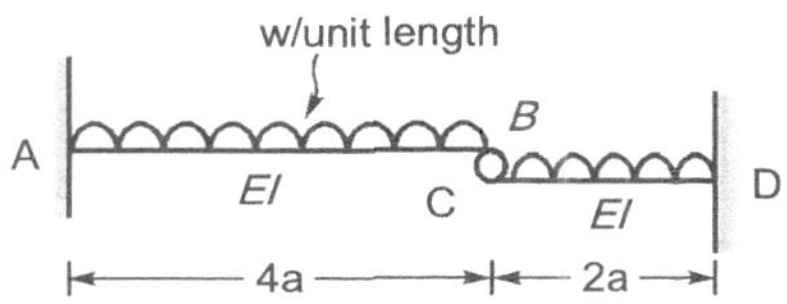

[10 marks : 2004]

Solution:

Let the reaction at the roller be R.

Now the vertical deflection at B and C will be equal

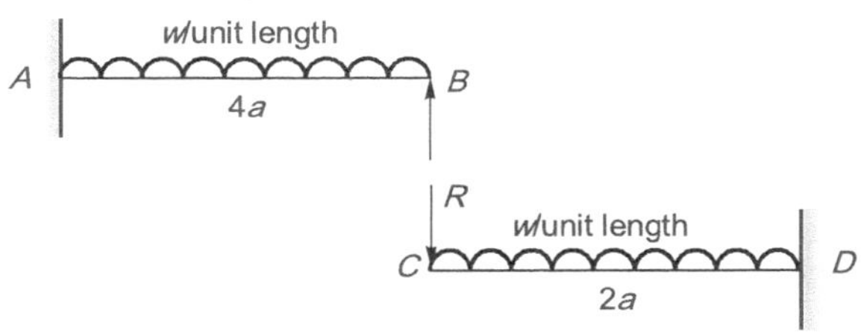

$$\Delta_B = \Delta_C$$

$$\frac{w(4a)^4}{8EI} - \frac{R(4a)^3}{3EI} = \frac{w(2a)^4}{8EI} + \frac{R(2a)^3}{3EI}$$

$$\frac{w}{8EI}\left[256a^4 - 16a^4\right] = \frac{R}{3EI}(8a^3 + 64a^3)$$

$$\frac{w}{8} \times 240a^4 = \frac{R}{3} \times 72a^3$$

$$R = \frac{3}{72} \times \frac{240}{8} wa$$

$\therefore$ $$R = \frac{5}{4} wa$$

Bending Moment Diagram (BMD)

Portion AB:

$$M_x(x \text{ from } B) = \frac{5}{4}wax - \frac{wx^2}{2} \qquad [0 \le x \le 4a]$$

$$M_x = \frac{5}{4}wax - \frac{w}{2}x^2 \quad \text{(Parabolic)}$$

at $x = 0$, $M_B = 0$

at $x = 4a$, $M_A = \frac{5}{4}wa(4a) - \frac{w(4a)^2}{2} = -3wa^2$

B.M. will be zero, if $\frac{5}{4}wax - \frac{w}{2}x^2 = 0$

$$\frac{5}{4}wax - \frac{wx^2}{2} = 0$$

$$x\left[\frac{5}{4}wa - \frac{wx}{2}\right] = 0$$

$\therefore$ $x = 0$ (Already known)

and $\frac{5}{4}wa - \frac{wx}{2} = 0$

$$x = 2.5\,a$$

Hence bending moment will be zero at a distance of $x = 0$ and $x = 2.5a$ where x is distance from point B in AB portion.

$$M_x = \frac{5}{4}wax - \frac{w}{2}x^2$$

$$\frac{\partial M_x}{\partial x} = \frac{5}{4}wa - wx = 0$$

$$x = \frac{5}{4}a$$

$$M_x\big|_{max} = \frac{5}{4}wa\left(\frac{5}{4}a\right) - \frac{w}{2}\left(\frac{5}{4}a\right)^2 = \frac{25}{32}wa^2$$

Hence maximumpositive bending moment is $\frac{25}{32}wa^2$, which is at a distance of $x = \frac{5}{4}wa$ from point B.

Portion CD:

$$M_x(x \text{ from } C) = -\frac{5}{4}wax - \frac{wx^2}{2} \qquad [0 \le x \le 2a]$$

$$M_x = -\frac{5}{4}wax - \frac{wx^2}{2} \quad \text{(Parabolic)}$$

at $x = 0$, $M_B = 0$

at $x = 2a$, $M_D = -\frac{5}{4}wa(2a) - \frac{w(2a)^2}{2}$

$$= -4.5\,wa^2$$

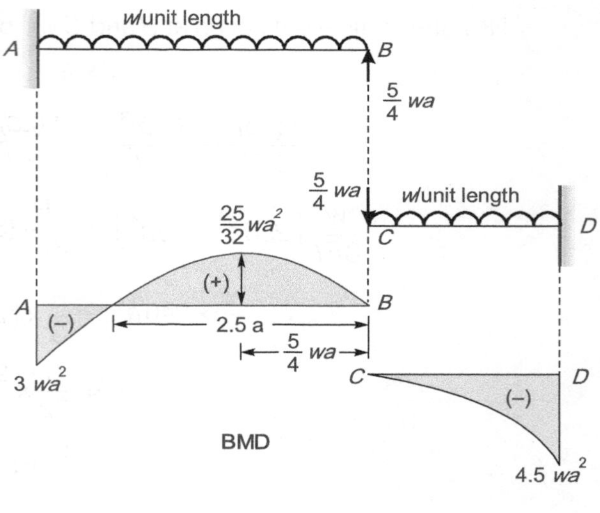

BMD

4.5 A stepped beam AB, as shown in the figure changes cross-section at C such that EI of AC is half that of CB. End A is simply supported and B is fixed. If on application of a moment 'M' at end A, Moment induced at B is $C_{BA} \times M$, determine the value of C_{BA}.

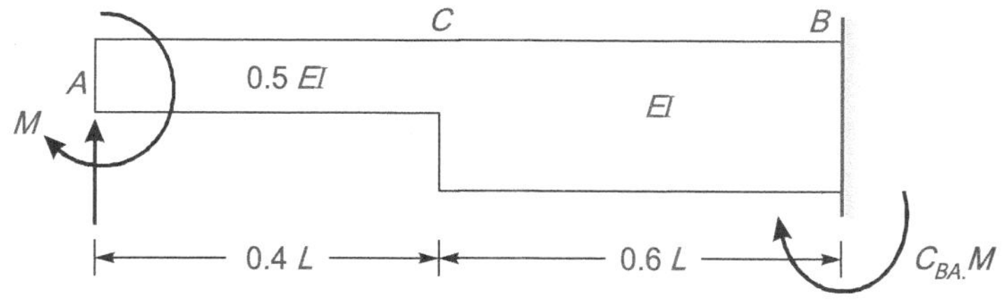

[20 marks : 2007]

Solution:

The given beam can be analysed by moment area method. Let the reaction induced at the simply supported end A be R.

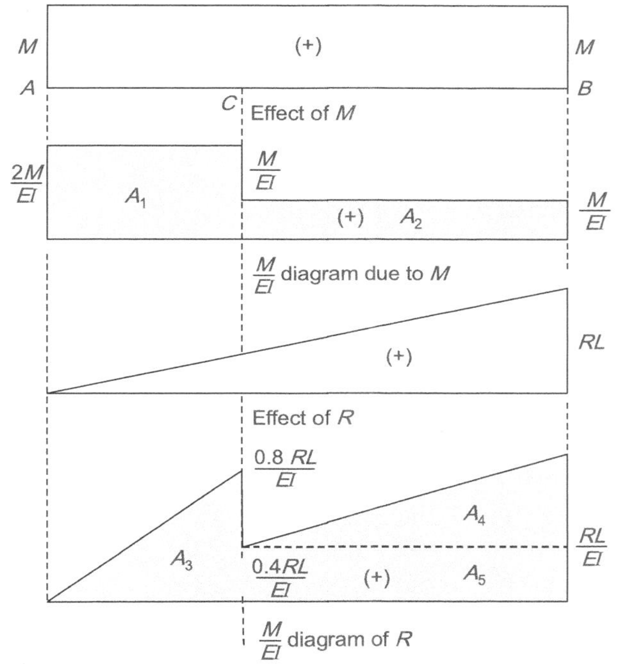

S.No.	Area	Distance of centroid from A
1.	$A_1 = 0.4L \times \dfrac{2M}{EI} = \dfrac{0.8ML}{EI}$	$\overline{X}_1 = 0.2L$
2.	$A_2 = 0.6L \times \dfrac{M}{EI} = \dfrac{0.6ML}{EI}$	$\overline{X}_2 = 0.4L + 0.3L = 0.7L$
3.	$A_3 = \dfrac{1}{2} \times 0.4L \times \dfrac{0.8RL}{EI} = 0.16\dfrac{RL^2}{EI}$	$\overline{X}_3 = 0.4L \times \dfrac{2}{3} = \dfrac{0.8L}{3}$
4.	$A_4 = \dfrac{1}{2} \times 0.6L \times \dfrac{0.6RL}{EI} = \dfrac{0.18RL^2}{EI}$	$\overline{X}_4 = 0.4L + \dfrac{2}{3} \times 0.6L = 0.8L$
5.	$A_5 = \dfrac{0.4RL}{EI} \times 0.6L = \dfrac{0.24RL^2}{EI}$	$\overline{X}_5 = 0.7L$

Deflection due to M at A = Moment of area of M/EI diagram between A and B about A

$$\delta_M = A_1\bar{X}_1 + A_2\bar{X}_2$$

$$= \frac{0.8ML}{EI}\times 0.2L + \frac{0.6ML}{EI}\times 0.7L = \frac{0.16ML^2}{EI} + \frac{0.42ML^2}{EI}$$

$$\delta_M = \frac{0.58ML^2}{EI}$$

Also, Deflection due to R at A = Moment of area of M/EI diagram between A and B about A

$$\delta_R = A_3\bar{X}_3 + A_4\bar{X}_4 + A_5\bar{X}_5$$

$$= \frac{0.16RL^2}{EI}\times\frac{0.8L}{3} + \frac{0.18RL^2}{EI}\times 0.8L + \frac{0.24RL^2}{EI}\times 0.7L = \frac{133}{375}\frac{RL^3}{EI}$$

At simply supported end B, net deflection due to M and R will be zero

$$\therefore \quad \delta_M + \delta_R = 0$$

$$\frac{0.58ML^2}{EI} + \frac{133RL^3}{375EI} = 0$$

$$\frac{133RL^3}{375EI} = -\frac{0.58ML^2}{EI}$$

$$R = -\frac{435M}{266L}$$

$\Sigma M_B = 0;$ $\quad M + RL + M_B = 0$

Hence, bending moment at B will be

$-M_B = M + RL;$ $\quad -M_B = M + \left(-\frac{435M}{266L}\right)L = M - \frac{435M}{266} = -\frac{169}{266}M$

$$M_B = \frac{169}{266}M \quad \text{(Sagging)}$$

But, $\quad M_B = C_{BA}M$

On comparing, we get $\quad C_{BA} = \frac{169}{266}$

4.6 **A simple frame *ABC* has circular cross-section with cross-sectional area '*a*' and moment of inertia *I*. The frame is supported and loaded as shown in the figure. Obtain the deflection under the load *P*, considering the effects of both the bending and axial forces. Young's modulus is *E* for the material.**

[15 marks : 2008]

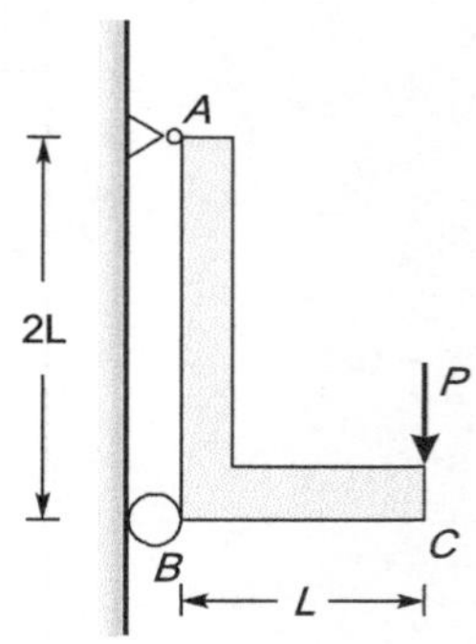

Solution:

Reactions:

Let H_A and H_B be the horizontal reactions at support A and B respectively. Also let R_A is vertical reaction at

(*i*) $\quad \Sigma F_y = 0$

$$R_A - P = 0$$

$$\therefore \quad R_A = P$$

(*ii*) $\Sigma F_x = 0$

$\Rightarrow$ $-H_A + H_B = 0$

$\Rightarrow$ $H_A = H_B$

(*iii*) Also, $\Sigma M_A = 0$

$\therefore$ $H_B \times 2L - P \times L = 0$

$\therefore$ $H_B = \dfrac{P}{2}$

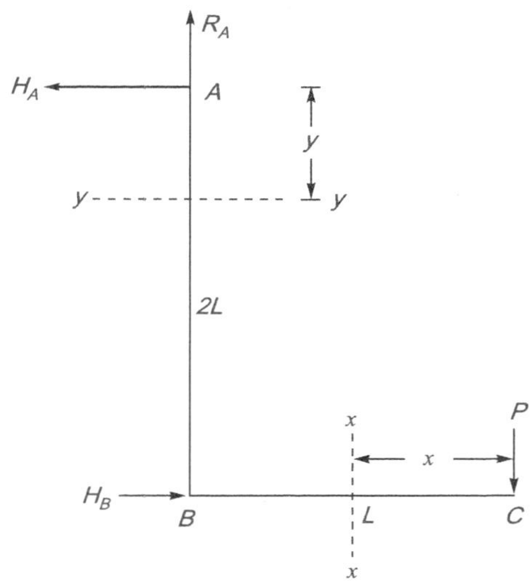

Strain energy stored:

Total strain energy stored,

$$U = U_{BC} + U_{AB}$$

Taking outer face as reference face.

Portion BC:

Consider a section *x-x* at a distance *x* from *C*.

$$U_{BC} = \int_0^L \frac{M_x^2\,dx}{2EI}$$

$$U_{BC} = \int_0^L \frac{(-Px)^2\,dx}{2EI}$$

$$U_{BC} = \int_0^L \frac{P^2x^2\,dx}{2EI}$$

$$U_{BC} = \frac{P^2}{2EI}\left[\frac{x^3}{3}\right]_0^L$$

$$U_{BC} = \frac{P^2L^3}{6EI}$$

Portion *AB*:

Consider a section *y-y* at a distance *y* from end *A*.

$$U_{AB} = \int_0^{2L} \frac{M_x^2\,dy}{2EI} + \int_0^{2L} \frac{P_x^2\,dy}{2AE}$$

$$U_{AB} = \int_0^{2L} \frac{\left(-\frac{P}{2}\cdot y\right)^2 dy}{2EI} + \int_0^{2L} \frac{P^2\,dy}{2aE}$$

$$U_{AB} = \frac{P^2}{8EI}\left[\frac{y^3}{3}\right]_0^{2L} + \frac{P^2}{2aE}[y]_0^{2L} = \frac{P^2(2L)^3}{24EI} + \frac{P^2(2L)}{2aE}$$

$$U_{AB} = \frac{P^2L^3}{3EI} + \frac{P^2L}{aE}$$

$\therefore$ Total strain energy, $U = U_{AB} + U_{BC}$

$$= \left(\frac{P^2L^3}{3EI} + \frac{P^2L}{aE}\right) + \left(\frac{P^2L^3}{6EI}\right) = \frac{P^2L^3}{2EI} + \frac{P^2L}{aE}$$

Deflection under load *P* is given by,

$$\frac{\partial U}{\partial P} = \frac{\partial}{\partial P}\left[\frac{P^2L^3}{2EI} + \frac{P^2L}{aE}\right]$$

$$\delta = \frac{2PL^3}{2EI} + \frac{2PL}{aE} = \frac{PL^3}{EI} + \frac{2PL}{aE}$$

4.7 A cantilever beam with circular cross-section of radius 100 mm is subjected to a uniformly distributed load over the entire span. It is given that the deflected shape of the beam has a maximum curvature of 1.018592×10^{-6} mm^{-1} and a maximum shear force of 1 kN. Find the intensity of loading on the beam and its span. $E = 2 \times 10^5$ MPa.

[10 marks : 2009]

Solution:

Let the intensity of loading and span be w per unit run and L respectively.

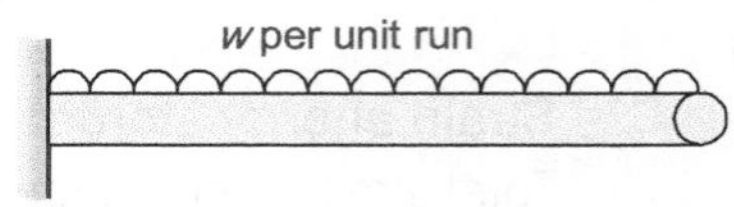

We know that the curvature of a beam is given by

$$\frac{d^2y}{dx^2} = \frac{M}{EI}$$

$$\therefore \quad \text{Maximum curvature} = \frac{M_{max}}{EI}$$

Given,

$$1.018592 \times 10^{-6} = \frac{M_{max}}{2 \times 10^5 \times \frac{\pi}{4} \times (100)^4}$$

$$\therefore \quad M_{max} = 16 \times 10^6 \text{ N-mm} = 16 \text{ kN-m}$$

But the maximum bending moment in a cantilever subjected to UDL . occurs at the fixed end i.e. $\frac{wL^2}{2}$

$$\therefore \quad \frac{wL^2}{2} = 16$$

$$\therefore \quad wL^2 = 32 \qquad \text{...(i)}$$

Also, the maximum shear force in a cantilever subjected to UDL occurs at the fixed end i.e.

$$(SF)_{max} = wL$$

But it is given that maximum SF = 1 kN

$$\therefore \quad wL = 1 \text{ kN} \qquad \text{...(ii)}$$

Substituting value of wL from (*ii*) in (*i*), we get

$$(wL) \times L = 32$$

$$\therefore \quad L = 32 \text{ m}$$

From equation (*ii*),

$$wL = 1$$

$$w \times 32 = 1$$

$$w = 0.03125 \text{ kN/m} = 31.25 \text{ N/m}$$

4.8 A stepped beam ABC, simply supported at A and fixed at C as shown in the figure carries a uniformly distributed load of intensity 'w' per unit length over BC. Determine the vertical reaction at A using moment area or energy method.

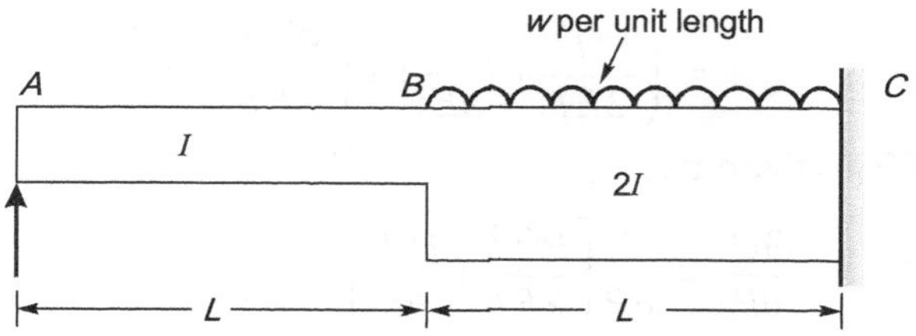

[20 marks : 2009]

Solution:

Moment area method: Let the reaction at A be R.

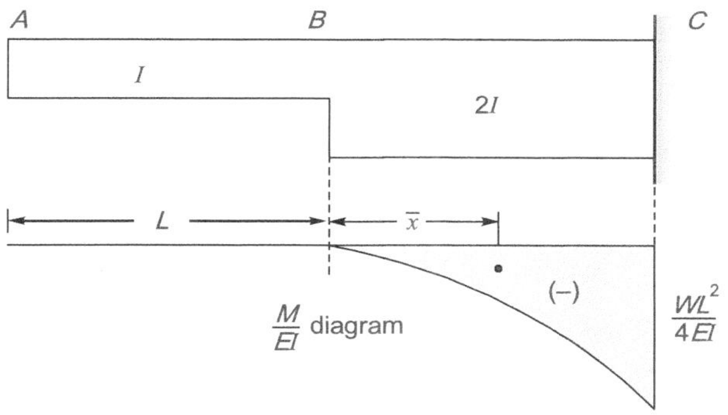

(i) Effect of UDL

Downward deflection at A due to UDL is given by

$(\Delta_A)_{UDL}$ = Moment of area of M/EI diagram between A and C about A

$$= A_1\bar{X}_1$$

where,
$$A_1 = -\frac{1}{3}\times L\times\frac{wL^2}{4EI} = -\frac{wL^3}{12EI}$$

$$\bar{X}_1 = (L+\bar{x}) = L+\frac{3}{4}L = \frac{7L}{4} \qquad \text{(From } A\text{)}$$

$$\therefore \quad (\Delta_A)_{Udl} = -\frac{wL^3}{12EI}\times\frac{7L}{4} = -\frac{7wL^4}{48EI} \quad \text{(Downward)} \qquad ...(i)$$

(ii) Effect of propped reaction R

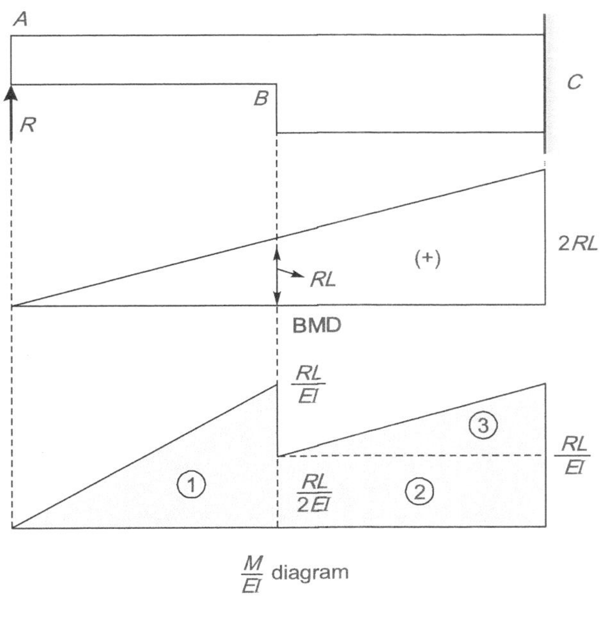

Upward deflection due to propped reaction

$(\Delta_A)_R$ = Moment of area of M/EI diagram between A and C about A

$$= A_1\bar{X}_1 + A_2\bar{X}_2 + A_3\bar{X}_3$$

$$= \left(\frac{1}{2}\times L\times\frac{RL}{EI}\right)\times\frac{2L}{3}+\left(L\times\frac{RL}{2EI}\right)\times\frac{3L}{2}+\left(\frac{1}{2}\times L\times\frac{RL}{2EI}\right)\times\frac{5L}{3}$$

$$= \frac{RL^3}{3EI}+\frac{3RL^3}{4EI}+\frac{5RL^3}{12}$$

$$(\Delta_A)_R = \frac{3}{2}\frac{RL^3}{EI} \quad ...(ii)$$

Since support A is unyielding. Hence net deflection at A will be zero

$$\Rightarrow \quad (\Delta_A)_{UDL} = (\Delta_A)_R$$

$$\frac{7}{48}\frac{wL^4}{EI} = \frac{3}{2}\frac{RL^3}{EI}$$

$$R = \frac{7\,wL}{72}(\uparrow)$$

Strain energy method:

We know that strain energy stored in the beam due to bending,

$$U = \int\frac{M_x^2\,dx}{2EI}$$

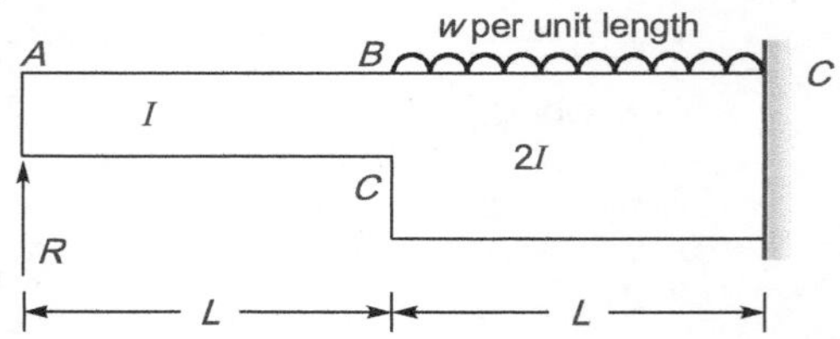

Deflection at A is given by

$$\delta_A = \frac{\partial U}{\partial R}$$

Since support is unyielding.

Hence,

$$\delta_A = \frac{\partial U}{\partial R} = \int\frac{M_x\frac{\partial M_x}{\partial R}dx}{EI} = 0$$

Portion AB:

$$M_x = R\cdot x \quad [0\le x\le L]$$

$$\frac{\partial M_x}{\partial R} = x$$

$$\therefore \quad \frac{\partial U_{AB}}{\partial R} = \frac{1}{EI}\int_0^L M_x\cdot\frac{\partial M_x}{\partial R}dx = \frac{1}{EI}\int(R\cdot x)\cdot x\,dx$$

$$\frac{\partial U_{AB}}{\partial R} = \frac{1}{EI}\int Rx^2dx \quad ...(i)$$

Portion BC:

$$M_x(x \text{ from } B) = R(L+x)-\frac{wx^2}{2} \quad [0\le x\le L]$$

$$\therefore \quad \frac{\partial M}{\partial R} = L+x$$

$$\therefore \quad \frac{\partial U_{BC}}{\partial R} = \frac{1}{2EI}\int_0^L\left[R(L+x)-\frac{wx^2}{2}\right]\times(L+x)dx$$

$$\frac{\partial U_{BC}}{\partial R} = \frac{1}{4EI}\int_0^L[2R(L+x)^2-wx^2(L+x)]dx \quad ...(ii)$$

But,

$$\frac{\partial U}{\partial R} = \frac{\partial U_{AB}}{\partial R}+\frac{\partial U_{BC}}{\partial R}$$

$$\therefore \quad \frac{1}{EI}\int_0^L Rx^2\,dx + \frac{1}{4EI}\int_0^L\left[2R(L+x)^2 - wx^2(L+x)\right]dx = 0$$

$$4R\left[\frac{x^3}{3}\right]_0^L + 2R\left[\frac{x^3}{3} + x^2L + L^2x\right]_0^L - wL\left[\frac{x^3}{3}\right]_0^L - w\left[\frac{x^4}{4}\right]_0^L = 0$$

$$\frac{4RL^3}{3} + 2R\left[\frac{L^3}{3} + L^3 + L^3\right] - \frac{wL^4}{3} - \frac{wL^4}{4} = 0$$

$$\frac{4RL^3}{3} + \frac{14RL^3}{3} = \frac{3wL^4 + 4wL^4}{12}$$

$$6RL^3 = \frac{7wL^4}{12}$$

$$R = \frac{7wL}{72}\ (\uparrow)$$

4.9 Find the reaction of B for the beam shown below. K is the stiffness of spring at B. Beam is of constant section. Use energy method of analysis.

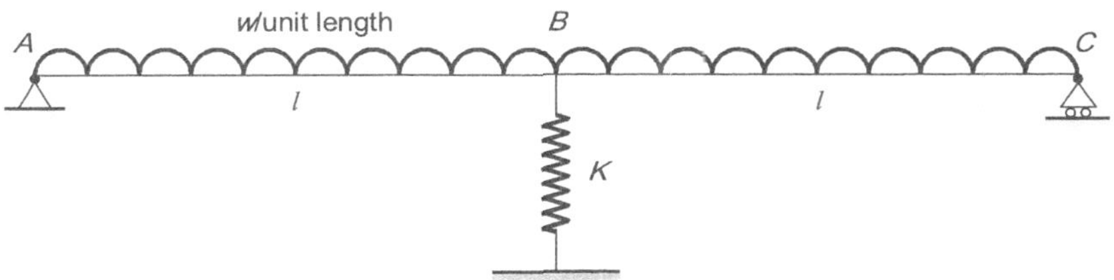

[10 marks : 2010]

Solution:

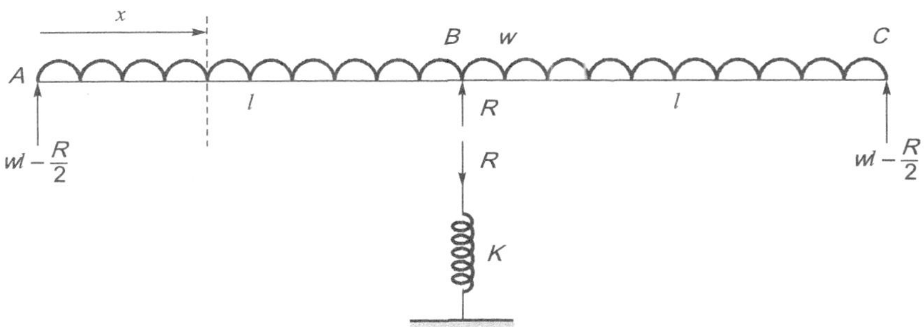

Let the reaction at B is R

$$\text{Downward deflection of beam } ABC \text{ at } B = \frac{-\partial U}{\partial R} = \frac{R}{K}$$

Strain energy of beam,

$$U = 2\int_0^l \frac{\left[(wl - \frac{R}{2})x - \frac{wx^2}{2}\right]^2 dx}{2EI}$$

$$\frac{R}{K} = -\frac{\partial U}{\partial R} = \frac{-4}{2EI}\int_0^l\left[(wl - \frac{R}{2})x - \frac{wx^2}{2}\right]\frac{(-x)}{2}dx$$

$$= \frac{+1}{EI}\left[(wl - \frac{R}{2})\frac{x^3}{3} - \frac{wx^4}{8}\Big|_0^l\right]$$

$$= \frac{1}{EI}\left[\frac{(wl - \frac{R}{2})l^3}{3} - \frac{wl^4}{8}\right] = \frac{1}{EI}\left[\frac{5wl^4}{24} - \frac{Rl^3}{6}\right]$$

$$R\left[\frac{1}{K} + \frac{l^3}{6EI}\right] = \frac{5wl^4}{24EI}$$

$$R = \frac{5wl^4}{24EI\left[\frac{1}{K} + \frac{1l^3}{6EI}\right]}$$

4.10 A beam *AB* of 6 m span carries a distributed load of varying intensity as shown in the figure below. Calculate the deflection at the centre of the beam. Load intensity at support B = 1.2 kN/m, $EI = 420 \times 10^6$ N-mm^2.

[16 marks : 2011]

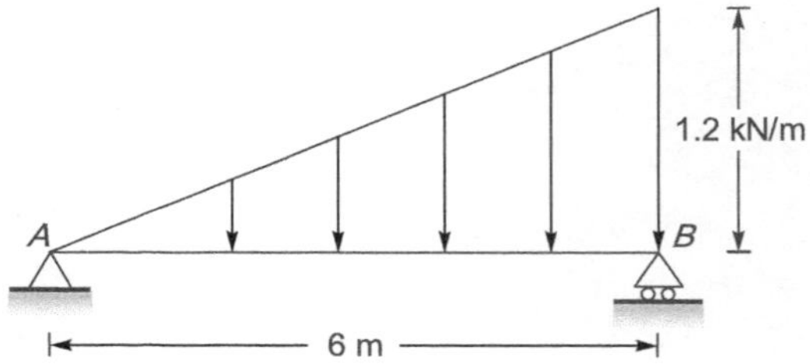

Solution:

Given:

Load intensity at support

$$B = 1.2 \text{ kN/m} = 1200 \text{ N/m}$$
$$EI = 420 \times 10^6 \text{ N-mm}^2$$
$$= 420 \times 10^6 \text{ N} - (10^{-3} \text{ m})^2$$
$$= 420 \times 10^6 \times 10^{-6} \text{ N-m}^2$$
$$= 420 \text{ N-m}^2$$

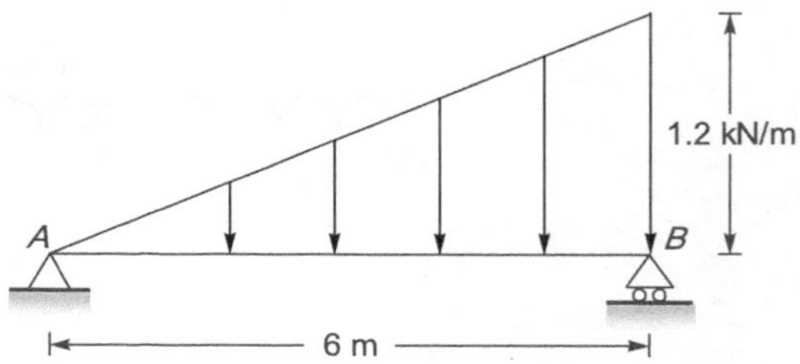

Let R_A and R_B be the react on at support A and B.

$$\Rightarrow \quad R_A + R_B = \frac{1}{2} \times 6 \times 1200 = 3600 \text{ N} \quad \text{...(i)}$$

$$\Sigma M_B = 0$$

$$\Rightarrow \quad R_A \times 6 - \frac{1}{2} \times 6 \times 1200 \times \frac{1}{3} \times 6 = 0$$

$$\Rightarrow \quad R_A = 1200 \text{ N}$$

$\therefore$ from (i) $\quad R_B = 3600 - 1200 = 2400$ N

Consider a section at a distance x from A

From similar triangle property

$$\frac{1200}{6} = \frac{w}{x}$$

$$\Rightarrow \quad w = 200x \text{ (equation of loading intensity)}$$

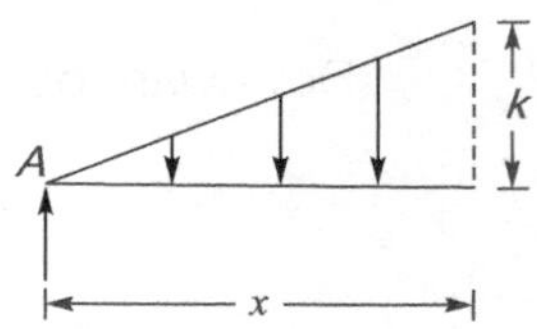

Using, $$M_x = EI\frac{d^2y}{dx^2}$$

or $$EI\frac{d^2y}{dx^2} = R_A x - \frac{1}{2} \times (x) \times (200x) \times \left(\frac{x}{3}\right)$$

$$= 1200\,x - 33.33\,x^3$$

On integration $$EI\frac{dy}{dx} = \frac{1200x^2}{2} - \frac{33.33x^4}{4} + C_1$$

Integrating again $$EIy = \frac{1200x^3}{2\times3} - \frac{33.33x^5}{4\times5} + C_1x + C_2$$

$$EIy = 200\,x^3 - 1.66\,x^5 + C_1x + C_2 \qquad \text{...(ii)}$$

At $x = 0, \quad y = 0$

Substituting in (2) $$0 = 0 - 0 + 0 + C_2 \quad \Rightarrow C_2 = 0$$

At $x = 6, \quad y = 0$

Substituting in (ii) $$0 = 200 \times 6^3 - 1.66 \times 6^5 + C_1 \times 6 + 0$$

$\Rightarrow$ $$C_1 = -5048.64$$

At $x = 3, \quad y = ?$

$$EIy = 200 \times 3^3 - 1.66 \times 3^5 - 5048.64 \times 3$$

$\Rightarrow$ $$y = \frac{5400 - 403.38 - 15145.92}{420} = -24.17 \text{ m}$$

Deflection at centre = 24.17 m (Downwards)

Remarks: *As the deflection obtained (24.17 m), at the centre is very big, it is not possible generally. It indicates that data given may have some error.*

4.11 A simply supported beam carries two point loads W each at its one-third sections as shown in figure. Determine the maximum deflection at its mid-span and slope at an end using the conjugate beam method.

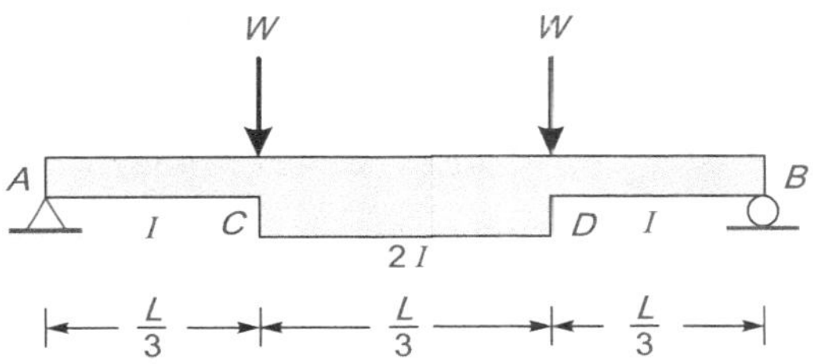

[15 marks : 2014]

Solution:

Reactions in real beam: By symmetry,

$$R_A = R_B = W$$

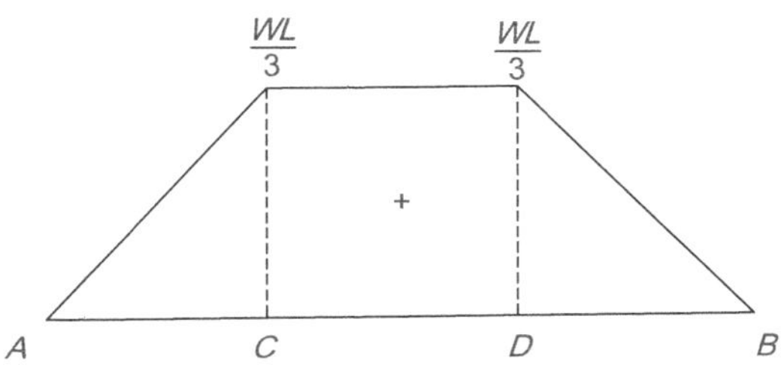

Fig. (i): BMD

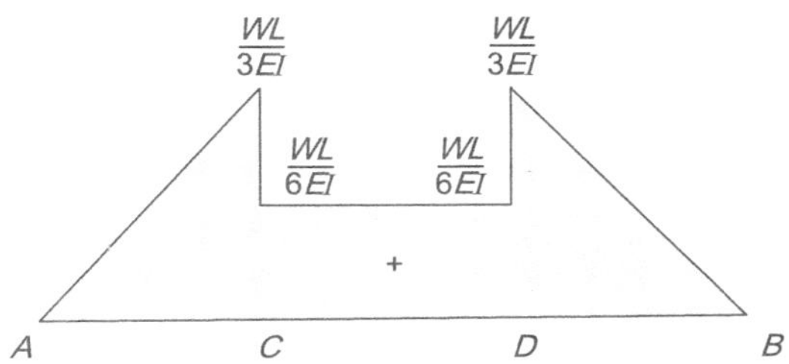

Fig. (ii): $\frac{M}{EI}$ diagram

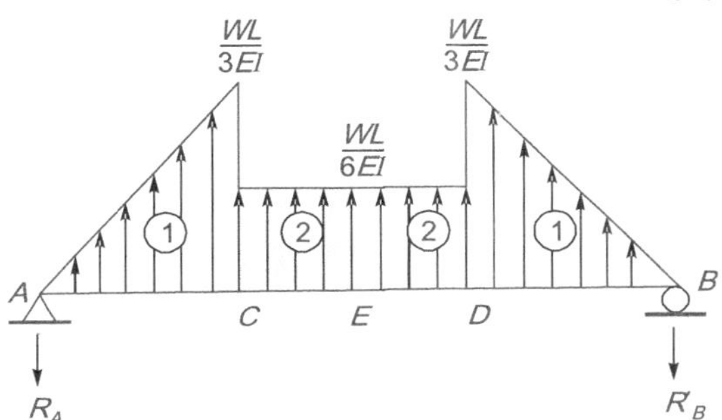

Fig. (iii): Conjugate beam

Reactions in conjugate beam:

By symmetry,

$$R_A' = R_B' = \frac{1}{2} \times \text{Area of } \frac{M}{EI} \text{ diagram}$$

$$= \frac{1}{2}\left[\frac{1}{2}\times\frac{L}{3}\times\frac{WL}{3EI}+\frac{L}{3}\times\frac{WL}{6EI}+\frac{1}{2}\times\frac{L}{3}\times\frac{WL}{3EI}\right]$$

$$= \frac{1}{2}\left[\frac{WL^2}{18EI}+\frac{WL^2}{18EI}+\frac{WL^2}{18EI}\right]$$

$$= \frac{WL^2}{12EI}(\downarrow)$$

S.N.	Part	Area	Centroid (from A)
1.	1	$A_1 = \frac{1}{2}\times\frac{L}{3}\times\frac{WL}{3EI} = \frac{WL^2}{18EI}$	$\bar{X}_1 = \frac{2}{3}\times\frac{L}{3} = \frac{2L}{9}$
2.	2	$A_2 = \frac{L}{6}\times\frac{WL}{6EI} = \frac{WL^2}{36EI}$	$\bar{X}_2 = \frac{L}{3}+\frac{L}{12} = \frac{5L}{12}$

According to conjugate beam method "the deflection at any section in real beam is equals to BM at that section in conjugate beam."

$\therefore$ **deflection at mid span,**

$$y_E = -R_A' \times \frac{L}{2} + A_1\left(\frac{L}{2}-\bar{X}_1\right) + A_2\left(\frac{L}{2}-\bar{X}_2\right)$$

$$= -\frac{WL^2}{12EI}\times\frac{L}{2}+\frac{WL^2}{18EI}\left(\frac{L}{2}-\frac{2L}{9}\right)+\frac{WL^2}{36EI}\left(\frac{L}{2}-\frac{5L}{12}\right)$$

$$= -\frac{WL^3}{24EI}+\frac{5WL^3}{324EI}+\frac{WL^3}{432EI} = -\frac{31WL^3}{1296EI}$$

$$\therefore \quad y_E = \frac{31}{1296}\frac{WL^3}{EI}(\downarrow)$$

According to conjugate beam method **"the slope at any section in real beam is equals to shear force at that section in conjugate beam."**

$$\therefore \quad \theta_A = \text{SF at } A \text{ in conjugate beam}$$

$$= -R_A' = -\frac{WL^2}{12EI}$$

$$\therefore \quad \theta_A = \frac{WL^2}{12EI}(\text{clockwise})$$

and by symmetry, $\theta_B = \theta_A = \frac{WL^2}{12EI}(\text{anticlockwise})$

4.12 **A beam fixed at one end and simply supported at the other end is having a hinge at *B* as shown in the figure. Determine the deflections (a) under the load and (b) at the hinge *B*. Use moment area method.**

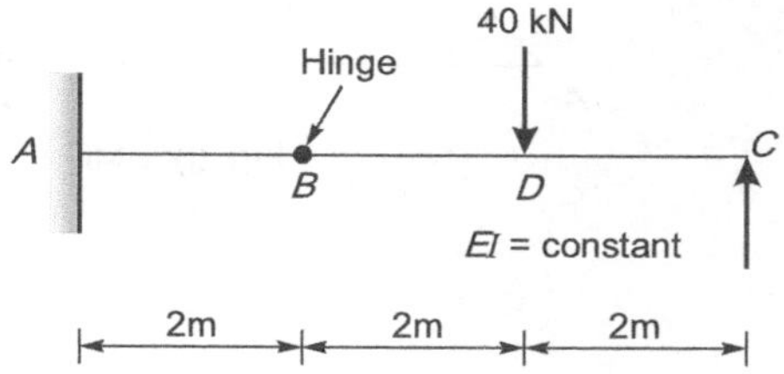

[10 marks : 2015]

Solution:

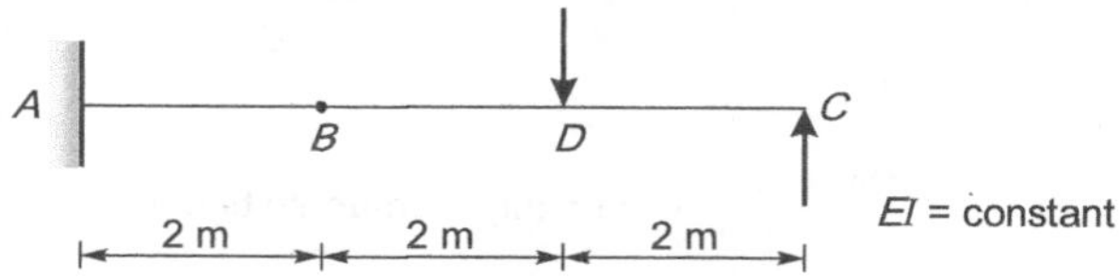

Taking member BC

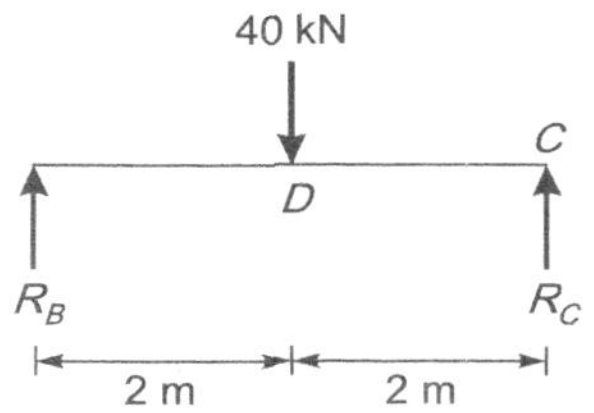

By symmetry $R_B = R_C = 20$ kN

Thus, FBD of the given beam

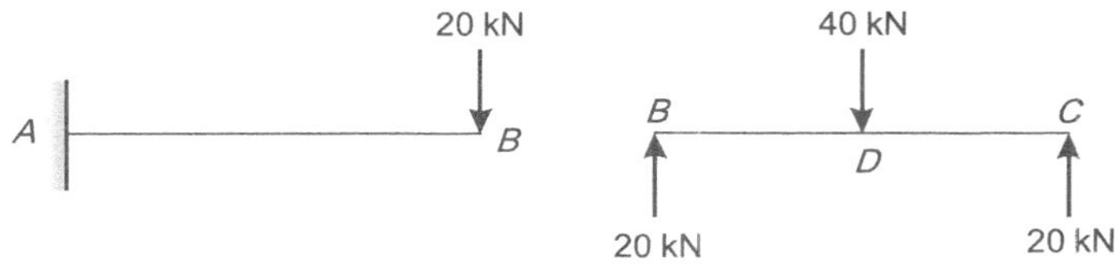

Thus, $\frac{BMD}{EI}$ of the given beam

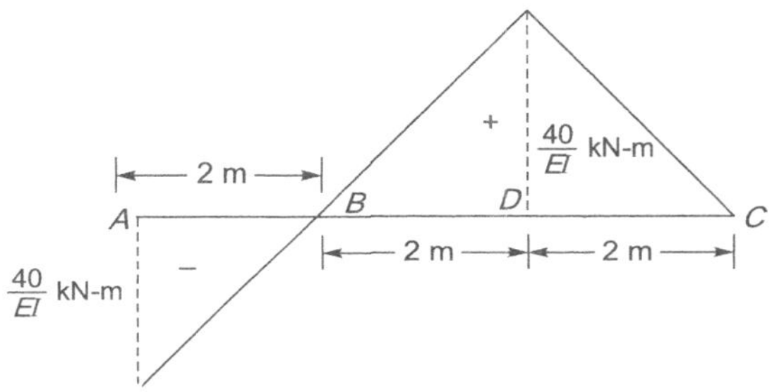

According to moment area method, **deflection at the hinge point B (δ_B)** is

$$\delta_B = \delta_A + \theta_A AB + \delta_{B/A}$$

$$= -\frac{1}{2}\times\frac{40}{EI}\times 2\times\frac{2}{3}\times 2 = -\frac{160}{3EI}(\downarrow)$$

Deflection under the load

$$\delta_C = \delta_B + \theta_{B_{BC}} BC + \delta_{C/B} = 0$$

$$\Rightarrow \quad -\frac{160}{3EI} + \theta_{B_{BC}} \times 4 + \left(\frac{1}{2}\times 4\times\frac{40}{EI}\times 2\right) = 0$$

$$\theta_{B_{BC}} \times 4 = \frac{160}{3EI} - \frac{160}{EI}$$

$$\therefore \quad \theta_{B_{BC}} = \frac{40}{3EI} - \frac{40}{EI} = -\frac{80}{3EI}$$

$$\therefore \quad \delta_D = \delta_B + \theta_{B_{BC}} BD + \delta_{D/B}$$

$$= -\frac{160}{3EI} - \left(\frac{80}{3EI}\times 2\right) + \frac{1}{2}\times 2\times\frac{40}{EI}\times\frac{2}{3}$$

$$= -\frac{160}{3EI} - \frac{160}{3EI} + \frac{80}{3EI}$$

$$= \frac{-320+80}{3EI} = -\frac{240}{3EI} = -\frac{80}{EI}(\downarrow)$$

Deflection shape of the beam will be

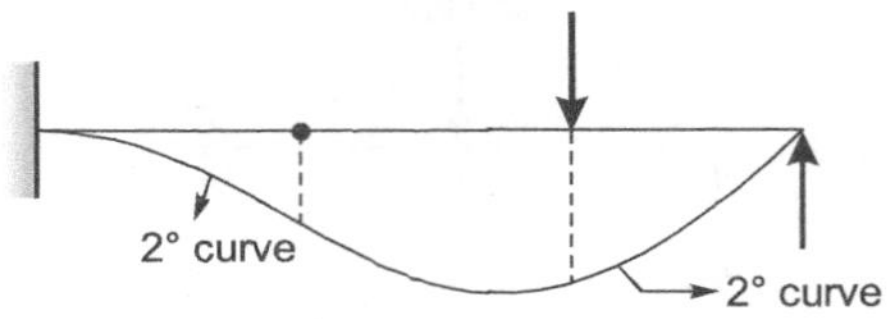

4.13 A propped cantilever beam of length 4 m is subjected to UDL of 30 kN/m over the entire length of the span. If the flexural rigidity of the beam of 2×10^4 kN-m², what would be the rotation at the propped support of the beam? Also determine the moment developed at the fixed support.

[5 marks : 2015]

Solution:

Given data: $EI = 2 \times 10^4$ kN-m², UDL $= w = 30$ kN/m

From compatibility method,

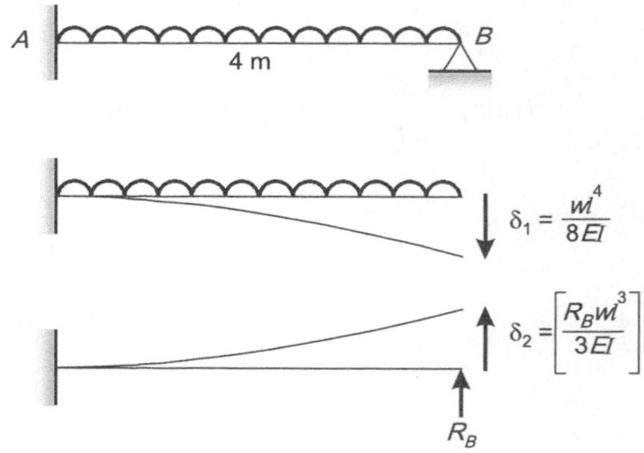

$$\delta_1 - \delta_2 = 0$$

$$\frac{wl^4}{8EI} - \frac{R_B l^3}{3EI} = 0$$

$$R_B = \frac{3wl}{8} = \frac{3 \times 30 \times 4}{8} = 45 \text{ kN}$$

Hence, rotation at free end $= \theta_1 - \theta_2$

$$\text{Rotation } (=\theta_1) \text{ due to reaction at } B = \frac{R_B l^2}{2EI} = \frac{45 \times 4^2}{2 \times 2 \times 10^4}$$

$$= 0.018 \text{ radian (anticlockwise)}$$

$$\text{Rotation } (=\theta_2) \text{ due to UDL} = \frac{wl^3}{6EI} = \frac{30 \times 4^3}{6 \times 2 \times 10^4}$$

$$= 0.016 \text{ radian (clockwise)}$$

Hence, net rotation at propped support of the beam

$$= \theta_1 - \theta_2$$

$$= 0.018 - 0.016$$

$$= 0.002 \text{ radian (anticlockwise)}$$

$$\Sigma M_A = 0$$

$$R_B(l) - 30 \times l \times \frac{l}{2} + M_A = 0$$

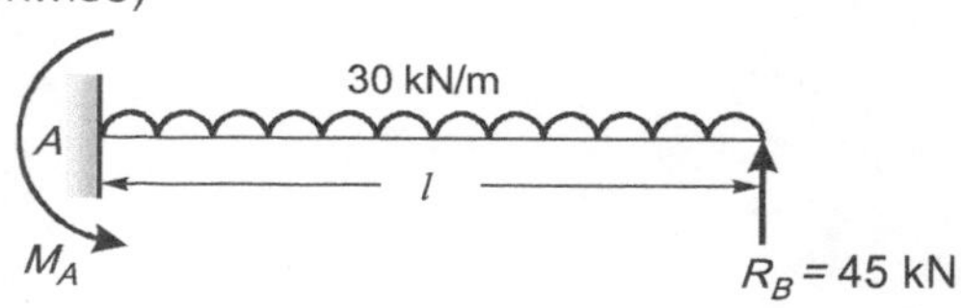

$$M_A = 30 \times 4 \times \frac{4}{2} - 45 \times 4$$

$$\text{Moment at fixed end } (=M_A) = 60 \text{ N-m (Hogging)}$$

4.14 Find the deflection at C for the beam as shown in Figure. EI = constant. Use conjugate beam method.

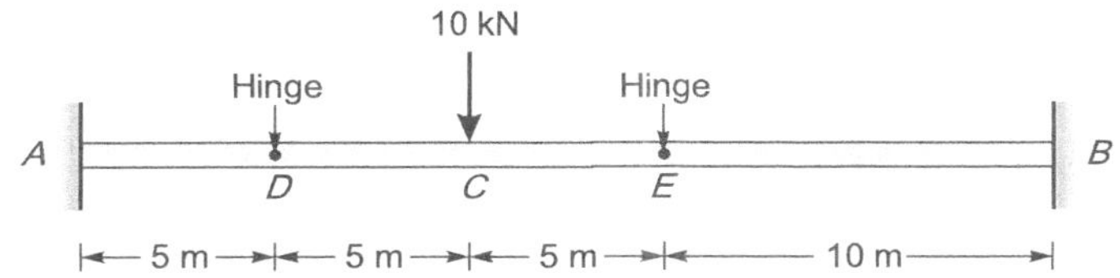

[20 marks : 2017]

Solution:

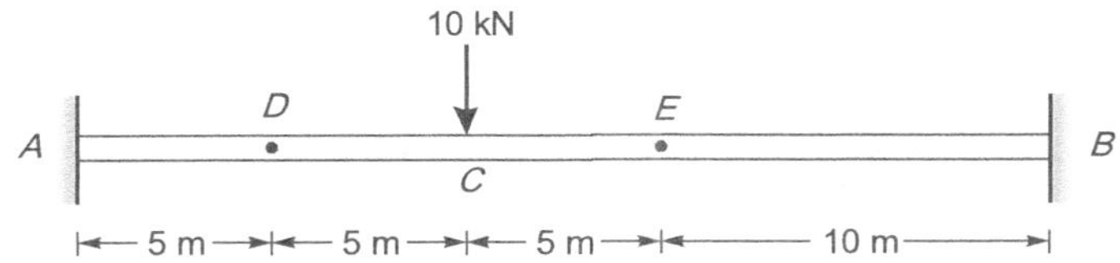

Free body diagram of above beam is shown below:

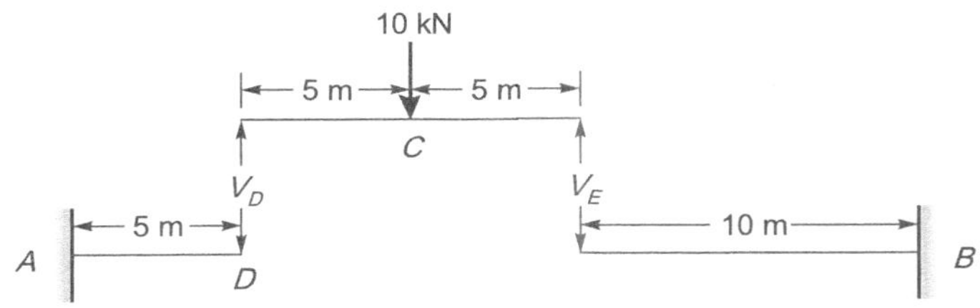

For portion DE,

Due to symmetry, $V_D + V_E = 10$ kN

$\Rightarrow$ $V_D = V_E = 5$ kN

$\therefore$ $M_A = V_D \times 5 = 5 \times 5 = 25$ kNm

$M_B = V_E \times 10 = 5 \times 10 = 50$ kNm

$M_C = V_E \times 5 = 5 \times 5 = 25$ kNm

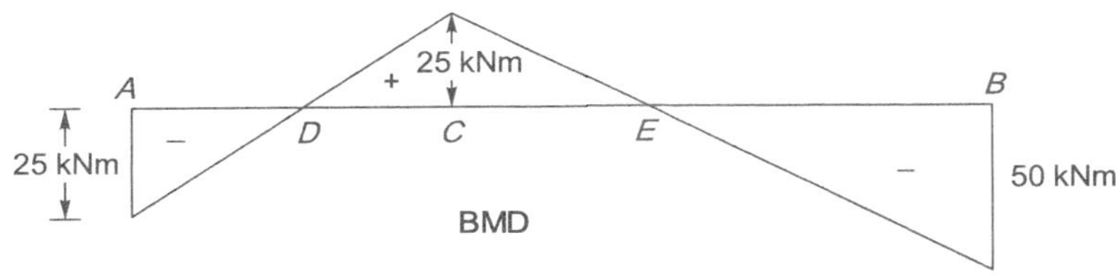

BMD

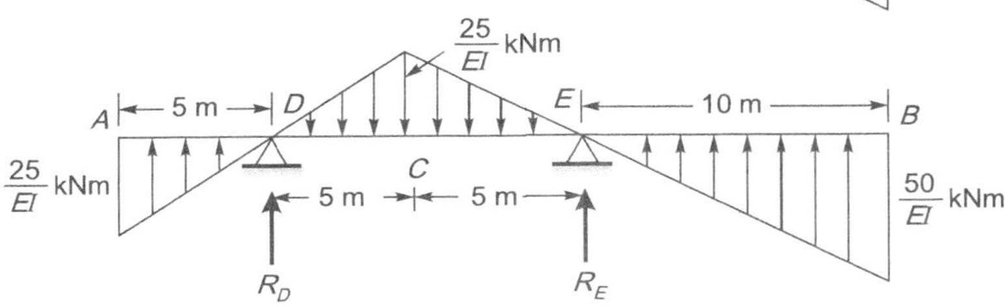

Conjugate Beam

Now $\Sigma F_y = 0$

$$\Rightarrow \quad R_D + R_E = \left[\frac{1}{2} \times 10 \times 25 + \left(-\frac{1}{2} \times 5 \times 25 - \frac{1}{2} \times 10 \times 50\right)\right]\frac{1}{EI}$$

$$\Rightarrow \quad R_D + R_E = \frac{125 - 62.50 - 250}{EI}$$

$$\Rightarrow \qquad R_D + R_E = \frac{-187.50}{EI}\text{kN} \qquad \text{...(i)}$$

$$\Sigma M_D = 0$$

$$\Rightarrow \qquad R_E \times 10 + \frac{1}{2} \times \frac{10 \times 50}{EI}\left(10 + \frac{2}{3} \times 10\right) = \frac{1}{2} \times \frac{10 \times 25}{EI} \times 5 + \frac{1}{2} \times \frac{5 \times 25}{EI} \times \frac{2}{3} \times 5$$

$$\Rightarrow \qquad R_E \times 10 + \frac{250 \times 16.67}{EI} = \frac{625}{EI} + \frac{625}{3EI}$$

$$\Rightarrow \qquad R_E \times 10 + \frac{4167.5}{EI} = \frac{625 + 208.33}{EI}$$

$$\Rightarrow \qquad R_E \times 10 = \frac{-3334.17}{EI}$$

or,

$$R_E = \frac{-333.417}{EI}(\downarrow) \qquad \text{...(ii)}$$

Putting this value in equation (i), we get

$$R_D = \frac{-187.50}{EI} + \frac{333.417}{EI} = \frac{145.917}{EI}(\uparrow)$$

Deflection under load at point C:

Let y_c = Deflection at C for the given beam

Now according to conjugate beam method.

$$y_c = \text{BM at } C \text{ for conjugate beam}$$

$$= R_D \times 5 + \frac{1}{2} \times \frac{5 \times 25}{EI} \times \left(5 + \frac{2}{3} \times 5\right) - \frac{1}{2} \times \frac{25 \times 5}{EI} \times \frac{5}{3}$$

$$= \frac{145.917 \times 5}{EI} + \frac{125}{2} \times \frac{8.333}{EI} - \frac{625}{6EI}$$

$$= \frac{729.585}{EI} + \frac{520.8125}{EI} - \frac{104.167}{EI}$$

$$= \frac{1146.23}{EI}(\downarrow) \text{ downwards at } C.$$

4.15 A beam of uniform cross-section and of length 2L is simply supported by rigid supports at its ends and by an elastic prop at its centre. If the prop deflects by an amount λ times the load it carries and if the beam carries a total uniformly distributed load of W find the load carried by the prop if EI is constant throughout the length of beam.

[12 marks : 2019]

Solution:

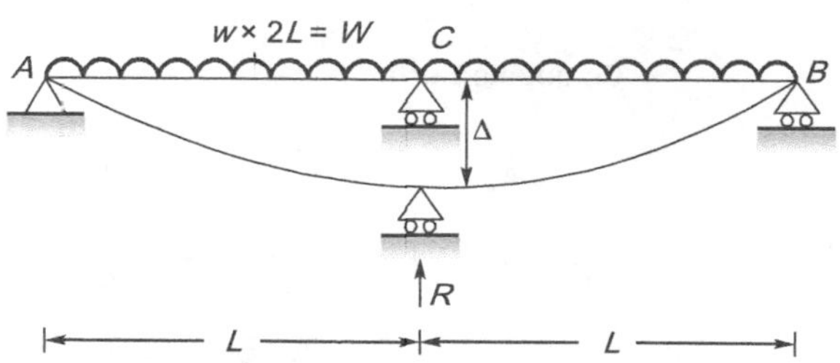

Prop at C sinks by amount Δ

If reaction developed at C is R then as per question

$$\Delta = \lambda R$$

$$w = \frac{W}{2L}$$

Now using compatibility condition

$$\frac{5}{384}\frac{w(2L)^4}{EI} - \frac{R(2L)^3}{48EI} = \Delta$$

$$\frac{5}{384}\frac{(W/2L)\times 16L^4}{EI} - \frac{8RL^3}{48EI} = \lambda R$$

$$\frac{5}{48}\frac{WL^3}{EI} = R\left[\lambda + \frac{L^3}{6EI}\right]$$

$$R = \frac{\frac{5}{48}\frac{WL^3}{EI}}{\left[\lambda + \frac{L^3}{6EI}\right]} = \frac{5WL^3}{8\left[L^3 + 6\lambda EI\right]}$$

4.16 Determine the vertical and horizontal deflections at the free end of the frame shown in figure. Support A is fixed, B is rigid joint and point C is free. Moment of inertia of BC = I and BA = 2I, E = 210 GPa and I = 300 × 10^{-6} m^4. Use Unit Load Method.

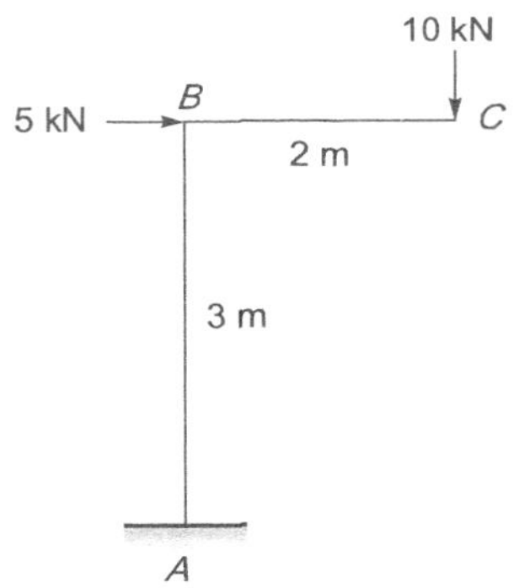

[20 marks : 2022]

Solution:

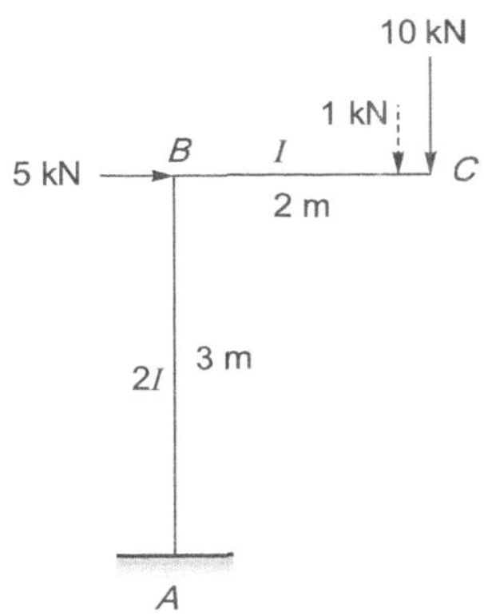

Vertical deflection at free end C.

$$E = 2.1 \times 10^5 \text{ N/mm}^2$$

$$= 2.1\times10^5 \times \frac{10^{-3}\text{kN}}{10^{-6}\text{ m}^2} = 2.1\times10^8 \text{ kN/m}^2$$

$$I = 300 \times 10^{-6} \text{ m}^4$$

So, $EI = 630 \times 10^2 \text{ kNm}^2$

By unit load method,

Portion	CB	BA
Origin	C	B
Limit (m)	0 to 2	0 to 3
M	$-10x$	$-(20 + 5x)$
m_1	$-x$	-2
MOI	I	$2I$

Here, m_1 is moment when unit load is applied in direction of deflection required as shown in figure.

$$\text{Deflection } (\delta_v) = \Sigma\int\frac{Mm_1\,dx}{EI} = \int_0^2\frac{-10x(-x)dx}{EI} + \int_0^3\frac{(20+5x)2dx}{2EI}$$

$$= \frac{26.67+82.5}{630\times10^2} = \frac{109.17}{630\times10^2}$$

$$= 1.733 \times 10^{-3} \text{ m} = 1.733 \text{ mm}$$

So, vertical deflection at free end = 1.733 mm

Horizontal deflection at point C,

Portion	CB	BA
Origin	C	B
Limit	0–2	0–3
M	$-10x$	$-(20 + 5x)$
m_2	0	$-x$
MOI	I	$2I$

$$(\delta_H) = \Sigma\int\frac{Mm_2dx}{EI} + \int_0^3\frac{-(20+5x)x - xdx}{2EI}$$

$$= \frac{1}{2EI}\int_0^3(20+5x)xdx$$

$$= \frac{135}{2EI} = \frac{67.5}{630\times10^2}$$

$$= 1.07 \times 10^{-3} \text{ m} = 1.07 \text{ mm}$$

So, horizontal deflection at free end = 1.07 mm

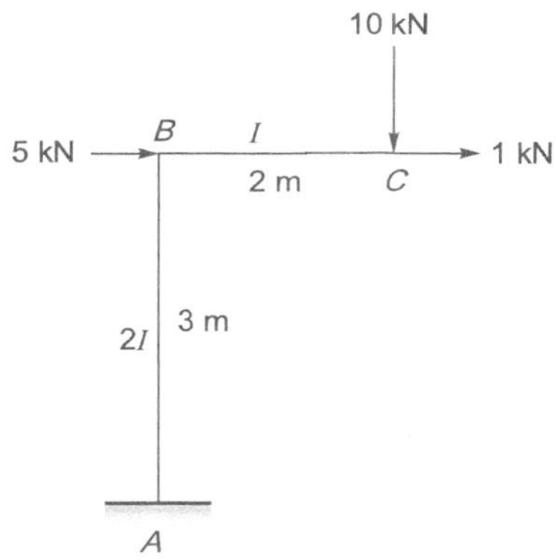

Alternate Solution:

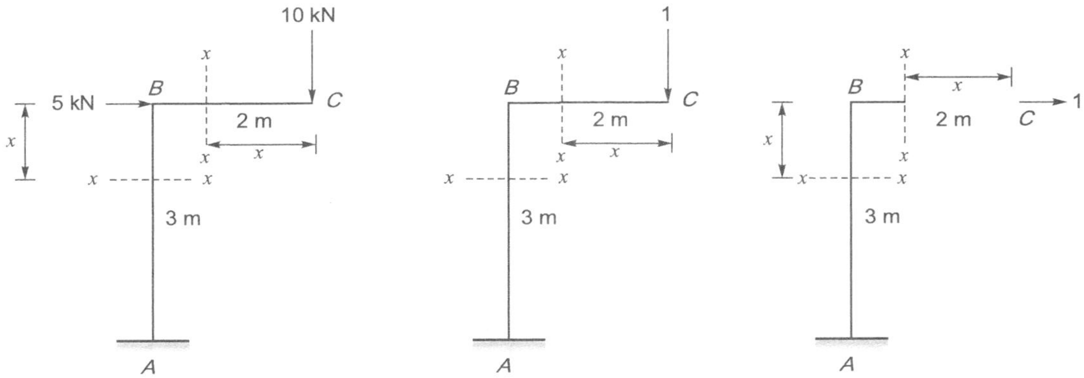

Member	M_x	m_1	m_2	$\delta_{V_c} = \int \frac{M_x m_1 dx}{EI}$	$\delta_{H_c} = \int \frac{M_x m_2 dx}{EI}$
CB	$-10x$	$-1x$	0	$\int_0^2 \frac{-10x(-x)dx}{EI} = \frac{10\times2^3}{3EI} = \frac{80}{3EI}$	0
BA	$-10\times2-5x$ $= -20-5x$	-1×2	$-1\times x$	$\int_0^3 \frac{(-20-5x)(-2)dx}{E(2I)}$ $= \frac{+120+\frac{10\times3^2}{2}}{2EI} = \frac{165}{2EI}$	$\int_0^3 \frac{(-20-5x)(-x)dx}{E(2I)}$ $= \frac{\frac{20\times3^2}{2}+\frac{5\times3^3}{3}}{2EI} = \frac{135}{2EI}$
				$= \frac{109.16}{EI}$	$= \frac{67.5}{EI}$

$$E = 210\times10^6 \text{ kN/m}^2$$
$$I = 300\times10^{-6} \text{ m}^4$$
$$EI = 630\times10^2 \text{ kNm}^2$$
$$\delta_{Vc} = \frac{109.16}{630\times10^2} = 1.73 \text{ mm}$$
$$\delta_{Hc} = \frac{67.5}{630\times10^2} = 1.07 \text{ mm}$$

5. Bending & Shear Stresses in Beams

5.1 A wire of 2 mm diameter is wound round a circular shaft of 1 m diameter. What are the maximum normal and shear stresses in the wire? Specify the planes of their occurence. $E = 1 \times 10^5$ N/mm^2. Assume no applied tensile pull on the wire.

[10 marks : 2002]

Solution:

Let the diameter of the shaft is $D = 1000$ mm
and the diameter of the wire is $d = 2$ mm
Modulus of Elasticity, $E = 1 \times 10^5$ N/mm^2

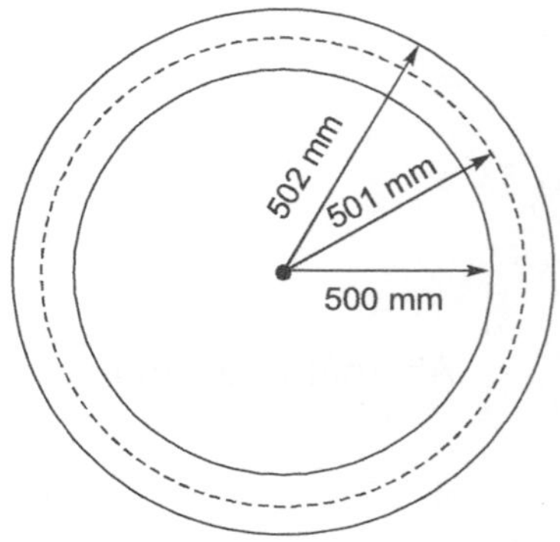

Now, normal stress in the wire is nothing but circumferential stress in the wire induced due to wrapping over the shaft. Hence the outer surface will be elongated and the central point of the wire will not undergo any change in length as it has no tension or compression.

$$\therefore \text{ Circumferential strain}, \epsilon = \frac{\pi D' - \pi D''}{\pi D''}$$

where D' = Diameter of the whole assembly.
D'' = Diameter of the central (neutral) circle

$$\epsilon = \frac{D' - D''}{D''} = \frac{1004 - 1002}{1002} = \frac{2}{1002}$$

Circumferential normal stress $\sigma_x = \epsilon \times E$

$$\sigma_x = \frac{2}{1002} \times 1 \times 10^5 = 199.60 \text{ N/mm}^2 \quad \textbf{(Ans.)}$$

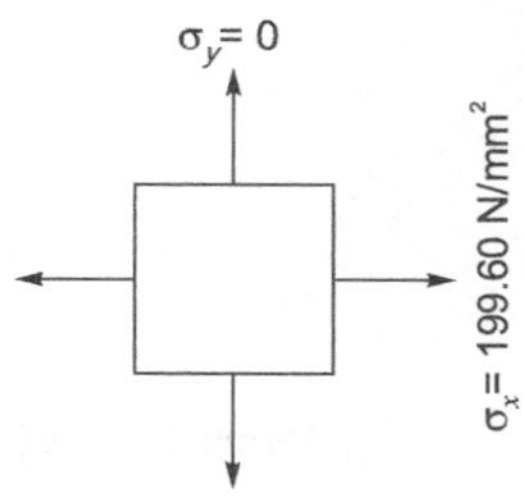

Maximum shear stress will be at a plane 45° to the plane of maximum normal stress.

$$\tau_{max} = \frac{\sigma_x - \sigma_y}{2} = \frac{199.600 - 0}{2} = 99.80 \text{ N/mm}^2 \quad \textbf{(Ans.)}$$

5.2 A cantilever beam AB of I-section has a span of 2 m and supports an inclined load of 10 kN at the free end B. The load passes through the centroid of the I-section and makes an angle 30° with the vertical in the plane of the cross-section as shown in figure. Find the normal stresses at the corners A, B, C and D of the cross section at the fixed end and shear stress component in the vertical direction at the centroid of cross-section. The dimensions of the cross-section are given in the figure.

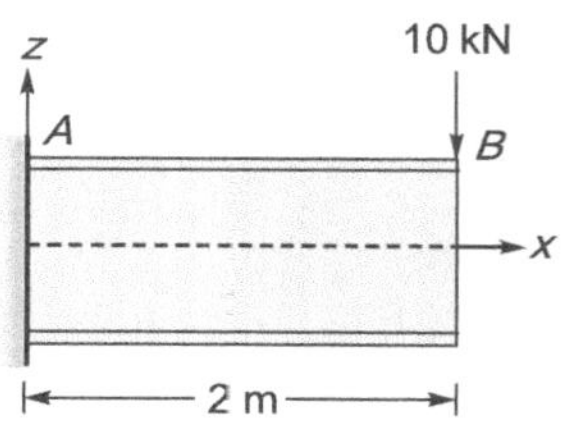

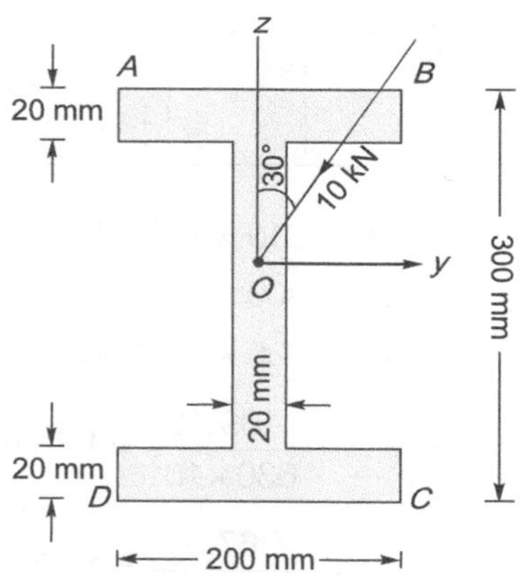

[20 marks : 2004]

Solution:

Resolving 10 kN in y and z directions, we get

$$P_z = 10\cos 30° = 8.66 \text{ kN}$$

$$P_y = 10\sin 30° = 5 \text{ kN}$$

Bending moment at fixed end at $x = 2$ m

$$M_z = 5 \times 2 = 10 \text{ kN-m}$$

$$M_y = 8.66 \times 2 = 17.32 \text{ kN-m}$$

Moment of inertia about y-axis

$$I_y = 2\left[\frac{200\times 20^3}{12} + 200\times 20\times 140^2\right] + \left[\frac{20\times 260^3}{12}\right] = 186.36\times 10^6 \text{ mm}^4$$

Moment of inertia about z axis,

$$I_z = 2\left[\frac{200^3\times 20}{12}\right] + \left[\frac{260\times 20^3}{12}\right] = 26.84\times 10^6 \text{ mm}^4$$

Since load passes through I and III quadrant. Hence NA will pass through II and IV quadrant.

Let Neutral axis is at β angle with y axis.

Let any point (y, z) in second or fourth quadrant

$$\sigma_x(y, z) = +\frac{M_y}{I_y}z - \frac{M_z}{I_z}y$$

at NA $\quad \sigma_x = 0$

$$\frac{M_y}{I_y}z = \frac{M_z}{I_z}y$$

$$\frac{z}{y} = \frac{M_z}{M_y}\times\frac{I_y}{I_z}$$

$$\tan\beta = \frac{10}{17.32}\times\frac{1.8636\times 10^8}{2.684\times 10^7}$$

$$\beta = 76°$$

Stress at A:

$$y = 100 \text{ mm}, z = 150 \text{ mm}$$

$$\sigma_A = +\frac{M_y}{I_y}z - \frac{M_z}{I_z}y = \frac{17.32\times 10^6\times 150}{1.8636\times 10^8} - \frac{10\times 10^6\times 100}{2.684\times 10^7} = +13.94 - 37.25$$

$$= -23.31 \text{ N/mm}^2 \text{ (compressive)}$$

Stress at B:

$$y = 100 \text{ mm}, z = 150 \text{ mm}$$

$$\sigma_B = +\frac{M_y}{I_y}z + \frac{M_z}{I_z}y = +\frac{17.32\times 10^6\times 150}{1.8636\times 10^8} + \frac{10\times 10^6\times 100}{2.684\times 10^7}$$

$$= +13.94 + 37.25$$

$$= +51.19 \text{ N/mm}^2 \text{ (tensile)}$$

Stress at C:

$$y = 100, z = 150 \text{ mm}$$

$$\sigma_C = -\frac{M_y}{I_y}z + \frac{M_z}{I_z}y = -13.94 + 37.25 = +23.31 \text{ N/mm}^2 \text{ (tensile)}$$

Stress at D:

$$\sigma_D = -\frac{M_y}{I_y}z - \frac{M_z}{I_z}y = -13.94 - 37.25 = -51.19 \text{ N/mm}^2 \text{ (compressive)}$$

$\therefore$ Hence,

$$\sigma_A = -23.31 \text{ N/mm}^2 \text{ (compressive)}$$
$$\sigma_B = +51.19 \text{ N/mm}^2 \text{ (tensile)}$$
$$\sigma_C = +23.31 \text{ N/mm}^2 \text{ (tensile)}$$
$$\sigma_D = -51.19 \text{ N/mm}^2 \text{ (compressive)}$$

Shear stress component in the vertical direction at the centroid of cross-section

$$\tau = \frac{VA\bar{y}}{I_{yy}b}$$

$$V = P_2 = 10 \cos 30° = 8.66 \text{ kN}$$
$$A\bar{y} = 200 \times 20 \times 140 + 130 \times 20 \times 65 = 729000 \text{ mm}^3$$
$$I_{yy} = 186.36 \times 10^6 \text{ mm}^4$$
$$b = 20 \text{ mm}$$

$$\tau = \frac{8.66 \times 10^3 \times 729000}{186.36 \times 10^6 \times 20} = 1.6938 \text{ N/mm}^2$$

5.3 A cantilever circular in section, has a uniform taper with the horizontal. The diameter of the cantilever at the free end is 'd' and it tapers at 't' units/m length on each side. A vertical load 'w' acts at the free end of the cantilever. Neglecting the self weight of the cantilever, find an expression for the distance of the section from the free end at which the bending stress is maximum. The length of the cantilever is 'h' units.

[10 marks : 2006]

Solution:

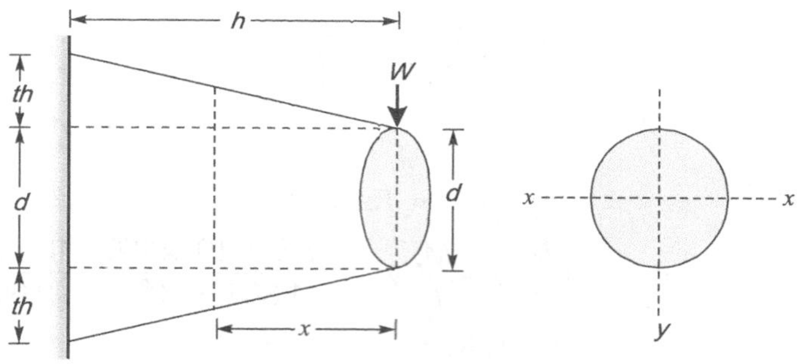

Let all the units be taken in SI system.
The diameter of the cantilever at fixed end = $d + 2th$
Assuming bending stress will be maximum at a distance x from free end.

Bending stress,
$$\sigma_x = \frac{M_x}{Z_x} \quad \text{...(i)}$$

Bending moment at distance $x \Rightarrow M_x = Wx$

Section modulus at a distance x from free end

$$Z_x = \frac{I_{xx}}{y_{max}}$$

$$I_{xx} = \frac{\pi}{64}(d_x)^4$$

where, $$d_x = d_1 + \left(\frac{d_2 - d_1}{L}\right)x$$

$\therefore$ $$d_x = d + \left[\frac{d + 2th - d}{h}\right]x$$

$$d_x = d + 2tx$$

and $$y_{max} = \frac{d_x}{2}$$

Hence $$Z_x = \frac{I_{xx}}{y_{max}} = \frac{\frac{\pi}{64}(d_x)^4}{\frac{d_x}{2}} = \frac{\pi}{32}d_x^3$$

Now from eq. (i) $$\sigma = \frac{M_x}{Z_x} = \frac{Wx}{\frac{\pi}{32}d_x^2}$$

$$\sigma = \frac{32}{\pi}\frac{Wx}{d_x^3} = \frac{32W}{\pi}\left[\frac{x}{d_x^3}\right]$$

or $$\sigma = \frac{32W}{\pi}\left[\frac{x}{(d + 2tx)^3}\right] \quad [\because d_x = d + 2tx]$$

For σ to be maximum $$\frac{d\sigma}{dx} = 0$$

$$\frac{32W}{\pi}\frac{d}{dx}\left[\frac{x}{(d + 2tx)^3}\right] = 0$$

$\Rightarrow$ $$\frac{(d + 2tx)^3 - x(3)(2t)(d + 2tx)^2}{(d + 2tx)^6} = 0$$

$$(d + 2tx)^2(d + 2tx - 6tx) = 0$$

either $$(d + 2tx) = 0$$

or $$(d - 4tx) = 0$$

$\therefore$ $$x = \frac{d}{4t}$$

Also, $$\sigma_{max} = \frac{32W}{\pi}\left[\frac{d/4t}{\left(d + \left(2t \times \frac{d}{4t}\right)\right)^3}\right]$$

$$= \frac{32W}{\pi}\left[\frac{d/4t}{(3d/2)^3}\right] = \frac{32W}{\pi}\left[\frac{8d}{4t \times 27d^3}\right] = \frac{64W}{27\pi t d^2}$$

5.4 A vertical flag staff has a height of 7.5 m above ground and is made up of square section throughout. Its dimensions being 75 mm × 75 mm at the top, changing uniformly to 150 mm × 150 mm at the ground. A horizontal force of 4000 N is applied at the top, location of this loading being along a diagonal of the section. Find the maximum stress due to this loading and its location.

[10 marks : 2006]

Solution:

Let the size of the tapered section at a distance x (in m) from the top be D_x

$$\therefore \quad D_x = 75 + \frac{150-75}{7.5}x$$

$$D_x = 75 + 10x$$

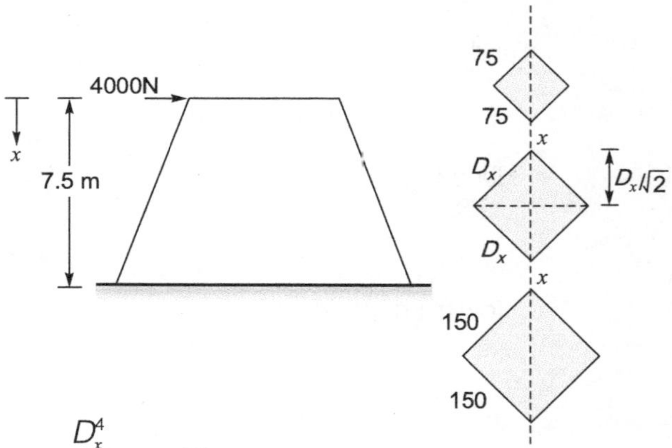

Moment of inertia of section,

$$I_x = \frac{2\sqrt{2}D_x \times \left(\frac{D_x}{\sqrt{2}}\right)^3}{12} = \frac{D_x^4}{12}$$

$$y_{max} \text{ from NA} = \frac{D_x}{\sqrt{2}}$$

$$\therefore \quad \text{Section modulus of section} = \frac{I_x}{y_{max}} = \frac{\frac{D_x^4}{12}}{D_x/\sqrt{2}} = \frac{D_x^3}{6\sqrt{2}}$$

Bending moment at section x-x = $4000x$ N-m or $4 \times 10^6 x$ N-mm

Maximum bending stress at section,

$$\sigma_x = \frac{M_x}{Z_x} = \frac{4\times10^6 x}{\frac{D_x^3}{6\sqrt{2}}} = \frac{4\times10^6 x\times 6\sqrt{2}}{(D_x)^3} = \frac{4\times10^6 x\times 6\sqrt{2}}{(75+10x)^3}$$

For σ_x to be maximum,

$$\frac{d\sigma_x}{dx} = 0$$

$$6\sqrt{2}\times4000\times10^3 \frac{d}{dx}\left[\frac{x}{(75+10x)^3}\right] = 0$$

$$\frac{(75+10x)^3 - \left[x\times3(75+10x)^2\times10\right]}{(75+10x)^6} = 0$$

$$(75 + 10x)^2\,[75 + 10x - 30x] = 0$$

$$(75 + 10x)^2[75 - 20x] = 0$$

Either $(75 + 10x)^2 = 0$

or $x = -7.5$ (not possible)

or $75 - 20x = 0$

$$x = 3.75 \text{ m}$$

$\therefore$ Maximum value of bending stress will be at $x = 3.75$ m

$$\therefore \quad \sigma_{max} = 6\sqrt{2}\times4\times10^6\times\frac{x}{(75+10x)^3}$$

$$= 6\sqrt{2} \times 4 \times 10^6 \times \frac{3.75}{(75 + 10 \times 3.75)^3}$$

$$= 89.39 \text{ N/mm}^2$$

5.5 A rectangular steel beam 180 mm × 260 mm in cross-section is used as a simply supported beam over a span of 8 m and is subjected to a UDL of 1 kN/m and also a concentrated load of 5 kN at 2 m from each support inclined at 30° to the vertical axis. Determine the bending stresses at four corners of the beam and location of the neutral axis of the cross-section.

[20 marks : 2006]

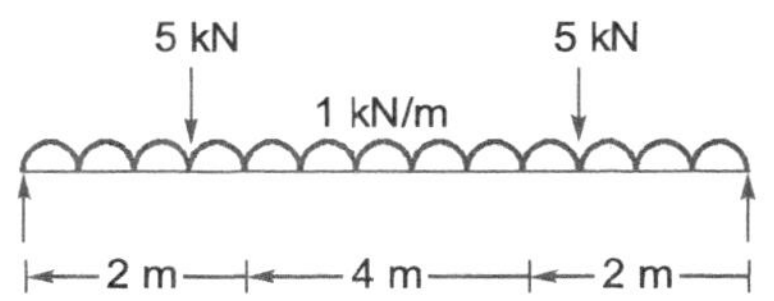

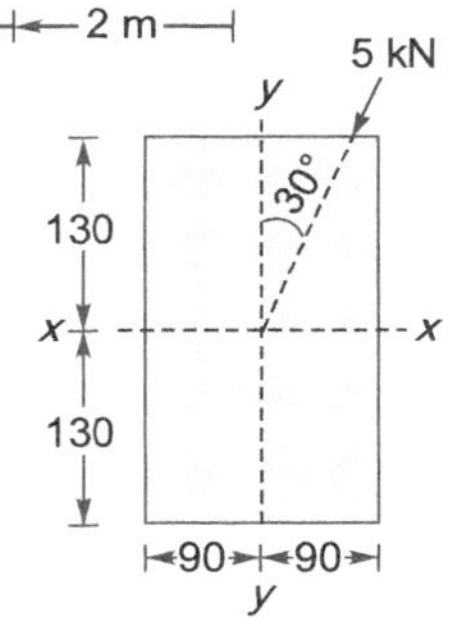

All dimensions are in mm.

Solution:

Resolve incline loads in the direction of y-axis and in the direction of x-axis.

Component of 5 kN in the direction of y-axis = 5 cos 30° = 4.33 kN

and component of 5 kN in x-direction = 5 sin 30° = 2.5 kN

Effect of vertical load:

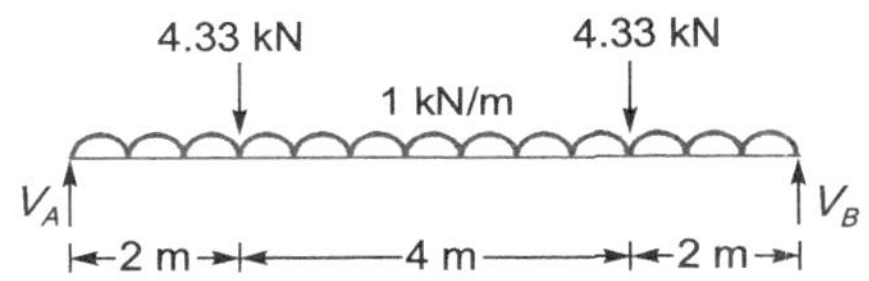

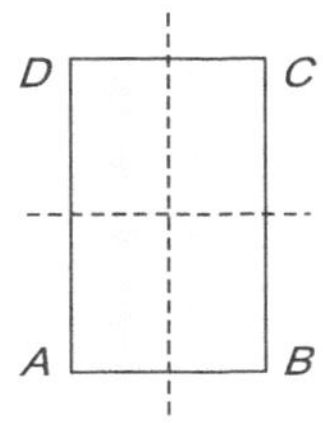

Since loading is symmetrical

Hence $$V_A = V_B = \frac{1}{2}[1 \times 8 + 2 \times 5 \cos 30°]$$

$$= \frac{1}{2}[8 + 2 \times 4.33] = 8.33 \text{ kN}$$

Since the loading is symmetrical, hence maximum bending moment will occur at the centre i.e., at x = 4 m from end.

$$(M_z)_c = V_A \times 4 - 4.33 \times 2 - 1 \times 4 \times 2$$

$$= 8.33 \times 4 - 4.33 \times 2 - 1 \times 4 \times 2 = 16.67 \text{ kN-m}$$

Effect of horizontal load:

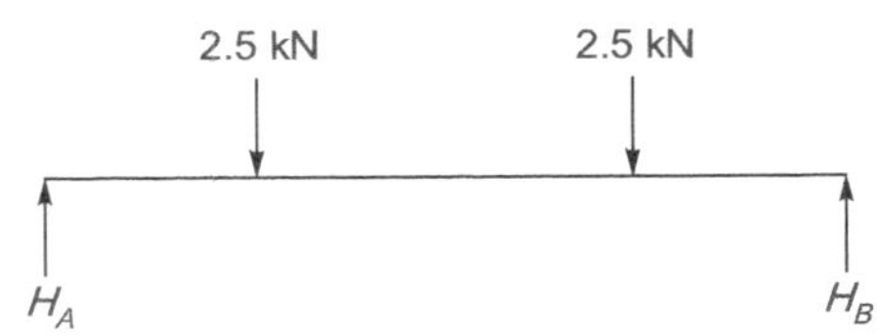

Since loading is symmetrical

$$\therefore \quad H_A = H_B = \frac{2.5 + 2.5}{2} = 2.5 \text{ kN}$$

Maximum bending moment due to horizontal load will be at centre of beam

$$\therefore \quad (M_y)_c = H_A \times 4 - 2.5 \times 2$$

$$= 2.5 \times 4 - 2.5 \times 2$$

$$= 5 \text{ kNm}$$

$$I_z = \frac{180 \times 260^3}{12} = 2636.4 \times 10^5 \text{ mm}^4$$

$$I_y = \frac{260 \times 180^3}{12} = 1263.6 \times 10^5 \text{ mm}^4$$

Since inclined load passes through quadrants I[st] and IIIrd. Hence NA with pass through quadrants IInd and IVth. Let NA make angle β will axis z-z.

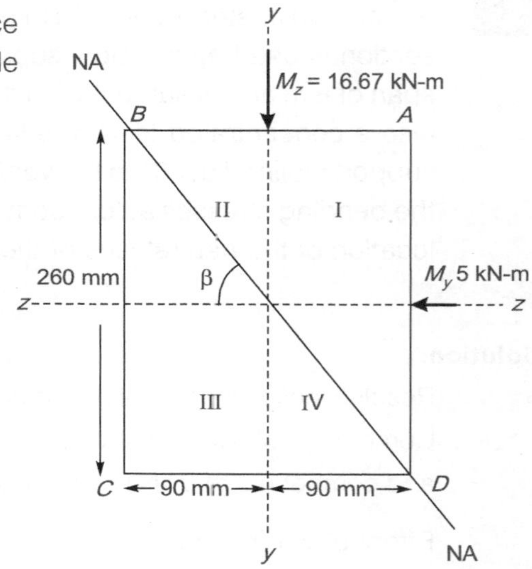

Then $$\tan\beta = \frac{y}{z} = \frac{M_y}{M_z} \times \frac{I_z}{I_y}$$

$$\tan\beta = \frac{5 \times 10^6}{16.67 \times 10^6} \times \frac{2636.4 \times 10^5}{1263.6 \times 10^5}$$

$$\tan\beta = 0.625$$

$$\therefore \quad \beta = 32^\circ$$

Stress at corner *A*:

$$z = 90 \text{ mm}$$

$$y = 130 \text{ mm}$$

$$\sigma_A = -\frac{M_y}{I_y} \times z - \frac{M_z}{I_z} \times y$$

$$= -\frac{5 \times 10^6}{1263.6 \times 10^5} \times 90 - \frac{16.67 \times 10^6}{2636.4 \times 10^5} \times 130$$

$$= -3.56 - 8.21$$

$$= -11.77 \text{ N/mm}^2 \text{ (compressive)}$$

Stress at corner *B*:

$$z = 90 \text{ mm}$$

$$y = 130 \text{ mm}$$

$$\sigma_B = -\frac{M_z}{I_z} \times y + \frac{M_y}{I_y} \times z = \frac{16.67 \times 10^6 \times 130}{2636.4 \times 10^5} + \frac{5 \times 10^6 \times 90}{1263.6 \times 10^5} = -8.21 + 3.56$$

$$= -4.65 \text{ N/mm}^2 \text{ (compressive)}$$

Stress at corner C:

$$z = 90 \text{ mm}$$

$$y = 130 \text{ mm}$$

$$\sigma_C = +\frac{M_z}{I_z} y + \frac{M_y}{I_y} \times z$$

$$= +\frac{16.67 \times 10^6 \times 130}{2636.4 \times 10^5} + \frac{5 \times 10^6 \times 90}{1263.6 \times 10^5} = +8.21 + 3.56$$

$$= +11.77 \text{ N/mm}^2 \text{ (tensile)}$$

Stress at corner *D*:

$$z = 90 \text{ mm}$$

$$y = 130 \text{ mm}$$

$$\sigma_D = +\frac{M_z}{I_z}\times 130 - \frac{M_y}{I_y}\times 90$$
$$= +8.21 - 3.56$$
$$= +4.65 \text{ N/mm}^2 \text{ (tensile)}$$

5.6 A cantilever beam of clear span 2 m is built into a wall of 30 cm thickness. The beam carries a concentrated load of 1 kN at the free end. Assume that the pressure exerted by the wall on the beam is linearly varying along the 30 cm thickness of the wall into which the beam is embedded. Find the pressure distribution exerted by the wall on the beam.

[15 marks : 2008]

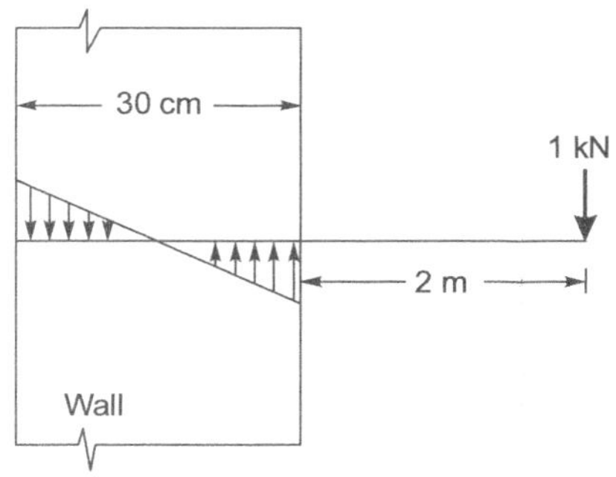

Solution:

Let moment at the centre of wall is M (hogging).

Taking moment about the centre of wall, we get

$$-1\times\left(2+\frac{30}{2}\times\frac{1}{100}\right)+M = 0$$

$$\Rightarrow \quad M = 2.15 \text{ kN-m}$$

Let the intensity of pressure at the ends be w per unit length.

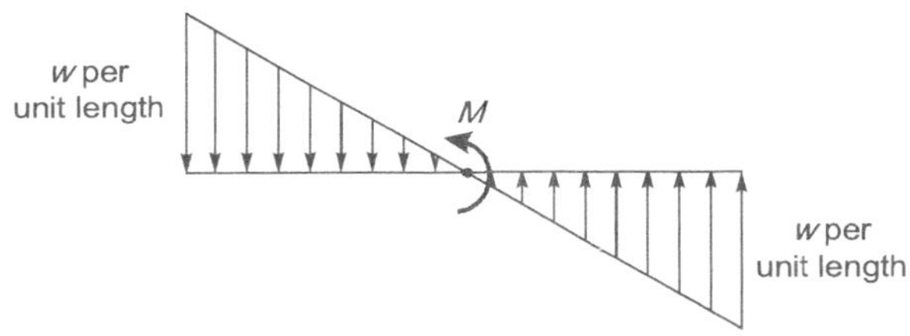

$$\therefore \quad M = \left[\frac{1}{2}\times w\times\left(\frac{30}{2}\times\frac{1}{100}\right)\times\left(\frac{2}{3}\times\frac{30}{2}\times\frac{1}{100}\right)\right]\times 2$$

$$\Rightarrow \quad 2.15 = w\times 0.15\times 0.1$$

$$\Rightarrow \quad w = \frac{2.15}{0.015} = 143.33 \text{ kN/m}$$

There is a vertical reaction induced on the wall due to the load at the free end.

$\therefore$ Vertical reaction = 1 kN

Thus, intensity of pressure due to vertical load $= \dfrac{1}{30/100} = 3.33$ kN/m

Location of zero pressure point: From similar triangular of resultant pressure distribution

$$\frac{140}{x} = \frac{146.66}{30-x} \Rightarrow \frac{30-x}{x} = \frac{146.66}{140} \Rightarrow \frac{30}{x} = \frac{146.66+140}{140}$$

$$x = 14.65 \text{ cm}$$

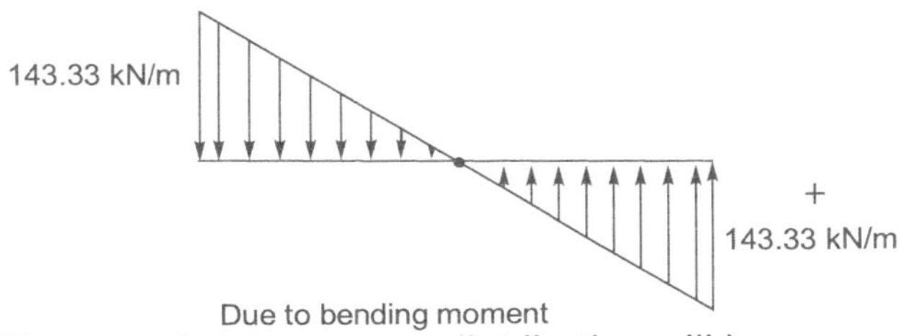

Due to bending moment

+

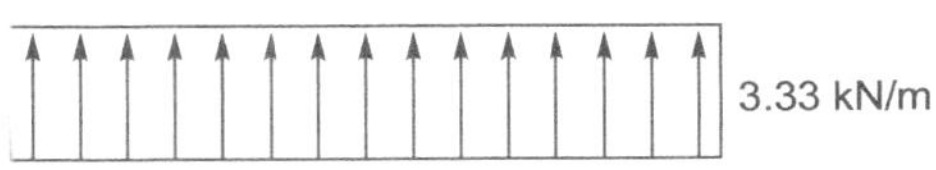

Due to vertical reaction

Thus resultant pressure distribution will be

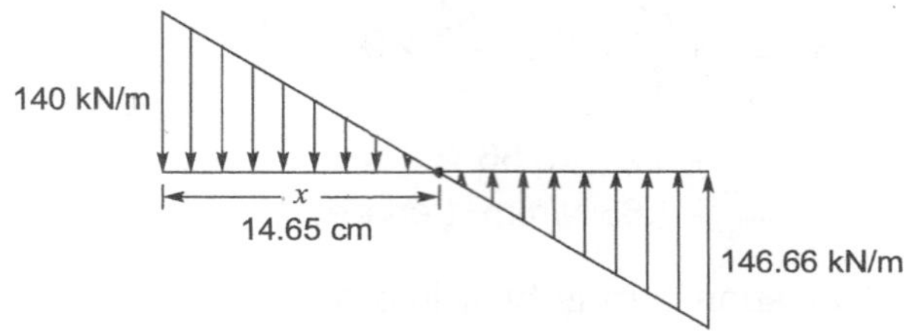

5.7 A simply supported beam of span 5 m and cross-section 200 mm × 600 mm is subjected to a uniformly distributed load of 30 kN/m including self-weight. A plane inclined at 60° to the axis of the beam is passing through a point *P* located on the central cross-section of the beam and 200 mm above the neutral axis. Find the normal stress and shear stress on the inclined plane at point *P*.

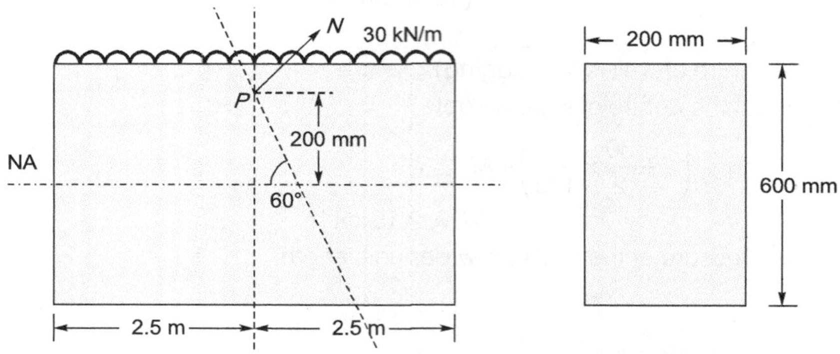

[10 marks : 2010]

Solution:

$$\text{BM at mid span} = \frac{wl^2}{8} = \frac{30(5)^2}{8} = 93.75 \text{ kNm}$$

$$\text{Bending stress at point } P, \quad \sigma_x = \frac{My}{I} = \frac{93.75 \times 10^6 \times 200}{\frac{200(600)^3}{12}} = 5.208 \text{ N/mm}^2$$

Shear force at mid span = 0

∴ Shear stress at point $P = 0$

Normal stress on the inclined plane

$$= \frac{\sigma_x}{2} + \frac{\sigma_x}{2}\cos 60° + 0$$

$$= \frac{-5.208}{2}\left(1 + \frac{1}{2}\right)$$

$$= -3.906 \text{ N/mm}^2$$

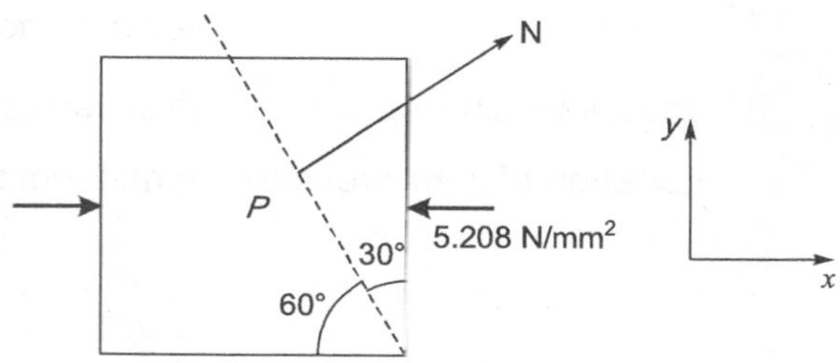

Shear stress on the inclined plane

$$= \frac{-\sigma_x}{2}\sin 60° = \frac{+5.208}{2} \times \frac{\sqrt{3}}{2}$$

$$= +2.255 \text{ N/mm}^2$$

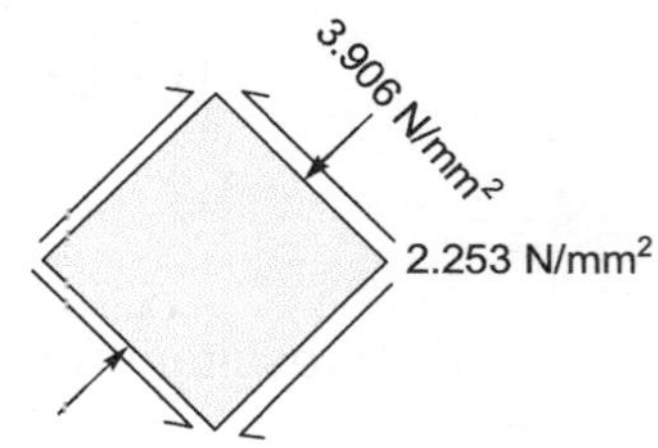

5.8 A circular log of timber has diameter D. Determine the dimension of the strongest rectangular section, one can cut from this log.

[10 marks : 2011]

Solution:

Let the dimension of the strongest rectangular section cut from the circular log of dia D be x and y.

Hence, $$D^2 = x^2 + y^2 \quad \text{...(i)}$$

The strength of any section will depend on its flexural strength i.e. Z

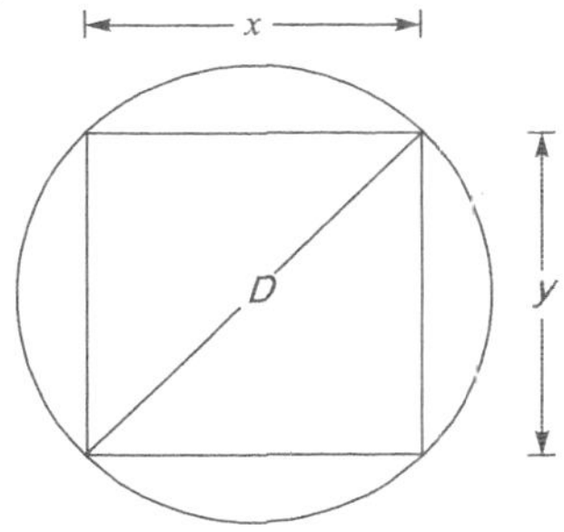

Flexural strength of rectangular section is given by

$$\Rightarrow \quad Z = \frac{I}{y} = \frac{\frac{xy^3}{12}}{\frac{y}{2}} = \frac{xy^2}{6}$$

for Z to be maximum differentiating it w.r.t. either x or y

$$\Rightarrow \quad \frac{dZ}{dx} = 0$$

$$\Rightarrow \quad \frac{d}{dx}\left(\frac{xy^2}{6}\right) = 0$$

From (i): $y^2 = D^2 - x^2$

$$\Rightarrow \quad \frac{d}{dx}\left\{\frac{1}{6} \times x \times \left(D^2 - x^2\right)\right\} = 0$$

$$\Rightarrow \quad \frac{1}{6}\frac{d}{dx}(D^2 x - x^3) = 0$$

$$\Rightarrow \quad \frac{1}{6}\left(D^2 - 3x^2\right) = 0$$

$$\Rightarrow \quad x^2 = \frac{D^2}{3}$$

or $$x = \frac{D}{\sqrt{3}}$$

Put value of x in equation (i)

$$D^2 = \frac{D^2}{3} + y^2$$

$$\Rightarrow \quad y = \sqrt{\frac{2}{3}}D$$

Dimension of strongest rectangular section from circular log of dia D will be

$$x = \frac{D}{\sqrt{3}} \quad \text{and} \quad y = \sqrt{\frac{2}{3}}D$$

5.9 A strip of copper 40 mm wide and 10 mm thick is bounded with another strip of aluminium of same size to form a bimetallic strip of 40 mm × 20 mm. The strip is subjected to a pure bending moment of 0.2 kNm as shown in the below figure. Calculate the radius of curvature of the strip and the maximum tensile and compressive stresses. $E_c = 1 \times 10^5$ MPa, $E_{al} = 0.6 \times 10^5$ MPa.

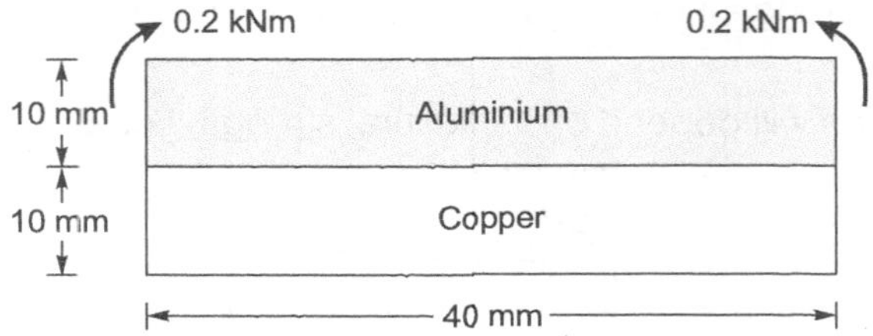

[20 marks : 2012]

Solution:

Since section of beam is composite whole section will bend about common neutral axis. In order to locate the common neutral axis, Let us convert the original section into an equivalent section in terms of copper as shown in the figure.

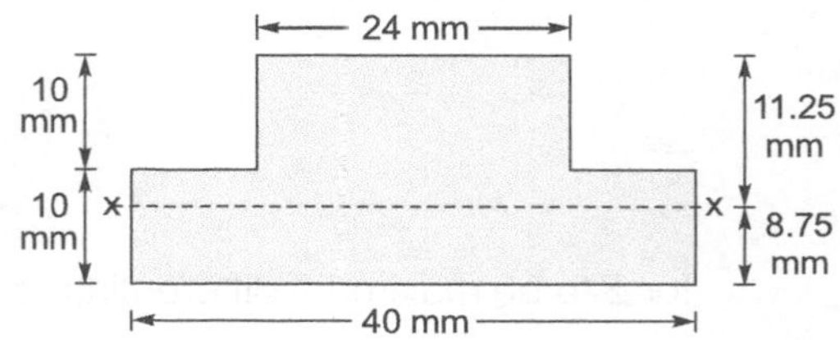

Modular ratio $m = \dfrac{E_c}{E_{al}} = \dfrac{1\times10^5}{0.6\times10^5} = 1.66$

Equivalent width of aluminium strip

$$= \frac{40}{m} = \frac{40}{1.66} = 24 \text{ mm}$$

Let neutral axis is $\bar{y}$ mm above the base of equivalent section

$$\therefore \quad \bar{y} = \frac{40\times10\times5+24\times10\times15}{40\times10+24\times10}$$

$$= 8.75 \text{ mm from the bottom}$$

Moment of inertia of equivalent section about NA is given by

$$I = \frac{40\times10^3}{12} + 40\times10\times(8.75-5)^2 + \frac{24\times10^3}{12} + 24\times10\times(15-8.75)^2$$

$$= 3333.33 + 5625 + 2000 + 9375$$

$$= 20333.33 \text{ mm}^4$$

Now, from the **equation of pure bending,**

$$\frac{M}{I} = \frac{E}{R}$$

$$\therefore \quad R = \frac{EI}{M}$$

$$= \frac{1\times10^5\times20333.33}{0.2\times10^6} \text{mm}$$

$$= 10166.66 \text{ mm} = 10.166 \text{ m}$$

Beam is subjected to sagging bending moment. Hence the stress developed at the top will be compressive and tensile at the bottom.

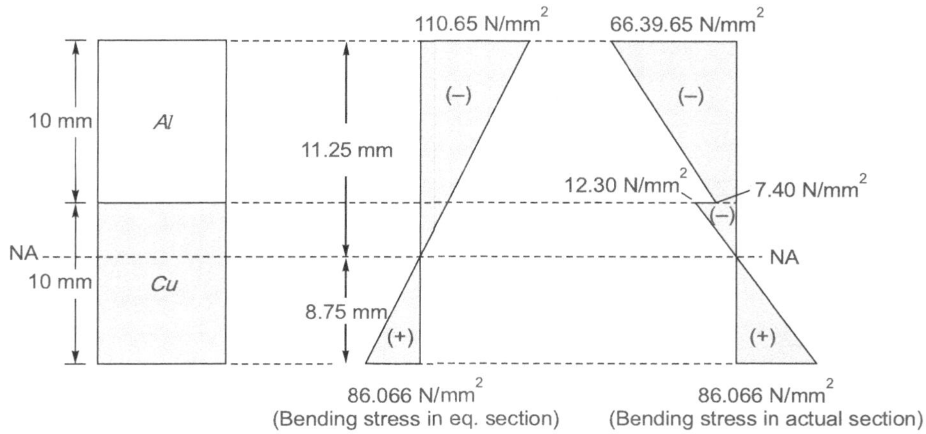

Stress in equivalent section.

$$(\sigma_c)_{top} = \frac{M}{I} \times y_{top} = \frac{0.2 \times 10^6 \times 11.25}{20333.33}$$

$$= -110.65 \text{ N/mm}^2 \text{ (compressive)}$$

and

$$(\sigma_c)_{bottom} = \frac{0.2 \times 10^6 \times 8.75}{20333.33}$$

$$= +86.066 \text{ N/mm}^2 \text{ (tensile)}$$

Stress in copper at the junction

$$\frac{110.65}{11.25} = \frac{(\sigma_c)_J}{1.25}$$

$$(\sigma_c)_J = 12.30 \text{ N/mm}^2$$

∴ Corresponding stress in aluminium are,

$$(\sigma_{Al})_T = \frac{(\sigma_c)_T}{m} = \frac{110.65}{1.66} = 66.65 \text{ N/mm}^2$$

$$(\sigma_{Al})_J = \frac{(\sigma_c)_J}{m} = \frac{12.30}{1.66} = 7.40 \text{ N/mm}^2$$

5.10 A simply supported rectangular beam of length L carries a UDL over its entire length. Determine the critical length at which the shearing stress 'τ' and flexural stress 'σ' reach their allowable values simultaneously. The breadth of beam section is 'b' and depth is 'd'.

[4 marks : 2013]

Solution:

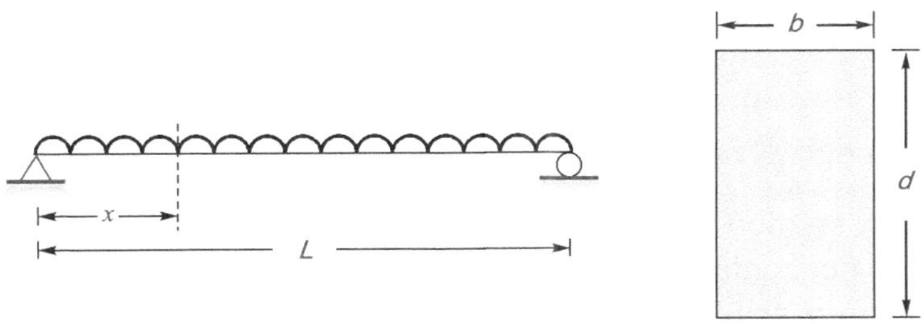

Let intensity of loading is w unit/length

(i) Position of neutral axis: $y = \frac{d}{2}$

(ii) Bending stress: $\sigma = \frac{My}{I} = \frac{M\frac{d}{2}}{bd^3/12} = \frac{6M}{bd^2}$

Maximum value of bending stress will occur at mid span

$$\sigma_{max} = \frac{wl^2}{8}$$

(iii) Shear stress

For rectangular section $\tau_{max} = \frac{3}{2}\tau_{av}$

$$\tau_{max} = \frac{3}{2}\times\frac{V}{bd} = \frac{3V}{2bd}$$

$$V = \frac{wL}{2} - wx$$

$$\therefore \quad \frac{M}{V} = \frac{Lx - x^2}{L - 2x}$$

$$\frac{Lx - x^2}{L - 2x} = \frac{\sigma_{max}\frac{bd^2}{6}}{\tau_{max}\frac{2}{3}bd}$$

Maximum shear stress will occur at support

$$\frac{3}{2}\times\frac{wl}{2bd} = \tau_{max}$$

$$\tau_{max} = \frac{2}{3bd} = \frac{wl}{L}$$

$$\frac{Lx - x^2}{L - 2x} = \frac{wl^2/8}{wl/L} = \frac{l}{4}$$

$$L^2 - 2\,Lx = 4\,Lx - 4x^2$$

$$Lx^2 - 6Lx + L^2 = 0$$

$$x = \frac{6L \pm \sqrt{36L^2 - 4\times 4\times L^2}}{8}$$

$$x = \frac{6L \pm \sqrt{20L^2}}{8} = \left(6 \pm 2\sqrt{5}\right)\frac{L}{8} = 1.309l, 0.191l$$

Hence critical length will be 0.191 l from left support.

5.11 A beam of square section of the side 'a' is placed such that (i) two sides are horizontal (ii) one diagonal is horizontal. Find the ratio of the moments of resistance of the section in two positions for same permissible bending stress.

[4 marks : 2013]

Solution:

$$MoR\,(M) = f_{max}Z \qquad \therefore M \propto Z$$

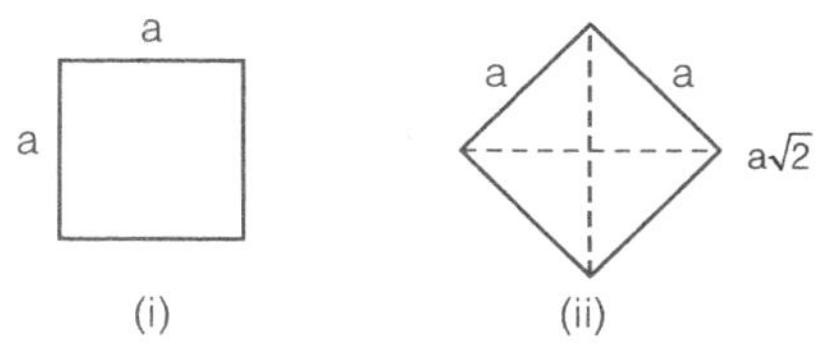

$$Z_1 = \frac{I_1}{a/2} = \frac{a.a^3/12}{a/2} = \frac{a^3}{6}$$

$$Z_2 = \frac{I_2}{a/\sqrt{2}} = \frac{2(a\sqrt{2})(a/\sqrt{2})^3}{12(a/\sqrt{2})} = \frac{(a\sqrt{2})(a/\sqrt{2})^2}{6}$$

$$= \frac{a^3\sqrt{2}}{2\times 6} = \frac{a^3}{6\sqrt{2}}$$

$$\therefore \quad \frac{M_1}{M_2} = \frac{MoR_1}{MoR_2} = \frac{f_{max}Z_1}{f_{max}Z_2} = \frac{Z_1}{Z_2} = \frac{a^3/6}{a^3/6\sqrt{2}} = \sqrt{2}$$

5.12 A rectangular beam 150 mm wide and 200 mm deep, is simply supported on a span of 6 m. Two loads of 5 kN each are applied to the beam, each load being 2 m from the supports as shown in the figures. The plane of loads make an angle of 30° with the vertical plane of symmetry. Find the direction of neutral axis and bending stress at a point marked '*X*'.

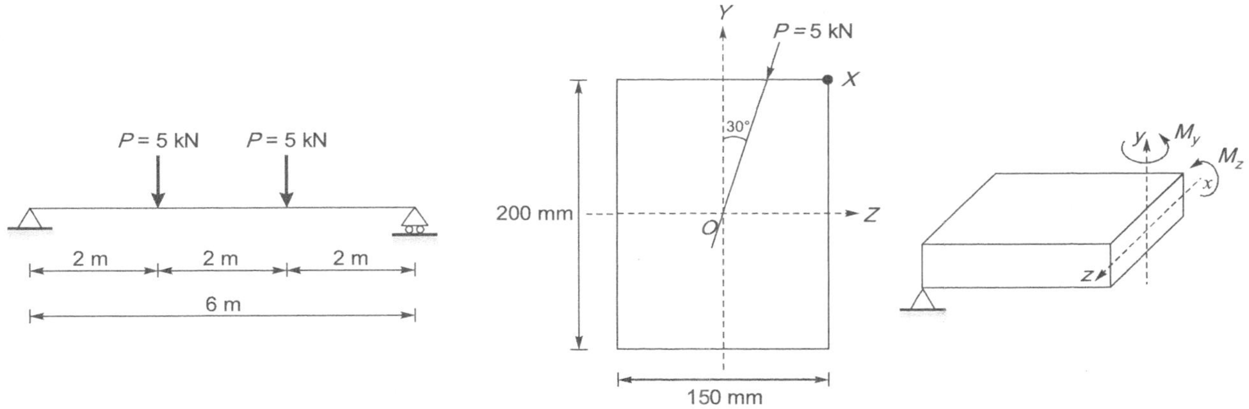

[20 marks : 2015]

Solution:

Since, at neutral axis, value of resultant stress due to moment M_z and M_y is zero.

Thus, **stress due to M_z,**

$$\sigma_z = \frac{M_z}{I_z}\times y$$

$$I_z = \frac{150\times 200^3}{12} = 100\times 10^6\,\text{mm}^4$$

$$M_z = 4.333\times 2 = 8.67\text{ kN-m}$$

$$\therefore \quad \sigma_z = \frac{8.67\times 10^6}{100\times 10^6}\times y = 0.0867\,y$$

Taking, σ_z as tensile.

Stress due to M_y,

$$\sigma_y = \frac{M_y}{I_y}\times z$$

$$\because \quad M_y = 2.5\times 2 = 5\text{ kN-m}$$

$$I_y = \frac{200 \times (150)^3}{12} = 56.25 \times 10^6 \text{ kN-m}$$

$$\therefore \quad \sigma_y = \frac{5 \times 10^6}{56.25 \times 10^6} \times z = 0.089z$$

Taking σ_y as compressive.

At neutral axis, σ_y and σ_z will be equal numerically.

Thus,
$$\sigma_y = \sigma_z$$
$$0.0867y = 0.089z$$
$$\Rightarrow \quad \frac{y}{z} = 1.025$$

Since,
$$\tan\theta = \frac{y}{z}$$
$$\therefore \quad \tan\theta = 1.025$$
$$\Rightarrow \quad \theta = 45.71°$$

Bending stress at 'X':

$$\sigma_z = 0.0867y = 0.0867 \times 100 = 8.67 \text{ N/mm}^2 \quad \text{(Compressive)}$$
$$\sigma_y = 0.089z = 0.089 \times 75 = 6.675 \text{ N/mm}^2 \quad \text{(Compressive)}$$

$\therefore$ **Resultant bending stress,**
$$\sigma = \sigma_z\,(\text{compressive}) + \sigma_y\,(\text{compressive})$$
$$= 8.67 + 6.675 = 15.345 \text{ N/mm}^2 \quad \text{(Compressive)}$$

5.13 **A bar with a square section of side 100 mm is used as a beam and placed in such a way that the diagonal is horizontal. When the section of the beam is subjected to a shear force of 600 kN, determine the maximum shear stress in the section and its location across its depth.**

[20 marks : 2017]

Solution:

The figure below shows a square beam $ABCD$, having diagonal AC horizontal.

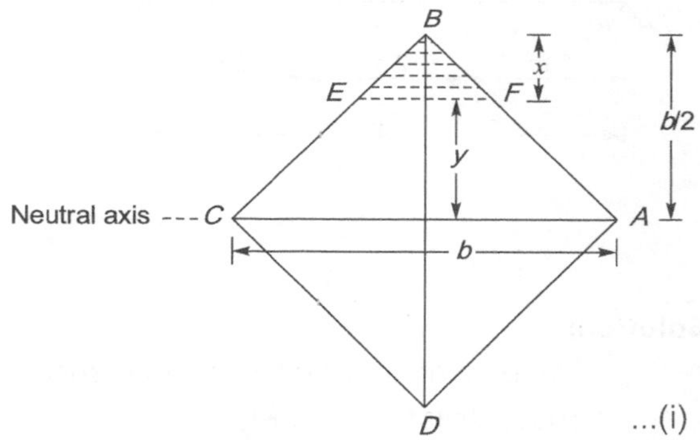

Let b = length of diagonal AC. This is also length of diagonal BD.

The N.A. of the beam shown in above figure passes through diagonal AC. Consider a level EF at a distance y from the NA. The shear stress at this level is given by

$$\tau = \frac{FA\bar{y}}{I \times b} \quad \text{...(i)}$$

where, $A\bar{y}$ = moment of shaded area about NA.

= Area of triangle BEF × Distance of C.G. of triangle BEF from N.A.

$$= \left(\frac{1}{2} \times EF \times x\right)\left(\frac{b}{2} - \frac{2}{3}x\right) = \left(\frac{1}{2} \times 2x \times x\right)\left(\frac{b}{2} - \frac{2}{3}x\right)$$

$$= x^2\left(\frac{b}{2} - \frac{2x}{3}\right)$$

I = Moment of inertial of the whole section about N.A.

$$= 2 \times \left\{\frac{b \times \left(\frac{b}{2}\right)^3}{12}\right\} = \frac{b^4}{48}$$

Substituting these values in equation (i), we get

$$\tau = \left\{\frac{Fx^2\left(\frac{b}{2}-\frac{2x}{3}\right)}{\frac{b^4}{48}\times 2x}\right\} = \frac{F\times 24x}{b^4}\left[\frac{3b-4x}{6}\right] = \frac{4F}{b^4}x(3b-4x)$$

At the top, $x = 0$, hence $\tau = 0$

At the NA, $x = \frac{b}{2}$ and hence $\tau = \frac{4F}{b^4}\times\frac{b}{2}\left(3b-\frac{4b}{2}\right)$

$$= \frac{4F}{b^4}\times\frac{b}{2}\times b = \frac{2F}{b^2}$$

Maximum shear stress:

Maximum shear stress will be obtained by differentiating equation (ii) w.r.t. x and equating it to zero.

$$\therefore \quad \frac{d}{dx}\left[\frac{4F}{b^4}(3bx-4x^2)\right] = 0$$

$$\Rightarrow \quad \frac{4F}{b^4}(3b-8x) = 0$$

$$\Rightarrow \quad x = \frac{3}{8}b \qquad \left[\because \frac{4F}{b^4}\text{ (cannot be zero)}\right]$$

Substituting this value of x in equation (ii), we get maximum shear stress.

$$\tau_{max} = \frac{4F}{b^4}\times\frac{3b}{8}\left(3b-4\times\frac{3b}{8}\right) = \frac{4F}{b^4}\times\frac{3b}{8}\times\frac{3b}{2} = \frac{9}{4}\frac{F}{b^2}$$

Here, $F = 600$ kN

and side of square section $= \sqrt{\left(\frac{b}{2}\right)^2+\left(\frac{b}{2}\right)^2} = \sqrt{\frac{b^2}{4}+\frac{b^2}{4}} = \frac{b}{\sqrt{2}} = 100$ mm

$$\Rightarrow \quad b = 100\sqrt{2}\text{ mm}$$

$$\therefore \quad \tau_{max} = \frac{9\times 600\times 10^3}{4\times(100\sqrt{2})^2} = 67.5\text{ N/mm}^2$$

$$\tau_{max}\text{ will occur at } x = \frac{3\times 100\sqrt{2}}{8} = 53.03\text{ mm from top.}$$

Alternative

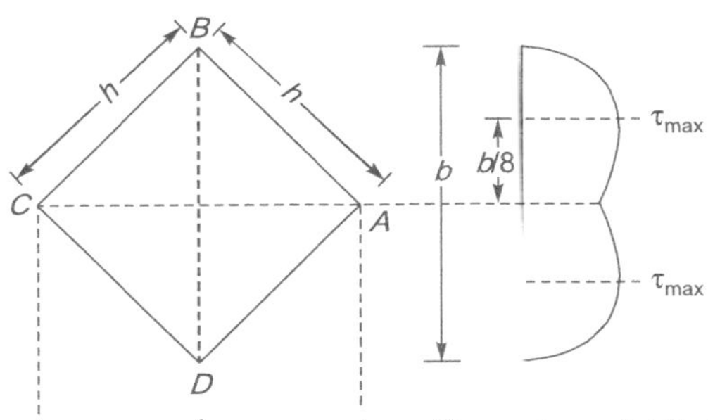

Shear stress distribution

$$\tau_{max} = \frac{9}{8}\tau_{av} = \frac{9}{8}\left(\frac{F}{h\times h}\right) = \frac{9}{8}\times\frac{600\times 10^3}{100\times 100}$$

$$\tau_{max} = 67.5\text{ N/mm}^2$$

From geometry,

$$\sqrt{2}\cdot h = b$$

$$\sqrt{2}\times 100 = b$$

$$b = 141.421 \text{ mm}$$

Maximum shear stress will occur at a distance of $D/8$ from the neutral axis

$$\text{Hence, distance from top} = \frac{b}{2}-\frac{b}{8} = \frac{3}{8}b = \frac{3}{8}\times 141.421 = 53.03 \text{ mm}$$

5.14 **A bracket has been loaded as shown in figure. Compute the top and bottom fibre stress at section A-B.**

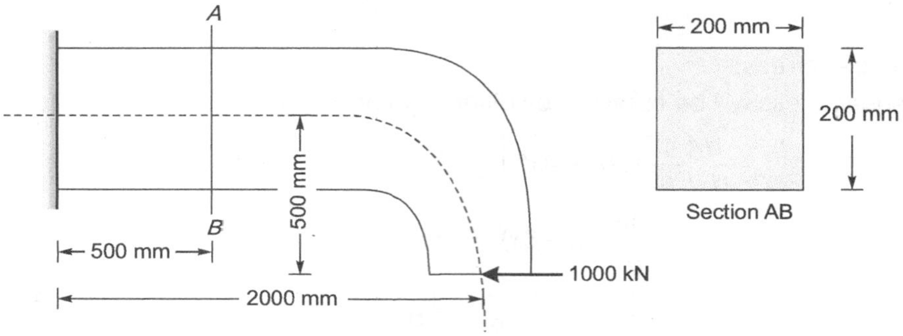

[12 marks : 2018]

Solution:

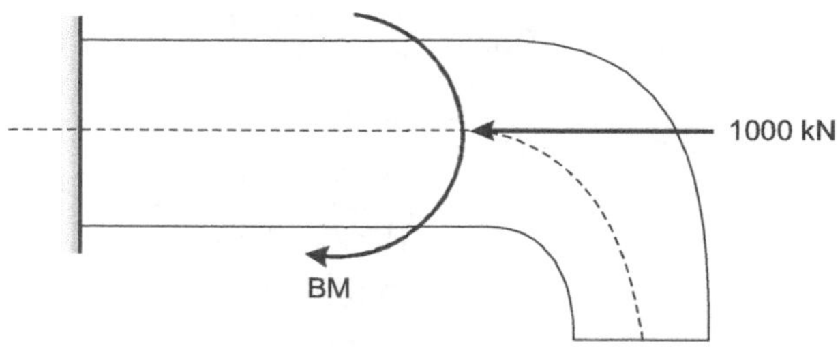

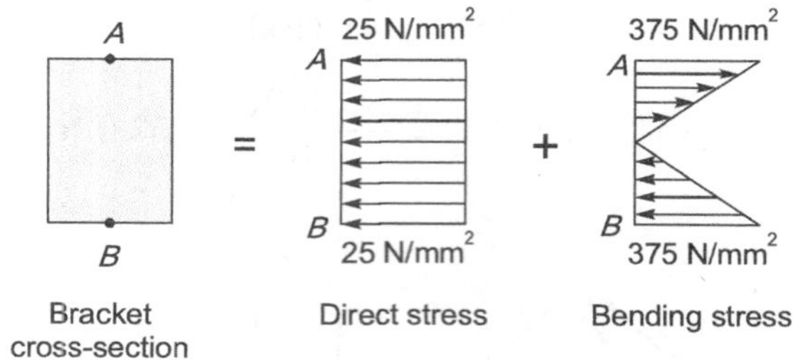

$$\text{BM} = (1000\times 0.5) = 500 \text{ kN-m}$$

Direct stress, $$\sigma_d = \frac{1000\times 10^3}{200\times 200} = 25 \text{ N/mm}^2 \text{(compressive)}$$

Bending stress, $$\sigma_b = \frac{My}{I} = \frac{500\times 10^6\times 100}{200\times \dfrac{200^3}{12}} \quad (y_{top} = y_{bottom} = 100 \text{ mm})$$

$\Rightarrow$ $\sigma_b = 375 \text{ N/mm}^2$ (Tensile at top and compressive at bottom)

(i) Stress at top fibre $= 375 - 25 = 350 \text{ N/mm}^2$ (Tensile)

(ii) Stress at bottom fibre $= 25 + 375 = 400 \text{ N/mm}^2$ (Compressive)

5.15 A beam of unsymmetrical I-section shown in figure is simply supported over a span of 8 m. It carries a uniformly distributed load of 20 kN/m over the entire span. Draw the sketch for shear stress variation across the depth of the cross-section located at a distance of 3 m from left end *A*.

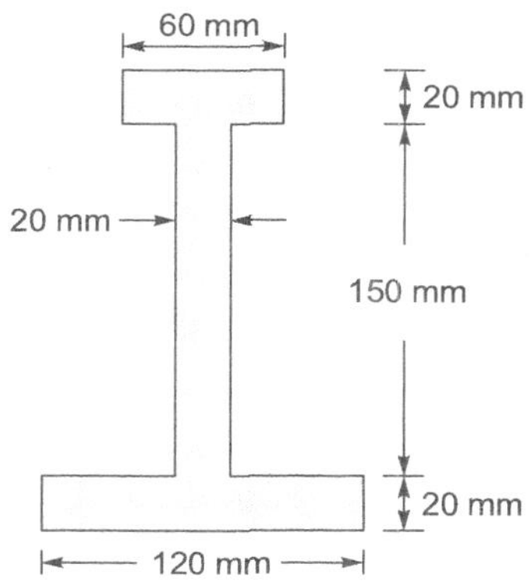

[20 marks : 2018]

Solution:

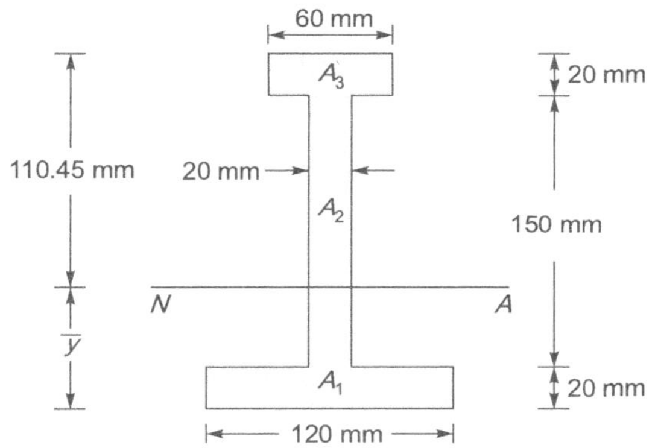

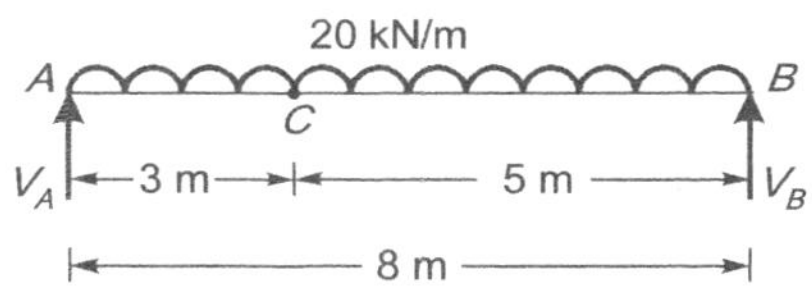

(i) Calculate shear force at *C*, $V_A = V_B = \dfrac{20 \times 8}{2} = 80$ kN

Shear force (SF) at C = 80 – (20 × 3) = 20 kN

(ii) Let neutral axis be at distance $\bar{y}$ from the bottom fibre.

So,
$$\bar{y} = \frac{A_1\bar{y}_1 + A_2\bar{y}_2 + A_3\bar{y}_3}{A_1 + A_2 + A_3}$$

$$= \frac{(120 \times 20 \times 10) + (20 \times 150) \times (20 + 75) + (60 \times 20)(170 + 10)}{(120 \times 20) + (150 \times 20) + (60 \times 20)}$$

$\Rightarrow$
$$\bar{y} = 79.55 \text{ mm}$$

(iii) Moment of inertia,
$$I = \left[\frac{60 \times 20^3}{12} + 60 \times 20 \times (110.45 - 10)^2\right]$$
$$+ \left[\left(\frac{20 \times 150^3}{12}\right) + \left(20 \times 150 \times (110.45 - 95)^2\right)\right]$$
$$+ \left[\frac{20^3 \times 120}{12} + 20 \times 120 \times (79.55 - 10)^2\right]$$

$$= 30178636.5 \text{ mm}^4$$

Now, shear stress at junction of top web and flange in flange portion is given by

$$\tau_1 = \frac{SA\bar{y}}{Ib} = \frac{20 \times 1000 \times 60 \times 20 \times (110.45 - 10)}{30178636.5 \times 60}$$

$$= 1.33 \text{ N/mm}^2$$

Shear stress at junction of web and top flange in web portion is given by,

$$\tau_1 b_1 = \tau_2 b_2$$

$$\Rightarrow \quad 1.33 \times 60 = \tau_2 \times 20$$

$$\Rightarrow \quad \tau = 1.33 \times \frac{60}{20} = 3.99 \text{ N/mm}^2$$

Shear stress at neutral axis

$$\tau = \frac{SA\bar{y}}{Ib} = \frac{20 \times 1000 \times \left[\left(20 \times 90.45 \times \frac{90.45}{2}\right) + (20 \times 60 \times 100.45)\right]}{30178636.5 \times 60}$$

$$\Rightarrow \quad \tau = 6.705 \text{ N/mm}^2$$

Shear stress at junction of web and bottom flange in flange portion is given by

$$\tau_3 = \frac{SA\bar{y}}{Ib} = \frac{20 \times 1000 \times 120 \times 20 \times (79.55 - 10)}{30178636.5 \times 120}$$

$$\Rightarrow \quad \tau_3 = 0.922 \text{ N/mm}^2$$

Shear stress at junction of web and bottom flange in web portion is given by

$$\tau_3 b_3 = \tau_4 b_4$$

$$\Rightarrow \quad 0.922 \times 120 = \tau_4 \times 20$$

$$\Rightarrow \quad \tau_4 = 0.922 \times \frac{120}{20} = 5.531 \text{ N/mm}^2$$

Shear stress variation across depth

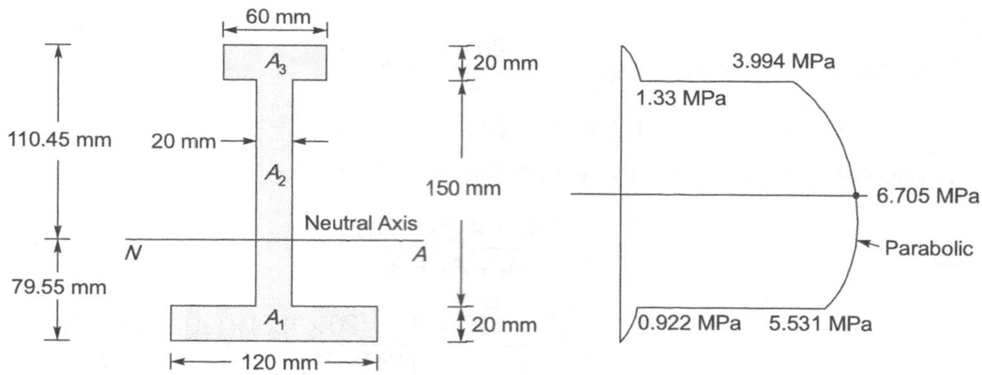

5.16 **A water main of 1200 mm internal diameter and 12 mm thick is running full. If the bending stress is not to exceed 56 MPa, find the longest span on which the pipe may be freely supported. Steel and water weigh 76.8 kN/m³ and 10 kN/m³ respectively.**

[12 marks : 2019]

Solution:

Given,

Internal diameter of water main = 1200 mm

Thickness, $t = 12$ mm

Given, bending stress is not to exceed 56 MPa

Unit weight of steel = 76.8 kN/m^3

Unit weight of water = 10 kN/m^3

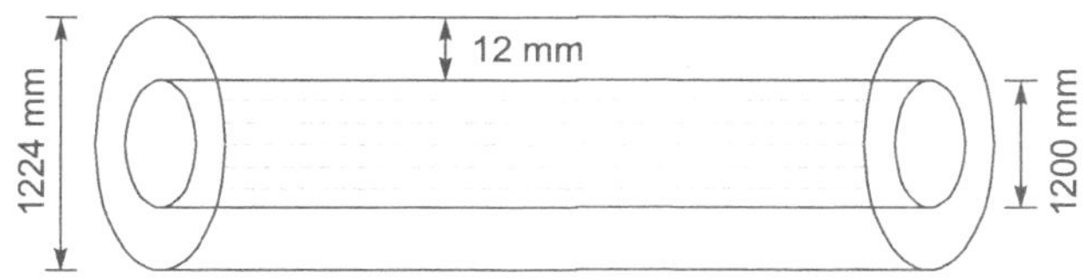

$$\text{Weight of water inside water main} = 10 \times \frac{\pi}{4} \times 1.2^2 = 11.3097 \text{ kN/m} \simeq 11.31 \text{ kN/m}$$

$$\text{Weight of steel tube} = 76.8 \times \frac{\pi}{4} \times \left(1.224^2 - 1.2^2\right)$$

$$= 3.509 \text{ kN/m} \simeq 3.51 \text{ kN/m}$$

Total uniformly distributed load,

$$w = 11.31 + 3.51 = 14.82 \text{ kN/m}$$

Max. BM,
$$M_{max} = \frac{wl^2}{8} = \frac{14.82 \times l^2}{8} \text{ kN/m} = 1.8525\, l^2 \text{ kNm}$$

Now, using bending equation,

$$\frac{M}{I} = \frac{\sigma}{y}$$

Here,
$$I = \frac{\pi}{64}\left(1224^4 - 1200^4\right) = 8.39 \times 10^9 \text{ mm}^4$$

By putting the values,
$$\sigma = \frac{M}{I} \times y$$

$$56 \text{ N/mm}^2 = \frac{1.8525 l^2 \times 10^6 \text{ Nmm} \times 612 \text{ mm}}{8.39 \times 10^9 \text{ mm}^4}$$

$$l = 20.36 \text{ m}$$

So, the maximum possible length of water main should be 20.457 m so that the bending stress is not exceed 56 MPa.

5.17

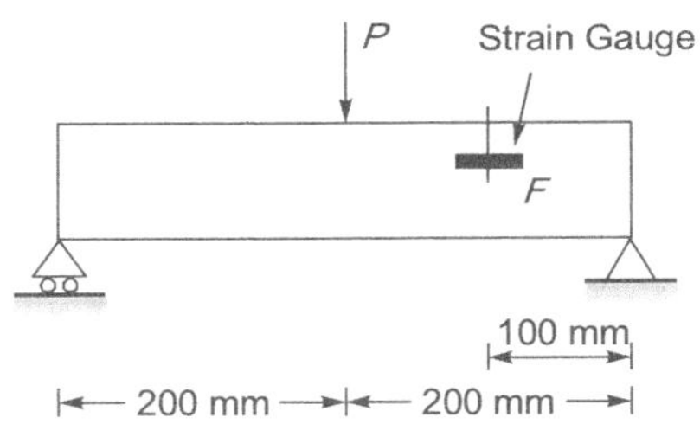

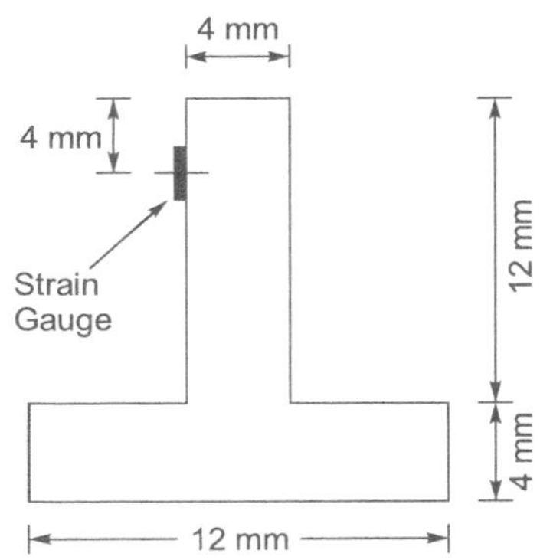

A small T-section is used in inverted position as a beam and is shown in figure over a span of 400 mm. If due to the application of forces shown, the longitudinal strain gauge at F registers a compressive strain of 1500 microstrains, determine the magnitude of P. Take E = 200 GPa.

[10 marks : 2019]

Solution:

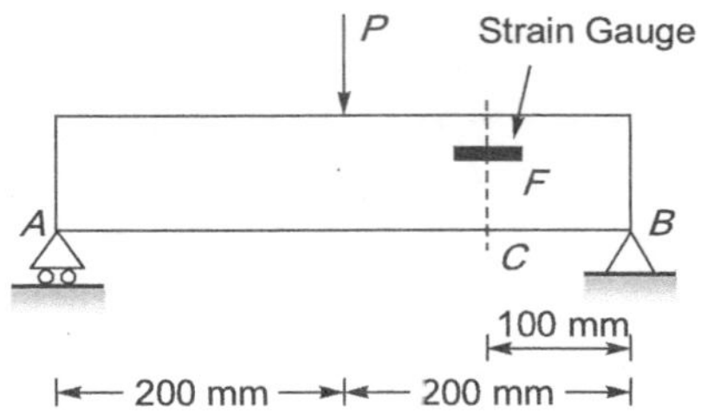

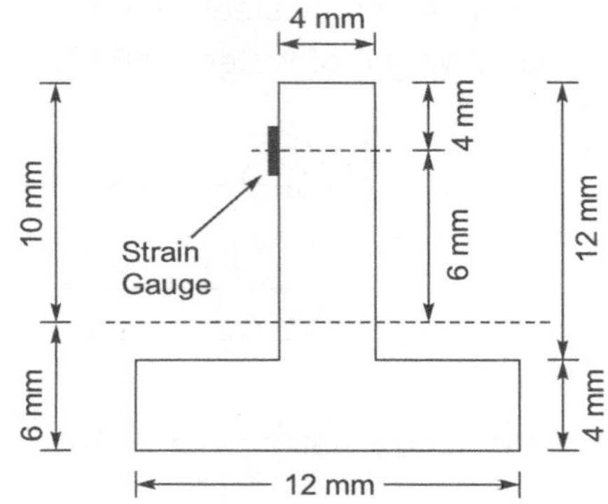

For CG of cross-section from bottom.

$$\bar{y} = \frac{12\times4\times2+12\times4\times10}{12\times4\times2} = 6 \text{ mm}$$

$$I = \frac{12\times4^3}{12}+12\times4\times4^2+\frac{4\times12^3}{12}+12\times4\times4^2$$

$$= 2176 \text{ mm}^4$$

Now as we know, $\frac{M}{I} = \frac{E}{R} = \frac{\sigma}{y}$

Stain at $F = 1500 \times 10^{-6}$

Stress at $F = 1500 \times 10^{-6} \times E = 300$ MPa

$$\frac{M}{I} = \frac{\sigma}{y}$$

$$\Rightarrow \quad M = \frac{300\times2176}{6} = 108800 \text{ Nmm}$$

Now moment at cross-section 'C' will be

$$= \frac{P}{2}\times100 = 50P$$

$$50\,P = 108800$$

$$P = 2176 \text{ N}$$

$$P = 2.176 \text{ kN}$$

6. Thick & Thin Cylinders and Spheres

6.1 A thin walled tube of circular cross-section with outer diameter 100 mm, thickness 2 mm and length 1000 mm is fixed at one end. It is subjected to a twisting moment of 1 kNm at the free end. Find the shear stress in the wall of the tube and the angle of twist at the free end. $E = 2 \times 10^5$ MPa and Poisson's ratio = 0.25.

What will be the shear stress in the wall of the tube if the cross-section of the tube is square with outside dimensions 100 mm × 100 mm and wall thickness 2 mm?

[10 marks : 2012]

Solution:

The shear stress in thin walled tube is given by

$$\tau = \frac{T}{2A_m t} \quad \text{...(i)}$$

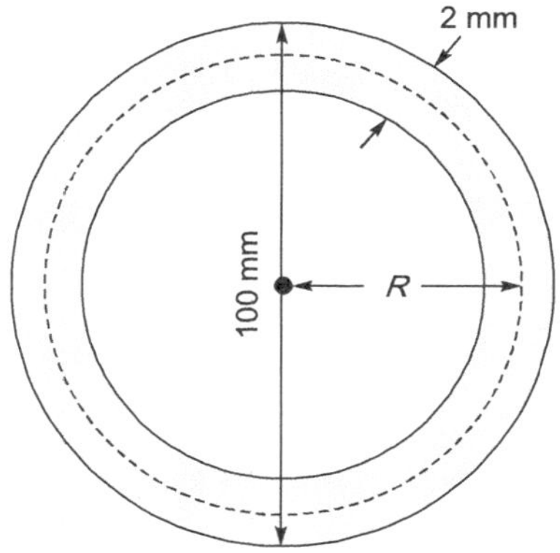

Where, A_m = area bounded by mean line

T = 1 kNm

Length of tube, L = 1000 mm

Mean radius, R = 50 – 1 = 49 mm

∴ Area bounded by mean line, $A_m = \pi R^2 = \pi \times 49^2\ \text{mm}^2$

∴ Shear stress in wall of tube, $\tau = \dfrac{1\times10^6}{2\times\pi\times49^2\times2} = 33.14\ \text{N/mm}^2$

Angle of twist for thin walled tube

$$\theta = \frac{TL}{4A_m^2 G}\Sigma\left(\frac{\Delta S}{t}\right) \quad \text{...(ii)}$$

In this question G is not given, but

$$E = 2G(1+\mu)$$

Hence, $G = \dfrac{E}{2(1+\mu)} = \dfrac{2\times10^5}{2(1+0.25)} = 0.8\times10^5\ \text{N/mm}^2$

From eq. (ii) $\theta = \dfrac{1\times10^6\times1000\times\left(\dfrac{S}{t}\right)}{4\times(\pi\times49^2)^2\times0.8\times10^5}$

$$= \frac{10^6\times1000\times2\pi\times49}{4\times(\pi\times49^2)^2\times0.8\times10^5\times2} = 8.455\times10^{-3}\ \text{radian.}$$

Or, θ can be found by the relation for circular tube

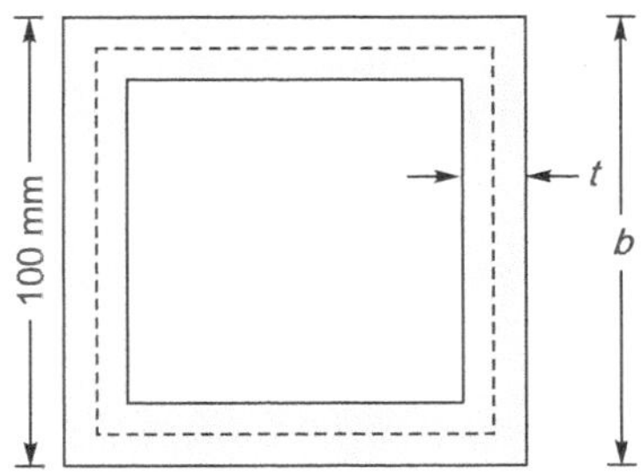

$$\theta = \frac{TL}{GI_P} = \frac{1\times10^6\times1000}{0.8\times10^5\times2\pi\times49^3\times2}$$

$$= 8.455\times10^{-3}\ \text{radian}$$

If section is square.

Mean dimension, b = 100 – 2 = 98 mm

The area bounded by mean line, $A_m = b\times b$

$= 98\times98 = 9604\ \text{mm}^2$

Shear stress, $\tau = \dfrac{T}{2A_m t} = \dfrac{1\times10^6}{2\times9604\times2} = 26.031\ \text{N/mm}^2$

6.2 **A vertical cylindrical steel storage tank has 30 m diameter and the same is filled up to a depth of 15 m with the gasoline of relative density 0.74. If the yield stress for steel is 250 MPa, find the thickness required for the wall plate. Adopt a factor of safety of 2.5 and neglect localized bending effects, if any.**

[10 marks : 2015]

Solution:

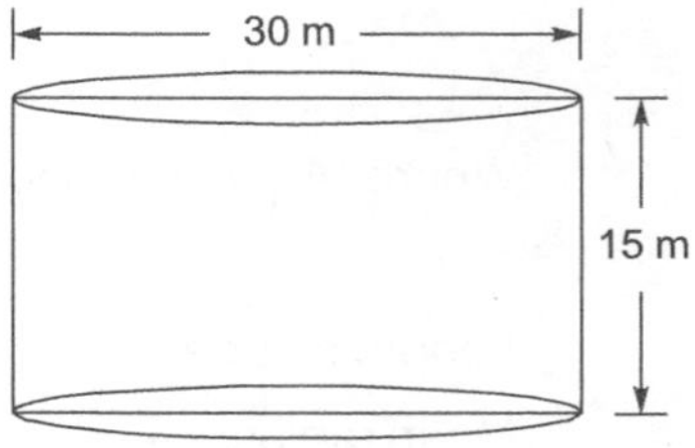

Density of gasoline $= 0.74 \times 1000 = 740 \text{ kg/m}^3$

$$\text{Hoop stress } (\sigma_h) = \frac{pd}{2t}$$

where $p = \rho g h = 740 \times 9.81 \times 15$

$= 1.089 \times 10^5 \text{ N/m}^2$

$= 0.1089 \text{ N/mm}^2$

$$\therefore \quad \sigma_h = \frac{0.1089 \times (30000)}{2t} \leq 250 \text{N/mm}^2$$

$\Rightarrow \quad t \geq 6.534 \text{ mm}$ ...(i)

$$\text{Longitudinal stress } (\sigma_t) = \frac{pd}{4t} = \frac{0.1089(30000)}{4t} \leq 250 \text{N/mm}^2$$

$\therefore \quad t \geq 3.267 \text{ mm}$ ...(ii)

From (i) and (ii), $t \geq 6.534 \text{ mm}$ (Maximum of above two thickness)

Using a FOS of 2.5, $t \geq 6.534 \times 2.5 = 16.335 \text{ mm}$

Adopt $t = 18 \text{ mm}$

7. Torsion in Shafts & Springs

7.1 Calculate the diameters of two solid shafts to transmit 200 horse power (metric) each without exceeding a shear stress of 700 kg/cm². One of these shafts operates at 20 rpm and the other at 20,000 rpm. Comment on the result with reference to savings of the material.

[10 marks : 2005]

Solution:

Given: $P \approx 200 \text{ metric HP} = 200 \times 0.735 \text{ kW} = 147 \text{ kW}$

... (1 mHP = 0.735 kW)

$\tau_{max} = 700 \text{ kg/cm}^2 = 70 \text{ MPa}, N_1 = 20, N_2 = 20000$

(i) For first shaft,
$$P = \frac{2\pi N_1 T_1}{60}$$

$$\Rightarrow \quad T_1 = \frac{60P}{2\pi N_1} = \frac{60 \times 147}{2\pi \times 20} = 70.187 \text{kN-m}$$

Let the diameter of the shaft is D_1 then, we know that $\frac{\tau_{max}}{R} = \frac{T}{I_P}$

$$\Rightarrow \quad \frac{\tau_{max}}{D_1/2} = \frac{T}{\frac{\pi}{32}D_1^4}$$

$$\Rightarrow \quad \frac{2\tau_{max}}{D_1} = \frac{32T}{\pi D_1^4}$$

$$\Rightarrow \quad D_1^3 = \frac{16T}{\pi\tau_{max}}$$

$$\Rightarrow \quad D_1^3 = \frac{16 \times 70.187 \times 10^6}{\pi \times 70}$$

$$\Rightarrow \quad D_1 = 172.204 \text{ mm}$$

(ii) For second shaft,

$$P = \frac{2\pi N_2 T_2}{60}$$

$$T_2 = \frac{60P}{2\pi N_2} = \frac{60 \times 147}{2\pi \times 20000} = 0.070187 \text{ kN-m}$$

and
$$D_2^3 = \frac{16 \times 0.070187 \times 10^6}{\pi \times 70}$$

$\Rightarrow$
$$D_2 = 17.2204 \text{ mm}$$

Now,
$$\text{Area of 1}^{st} \text{ bar (shaft)} = \frac{\pi}{4} \times D_1^2 = \frac{\pi}{4} \times 172.204^2 = 23290.36 \text{ mm}^2$$

$$\text{Area of 2}^{nd} \text{ bar (shaft)} = \frac{\pi}{4} \times D_2^2 = \frac{\pi}{4} \times 17.2204^2 = 232.9036 \text{ mm}^2$$

Now,
$$\text{Ratio of two areas} = \frac{A_2}{A_1} = \frac{232.9036}{23290.36} \times 100 = 1.00\%$$

From the above calculation, it is evident that when the rpm of the shaft is increased from 20 to 20,000, then diameter of the shaft is reduced enormously, which results into the saving of the material of the shaft which is as high as 99%.

7.2 **Two circular shafts *AB* and *BC* of the same material but different diameter are welded together at point *B* as shown in the figure. Ends *A* and *C* are fixed. An external torque with magnitude T_o is applied to the shafts at point *B*. Find the torques exerted on the ends of the shaft at *A* and *C*.**

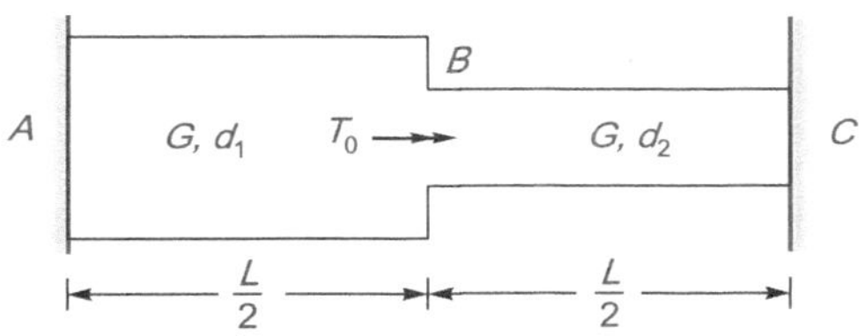

[15 marks : 2006]

Solution:

Since the material is same, i.e., *E* is same, therefore modulus of rigidity, *G* will also be same.
Let T_A and T_C be the torsional reaction at end *A* and *C* respectively.
Since shaft is in equilibrium

$$T_A + T_C = T_0 \quad \text{...(i)}$$

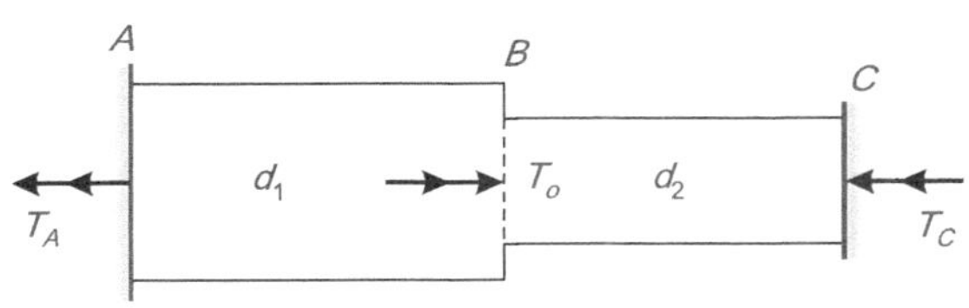

Drawing *FBD* of the whole assembly in parts, we have

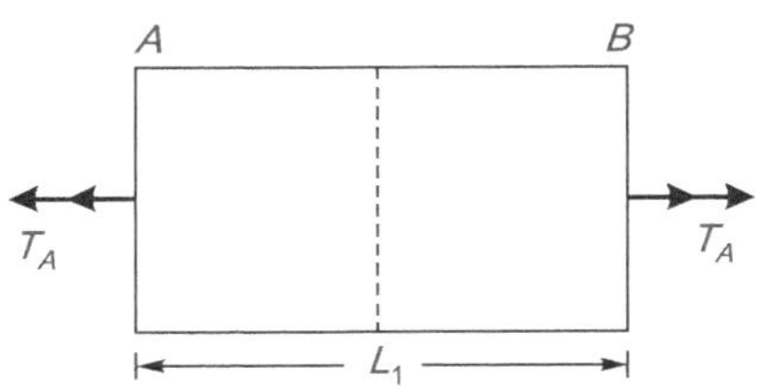

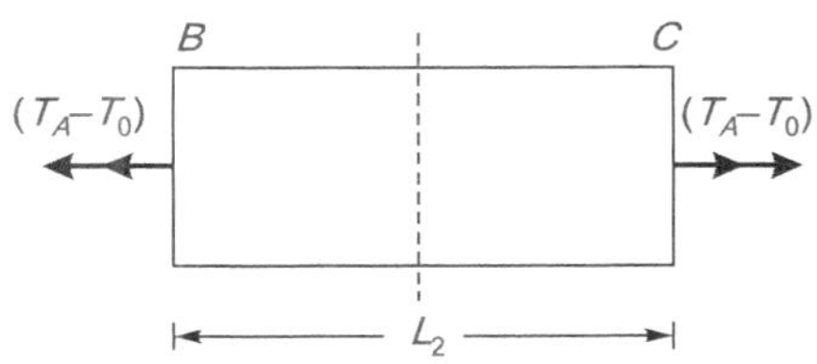

$$T_{AB} = -T_A$$

and $$T_{BC} = -(T_A - T_0)$$

Since the ends are fixed. Hence total angle of twist from *A* to *C* will be zero.

$\therefore$ $$\theta_{AC} = \theta_{AB} + \theta_{BC} = 0$$

$$\frac{-T_A L_1}{GI_{P1}} + \frac{-(T_A - T_o)L_2}{GI_{P2}} = 0$$

$$\frac{-T_A L_1}{GI_{P1}} + \frac{(T_0 - T_A)L_2}{GI_{P2}} = 0$$

$\therefore$ $$\frac{T_A L_1}{(T_0 - T_A)L_2} = \frac{I_{P1}}{I_P} = \frac{\frac{\pi d_1^4}{32}}{\frac{\pi d_2^4}{32}}$$

$$\frac{T_A L_1}{(T_0 - T_A)L_2} = \left(\frac{d_1}{d_2}\right)^4$$

$$T_A L_1 = (T_0 - T_A)L_2\left(\frac{d_1}{d_2}\right)^4$$

$\therefore$ $$T_A = \frac{T_0 L_2 d_1^4}{L_1 d_2^4 + L_2 d_1^4}$$

and from eq. (i), we get

$$T_C = \frac{T_0 L_1 d_2^4}{L_1 d_2^4 + L_2 d_1^4}$$

7.3 **The internal diameter of a shaft of steel is 70% of the external diameter. The shaft is to transmit 3500 kW at 200 rpm. If the maximum allowable stress in the shaft material is 50 MPa, calculate the diameter of the shaft. Find also the maximum twist of the shaft when it is stressed to the maximum permissible value. The length of the shaft is 4 m. Take *G* = 80 MPa.**

[10 marks : 2006]

Solution:

Let the outer diameter of shaft = D_o, Internal diameter of shaft = $0.7D_o$, Power, P = 3500 kW, N = 200 rpm, τ_{max} = 50 MPa, Length of shaft, L = 4 m = 4000 mm, Modulus of rigidity, G = 80 MPa.

We know that, $$P = \frac{2\pi NT}{60}$$ (Where, T = mean torque)

$\therefore$ $$T = \frac{60P}{2\pi N} = \frac{60\times 3500}{2\pi\times 200} = 167.11 \text{ kNm}$$

Assuming $$T_{max} = T_{mean}$$

$\therefore$ $$T_{max} = 167.11 \text{ kNm}$$

We know, $$T_{max} = \tau_{max} Z_p$$

$\therefore$ $$167.11\times 10^6 = 50\times\frac{\pi(D_o^4 - D_i^4)}{16D_o}$$

$$\frac{167.11\times 16\times 10^6}{50\pi} = \frac{D_o^4 - (0.7)^4 D_o^4}{D_o}$$

$$\frac{167.11\times16\times10^6}{50\pi} = 0.7599\,D_0^{\,3}$$

$$D_o = \left(\frac{167.11\times16\times10^6}{50\times0.7599\times\pi}\right)^{1/3}$$

$$D_o = \mathbf{281.89\ mm}$$

$$D_i = 0.7D_o = \mathbf{197.32\ mm}$$

Maximum twist,

$$\theta = \frac{TL}{GI_P} = \frac{167.11\times10^6\times4000}{80\times\dfrac{\pi\left(281.89^4-197.32^4\right)}{32}}$$

$$\theta = 17.737\text{ radian}$$

7.4 A solid shaft transmits 250 kW at 100 rpm. If the shear stress is not to exceed 75 N/mm², what should be the diameter of shaft? If this shaft is to be replaced by a hollow shaft whose internal diameter shall be 0.6 times the outer diameter, determine the size and percentage saving in weight, maximum stress being same.

[10 marks : 2011]

Solution:

Given, τ_{max} = 75 N/mm², N = 100 rpm, Power, P = 250 kW

We know that,

$$P = \frac{2\pi NT}{60}$$

$$\therefore \quad T = \frac{60P}{2\pi N} = \frac{60\times250}{2\pi\times100} = 23.87\text{ kNm}$$

(i) Diameter of shaft based on strength criteria

We know,

$$T_{max} = \tau_{max}Z_P$$

Assuming

$$T_{max} = T_{mean} = 23.87\text{ kNm}$$

$$\therefore \quad 23.87\times10^6 = 75\times\frac{\pi\times D^3}{16}$$

$$D^3 = \frac{16\times23.87\times10^6}{75\pi}$$

$$\therefore \quad D = \left(\frac{16\times23.87\times10^6}{75\pi}\right)^{1/3}$$

$$D = 117.468\text{ mm}$$

(ii) Now this shaft is replaced by a hollow shaft whose

$$\text{Internal dia} = D_i$$

$$\text{Outer dia} = D_o$$

Given,

$$D_i = 0.6D_o$$

Also,

$$T_{max} = \tau_{max}Z_P$$

$$23.87\times10^6 = 75\times\frac{\pi\left(D_o^4-D_i^4\right)}{16D_o}$$

$$\frac{D_o^4-(0.6D_o)^4}{D_o} = \frac{16\times23.87\times10^6}{75\pi}$$

$$\frac{D_o^4 - 0.1296 D_o^4}{D_o} = \frac{16 \times 23.87 \times 10^6}{75\pi}$$

$$0.8704\, D_o^3 = \frac{16 \times 23.87 \times 10^6}{75\pi}$$

$$D_o = 123.03 \text{ mm}$$

$$\therefore \quad D_i = 0.6 D_o = 73.82 \text{ mm}$$

(iii) Percentage saving in weight

$$= \frac{\text{c/s area of solid shaft} - \text{c/s area of hollow shaft}}{\text{c/s area of solid shaft}}$$

$$= \frac{\frac{\pi D^2}{4} - \frac{\pi}{4}\left(D_o^2 - D_i^2\right)}{\frac{\pi}{4} D^2} = \frac{D^2 - \left(D_c^2 - D_i^2\right)}{D^2}$$

$$= \frac{(117.468)^2 - \left[(123.03)^2 - (73.82)^2\right]}{(117.468)^2} = 0.2979$$

Hence, 29.79% of weight is saved.

7.5 An aluminium shaft with a constant diameter of 50 mm is loaded by torques applied to gears attached to it as shown in figure below. Using G = 28 GPa, determine the relative angle of twist of gear D relative to gear A.

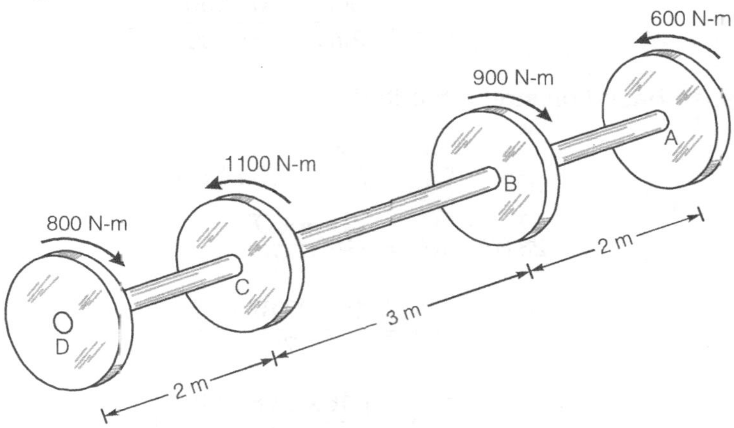

[4 marks : 2013]

Solution:

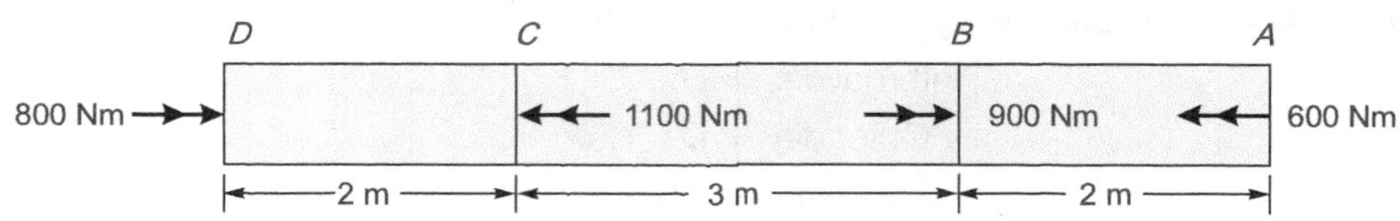

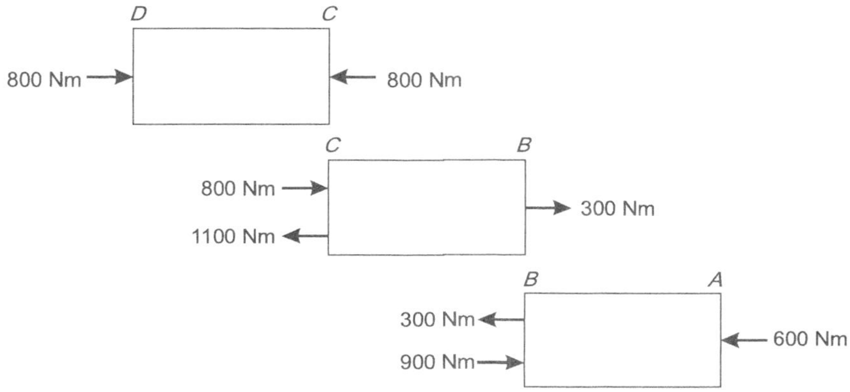

Torques in shafts

$$T_{DC} = +800 \text{ Nm}$$
$$T_{CB} = +800 - 1100 = -300 \text{ Nm}$$
$$T_{BA} = +800 - 1100 + 900 = +600 \text{ Nm}$$

Polar moment of inertia

$$(I_P)_{DC} = (I_P)_{CB} = (I_P)_{BA} = \frac{\pi \times 50^4}{32}$$
$$I_P = 61.36 \times 10^4 \text{ mm}^4$$

Since shafts are in series,

$\therefore$
$$\theta_{DA} = \theta_{DC} + \theta_{CB} + \theta_{BA}$$
$$= \left(\frac{TL}{GI_P}\right)_{DC} + \left(\frac{TL}{GI_P}\right)_{CB} + \left(\frac{TL}{GI_P}\right)_{BA}$$
$$= \frac{800 \times 10^3 \times 2000 - 300 \times 10^3 \times 3000 + 600 \times 10^3 \times 2000}{28 \times 10^3 \times 61.36 \times 10^4}$$
$$= 0.1106 \text{ radian} = 6.34°$$

$\therefore$ Relative angle of twist of gear D with respect to A is 6.34°.

7.6 **A solid circular shaft has a uniform taper from one end to the other. The ratio of radius at the larger end to that at the smaller end is 1.2. Determine the error that occurs if the angle of twist for a given length is calculated using the mean radius of the shaft.**

[20 marks : 2017]

Solution:

Step-1: Calculation of radius :

Let us set up a coordinate system with variable x denoting the distance from small end of the shaft. The radius at a section at the distance x from the small end is

$$r = a + \frac{(b-a)x}{L}$$

where a = radius at small end
b = radius at large end
L = length of the bar

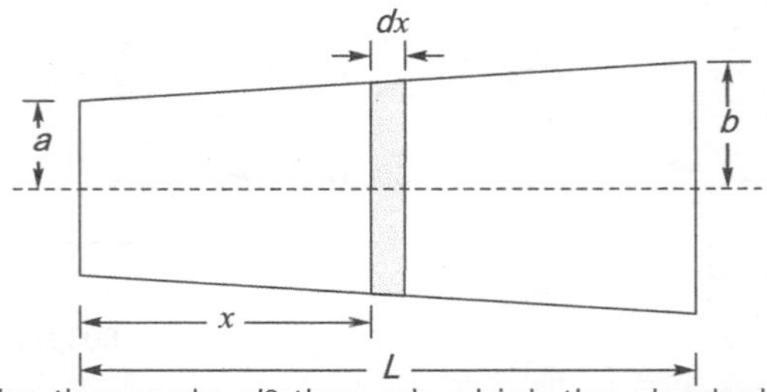

Step-2 : Calculation of polar moment of inertia:

Provided the angle of taper is small, it is sufficient to consider the angle $d\theta$ through which the shaded element of length dx is twisted. This is obtained by applying expression $\theta = \frac{Tl}{GJ}$ to the element of length dx and radius $r = a + \left\{\frac{(b-a)x}{L}\right\}$.

For such an element, the polar moment of inertia is

$$J = \frac{\pi D^4}{32} = \frac{\pi r^4}{2} = \frac{\pi}{2}\left[a + \frac{(b-a)x}{L}\right]^4$$

Thus,

$$d\theta = \frac{Tdx}{G\frac{\pi}{2}\left[a + \frac{(b-a)x}{L}\right]^4} \quad \text{...(i)}$$

Step 3 : Calculation of angle of twist:

The angle of twist in the length L is found by integrating the eq. (i). Thus,

$$\theta = \frac{2T}{G\pi}\int_0^L \frac{dx}{\left[a + \frac{(b-a)x}{L}\right]^4}$$

$$= \frac{2T}{G\pi}\left(\frac{-1}{3}\right)\left(\frac{L}{b-a}\right)\left[\frac{1}{\left\{a + \frac{(b-a)x}{L}\right\}^3}\right]_0^L$$

$$= \frac{2TL}{3G\pi(b-a)}\left\{\frac{-1}{b^3} + \frac{1}{a^3}\right\} \quad \text{...(ii)}$$

Given, $b = 1.2a$

Put this value in eq. (ii), we get

$$\theta = \frac{1.40433TL}{G\pi a^4}$$

$$\text{Mean radius of solid shaft} = \frac{b+a}{2} = \frac{2.2a}{2} = 1.1a$$

$$\therefore \quad \theta_m = \frac{TL}{G \times \frac{\pi}{2}(1.1a)^4} = \frac{1.36603TL}{G\pi a^4}$$

Step 4 : Calculation of percent error,

Using these values of θ and θ_m, we find

$$\text{Percent error} = \left(\frac{1.40433 - 1.36603}{1.40433}\right) \times 100$$

$$= \left(\frac{0.03831}{1.40433}\right) \times 100 \simeq 2.73\%$$

7.7 A single solid circular shaft 400 mm diameter running at 200 RPM, is to be replaced by two hollow circular shafts of equal size running at 100 RPM and developing 50% additional power. The internal diameter of hollow shaft may be taken as one third of their external diameters. If the working stress of the new shaft is 30% greater than that of the former, find the external and internal diameters of the hollow shafts.

[20 marks : 2018]

Solution:

Power, $$P = \frac{2\pi NT}{60} \quad ...(i)$$

and $$\frac{T}{J} = \frac{\tau}{R}$$

$\Rightarrow$ $$T = \frac{\tau J}{R} \quad ...(ii)$$

From (i) and (ii), we get

$$P = \frac{2\pi N}{60}\left(\frac{\tau J}{R}\right) \quad ...(iii)$$

Now, for solid shaft,

$D = 400$ mm, $N = 200$ rpm

So, $$P_1 = \frac{2\pi \times 200}{60} \times \tau_1 \times \frac{\pi}{32} \times \frac{400^4}{200}$$

$\Rightarrow$ $$P_1 = 263.189 \times 10^6 \tau_1 \quad ...(iv)$$

Now, for hollow shafts,

$N = 100$ rpm, $P_2 = 1.5P_1$, $\tau_2 = 1.3\tau_1$

Let external diameter be 'D' and internal diameter be 'd'. So, $d = \left(\frac{D}{3}\right)$

Now, $$P_2 = 2 \times \frac{2\pi N_2}{60} \times \frac{\tau_2 J_2}{R_2}$$

$\Rightarrow$ $$1.5P_1 = 2 \times \frac{2\pi \times 100}{60} \times 1.3\tau_1 \times \frac{\pi}{32} \times \frac{\left[D^4 - \left(\frac{D}{3}\right)^4\right]}{\frac{D}{2}}$$

$\Rightarrow$ $$P_1 = 3.564\tau_1 \frac{D^4}{D}\left(1 - \frac{1}{3^4}\right) \quad ...(v)$$

Dividing (iv) and (v) we get

$$\frac{263.189}{3.564} = \frac{D^4}{D}\left(1 - \frac{1}{3^4}\right)$$

On solving,

$D = 421.28$ mm = External diameter of hollow shaft

And, Internal diameter of hollow shaft $= d = \frac{D}{3} = \frac{421.28}{3}$

$\Rightarrow$ $$d = 140.43 \text{ mm}$$

7.8 A solid steel shaft has to transmit 75 kW at 200 rpm. Taking allowable shear stress as 70 N/mm^2, find suitable diameter for the shaft, if the maximum torque transmitted at each revolution exceeds the mean by 30%.

[8 marks : 2019]

Solution:

The power transmitted, $P = 75$ kW

Speed, $N = 200$ rpm

Allowable shear stress, $\tau = 70\ \text{N/mm}^2$

Using equation, $$P = \frac{2\pi NT}{60}$$

$$75 = \frac{2\pi \times 200 \times T}{60}$$

$$T = 3.581\ \text{kNm}$$

If the maximum torque transmitted at each revolution exceeds the mean by 30%.

So, the modified torque, $T' = 1.3\ T$

$$T' = 1.3 \times 3.581$$

$$T' = 4.6553\ \text{kNm}$$

Using torque equation, $$\frac{\tau}{R} = \frac{T}{I_P}$$

$$\tau = \frac{T}{I_P} \times R = \frac{T}{Z_p}$$

$$70 = \frac{4.6553 \times 10^6}{\frac{\pi D^3}{16}}$$

We get, $D = 69.706$ mm

So, the diameter of solid shaft,

$$D = 69.706\ \text{mm}$$

7.9 **The electric motor exerts a torque of 800 N on the steel shaft ABCD when it is rotating at constant speed. Design specifications require that the diameter of the shaft be uniform from A to D and that the angle of twist between A and D not exceed 1.5°. Knowing that $\tau_{max} \geq 60$ MPa and G = 77 GPa, determine the diameter of the shaft that my be used.**

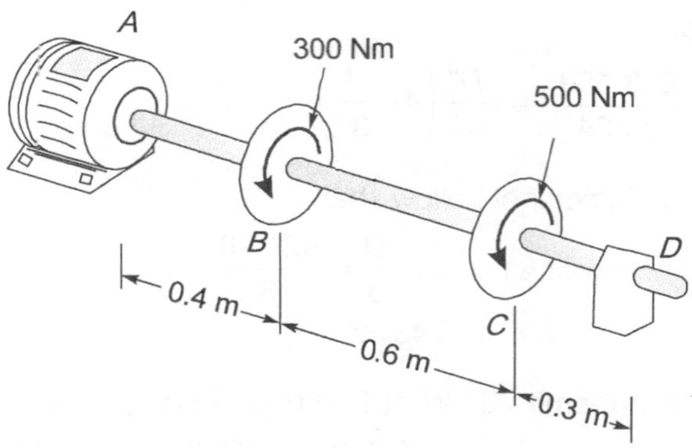

[12 Marks : 2021]

Solution:

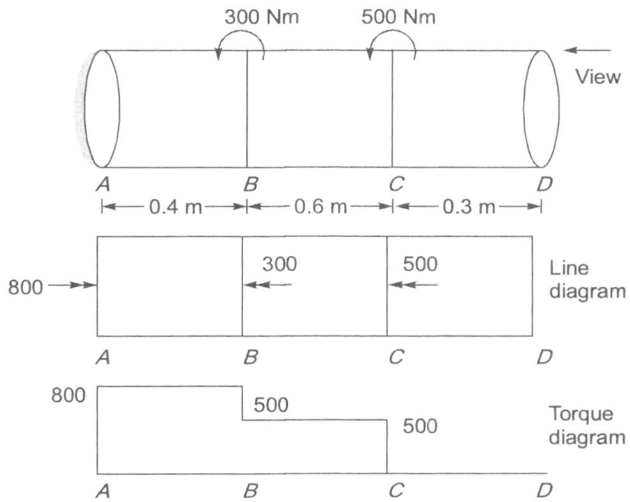

Crieteria: 1

Maximum shear stress crieteria:

$$\tau_{max} = \frac{16T_{max}}{\pi D^3}$$

$$60 = \frac{16(800 \times 10^3)}{\pi D^3}$$

$$D = 40.79 \text{ mm} = 40.8 \text{ mm}$$

Crieteria: 2

angle of twist crieteria:

$$\theta_{DA} = \theta_D - \theta_A = \theta_{AB} + \theta_{BC} + \theta_{CD} = \left(1.5 \times \frac{\pi}{180}\right) \quad \text{... (i)}$$

$$\theta_{DC} = 0$$

$$\theta_{BC} = \frac{(500 \times 10^3)(0.6 \times 10^3)}{(77 \times 10^3)\frac{\pi}{32}D^4}$$

$$\theta_{AB} = \frac{(800 \times 10^3)(0.4 \times 10^3)}{(77 \times 10^3)\left(\frac{\pi}{32}D^4\right)}$$

Using equation (i),

$$\frac{(500 \times 10^3)(0.6 \times 10^3)}{(77 \times 10^3)\frac{\pi}{32}D^4} + \frac{(800 \times 10^3)(0.4 \times 10^3)}{(77 \times 10^3)\left(\frac{\pi}{32}D^4\right)} + 0 = 1.5 \times \frac{\pi}{180}$$

$$D = 42.07 \text{ mm}$$

So, dia. of shaft that may be used will be maximum of above two crieteria, i.e.

$$D = 42.07 \text{ mm} \simeq 42 \text{ mm}$$

7.10 Determine the maximum principal stress developed in a cylindrical shaft 10 cm in diameter, subjected to a BM of 3.0 kN-m and twisting moment of 4.50 kN m. If the yield stress of the shaft material is 230 MN /m^2, determine the factor of safety according to the maximum shearing stress theory of failure.

[12 marks : 2023]

Solution:

Bending moment, M = 3 kN-m, Twisting moment, T = 4.5 kN-m

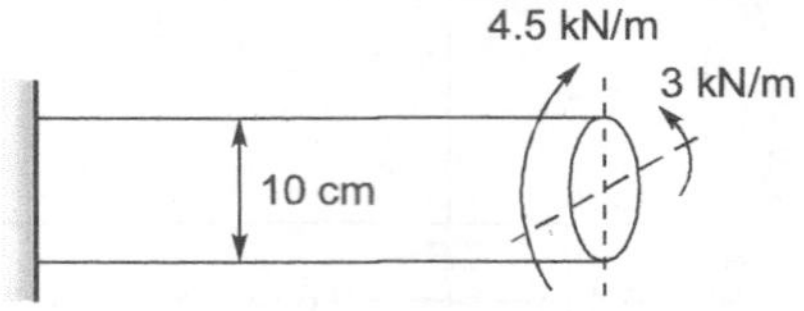

Now, maximum principal stress, $\sigma_{P_1} = \frac{16}{\pi D^3}\left[M+\sqrt{M^2+T^2}\right]$

where D is diameter of shaft

So, $$\sigma_{P_1} = \frac{16}{\pi\times(100)^3}\left[3+\sqrt{3^2+4.5^2}\right]\times 10^6 = 42.82 \text{ N/mm}^2$$

Minor principal stress, $$\sigma_{P_2} = \frac{16}{\pi D^3}\left[M-\sqrt{M^2+T^2}\right]$$

$$= \frac{16}{\pi\times(100)^3}\left[3-\sqrt{3^2+4.5^2}\right]\times 10^6 = -12.26 \text{ N/mm}^2$$

Therefore, maximum shear stress,

$$\tau_{max} = \frac{\sigma_{P_1}-\sigma_{P_2}}{2} = \frac{42.82-(-12.26)}{2} = 27.54 \text{ N/mm}^2$$

Now, as per maximum shear stress theory,

$$\tau_{max} \le \left(\frac{\tau_y}{FOS}\right)$$

$$\Rightarrow \quad FOS = \frac{\sigma_y}{2\tau_{max}} = \frac{230}{2\times 27.54} = 4.17$$

8. Theory of Columns

8.1 A column of length L, with the ends fixed in position but free to rotate, is provided with a bracket which carries a vertical load P at a distance 'a' from the axis of the column. Due to improper connection, the bracket slides down. Draw thrust, bending and shear force diagrams when the bracket is at a height 'x' from the base.

[10 marks : 2002]

Solution:

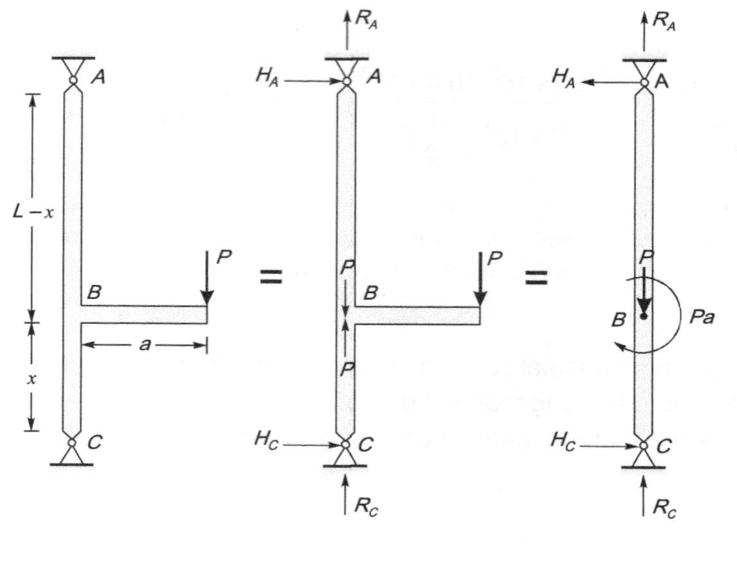

Reactions:

$\Sigma F_x = 0,\qquad H_A + H_C = 0 \qquad \text{...(i)}$

$\Sigma f_y = 0,\qquad R_A + R_C = P \qquad \text{...(ii)}$

$\Sigma M_A = 0,\qquad H_C \times L = Pa$

$$\therefore \qquad H_C = \frac{Pa}{L}(\rightarrow)$$

Hence,
$$H_A = -H_C = -\frac{Pa}{L} \text{ or } \frac{Pa}{L}(\leftarrow)$$

By the compatibility condition,

$$\Delta_{AB} + \Delta_{BC} = 0$$

$$\frac{R_A(L-x)}{AE} + \frac{(R_A-P)x}{AE} = 0$$

$$R_A L - R_A x + R_A x - Px = 0$$

$$R_A = \frac{Px}{L} \text{ (Tension)}$$

From eq. (ii),
$$R_C = \frac{P(L-x)}{L} \text{ (Compression)}$$

Axial thrust diagram: Taking tensile force positive and compressive force negative

Portion ***AC*:**

$$P_y(y \text{ from } A) = +R_A \qquad [0 \le y < L-x]$$

$$P_y = \frac{Px}{L}(\text{ constant})$$

Portion ***BC*:**

$$P_y(y \text{ from } A) = +R_A - P \qquad [L-x < y < L]$$

$$P_y = \frac{Px}{L} - P = -\frac{P(L-x)}{L}(\text{constant})$$

For SFD:

$$S_y(y \text{ from } A) = -H_A \qquad [0 < y < L]$$

$$S_y = -\frac{Pa}{L}(\text{constant})$$

Since there is no force any where in the column, the SF remain constant throughout the length of the column of magnitude $-\frac{Pa}{L}$

For BMD: Taking left hand side of column as reference face.

Portion CB:

$$M_y(y \text{ from } C) = -\frac{Pa}{L}y \qquad [0 \le y \le x]$$

$$M_C = 0$$

$$M_B = -\frac{Pax}{L}$$

Portion AB:

$$M_y(y \text{ from } A) = +\frac{Pa}{L}y \qquad [0 \le y \le (L-x)]$$

$$M_A = 0$$

$$M_B = +\frac{Pa}{L}(L-x)$$

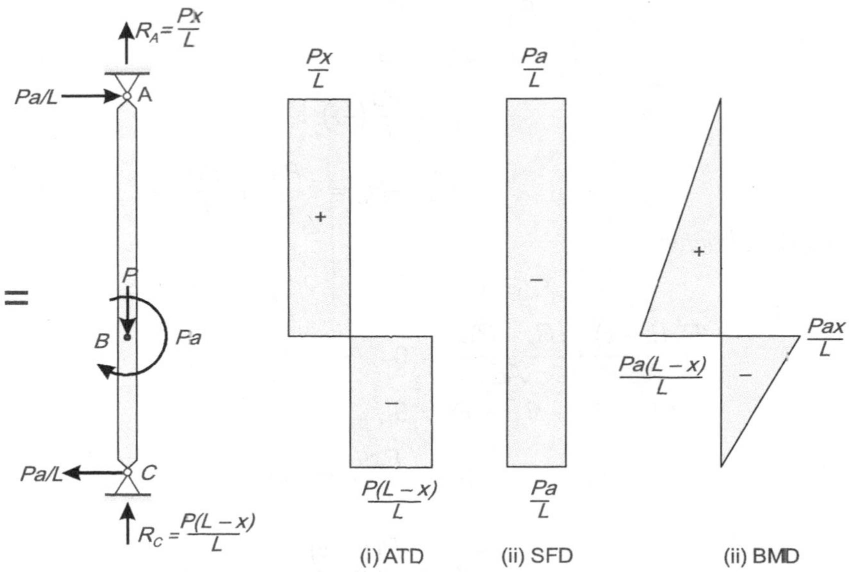

9. Shear Centre, Moment of Inertia & Principal Axes

9.1 The combined angle and channel section in figure below forms part of a runway beam. Calculate:

(a) Coordinate of the centroid

(b) Second moment of area about X-X

(c) Second moment of area about Y-Y

(d) Product of inertia about O

[10 × 4 = 40 marks : 1999]

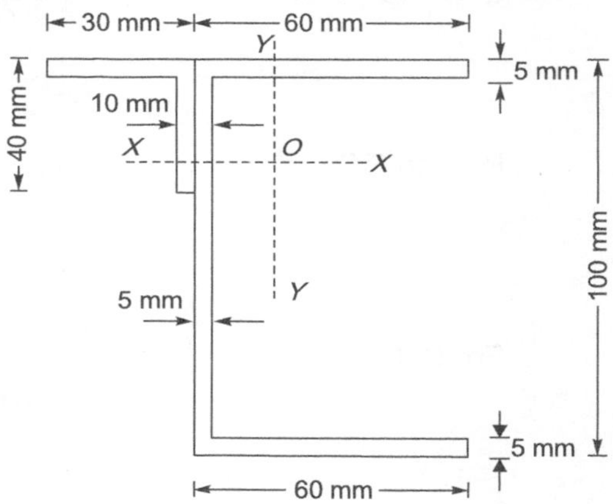

Solution:

Dividing the whole arrangement into five parts of area A_1, A_2, A_3, A_4 and A_5 and taking O' as reference point (origin)

Area $A_1 = 30 \times 5 = 150 \text{ mm}^2$

$\bar{x}_1 = 15$ mm and $\bar{y}_1 = 2.5$ mm

Area $A_2 = 35 \times 5 = 175 \text{ mm}^2$

$\bar{x}_2 = 27.5$ mm and $\bar{y}_2 = 22.5$ mm

Area $A_3 = 60 \times 5 = 300 \text{ mm}^2$

$\bar{x}_3 = 60$ mm and $\bar{y}_3 = 2.5$ mm

Area $A_4 = 90 \times 5 = 450 \text{ mm}^2$

$\bar{x}_4 = 32.5$ mm and $\bar{y}_4 = 50$ mm

Area $A_5 = 60 \times 5 = 300 \text{ mm}^2$

$\bar{x}_5 = 60$ mm and $\bar{y}_5 = 97.5$ mm

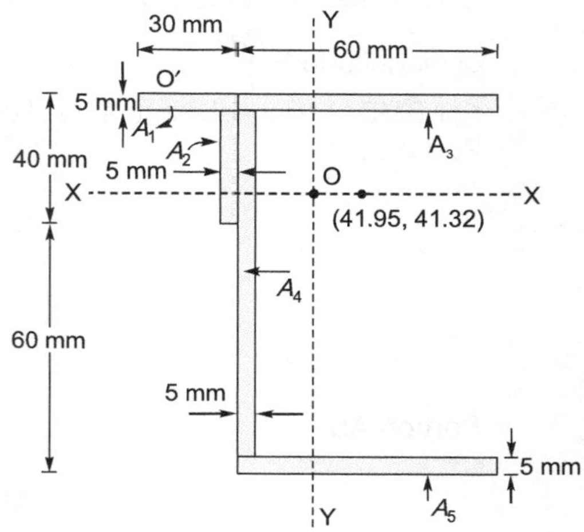

(i) **Coordinate of centroid**

$$\bar{x} = \frac{A_1\bar{x}_1 + A_2\bar{x}_2 + A_3\bar{x}_3 + A_4\bar{x}_4 + A_5\bar{x}_5}{A_1 + A_2 + A_3 + A_4 + A_5}$$

$$= \frac{(150\times15)+(175\times27.5)+(300\times60)+(450\times32.5)+(300\times60)}{150+175+300+450+300} = 41.95 \text{ mm}$$

$$\bar{y} = \frac{A_1\bar{y}_1 + A_2\bar{y}_2 + A_3\bar{y}_3 + A_4\bar{y}_4 + A_5\bar{y}_5}{A_1 + A_2 + A_3 + A_4 + A_5}$$

$$= \frac{(150\times2.5)+(175\times22.5)+(300\times2.5)+(450\times50)+(300\times97.5)}{150+175+300+450+300} = 41.32 \text{ mm}$$

$\therefore$ Coordinate of the centroid of the area are 41.95 mm and 41.32 mm

(ii) **Second moment of area about x-x**

Coordinate of centroids of each area with respect to new origin O.

$$A_1 \rightarrow \begin{cases} \bar{x}_1 = -(41.95-15) = -26.95 \text{ mm} \\ \bar{y}_1 = +(41.32-2.5) = +38.82 \text{ mm} \end{cases}$$

$$A_2 \rightarrow \begin{cases} \bar{x}_2 = -(41.95-27.5) = -14.45 \text{ mm} \\ \bar{y}_2 = +(41.32-22.5) = +18.82 \text{ mm} \end{cases}$$

$$A_3 \rightarrow \begin{cases} \bar{x}_3 = +(60-41.95) = +18.05 \text{ mm} \\ \bar{y}_3 = +(41.32-2.5) = +38.82 \text{ mm} \end{cases}$$

$$A_4 \rightarrow \begin{cases} \bar{x}_4 = -(41.95-32.5) = -9.45 \text{ mm} \\ \bar{y}_4 = -(50-41.32) = -8.68 \text{ mm} \end{cases}$$

$$A_5 \rightarrow \begin{cases} \bar{x}_5 = +(60-41.95) = +18.05 \text{ mm} \\ \bar{y}_5 = +(97.5-41.32) = 56.18 \text{ mm} \end{cases}$$

We know, $I_x = \Sigma I_{x\text{self}} + \Sigma A\bar{y}^2$

$$\Sigma I_{x\text{ self}} = \frac{30\times5^3}{12} + \frac{5\times35^3}{12} + \frac{60\times5^3}{12} + \frac{5\times90^3}{12} + \frac{60\times5^3}{12} = 0.323 \times 10^6 \text{ mm}^4$$

$$\Sigma A\bar{y}^2 = (150 \times 38.82^2) + (175 \times 18.82^2) + (300 \times 38.82^2) + [450 \times (-8.68)^2] + [300 \times (56.18)^2]$$

$$= 1.72 \times 10^6 \text{ mm}^4$$

Hence, $I_x = \Sigma I_{x\text{self}} + \Sigma A\bar{y}^2$

$$= 0.323 \times 10^6 + 1.72 \times 10^6 \text{ mm}^4 = 2.043 \times 10^6 \text{ mm}^4$$

(iii) **Second moment of area about y-y**

$$I_y = \Sigma I_{y\text{self}} + \Sigma A\bar{x}^2$$

$$\Sigma I_{y\ self} = \frac{5\times30^3}{12}+\frac{35\times5^3}{12}+\frac{5\times60^3}{12}+\frac{90\times5^3}{12}+\frac{5\times60^3}{12} = 1.93\times10^5\ mm^4$$

$$\Sigma A\bar{x}^2 = [150\times(-26.95)^2]+[175\times(-14.45)^2]+(300\times18.05^2)+[450\times(-9.45)^2]+(300\times18.05^2)$$

$$= 3.81\times10^5\ mm^4$$

Hence, $I_y = 1.93\times10^5 + 3.81\times10^5$

$I_y = 5.74\times10^5\ mm^4$

(Here $\bar{x}$ & $\bar{y}$ are distance of C.G. of individual areas from centroid of the whole section)

(iv) **Product of inertia:**

$$I_{xy} = \Sigma(I_{xy})_{self} + \Sigma A\bar{x}\bar{y}$$

Hence $(I_{xy})_{self}$ will be zero for each area because each area have symmetrical about their centroidal axes.

$$I_{xy} = \Sigma A\bar{x}\bar{y}$$

$$= [150\times(-26.95)\times38.82]+[175\times(-14.45)\times18.82)]+[300\times18.05\times38.82]$$
$$+[450\times(-9.45)\times(-8.68)]+[300\times18.05\times(-56.18)]$$

$$I_{xy} = -26.16\times10^4\ mm^4$$

9.2 **A steel tube is to be used as a post for a road sign board as shown in figure. The maximum wind pressure on the sign board is 1960 N/m². The angle of rotation of the tube at the bottom of the sign board marked as A must not exceed 4° and the maximum shear stress (due to torsion only) must not be greater than 38 MPa. Determine the mean diameter of the tube if the wall thickness is 4.2 mm. Take G = 70 GPa. Assume wind is transmitting only over the sign board portion.**

[20 marks : 2020]

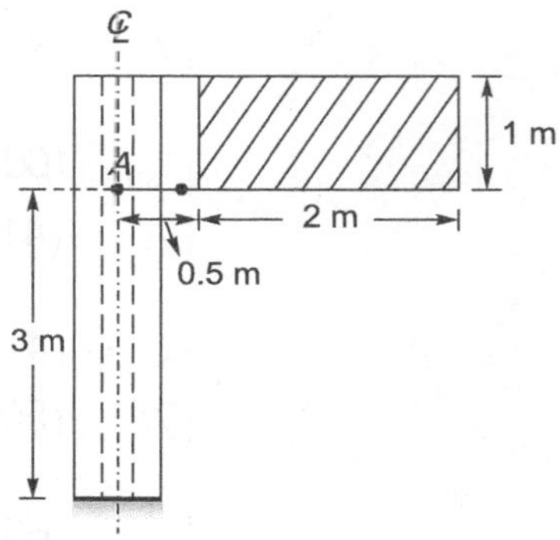

Solution:

Wind pressure, $p = 1960\ N/m^2$

Force on sign board, $pA = 1960\times1\times2 = 3920\ N$

Torsion on tube, $T = Fr = 3920\times\left(0.5+\frac{2}{2}\right) = 3920\times1.5 = 5880\ Nm$

At a point A, rotation due to torsion,

$$\phi_A = \frac{Tl_A}{GJ}$$

Assuming the tube to be thin cylinder with mean diameter d_m.

Polar MOI, $J = \pi d_m \times t\times\left(\frac{d_m}{2}\right)^2 = \frac{\pi d_m^3 t}{4}$

For $\phi_A \le 4°$

$$\frac{Tl_A}{GJ} \le \frac{4\pi}{180}$$

$$\frac{5880\times10^3\times3000}{70\times10^3\times J} \le \frac{4\pi}{180}$$

$$J \ge 3.609\times10^6\ mm^4$$

$$\frac{\pi d_m^3 \times t}{4} \geq 3.609 \times 10^6$$

So, $d_m \geq 103.048$ mm ...(i)

Also, $\tau_{max} \leq 38$ MPa

$$\frac{T}{2A_m t} \leq 38 \text{ MPa}$$

$$\Rightarrow \quad \frac{5880 \times 10^3}{2A_m \times 4.2} \leq 38 \text{ MPa}$$

$$A_m \geq 18421.053 \text{ mm}^2$$

$$\frac{\pi d_m^2}{4} \geq 18421.053 \text{ mm}^2$$

So, $d_m \geq 153.148$ mm ...(ii)

From eq. (i) and (ii), $d_m \geq 153.1448$ mm

For calculated mean dia., $\frac{d_m}{t} = \frac{153.148}{4.2} = 36.46 > 20$

So, our assumption of thin cylinder is correct.

■■■■

3 Structural Analysis

Revised Syllabus of ESE : *Basics of strength of materials, Types of stresses and strains, Bending moments and shear force, concept of bending and shear stresses; Analysis of determinate and indeterminate structures; Trusses, beams, plane frames; Rolling loads, Influence Lines, Unit load method & other methods; Free and Forced vibrations of single degree and multi degree freedom system; Suspended Cables; Concepts and use of Computer Aided Design.*

---

1. Determinacy and Indeterminacy

1.1 What do you understand by static indeterminacy and kinematic indeterminacy of a 2-*D* framed structure? Explain with an example of a fixed end beam.

[5 marks : 2014]

Solution:

Static indeterminacy (D_S): Those structures which can not be analysed by using condition of static equilibrium alone are called indeterminate structures. To analyse these indeterminate structures extra equilibrium condition are required, called compatibility conditions and numbers of compatibility conditions needed to analyse structure is known as degree of indeterminacy.

D_S = Total no. of reaction present (Both internal and external) – No. of available equilibrium equations.

For 2-*D* Rigid Frame: In two dimensional rigid member, each member has three internal reactions (viz. R_x, R_y and M_z) and at each joint three equilibrium condition (viz. $\Sigma F_x = 0$, $\Sigma F_y = 0$ and $\Sigma M_x = 0$) are available

Let there are r_e number of external support conditions.

∴ Total no. of reaction present,

$$R = \text{External reaction} + \text{Internal reaction}$$

$$R = r_e + 3m$$

and total no. of available equilibrium conditions,

$$E = 3j$$

∴ $$D_S = R - E$$

$$D_S = r_e + 3\text{ m} - 3j \quad \text{... when all joint are rigid}$$

$$D_S = r_e + 3\text{m} - 3j - r_r \quad \text{... when some joints are hybrid}$$

where r_r = Number of released reactions

Example:

Here,

$$r_e = 3 + 1 = 4$$

$$m = 1$$

$$j = 2$$

$$D_S = r_e + 3\text{m} - 3j$$

$$= 4 + 3 \times 1 - 3 \times 2$$

$$= 1 \text{ (indeterminate to 1}^{\text{st}}\text{ degree)}$$

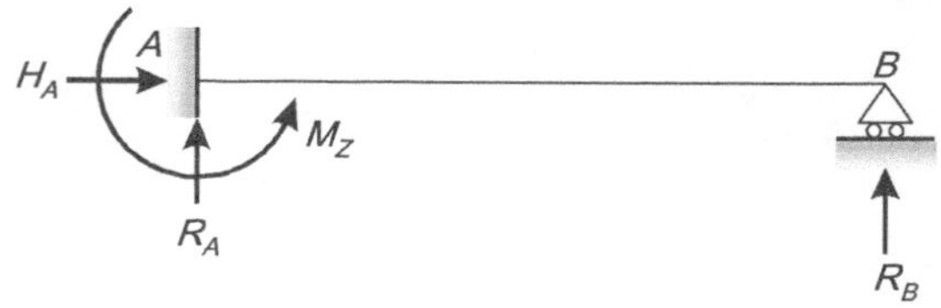

Kinematic Indeterminacy (D_k): It refers to the total no. of available degree of freedom at all joints. It is equal to total no. of unrestrained displacement component at all joints.

D_k = Total degree of freedom at all joints – degree of freedom restrained by supports

2-*D* Rigid Frames: At each joint there are three degree of freedom (*viz.* Δ_x, Δ_y and θ_z). Hence at all joint there will be $3j$ degree of freedoms. But at supports displacements are not available in the direction of reaction component.

$\therefore \quad D_k$ = unrestrained displacement component

$\Rightarrow \quad D_k = 3j - r_e$ ... members are axially flexible

$D_k = 3j - r_e - m''$ ...m'' member are axially rigid

Example:

$D_k = 3j - r_e$

Here, $j = 2$, $r_e = 3$

$\therefore \quad D_k = 3 \times 2 - 3$

$\Rightarrow \quad D_k = 3$ (i.e., θ_B, Δ_{HB} & Δ_{VB})

1.2 **Find the total degree of statical indeterminacy (both internal and external) for the bridge truss shown in the figure.**

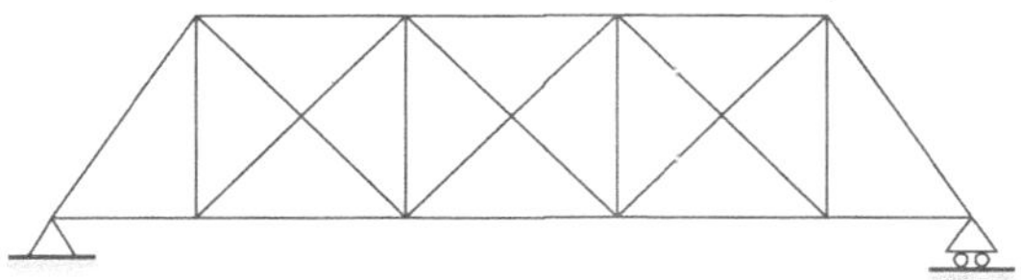

[5 marks : 2015]

Solution:

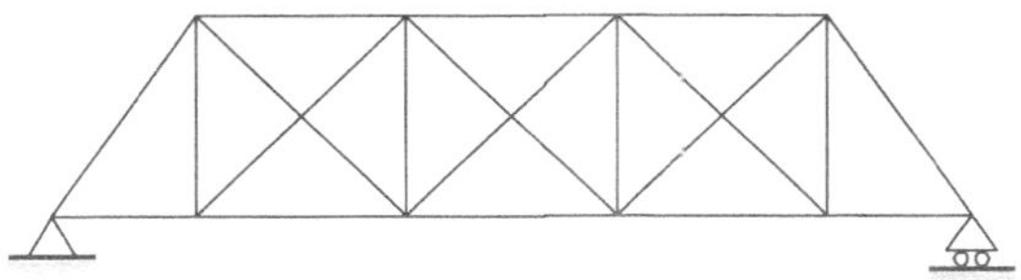

Total number of members, $m = 20$

Total number of joints, $j = 10$

Total number of external reactions, $r_e = 3$

Total degree of external indeterminacy, $E = r_e - 3 = 3 - 3 = 0$

Total degree of internal indeterminacy, $I = m - (2j - 3)$

$= 20 - (2 \times 10 - 3)$

$= 20 - 17 = 3$

$\therefore$ **Total indeterminacy** $= E + I$

$= 0 + 3 = 3$

Alternatively,

Total indeterminacy $= (r_e - 3) + (m - (2j - 3))$

$= m + r_e - 2j$

$= 20 + 3 - 2(10) = 3$

2. Influence Line Diagram and Rolling Loads

2.1 State Muller Breslau principle : Derive the equation for influence line for moment M_B at the joint B of the rigid frame ABC shown in the below figure, when a unit load travels over the beam AB.

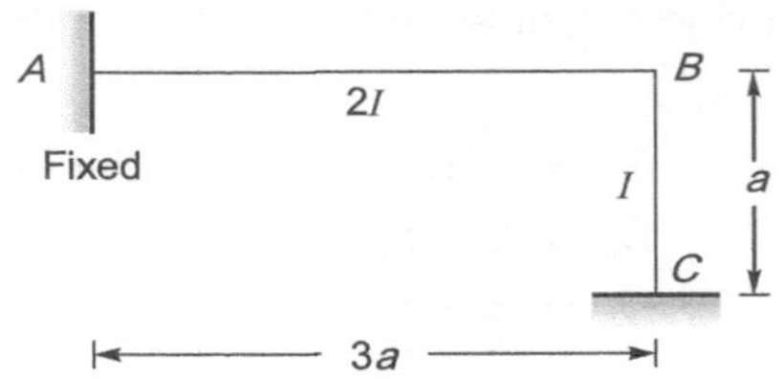

[20 marks : 2000]

Solution:

Muller-Breslau Principle : "The ILD for any stress function (SF, BM and Reaction) in a structure is represented by it's deflected shape obtained by removing the restraint offered by that stress function and introducing a directly generalized unit displacement in the positive direction of that stress function."

Assume unit load travels from A, now unit load is at x distance from A.

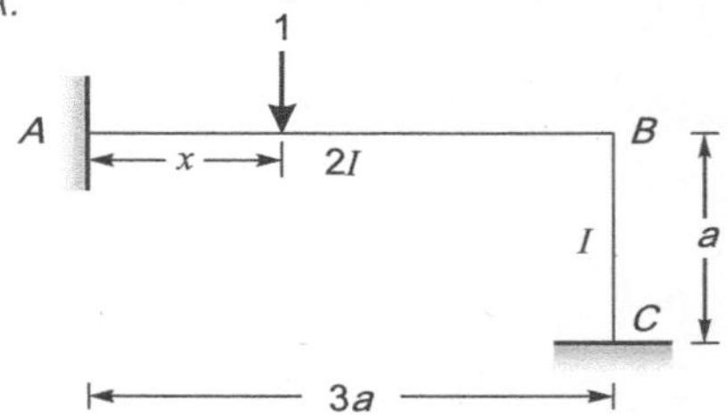

$$\overline{M}_{AB} = \frac{-1\times x\times(3a-x)^2}{(3a)^2}$$

$$\overline{M}_{BA} = \frac{1\times(3a-x)x^2}{(3a)^2}$$

$$\overline{M}_{BC} = 0$$

$$\overline{M}_{CB} = 0$$

Analysis by slope deflection method

$$M_{BA} = \overline{M}_{BA} + \frac{2E(2I)}{L_{AB}}\left(2\theta_B + \theta_A - \frac{3\delta}{l}\right)$$

[$\theta_A = 0$ since end A is fixed and settlement = 0]

$$= \frac{1\times(3a-x)x^2}{9a^2} + \frac{4EI}{3a}(2\theta_B)$$

$$M_{BA} = \frac{(3a-x)x^2}{9a^2} + \frac{8EI}{3}(\theta_B) \quad \text{...(i)}$$

$$M_{BC} = \overline{M}_{BC} + \frac{2EI}{L_{BC}}\left(2\theta_B + \theta_C - \frac{3\delta}{l}\right)$$

[$\theta_A = 0$ since end C is fixed and settlement = 0]

$$M_{BC} = 0 + \frac{4EI\theta_B}{a} \quad \text{...(ii)}$$

By joint equilibrium equation at B

$$M_{EA} + M_{BC} = 0$$

$$\frac{(3a-x)x^2}{9a^2} + \frac{8EI}{3}(\theta_B) + \frac{4EI\theta_B}{a} = 0 \quad \text{[from (i) and (ii)]}$$

$$\theta_B = -\frac{(3a-x)x^2}{60aEI}$$

Hence,
$$M_{BA} = \frac{(3a-x)x^2}{9a^2} - \frac{8EI}{3} \times \frac{(3a-x)x^2}{60aEI} = \frac{(3a-x)x^2}{15a^2}$$

Hence, variation of moment at joint B can be calculated with the equations of M_{BA} or M_{BC}.

2.2 **State muller Breslau principle. Derive the equation for influence line for the reaction R_B for the beams shown in the figure. *EI* is constant throughout.**

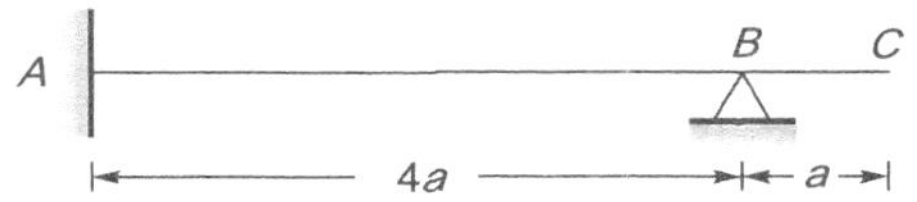

[12 marks : 2003]

Solution:

Muller Breslau Principle: "The ILD for any stress function in a structure is represented by it's deflected shape obtained by removing the restraint offered by the stress function (SF, BM and reaction) and introducing a directly generalized unit displacement in the positive direction of that stress function".

Assume unit load travels from *A*, now unit load is at *x* distance from *A*

Case (i) $0 \le x \le 4a$

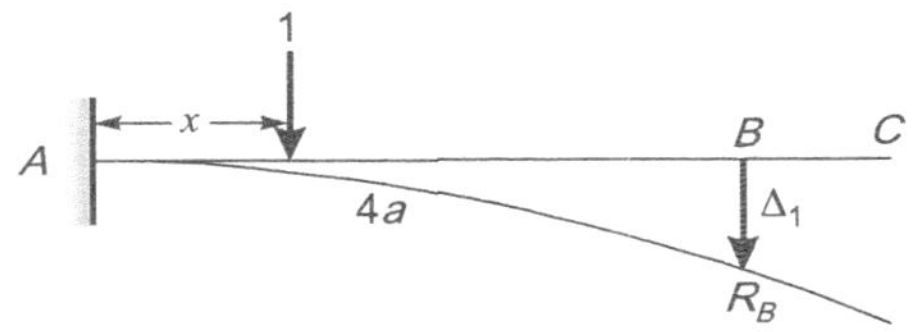

$$\Delta_1 = \frac{1 \cdot x^3}{3EI} + \frac{1 \cdot x^2}{2EI}(4a - x) \qquad \text{\{Downward\}}$$

Where Δ_1 is deflection at *B* due to unit load.

$$\Delta_2 = \frac{R_B(4a)^3}{3EI} = \frac{64R_Ba^3}{3EI}$$

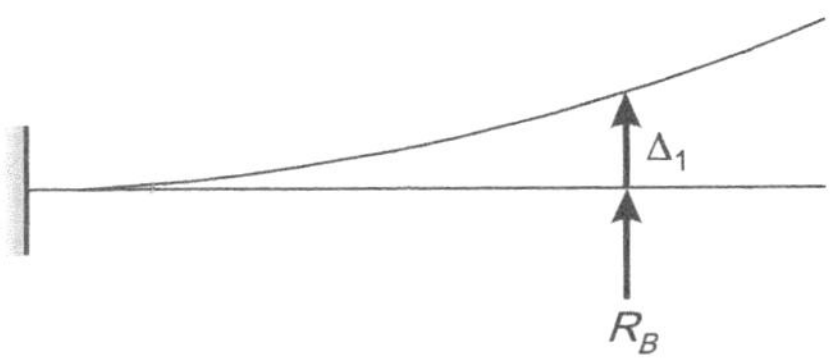

Since joint *B* is hinged, hence net deflection is zero

$$\Delta_1 = \Delta_2$$

$$\frac{x^3}{3EI} + \frac{x^2(4a-x)}{2EI} = \frac{64R_Ba^3}{3EI}$$

$$R_B = \frac{3EI}{64a^3}\left[\frac{x^3}{3EI} + \frac{2ax^2}{EI} - \frac{x^3}{2EI}\right]$$

$$= \frac{3EI}{64a^3}\left[\frac{2ax^2}{EI} - \frac{x^3}{2EI}\right] = \frac{3x^2}{32a^2} - \frac{1}{128}\frac{x^3}{a^3}$$

$$R_B = \frac{x^2(12a - x)}{128a^3} \qquad \text{(when } 0 \le x \le 4a\text{)}$$

Case (ii) $4a \le x \le 5a$

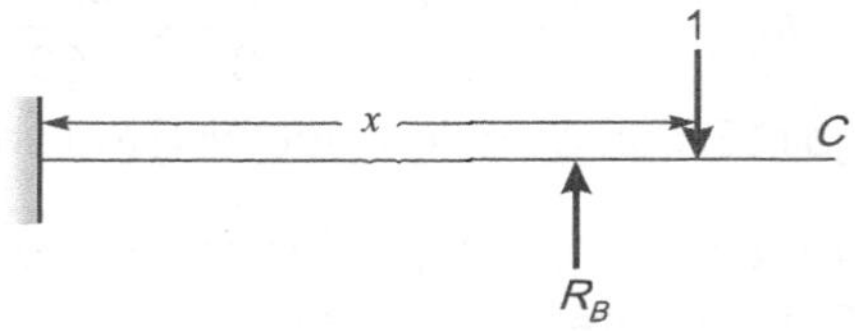

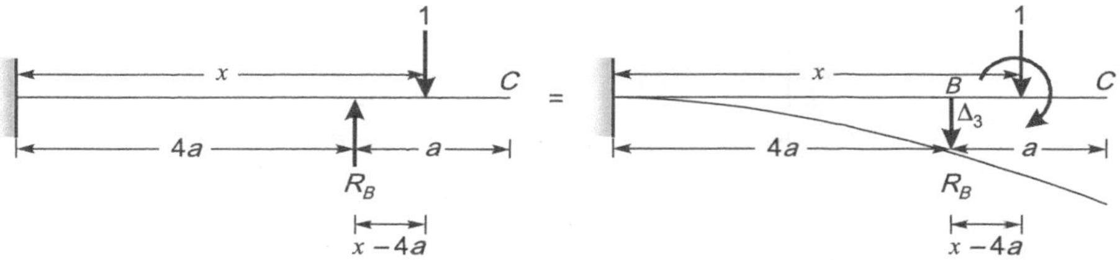

$$\Delta_3 = \frac{1\cdot(4a)^3}{3EI} + \frac{(4a)^2(x - 4a)\cdot 1}{2EI}$$

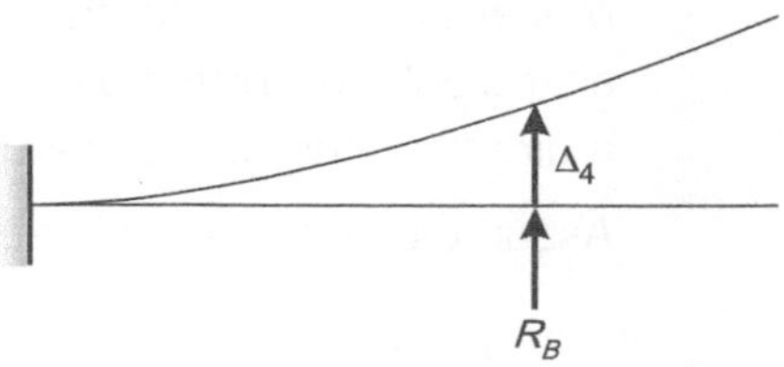

$$\Delta_4 = \frac{R_B(4a)^3}{3EI} = \frac{64R_B a^3}{3EI}$$

Since joint B is hinged, hence net deflection is zero.

$$\Delta_3 = \Delta_4$$

$$\frac{1\cdot(4a)^3}{3EI} + \frac{1\cdot(x - 4a)(4a)^2}{2EI} = \frac{64R_B a^3}{3EI}$$

$$R_B = \frac{3EI}{64a^3}\left[\frac{64a^3}{3EI} + \frac{16a^2(x - 40)}{2EI}\right]$$

$$= 1 + \frac{3EI \times 16a^2(x - 4a)}{64a^2 \times 2EI}$$

$$= 1 + \frac{3}{8a}(x - 4a)$$

$$= 1 + \frac{3x}{8a} - \frac{3}{2}$$

$$R_B = \frac{3x}{8a} - 0.5 \qquad \text{when } 4a \le x \le 5a$$

2.3 **State, explain and prove 'Muller-Breslau' influence line theorem for a statically determinate beam. State whether the theorem is also applicable to statically indeterminate beam.**

[10 marks : 2005]

Solution:

Muller Breslau Principle: "The ILD for any stress function in a structure is represented by it's deflected shape obtained by removing the restraint offered by that stress function (SF, BM and Reaction) and introducing a directly generalized unit displacement in the positive direction of that stress function."

Let us consider a simply supported beam AB and we have to find out influence line for vertical reaction at support B.

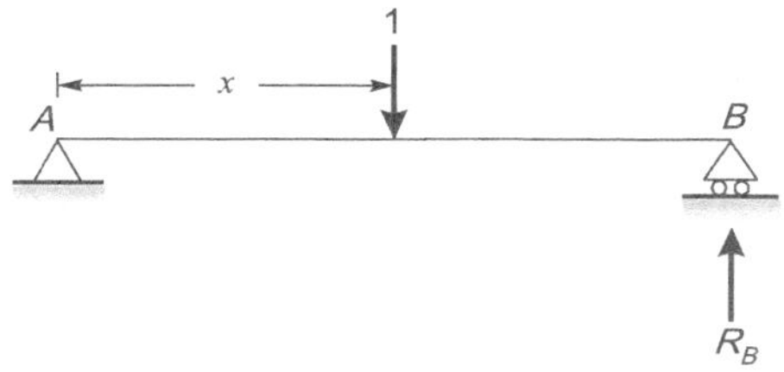

Case (i)

By Muller Breslau Principle: Remove the restraint (support B, reaction) and apply unit displacement in the direction of restraint.

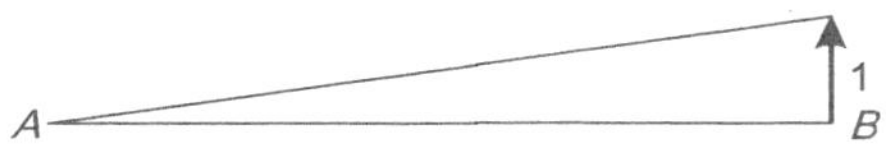

[ILD for vertical reaction at B]

Case (ii)

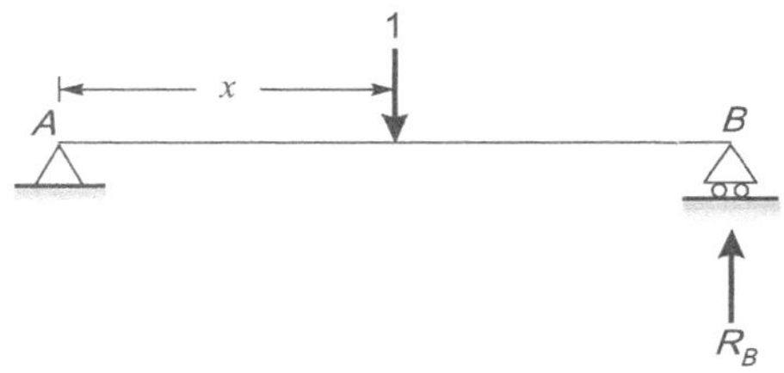

$$\Sigma M_A = 0$$

$$R_B(L) - 1 \cdot x = 0$$

$$R_B = \frac{x}{L} \text{(Linear)}$$

at
$$x = 0, \quad R_B = 0$$
$$x = L \quad R_B = 1$$

Hence, ILD for R_B (Vertical reaction at B)

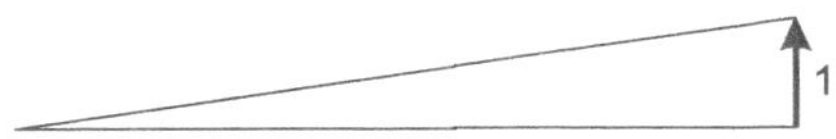

[ILD for R_B]

Remark: Hence it is clear that ILD for vertical reaction at B is same, as found in muller breslau principle. Hence proved.

Muller Breslau principle is also applicable for statically indeterminate beams. ILD for statically determinate beams is linear but ILD for statically indeterminate beam is non linear.

2.4 **A uniformly distributed load of 2000 kg/m, 6 m long crosses over a girder simply supported at ends over a span of 10 m from left to right. Calculate maximum bending moment in girder at a point 4.5 m from left hand end using influence line.**

[20 marks : 2009]

Solution:

Assumptions

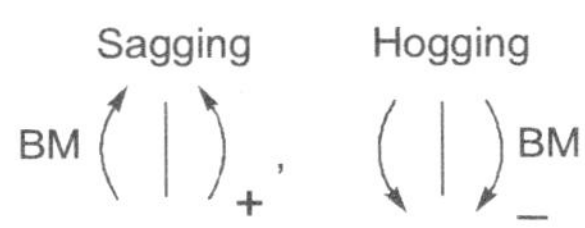

, 2000 kg/m = 20 kN/m

To get maximum benging moment, we need to determine or draw ILD for bending moment at C

Taking top face as reference face

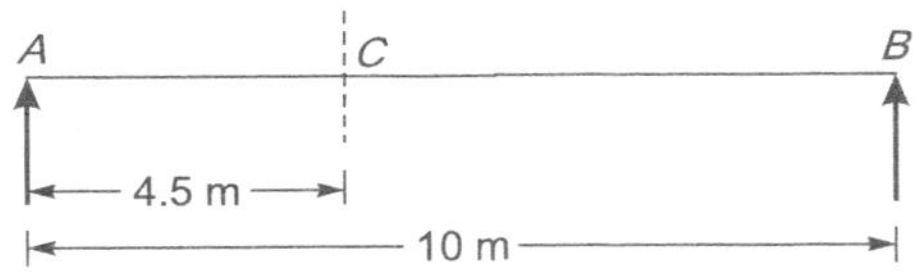

ILD: Response of certain quantity (BM, Shear force, Stress) at a section on span when unit load moves on span.

Case-1: Taking 1 kN load at left of C

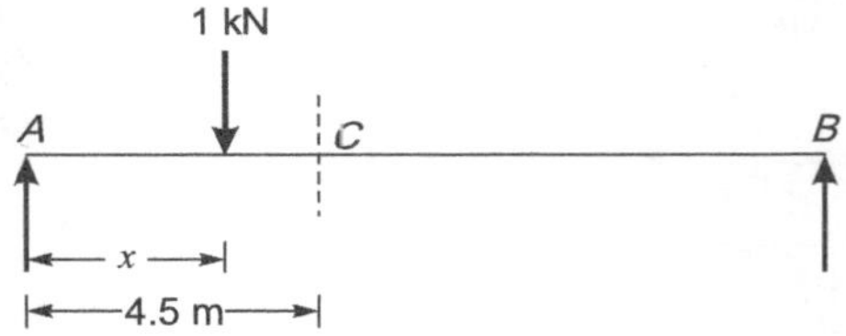

For $0 \le x \le 4.5$ m

Reaction at B $\quad R_B = \dfrac{x}{10}$

BM at C $\quad M_{cx} = \dfrac{5.5x}{10}$

Case-2: Load on right of C

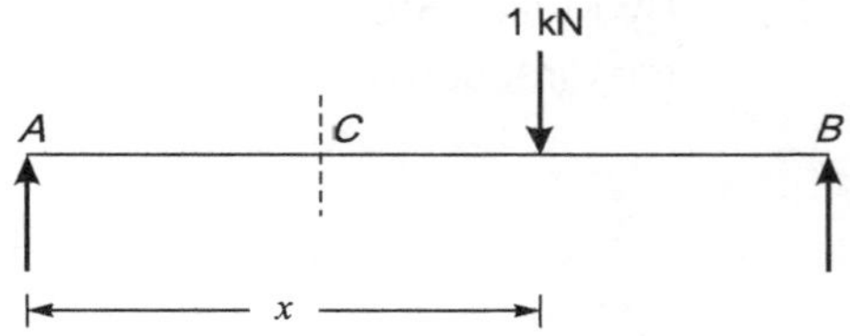

$4.5 \le x \le 10$ m

Reaction at B $\quad R_B = \dfrac{x}{10}$

Reaction at A $\quad R_A = \left(1-\dfrac{x}{10}\right)$

BM at C $\quad M_{cx} = 4.5\left(1-\dfrac{x}{10}\right)$

Drawing ILD

When a UDL moves over a girder then BM at a section is the area of ILD considering the portion of girder which is loaded by UDL.

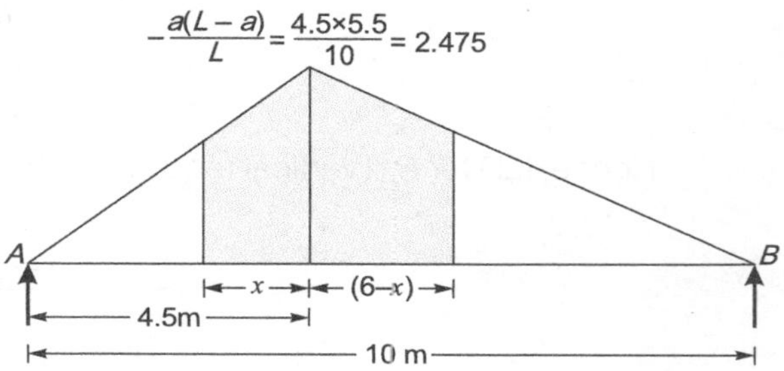

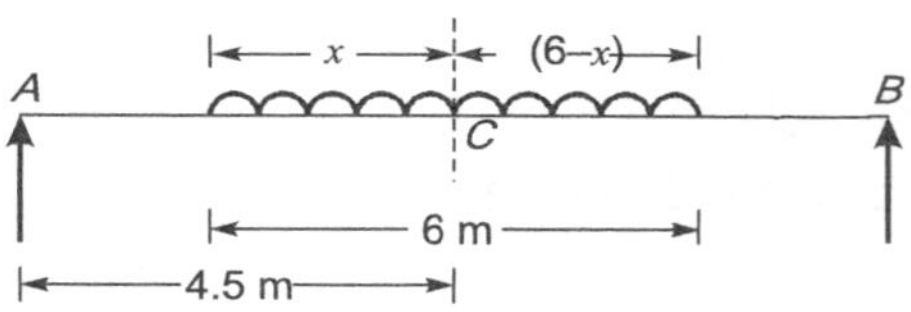

Itis a case of UDL shorter than span, hence maximum BM at a particular section will occur if that section divides the length of UDL in the same ratio as it (section) divdes the span.

Hence,
$$\frac{x}{6-x} = \frac{4.5}{10-4.5}$$
$$x = 2.7 \text{ m}$$

Then ordinates in the ILD under UDL position will be α, β

$$\alpha = \frac{2.475}{4.5}\times(4.5-2.7) = 0.99 = \beta$$

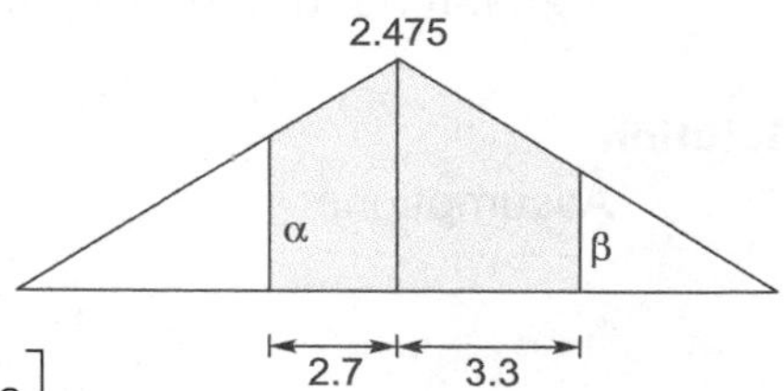

So maximum bending moment at C

$$= 20\times\left[\frac{1}{2}\times(2.475-0.99)\times6+0.99\times6\right]$$
$$= 207.9 \text{ kNm}$$

2.5 **A simply supported beam has a span of 10 m, a 7 m long UDL of 10 kN/m intensity crosses the beam from left to right. When the head of the load is 1 m from the right support, find the support reactions, BM and SF at the mid-span using the influence line diagram.**

[10 marks : 2014]

Solution:

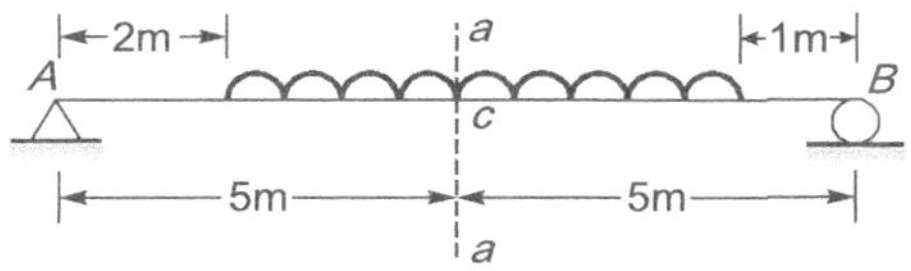

1. Support reactions:

$$R_A = \text{Area of ILD below UDL} \times \text{Loading rate}$$

$$= 7 \times \frac{(0.8 + 0.1)}{2} \times 10 \text{ kN} = 31.5 \text{ kN}$$

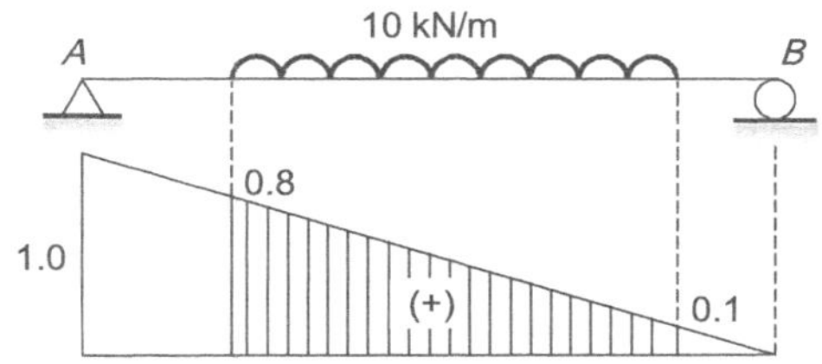

Fig. (i) ILD for R_A

$$R_B = \text{Area of ILD below UDL} \times \text{Loading rate}$$

$$= 7 \times \frac{(0.2 + 0.9)}{2} \times 10 \text{ kN} = 38.5 \text{ kN}$$

Check:

$$\Sigma F_y = R_A + R_B - 10 \times 7$$

$$= 31.5 + 38.5 - 70$$

$$= 0 \text{ (OK)}$$

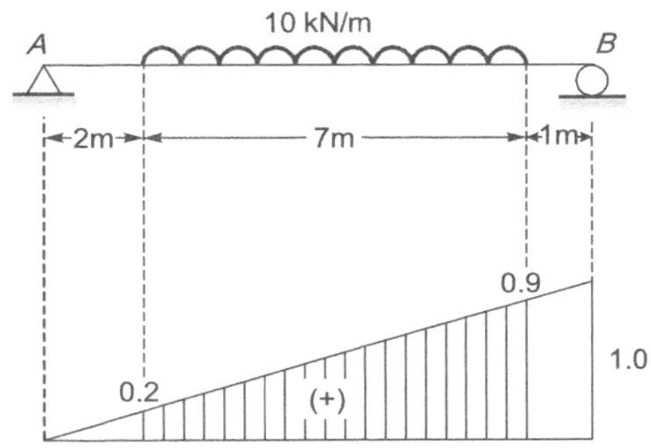

Fig. (ii) ILD for R_B

2. *BM* at mid span (M_c)

$$M_C = \text{Area of ILD below UDL} \times \text{Loading rate}$$

$$= \left[3 \times \frac{(1 + 2.5)}{2} + 4 \times \frac{(2.5 + 0.5)}{2}\right] \times 10$$

$$= 112.5 \text{ kN-m (Sagging)}$$

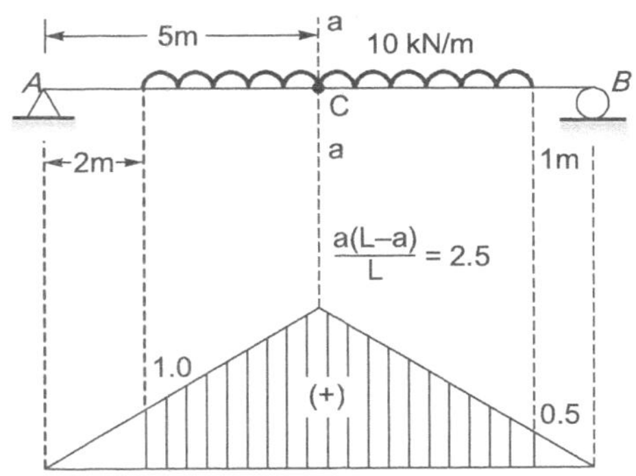

Fig. (iii) ILD for M_C

3. *SF* at mid span (S_C)

$$S_C = \text{Area of ILD below UDL} \times \text{Loading rate}$$

$$= \left[-3 \times \frac{(0.2 + 0.5)}{2} + 4 \times \frac{(0.5 + 0.1)}{2}\right] \times 10$$

$$= [-1.05 + 1.2] \times 10 = 1.5 \text{ kN}$$

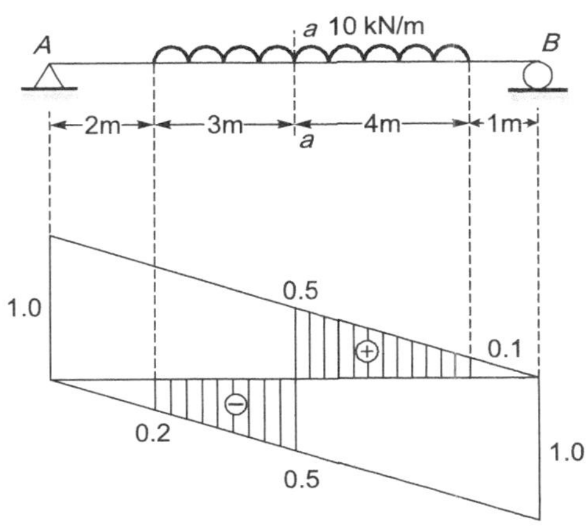

Fig. (iv) ILD for S_C

2.6 A continuous beam ABCDE is subjected to Dead Load and Live Load. Draw the arrangements of Live Load that produce maximum bending moment at support B and span BC of the beam.

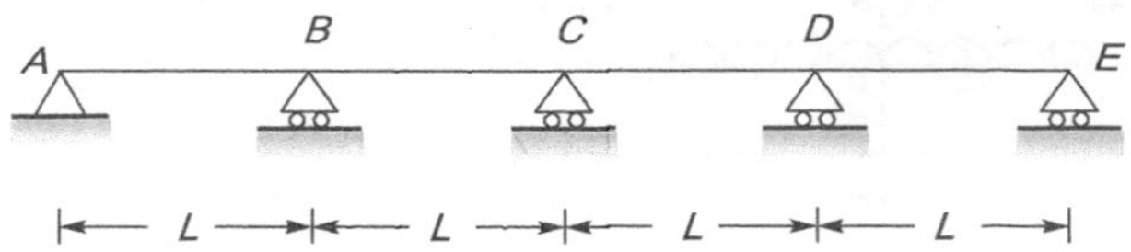

[8 marks : 2016]

Solution:

Loading arrangement for maximum BM at support *B*.

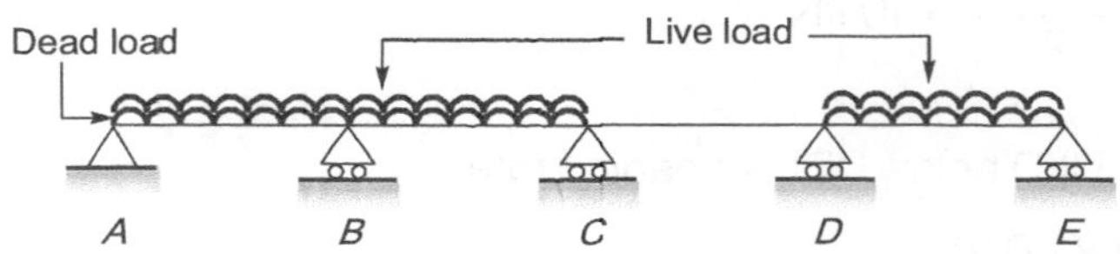

Loading arrangement for maximum BM of span BC.

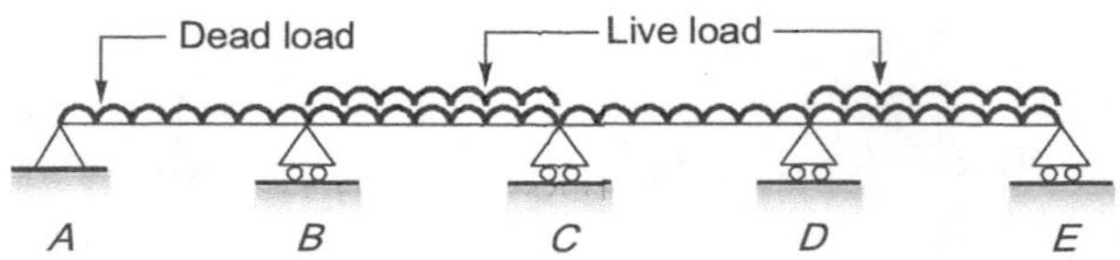

2.7 State the Muller-Breslau principle. Hence draw influence line for moment at D in the beam shown in figure.

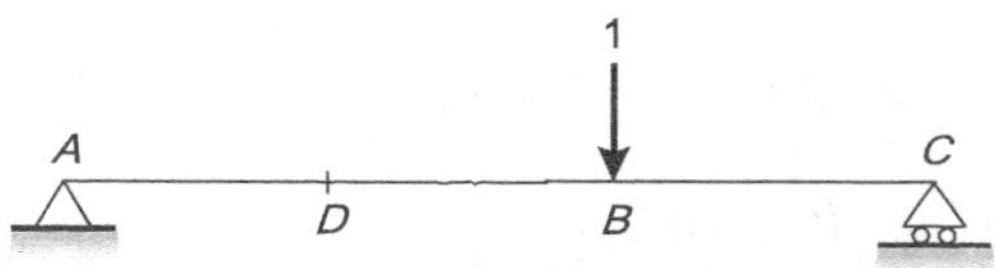

[8 marks : 2016]

Solution:

Muller-Breslau principle states that the influence lines of an action (force, moment etc.) assumes the scaled form of the deflection that the structure displays after removing the restraint on the point where the action is evaluated and applying a point load at that point which causes a unit displacement.

Apply unit rotation at *D* i.e., $\theta_1 + \theta_2 = 1$

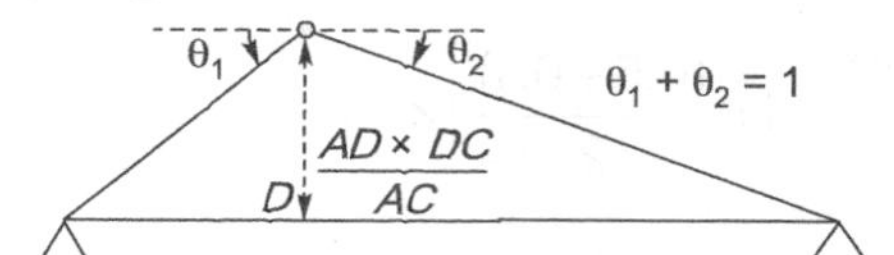

ILD for BM at *D*

2.8 A beam AB is simply supported over a span of 15 m. An udl of 25 kN/m intensity and 5 m length moves over the beam from end A to B. Draw the influence line diagram for bending moment and shear force at section C located at 6 m from end A. Hence calculate the maximum bending moment and shear force at section C.

[12 marks : 2018]

Solution:

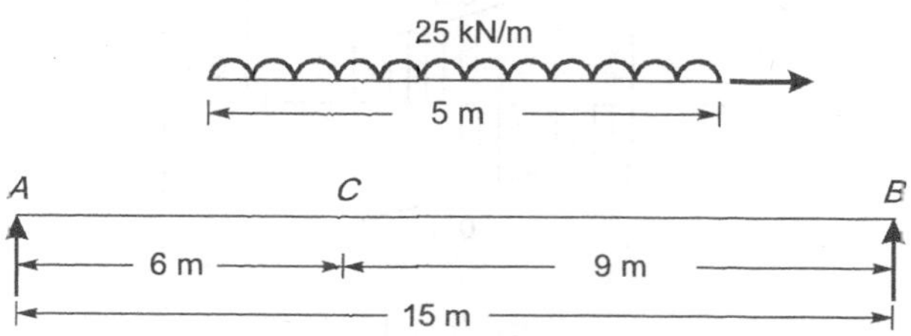

(i) Bending moment at 'C'

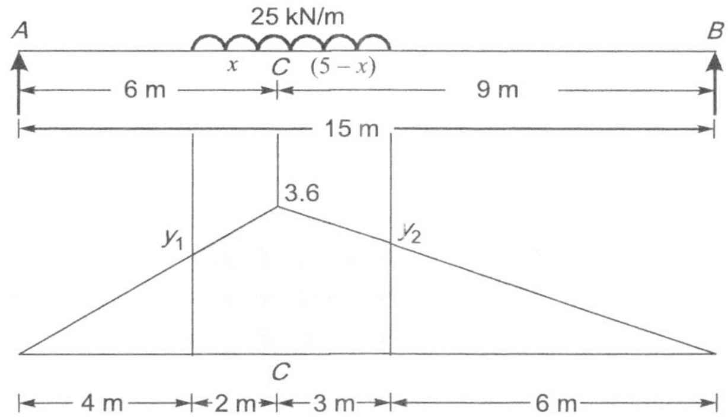

(Influence line diagram for BM at C)

Load should be divided by the section in the same ratio as the span.

$$\therefore \quad \frac{x}{5-x} = \frac{6}{9}$$

$$\Rightarrow \quad x = 2 \text{ m}$$

$$\text{Now,} \quad \frac{3.6}{6} = \frac{y_1}{4}$$

$$\Rightarrow \quad y_1 = 2.4$$

$$\text{Now,} \quad \frac{3.6}{9} = \frac{y_2}{6}$$

$$\Rightarrow \quad y_2 = 2.4$$

Maximum BM at C = (Shaded area of ILD) × Loading intensity

$= 0.5 \times (2.4 + 3.6) \times 5 \times 25$

$= 375$ kN-m

(ii) ILD for shear force at C

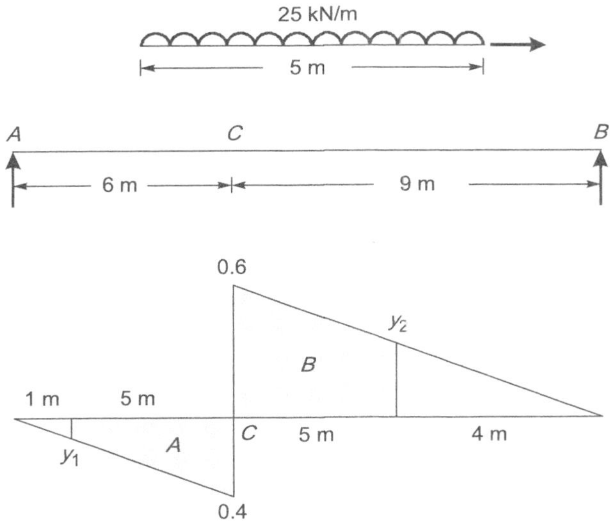

(Influence line diagram)

(A) For maximum negative shear force at C.

Load is placed such that head of the load touches 'C' and tail is at 5 m from C.

$$\frac{0.4}{6} = \frac{y_1}{1}$$

$$\Rightarrow \qquad y_1 = \left(\frac{1}{15}\right)$$

$$\text{Maximum negative SF} = \text{Loading intensity} \times \text{Area of shaded region (A)}$$

$$= 25 \times 0.5 \times \left(0.4 + \frac{1}{15}\right) \times 5 = 29.167 \text{ kN}$$

(B) For maximum positive shear force at C
Keep tail of load at C

$$\frac{0.6}{9} = \frac{y_2}{4}$$

$$\Rightarrow \qquad y_2 = \left(\frac{4}{15}\right)$$

$$\text{Maximum positive SF} = \text{Loading intensity} \times \text{Area of region (B)}$$

$$= 25 \times 0.5 \times \left(0.6 + \frac{4}{15}\right) \times 5 = 54.167 \text{ kN}$$

2.9 **A uniformly distributed load of 40 kN/m and 5 m long crosses a simply supported beam of span 15 from left to right. Draw the influence line diagram for shear force and bending moment at a section 6 m from left end. Use these diagrams to get the maximum shear force and bending moment at this section.**

[20 marks : 2019]

Solution:

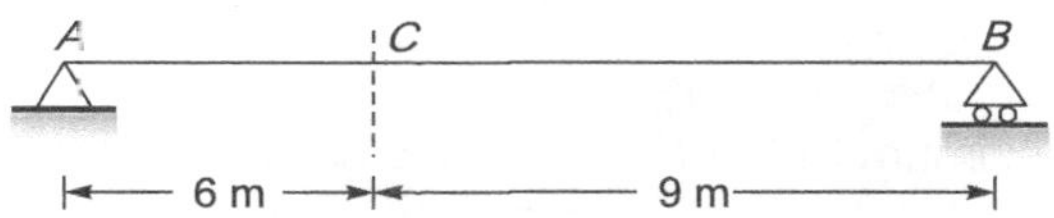

ILD for shear force at C
Using Muller Breslau's principle, ILD for shear at C is given as

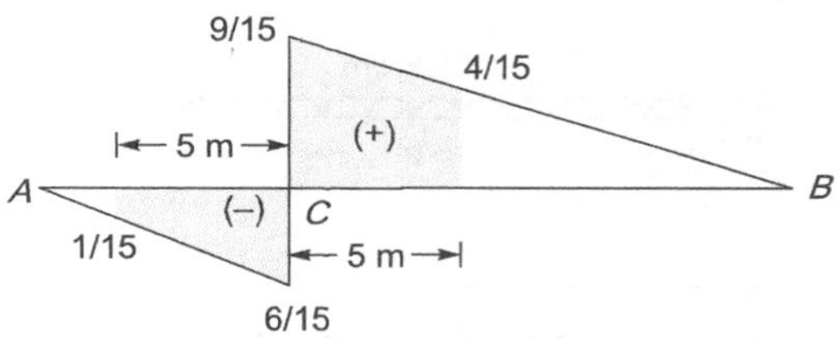

For max (+) SF at 'C', load should be placed at section CB

$$(+)\ SF_{max} = \left[\frac{\frac{9}{15} + \frac{4}{16}}{2}\right] \times 5 \times 40 = 86.67 \text{ kN}$$

For max (–) SF at 'C', loac should be placed at section AC

$$(-)\ SF_{max} = \left[\frac{\frac{1}{15} + \frac{6}{16}}{2}\right] \times 5 \times 40 = -46.67 \text{ kN}$$

ILD for Bending moment at C
Using Muller Breslau's principle, ILD of BM is given as

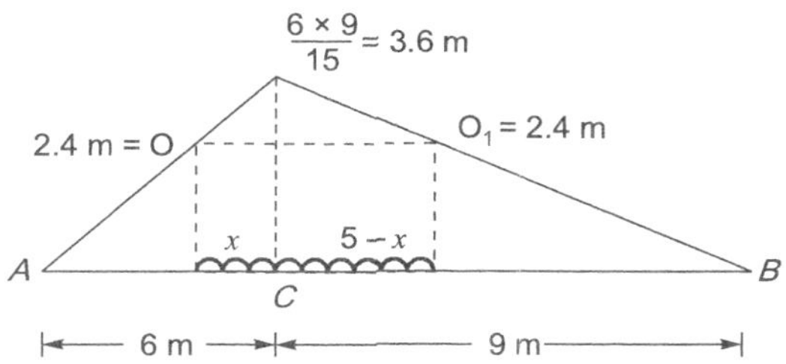

For max BM, point C should divide load in same proportion as $AC : CB$

So,
$$\frac{x}{5-x} = \frac{6}{9}$$

$\Rightarrow$
$$x = 2 \text{ m}$$

$$\text{Ordinate at } O \text{ and } O_1 = \frac{3.6}{6} \times 4 = 2.4 \text{ m}$$

So,
$$\text{Max BM} = \left[2.4 \times 5 + (3.6 - 2.4) \times \frac{1}{2} \times 5\right] \times 40$$

$$BM_{max} = 600 \text{ kNm}$$

2.10 **Determine the absolute bending moment in a 16 m long beam AB due to the truck loading shown in the figure.**

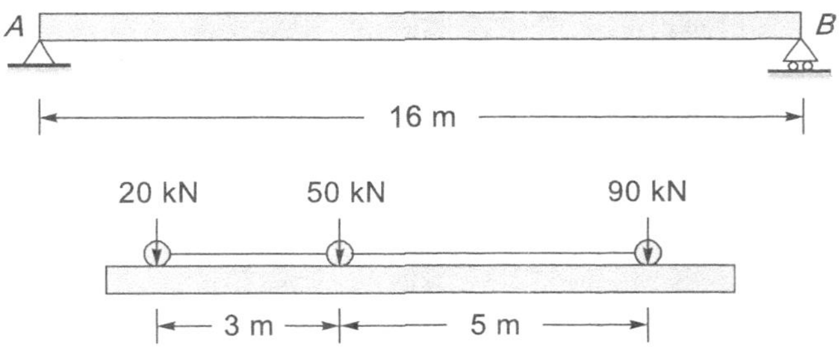

[20 Marks : 2021]

Solution:

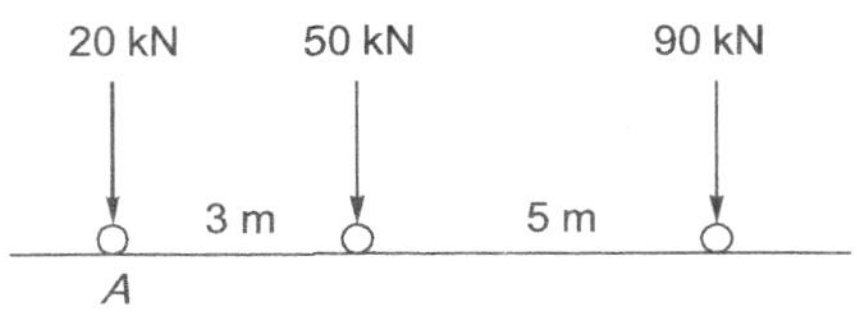

Absolute maximum BM occurs near centre.

Resultant of load system

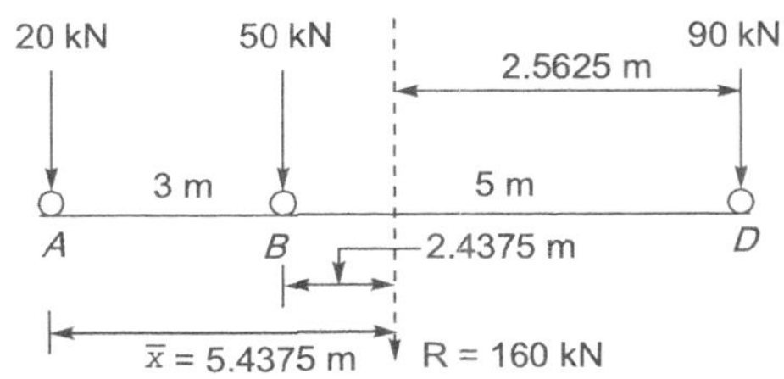

$$R \times \bar{x} = 20 \times 0 + (50 \times 3) + 90 \times 8$$

$$\bar{x} = \frac{150 + 120}{160} = 5.4375 \text{m}$$

I[st]-possibility: Let us assume that maximum BM occur under the 50 kN load:

To have a maximum bending moment under any point load, load position should be such that, redultant and the load under which pcint load is maximum is equidistant from mid-span

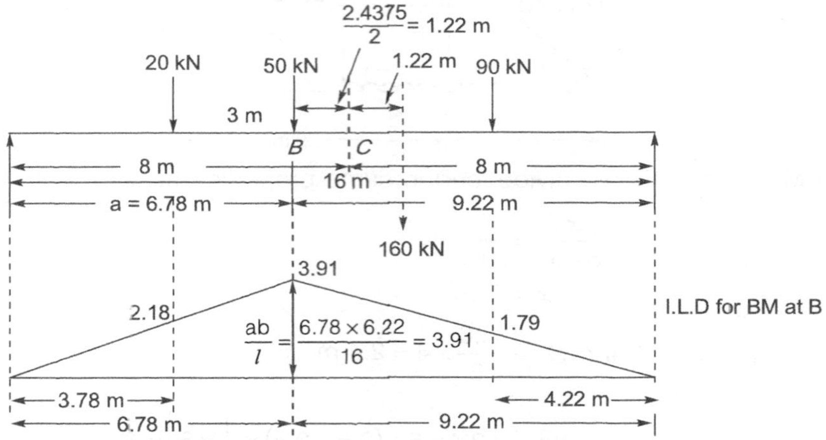

Maximum BM under 50 kN load = 20 × 2.18 + 50 × 3.91 + 90 × 1.79 = 400.2 kN-m

2nd possibility: Let us assume that maximum BM occur under the 90 kN load:

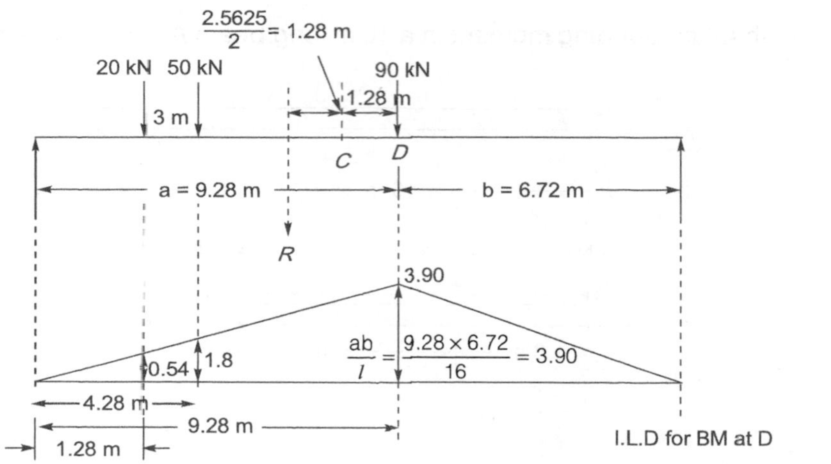

Maximum BM at D = 20 × 0.54 + 50 × 1.8 + 90 × 3.9 = 451.8 kNm

∴ Absolute maximum bending moment = Maximum of above two values = 451.8 kNm

2.11 **Two wheel loads of 8000 N and 4000 N at a fixed distance of 2000 mm, cross a beam of 20 m span. Draw the influence line diagram for bending moment and shear force for a point 8 m from the left side and also determine the maximum bending moment and shear force at that point. Also, evaluate the absolute maximum bending moment due to the given loading system. The loads cross the beam from left to right.**

[20 marks : 2022]

Solution:

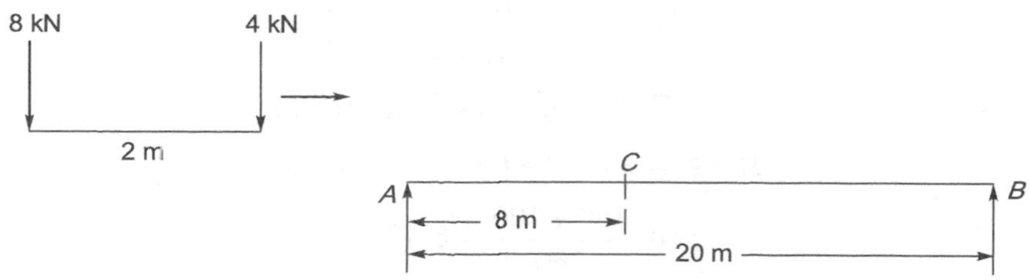

Influence line diagram for $(S.F)_C$: Using Muller Breslau's principle, ILD of stress function can be drawn by removing the restraint offered by that stress function and introducing directly a unit displacement. So, giving unit displacement at C and removing the restrain offered by S.F at C, we get

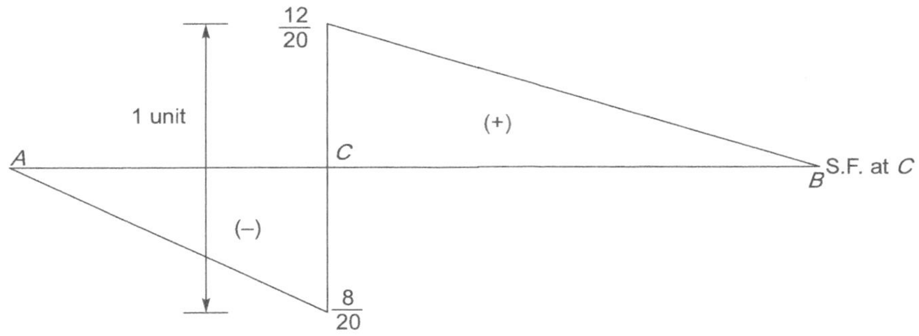

Similarly for Bending Moment at C, provide hinge at C and apply unit displacement in positive direction of moment.

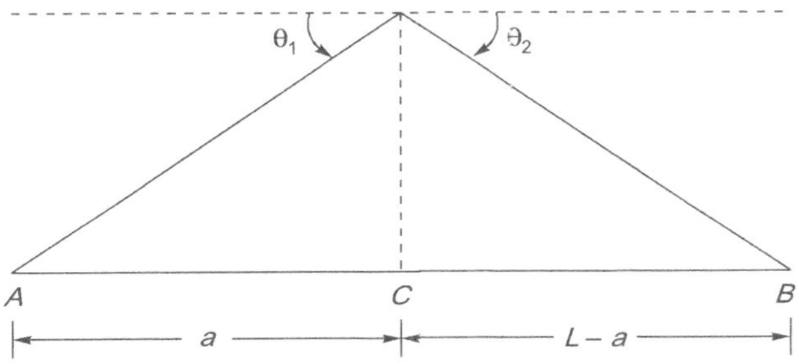

$$\theta_1 + \theta_2 = 1$$

$$\therefore \qquad \theta_1 = \frac{L-a}{L}; \quad \theta_2 = \frac{a}{L}$$

Putting values of a and L, we get ILD for B.M. at C,

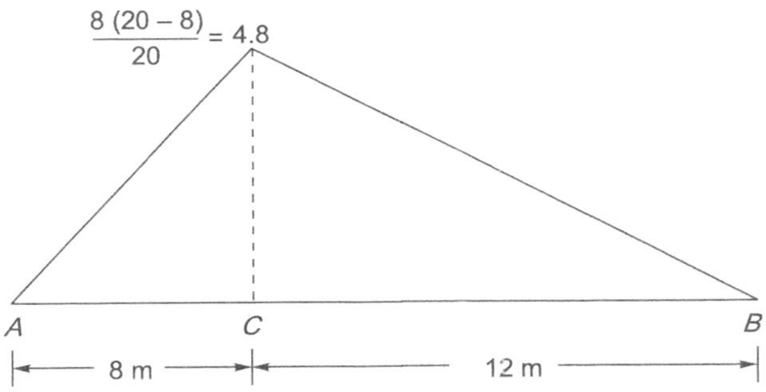

When given load moves on the span AB

(i)To calculate maximum shear force at C ILD for $(S.F.)_C$.

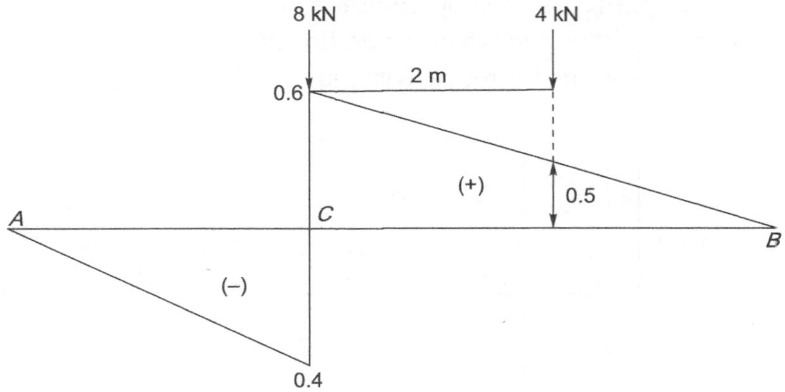

To have maximum shear force, place 8 kN load on just right of *C* as shown in figure.

The value of ordinates of ILD at 8 kN and 4 kN will be 0.6 and 0.5 respectively.

So, maximum shear force $= 8 \times 0.6 + 4 \times 0.5 = 6.8$ kN

(ii) To calculate maximum bending moment at C.

To have maximum shear force, place 8 kN load on just right of *C* as shown in figure.

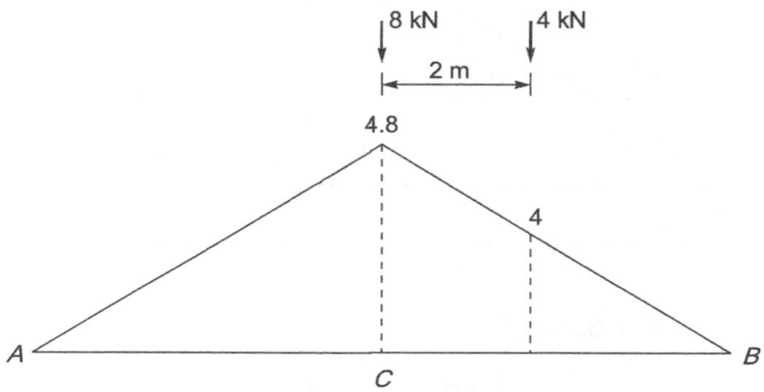

The value of ordinates of ILD at 8 kN and a 4 kN will be 4.8 and 4 respectively.

So, maximum B.M at C $= 8 \times 4.8 + 4 \times 4 = 54.4$ kNm

To calculate maximum bending moment any where in beam we have to place the loading on beam in such a way that centre of span is mid way between C.G of load system and load under consideration. Since 8 kN is nearer to C.G. Therefore maximum bending moment will occur below 8 kN.

C.G. of load system:

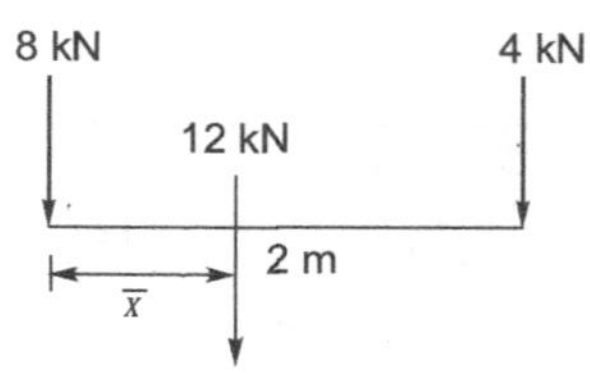

$$\bar{x} = \frac{8 \times 0 + 4 \times 2}{12} = 0.67 \text{m}$$

So, place the load as shown in figure.

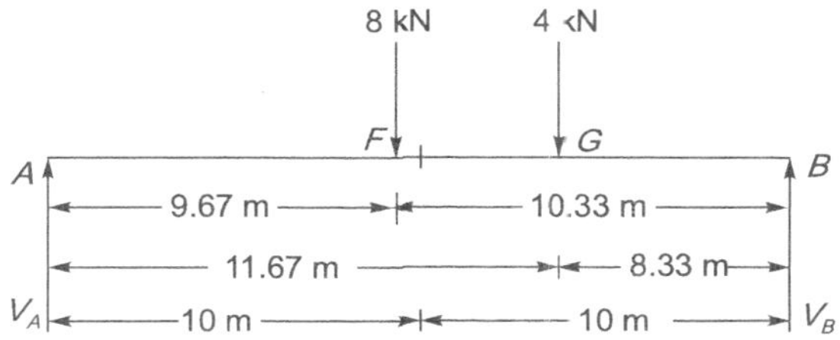

$$\Sigma M_B = 0$$

$\Rightarrow V_A \times 20 - 8 \times 10.33 - 4 \times 8.33 = 0$

$$\Rightarrow \quad V_A = 5.798 \text{ kN}$$

$$\therefore \quad V_B = 12 - V_A = 6.202 \text{ kN}$$

Maximum bending moment will be at F.

$$\therefore \quad (\text{B.M})_F = V_A \times 9.67$$

$$= 5.798 \times 9.67 = 56.067 \text{ kNm}$$

2.12 **A uniformly distributed load of 45 kN/m longer than the span rolls over a simply supported girder of 35 m span. Using influence line diagram for shear force and bending moment, determine the maximum shear force and maximum bending moment at a section 14 m from left-hand support.**

[20 marks : 2023]

Solution:

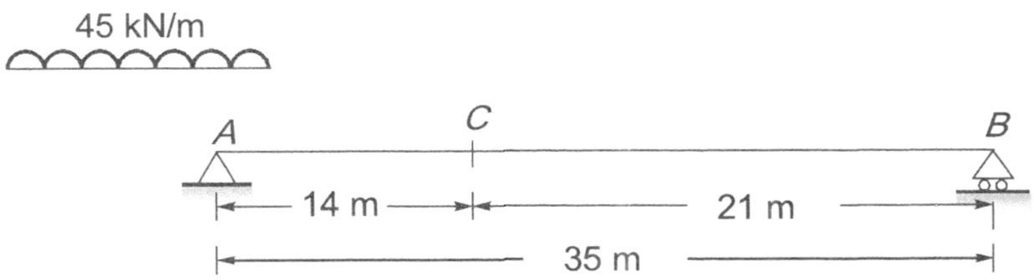

Influence line diagram of shear force at C using Muller breaslau's principle is shown below.

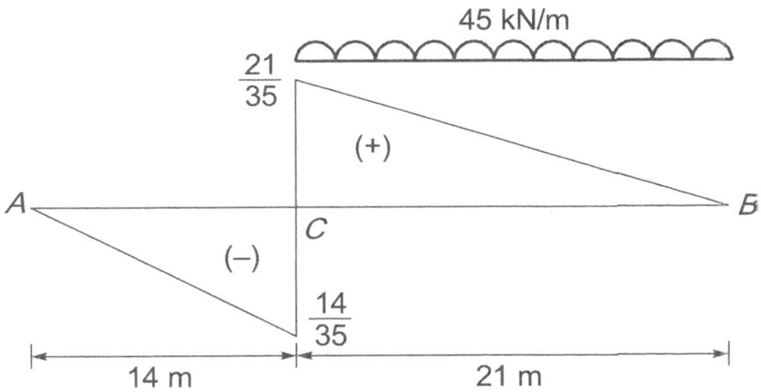

Now, when load moves from A to C, then shear force at C will be maximum when load covers the whole portion BC as shown in figure above.

So, $$(\text{S.F.})_C = \frac{1}{2} \times 21 \times \frac{21}{35} \times 45 = 283.5 \text{ kN}$$

Influence line diagram for bending moment at *C*.

Influence line diagram of bending moment at C using Muller breaslau's principle is shown below.

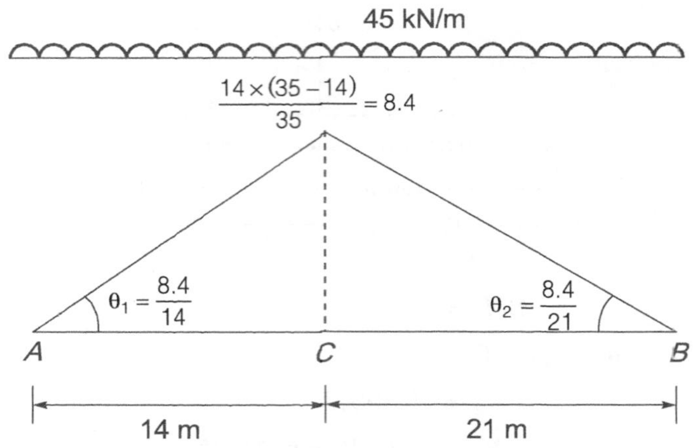

Now, when load moves from A to C, bending moment will be maximum at C when load covers the entire span.

So,
$$(\text{B.M.})_C = \frac{1}{2} \times 35 \times 8.4 \times 45$$
$$= 6615 \text{ kN-m}$$

3. Arches and Suspended Cables

3.1 **A symmetrical triangular shaped arch ACB is hinged at supports *A* and *B*. The two members of the arch are of the same cross-section. A vertical load '*W*' is placed as shown in the figure. Joint *C* is rigid.**

Determine the components of reactions and draw bending moment diagram indicating maximum positive and negative magnitudes.

[20 marks : 2007]

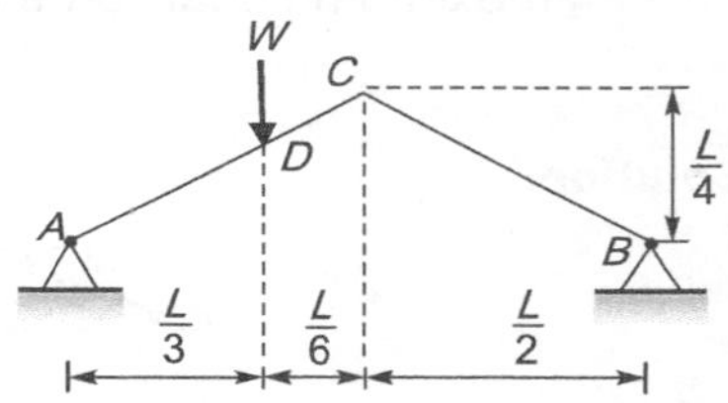

Solution:

The figure below shows the triangular arch carrying the load '*W*' at a distance *L*/3 from end *A*. Let the vertical reactions at *A* and *B* be V_1 and V_2 respectively. Let *H* be the horizontal thrust at each support.

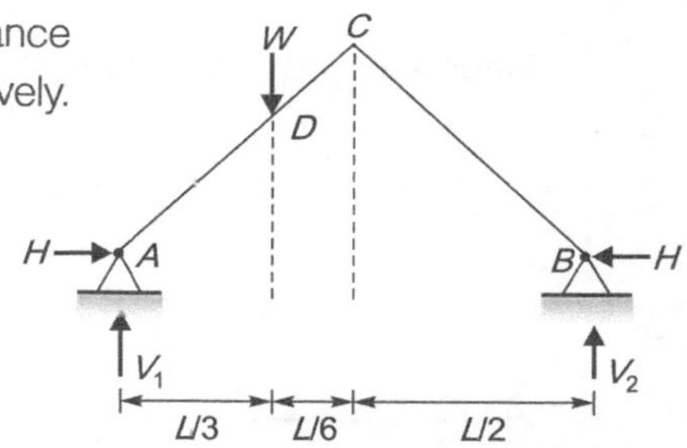

From the diagram above, we have, $V_1 + V_2 = W$

Taking moment about hinge *A*, we get

$$V_2 \times L - W \times \frac{L}{3} = 0$$

$$\Rightarrow \quad V_2 = \frac{W}{3}$$

$$\therefore \quad V_1 = W - \frac{W}{3} = \frac{2W}{3}$$

Now taking the same arch, but placing the load *W* at a distance *L*/3 from *B*. In this case the vertical reactions will interchange i.e. the reaction at *A* and *B* will be respectively V_2 and V_1.

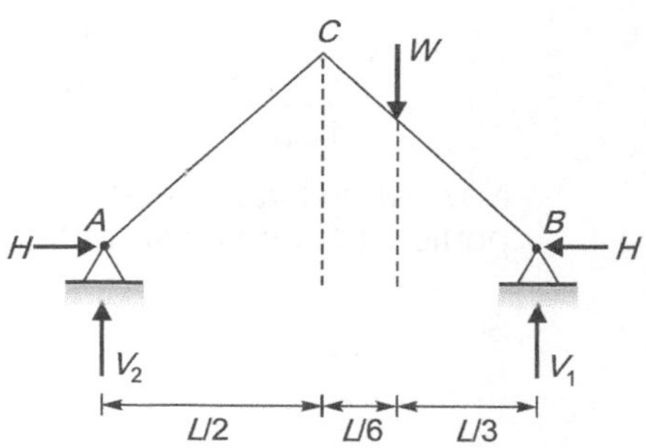

But the horizontal thrust will remain same.

If both the above arrangement are clubbed together, we get

For this load system, the horizontal thrust will be 2H at each support. The vertical reaction at each support will be $V_1 + V_2$

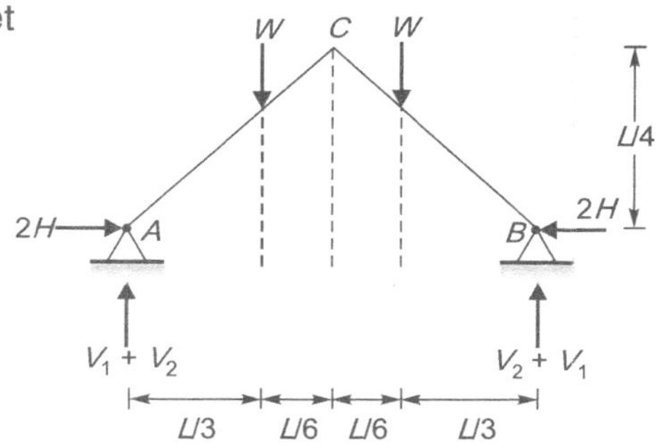

But $\quad V_1 + V_2 = \dfrac{2W}{3} + \dfrac{W}{3} = W$

At any section x from A, $\left(x < \dfrac{L}{3}\right)$ the beam moment at the section $= M = Wx$

But at any section x from A $\left(\dfrac{L}{3} < x < \dfrac{L}{2}\right)$, the beam moment at the section

$$= M = Wx - W\left(x - \frac{L}{3}\right) = Wx - Wx + \frac{WL}{3} = \frac{WL}{3} \text{ (constant)}$$

If x and y be the coordinates of a point on the arch, then we have

$$\Rightarrow \quad \frac{y}{x} = \frac{L/4}{L/2} = \frac{L \times 2}{L \times 4}$$

$$y = \frac{x}{2}$$

Now we know that the **horizontal thrust** is given by

$$\Rightarrow \quad 2H = \frac{\int My\,dx}{\int y^2\,dx} = \frac{2\left[\int_0^{L/3} Wx \times \frac{x}{2}\,dx + \int_{L/3}^{L/2} \frac{WL}{3} \cdot \frac{x}{2}\,dx\right]}{2\int_0^{L/2} \frac{x^2}{4}\,dx}$$

$$\Rightarrow \quad 2H = \frac{\frac{W}{2}\int_0^{L/3} x^2\,dx + \frac{WL}{6}\int_{L/3}^{L/2} x\,dx}{\frac{1}{4}\int_0^{L/2} x^2\,dx} = \frac{\frac{W}{2} \times \left[\frac{x^3}{3}\right]_0^{L/3} + \frac{WL}{6}\left[\frac{x^2}{2}\right]_{L/3}^{L/2}}{\frac{1}{4}\left[\frac{x^3}{3}\right]_0^{L/2}}$$

$$\Rightarrow \quad 2H = \frac{\frac{W}{6} \times \frac{L^3}{27} + \frac{WL}{12} \times \left[\frac{L^2}{4} - \frac{L^2}{9}\right]}{\frac{1}{12} \times \frac{L^3}{8}} = \frac{\frac{WL^3}{162} + \frac{WL}{12} \times \frac{5L^2}{36}}{L^3/96} = \frac{\frac{WL^3}{162} + \frac{5WL^3}{432}}{L^3/96}$$

$$\Rightarrow \quad 2H = \frac{\frac{8WL^3 + 15WL^3}{1296}}{L^3/96} = \frac{23WL^3 \times 96}{L^3 \times 1296}$$

$$\Rightarrow \quad H = \frac{23 \times 96 \times W}{1296 \times 2} = \frac{23}{27}W$$

Assuming the inner face of the arch as reference face.

Portion *AD*

$$M_x(x \text{ from } A) = V_1 x - Hy\left(x < \frac{L}{3}\right) = \frac{2W}{3} \times \frac{L}{3} - \frac{23}{27}W \times \frac{x}{2}$$

$$\Rightarrow \qquad M_D = \frac{2WL}{9} - \frac{23}{27}W \times \frac{L}{6} = \frac{13WL}{162}$$

Portion *DC*

$$M_x(x \text{ from } A) = V_1 x - Hy - W\left(x - \frac{L}{3}\right) \quad \left[\frac{L}{3} < x < \frac{L}{2}\right]$$

$$= \frac{2Wx}{3} - \frac{23}{27}W \times \frac{x}{2} - Wx + \frac{WL}{3}$$

$$= \frac{2Wx}{3} - \frac{23Wx}{54} - Wx + \frac{WL}{3}$$

$$= \frac{36Wx - 23Wx - 54Wx + 18WL}{54}$$

$$M_C = \frac{18WL - 41W \times \frac{L}{2}}{54} = -\frac{5WL}{108}$$

Portion *BC*

$$M_x(x \text{ from } B) = V_2 x - Hy \left[x < \frac{L}{2}\right]$$

$$= \frac{W}{3}x - \frac{23Wx}{27 \times 2}$$

$$= \frac{18Wx - 23Wx}{54}$$

$$= -\frac{5Wx}{54}$$

$$M_C = -\frac{5W}{54} \times \frac{L}{2} = -\frac{5WL}{108}$$

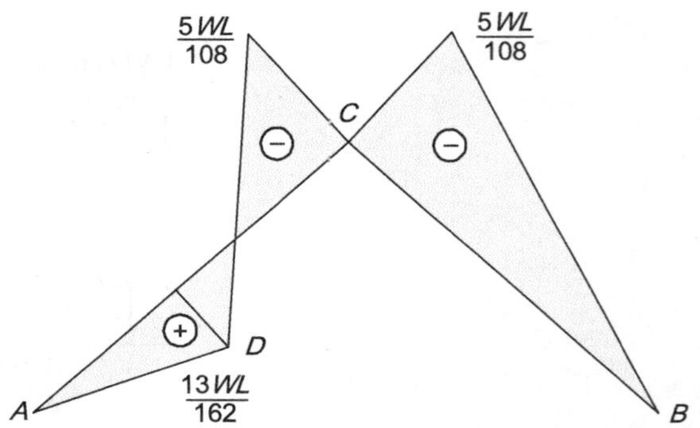

Bending Moment Diagram

3.2 A cable suspends across a gap of 250 m and carries a UDL of 10 kN/m horizontally. Calculate the maximum tension if the maximum sag is 1/25. Also compute sag at 50 m from one end.

[10 marks : 2014]

Solution:

Since load is uniform along the horizontal span and supports are at the same level

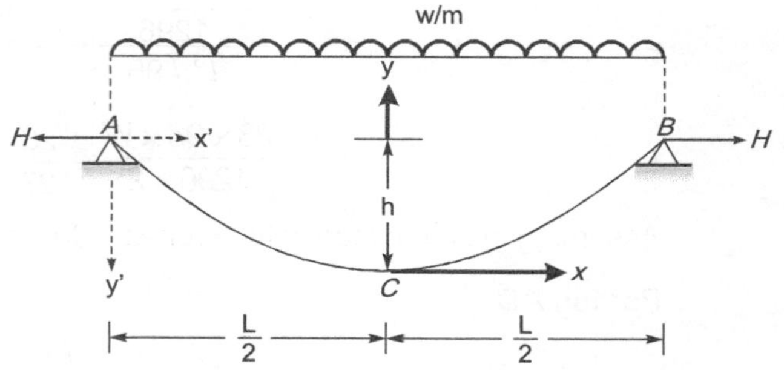

$$\therefore \qquad \frac{d^2y}{dx^2} = \frac{w}{H} \qquad \text{...(i)}$$

Integrating (i), we get

$$\frac{dy}{dx} = \frac{w}{H}x + C_1 \qquad \text{...(ii)}$$

Again integrating (ii), we get

$$y = \frac{w}{H} \times \frac{x^2}{2} + C_1 x + C_2$$

Using boundary condition,

With origin at C, at $x = 0$, $y = 0$

Hence from eq. (iii) $C_2 = 0$

From eq. (ii) $C_1 = 0$

at $x = 0$; $\frac{dy}{dx} = 0$ (due to symmetry)

The equation of cable becomes,

$$y = \frac{w}{2H}x^2 \quad \text{...(iii)}$$

at $x = \frac{L}{2}$, $y = h$

$$\therefore \quad h = \frac{w}{2H}\left(\frac{L}{2}\right)^2$$

or $$H = \frac{wL^2}{8h}$$

Substituting value of H in eq. (iii), we get

$$y = \frac{8h.H}{L^2} \times \frac{1}{2H} \times x^2 = \frac{4hx^2}{L^2} \quad \text{...(iv)}$$

When origin is shifted to A, in order to find the shape of the cable, moment about a point P with coordinate x and y as shown in figure will be

$$M_p = V_A(x) - \frac{wx^2}{2} - Hy = 0$$

$$\Rightarrow \quad \frac{wL}{2}x - \frac{wx^2}{2} = \frac{wL^2}{8h}y$$

On rearranging, we get $$y = \frac{4h}{L^2}x(L - x)$$

given, $$h = \frac{L}{25} = 10\text{m}$$

$$L = 250 \text{ m}$$

$$w = 10 \text{ kN/m}$$

(i) **Maximum tension**

By symmetry, $$R_A = R_B = \frac{wL}{2} = \frac{250 \times 10}{2} = 1250 \text{ kN}$$

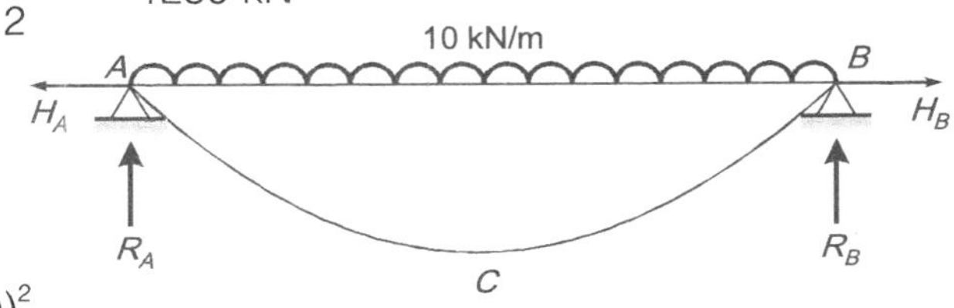

Horizontal thrust,

$$H_A = H_B = H = \frac{wL^2}{8h}$$

$$H = \frac{10 \times (250)^2}{8 \times 10} = 7812.5 \text{ kN}$$

Maximum tension occurs at higher support, here both supports are at same level

$$\therefore \quad T_{max} = \sqrt{R_A^2 + H_A^2} = \sqrt{(1250)^2 + (7812.5)^2}$$

$$= 7911.86 \text{ kN}$$

(ii) **Sag at 50 m from one end**

Equation of cable, $$y = \frac{4h}{L^2}x(L - x)$$

at $x = 50$ m $\qquad y = \dfrac{4\times10}{(250)^2}\times50\times(250-50)$

$\Rightarrow \qquad y = 6.4$ m

3.3 A 3-hinged parabolic arch of 16 m span has its abutments *A* and *B* at a depth of 4 m and 8 m respectively below the crown *C*. It is loaded as shown in figure. Determine the horizontal thrust and the vertical reactions at the supports.

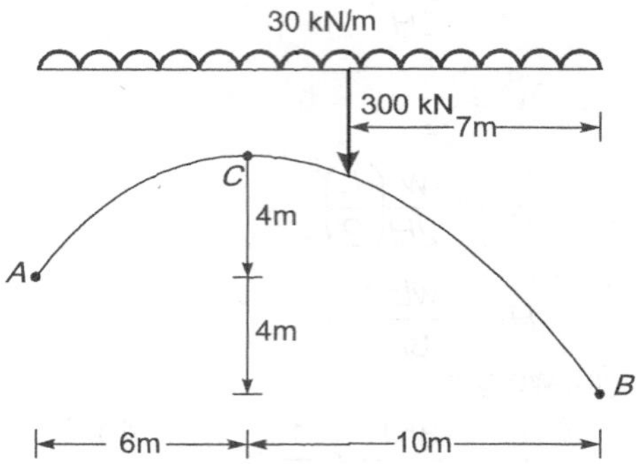

[15 marks : 2014]

Solution:

Let R_A and R_B be the vertical reactions at the supports *A* and *B*. Let *H* be the equal horizontal reaction at both supports *A* and *B*

$\Sigma F_y = 0; \qquad R_A + R_B = 30 \times 16 + 300 = 780 \qquad$ ...(i)

Since there is a hinge at crown. Hence *BM* at crown will be zero.

So taking moment about *C* from left side of *C*, $M_C = 0$

$$R_A\times6-H\times4-\frac{w\times6^2}{2} = 0$$

$$6R_A = 4H+\frac{30\times6^2}{2}$$

$$6R_A = 4H+540 \qquad \text{...(ii)}$$

Taking moment about *C* from right side of *C*, $M_C = 0$

$$-R_B\times10+H\times8+\frac{30\times10^2}{2}+300\times3=0$$

$$-10R_B + 8H + 2400 = 0 \qquad \text{...(iii)}$$

From (i) and (ii), we get

$$6(780 - R_B) = 4H + 540$$

$$4680 - 6R_B = 4H + 540$$

$$4H + 6R_B = 4140 \qquad \text{...(iv)}$$

On solving eq. (iii) and (iv), we get

$$R_B = 485.45 \text{ kN}$$

$$H = 306.81 \text{ kN}$$

From eq. (i), $\qquad R_A = 780 - R_B = 294.55$ kN

Thus, $\qquad R_A = 294.55$ kN

$$R_B = 485.45 \text{ kN}$$

$$H = 306.81 \text{ kN}$$

Check: $\Sigma F_y = R_A + T_B - 30 \times 16 - 300$

$$\Sigma F_y = 294.55 + 485.45 - 480 - 300$$

$$\Sigma F_y = 0 \quad \text{(OK)}$$

3.4 A symmetrical three hinged parabolic arch has 40 m span and 5 m rise. A vertical downward load of 30 kN and a horizontal load of 20 kN (acting in the right hand side direction) act at one quarter span form left hand support. Determine reactions at the supports.

[5 marks : 2015]

Solution:

$$R_A + R_B = 30 \text{ kN} \quad \text{...(i)}$$

$$H_A + 20 = H_B \quad \text{...(ii)}$$

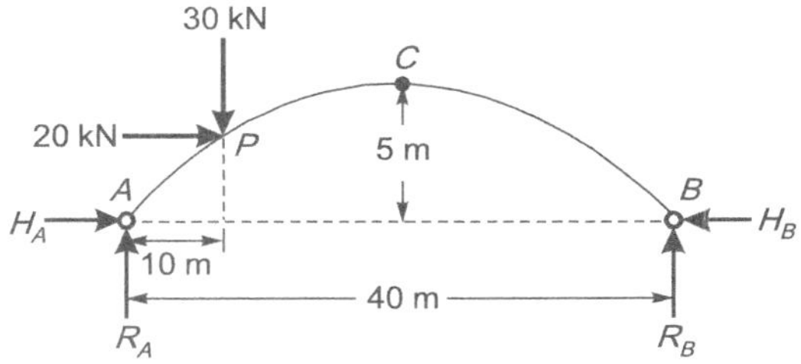

Equation of arch is, $y = \frac{4y_c}{L^2}(Lx - x^2)$

$\Rightarrow$ $y = \frac{4(5)}{40^2}(40x - x^2)$

$\Rightarrow$ $y = \frac{1}{80}(40x - x^2)$

At P, $x = 10$ m,

$\therefore$ $y = \frac{1}{80}(40 \times 10 - 10^2) = 3.75\text{m}$

Taking moments about B, $R_A(40) = 30(30) - 20(3.75)$

$\Rightarrow$ $R_A = 20.625 \text{ kN} (\uparrow)$

$\therefore$ $R_B = 30 - R_A$

$\Rightarrow$ $R_B = 9.375 \text{ kN} (\uparrow)$

Taking moments about C of the right hand portion of arch,

$$R_B(20) = H_B(5)$$

$\Rightarrow$ $H_B = \frac{9.375 \times 20}{5} = 37.5\text{kN}$

$\therefore$ $H_A = H_B - 20 = 17.5 \text{ kN}$

3.5 A two hinged symmetrical parabolic arch of span 30 m and central rise 6 m carries a point of load of 40 kN at a distance of 5 m from the left support. Find the horizontal thrust at each support. Also find the maximum bending moment.

[10 marks : 2016]

Solution:

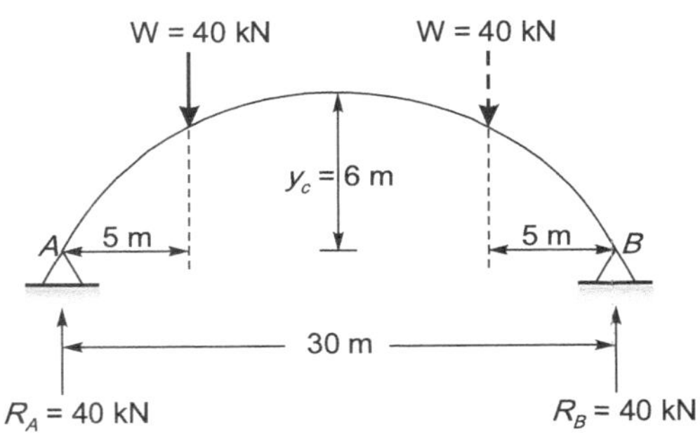

Horizontal thrust $(H) = \frac{\int M'y\,dx}{\int y^2 dx}$ where M' = Beam moment

Apply a fictitious load of 40 kN at 5 m from right support B to take advantage of symmetry.

$\therefore$ Horizontal thrust will become twice i.e. $2H$.

$$\therefore \quad 2H = \frac{\int M'y\,dx}{\int y^2 dx}$$

and $R_A = W = 40$ kN due to symmetry

Equation of parabolic arch is $y = \frac{4y_c}{L^2}(Lx - x^2) = \frac{4\times 6}{30^2}(30x - x^2) = \frac{2}{75}(30x - x^2)$

$$\text{Denominator} = \int y^2 dx = \int_0^{30}\left(\frac{2}{75}\right)^2 (30x - x^2)^2 dx$$

$$= \left(\frac{2}{75}\right)^2 \int_0^{30} (900x^2 + x^4 - 60x^3)dx$$

$$= \left(\frac{2}{75}\right)^2 \left[900\frac{x^3}{3} + \frac{x^5}{5} - 60\times\frac{x^4}{4}\right]_0^{30}$$

$$= \left(\frac{2}{75}\right)^2 \left[300(30^3 - 0^3) + \frac{1}{5}(30^5 - 0^5) - 15(30^4 - 0^4)\right] = 576$$

Short tip: For parabolic arch $y = \frac{4y_c}{L^2}(Lx - x^2), \int_0^L y^2 dx = \frac{8}{15}y_c^2 L = 576$

Beam moment for $(0 < x < 5)$

$$M' = R_A x$$

Beam moment for $(5 < x < 15)$

$$M' = R_A x - W(x-5)$$
$$= Wx - W(x-5)$$
$$= Wx - Wx + W(5)$$
$$M' = W(5)$$

$$\text{Numerator} = \int M'y\,dx = 2\left[\int_0^5 R_A \cdot xy\,dx + \int_5^{15} W(5)\,y\,dx\right]$$

$$= 2\left[\int_0^5 W\cdot x\cdot\frac{4y_c}{L^2}(Lx - x^2)dx + \int_5^{15} 5W\cdot\frac{4y_c}{L^2}(Lx - x^2)dx\right]$$

$$= 2\left[\frac{4Wy_c}{L^2}\left\{\frac{L}{3}(5^3) - \frac{1}{4}(5^4)\right\} + \frac{20Wy_c}{L^2}\left\{\frac{L}{2}(15^2 - 5^2) - \frac{1}{3}(15^3 - 5^3)\right\}\right]$$

$$= 2\left[\frac{4\times 40\times 6}{30^2}\left(\frac{30\times 5^3}{3} - \frac{5^4}{4}\right) + \frac{20\times 40\times 6}{30^2}\left(\frac{30}{2}(200) - \frac{1}{3}(3250)\right)\right]$$

$$= 2[1166.67 + 10222.22] = 22777.78$$

$$\therefore \quad 2H = \frac{\int M'y\,dx}{\int y^2 dx} = \frac{22777.78}{576} = 39.545$$

$\therefore \quad H = 19.772$ kN

$\therefore$ Horizontal thrust at supports = 19.772 kN

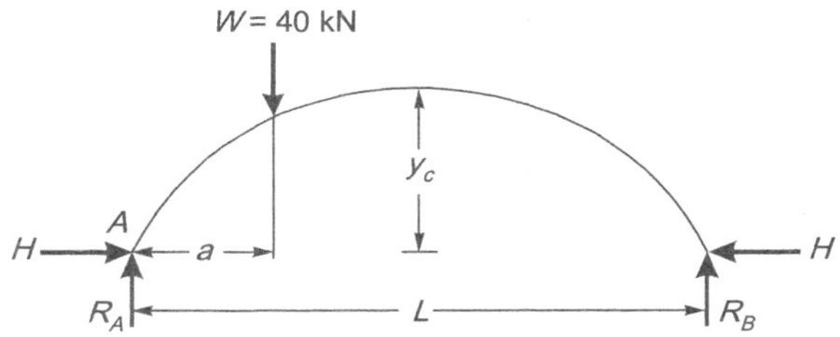

Short tip for check:

Use: $$H = \frac{5}{8}\frac{Wa}{y_c L^3}(L-a)(L^2 + aL - a^2)$$

$$= \frac{5}{8}\times\frac{40\times5}{6\times30^3}(30-5)\left(30^2+5\times30-5^2\right) = 19.772 \text{ kN}$$

Taking BM about A

$$R_B = (30) - 40\times5 = 0$$

$$R_B = 6.67 \text{ kN}$$

$\therefore \quad M_{-ve} = R_B\cdot x - H_y$ where x is measured from B

$$= 6.67x - 19.772\times\frac{2}{75}(30x - x^2)$$

For M_{-ve} to be maximum, $\dfrac{dM_{-ve}}{dx} = 0$

$$\Rightarrow \quad 6.67 - 19.772\times\frac{2}{75}(30-2x) = 0$$

$$\Rightarrow \quad 12.65 = 30 - 2x$$

$$\Rightarrow \quad x = 8.675 \text{ m from } B$$

$$\therefore \quad M_{-ve\,max} = 6.67(8.675) - 19.772\times\frac{2}{75}(30\times8.675 - 8.675^2)$$

$$= -39.68 \text{ Nm}$$

3.6 A suspension cable of 160 m span and 16 m central dip carries a load of 1/2 kN per linear horizontal metre. Calculate the maximum and minimum tension in the cable. Also find horizontal and vertical forces in each pier under the following alternate conditions:

(i) If the cable passes over the frictionless pulley on the top of the piers.

(ii) If the cable is firmly clamped to saddles carried on frictionless roller on the top of the piers. In each case the backstay is inclined at 30° to the horizontal.

[20 marks : 2019]

Solution:

Given,

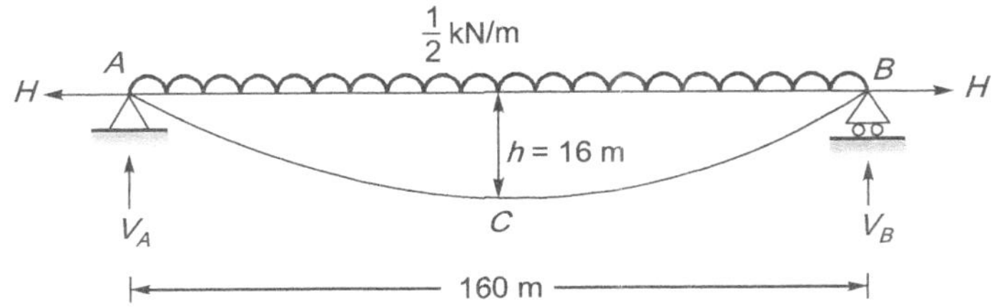

The reactions, $$V_A = V_B = \frac{\frac{1}{2}\times160}{2} = 40 \text{ kN}$$

The horizontal reaction, $H = \frac{wl^2}{8h} = \frac{\frac{1}{2} \times 160^2}{8 \times 16} = 100 \text{ kN}$

The minimum reaction will occur at mid span, which is equal to the value of H.

So, $T_{min} = H = 100 \text{ kN}$

At support,

Maximum tension, $T_{max} = \sqrt{100^2 + 40^2} = 107.703 \text{ kN}$

Assuming the shape of cable to be parabola,

So, the dip, $y = \frac{4h}{l^2} x(l - x)$

$$\text{Slope of cable} = \frac{dy}{dx} = \frac{4h}{l^2}(L - 2x)$$

At support, ($x = 0$)

$$\text{Slope} = \frac{4h}{l} = \frac{4 \times 16}{160} = 0.4$$

$$\tan\theta = 0.4$$

So, $\theta = 21.80°$

(i) **If the cable passes over the frictionless pulley on the top of the piers**

In this case, tension in cable on two sides will be equal

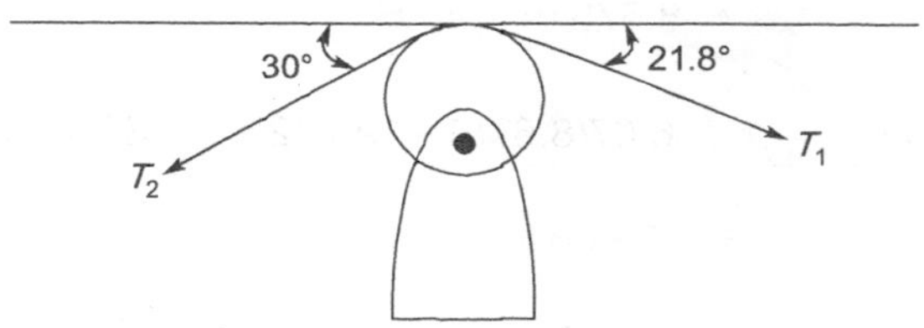

Here, $T_1 = T_2 = T_{max} = 107.703 \text{ kN}$

So, vertical pressure on pier, $V = T_{max}(\sin 21.8° + \sin 30°)$

$= 107.703(\sin 21.8° + \sin 30°)$

$= 93.85 \text{ kN}$

Horizontal thrust on pier, $H = T_{max}(\cos 21.8° - \cos 30°)$

$= 107.703(\cos 21.8° - \cos 30°)$

$= 6.73 \text{ kN } (\rightarrow)$

(ii) **If the cable is firmly clamped to saddles carried on frictionless roller on the top of the piers.**

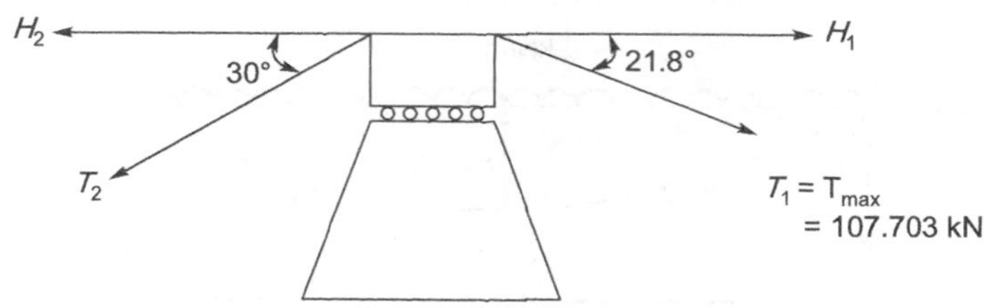

Here, $H_1 = H_2$

$$T_1 \cos 21.8° = T_2 \cos 30°$$

$$T_2 = \frac{T_1 \cos 21.8°}{\cos 30°} = \frac{107.703 \times \cos 21.8°}{\cos 30°} = 115.47 \text{ kN}$$

Vertical pressure on pier, $V = T_1 \sin 21.8° + T_2 \sin 30°$

$= 107.703 \sin 21.8° + 115.45 \sin 30°$

$= 97.72$ kN

3.7 A cable of uniform cross-section hangs between two points A and B, which are 150 m apart. The end 'A' of the cable is 3 m above the other end of the cable. The sag of the cable measured from 'B' = 2 m. If the cable carries a UDL of 12 kN/m, determine the maximum tension in the cable. Also find the horizontal pull.

[12 marks : 2020]

Solution:

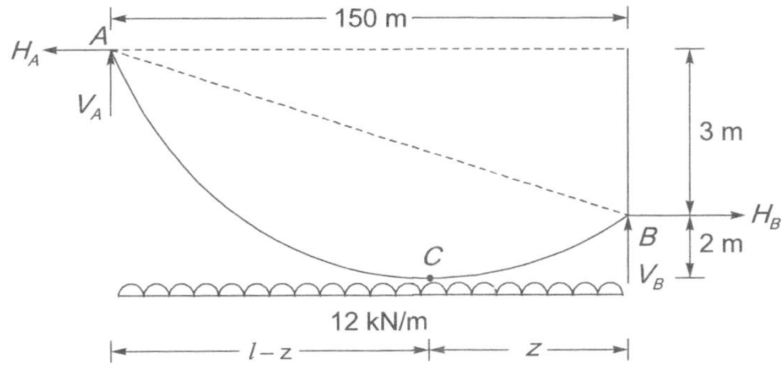

w = UDL = 12 kN/m, l = 150 m, d = 3 m, h = 3 + 2 = 5 m

Lowest point, $y_c = 2$ m

Location of lowest point:

$$\frac{l-z}{z} = \sqrt{\frac{y_C + d}{y_C}} = \sqrt{\frac{5}{2}} = 1.581$$

$$l - z = 1.581 z$$

So $z = 58.117$ m

$$\Sigma F_x = 0$$

$\Rightarrow$ $-H_A + H_B = 0$ ...(i)

$$\Sigma F_y = 0$$

$\Rightarrow$ $V_A + V_B - 12 \times 150 = 0$ ...(ii)

$$\Sigma M_A = 0$$

$\Rightarrow$ $-V_B \times 150 - H_B \times 3 + 12 \times 150 \times 75 = 0$

$\Rightarrow$ $150 V_B + 3H_B = 1800 \times 75$ ...(iii)

$$M_C = 0 \text{ (RHS)}$$

$\Rightarrow$ $-V_B \times z + H_B \times 2 + \frac{12z^2}{2} = 0$

$\Rightarrow$ $-58.117V_B + 2H_B = -6(58.117)^2$ ...(iv)

From eq. (iii) and (iv) $V_B = 697.366$ kN

$$H_B = 10131.670 \text{ kN} = H_A$$

$$V_A = 1102.634 \text{ kN} \quad \text{From eq. (ii)}$$

Max tension in the cable:

$$T_{max} = \sqrt{H^2 + V_A^2}$$

$$= \sqrt{1102.634^2 + 10131.670^2}$$

$$= 10191.493 \text{ kN}$$

3.8 **Sketch influence line diagram for the bending moment at a point 'C' located 6 m from one of the supports of a three hinged symmetrical parabolic arch having span of 18 m and central rise 2.5 m. Locate the point from where the moving load changes the sign of bending moment at C.**

[20 marks : 2020]

Solution:

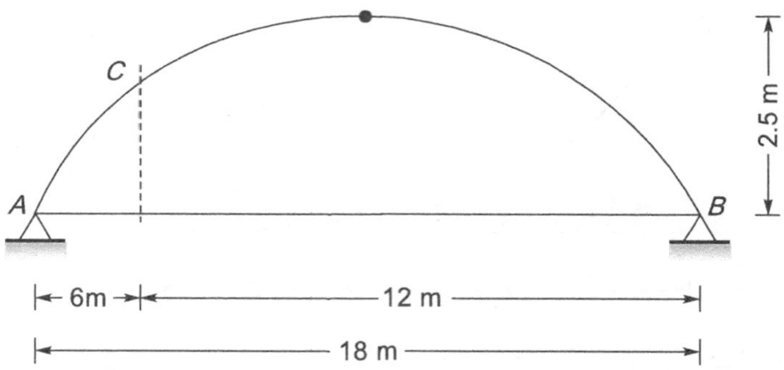

ILD for bending moment at C (M_c)

Bending moment at any section= BM in simply supported beam – Hy

∴ ILD can be drawn by superimposing beam ILD and (Hy) ILD

We know, for parabolic arch

$$y = \frac{4h}{l^2}x(l-x)$$

$$y \text{ at C } (x = 6 \text{ m}) = \frac{4\times 2.5}{(18)^2}\times 6\times(18-6) = \frac{20}{9}$$

ILD for bending moment,

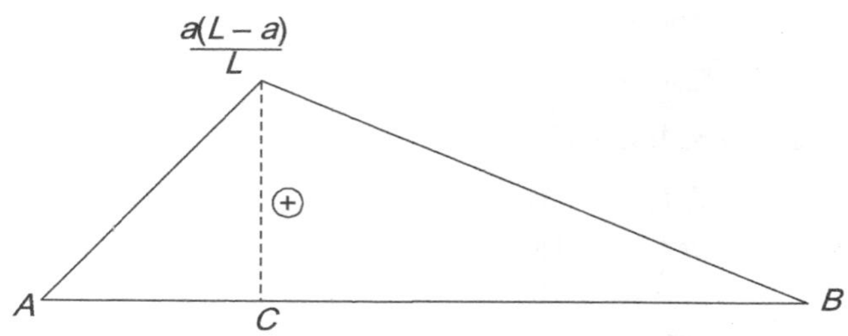

Beam ILD for M_C

$$\frac{a(L-a)}{L} = \frac{6\times 12}{18} = 4$$

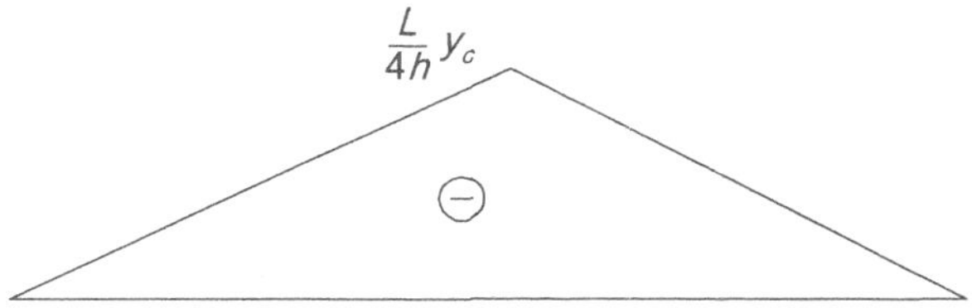

ILD for Hy_c

$$\frac{Ly_C}{4h} = \frac{18}{4\times2.5}y_C$$

$$\frac{18}{4\times2.5}\times\frac{20}{9} = 4$$

Superimposing both ILD

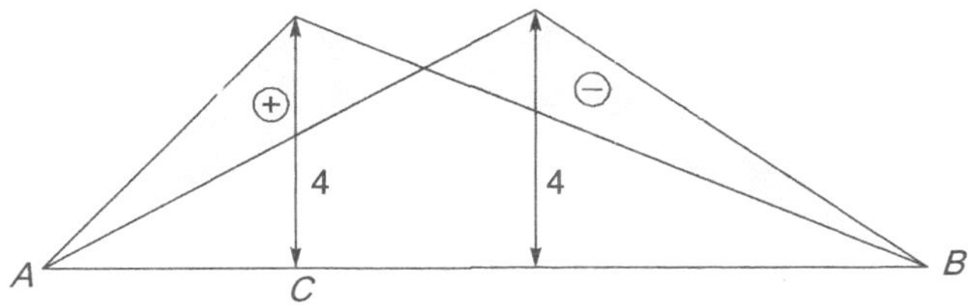

ILD for bending moment at C

Assuming load is placed at Z than BM at C would be zero because at Z distance from A. ILD changes its sign.

$$\Sigma M_B = 0$$

$$\Rightarrow \quad V_A = \frac{18-Z}{18}$$

$$M_D = 0 \text{ (LHS)}$$

$$\Rightarrow \quad H_A = \frac{Z}{5}$$

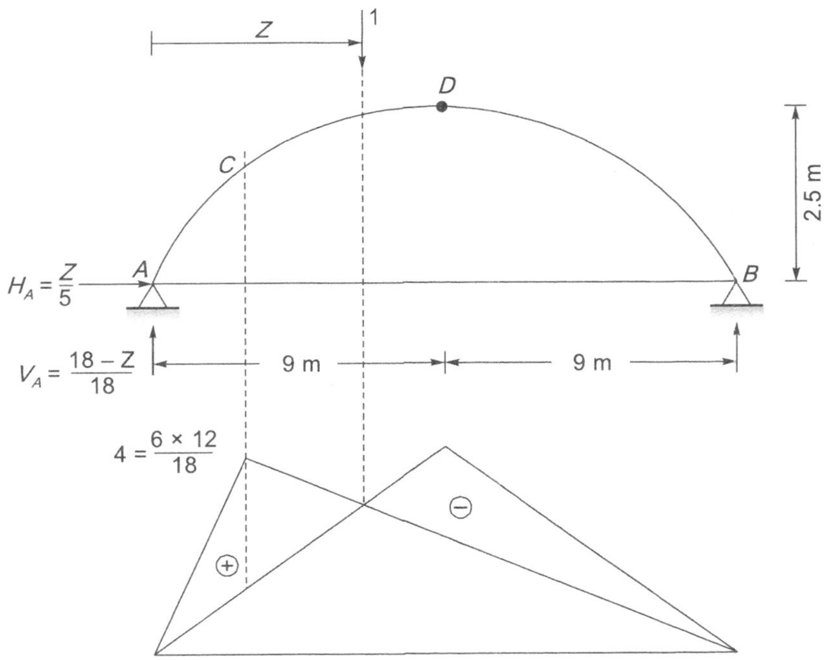

Now, $BM_C = 0$

$$\Rightarrow \quad \frac{18-Z}{18}\times 6-\frac{Z}{5}\times y_c = 0$$

$$\Rightarrow \quad \frac{18-Z}{18}\times 6-\frac{Z}{5}\left\{\frac{4\times 2.5}{18^2}\times 6(18-6)\right\} = 0$$

$$\Rightarrow \quad Z = 7.714\text{ m}$$

Alternative Solution:

Location where bending moment changes its sign is the intersection point of both triangles i.e., E

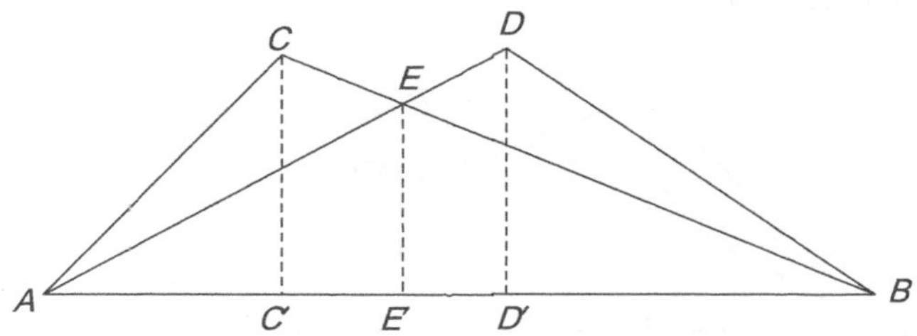

$\Delta ADD'$ and $\Delta AEE'$

$$\frac{EE'}{AE'} = \frac{DD'}{AD'}$$

$$EE' = \frac{4}{9}\times AE' \qquad \text{...(i)}$$

$\Delta BEE'$ and $\Delta BC'C$

$$\frac{EE'}{BE'} = \frac{CC'}{BC'}$$

$$EE' = \frac{4}{12}\times BE' \qquad \text{...(ii)}$$

Equating eq. (i) and (ii)

$$\frac{4}{9}AE' = \frac{4}{12}BE'$$

$$\Rightarrow \quad AE' = \frac{3}{4}BE'$$

$$AE' + BE' = 18$$

$$\frac{7}{4}BE' = 18$$

$$BE' = 10.285\text{ m}$$

$$AE' = 7.714\text{ m}$$

Bending moment changes, its sign at point 7.714 m from A.

39 A cable is suspended from the points (1) and (2) which are 90 m apart horizontally and are at different levels, the point (1) being 5.50 m vertically higher than the point (2) and the lowest point in the cable is 11 m below point (1). The cable is subjected to a UDL of 35 kN/m over the horizontal span. Determine the horizontal and vertical reactions at each end and also the maximum tension in the cable.

[12 marks : 2022]

Solution:

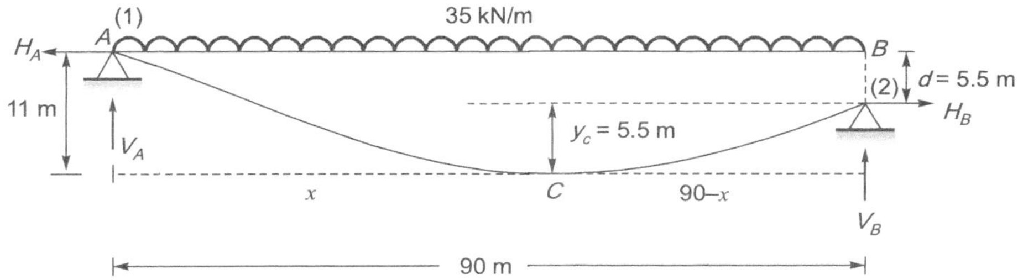

$$\Sigma M_B = 0$$

$$\Rightarrow \quad V_A \times 90 - H_A \times 5.5 - 35 \times 90 \times \frac{90}{2} = 0$$

$$\Rightarrow \quad 90\,V_A - 5.5\,H_A = 35 \times 90 \times 45 \qquad \text{...(i)}$$

BM at $C = 0$

Location of lowest point (C)

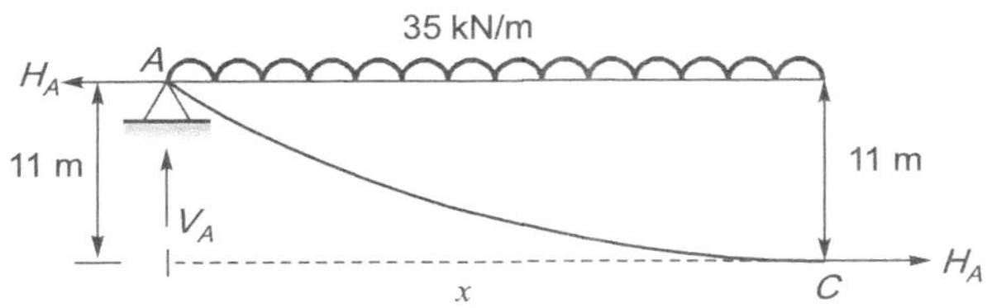

$$\Sigma F_y = 0$$

$$\Rightarrow \quad V_A - 35x = 0$$

$$\Rightarrow \quad V_A = 35x \qquad \text{...(ii)}$$

$$\Sigma M_A = 0$$

$$\Rightarrow \quad H_A \times 11 - 35\frac{x^2}{2} = 0$$

$$\Rightarrow \quad H_A = 1.591\,x^2 \qquad \text{...(iii)}$$

Put values from (ii) and (iii) into (i), we get

$$\Rightarrow \quad 90 \times 35x - 5.5 \times 1.591\,x^2 = 35 \times 90 \times 45$$

$$\therefore \quad 8.751\,x^2 - 3150x + 141750 = 0$$

$$x = 52.72 \text{ m}$$

$$\therefore \quad V_A = 35 \times 52.72$$

$$= 1845.2 \text{ kN}$$

$$H_A = 1.591 \times 52.72^2$$

$$= 4422 \text{ kN}$$

Now,

$$V_B = 35 \times 90 - V_A$$

$$= 35 \times 90 - 1845.2 = 1304.8 \text{ kN}$$

$$H_A = H_B = 4422 \text{ kN}$$

Maximum tension will be at support A since $V_A > V_B$.

$$\therefore \quad T_{max} = \sqrt{V_A^2 + H_A^2} = \sqrt{(1845.2)^2 + (4422)^2}$$

$$= 4791.54 \text{ kN}$$

3.10 A steel cable of 12 mm diameter is stretched across two poles 80 m apart. If the central dip is 1.10 m at normal temperature, determine the stress intensity in the cable. Also, determine the change in temperature necessary to raise the stress to 80 MPa. Take unit weight of steel γ = 78 kN/m³ and α = 12 × 10^{-6}/°C.

[12 marks : 2023]

Solution:

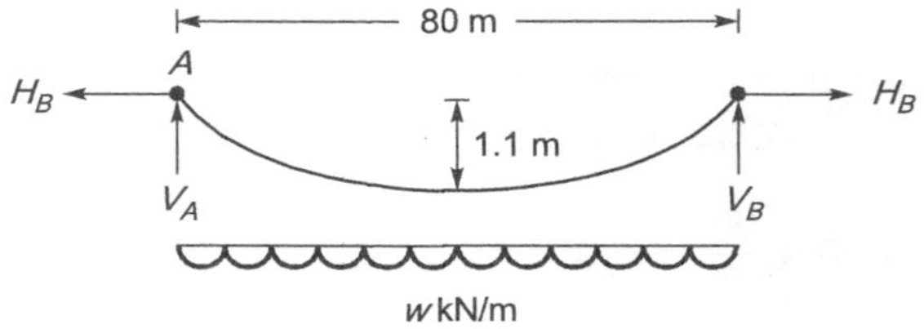

Let, V_A and V_B be the vertical reactions developed at A and B and H_A and H_B be the horizontal reactions developed at A and B respectively.

Now, Length of cable, $L = l\left[1+\frac{8}{3}\times\frac{h^2}{l^2}\right] = 80\left[1+\frac{8}{3}\times\frac{1.1^2}{80^2}\right] = 80.04\text{m}$

So, Self weight of cable, $= \gamma AL$

where γ is density of cable and A is cross-sectional area of cable.

$$= 78\times\frac{\pi}{4}\times(0.012)^2\times 80.04 = 706.1\times 10^{-3}\text{ kN}$$

$$= 706.1\text{ N}$$

Now, $$H = \frac{WL}{8h} = \frac{706.1\times 80}{8\times 1.1} = 6419.09\text{ N}$$

Also, $$V_A = V_B = \frac{W}{2} = \frac{706.1}{2} = 353.05\text{ N}$$

Now, maximum tension in cable, $T_{max} = \sqrt{V_A^2+H^2} = \sqrt{353.05^2+6419.09^2} = 6428.79\text{ N}$

Therefore, stress in cable, $f = \frac{T_{max}}{A} = \frac{6428.79\text{ N}}{\frac{\pi}{4}\times(12\text{ mm})^2} = 56.84\text{ N/mm}^2$

Now, $$\frac{\delta f}{f} = \frac{-\delta h}{h}$$

where δf and δh are change in stress and dip respectively due to temperature change.

$$\frac{\delta h}{h} = \frac{3}{16}\alpha t\frac{l^2}{h^2}$$

where α is coefficient of thermal expression and t is temperature change

So, $$\frac{\delta f}{f} = \frac{-3}{16}\alpha t\frac{l^2}{h^2}$$

$$\Rightarrow \quad \frac{(80-56.84)}{56.84} = \frac{-3}{16}\times 12\times 10^{-6}\times t\times\frac{80^2}{1.1^2}$$

$$\Rightarrow \quad t = -34.24°$$

(–ve) indicates that temperature should be decreased than normal temperature.

4. Methods of Structural Analysis

4.1 The rigid jointed portal frame shown in figure below is fully fixed at point *A* and *D* and is to be analysed by the moment distribution method, for the load system shown. It can be assumed that $EI = 43500$ kNm^2 for all members.

A. Draw the BMD throughout the structure.

B. Estimate the horizontal deflection at *B*.

[10 + 30 marks : 1999]

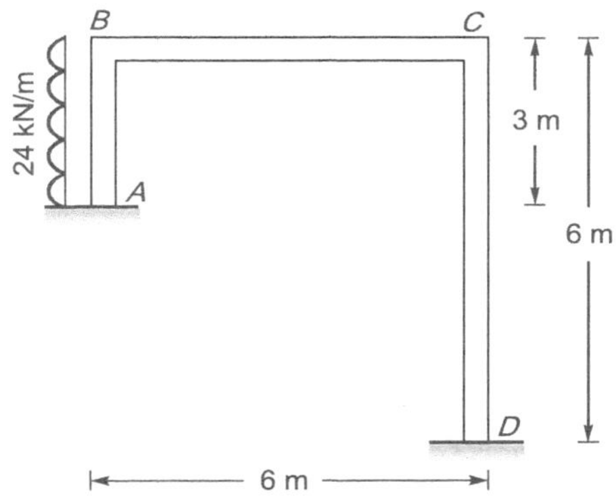

Solution:

A. Distribution Factors

Joint	Member	Relative Stiffness	Total Relative Stiffness	Distribution Factors
B	BA	$\frac{I}{3} = \frac{2I}{6}$	$\frac{3I}{6}$	$\frac{2}{3}$
	BC	$\frac{I}{6}$		$\frac{1}{3}$
C	CB	$\frac{I}{6}$	$\frac{2I}{6}$	$\frac{1}{2}$
	CD	$\frac{1}{6}$		$\frac{1}{2}$

(i) Non sway analysis

Fixed end moments

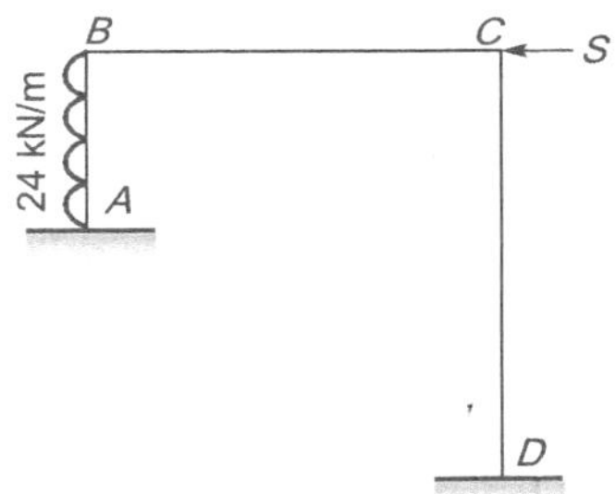

$$\bar{M}_{AB} = \frac{-24 \times 3^2}{12} = -18 \text{ kN-m}$$

$$\bar{M}_{BA} = +\frac{24 \times 3^2}{12} = +18 \text{ kN-m}$$

$$\bar{M}_{BC} = \bar{M}_{CB} = \bar{M}_{CD} = \bar{M}_{DC} = 0$$

A	B		C		D
	$\frac{2}{3}$	$\frac{1}{3}$	$\frac{1}{2}$	$\frac{1}{2}$	
–18	+18	0	0	0	0
	–12	–6	0	0	
–6		0	–3		0
	0	0	+1.5	+1.5	
0		+0.75	0		+0.75
	–0.50	–0.25	0	0	
–0.25		0	–0.125		0
	0	0	+0.06	+0.06	
0		+0.03	0		+0.03
	–0.02	–0.01	0	0	
–0.01			–0.005		0
–24.26	+5.48	–5.48	–1.57	+1.57	+0.78

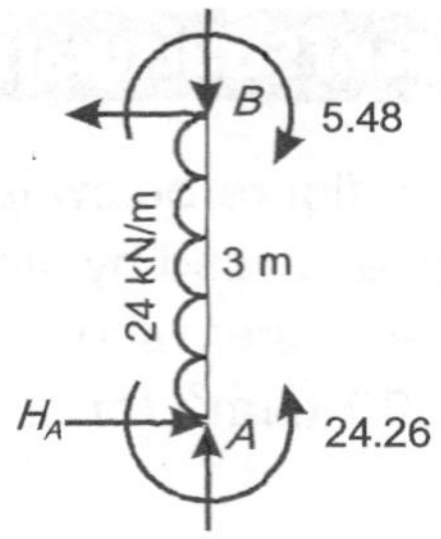

Taking moment about B,

$$3H_A + 24.26 - 5.48 + 24 \times 3 \times 1.5 = 0$$

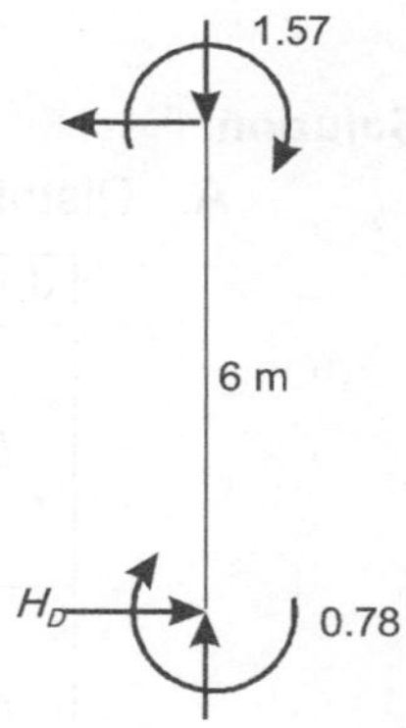

$$\text{Horizontal reaction at } A = \frac{-24.26 + 5.48 - 24 \times 3 \times 1.5}{3}$$

$$= -42.26 \text{ kN } (\rightarrow) = 42.26 \text{ kN } (\leftarrow)$$

Taking moment about C,

$$6H_D - 1.57 - 0.78 = 0$$

$$\text{Horizontal reaction at } D = \frac{1.57 + 0.78}{6} = 0.39 \text{ kN } (\rightarrow)$$

Force acting from left to right = Total UDL + Horizontal reaction at D

$= (24 \times 3) + 0.39 = 72.39$ kN

Force acting from right to left = 42.26 kN

$\therefore$ Sway force, $S = 72.39 - 42.26 = 30.13$ kN $(\rightarrow)$

(ii) Sway analysis

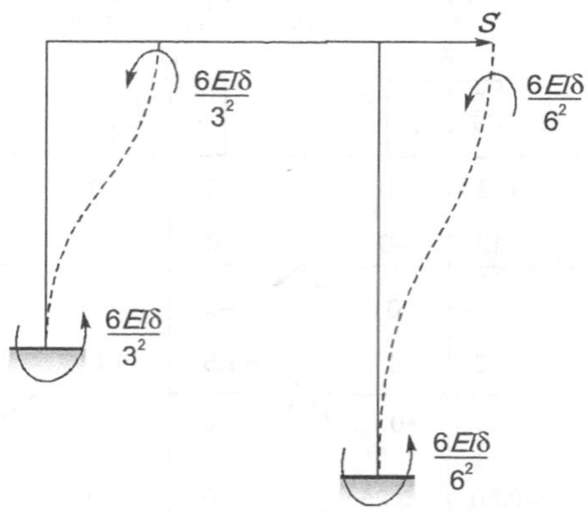

Since sway of the frame is from left to right, the initial fixing moments due to deflection Δ will be in anticlockwise direction

$$\bar{M}_{AB} = \bar{M}_{BA} = \frac{-6EI\Delta}{L^2} = \frac{-6EI\Delta}{3^2} = \frac{-6EI\Delta}{9}$$

$$\bar{M}_{BC} = \bar{M}_{CB} = 0$$

$$\bar{M}_{CD} = \bar{M}_{DC} = \frac{-6EI\Delta}{L^2} = \frac{-6EI\Delta}{6^2} = \frac{-6\,EI\Delta}{36}$$

∴	$\bar{M}_{AB}$	:	$\bar{M}_{BA}$	:	$\bar{M}_{BC}$	:	$\bar{M}_{CB}$	:	$\bar{M}_{CD}$	:	$\bar{M}_{DC}$
	$\frac{-1}{9}$	:	$\frac{-1}{9}$	:	0	:	0	:	$\frac{-1}{36}$	:	$\frac{-1}{36}$
	−4	:	−4	:	0	:	0	:	−1	:	−1
	−36	:	−36	:	0	:	0	:	−9	:	−9

	A	B ($\frac{2}{3}$)	B ($\frac{1}{3}$)	C ($\frac{1}{2}$)	C ($\frac{1}{2}$)	D
	−36	−36	0	0	−9	−9
		+24	+12	+4.5	+ 4.5	
	+12		+2.25	+6		+2.25
		−1.50	−0.75	−3	−3	
	−0.75		−1.50	−0.38		−1.50
		+1.00	+0.5	+0.19	+0.19	
	+0.5		+0.09	+0.25		+0.10
		−0.06	−0.03	−0.121	−0.13	
	−0.03		−0.06	−0.02		−0.07
		+0.04	+0.02	+0.01	+0.01	
	+0.02		0	+0.01		0
		0	0	0	0	
Col. (a)	−24.26	−12.52	+12.52	+7.43	−7.43	−8.22

Let the moments shown in col. (a) be due to a sway force S'.

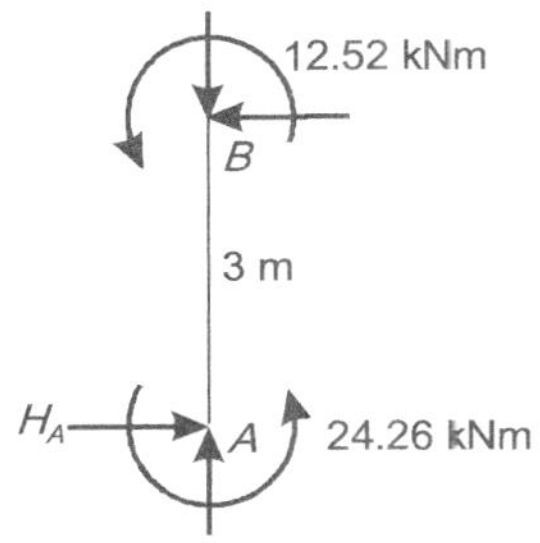

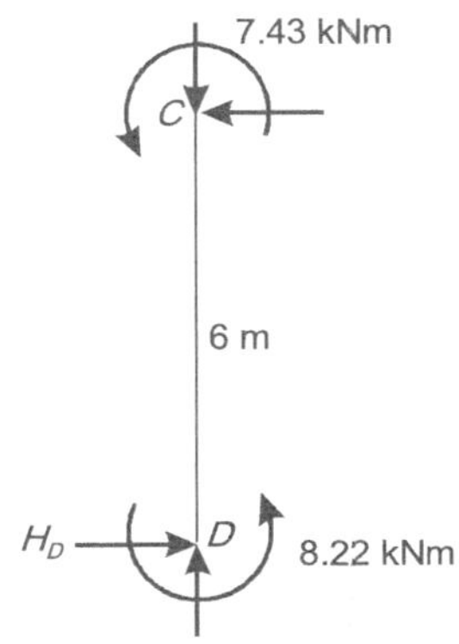

Taking moment about B,

$$3H_A + 24.26 + 12.52 = 0$$

$$\text{Horizontal reaction at } A = \frac{-24.26 - 12.52}{3}$$

$$= -12.26 \text{ kN} (\rightarrow) = 12.26 \text{ kN} (\leftarrow)$$

Taking moment about C,

$$6H_D + 8.22 + 7.43 = 0$$

$$\text{Horizontal reaction at } D = \frac{-7.43 - 8.22}{6}$$

$$= -2.61 \text{ kN} (\rightarrow) = 2.61 \text{ kN} (\leftarrow)$$

∴ Sway force, $S' = 12.26 + 2.61 = 14.87$ kN (→)

Thus, the moments shown in col. (a) are due to a sway force of 14.87 kN. Hence, for the actual sway force of 30.13 kN the actual sway moments will be $\frac{30.13}{14.87} \times$ col. (a) moments.

	M_{AB}	M_{BA}	M_{BC}	M_{CB}	M_{CD}	M_{DC}
col. (a)	−24.26	−12.52	+12.52	+7.43	−7.43	−8.22
Actual Sway moments	−49.16	−25.37	+25.37	+15.06	−15.06	−16.66
Non-Sway moments	−24.26	+5.48	−5.48	−1.57	+1.57	+0.78
Final moments	−73.42	−19.89	+19.89	+13.50	−13.50	−15.88

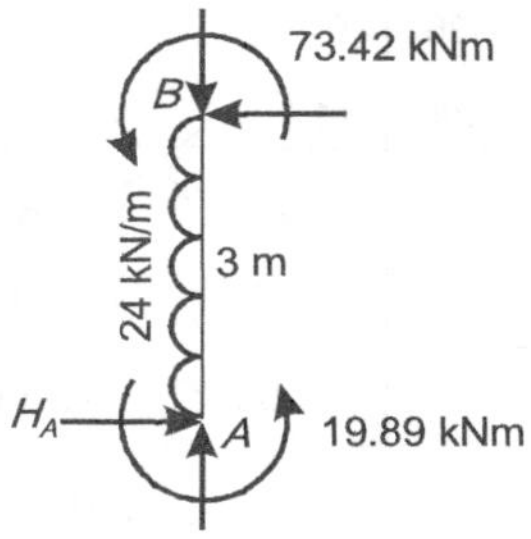

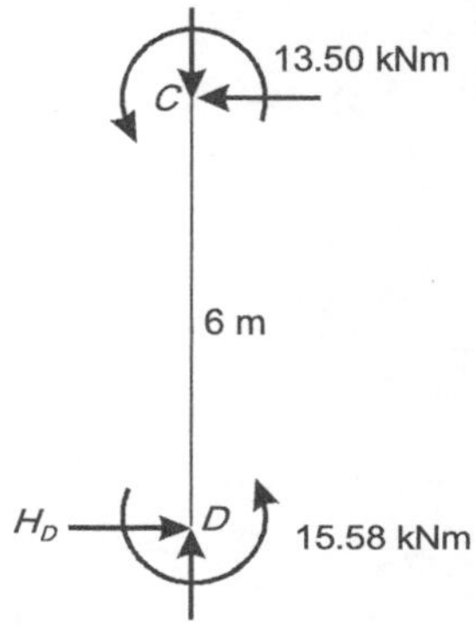

Taking moment about B,

$$3H_A + 19.89 + 73.42 + 24 \times 3 \times 1.5 = 0$$

Actual horizontal reaction at $A = \frac{-73.42 - 19.89 - 24 \times 3 \times 1.5}{3}$

$$= -67.10 \text{ kN} (\rightarrow) = 67.10 \text{ kN} (\leftarrow)$$

Taking moment about C,

$$6H_D + 15.58 + 13.50 = 0$$

Actual horizontal reaction at $D = \frac{-13.50 - 15.88}{6} = -4.90 \text{ kN} (\rightarrow) = 4.91 \text{ kN} (\leftarrow)$

Vertical reaction at $A = -\left(\frac{19.89 + 13.50}{6}\right) = -5.57 \text{ kN}(\uparrow) = 5.57 \text{ kN} (\downarrow)$

∴ Vertical reaction at $D = 5.57$ kN (↑)

Bending moment diagram

Taking outer face of the portal frame as reference face.

Simply supported moment for

$$AB = \frac{24 \times 3^2}{8} = 27 \text{ kN-m}$$

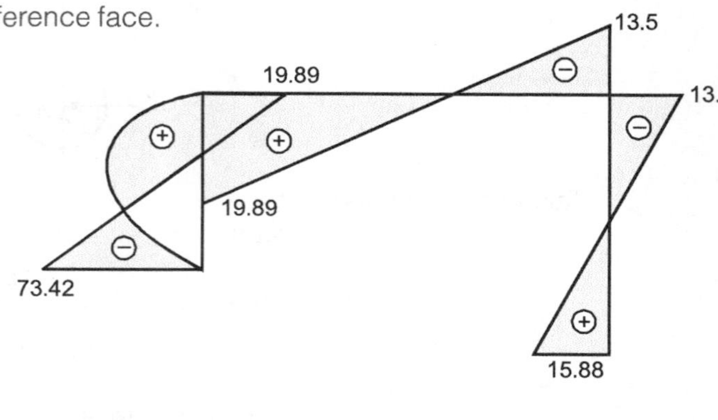

B. We have

$$M_{AB} = -73.42 \text{ kN-m}$$

and $$\bar{M}_{AB} = -18 \text{ kN-m}$$

$$M_{BA} = -19.89 \text{ kN-m}$$

and $$\bar{M}_{BA} = +18 \text{ kN-m}$$

The sway force is from left to right, hence, the deflection will be positive.

Writing slope deflection equations for M_{AB} and M_{BA}, we get

$$M_{AB} = \bar{M}_{AB} + \frac{2EI}{L}\left(2\theta_A + \theta_B - \frac{3\delta}{L}\right)$$

$\Rightarrow$ $$-73.42 = -18 + \frac{2EI}{3}\left(0 + \theta_B - \frac{3\delta}{3}\right)$$

$\Rightarrow$ $$EI(\delta - \theta_B) = \frac{55.42 \times 3}{2}$$

$$\delta - \theta_B = \frac{83.13}{EI} \quad \text{... (i)}$$

and $$M_{BA} = \bar{M}_{BA} + \frac{2EI}{L}\left(2\theta_B + \theta_A - \frac{3\delta}{L}\right)$$

$\Rightarrow$ $$-19.89 = 18 + \frac{2EI}{3}\left(2\theta_B + 0 - \frac{3\delta}{3}\right)$$

$\Rightarrow$ $$(\delta - 2\theta_B)\,EI = \frac{37.89 \times 3}{2}$$

$$\delta - 2\theta_B = \frac{56.835}{EI} \quad \text{...(ii)}$$

Solving (i) and (ii) we get

$$\theta_B = \frac{26.295}{EI} \text{ and } \delta = \frac{109.425}{EI}$$

But $$EI = 43500 \text{ kN-m}^2$$

$\therefore$ $$\delta = \frac{109.425}{43500}$$

$\Rightarrow$ $$\delta = 2.52 \text{ mm}$$

Thus the horizontal deflection at B is 2.52 mm from left to right.

4.2 The continuous beam $ABCD$ shown in the figure below is fixed at end A, and simply supported at B and C and free at the end D. The beam carries a concentrated load P at the free end. Analyse the beam by using (i) Slope deflection method (ii) Moment distribution method. Also draw the BMD and SFD.

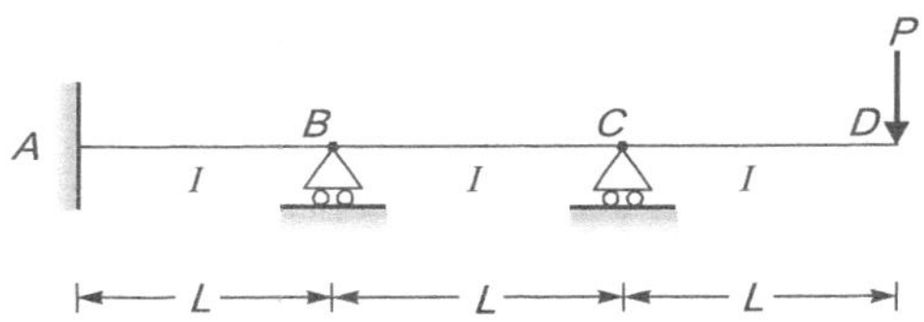

[20 marks : 2000]

Solution:

(i) Slope deflection method

Fixed end moments

$$\bar{M}_{AB} = 0; \qquad \bar{M}_{BA} = 0$$

$$\bar{M}_{BC} = 0; \qquad \bar{M}_{CB} = 0$$

$$\bar{M}_{CD} = M_{CD} = -PL$$

For member AB,

$$M_{AB} = \bar{M}_{AB} + \frac{2EI}{L}(2\theta_A + \theta_B) = 0 + \frac{2EI}{L}(0 + \theta_B)$$

$\therefore$ $$M_{AB} = \frac{2EI}{L}\theta_B \quad \text{...(i)}$$

$$M_{BA} = \bar{M}_{BA} + \frac{2EI}{L}(2\theta_B + \theta_A) = 0 + \frac{2EI}{L}(2\theta_B + 0)$$

$$\therefore \quad M_{BA} = \frac{4EI}{L}\theta_B \quad \text{...(ii)}$$

For member BC

$$M_{BC} = \bar{M}_{BC} + \frac{2EL}{L}(2\theta_B + \theta_C) = 0 + \frac{2EI}{L}(2\theta_B + \theta_C)$$

$$M_{BC} = \frac{2EI}{L}(2\theta_B + \theta_C) \quad \text{...(iii)}$$

$$M_{CB} = \bar{M}_{CB} + \frac{2EI}{L}(2\theta_C + \theta_B) = 0 + \frac{2EI}{L}(2\theta_C + \theta_B)$$

$$\therefore \quad M_{CB} = \frac{2EI}{L}(2\theta_C + \theta_B) \quad \text{...(iv)}$$

Equilibrium condition at B,

$$M_{BA} + M_{BC} = 0$$

$$\Rightarrow \quad \frac{4EI}{L}\theta_B + \frac{2EI}{L}(2\theta_B + \theta_C) = 0$$

$$\Rightarrow \quad \frac{8EI}{L}\theta_B + \frac{2EI}{L}\theta_C = 0$$

$$\Rightarrow \quad \frac{2EI}{L}(4\theta_B + \theta_C) = 0$$

$$\Rightarrow \quad \theta_C = -4\theta_B \quad \text{...(v)}$$

Equilibrium condition at C,

$$M_{CB} + M_{CD} = 0$$

$$\Rightarrow \quad \frac{2EI}{L}(2\theta_C + \theta_B) - PL = 0 \quad \text{[from (v)]}$$

$$\Rightarrow \quad \frac{4EI}{L}(-4\theta_B) + \frac{2EI}{L}\theta_B - PL = 0$$

$$\Rightarrow \quad -\frac{16EI}{L}\theta_B + \frac{2EI}{L}\theta_B = PL$$

$$\Rightarrow \quad -\frac{14EI}{L}\theta_B = PL$$

$$\Rightarrow \quad \theta_B = -\frac{PL^2}{14EI}$$

$$\therefore \quad \theta_C = -4\theta_B = -4 \times \left(\frac{-PL^2}{14EI}\right) = \frac{2PL^2}{7EI}$$

$$M_{AB} = \frac{2EI}{L}\theta_B = \frac{2EI}{L} \times \left(\frac{-PL^2}{14EI}\right) = \frac{-PL}{7}$$

$$M_{BA} = \frac{4EI}{L}\theta_B = \frac{4EI}{L} \times \left(\frac{-PL^2}{14EI}\right) = \frac{-2PL}{7}$$

$$M_{BC} = \frac{2EI}{L}(2\theta_B + \theta_C) = \frac{2EI}{L} \times \left[2 \times \left(\frac{-PL^2}{14EI}\right) + \frac{2PL^2}{7EI}\right] = \frac{2PL}{7}$$

$$M_{CB} = \frac{2EI}{L}(2\theta_C + \theta_B) = \frac{2EI}{L} \times \left[2 \times \frac{2PL^2}{7EI} + \frac{-PL^2}{14EI}\right] = PL$$

(ii) Moment distribution method:

Distribution Factors

Joint	Member	Relative Stiffness	Total Relative Stiffness	Distribution Factors
B	BA	$\frac{I}{L}=\frac{4I}{4L}$	$\frac{7I}{4L}$	$\frac{4}{7}$
	BC	$\frac{3}{4}\times\frac{I}{L}=\frac{3I}{4L}$		$\frac{3}{7}$

Fixed end moments

$$\bar{M}_{AB} = 0;\ \bar{M}_{BA} = 0$$

$$\bar{M}_{BC} = 0;\quad \bar{M}_{CB} = 0$$

$$\bar{M}_{CD} = M_{CD} = -PL$$

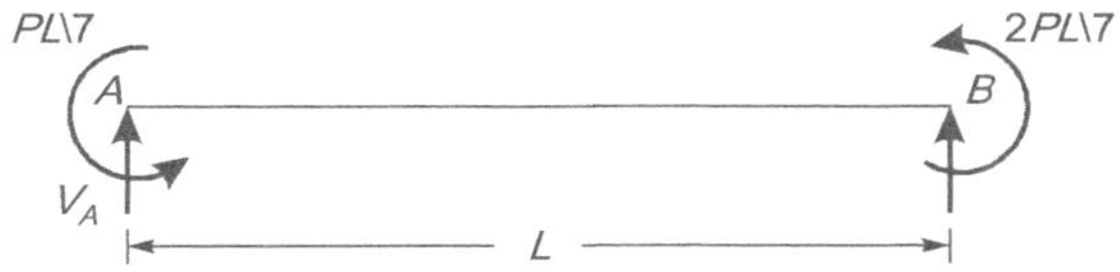

Taking moment about B,

$$V_A(L) - \frac{PL}{7} - \frac{2PL}{7} = 0$$

A	B: $\frac{4}{7}$	B: $\frac{3}{7}$	C	D
		$+\frac{PL}{2}$	$+PL$	$-PL$
0	0	0	0	
0	0	$+\frac{PL}{2}$	$+PL$	$-PL$
	$\frac{-2PL}{7}$	$\frac{-1.5PL}{7}$		
$\frac{-PL}{7}$				
$-\frac{PL}{7}$	$-\frac{2PL}{7}$	$+\frac{2PL}{7}$	$+PL$	$-PL$

Vertical reaction at

$$A = -\frac{\left(\frac{-PL}{7} - \frac{2PL}{7}\right)}{L} = \frac{3P}{7}\ (\uparrow)$$

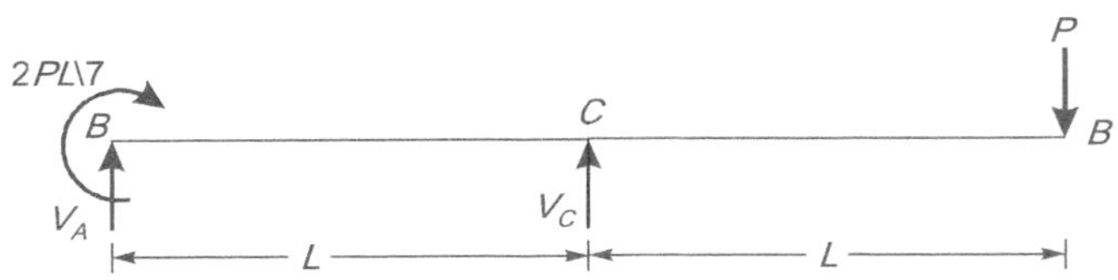

Taking moment about B,

$$LV_C - P(2L) - \frac{2PL}{7} = 0$$

Vertical reaction at C, $V_C = \dfrac{\dfrac{2PL}{7}+2PL}{L} = \dfrac{16P}{7}\ (\uparrow)$

Vertical reaction at B, $V_B = P-(V_A+V_C)$

$$= P-\left(\frac{3P}{7}+\frac{16P}{7}\right) = -\frac{12P}{7}(\uparrow) = \frac{12P}{7}(\downarrow)$$

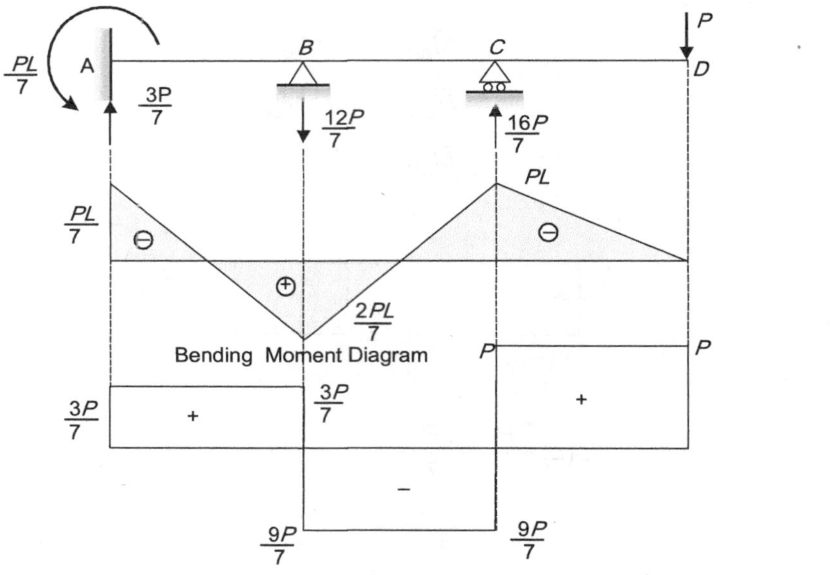

Bending Moment Diagram

Shear Force Diagram

4.3 Analyse the structure shown in figure below by stiffness method. Draw BMD and deflected shape.

[10 marks : 2000]

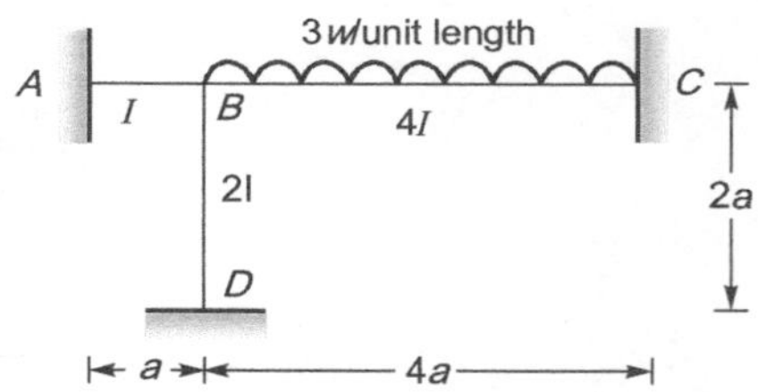

Solution:

The given portal frame can be easily analysed by slope deflection method which is a stiffness method

Fixed end moments

$$\bar{M}_{AB} = 0;\ \bar{M}_{BA} = 0$$

$$\bar{M}_{BC} = -\frac{3w\times(4a)^2}{12} = -4wa^2;\ \bar{M}_{CB} = +4wa^2$$

$$\bar{M}_{BD} = 0 \text{ and } \bar{M}_{DB} = 0$$

Slope deflection equations

$$M_{AB} = \bar{M}_{AB}+\frac{2EI}{a}(0+\theta_B) = 0+\frac{2EI\theta_B}{a} = \frac{2EI\theta_B}{a}$$

$$M_{BA} = \bar{M}_{BA}+\frac{2EI}{a}(2\theta_B+0) = 0+\frac{4EI\theta_B}{a} = \frac{4EI\theta_B}{a}$$

$$M_{BC} = \bar{M}_{BC}+\frac{2E(4I)}{4a}(2\theta_B+0) = -4wa^2+\frac{4EI\theta_B}{a}$$

$$M_{CB} = \bar{M}_{CB} + \frac{2E(4I)}{4a}(0+\theta_B) = 4wa^2 + \frac{2EI\theta_B}{a}$$

$$M_{BD} = \bar{M}_{BD} + \frac{2E(2I)}{2a}(2\theta_B+0) = 0 + \frac{4EI\theta_B}{a} = \frac{4EI\theta_B}{a}$$

$$M_{DB} = \bar{M}_{DB} + \frac{2E(2I)}{2a}(0+\theta_B) = 0 + \frac{2EI\theta_B}{a} = \frac{2EI\theta_B}{a}$$

Joint equilibrium at B

$$M_{BA} + M_{BC} + M_{BD} = 0$$

$$\Rightarrow \quad \frac{4EI\theta_B}{a} + \frac{4EI\theta_B}{a} - 4wa^2 + \frac{4EI\theta_B}{a} = 0$$

$$\Rightarrow \quad \frac{12\,EI\theta_B}{a} = 4wa^2$$

$$\Rightarrow \quad \theta_B = \frac{4wa^3}{12EI}$$

$$\Rightarrow \quad \theta_B = \frac{wa^3}{3EI}$$

$$\Rightarrow \quad EI\theta_B = \frac{wa^3}{3}$$

Substituting the value of $EI\theta_B$ in slope deflection equations, we get

$$M_{AB} = \frac{2}{a}\times\frac{wa^3}{3} = \frac{2}{3}wa^2$$

$$M_{BA} = \frac{4}{a}\times\frac{wa^3}{3} = \frac{4}{3}wa^2$$

$$M_{BC} = -4wa^2 + \frac{4}{a}\times\frac{wa^3}{3} = -\frac{8}{3}wa^2$$

$$M_{CB} = 4wa^2 + \frac{2}{a}\times\frac{wa^3}{3} = \frac{14}{3}wa^2$$

$$M_{BD} = \frac{4}{a}\times\frac{wa^3}{3} = \frac{4}{3}wa^2$$

$$M_{DB} = \frac{2}{a}\times\frac{wa^3}{3} = \frac{2}{3}wa^2$$

Maximum positive BM

For BC portion

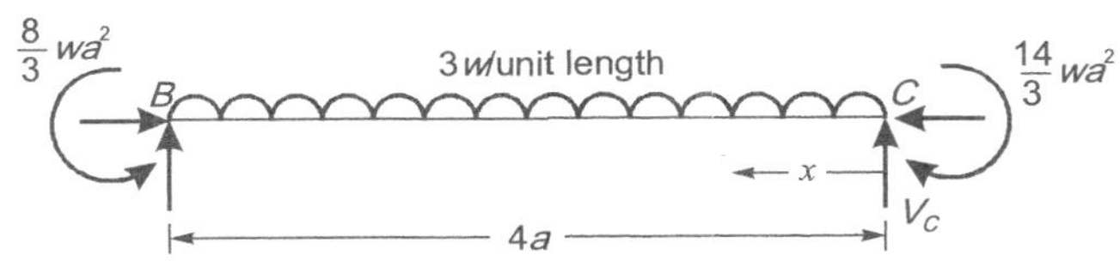

Taking moment about B,

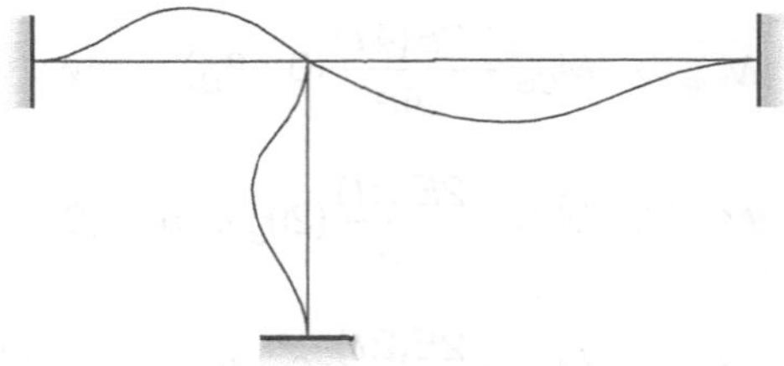

Deflected shape

$$V_C(4a) - \frac{14}{3}wa^2 - 3w \times 4a \times 2a + \frac{8}{3}wa^2 = 0$$

$$V_C = 6.5\,wa$$

$$M_x = V_C x - \frac{14}{3}wa^2 - 3wx\frac{x}{2}$$

$$M_x = 6.5wax - \frac{14}{3}wa^2 - 1.5wx^2$$

For maximumbending moment

$$\frac{d}{dx}(M_x) = 0$$

$$6.5wa - 3wx = 0$$

$$x = \frac{13}{6}a$$

$$M_x|_{max} = \frac{19}{8}wa^2 = \frac{19}{8}wa^2 = M_{max}$$

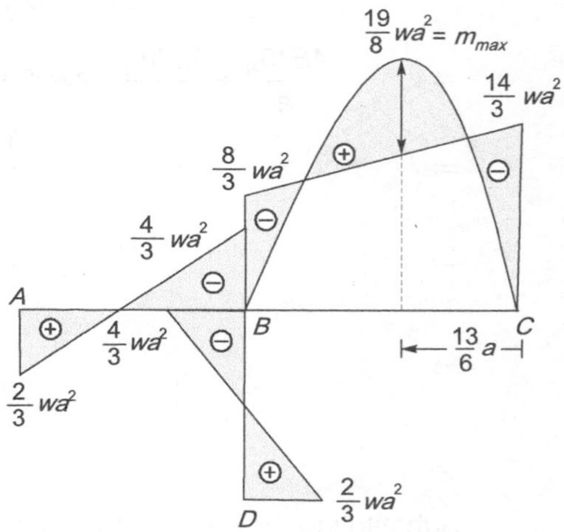

BMD

4.4 Using Betti's theorem, determine the elongation of a bar 10 mm in diameter and 1 m long when it is subjected to a force of 10 kN at diametrically opposite points at midpoint of its length. $E = 2 \times 10^5$ MPa, $\mu = 0.3$.

[8 marks : 2001]

Solution:

The loading on the bar may be shown as

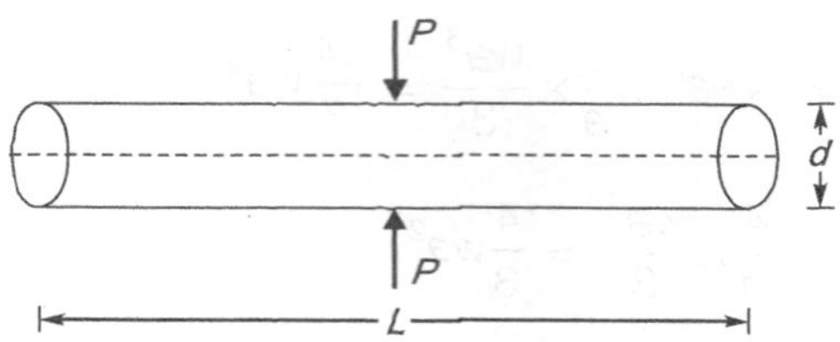

Bettis' theorem says that the deflection along the axis of the bar due to force P at centre is equal to the deflection at the centre due to force P at the ends (axis).

The elongation along x - direction is given as

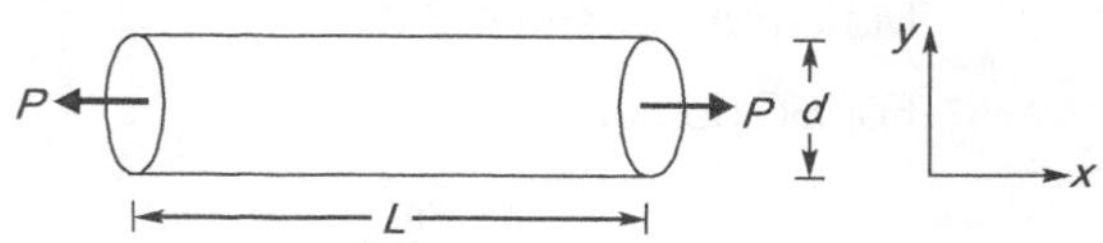

$$\Delta_x = \frac{PL}{AE}$$

$$\frac{\Delta_x}{L} = \frac{P}{AE}$$

$$\Rightarrow \quad \varepsilon_x = \frac{P}{AE}$$

∴ Strain in y - direction will be

$$\varepsilon_y = \mu\varepsilon_x$$

$$\Rightarrow \qquad \varepsilon_y = \frac{\mu P}{AE}$$

But we know that

$$\frac{\Delta_y}{D} = \varepsilon_y$$

$$\Rightarrow \qquad \Delta_y = \frac{\mu P}{AE} \times D \qquad \left[\because A = \frac{\pi D^2}{4}\right]$$

$$\Rightarrow \qquad \Delta_y = \frac{\mu P \times 4}{\pi DE}$$

Now if Δ is the deflection along the ends due to force P at the centre, then by Betti's theorem

$$\Delta = \Delta_y = \frac{4\mu P}{\pi DE} = \frac{4 \times 0.3 \times 10 \times 10^3}{\pi \times 10 \times 2 \times 10^5} = 1.91 \times 10^{-3}\ \text{mm}$$

4.5 Analyse the portal frame shown in figure by strain energy method and draw BMD, EI is constant.

[18 marks : 2001]

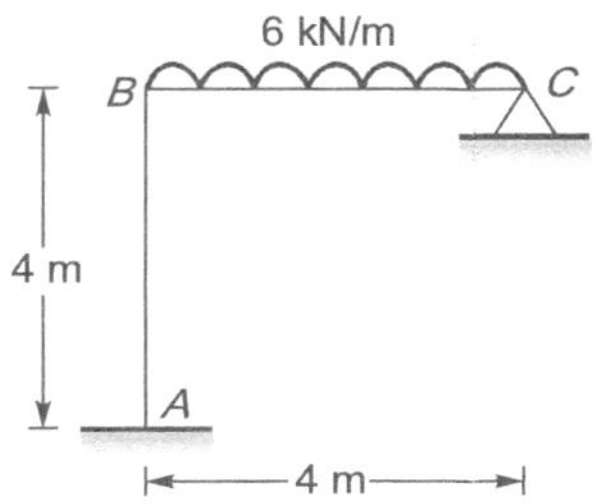

Solution:

(i) Analysing the portal frame by moment distribution method.

Distribution Factors.

Joint	Member	Relative Stfiffness	Total Relative Stiffness	Distribution Factors
B	BA	$\frac{I}{4} = \frac{4I}{16}$	$\frac{7I}{16}$	$\frac{4}{7}$
	BC	$\frac{3}{4} \times \frac{I}{4} = \frac{3I}{16}$		$\frac{3}{7}$

Fixed End Moments:

$$\bar{M}_{AB} = 0$$

$$\bar{M}_{BA} = 0$$

$$\bar{M}_{BC} = -\frac{6 \times 4^2}{12} = -8\ \text{kN-m}$$

$$\bar{M}_{CB} = +8\ \text{kN-m}$$

A	B	B	C
	$\frac{4}{7}$	$\frac{3}{7}$	
0	0	−8	+8
		−4	−8
0	0	−12	0
	+6.857	+5.143	
+3.429			
+3.429	+6.857	−6.857	0

Reactions

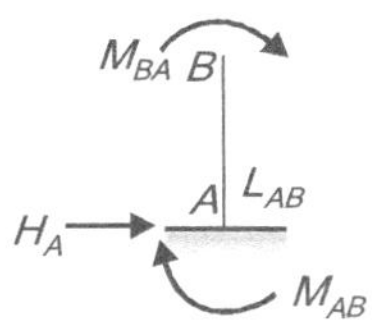

$$H_A = \frac{M_{AB} + M_{BA}}{L_{AB}} = \frac{3.429 + 6.857}{4} = 2.572\ \text{kN}\ (\rightarrow)$$

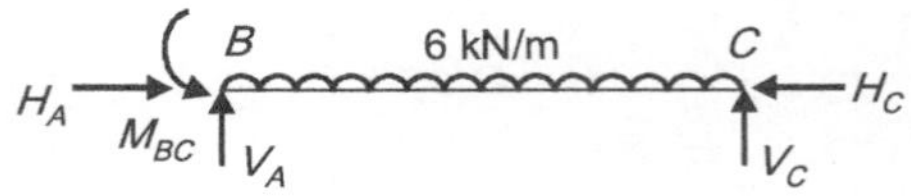

$$\Sigma M_C = 0$$

$$V_A \times 4 - 6.857 - 6 \times 4 \times 2 = 0$$

$\Rightarrow$ $$V_A = 13.71 \text{ kN}$$

Also $$H_A - H_C = 0$$

$$H_A = H_C = 2.572 \text{ kN } (\leftarrow)$$

$$V_C = 6 \times 4 - V_A = 24 - 13.71 = 10.29 \text{ kN}$$

$$\text{Free Bending moment in } BC = \frac{6 \times 4^2}{8} = 12 \text{ kN-m}$$

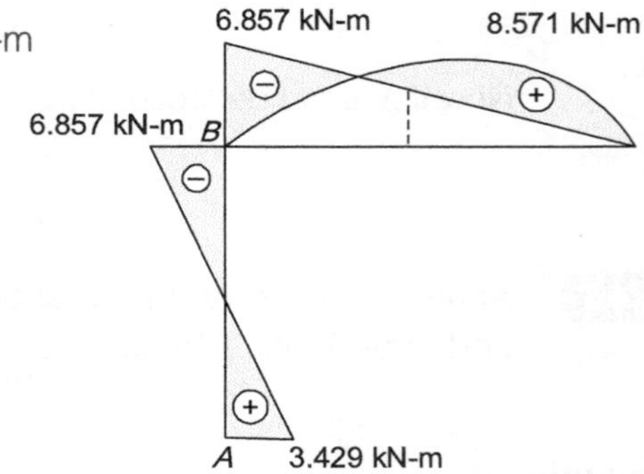

Taking outer face as reference face.

Net Bending moment at the centre of BC

$$= 12 - \frac{6.857}{2}$$

$$= 12 - 3.429$$

$$= 8.571 \text{ kN-m}$$

(ii) Analysing by strain energy method

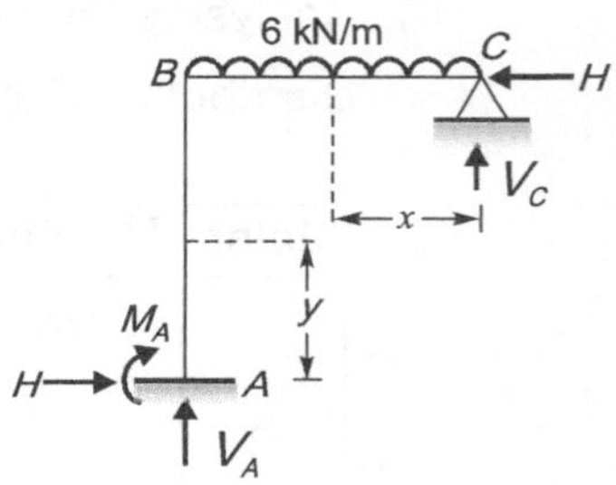

$$\Sigma F_y = 0$$

$\Rightarrow$ $$V_A + V_C - 6 \times 4 = 0$$

$\Rightarrow$ $$V_A = 24 - V_C$$

or $$V_C = 24 - V_A$$

Also $$\Sigma M_C = 0$$

$\Rightarrow$ $$V_A \times 4 - H \times 4 + M_A - 6 \times 4 \times 2 = 0$$

$\Rightarrow$ $$4V_A - 4H + M_A - 48 = 0$$

$\Rightarrow$ $$V_A = H + 12 - \frac{M_A}{4}$$

$\therefore$ $$V_C = 24 - V_A = 24 - H - 12 + \frac{M_A}{4}$$

$\Rightarrow$ $$V_C = 12 - H + \frac{M_A}{4}$$

Now, we have

$$U = \int \frac{M^2 ds}{2EI}$$

The value of H will be such that the total strain energy is minimum i.e.

$\Rightarrow$ $$\frac{\partial U}{\partial H} = 0$$

$\Rightarrow$ $$\frac{\partial U}{\partial H} = \int \frac{2M \cdot \frac{\partial M}{\partial H} \cdot ds}{2EI} = 0$$

$$\Rightarrow \quad \frac{\partial U}{\partial H} = \int \frac{M \cdot \frac{\partial M}{\partial H} \cdot ds}{EI} = 0 \qquad \text{... (i)}$$

Similarly $\quad \frac{\partial U}{\partial M_A} = 0$

$$\frac{\partial U}{\partial M_A} = \int \frac{M \cdot \frac{\partial M}{\partial M_A} \cdot ds}{EI} = 0 \qquad \text{...(ii)}$$

Member	M	$\frac{\partial M}{\partial H}$	$\frac{\partial M}{\partial M_A}$	Limit of integration
AB	$M_A - Hy$	$-y$	1	0 *to* 4
BC	$12x - Hx + \frac{M_A x}{4} - 3x^2$	$-x$	$\frac{x}{4}$	0 *to* 4

From (i), we get

$$\Rightarrow \quad \int_0^4 \frac{[(M_A - Hy)(-y)dy]}{EI} \pi + \int_0^4 \frac{\left[\left(12x - Hx + \frac{M_A}{4}x - 3x^2\right)(-x)dx\right]}{EI} = 0$$

$$\Rightarrow \quad \int_0^4 \left(-M_A y + Hy^2\right) dy + \int_0^4 \left(-12x^2 + Hx^2 - \frac{M_A}{4}x^2 + 3x^3\right) dx = 0$$

$$\Rightarrow \quad \left[-\frac{M_A y^2}{2} + \frac{Hy^3}{3}\right]_0^4 + \left[-\frac{12x^3}{3} + \frac{Hx^3}{3} - \frac{M_A x^3}{12} + \frac{3x^4}{4}\right]_0^4 = 0$$

$$\Rightarrow \quad \left[-8M_A + \frac{64H}{3}\right] + \left[-256 + \frac{64H}{3} - \frac{16M_A}{3} + 192\right] = 0$$

$$\Rightarrow \quad -8M_A + \frac{64H}{3} - 256 + \frac{64H}{3} - \frac{16M_A}{3} + 192 = 0$$

$$\Rightarrow \quad \frac{-40M_A}{3} + \frac{128H}{3} - 64 = 0$$

$$\Rightarrow \quad -40M_A + 128H - 192 = 0$$

$$16H - 5M_A = 24 \qquad \text{...(iii)}$$

From (ii), we get

$$\int_0^4 \frac{(M_A - Hy) \times 1 \times dy}{EI} + \int_0^4 \frac{\left(12x - Hx + \frac{M_A x}{4} - 3x^2\right) \times \frac{x}{4} \times dx}{EI} = 0$$

$$\Rightarrow \quad \left[M_A y - \frac{Hy^2}{2}\right]_0^4 + \frac{1}{4}\left[\frac{12x^3}{3} - \frac{Hx^3}{3} + \frac{M_A x^3}{12} - \frac{3x^4}{4}\right]_0^4 = 0$$

$$\Rightarrow \quad 4M_A - 8H + \frac{1}{4}\left[4\times 4^3 - \frac{4^3 H}{12} + \frac{4^3 \times M_A}{12} - \frac{3\times 4^4}{4}\right] = 0$$

$$\Rightarrow \quad 4M_A - 8H + 64 - \frac{16\,H}{3} + \frac{4\,M_A}{3} - 48 = 0$$

$$\Rightarrow \quad \frac{16}{3}M_A - \frac{40\,H}{3} + 16 = 0$$

$$\Rightarrow \quad -40H + 16\,M_A = -48$$

$$\Rightarrow \quad -5H + 2M_A = -6 \quad \text{... (iv)}$$

Solving (iii) and (iv), we get

$$H = 2.571\text{ kN}$$

$$M_A = 3.429\text{ kN-m}$$

$$\therefore \quad V_A = H + 12 - \frac{M_A}{4} = 2.571 + 12 - \frac{3.429}{4}$$

$$= 13.71\text{ kN}$$

and $V_C = 24 - V_A = 24 - 13.71 = 10.29\text{ kN}$

Assuming outer face as reference face

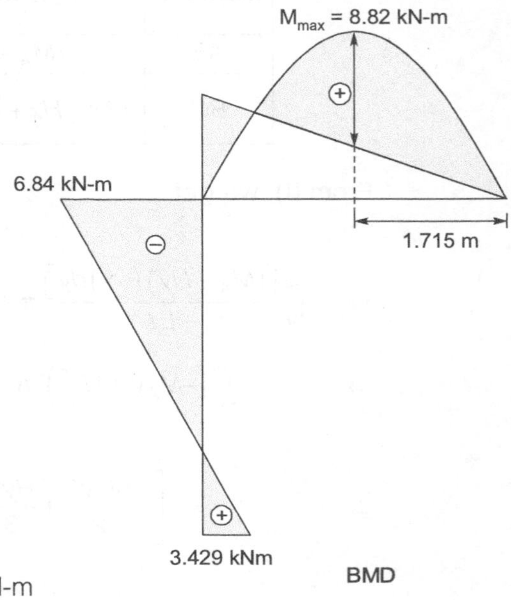

Bending moment in *BC*

$$M_x(x\text{ from }C) = 10.29\,x - \frac{6x^2}{2} \quad [0 \le x \le 4]$$

$$M_x = 10.29x - 3x^2$$

$$M_C = 0$$

$$M_B = 10.29\times 4 - 3\times(4)^2 = -6.84\text{ kN-m}$$

For M_{max}, $\dfrac{dM_x}{dx} = 0$

$$\Rightarrow \quad 10.29 - 6x = 0$$

$$x = \frac{10.29}{6} = 1.715\text{ m}$$

$$\therefore \quad M_{max} = 10.29\times 1.715 - 3\times(1.715)^2 = 8.82\text{ kN-m}$$

Bending moment in *AB*

$$M_y(y\text{ from }A) = M_A - Hy \quad [0 \le y \le 4]$$

$$M_y = 3.429 - 2.571y$$

$$M_A = 3.429\text{ kN-m}$$

$$M_B = 3.429 - 2.571\times 4 = -6.85\text{ kN-m}$$

For $M_y = 0$, we have

$$3.429 - 3.571y = 0$$

$$\Rightarrow \quad y = 1.33\text{ m}$$

4.6 Analyse the rigid frame shown in figure below by moment distribution method.

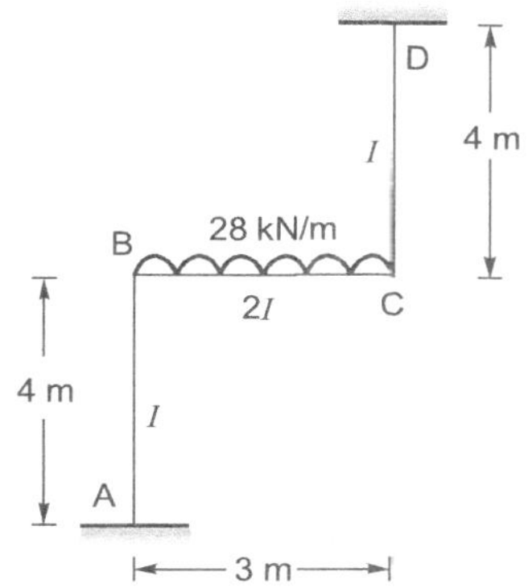

[40 marks : 2001]

Solution:

(i) Non sway analysis.

Distribution Factors

Joint	Member	Relative Stiffness	Total Relative Stiffness	Distribution Factors
B	BA	$\frac{I}{4}=\frac{3I}{12}$	$\frac{11I}{12}$	$\frac{3}{11}$
	BC	$\frac{2I}{3}=\frac{8I}{12}$		$\frac{8}{11}$
C	CB	$\frac{2I}{3}=\frac{8I}{12}$	$\frac{11I}{12}$	$\frac{8}{11}$
	CD	$\frac{I}{4}=\frac{3I}{12}$		$\frac{3}{11}$

Fixed End Moments

$$\bar{M}_{AB} = \bar{M}_{BA} = 0$$

$$\bar{M}_{BC} = -\frac{28 \times 3^2}{12} = -21 \text{ kN-m}$$

$$\bar{M}_{CB} = +\frac{28 \times 3^2}{12} + 21 \text{ kN-m}$$

$$\bar{M}_{CD} = \bar{M}_{DC} = 0$$

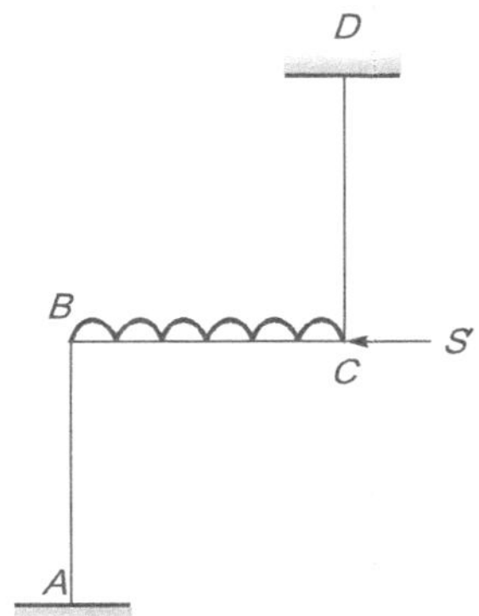

A	B		C		D
	$\frac{3}{11}$	$\frac{8}{11}$	$\frac{8}{11}$	$\frac{3}{11}$	
0	0	–21	+21	0	0
	+5.73	+15.27	–15.27	–5.73	
+2.86		–7.63	+7.63		–2.86
	+2.08	+5.55	–5.55	–2.08	
+1.04		–2.78	+2.78		–1.04
	+0.76	+2.02	–2.02	–0.76	
+0.38		–1.01	+1.01		–0.38
	+0.27	+0.74	–0.74	–0.27	
+0.14		–0.37	+0.37		–0.14
	+0.1	+0.27	–0.27	–0.1	
+0.05		–0.14	+0.14		–0.05
	+0.04	+0.1	–0.1	–0.04	
+4.47	**+8.98**	**–8.98**	**+8.98**	**–8.98**	**–4.47**

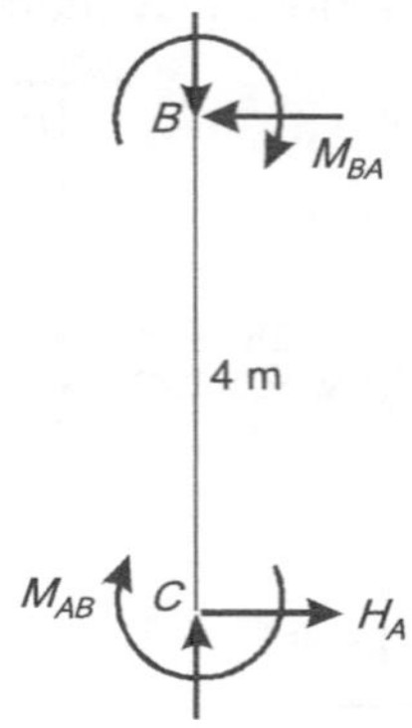

Taking moment about B,

$$\therefore \quad 4H_A - M_{AB} - M_{BA} = 0$$

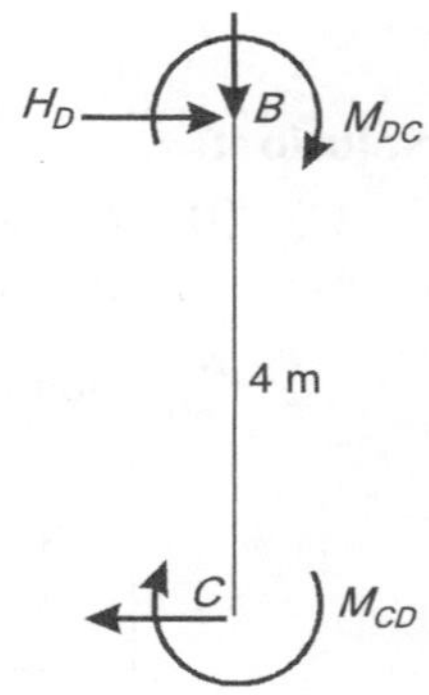

$$\text{Horizontal reaction at } A = \frac{M_{AB} + M_{BA}}{4}$$

$$= \frac{4.47 + 8.98}{4} = 3.3625 \text{ kN } (\rightarrow)$$

Taking moment about C,

$$4H_D + M_{DC} + M_{CD} = 0$$

$$\text{Horizontal reaction at } D = -\left(\frac{M_{CD} + M_{DC}}{4}\right)$$

$$= -\frac{(-8.98 - 4.47)}{4} = 3.3625 \text{ kN } (\rightarrow)$$

$$\therefore \quad \text{Total sway force, } S' = 3.3625 + 3.3625 = 6.725 \text{ kN } (\rightarrow)$$

(ii) Sway Analysis

The rigid frame will now be analysed for a sway force of 6.725 kN acting from left to right.

The initial moments for AB will be negative while for CD, they are positive.

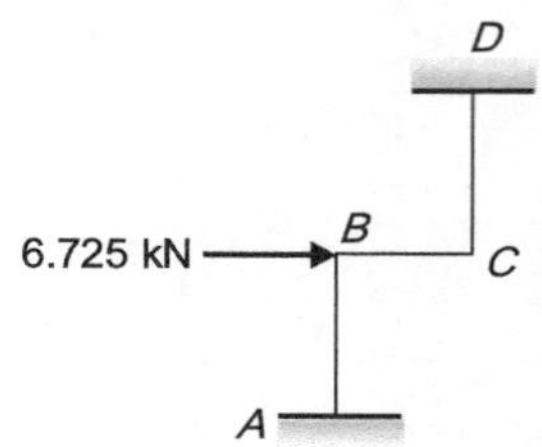

$$\bar{M}_{AB} = \bar{M}_{BA} = -\frac{6EI\Delta}{L^2} = -\frac{6EI\Delta}{4^2} = -\frac{6EI\Delta}{16}$$

$$\bar{M}_{BC} = \bar{M}_{CB} = 0$$

$$\bar{M}_{CD} = \bar{M}_{DC} = +\frac{6EI\Delta}{L^2} = +\frac{6EI\Delta}{4^2} = +\frac{6EI\Delta}{16}$$

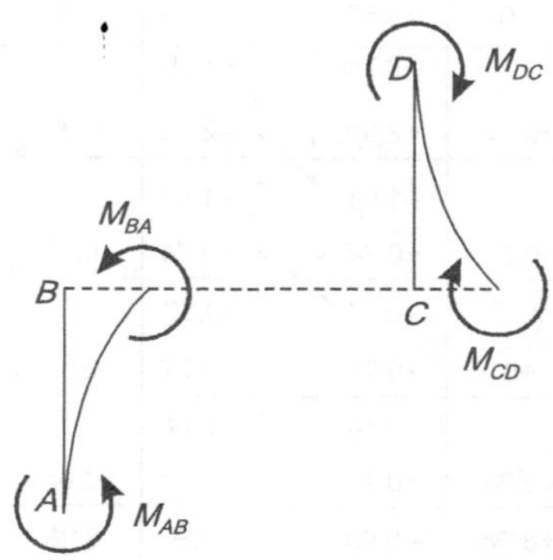

	$\bar{M}_{AB}$	:	$\bar{M}_{BA}$	:	$\bar{M}_{BC}$	:	$\bar{M}_{CB}$	:	$\bar{M}_{CD}$	:	$\bar{M}_{DC}$
	$-\dfrac{6EI\Delta}{16}$	:	$-\dfrac{6EI\Delta}{16}$	:	0	:	0	:	$+\dfrac{6EI\Delta}{16}$	:	$+\dfrac{6EI\Delta}{16}$
∴	−1	:	−1	:	0	:	0	:	+1	:	+1
	−10	:	−10	:	0	:	0	:	+10	:	+10

Joint	A	B	B	C	C	D
DF		3/11	8/11	8/11	3/11	
	−10	−10	0	0	+10	+10
		+2.73	+7.27	−7.27	−2.73	
	+1.36		−3.64	+3.64		−1.36
		+0.99	+2.65	−2.65	−0.99	
	+0.49		−1.33	+1.33		−0.49
		+0.36	+0.97	−0.97	−0.36	
	+0.18		−0.48	+0.48		−0.18
		+0.13	+0.35	−0.35	−0.13	
	+0.06		−0.18	+0.18		−0.06
		+0.05	+0.13	−0.13	−0.05	
	+0.02		−0.07	+0.07		−0.02
		+0.02	+0.05	−0.05	−0.02	
col. (a)	−7.89	−5.72	+5.72	−5.72	+5.72	+7.89

Let S be the sway force for which the moments are given in col. (a).

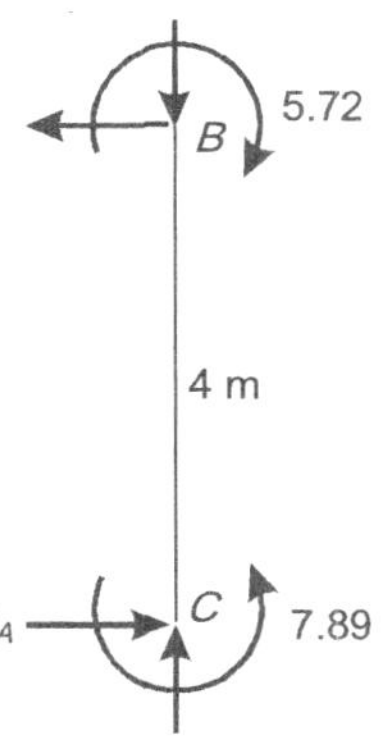

$\Sigma M_B = 0,$ $\quad 4H_A + 7.89 + 5.72 = 0$

Horizontal reaction at $A = \dfrac{-7.89 - 5.72}{4}$

$= -3.4025 \text{ kN} (\rightarrow) = 3.4025 \text{ kN} (\leftarrow)$

$\Sigma M_C = 0,$ $\quad 4H_D + 7.89 + 5.72 = 0$

Horizontal reaction at $D = -\left(\dfrac{5.72 + 7.89}{4}\right)$

$= -3.4025 \text{ kN} (\rightarrow) = 3.4025 \text{ kN} (\leftarrow)$

∴ Sway force, $\quad S = 3.4025 + 3.4025 = 6.805 \text{ kN} (\rightarrow)$

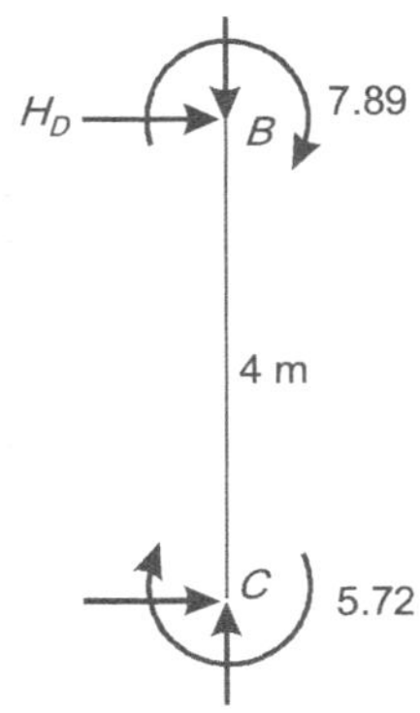

Thus for a sway force of 6.805 kN, the sway moments are as per col. (a). Hence for actual sway force of 6.725 kN, the corresponding sway moments will be $\frac{6.725}{6.805} \times$ col. (a).

	M_{AB}	M_{BA}	M_{BC}	M_{CB}	M_{CD}	M_{DC}
col. (a)	– 7.89	– 5.72	+ 5.72	– 5.72	+ 5.72	+ 7.89
Sway moments	– 7.80	– 5.65	+ 5.65	– 5.65	+ 5.65	+ 7.80
Non sway moments	+ 4.47	+ 8.98	– 8.98	+ 8.98	– 8.98	– 4.47
Final moments	– 3.33	+ 3.33	– 3.33	+ 3.33	– 3.33	+ 3.33

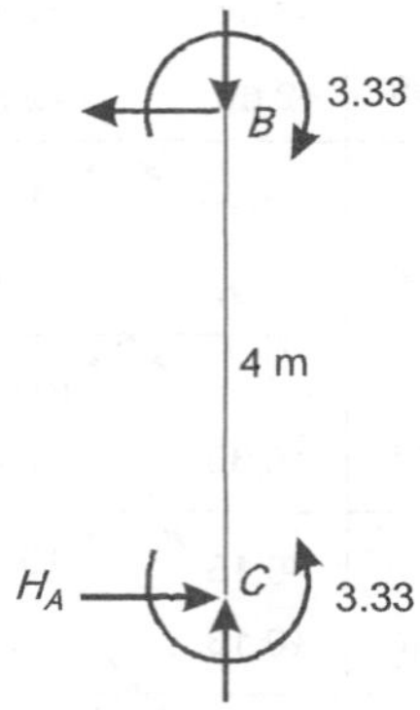

$\Sigma M_B = 0, \qquad 4H_A + 3.33 - 3.33 \approx 0$

$\therefore$ Actual horizontal reaction at $A = \frac{-3.33 + 3.33}{4} = 0$

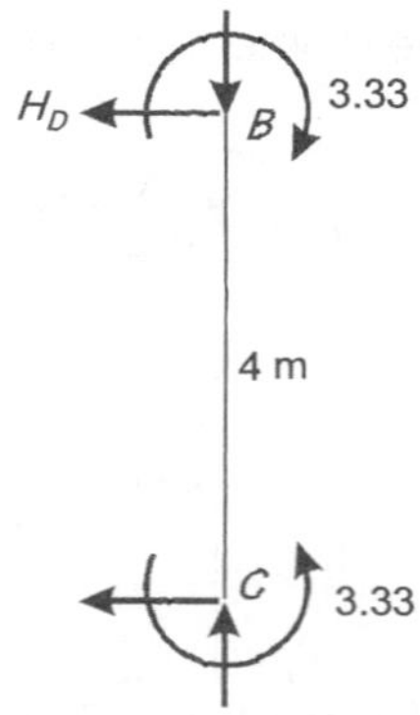

$\Sigma M_C = 0, \qquad 4H_D + 3.33 - 3.33 = 0$

Actual horizontal reaction at $D = -\frac{(-3.33 + 3.33)}{4} = 0$

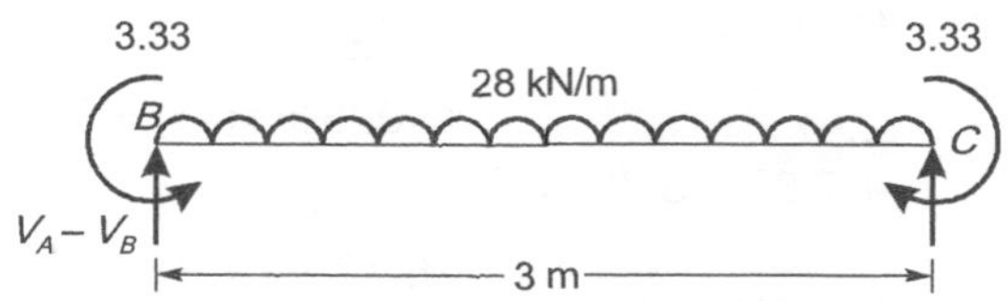

$\Sigma M_C = 0,$

$3V_A - 3.33 - 28 + 3 \times 1.5 + 3.33 = 0$

$$\text{Vertical reaction at } A = -\left(\frac{-3.33 + 3.33 - 28 \times 3 \times 1.5}{3}\right) = 42 \text{ kN } (\uparrow)$$

$$\text{Vertical reaction at } D = 28 \times 3 - 42 = 42 \text{ kN } (\uparrow)$$

Bending moment Diagram

Taking right hand side face of columns as reference face and top face of the beam as reference face.

$$\text{Maximum simply supported moment in } BC = \frac{28 \times 3^2}{8} = 31.5 \text{ kN-m (Sagging)}$$

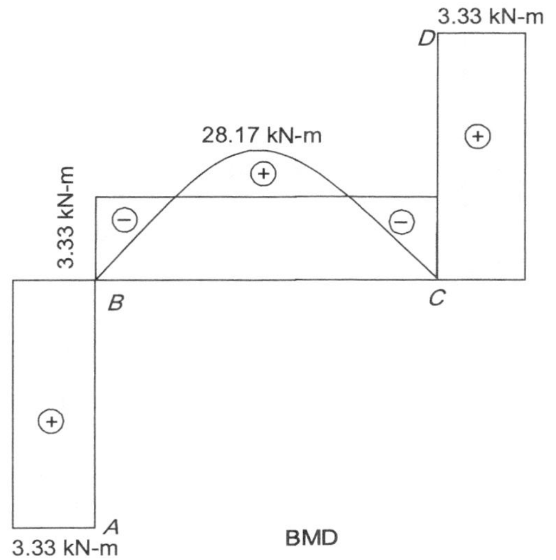

4.7 Two prismatic beams of same area of cross section *AB* and *BD* have fixed ends at *A* and *D* forming a hinge at *B*. Beam *BD* carries a single vertical load '*W*' at mid point *C*. *AB* = *BC* = *CD* = *L*. If *A*, *B*, *C* and *D* lie in one horizontal line with *AD* = 3*L*, analyse the beam and draw BMD and SFD.

[20 marks : 2002]

Solution:

The diagram can be shown:

Analysing the given beam by **slope-deflection method.**

Fixed end Moments

$$\bar{M}_{AB} = 0; \; \bar{M}_{BA} = 0$$

$$\bar{M}_{BD} = -\frac{W \times 2L}{8} = -\frac{WL}{4}$$

$$\bar{M}_{DB} = +\frac{WL}{4}$$

Let the deflection at the hinge be δ,

Slope deflection equations

$$M_{AB} = \bar{M}_{AB} + \frac{2EI}{L}\left(0 + \theta_{BA} - \frac{3\delta}{L}\right) = 0 + \frac{2EI}{L}\left(\theta_{BA} - \frac{3\delta}{L}\right)$$

$$M_{BA} = \bar{M}_{BA} + \frac{2EI}{L}\left(2\theta_{BA} + 0 - \frac{3\delta}{L}\right) = 0 + \frac{2EI}{L}\left(2\theta_{BA} - \frac{3\delta}{L}\right)$$

$$M_{BD} = \bar{M}_{BD} + \frac{2EI}{2L}\left(2\theta_{BD} + 0 + \frac{3\delta}{2L}\right) = \frac{EI}{L}\left(2\theta_{BD} + \frac{3\delta}{2L}\right) - \frac{WL}{4}$$

$$M_{DB} = \bar{M}_{DB} + \frac{2EI}{2L}\left(0 + \theta_{BD} + \frac{3\delta}{2L}\right) = \frac{EI}{L}\left(\theta_{BD} + \frac{3\delta}{2L}\right) + \frac{WL}{4}$$

Joint equilibrium conditions

From the above figure it is clear that both M_{BA} and M_{BD} are equal to zero

$$M_{BA} = 0 \quad \text{...(i)}$$

$$M_{BD} = 0 \quad \text{...(ii)}$$

Shear equation:

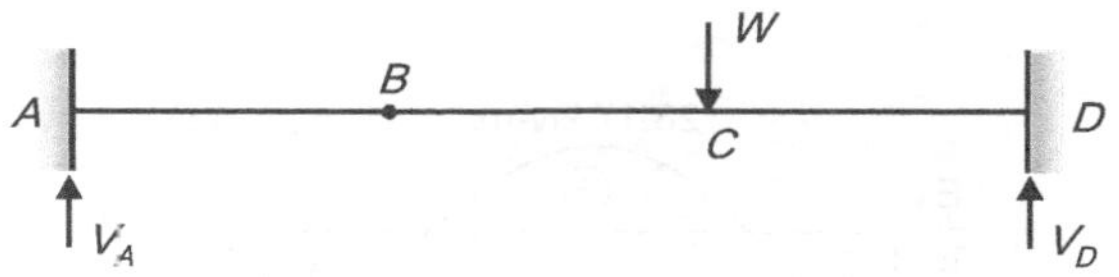

$$\Sigma F_y = 0$$

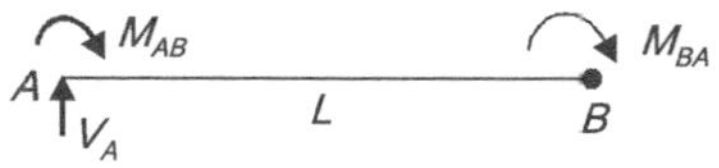

$$V_A + V_D - W = 0 \quad \text{...(iii)}$$

$$V_A = -\left(\frac{M_{AB} + M_{BA}}{L}\right) \qquad [\because M_{BA} = 0]$$

$$\Rightarrow \quad V_A = -\frac{M_{AB}}{L}$$

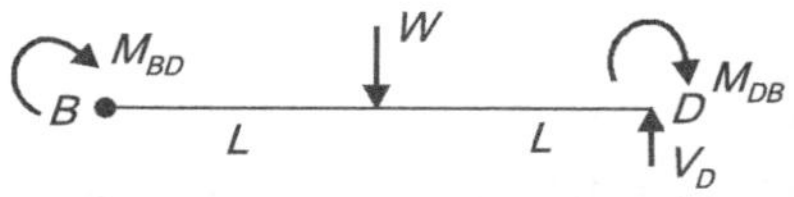

$$V_D = \frac{M_{DB} + M_{BD} + WL}{2L} \qquad [\because M_{BD} = 0]$$

$$\Rightarrow \quad V_D = \frac{M_{DB} + WL}{2L}$$

Putting values of V_A and V_D in (iii), we get

$$\frac{-M_{AB}}{L} + \frac{M_{DB} + WL}{2L} - W = 0$$

$$\Rightarrow \quad -2M_{AB} + M_{DB} + WL - 2WL = 0$$

$$\Rightarrow \quad -2M_{AB} + M_{DB} - WL = 0$$

$$\Rightarrow \quad -2 \times \frac{2EI}{L}\left(\theta_{BA} - \frac{3\delta}{L}\right) + \frac{EI}{L}\left(\theta_{BD} + \frac{3\delta}{2L}\right) + \frac{WL}{4} = WL$$

$$\Rightarrow \quad \frac{-4EI}{L}\left(\theta_{BA} - \frac{3\delta}{L}\right) + \frac{EI}{L}\left(\theta_{BD} + \frac{3\delta}{2L}\right) = \frac{3WL}{4} \quad \text{...(iv)}$$

From (i), we have

$$M_{BA} = 0$$

$$\Rightarrow \quad \frac{2EI}{L}\left(2\theta_{BA} - \frac{3\delta}{L}\right) = 0$$

$$\Rightarrow \quad \frac{\delta}{L} = \frac{2}{3}\theta_{BA}$$

Substituting this value in (iv), we get

$$\frac{-4EI}{L}\left(\theta_{BA} - 3 \times \frac{2}{3}\theta_{BA}\right) + \frac{EI}{L}\left(\theta_{BD} + \frac{3}{2} \times \frac{2}{3}\theta_{BA}\right) = \frac{3WL}{4}$$

$$\Rightarrow \quad \frac{4EI\,\theta_{BA}}{L} + \frac{EI}{L}(\theta_{BD}+\theta_{BA}) = \frac{3WL}{4}$$

$$\Rightarrow \quad 4EI\theta_{BA} + EI\theta_{BD} + EI\theta_{BA} = \frac{3WL^2}{4}$$

$$5\theta_{BA} + \theta_{BD} = \frac{3WL^2}{4EI} \quad \text{... (v)}$$

From (ii), we get

$$\Rightarrow \quad M_{BD} = 0$$

$$\Rightarrow \quad \frac{EI}{L}\left(2\theta_{BD} + \frac{3\delta}{2L}\right) - \frac{WL}{4} = 0$$

$$\Rightarrow \quad \frac{EI}{L}\left(2\theta_{BD} + \frac{3}{2}\times\frac{2}{3}\theta_{BA}\right) = \frac{WL}{4}$$

$$\Rightarrow \quad \theta_{BA} + 2\theta_{BD} = \frac{WL^2}{4EI} \quad \text{...(vi)}$$

Solving (v) and (vi), we get

$$\theta_{BA} = \frac{5}{36}\frac{WL^2}{EI} \text{ and } \theta_{BD} = \frac{WL^2}{18EI}$$

$$\therefore \quad M_{AB} = \frac{2EI}{L}\left(\theta_{BA} - \frac{3\delta}{L}\right)$$

$$= \frac{2EI}{L}\left(\theta_{BA} - 3\times\frac{2}{3}\theta_{BA}\right)$$

$$= \frac{-2EI}{L}\times\theta_{BA} = \frac{-2EI}{L}\times\frac{5}{36}\times\frac{WL^2}{EI}$$

$$M_{AB} = \frac{-5}{18}WL = \frac{-10}{36}WL$$

and

$$M_{DB} = \frac{EI}{L}\left(\theta_{BD} + \frac{3\delta}{2L}\right) + \frac{WL}{4}$$

$$= \frac{EI}{L}\left(\theta_{BD} + \frac{3}{2}\times\frac{2}{3}\theta_{BA}\right) + \frac{WL}{4}$$

$$= \frac{EI}{L}(\theta_{BD} + \theta_{BA}) + \frac{WL}{4}$$

$$= \frac{EI}{L}\left(\frac{WL^2}{18EI} + \frac{5}{36}\frac{WL^2}{EI}\right) + \frac{WL}{4}$$

$$= \frac{7EI}{L}\times\frac{WL^2}{36EI} + \frac{WL}{4} = \frac{16}{36}WL$$

$$\Rightarrow \quad M_{BD} = \frac{4}{9}WL$$

Also,

$$V_A = \frac{-M_{AB}}{L} = -\left(\frac{-5}{18}\right)\frac{WL}{L} = \frac{5}{18}W$$

$$\therefore \qquad V_D = W - \frac{5}{18}W = \frac{13}{18}W$$

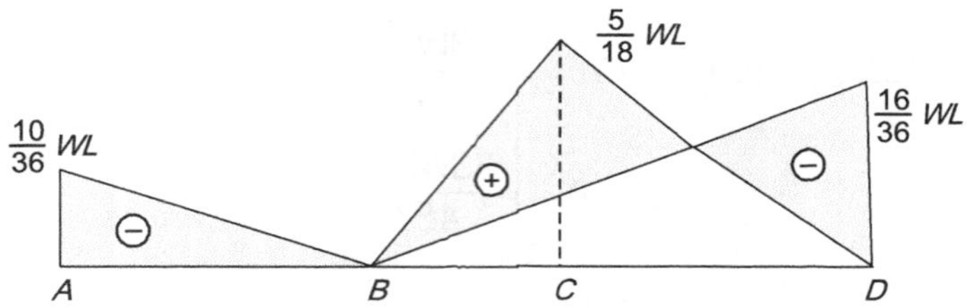

Bending Moment Diagram

$$\text{Maximum simply supported } BM \text{ in } BD = \frac{W(2L)}{4} = \frac{2WL}{4} = \frac{WL}{2}$$

$$\text{Net positive } BM \text{ at } C = \frac{WL}{2} - \frac{16WL}{36 \times 2} = \frac{10WL}{36} = \frac{5WL}{18}$$

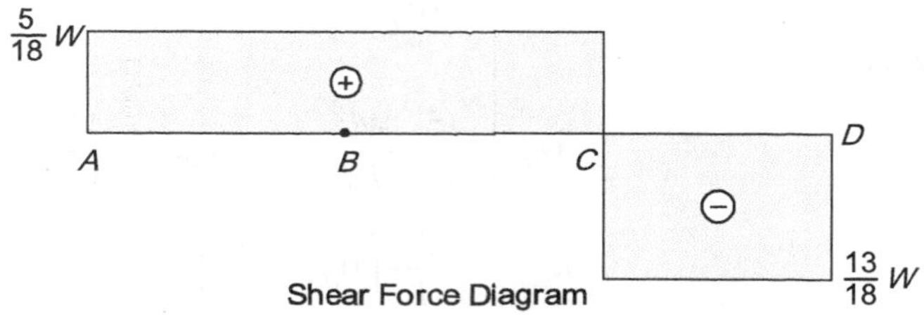

Shear Force Diagram

4.16 Use stiffness method to analyse the rigid frame shown in the figure below. Draw neatly the BMD.

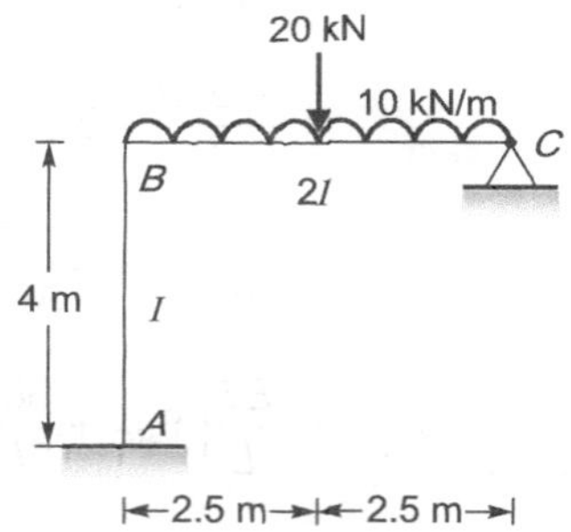

[20 marks : 2003]

Solution:

Analysing the given rigid frame by **slope deflection method** which is a stiffness method.

Fixed end moments

$$\bar{M}_{AB} = 0; \; \bar{M}_{BA} = 0$$

$$\bar{M}_{BC} = -\left(\frac{10 \times 5^2}{12} + \frac{20 \times 5}{8}\right) = -33.33 \text{ kN-m}$$

$$\bar{M}_{CB} = +33.33 \text{ kN-m}$$

Slope deflection equations

$$M_{AB} = \bar{M}_{AB} + \frac{2EI}{L}\left(2\theta_A + \theta_B - \frac{3\delta}{L}\right)$$

$$\Rightarrow \quad M_{AB} = 0 + \frac{2EI}{4}(0 + \theta_B - 0)$$

$$\Rightarrow \quad M_{AB} = 0.5EI\theta_B \qquad \text{... (i)}$$

and $$M_{BA} = \bar{M}_{BA} + \frac{2EI}{L}\left(2\theta_B + \theta_A - \frac{3\delta}{L}\right)$$

$$\Rightarrow \quad M_{BA} = 0 + \frac{2EI}{4}(2\theta_B + 0 - 0)$$

$$\Rightarrow \quad M_{BA} = EI\theta_B \qquad \text{... (ii)}$$

and $$M_{BC} = \bar{M}_{BC} + \frac{2E(2I)}{L}\left(2\theta_B + \theta_C - \frac{3\delta}{L}\right)$$

$$\Rightarrow \quad M_{BC} = -33.33 + \frac{4EI}{5}(2\theta_B + \theta_C)$$

$$\Rightarrow \quad M_{BC} = 0.8\,EI(2\theta_B + \theta_C) - 33.33 \qquad \text{... (iii)}$$

and $$M_{CB} = \bar{M}_{CB} + \frac{2E(2I)}{L}\left(2\theta_C + \theta_B - \frac{3\delta}{L}\right)$$

$$\Rightarrow \quad M_{CB} = +33.33 + \frac{4EI}{5}(2\theta_C + \theta_B)$$

$$M_{CB} = 0.8EI(\theta_B + 2\theta_C) + 33.33 \qquad \text{... (iv)}$$

Equilibrium equations

(a) $$M_{BA} + M_{BC} = 0$$

$$\Rightarrow \quad EI\theta_B + 0.8EI(2\theta_B + \theta_C) - 33.33 = 0$$

$$\Rightarrow \quad 2.6\theta_B + 0.8\theta_C = \frac{33.33}{EI} \qquad \text{... (v)}$$

(b) $$M_{CB} = 0$$

$$\Rightarrow \quad 0.8\,EI\,(\theta_B + 2\theta_C) + 33.33 = 0$$

$$\Rightarrow \quad 0.8\theta_B + 1.6\theta_C = \frac{-33.33}{EI} \qquad \text{... (vi)}$$

Solving (v) and (vi), we get

$$\theta_B = \frac{22.725}{EI};\ \theta_C = \frac{-32.194}{EI}$$

$\therefore$ From (i), $$M_{AB} = 0.5\,EI \times \frac{22.725}{EI} = 11.36 \text{ kN-m}$$

From (ii), $$M_{BA} = EI \times \frac{22.725}{EI} = 22.725 \text{ kN-m}$$

From (iii), $$M_{BC} = 0.8\ EI\left(2 \times \frac{22.725}{EI} - \frac{32.194}{EI}\right) - 33.33 = -22.725 \text{ kN-m}$$

From (iv), $$M_{CB} = 0$$

Reactions

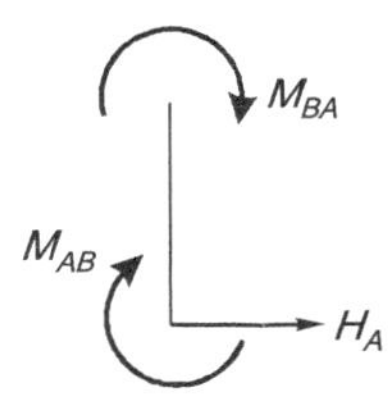

$$H_A = \frac{M_{AB} + M_{BA}}{L_{AB}}$$

$$= \frac{11.36 + 22.725}{4} = 8.52 \text{ kN}$$

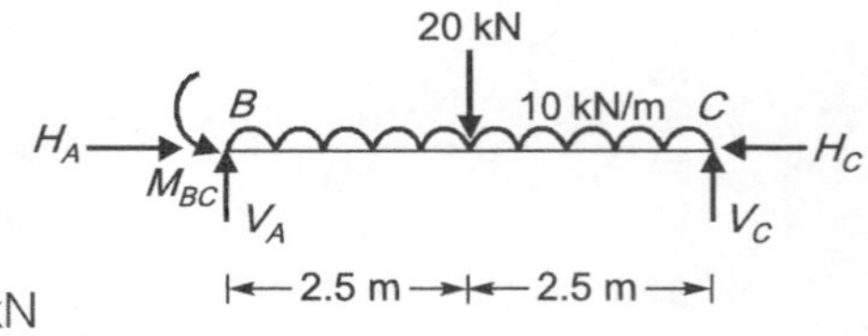

$$\Sigma M_C = 0$$

$\Rightarrow$ $V_A \times 5 - 22.725 - 20 \times 2.5 - 10 \times 5 \times 2.5 = 0$

$\Rightarrow$ $5V_A = 197.725$

$\Rightarrow$ $V_A = 39.545 \text{ kN}$

and $V_C = 20 + (10 \times 5) - V_A$

$\Rightarrow$ $V_C = 70 - 39.545$

$\Rightarrow$ $V_C = 30.455 \text{ kN}$

Also $H_A - H_C = 0$

$\Rightarrow$ $H_A = H_C = 8.52 \text{ kN}$

Free BMD of portion *BC*

$$M_x(x \text{ from } B) = 35x - 5x^2 \quad [0 \le x \le 2.5]$$

$$M_{\text{centre}} = 35 \times 2.5 - 5 \times (2.5)^2 = 56.25 \text{ kN-m}$$

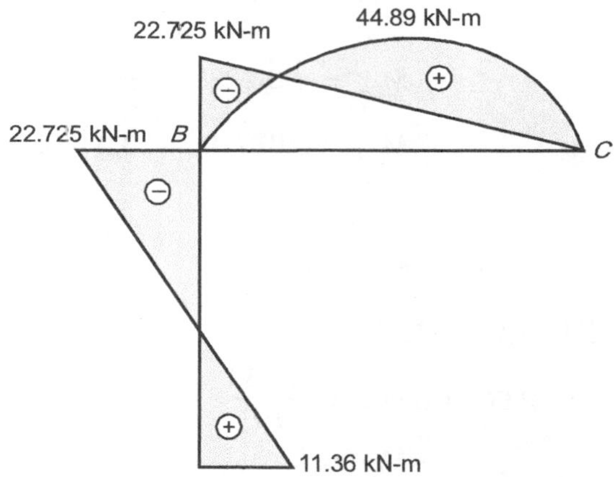

Maximum sagging moment at the centre of beam *BC*

$$= 56.25 - \frac{22.75}{2} = 44.89 \text{ kN-m}$$

4.8 A rigid frame *ABCD* of uniform *EI* shown in the following figure is subjected to horizontal and vertical forces. Determine horizontal displacement at the support *D*.

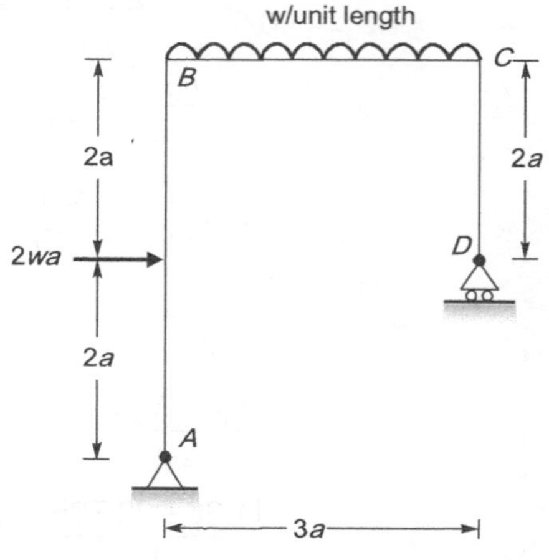

[15 marks : 2003]

Solution:

Let the reactions at A be H_A and V_A and the reaction at D be V_D in the direction as shown in figure below. The displacement in the horizontal direction at D is required. Hence, let us apply a fictitious force Q at D.

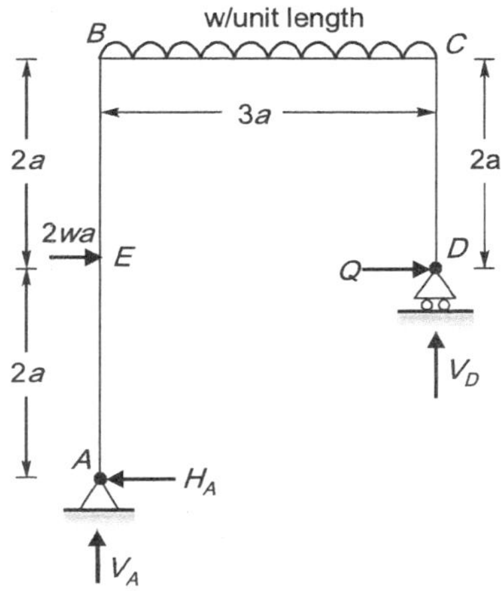

$$\therefore \quad \Sigma F_y = 0$$
$$\Sigma F_x = 0$$
$$\Rightarrow \quad V_D + V_A = w \times 3a$$
$$2wa + Q - H_A = 0$$
$$\Rightarrow \quad V_A + V_D = 3wa$$
$$\Rightarrow \quad H_A = 2wa + Q$$
and
$$\Sigma M_D = 0$$
$$\Rightarrow \quad V_A \times 3a + H_A \times 2a - w \times 3a \times 1.5a = 0$$
$$\Rightarrow \quad 3aV_A + (2wa + Q)2a - 4.5wa^2 = 0 \qquad [\because H_A = 2wa + Q]$$
$$\Rightarrow \quad 3aV_A + 2aQ - 0.5wa^2 = 0$$
$$\Rightarrow \quad V_A = \frac{-2Q + 0.5wa}{3}$$
$$\therefore \quad V_D = 3wa - \left(\frac{-2Q + 0.5wa}{3}\right)$$
$$V_D = \frac{8.5\,wa + 2Q}{3}$$

Now, the strain energy stored by the whole frame is given by

$$U = \Sigma\int \frac{M^2 ds}{2EI}$$

i.e.

$$U = U_{AE} + U_{EB} + U_{BC} + U_{CD}$$

Strain energy stored in portion AE

$$M_y = H_A y \qquad [0 < y < 2a]$$
$$M_y = (2wa + Q)\,y$$
$$\therefore \quad U_{AE} = \int_0^{2a} \frac{(2wa + Q)^2 y^2 dy}{2EI}$$
$$\Rightarrow \quad \frac{\partial U_{AE}}{\partial Q} = \int_0^{2a} \frac{2(2wa + Q)y^2 dy}{2EI}$$
$$\Rightarrow \quad \frac{\partial U_{AE}}{\partial Q} = \int_0^{2a} \frac{2wa\,y^2 dy}{EI} \qquad [\because Q = 0]$$
$$\Rightarrow \quad \frac{\partial U_{AE}}{\partial Q} = \frac{2wa}{EI}\left[\frac{y^3}{3}\right]_0^{2a}$$
$$\Rightarrow \quad \frac{\partial U_{AE}}{\partial Q} = \frac{2wa}{3EI} \times (2a)^3$$
$$\Rightarrow \quad \frac{\partial U_{AE}}{\partial Q} = \frac{16}{3}\frac{wa^4}{EI} \qquad \text{...(i)}$$

Strain energy stored in portion *EB*

$$M_y(y \text{ from } A) = H_A y - 2wa(y-2a) \qquad [2a < y < 4a]$$

$$\Rightarrow \quad M_y = (2wa + Q)y - 2way + 4wa^2$$

$$\Rightarrow \quad M_y = 2way + Qy - 2way + 4\,wa^2$$

$$\Rightarrow \quad M_y = 4wa^2 + Qy$$

$$\therefore \quad U_{EB} = \int_{2a}^{4a} \frac{\left(4wa^2 + Qy\right)^2 dy}{2EI}$$

$$\Rightarrow \quad \frac{\partial U_{EB}}{\partial Q} = \int_{2a}^{4a} \frac{2\left(4wa^2 + Qy\right) y\, dy}{2EI} \qquad [\because Q = 0]$$

$$\Rightarrow \quad \frac{\partial U_{EB}}{\partial Q} = \frac{4wa^2}{EI}\int_{2a}^{4a} y\, dy$$

$$\Rightarrow \quad \frac{\partial U_{EB}}{\partial Q} = \frac{4wa^2}{EI}\left[\frac{y^2}{2}\right]_{2a}^{4a}$$

$$\Rightarrow \quad \frac{\partial U_{EB}}{\partial Q} = \frac{2wa^2}{EI}(16a^2 - 4a^2)$$

$$\Rightarrow \quad \frac{\partial U_{EB}}{\partial Q} = \frac{24wa^4}{EI} \qquad \text{...(ii)}$$

Strain energy stored in portion *BC*

$$M_x(x \text{ from } C) = 2Qa + V_D x - \frac{wx^2}{2} \qquad [0 < x < 3a]$$

$$\Rightarrow \quad M_x = 2Qa + \left(\frac{8.5wa + 2Q}{3}\right) x - \frac{wx^2}{2}$$

$$\Rightarrow \quad M_x = \frac{12Qa + 17wax + 4Qx - 3wx^2}{6}$$

$$\therefore \quad U_{BC} = \int_0^{3a} \frac{\left(12Qa + 17wax + 4Qx - 3wx^2\right)^2 dx}{36 \times 2EI}$$

$$\Rightarrow \quad \frac{\partial U_{BC}}{\partial Q} = \int_0^{3a} \frac{\left(12Qa + 4Qx + 17wax - 3wx^2\right)(12a + 4x)\, dx}{36EI}$$

$$\Rightarrow \quad \frac{\partial U_{BC}}{\partial Q} = \int_0^{3a} \frac{\left(204wa^2 x + 32wax^2 - 12wx^3\right) dx}{36\,EI} \qquad [\because Q = 0]$$

$$\Rightarrow \quad \frac{\partial U_{BC}}{\partial Q} = \frac{1}{36EI}\left[\frac{204wa^2}{2} \times (3a)^2 + \frac{32wa}{3} \times (3a)^3 - \frac{12w}{4} \times (3a)^4\right]$$

$$\Rightarrow \quad \frac{\partial U_{BC}}{\partial Q} = \frac{1}{36EI}\left[918\,wa^4 + 288wa^4 - 243\,wa^4\right]$$

$$\Rightarrow \quad \frac{\partial U_{BC}}{\partial Q} = \frac{963\,wa^4}{36\,EI}$$

$$\Rightarrow \quad \frac{\partial U_{BC}}{\partial Q} = \frac{107\, wa^4}{4\, EI} \quad \text{...(iii)}$$

Strain energy stored in portion CD

$$M_y (y \text{ from } D) = Qy \quad [0 < y < 2a]$$

$$\Rightarrow \quad M_y = Qy$$

$$\therefore \quad U_{CD} = \int_0^{2a} \frac{Q^2 y^2\, dy}{2EI}$$

$$\frac{\partial U_{CD}}{\partial Q} = \int_0^{2a} \frac{2Q\, y^2\, dy}{2\, EI} \quad [\because Q = 0]$$

$$\frac{\partial U_{CD}}{\partial Q} = 0 \quad \text{... (iv)}$$

Now, we have

$$U = U_{AE} + U_{EB} + U_{BC} + U_{CD}$$

$$\Rightarrow \quad \frac{\partial U}{\partial Q} = \frac{\partial U_{AE}}{\partial Q} + \frac{\partial U_{EB}}{\partial Q} + \frac{\partial U_{BC}}{\partial Q} + \frac{\partial U_{CD}}{\partial Q}$$

$$\Rightarrow \quad \frac{\partial U}{\partial Q} = \frac{16wa^4}{3\, EI} + \frac{24wa^4}{EI} + \frac{107wa^4}{4\, EI} + 0$$

$$\Rightarrow \quad \frac{\partial U}{\partial Q} = \frac{64wa^4 + 288wa^4 + 321wa^4}{12\, EI}$$

$$\Rightarrow \quad \frac{\partial U}{\partial Q} = \frac{673wa^4}{12EI} = \frac{56.08\, wa^4}{EI}$$

$$\Rightarrow \quad \frac{\partial U}{\partial Q} = \text{Horizontal displacement at the support } D = \delta_H$$

$$\therefore \quad \delta_H = \frac{56.08\, wa^4}{EI}$$

4.9 A cantilever beam AB of length '$2a$' carries a UDL of intensity 'w' over half the span adjoining fixed support. The beam is strengthened by supporting its free end over the midpoint of transverse simply supported beam of length 'a'. EI is same for both the beams. Draw bending moment diagram for the cantilever beam.

[15 marks : 2003]

Solution:

The given arrangement can be shown as given below.

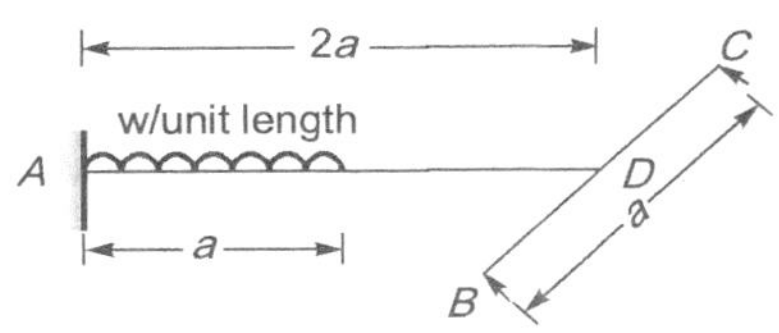

Let the reaction at D be R, then

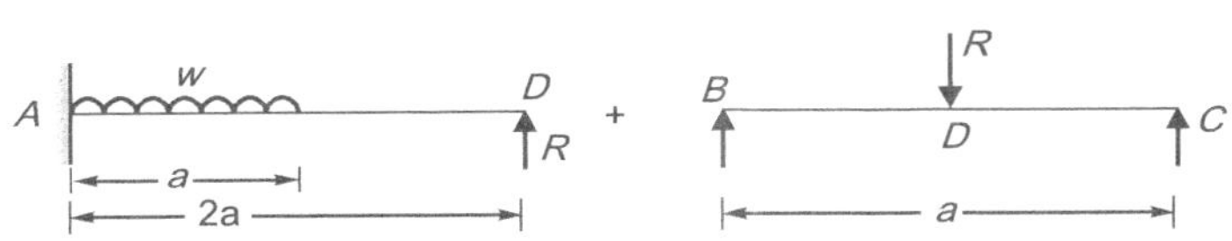

The deflection at D in cantilever may be given as

$$\Delta_{D,\,\text{cantilever}} = \frac{wa^4}{8EI} + \frac{wa^3}{6EI} \times a - \frac{R(2a)^3}{3EI}$$

$$= \frac{wa^4}{8EI} + \frac{wa^4}{6EI} - \frac{8Ra^3}{3EI}$$

$$= \frac{7wa^4}{24EI} - \frac{8Ra^3}{3EI} \quad \text{...(i)}$$

The deflection at D in simply supported beam may be given as

$$\Delta_{D,\,\text{ssb}} = \frac{Ra^3}{48EI} \quad \text{...(ii)}$$

But $\Delta_{D,\,\text{cantilever}} = \Delta_{D,\,\text{ssb}}$

$$\therefore \quad \frac{7wa^4}{24EI} - \frac{8Ra^3}{3EI} = \frac{Ra^3}{48EI}$$

$$\Rightarrow \quad \frac{7wa^4}{24EI} = \frac{Ra^3}{48EI} + \frac{8Ra^3}{3EI}$$

$$\Rightarrow \quad \frac{7wa}{24} = \frac{R}{48} + \frac{8R}{3}$$

$$\Rightarrow \quad 129R = 14wa$$

$$\Rightarrow \quad R = \frac{14wa}{129}$$

Bending Moment Diagram

Taking top face of the cantilever as reference face.

Portion *DE*

$$M_x(x \text{ from } D) = \frac{14wa}{129}x \quad [0 \le x \le a]$$

$$M_D = 0$$

$$M_E = \frac{14wa}{129} \times a = \frac{14wa^2}{129}$$

Portion *EA*

$$M_x(x \text{ from } D) = \frac{14wa}{129}x - w(x-a)\frac{(x-a)}{2} \quad [a \le x \le 2a]$$

At $x = 2a$,

$$M_A = \frac{14wa}{129} \times 2a - w(2a-a)\frac{(2a-a)}{2}$$

$$\Rightarrow \quad M_A = \frac{28wa^2}{129} - \frac{wa^2}{2}$$

$$\Rightarrow \quad M_A = \frac{56wa^2 - 129wa^2}{258} = \frac{-73wa^2}{258} = -\frac{-36.5wa^2}{129}$$

For $M_x = 0$, we have

$$\frac{14wa}{129}x - w\frac{(x-a)^2}{2} = 0$$

$$\Rightarrow \quad 28wax - 129w(x-a)^2 = 0$$

$$\Rightarrow \quad 28ax - 129(x^2 + a^2 - 2ax) = 0$$

$$\Rightarrow \quad 28ax - 129x^2 - 129a^2 + 258ax = 0$$

$$\Rightarrow \quad -129x^2 + 286ax - 129a^2 = 0$$

$$\Rightarrow \quad 129x^2 - 286ax + 129a^2 = 0$$

$$\Rightarrow \quad x = \frac{286a \pm \sqrt{(286a)^2 - 4 \times 129 \times 129a^2}}{2 \times 129}$$

$$\Rightarrow \quad x = \frac{286a \pm 123.42a}{258} = 1.587a$$

For maximum bending moment,

$$M_x \text{ (x from } D) = \frac{14wa}{129} - \frac{w(x-a)^2}{2}$$

$$\frac{dM_x}{dx} = 0$$

$$\frac{14wa}{129} - \frac{2w(x-a)}{2} = 0$$

$$x = \left(1 + \frac{14}{129}\right)a = 1.1085a$$

$$M_x|_{max} = 0.1144\ wa^2$$

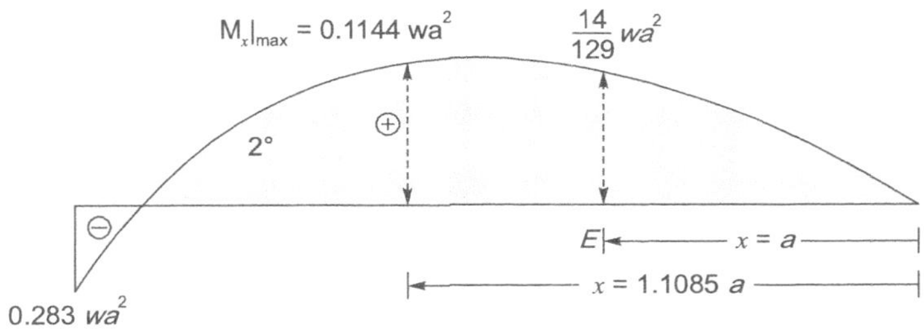

4.10 Analyse the frame *ABCD* shown in figure using moment distribution method. The properties of the sections geometry of the frame and the loading are given in figure. Draw the BM and axial force diagrams.

[40 marks : 2004]

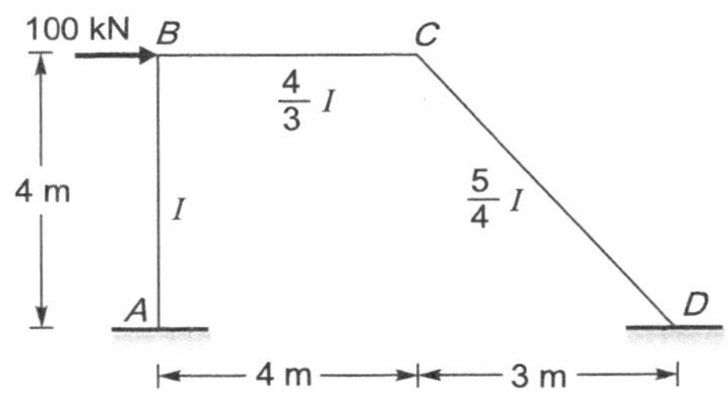

Solution:

There are no external loads acting on the portal frame except the sway force of 100 kN from left to right. So only the sway analysis will be carried out.

Distribution Factors:

Joint	Member	Relative Stiffness	Total Relative Stiffness	Distribution Factor
B	BA	$\frac{I}{4} = \frac{I}{4}$	$\frac{7I}{12}$	$\frac{3}{7}$
	BC	$\frac{4}{3}I \times \frac{1}{4} = \frac{I}{3}$		$\frac{4}{7}$
C	CB	$\frac{4}{3}I \times \frac{1}{4} = \frac{I}{3}$	$\frac{7I}{12}$	$\frac{4}{7}$
	CD	$\frac{5}{4}I \times \frac{1}{5} = \frac{I}{4}$		$\frac{3}{7}$

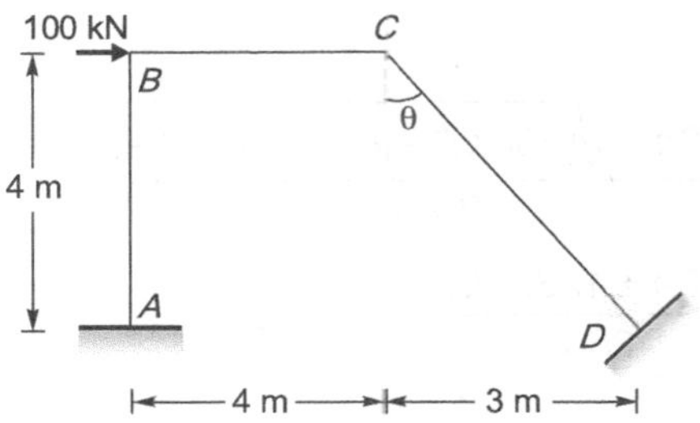

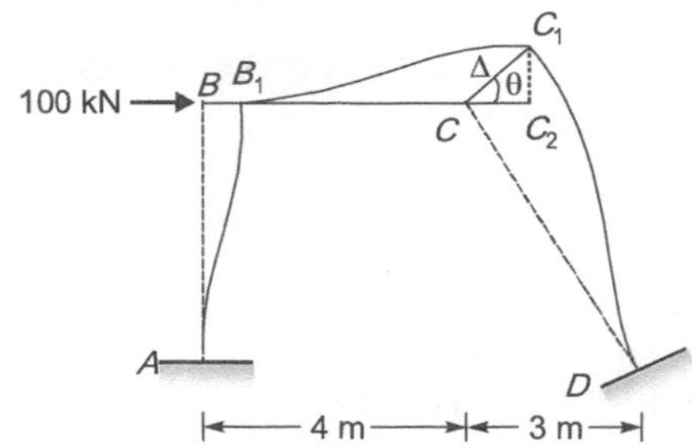

Let the frame $ABCD$ deflect to the position $AB_1\ C_1\ D$ due to the sway force.

Let, $$CC_1 = \Delta$$

In ΔCC_1C_2, $$\cos\theta = \frac{CC_2}{CC_1}$$

$$BB_1 = CC_2 = \Delta\cos\theta = \frac{4}{5}\Delta$$

$$C_1C_2 = \Delta\sin\theta = \frac{3}{5}\Delta$$

$$M_{FAB} = -\frac{6EI\Delta_{BB_1}}{l^2} = \frac{-6EI\frac{4}{5}\Delta}{16} = M_{FBA}$$

$$M_{FBC} = \frac{6E\frac{4}{3}I\,\Delta_{C_1C_2}}{l^2} = \frac{6E\frac{4}{3}I\times\frac{3}{5}\Delta}{16} = M_{FBC} = \frac{24}{80}EI\Delta$$

$$M_{FCD} = M_{FDC} = \frac{-6E\times\frac{5I\Delta}{4}}{25} = -\frac{6EI\Delta}{20} = -\frac{24EI\Delta}{80}$$

Ratio of fixed end moments for AB, BA, BC, CB, CD and DC

$$-\frac{24EI\Delta}{80} : -\frac{24EI\Delta}{80} :: \frac{24EI\Delta}{80} : \frac{24EI\Delta}{80} :: -\frac{24EI\Delta}{80} : -\frac{24EI\Delta}{80}$$

$-50 : -50 :: 50 : 50 :: -50 : -50$

	A	B		C		D
		$\frac{3}{7}$	$\frac{4}{7}$	$\frac{4}{7}$	$\frac{3}{7}$	
Col.(a)	–50	–50	50	50	–50	–50

Since the joints are balanced so there is no need to further distribution.

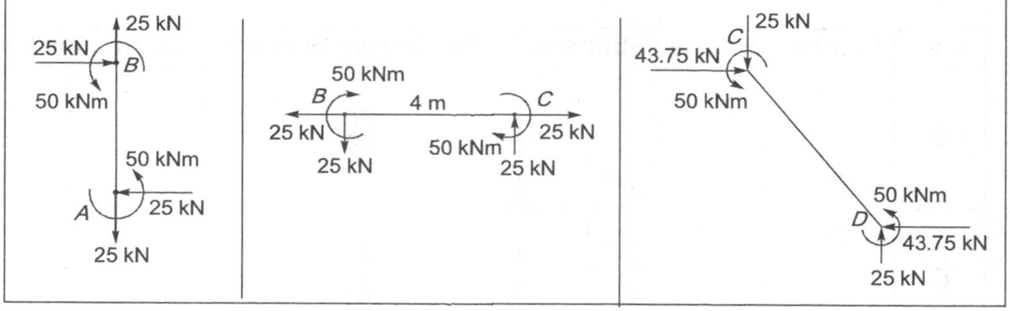

$$\text{Vertical reaction at } D = \frac{M_{BC} + M_{CB}}{L_{BC}} = \frac{50+50}{4} = 25\ \text{kN}(\uparrow)$$

$\therefore$

$$\text{Horizontal reaction at } A = \frac{M_{AB} + M_{BA}}{L_{AB}} = \frac{-50-50}{4} = -25\ \text{kN}(\rightarrow) = 25\ \text{kN}(\leftarrow)$$

$$\text{Horizontal reaction at } D = \frac{M_{CD} + M_{DC} - V_D \times 3}{4} = \frac{-50-50-25\times 3}{4}$$

$$= -43.75(\rightarrow) = 43.75\ \text{kN}\ (\leftarrow)$$

Let the sway force S be acting from left to right and then

$$H_A + H_D + S = 0$$

$$-25 - 43.75 + S = 0$$

$$\Rightarrow \quad S = 68.75\ \text{kN}\ (\rightarrow)$$

For a sway force of 68.75 kN, the sway moment are as per Col. (a)

For the actual sway force of 100 kN, the corresponding sway moments will be

$$\frac{100}{68.75} \times \text{Col. (a) moments}$$

	A		B		C	D
col. (a)	–50	–50	50	50	–50	–50
Actual sway moments $\frac{100}{68.75}\times$ col. (a)	–72.72	–72.72	72.72	72.72	–72.72	–72.72

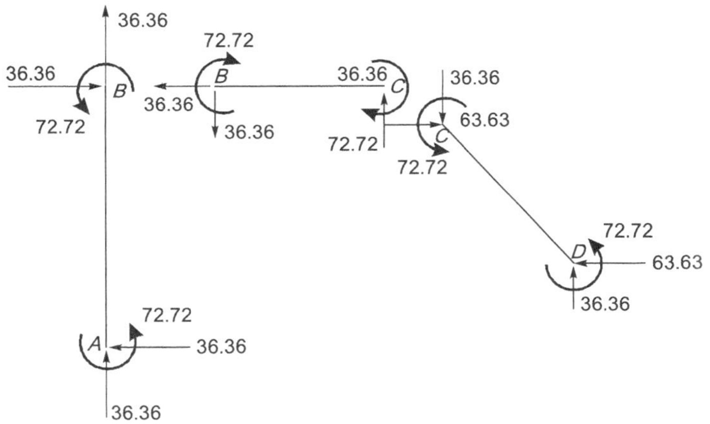

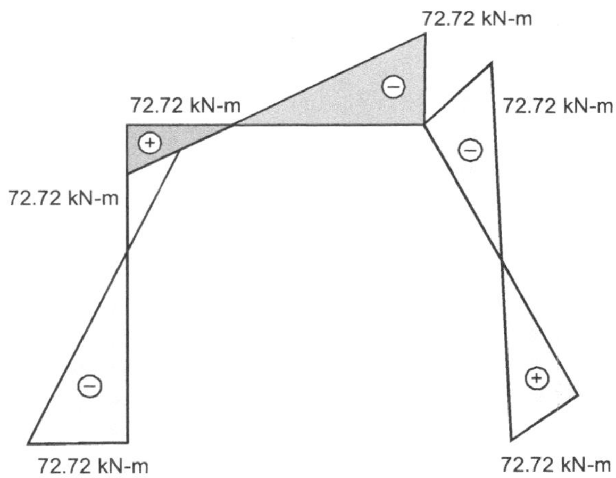

Bending Moment Diagram

4.11 Analyse the rigid jointed frame shown in the figure using slope deflection method. Find all the reaction components at the supports A and D. Draw BMD and SFD for all the members.

[30 marks : 2005]

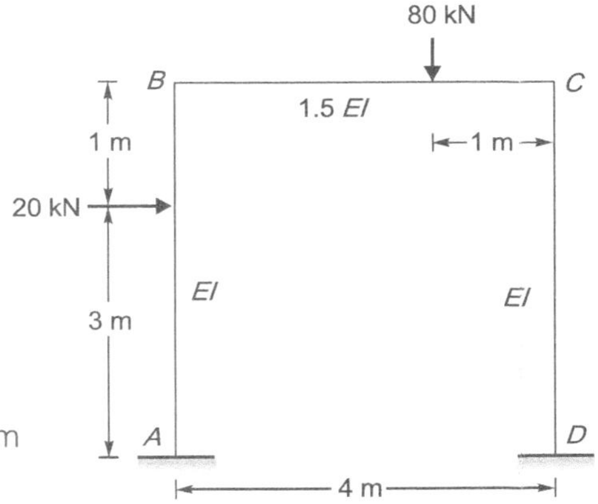

Solution:

Fixed end moments:

$$\bar{M}_{AB} = -\frac{20\times 3\times 1^2}{4^2} = -3.75\ \text{kN-m}$$

$$\bar{M}_{BA} = +\frac{20\times 1\times 3^2}{4^2} = +11.25\ \text{kN-m}$$

$$\bar{M}_{BC} = -\frac{80\times 3\times 1}{4^2} = -15 \text{ kN-m}$$

$$\bar{M}_{CB} = +\frac{80\times 1\times 3^2}{4^2} = 45 \text{ kN-m}$$

$$\bar{M}_{CD} = \bar{M}_{DC} = 0$$

Slope deflection equations

Member *AB*

$$M_{AB} = \bar{M}_{AB} + \frac{2EI}{L}\left(2\theta_A + \theta_B - \frac{3\delta}{L}\right)$$

$$\Rightarrow \quad M_{AB} = -3.75 + \frac{2EI}{4}\left(\theta_B - \frac{3\delta}{4}\right)$$

$$\Rightarrow \quad M_{AB} = -3.75 + \frac{EI}{2}\theta_B - \frac{3EI\delta}{8}$$

and

$$M_{BA} = \bar{M}_{BA} + \frac{2EI}{L}\left(2\theta_B + \theta_A - \frac{3\delta}{L}\right)$$

$$\Rightarrow \quad M_{BA} = +11.25 + \frac{2EI}{4}\left(2\theta_B - \frac{3\delta}{4}\right)$$

$$\Rightarrow \quad M_{BA} = 11.25 + EI\theta_B - \frac{3EI\delta}{8}$$

Member *BC*

$$M_{BC} = \bar{M}_{BC} + \frac{2EI}{L}(2\theta_B + \theta_C)$$

$$\Rightarrow \quad M_{BC} = -15 + \frac{2\times 1.5\,EI}{4}(2\theta_B + \theta_C)$$

$$\Rightarrow \quad M_{BC} = -15 + 1.5\,EI\theta_B + 0.75\,EI\theta_C$$

$$M_{CB} = \bar{M}_{CB} + \frac{2EI}{L}(2\theta_C + \theta_B)$$

$$\Rightarrow \quad M_{CB} = 45 + \frac{2\times 1.5\,EI}{4}(2\theta_C + \theta_B)$$

$$\Rightarrow \quad M_{CB} = 45 + 0.75\,EI\theta_B + 1.5\,EI\,\theta_C$$

Member *CD*

$$M_{CD} = \bar{M}_{CD} + \frac{2EI}{L}\left(2\theta_C + \theta_D - \frac{3\delta}{L}\right)$$

$$\Rightarrow \quad M_{CD} = 0 + \frac{2EI}{4}\left(2\theta_C - \frac{3\delta}{4}\right)$$

$$\Rightarrow \quad M_{CD} = EI\theta_C - \frac{3EI\delta}{8}$$

$$M_{DC} = \bar{M}_{DC} + \frac{2EI}{4}\left(2\theta_D + \theta_C - \frac{3\delta}{L}\right)$$

$$\Rightarrow \quad M_{DC} = 0 + \frac{2EI}{4}\left(\theta_C - \frac{3\delta}{4}\right)$$

$$\Rightarrow \quad M_{DC} = 0.5\,EI\theta_C - \frac{3EI\delta}{8}$$

Equilibrium condition at B,

$$M_{BA} + M_{BC} = 0$$

$$\Rightarrow \quad 11.25 + EI\theta_B - \frac{3EI\delta}{8} + 1.5\,EI\theta_B + 0.75EI\theta_C - 15 = 0$$

$$\Rightarrow \quad 0.75\,EI\theta_C + 2.5\,EI\theta_B - \frac{3EI\delta}{8} = 3.75 \qquad \text{... (i)}$$

Equilibrium condition at C,

$$M_{CB} = M_{CD} = 0$$

$$\Rightarrow \quad (45 + 0.75\,EI\theta_B + 1.5\,EI\theta_C) + \left(EI\theta_C - \frac{3EI\delta}{8}\right) = 0$$

$$\Rightarrow \quad 0.75\,EI\theta_B + 2.5\,EI\theta_C - \frac{3EI\delta}{8} = -45 \qquad \text{...(ii)}$$

For **horizontal equilibrium**, we have

$$H_A + H_D + 20 = 0$$

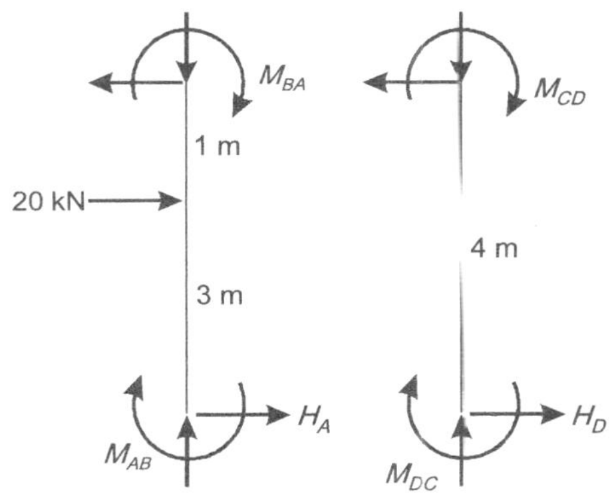

$$\Rightarrow \quad \frac{M_{AB} + M_{BA} - 20 \times 1}{4} + \frac{M_{CD} + M_{DC}}{4} + 20 = 0$$

$$\Rightarrow \quad M_{AB} + M_{BA} - 20 + M_{CD} + M_{DC} + 80 = 0$$

$$\Rightarrow \quad -3.75 + \frac{EI}{2}\theta_B - \frac{3EI\delta}{8} + 11.25 + EI\theta_B - \frac{3EI\delta}{8} + EI\theta_C - \frac{3EI\delta}{8}$$

$$+ 0.5\,EI\theta_C - \frac{3EI\delta}{8} + 60 = 0$$

$$\Rightarrow \quad 1.5\,EI\theta_B + 1.5EI\theta_C - \frac{12EI\delta}{8} = -67.5$$

$$\Rightarrow \quad EI\theta_B + EI\theta_C - EI\delta = -45 \qquad \text{...(iii)}$$

Solving (i), (ii) and (iii), we get

$$EI\theta_B = 12.43;\ EI\theta_C = -15.43;\ EI\delta = 42$$

Final moments

$$M_{AB} = -3.75 + 0.5\ EI\theta_B - 0.375\ EI\delta$$
$$= -3.75 + 0.5 \times 12.43 - 0.375 \times 42$$

$\Rightarrow$ $$M_{AB} = -13.285 \text{ kN-m}$$
$$M_{BA} = 11.25 + 12.43 - 0.375 \times 42 = +7.93 \text{ kN-m}$$
$$M_{BC} = -15 + 1.5 \times 12.43 - 0.75 \times 15.43 = -7.93 \text{ kN-m}$$
$$M_{CB} = 45 + 0.75 \times 12.43 - 1.5 \times 15.43 = +31.18 \text{ kN-m}$$
$$M_{CD} = -15.43 - 0.375 \times 42 = -31.18 \text{ kN-m}$$
$$M_{DC} = -0.5 \times 15.43 - 0.375 \times 42 = -23.465 \text{ kN-m}$$

$$\text{Horizontal reaction at } A = \frac{M_{AB} + M_{BA} - 20 \times 1}{4} = \frac{-13.285 + 7.93 - 20}{4}$$
$$= -6.34 \text{ kN } (\rightarrow) = 6.34 \text{ kN } (\leftarrow)$$

$$\text{Horizontal reaction at } D = \frac{M_{CD} + M_{DC}}{4} = \frac{-31.18 - 23.465}{4}$$
$$= -13.66 \text{ kN } (\rightarrow) = 13.66 \text{ kN } (\leftarrow)$$

$$\text{Vertical reaction at } A = -\left(\frac{M_{BC} + M_{CB} - 80 \times 1}{4}\right) = -\left(\frac{-7.93 + 31.18 - 80}{4}\right)$$
$$= 14.1875 \text{ kN } (\uparrow)$$

$$\text{Vertical reaction at } D = \frac{M_{BC} + M_{CB} + 80 \times 3}{4} = \frac{-7.93 + 31.18 + 240}{4} = 65.8125 \text{ kN } (\uparrow)$$

Alternatively $$V_D = 80 - V_A = 80 - 14.1875 = 65.8125 \text{ kN}$$

Shear force and bending moment diagrams:

Taking outer surface of the portal frame as reference face.

$$\text{Maximum simply supported moment in } AB = \frac{20 \times 3 \times 1}{4} = 15 \text{ kN-m}$$

$$\text{Maximum simply supported moment in } BC = \frac{80 \times 3 \times 1}{4} = 60 \text{ kN-m}$$

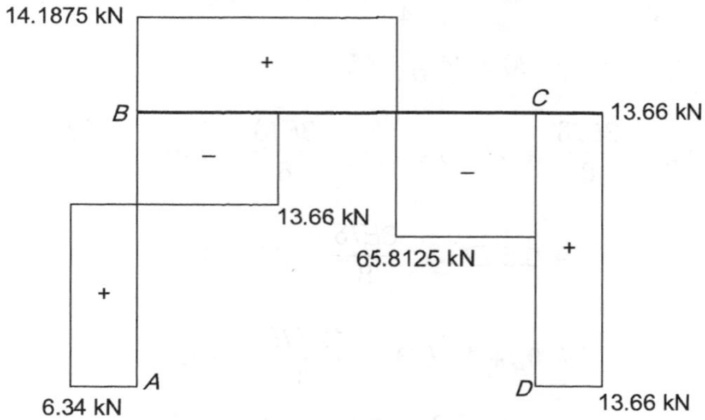

Shear Force Diagram

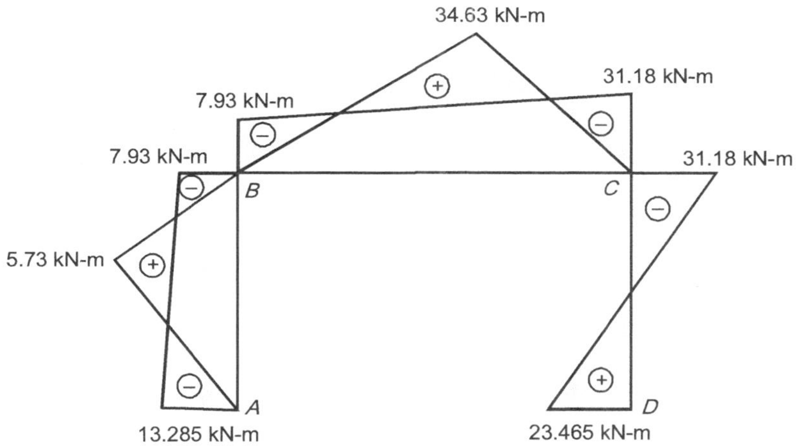

Bending Moment Diagram

4.12 Compare the force and displacement methods of structural analysis.

[10 marks : 2006]

Solution:

The points of comparison between force and displacement methods are:

Force Methods	Displacement Methods
1. In these methods forces (i.e. SF, BM, reactions) are taken as unknowns and equations are expressed in terms of these unknown forces.	1. In these methods displacements (i.e. Δ, θ) are taken as unknowns and equations are expressed in terms of these unknown forces.
2. Additional compatibility equations are required. Hence this method is also called compatibility method.	2. Additional equilibrium equations are written to find displacement components. Hence this method is also called equilibrium method
3. In the compatibility equations flexibilities of members appear, hence this method is also called flexibility method.	3. In this method stiffnesses of members appear, hence this method is also called stiffness method.
4. When degree of kinematic indeterminacy (D_k) is greater than degree of static indeterminacy (D_s), then the force methods are better.	4. When degree of kinematic indeterminacy (D_k) is less than degree of static indeterminacy (D_s), then displacement methods are better.
5. Examples of force methods are virtual work method. Clapeyron's three moment theorem, Castigliano's theorem, strain energy method, etc.	5. Examples of displacement methods are moment distribution method, slope deflection method, Kani's method etc.

4.13 Using the force (flexibility/compatibility) method, analyse the structure in the figure below :

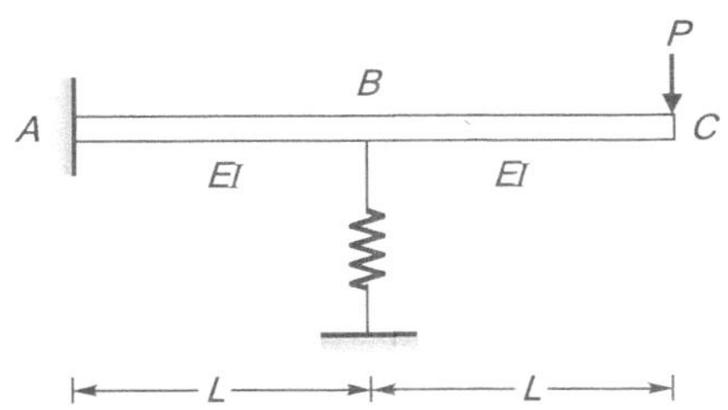

[20 marks : 2006]

Solution:

Let the stiffness in the spring be K.

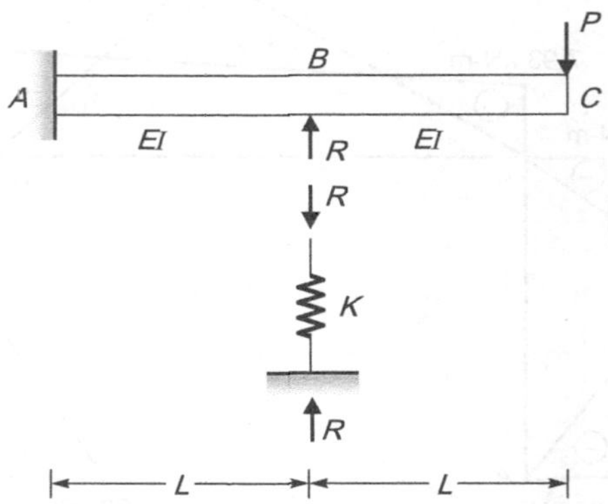

Deflection in spring, (compression) $\Delta_S = \dfrac{R}{K}$

Using compatibility condition,
Downward deflection of beam ABC at B = Deflection in spring. ...(i)

Downward deflection of beam ABC at B,

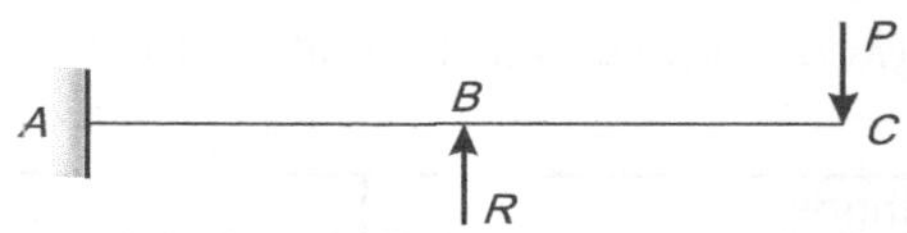

(Using Maxwell reciprocal theorem($\delta_C = \delta_B$)

$$\delta_{BB} = \frac{PL^3}{3EI} + \frac{PL^2}{2EI} \times L - \frac{RL^3}{3EI}$$

$$= \frac{PL^3}{3EI} + \frac{PL^3}{2EI} - \frac{RL^3}{3EI} = \frac{5PL^3}{6EI} - \frac{RL^3}{3EI}$$

By equation (i) from **compatibility condition,**

$$\frac{5PL^3}{6EI} - \frac{RL^3}{3EI} = \frac{R}{K}$$

$$\Rightarrow \quad \frac{5PL^3}{6EI} = \frac{R}{K} + \frac{RL^3}{3EI} = R\left[\frac{1}{K} + \frac{L^3}{3EI}\right] = \frac{R}{6EI}\left[\frac{6EI}{K} + 2L^3\right]$$

$$\Rightarrow \quad R = \frac{5PL^3}{\left(2L^3 + \dfrac{6EI}{K}\right)}$$

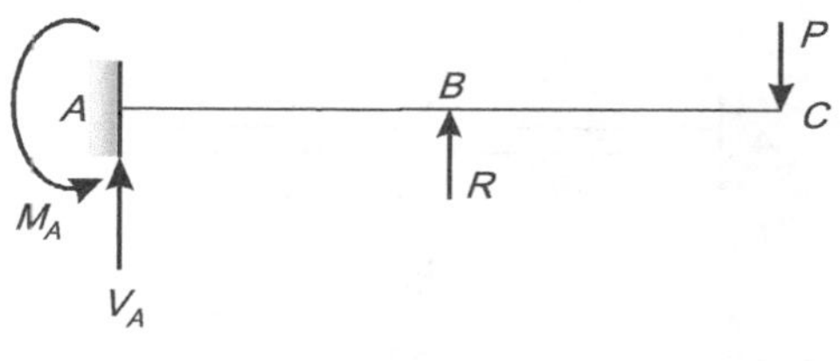

$\Sigma F_y = 0;$ $$V_A = P - R = P - \frac{5PL^3}{\left(\dfrac{6EI}{K} + 2L^3\right)}$$

$$\Sigma M_A = 0; \quad -M_A - R \times L + P \times 2L = 0$$

$$M_A = 2PL - RL$$

$$M_A = 2PL - \left[\frac{5PL^3}{\frac{6EI}{K} + 2L^3}\right]^2$$

$$M_A = \frac{\frac{12PLEI}{K} - PL^4}{\frac{6EI}{K} + 2L^3}$$

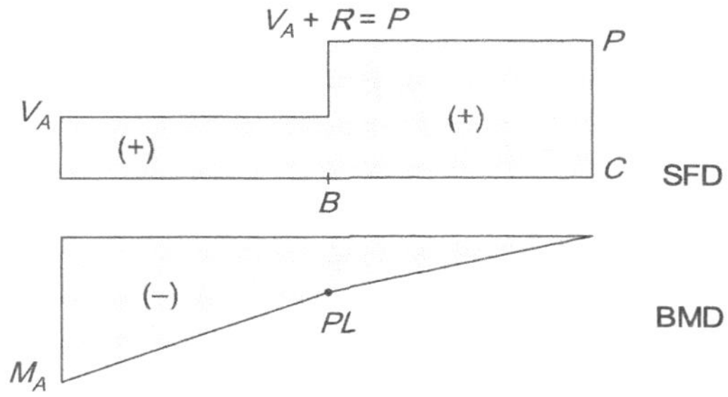

4.14 A rigid uniform straight rod *BC* of 2 kN weight and 2 m length is held in horizontal position by two inextensible flexible strings *AB* and *CD* as shown in the figure. Another 1 kN weight is also suspended on the string *AB* at *O*. Calculate the horizontal and vertical reaction components at the supports *A* and *D*, and the sag at *O*.

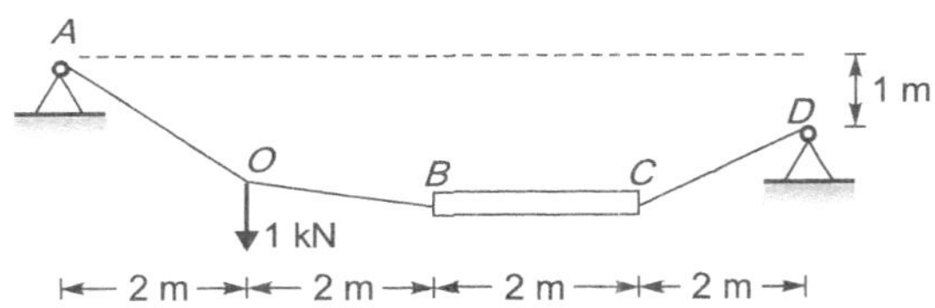

[10 marks : 2008]

Solution:

Let V_A and V_D be the vertical reactions at the supports *A* and *D*. Let *H* be the horizontal reaction at each support. Let θ_1, θ_2 and θ_3 be the inclinations of the different parts of the cable with the horizontal.

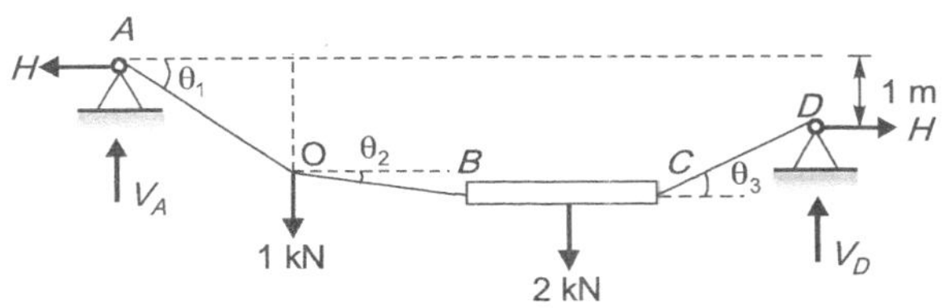

Calculation of reactions:

$$\Sigma F_y = 0$$

$$\Rightarrow \quad V_A + V_D - 1 - 2 = 0$$

$$\Rightarrow \quad V_A + V_D = 3 \quad \text{...(i)}$$

Taking moments about D, we get

$$V_A \times 8 - H \times 1 - 1 \times 6 - 2 \times 3 = 0$$

$$\Rightarrow \quad V_A = \frac{12 + H}{8} \quad \text{...(ii)}$$

Since bending moment at O is zero, taking moments about O from left, we get

$$V_A \times 2 - H \times 2\tan\theta_1 = 0$$

$$\Rightarrow \left(\frac{12+H}{8}\right) \times 2 - 2H\tan\theta_1 = 0 \qquad \text{[from (ii)]}$$

$$\Rightarrow \tan\theta_1 = \frac{12+H}{8H} \qquad \text{...(iii)}$$

Similarly, taking moments about C from right, we get

$$V_D \times 2 - H \times 2\tan\theta_3 = 0$$

$$\Rightarrow (3 - V_A) \times 2 - 2H\tan\theta_3 = 0 \qquad \text{[from (ii)]}$$

$$\Rightarrow \left[3 - \left(\frac{12+H}{8}\right)\right] \times 2 - 2H\tan\theta_3 = 0 \qquad \text{[from (ii)]}$$

$$\Rightarrow \left(\frac{12-H}{8}\right) \times 2 - 2H\tan\theta_3 = 0$$

$$\Rightarrow \tan\theta_3 = \frac{12-H}{8H} \qquad \text{...(iv)}$$

Similarly, taking moments about B from left, we get

$$V_A \times 4 - H \times (2\tan\theta_1 + 2\tan\theta_2) - 1 \times 2 = 0$$

$$\Rightarrow \left(\frac{12+H}{8}\right) \times 4 - 2H(\tan\theta_1 + \tan\theta_2) - 2 = 0$$

$$\Rightarrow \tan\theta_1 + \tan\theta_2 = \frac{12+H}{4H} - \frac{1}{H}$$

$$\Rightarrow \tan\theta_2 = \frac{12+H}{4H} - \frac{1}{H} - \frac{12+H}{8H} \qquad \text{[From (iii)]}$$

$$\Rightarrow \tan\theta_2 = \frac{12+H}{8H} - \frac{1}{H} \qquad \text{...(v)}$$

Now, we have

$$\Rightarrow 2\tan\theta_1 + 2\tan\theta_2 - 2\tan\theta_3 = 1$$

$$\Rightarrow 2 \times \left[\frac{12+H}{8H} + \frac{12+H}{8H} - \frac{1}{H} - \frac{12-H}{8H}\right] = 1$$

$$\Rightarrow 2 \times \left[\frac{12+H+12+H-8-12+H}{8H}\right] = 1$$

$$\Rightarrow 3H + 4 = 4H$$

$$\Rightarrow H = 4 \text{ kN}$$

From (ii) $$V_A = \frac{12+H}{8} = \frac{12+4}{8} = 2 \text{ kN}$$

From (i) $$V_D = 3 - V_A = 3 - 2 = 1 \text{ kN}$$

$$\text{The sag at } O = 2\tan\theta_1 = 2 \times \frac{12+H}{8H} = 2 \times \frac{12+4}{8 \times 4} = 1 \text{ m}$$

4.15 Analyse the frame *ABCD* shown in figure using moment distribution method. The geometry of the frame, loading and properties are shown in the figure. Also draw the bending moment diagram.

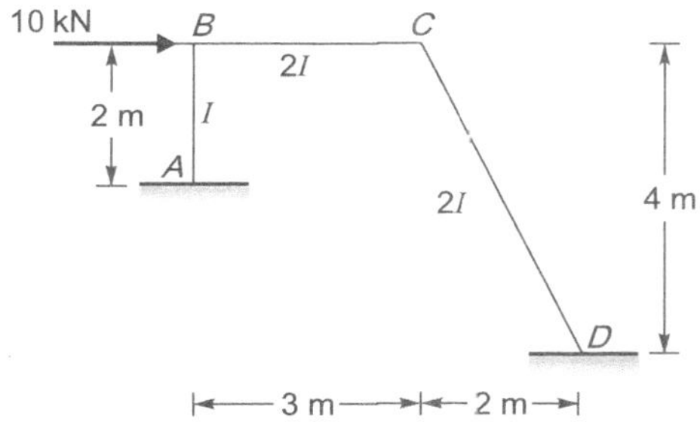

[20 marks : 2008]

Solution:

There are no external loads acting on the portal frame except the sway force of 10 kN from left to right. Hence only the sway analysis is to be carried out.

Distribution Factors

Joint	Member	Relative Stiffness	Total Relative Stiffness	Distribution Factor
B	BA	$\frac{I}{2}$	$\frac{7I}{6}$	$\frac{3}{7} = 0.43$
	BC	$\frac{2I}{3}$		$\frac{4}{7} = 0.57$
C	CB	$\frac{2I}{3}$	$\frac{(2\sqrt{5}+3)}{3\sqrt{5}}$	0.6
	CD	$\frac{2I}{2\sqrt{5}}$		0.4

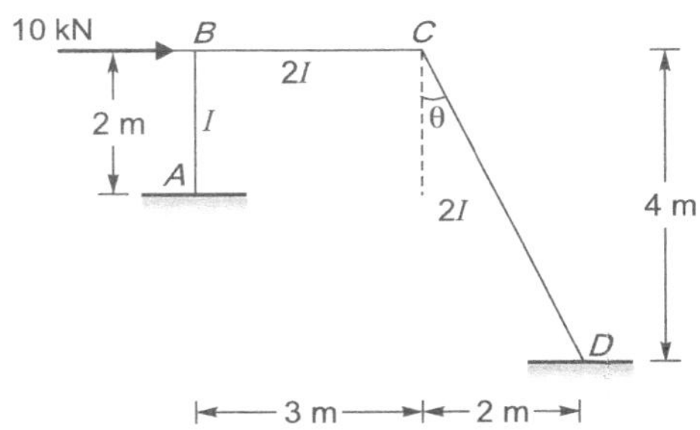

Let member *CD* makes an angle θ with the vertical

$$\therefore \quad \text{Length of } CD = \sqrt{2^2+4^2} = 2\sqrt{5}\text{ m}$$

$$\sin\theta = \frac{2}{2\sqrt{5}} = \frac{1}{\sqrt{5}}$$

$$\cos\theta = \frac{4}{2\sqrt{5}} = \frac{2}{\sqrt{5}}$$

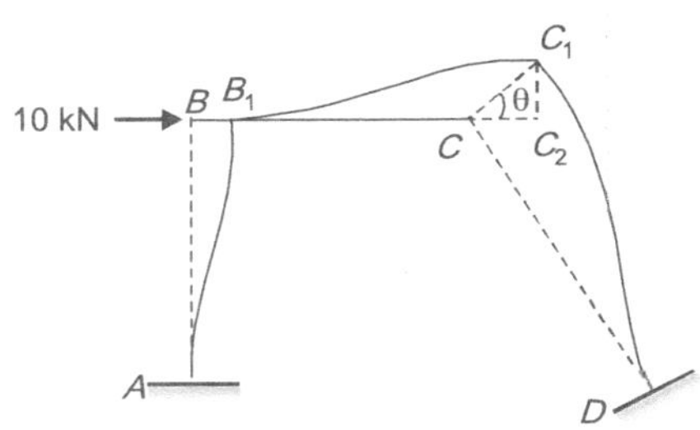

Let the frame *ABCD* deflect to the position AB_1C_1D due to the sway force.

Let $\quad BB_1 = \delta$

$\therefore \quad CC_2 = \delta$

In ΔCC_1C_2, $\quad \cos\theta = \frac{CC_2}{CC_1}$

$$\Rightarrow \quad CC_1 = \frac{\delta}{2/\sqrt{5}}$$

$$\Rightarrow \quad CC_1 = \frac{\sqrt{5}\delta}{2}$$

Again, in $\Delta\, CC_1C_2$, $\quad \sin\theta = \dfrac{C_1C_2}{CC_1}$

$\Rightarrow \quad C_1C_2 = \dfrac{1}{\sqrt{5}} \times \dfrac{\sqrt{5}\delta}{2}$

$\Rightarrow \quad C_1C_2 = \dfrac{\delta}{2}$

Fixed end Moments, $\quad \bar{M}_{BA} = \dfrac{-6EI\delta}{AB^2} = \dfrac{-6EI\delta}{(2)^2} = \dfrac{-6EI\delta}{4} = \bar{M}_{BA}$

$$\bar{M}_{BC} = \frac{6E(2I)\times\frac{\delta}{2}}{BC^2} = \frac{6E(2I)\times\frac{\delta}{2}}{(3)^2} = \frac{6EI\delta}{9} = \bar{M}_{CB}$$

$$\bar{M}_{CD} = -\frac{6E(2I)\times\frac{\sqrt{5}\delta}{2}}{CD^2} = \frac{-6\sqrt{5}\,EI\delta}{(2\sqrt{5})^2} = \frac{-6\sqrt{5}\,EI\delta}{20} = \bar{M}_{DC}$$

$\therefore$ Ratio of fixed end moments for AB, BC and CD

$$= \frac{-6}{4} : +\frac{6}{9} : -\frac{6\sqrt{5}}{20}$$

$$= -45 : +20 : -9\sqrt{5}$$

$$= -45 : +20 : -20.12$$

	A	B	B	C	C	D
		0.43	0.57	0.6	0.4	
	–45	–45	+20	+20	–20.12	–20.12
		+10.75	+14.25	+0.072	+0.048	
	+5.375		+0.036	+7.125		+0.024
		–0.015	–0.021	–4.275	–2.85	
	–0.007		–2.14	–0.011		–1.425
		+0.92	+1.22	+0.0066	+0.0044	
	+0.46		+0.0033	+0.61		+0.0022
		–0.0014	–0.0019	–0.366	–0.244	
col. (a)	–39.17	–33.35	+33.35	+23.16	–23.16	–21.52

Let the moments in col. (a) are induced due to a sway force S

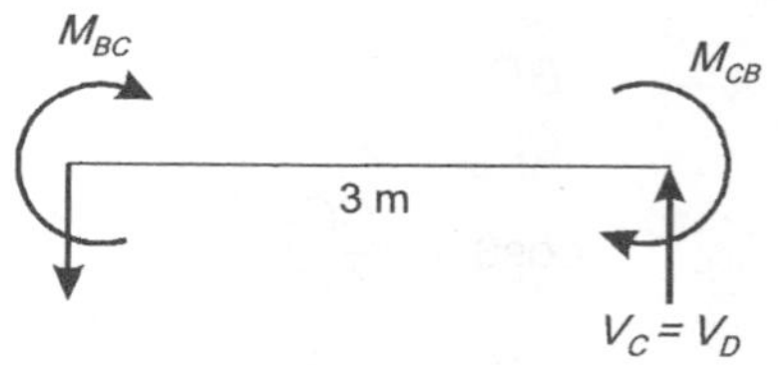

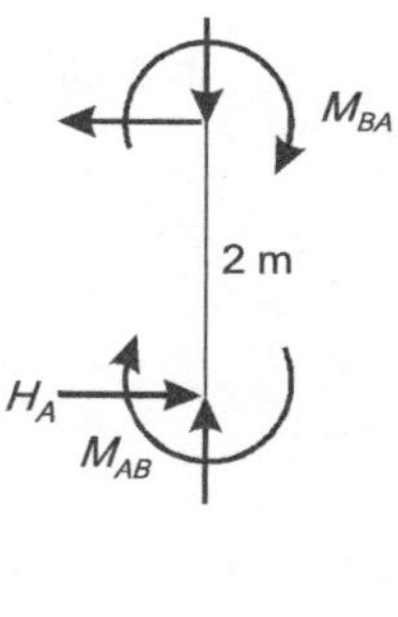

$$\text{Vertical reaction at } D = \frac{M_{BC} + M_{CB}}{L_{BC}}$$

$$= \frac{33.35 + 23.16}{3} = 18.84 \text{ kN } (\uparrow)$$

$$\text{Horizontal reaction at } A = \frac{M_{BA} + M_{AB}}{L_{AB}} = \frac{-39.17 - 33.35}{2}$$

$$= -36.26 \text{ kN } (\rightarrow) = 36.26 \text{ kN } (\leftarrow)$$

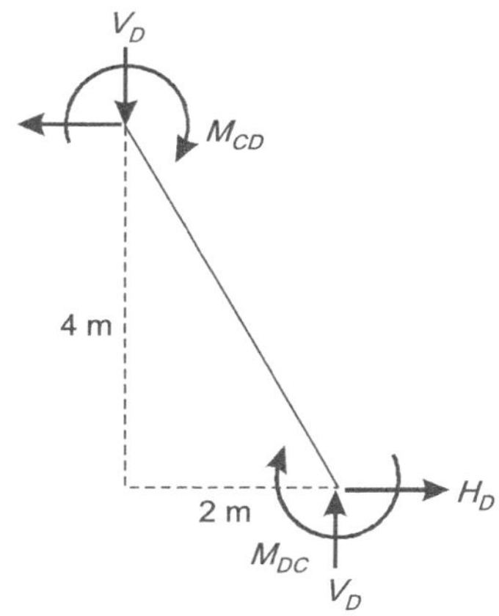

$$\text{Horizontal reaction at } D = \frac{M_{CD} + M_{DC} - V_D \times 2}{4}$$

$$= \frac{-23.16 - 21.52 - 18.84 \times 2}{4}$$

$$= -20.59 \text{ kN } (\rightarrow) = 20.59 \text{ kN } (\leftarrow)$$

Let the sway force S be acting from left to right, then

$$H_A + H_D + S = 0$$

$$\Rightarrow \quad -36.26 - 20.59 + S = 0$$

$$\Rightarrow \quad S = 56.85 \text{ kN } (\rightarrow)$$

Thus for a sway force of 56.85 kN, the sway moments are as in col. (a)

$\therefore$ For the actual sway force of 10 kN, the corresponding sway moments will be

$$= \frac{10}{56.85} \times \text{col. (a) moments}$$

Actual Sway Moments

$$M_{AB} = \frac{10}{56.85} \times (-39.17) = -6.89 \text{ kN-m}$$

$$M_{BA} = \frac{10}{56.85} \times (-33.35) = -5.87 \text{ kN-m}$$

$$M_{BC} = \frac{10}{56.85} \times 33.35 = +5.87 \text{ kN-m}$$

$$M_{CB} = \frac{10}{56.85} \times 23.16 = +4.07 \text{ kN-m}$$

$$M_{CD} = \frac{10}{56.85} \times (-23.16) = -4.07 \text{ kN-m}$$

$$M_{DC} = \frac{10}{56.85} \times (-21.52) = -3.78 \text{ kN-m}$$

$$\text{Vertical reaction at } D = \frac{5.87 + 4.07}{3} = 3.31 \text{ kN } (\uparrow)$$

$$\text{Vertical reaction at } A = -\left(\frac{5.87 + 4.07}{3}\right) = -3.31 \text{ kN } (\uparrow) = 3.31 \text{ kN } (\downarrow)$$

$$\text{Horizontal reaction at } A = \frac{-6.89 - 5.87}{2} = -6.38 \text{ kN } (\rightarrow) = 6.38 \text{ kN } (\leftarrow)$$

$$\text{Horizontal reaction at } D = \frac{-4.07 - 3.78 - 3.31 \times 2}{4} = -3.62 \text{ kN } (\rightarrow) = 3.62 \text{ kN } (\leftarrow)$$

Bending Moment Diagram

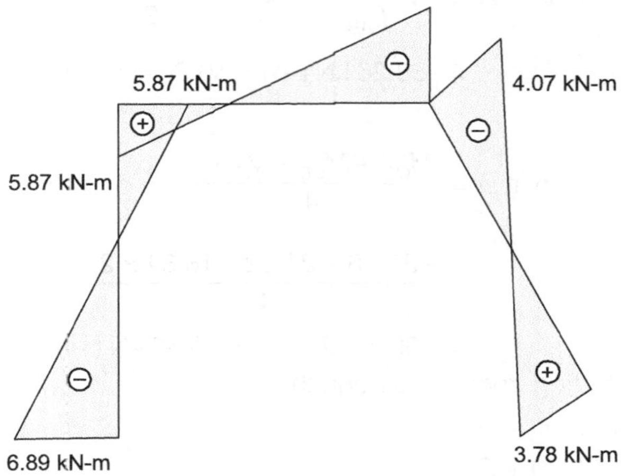

4.16 Employing slope-deflection method, evaluate BM at salient points of the frame shown below. Draw BM diagram indicating BM values on it.

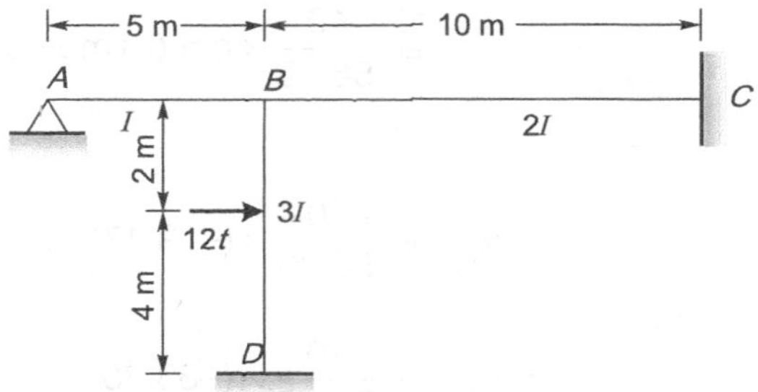

[18 marks : 2010]

Solution:

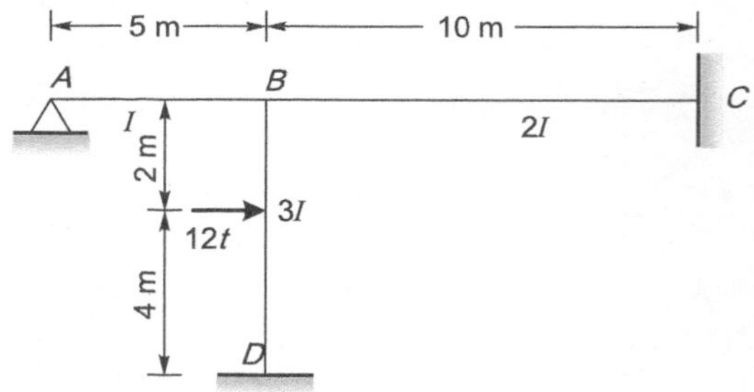

D_K = Kinetic degree of indeterminacy

$= 2\ (\theta_A, \theta_B)$

Assumptions:

1. 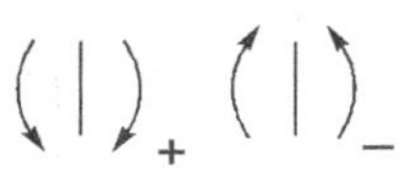
2. Top face is the reference face
3. All members are inextensible axially from figure as it is clear that sway will be zero.

Fixed end moments

$$M_{fAB} = 0 = M_{fBA}$$

$$M_{fBC} = 0 = M_{fCB}$$

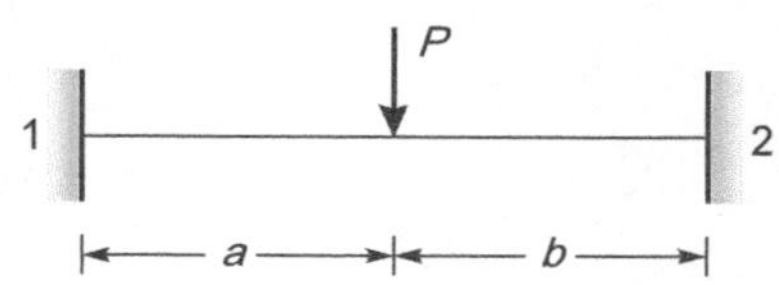

$$\therefore \quad M_{12} = -\frac{Pab^2}{L^2} \ (\circlearrowleft)$$

and $$M_{21} = \frac{Pba^2}{L^2} \ (\circlearrowright)$$

where $L = a + b$

$$M_{fBD} = -\frac{12 \times 2 \times 4^2}{6^2} = -10.67 \text{ kNm}$$

and
$$M_{fDB} = \frac{12 \times 4 \times 2^2}{6^2} = 5.33 \text{ kNm}$$

$$M_{AB} = \frac{4EI}{5}\theta_A + \frac{2EI}{5}\theta_B \quad \text{...(i)}$$

[Rotation are all assumed in (↻) direction]

$$M_{BA} = \frac{2EI}{5}\theta_A + \frac{4EI}{5}\theta_B \quad \text{...(ii)}$$

$$M_{BC} = \frac{8EI}{10}\theta_B + \frac{4EI}{10}\theta_C \quad \text{...(iii)}$$

$$M_{CB} = \frac{4EI}{10}\theta_B \quad (\because \theta_c = 0) \quad \text{...(iv)}$$

$$M_{BD} = -10.67 + \frac{12EI}{6}\theta_B \quad \text{...(v)}$$

$$M_{DB} = +5.33 + \frac{6EI}{6}\theta_B \quad \text{...(vi)}$$

Now applying joint equilibrium equation

$$M_{BD} + M_{BA} + M_{BC} = 0$$

$$\frac{2EI}{5}\theta_A + \frac{18}{5}EI\theta_B = 10.67 \quad \text{...(vii)}$$

Also
$$M_{AB} = 0 \text{ (simple pin support)}$$

$$\theta_B = -2\theta_A \quad \text{...(viii)}$$

$$\frac{17}{5}EI\theta_B = 10.67$$

$\Rightarrow$
$$\theta_B = \frac{3.1382}{EI}$$

We get,
$M_{BA} = 1.883$ kNm $M_{BC} = 2.51$ kNm

$M_{BD} = -4.3936$ kNm $M_{CB} = 1.2553$ kNm

$M_{DB} = 8.468$ kNm

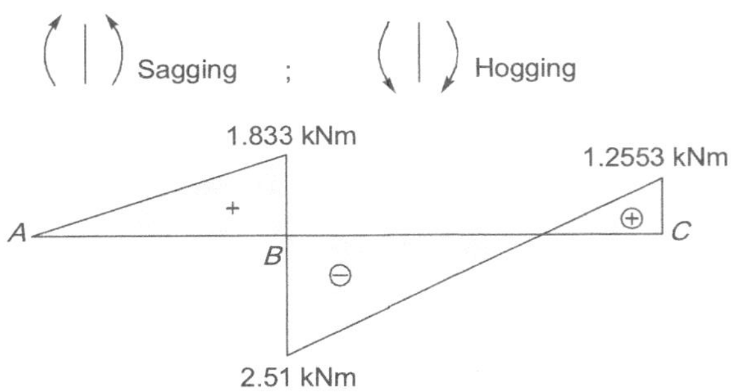

In span *BD* assuming simple supported maximum moment will be at loading point

$$M_{max} = \frac{Pab}{L} = \frac{12 \times 2 \times 4}{6} = 16 \text{ kNm}$$

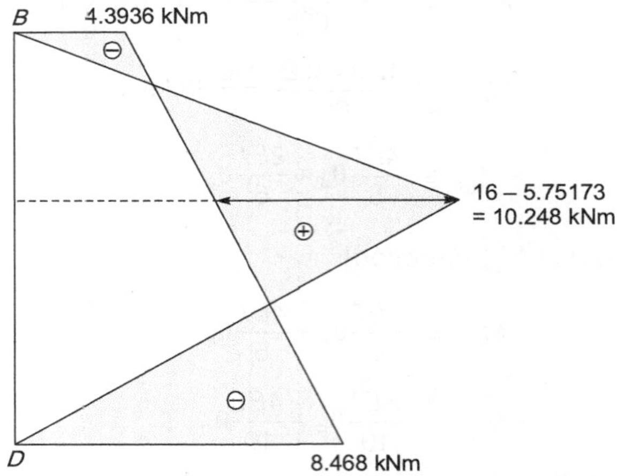

4.17 Analyse the continuous beam shown in the figure below by slope deflection method. Support B settles down by 5 mm.
$E = 2 \times 10^5$ N/mm², $I = 36 \times 10^6$ N/mm⁴

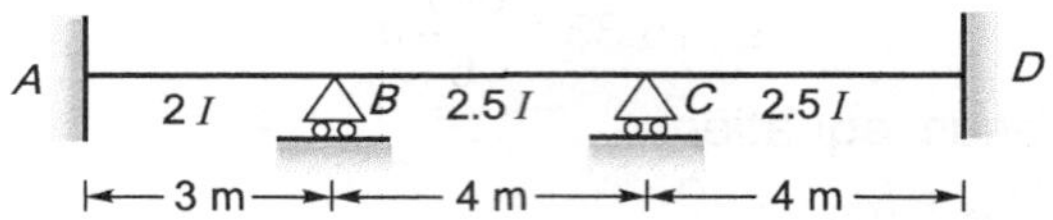

[10 marks : 2011]

Solution:

Given:

Support B settles down by 5 mm

$$E = 2 \times 10^5 \text{ N/mm}^2$$

$$I = 36 \times 10^6 \text{ mm}^4$$

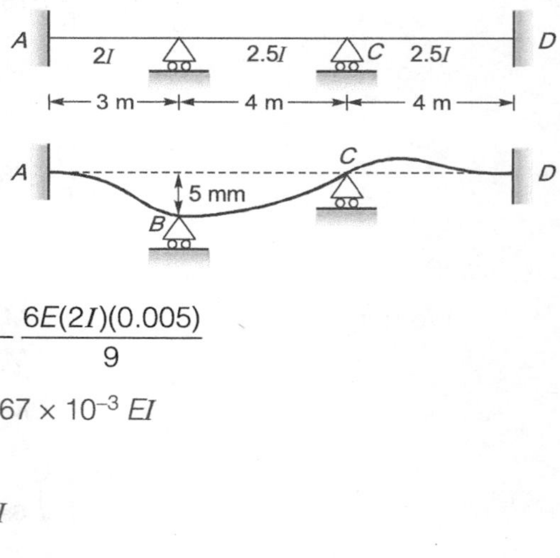

Using slope-deflection equations:

For AB

$$M_{AB} = \frac{2E(2I)(2\theta_A)}{3} + \frac{2E(2I)\theta_B}{3} - \frac{6E(2I)(0.005)}{9}$$

$$= 2.67\theta_A EI + 1.33\, EI\theta_B - 6.67 \times 10^{-3}\, EI$$

Since $\theta_A = 0$

$$M_{AB} = 1.33\, EI\theta_B - 6.67 \times 10^{-3}\, EI$$

For BA

$$M_{BA} = \frac{2E(2I)(2\theta_B)}{3} + \frac{2E(2I)\theta_A}{3} - \frac{6E(2I)(0.005)}{9}$$

$$= 2.67\, \theta_B EI - 6.67 \times 10^{-3}\, EI$$

For BC

$$M_{BC} = \frac{2E \times 2.5I(2\theta_B)}{4} + \frac{2E \times 2.5I\, \theta_C}{4} + \frac{6E(2.5I)(0.005)}{16}$$

$$= 2.5EI\theta_B + 1.25\, EI\theta_C + 4.6875\, EI \times 10^{-3}$$

$$\Sigma M_B = M_{BA} + M_{BC}$$

$$= 5.17EI\theta_B + 1.25\, EI\theta_C - 1.9825 \times 10^{-3}\, EI = 0 \quad \text{...(i)}$$

Applying, compatibility condition

For *CB*
$$M_{CB} = \frac{2E \times 2.5I(2\theta_C)}{4} + \frac{2E(2.5I)\theta_B}{4} + 4.6875 \times 10^{-3} EI$$
$$= 2.5\, EI\theta_C + 1.25\, EI\theta_B + 4.6875 \times 10^{-3} EI$$

For *CD*
$$M_{CD} = \frac{2E(2.5I)(2\theta_C)}{4} = 2.5\, EI\theta_C$$
$$M_{DC} = \frac{2E(2.5I)\theta_C}{4} = 1.25\, EI\theta_C$$
$$\Sigma M_C = M_{CB} + M_{CD} = 0$$
$$\Sigma M_C = 5EI\theta_C + 1.25\, EI\theta_B + 4.6875 \times 10^{-3}\, EI = 0 \quad \text{...(ii)}$$

By solving equation (i) and (ii)
$$\theta_B = 6.49 \times 10^{-4}$$
$$\theta_C = -1.099 \times 10^{-3}$$

by putting the value of θ_B and θ_C in the equation of moments we get
$$M_{AB} = -5.8 \times 10^{-3}\, EI$$
$$= -5.8 \times 10^{-3} \times 2 \times 10^5 \text{ N/mm}^2 \times 36 \times 10^6 \text{ mm}^4\text{/m}$$
$$= -5.8 \times 10^{-3} \times 2 \times 10^5 \times 36 \times 10^6 \times 10^{-6} \text{ N-m}$$
$$= -41.76 \text{ kN-m}$$
$$M_{BA} = -4.93 \times 10^3 \times 2 \times 10^5 \text{ N/mm}^2 \times 36 \times 10^6 \text{ mm/m}$$
$$= -4.93 \times 10^{-3} \times 2 \times 10^5 \times 36 \times 10^6 \times 10^{-6} \text{ N-m}$$
$$= -35.49 \text{ kN-m}$$
$$M_{BC} = 35.49 \text{ kN-m} \qquad M_{CB} = 19.8 \text{ kN-m}$$
$$M_{CD} = -19.8 \text{ kN-m} \qquad M_{DC} = -9.864 \text{ kN-m}$$

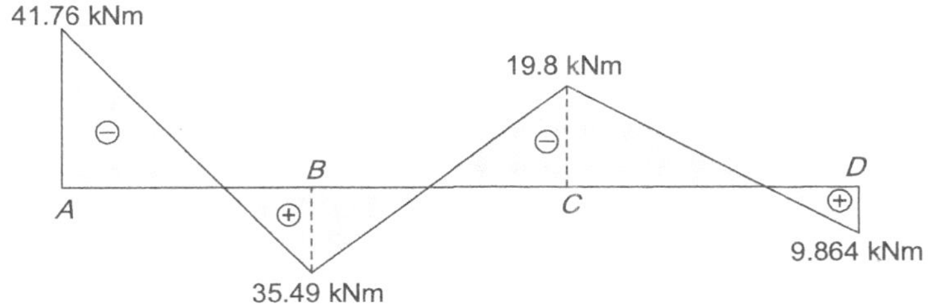

Bending Moment Diagram

4.18 Analyse the frame shown above by compatibility method. *EI* is constant. Draw *BM* and *SF* diagrams.

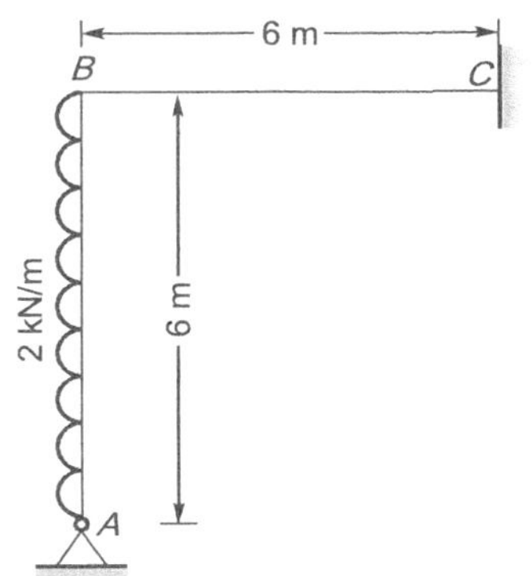

[25 marks : 2012]

Solution:

Let the vertical reactions at A and C be $V_A(\uparrow)$ and $V_C(\uparrow)$

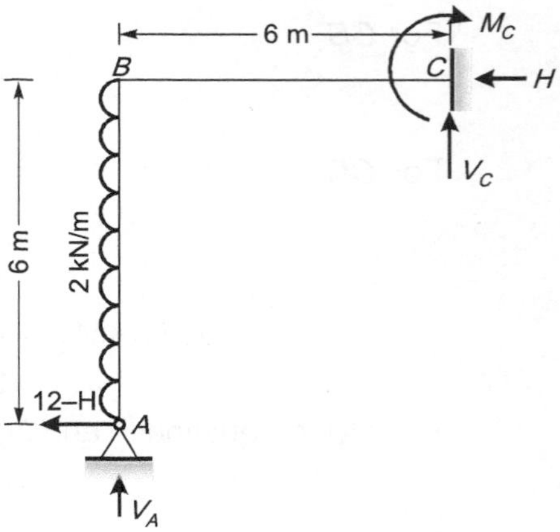

Calculation of reactions

$$\Sigma F_y = 0$$

$$\Rightarrow \quad V_A + V_C = 0$$

$$\Rightarrow \quad V_A = -V_C$$

$$\Sigma F_x = 0$$

Let $\quad H_C = H$

$$\Rightarrow \quad H_A - 2 \times 6 + H = 0$$

$$\Rightarrow \quad H_A = 12 - H$$

$$\Sigma M_A = 0$$

$$\Rightarrow \quad M_C - 6 \times V_C - H \times 6 + 2 \times 6 \times 3 = 0$$

$$\Rightarrow \quad M_C - 6\,V_C - 6H + 36 = 0$$

$$\Rightarrow \quad M_C + 36 - 6H = 6\,V_C$$

$$\Rightarrow \quad V_C = 6 - H + \frac{M_C}{6}$$

$$\therefore \quad V_A = -V_C = H - \frac{M_C}{6} - 6$$

Let redundants are H and M_C

Now as we know $\quad U = \int \frac{M^2 dS}{2EI}$

The value of H and M_C will be such that the total strain energy is minimum i.e.,

$$\Rightarrow \quad \frac{\partial U}{\partial H} = 0$$

$$\Rightarrow \quad \frac{\partial U}{\partial H} = \int \frac{2M \frac{\partial M}{\partial H} dS}{2EI} = 0$$

$$\Rightarrow \quad \frac{\partial U}{\partial H} = \Sigma \int \frac{M \frac{\partial M}{\partial H} dS}{EI} = 0 \quad \text{...(i)}$$

Similarly $\quad \frac{\partial U}{\partial M_C} = 0$

$$\Rightarrow \quad \Sigma \int \frac{M \frac{\partial M}{\partial M_C} dS}{EI} = 0 \quad \text{...(ii)}$$

The bending moment M at any section and the corresponding values of $\frac{\partial M}{\partial H}$ and $\frac{\partial M}{\partial M_C}$ of the frame are tabulated below:

Member	M	$\frac{\partial M}{\partial H}$	$\frac{\partial M}{\partial M_C}$	Limit of integration
AB	$(12-H)y - 2y\frac{y}{2}$	$-y$	0	0 *to* 6
CB	$\left(6 - H + \frac{M_C}{6}\right)x - M_C$	$-x$	$\frac{x}{6} - 1$	0 *to* 6

From eq. (i) we get,

$$\Rightarrow \int_0^6 \frac{\left[(12-H)y - 2y\frac{y}{2}\right](-y)dy}{EI} + \frac{\int_0^6\left\{\left(6-H+\frac{M_C}{6}\right)x - M_C\right\}(-x)(dx)}{EI} = 0$$

$$\Rightarrow \int_0^6\left(12y - Hy - y^2\right)(-y)dy + \int_0^6\left(6x - Hx + \frac{M_c x}{6} - M_c\right)(-x)dx = 0$$

$$\Rightarrow \int_0^6\left(-12y^2 + Hy^2 + y^3\right)dy + \int_0^6\left(-6x^2 + Hx^2 - \frac{M_C x^2}{6} + M_c.x\right)(dx) = 0$$

$$\Rightarrow \left[-12\frac{y^3}{3} + \frac{Hy^3}{3} + \frac{y^4}{4}\right]_0^6 + \left[-6\frac{x^3}{3} + \frac{Hx^3}{3} - \frac{M_c x^3}{18} + \frac{M_C x^2}{2}\right]_0^6 (dx) = 0$$

$$\Rightarrow -12\times\frac{6^3}{3} + H\times\frac{6^3}{3} + \frac{6^4}{4} - \frac{6\times 6^3}{3} + \frac{H\times 6^3}{3} - \frac{M_C\times 6^3}{18} + \frac{M_C\times 6^2}{2} = 0$$

$$\Rightarrow -864 + 72H + 324 - 432 + 72H - 12M_C + 18M_C = 0$$

$$\Rightarrow 144H + 6M_C = 972$$

$$\Rightarrow 24H + M_C = 162 \quad \text{...(iii)}$$

From eq. (ii) we get,

$$\Rightarrow \int_0^6 \frac{\left[(12-H)y - 2y\frac{y}{2}\right](0)dy}{EI} + \frac{\left[\int_0^6\left\{\left(6-H+\frac{M_C}{6}\right)x - M_C\right\}\left(\frac{x}{6}-1\right)(dx)\right]}{EI} = 0$$

$$\Rightarrow \int_0^6 \frac{\left(6x - Hx + \frac{M_C x}{6} - M_C\right)\left(\frac{x}{6}-1\right)}{EI}dx = 0$$

$$\Rightarrow \int_0^6\left(x^2 - \frac{Hx^2}{6} + \frac{M_C x^2}{36} - \frac{M_C x}{6} - 6x + Hx - \frac{M_C x}{6} + M_C\right)dx = 0$$

$$\Rightarrow \left[\frac{x^3}{3} - \frac{Hx^3}{18} + \frac{M_c x^3}{108} - \frac{M_c x^2}{12} - \frac{6x^2}{2} + \frac{Hx^2}{2} - \frac{M_C x^2}{12} + M_C x\right]_0^6 = 0$$

$$\Rightarrow \left[\frac{6^3}{3} - \frac{H\times 6^3}{18} + \frac{M_C\times 6^3}{108} - \frac{M_C\times 6^2}{12} - \frac{6\times 6^2}{2} + \frac{H\times 6^2}{2} - \frac{M_C\times 6^2}{12} + 6M_C\right] = 0$$

$$\Rightarrow 72 - 12H + 2M_C - 3M_C - 108 + 18H - 3M_C + 6M_C = 0$$

$$\Rightarrow -36 + 6H + 2M_C = 0$$

$$\Rightarrow M_C + 3H = 18 \quad \text{...(iv)}$$

Solving (iii) and (iv) we get

$$H = 6.8571 \text{ kN}$$

$$M_C = -2.571 \text{ kN-m}$$

$$\therefore \quad H_C = H = 6.8571 \text{ kN}$$

$$H_A = 12 - H = 5.1429 \text{ kN}$$

$$V_A = H - \frac{M_C}{6} - 6 = 6.8571 + \frac{2.571}{6} - 6 = 1.2856 \text{ kN}$$

$$V_C = 6 - H + \frac{M_C}{6} = 6 - 6.8571 + \frac{(-2.571)}{6} = -1.2856 \text{ kN}$$

Bending moment diagram

Portion *AB*

$$M_y (y \text{ from } A) = H_A y - 2y\frac{y}{2}$$

$$= 5.1429\,y - y^2$$

At y = 0, $M_y = M_A = 0$

$$M_B = 5.1429 \times 6 - 6^2 = -5.143 \text{ kNm}$$

For M_{max}, $\dfrac{dM_y}{dy} = 0$

$\Rightarrow$ $5.1429 - 2y = 0$

$\Rightarrow$ $y = 2.571$ m

$\therefore$ $M_{max} = 5.1429 \times 2.571 - (2.571)^2 = 6.61$ kN-m

For $M_y = 0$, we have

$$5.1429y - y^2 = 0$$

$\Rightarrow$ $y(5.1429 - y) = 0$

$\Rightarrow$ $y = 0$ and $y = 5.1429$ m

Portion *CB*

$$M_x = V_C x - M_C$$

$$= -1.2856\,x + 2.571$$

$$M_C = 2.571 \text{ kN-m}$$

$$M_B = -5.143 \text{ kN-m}$$

For $M_x = 0$, we have

$$-1.2856\,x + 2.571 = 0$$

$\Rightarrow$ $x = 1.999 \text{ m} \simeq 2\text{m}$

Taking outer face as reference face

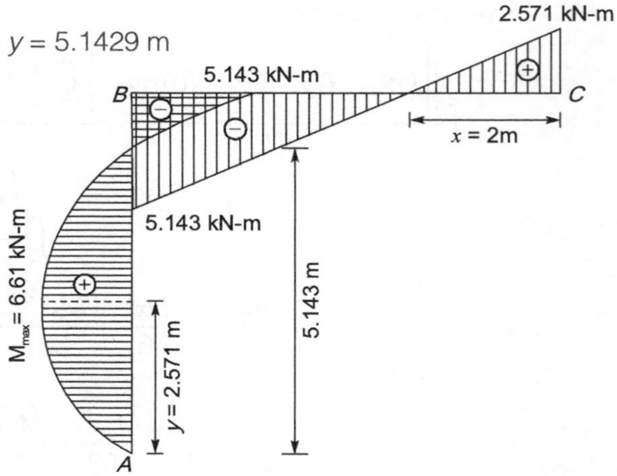

BMD

Shear force diagram

Portion *AB*

Shear force in $AB = H_A - 2y$

$$= 5.1429 - 2y$$

At $y = 0$, $SF_A = 5.1429$ kN

At $y = 6$ m, $SF_B = -6.8571$ kN

For *SF* to be zero

$\Rightarrow$ $5.1429 - 2y = 0$

$\Rightarrow$ $y = 2.571$ m

Portion *CB*

$$SF = -V_C$$

$$= 1.2856 \text{ kN}$$

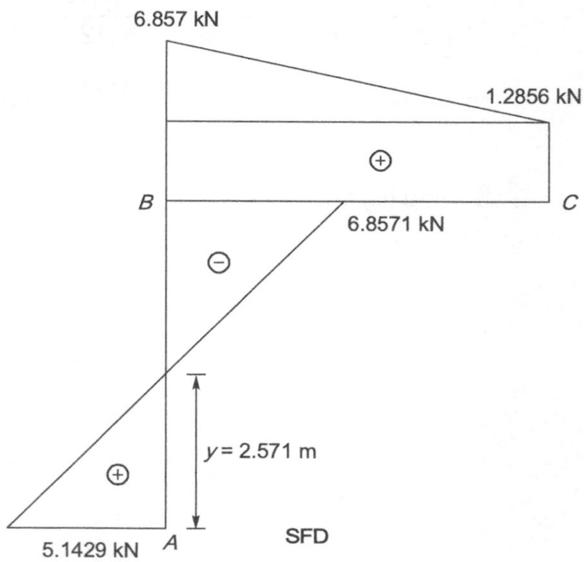

SFD

4.18 Analyse the rigid frame shown in the below figure by moment distribution method, taking flexural rigidity EI to be uniform for all members.

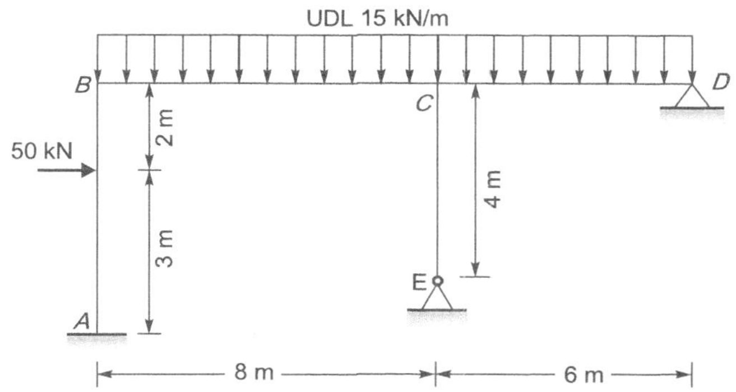

[20 marks : 2012]

Solution:

Analysing the given frame using moment distribution method.

(i) Distribution factors

Joint	Member	Relative Stiffness	Total R.S.	DF
B	BA	$\frac{I}{5}=\frac{8I}{40}$	$\frac{13I}{40}$	$\frac{8}{13}$
	BC	$\frac{I}{8}=\frac{5I}{40}$		$\frac{5}{13}$
C	CB	$\frac{I}{8}=\frac{2I}{16}$	$\frac{7I}{16}$	$\frac{2}{7}$
	CD	$\frac{3}{4}\frac{I}{6}=\frac{2I}{16}$		$\frac{2}{7}$
	CE	$\frac{3}{4}\frac{I}{4}=\frac{3I}{16}$		$\frac{3}{7}$

(ii) Fixed end moments

$$M_{\overline{AB}} = -\frac{50\times3\times2^2}{5^2} = -24 \text{ kNm}$$

$$M_{\overline{BA}} = \frac{50\times2\times3^2}{5^2} = 36\text{kNm}$$

$$M_{\overline{BC}} = -\frac{15\times8^2}{12} = -80\text{kNm}$$

$$M_{\overline{CB}} = \frac{15\times8^2}{12} = 80 \text{ kNm}$$

$$M_{\overline{CD}} = -\frac{15\times6^2}{12} = -45\text{kNm}$$

$$M_{\overline{DC}} = \frac{15\times6^2}{12} = 45\text{kNm}$$

A	B (8/13)	B (5/13)	C (2/7)	C (2/7)	D	C (3/7)	E
–24	+36	–80	+80	–45	45		
					–45		
				–22.5			
–24	+36	–80	+80	–67.5	0		
	27.08	+16.92	–3.57	–3.57		–5.36	
13.54		–1.785	8.46		0		0
	1.098	0.687	–2.417	–2.417		–3.626	
0.549		–1.2085	0.3435		0		0
	0.744	0.464	–0.098	–0.098		–0.1472	
0.372					0		0
–9.54	64.922	–64.922	82.71	–73.58	0	–9.13	

(iii) Bending moment diagram

Taking top face of the beam as reference face and outer face of the columns as reference face.

Taking simply supported span *AB*

$$\text{Maximum moment under the load} = \frac{50 \times 3 \times 2}{5} = 60 \text{ kNm(sagging)}$$

$$\text{Fixed moment under the load} = 9.54 + \left(\frac{64.922 - 9.54}{5}\right) \times 3 = 42.76925 \text{ kNm (hogging)}$$

$\therefore$ Resultant moment under the load = 60 – 42.7692 = 17.2308 kNm (sagging)

Taking simply supported span *BC*

$$\text{Maximum moment} = \frac{15 \times 8^2}{8} = 120 \text{ kNm(sagging)}$$

$$\text{Fixed moment at centre} = 64.922 + \left(\frac{82.71 - 64.922}{8}\right) \times 4 = 73.816 \text{ (hogging)}$$

$\therefore$ Resultant moment at the centre of span *BC* = 120 – 73.816 = 46.184 kNm (sagging)

Taking simply supported span *CD*

Maximum moment = 67.5 kNm (sagging)

$$\text{Fixed moment at the centre} = 73.58 - \frac{73.58}{2} = 36.79 \text{ kNm(hogging)}$$

$\therefore$ Resultant moment at the centre of span *CD* = 67.5 – 36.79 = 30.71 kNm (sagging)

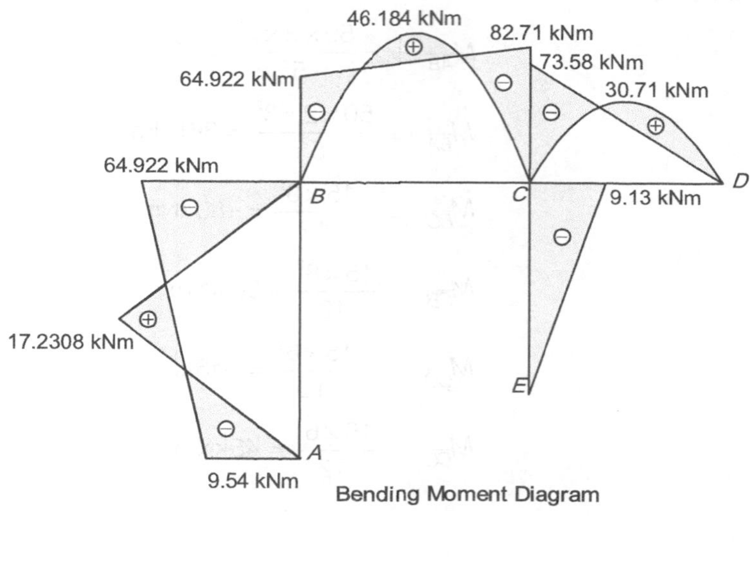

Bending Moment Diagram

4.19 Find the end moments of the frame shown below. Use Moment Distribution method. Also draw BMD.

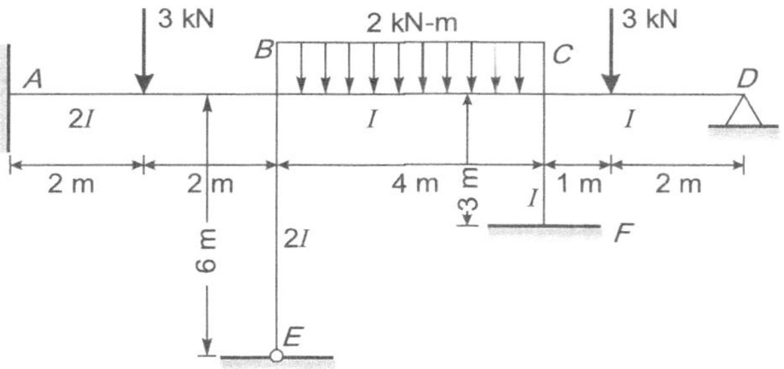

[20 marks : 2013]

Solution:

(i) Fixed End Moment

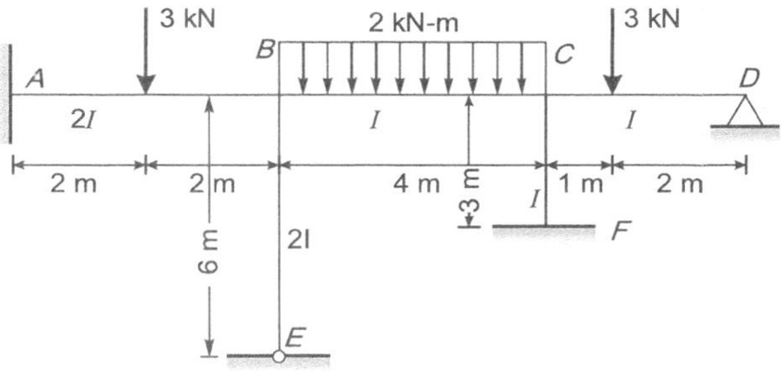

$$M_{FAB} = \frac{-3\times 4}{8} = -1.5 \text{ kNm}$$

$$M_{FBA} = 1.5 \text{ kNm}$$

$$M_{FBE} = 0$$

$$M_{FBC} = -\frac{wl^2}{12} = \frac{-2\times 4^2}{12} = -2.67 \text{ kNm}$$

$$M_{FCB} = 2.67 \text{ kNm}$$

$$M_{FCD} = \frac{-3\times 1\times 2^2}{3^2} = -1.33 \text{ kNm}$$

$$M_{FDC} = \frac{3\times 2\times 1^2}{3^2} = +0.67 \text{ kNm}$$

$$M_{FCF} = M_{FFC} = 0$$

Distribution Factor

Joint	Member	Relative stiffness	Total Relative stiffness	D.F
B	BA	$\frac{2I}{4}=\frac{I}{2}$	I	$\frac{1}{2}$
	BE	$\frac{3}{4}\times\frac{2I}{6}=\frac{I}{4}$		$\frac{1}{4}$
	BC	$\frac{I}{4}$		$\frac{1}{4}$
C	CB	$\frac{I}{4}$	$\frac{10I}{12}$	$\frac{3}{10}$
	CF	$\frac{I}{3}$		$\frac{4}{10}$
	CD	$\frac{3}{4}\times\frac{I}{3}=\frac{I}{4}$		$\frac{3}{10}$

A	B			C			D	E	F
	1/2	1/4	1/4	3/10	3/10	4/10			
AB	*BA*	*BC*	*BE*	*CB*	*CD*	*CF*	*DC*	*EB*	*FC*
−1.5 0.292	1.5 0.585	−2.67 0.292	0 0.292	2.67 0.146	−1.33	0	0.67	0 0	0
−1.207	2.085	−2.378 −0.223	0.292	2.816 −0.446	−1.33 −0.446	0 −0.594	0.67 −0.223	0	0 −0.297
−1.207 0.055	2.085 0.111	−2.601 0.055	0.292 0.055	2.37 0.028	−1.776 −0.224	−0.594	0.447 −0.447	0	−0.297
−1.152	2.197	−2.546 0.029	0.347	2.398 0.058	−2.0 0.058	−0.594 0.078	0	0 0	−0.297 0.039
−1.152 −0.007	2.197 −0.014	−2.517 −0.006	0.347 −0.006	2.456 −0.003	−1.942 – 0.003	−0.516	0	0	−0.258
−1.159	**2.183**	**−2.523**	**0.341**	**2.453**	**−1.942**	**−0.516**	**0**	**0**	**−0.258**

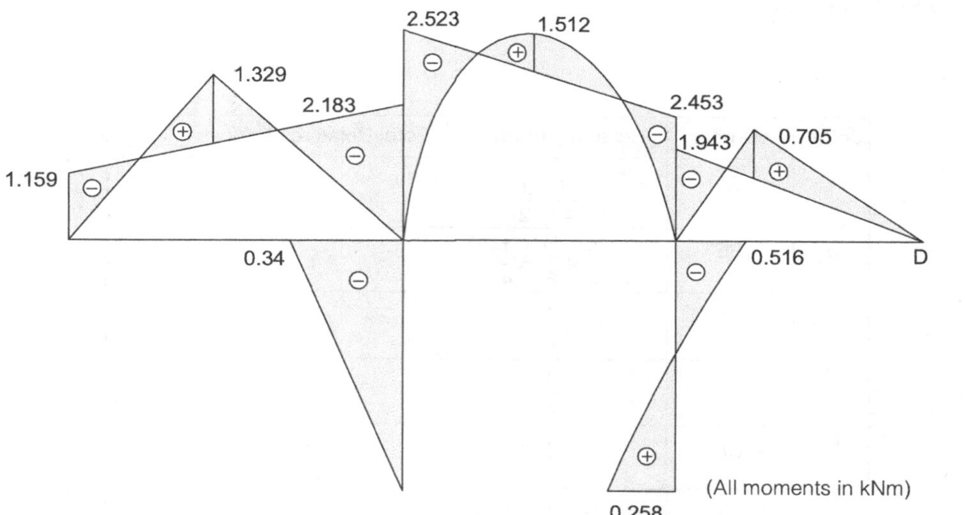

BMD

[Outer face is reference face]

4.20 Sketch the elastic curve for the frame as shown.

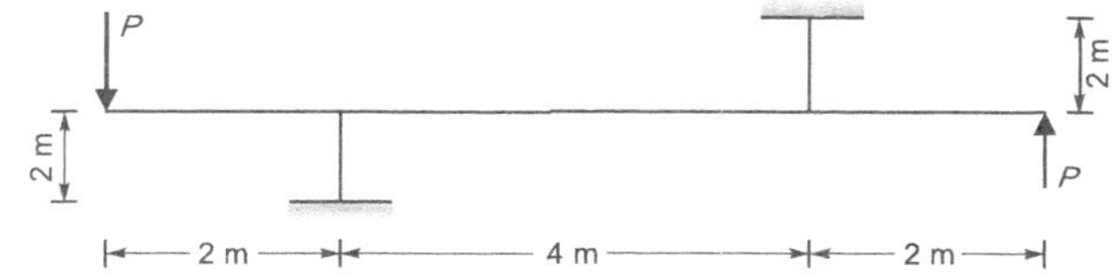

[4 marks : 2013]

Solution:

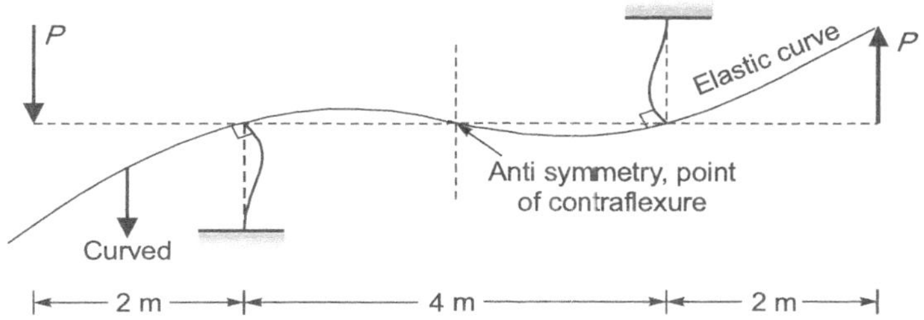

4.21 For the rigid frame shown in figure below, find by unit load method, the slopes and deflection at *A*, *C*.

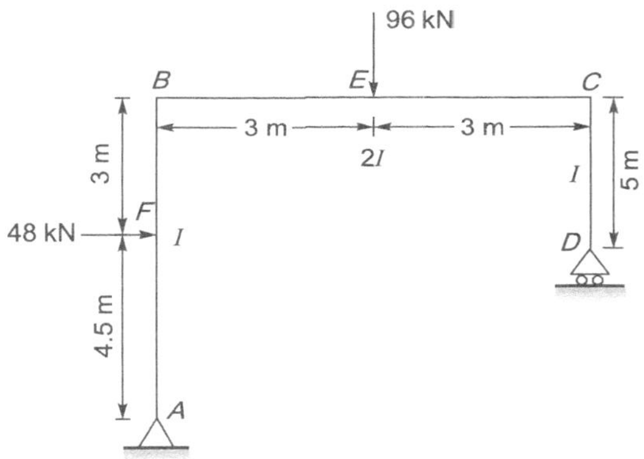

[20 marks : 2013]

Solution:

Since support *D* is a roller support, there will be no horizontal reaction at *D*, hence no BM in the member *CD*. Thus the structure can be represented in another way

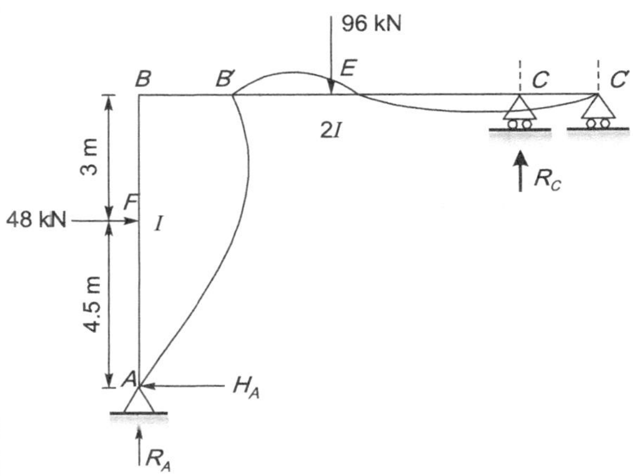

Equilibrium condition

$\Sigma F_x = 0;$ $\qquad H_A = 48$ kN

$\Sigma M_C = 0$

$$R_A \times 6 + 48 \times 7.5 - 48 \times 3 - 96 \times 3 = 0$$

$$R_A = 12 \text{ kN}$$

$\Sigma F_y = 0$ $\qquad R_C = 84$ kN

From modified diagram, support *C* has roller support it can not be deflected in vertical direction. So vertical deflection at *C* = 0

Since support *A* is hinged support, so deflection at *A* zero (0).

Thus we have to calculate slope at A and slope and horizontal deflection at C.

Using unit load method:

$$\Delta = \int \frac{Mmdx}{EI}, \theta = \int \frac{Mm'dx}{EI}$$

where, $M \rightarrow$ BM due to external loading

$m \rightarrow$ BM when external loads are removed and unit load is applied at that point where deflection is required.

$m' \rightarrow$ BM when external loads are removed and unit moment is applied at that point where slope is required.

Segment	Limit	M	m_C	$m_{\theta C}$	$m_{\theta A}$
CE (2I)	0-3	$84x$	$1.25x$	$\left(1-\frac{x}{6}\right)$	$\frac{x}{6}$
EB (2I)	3-6	$84x - 96(x-3)$ $=(-12x+288)$	$1.25x$	$\left(1-\frac{x}{6}\right)$	$\frac{x}{6}$
BF (I)	0-3	$84 \times 6 - 96 \times 3 = 216$	$1.25 \times 6 - 1 \times x = 7.5 - x$	$1-\frac{1}{6}\times 6 = 0$	1
AF (I)	0-4.5	$48x$	x	0	1

$m_c \rightarrow$ BM due to horizontal unit load at C

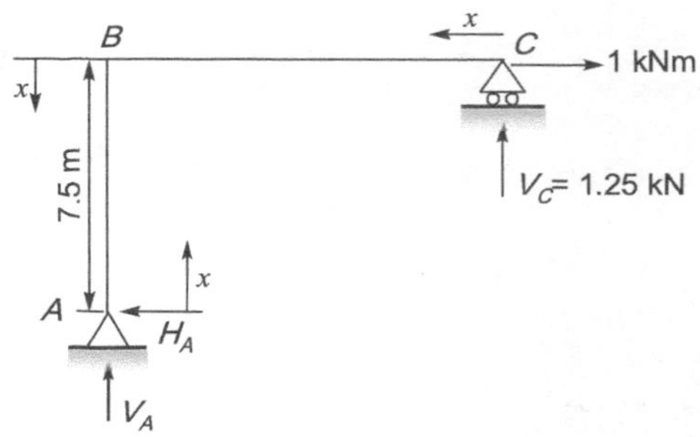

$\Sigma F_x = 0;$ $H_A = 1$ kN

$\Sigma M_A = 0;$ $V_C \times 6 = 1 \times 7.5$

$V_C = 1.25$ kN

$\Sigma F_y = 0;$ $V_A = -1.25$ kN

$m_{\theta_C} \rightarrow$ BM due to unit anticlockwise moment at C

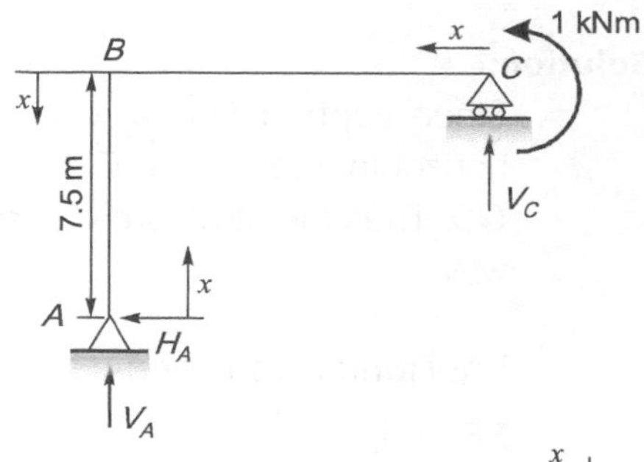

$\Sigma F_x = 0;$ $H_A = 0$

$\Sigma M_A = 0;$ $V_C \times 6 + 1 = 0$

$$V_C = -\frac{1}{6}\text{KN}$$

$\Sigma F_y = 0;$ $V_A = \frac{1}{6}$KN

$m_{\theta A} \rightarrow$ BM due to unit clockwise moment at A

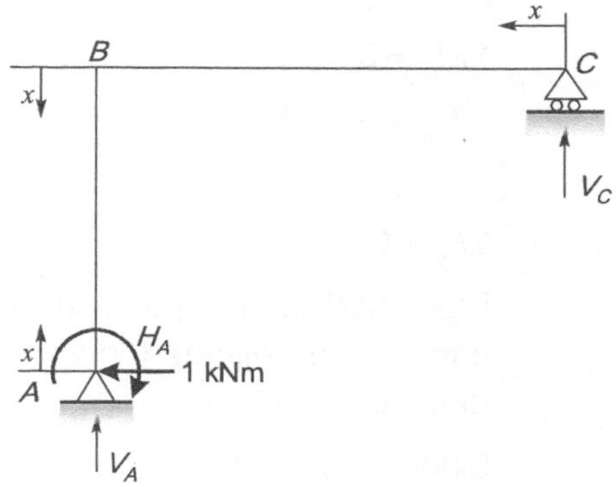

$\Sigma F_x = 0;$ $H_A = 0$

$\Sigma M_C = 0;$ $V_A \times 6 + 1 = 0$

$\Sigma f_y = 0;$ $V_A = -\frac{1}{6}$KN

$$V_C = \frac{1}{6}\text{KN}$$

$$\Delta_C = \int \frac{Mm_C dx}{EI} = \int_0^3 \frac{84x \times 1.25x dx}{2EI} + \int_3^6 \frac{(-12x+288)1.25x dx}{2EI} + \int_0^3 \frac{216(7.5-x)dx}{EI} + \int_0^{4.5} \frac{48x^2}{EI} dx$$

$$\Delta_C = \frac{472.5}{EI} + \frac{1957.5}{EI} + \frac{3888}{EI} + \frac{1458}{EI} = \frac{7776}{EI}(\rightarrow)$$

$$\theta_C = \int \frac{Mm_{\theta_c} dx}{EI} = \int_0^3 \frac{84x\left(1-\frac{x}{6}\right)dx}{2EI} + \int_3^6 \frac{(-12x+288)\left(1-\frac{x}{6}\right)dx}{2EI} + 0 + 0$$

$$\theta_C = \frac{126}{EI} + \frac{90}{EI} = \frac{216}{EI}(\text{Anticlockwise})$$

$$\dot{\theta}_A = \int \frac{Mm_{\theta_A} dx}{EI} = \int_0^3 \frac{84x \times \frac{x}{6}dx}{2EI} + \int_3^6 \frac{(-12x+288)\frac{x}{6}dx}{2EI} + \int_0^3 \frac{216dx}{EI} + \int_0^{4.5} \frac{48x \times 1dx}{EI}$$

$$\theta_A = \frac{63}{EI} + \frac{261}{EI} + \frac{648}{EI} + \frac{486}{EI}$$

$$\theta_A = \frac{1458}{EI}(\text{clockwise})$$

At A, $\theta_A = \frac{1458}{EI}(\text{clockwise}),\ \Delta_A = 0$

At C, $\theta_C = \frac{216}{EI}(\text{anticlockwise}),\ \Delta_{C_H} = \frac{7776}{EI}(\rightarrow)$

$\Delta_{C_V} = 0$

4.22 Analyze the beam shown in figure using the strain energy method and draw bending moment diagram and shear force diagram.

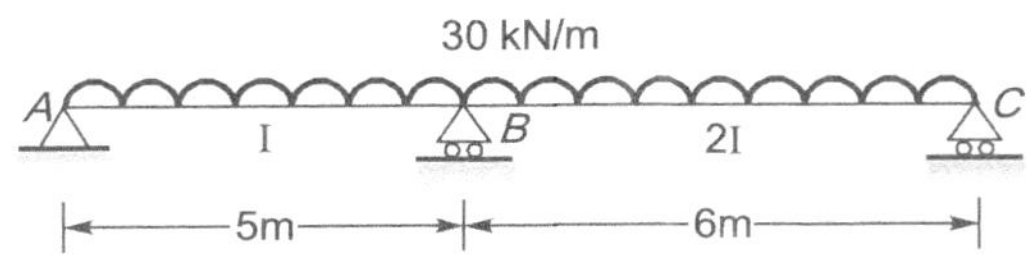

[15 marks : 2014]

Solution:

$D_S = R_e - 3$

Here, $R_e = 2 + 1 + 1 = 4$

$\therefore$ $D_S = 4 - 3 = 1$

Let reaction R_A is redundant, say $R_A = R$

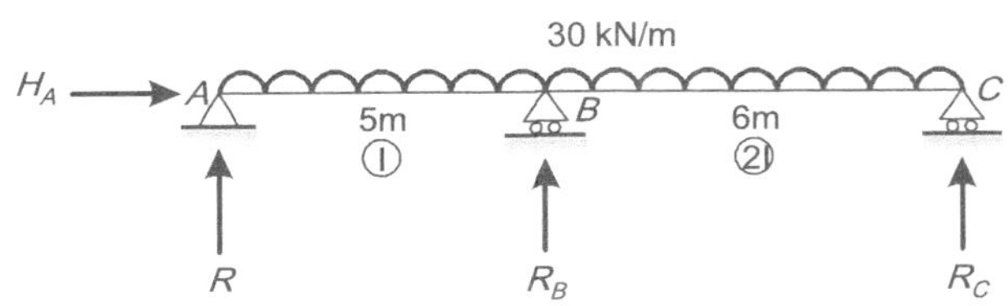

$\Sigma F_x = 0;$

$$\therefore \quad H_A = 0$$

$\Sigma F_y = 0;$

$$R + R_B + R_C = 30 \times 11$$

$$R_B + R_C = 330 - R \quad \text{...(i)}$$

$\Sigma M_C = 0;$

$$R \times 11 + R_B \times 6 - 30 \times 11 \times 5.5 = 0$$

$$6R_B = 1815 - 11\,R$$

$$R_B = \frac{1815 - 11R}{6} (\uparrow)$$

From eq. (i)

$$R_C = 330 - R - R_B$$

$$R_C = \frac{165 + 5R}{6} (\uparrow)$$

The true value of redundant R will be that for which total strain energy of the system is minimum i.e.,

$$\frac{\partial U}{\partial R} = 0$$

$$U = U_{AB} + U_{BC}$$

$$= \int_0^5 \frac{M_x^2 dx}{2EI} + \int_0^6 \frac{M_x^2 dx}{2 \times 2EI}$$

$$= \int_0^5 \frac{\left(Rx - \frac{wx^2}{2}\right)^2}{2EI} dx + \int_0^6 \frac{\left[\left(\frac{165 + 5R}{6}\right).x - \frac{wx^2}{2}\right]^2 dx}{4EI}$$

$$\frac{\partial U}{\partial R} = \int_0^5 \frac{2\left(R.x - \frac{wx^2}{2}\right)x}{2EI} dx + \int_0^6 \frac{2\left[\left(\frac{165 + 5R}{6}\right)x - \frac{wx^2}{2}\right]\frac{5}{6}xdx}{4EI}$$

$$\frac{\partial U}{\partial R} = \int_0^5 \frac{\left(R.x - \frac{wx^2}{2}\right)x}{EI} dx + \frac{\int_0^6 \frac{5}{12}\left[\left(\frac{165 + 5R}{6}\right)x - \frac{wx^2}{2}\right]xdx}{EI}$$

$$\therefore \quad \int_0^5 12\left(R.x^2 - \frac{wx^3}{2}\right) dx + \int_0^6 5\left[\left(\frac{165 + 5R}{6}\right)x^2 - \frac{wx^3}{2}\right]dx = 0$$

$$12\left[R\frac{x^3}{3} - \frac{wx^4}{2 \times 4}\right]_0^5 + 5\left[\left(\frac{165 + 5R}{6}\right)\frac{x^3}{3} - \frac{wx^4}{2 \times 4}\right]_0^6 = 0$$

$$12\left[\frac{125R}{3} - 2343.75\right] + 5\left[\left(\frac{165 + 5R}{6}\right) \times 72 - 4860\right] = 0$$

$$500R - 28125 + 9900 + 300R - 24300 = 0$$

$$800\,R = 42525$$

$$R = 53.15 \text{ kN} (\uparrow)$$

Hence,

$$R_B = \frac{1815 - 11 \times 53.15}{6} = 205.05 \text{ kN}$$

$$R_C = \frac{165 + 5 \times 53.15}{6} = 71.80 \text{ kN}$$

Check:

$$\Sigma F_y = R_A + R_B + R_C - 30 \times 11$$
$$= 53.15 + 205.05 + 71.80 - 330$$
$$= 0 \qquad \text{(OK)}$$

For shear force diagram:

Portion *AB*:

$$S_x \,(x \text{ from } A) = +R_A - wx \qquad [0 < x < 5]$$
$$S_x = 53.15 - 30x \qquad \text{(linear)}$$

at $x = 0$, $S_A = 53.15$ kN

at $x = 5$ m, S_B (Just left of *B*) $= 53.15 - 30 \times 5 = -96.85$ kN

$$M = Rx - wx \times \frac{x}{2} = 53.15x - \frac{30x^2}{2} = 53.15x - 15x^2$$

For M_{max}: $\frac{dM}{dx} = 0$

$$53.15 - 15 \times 2x = 0$$
$$x = 1.77 \text{ m (from A)}$$

$\therefore$ M_{max} (at $x = 1.77$ m) $= 53.15 \times 1.77 - \frac{30 \times (1.77)^2}{2} = 47.08$ kN-m

Portion *BC*:

$$S_x \,(x \text{ from } C) = -R_C + wx \qquad [0 < x < 6]$$
$$S_x = -71.80 + 30x \qquad \text{(linear)}$$

at $x = 0$, $S_C = -71.80$ kN

at $x = 6$ m, S_B (Just right of *B*) $= 108.2$ kN

For bending moment diagram:

Portion *AB*:

$$M_x \,(x \text{ from } A) = R_A x - \frac{wx^2}{2} \qquad [0 \le x \le 5]$$
$$M_x = 53.15x - \frac{30x^2}{2} \qquad \text{(Parabolic)}$$

at $x = 0$, $M_A = 0$

at $x = 5$ m, $M_B = 53.15 \times 5 - \frac{30 \times 5^2}{2} = -109.25$ kN-m

Portion *BC*:

$$M_x (x \text{ from } C) = R_C x - \frac{wx^2}{2} \qquad [0 \le x \le 6]$$
$$M_x = 71.80x - \frac{30x^2}{2} \qquad \text{(Parabolic)}$$

at $x = 0$, $M_C = 0$

at $x = 6$ m, $M_B = 71.80 \times 6 - \frac{30 \times 6^2}{2} = -109.20$ kN-m

Form $M_{max:}$

$$\frac{dM}{dx} = 0$$

$$71.80 - 15 \times 2x = 0$$

$$x = 2.39 \text{ m (from } C)$$

$$\therefore \quad M_{max} \text{ (at } x = 2.39 \text{ m)} = 71.80 \times 2.39 - 15 \times (2.39)^2 = 85.92 \text{ kN-m}$$

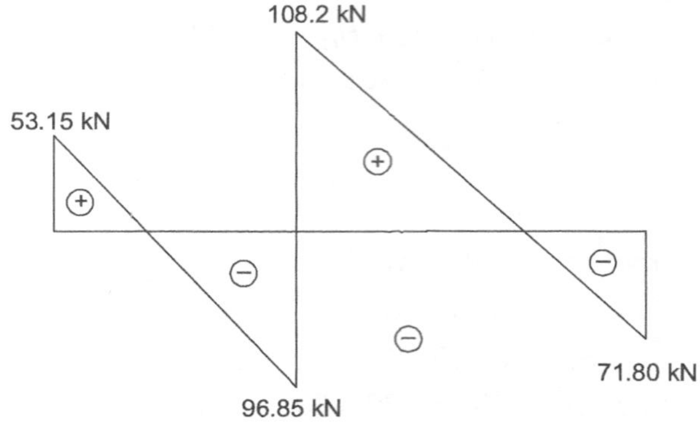

SFD

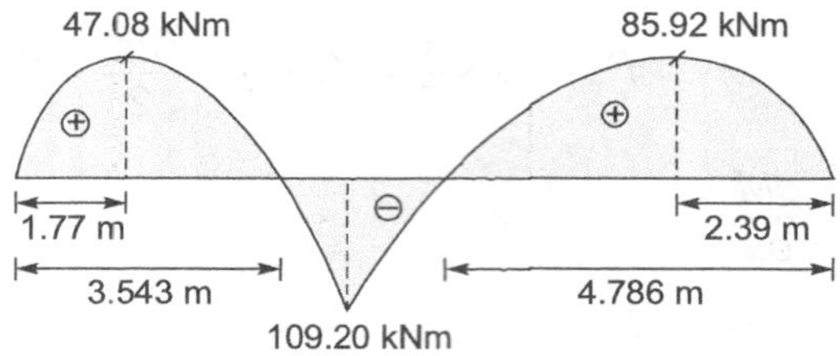

BMD

4.23 Determine the vertical deflection at the free end of a circular cantilever frame shown in figure using the unit load method. Take EI = constant.

[5 marks : 2014]

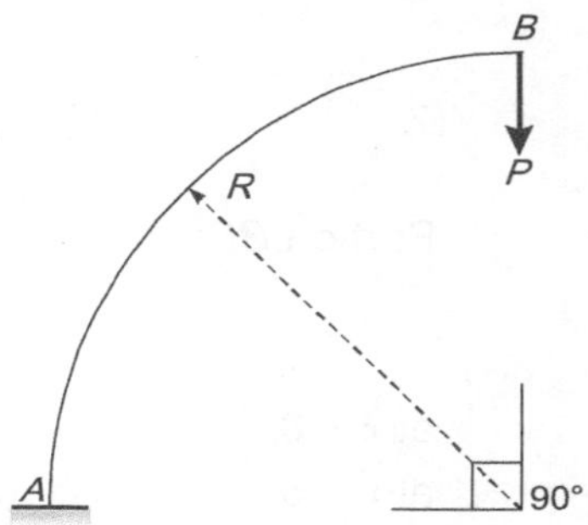

Solution:

According to **unit load method**, the deflection at a point in desired direction is given by

$$\Delta = \int \frac{Mm.ds}{EI}$$

Where, $M = BM$ at any section due to given loading

$m = BM$ at any section when unit load is applied in the direction of desired deflection

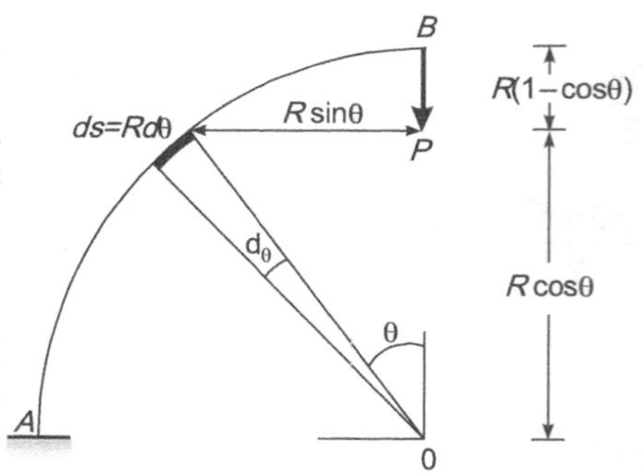

Portion	AB
Origin	B
Limit	0-90° or $0 - \frac{\pi}{2}$
M	$-PR\sin\theta$
m	$-R\sin\theta$
EI	EI

Vertical deflection at B,

$$\Delta_{BV} = \int \frac{Mmds}{EI} = \int_0^{\frac{\pi}{2}} \frac{Mm(Rd\theta)}{EI}$$

$$= \int_0^{\frac{\pi}{2}} \frac{(-PR\sin\theta)(-R\sin\theta)Rd\theta}{EI}$$

$$= \int_0^{\frac{\pi}{2}} \frac{PR^3 \sin^2\theta d\theta}{EI} = \frac{PR^3}{EI}\int_0^{\frac{\pi}{2}} \sin^2\theta d\theta = \frac{PR^3}{EI}\times\frac{\pi}{4}$$

$$\therefore \quad \Delta_{BV} = \frac{\pi}{4}\frac{PR^3}{EI}(\downarrow)$$

4.24 Analyze the plane box frame shown in figure using the moment distribution method and making use of symmetry. Also, draw bending moment diagram.

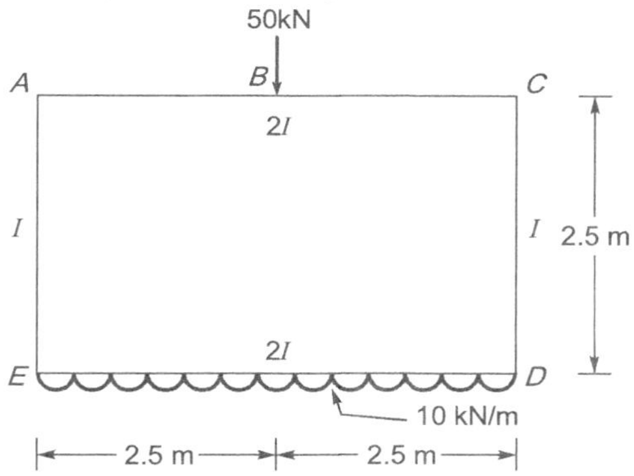

[15 marks : 2014]

Solution:

Fixed end moment:

$$\bar{M}_{AC} = -\frac{PL}{8} = -\frac{50\times5}{8} = -31.25 \text{ kN-m}$$

$$\bar{M}_{CA} = +\frac{PL}{8} = +31.25 \text{ kN-m}$$

$$\bar{M}_{AE} = \bar{M}_{EA} = 0$$

$$\bar{M}_{CD} = \bar{M}_{DC} = 0$$

$$\bar{M}_{DE} = -\frac{wL^2}{12} = -\frac{10\times5^2}{12} = -20.83 \text{ kN-m}$$

$$\bar{M}_{ED} = +\frac{wL^2}{12} = +20.83 \text{ kN-m}$$

Distribution factors:

Joint	Member	R.S	Total R.S	D.F
A	AE	$\frac{I}{2.5} = 0.4I$	0.8 I	0.5
	AC	$\frac{2I}{5} = 0.4I$		0.5
C	CA	$\frac{2I}{5} = 0.4I$	0.8 I	0.5
	CD	$\frac{I}{2.5} = 0.4I$		0.5
D	DC	$\frac{I}{2.5} = 0.4I$	0.8 I	0.5
	DE	$\frac{2I}{5} = 0.4I$		0.5
E	ED	$\frac{2I}{5} = 0.4I$	0.8 I	0.5
	EA	$\frac{I}{2.5} = 0.4I$		0.5

Moment distribution:

E	A		C		D		E		A
	0.5	0.5	0.5	0.5	0.5	0.5	0.5	0.5	
	0	–31.25	31.25	0	0	–20.83	+20.83	0	
	15.625	15.625	–15.625	–15.625	10.415	10.415	–10.415	–10.415	
	–5.207	–7.812	7.812	5.207	–7.812	–5.207	5.207	7.812	
	6.509	6.509	–6.509	–6.509	6.509	6.509	–6.509	–6.509	
	–3.254	–3.254	3.254	3.254	–3.254	–3.254	3.254	3.254	
	+3.254	+3.254	–3.254	–3.254	3.254	3.254	–3.254	–3.254	
	-----	-----	-----	-----	-----	-----	-----	-----	
Final End Moment	16.927	–16.927	16.927	–16.927	9.112	–9.112	9.112	–9.112	

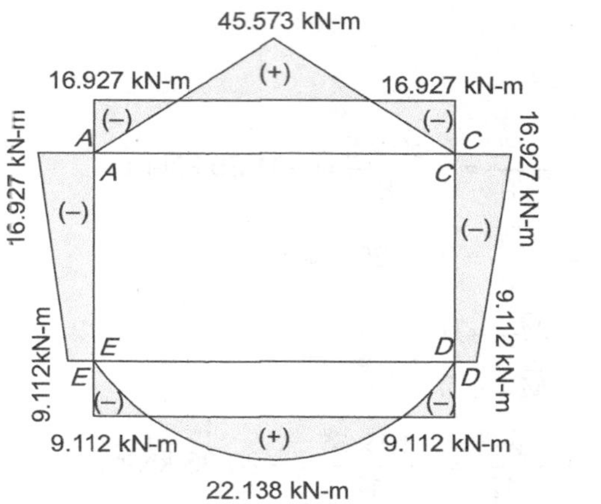

BMD

4.25 Analyse the box-frame shown in figure below by moment distribution method.

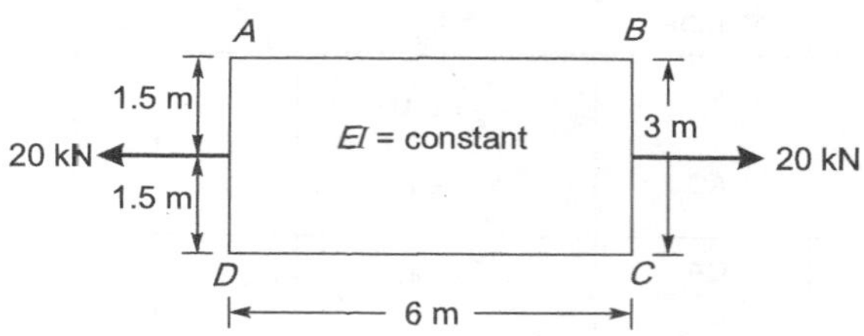

Draw bending moment diagram with relevant values.

[20 marks : 2015]

Solution:

Fixed end moments,

$$\overline{M}_{DA} = +\frac{WL}{8} = +\frac{20 \times 3}{8} = +7.5\text{kN-m}$$

$$\overline{M}_{AD} = -\frac{WL}{8} = -\frac{20\times 3}{8} = -7.5\text{kN-m}$$

$$\overline{M}_{BC} = +7.5\text{ kN-m}$$

$$\overline{M}_{CB} = -7.5\text{ kN-m}$$

$$\overline{M}_{AB} = \overline{M}_{BA} = \overline{M}_{CD} = \overline{M}_{DC} = 0$$

Distribution factors,

Joint	Member	Stiffness	Relative stiffness	Distribution Factor
A	*AB*	*I*/6	*I*/2	1/3
	AD	*I*/3		2/3
B	*BA*	*I*/6	*I*/2	1/3
	BC	*I*/3		2/3

Similarly for joints *C* and *D* also.

Moment distribution table,

	A (*D′*)	*A*	*B*	*B*	*C*	*C*	*D*	*D* (*A′*)
	2/3	1/3	1/3	2/3	2/3	1/3	1/3	2/3
	–7.5	0	0	7.5	–7.5	0	0	7.5
	5	2.5	–2.5	–5	5	2.5	–2.5	–5
	–2.5	–1.25	1.25	2.5	–2.5	–1.25	1.25	2.5
	+2.5	+1.25	–1.25	–2.5	+2.5	+1.25	–1.25	–2.5
	–1.25	–0.625	0.625	1.25	–1.25	–0.625	0.625	1.25
	+1.25	+0.625	–0.625	–1.25	–1.25	+0.625	–0.625	–1.25
	.	.	.	.	.	.	.	.
	.	.	.	.	.	.	.	.
	.	.	.	.	.	.	.	.
		[All the terms below will further cancel out]						
Final end moments	–2.5	2.5	–2.5	2.5	–2.5	2.5	–2.5	2.5

Taking inner face as the reference face, hence drawing SAGGING of fixed end moment diagram outwards and also HOGGING of simply supported moment diagram outwards.

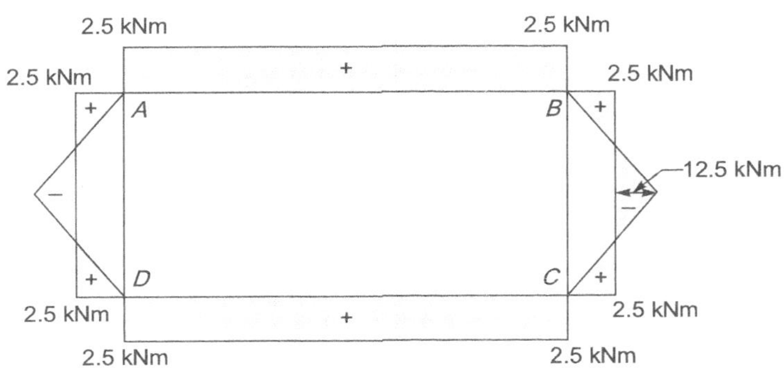

BMD

$$\text{Moment at the centre} = -\frac{WL}{4} + 2.5 = -\frac{20\times 3}{4} + 2.5 = -12.5\text{kN-m}$$

4.26 In the rigid frame shown in figure, a central concentrated load is acting on "*BC*". Find slopes θ_A, θ_B, θ_C, θ_D.

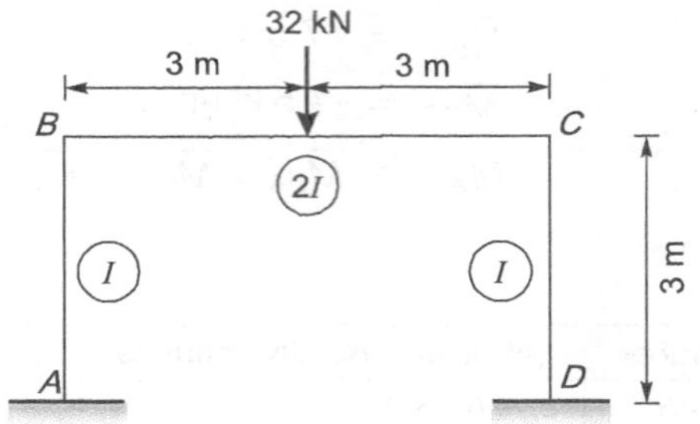

[10 marks : 2015]

Solution:

Due to symmetry in geometry and loading of figure, there will not be any sway in the frame.

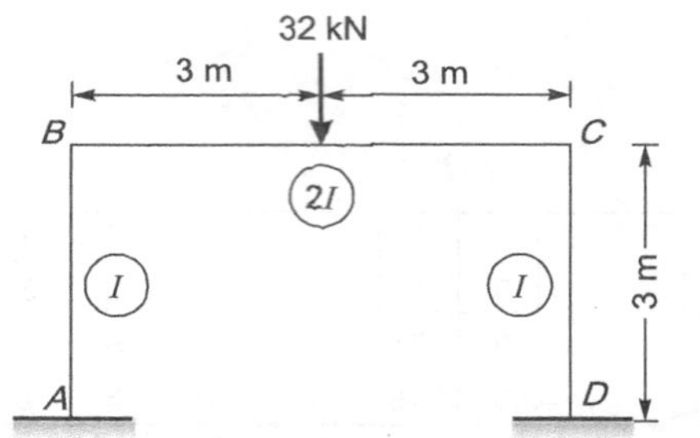

Ends *A* and *D* are fixed

$$\therefore \quad \theta_A = \theta_D = 0$$

Slope-deflection equations,

$$M_{FAB} = M_{FBA} = M_{FCD} = M_{FDC} = 0$$

$$M_{FBC} = -\frac{32 \times 6}{8} = -24 \text{ kNm}$$

$$M_{FCB} = +\frac{32 \times 6}{8} = +24 \text{ kNm}$$

Joint *B*,

$$M_{BA} = M_{FBA} + \frac{2EI}{3}(2\theta_B + 0) = \frac{4}{3}EI\theta_B$$

$$M_{BC} = M_{FBC} + \frac{2E(2I)}{6}(2\theta_B + \theta_C)$$

$$= -24 + \frac{2EI}{3}(2\theta_B + \theta_C)$$

$$= \frac{4}{3}EI\theta_B + \frac{2}{3}EI\theta_C - 24$$

But

$$M_{BA} + M_{BC} = 0$$

$$\Rightarrow \quad \frac{4}{3}EI\theta_B + \frac{2}{3}EI\theta_C - 24 + \frac{4}{3}EI\theta_B = 0$$

$\Rightarrow \quad 4EI\theta_B + EI\theta_C = 36$ ...(i)

Joint *C*,

$\Rightarrow \quad \theta_B = \frac{12}{EI}$ (Clockwise) ($\because \theta_B = -\theta_C$)

Hence, $\quad \theta_C = -\frac{12}{EI}$ (Anticlockwise)

$\therefore \quad \theta_A = 0,\ \theta_B = \frac{12}{EI},\ \theta_C = -\frac{12}{EI},\ \theta_D = 0$

4.27 Analyze the frame shown in figure by the moment distribution method and draw bending moment diagram.

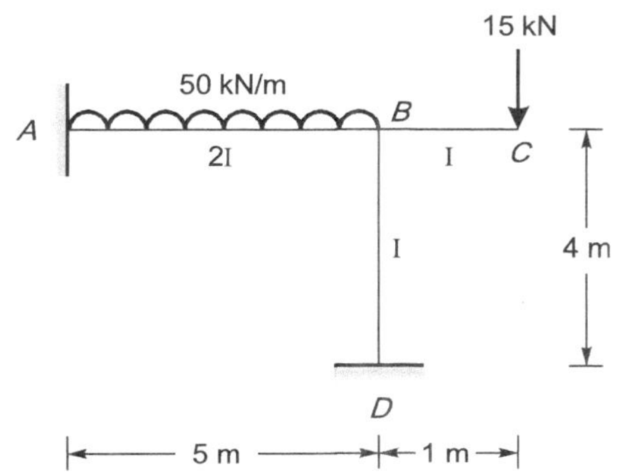

[15 marks : 2016]

Solution:

Removing cantilever portion *BC* to reduce kinetic Indeterminacy of structure.

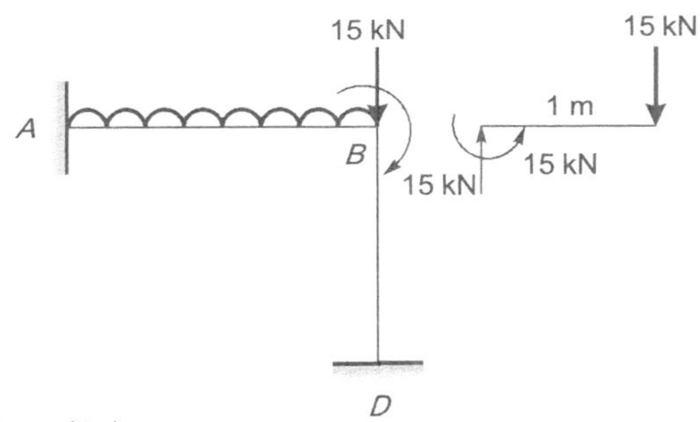

$D_k = 1$ i.e., (θ_B)

$$M_{FAB} = \frac{-wL^2}{12} = -\frac{50 \times 5^2}{12} = -104.17 \text{ kNm}$$

$$M_{FBA} = \frac{wL^2}{12} = 104.17 \text{ kNm}$$

$$M_{FBD} = M_{FDB} = 0,\ M_{BC} = -15 \text{ kNm}$$

Distribution factor

Joint	Member	Stiffness	Total Stiffness of Joint	Distribution Factors
B	BA	$\frac{4E(2I)}{5}$	2.6*EI*	0.62
	BD	$\frac{4EI}{4}$		0.38

Distribution table

$$M_{BA} + M_{BC} + M_{BD} = 0 \quad \text{(For joint B)}$$

Hence unbalanced moment at joint B $= 104.17 - 15 = 89.17$ kNm

Assuming reference face upwards for horizontal members and right face for vertical members.

Joint	A	B		D	
Member	AB	BA	BC	BD	DC
DF	—	0.62	—	0.38	–
FEM	–104.17	104.17	–15	0	0
Bal.	—	–55.28	—	–33.88	–
CO	–27.64	—	—	—	–16.94
Final	–131.84	48.89	–15	–33.88	–16.94

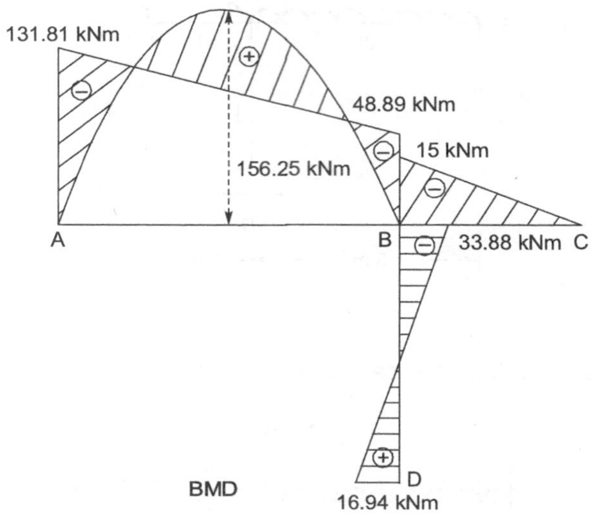

4.28 Analyze a 2 span beam shown in figure using the method of consistent deformations and draw shear force and bending moment diagrams. Treat reaction at A as redundant.

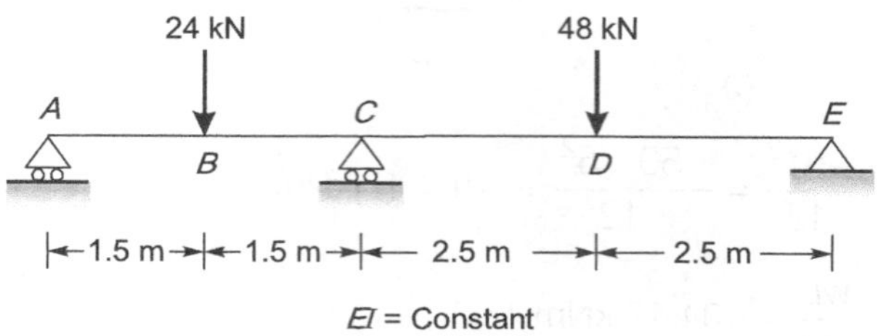

[20 marks : 2016]

Solution:

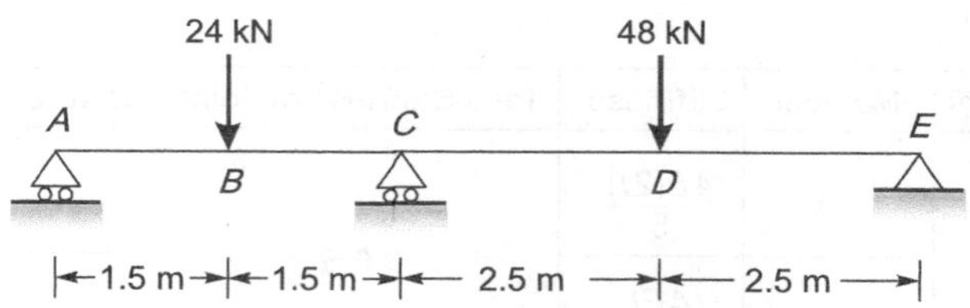

Assuming reaction at *A* as redundant, then above structure can be divided into following structures using principle of superposition.

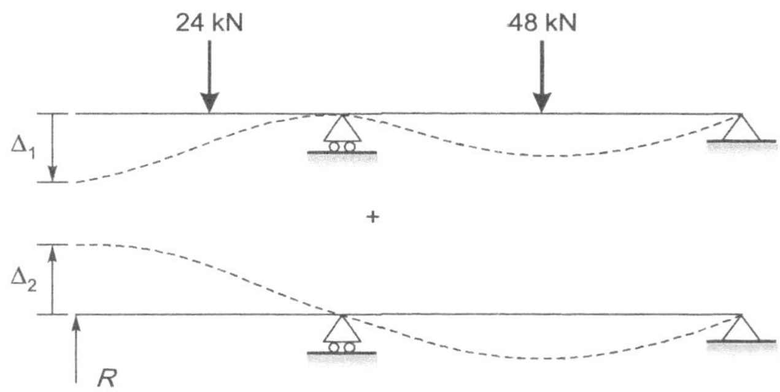

Compatibility condition at *A*:

$$-\Delta_1 + \Delta_2 = 0$$

Calculation of Δ_1:

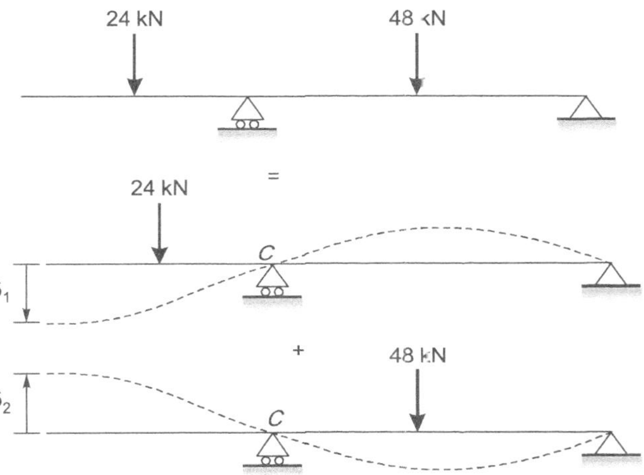

$$\Delta_1 = -\delta_1 + \delta_2$$

$$= -\left[\frac{24 \times BC^3}{3EI} + \theta_B \times BA + \theta_C \times AC\right] + [\theta_C \times AC]$$

$$= -\left[\frac{24 \times BC^3}{3EI} + \frac{24 \times BC^2}{2EI} \times BA + \frac{(24 \times BC)CE}{3EI} \times AC\right] + \left[\frac{48 \times CE^2}{16EI} \times AC\right]$$

$$= -\left[\frac{24 \times 1.5^3}{3EI} + \frac{24 \times 1.5^2 \times 1.5}{2EI} + \frac{24 \times 1.5 \times 5}{3EI} \times 3\right] + \left[\frac{48 \times 5^2}{16EI} \times 3\right]$$

$$= -\left[\frac{27}{EI} + \frac{40.5}{EI} + \frac{180}{EI}\right] + \left[\frac{225}{EI}\right]$$

$$= \frac{-22.5}{EI}$$

Calculation of Δ_2 :

$$\Delta_2 = \frac{R \times AC^3}{3EI} + \theta_C \times AC$$

$$= \frac{R \times AC^3}{3EI} + \frac{(R \times AC)CE}{3EI} \times AC$$

$$= \frac{R \times 3^3}{3EI} + \frac{(R \times 3)5}{3EI} \times 3 = \frac{24R}{EI}$$

Note: Δ_1 and Δ_2 can be calculated using either conjugate beam method or moment area method also.

From compatibility condition:

$$-\Delta_1 + \Delta_2 = 0$$

$$\Rightarrow \quad \frac{-22.5}{EI} + \frac{24R}{EI} = 0$$

$$\Rightarrow \quad R = 0.94 \text{ kN}$$

Calculation of other support, reaction:

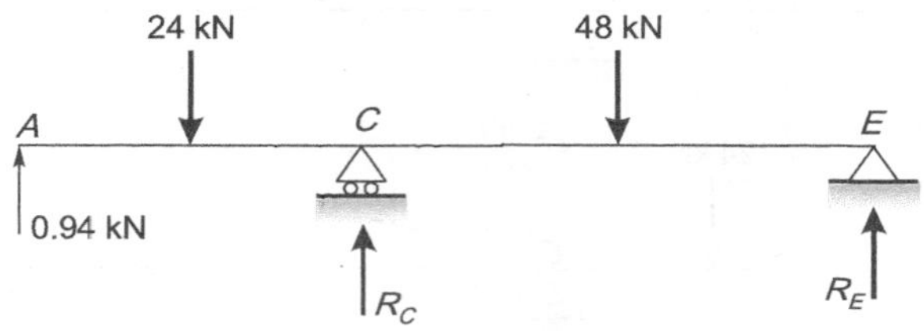

$$\Sigma M_E = 0$$

$$\Rightarrow \quad 0.94 \times 8 - 24 \times 6.5 + R_C \times 5 - 48 \times 2.5 = 0$$

$$\Rightarrow \quad R_C = 53.7 \text{ kN}$$

$$\Sigma F_y = 0$$

$$\Rightarrow \quad 0.94 + R_C + R_E - 24 - 48 = 0$$

$$\Rightarrow \quad R_E = 17.36 \text{ kN}$$

Now, $BM_C = 0.94 \times 3 - 24 \times 1.5 = -33.18$ kN

BMD using superposition:

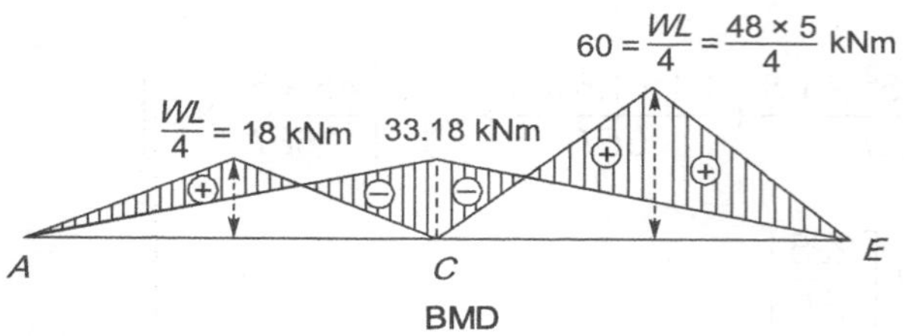

BMD

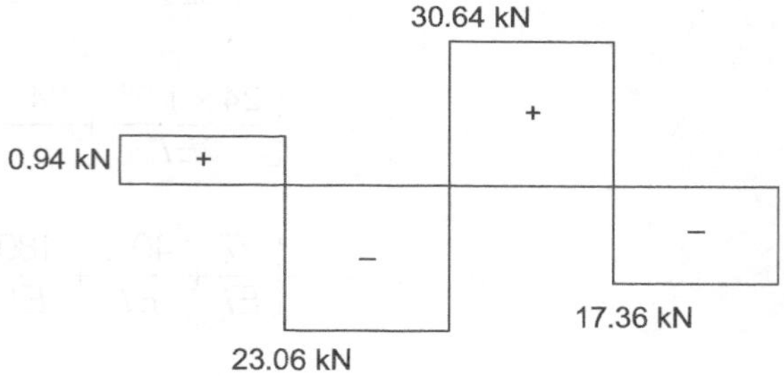

SFD

4.29 Determine the horizontal deflection at F for the frame shown in Figure. Take EI = constant.

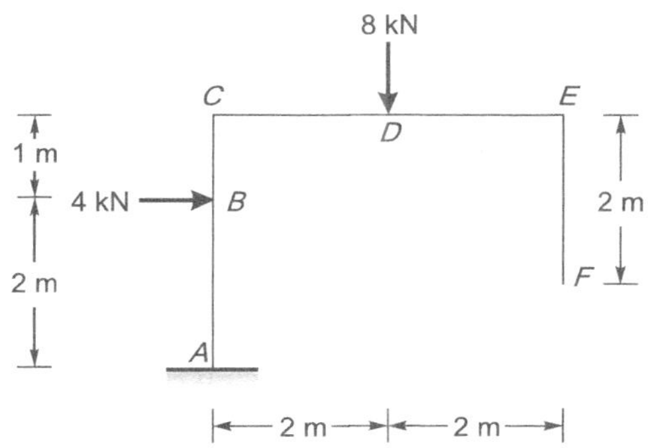

[20 marks : 2017]

Solution:

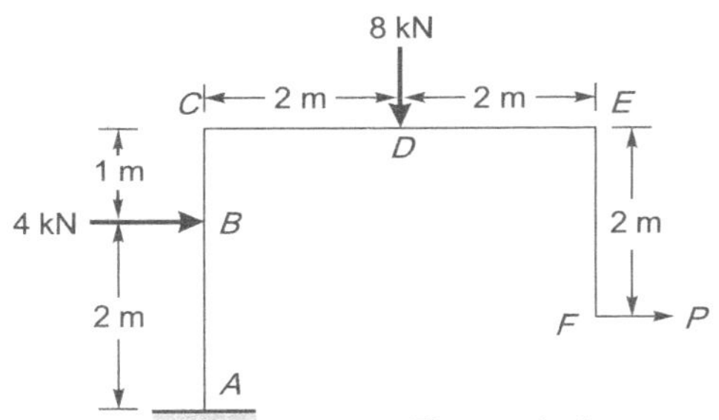

EI = constant

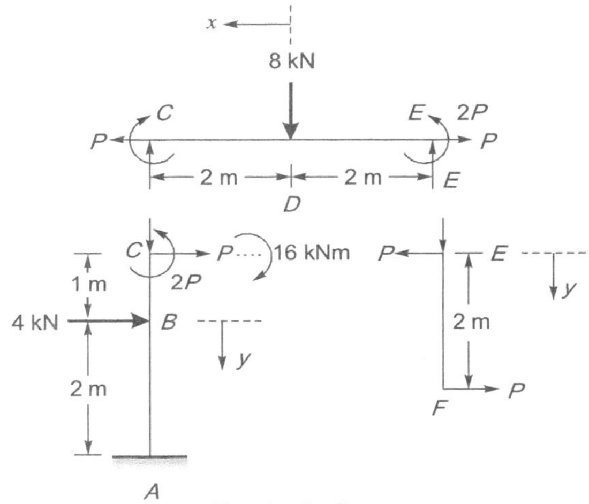

Free body diagram

Let us apply a fictitious horizontal load P at F.

Strain energy stored in frame = $U_{AB} + U_{BC} + U_{CD} + U_{DE} + U_{EF}$

$$\Rightarrow \qquad U = \Sigma\int \frac{M^2 ds}{2EI}$$

$$U_{EF} = \int_0^2 \frac{(Py)^2 dy}{2EI}$$

$$U_{CB} = \int_0^1 \frac{(Py + 16 - 2P)^2 dy}{2EI}$$

$$U_{DE} = \int_0^2 \frac{(2P)^2 dx}{2EI}$$

and

$$U_{BA} = \int_0^2 \frac{\{P(y+1) + 4y + 16 - 2P\}^2 dy}{2EI}$$

$$U_{CD} = \int_0^2 \frac{(2P - 8x)^2 dx}{2EI}$$

Horizontal deflection at F,

$$\Rightarrow \quad \delta_F = \frac{\partial U}{\partial P} = \int_0^2 \frac{2Py^2 dy}{2EI} + \int_0^2 \frac{2\times 2P \times 2dx}{2EI} + \int_0^2 \frac{2(2P-8x)\times 2dx}{2EI}$$

$$+\int_0^1 \frac{2(Py+16-2P)\times(y-2)dy}{2EI} + \int_0^2 \frac{2\{P(y+1)+4y+16-2P\}\times\{(y+1)-2\}dy}{2EI}$$

Putting $P = 0$, we get

$$\delta_F = 0+0+\int_0^2 \frac{2\times(-8x)\times 2dx}{2EI} + \int_0^1 \frac{2(16)\times(y-2)dy}{2EI} + \int_0^2 \frac{2\{4y+16\}(y-1)dy}{2EI}$$

$$\Rightarrow \quad \delta_F = \frac{1}{EI}\left\{-16\left[\frac{x^2}{2}\right]_0^2 + 16\left[\frac{y^2}{2}-2y\right]_0^1 + \int_0^2 4(y+4)(y-1)dy\right\}$$

$$= \frac{1}{EI}\left\{-16\times\left[\frac{2^2}{2}-0\right]+16\left[\frac{1}{2}-2-0\right]+4\int_0^2(y^2+3y-4)dy\right\}$$

$$= \frac{1}{EI}\left\{(-16\times 2)+16\times\left(-\frac{3}{2}\right)+4\left[\frac{y^3}{3}+\frac{3y^2}{2}-4y\right]_0^2\right\}$$

$$= \frac{1}{EI}\left\{-32-24+4\left[\frac{(2)^3}{3}+\frac{3\times 2^2}{2}-4\times 2\right]\right\}$$

$$= \frac{1}{EI}\left\{-56+4\left[\frac{8}{3}+6-8\right]\right\} = \frac{1}{EI}\left\{-56+4\times\frac{2}{3}\right\}$$

$$= \frac{1}{EI}\{-56+2.667\} = \frac{-53.33}{EI}(\rightarrow)$$

$$\therefore \quad \delta_{HF} = \frac{53.33}{EI}(\leftarrow) \text{ towards left in horizontal direction at } F.$$

4.30 Analyse the frame shown in figure by slope deflection method and draw the BMD and SFD.

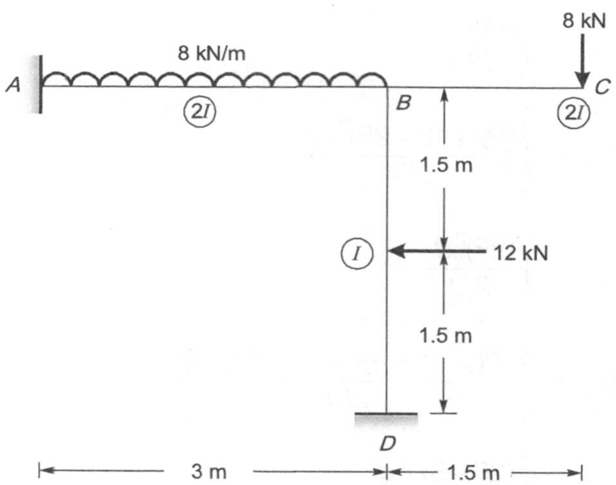

[20 marks : 2018]

Solution:

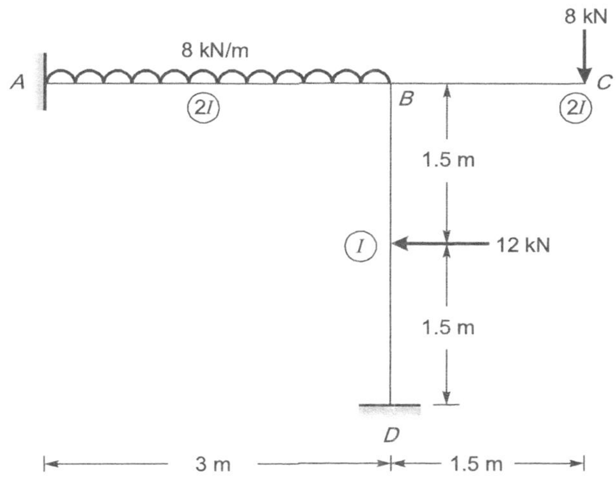

(i) Fixed end moments

$$M_{\overline{AB}} = \frac{-wl^2}{12} = -6 \text{ kN-m} \qquad M_{\overline{BA}} = \frac{-wl^2}{12} = +6 \text{ kN-m}$$

$$M_{\overline{BD}} = \frac{-Wl}{8} = -4.5 \text{ kN-m} \qquad M_{\overline{DB}} = \frac{+Wl}{8} = +4.5 \text{ kN-m}$$

$$M_{\overline{BC}} = M_{BC} = -8 \times 1.5 = -12 \text{ kN-m}$$

(ii) Slope deflection equations:

$$M_{AB} = M_{\overline{AB}} + \frac{2E(2I)}{L}\left[2\theta_A + \theta_B - \frac{3\delta}{L}\right]$$

$$\Rightarrow \qquad M_{AB} = -6 + \frac{4EI(\theta_B)}{3} \qquad \text{...(i)}$$

$$M_{BA} = M_{\overline{BA}} + \frac{2EI(2I)}{L}\left[2\theta_B + \theta_A - \frac{3\delta}{L}\right]$$

$$\Rightarrow \qquad M_{BA} = 6 + \frac{4EI}{3}(2\theta_B) \qquad \text{...(ii)}$$

$$M_{BD} = M_{\overline{BD}} + \frac{2EI}{L}\left[2\theta_B + \theta_D - \frac{3\delta}{L}\right]$$

$$\Rightarrow \qquad M_{BD} = -4.5 + \frac{2EI}{3}(2\theta_B) \qquad \text{...(iii)}$$

$$M_{DB} = M_{\overline{DB}} + \frac{2EI}{L}\left[2\theta_D + \theta_B - \frac{3\delta}{L}\right]$$

$$\Rightarrow \qquad M_{DB} = 4.5 + \frac{2EI}{3}(\theta_B) \qquad \text{...(iv)}$$

Now, at joint B,

$$M_{BA} + M_{BC} + M_{BD} = 0$$

$$\Rightarrow \quad 6 + \frac{4EI}{3}(2\theta_B) - 12 - 4.5 + \frac{2EI}{3}(2\theta_B) = 0$$

$$\Rightarrow \quad 4EI\theta_B = 10.5$$

$$\Rightarrow \quad EI\theta_B = 2.625$$

On substituting the value of $EI\,\theta_B$ in above equations,

$$M_{AB} = -6 + \frac{4}{3}(2.625) = -2.5 \text{ kN-m} \qquad M_{BA} = 6 + \frac{8}{3}(2.625) = 13 \text{ kN-m}$$

$$M_{BC} = -12 \text{ kN-m}$$

$$M_{BD} = -4.5 + \frac{4}{3}(2.625) = -1 \text{ kN-m} \qquad M_{DB} = 4.5 + \frac{2}{3}(2.625) = 6.25 \text{ kN-m}$$

Now,

Free Body diagrams,

(i) Taking moment about B,

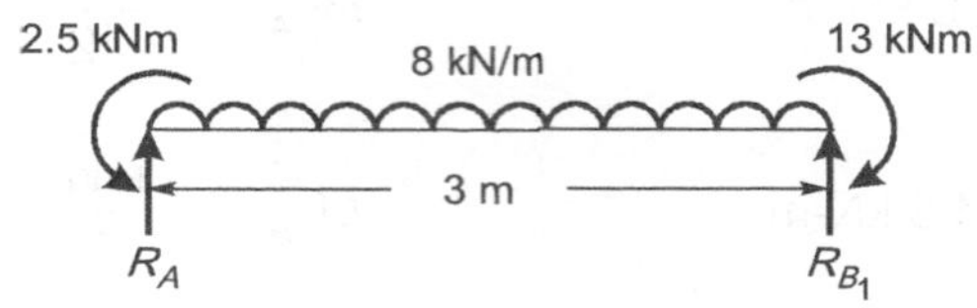

$\Sigma M_B = 0; \quad 3R_A + 13 - 2.5 = (8 \times 3 \times 1.5)$

$\Rightarrow \quad R_A = 8.5 \text{ kN}$

$\therefore \quad R_{B_1} = 24 - R_A = 5.5 \text{ kN}$

(ii)

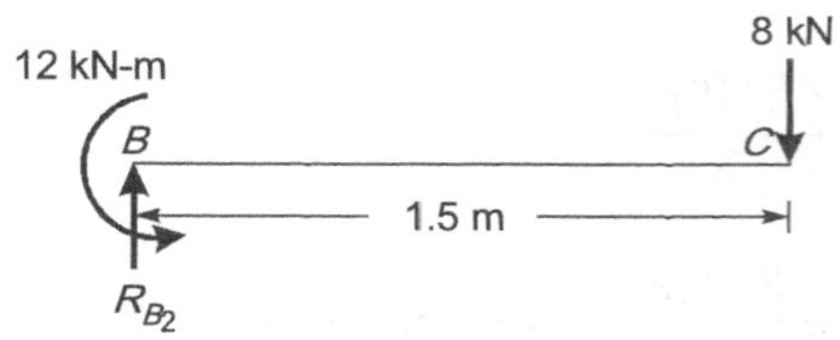

$\Sigma F_y = 0; \quad R_{B_2} = 8 \text{ kN}$

(iii)

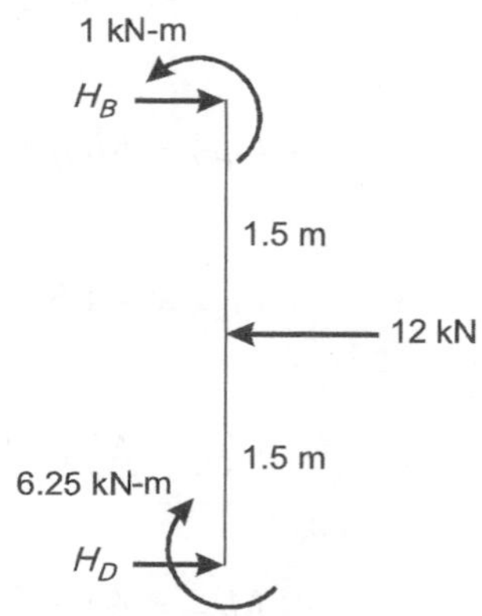

Taking moment about B,

$$3H_D + 1 = 6.25 + (12 \times 1.5)$$

$$\Rightarrow \quad H_D = 7.75 \text{ kN}$$

$$\therefore \quad H_B = 12 - 7.75 = 4.25 \text{ kN}$$

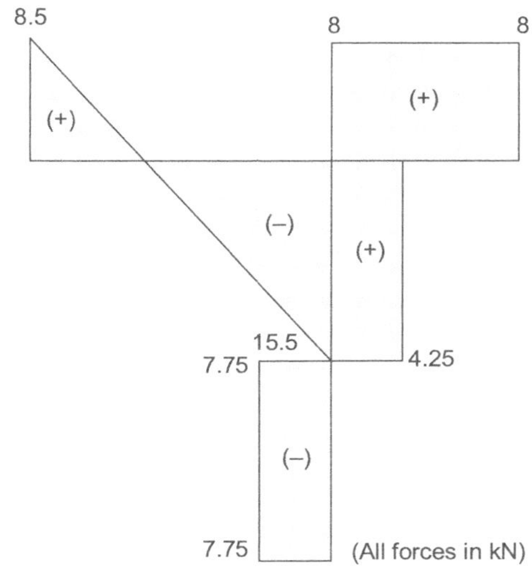

Shear force diagram

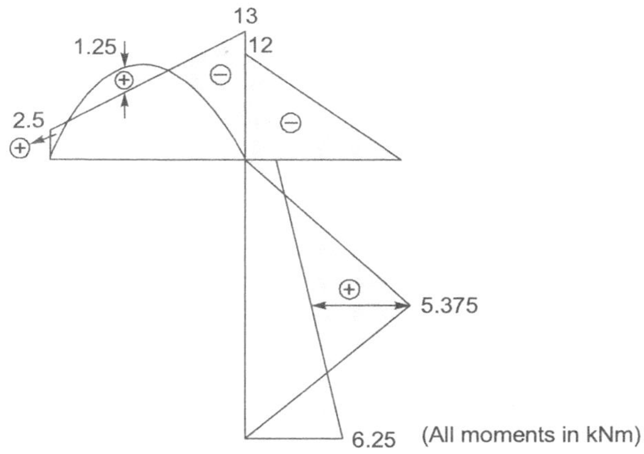

Bending Moment Diagram

4.31 A beam of span L carries a uniformly distributed load ω per unit length on its whole span. It has one simple support at its left end and other support is at a distance of a from the other end. Find the value of a so that the maximum bending moment for the beam is as small as possible. Find also the maximum bending moment for this condition.

[20 marks : 2019]

Solution:

For the given arrangement:

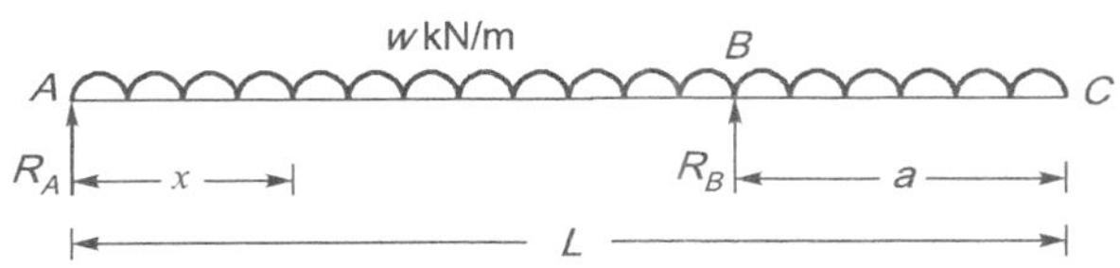

$$R_A + R_B = wL \quad \ldots(i)$$

Taking moment about A,
$$w \times L \times \frac{L}{2} = R_B \times (L - a)$$

$$R_B = \frac{wL^2}{2(L-a)}$$

From equation (i),
$$R_A = \frac{wL(L-2a)}{2(L-a)}$$

Moment in span AB,
$$M_x = \frac{wL(L-2a)}{2(L-a)} \times x - \frac{wx^2}{2}$$

For maximum bending moment in span *AB*,

$$\frac{dM_x}{dx} = 0$$

$$\frac{wL(L-2a)}{2(L-a)} - \frac{2wx}{2} = 0$$

So, $$x = \frac{L(L-2a)}{2(L-a)}$$

The BMD of the beam will be

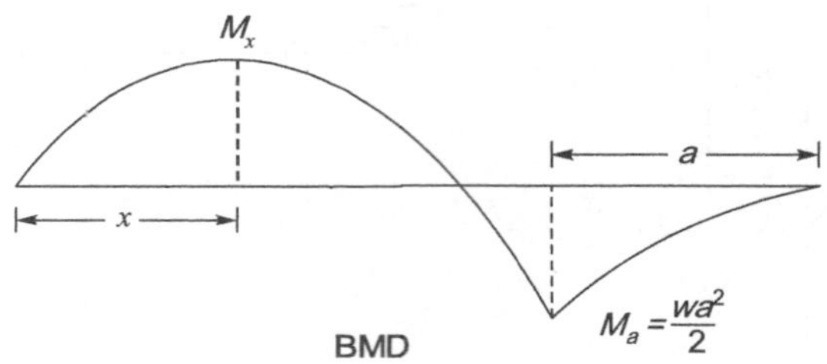

If the maximum bending moment for the beam is as small as possible.
So, the maximum +ve BM = Maximum –ve BM
i.e., $$M_x = M_a$$

$$\frac{wL^2(L-2a)^2}{4(L-a)^2} - \frac{wL^2(L-2a)^2}{2\times 4(L-a)^2} = \frac{wa^2}{2}$$

After solving, $$L^2 - 2aL = 2aL - 2a^2$$

$$a^2 - 2aL + 0.5\,L^2 = 0$$

$$a = 1.707\,L,\ 0.292\,L;\quad a = 1.7072 \text{ is not possible}$$

So, $$a = 0.292\,L$$

So, Maximum BM, $$M_{max} = \frac{wa^2}{2} = \frac{w(0.292L)^2}{2} = 0.042\,wL^2$$

4.32 **Frame ABCD shown in figure, is acted upon by a UDL of intensity 20 kN/m on the horizontal span. What should be value of horizontal force '*P*' applied at C, that will prevent sway of the frame? Draw BMD.**

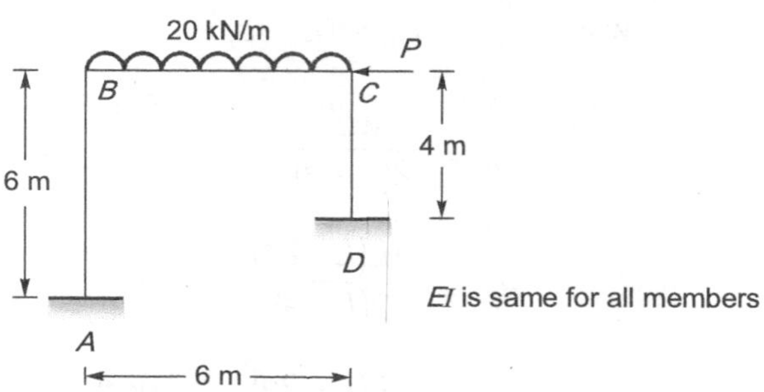

[20 marks : 2020]

Solution:

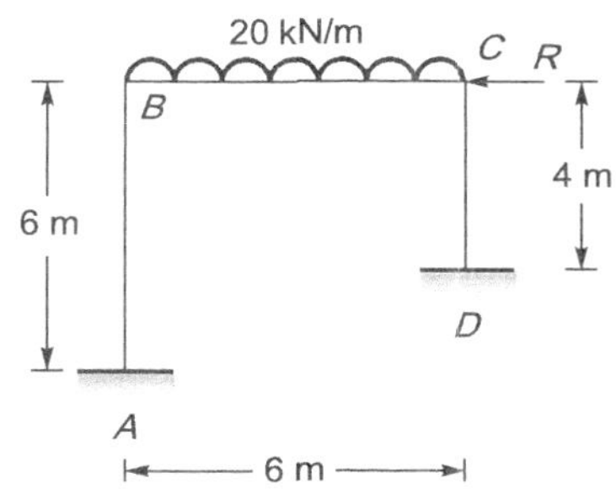

Joint	Member	Relative stiffness	Total stiffness	Distribution factor
B	BA	$\frac{I}{6}$	$\frac{I}{3}$	$\frac{1}{2}$
	BC	$\frac{I}{6}$		$\frac{1}{2}$
C	CB	$\frac{I}{6}$	$\frac{5I}{12}$	$\frac{2}{5}$
	CD	$\frac{I}{4}$		$\frac{3}{5}$

$$(M_F)_{BA} = (M_F)_{AB} = (M_F)_{CD} = (M_F)_{DC} = 0$$

$$(M_F)_{BC} = -60 \text{ kNm}$$

$$(M_F)_{CB} = +60 \text{ kNm}$$

	$\frac{1}{2}$	$\frac{1}{2}$	$\frac{2}{5}$	$\frac{3}{5}$	
0	0	–60	60	0	0
	30	30	–24	–36	
15		–12	+15		–18
	+6	+6	–6	–9	
+3		–3	+3		–4.5
	+1.5	+1.5	–1.2	–1.8	
0.75		–0.6	+0.75		–0.9
	+0.3	+0.3	–0.3	–0.45	
18.75	37.8	–37.8	47.25	–47.25	–23.4

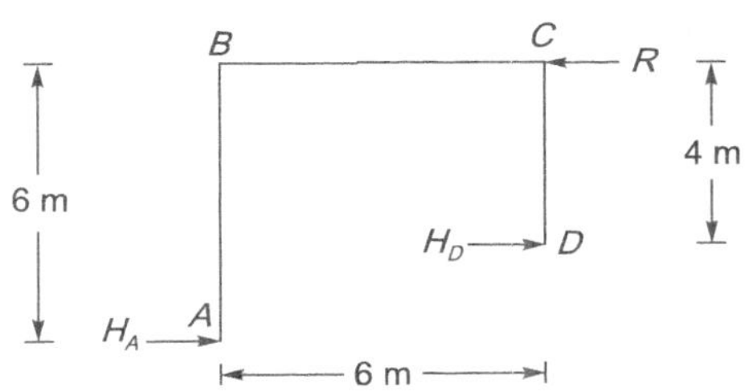

$$\Sigma F_x = 0$$

$$\Rightarrow \quad H_A + H_D - R = 0$$

$$\Rightarrow \quad R = H_A + H_D$$

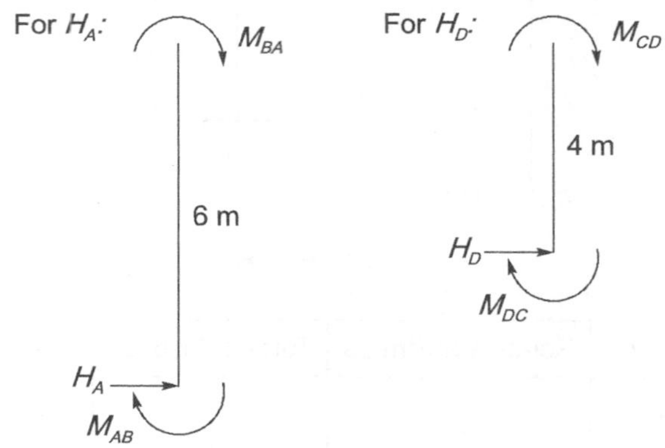

$$\text{Horizontal reaction at } A = \frac{M_{AB} + M_{BC}}{6} = \frac{56.55}{6} = 9.425 \text{ kN}$$

$$(H)_D = \frac{M_{DC} + M_{CD}}{4} = -17.6625 \text{ kN}$$

Sway force (unbalanced horizontal force) = –8.2375 kN

–ve sign represents R is towards right.

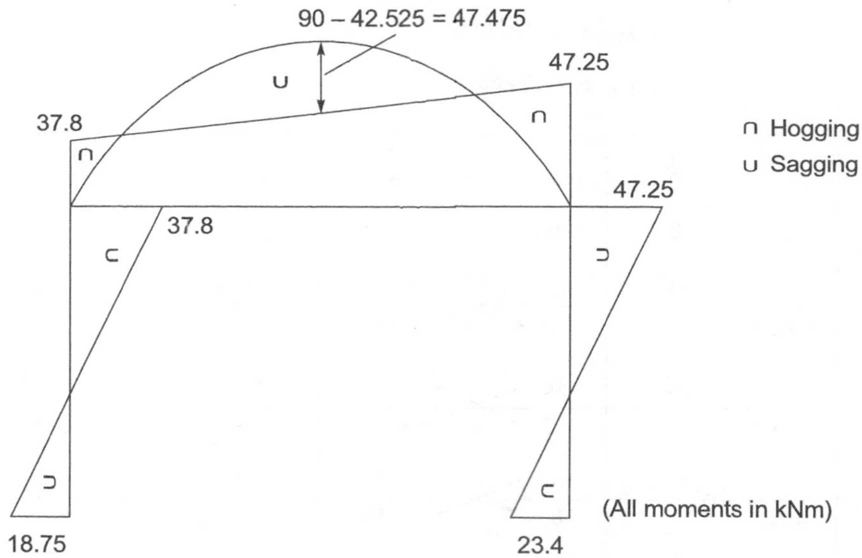

4.33 Analyse a continuous beam shown in the figure. During loading the support B sinks by 12 mm. E = 210 GPa, I = 5131.6 × 10^4 mm^4. EI is constant. Draw BMD and Elastic curve.

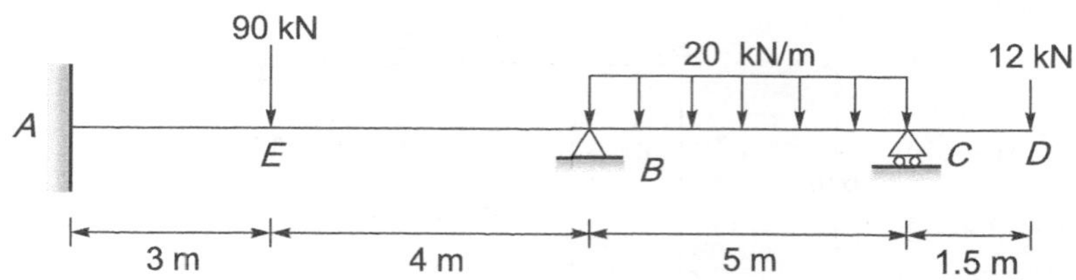

[20 Marks : 2021]

Solution:

Let us do calculation by using moment-distribution method:

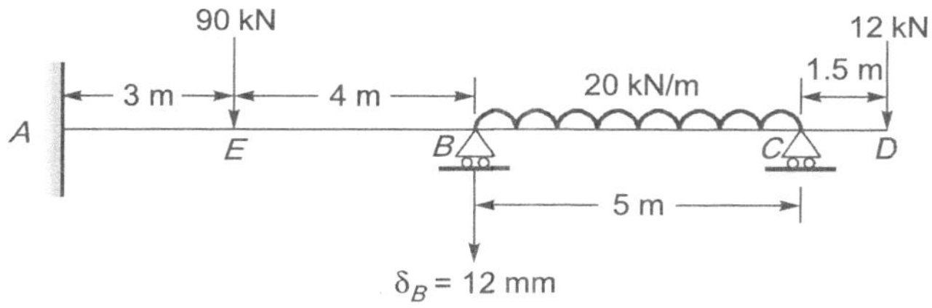

$E = 210$ GPa, $I = 5131.6 \times 10^4$ mm^4 $= \dfrac{5131.6\times10^4}{10^{12}}$ m^4

$$EI = 210\times10^6\times\frac{5131.6\times10^4}{10^{12}} = 10776.36 \text{ kN-m}^2$$

Step-1: Fixed end moments; due to applied loads and sinking of supports.

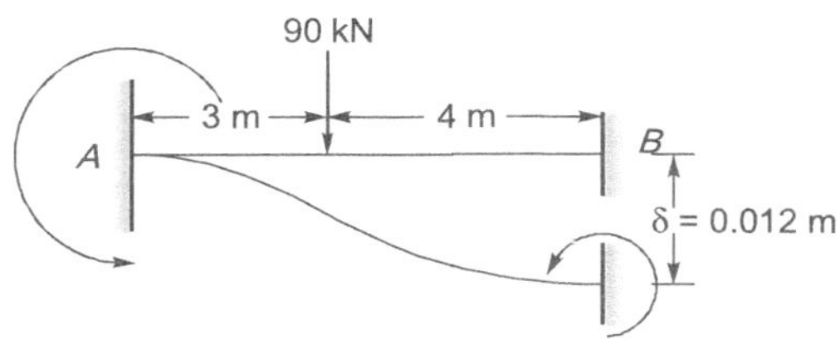

$$M_{FAB} = -\frac{90\times3\times4^2}{7^2}-\frac{6\times10776.36\times0.012}{7^2} = -104 \text{ kNm}$$

$$M_{FBA} = +\frac{90\times3^2\times4}{7^2}-\frac{6\times10776.36\times0.012}{7^2} = 50.29 \text{ kNm}$$

$$M_{FBC} = -\frac{20\times5^2}{12}+\frac{6\times10776.36\times0.012}{5^2} = -10.63 \text{ kNm}$$

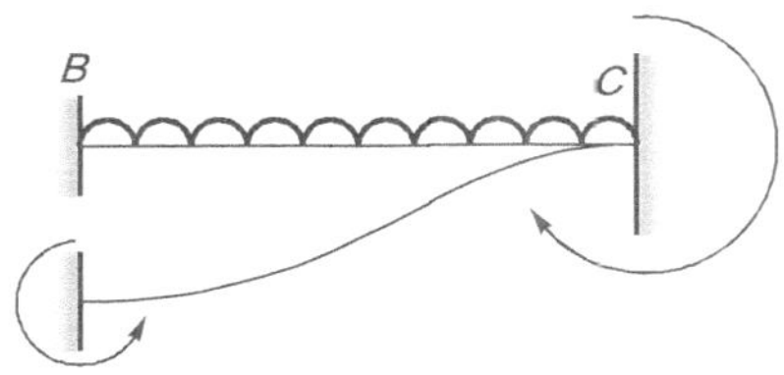

$$M_{FCB} = +\frac{20\times5^2}{12}+\frac{6\times10776.36\times0.012}{5^2} = 72.7 \text{ kNm}$$

$$M_{FCD} = -18 \text{ kNm}$$

Step-2: Distribution factors

Joint	Member	k	Σk	$DF=\dfrac{k}{\Sigma k}$
	BA	$I/7$		0.49
B			$0.293I$	
	BC	$\dfrac{3}{4}\times\dfrac{I}{5}=\dfrac{3I}{20}$		0.51

Step-3: End Moment distribution

Joint	A	B		C		D
DF	0	0.49	0.51	1	0	–
FEMS	–104	50.29	–10.63	72.7	–18	–
Release C COM			–27.35	–54.7	0	
Net FEMS	–104	50.29	–37.98	+18.0	–18	
Bal B		–6.03	–6.28			
COM	–3.02			0		
Final and moments	–107.02	44.26	–44.26	+18	–18	–

Step-4: BMD

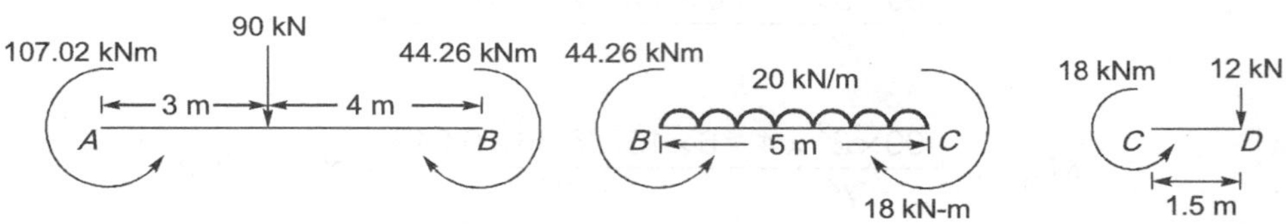

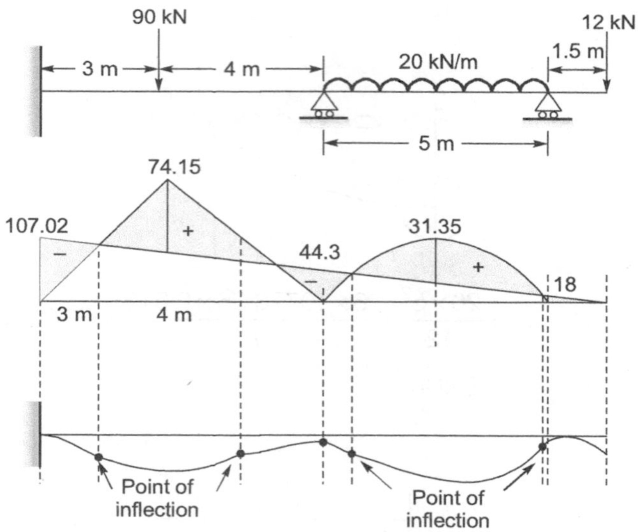

4.34 A rigid jointed portal frame *ABCD* has the horizontal beam member *BC* of length 6 m and moment of inertia. 2I. The left vertical column *AB* of height 6 m and moment of inertia 2I and the right vertical column *DC* of length 4 m and moment of inertia is 1.5I. The supports *A* and *D* are hinged. A uniformly distributed load of 10 kN/m acts on the full length of the beam *BC*. Analyse the portal and draw the bending moment diagram. The analysis of the frame for a sway load of 1 kN at *B* in the direction of $\overrightarrow{BC}$ gives the moment M_{BC} = +2 kN-m and M_{CB} = +2.66 kN-m both clockwise. The *B* and *C* are at the same level.

[20 marks : 2022]

Solution:

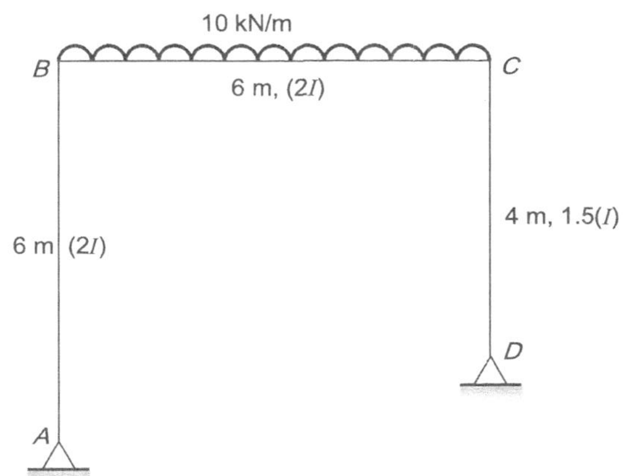

$$M_{FBC} = -\frac{10 \times 6^2}{12} - 30 \text{ kNm}$$

$$M_{FCB} = 30 \text{ kNm}$$

Due to 1 kN sway force,

$$M_{BC} = +2 \text{ kNm}$$
$$M_{CB} = 2.66 \text{ kNm}$$

(i) Distribution factors

Joint	Member	k	Σk	$DF = \frac{k}{\Sigma k}$
B	BA	$\frac{3}{4}\left(\frac{2I}{6}\right)$	$\frac{14I}{24}$	$\frac{6}{14} = 0.43$
	BC	$\frac{2I}{6} = \frac{8I}{24}$		$\frac{8}{14} = 0.57$
C	CB	$\frac{2I}{6} = \frac{I}{3} = 0.33I$	$0.61I$	0.54
	CD	$\frac{3}{4}\left(\frac{1.5I}{4}\right) = \frac{9I}{32} = 0.28I$		0.46

(ii) BMD

Joint	A		B		C	D
D.F.	1	0.43	0.57	0.54	0.46	1
FEMs	0	0	–30	+30		
Balance		+12.9	+17.1	–16.2	–13.8	
COM	0		–8.1	8.55		0
Balance		+3.48	+4.62	–4.62	–3.93	
COM	0		–2.31	2.31		0
Balance		+1.0	+1.31	–1.25	–1.06	
Final end moment	0	17.38	–17.38	+18.79	–18.79	

To find sway force

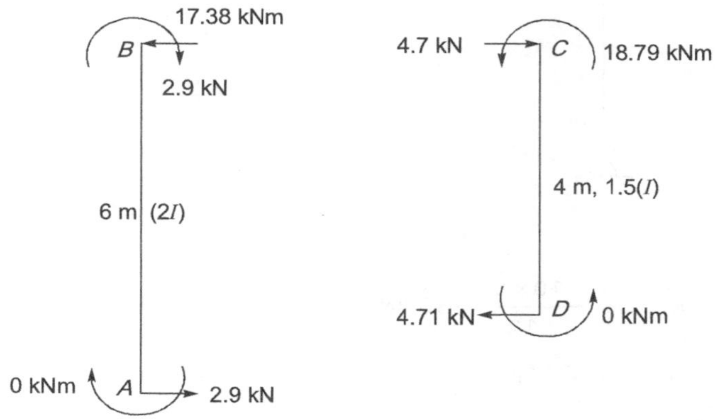

$$\Sigma x \text{ at beam level} = -2.9 + 4.7 = 1.8 \text{ kN } (\rightarrow)$$

So, sway force at beam level = 1.8 kN (←)

For a sway force of 1 kN (→)

$$M_{BC} = +2 \text{ kNm}$$

$$M_{CB} = 2.66 \text{ kNm}$$

Correction factor, $\dfrac{\overleftarrow{1.8}}{\overrightarrow{1}} = -1.8$

Final moment

Joint	A		B	C		D
1. Non-sway end moments	0	17.38	–17.38	+18.79	–18.79	0
2. Corrected sway moments	0	+1.8 × 2	–1.8 × 2	+1.8 × 2.66	1.8 × 2.66	0
		= 3.6	= –3.6	= +4.79	= +4.79	0
Final end moments	0	20.98	20.98	14	–14	0

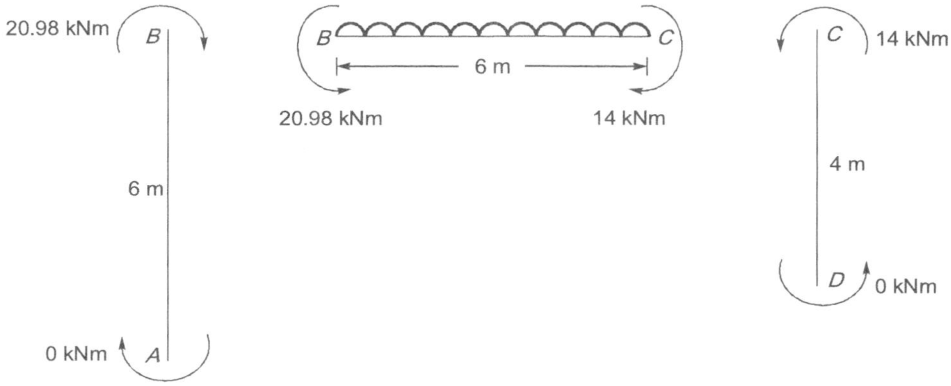

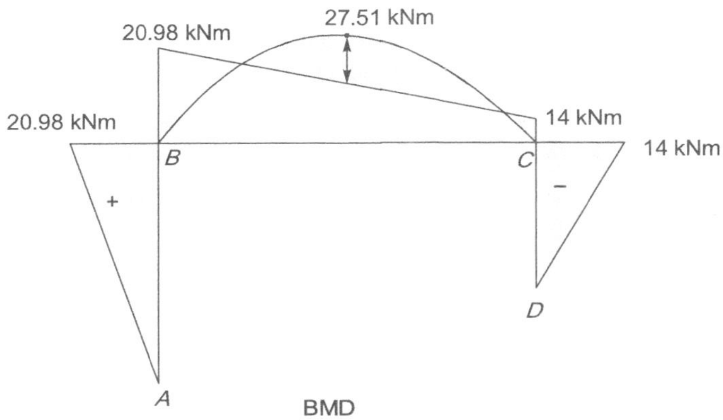

BMD

4.35 A continuous beam *ABC* consists of two spans *AB* and *BC* of length 15.50 and 11.50 m respectively. The span *AB* carries a UDL of 28 kN/m and span *BC* carries a point load of 100 kN at 8 m from support *B*. EI is constant for both the span. The end supports *A* and *C* are fixed, while support *B* is simply supported. Determine the support moments and reactions. And also draw the shear force and bending moment diagrams.

[20 marks : 2022]

Solution:

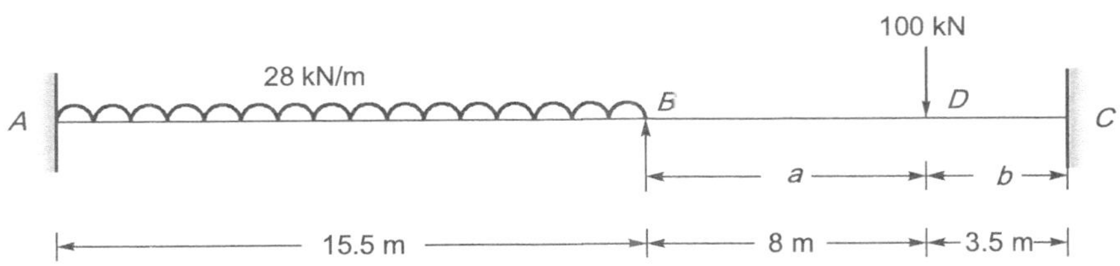

First calculating fixed end moment

$$M_{FAB} = -\frac{wl^2}{12} = -\frac{28 \times 15.5^2}{12} = -560.58 \text{ kNm}$$

$$M_{FBA} = \frac{wl^2}{12} = \frac{28 \times 15.5^2}{12} = 560.58 \text{ kNm}$$

$$M_{FBC} = -\frac{Wab^2}{l^2} = -\frac{100 \times 8 \times 3.5^2}{11.5^2} = -74.102 \text{ kNm}$$

$$M_{FCB} = \frac{Wab^2}{l^2} = \frac{100 \times 3.5 \times 8^2}{11.5^2} = 169.376 \text{ kNm}$$

Now distribution factors for joint *B*.

Joint	Member	Stiffness	Total Stiffness	D.F.
B	*BA*	$\frac{4EI}{15.5}$	$\frac{432EI}{713}$	0.43
	BC	$\frac{4EI}{11.5}$		0.57

By using moment distribution method.

	A	B	B	C
D.F.		0.43	0.57	
F.E.M.	–560.58	560.58	–74.102	169.376
B.M.		–209.2	–277.3	
C.O.M	–104.6			–138.65
Final Moment	–665.2	351.4	–351.34	30.73

Taking individual spans *AB* and *BC*.

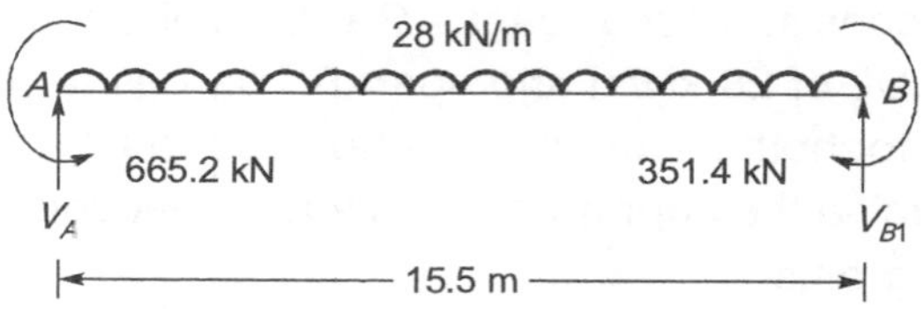

$$\Rightarrow \qquad V_A + V_{B1} = 434 \text{ kN} \qquad \text{...(i)}$$

Taking $\Sigma M_B = 0$

$$\Rightarrow V_A \times 15.5 - 665.2 - 28 \times 15.5 \times \frac{15.5}{2} + 351.4 = 0$$

$$\Rightarrow \qquad V_A = 237.25 \text{ kN}$$

Putting value in eq. (i)

$$V_{B1} = 196.75 \text{ kN}$$

Similarly for span *BC*.

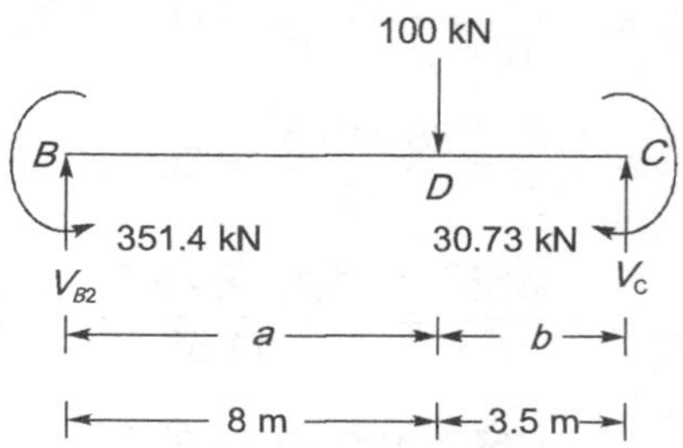

$\Rightarrow \quad V_{B2} + V_C = 100 \text{ kN}$ ...(ii)

Taking $\Sigma M_C = 0$

$\Rightarrow \quad V_{B2} \times 11.5 - 100 \times 3.5 - 351.4 + 30.73 = 0$

$\Rightarrow \quad V_{B2} = 58.30 \text{ kN}$

Putting value in eq. (ii)

$V_C = 41.70 \text{ kN}$

Hence, $V_A = 237.25 \text{ kN}$

$V_B = V_{B1} + V_{B2}$

$V_B = 196.75 + 58.30 = 255.05 \text{ kN}$

$V_C = 41.70 \text{ kN}$

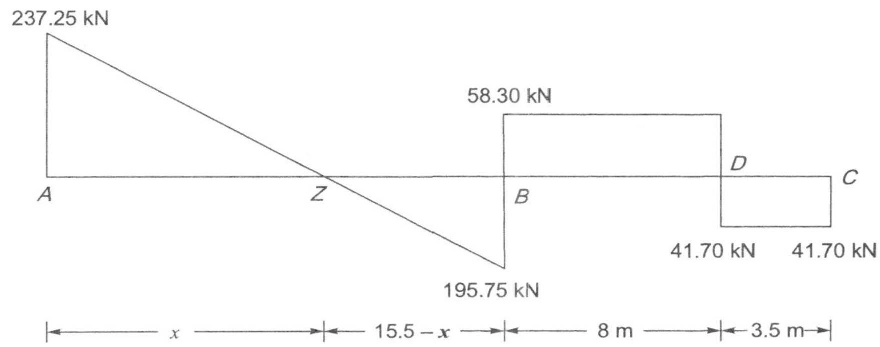

SFD

Point Z can be found out by similar triangle property

Then $$\frac{x}{15.5 - x} = \frac{237.25}{196.75}$$

$\Rightarrow \quad 196.75x = 3677.375 - 237.25x$

$\Rightarrow \quad x = 8.47$

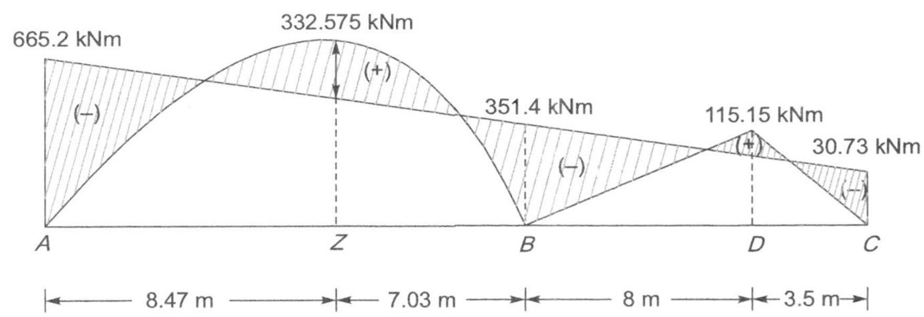

BMD

4.36 Analyze the portal frame shown in the figure below by moment distribution method. The frame is fixed at A and D, and has rigid joints at B and C. Draw the bending moment diagram and sketch the deflected shape of the structure. Take EI as constant:

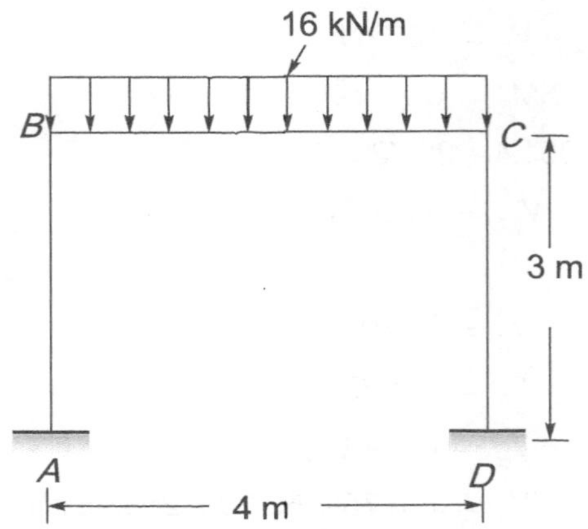

[20 marks : 2023]

Solution:

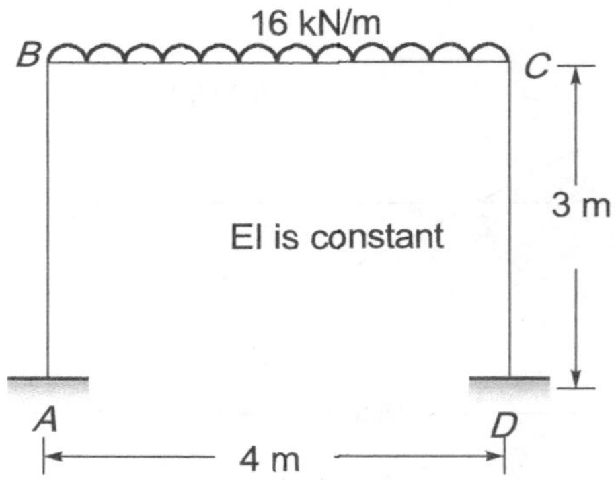

Fixed end moment,

$$M_{FAB} = M_{FBA} = 0$$

$$M_{FBC} = \frac{-wL^2}{12}$$

$$= \frac{-16 \times 4^2}{12} = -21.33 \text{ kN-m}$$

$$M_{FCB} = \frac{wL^2}{12} = \frac{16 \times 4^2}{12} = 21.33 \text{ kN-m}$$

$$M_{FCD} = M_{FDC} = 0$$

Distribution factor:

Joint	Member	Stiffness	Total stiffness	Distribution factor
B	BA	$\frac{4EI}{3}$	$\frac{7EI}{3}$	$\frac{4}{7}$
	BC	$\frac{4EI}{4}$		$\frac{3}{7}$
C	CB	$\frac{4EI}{4}$	$\frac{7EI}{3}$	$\frac{3}{7}$
	CD	$\frac{4EI}{3}$		$\frac{4}{7}$

Moment distribution table

Joint	A	B		C		D
Distribution factor		$\frac{4}{7}$	$\frac{3}{7}$	$\frac{3}{7}$	$\frac{4}{7}$	
Fixed end moment	0	0	−21.33	21.33	0	0
Balancing moment		12.19	9.14	−9.14	−12.19	
Carry over moment	6.095		−4.57	4.57		−6.095
Balancing moment		2.61	1.96	−1.96	−2.61	
Carry over moment	1.305		−0.98	0.98		−1.305
Balancing moment		0.56	0.42	−0.42	−0.56	
Carry over moment	0.28		−0.21	0.21		−0.28
Final end moment	7.7	15.5	−15.5	−15.5	−15.5	−7.7

Now, B.M.D. can be drawn as shown.

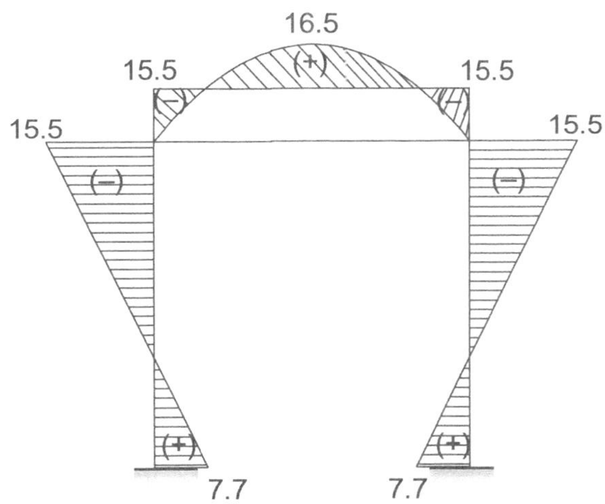

Deflected shape is shown below.

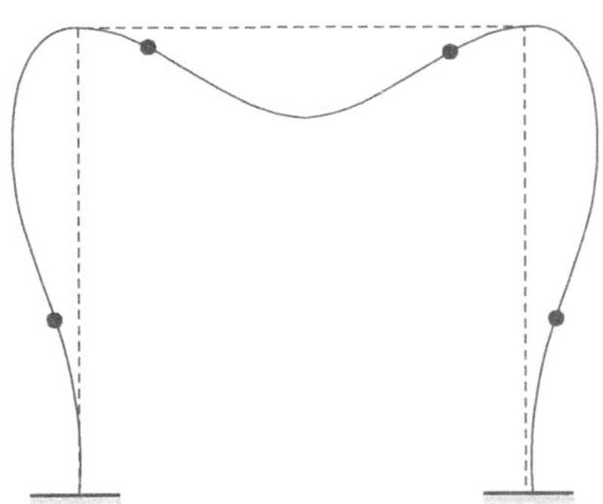

4.37 Determine the vertical and horizontal deflections at the free end of the frame shown in the figure below. Take $EI = 12 \times 10^4$ kN-m^2:

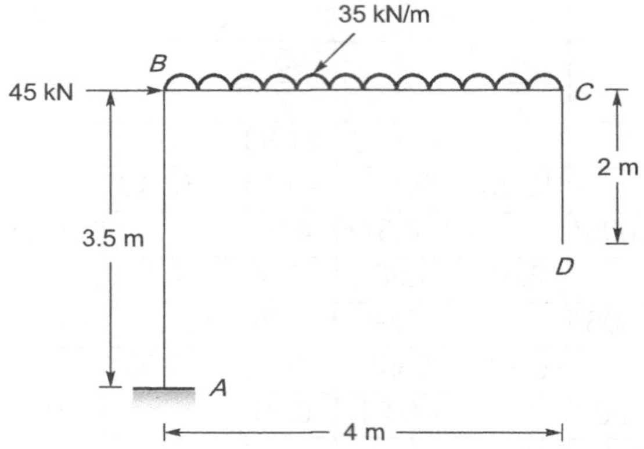

[20 marks : 2023]

Solution:

Vertical deflection of *D* :

For vertical deflection at *D*, Apply a unit load at *D* in downward direction as shown in figure below.

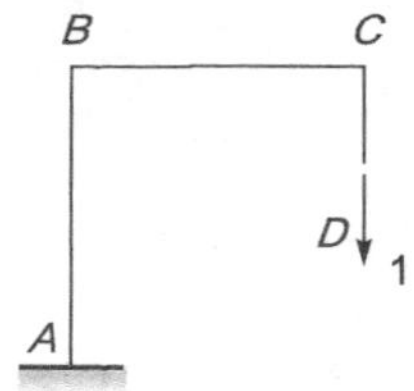

As per unit load method calculations are done below:

Portion	*DC*	*CB*	*BA*
Origin	*D*	*C*	*B*
Limit	0 - 2	0 - 4	0 - 3.5
M	0	$\frac{-35x^2}{2}$	$-45x - 280$
m_1	0	$-x$	-4

Therefore vertical deflection of *C*, $\Delta_{VC} = \sum \frac{Mm, dx}{EI}$

So,
$$\Delta_{VC} = 0 + \int_0^4 \frac{\frac{-35x^2}{2} \times -x \; dx}{EI} + \int_0^{3.5} \frac{(-4)\times(-45x - 280)}{EI}$$

$$= \frac{1120}{EI} + \frac{10045}{2EI} = \frac{6142.5}{EI} = \frac{6142.5}{12 \times 10^4} = 51 \times 10^{-3} \text{ m} = 51 \text{ mm}$$

Horizontal deflection at *D* :

For horizontal deflection at *D*, Apply a unit load at *D* in horizontal direction as shown in figure below.

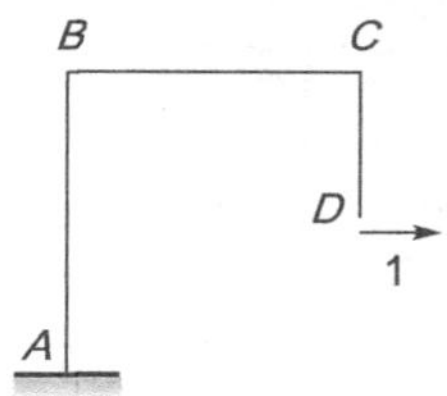

As per unit load method, calculations are done below:

Portion	DC	CB	BA
Origin	D	C	B
Limit	0 - 2	0 - 4	0 - 3.5
M	0	$\frac{-35x^2}{2}$	$-45x-280$
m_2	x	2	$2-x$

So, $$\Delta_{HC} = \int \frac{Mm_2dx}{EI}$$

$$= 0 + \int_0^4 \frac{\frac{-35x^2}{2} \times 2dx}{EI} + \int_0^{3.5} \frac{(-45x-280)(2-x)}{EI}$$

$$= \frac{-2240}{3EI} + \frac{-1225}{8EI} = \frac{-899.79}{EI} = \frac{-899.79}{12\times10^4} = -7.5 \text{ mm}$$

(–ve) sign indicates that deflection will be in opposite direction to that of load applied.

5. Trusses

5.1 Figure below shows a plane pinjointed truss *BCDEFG* of the given dimensions and carrying the loading shown. All member have the same axial rigidity.

(i) Calculate the axial force in member *FG*.

(ii) Evaluate the force in the same member if it is fabricatered 0.1% too short, and the structure carries no external load.

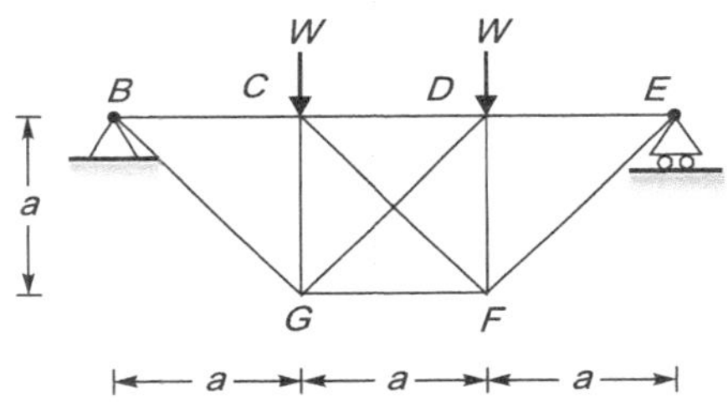

[30 + 10 = 40 marks : 1999]

Solution:

(i) The degree of determinacy may be given by

$$D_S = m + r_e - 2j$$

Here $m = 10, r_e = 3, j = 6$

$\therefore$ $D_S = 10 + 3 - 2 \times 6 = +1$

After removing *GF*, the static indeterminacy will be

$$D_S = m + r_e - 2j$$

$\Rightarrow$ $D_S = (10-1) + 3 - 2 \times 6$

$\Rightarrow$ $D_S = 0$

P-system of forces

After removing *GF*, the resultant truss will be as in the given figure.

$$\Sigma F_y = 0$$

$\Rightarrow$ $V_B + V_E = 2W$

and $\Sigma M_B = 0$

$\Rightarrow$ $V_E \times 3a - W \times 2a - W \times a = 0$

$\Rightarrow$ $V_E = \frac{3Wa}{3a}$

$\Rightarrow$ $V_E = W$

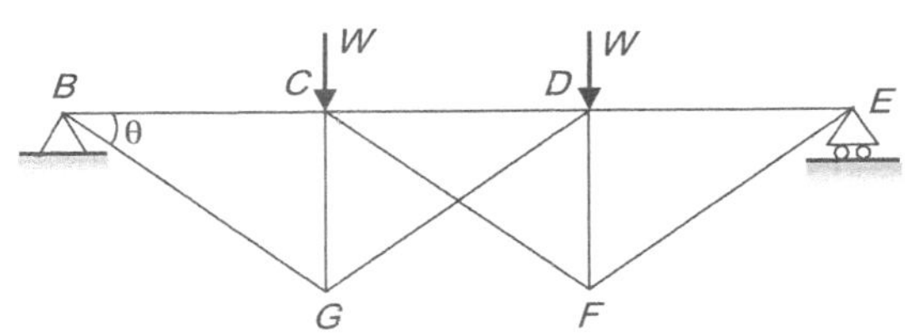

$\therefore \quad V_B = W$

Also $\quad \tan\theta = \dfrac{a}{a}$

$\Rightarrow \quad \tan\theta = 1, \quad \sin\theta = \dfrac{1}{\sqrt{2}}$ and $\cos\theta = \dfrac{1}{\sqrt{2}}$

Assuming tensile forces as positive and compressive forces as negative

Considering joint *B*

$\Sigma F_y = 0$	AND $\Sigma F_x = 0$
$\Rightarrow W - F_{BG}\sin\theta = 0$	$\Rightarrow F_{BC} + F_{BG}\cos\theta = 0$
$\Rightarrow F_{BG} = \dfrac{W}{\sin\theta}$	$\Rightarrow F_{BC} = -\sqrt{2}\,W \times \dfrac{1}{\sqrt{2}}$
$\Rightarrow F_{BG} = \sqrt{2}\,W$ (Tension)	$\Rightarrow F_{BC} = -W$ (Compression)

Considering joint *G*

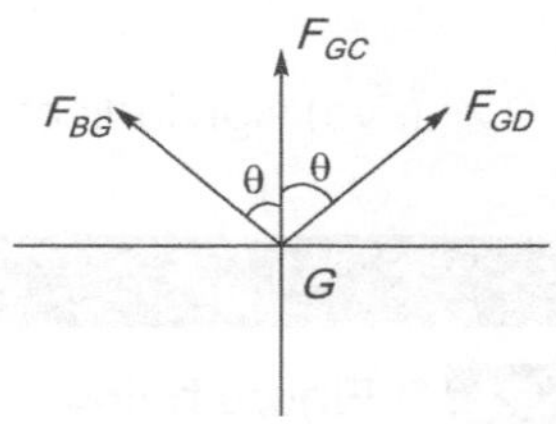

$\Sigma F_x = 0$	AND $\Sigma F_y = 0$
$\Rightarrow F_{GD}\sin\theta - F_{BG}\sin\theta = 0$	$\Rightarrow F_{GD}\cos\theta + F_{BG}\cos\theta + F_{GC} = 0$
$\Rightarrow F_{GD} = F_{BG}$	$\Rightarrow F_{GC} = -\sqrt{2}\,W \times \dfrac{1}{\sqrt{2}} - \sqrt{2}\,W \times \dfrac{1}{\sqrt{2}}$
$\Rightarrow F_{GD} = \sqrt{2}\,W$ (Tension)	$F_{GC} = -2\,W$ (Compression)

Considering joint *C*

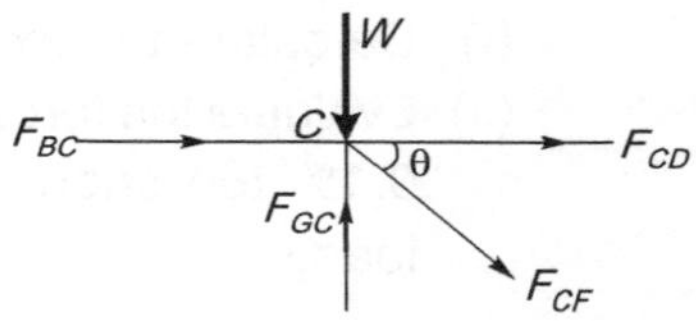

$\Sigma F_y = 0$	AND $\Sigma F_x = 0$
$\Rightarrow F_{GC} - W - F_{CF}\sin\theta = 0$	$\Rightarrow F_{CD} + F_{CF}\cos\theta + F_{BC} = 0$
$\Rightarrow 2W - W = F_{CF}\sin\theta$	$\Rightarrow F_{CD} = -W\sqrt{2} \times \dfrac{1}{\sqrt{2}} - W$
$\Rightarrow F_{CF} = W\sqrt{2}$ (Tension)	$\Rightarrow F_{CD} = -2W$ (Compression)

Now, as the truss is symmetrical, the forces in remaining members will be corresponding to the forces in the above members.

i.e.,

$$F_{BG} = F_{EF} = \sqrt{2}\,W \text{(Tension)}$$
$$F_{BC} = F_{ED} = -W \text{(Compression)}$$
$$F_{CG} = F_{FD} = -2W \text{(Compression)}$$

***K*- system of forces**

The truss is now analysed by removing all external loads and applying unit tensile forces at *G* and *F* respectively.

The only forces acting on the truss are unit tensile forces at *G* and *F* respectively which are opposite in nature. Therefore the vertical reaction will be zero at *B* and *E* respectively.

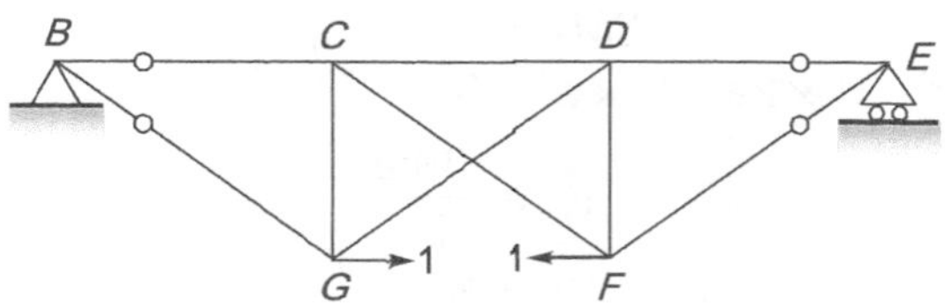

$\therefore \quad K_{BC} = 0$ and $K_{BG} = 0$

Also $\quad K_{ED} = 0$ and $K_{EF} = 0$ and $K_{GF} = +1$

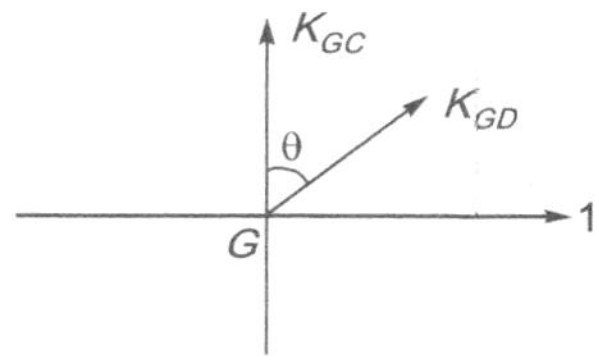

Considering joint G

$\Sigma F_x = 0$ AND $\Sigma F_y = 0$

$\Rightarrow \quad K_{GD}\sin\theta + 1 = 0$

$\Rightarrow \quad K_{GD} = -\sqrt{2}$ (Compression)

$K_{GC} + K_{GD}\cos\theta = 0$

$\Rightarrow \quad K_{GC} = -(-\sqrt{2}) \times \dfrac{1}{\sqrt{2}}$

$\Rightarrow \quad K_{GC} = 1$ (Tension)

By symmetry

$K_{CF} = -\sqrt{2}$ (Compression)

$K_{FD} = 1$ (Tension)

Considering joint C

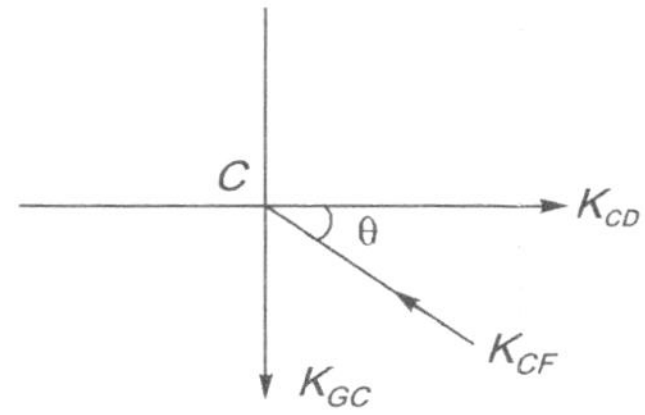

$\Sigma F_x = 0$

$\Rightarrow \quad K_{CD} - K_{CF}\cos\theta = 0$

$\Rightarrow \quad K_{CD} = \sqrt{2} \times \dfrac{1}{\sqrt{2}}$

$\Rightarrow \quad K_{CD} = 1$

Member	*P*	*K*	*L*	*PKL*	K^2L	*KX′*
BC	$-W$	0	a	0	0	0
BG	$\sqrt{2}W$	0	$a\sqrt{2}$	0	0	0
CD	$-2W$	1	a	$-2Wa$	a	X'
CF	$\sqrt{2}W$	$-\sqrt{2}$	$\sqrt{2}a$	$-2\sqrt{2}Wa$	$2\sqrt{2}a$	$-\sqrt{2}X'$
GC	$-2W$	1	a	$-2Wa$	a	X'
ED	$-W$	0	a	0	0	0
EF	$\sqrt{2}W$	0	$\sqrt{2}a$	0	0	0
FD	$-2W$	1	a	$-2Wa$	a	X'
GF	0	1	a	0	a	X'
GD	$\sqrt{2}W$	$-\sqrt{2}$	$\sqrt{2}a$	$-2\sqrt{2}Wa$	$2\sqrt{2}a$	$-\sqrt{2}X'$

$$\Sigma PKL = -(6Wa + 4\sqrt{2}\,Wa)$$

$$\Sigma K^2L = 4a + 4\sqrt{2}\,a$$

We know that,

$$X = -\frac{\Sigma \dfrac{PKL}{AE}}{\Sigma \dfrac{K^2L}{AE}}$$

Axial rigidity of members is constant i.e. AE is constant.

$$\therefore \quad X = -\frac{\Sigma PKL}{\Sigma K^2L} - \frac{-(6Wa + 4\sqrt{2}\,Wa)}{4a + 4\sqrt{2}\,a}$$

$$\Rightarrow \quad X = \left(\frac{3 + 2\sqrt{2}}{2 + 2\sqrt{2}}\right) W = 1.207\ W$$

Force in $GF = P + KX = 0 + 1 \times 1.207\ W$

$$= 1.207\ W \text{(Tension)}$$

(ii) *GF* is 0.1 % too short

$$\Delta = \frac{0.1}{100} \times a = \frac{a}{1000}$$

Let forces in all the members are expressed in terms of X'.

$$\therefore \quad S = P + KX'$$

But the structure carries no external load i.e. $P = 0$

$$\therefore \quad S = KX'$$

The total strain energy stored in the structure may be given by

$$U = U_{BG} + U_{BC} + U_{CD} + U_{CF} + U_{GC} + U_{ED} + U_{EF} + U_{FD} + U_{GF} + U_{GD}$$

$$\Rightarrow \quad U = 0 + 0 + \frac{(X')^2 a}{2AE} + \frac{\left(-\sqrt{2}X'\right)^2 \sqrt{2}a}{2AE} + \frac{(X')^2 a}{2AE} + 0 + 0 + \frac{(X')^2 a}{2AE}$$

$$+ \frac{(X')^2 a}{2AE} + \frac{\left(-\sqrt{2}X'\right)^2 \times \sqrt{2}a}{2AE}$$

$$\Rightarrow \quad U = \frac{4a(X')^2}{2AE} + \frac{4\sqrt{2}\,a(X')^2}{2AE}$$

$$\Rightarrow \quad U = \frac{4a(X')^2}{2AE}\left(\sqrt{2} + 1\right)$$

$$\Rightarrow \quad \frac{\partial U}{\partial X'} = \frac{8a\,X'\left(\sqrt{2}+1\right)}{2AE} = \Delta$$

$$\therefore \quad \frac{4\left(\sqrt{2}+1\right)a\,X'}{AE} = \frac{a}{1000}$$

$$\Rightarrow \quad X' = \frac{AE}{4000\left(\sqrt{2}+1\right)}$$

Thus force in *GF* due to lack of fit = KX'

$$= 1 \times \frac{AE}{4000\left(\sqrt{2}+1\right)} = \frac{AE}{4000\left(\sqrt{2}+1\right)}$$

5.2 When fabricating the pinjointed plane frame shown in the figure below, it was found that the member *AC* is fabricated 'Δ' too long. Determine the forces in the members after assembly. All members have same axial rigidity (*AE*).

[20 marks : 2000]

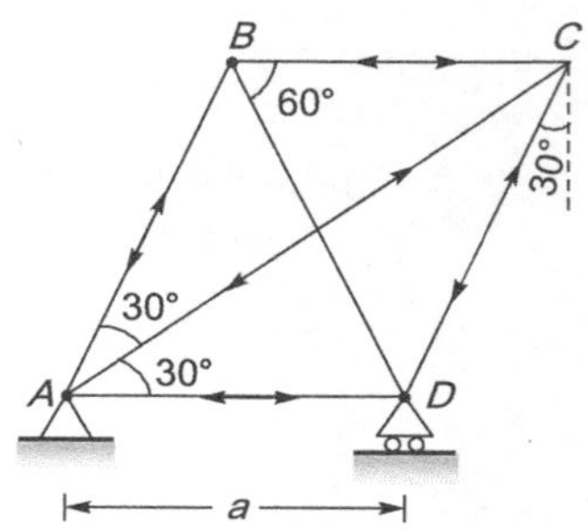

Solution:

The degree of indeterminacy of the plane frame may be given as

$$D_S = m + r_e - 2j$$

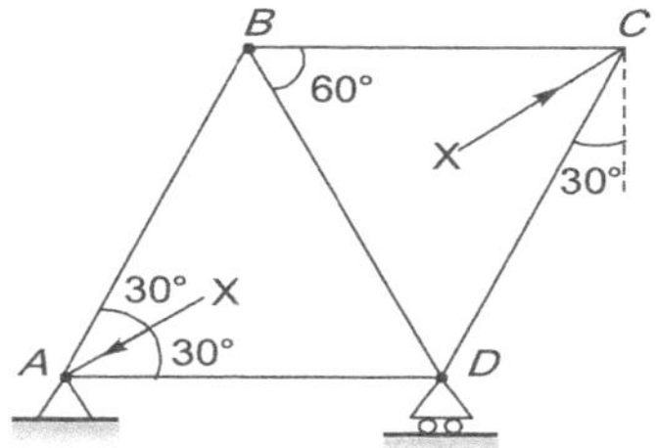

Here $m = 6, r_e = 3, j = 4$

$\therefore$ $D_S = 6 + 3 - 2 \times 4$

$\Rightarrow$ $D_S = 1$

After removing AC, the degree of indeterminacy will be

$$D_S = (6 - 1) + 3 - 2 \times 4$$

$\Rightarrow$ $D_S = 0$

Since the member AC is 'Δ' too long, therefore, the forces at A and C will be compressive in nature. Let it be X.

Reaction at A and D will be zero.

Considering joint A

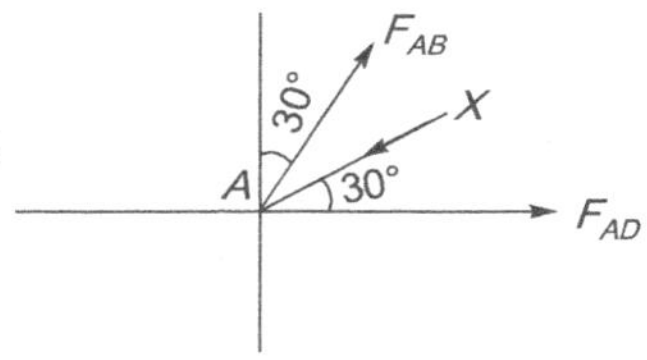

$\Sigma F_x = 0$

$\Rightarrow$ $F_{AD} - X\cos 30° + F_{AB}\sin 30° = 0$

$\Rightarrow$ $F_{AD} = \dfrac{\sqrt{3}}{2}X - \dfrac{0.5X}{\sqrt{3}}$

$\Rightarrow$ $F_{AD} = \dfrac{X}{\sqrt{3}}$ (Tension)

AND

$\Sigma F_y = 0$

$\Rightarrow$ $F_{AB}\cos 30° - X\sin 30° = 0$

$\Rightarrow$ $F_{AB} \times \dfrac{\sqrt{3}}{2} = \dfrac{X}{2}$

$\Rightarrow$ $F_{AB} = \dfrac{X}{\sqrt{3}}$ (Tension)

Considering joint B

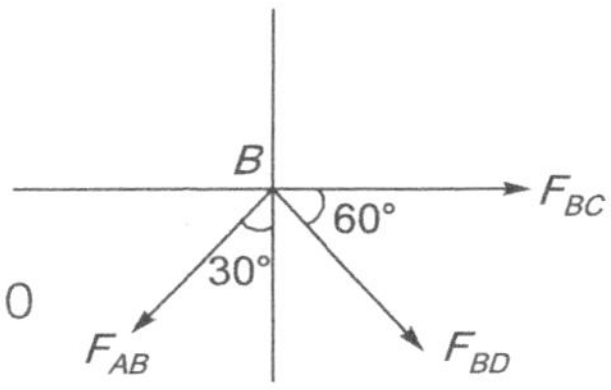

$\Sigma F_y = 0$

$\Rightarrow$ $F_{AB}\cos 30° + F_{BD}\sin 60° = 0$

$\Rightarrow$ $F_{BD}\dfrac{\sqrt{3}}{2} = -\dfrac{X}{\sqrt{3}} \times \dfrac{\sqrt{3}}{2}$

$\Rightarrow$ $F_{BD} = -\dfrac{X}{\sqrt{3}}$ (Compression)

AND

$\Sigma F_x = 0$

$\Rightarrow$ $F_{BC} + F_{BD}\cos 60° - F_{AB}\sin 30° = 0$

$\Rightarrow$ $F_{BC} = \dfrac{X}{\sqrt{3}} \times \dfrac{1}{2} - \left(-\dfrac{X}{\sqrt{3}} \times \dfrac{1}{2}\right)$

$\Rightarrow$ $F_{BC} = \dfrac{X}{2\sqrt{3}} + \dfrac{X}{2\sqrt{3}}$

$F_{BC} = \dfrac{X}{\sqrt{3}}$ (Tension)

Considering joint D

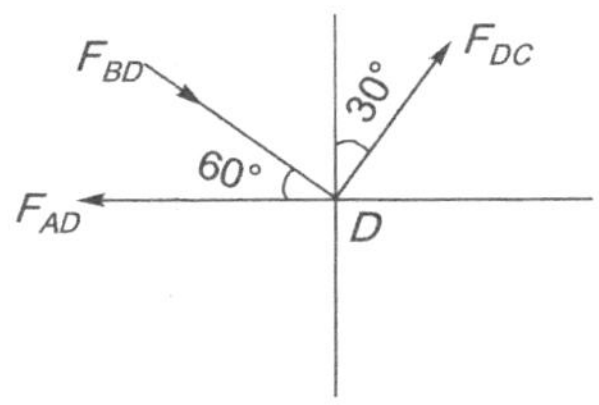

$\Sigma F_x = 0$

$\Rightarrow$ $F_{DC}\sin 30° + F_{BD}\cos 60° - F_{AD} = 0$

$\Rightarrow$ $\dfrac{F_{DC}}{2} = \dfrac{X}{\sqrt{3}} - \dfrac{X}{\sqrt{3}} \times \dfrac{1}{2}$

$\Rightarrow$ $\dfrac{F_{DC}}{2} = \dfrac{2X - X}{2\sqrt{3}}$

$\Rightarrow$ $F_{DC} = \dfrac{X}{\sqrt{3}}$ (Tension)

We know that diagonals of a rhombus bisect each other at right angles.

$\therefore$ $\dfrac{BD/2}{a} = \cos 60°$

$\Rightarrow$ $BD = a$

Also $\dfrac{AC/2}{a} = \sin 60°$

$\Rightarrow$ $AC = \sqrt{3}a$

Now,

$$U = U_{AB} + U_{AD} + U_{BC} + U_{BD} + U_{DC} + U_{AC}$$

$$\Rightarrow \quad U = \frac{\left(\frac{X}{\sqrt{3}}\right)^2 \times a}{2AE} + \frac{\left(\frac{X}{\sqrt{3}}\right)^2 \times a}{2AE} + \frac{\left(\frac{X}{\sqrt{3}}\right)^2 \times a}{2AE} + \frac{\left(-\frac{X}{\sqrt{3}}\right)^2 \times a}{2AE} + \frac{\left(\frac{X}{\sqrt{3}}\right)^2 \times a}{2AE} + \frac{X^2 \times \sqrt{3}a}{2AE}$$

$$U = \frac{5X^2a}{6AE} + \frac{\sqrt{3}X^2a}{2AE}$$

$$\Rightarrow \quad \frac{\partial U}{\partial X} = \frac{10Xa}{6AE} + \frac{2\sqrt{3}Xa}{2AE} \qquad \left[\because \frac{\partial U}{\partial X} = \Delta\right]$$

$$\Rightarrow \quad \frac{Xa}{AE}\left(\frac{5}{3} + \sqrt{3}\right) = \Delta$$

$$\Rightarrow \quad X = \left(\frac{3}{5+3\sqrt{3}}\right)\frac{AE\Delta}{a}$$

Forces in the members of the plane frame

$$\text{Force in } AB = \frac{X}{\sqrt{3}} = \left(\frac{3}{5+3\sqrt{3}}\right)\frac{AE\Delta}{a} \times \frac{1}{\sqrt{3}} = \left(\frac{\sqrt{3}}{5+3\sqrt{3}}\right)\frac{AE\Delta}{a} \text{(Tension)}$$

$$\text{Force in } AD = \frac{X}{\sqrt{3}} = \left(\frac{3}{5+3\sqrt{3}}\right)\frac{AE\Delta}{a} \times \frac{1}{\sqrt{3}} = \left(\frac{\sqrt{3}}{5+3\sqrt{3}}\right)\frac{AE\Delta}{a} \text{(Tension)}$$

$$\text{Force in } BC = \frac{X}{\sqrt{3}} = \left(\frac{3}{5+3\sqrt{3}}\right)\frac{AE\Delta}{a} \times \frac{1}{\sqrt{3}} = \left(\frac{\sqrt{3}}{5+3\sqrt{3}}\right)\frac{AE\Delta}{a} \text{(Tension)}$$

$$\text{Force in } DC = \frac{X}{\sqrt{3}} = \left(\frac{3}{5+3\sqrt{3}}\right)\frac{AE\Delta}{a} \times \frac{1}{\sqrt{3}} = \left(\frac{\sqrt{3}}{5+3\sqrt{3}}\right)\frac{AE\Delta}{a} \text{(Tension)}$$

$$\text{Force in } BD = -\frac{X}{\sqrt{3}} = -\left(\frac{3}{5+3\sqrt{3}}\right)\frac{AE\Delta}{a} \times \frac{1}{\sqrt{3}} = -\left(\frac{\sqrt{3}}{5+3\sqrt{3}}\right)\frac{AE\Delta}{a} \text{(Compression)}$$

$$\text{Force in } AC = -X = -\left(\frac{3}{5+3\sqrt{3}}\right)\frac{AE\Delta}{a} \text{(Compression)}$$

5.3 All members of the truss shown in the below figure are pin jointed. Calculate the reactions and forces in all the members.

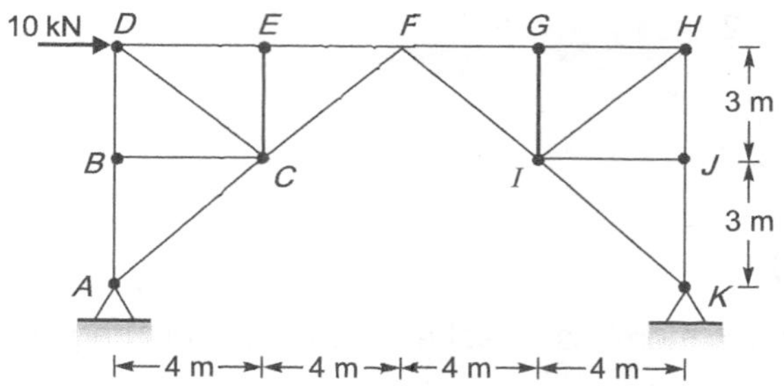

[20 marks : 2002]

Solution:

The degree of indeterminacy of the given truss may be given as

$$D_S = m + r_e - 2j$$

Here $m = 18,\ r_e = 4,\ j = 11$

$\therefore$ $D_S = 18 + 4 - 2 \times 11$

$\Rightarrow$ $D_S = 0$

The given truss is statically determinate.

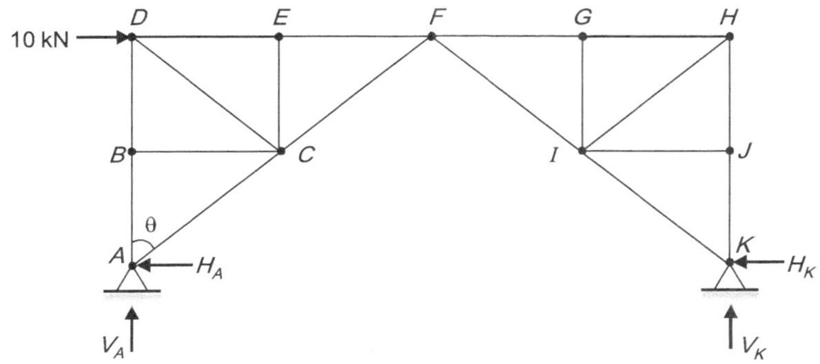

Reactions

$$\Sigma F_x = 0$$

$\Rightarrow$ $H_A + H_K = 10$ ... (i)

$$\Sigma F_y = 0$$

$\Rightarrow$ $V_A + V_K = 0$ ... (ii)

$$\Sigma M_A = 0$$

$\Rightarrow$ $V_K \times 16 - 10 \times 6 = 0$

$\Rightarrow$ $V_K = 3.75 \text{ kN}$

$\therefore$ From (ii), we have $V_A = -V_K = -3.75 \text{ kN } (\downarrow)$

Compatibility condition

$\Sigma M_F = 0$ from right

Also, $\tan\theta = \frac{4}{3}$; $\sin\theta = \frac{4}{5}$; $\cos\theta = \frac{3}{5}$

$\Rightarrow$ $V_K \times 8 - H_K \times 6 = 0$

$\Rightarrow$ $H_K = \frac{3.75 \times 8}{6}$

$\Rightarrow$ $H_K = 5 \text{ kN}$

Assume tension as positive and compression as negative

$\therefore$ From (i), we get

$H_A = 10 - 5$

$\Rightarrow$ $H_A = 5 \text{ kN}$

Considering joint A

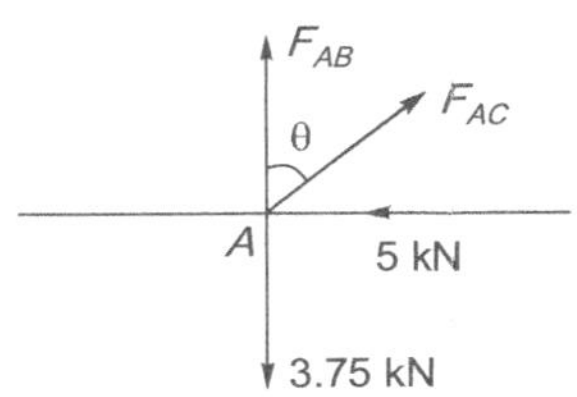

$\Sigma F_x = 0$ AND $\Sigma F_y = 0$

$\Rightarrow$ $F_{AC}\sin\theta - 5 = 0$	$\Rightarrow$ $F_{AC}\cos\theta + F_{AB} - 3.75 = 0$
$\Rightarrow$ $F_{AC} \times \frac{4}{5} = 5$	$\Rightarrow$ $F_{AB} = 3.75 - 6.25 \times \frac{3}{5}$
$F_{AC} = 6.25 \text{ kN (Tension)}$	$\Rightarrow$ $F_{AB} = 0$

Since $F_{AB} = 0$, therefore *BC* and *BD* will also carry zero force.

Considering joint *D*

$\Sigma F_y = 0$ AND $\Sigma F_x = 0$

$\Rightarrow F_{DC}\cos\theta = 0$ | $\Rightarrow F_{DC}\sin\theta + F_{DE} + 10 = 0$

$\Rightarrow F_{DC} = 0$ | $\Rightarrow F_{DE} = -10$ kN (Compression)

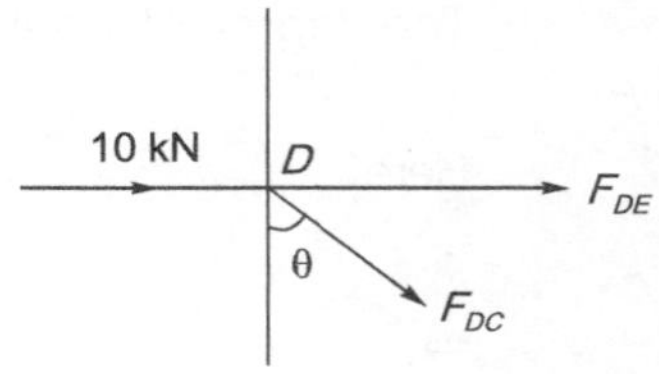

Considering joint *E*

$\Sigma F_x = 0$ AND $\Sigma F_y = 0$

$\Rightarrow F_{EF} + F_{DE} = 0$ | $\Rightarrow F_{EC} = 0$

$\Rightarrow F_{EF} + 10 = 0$

$\Rightarrow F_{EF} = -10$ kN (Comp.)

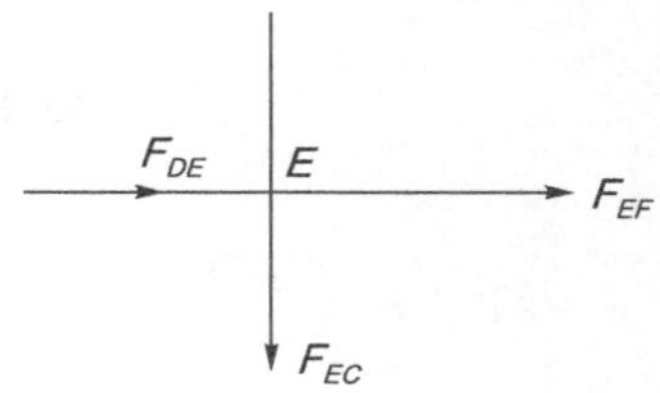

Considering joint *C*

$\Sigma F_x = 0$

$\Rightarrow F_{CF}\sin\theta - F_{AC}\sin\theta = 0$

$\Rightarrow F_{CF} = F_{AC}$

$\Rightarrow F_{CF} = 6.25$ kN (Tension)

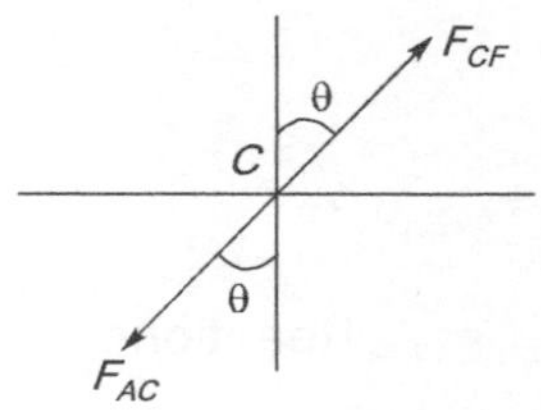

Considering joint *K*

$\Sigma F_x = 0$ AND $\Sigma F_y = 0$

$\Rightarrow F_{KI}\sin\theta + 5 = 0$ | $\Rightarrow F_{KI}\cos\theta + F_{KJ} + 3.75 = 0$

$\Rightarrow F_{KI} \times \frac{4}{5} = -5$ | $\Rightarrow F_{KJ} = -3.75 - (-6.25) \times \frac{3}{5}$

$\Rightarrow F_{KI} = -6.25$ kN (Comp.) | $\Rightarrow F_{KJ} = 0$

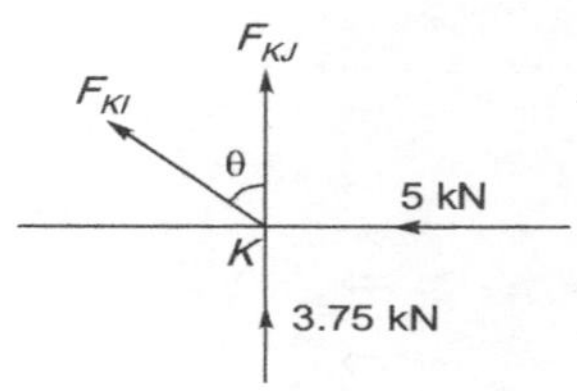

Since $F_{KJ} = 0$, therefore *JI* and *JH* will carry zero force. As a result *HI* and *HG* will also carry zero force.

Considering joint *I*

$\Sigma F_x = 0$ AND $F_y = 0$

$\Rightarrow -F_{KI}\sin\theta - F_{IF}\sin\theta = 0$ | $\Rightarrow F_{IG} + F_{KI}\cos\theta + F_{IF}\cos\theta = 0$

$\Rightarrow F_{IF} + F_{KI} = 0$

$\Rightarrow F_{IF} = -6.25$ kN (Comp.) | $\Rightarrow F_{IG} = -\frac{3}{5}(6.25 - 6.25)$

$\Rightarrow F_{IG} = 0$

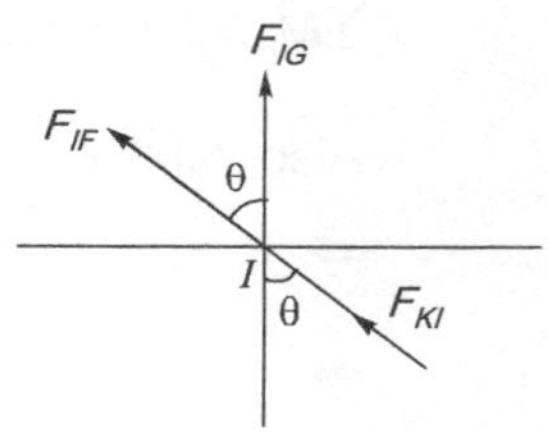

Considering joint *F*

$$\Sigma F_x = 0$$

$$\Rightarrow F_{FG} - F_{IF}\sin\theta + F_{EF} - F_{CF}\sin\theta = 0$$

$$\Rightarrow F_{FG} - 6.25 \times \frac{4}{5} + 10 - 6.25 \times \frac{4}{5} = 0$$

$$\Rightarrow F_{FG} = 0$$

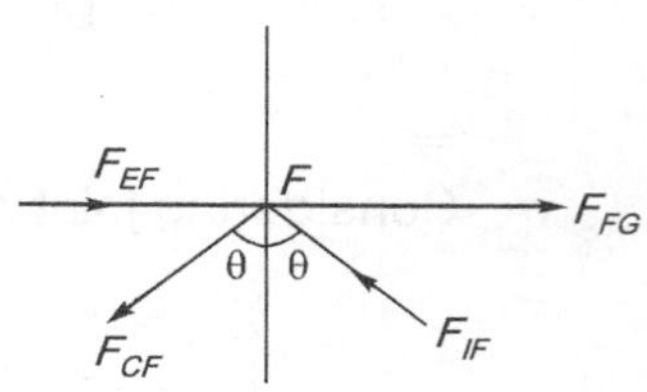

$\therefore$ $F_{AB} = 0$

$F_{AC} = 6.25$ kN (Tension) $\quad F_{BC} = 0$

$F_{BD} = 0$	$F_{DC} = 0$
$F_{DE} = -10$ kN (Compression)	$F_{EC} = 0$
$F_{EF} = -10$ kN (Compression)	$F_{CF} = 6.25$ kN (Tension)
$F_{KJ} = 0$	$F_{KI} = -6.25$ kN (Compression)
$F_{JI} = 0$	$F_{JH} = 0$
$F_{HI} = 0$	$F_{HG} = 0$
$F_{IF} = -6.25$ kN (Compression)	$F_{IG} = 0$

5.4 The square truss *ABCD* shown below carries a load of 10 kN attached to a string *GF* and passing over a frictionless pulley at *A*. *GF* and *AB* are parallel. The truss is supported on rollers at *C* and *D* and by a link *BE* at *B*. Find the forces in the member of the truss.

[15 marks : 2003]

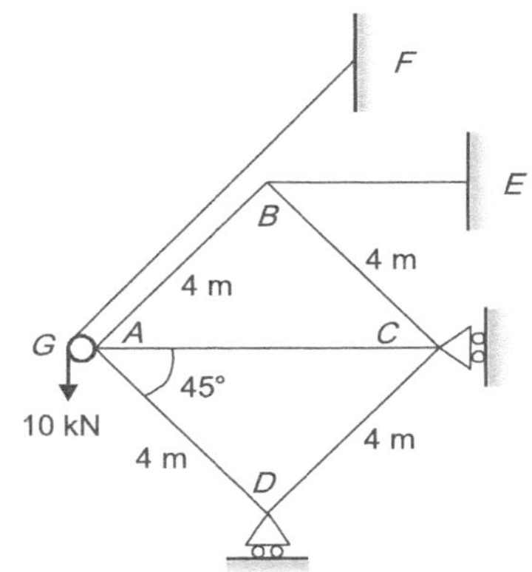

Solution:

The degree of indeterminacy of the square truss may be given as

$$D_S = m + r_e - 2j$$

Here $m = 6,\ r_e = 4,\ j = 5$

$\therefore$ $D_S = 6 + 4 - 2 \times 5$

$\Rightarrow$ $D_S = 0$

Thus, the given square truss is statically determinate.

Let the tension in the string be *T*, but *T* will be equal to 10 kN as the pulley is frictionless.

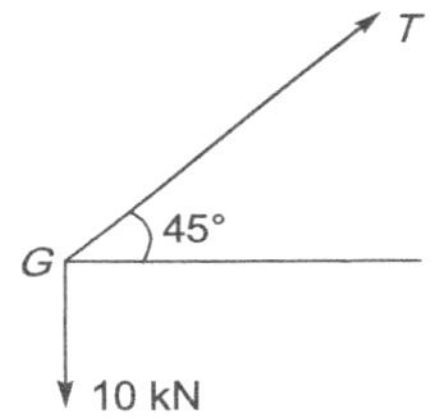

$$\therefore \text{ Total downward load} = -T\sin 45° + 10 = -\frac{10}{\sqrt{2}} + 10 = 2.93 \text{ kN } (\downarrow)$$

Total load acting on the truss from left to right $= T\cos 45°$

$$= \frac{10}{\sqrt{2}} = 7.07 \text{ kN } (\rightarrow)$$

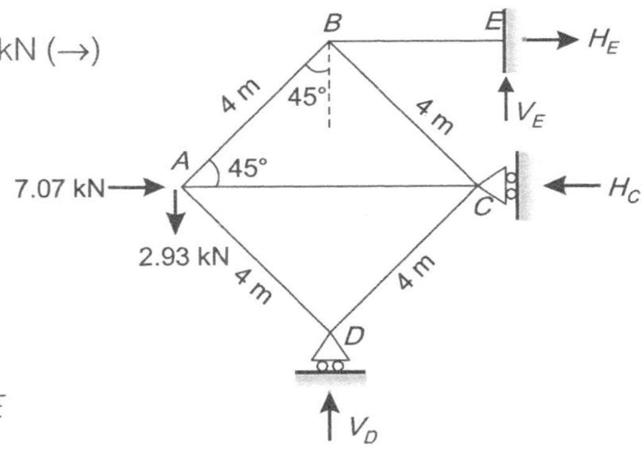

$$\cos 45° = \frac{BE}{4}$$

$\Rightarrow$ $BE = \dfrac{4}{\sqrt{2}}$

Reactions

$\Rightarrow$ $\Sigma M_B = 0$ for link *BE*

$\Rightarrow$ $V_E \times BE = 0$

$\Rightarrow$ $V_E \times \dfrac{4}{\sqrt{2}} = 0$

$\Rightarrow$ $V_E = 0$

$\Sigma F_y = 0$

$\Rightarrow$ $V_E + V_D - 2.93 = 0$

$V_D = 2.93$ kN

Also $\Sigma M_B = 0$ for the whole truss

$$\Rightarrow \quad 7.07 \times \frac{4}{\sqrt{2}} + 2.93 \times \frac{4}{\sqrt{2}} - H_C \times \frac{4}{\sqrt{2}} = 0$$

$$\Rightarrow \quad H_C = 10 \text{ kN}$$

and $\quad \Sigma F_x = 0$

$$\Rightarrow \quad H_E - H_C + 7.07 = 0$$

$$\Rightarrow \quad H_E = 10 - 7.07$$

$$\Rightarrow \quad H_E = 2.93 \text{ kN}$$

Assuming tension as positive and compression as negative.

Considering joint *E*

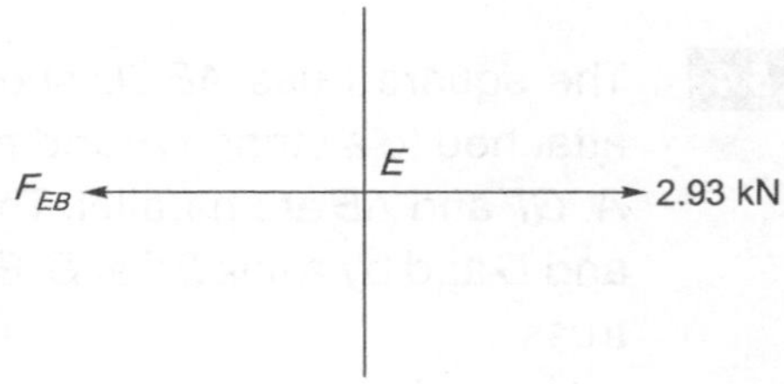

$$\Sigma F_x = 0$$

$$\Rightarrow \quad 2.93 - F_{EB} = 0$$

$$F_{EB} = 2.93 \text{ kN (Tension)}$$

Considering joint *B*

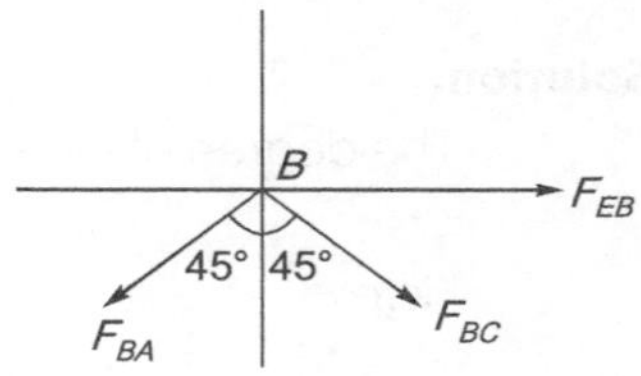

$$\Sigma F_x = 0$$

$$\Rightarrow \quad F_{BC} \sin 45° + F_{EB} - F_{BA} \sin 45° = 0$$

$$\Rightarrow \quad \frac{F_{BC}}{\sqrt{2}} - \frac{F_{BA}}{\sqrt{2}} = -2.93$$

$$\Rightarrow \quad F_{BC} - F_{BA} = -2.93\sqrt{2} \qquad \text{... (i)}$$

$$\Sigma F_y = 0$$

$$\Rightarrow \quad F_{BC} \cos 45° + F_{BA} \cos 45° = 0$$

$$\Rightarrow \quad F_{BC} + F_{BA} = 0 \qquad \text{... (ii)}$$

Solving (i) and (ii), we get

$$F_{BC} = -2.07 \text{ kN (Compression)}$$

$$F_{BA} = 2.07 \text{ kN (Tension)}$$

Considering joint *C*

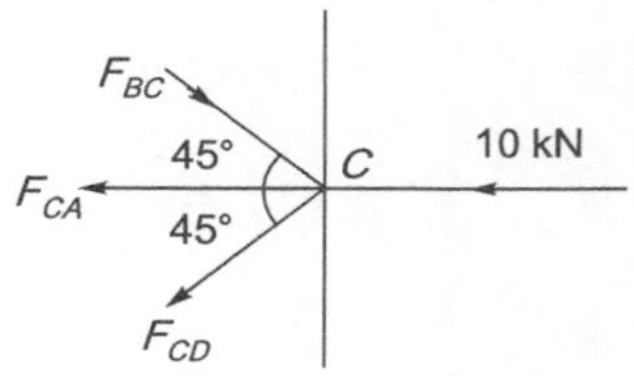

$$\Sigma F_y = 0$$

$$\Rightarrow \quad F_{CD} \sin 45° + F_{BC} \sin 45° = 0$$

$$\Rightarrow \quad F_{CD} = -F_{BC}$$

$$\Rightarrow \quad F_{CD} = -2.07 \text{ kN (Compression)}$$

$$\Sigma F_x = 0$$

$$\Rightarrow \quad F_{BC} \cos 45° - F_{CD} \cos 45° - F_{CA} - 10 = 0$$

$$\Rightarrow \quad 2.07 \times \frac{1}{\sqrt{2}} - (-2.07) \times \frac{1}{\sqrt{2}} - 10 = F_{CA}$$

$$\Rightarrow \quad F_{CA} = -7.07 \text{ kN (Compression)}$$

Considering joint *D*

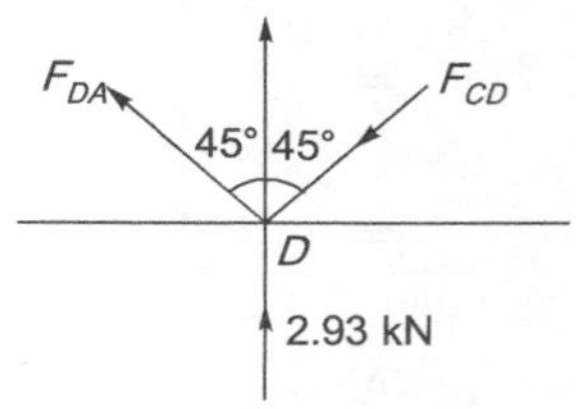

$$\Sigma F_x = 0$$

$$\Rightarrow \quad F_{DA} \sin 45° + F_{CD} \sin 45° = 0$$

$$\Rightarrow \quad \frac{F_{DA}}{\sqrt{2}} = \frac{-2.07}{\sqrt{2}}$$

$\Rightarrow$ $F_{DA} = -2.07$ kN (Compression)

Forces in members

$F_{EB} = 2.93$ kN (Tension)

$F_{BC} = -2.07$ kN (Compression)

$F_{BA} = 2.07$ kN (Tension)

$F_{CD} = -2.07$ kN (Compression)

$F_{CA} = -7.07$ kN (Compression)

$F_{DA} = -2.07$ kN (Compression)

5.5 Find the forces in the members of the truss shown in figure. The truss is hinged at A and simply supported at C and E. It carries 2 kN, 4 kN and 2 kN at nodes G, H and I respectively. The diagonals make 45° with the horizontal and vertical members.

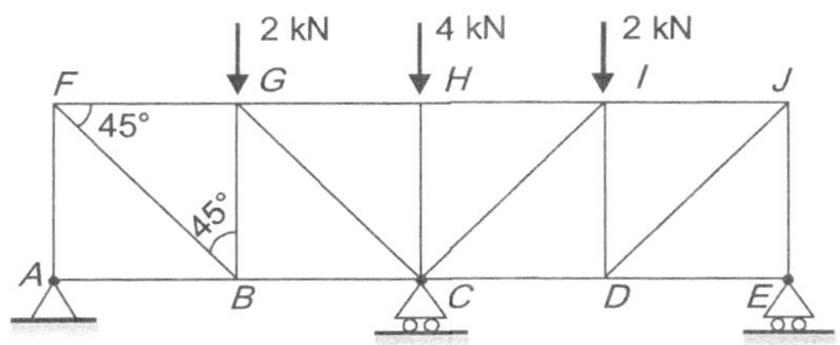

[30 marks : 2004]

Solution:

The degree of indeterminacy of the given truss may be given as

$$D_S = m + r_e - 2j$$

Here $m = 17,\ r_e = 4,\ j = 10$

$\therefore$ $D_S = 17 + 4 - 2 \times 10$

$\Rightarrow$ $D_S = 1$

However, this truss is internally determinate. It is externally indeterminate because total number of external reactions are more than the available equilibrium equations.

P system of forces

Removing the support at C, the truss will become statically determinate.

$\therefore$ $D_S = 17 + 3 - 2 \times 10$

$\Rightarrow$ $D_S = 0$

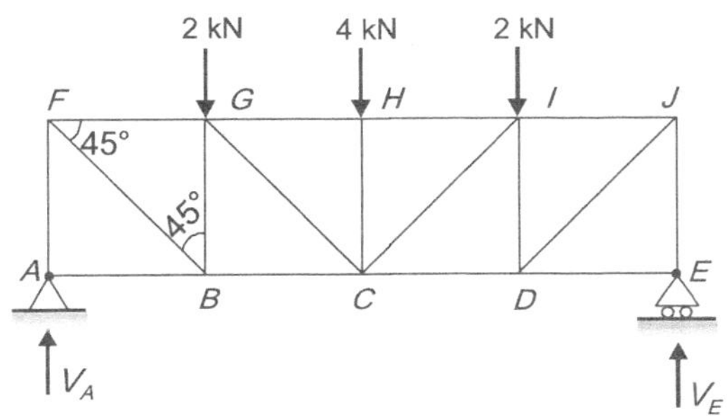

Since the truss is loaded symmetrically, the reactions at the supports will be equal

$\therefore \quad V_A = V_E = 4 \text{ kN}$

Assuming tension as positive and compression as negative.

Considering joint *A*

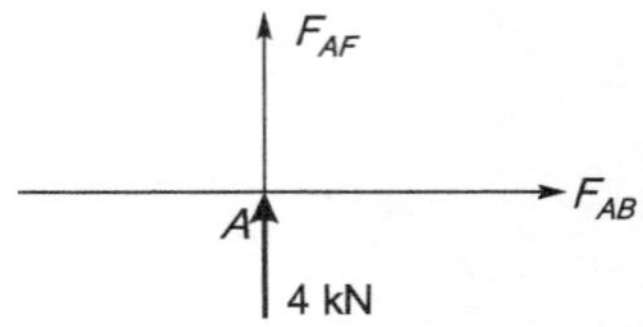

$\Sigma F_x = 0$ AND $\Sigma F_y = 0$

$\Rightarrow \quad F_{AB} = 0$

$\Rightarrow \quad F_{AF} + 4 = 0$

$\Rightarrow \quad F_{AF} = -4 \text{ kN (Compression)}$

Considering joint *F*

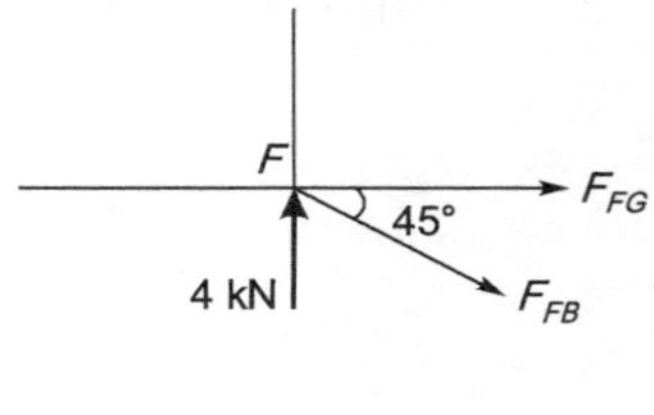

$\Sigma F_y = 0$ AND $\Sigma F_x = 0$

$\Rightarrow \quad 4 - F_{FB}\sin 45° = 0$

$\Rightarrow \quad \dfrac{F_{FB}}{\sqrt{2}} = 4$

$\Rightarrow \quad F_{FB} = 4\sqrt{2} \text{ kN (Tension)}$

$\Rightarrow \quad F_{FG} + F_{FB}\cos 45° = 0$

$\Rightarrow \quad F_{FG} = -4\sqrt{2} \times \dfrac{1}{\sqrt{2}}$

$\Rightarrow \quad F_{FG} = -4 \text{ kN (Compression)}$

Considering joint *B*

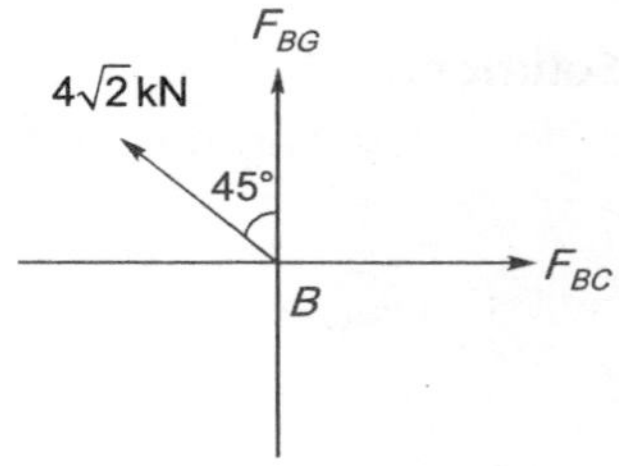

$\Sigma F_x = 0$ AND $\Sigma F_y = 0$

$\Rightarrow \quad F_{BC} - 4\sqrt{2}\sin 45° = 0$

$\Rightarrow \quad F_{BC} = 4 \text{ kN (Tension)}$

$\Rightarrow \quad F_{BG} + 4\sqrt{2}\cos 45° = 0$

$\Rightarrow \quad F_{BG} = -4 \text{ kN (Compression)}$

Considering joint *G*

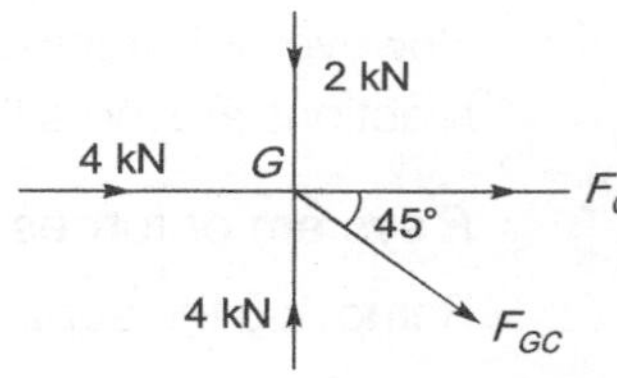

$\Sigma F_y = 0$ AND $\Sigma F_x = 0$

$\Rightarrow \quad 4 - 2 - F_{GC}\sin 45° = 0$

$\Rightarrow \quad \dfrac{F_{GC}}{\sqrt{2}} = 2$

$\Rightarrow \quad F_{GC} = 2\sqrt{2} \text{ kN (Tension)}$

$\Rightarrow \quad F_{GH} + F_{GC}\cos 45° + 4 = 0$

$\Rightarrow \quad F_{GH} = -2\sqrt{2} \times \dfrac{1}{\sqrt{2}} - 4$

$\Rightarrow \quad F_{GH} = -6 \text{ kN (Compression)}$

Considering joint *H*

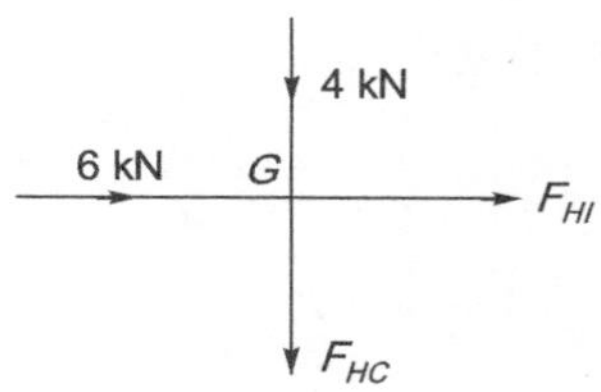

$\Sigma F_x = 0$ AND $\Sigma F_y = 0$

$\Rightarrow \quad F_{HI} + 6 = 0$

$\Rightarrow \quad F_{HI} = -6 \text{ kN (Comp.)}$

$\Rightarrow \quad F_{HC} + 4 = 0$

$\Rightarrow \quad F_{HC} = -4 \text{ kN (Compression)}$

Since, the truss is symmetrically loaded, the forces in the members which are right of *H* will be same as their corresponding members which are left of *H*.

$$\therefore \quad F_{AB} = F_{ED} = 0$$

$$F_{AF} = F_{EJ} = -4 \text{ kN (Compression)}$$

$$F_{FG} = F_{JI} = -4 \text{ kN (Compression)}$$

$$F_{FB} = F_{JD} = -4\sqrt{2} \text{ kN (Tension)}$$

$$F_{BG} = F_{DI} = -4 \text{ kN (Compression)}$$

$$F_{BC} = F_{DC} = 4 \text{ kN (Tension)}$$

$$F_{GH} = F_{HI} = -6 \text{ kN (Compression)}$$

$$F_{GC} = F_{IC} = 2\sqrt{2} \text{kN (Tension)}$$

$$F_{HC} = -4 \text{ kN (Compression)}$$

K-system of forces: Now removing all the external loads on the truss and applying a unit force in the upward direction at C.

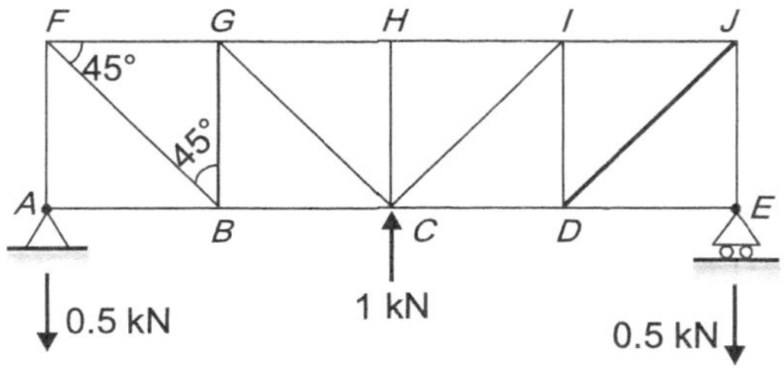

Considering Joint A

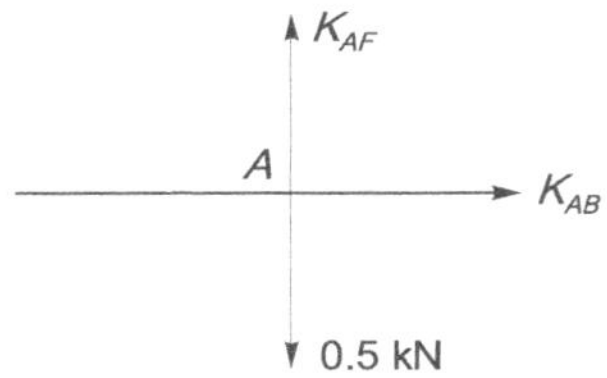

$\Sigma F_x = 0$ AND $\Sigma F_y = 0$

$\Rightarrow K_{AB} = 0$

$\Rightarrow K_{AF} - 0.5 = 0$

$\Rightarrow K_{AF} = 0.5$ kN (Tension)

Considering joint F

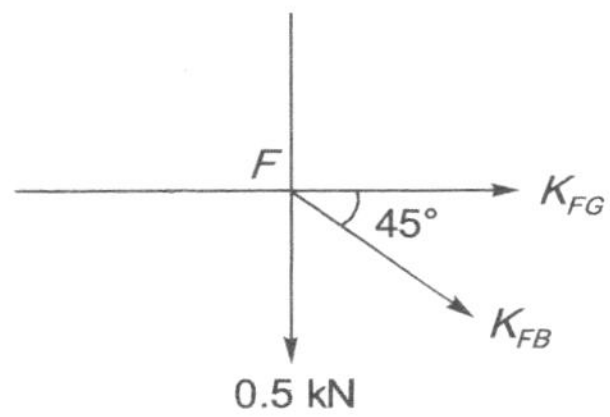

$\Sigma F_y = 0$ AND $\Sigma F_x = 0$

$\Rightarrow K_{FB} \sin 45° + 0.5 = 0$

$\Rightarrow \dfrac{K_{FB}}{\sqrt{2}} = -0.5$

$\Rightarrow K_{FB} = -0.5\sqrt{2}$ (Comp.)

$\Rightarrow K_{FG} + K_{FB} \cos 45° = 0$

$\Rightarrow K_{FG} - 0.5\sqrt{2} \times \dfrac{1}{\sqrt{2}} = 0$

$\Rightarrow K_{FG} = 0.5$ kN(Tension)

Considering joint B

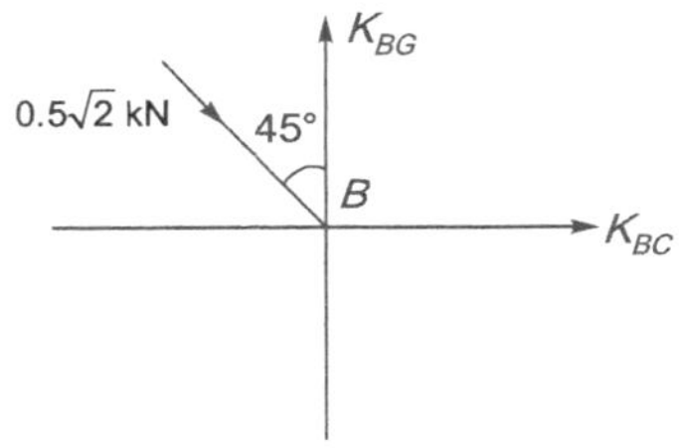

$\Sigma F_y = 0$ AND $\Sigma F_x = 0$

$\Rightarrow K_{BG} - 0.5\sqrt{2} \cos 45° = 0$

$\Rightarrow K_{BG} = 0.5$ kN (Tension)

$\Rightarrow K_{BC} + 0.5\sqrt{2} \sin 45° = 0$

$\Rightarrow K_{BC} = -0.5$ kN (Compression)

Considering joint G

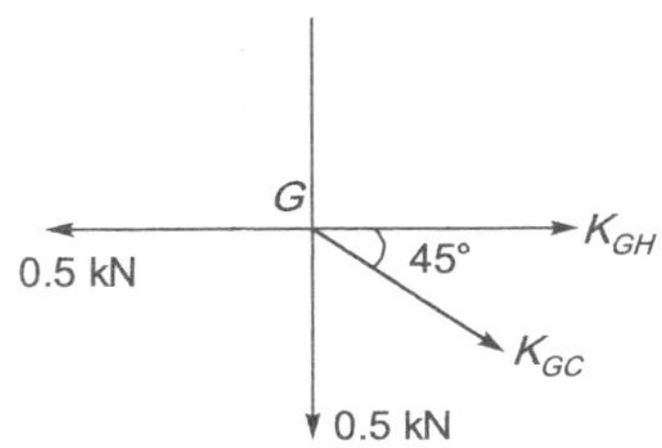

$\Sigma F_y = 0$ AND $\Sigma F_x = 0$

$\Rightarrow K_{GC} \sin 45° + 0.5 = 0$

$\Rightarrow K_{GC} = -0.5\sqrt{2}$ kN (Comp.)

$\Rightarrow K_{GH} + K_{GC} \cos 45° - 0.5 = 0$

$\Rightarrow K_{GH} = 0.5 - \left(-0.5\sqrt{2} \times \dfrac{1}{\sqrt{2}}\right)$

$\Rightarrow K_{GH} = 1$ kN (Tension)

Considering joint H

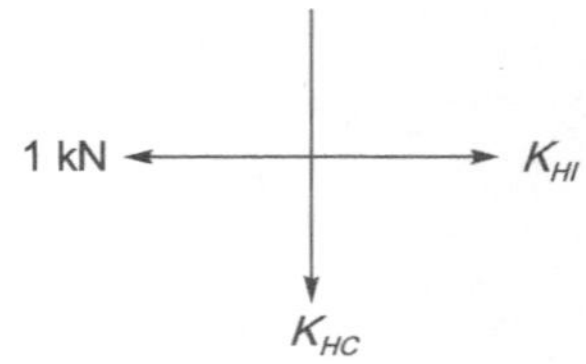

$\Sigma F_x = 0$ AND $\Sigma F_y = 0$

$\Rightarrow \quad K_{HI} - 1 = 0 \qquad \Rightarrow \quad K_{HC} = 0$

$\Rightarrow \quad K_{HI} = 1$ kN (Tension)

Similarly, forces in other members will be given as

$K_{ED} = K_{AB} = 0$ kN

$K_{EJ} = K_{AF} = 0.5$ kN (Tension)

$K_{JI} = K_{FG} = 0.5$ kN (Tension)

$K_{JD} = K_{FB} = -0.5\sqrt{2}$ kN (Compression)

$K_{DI} = K_{BG} = 0.5$ kN (Tension)

$K_{DC} = K_{BC} = -0.5$ (Compression)

$K_{HI} = K_{GH} = 1$ kN (Tension)

$K_{IC} = -K_{GC} = -0.5\sqrt{2}$kN (Compression)

$K_{HC} = 0$

Taking length of vertical and horizontal members as 'a' and 'a' respectively.

Members	P	K	L	PKL	K^2L	$P+KX$
AB	0	0	a	0	0	0
AF	–4	0.5	a	$-2a$	$0.25a$	–0.71
FG	–4	0.5	a	$-2a$	$0.25a$	–0.71
FB	$4\sqrt{2}$	$-0.5\sqrt{2}$	$\sqrt{2}a$	$-4\sqrt{2}a$	$0.5\sqrt{2}a$	1
BG	–4	0.5	a	$-2a$	$0.25a$	–0.71
BC	4	–0.5	a	$-2a$	$0.25a$	0.71
GH	–6	1	a	$-6a$	a	0.58
GC	$2\sqrt{2}$	$-0.5\sqrt{2}$	$\sqrt{2}a$	$-2\sqrt{2}a$	$0.5\sqrt{2}a$	–1.82
ED	0	0	a	0	0	0
EJ	–4	0.5	a	$-2a$	$0.25a$	–0.71
JI	–4	0.5	a	$-2a$	$0.25a$	–0.71
JD	$4\sqrt{2}$	$-0.5\sqrt{2}$	$\sqrt{2}a$	$-4\sqrt{2}a$	$0.5\sqrt{2}a$	1
DI	–4	0.5	a	$-2a$	$0.25a$	–0.71
DC	4	–0.5	a	$-2a$	$0.25a$	0.71
HI	–6	1	a	$-6a$	a	0.58
IC	$2\sqrt{2}$	$-0.5\sqrt{2}$	$\sqrt{2}a$	$-2\sqrt{2}a$	$0.5\sqrt{2}a$	–1.82
HC	–4	0	a	0	0	–4

$\therefore \qquad \Sigma PKL = -\left(28a + 12\sqrt{2}\,a\right)$

and $\qquad \Sigma K^2L = 4a + 2\sqrt{2}a$

The reaction at C may be given as

$$X = -\frac{\Sigma \dfrac{PKL}{AE}}{\Sigma \dfrac{K^2L}{AE}}$$

If the truss members have same axial stiffness, then

$$X = -\frac{\frac{1}{AE}\Sigma PKL}{\frac{1}{AE}\Sigma K^2L}$$

$\Rightarrow$ $$X = -\frac{\Sigma PKL}{\Sigma K^2L} = -\frac{-\left(28a + 12\sqrt{2}a\right)}{4a + 2\sqrt{2}a}$$

$\Rightarrow$ $$X = \frac{28 + 12\sqrt{2}}{4 + 2\sqrt{2}} = 6.58 \text{ kN}$$

The forces in each member can now be given as

$$S = P + KX \text{ (shown in table)}$$

5.6 **Calculate the magnitude and nature of forces in members marked 1, 2, 3 and 4. Assume all member to be pinjointed.**

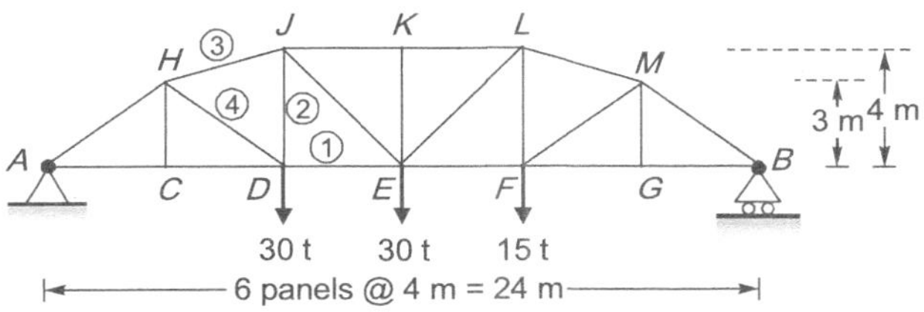

[20 marks : 2005]

Solution:

Degree of static determinacy (D_S)

$$D_S = m + r_e - 2j$$

Here $m = 21,\ r_e = 3,\ j = 12$

$\therefore$ $D_S = 21 + 3 - 2 \times 12 = 0$

Hence the truss is statically determinate

Reactions:

$$\Sigma F_y = 0$$

$\Rightarrow$ $V_A + V_B = 30 + 30 + 15$

$\Rightarrow$ $V_A + V_B = 75 \text{ t}$ ... (i)

and $\Sigma M_A = 0$

$\Rightarrow$ $V_B \times 24 = 30 \times 8 + 30 \times 12 + 15 \times 16$

$\Rightarrow$ $V_B = 35 \text{ t}$

$V_A = 75 - 35 = 40 \text{ t}$

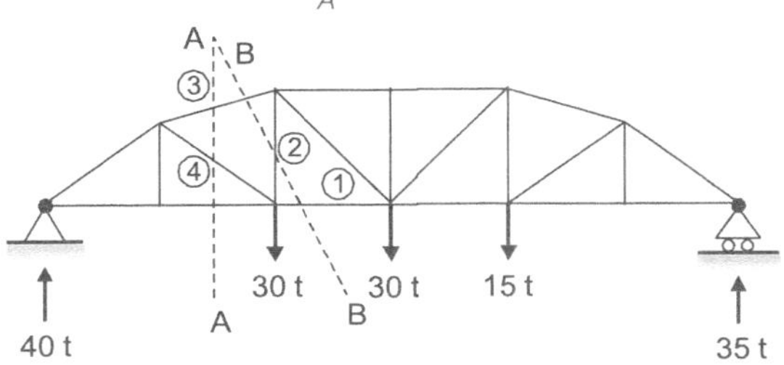

From section A-A

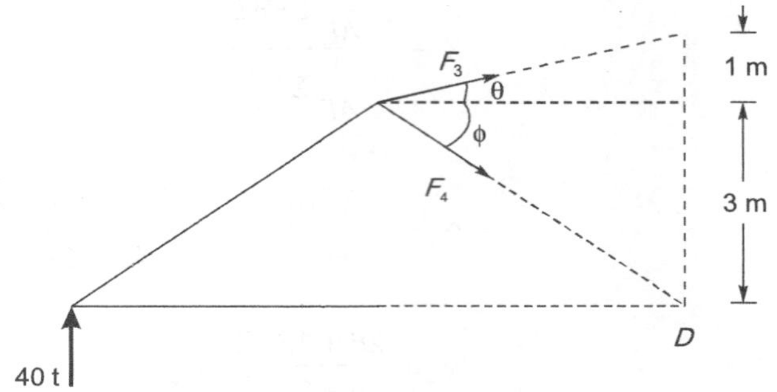

$$\sin\theta = \frac{1}{\sqrt{17}} \qquad \sin\phi = \frac{3}{5}$$

$$\cos\theta = \frac{4}{\sqrt{17}} \qquad \cos\phi = \frac{4}{5}$$

$$\tan\theta = \frac{1}{4} \qquad \tan\phi = \frac{3}{4}$$

From $\Sigma F_y = 0$

$\Rightarrow$ $F_3 \sin\theta - F_4 \sin\phi + 40 = 0$

$\Rightarrow$ $\frac{F_3}{\sqrt{175}} + 40 - \frac{3F_4}{5} + 40 = 0$ ...(i)

From $\Sigma M_D = 0$

$\Rightarrow$ $F_3 \cos\theta \times 3 + F_3 \sin\phi \times 4 + 40 \times 8 = 0$

$\Rightarrow$ $\frac{12F_3}{\sqrt{17}} + \frac{4F_3}{\sqrt{17}} + 320 = 0$

$\Rightarrow$ $F_3 = -\frac{320 \times \sqrt{17}}{16} = -20\sqrt{17} = -82.46\ \text{t}$

Hence, $F_3 = 82.46\ \text{t}$ (Compressive)

From (i) $F_4 = \left(\frac{F_3}{\sqrt{17}} + 40\right) \times \frac{5}{3} = (-20 + 40) \times \frac{5}{3} = 33.33t$

Hence, $F_3 = 33.33\ \text{t}$ (tensile)

From section B-B

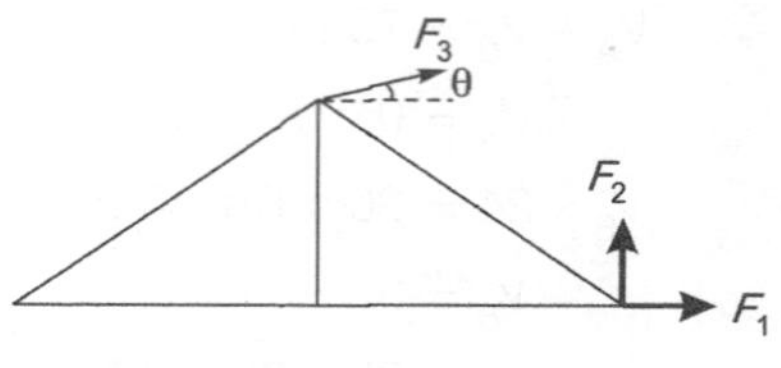

$$\Sigma F_x = 0$$

$$F_3 \cos\theta + F_1 = 0$$

$$F_1 = -F_3 \cos\theta$$

$$F_1 = 20\sqrt{17} \times \frac{4}{\sqrt{17}} = 80t$$

$$F_1 = 80\ \text{t (tensile)}$$

From $\Sigma F_y = 0$

$$F \sin\theta + F_2 + 40 = 0$$

$$F_2 = -F_3 \sin\theta - 40 = 20\sqrt{17} \times \frac{1}{\sqrt{17}} - 40 = 20 - 40 = -20$$

$$F_2 = 20 \text{ t (compressive)}$$

Hence,

$$F_1 = 80 \text{ t (tensile)}$$
$$F_2 = 20 \text{ t (compressive)}$$
$$F_3 = 82.46 \text{ t (compressive)}$$
$$F_4 = 3.33 \text{ t (tensile)}$$

5.7 A small truss as shown in figure has all the pinjointed members having 20 cm^2 cross sectional area. Plane of roller is inclined and parallel to member *AD*. Determine the movement of joint '*C*' on application of a vertical load of 4t at joint *B*. $E = 2 \times 10^6$ kg/cm^2.

[20 marks : 2005]

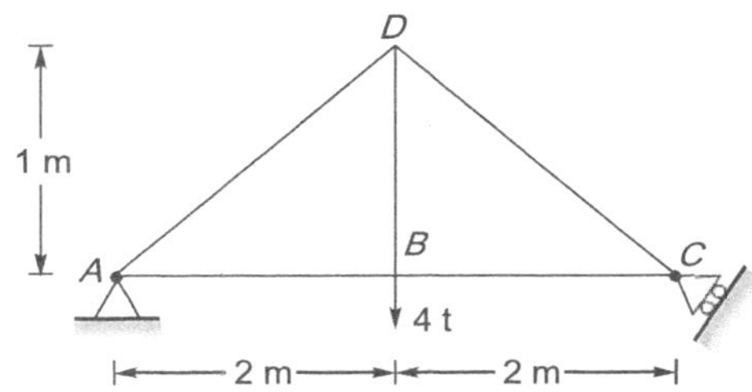

Solution:

The static determinacy of the truss may be given as

$$D_S = m + r_e - 2j$$

Here $m = 5,\ r_e = 3,\ j = 4$

$\therefore$ $D_S = 5 + 3 - 2 \times 4$

$\Rightarrow$ $D_S = 0$

Thus, the truss is statically determinate.

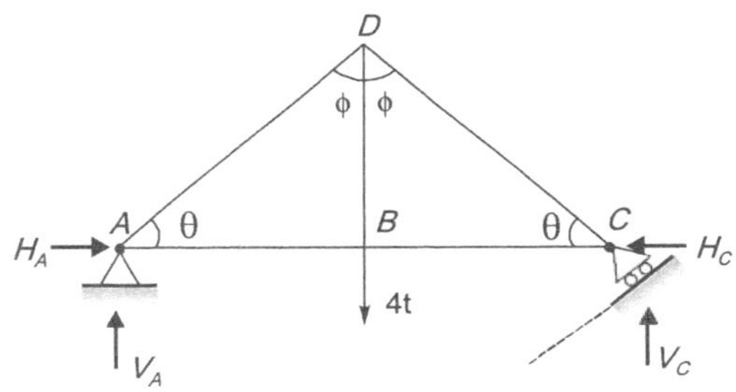

Resolving R_C into horizontal and vertical components respectively as H_C and V_C.

$\therefore$ $H_C = R_C \sin\theta$ and $V_C = R_C \cos\theta$

Also $\tan\theta = \frac{1}{2};\ \sin\theta = \frac{1}{\sqrt{5}};\ \cos\theta = \frac{2}{\sqrt{5}}$

and $\tan\phi = 2;\ \sin\phi = \frac{2}{\sqrt{5}}$ and $\cos\phi = \frac{1}{\sqrt{5}}$

Reactions

$$\Sigma F_y = 0$$

$$V_A + V_C = 4 \quad \text{... (i)}$$

$$\Sigma F_x = 0$$

$\Rightarrow$ $H_A - H_C = 0$

$\Rightarrow$ $H_A = H_C$ ... (ii)

$$M_A = 0$$

$\Rightarrow$ $V_C \times 4 - 4 \times 2 = 0$

$\Rightarrow$ $V_C = \frac{8}{4} = 2 \text{ t}$

From (i), we get

$$V_A = 4 - V_C$$

$$V_A = 4 - 2$$

$$V_A = 2\text{ t}$$

Now, we have

$$\tan\theta = \frac{H_C}{V_C} \qquad \left[\because \tan\theta = \frac{1}{2}\right]$$

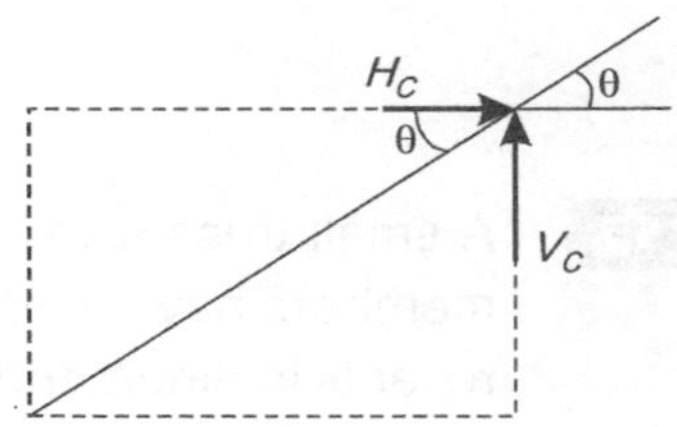

$$\therefore \quad \frac{1}{2} = \frac{H_C}{V_C}$$

$$\Rightarrow \quad H_C = \frac{V_C}{2}$$

$$\Rightarrow \quad H_C = \frac{2}{2}$$

$$\Rightarrow \quad H_C = 1\text{ t}$$

$\therefore$ From (ii), we get

$$H_A = H_C = 1\text{ t}$$

P-system of forces

Considering joint *A*

$\Sigma F_y = 0$ AND $\Sigma F_x = 0$

$\Rightarrow \quad F_{AD}\sin\theta + 2 = 0$ | $\Rightarrow \quad F_{AD}\cos\theta + F_{AB} + 1 = 0$

$\Rightarrow \quad F_{AD} \times \frac{1}{\sqrt{5}} = -2$ | $\Rightarrow \quad F_{AB} = -1 - \left(-2\sqrt{5}\right) \times \frac{2}{\sqrt{5}}$

$\Rightarrow \quad F_{AD} = -2\sqrt{5}$ t (Comp.) | $\Rightarrow \quad F_{AB} = 3$ t (Tension)

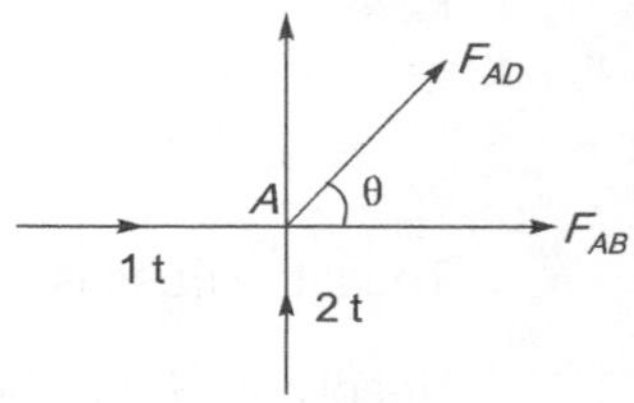

Considering joint *B*

$\Sigma F_x = 0$ AND $\Sigma F_y = 0$

$\Rightarrow \quad F_{BC} - 3 = 0$ | $\Rightarrow \quad F_{BD} - 4 = 0$

$\Rightarrow \quad F_{BC} = 3$ t (Tension) | $\Rightarrow \quad P_{BD} = 4$ t (Tension)

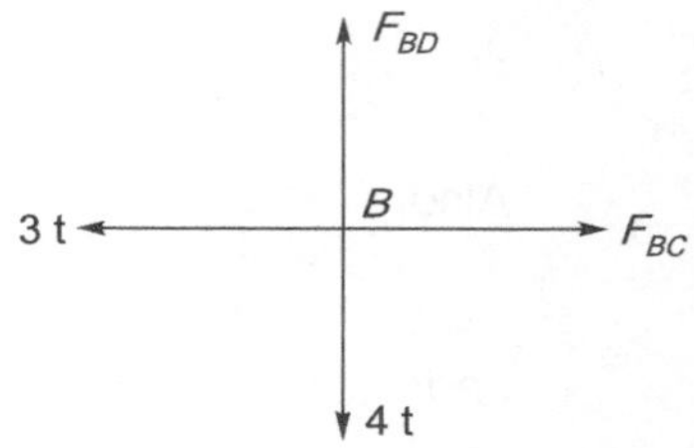

Considering joint *D*

$$\Sigma F_x = 0$$

$$\Rightarrow \quad F_{DC}\sin\phi + 2\sqrt{5}\sin\phi = 0$$

$$\Rightarrow \quad F_{DC} = -2\sqrt{5}\text{ t (Compression)}$$

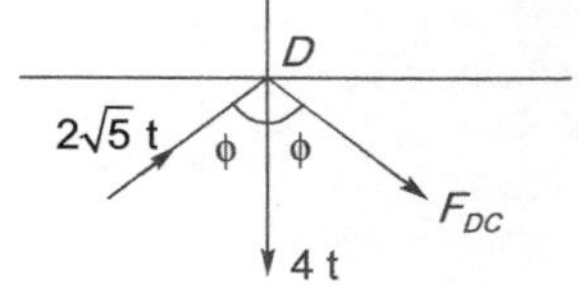

k-system of forces

Reactions

$$\Sigma F_x = 0$$

$$\Rightarrow \quad H_A - H_C + \cos\theta = 0$$

$$\Rightarrow \quad H_A - H_C = -\frac{2}{\sqrt{5}} \quad \text{... (i)}$$

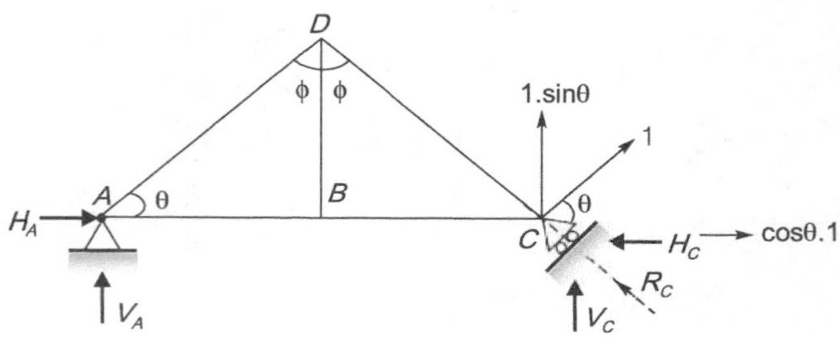

$$\Sigma F_y = 0$$

$$\Rightarrow \quad V_A + V_C + \sin\theta = 0$$

$$\Rightarrow \quad V_A + V_C = -\frac{1}{\sqrt{5}} \qquad \text{... (ii)}$$

$$\Sigma M_A = 0$$

$$\Rightarrow \quad V_C \times 4 + 4\sin\theta = 0$$

$$\Rightarrow \quad V_C = -\sin\theta$$

$$\Rightarrow \quad V_C = -\frac{1}{\sqrt{5}}\,t\,(\uparrow) = \frac{1}{\sqrt{5}}\,t(\downarrow)$$

From (ii), we get

$$V_A = -\frac{1}{\sqrt{5}} - V_C$$

$$\Rightarrow \quad V_A = -\frac{1}{\sqrt{5}} - \left(-\frac{1}{\sqrt{5}}\right)$$

$$\Rightarrow \quad V_A = 0$$

Also $\tan\theta = \dfrac{H_C}{V_C}$

$$\Rightarrow \quad \frac{1}{2} = \frac{H_C}{V_C}$$

$$\Rightarrow \quad H_C = \frac{V_C}{2}$$

$$\therefore \quad H_C = -\frac{1}{2\sqrt{5}}\,t\,(\leftarrow) = \frac{1}{2\sqrt{5}}\,t\,(\rightarrow)$$

From (i), we get

$$\Rightarrow \quad H_A = -\frac{2}{\sqrt{5}} + H_C$$

$$\Rightarrow \quad H_A = -\frac{2}{\sqrt{5}} - \frac{1}{2\sqrt{5}}$$

$$\Rightarrow \quad H_A = -0.5\sqrt{5}\,t\,(\rightarrow) = 0.5\sqrt{5}\,t\,(\leftarrow)$$

Considering joint A

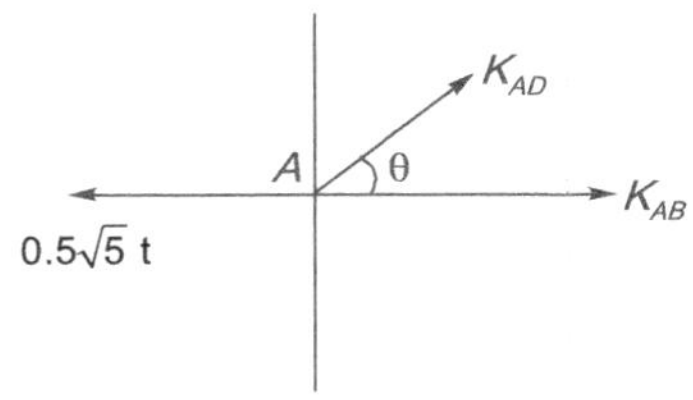

$\Sigma F_y = 0$ AND $\Sigma F_x = 0$

$\Rightarrow \quad K_{AD}\sin\theta = 0$ | $\Rightarrow \quad K_{AD}\cos\theta + K_{AB} - 0.5\sqrt{5} = 0$

$\Rightarrow \quad K_{AD} = 0$ | $\Rightarrow \quad K_{AB} = 0.5\sqrt{5}\,t$ (Tension)

Considering joint B

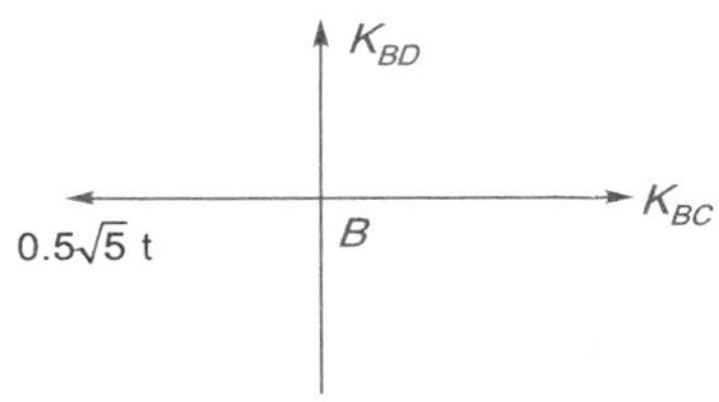

$\Sigma F_x = 0$ AND $\Sigma F_y = 0$

$\Rightarrow \quad K_{BC} - 0.5\sqrt{5} = 0$ | $K_{BD} = 0$

$\Rightarrow \quad K_{BC} = 0.5\sqrt{5}$ t (Tension)

As there is no force in *BD* and *AD*, hence *DC* also does not carry any force. Therefore $K_{DC} = 0$.

$$\Sigma PKL = 13.416$$

Member	*P*	*K*	*L*	*PKL*
AB	3	$0.5\sqrt{5}$	2	6.708
AD	$-2\sqrt{5}$	0	$\sqrt{5}$	0
BC	3	$0.5\sqrt{5}$	2	6.708
BD	4	0	1	0
DC	$-2\sqrt{5}$	0	$\sqrt{5}$	0

The deflection at *C* will be given as

$$\delta_C = \frac{\Sigma PKL}{AE}$$

$$\Rightarrow \quad \delta_1 = \frac{13.416 \times 10^3 \times 100}{20 \times 2 \times 10^6}$$

$$\Rightarrow \quad \delta_C = 0.0335 \text{ cm}$$

$$\Rightarrow \quad \delta_C = 0.335 \text{ mm}$$

5.8 **For a three bar truss shown in the figure, compute the vertical displacement of node 1 by displacement (stiffness/equilibrium) method.**

[20 marks : 2006]

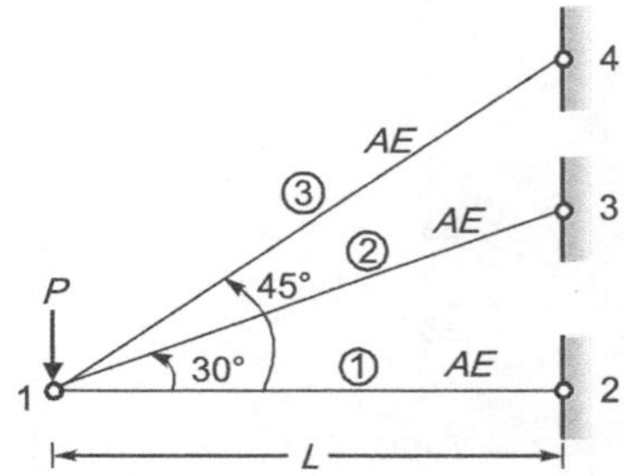

Solution:

Step-1: Find D_k and give coordinate no. to the displacement $D_k = 2\,(\Delta A_x, \Delta A_y)$

Give coordinate number (1) and (2) for ΔA_x and ΔA_y

{Always give (1) and (2) along +ve x-axis and +y-axis respectively}

Step-2: Find $[P]_L \Rightarrow$ {There are no fixed end moment in truss so $[P]_L$ is always zero}

Step-3: Stiffness matrix [k]

Member	$\frac{AE}{l}$	θ	cosθ	sinθ	$\frac{AE}{l}\cos^2\theta$	$\frac{AE}{l}\sin^2\theta$	$\frac{AE}{l}\sin\theta\cos\theta$
AB	$\frac{AE}{1}$	0°	1	0	$\frac{AE}{l}$	0	0
AC	$\frac{AE}{1.15\,l}$	30°	0.866	0.5	$\frac{0.65AE}{l}$	$\frac{0.22\,AE}{l}$	$\frac{0.38\,AE}{l}$
AD	$\frac{AE}{1.414\,l}$	45°	0.707	0.707	$\frac{0.35AE}{l}$	$\frac{0.35\,AE}{l}$	$\frac{0.35\,AE}{l}$
					$k_{11}=\frac{2AE}{l}$	$k_{22}=\frac{0.57AE}{l}$	$k_{12}=k_{21}=\frac{0.73AE}{l}$

Note: θ is always measured in anticlockwise direction from +ve *x*-axis.

$$\cos 30° = \frac{l}{l_{AC}}$$

$$l_{AC} = \frac{l}{\cos 30°} = 1.15l$$

$$l_{AD} = \frac{l}{\cos 45°} = \sqrt{2}\,l = 1.414l$$

So,

$$[k] = \begin{bmatrix} \frac{2AE}{l} & 0.73\frac{AE}{l} \\ 0.73\frac{AE}{l} & 0.570\frac{AE}{l} \end{bmatrix} = \frac{AE}{l}\begin{bmatrix} 2 & 0.73 \\ 0.73 & 0.57 \end{bmatrix}$$

Step-4: Use joint equilibrium to find uknown displacements

$$[P] = [P]_L + [P]_\Delta$$

$$[P] = \begin{bmatrix} P_1 \\ P_1 \end{bmatrix} = \begin{bmatrix} 0 \\ -P \end{bmatrix}$$

$$[P]_\Delta = [P] - [P]_L = \begin{bmatrix} 0 \\ -P \end{bmatrix} - \begin{bmatrix} 0 \\ 0 \end{bmatrix} = \begin{bmatrix} 0 \\ -P \end{bmatrix}$$

We know that $[P]_\Delta = [k][\Delta]$

$$[\Delta] = [k]^{-1}[P]_\Delta = \frac{l}{AE(2\times 0.27 - 0.73\times 0.73)}\begin{bmatrix} 0.57 & -0.73 \\ -0.73 & 2 \end{bmatrix}\begin{bmatrix} 0 \\ -P \end{bmatrix}$$

$$[\Delta] = \begin{bmatrix} \Delta_1 \\ \Delta_2 \end{bmatrix} = \begin{bmatrix} 1.19\dfrac{Pl}{AE} \\ -3.3\dfrac{El}{AE} \end{bmatrix} = \begin{bmatrix} \Delta A_x \\ \Delta A_y \end{bmatrix}$$

Step-5: Find forces in all members

$$F_{AB} = -\frac{AE}{l}\left[(\Delta A_x - \Delta B_x)\cos\theta_{AB} + (\Delta A_y - \Delta B_y)\sin\theta_{AB}\right]$$

$$= -\frac{AE}{l}\left[\left(\frac{1.19Pl}{AE} - 0\right)\cos 0 + \left(\frac{-3.3Pl}{AE} - 0\right)\sin 0\right]$$

$F_{AB} = -1.19P$ {–ve sing indicates F_{AB} is in compression}

$$F_{AC} = -\frac{AE}{1.5l}\left[\left(\frac{1.19Pl}{AE} - 0\right)\cos 30° + \left(\frac{-3.3Pl}{AE} - 0\right)\sin 30°\right]$$

$= 0.54$ P (Tensile)

$$F_{AD} = -\frac{AE}{1.414l}\left[\left(\frac{1.19Pl}{AE} - 0\right)\cos 45° + \left(\frac{-3.3Pl}{AE} - 0\right)\sin 45°\right]$$

$= 1.05$ P (Tensile)

Alternate :

The axial compression (Δ) due to movement of node 2 = $\Delta_{2x}\cos\theta + \Delta_{2y}\sin\theta$

$\Delta_{2x} \rightarrow$ displacement of node 2 in x–direction

The axial elongation (Δ') due to movement of node 1 = $\Delta_{1x}\cos\theta + \Delta_{1y}\sin\theta$

$\therefore$ Net elongation of 12 = $(\Delta_{1x} - \Delta_{2x})\cos\theta + (\Delta_{1y} - \Delta_{2y})\sin\theta$

But net elogation of 12 = $\dfrac{P_{12}L}{AE}$

$$\Rightarrow \quad P_{12} = \frac{AE}{L}\left[(\Delta_{1x} - \Delta_{2x})\cos\theta + (\Delta_{1y} - \Delta_{2y})\sin\theta\right]$$

Now from the above figure we have

$$\cos 30° = \frac{L_{12}}{L_{13}}$$

and $$\cos 45° = \frac{L_{12}}{L_{14}}$$

$$\Rightarrow \quad L_{13} = \frac{2L_{12}}{\sqrt{3}}$$

$$\Rightarrow \quad L_{14} = \frac{\sqrt{2}L_{12}}{1}$$

$$\Rightarrow \quad L_{13} = \frac{2L}{\sqrt{3}} \quad [\because L_{12} = L]$$

$$\Rightarrow \quad L_{14} = \sqrt{2}L$$

For member 12, the value of $\theta = 0°$

$$\therefore \quad P_{12} = \frac{AE}{L}\left[(\Delta_{1x} - 0)\cos 0° + (\Delta_{1y} - 0)\sin 0°\right]$$

$$\Rightarrow \quad P_{12} = \frac{AE}{L}\Delta_{1x} \quad \text{... (i)}$$

Similarly $$P_{13} = \frac{AE}{2L} \times \sqrt{3}\,[(\Delta_{1x} - \Delta_{3x})\cos 30° + (\Delta_{1y} - \Delta_{3y})\sin 30°]$$

$$\Rightarrow \quad P_{13} = \frac{\sqrt{3}}{2}\frac{AE}{L}\left[\Delta_{1x} \times \frac{\sqrt{3}}{2} + \Delta_{1y} \times \frac{1}{2}\right]$$

$$\Rightarrow \quad P_{13} = \frac{\sqrt{3}}{4}\frac{AE}{L}\left[\sqrt{3}\,\Delta_{1x} + \Delta_{1y}\right] = \frac{\sqrt{3}}{4}\frac{AE}{L}\left(\sqrt{3}\Delta_{1x} + \Delta_{1y}\right) \quad \text{... (ii)}$$

$$\Rightarrow \quad P_{14} = \frac{AE}{\sqrt{2}L}\left[(\Delta_{1x} - \Delta_{4x})\cos 45° + (\Delta_{1y} - \Delta_{4x})\sin 45°\right]$$

$$\Rightarrow \quad P_{14} = \frac{AE}{\sqrt{2}L}\left[\Delta_{1x} \times \frac{1}{\sqrt{2}} + \Delta_{1y} \times \frac{1}{\sqrt{2}}\right]$$

$$\Rightarrow \quad P_{14} = \frac{AE}{\sqrt{2}L} \times \frac{1}{\sqrt{2}}\,[\Delta_{1x} + \Delta_{1y}]$$

$$\Rightarrow \quad P_{14} = \frac{AE}{2L}\,[\Delta_{1x} + \Delta_{1y}] \quad \text{... (iii)}$$

Equilibrium equation at node 1

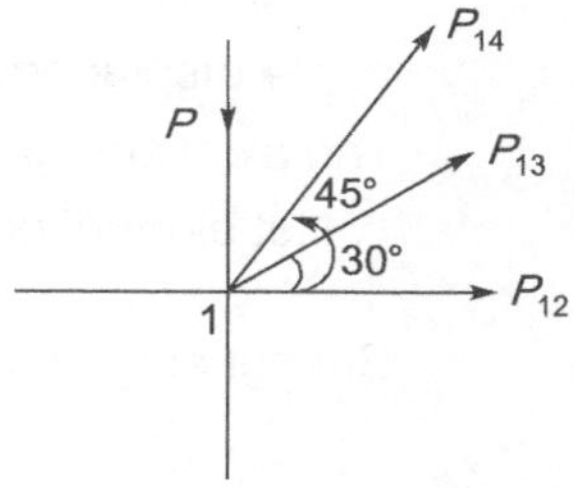

$$\Sigma F_x = 0$$

$$\Rightarrow \quad P_{12} + P_{13}\cos 30° + P_{14}\cos 45° = 0$$

$$\Rightarrow \quad \frac{AE}{L}\Delta_1 x + \frac{\sqrt{3}}{4}\left(\sqrt{3}\,\Delta_{1x} + \Delta_{1y}\right)\cos 30° + \frac{AE}{2L}\left(\Delta_{1x} + \Delta_{1y}\right)\cos 45° = 0$$

$$\Rightarrow \quad \frac{AE}{L}\left[\Delta_1 x + \frac{3}{8}\left(\sqrt{3}\Delta_{1x} + \Delta_{1y}\right) + \frac{1}{2\sqrt{2}}\left(\Delta_{1x} + \Delta_{1y}\right)\right] = 0$$

$$\Rightarrow \quad \Delta_{1x} + \frac{3\sqrt{3}}{8}\Delta_{1x} + \frac{\Delta_{1x}}{2\sqrt{2}} + \frac{3}{8}\Delta_{1y} + \frac{\Delta_{1y}}{2\sqrt{2}} = 0$$

$$\Rightarrow \quad 2\Delta_{1x} + 0.73\,\Delta_{1y} = 0 \qquad \text{... (iv)}$$

$$\Sigma F_y = 0$$

$$\Rightarrow \quad P_{13}\sin 30^\circ + P_{14}\sin 45^\circ - P = 0$$

$$\Rightarrow \quad \frac{\sqrt{3}}{8}\frac{AE}{L}\left(\sqrt{3}\,\Delta_{1x} + \Delta_{1y}\right) + \frac{1}{2\sqrt{2}}\frac{AE}{L}\left(\Delta_{1x} + \Delta_{1y}\right) = P$$

$$\Rightarrow \quad \frac{3}{8}\Delta_{1x} + \frac{1}{2\sqrt{2}}\Delta_{1x} + \frac{\sqrt{3}}{8}\Delta_{1y} + \frac{1}{2\sqrt{2}}\Delta_{1y} = \frac{PL}{AE}$$

$$\Rightarrow \quad 0.73\,\Delta_{1x} + 0.57\,\Delta_{1y} = \frac{PL}{AE}$$

From (iv), we have

$$\therefore \quad \Delta_{1x} = -\frac{0.73\Delta_{1y}}{2}$$

Substituting in (v), we get

$$\left(\frac{-0.73\Delta_{1y}}{2}\right) + 0.57\,\Delta_{1y} = \frac{PL}{AE}$$

$$\Rightarrow \quad 0.30355\,\Delta_{1y} = \frac{PL}{AE}$$

$$\Rightarrow \quad \Delta_{1y} = \frac{3.29\,PL}{AE}$$

5.9 Calculate forces in all the pin-jointed members of the truss as shown in the figure.

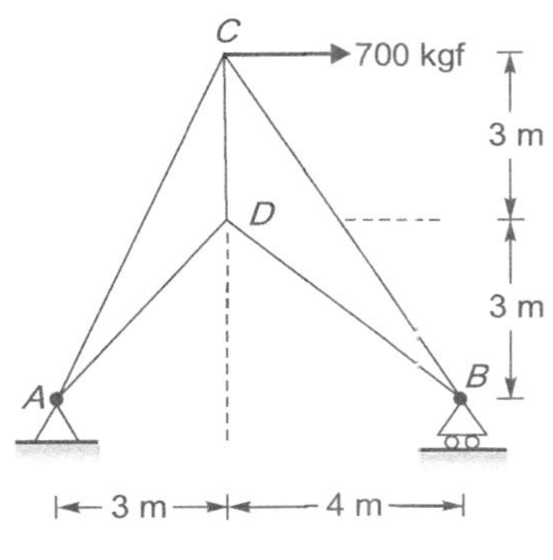

Note that the truss is not symmetrical. Show the magnitude and nature of forces on the truss diagram.

[20 marks : 2007]

Solution:

Let us take 100 kgf = 1 kN and analysing the **truss by method of joints.**

Let the vertical reactions at A and B be V_A and V_B respectively.

Let the horizontal reaction at A be H_A.

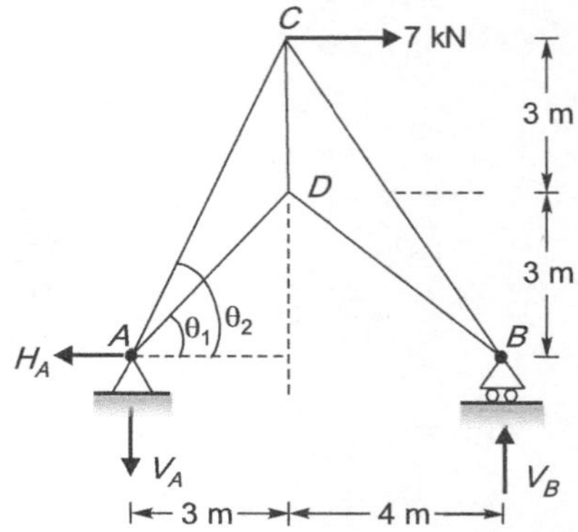

(i)
$$\Sigma F_y = 0$$
$$-V_A + V_B = 0$$
$$\Rightarrow \quad V_A = V_B$$

(ii)
$$\Sigma M_A = 0$$
$$\Rightarrow \quad V_B \times 7 - 7 \times 6 = 0$$
$$\Rightarrow \quad V_B = 6 \text{ kN}$$
$$\therefore \quad V_A = 6 \text{ kN}$$

(iii)
$$\Sigma F_x = 0$$
$$\Rightarrow \quad -H_A + 7 = 0$$
$$\Rightarrow \quad H_A = 7 \text{ kN}$$

Considering joint A and assuming tensile force in both members AD and AC

Assuming tensile force as positive and compressive force as negative.

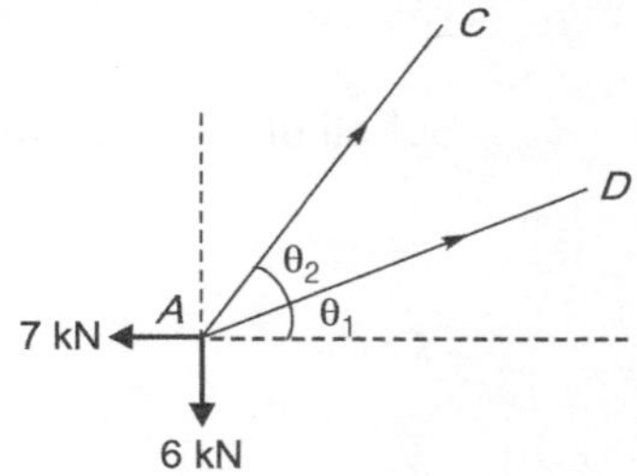

$$\tan\theta_1 = \frac{3}{3} = 1 \quad \text{and} \quad \tan\theta_2 = \frac{6}{3} = 2$$
$$\sin\theta_1 = \frac{1}{\sqrt{2}} \qquad \sin\theta_2 = \frac{2}{\sqrt{5}}$$
$$\cos\theta_1 = \frac{1}{\sqrt{2}} \qquad \cos\theta_2 = \frac{1}{\sqrt{5}}$$

$$\Sigma F_x = 0$$
$$\Rightarrow \quad F_{AD}\cos\theta_1 + F_{AC}\cos\theta_2 - 7 = 0$$
$$\Rightarrow \quad F_{AD} \times \frac{1}{\sqrt{2}} + \frac{F_{AC}}{\sqrt{5}} - 7 = 0 \qquad \text{...(i)}$$

and
$$\Sigma F_y = 0$$
$$F_{AD}\sin\theta_1 + F_{AC}\sin\theta_2 - 6 = 0$$
$$\Rightarrow \quad \frac{F_{AD}}{\sqrt{2}} + \frac{2F_{AC}}{\sqrt{5}} = 6 \qquad \text{...(ii)}$$

Solving (i) and (ii), we get

$$F_{AD} = 11.314 \text{ kN (Tensile)}$$
$$F_{AC} = -2.236 \text{ kN (Compressive)}$$

Considering joint B and assuming tensile force in members BD and BC.

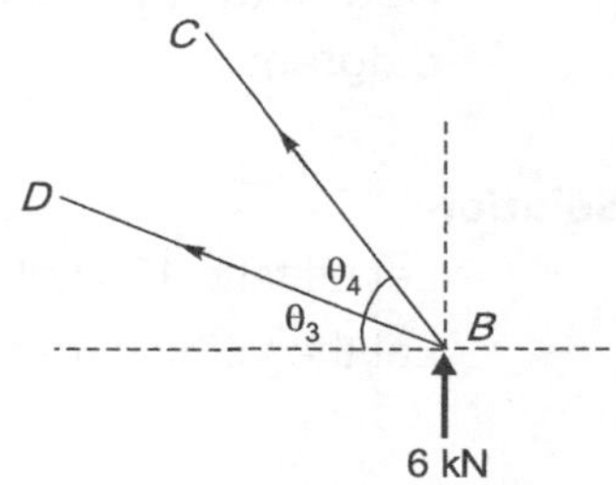

$$\tan\theta_3 = \frac{3}{4} \quad \text{and} \quad \tan\theta_4 = \frac{6}{4} = \frac{3}{2}$$
$$\sin\theta_3 = \frac{3}{5} \qquad \sin\theta_4 = \frac{3}{\sqrt{13}}$$
$$\cos\theta_3 = \frac{4}{5} \qquad \cos\theta_4 = \frac{2}{\sqrt{13}}$$

$$\Sigma F_x = 0$$

$$\Rightarrow \quad -F_{BD}\cos\theta_3 - F_{BC}\cos\theta_4 = 0$$

$$\Rightarrow \quad \frac{4}{5}F_{BD} + \frac{2}{\sqrt{13}}F_{BC} = 0 \qquad \text{... (iii)}$$

$$\Sigma F_y = 0$$

$$\Rightarrow \quad F_{BD}\sin\theta_3 + F_{BC}\sin\theta_4 + 6 = 0$$

$$\Rightarrow \quad \frac{3}{5}F_{BD} + \frac{3}{\sqrt{13}}F_{BC} = -6 \qquad \text{... (iv)}$$

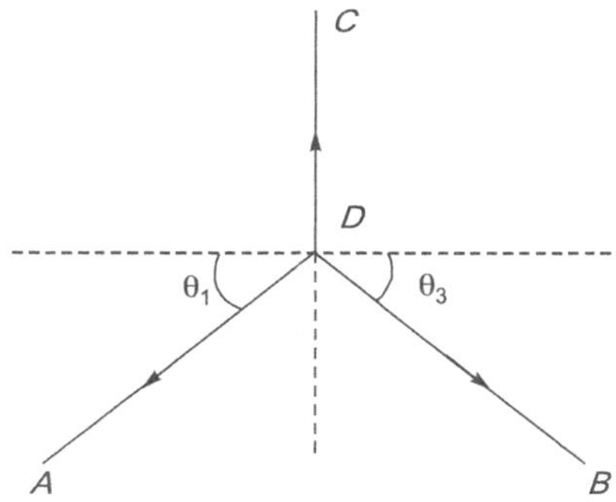

Solving (iii) and (iv), we get

$$\Rightarrow \quad F_{BD} = 10 \text{ kN (Tensile)}$$

$$\Rightarrow \quad F_{BC} = -14.422 \text{ kN (Compressive)}$$

Considering joint D, and assuming tensile force in CD

$$\Sigma F_y = 0$$

$$\Rightarrow \quad -F_{AD}\sin\theta_1 - F_{BD}\sin\theta_3 + F_{DC} = 0$$

$$\Rightarrow \quad F_{DC} = 11.314 \times \frac{1}{\sqrt{2}} + 10 \times \frac{3}{5}$$

$$\Rightarrow \quad F_{DC} = 8 + 6$$

$$\Rightarrow \quad F_{DC} = 14 \text{ kN (Tensile)}$$

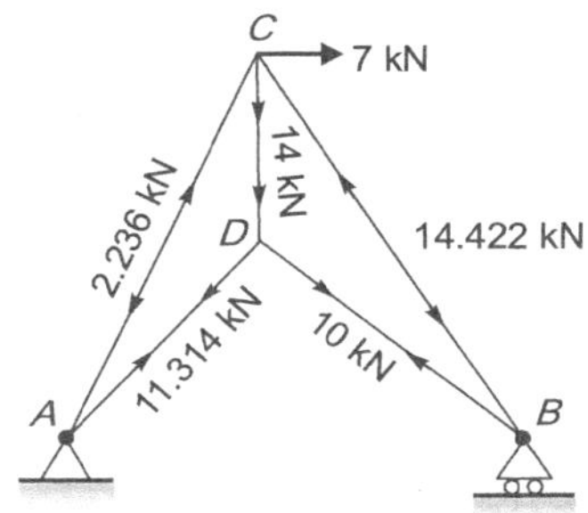

5.10 **Find the magnitude and the nature of forces in the members of the truss shown in figure. Find also the reaction components at the supports. Tabulate the results.**

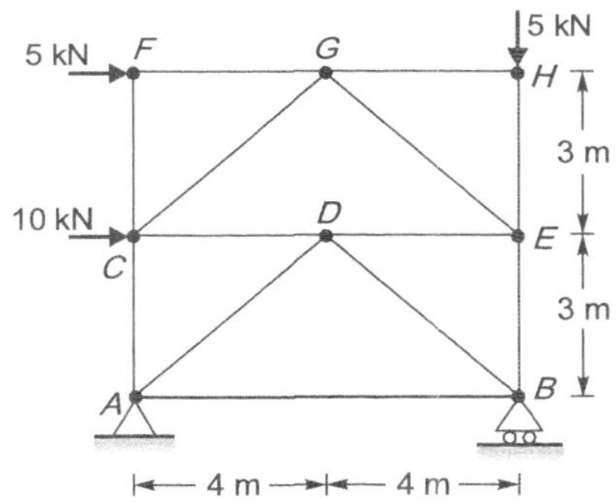

[25 marks : 2008]

Solution:

Let the reactions at A be $V_A(\uparrow)$ and $H_A(\leftarrow)$. Also let the reaction at B be $V_B(\uparrow)$.
Solving the problem by method of joints and starting from joint F, we get

Joint F: Assuming tensile forces as positive and compressive forces as negative.

$\Sigma F_x = 0$ AND $\Sigma F_y = 0$

$\Rightarrow F_{FG} + 5 = 0$ | $\Rightarrow F_{FC} = 0$

$\Rightarrow F_{FG} = -5$ kN (Comp.)

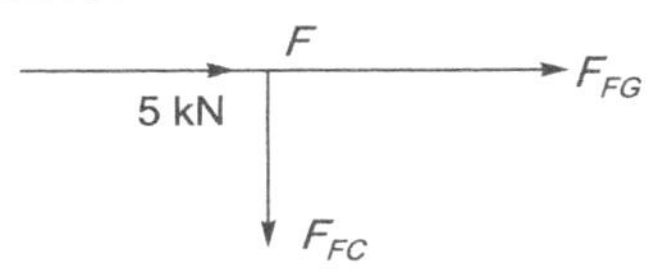

Joint *H*:

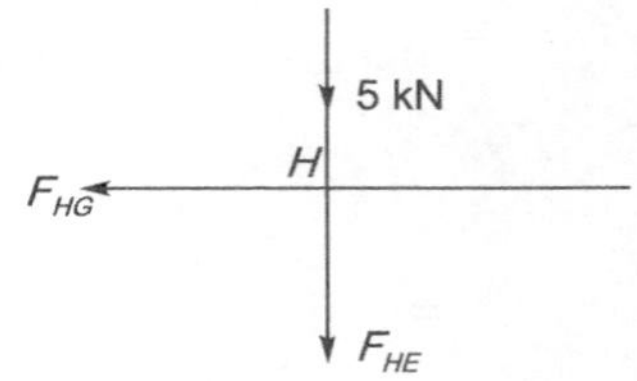

$\Sigma F_x = 0$

$\Rightarrow \quad F_{HG} = 0$

AND

$\Sigma F_y = 0$

$\Rightarrow \quad F_{HE} + 5 = 0$

$\Rightarrow \quad F_{HE} = -5$ kN (Comp.)

Joint *G*:

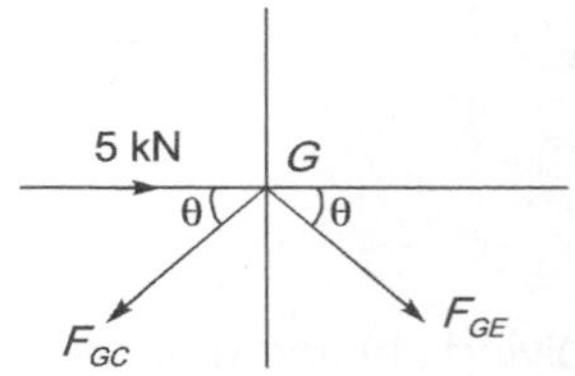

$\tan\theta = \frac{3}{4}$

$\sin\theta = \frac{3}{5}$

$\cos\theta = \frac{4}{5}$

$\Sigma F_x = 0$

$\Rightarrow \quad 5 - F_{GC}\cos\theta + F_{GE}\cos\theta = 0$

$\Rightarrow \quad \frac{4}{5}F_{GC} - \frac{4}{5}F_{GE} = 5$

$\Rightarrow \quad F_{GC} - F_{GE} = 6.25$ ...(i)

AND

$\Sigma F_y = 0$

$\Rightarrow \quad F_{GC}\sin\theta + F_{GE}\sin\theta = 0$

$\Rightarrow \quad F_{GC} + F_{GE} = 0$ ...(ii)

Solving (i) and (ii), we get

$F_{GC} = 3.125$ kN (Tension) and $F_{GE} = -3.125$ kN (Compression)

Joint *C*:

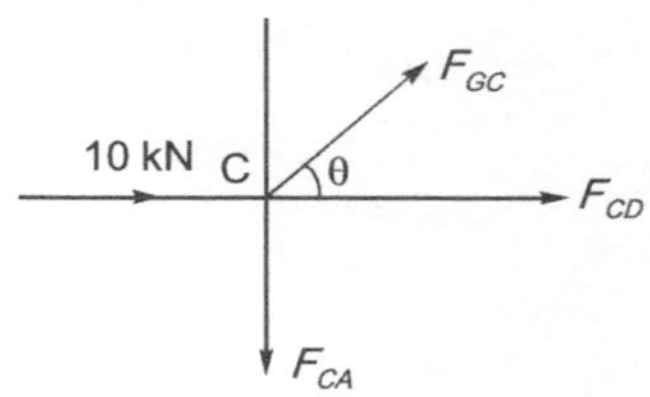

$\Sigma F_x = 0$

$\Rightarrow \quad 10 + F_{CD} + 3.125\cos\theta = 0$

$\Rightarrow \quad F_{CD} = -10 - 3.125 \times \frac{4}{5}$

$\Rightarrow \quad F_{CD} = -12.5$ kN (Comp.)

AND

$\Sigma F_y = 0$

$F_{GC}\sin\theta - F_{CA} = 0$

$\Rightarrow \quad 3.125\sin\theta - F_{CA} = 0$

$\Rightarrow \quad F_{CA} = 3.125 \times \frac{3}{5}$

$\Rightarrow \quad F_{CA} = 1.875$ kN (Tension)

Joint *E*:

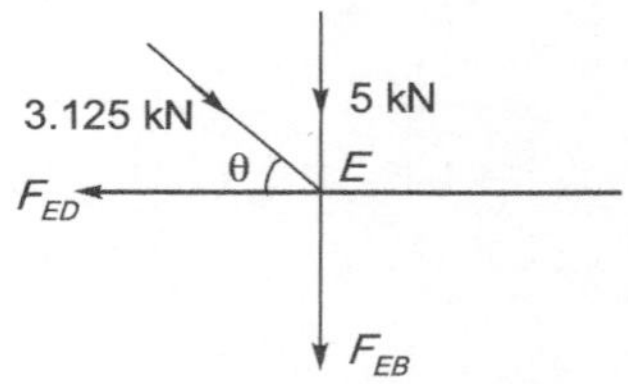

$\Sigma F_x = 0$

$\Rightarrow \quad 3.125\cos\theta - F_{ED} = 0$

$\Rightarrow \quad F_{ED} = 3.125 \times \frac{4}{5}$

$\Rightarrow \quad F_{ED} = 2.5$ kN (Tension)

AND

$\Sigma F_y = 0$

$\Rightarrow \quad 3.125\sin\theta + 5 + F_{EB} = 0$

$\Rightarrow \quad F_{EB} = -5 - 3.125 \times \frac{3}{5}$

$\Rightarrow \quad F_{EB} = -6.875$ kN (Comp.)

Joint *D*:

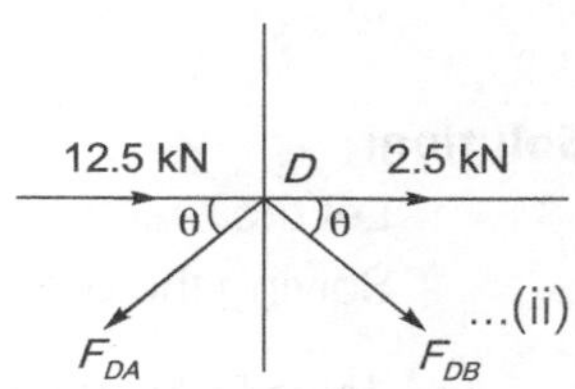

$\Sigma F_x = 0$

$\Rightarrow \quad F_{DB}\cos\theta - F_{DA}\cos\theta + 12.5 + 2.5 = 0$

$\Rightarrow \quad (F_{DA} - F_{DB}) \times \frac{4}{5} = 15$

$\Rightarrow \quad F_{DA} - F_{DB} = 18.75$ ...(i)

AND

$\Sigma F_y = 0$

$\Rightarrow \quad F_{DA}\sin\theta + F_{DB}\sin\theta = 0$

$\Rightarrow \quad F_{DA} + F_{DB} = 0$ ...(ii)

Solving (i) and (ii), we get

$F_{DA} = 9.375$ kN (Tension) and $F_{DB} = -9.375$ kN (Compression)

Joint *A*:

	$\Sigma F_x = 0$	AND	$\Sigma F_y = 0$
$\Rightarrow$	$9.375 \cos\theta + 7.5 - H_A = 0$	$\Rightarrow$	$V_A + 1.875 + 9.375 \sin\theta = 0$
$\Rightarrow$	$H_A = 9.375 \times \frac{4}{5} + 7.5$	$\Rightarrow$	$V_A = -1.875 - 9.375 \times \frac{3}{5}$
$\Rightarrow$	$H_A = 15 \text{ kN} (\leftarrow)$	$\Rightarrow$	$V_A = -7.5 \text{ kN} (\uparrow) = 7.5 \text{ kN} (\downarrow)$

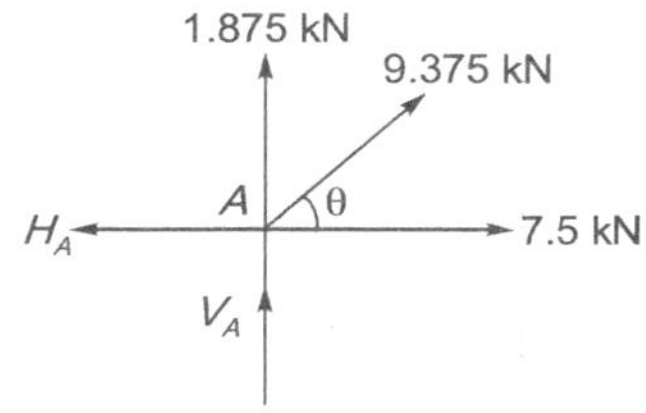

Joint *B*:

	$\Sigma F_x = 0$	AND	$\Sigma F_y = 0$
$\Rightarrow$	$9.375 \cos\theta - F_{AB} = 0$	$\Rightarrow$	$V_B - 9.375 \sin\theta - 6.875 = 0$
$\Rightarrow$	$F_{AB} = 9.375 \times \frac{4}{5}$	$\Rightarrow$	$V_B = 9.375 \times \frac{3}{5} + 6.875$
$\Rightarrow$	$F_{AB} = 7.5 \text{ kN (Tension)}$	$\Rightarrow$	$V_B = 12.5 \text{ kN} (\uparrow)$

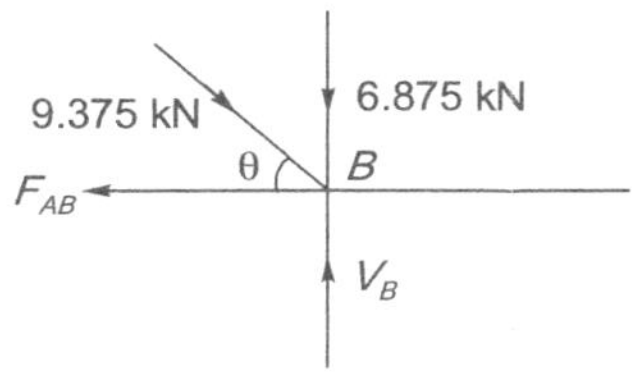

Member	Force (kN)	Nature
AB	7.5	tensile
DA	9.375	tensile
CA	1.875	tensile
DB	9.375	compressive
EB	6.875	compressive
CD	12.5	compressive
GC	3.125	tensile
FC	0	–
ED	2.5	tensile
GE	3.125	compressive
HE	5	compressive
FG	5	compressive
HG	0	–

5.11 All the members of steel truss shown in the figure are pin-jointed and have same area of cross-section. Calculate the vertical and horizontal components of the reaction at *A* and *B*.

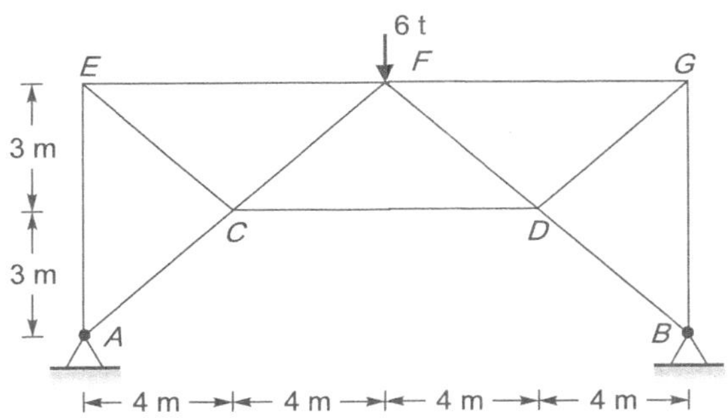

[20 marks : 2009]

Solution:

Let the vertical reactions at *A* and *B* be $V_A (\uparrow)$ and $V_B (\uparrow)$ respectively. Also let the horizontal thrust at *A* and *B* be $H (\rightarrow)$ and $H (\leftarrow)$ respectively.

The static indeterminacy of the steel truss may be given as

$$D_S = m + r_e - 2j$$

$$m = 11,\ r_e = 4,\ j = 7$$

Hence $\quad D_S = 11 + 4 - 2 \times 7 = 1$

Hence given steel truss is statically indeterminate by one degree.

$$D_{Se} = r_e - 3 = 4 - 3 = 1$$

But, $\quad \Sigma M_A = 0; \quad V_B \times 16 - 6 \times 8 = 0$

$\Rightarrow \quad V_B = 3\text{ t}$

$\therefore \quad V_A = 6 - V_B = 6 - 3 = 3\text{ t}$

***P* system of forces:**

The truss is indeterminate externally by one degree. Analysising the truss considering one of the supports, say the support at B to be a roller support.

Now this truss is a determinate one and such a truss can be analysed by method of joints. Assuming tensile forces in members to be positive and compressive forces to be negative.

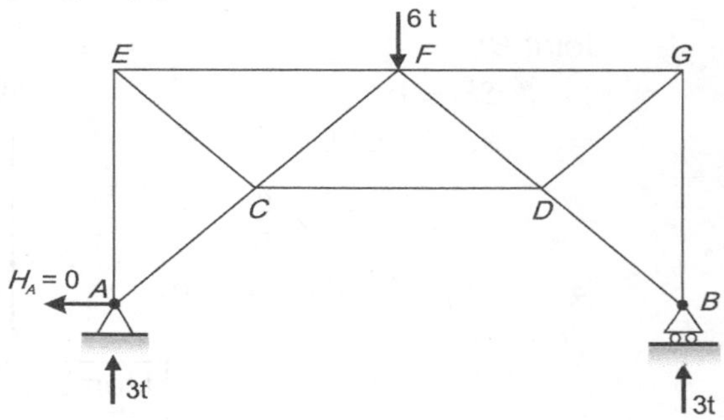

Joint *A*

$$\tan\theta = \frac{4}{3};\ \sin\theta = \frac{4}{5};\ \cos\theta = \frac{3}{5}$$

$\Sigma F_x = 0$ AND $\Sigma F_y = 0$

$\Rightarrow \quad F_{AC} \sin\theta = 0 \qquad \Rightarrow \quad F_{AE} + F_{AC}\cos\theta + 3 = 0$

$\Rightarrow \quad F_{AC} = 0 \qquad \Rightarrow \quad F_{AE} = -3\text{ t (Compressive)}$

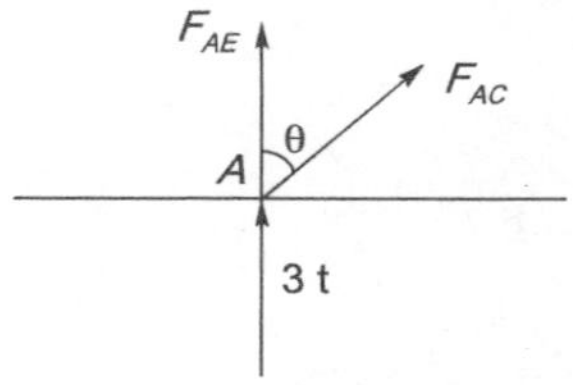

Joint *E*

$$\tan\theta = \frac{3}{4};\ \sin\theta = \frac{3}{5};\ \cos\theta = \frac{4}{5}$$

$\Sigma F_y = 0$ AND $\Sigma F_x = 0$

$\Rightarrow \quad -F_{EC}\sin\theta + 3 = 0 \qquad \Rightarrow \quad F_{EF} + F_{EC}\cos\theta = 0$

$\Rightarrow \quad F_{EC} = \dfrac{3}{\sin\theta} \qquad \Rightarrow \quad F_{EF} = -5 \times \dfrac{4}{5}$

$\Rightarrow \quad F_{EC} = 3 \times \dfrac{5}{3} \qquad \Rightarrow \quad F_{EF} = -4\text{ t (Compressive)}$

$\Rightarrow \quad F_{EC} = 5\text{ t (Tensile)}$

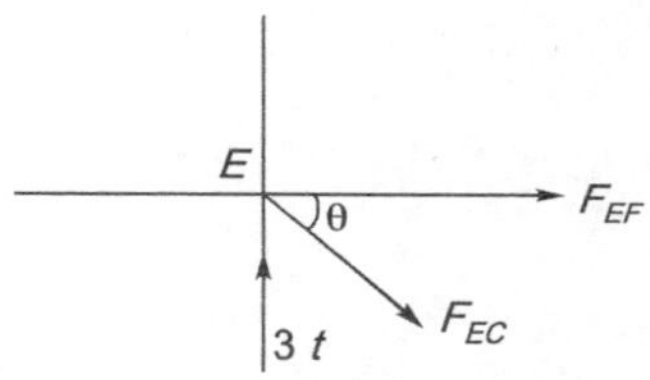

Joint *C*

$$\tan\phi = \frac{3}{4};\ \sin\phi = \frac{3}{5};\ \cos\phi = \frac{4}{5},\ \tan\theta = \frac{4}{3};\ \sin\theta = \frac{4}{5};\ \cos\theta = \frac{3}{5}$$

$\Sigma F_x = 0$ AND $\Sigma F_y = 0$

$\Rightarrow \quad F_{CD} + F_{CF}\cos\phi - 5\sin\theta = 0 \qquad \Rightarrow \quad F_{CF}\sin\phi + 5\cos\theta = 0$

$\Rightarrow \quad F_{CD} + F_{CF} \times \dfrac{4}{5} - 5 \times \dfrac{4}{5} = 0 \qquad \Rightarrow \quad F_{CF} \times \dfrac{3}{5} + 5 \times \dfrac{3}{5} = 0$

$\Rightarrow \quad F_{CD} + \dfrac{4}{5} \times F_{FC} = 4 \qquad \Rightarrow \quad F_{CF} = -3 \times \dfrac{5}{3}$

$\Rightarrow \quad F_{CD} + \dfrac{4}{5} \times (-5) = 4 \qquad \Rightarrow \quad F_{CF} = -5\text{ t (Compression)}$

$\Rightarrow \quad F_{CD} + \dfrac{4}{5} \times (-5) = 4$

$\Rightarrow \quad F_{CD} = 8\text{ t (Tensile)}$

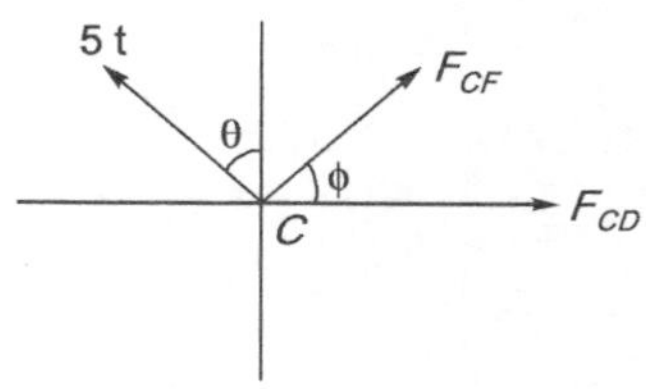

Since the truss is symmetrical, the force in other members will be

$F_{BG} = F_{AE} = -3$ t (Compressive)
$F_{BD} = F_{AC} = 0$
$F_{GD} = F_{EC} = 5$ t (Tensile)
$F_{GF} = F_{EF} = -4$ t (Compressive)
$F_{DF} = F_{CF} = -5$ t (Compressive)
$F_{CD} = 8$ t (Tension)

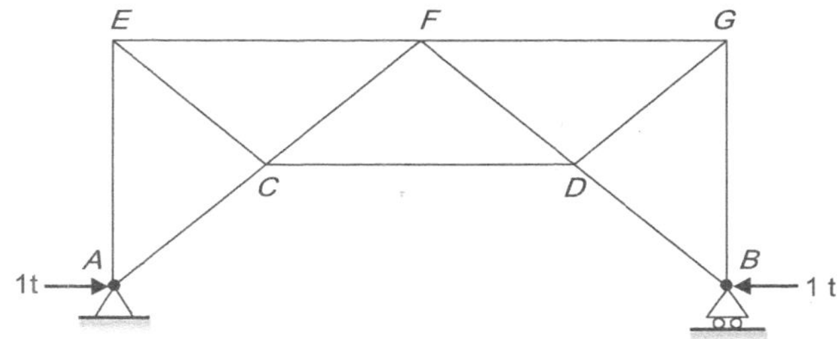

K System of Forces

Now the external load is removed and a pair of unit horizontal loads are applied at *A* and *B*.

Joint *A*

$$\tan\theta = \frac{4}{3};\ \sin\theta = \frac{4}{5};\ \cos\theta = \frac{3}{5}$$

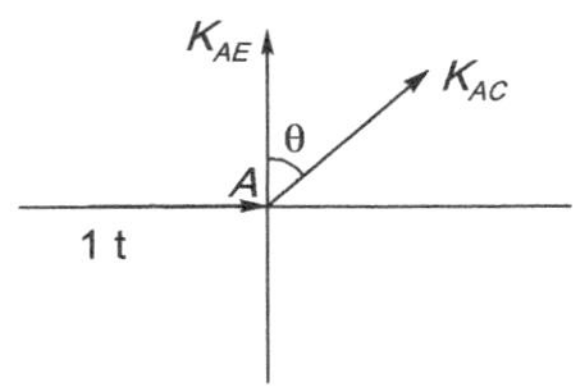

$\Sigma F_x = 0$

$\Rightarrow \quad K_{AC}\sin\theta + 1 = 0$

$\Rightarrow \quad K_{AC} = -\dfrac{1}{\sin\theta}$

$\Rightarrow \quad K_{AC} = -1\times\dfrac{5}{4}$

$\Rightarrow \quad K_{AC} = -1.25$ t

AND

$\Sigma F_y = 0$

$\Rightarrow \quad K_{AC}\cos\theta + K_{AE} = 0$

$\Rightarrow \quad K_{AE} = -\left(-1.25\times\dfrac{3}{5}\right)$

$\Rightarrow \quad K_{AE} = 0.75$ t (Tensile)

Joint *E*

$$\tan\theta = \frac{3}{4};\ \sin\theta = \frac{3}{5};\ \cos\theta = \frac{4}{5}$$

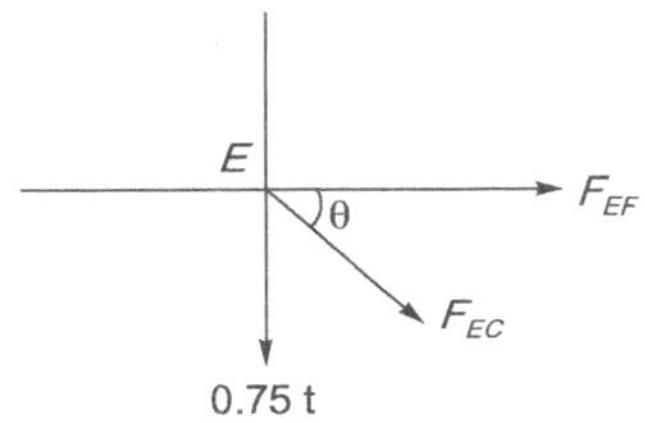

$\Sigma F_x = 0$

$\Rightarrow \quad K_{EF} + K_{EC}\cos\theta = 0$

$\Rightarrow \quad K_{EF} = -(-1.25)\times\dfrac{4}{5}$

$\Rightarrow \quad K_{EF} = 1$ t (Tensile)

AND

$\Sigma F_y = 0$

$\Rightarrow \quad -K_{EC}\sin\theta - 0.75 = 0$

$\Rightarrow \quad K_{EC} = -\dfrac{0.75}{\sin\theta}$

$\Rightarrow \quad K_{EC} = -0.75\times\dfrac{5}{3}$

$\Rightarrow \quad K_{EC} = -1.25$ t (Compressive)

Joint *C*

$$\tan\phi = \frac{3}{4};\ \sin\phi = \frac{3}{5};\ \cos\phi = \frac{4}{5}\ \text{ and }\ \tan\theta = \frac{4}{3};\ \sin\theta = \frac{4}{5};\ \cos\theta = \frac{3}{5}$$

$\Sigma F_x = 0$

$\Rightarrow \quad K_{CD} + K_{CF}\cos\phi + 1.25\sin\theta + 1.25\cos\phi = 0$

$\Rightarrow \quad K_{CD} + 1.25\times\dfrac{4}{5} + 1.25\times\dfrac{4}{5} = 0$

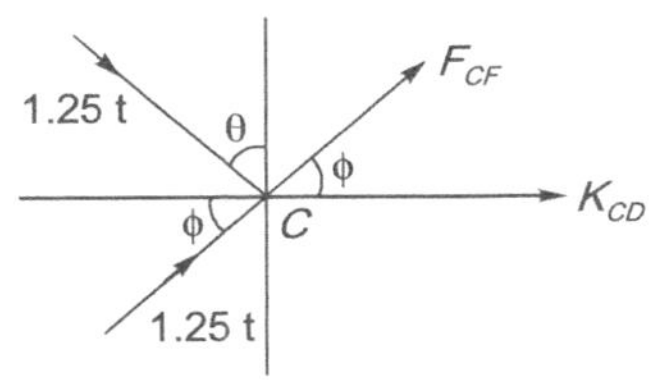

$\Rightarrow \quad K_{CD} = -2\times 1.25\times\dfrac{4}{5}$

$\Rightarrow \quad K_{CD} = -2$ t (Compressive)

$\Sigma F_y = 0$

$\Rightarrow \quad K_{CF}\sin\phi + 1.25\sin\phi - 1.25\cos\theta = 0$

$\Rightarrow \quad K_{CF}\sin\phi + 1.25\times\dfrac{3}{5} - 1.25\times\dfrac{3}{5} = 0$

$\Rightarrow \quad K_{CF} = 0$

K-force in remaining members

$\therefore$ $K_{BG} = K_{AE} = 0.75$ t (Tensile) $K_{BD} = K_{AC} = -1.25$ t (Compressive)

$K_{GD} = K_{EC} = -1.25$ t (Compressive) $K_{GF} = K_{EF} = 1$ t (Tensile)

$K_{DF} = K_{CF} = 0$ $K_{CD} = -2$ t (Compressive)

Length of members of truss

$$L_{AE} = L_{BG} = 6 \text{ m}$$

$$L_{AC} = L_{BD} = \sqrt{4^2 + 3^2} = 5 \text{ m}$$

$$L_{EC} = L_{GD} = \sqrt{4^2 + 3^2} = 5 \text{ m}$$

$$L_{EF} = L_{GF} = 8 \text{ m}$$

$$L_{CF} = L_{DF} = \sqrt{4^2 + 3^2} = 5 \text{ m}$$

$$L_{CD} = 8 \text{ m}$$

The quantities which have been calculated are tabulated:

Member	*P*	*K*	*L*	*PKL*	K^2L
AE	–3	0.75	6	–13.5	3.375
AC	0	–1.25	5	0	7.8125
EC	5	–1.25	5	–31.25	7.8125
EF	–4	1	8	–32	8
CF	–5	0	5	0	0
CD	8	–2	8	–128	32
BG	–3	0.75	6	–13.5	3.375
BD	0	–1.25	5	0	7.8125
GD	5	–1.25	5	–31.25	7.8125
GF	–4	1	8	–32	8
DF	–5	0	5	0	0
				$\Sigma PKL = -281.5$	$\Sigma K^2L = 86$

The horizonal thrust is given by

$$H = -\frac{\Sigma \frac{PKL}{AE}}{\Sigma \frac{K^2L}{AE}}$$

Assuming material of all the truss members to be same. Also, it is given that all the members have same area of cross-section.

$$\therefore \quad H \approx -\frac{\Sigma PKL}{\Sigma K^2L} = -\frac{(-281.5)}{86} = 3.27 \text{ t}$$

$$V_A = 3\text{t} (\uparrow),\ V_B = 3\text{t} (\uparrow),\ H_A = 3\text{t} (\rightarrow),\ H_B = 3\text{t} (\leftarrow)$$

5.12 **Calculate the bar forces in all the members of the plane truss shown below which is used as a tower. All inclined members have a slope of 45° to the horzional planes. Also find reaction at supports *D*, *E* and *F*.**

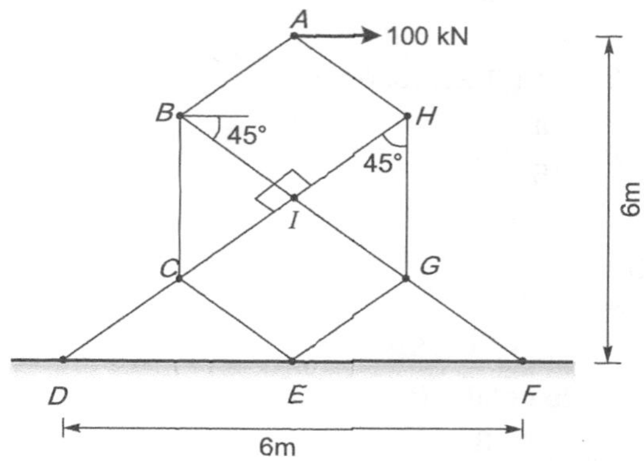

[20 marks : 2010]

Solution:

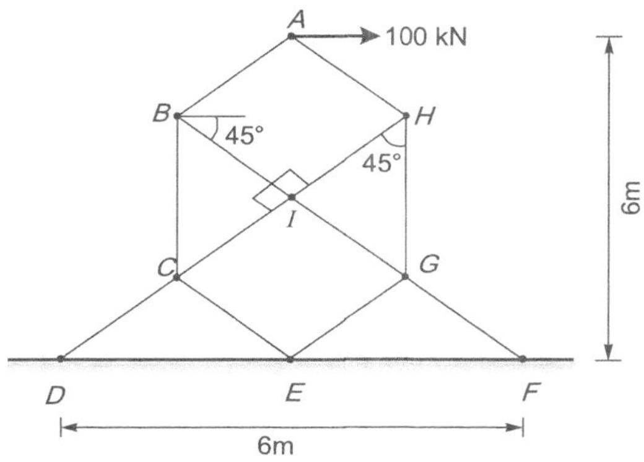

Given: All inclined members have a slope of 45°.

Assumption: All member forces are assumed to be tensile

(+) Tensile

(–) Compression

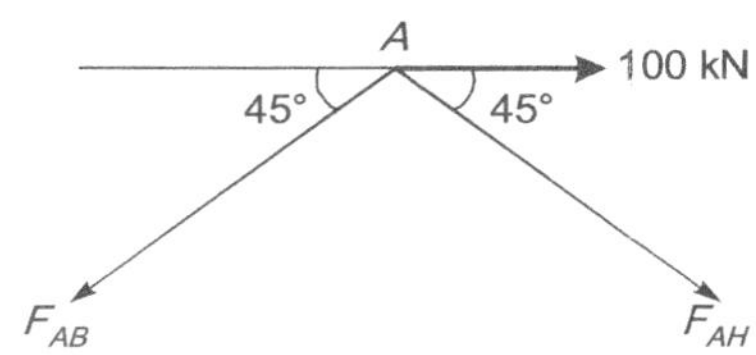

Considering joint *A*

$\Sigma F_x = 0$

$\Sigma F_y = 0$

$$F_{AB} \sin 45° + F_{AH} \sin 45° = 0$$

$$F_{AB} + F_{AH} = 0 \quad \text{...(i)}$$

and,
$$F_{AB} \cos 45° = 100 + F_{AH} \cos 45° \quad \text{...(ii)}$$

$$F_{AB} = \frac{100}{\sqrt{2}} \text{ kN}$$

$\Rightarrow$
$$F_{AB} = 70.71 \text{ kN}$$

$$F_{AH} = -70.71 \text{ kN}$$

Considering joint *B*

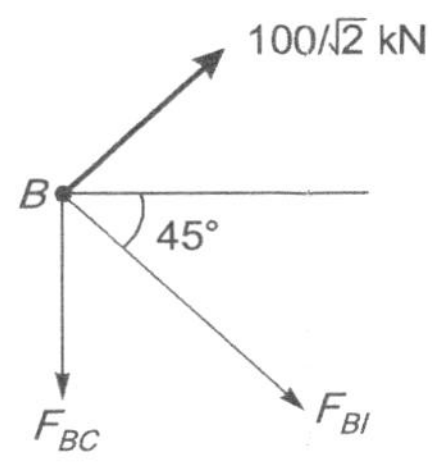

$$F_{BI} \cos 45° + \frac{100}{\sqrt{2}} \cos 45° = 0$$

$$F_{BI} = -70.71 \text{ KN}$$

$$F_{BC} = \frac{100}{\sqrt{2}} \sin 45° + F_{BI} \cos 45°$$

$$F_{BC} = 100 \text{ KN}$$

Considering joint *I*

At joint *I*, directions *BG* and *CH* are mutually perpendicular.

Thus summation of forces along both these direction should also be zero and as F_{CI} and F_{CH} will not have any component in BG direction so we can say that

$$[F_{BI} = F_{IG}]; \qquad [F_{CI} = F_{IH}]$$

$$F_{IG} = -70.71 \text{ kN}$$

$$F_{CI} = 70.71 \text{ kN}$$

At joint *H*

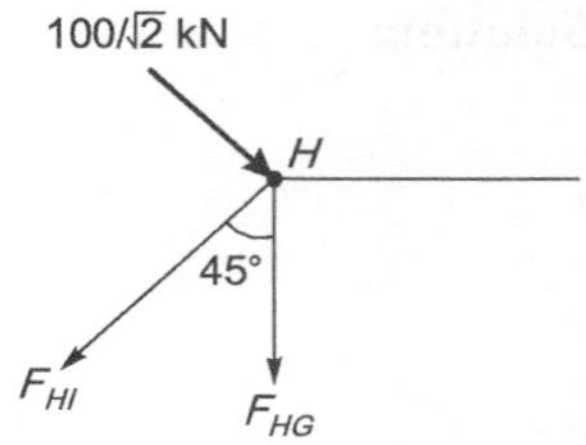

$$F_{HI} = 70.71 \text{ kN}$$
$$F_{HG} = -100 \text{ kN}$$

Considering joint *C*

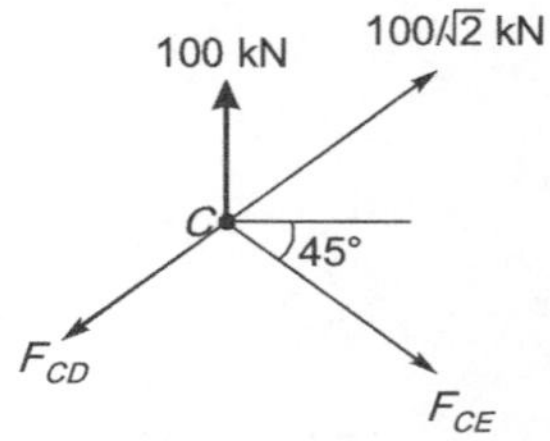

$\Sigma F_x = 0;\quad F_{CE}\cos 45° + 50 = F_{CD}\cos 45°$ ...(iii)

$\Sigma F_y = 0;\quad F_{CD}\cos 45° + F_{CE}\cos 45° = 150$ ...(iv)

$$F_{CE} = 70.71 \text{ kN}$$
$$F_{CD} = 141.42 \text{ kN}$$

Similarly at Joint *G*

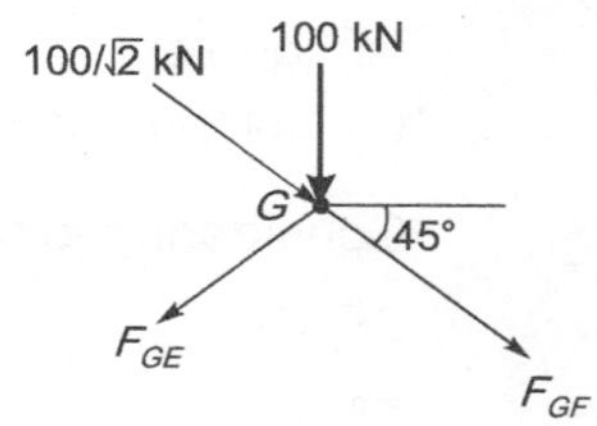

We get

$$F_{GE} = -70.71 \text{ kN}$$
$$F_{GF} = -141.42 \text{ kN}$$

At joint *E*

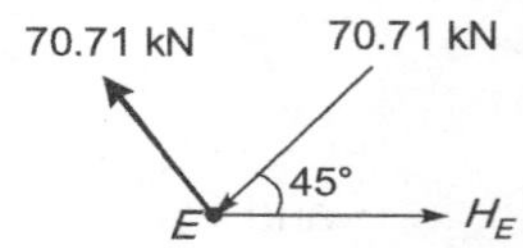

$$H_E = (70.71 \cos 45°) \times 2$$
$$H_E = 100 \text{ kN}$$

At joint *D*

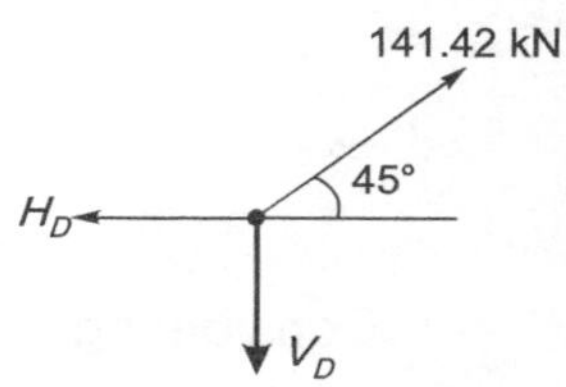

$$V_D = 100 \text{ kN}$$
$$H_D = 100 \text{ kN}$$

At joint *F*

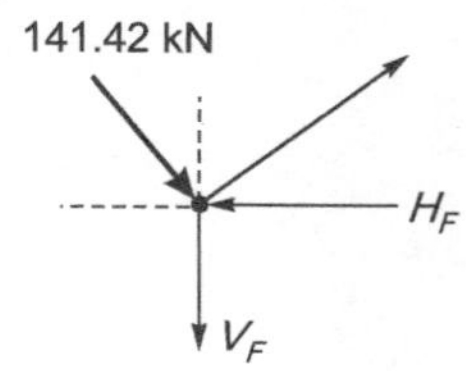

$$H_F = 141.42 \cos 45° = 100 \text{ kN}$$

$\Sigma F_x = 0,\quad 141.42\cos 45° + V_F = 0,\quad V_F = -100 \text{ kN}$

5.13 Three members of the truss shown in the figure below have been cut either too long (+ve) or too short (–ve) as per details given below. Calculate the vertical displacement of joint *D* due to discrepancies in the length of these members by unit load method.

$\delta_{AB} = -3$ mm, $\delta_{BC} = 2$ mm; $\delta_{EF} = 4$ mm

All members have the same area of cross-section.

[10 marks : 2011]

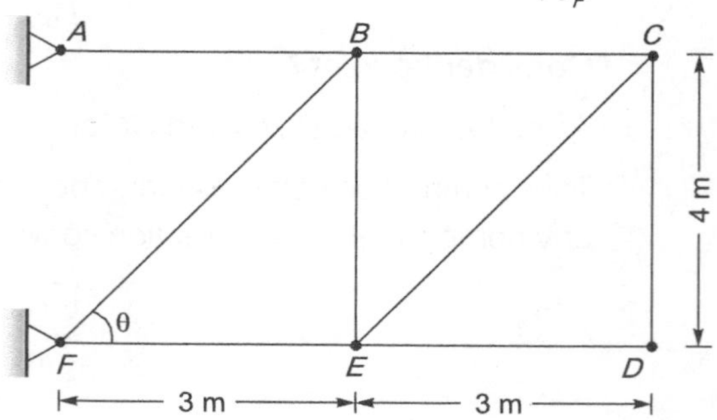

Solution:

All members have same area of cross-section

Given: $\delta_{AB} = -3$ mm, $\delta_{BC} = 2$ mm, $\delta_{EF} = 4$ mm,

$\tan\theta = 4/3$

We have to determine vertical displacement at joint D.

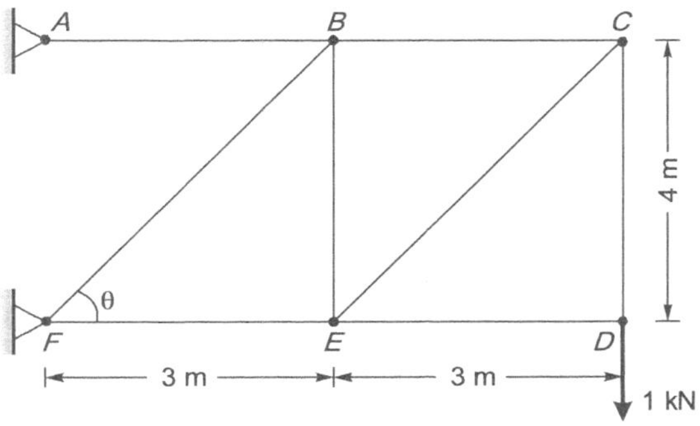

By unit load method:

Apply unit vertical load at joint D.

k_j = Member force in j^{th} member due to application of unit vertical load at D.

$$k_{CD} = 1 \text{ kN}; \quad k_{DE} = 0 \text{ kN}$$

$$k_{CE} = -\frac{5}{4} \text{ kN}; \quad k_{BC} = \frac{3}{4} \text{ kN}$$

$$k_{BE} = 1 \text{ kN}; \quad k_{EF} = -\frac{3}{4} \text{ kN}$$

$$k_{BF} = -\frac{5}{4} \text{ kN}; \quad k_{AB} = \frac{3}{2} \text{ kN}$$

Member	δ	k	L	δk
AB	–3	3/2	3 m	–9/2
BC	2	3/4	3 m	6/4
CD	—	1	4 m	—
DE	—	0	3 m	—
EF	4	–3/4	3 m	–3
FB	—	–5/4	5 m	—
EC	—	–5/4	5 m	—
BE	—	1	4 m	—
				$\Sigma\delta k = -6$

$$\Delta_{DV} = \frac{\Sigma PkL}{AE}$$

Here, $\frac{P_i L_i}{A_i E_i} = \delta_i$

$$\Delta_{DV} = \Sigma\delta k$$

$$\Delta_{DV} = -6 \text{ mm (upward)}$$

5.14 Analyse the truss shown in the figure below and list out the forces in the members.

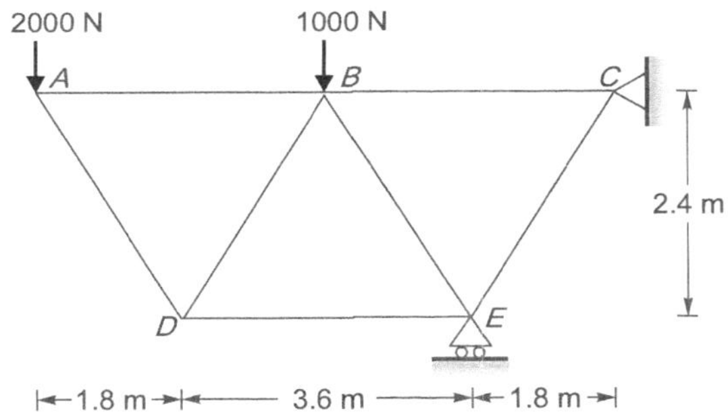

[14 marks : 2011]

Solution:

Using method of joints

At support E

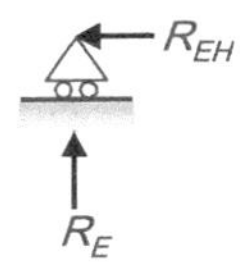

$$R_{EH} = 0 \quad \text{(Roller support, } \Sigma H = 0\text{)}$$

At support C

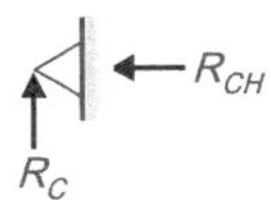

$$R_{CH} = 0 \;(\because \text{Because there is no external horizontal force})$$

$\Rightarrow \quad \Sigma V = 0$

$$R_E + R_C = 2000 + 1000 = 3000 \quad \text{...(i)}$$

$$\Sigma M_c = 0$$

$\Rightarrow \quad R_E \times 1.8 - 1000 \times 3.6 - 2000 \times 7.2 = 0$

or $\quad R_E = 10000$ N

From (i) $\quad R_C = 3000 - 10000 = -7000$ N

At Joint *A*

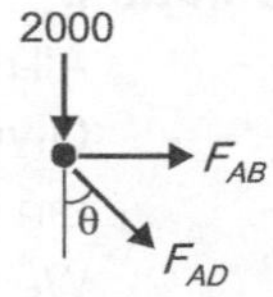

$$\tan\theta = \frac{1.8}{2.4} = \frac{3}{4}$$

$$\cos\theta = \frac{2.4}{\sqrt{2.4^2 + 1.8^2}} = \frac{2.4}{3} = \frac{4}{5}$$

$$\sin\theta = \frac{1.8}{\sqrt{2.4^2 + 1.8^2}} = \frac{1.8}{3} = \frac{3}{5}$$

$\Sigma F_y = 0$ $\quad 2000 + F_{AD}\cos\theta = 0$

$\Rightarrow$ $\quad 2000 + F_{AD} \times \frac{4}{5} = 0$

$\Rightarrow$ $\quad F_{AD} = F_{DA} = -2500$ N (Compressive)

$\Sigma F_x = 0$ $\quad F_{AB} + F_{AD}\sin\theta = 0$

$$F_{AB} = -F_{AD}\sin\theta = -(-2500) \times \frac{3}{5} = 1500 \text{ N}$$

$\Rightarrow$ $\quad F_{AB} = F_{BA} = 1500$ N (Tensile)

At Joint *D*

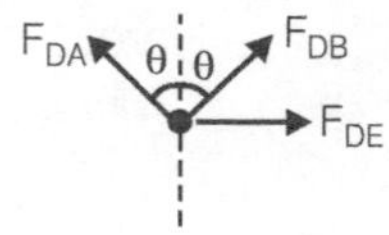

$$\cos\theta = \frac{4}{5} \text{ and } \sin\theta = \frac{3}{5}$$

$\Sigma F_y = 0$ $\quad F_{DA}\cos\theta + F_{DB}\cos\theta = 0$

$$-2500 \times \frac{4}{5} + F_{DB} \times \frac{4}{5} = 0$$

$\Rightarrow$ $\quad F_{DB} = F_{BD} = 2500$ N (Tensile)

$\Sigma F_x = 0$ $\quad -F_{DA}\sin\theta + F_{DB}\sin\theta + F_{DE} = 0$

$$-(-2500) \times \frac{3}{5} + 2500 \times \frac{3}{5} + F_{DE} = 0$$

$\Rightarrow$ $\quad F_{DE} = F_{ED} = -3000$ N (Compressive)

At Joint *C*

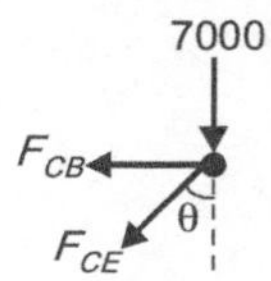

$$\cos\theta = \frac{4}{5}, \quad \sin\theta = \frac{3}{5}$$

$\Sigma F_y = 0$ $\quad 7000 + F_{CE}\cos\theta = 0$

$\Rightarrow$ $\quad F_{CE} = F_{EC} = -8750$ N (Compressive)

$\Sigma F_x = 0$ $\quad F_{CB} + F_{CE}\sin\theta = 0$

$\Rightarrow$ $\quad F_{CB} = -(-8750) \times 3/5 = 5250$ N

$\Rightarrow$ $\quad F_{CB} = F_{BC} = 5250$ N (Tensile)

At Joint *B*

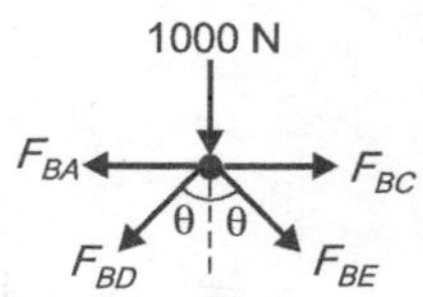

$$\cos\theta = \frac{4}{5}, \quad \sin\theta = \frac{3}{5}$$

$\Sigma F_y = 0$ $\quad 1000 + F_{BD}\cos\theta + F_{BE}\cos\theta = 0$

$\Rightarrow$ $\quad 1000 + 2500 \times \frac{4}{5} + F_{BE} \times \frac{4}{5} = 0$

$\Rightarrow$ $\quad F_{BE} = F_{EB} = -3750$ N (Compressive)

Member	Force value	Force Nature
AD	2500 N	Compressive
AB	1500 N	Tensile
DB	2500 N	Tensile
DE	3000 N	Compressive
BE	3750 N	Compressive
BC	5250 N	Tensile
EC	8750 N	Compressive

5.15 Find the member forces in the member marked *A*, *B* and *C* for the truss as show.

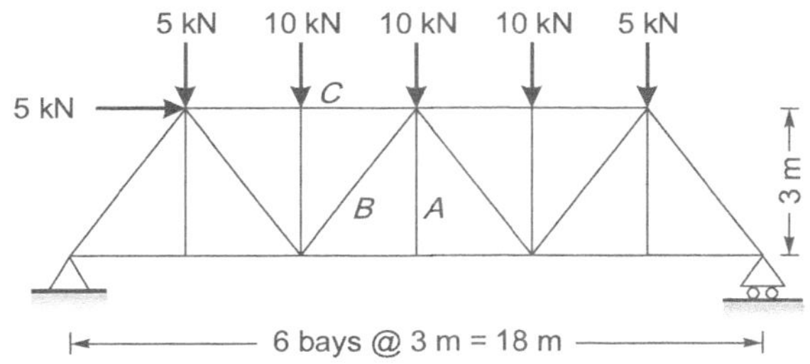

[10 marks : 2013]

Solution:

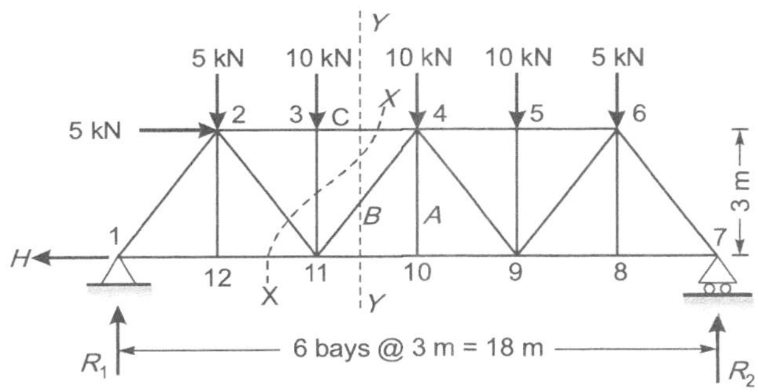

$$H = 5 \text{ kN}$$

$$R_1 + R_2 = 40 \text{ kN}$$

Taking moment about joint 1, $R_2(18) = 5(3) + 5(3) + 10(6) + 10(9) + 10(12) + 5(15)$

$$\Rightarrow \quad R_2 = 20.833 \text{ kN}$$

$$\therefore \quad R_1 = 19.1667 \text{ kN}$$

Section *X*-*X* is as shown above.

Considering the equilibrium of LHS of section *x*-*x* and taking moment about joint 11,

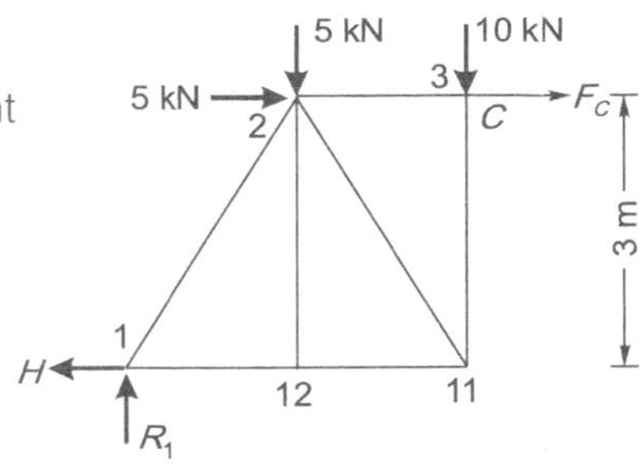

$$F_c(3) + 5(3) + R_1(6) = 5(3)$$

$$\Rightarrow \quad F_c(3) + 15 + 115 = 15$$

$$\Rightarrow \quad F_c = -\frac{115}{3} = -38.333 \text{kN}$$

$$\therefore \quad F_c = 38.333 \text{ kN (compression)}$$

$$F_A = 0 \text{ (due to eq. of joint 10)}$$

Consider section *y-y* as shown in figure

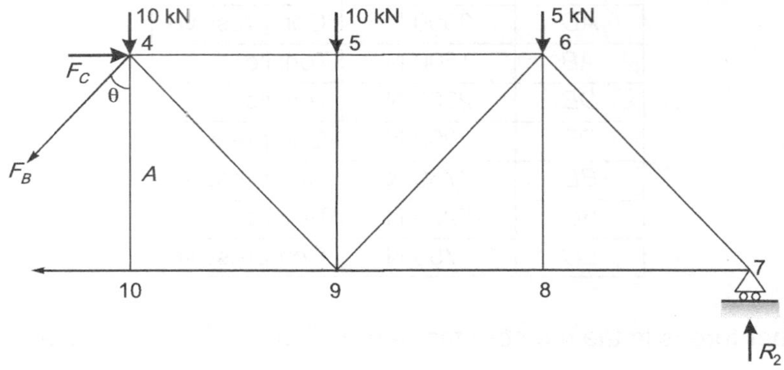

$$\theta = 45°$$

Considering the vertical equilibrium of RHS of section *y-y*

$$F_B \cos 45° + 10 + 10 + 5 = R_2$$

$$\Rightarrow \quad \frac{F_B}{\sqrt{2}} + 25 = 20.8333$$

$$\Rightarrow \quad F_B = -5.8926 \text{ kN}$$

$$\therefore \quad F_B = 5.8926 \text{ kN (Compression)}$$

∴ Force in member, $A = 0$

Force in member, $B = 5.8926$ kN (Compression)

Force in member, $C = 38.333$ kN (Compression)

5.16 Determine the vertical deflection of joint L_2 of the truss shown in figure. The area of cross-section of the members is 20 cm² each. Take $E = 200$ GPa.

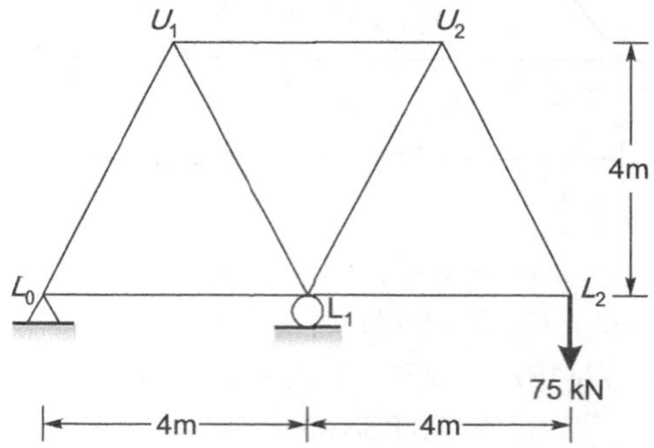

[15 marks : 2014]

Solution:

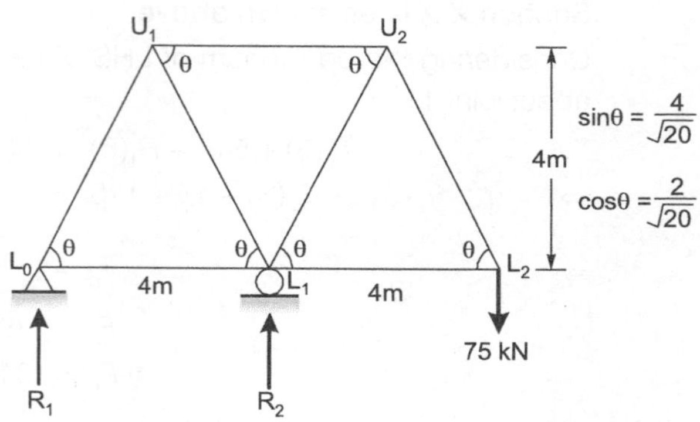

$\Sigma F_x = 0;$ $\quad H_1 = 0$

$\Sigma F_y = 0;$ $\quad R_1 + R_2 = 75$ kN ...(i)

$\Sigma M_A = 0;$ $\quad -R_2 \times 4 + 75 \times 8 = 0$

$$R_2 = 150 \text{ kN } (\uparrow)$$

From eq. (i), we get

$$R_1 = 75 - R_2 = 75 - 150$$

$$R_1 = -75 \text{ kN or } 75 \text{ kN } (\downarrow)$$

P-system of forces:
Consider joint L_0:

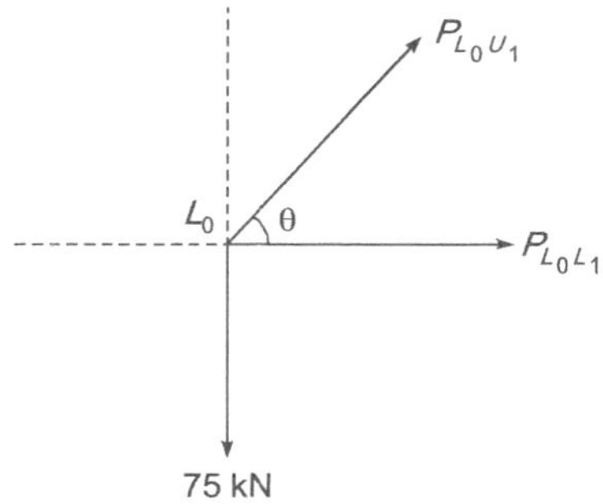

$$\Sigma F_y = 0; \quad P_{L_0U_1}\sin\theta = 75$$

$$P_{L_0U_1} = \frac{75\sqrt{20}}{4} = 83.85 \text{ kN (Tension)}$$

$$\Sigma F_x = 0; \quad P_{L_0L_1} + P_{L_0U_1}\cos\theta = 0$$

$$P_{L_0L_1} = -83.85\times\frac{2}{\sqrt{20}} = -37.5 \text{ kN (Compression)}$$

Consider joint L_2;

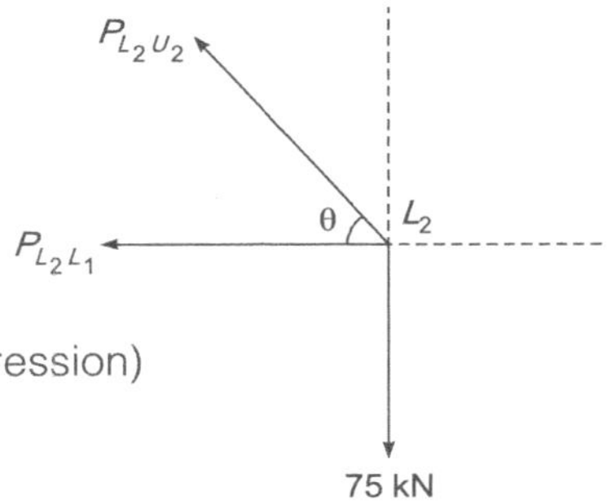

$$\Sigma F_y = 0; \quad P_{L2U2}\sin\theta = 75$$

$$P_{L2U2} = 83.85 \text{ kN (Tension)}$$

$$\Sigma F_x = 0; \quad P_{L2L1} + P_{L2U2}\cos\theta = 0$$

$$P_{L2L1} = -P_{L2U2}\cos\theta = -37.5 \text{ kN (Compression)}$$

Consider joint U_1:

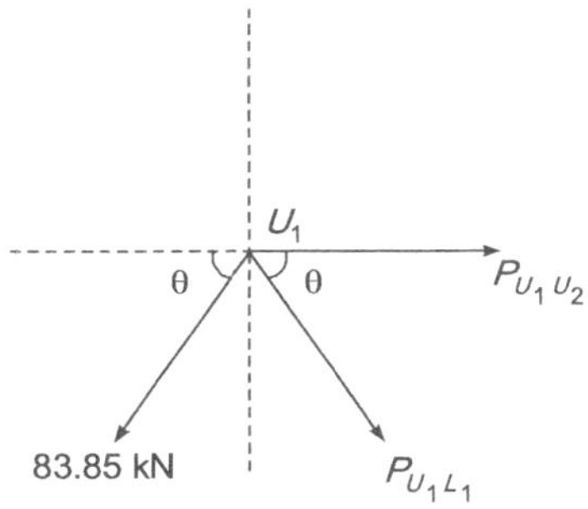

$$\Sigma F_y = 0; \quad P_{U1L1}\sin\theta + 83.85\sin\theta = 0$$

$$P_{U1L1} = -83.85 \text{ kN (Compression)}$$

$$\Sigma F_x = 0; \quad P_{U1U2} + P_{U1L1}\cos\theta = 83.85\cos\theta$$

$$P_{U1U2} = 83.85\cos\theta - P_{U1L1}\cos\theta$$

$$= (83.85 + 83.85)\cos\theta$$

$$= 75 \text{ kN (Tension)}$$

By symmetry,

$$P_{U2L1} = P_{U1L1}$$

$$= 83.85 \text{ kN (Compression)}$$

k-system of forces: Remove all external loading and apply unit load at L_2 in vertically downward direction

Hence *k*-forces will be equal to $\frac{P\text{ - force}}{75}$ kN

Member	P (kN)	k (kN)	L (kN)	PkL
L_0L_1	–37.5	–0.5	4m	75
L_0U_1	83.85	1.118	$\sqrt{20}$ m	419.23
U_1U_2	75	1	4m	300
U_1L_1	–83.85	–1.118	$\sqrt{20}$ m	419.23
U_2L_1	–83.85	–1.118	$\sqrt{20}$ m	419.23
U_2L_2	83.85	1.118	$\sqrt{20}$ m	419.23
L_1L_2	–37.5	–0.5	4m	75
				ΣPkL = 2126.92

$$\therefore \quad \Delta_{L2} = \frac{\Sigma PkL}{AE} = \frac{2126.92 \times 10^6}{(20 \times 10^2) \times 2 \times 10^5} \text{ mm} = 5.32 \text{ mm (downward)}$$

5.17 If member U_1L_1 is fabricated 10 mm too short, determine the vertical deflection at L_2.

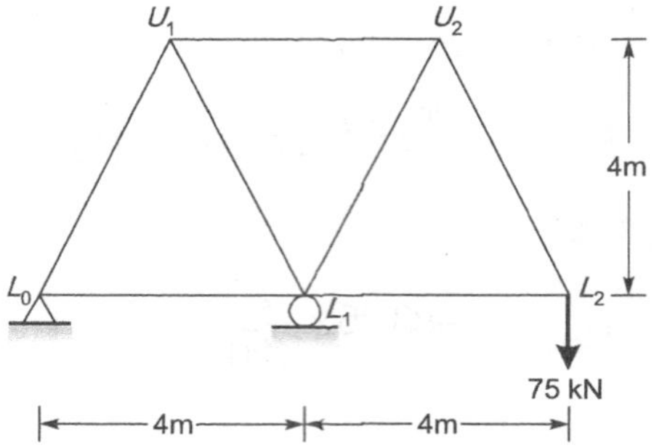

[5 marks : 2014]

Solution:

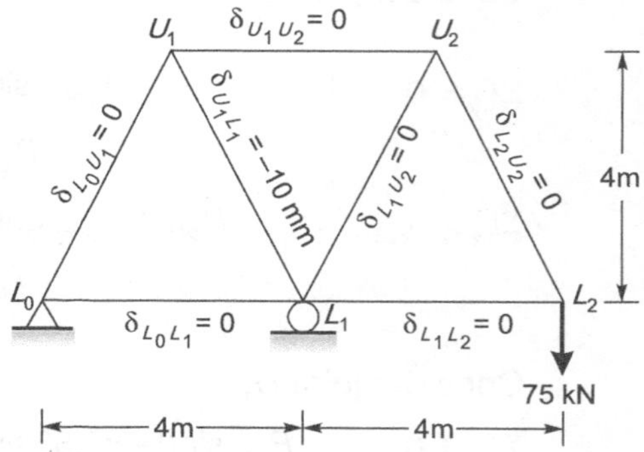

$$\Delta_{L2} = \Sigma k\left[\frac{PL}{AE} + \delta\right] = \frac{\Sigma PkL}{AE} + \Sigma k.\delta$$

$$= 5.32 + (-1.118) \times (-10)$$

$$= 5.32 + 11.180$$

$$= 16.50 \text{ mm } (\downarrow)$$

5.18 Determine the forces in all the members of the truss as shown in figure, due to a fall of temperature of the member BD by 30°C. Cross-sectional area of all the members = 1500 mm², α = 0.00001/°C.

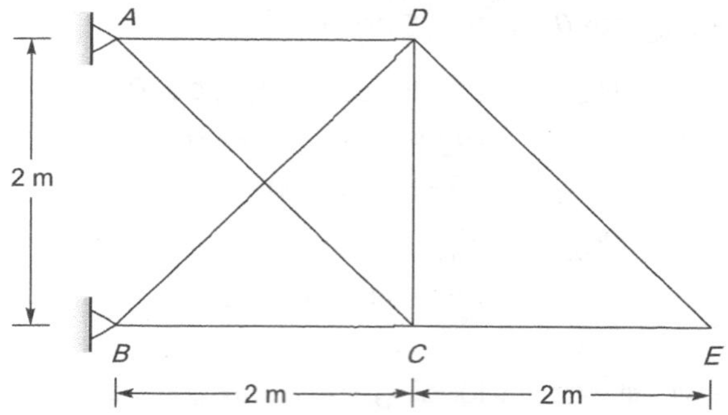

[20 marks : 2017]

Solution:

Assumption:

(i) $E = 2 \times 10^5$ N/mm², for all members

(ii) Taking tension as positive and compressive force as negative.

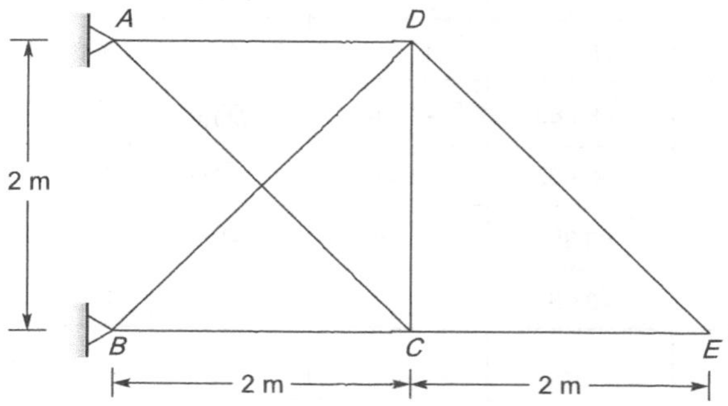

Given ΔT in member BD = –30°C

Cross-sectional area of all members, A = 1500 mm²

α = 0.00001/°C

$\because$ $D_S = (m + r_e) - 2j = (7 + 4) - 5 \times 2 = 1.0$

Let us consider one of diagonal members say BD as redundant.

Let us remove this member and analyse the structure.

P-system of forces

$\because$ No external load acts on the member. All forces in members due to external load is zero.

K-system of force:

Free compression of member BD, $\Delta_T = -L\alpha\Delta T$

$$= -2\sqrt{2} \times 1 \times 10^{-5} \times 30°$$

$$= -8.48528 \times 10^{-4} \text{ m}$$

As free compression is prevented, a tensile force develops in the member BD. Let this force be X.

$\therefore$ Apply unit force 1 kN at each B and D in place of member BD. The determinate truss obtained after removing member BD is as shown below.

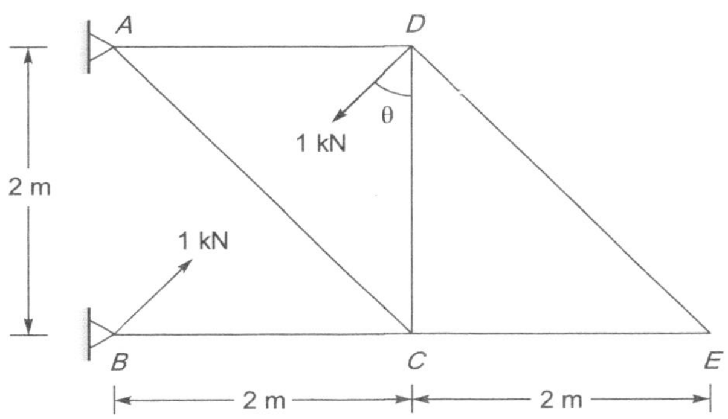

Since these forces balances themselves there will be no reactions at supports.

At joint E,

$$\Sigma F_x = 0 \quad \text{and} \quad \Sigma F_y = 0$$

$\Rightarrow$ $k_{CE} = 0$ and $k_{DE} = 0$

At joint D,

Resolving vertically at D,

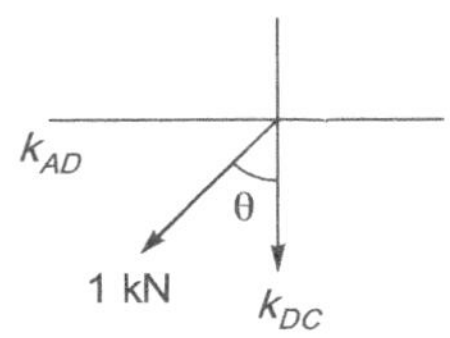

$$\tan\theta = \frac{2}{2} \Rightarrow 1$$

$\Rightarrow$ $\theta = 45°$

$$\Sigma F_y = 0$$

$\Rightarrow$ $1\cos 45° + k_{DC} = 0$

$\Rightarrow$ $k_{DC} = -\frac{1}{\sqrt{2}}$ kN (compressive)

$\Sigma F_x = 0$, Resolving horizontally at D.

$$1\sin 45° + k_{AD} = 0$$

$\Rightarrow$ $k_{AD} = -\frac{1}{\sqrt{2}}$ kN

At joint C,

Resolving horizontally at C,

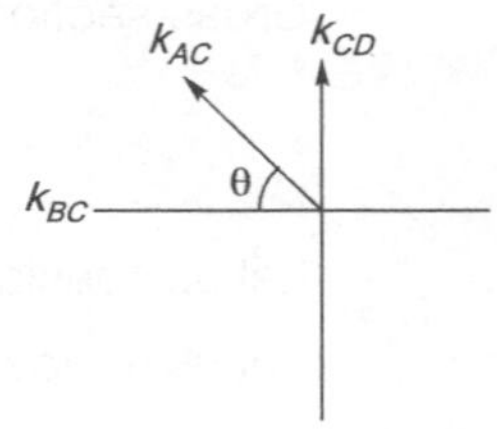

$$\Sigma F_x = 0$$

$$\Rightarrow \quad k_{BC} + k_{AC}\cos 45° = 0$$

$$\Sigma F_y = 0$$

$$\Rightarrow \quad k_{AC}\sin\theta + k_{DC} = 0$$

$$\Rightarrow \quad k_{AC} \times \frac{1}{\sqrt{2}} - \frac{1}{\sqrt{2}} = 0$$

$$\Rightarrow \quad k_{AC} = 1 \text{ kN}$$

$$\therefore \quad k_{BC} = -1 \times \frac{1}{\sqrt{2}} = \frac{-1}{\sqrt{2}} \text{ kN}$$

Member	P(kN)	K(kN)	L(m)	K^2L	$S = P + Kx = Kx$ (kN)
AD	0	$-\frac{1}{\sqrt{2}}$	2	1	–20.79
DE	0	0	$2\sqrt{2}$	0	0
CE	0	0	2	0	0
CB	0	$-\frac{1}{\sqrt{2}}$	2	1	–20.79
AC	0	1	$2\sqrt{2}$	$2\sqrt{2}$	29.4
BD	0	1	$2\sqrt{2}$	$2\sqrt{2}$	29.4
CD	0	$-\frac{1}{\sqrt{2}}$	2	1	–20.79
				$\Sigma K^2L = 3 + 4\sqrt{2}$	

$$x = -\left\{\frac{\Sigma \frac{pKL}{AE} + \Sigma K\alpha L\Delta T}{\Sigma \frac{K^2L}{AE}}\right\}$$

$\because$ $P = 0$ for all members and $\Delta T = -30°C$ (temperature decrease is only for member BD)

$$\therefore \quad X = -\frac{K\alpha L\Delta T}{\Sigma \frac{K^2L}{AE}} = -\frac{K\alpha L\Delta T}{\Sigma K^2L}(AE)$$

$$= -\frac{1 \times \left(-8.48528 \times 10^{-4}\right)}{3 + 4\sqrt{2}} \times 1500 \times 2 \times 10^5 \times 10^{-3}$$

$$= 29.4 \text{ kN}$$

Hence final forces in the members, are shown below:

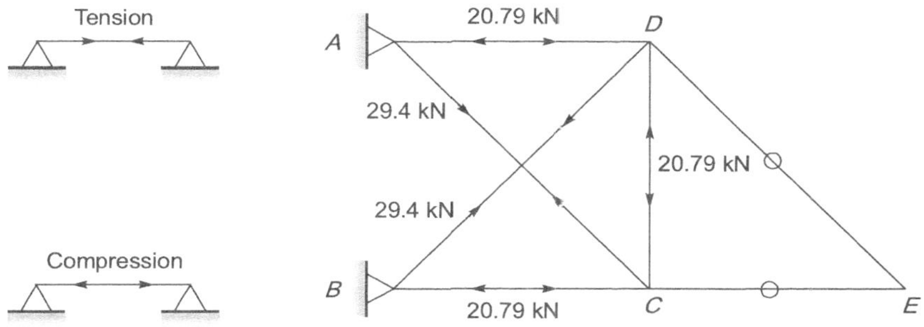

5.19 The frame shown in figure is fabricated using members with cross-sectional areas as given below:

Diagonal members = 2000 sq.mm

Other members = 1000 sq.mm

E for the material of members = 2 × 10^5 N/sq.mm

Member AC was fitted last and the length of the member was 1 mm short. Determine the forces developed in the members when AC was pulled and fitted in position.

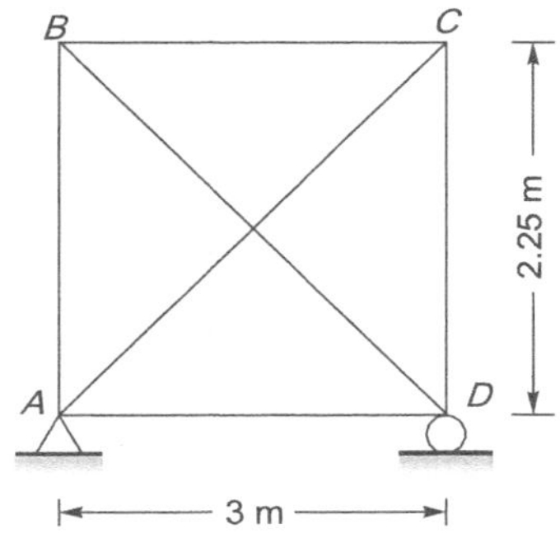

[20 marks : 2018]

Solution:

Given data: Diagonal members c/s area = 2000 mm^2; other members c/s area = 1000 mm^2 and $E = 2 \times 10^5$ N/mm^2.

External static indeterminacy,

$$D_{se} = \text{No. of reactions} - \text{No. of equilibrium equations}$$
$$= 3 - 3 = 0$$

Internal static indeterminacy,

$$D_{si} = m - (2j - 3)$$
$$= 6 - (2 \times 4 - 3)$$
$$= 1$$

∴ Structure is statically indeterminate internally.

Hence, forces are induced in the members due to lack of fit of a member.

Let the force in member *AC* due to lack of fit be '*X*'.

Strain energy,
$$U = \frac{\Sigma (kX)^2 L}{2AE}$$

$$\frac{\partial U}{\partial X} = \delta$$

$$\Rightarrow \quad \Sigma\left(\frac{2k^2 XL}{2AE}\right) = \delta$$

$$\Rightarrow \quad X = \left(\frac{\delta}{\frac{\Sigma k^2 L}{AE}}\right)$$

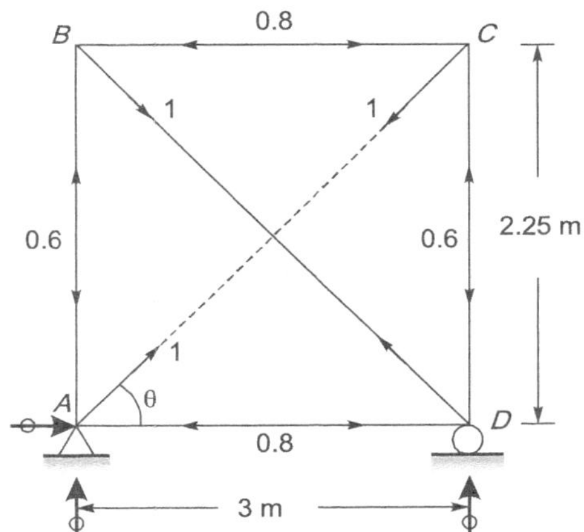

k-forces

k is the force induced in all the members due to unit force applied at joints A and C.

$$\cos\theta = \frac{3}{\sqrt{3^2 + 2.25^2}} = 0.8$$

$$\sin\theta = \frac{2.25}{\sqrt{3^2 + 2.25^2}} = 0.6$$

Sign convention:

Compression : –ve Tension : +ve

Member	k	L(cm)	$\frac{k^2L}{AE}$	Force in members due to lack of fit = kX (in kN)
AB	–0.6	2.25	4.05×10^{-6}	–13.029
BC	–0.8	3	9.6×10^{-6}	–17.372
CD	–0.6	2.25	4.05×10^{-6}	–13.029
DA	–0.8	3	9.6×10^{-6}	–17.372
AC	1	3.75	9.375×10^{-6}	21.715
BD	1	3.75	9.375×10^{-6}	21.715

$$X = \frac{\delta}{\Sigma\frac{k^2L}{AE}} = \frac{10^{-3}}{4.605\times10^{-5}} = 21.715 \text{ kN}$$

5.20 **Determine the forces in the members of the truss shown in figure. All members have same axial rigidity.**

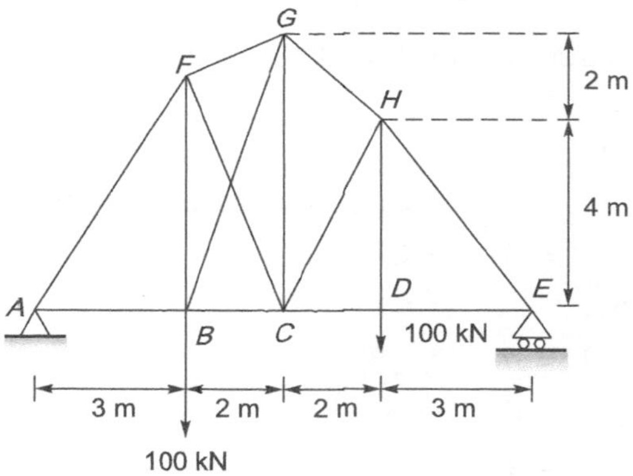

[20 marks : 2020]

Solution:

Note: Since data of BF is not given so as per diagram assuming BF = 5 m.

Truss is indeterminate with 1-degree, so considering member force BG as redundant "R".

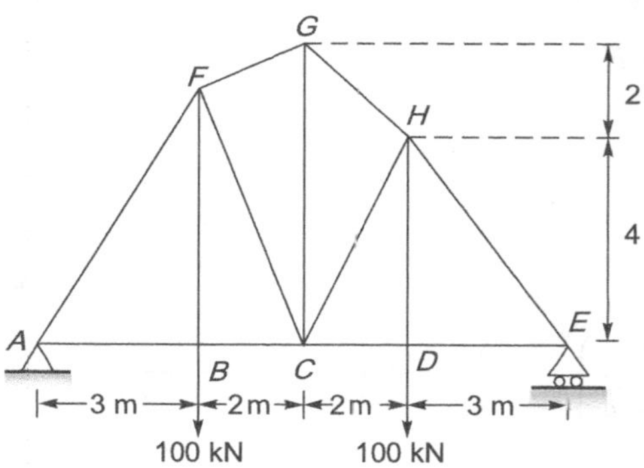

P-Force

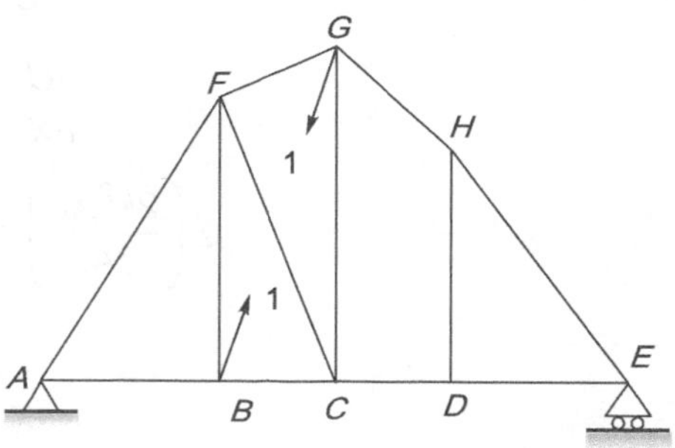

K-forces

$$R = -\frac{\frac{\Sigma P_i K_i l_i}{AE}}{\frac{\Sigma K_i^2 l}{AE}}$$

P-force calculation

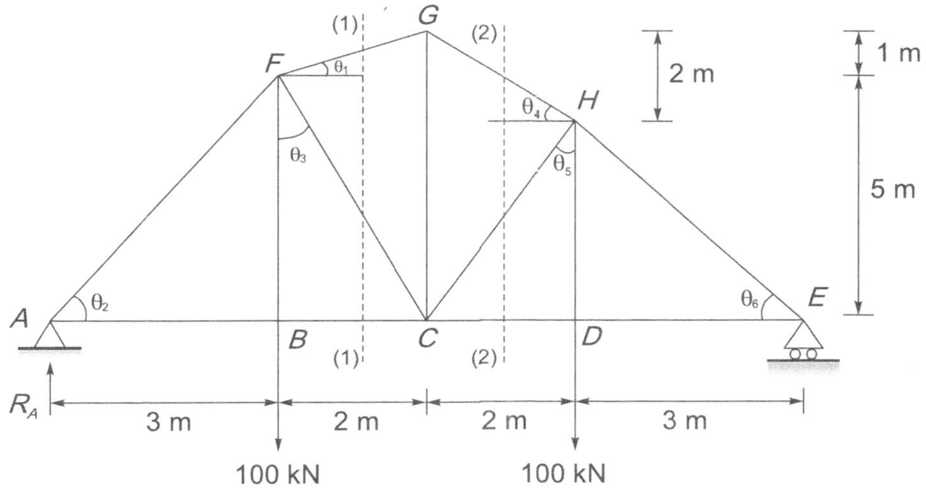

$\Sigma M_E(\circlearrowright) = 0$

$$R_A \times 10 - 100 \times 7 - 100 \times 3 = 0$$

$\Sigma F_y\,(\uparrow) = 0$

$$R_A + R_E = 200 \text{ kN}$$
$$R_E = 100 \text{ kN}$$

Take joint A,

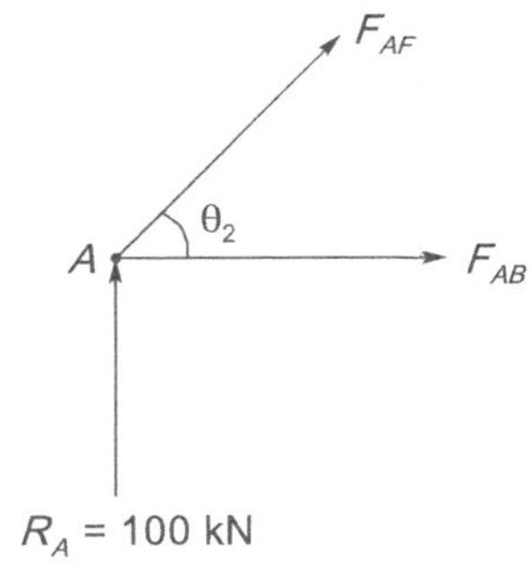

$$\tan\theta_2 = \frac{5}{3}$$

$$\theta_2 = 59.036°$$

$$F_{AF} = -\frac{100}{\sin\theta_2} = -\frac{100}{\sin 59.036}$$

$$= -116.619 \text{ kN}$$

$$F_{AB} = 100 \cot\theta_2$$

$$= 100 \cot 59.036 = 60 \text{ kN}$$

Take joint B,

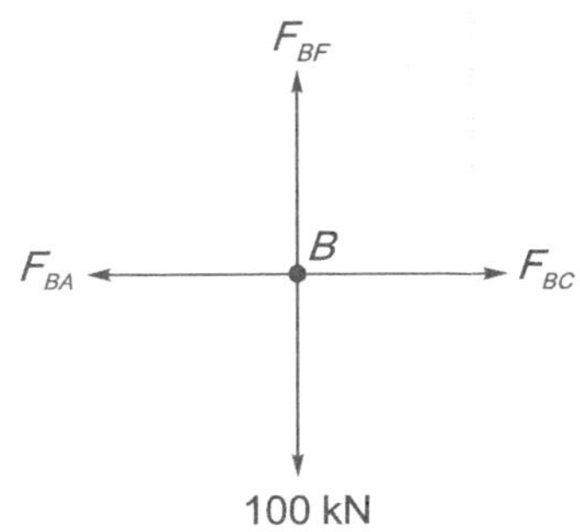

$\Sigma F_x = 0$

$\Rightarrow$ $$F_{BC} = F_{BA} = 60 \text{ kN}$$

$\Sigma F_y = 0$

$\Rightarrow$ $$F_{BF} = 100 \text{ kN}$$

Take section (1)-(1)

$$\tan\theta_1 = \frac{1}{2}$$

$$\theta_1 = 26.565°$$

$$\tan\theta_3 = \frac{2}{5}$$

$$\theta_3 = 21.801°$$

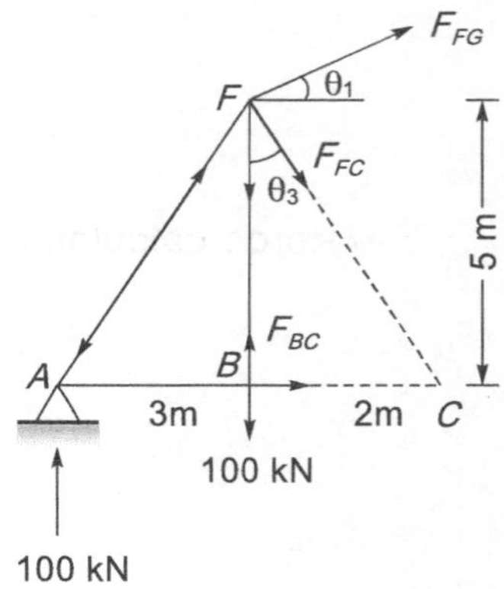

$\Sigma M_C(\curvearrowright) = 0$

$$100 \times 5 - 100 \times 2 + F_{FG}\sin\theta_1 \times 2 + F_{FG}\cos\theta_1 \times 5 = 0$$

$$-\frac{300}{[2\sin 26.565° + 5\cos 26.565°]} = F_{FG}$$

$$F_{FG} = -55.9 \text{ kN}$$

$\Sigma F_y(\uparrow) = 0$

$$-F_{FC}\cos\theta_3 + F_{FG}\sin\theta_1 = 0$$

$$F_{FC} = -55.901 \times \frac{\sin 26.565°}{\cos 21.801°} = -26.925 \text{ kN}$$

Take joint E,

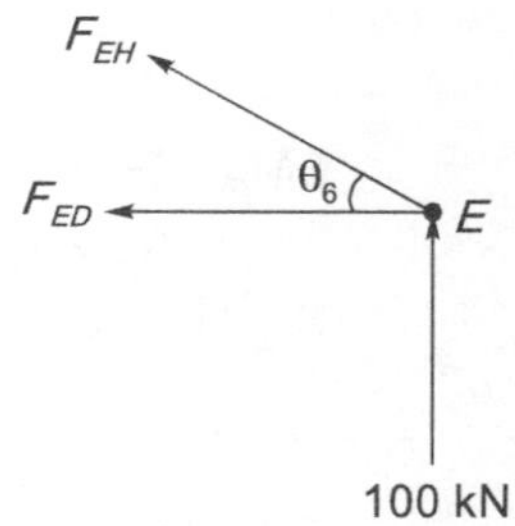

$$\tan\theta_6 = \frac{4}{3}$$

$$\theta_6 = 53.13°$$

$$F_{EH} = -\frac{100}{\sin 53.13°} = -125 \text{ kN}$$

$$F_{ED} = 125\cos 53.13° = 75 \text{ kN}$$

Take joint D,

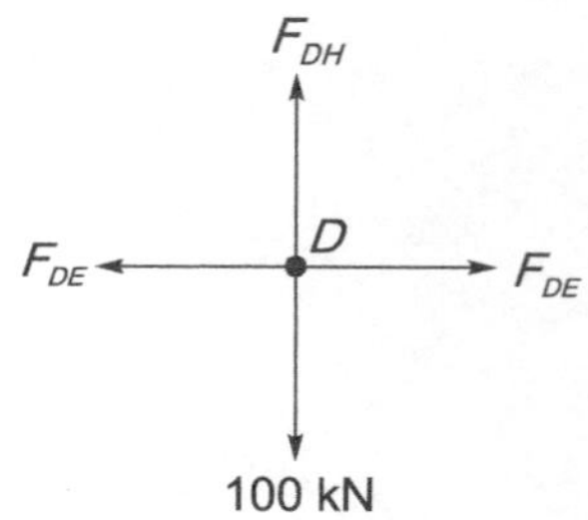

$\Sigma F_y(\uparrow) = 0$

$$F_{DH} = 100 \text{ kN}$$

$\Sigma F_x = 0$

$$F_{DC} = F_{DE} = 75 \text{ kN}$$

Take section (2)-(2)

$$\tan\theta_4 = \frac{2}{2}$$

$$\theta_4 = 45°$$

$$\tan\theta_5 = \frac{2}{4}$$

$$\theta_5 = 26.565°$$

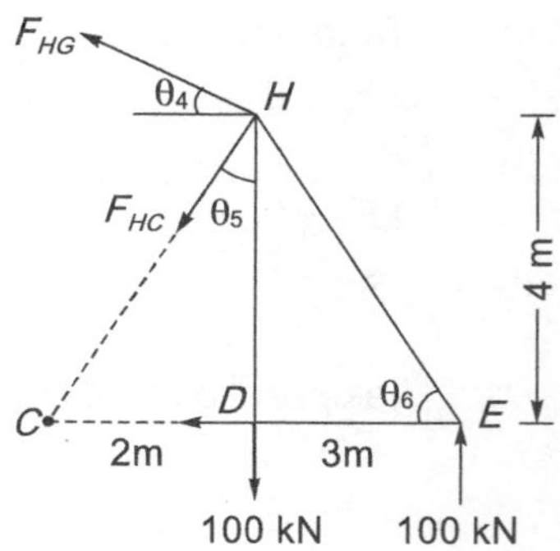

$\Sigma M_C(\curvearrowright) = 0$

$$-100 \times 5 + 100 \times 2 - F_{HG}\cos\theta_4 \times 4 - F_{HG}\sin\theta_4 \times 2 = 0$$

$$300 + F_{HG} \times \cos 45° + 4 + F_{HG}\sin 45° \times 2 = 0$$

$$F_{HG} = -\frac{300}{[4\cos 45° + 2\sin 45°]} = -70.7106 \text{ kN}$$

$\Sigma F_y (\uparrow) = 0$

$$F_{HG} \sin\theta_4 - F_{HC} \cos\theta_5 = 0$$

$$-70.7106 \times \frac{\sin 45°}{\cos 26.565°} = F_{HC}$$

$$F_{HC} = -55.901 \text{ kN}$$

Take joint G,

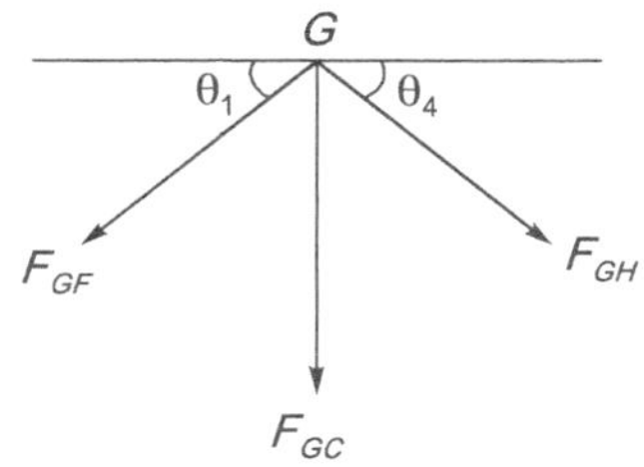

$\Sigma F_y = 0$

$$F_{GF} \sin\theta_1 + F_{GH} \sin\theta_4 + F_{GC} = 0$$

$$F_{GC} - 55.901 \times \sin 26.565° - 70.7106 \times \sin 45° = 0$$

$$F_{GC} = 75 \text{ kN}$$

K-force calculation

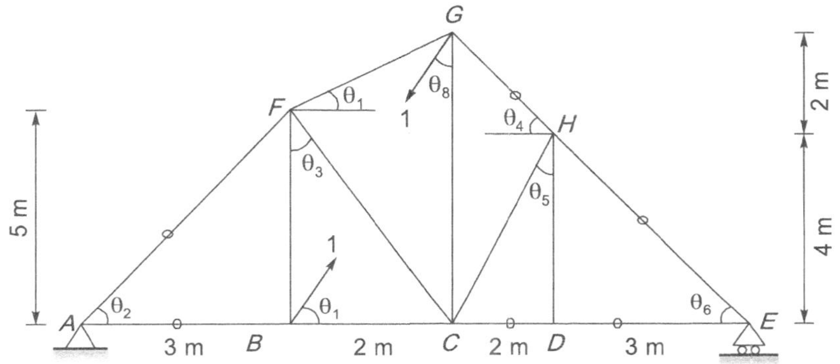

Take joint B,

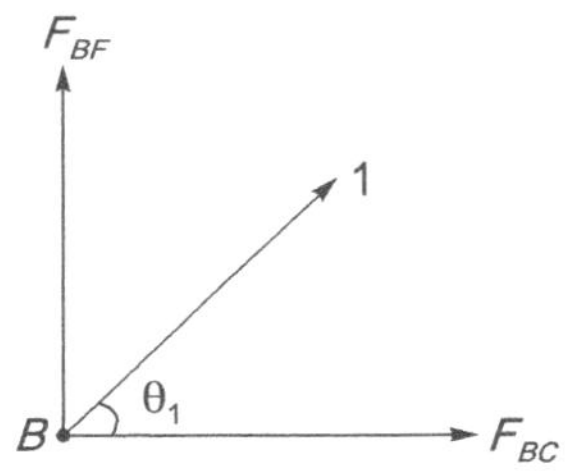

$$\tan\theta_7 = \frac{6}{2}$$

$$\theta_7 = 71.565°$$

$\Sigma F_y = 0,$

$$F_{BF} = 1 \sin\theta_7 = 0$$

$$F_{BF} = -\sin 71.565 = -0.949$$

$\Sigma F_x = 0$

$$F_{BC} = -\cos\theta_7 = -\cos 71.565 = -0.3162$$

Take joint C,

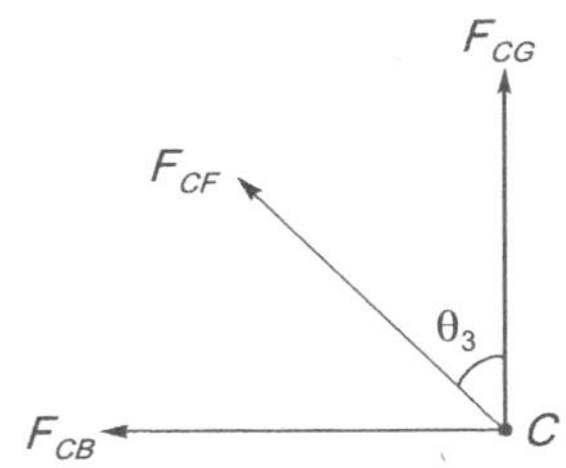

$$\tan\theta_3 = \frac{2}{5}$$

$$\theta_3 = 21.801°$$

$\Sigma F_x = 0$

$$F_{CF} \sin 21.801° + F_{CB} = 0$$

$$F_{CF} = -\frac{F_{CB}}{\sin 21.801°} = -\left(\frac{-0.3162}{\sin 21.801}\right) = 0.8514$$

$\Sigma F_y = 0,$

$$F_{CF} = \cos\theta_3 + F_{CG} = 0$$

$$F_{CF} = -0.8514 \cos 21.801°$$

$$F_{CG} = -0.791$$

Take joint F,

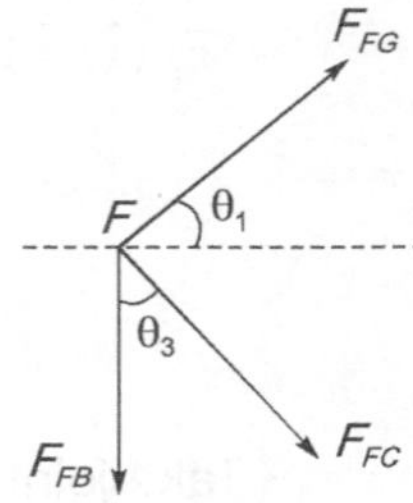

$$\tan\theta_1 = \frac{1}{2}$$

$$\theta_1 = 26.565°$$

$\Sigma F_x = 0$

$$F_{FG}\cos\theta_1 + F_{FC}\sin\theta_3 = 0$$

$$F_{FG} = -\frac{0.8514 \times \sin 21.801°}{\cos 26.565°} = -0.354$$

Member	P	K	L	KPL	K^2L	P + RK
AB	60	0	3	0	0	60.0
BC	60	–0.316	2	–37.920	0.200	44.2
CD	75	0	2	0.000	0	75.0
DE	75	0	3	0	0	75.0
AF	–116.6	0	5.83	0	0	–116.6
FG	–55.9	–0.354	2.236	44.247	0.280	–73.6
GH	–70.71	0	283	0.000	0	–70.7
HE	–125	0	5	0	0	–125.0
FB	100	–0.949	5	–474.500	4.503	52.6
FC	–26.926	0.851	5.385	–123.392	3.900	15.6
GB	0	1	6.24	0	6.324	50.0
GC	75	–0.791	6	–355.950	3.754	35.5
HC	–55.9	0	4.472	0	0	–55.9
HD	100	0	4	0	0	100.0
				Σ = –947.515	18.961	

$$\Sigma PKL = -947.45$$

$$\Sigma K^2L = 18.96$$

So,

$$R = -\left(\frac{\Sigma PKL}{\Sigma K^2L}\right) = -\left(-\frac{947.45}{18.96}\right) = 49.97 \text{ kN}$$

5.21 Determine the vertical displacement of joint A of the metal truss shown in figure, due to heating from the wall member *BC* is subjected to an increase in temperature of $\Delta T = 70°C$. Take coefficient of thermal expansion of member $\alpha = 1.05 \times (10^{-5})/°C$ and $E = 210 \times 10^3$ N/mm^2. The cross-sectional area of each member is indicated in the figure in mm^2.

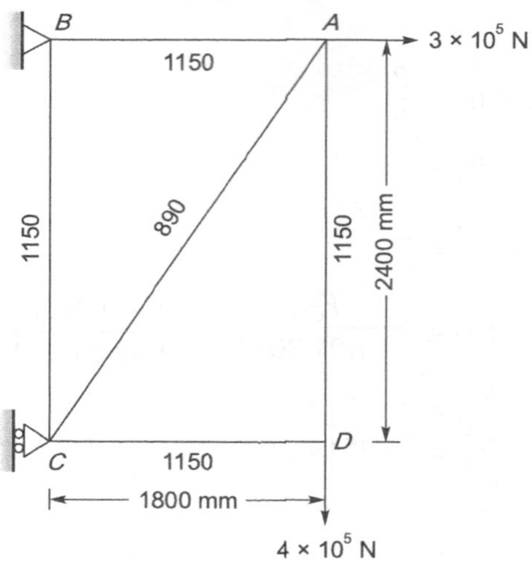

[20 marks : 2022]

Solution:

Deflection at A due to external loads only:

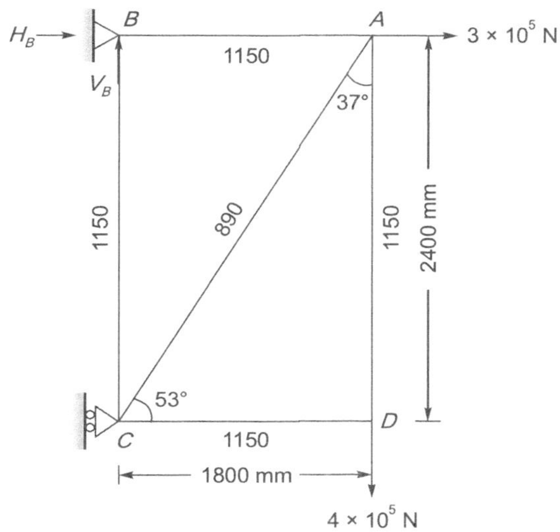

Reaction at B are H_B and V_B

Reaction at C is H_C

Taking, $\Sigma M_C = 0$

$\Rightarrow H_B \times 2.4 + 300 \times 2.4 + 400 \times 1.8 = 0$

$\Rightarrow \quad H_B = -600$ kN i.e., 600 kN ($\leftarrow$)

$\therefore \quad H_C = 300$ kN ($\rightarrow$)

$V_B = 400$ kN

Considering joint B,

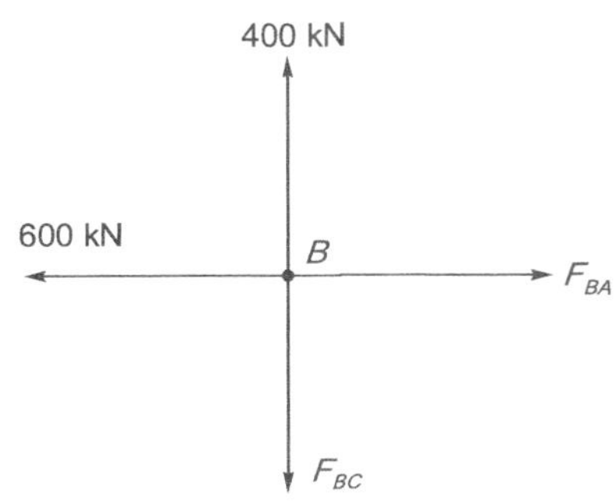

So, $\quad F_{BA} = 600$ kN (Tension)

$F_{BC} = 400$ kN (Tension)

Considering joint A,

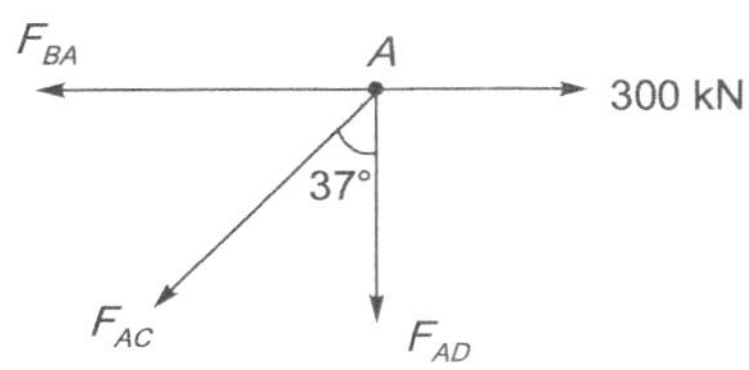

$\Rightarrow \quad F_{BA} + F_{AC} \cos 53° = 300$ kN

$$\Rightarrow \quad 600 + F_{AC} \times 0.6 = 300 \text{ kN}$$

$$\Rightarrow \quad F_{AC} = -500 \text{ kN (compression)}$$

$$F_{AC} \cos 37° + F_{AD} = 0$$

$$\Rightarrow \quad F_{AD} = 400 \text{ kN (Tension)}$$

$$F_{CD} = 0$$

Now remove external load and apply unit load at A in the vertical direction,

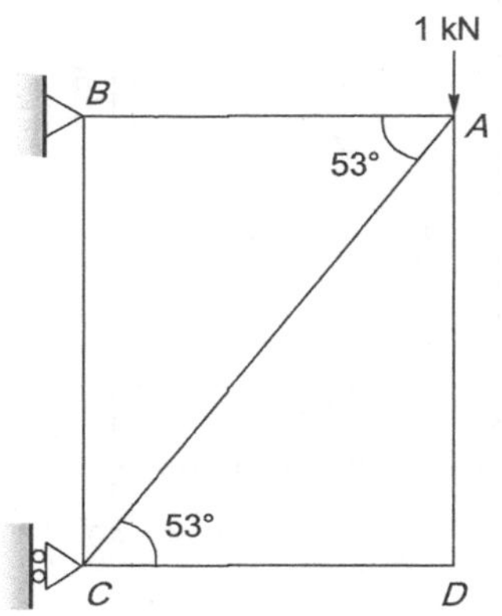

$$k_{DC} = k_{AD} = 0$$

Considering joint A,

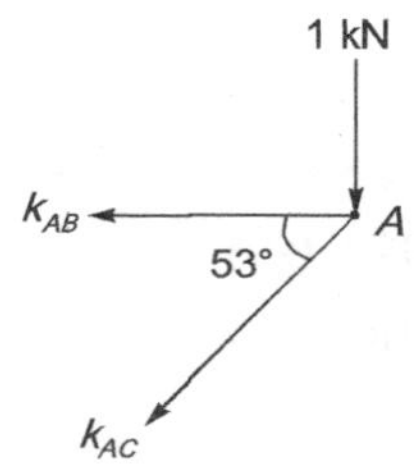

$$k_{AC} \cos 53° + k_{AB} = 0$$

and
$$k_{AC} \sin 53° + 1 = 0$$

$$\Rightarrow \quad k_{AC} = -1.25$$

$$\therefore \quad k_{AB} = -k_{AC} \cos 53° = 0.75$$

Considering joint B,

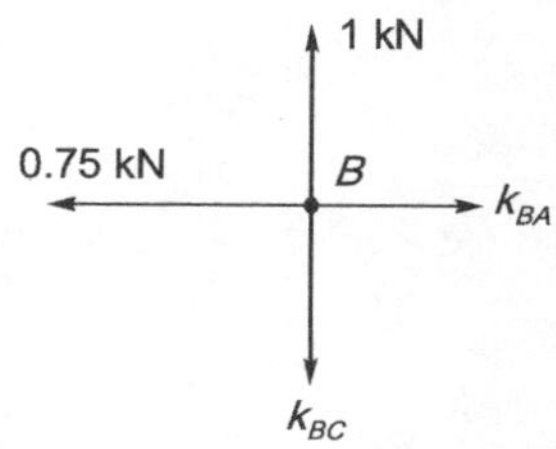

$$k_{BC} = k_{AC} \cos 37° = 1$$

Now,

Member	P(kN)	k (kN)	L (m)	PkL (kNm)	$\frac{PkL}{AE}$ (m)
AB	600	0.75	1.8	810	3.354×10^{-3}
BC	400	1	2.4	960	3.975×10^{-3}
CD	0	0	1.8	0	0
DA	400	0	2.4	0	0
AC	–500	–1.25	3	1875	0.01

$$\Delta = \sum \frac{PkL}{AE} \text{ (Due to external load only)}$$

$$\Rightarrow \quad \Delta = (3.354 + 3.975 + 10) \text{ mm}$$

$$= 17.329 \text{ mm}$$

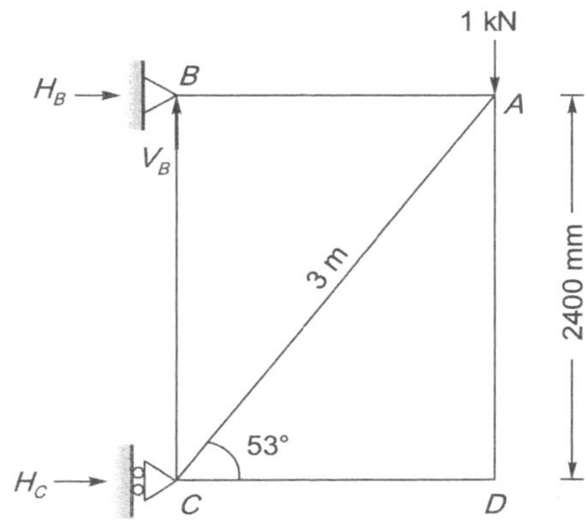

Now, deflection at A due to temperature change.

We need to calculate force in members BC only because no temperature change is given in other members. Apply unit load in the vertical direction of at A.

$$V_B = 1 \text{ kN}$$

Taking $\Sigma M_C = 0$

$$\Rightarrow \quad H_B \times 2.4 + 1 \times 1.8 = 0$$

$$\Rightarrow \quad H_B = -0.75 \text{ kN}$$

$$\Rightarrow \quad H_B = 0.75 \text{ kN} (\leftarrow)$$

Considering joint B,

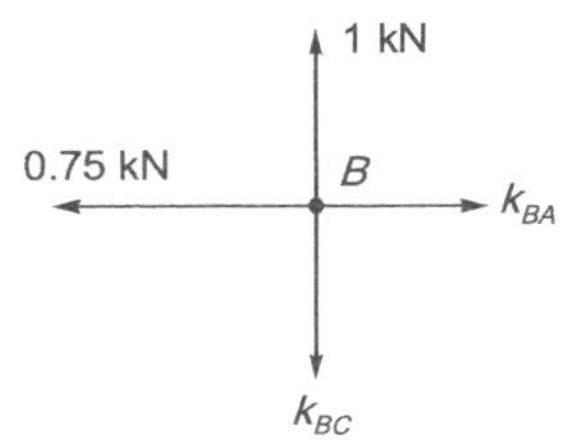

$$k_{BC} = 1$$

Therefore, $\Delta_A = k_i(L\alpha\Delta T)$

$$= 1 \times (2400 \times 1.05 \times 10^{-5} \times 70)$$

$$= 1.764 \text{ mm}$$

So, Total deflection at $A = 17.329 + 1.764 = 19.093$ mm

5.22 Determine the forces in the member of the braced frame as shown in the figure below. Also, determine the drift due to shear in each storey. Areas of diagonals and horizontal girders are shown in brackets and they are in mm^2.
Take $E = 205$ kN/mm^2 :

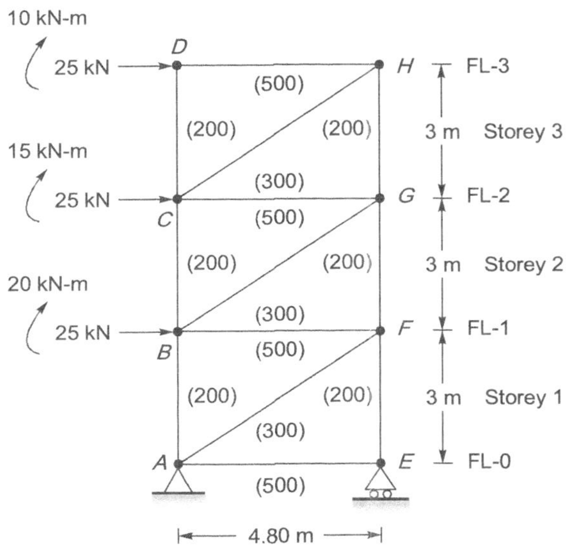

[20 marks : 2023]

Solution:

Let, V_A and V_E be the vertical reactions developed at A and E respectively and H_A be the horizontal reaction developed at A.

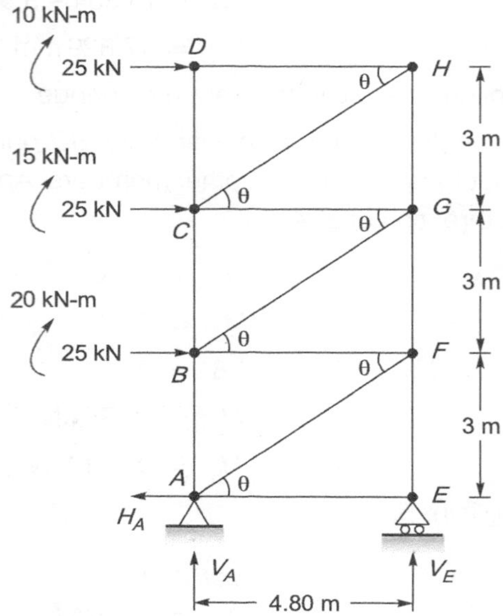

Now, $\Sigma F_y = 0$

$\Rightarrow \quad V_A + V_E = 0 \quad ...(i)$

$\Sigma F_x = 0$

$\Rightarrow \quad 25 + 25 + 25 - H_A = 0$

$\Rightarrow \quad H_A = 75\text{ kN} (\leftarrow)$

$\Sigma M_E = 0$

$\Rightarrow \quad 25 \times 9 + 25 \times 6 + 25 \times 3 + V_A \times 4.8 = 0$

$\Rightarrow \quad V_A = 93.75\text{ kN} (\downarrow)$

So using (i) $\quad V_E = 93.75\text{ kN} (\uparrow)$

Note: Moments applied will not be considered in analysis of truss.

Now, number of numbers, $m = 13$

Number of external reactions, $r_e = 3$

Number of joints, $j = 8$

So, degree of static indeterminacy,

$$D_s = m + r_e - 2j$$
$$= 13 + 3 - 2 \times 8 = 0$$

Force in member DC and AE will be zero because

(i) At D, two forces are collinear and one is non-collinear.

(ii) At E, two forces are collinear and one is non-collinear.

In figure,

$$\tan\theta = \frac{3}{4.8} \Rightarrow \theta = 32°$$

Now, $\quad F_{DH} = 25\text{ kN (compressive)}$

Consider joint H

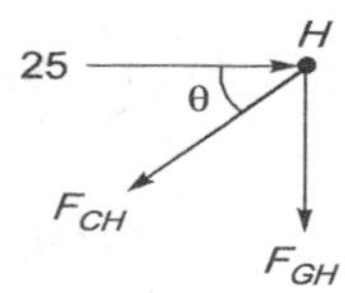

$\Sigma F_x = 0$

$\Rightarrow \quad F_{CH}\cos\theta - 25 = 0$

$\Rightarrow \quad F_{CH} = \dfrac{25}{\cos\theta} = 29.5\text{ kN (Tension)}$

$\Sigma F_y = 0$

$\Rightarrow \quad F_{CH}\sin\theta + F_{GH} = 0$

$\Rightarrow \quad 29.5 \times \sin 32° + F_{GH} = 0$

$\Rightarrow \quad F_{GH} = -15.63\text{ kN} = 15.63\text{ kN (compressive)}$

Forces in other members are shown below in figure.

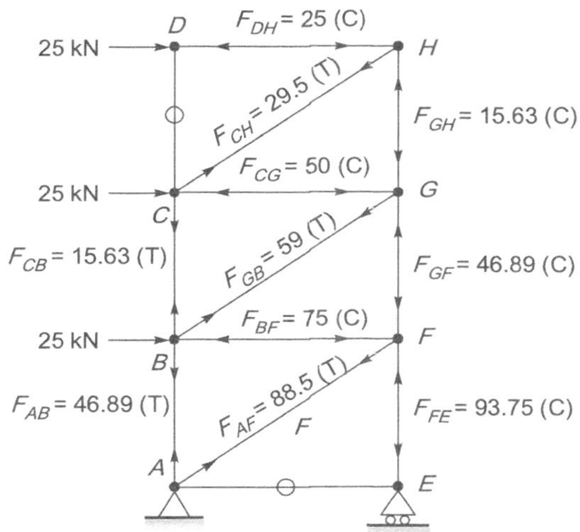

Drift due to shear:

Let, a unit load is applied at joint F as shown in figure. Due to this unit load, there will be force in only AF and EF. In all other members, force will be zero.

So, consider joint F

Now, $\Sigma F_x = 0$

$\Rightarrow$ $K_{AF} \cos\theta - 1 = 0$

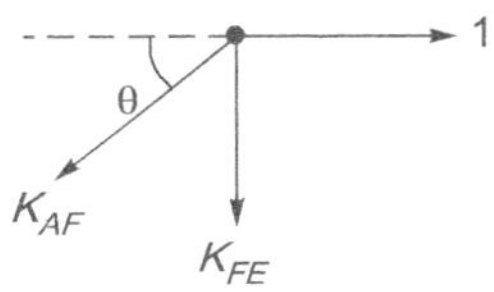

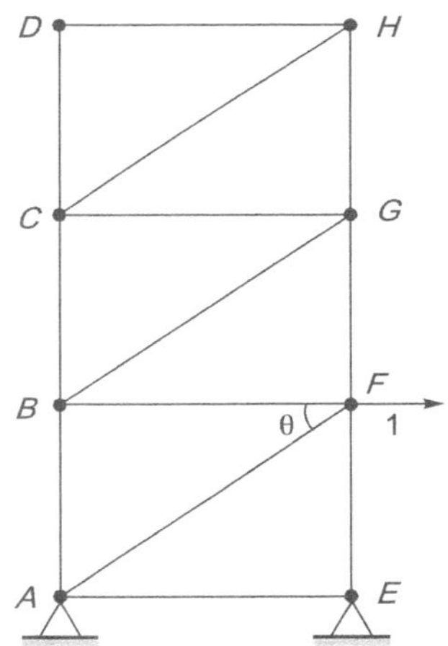

$\Rightarrow$ $K_{AF} = \dfrac{1}{\cos\theta} = \dfrac{1}{\cos 32^\circ} = 1.18$

Also, $\Sigma F_y = 0$

$\Rightarrow$ $K_{AF} \sin\theta + K_{FE} = 0$

$\Rightarrow$ $1.18 \times \sin 32^\circ + K_{FE} = 0$

$\Rightarrow$ $K_{FE} = -0.625$

Also, $L_{AF} = \sqrt{AE^2 + EF^2}$

$= \sqrt{4.8^2 + 3^2} = 5.66$ m

Now, Deflection of F, $\Delta_F = \dfrac{P_{AF} \times k_{AF} \times L_{AF}}{A_{AF} \times E} + \dfrac{P_{EF} \times k_{EF} \times L_{EF}}{A_{EF} \times E}$

$$= \frac{88.5 \times 1.18 \times 5660}{300 \times 205} + \frac{(-93.75) \times (-0.625) \times 3000}{200 \times 205} = 13.9 \text{ mm}$$

By linear interpolation,

Deflection of G, $\Delta_G = 2 \times 13.9 = 27.8$ mm

Deflection of H, $\Delta_H = 3 \times 13.9 = 41.7$ mm

6. Matrix Method of Structural Analysis

6.1 Develop the flexibility matrix for the cantilever beam for the degrees of freedom shown below:

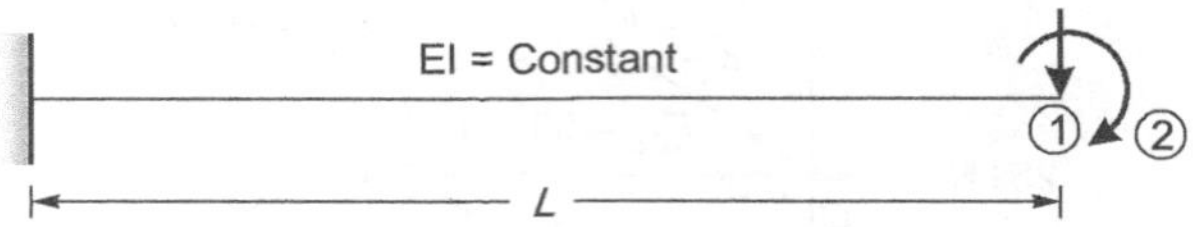

[10 marks : 2010]

Solution:

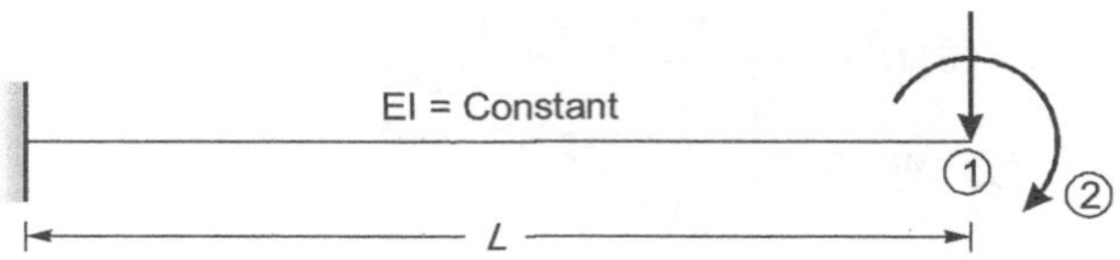

Flexibility matrix for given beam will be 2 × 2 matrix because degree of freedom is 2. (as given)

$$A = \begin{bmatrix} f_{11} & f_{12} \\ f_{21} & f_{22} \end{bmatrix}$$

It will be a symmetric matrix

$f_{12} = f_{21}$ (Maxwell's unit load theory,

f_{11} = Vertical deflection at (1) due to application of unit vertical load at (1)

f_{12} = Vertical deflection at (1) due to application of unit moment at (2)

f_{21} = Rotation at (2) (in ↻) due to application of unit vertical load

f_{22} = Rotation at (2) (in ↻) due to application of unit clockwise moment at (2)

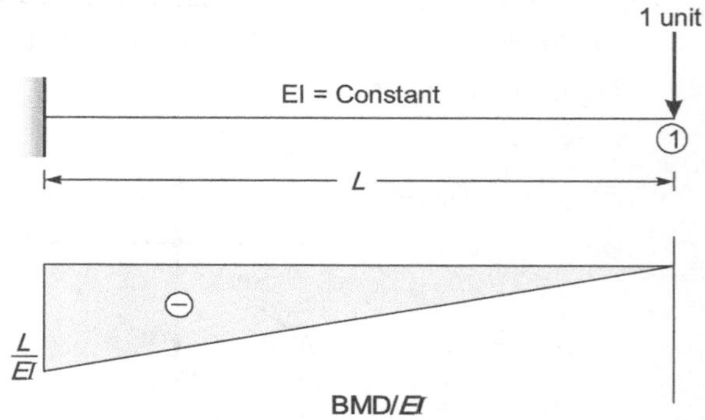

f_{11} = Deflection by moment area theory
= Moment of area about the far end

$$= \frac{1}{2} \times L \times \frac{L}{EI} \times \frac{2L}{3} = \frac{L^3}{3EI}$$

f_{21} = By moment area theorem (1)

$$= \text{Area under } \frac{\text{BMD}}{EI} = \frac{1}{2} \times L \times \frac{L}{EI} = \frac{L^2}{2EI}$$

$$f_{12} = \frac{L^2}{2EI}$$

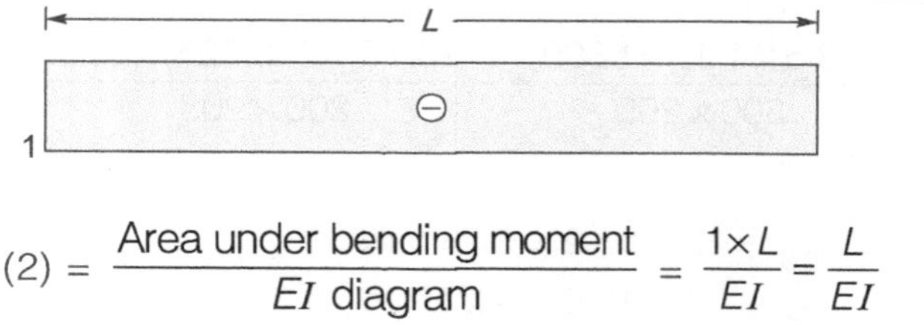

$$\text{Rotation at (2)} = \frac{\text{Area under bending moment}}{EI \text{ diagram}} = \frac{1 \times L}{EI} = \frac{L}{EI}$$

Thus, $$A = \begin{bmatrix} \frac{L^3}{3EI} & \frac{L^2}{2EI} \\ \frac{L^2}{2EI} & \frac{L}{EI} \end{bmatrix}$$

6.2 Develop the stiffness matrix for the frame shown in the figure below with coordinates 1 and 2 indicated.

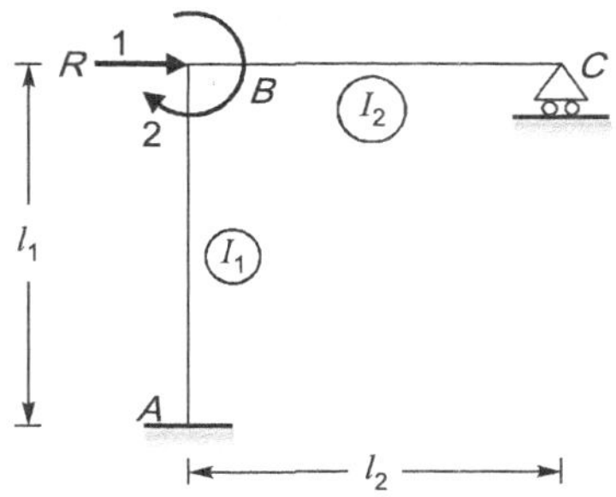

[10 marks : 2011]

Solution:

The form of the stiffness matrix will be

$$K = \begin{bmatrix} K_{11} & K_{12} \\ K_{21} & K_{22} \end{bmatrix}$$

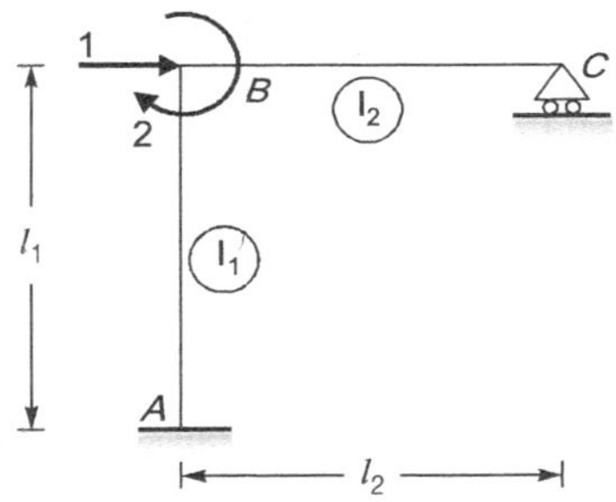

Column 1: K_{11} and K_{21}

Giving a unit displacement along the cordinate ① i.e. in horizontal reaction.

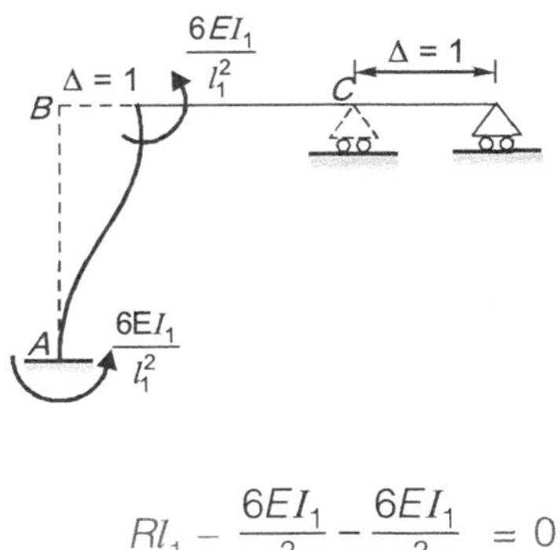

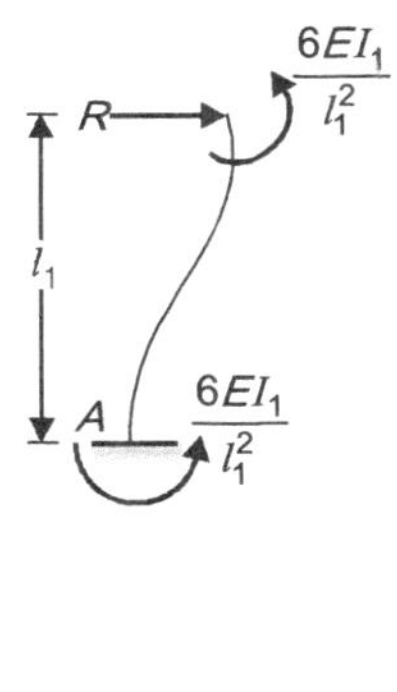

$\Sigma M = 0$ at $A \Rightarrow$ $$Rl_1 - \frac{6EI_1}{l_1^2} - \frac{6EI_1}{l_1^2} = 0$$

$\Rightarrow$ $$R = \frac{12EI_1}{l_1^3}$$

$\therefore$ $$K_{11} = R_1 = \frac{12EI_1}{l_1^3}$$

and $$K_{21} = \frac{-6EI_1}{l_1^2}$$ $\left(\because \text{Induced moment is opposite to the direction of coordinate (2).}\right)$

Column 2: Giving unit rotation (i.e. θ = 1) along coordinate ②.

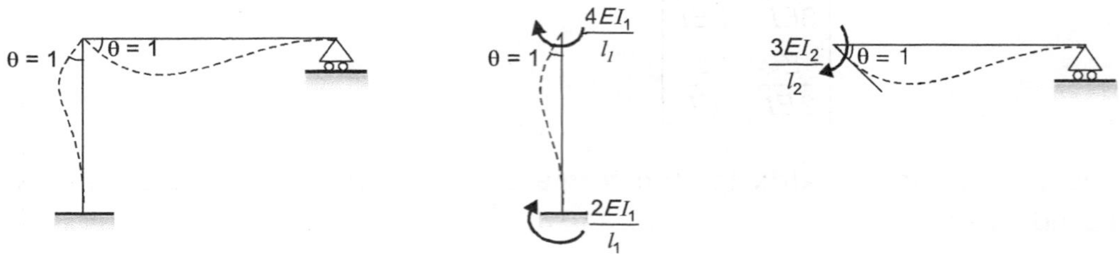

or,
$$K_{22} = \frac{4EI_1}{l_1} + \frac{3EI_2}{l_2}$$

and $K_{12} = K_{21} = \frac{-6EI_1}{l_1^2}$ $[K_{ij} = K_{ji}$ for stiffnes matrix]

∴
$$K = \begin{bmatrix} \frac{12EI_1}{l_1^3} & \frac{-6EI_1}{l_1^2} \\ \frac{-6EI_1}{l_1^2} & \frac{4EI_1}{l_1} + \frac{3EI_2}{l_2} \end{bmatrix}$$

6.3 **Derive the flexibility matrix of the plane beam shown in figure with respect to the degrees of freedom shown. Take EI = constant.**

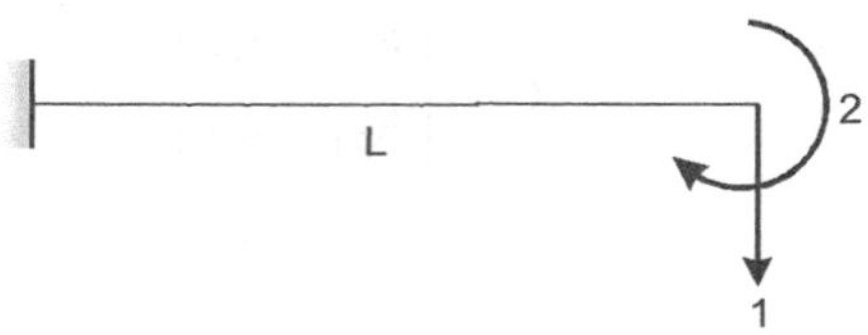

[15 marks : 2014]

Solution:

First Column: To generate first column of flexibility matrix, apply unit load in the direction of coordinate 1.

f_{ab} = deflection in the direction of coordinate 'a' when unit load is the direction of coordinate 'b'

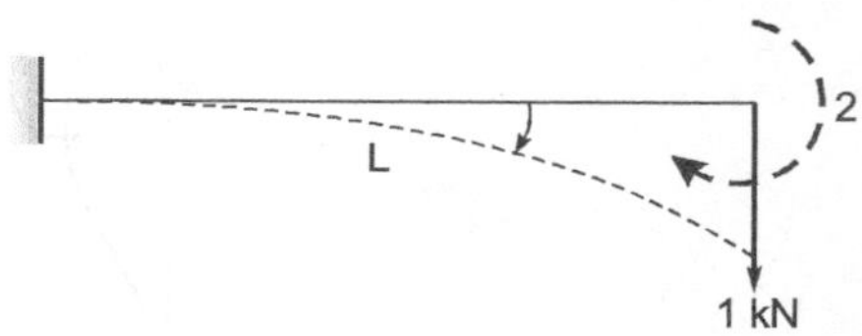

∴
$$f_{11} = \frac{1 \times L^3}{3EI} = \frac{L^3}{3EI}$$

∴
$$f_{21} = \frac{1 \times L^2}{2EI} = \frac{L^2}{2EI}$$

Also from the **Maxwell's reciprocal theorem,**

$$f_{12} = f_{21} = \frac{L^2}{2EI}$$

Second column: To generate second column of flexibility matrix, apply unit load (i.e., moment) in the direction of coordinate (2)

$$f_{12} = \frac{ML^2}{2EI} = \frac{L^2}{2EI}$$

$$f_{22} = \frac{ML}{EI} = \frac{L}{EI}$$

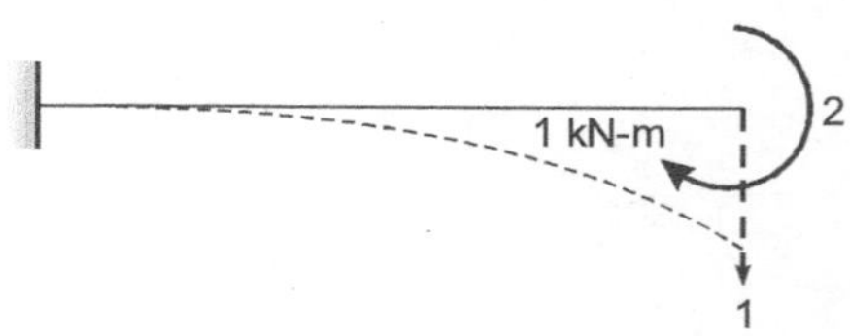

Hence **flexibility matrix** for given coordinate is given by

$$[f] = \begin{bmatrix} \frac{L^3}{3EI} & \frac{L^2}{2EI} \\ \frac{L^2}{2EI} & \frac{L}{EI} \end{bmatrix}$$

6.4 A plane frame is shown in figure. Determine the size of the reduced stiffness matrix with axial deformations and without axial deformations after introducing the boundary conditions. Also, show the active degrees of freedom in both cases. Node number is shown in circles.

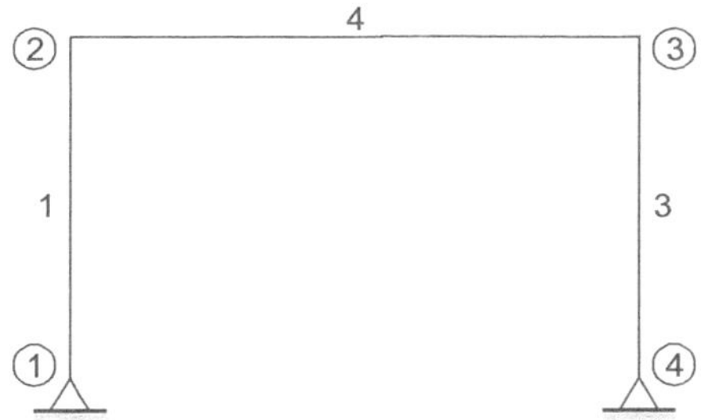

[10 marks : 2014]

Solution:

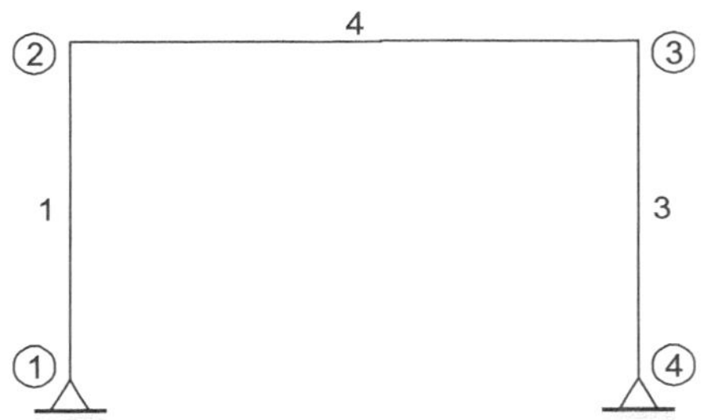

Plane size

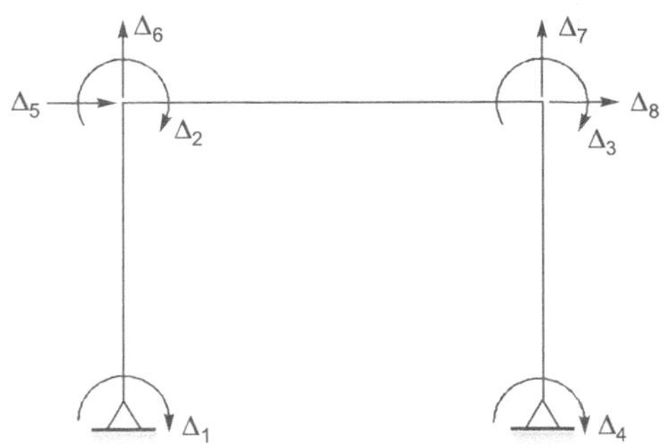

Size of the reduced stiffness matrix with axial deformations

$$D_k = 3J - C$$

Where, J = No. of joints

C = No. of constraints

$D_k = 4 \times 3 - 4 = 8$

Hence size of stiffness matrix will be [8 × 8]

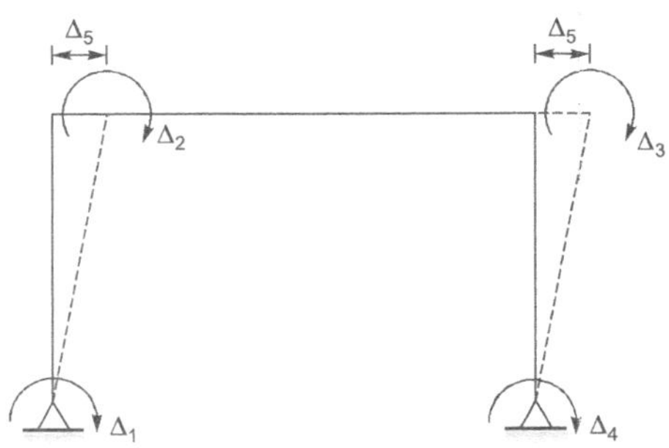

Size of the reduced stiffness matrix without axial deformation

$$D_k = 3J - C - m$$

Where, m = No. of axial rigid member

$$D_k = 3 \times 4 - 4 - 3$$

$$D_k = 5$$

Hence size of stiffness matrix will be [5 × 5]

6.5 A plane frame is shown in figure. The stiffness matrix of elements 1 and 2 with respect to global axes are also shown. Assemble the global stiffness matrix of the frame. Node number is shown in circles.

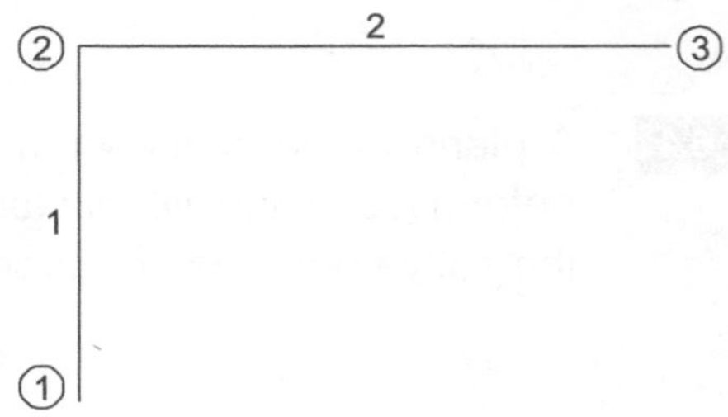

Member 1, $$K_{6\times 6} = \begin{bmatrix} K_{11} & K_{12} \\ K_{21} & K_{22} \end{bmatrix}_{6\times 6}$$

Member 2, $$K_{6\times 6} = \begin{bmatrix} K'_{11} & K'_{12} \\ K'_{21} & K'_{22} \end{bmatrix}_{6\times 6}$$

where, each of the K_{ij} is a 3 × 3 sub-matrix.

For Member 1,

$$\text{sub-matrix } K_{11} = \begin{bmatrix} K_{11} & K_{12} & K_{13} \\ K_{21} & K_{22} & K_{23} \\ K_{31} & K_{32} & K_{33} \end{bmatrix}$$

For Member 2,

$$\text{sub-matrix } K'_{11} = \begin{bmatrix} K_{41} & K_{42} & K_{43} \\ K_{51} & K_{52} & K_{53} \\ K_{61} & K_{62} & K_{63} \end{bmatrix}$$

[8 marks : 2014]

Solution:

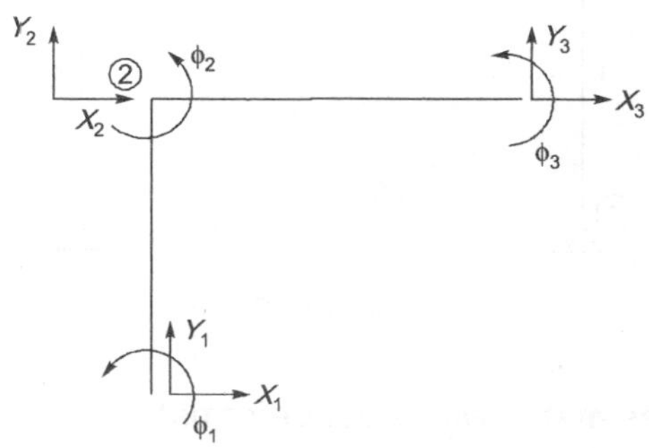

$D' \rightarrow$ Displacement matrix for frame X, Y, ϕ, are degree of freedom for frame.

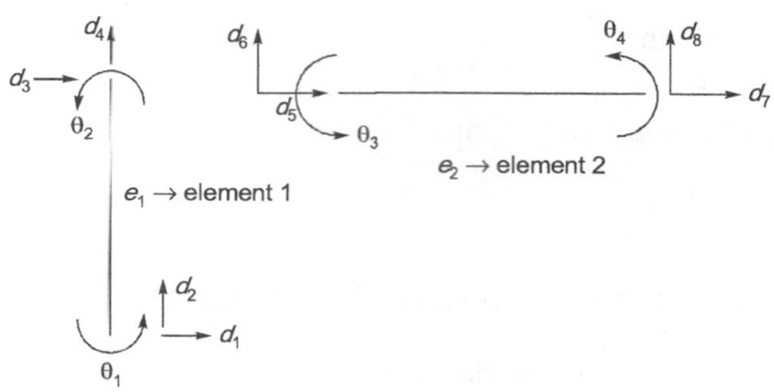

$d' \rightarrow$ Displacement matrix for unconnected element

Q, D are DOF of elements e_1 and e_2

Member 1, e_1, K_{e_1}, $$K_{6\times 6} = \begin{bmatrix} K_{11_{3\times 3}} & K_{12_{3\times 3}} \\ K_{21_{3\times 3}} & K_{22_{3\times 3}} \end{bmatrix}_{6\times 6}$$

Member 2, e_2, K_{e_2} , $K_{6\times 6} = \begin{bmatrix} K'_{11_{3\times3}} & K'_{12_{3\times3}} \\ K'_{21_{3\times3}} & K'_{22_{3\times3}} \end{bmatrix}_{6\times6}$

To define relationship between d' and D'. Use matrix B where $B \rightarrow$ transformation Matrix

At each joint displacement of element (1) and element (2) should be equal to the displacement of frame at the corresponding joint of frame.

Ex. $\theta_2 = \theta_3 = \phi_2$, $X_3 = d_7$, $\theta_1 = \phi_1$.

$$d' = BD'$$

Similarly all such relationships between the element and the frame are developed and the relationship is developed based on above relations.

$$\begin{bmatrix} \theta_1 \\ d_1 \\ d_2 \\ \theta_2 \\ d_3 \\ d_4 \\ \theta_3 \\ d_5 \\ d_6 \\ \theta_4 \\ d_7 \\ d_8 \end{bmatrix}_{12\times1} = \begin{bmatrix} 1 & 0 & 0 & 0 & 0 & 0 & 0 & 0 & 0 \\ 0 & 1 & 0 & 0 & 0 & 0 & 0 & 0 & 0 \\ 0 & 0 & 1 & 0 & 0 & 0 & 0 & 0 & 0 \\ 0 & 0 & 0 & 1 & 0 & 0 & 0 & 0 & 0 \\ 0 & 0 & 0 & 0 & 1 & 0 & 0 & 0 & 0 \\ 0 & 0 & 0 & 0 & 0 & 1 & 0 & 0 & 0 \\ 0 & 0 & 0 & 1 & 0 & 0 & 0 & 0 & 0 \\ 0 & 0 & 0 & 0 & 1 & 0 & 0 & 0 & 0 \\ 0 & 0 & 0 & 0 & 0 & 1 & 0 & 0 & 0 \\ 0 & 0 & 0 & 0 & 0 & 0 & 1 & 0 & 0 \\ 0 & 0 & 0 & 0 & 0 & 0 & 0 & 1 & 0 \\ 0 & 0 & 0 & 0 & 0 & 0 & 0 & 0 & 1 \end{bmatrix}_{12\times9} \begin{bmatrix} \phi_1 \\ X_1 \\ Y_1 \\ \phi_2 \\ X_2 \\ Y_2 \\ \phi_3 \\ X_3 \\ Y_3 \end{bmatrix}_{9\times1}$$

To form stiffness transformation matrix, use stiffness matrix of frame.

$$K = B^T kB$$

where, $k = \begin{bmatrix} k_{e1} & 0 \\ 0 & k_{e2} \end{bmatrix}_{12\times12}$

So, $K_{9\times9} = [B^T]_{9\times12} \times [k]_{12\times12} \times [B]_{12\times9}$

Using the relationship, global stiffness matrix for frame can be assembled.

6.6 **A member in the plane truss is having the following data : L = length of the member, A = area of cross-section, E = Young's modulus of elasticity and α = angle in the first quadrant from x-axis in anticlockwise direction.**
Derive the element stiffness matrix of the plane truss member in local and global coordinate system.
[10 marks : 2015]

Solution:

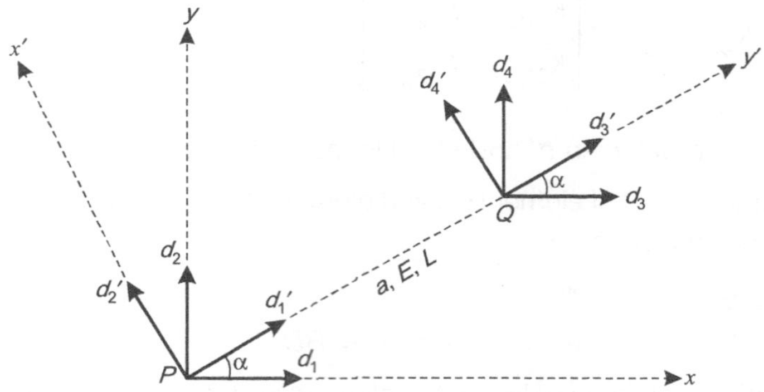

Let x and y are global coordinate and, x' and y' are local coordinate of element PQ of plane truss.

Where d_1, d_2, d_3 and d_4 are degrees of freedom in global coordinate and d_1', d_2', d_3' and d_4' are degrees of freedom in local coordinate.

Element stiffness matrix in local coordinate (K') will be found by providing unit deflection in the corresponding direction.

$$k' = \begin{bmatrix} +\frac{aE}{L} & 0 & -\frac{aE}{L} & 0 \\ 0 & 0 & 0 & 0 \\ -\frac{aE}{L} & 0 & +\frac{aE}{L} & 0 \\ 0 & 0 & 0 & 0 \end{bmatrix}$$ (Tension negative and compression positive)

Rotation matrix,

$$R = \begin{bmatrix} \cos x & \sin x & 0 & 0 \\ -\sin x & \cos x & 0 & 0 \\ 0 & 0 & \cos x & \sin x \\ 0 & 0 & -\sin x & \cos x \end{bmatrix}$$

Now element stiffness matrix in global coordinate will by given

$$k = R^T K' R$$

$$k = \frac{aE}{L}\begin{bmatrix} \cos x & -\sin x & 0 & 0 \\ \sin x & \cos x & 0 & 0 \\ 0 & 0 & \cos x & -\sin x \\ 0 & 0 & \sin x & \cos x \end{bmatrix}\begin{bmatrix} +1 & 0 & -1 & 0 \\ 0 & 0 & 0 & 0 \\ -1 & 0 & +1 & 0 \\ 0 & 0 & 0 & 0 \end{bmatrix}\begin{bmatrix} \cos x & \sin x & 0 & 0 \\ -\sin x & \cos x & 0 & 0 \\ 0 & 0 & \cos x & \sin x \\ 0 & 0 & -\sin x & \cos x \end{bmatrix}$$

$$k = \frac{aE}{L}\begin{bmatrix} +\cos x & 0 & -\cos x & 0 \\ +\sin x & 0 & -\sin x & 0 \\ -\cos x & 0 & +\cos x & 0 \\ -\sin x & 0 & +\sin x & 0 \end{bmatrix}\begin{bmatrix} \cos x & \sin x & 0 & 0 \\ -\sin x & \cos x & 0 & 0 \\ 0 & 0 & \cos x & \sin x \\ 0 & 0 & -\sin x & \cos x \end{bmatrix}$$

$$k = \frac{aE}{L}\begin{bmatrix} +\cos^2 x & +\sin x\cos x & -\cos^2 x & -\sin x\cos x \\ +\sin x\cos x & +\sin^2 x & -\sin x\cos x & -\sin^2 x \\ -\cos^2 x & -\sin x\cos x & +\cos^2 x & \sin x\cos x \\ -\sin x\cos x & -\sin^2 x & \sin x\cos x & \sin^2 x \end{bmatrix}$$

6.7 A propped cantilever beam is shown in figure. Analyze it using the stiffness matrix method and find the reactions in the spring. Neglect axial deformation in the beam. Show the degree of freedom.

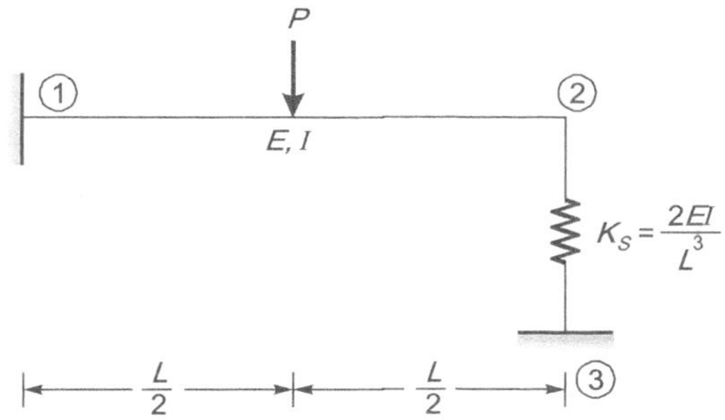

[12 marks : 2017]

Solution:

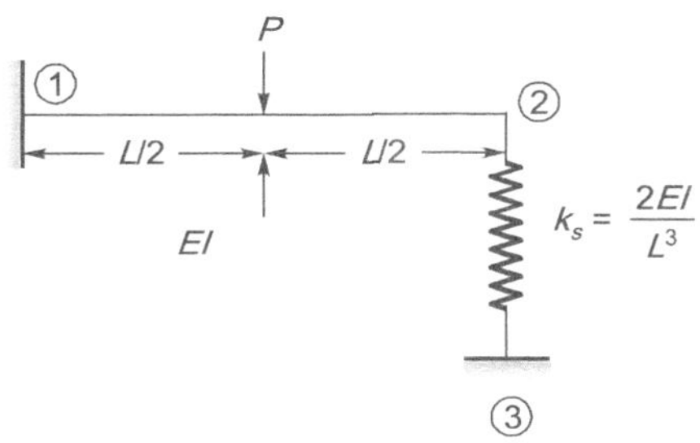

The degree of freedom is 2. As there are two independent displacement components at node 2 of fixed end beam at one end resting on elastic spring support at another end.

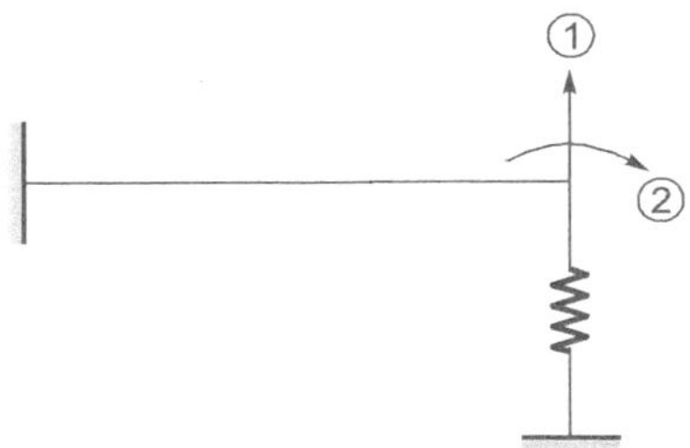

The stiffness matrix with reference to the chosen coordinates may be developed as;

$$\begin{bmatrix} \Delta_1 \\ \Delta_2 \end{bmatrix} = \begin{bmatrix} \Delta \\ \theta \end{bmatrix} \text{(Displacement components)}$$

$$\Rightarrow \qquad k_{ij} = \begin{bmatrix} k_{11} & k_{12} \\ k_{21} & k_{22} \end{bmatrix}$$

$$\therefore \qquad k_{11} = \frac{12EI}{L^3} \qquad k_{21} = \frac{6EI}{L^2}$$

$$k_{12} = \frac{6EI}{L^2} \qquad k_{22} = \frac{4EI}{L}$$

$$\therefore \qquad [k] = \begin{bmatrix} \frac{12EI}{L^3} & \frac{6EI}{L^2} \\ \frac{6EI}{L^2} & \frac{4EI}{L} \end{bmatrix}$$

Given, k_s = Stiffness of spring = $\frac{2EI}{L^3}$

$\therefore$ $[k_M]$ = Modified stiffness matrix = $\begin{bmatrix} \left(\frac{12EI}{L^3}+\frac{2EI}{L^3}\right) & \frac{6EI}{L^2} \\ \frac{6EI}{L^2} & \frac{4EI}{L} \end{bmatrix}$

$$\Rightarrow \quad [k_s] = \begin{bmatrix} \frac{14EI}{L^3} & \frac{6EI}{L^2} \\ \frac{6EI}{L^2} & \frac{4EI}{L} \end{bmatrix}$$

We know, $\begin{bmatrix} \Delta_1 \\ \Delta_2 \end{bmatrix} = [k_M]^{-1}\{[P_M]-[P']\}$ ...(i)

$[P_M]$ = No external loads at nodes.

$$\Rightarrow \quad [P'] = \begin{bmatrix} P_1' \\ P_2' \end{bmatrix}$$

Matrix is determined by considering the span. *AB* as fixed. Determining the forces generated at nodes due to central point load *P* on beam *AB*.

$$\therefore \quad \begin{bmatrix} P_1' \\ P_2' \end{bmatrix} = \begin{bmatrix} \frac{P}{2} \\ \frac{PL}{8} \end{bmatrix}$$

Put these value in equation (i);

$$\Rightarrow \quad \begin{bmatrix} \Delta_1 \\ \Delta_2 \end{bmatrix} = \begin{bmatrix} \frac{14EI}{L^3} & \frac{6EI}{L^2} \\ \frac{6EI}{L^2} & \frac{4EI}{L} \end{bmatrix}^{-1} \left\{0 - \begin{bmatrix} \frac{P}{2} \\ \frac{PL}{8} \end{bmatrix}\right\}$$

$$\Rightarrow \quad \begin{bmatrix} \Delta_1 \\ \Delta_2 \end{bmatrix} = \frac{-1}{\left\{\frac{56(EI)^2}{L^4} - \frac{36(EI)^2}{L^4}\right\}} \begin{bmatrix} \frac{4EI}{L} & \frac{-6EI}{L^2} \\ \frac{-6EI}{L^2} & \frac{14EI}{L^3} \end{bmatrix} \begin{bmatrix} \frac{P}{2} \\ \frac{PL}{8} \end{bmatrix}$$

$$= \frac{-L^4}{20(EI)^2} \begin{bmatrix} \frac{4PEI}{2L} - \frac{3EI \times P}{4L} \\ \frac{-3EIP}{L^2} + \frac{7PEI}{4L^2} \end{bmatrix} = \frac{-L^4}{20(EI)^2} \begin{bmatrix} \frac{5PEI}{4L} \\ \frac{-5PEI}{4L^2} \end{bmatrix}$$

$$\therefore \quad \Delta_1 = \frac{-L^4}{20(EI)^2} \times \frac{5PEI}{4L} = \frac{-PL^3}{16EI}$$

$$\Delta_2 = \frac{-L^4}{20(EI)^2} \times \left(\frac{-5}{4}\right) \times \left(\frac{PEI}{L^2}\right) = \frac{PL^2}{16EI}$$

$\therefore$ Let *R* be reaction is spring.

Hence, $R = k_s\Delta_1 = -\frac{2EI}{L^3} \times \frac{PL^3}{16EI} = \frac{-P}{8}(\downarrow)$

$$\therefore \qquad \text{BM at support } A = R \times L - \frac{PL}{2}$$

$$= \frac{P}{8} \times L - \frac{PL}{2} = \left(\frac{PL - 4PL}{8}\right) = \frac{-3PL}{8}$$

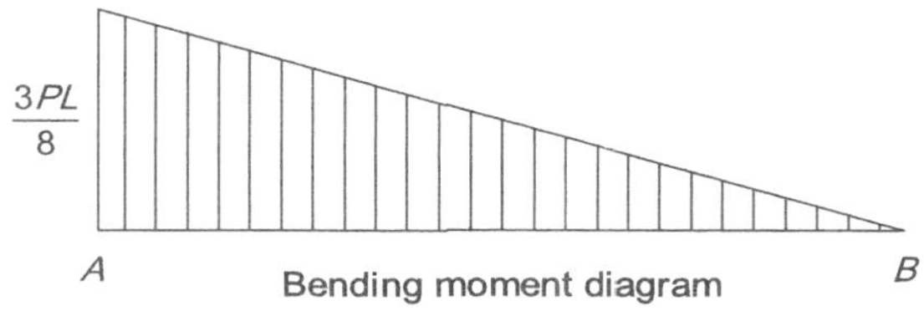

Bending moment diagram

6.8 Distinguish between Flexibility Method and Stiffness Method used for analysis of structures.

[4 marks : 2018]

Solution:

Flexible Method	Stiffness Method
1. This method is also known as compatibility method of force method.	1. This method is also known as equilibrium method or deformation method.
2. Redundant forces are chosen as unknown.	2. Displacements of joints are taken as unknown.
3. Additional equations are obtained by considering the geometrical conditions imposed on the deformations of the structures.	3. Equilibrium equations are expressed in terms of these displacements and the external loads are solved to give the actual joint displacements from which redundant forces can be computed.
4. It is used when degree of static indeterminacy is less.	4. It is used when degree of kinematic indeterminacy is less.
5. Examples : Method of consistent deformation, three moment theorem, column analogy method, Castigliano's theorem of strain energy, etc.	5. Examples : Slope-deflection method, moment distribution method, minimum potential energy method, etc.

6.9 Briefly explain the procedure and then develop the stiffness matrix for the beam element shown in figure with respect to the degrees of freedom 1, 2 and 3. The cross-sectional area A and flexural rigidity EI are constant for the beam.

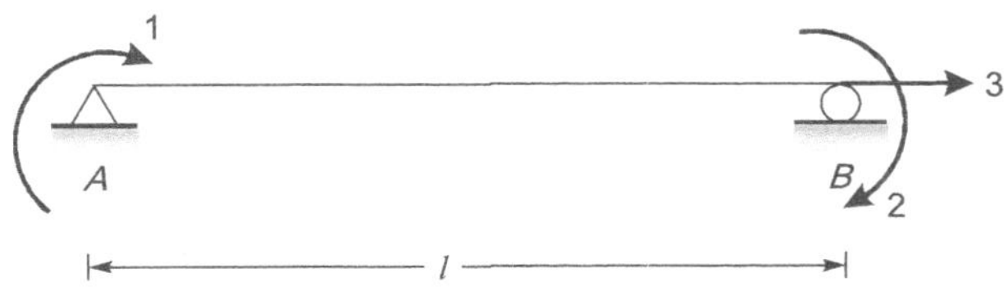

[16 marks : 2018]

Solution:

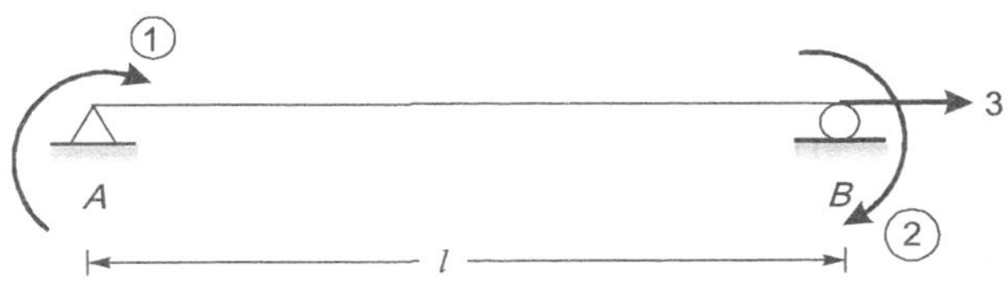

Steps to develop stiffness matrix are:

- Determine degree of kinematic indeterminacy (D_k).
- Assign the co-ordinate numbers to the unknown displacements.

- For 'n' degrees of freedom, stiffness matrix of order $n \times n$ will be obtained.
- The elements of the stiffness matrix indicate the forces exerted on the member by the restraints at the ends of the member when unit displacements are imposed at each end of the member.
  Hence, apply unit displacement along each degree of freedom, holding all other displacements to zero.
- The displacement causes both shear and moment in the beam. And these restraint actions are the elements of the stiffness matrix of the members.

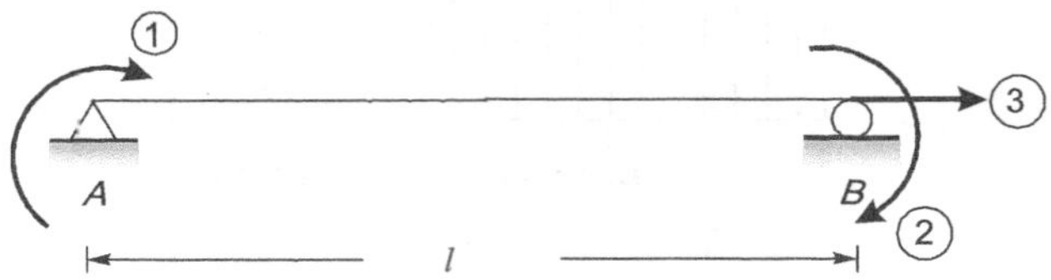

Stiffness matrix, $k = \begin{bmatrix} k_{11} & k_{12} & k_{13} \\ k_{21} & k_{22} & k_{23} \\ k_{31} & k_{32} & k_{33} \end{bmatrix}$

- Applying unit displacement in co-ordinate (1)

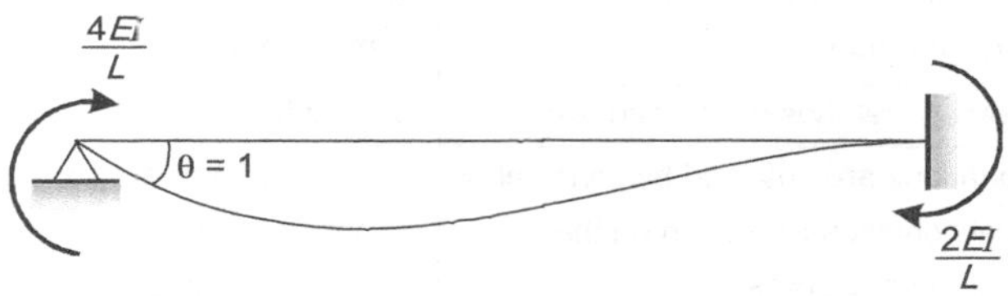

So, $k_{11} = \frac{4EI}{L};\ k_{21} = \frac{2EI}{L};\ k_{31} = 0$

- Applying unit displacement in co-ordinate (2)

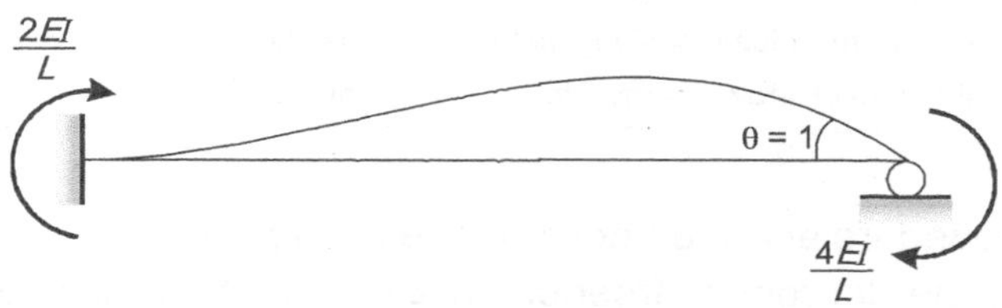

$$k_{12} = \frac{2EI}{L};\ k_{22} = \frac{4EI}{L};\ k_{32} = 0$$

- Applying unit displacement in coordinate (3)

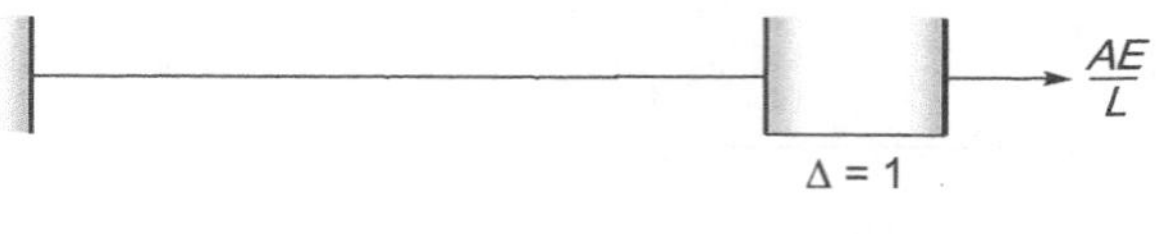

So, $k_{13} = 0;\ k_{23} = 0;\ k_{33} = \frac{AE}{L}$

So, Stiffness matrix, $k = \begin{bmatrix} \frac{4EI}{L} & \frac{2EI}{L} & 0 \\ \frac{2EI}{L} & \frac{4EI}{L} & 0 \\ 0 & 0 & \frac{AE}{L} \end{bmatrix}$

6.10 Develop the flexibility matrix for the beam shown in figure, with respect to the generalized coordinates mentioned. EI is constant for all members.

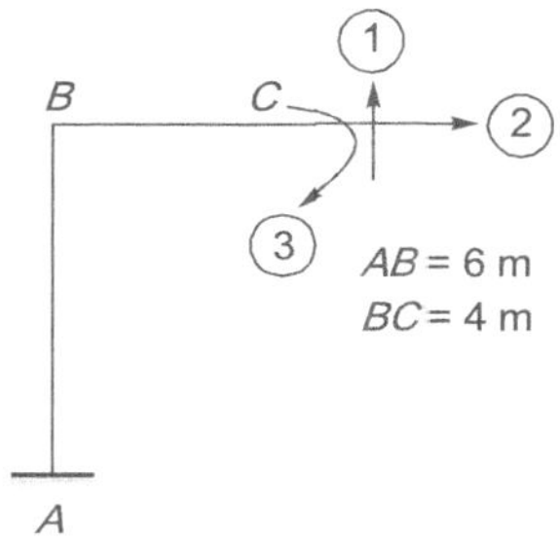

[20 marks : 2020]

Solution:

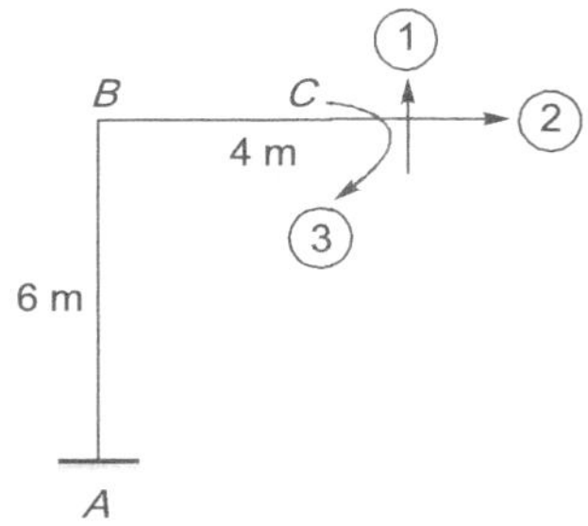

Order of flexibility matrix is 3.

The required deflection may be calculated from unit load method.

Portion	CB	BA
Origin	C	B
Limit	0-4	0-6
m_1	x	4
m_2	0	$-x$
m_3	−1	−1

m_1 = Bending moment at any section when unit load is in the direction of coordinate 1.
m_2 = Bending moment at any section when unit load is in the direction of coordinate 2.
m_3 = Bending moment at any section when unit load is in the direction of coordinate 3.

First column of flexibility matrix

$$f_{11} = \int \frac{m_1 m_1 \, dx}{EI} = \int_0^4 \frac{x^2 dx}{EI} + \int_0^6 \frac{(4)^2 \, dx}{EI} = \frac{117.33}{EI}$$

$$f_{21} = \int \frac{m_1 m_2 \, dx}{EI} = \int_0^4 \frac{x(0) dx}{EI} + \int_0^6 \frac{-4x \, dx}{EI} = -\frac{72}{EI}$$

$$f_{31} = \int \frac{m_1 m_3 \, dx}{EI} = \int_0^4 \frac{-x \, dx}{EI} + \int_0^6 \frac{-4 \, dx}{EI} = -\frac{32}{EI}$$

Second column of flexibility matrix

$$f_{12} = \int \frac{m_1 m_2 \, dx}{EI}$$

Also, $f_{21} = f_{12} = -\dfrac{72}{EI}$ (Property of flexibility matrix by reciprocal theorem)

$$f_{22} = \int \frac{m_2 m_2\, dx}{EI} = \int_0^4 \frac{0\,dx}{EI} + \int_0^6 \frac{x^2 dx}{EI} = \frac{72}{EI}$$

$$f_{32} = \int \frac{m_2 m_3\, dx}{EI} = \int_0^4 \frac{0(-1)\,dx}{EI} + \int_0^6 \frac{x\,dx}{EI} = \frac{18}{EI}$$

Third column of flexibility matrix

$$f_{13} = \int \frac{m_1 m_3\, dx}{EI}$$

$$f_{13} = f_{31} = -\frac{32}{EI}$$

$$f_{23} = \int \frac{m_2 m_3\, dx}{EI}$$

$$f_{23} = f_{32} = \frac{18}{EI}$$

$$f_{33} = \int \frac{m_3 m_3\, dx}{EI} = \int_0^4 \frac{(-1)^2\, dx}{EI} + \int_0^6 \frac{(-1)^2\, dx}{EI} = \frac{10}{EI}$$

$$[f] = \frac{1}{EI}\begin{bmatrix} 117.33 & -72 & -32 \\ -72 & 72 & 18 \\ -32 & 18 & 10 \end{bmatrix}$$

7. Structural Dynamics

7.1 A motor of 150 kg mass is supported by four springs as shown in figure. Each of the springs has a stiffness of 120 kN/m. The unbalance of the rotor is equivalent to a mass of 40 g located at 150 mm from the axis of rotation. The motor is constrained to move vertically. Find
(i) the speed of motor at which resonance will occur.
(ii) the amplitude of vibration of the motor when the speed is 1000 rev./m.

[12 marks : 2018]

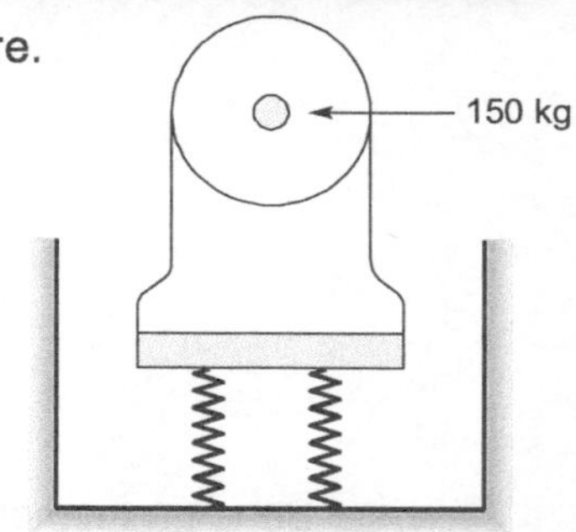

$k = 120$ kN/m

Solution:

Since deflection in all 4 springs are equal, hence springs are in parallel.

$$\therefore \quad k_{eq} = k_1 + k_2 + k_3 + k_4 = 4k = 480 \text{ kN/m}$$

Given $M = 150$ kg

(i) w_n = undamped natural frequency

$$= \sqrt{\frac{k_{eq}}{M}} = \sqrt{\frac{480000}{150}} = 56.569 \text{ rad/s}$$

Resonance will occur at natural frequency of the system i.e. w_n

$$\frac{2\pi N}{60} = w_n$$

$$\therefore \quad N = 540.2 \text{ rpm}$$

(ii)

$$\text{Force exerted, } f_0 = mrw^2$$

Here,

$$\overline{w} = \frac{2\pi N}{60} = 104.67 \text{ rad/sec}$$

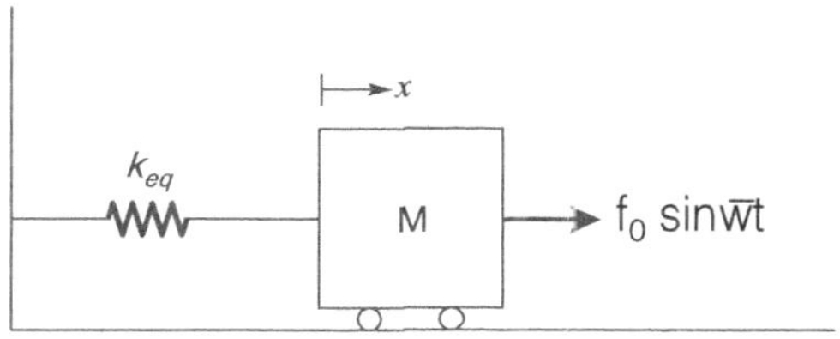

$$m\ddot{x} + kx = f_0 \sin\bar{w}t \quad ...(i)$$

The solution can be expressed as $x(t) = x_c(t) + x_p(t)$

$$x_c(t) = \text{complementary solution}$$

$$x_p(t) = \text{particular solution}$$

$$x_c(t) = A\cos wt + B\sin wt,\ w = \sqrt{\frac{k}{m}}$$

$$x_p(t) = x\sin\bar{w}t$$

$$x = \text{peak value}$$

∴ Putting this in equation (i)

$$-mx\bar{w}^2 \sin wt + kx\sin\bar{w}t = f_0 \sin\bar{w}t$$

$$-mx\bar{w}^2 + kx = f_0$$

$$x = \left[\frac{f_0}{k - m\bar{w}^2}\right]$$

Hence placing $f_0 = mrw^2$

$$= 0.04 \times 0.15 \times 104.67^2 = 65.735 \text{ N}$$

$$k = 48 \times 10^4 \text{ N/m}$$

$$m\bar{w}^2 = 150 \times 104.67^2 = 1644941.7$$

$$\therefore \quad x = \left|\frac{65.735 \text{ N}}{480000 - 1644941.7}\right| = 5.65 \times 10^{-5} \text{ m}$$

Hence, amplitude of vibration of motor = 0.0565 mm.

7.2 A machine is mounted at the centre of a simple supported beam that can exert a harmonic load $F(t) = 20 \sin (0.12t)$ kN in vertical direction. The length of beam is 4 m and its cross-section is uniform throughout. Cross-section of beam width 20 mm and depth 40 mm. A weight $W = 200$ N is suspended from the centre of the beam by a spring of spring constant $K_S = 40$ N/m. Determine the natural frequency of the weight W. Neglect mass of the beam and weight of machine. $E = 2 \times 10^5$ MPa.

[12 marks : 2020]

Solution:

The arrangement of machine and spring system could be simplified as shown below:

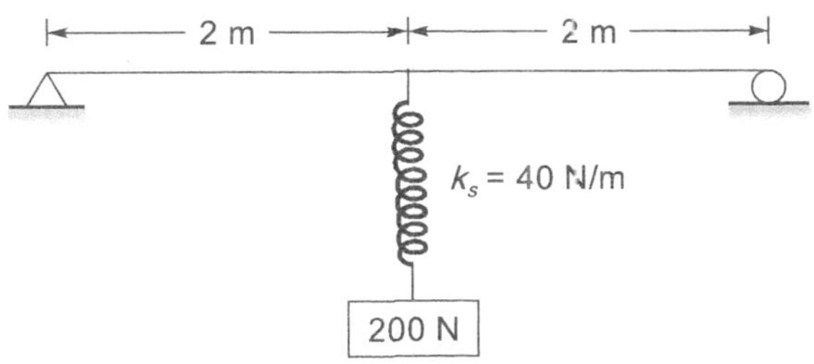

$$\text{Cross-section of beam} = 20 \text{ mm} \times 40 \text{ mm}$$

So, MOI of beam, $$I = \frac{bd^3}{12} = \frac{20 \times 40^3}{12}$$

Stiffness of beam loaded at midpoint

$$= \frac{48EI}{l^3} = \frac{48 \times 2 \times 10^5 \times 20 \times 40^3}{12 \times (4000)^3}$$

$$k_{beam} = 16 \text{ N/mm} = 16000 \text{ N/m}$$

Now, spring and beam are in series as shown below:

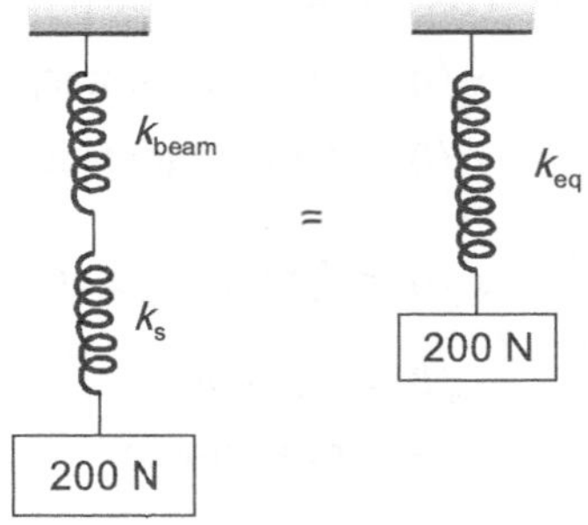

$$k_{eq} = \frac{k_s \times k_{beam}}{k_s + k_{beam}} = \frac{16000 \times 40}{16000 + 40} = 39.90 \text{ N/m}$$

Natural circular frequency of system,

$$\omega_n = \sqrt{\frac{k}{m}} = \sqrt{\frac{39.90}{\frac{200}{9.81}}} = 1.399 \text{ rad/s}$$

$$\text{Natural frequency} = \frac{1}{T} = \frac{\omega}{2\pi} = \frac{1.399}{2\pi} = 0.22 \text{ Hz}$$

7.3 **Derive the expression for displacement of an undamped free vibration of motion for a single degree of freedom system from first principles. Plot the undamped free vibration response.**

[12 Marks : 2021]

Solution:

Displacement of an undamped free vibration of motion for a single degree of freedom system.

(i) Differential equation for motion:

$$a = \ddot{y} = \frac{d^2y}{dt^2}$$

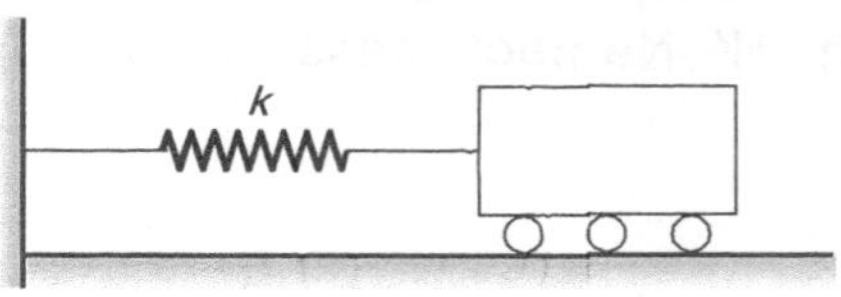

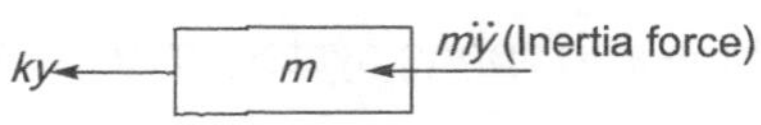

Body in dynamic equilibrium

$\Sigma x = 0 \Rightarrow$ $$-ky - m\ddot{y} = 0$$

$\Rightarrow$ $$m\ddot{y} + ky = 0 \quad \text{... (A)}$$

(ii) Solutions to differential equation of motion:

Since it is a second order linear differential homogeneous equation with constant coefficients, a trial solution can be

$$y = A\cos\omega t \quad \text{... (i)}$$

$$y = B\sin\omega t \quad \text{... (ii)}$$

Substituting equation (i) into equation (A) gives,

$$(-m\omega^2 + k)A\cos\omega t = 0$$

To satisfy this equation at anytime,

$$-m\omega^2 + k = 0$$

i.e. $\omega^2 = \dfrac{k}{m} = \sqrt{\dfrac{k}{m}}$ is called natural frequency of the system.

Since the differential equation is linear,

$y = A\cos\omega t + B\sin\omega t$ is also a solution

So, $\dot{y} = -A\omega\sin t + B\omega\cos\omega t$

At $t = 0$, $y = y_o$, $v = v_o$, $\dot{y} = v = v_o$

$$y_o = A\cos 0 + B\sin o \Rightarrow A = y_o$$

$$v_o = -A\omega\sin 0 + B\omega\cos 0 \Rightarrow v_o = B\omega \Rightarrow B = \frac{v_o}{\omega}$$

Substituting A and B in the equation of motion, we get,

$$y = y_o\cos\omega t + \frac{v_o}{\omega}\sin\omega t$$

Free vibration response:

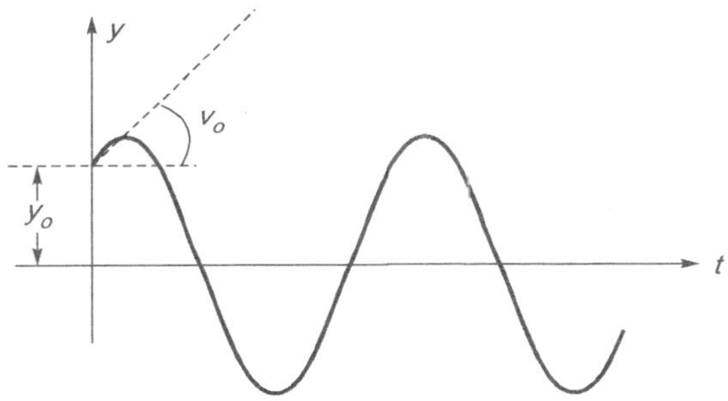

Undamped free vibration response

7.4 **For the vibrating system shown in the figure below, determine the following parameters**

(1) Natural frequency of the vibrating system

(2) Critical damping of the vibrating system

(3) Damping ratio

(4) Damped natural frequency of the vibrating system

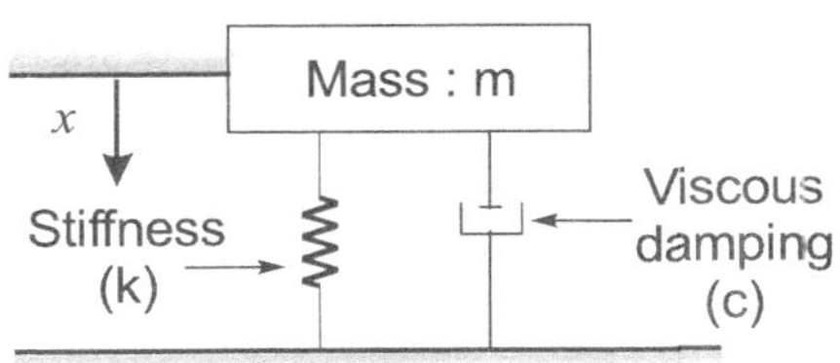

$m = 10$ kg
$k = 2400$ N/m
$c = 77$ N-s/m

[10 marks : 2023]

Solution:

1. Natural frequency of system,

$$\omega_n = \sqrt{\frac{k}{m}} \quad \text{...(i)}$$

where k is stiffness and m is mass.

So, $\omega_n = \sqrt{\dfrac{2400}{10}} = 15.49$ rad/s

2. Critical damping coefficient,

$$C_{cr} = \frac{2k}{\omega_n}$$

$$= \frac{2 \times 2400}{15.49} = 309.88 \text{ N-s/m}$$

3. Damping ratio, $\xi = \dfrac{C}{C_{cr}}$

$$= \frac{77}{309.88} = 0.25$$

As the ratio is less than 1, system is underdamped.

4. Damped natural frequency of system,

$$\omega_D = W_n\sqrt{1-\xi^2}$$

$$= 15.49\sqrt{1-0.25^2} \simeq 15 \text{ rad/s}$$

7.5 **A mass of 1000 kg is placed at the free end of a cantilever beam of span 3 m. Assume that the beam is massless compared to applied mass. The flexural rigidity of the beam is 10^3 kN-m². Determine the natural time period of the system. (Assume g =10 m/s²)**

[10 marks : 2023]

Solution:

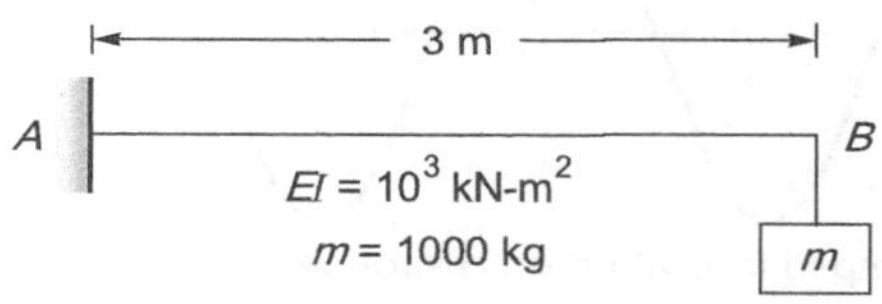

Stiffness of beam,

$$k_b = \frac{3EI}{L^3} = \frac{3 \times 10^3 \text{ kN-m}^2}{(3\text{ m})^3}$$

$$= 111.11 \text{ kN-m} = 111.11 \times 10^3 \text{ N-m}$$

Now, Natural time period, $T = 2\pi\sqrt{\dfrac{m}{k}} = 2\pi\sqrt{\dfrac{1000}{111.11 \times 10^3}} = 0.596$ seconds

■■■■

4 Design of Steel Structures

Revised Syllabus of ESE : *Principles of Working Stress methods, Design of tension and compression members, Design of beams and beam column connections, built-up sections, Girders, Industrial roofs, Principles of Ultimate load design.*

---

1. Structural Fasteners

1.1 The figure below shows a steel bracket connection using 22 mm diameter (12 No.) rivets. The connection transmits an eccentric load of 200 kN, at a distance of 250 mm from the centre of group of rivets. Find the maximum force in the rivets. Check whether the rivets are safe if the total thickness of plates connecting the rivets is 20 mm.

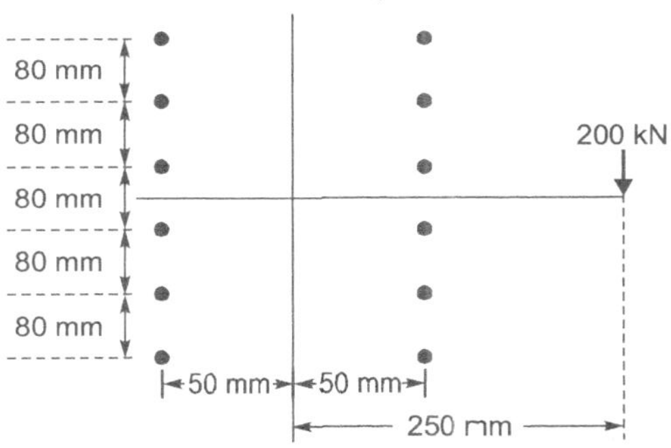

[25 marks : 2002]

Solution :

Assuming power driven shop rivets are used.

Nominal diameter of rivets = 22 mm

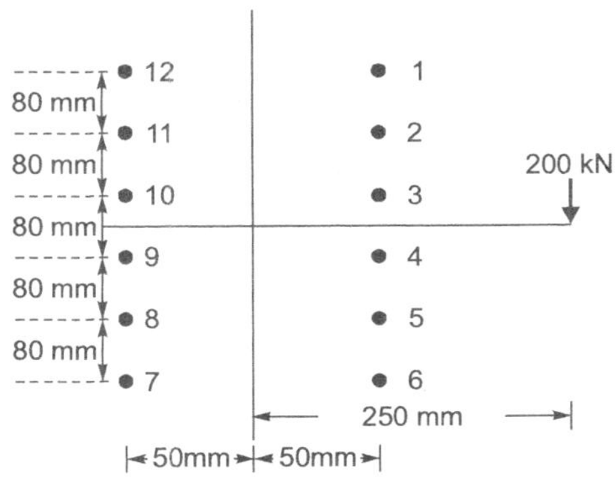

Gross diameter = 22 + 1.5 = 23.5 mm

Strength of rivet in shearing

$$= \frac{\pi}{4}\times(23.5)^2\times\frac{100}{1000} = 43.37 \text{ kN}$$

Strength of rivet in bearing

$$= 23.5\times 20\times\frac{300}{1000} = 141 \text{ kN}$$

Hence, the rivet value, R_v = 43.37 kN

Force in each rivet due to direct shear,

$$F_D = \frac{200}{12} = 16.67 \text{ kN}$$

Let the distance of rivets 1, 2, 3 ..... from the CG of rivet group be r_1, r_2, ...

$$r_1 = \sqrt{(200)^2 + (50)^2} = 206.16 \text{ mm}$$

$$r_2 = \sqrt{(120)^2 + (50)^2} = 130 \text{ mm}$$

$$r_3 = \sqrt{(40)^2 + (50)^2} = 64.03 \text{ mm}$$

Now

$r_1 = r_6 = r_7 = r_{12}$; $r_2 = r_5 = r_8 = r_{11}$; $r_3 = r_4 = r_9 = r_{10}$

Force in a rivet due to torsional moment,

$$F_T = \frac{(Pe)r_i}{\sum_{i=1}^{n} r_i^2}$$

Here $P = 200$ kN, $e = 250$ mm

$$\sum_{i=1}^{n} r_i^2 = 4[(206.16)^2 + (130)^2 + (64.03)^2] = 254007.15 \text{ mm}^2$$

$$\therefore \quad F_{T_1} = \frac{200 \times 250 \times 206.16}{254007.15} = 40.58 \text{ kN}$$

$$F_{T_2} = \frac{200 \times 250 \times 130}{254007.15} = 25.59 \text{ kN}$$

$$F_{T_3} = \frac{200 \times 250 \times 64.03}{254007.15} = 12.60 \text{ kN}$$

The most suspected rivet is rivet 1. Thus checking the safety of the joint for rivet 1.

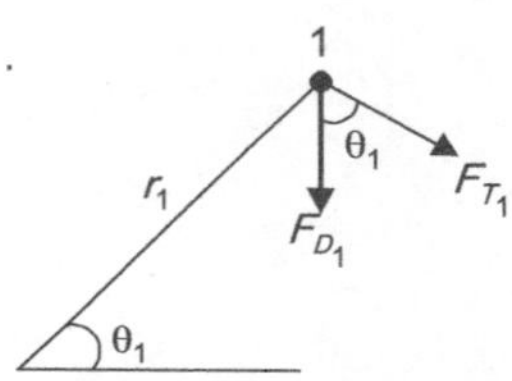

$$\cos\theta_1 = \frac{50}{206.16} = 0.243$$

Resultant force in rivet 1,

$$F_{r_1} = \sqrt{(F_{D_1})^2 + (F_{T_1})^2 + 2 \times F_{D_1} \times F_{T_1} \times \cos\theta_1}$$

$$= \sqrt{(16.67)^2 + (40.58)^2 + 2 \times 16.67 \times 40.58 \times 0.243}$$

$$= 47.47 \text{ kN}$$

Since $F_{r_1} > R_V$, hence the connection is unsafe.

Now $$\cos\theta_2 = \frac{50}{130} = 0.385$$

$$\therefore \quad F_{r_2} = \sqrt{(F_{D_2})^2 + (F_{T_2})^2 + 2 \times F_{D_2} \times F_{T_2} \times \cos\theta_2}$$

$$= \sqrt{(16.67)^2 + (25.59)^2 + 2 \times 16.67 \times 25.59 \times 0.385}$$

$$= 35.51 \text{ kN}$$

$$\cos\theta_3 = \frac{50}{64.03} = 0.781$$

$$\therefore \quad F_{r_3} = \sqrt{(F_{D_3})^2 + (F_{T_3})^2 + 2 \times F_{D_3} \times F_{T_3} \times \cos\theta_3}$$

$$= \sqrt{(16.67)^2 + (12.60)^2 - 2 \times 16.67 \times 12.60 \times 0.781}$$

$$= 27.65 \text{ kN}$$

$$\cos\theta_3 = \cos\theta_4 = 0.781$$

$$\therefore \quad F_{r_3} = F_{r_4} = 27.65 \text{ kN}$$

$$\cos\theta_2 = \cos\theta_5 = 0.385$$

$$\therefore \quad F_{r_2} = F_{r_5} = 35.51 \text{ kN}$$

$$\cos\theta_1 = \cos\theta_6 = 0.243$$

$$\therefore \quad F_{r_1} = F_{r_6} = 47.47 \text{ kN}$$

$$\cos\theta_7 = \cos(180 - \theta_1) = -0.243 = \cos\theta_{12}$$

$$\therefore \quad F_{r_7} = \sqrt{(F_{D_7})^2 + (F_{T_7})^2 + 2 \times F_{D_7} \times F_{T_7} \times \cos\theta_7}$$

$$= \sqrt{(16.67)^2 + (40.58)^2 + 2 \times 16.67 \times 40.58 \times (-0.243)}$$

$$= 39.95 \text{ kN} = F_{r_{12}}$$

$$\cos\theta_8 = \cos(180 - \theta_2) = -0.385 = \cos\theta_{11}$$

$$\therefore \quad F_{r_8} = \sqrt{(F_{D_8})^2 + (F_{T_8})^2 + 2 \times F_{D_8} \times F_{T_8} \times (\cos\theta_8)}$$

$$= \sqrt{(16.67)^2 + (25.59)^2 + 2 \times 16.67 \times 25.59 \times (-0.385)}$$

$$= 24.58 \text{ kN} = F_{r_{11}}$$

$$\cos\theta_9 = \cos(180 - \theta_3) = -0.781 = \cos\theta_{10}$$

$$\therefore \quad F_{r_9} = \sqrt{(F_{D_9})^2 + (F_{T_9})^2 + 2 \times F_{D_9} \times F_{T_9} \times \cos\theta_9}$$

$$= \sqrt{(16.67)^2 + (12.60)^2 + 2 \times 16.67 \times 12.60 \times (-0.781)}$$

$$= 10.42 \text{ kN} = F_{r_{10}}$$

Thus

$F_{r_1} = F_{r_6} = 47.47$ kN $\qquad F_{r_3} = F_{r_4} = 27.65$ kN $\qquad F_{r_8} = F_{r_{11}} = 24.58$ kN

$F_{r_2} = F_{r_5} = 35.51$ kN $\qquad F_{r_7} = F_{r_{12}} = 39.95$ kN $\qquad F_{r_9} = F_{r_{10}} = 10.42$ kN

1.2 **In an industrial shade on edge support consisting of two angle 110 mm × 100 mm is to be connected to a 16 mm gusset plate for a tensile load of 650 kN. Design the moment free welded connection. The distances of centroid of angle from the back side of legs are $C_x = C_y = 30.9$ mm. The strength of the weld per mm thickness per mm length is 76 MPa.**

[20 marks : 2003]

Solution :

The edge support is consisted of two angles i.e. the load taken by them is 650 kN. Thus each angle will take $\frac{650}{2} = 325$ kN

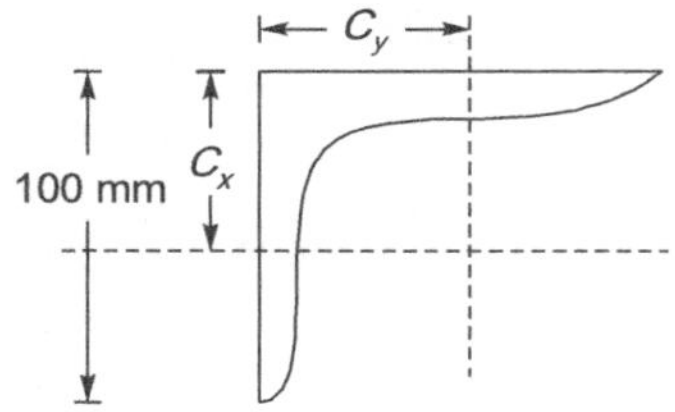

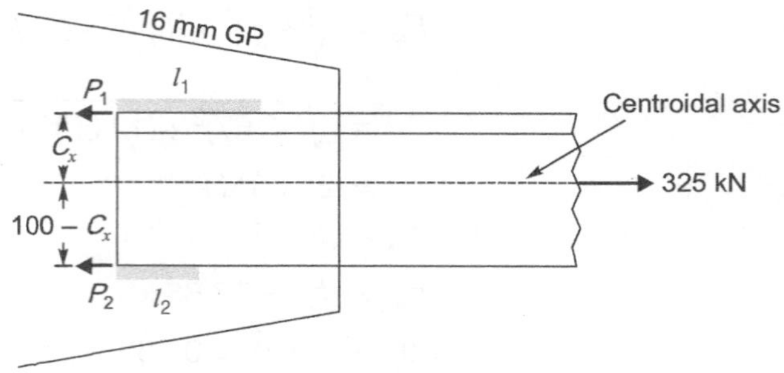

Assuming the thickness of angles as 8 mm, hence minimum size will be 5 mm, because the minimum thickness of thicker part governs the minimum size of weld.

From the above figure, we have

$$P_1 + P_2 = 325 \quad \text{...(i)}$$

Taking moment about the line of action of load of 325 kN

$$\therefore \quad P_1 \times C_x - P_2 \times (100 - C_x) = 0$$

$$\Rightarrow \quad P_1 \times 30.9 - P_2 \times (100 - 30.9) = 0$$

$$\Rightarrow \quad P_1 = \frac{69.1}{30.9} P_2$$

$$\Rightarrow \quad P_1 = 2.24\, P_2 \quad \text{...(ii)}$$

Substituting (ii) in (i), we get

$$P_2 = \frac{325}{3.24} = 100.309 \text{ kN}$$

and $P_1 = 325 - 100.309 = 224.691$ kN

For l_1 length of weld, $P_1 = l_1 \times$ throat thickness $\times$ permissible stress in weld

$$\Rightarrow \quad 224.691 \times 10^3 = l_1 \times \frac{5}{\sqrt{2}} \times 76$$

$$\Rightarrow \quad l_1 = 836.21 \text{ mm} \approx 837 \text{ mm}$$

For l_2 length of weld $\quad P_2 = l_2 \times$ throat thickness $\times$ permissible stress in weld

$$\Rightarrow \quad 100.309 \times 10^3 = l_2 \times \frac{5}{\sqrt{2}} \times 76$$

$$\Rightarrow \quad l_2 = 373.311 \text{ mm} \approx 374 \text{ mm}$$

Thus adopting $l_1 = 837$ mm and $l_2 = 374$ mm

1.3 **A circular shaft of diameter 150 mm is welded to a rigid plate by an external alround fillet weld of size 10 mm. If a torque of 10 kN-m is applied to the shaft, find the maximum stress in the weld.**

[10 marks : 2004]

Solution :

Let '*d*' be the diameter of the shaft and *S* be the size of weld. Let *t* be the throat thickness, then

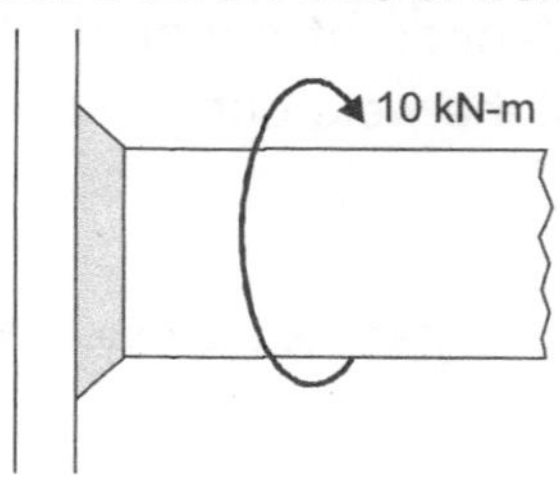

$$t_t = 0.7s = 0.7 \times 0 = 7 \text{ mm}$$

$$f_s = \text{Max. shear stress in weld} = \frac{T}{Z_P}$$

$$Z_p = \text{Polar modulus of weld area}$$

$$= A \times r = \pi dt \times \frac{d}{2} = \pi d^2 \frac{t}{2}$$

$$f_s = \frac{T}{Z_P} = \frac{10 \times 10^6 \text{ Nmm}}{\pi \times 150^2 \times \frac{7}{2}} = 40.4 \text{ N/mm}^2$$

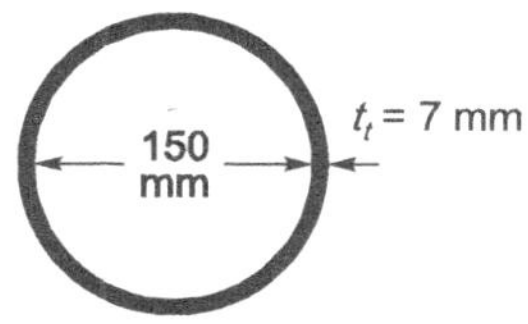

Effective area of weld

1.4 Determine the number and pattern of 20 mm diameter power driven shop rivets for the truss connection as shown in the figure below:

Sketch the arrangement of rivets.
Assume the stresses as follows:
Shear stress = 1025 kgf/cm²
Bearing stress = 2360 kgf/cm²
1 kN = 100 kgf
1 N/mm² = 10 kgf/cm²

[12 marks : 2007]

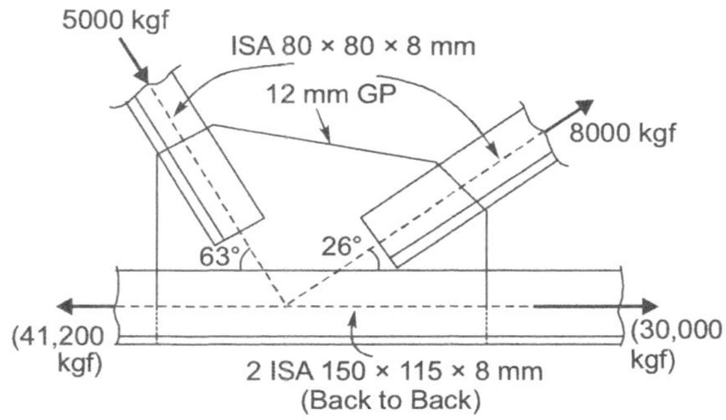

Solution :

Nominal diameter of power driven stop rivets = 20 mm
Gross diameter = 20 + 1.5 = 21.5 mm

(i) Taking the member subjected to an axial load (compression) of 5000 kgf = 50 kN

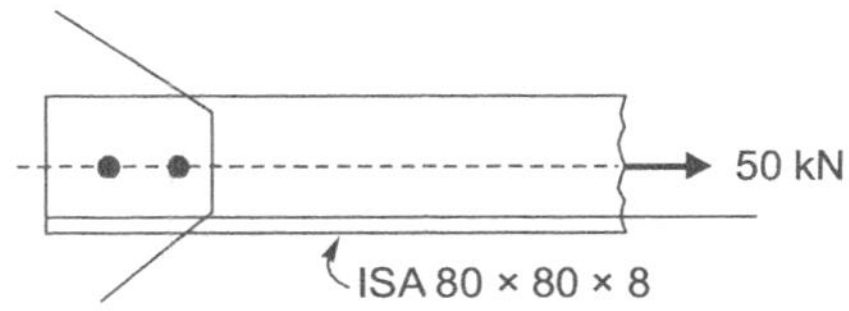

$$\text{Strength of rivet in shearing} = \frac{\pi}{4} \times (21.5)^2 \times \frac{102.5}{1000} = 37.21 \text{ kN}$$

$$\text{Strength of rivet in bearing} = 21.5 \times 8 \times \frac{236}{1000} = 40.59 \text{ kN}$$

$$\therefore \quad \text{Rivet value, } R_v = 37.21 \text{ kN}$$

$$\text{Number of rivets, } n = \frac{P}{R_v} = \frac{50}{37.21} = 1.34 \approx 2$$

Provide 2 rivets.

(ii) Member subjected to an axial tensile load of 80 kN

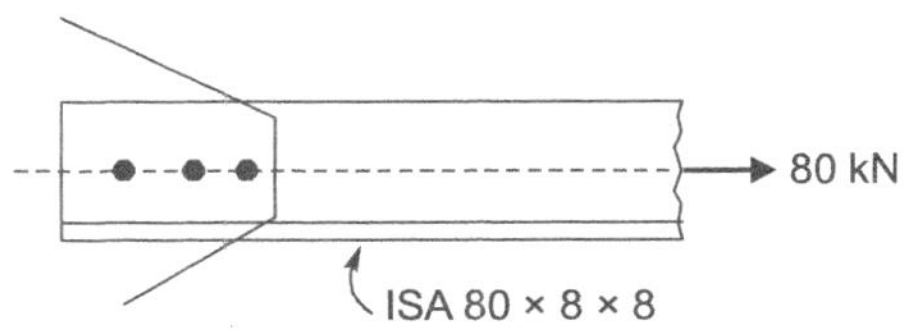

$$\text{Rivet value, } R_v = 37.21 \text{ kN}$$

$$\text{Number of rivets, } n = \frac{P}{R_v} = \frac{80}{37.21} = 2.15 \approx 3 \text{ rivets}$$

Providing 3 rivets in a single line.

(iii) Members subjected to an axial tensile load of 300 kN

Strength of rivet in double shear

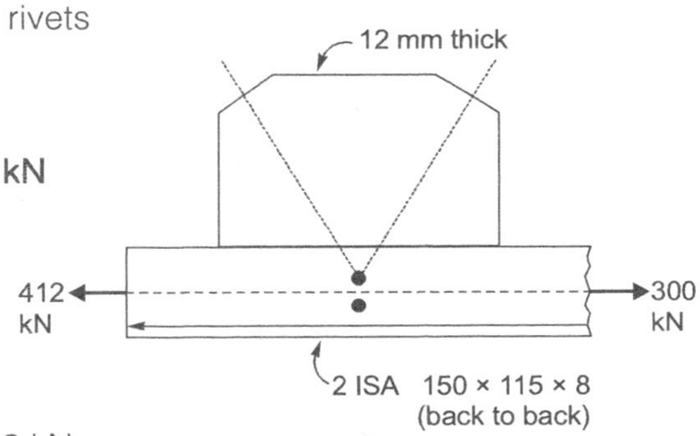

$$= \frac{\pi}{4} \times (21.5)^2 \times \frac{102.5}{1000} \times 2$$

$$= 74.43 \text{ kN}$$

$$\text{Strength of rivet in bearing} = 21.5 \times 12 \times \frac{236}{1000} = 60.89 \text{ kN}$$

$\therefore$ Rivet value, $R_v = 60.89$ kN

Rivets in the continuous leg should be designed for balanced force i.e

$$412 - 300 = 112 \text{ kN}$$

$$\text{Number of rivets, } n = \frac{P}{R_v} = \frac{112}{60.89} = 1.84 \approx 2 \text{ rivets}$$

Providing 2 rivets one in each row is connected leg

pitch of rivets $= 2.5 \times 20 = 50$ mm

edge distance $= 1.5 \times 21.5 = 32.25$ mm

1.5 **A plate of 150 mm width and 20 mm thick is welded to another plate by fillet weld as shown in figure below. The size of the weld is 12 mm throughout. Compute the average shear stress produced in the weld for the full strength of the plate if the allowable stress is 150 N/mm².**

[10 marks : 2008]

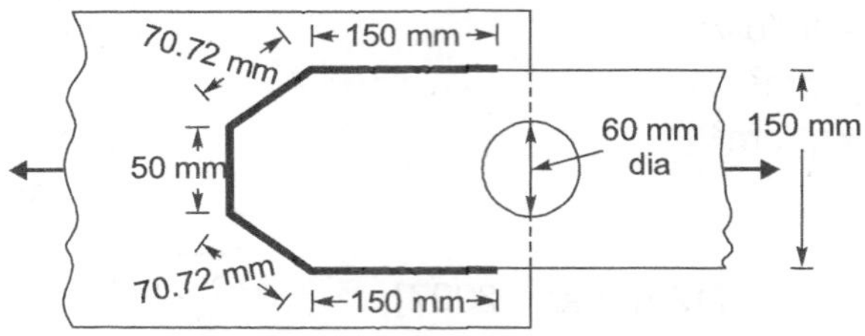

Solution :

The full strength of the plate is given by

$$P = \text{Area of cross-section of plate} \times \text{permissible stress in tension}$$

$$= (150 - 60) \times 20 \times \frac{150}{10^3} = 270 \text{ kN}$$

This force of 450 kN will be resisted by the fillet weld at the edges and slot weld.

Let the average shear stress produced in the welds be f_s, then

$$P = (\text{Total effective length of fillet weld at the edges} \times \frac{s}{\sqrt{2}} \times f_s)$$

$$+ (\text{Circumference of the slot} \times \frac{s}{\sqrt{2}} \times f_s)$$

$$\Rightarrow \quad 270 \times 10^3 \text{ N} = (2 \times 150 + 2 \times 70.72 + 50) \times f_s \times 0.7 \times 12$$

$$\Rightarrow \quad f_s = 65.4 \text{ N/mm}^2$$

1.6 **Explain the importance of welded connection in building connection.**

[5 marks : 2010]

Solution:

1. Welded connection leads to saving in expenditure on material and labour.
2. Welding permits architects and structural engineer complete freedom of design.
3. Welded joints are better for fatigue, impact and vibration as compared to bolted joint.
4. Rigidity of joint also leads to reduce deflections.
5. The compactness and rigidity of welded joints permits design assumption to be realised more accurately.
6. It leads to rapid construction, thus saving in time of project.
7. It leads to rigid connection which in turn leads to reduced beam depth and weight.

1.7 (i) List out five merits and three demerits of welded joints.

(ii) Determine the size of the fillet weld required to join a bracket plate with the flange of a column section shown in the figure below. Permissible stress in weld = 108 MPa.

[4 + 4 = 8 marks : 2011]

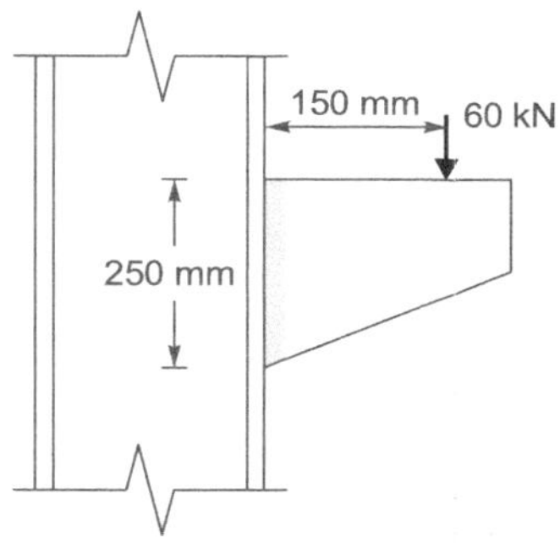

Solution:

(i) Advantages of welded joints:

(i) As no holes are required for welding, the structural members are more effective in taking load.

(ii) The overall weight of structural steel required is reduced by the use of welded joints.

(iii) Welded joints are often economical as less labour and material are required for a joint.

(iv) The speed of fabrication is higher with the welding process.

(v) Any shape of joint can be made with ease.

(vi) Complete rigid joints can be provided with the welding process.

Disadvantages of welded joints:

(i) Skilled labour and electricity are required for welding.

(ii) Internal stresses and warping are produced due to uneven heating and cooling.

(iii) Welded joints are more brittle and therefore their fatigue strength is less than the members joined.

(iv) Defects like internal air pockets, slag inclusion and incomplete penetration are diffcult to detect.

(ii) Let, thickness of weld throat = t mm

Vertical shear stress at weld,

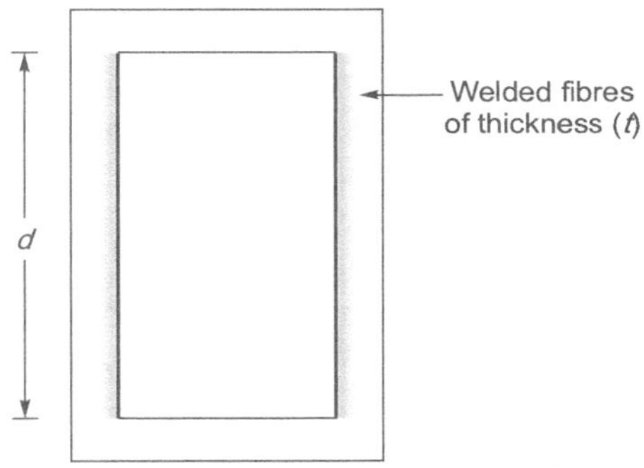

$$p_s = \frac{W}{2 \times d \times t} = \frac{60 \times 10^3}{2 \times 250 \times t}$$

$$= \frac{60 \times 10^3}{500 \times t} = \frac{120}{t} \text{MPa}$$

Horizontal shear stress due to bending at extreme fibre,

$$p_b = \frac{W \times e \times \dfrac{d}{2}}{\left(t \times \dfrac{d^3}{12}\right) \times 2} = \frac{6 \times W \times e}{2 \times t \times d^2}$$

$$= \frac{6 \times 60 \times 10^3 \times 150}{2 \times t \times 250^2} = \frac{432}{t} \text{MPa}$$

Resultant stress at extreme fibre,

$$p_r = \sqrt{p_s^2 + p_b^2} = \sqrt{\left(\frac{120}{t}\right)^2 + \left(\frac{432}{t}\right)^2}$$

$$= \frac{448.357}{t} \text{MPa} = \frac{448.4}{t} \text{MPa}$$

Since, $\quad p_r$ = Permissible stress

$$\therefore \quad \frac{448.4}{t} = 108$$

or $$t = 4.15 \text{ mm}$$

Therefore, size of weld $$s = \frac{t}{0.7} = 5.9 \text{ mm} = 6 \text{ mm (say)}$$

1.8 A plate is connected to the flange of an ISMB as shown. The factored load is 100 kN. Find the size of the weld. Assume shop weld and ultimate strength of weld as 410 MPa.

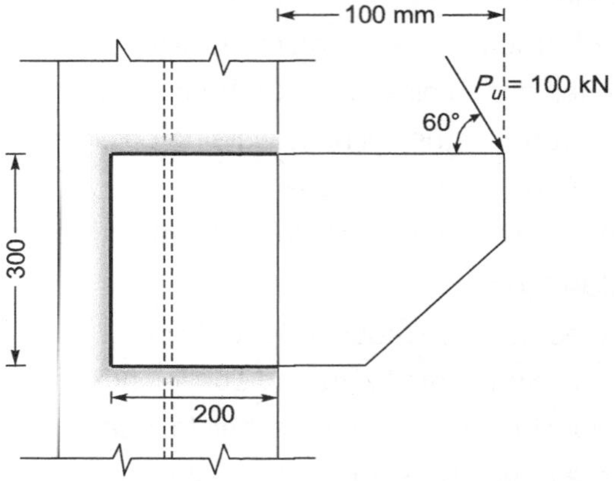

[20 marks : 2013]

Solution:

Let $\bar{x}$ = distance of centroid of weld group from left edge.

s = size of weld

t = effective throat thickness of weld

$\therefore \quad t = 0.7\ s$

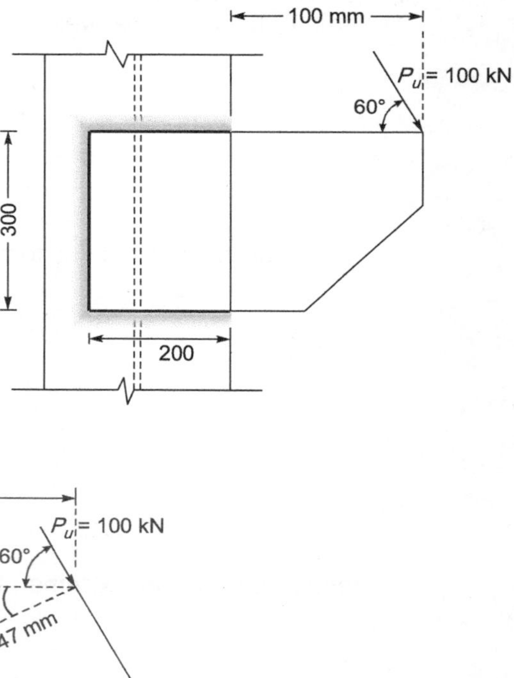

$$\bar{x} = \frac{(2\times200\times t)100+(300\times t)0}{(2\times200\times t)+(300\times t)}$$

$$\bar{x} = \frac{(400\times100)t}{700t} = 57.143 \text{ mm from left edge}$$

$$\tan\theta_1 = \frac{150}{200-\bar{x}} = \frac{150}{200-57.143} = 1.05$$

$$\theta_1 = 46.397°$$

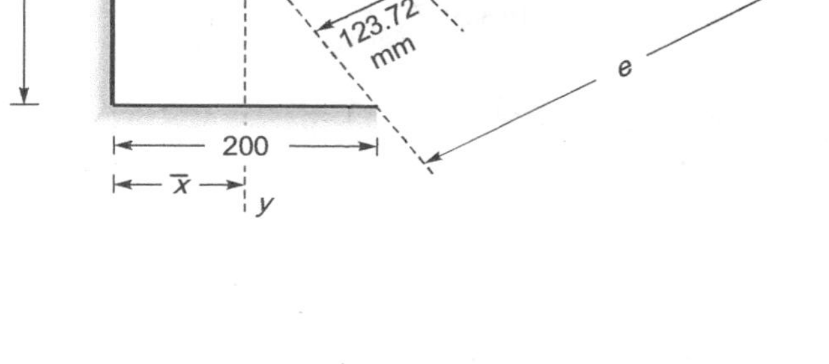

$$I_{xx} = \frac{t \times 300^3}{12} + 2\left[\frac{200t^3}{12} + 200t \times (150)^2\right]$$

$$= (2.25\, t \times 10^6) + 33.33\, t^3 + (9t \times 10^6) \simeq 11.25\, t \times 10^6 \text{ mm}^4 \text{[Neglecting } 33.33\, t^3]$$

$$I_{yy} = \left\{\frac{300t^3}{12} + 300t(57.143)^2\right\} + 2\left\{\frac{t \times 200^3}{12} + 200t(100 - 57.143)^2\right\}$$

$$= \{25t^3 + 0.9796t \times 10^6\} + \{1.333t \times 10^6 + 0.7347t \times 10^6\}$$

$$= 3.0473\, t \times 10^6 \text{ mm}^4 \quad \text{[Neglecting } 25\, t^3]$$

$\therefore$ Polar MOI $= J = I_{xx} + I_{yy} = (11.25\, t + 3.0473\, t)10^6 = 14.2973\, t \times 10^6$ mm^4

Calculating 'e' of load P

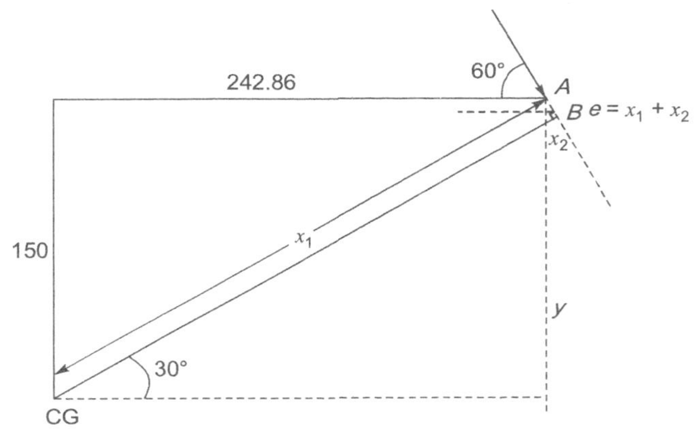

$$\frac{242.86}{x_1} = \cos 30°$$

$\Rightarrow$ $x_1 = 280.43$ mm

$\Rightarrow$ $\dfrac{y}{242.86} = \tan 30°$

$\Rightarrow$ $y = 140.215$ mm

$(150 - y) = 9.78$ mm

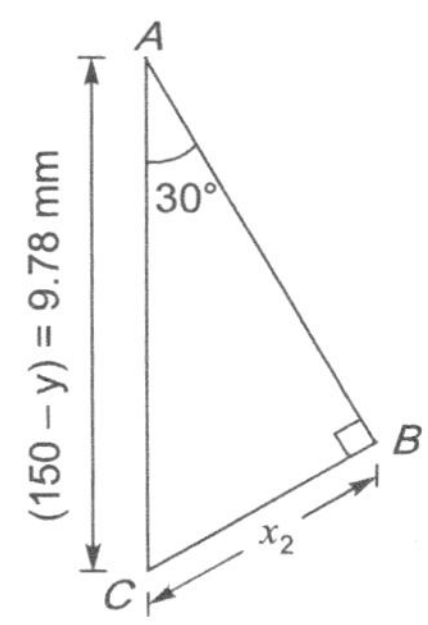

In ΔABC, $\dfrac{x_2}{9.78} = \sin 30°$

$x_2 = 4.89$ mm

$e = x_1 + x_2 = 280.43 + 4.89 = 285.32$ mm

Direct shear stress, $\tau_{vf} = \dfrac{142.857}{t}$ N/mm^2

$\theta = 46.397° - 30° = 16.397°$

Where $\theta = 16.397°$ is angle between direct shear stress and torsional shear stress.

Shear stress due to twisting/torsion $= \dfrac{\text{Per}}{J}$

$$\tau_{vr} = \frac{(100 \times 10^3)(285.32)(207.143)}{14.2973t \times 10^6} = \frac{413.379}{t}$$

$$\text{Resultant shear stress} = \sqrt{\tau_{vf}^2 + \tau_{vr}^2 + 2\tau_{vf}.\tau_{vr}\cos\theta}$$

$$f_R = \frac{551.9}{t}\ \text{N/mm}^2$$

Ultimate strength of weld = f_u = 410 N/mm²

$$f_{wd},\ \text{Design shear strength of weld} = \left(\frac{f_u}{\sqrt{3}}\right)\frac{1}{\gamma_{mW}}$$

$$\gamma_{mW} = 1.25$$

$$f_{wd} = 189.37\ \text{N/mm}^2$$

Hence, $f_R \leq f_{wd}$

$$\Rightarrow \quad \frac{551.9}{t} \leq 189.37$$

$$t \geq 2.914\ \text{mm}$$

$$\Rightarrow \quad s \geq 2.914 \times \sqrt{2} = 4.12\ \text{mm} \approx 5\ \text{mm}$$

1.9 **(i) Name 10 commonly used hot rolled structural section for steel construction.**
(ii) Give 6 reasons for failure of riveted joints.

[2 + 2 = 4 marks : 2013]

Solution:

(i) The largest category of standard shapes of steel sections that is available for steel construction are those which are produced by hot rolling.

Commonly used hot rolled sections are:

• Rolled steel I-section 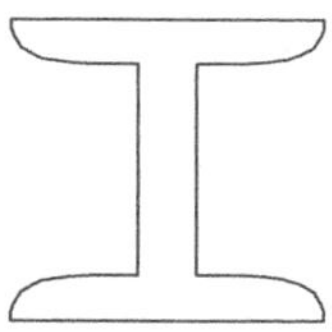	• Rolled steel channel sections 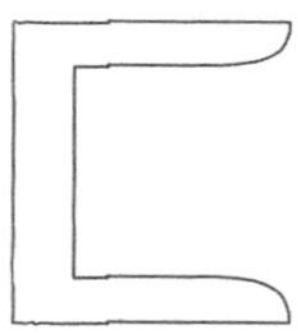	• Rolled steel T-section 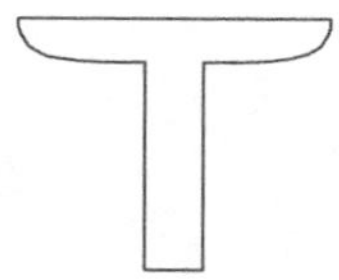
• Rolled steel angle sections 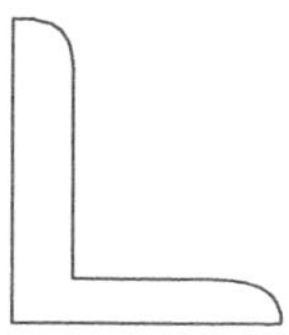	• Rolled steel tube sections 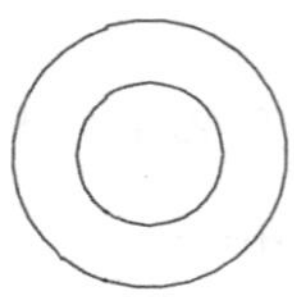	• Rolled steel bars 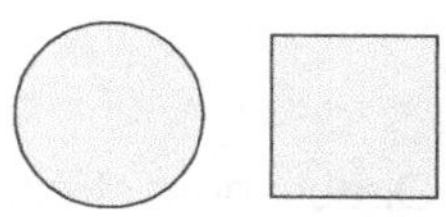
• Rolled steel flats 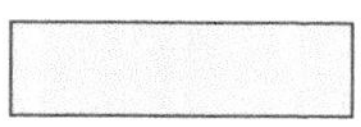	• Rolled steel plates	• Rolled steel sheets
• Rolled steel strips		

(ii) Failure of riveted joints.

The six (06) modes of failure of riveted joints are as follows:

1. **Shear failure of rivets:** When shear stress in rivet exceeds the permissible shear stress of rivet.

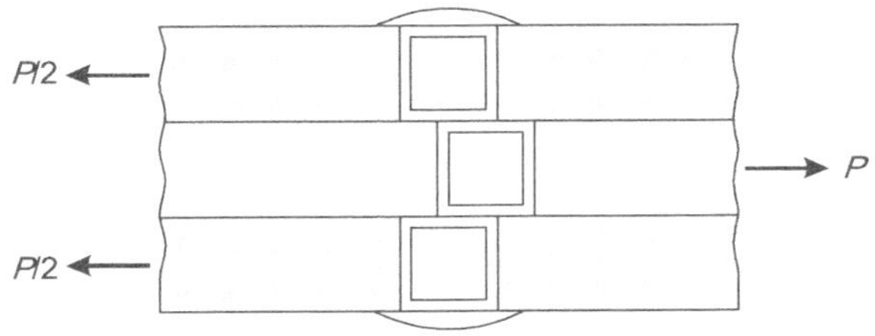

2. **Shear failure of plate:** When rivets are placed very near to the edge of plate, then this failure takes place. This type of failure can be avoided by providing adequate edge distance.

3. **Tearing/tension failure of plates:** Here in this failure, at the location of net cross-section, tensile stress in plate exceeds the permissible limit.

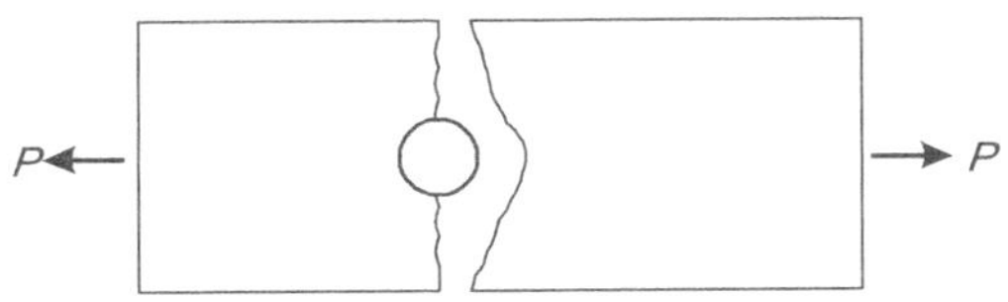

4. **Bearing failure of plate:** When bearing stress in plate exceeds the permissible limit, then crushing of plates takes place.

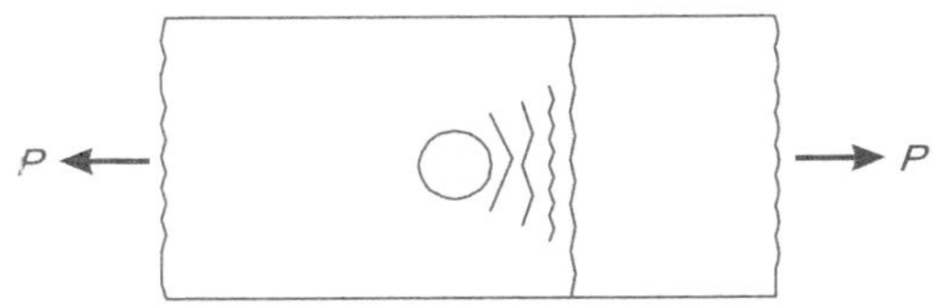

5. **Splitting of plate:** When rivets are placed at very small edge distance then the plate tends to split or shear out.

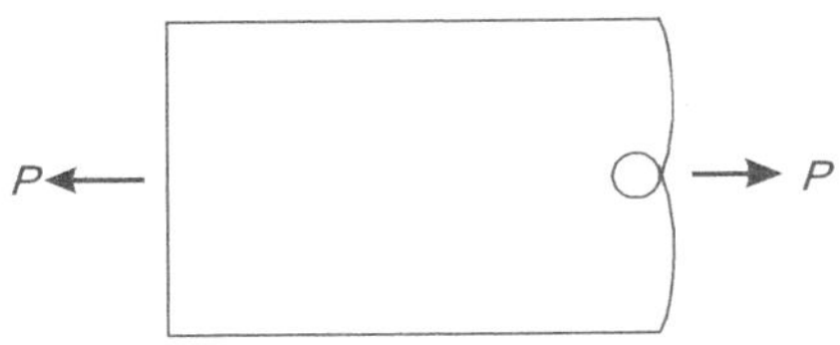

6. **Bearing failure of rivets:** When plates are stronger than rivets in bearing then there is a possibility that rivets may get crushed around the half circumference when bearing stress of rivet exceeds the allowable limit.

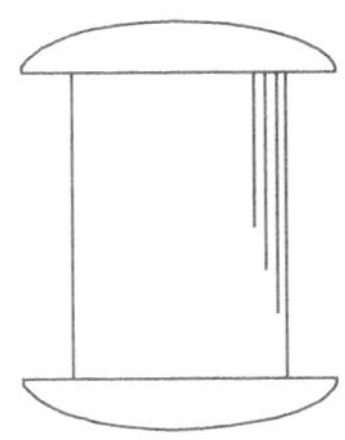

1.10 What are the various limit states of design for a steel structure as per IS : 800-2007?

[5 marks : 2014]

Solution:

Limit state are the states beyond which the structure no longer satisfies the specified performance requirement. **As per IS 800: 2007, the limit states are generally grouped under.**

1. **Limit State of Strength:** Limit state of strength are associated with failure of structure under the worst combination of loading including appropriate partial factor of safety. The limit state of strength include
   (a) Loss of stability/equilibrium of structure (including the effect of sway) or any of parts including supports and foundation.
   (b) Strength limit (general yielding, formation of mechanism, rupture of structure or any of its parts of components)

(c) Fatigue and brittle failure.

2. **Limit state of serviceability:** There are limit states beyond which specified service criteria are no longer met. These include
   (a) Deformation and deflections
   (b) Vibrations in the structure or any of its component causing discomfort to people or damages to the structure
   (c) Ponding of structures
   (d) Corrosion and durability
   (e) repairable damage due to fatigue

1.11 **State the assumptions made for designing riveted connections in steel.**

[5 marks : 2014]

Solution:

Assumptions in riveted connections:

1. Friction between the plates is neglected.
2. Shearing stresses are assumed to be distributed uniformly (but actually shear stress varies parabolically over the cross-section).

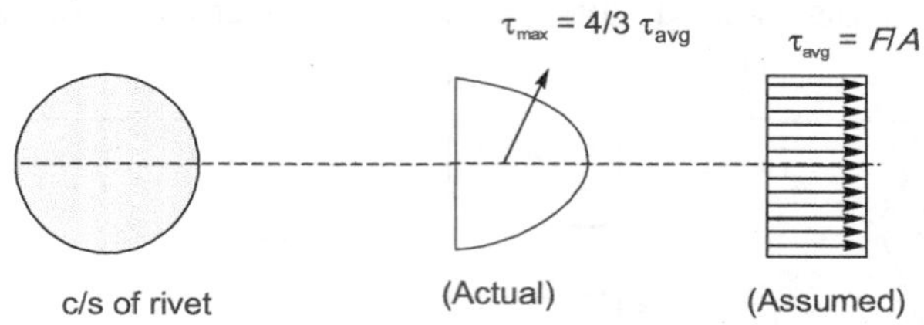

3. Bearing stress distribution is also assumed uniform and contact area is $d \times t$, where 'd' is the diameter and 't' is the thickness of the plate.
4. Bending stresses in rivets are neglected.
5. The rivets fill the hole completely.
6. The distribution of direct stresses on the portion between the rivets hole is assumed uniform.

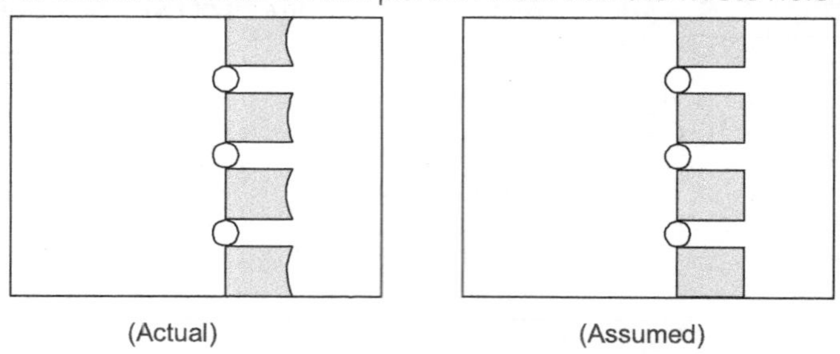

7. Rivets in group subjected to direct load, share the direct load equally.

1.12 **2 ISA 75 × 75 × 8 carry a load of 150 kN and are placed back to back through a 6 mm gusset plate. The permissible shear stress is 100 MPa and bearing stress is 300 MPa. Design the riveted connection and show the arrangement with a neat sketch.**

[10 marks : 2014]

Solution:

Diameter of rivet by Unwin's formula

$$d = 6.04\sqrt{t_{min}} = 6.04\sqrt{6} = 14.795 \text{ mm} \simeq 16 \text{ mm}$$

Hence use 16 mm dia. rivet in connection

Diameter of hole, $d' = 16 + 1.5 = 17.5$ mm

Strength of rivet in shearing, $P_s = 2\times\frac{\pi d'^2}{4}\times f_s$ [∵ rivets are in double shear]

$$= \frac{2\times\pi\times 17.5^2\times 100}{4}\text{ N} = 48.10\text{ kN}$$

Strength of rivet in bearing, $P_b = d'\times t_{min}\times f_b$

where t_{min} = Minimum of $\begin{cases}\text{(a) Sum of cover plates, } t = 8+8 = 16\text{ mm}\\ \text{(b) Main plate, } t = 6\text{ mm}\end{cases}$

$$\therefore \quad t_{min} = 6\text{ mm}$$

$$\therefore \quad P_b = 17.5\times 6\times 300 = 31.5\text{ kN}$$

So rivet value, R_v = Minimum of $\begin{cases}\text{(a) } P_s = 48.10\text{ kN}\\ \text{(b) } P_b = 31.5\text{ kN}\end{cases}$

$$\therefore \quad R_v = 31.5\text{ kN}$$

No. of rivets required for connection is given by,

$$N = \frac{P}{R_v} = \frac{150}{31.5}$$

$$\Rightarrow \quad N = 4.76 \simeq 5\text{ Nos}$$

Pitch distance, $p = 2.5\times d$ [where d is nominal dia.]

$$= 2.5\times 16 = 40\text{ mm}$$

Minimum edge distance according to no tearing criteria is given by,

$$e = 2\times d' \quad \text{[where } d' \text{ is hole dia.]}$$

$$= 2\times 17.5 = 35\text{ mm}$$

Detail of connection shown in figure.

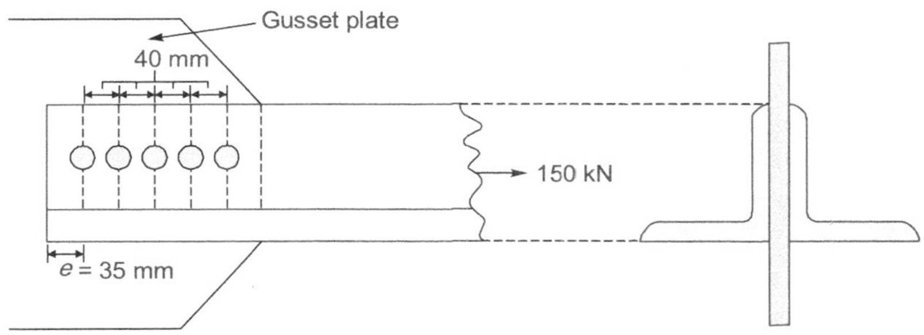

1.13 The vertical member of a triangular pratt truss is composed of 2 Nos. ISA 75 × 75 × 6 (connected back to back on each side of the gusset of 10 mm thickness). The factored forces in the member are: 107 kN (Compression)
79 kN (Tension)

Design the fillet weld connection. Assume f_u = 410 MPa and γ_{mw} = 1.25. Welding shall be done along the length of the member.

[10 marks : 2015]

Solution:

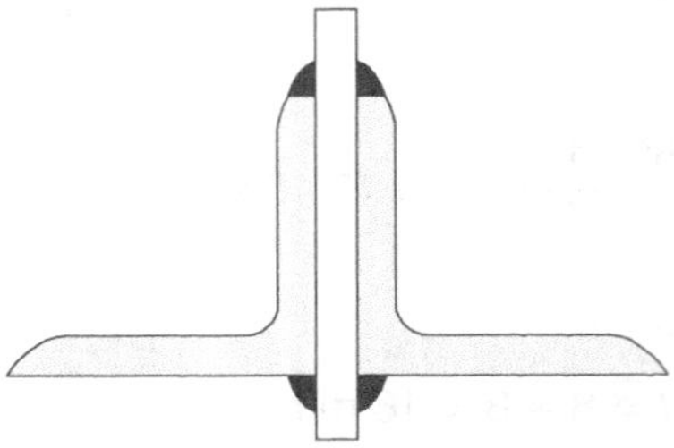

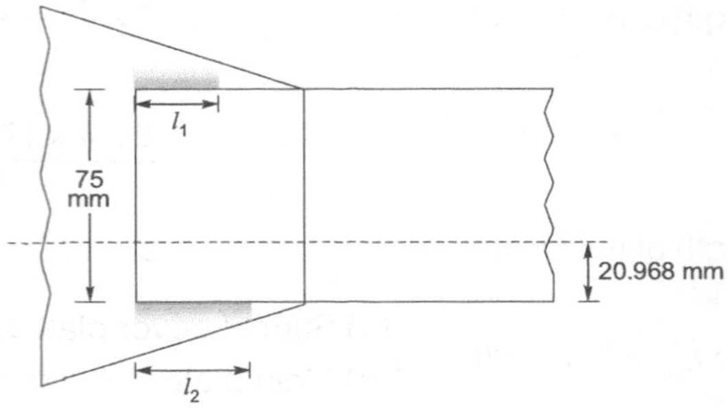

Whether a member is in compression or is in tension, force transfer in fillet weld occurs through shear only.

∴ Design load = Maximum of compression and tension = 107 kN

Let l_1, l_2 = Lengths of weld on upper and lower portion of angle respectively.

∴ Total weld length = $l_1 + l_2$ (∵ Welding is required along the length of member only)

Strength of weld required = 107 kN = P

$$\text{Size of weld} = 6 - \frac{6}{4} = 4.5\text{mm}$$

So, providing 4 mm size for fillet weld.

$$\therefore \quad \text{Strength of weld } (T_{dw}) = t_t \frac{f_u}{\sqrt{3}\gamma_{mw}} = (0.7 \times 4)\frac{410}{1.25\sqrt{3}} = 530.238 \text{ N/mm}$$

$$2(P_1 + P_2) = 107 \times 10^3$$

Here 2 is multiplied on strength side because there are two welds on upper side and two welds on lower side because of two angles.

$$\therefore \quad 2(l_1 + l_2)530.238 = 107 \times 10^3$$

$$\Rightarrow \quad l_1 + l_2 = 100.897 \text{ mm}$$

$$A_1 = 75 \times 6 = 450 \text{ mm}^2,$$

$$A_2 = (75 - 6)6 = 414 \text{ mm}^2$$

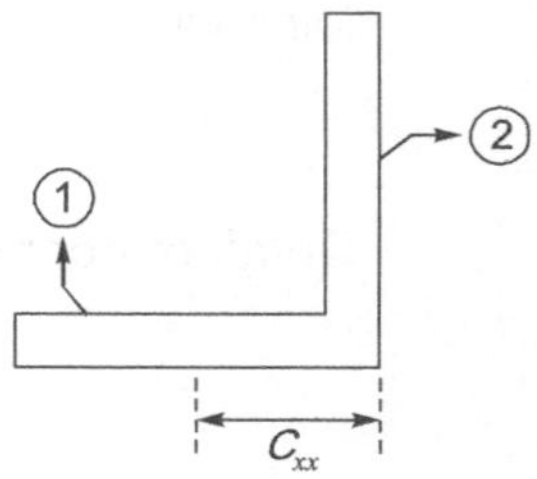

$$C_{xx} = \frac{450 \times \frac{75}{2} + 414 \times 3}{450 + 414} = 20.968\text{mm}$$

Taking moment about bottom weld length,

$$2P_1 \times 75 = P \times 10^3 \times C_{xx}$$

$$2 \times 530.238\, l_1 (75) = 107 \times 10^3 \times 20.968$$

$$\Rightarrow \quad l_1 = 28.21 \text{ mm}, \quad l_2 = 100.897 - 28.21 = 72.687 \text{ mm}$$

1.14 Two steel plates in axial tension are to be connected by bolts as shown in figure. Determine its tensile strength. Bolt dia = 25 mm, Hole dia = 28 mm, Plate thickness = 16 mm each, f_u= 440 MPa, f_y = 290 MPa

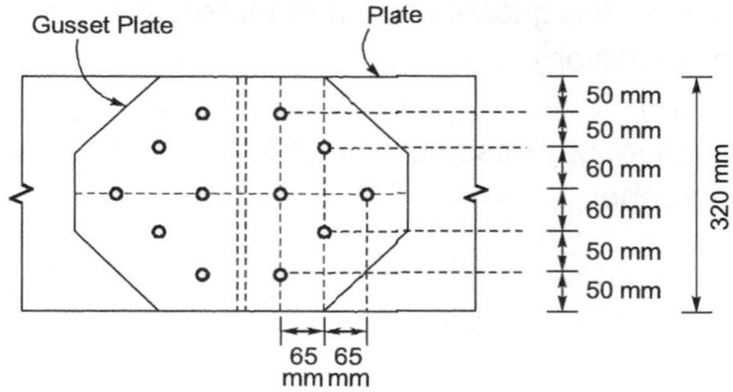

[20 marks : 2016]

Solution:

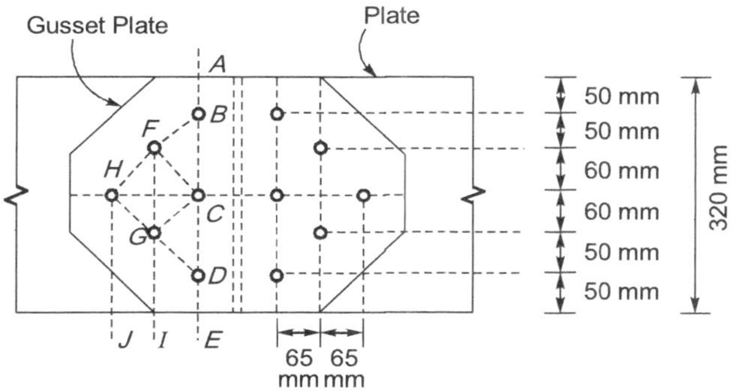

Tensile strength of connection or (joint) = Least value of P_s, P_b and P_t

P_s = Shear strength of all bolts in the joint

$$= n \times \left[0.78 \times \frac{\pi}{4} \times d^2\right] \times f_s \qquad \left(f_s = \frac{f_u}{\sqrt{3} \times 1.25}\right)$$

$$= 6 \times \left[0.78 \times \frac{\pi}{4} \times (25)^2\right] \times \frac{440}{\sqrt{3} \times 1.25} = 446.6 \text{ kN}$$

P_b = Bearing strength of all bolts in the joint

$= n \times (d \times t) \times f_b$

t = Thickness of cover plate = 16 mm
or Thickness of thinner main plate = 16 mm } whichever is less

f_b = Design bearing stress

$$= 2.5 \times k_b \times \frac{f_u p}{1.25}$$

$$k_b = \frac{p}{3d_0} - 0.25 = \frac{65}{3 \times 25} - 0.25 = 0.52$$

$$\frac{f_{u,}b}{f_u,p} = \frac{440}{440} = 1$$

$k_b = \{0.52, 0.595, 1\}$

$k_b = 0.52$

$P_b = n \times (d \times t)\, f_b$

$$= 6 \times 25 \times 16 \times 2.5 \times 0.52 \times \frac{440}{1.25} = 1098.2 \text{ kN}$$

P_T = Critical section for cover plate is at (3)-(3) and for main plate at (1)-(1)

$$(P_T)_{xx} = A_g \times \frac{f_u}{1.1}$$

$$= (320 \times 0.16) \times \frac{290}{1.1} = 1349.8 \text{ kN}$$

$$(P_T)_{(1)\text{-}(1)} = A_{net} \times \frac{0.9 f_u}{1.25} = (8 - d) \times t \times \frac{0.9 f_u}{1.25}$$

$$\text{For main plate} = (320 - 28) \times 16 \times 0.9 \times \frac{440}{1.25}$$

$(P_T)_{(3)-(3)}$ For cover plate

$$= (B - 3d) \times t \times \frac{0.9 f_u}{1.25}$$

$$= (320 - 3 \times 28) \times 16 \times 0.9 \times \frac{440}{1.25} = 1196.6 \text{ kN}$$

So, tensile strength of connection = 466.8 kN

1.15 A welded bracket connection is shown in figure. It supports a factored load of 150 kN at a distance of 120 mm from the face of column. Design the fillet weld on two sides. Grade of steel = Fe410, $f_y = 250$ MPa.

[10 marks : 2016]

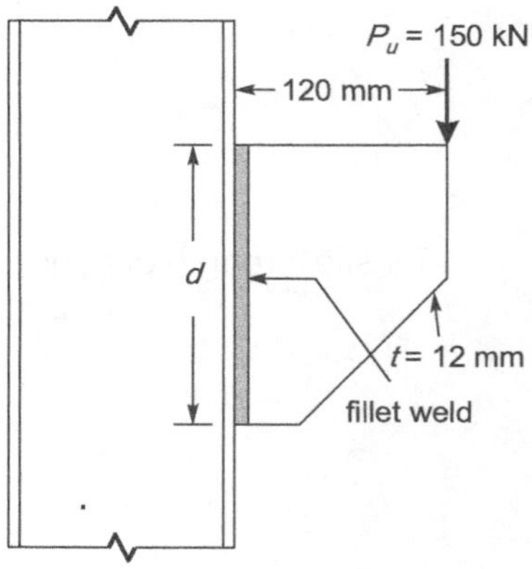

Solution:

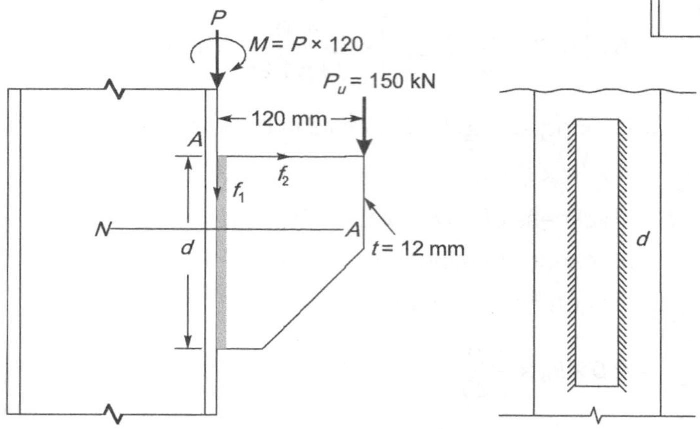

No. of unknown = *l* [*d* and size *S*] so assume *S* and find '*d*'

Using, $$f_e = \sqrt{f_2^2 + 3f_1^2} \leq f_s = \frac{410}{\sqrt{3} \times 1.25}$$

For a thickness plate of 12 mm, minimum size of weld, is 5 mm, so provide S = 5 mm

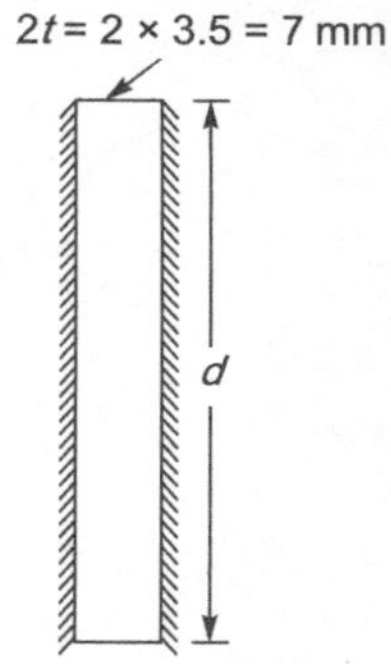

Eff. area of weld (treated as thin area)

1. f_1 = Direct shear stress at $A = \dfrac{P}{\text{weld area}}$

$$= \frac{150 \times 10^3}{7 \times d} = \frac{21.43 \times 10^3}{d} \text{ N/mm}^2$$

2. f_2 = Bending tensile stress at $A = \dfrac{M}{Z}$

$$= \frac{150 \times 10^3 \text{N} \times 120}{\left(\dfrac{7 \times d^2}{6}\right)} = \frac{15.42 \times 10^6}{d^2} \text{ N/mm}^2$$

$$Z = \frac{I_{xx}}{y} = \frac{bd^2}{6} = \frac{7 \times d^2}{6}$$

3. f_e = Equivalent shear stress = $\sqrt{f_2^2 + 3f_1^2}$

$$= \sqrt{\left(\frac{15.4^2 \times 10^6}{d^2}\right)^2 + 3\left(\frac{21.43 \times 10^3}{d}\right)^2} = \frac{410}{\sqrt{3} \times 1.25}$$

d = depth of bracket or length of weld = 320.5 mm ≃ 325 mm

1.16 **A rafter member of a truss consists of double angle ISA, 75 × 75 × 6 welded on the opposite sides of a 10 mm gusset plate. Design the fillet weld (shop) for the member. The factored compressive force due to DL + LL = 240 kN and the factored tensile force due to DL + WL = 200 kN. Assume E 250 grade steel. $C_{xx} = C_{yy}$ = 20.6 mm, γ_{mw} = 1.25.**

[12 marks : 2017]

Solution :

Given, E250, f_u = 410 MPa and f_y = 250 MPa, $C_{xx} = C_{yy}$ = 20.5 mm, y_{mw} (for shop welding) = 1.25

Factored compressive force = 240 kN (DL + LL)

Factored tensile force = 200 kN (DL + WL)

The maximum axial force governs design of fillet weld connection.

Let the length of welds at top and bottom of angle section be L_{w1} and L_{w2} respectively for each angle.

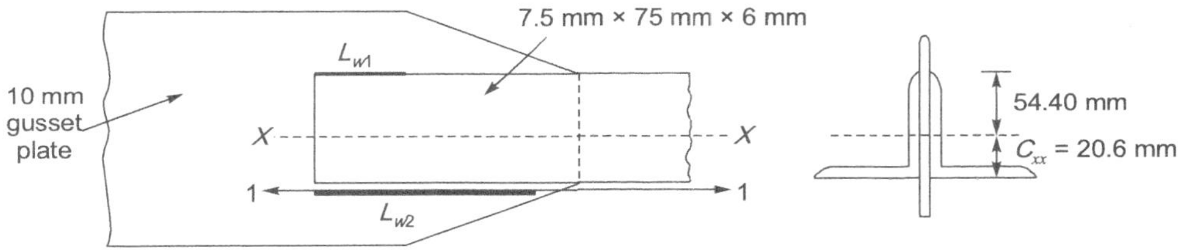

∴ Total length of weld = $2(L_{w1} + L_{w2})$

Minimum size of weld = 3 mm (for 10 mm gusset (thicker member))

Maximum size of weld = 6 – 1.5 = 4.5 mm

Hence adopt side of weld = 4.0 mm

Effective throat thickness = kS = 0.7 × 4.0 = 2.8 mm

$$\text{Design strength of weld} = P_{dw} = L_w t_t \times \frac{f_u}{\sqrt{3} y_{mw}}$$

Design strength of weld per mm length of weld,

$$= 1 \times 2.8 \times \frac{410}{\sqrt{3} \times 1.25} = 530.24 \text{ N/mm}$$

Equating the design strength of the weld to factored compressed force.

$$530.24\,(L_{w1} + L_{w2}) \times 2 = 240 \times 10^3$$

$$\Rightarrow \quad L_{w1} + L_{w2} = 226.313 \text{ mm}$$

Taking moment about 1-1 axis

$$530.24 \times 2L_{w1} \times 75 = 240 \times 10^3 \times 20.60$$

$$L_{w1} = 62.16 \simeq 65 \text{ mm}$$

$$\therefore \quad L_{w2} = 226.313 - 65 = 161.313 \simeq 165 \text{ mm}$$

1.17 A tie member of a truss consisting of an angle section ISA 65 × 65 × 6 is welded to a gusset plate. Design a fillet weld to transmit a load equal to full tensile strength of the plate. Assume shop weld. Take grade of steel E 250 (Fe410). Also sketch the weld length.
Properties of ISA 65 × 65 × 6
A = 744 mm², C_z = 18.1 mm
Thickness of gusset plate is 10 mm.

[12 marks : 2020]

Solution:

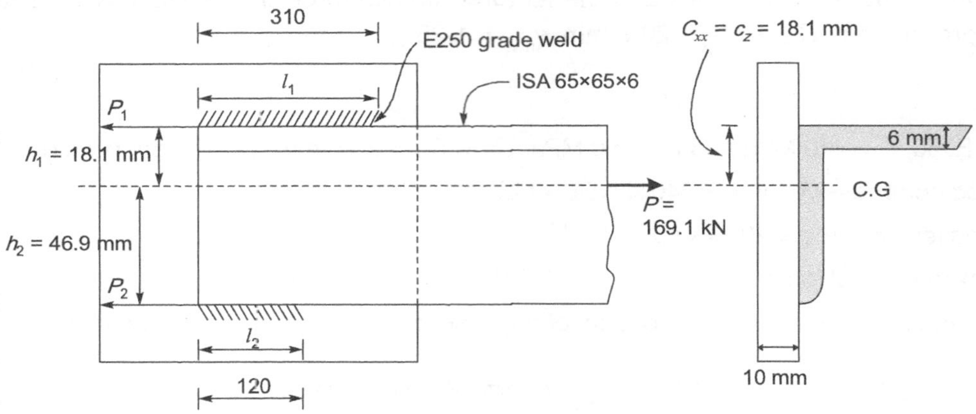

ISA 65 × 65 × 6
A = 744 mm²
C_z = 181 mm

$$\text{Max.size of weld} = \frac{3}{4}\times 6 = 4.5 \text{ mm}$$

Min. size of weld = for a thicker plate of 10 mm
Min. size of weld = 3 mm

So, provide. s = size of weld = 3 mm [Always provide minimum size of weld]

P_T = Factored tensile strength of angle section

$$= A_g \times \frac{f_y}{1.1} = 744\times\frac{250}{1.1} = 169.1 \text{ kN}$$

P_T = 169.1 kN

[Since there are no holes in the angle, no net area cracking]

Analysis : Finding P_1 and P_2

$\Sigma X = 0$

$\Rightarrow \quad -P_1 - P_2 + P = 0$

$\Rightarrow \quad P_1 + P_2 = 169.1 \text{ kN}$ ...(i)

$\Sigma M_{\text{C.G. line}} = 0$

$\Rightarrow \quad -P_1 \times h_1 + P_2 \times h_2 = 0$ {A characteristic of any moment free connection}

$\Rightarrow \quad -P_1 \times 18.1 = -P_2 \times 46.9$

$P_1 = 2.59\, P_2$ ...(ii)

Eq. (ii) in (i)

$\Rightarrow \quad 2.59\, P_2 + P_2 = 169.1$

$\Rightarrow \quad P_1 = 2.59 \times 47.1 = 122 \text{ kN}$

Design : Finding l_1 and l_2

$$P_1 = f_s \times l_1 \times t_t$$

$$\Rightarrow \quad 122 \times 10^3 \text{ N} = \frac{410}{\sqrt{3} \times 1.25} \times l_1 \times 0.7 \times 3$$

$$f_s = \frac{f_u}{\sqrt{3} \times 1.25} \text{ for E 250 grades, } f_u = 410 \text{ MPa}$$

$$t_t = 0.7s = 0.7 \times 3 = 2.1 \text{ mm = throat the thickness}$$

$$l_1 = 306.7 \text{ mm}$$

Provide effective length

$$l_1 = 310 \text{ mm}$$

$$P_2 = f_s \times l_2 \times t_t$$

$$47.1 \times 10^3 \text{ N} = \frac{410}{\sqrt{3} \times 1.25} \times l_2 \times 0.7 \times 3$$

$$l_2 = 118.4 \text{ mm}$$

Provide effective length,

$$l_2 = 120 \text{ mm as shown in figure}$$

Note : Always welds must begin from the end of angle section as shown in figure. (Gaylord and gaylord, design of steel structures)

1.18 A bracket plate is connected to a flange of ISMB 500 as shown in figure. Find the safe load P carried by the joint. M16 bolts of grade 4.6 are provided at a pitch of 50 mm and end distance 30 mm.

Given:

Thickness of bracket plate = 10 mm

Width of flange b_f = 180 mm

Thickness of flange t_f = 17.2 mm

Steel grade of bracket plate Fe410

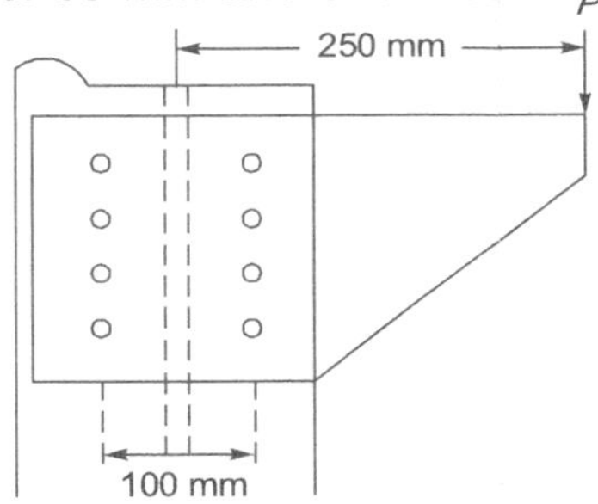

[20 marks : 2020]

Solution:

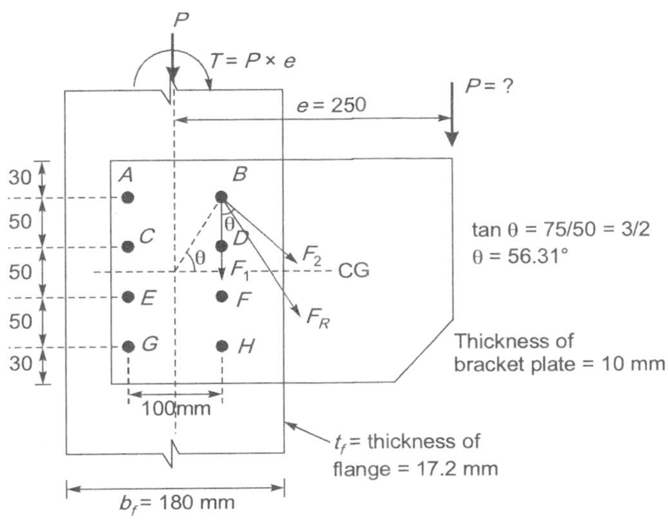

Diameter of bolt, $\phi = 16$ mm

Diameter of bolt hole in the plate,

$$d = \phi + 2 = 18 \text{ mm}$$

Note : The most critically stressed bolt is the bolt for which 'r' is maximum and θ is minimum. In this case, bolts B and H are most critically stressed because, r is maximum and θ is minimum.

Analysis :

1. $$F_1 = \frac{P}{n} = \frac{P}{8} = 1.125\, P$$

2. $$F_2 = \frac{P.e}{\Sigma r^2} \times r_B$$

$$e = 250 \text{ mm}$$

$$r_A = r_B = r_G = r_H = \sqrt{75^2 + 50^2} = 90.14 \text{ mm}$$

$$r_C = r_D = r_E = r_F = \sqrt{25^2 + 50^2} = 55.9 \text{ mm}$$

$$F_2 = \frac{P \times 250}{4[90.14^2 + 55.9^2]} \times 90.14 = 0.5P$$

3. $$F_R = \sqrt{F_1^2 + F_2^2 + 2F_1 F_2 \cos\theta}$$

$$= \sqrt{(0.125P)^2 + (0.5P)^2 + 2 \times 0.125P \times 0.5P \times \cos 56.3°} = 0.579\, P$$

4. $F_R \le R_V$

Bolt value (R_V)

Bolt is in single shear

$$P_s = \text{shear strength of bolt} = \left(0.78 \times \frac{\pi}{4} \times d^2\right) \times f_s$$

$$= 0.78 \times \frac{\pi}{4} \times 16^2 \times \frac{400}{\sqrt{3} \times 1.25} = 28.975 \text{ kN}$$

$$t = \text{t of bracket} = 10 \text{ mm}$$

$$P_b = (d \times t)\, f_b$$

$$k_b = \min \begin{cases} = \dfrac{p}{3d_0} - 0.25 = \dfrac{50}{3 \times 18} - 0.25 = 0.675 \\ = \dfrac{e}{3d_0} = \dfrac{30}{3 \times 18} = 0.555 \\ = \dfrac{f_{ub}}{f_{up}} = \dfrac{400}{410} \\ = 1.0 \end{cases}$$

$$f_b = 2.5 \times k_b \times \frac{f_{up}}{1.25} = 2.5 \times 0.555 \times \frac{410}{1.25}$$

$$P_b = (16 \times 10) \times 2.5 \times 0.555 \times \frac{410}{1.25} \times 10^{-3} = 72.816 \text{ kN}$$

So, Bolt value $R_V = 28.975$ kN

$$0.579\, P \le 28.975 \text{ kN}$$

$\Rightarrow$ $P \le 50.043$ kN

So, P = Max factored load that can be applied

$= 50.04$ kN

$$P_{safe} = \text{safe load} = \frac{P}{1.5} = \frac{50.04}{1.5} = 33.36 \text{ kN}$$

1.19 Determine the maximum load P the joint can carry if bolts of M 20, grade 4.6 are used.

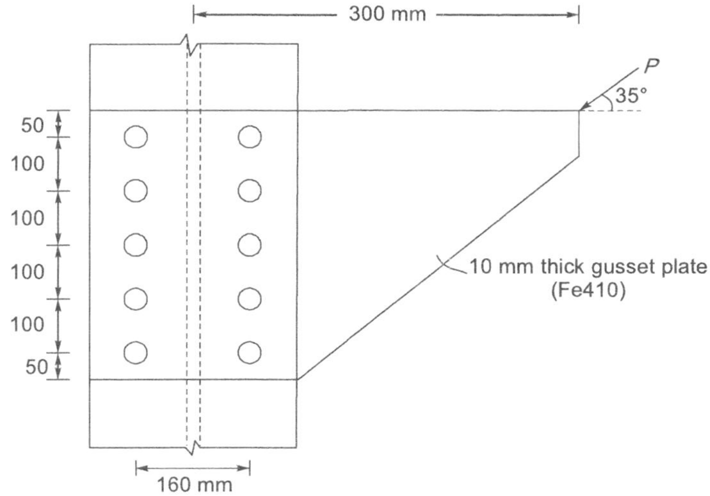

Design shear strength of bolt $V_{nsb} = \frac{f_{ub}}{\sqrt{3}} \cdot \frac{A_{nb}}{\gamma_{mb}}$

Bearing strength of bolt $= \frac{1}{\gamma_{mb}} 2.5\, k_b \cdot d \cdot t \cdot f_u$

Where k_b = Least of $\left[\frac{e}{3d_0}, \left(\frac{p}{3d_0} - 0.25\right), \frac{f_{ub}}{f_u}, 1.0\right]$

[20 Marks : 2021]

Solution:

Given: M20 bolts of grade 4.6

$\Rightarrow$ $d = 20$ mm, $d_o = 22$ mm, $f_{ub} = 400$ N/mm^2; $f_{yb} = 240$ N/mm^2

Fe 410 plate

$\Rightarrow$ $f_y = 250$ N/mm^2

$f_u = 410$ N/mm^2

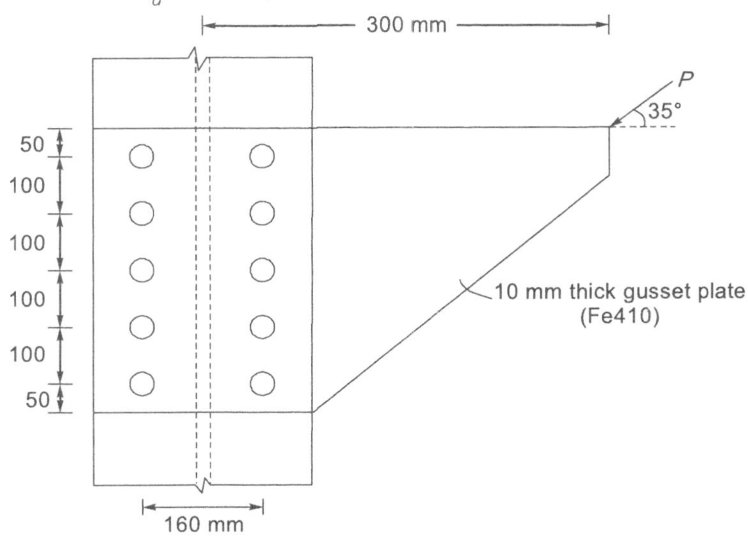

(1) Bolt strength (V_{db})

(a) Design shear capacity of Bolt, V_{dsb},

$$V_{dsb} = \frac{f_{ub}}{\sqrt{3}\gamma_{mb}}(n_s A_{sb} + n_n A_{nb}) \qquad \text{(Here } n_s = 0 \text{ given in question)}$$

$$V_{dsb} = \frac{f_{ub}}{\sqrt{3}\gamma_{mb}} n_n A_{nb}$$

$$= \frac{400}{\sqrt{3}\times 1.25}\times 1\times 0.78\times\frac{\pi}{4}\times 20^2 = 45.27 \text{ kN}$$

(b) Design bearing capacity of Bolt,

$$k_b = \frac{e}{3d_0} = \frac{50}{3(20+2)} = 0.757$$

$$= \frac{p}{3d_0} - 0.25 = \frac{100}{3(20+2)} - 0.25 = 1.265$$

$$= \frac{f_{ub}}{f_u} = \frac{400}{410} = 0.97$$

$$= 1$$

k_b is minimum of all above $k_b = 0.757$

$$V_{dpb} = \frac{2.5\, k_b d t f_u}{\gamma_{mb}} = \frac{2.5\times 0.757\times 20\times 10\times 410}{1.25} = 124.148 \text{ kN}$$

Bolt strength is minimum out of V_{dsb} and V_{dpb}

$\therefore$ $V_{db} = 45.27$ kN

(2) Maximum resultant shear force (F_{Rmax})

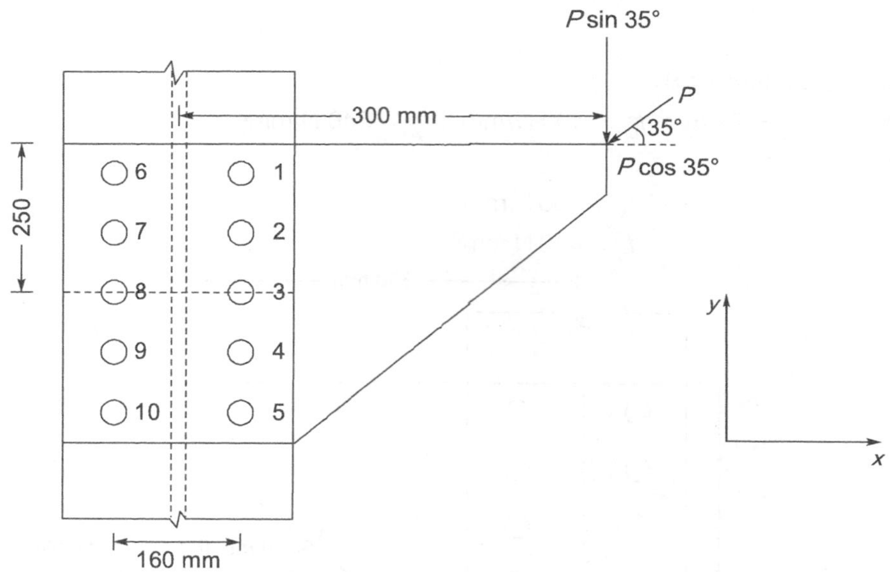

Due to inclined loading, two direct force/loads will be induced in x and y direction. Along with a torsional moment shear force.

Direct shear force:

(a) Direct shear force in x direction.

$$F_{D_1} = \frac{P\cos 35°}{10} = 0.082\text{P kN}$$

(b) Direct shear force in *y* direction.

$$F_{D_2} = \frac{P\sin 35°}{10} = 0.05736 \text{ P kN}$$

Torsional shear force:

Torsional moment on bolted connection.

Torsional moment = P sin 35 × 0.3 – P cos 35° × 0.25 = 0.0327 P.

Net torsional moment is 0.0327 P kNm in anticlockwise direction.

∴ Maximum radial distance (r_{max})

$$r_{max} = \sqrt{(100+100)^2 + 80^2} = 215.40 \text{ mm}$$

Radial distance for bolts,

$$r_1 = r_5 = r_6 = r_{10} = r_{max} = 215.40 \text{ mm}$$

$$r_2 = r_4 = r_7 = r_9 = \sqrt{100^2 + 80^2} = 128.06 \text{ mm}$$

$$r_3 = r_8 = 80 \text{ mm}$$

Now, Maximum resultant shear force will be generated at the bolt on which radial distance from C. G is maximum and inclination between direct shear force and torsional shear force is Minimum.

Maximum torsional shear force,

$$F_2 = \frac{(TM)\times r_{max}}{\sum_{i=1}^{n} r_i^2} = \frac{0.0327\,P\times 10^3 \times 215.40}{4\times 215.40^2 + 4\times 128.06^2 + 2\times 80^2}$$

$$= 0.02668 \text{ P kN}$$

(e) Total force acting along on bolt 6

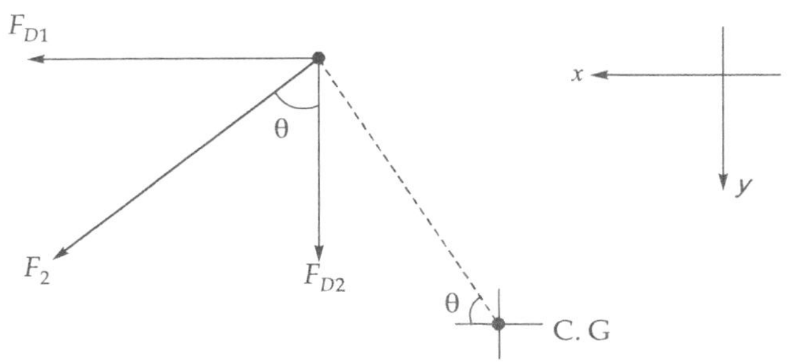

$$\theta = \tan^{-1}\frac{200}{80} = 68.2°$$

$$F_x = F_{D1} + F_2 \sin\theta = 0.082 \text{ P} + 0.02668 \text{ P} \sin 68.2$$

$$= 0.10678 \text{ P kN}$$

$$F_y = F_{D2} + F_2 \cos\theta$$

$$= 0.05736 \text{ P} + 0.02668 \text{ P} \times \cos 68.2°$$

$$= 0.0673 \text{ P kN}$$

$$F_R = \sqrt{F_x^2 + F_y^2} = \sqrt{(0.10678P)^2 + (0.0673P)^2}$$

$$= 0.12622 \text{ P kN}$$

3. For safe design, $F_R \ngtr$ Bolt strength (V_{db})

$$\Rightarrow \quad 0.12622\,P_{max} = 45.27 \text{ kN}$$

$$P_{max} = 358.662 \text{ kN}$$

Safe load, $$P_{safe} = \frac{P_{max}}{\gamma_f} = \frac{358.662}{1.5} = 239.108 \text{ kN}$$

1.20 **The rafter member of a truss consists of two angles ISA 75 × 75 × 8 placed (back-to-back) both sides of the gusset of thickness 10 mm. It carries factored axial compressive force of 200 kN. Determine the number of 16 mm diameter, 4.6 grade ordinary bolts for the joint. Assume E250 grade of steel and the cross-sectional area in the threaded part for 16 mm diameter bolt is 157 mm^2. Use K_b = 0.49, γ_{mb} = 1.25. Use limit state method of design.**

[12 marks : 2023]

Solution:

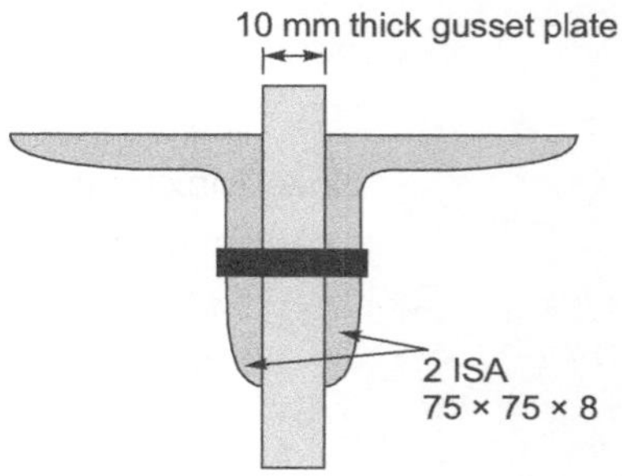

Factored compressive force, P_u = 200 kN,

Diameter of bolts, d = 16 mm

Net area of bolts, A_{nb} = 157 mm^2

$k_b = 0.49$; $f_{ub} = 400$ N/mm^2 ; $f_{up} = 410$ N/mm^2

Strength of bolt in shearing (double shear):

$$V_{dsb} = 2 \times A_{nb} \times \frac{f_{ub}}{\sqrt{3} \times \gamma_{mb}}$$

$$= 2 \times 157 \times \frac{400}{\sqrt{3} \times 1.25} = 58012.155 \text{ N}$$

Strength of bolt in bearing:

$$t_{min} = \begin{Bmatrix} 10 \text{ mm} \\ 8+8=16 \text{ mm} \end{Bmatrix} \text{minimum}$$

$$= 10 \text{ mm}$$

$$\therefore \quad V_{dpb} = 2.5 k_b d t_{min} \times \frac{f_{up}}{\gamma_{mb}}$$

$$= 2.5 \times 0.49 \times 16 \times 10 \times \frac{410}{1.25} = 64288 \text{ N}$$

So, strength of each bolt = minimum {58012.155, 64288}

= 58012.155 N

$$\therefore \quad \text{No. of bolts required} = \frac{200 \times 10^3}{58012.155} = 3.44 \approx 4 \text{ bolts}$$

1.21 **A single angle ISA 100 × 100 × 10 is connected to a gusset plate of thickness 10 mm by weld along two parallel edges. The size of weld (fillet) is 6 mm. The member is subjected to an axial compressive load of 150 kN (factored). Find the weld length along two parallel edges. Assume E250 grade of steel and shop welded. For ISA 100 × 100 × 10, $c_y = c_z$ = 27.6 mm. Use limit state method.**

[12 marks : 2023]

Solution:

Given

Factored load, P_u = 150 kN

Size of weld, s = 6 mm,

$f_y = 250$ MPa, $f_u = 410$ MPa

For ISA 100 × 100 × 10, $C_y = C_z = 27.6$ mm

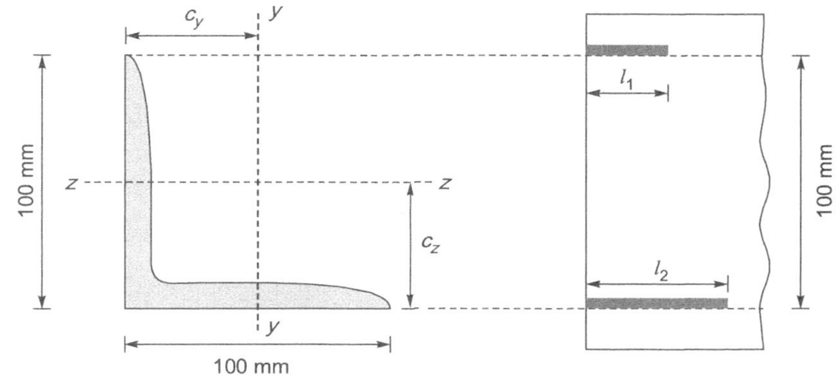

Let, force resisted by upper portion $l_1 = P_1$

$$\therefore \quad P_1 = \frac{150 \times C_z}{100} = \frac{150 \times 27.6}{100}$$

$$= 41.4 \text{ kN}$$

Force resisted by lower portion, $l_2 = P_2$

$$\therefore \quad P_2 = \frac{150(100 - C_z)}{100} = \frac{150(100 - 27.6)}{100} = 108.6 \text{ kN}$$

∴ Throat thickness, $t = 0.7\ s = 0.7 \times 6 = 4.2$ mm

Length of weld l_1 had to resist a force of P_1

$$\therefore \quad P_1 = l_1 t \frac{f_u}{\sqrt{3}\,\gamma_{mw}}$$

$$\Rightarrow \quad 41.4 \times 10^3 = \frac{l_1 \times 4.2 \times 410}{\sqrt{3} \times 1.25} \qquad (\because \text{For shop weld, } \gamma_{mw} = 1.25)$$

$$\therefore \quad l_1 = 52.05 \text{ mm} \simeq 53 \text{ mm}$$

Similarly, length of weld l_2 has to resist a force of P_2

$$\therefore \quad P_2 = \frac{l_2\, t f_u}{\sqrt{3}\,\gamma_{mw}}$$

$$108.6 \times 10^3 = \frac{l_2 \times 4.2 \times 410}{\sqrt{3} \times 1.25}$$

$$\therefore \quad l_2 = 136.54 \text{ mm} \simeq 137 \text{ mm}$$

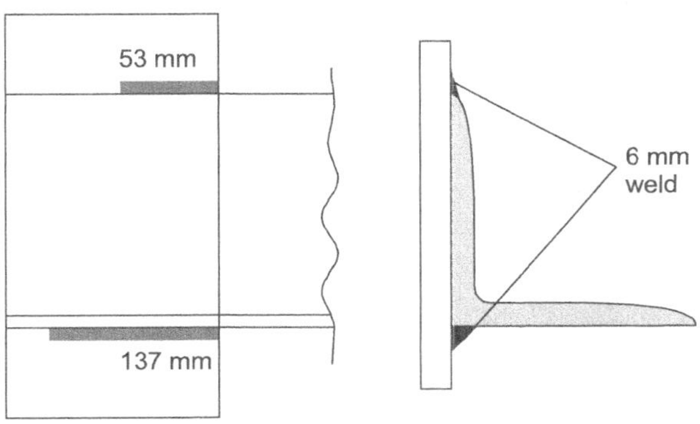

1.22 A cantilever beam (ISMB 500) in connected to the flange of the column (ISHB 450) by fillet weld of size 5 mm. The beam is subjected to a vertical load P and a horizontal load $\frac{P}{2}$ at a distance of 200 mm from the flange of the column as shown in the figure below. Find the factored P that can be applied for the joint. Assume E250 grade of steel, site weld. Given, $f_e = \sqrt{f_a^2 + 3q^2}$. Use limit state method. Assume that the beam section is safe :

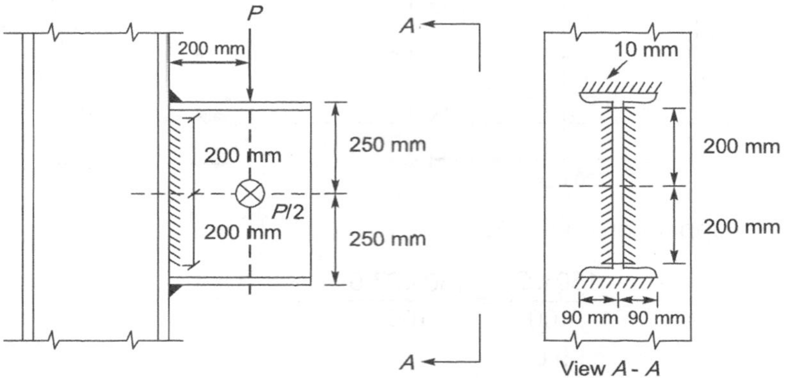

[20 marks : 2023]

Solution:

Concept :

Welding will be subjected to direct shear force in y and z direction and bending moment about both z and y axis.

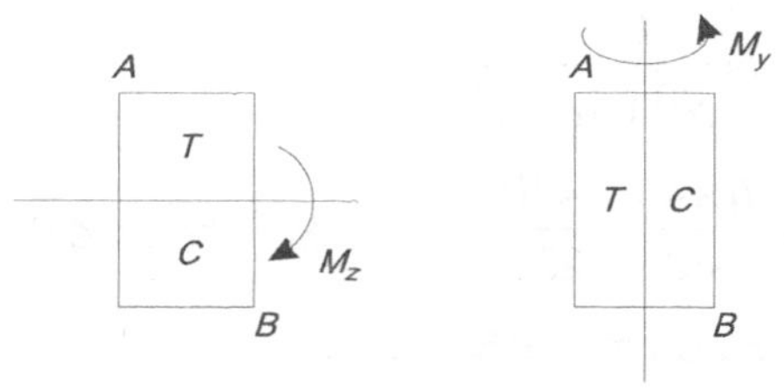

Critical points will be A and B as the effect of stress will be added

Welded Area

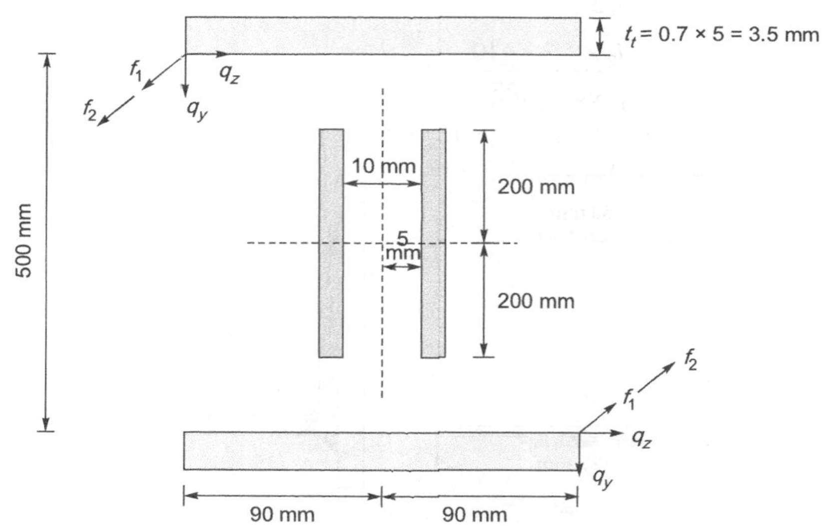

Assuming welding as thin line element t_t^2 and t_t^3 will be neglected.

Calculation of stress:

1. Direct shear stress in *y*-direction

$$q_y = \frac{P}{A_w} = \frac{P \times 10^3 \text{ N}}{l_w \times t_t}$$

$$= \frac{P \times 10^3}{4060} = 0.246 \text{ P}$$

2. Direct shear stress in *z*-direction

$$q_z = \frac{P/2}{A_w}$$

$$\Rightarrow \quad = \frac{P}{2} \times \frac{10^3}{l_w t_t} = 0.123 \text{ P}$$

3. Maximum flexural stress at point *A* or *B* due to M_{zz}

$$f_1 = \frac{M_{zz}}{I_{zz}} \times y_{max}$$

where $$I_{zz} = \frac{3.5 \times 400^3}{12} \times 2 + \left(180 \times 3.5 \times 250^2\right) \times 2$$

$$= 116.083 \times 10^6 \text{ mm}^4$$

$$f_1 = \frac{P \times 10^3 \times 200 \times 250}{116.083 \times 10^6} = 0.4307 \text{ P}$$

4. Maximum flexural stress of point *A* or *B* due to M_{yy}

$$f_2 = \frac{M_{yy}}{I_{yy}} \times z_{max}$$

where $$I_{yy} = \left(\frac{3.5 \times 180^3}{12} \times 2\right) + \left[400 \times 3.5 \times 5^2\right] \times 2$$

$$= 3.472 \times 10^6 \text{ mm}^4$$

$$f_2 = \frac{\frac{P}{2} \times 200 \times 10^3 \times 90}{3.472 \times 10^6} = 2.592 \text{ P}$$

Resultant direct shear stress at point *A* or *B*

$$q = \sqrt{q_y^2 + q_z^2} = \sqrt{(0.246P)^2 + (0.123P)^2}$$

$$= 0.275 \text{ P}$$

Resultant flexural stress at point *A* or *B*

$$f = f_1 + f_2 = 0.4307 \text{ P} + 2.592 \text{ P}$$

$$= 3.0227 \text{ P}$$

Equivalent stress, $f_e = \sqrt{3q^2 + f^2}$

$$f_e = \sqrt{3(0.275P)^2 + (3.0227P)^2} = 3.06\text{ P}$$

For safety $f_e \le \dfrac{f_u}{\sqrt{3}\,\gamma_{M_u}}$, $\quad 3.06P \le \dfrac{410}{\sqrt{3}\times 1.50}$, $\quad P \le 51.57$ kN

So, required P value = 51.57 kN

2. Tension Member

2.1 An angle section ISA 75 × 75 × 10 mm is used as a tension member and is connected to the gusset plates at the joint with 16 mm diameter rivets in single line on both the legs each at a pitch of 50 mm. The rivets on one leg are staggered by 25 mm relative to the rivets on another leg. The rivet line is at a distance of 35 mm centrally from the toe edge of each leg.

(i) Find the net area of the tension member after making the sketch of it at the joint.

(ii) Find the maximum tensile force in the member if the permissible tension stress $\sigma_{st} = 150$ MPa

[6 + 4 marks : 2004]

Solution:

(i) The sketch of the joint may be shown as given below :

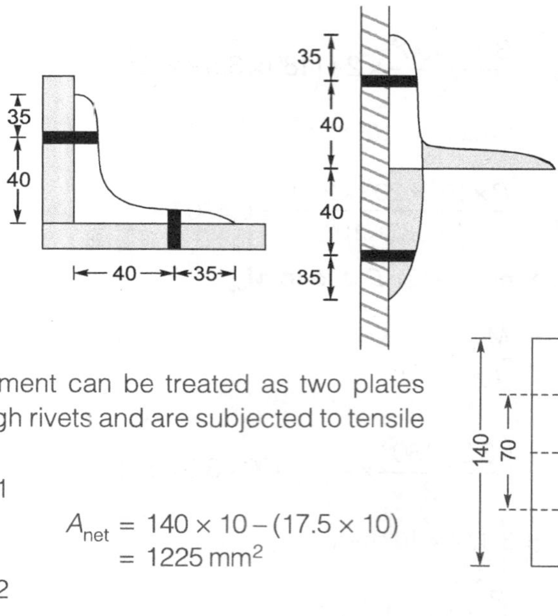

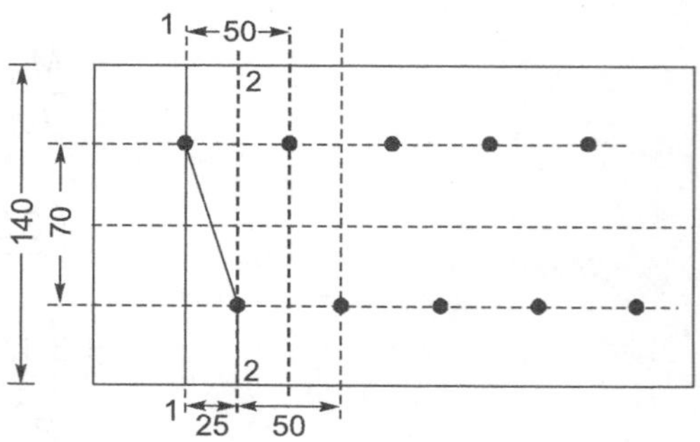

Now the arrangement can be treated as two plates connecting through rivets and are subjected to tensile loads.

A_{net} for section 1-1

$$A_{net} = 140 \times 10 - (17.5 \times 10) = 1225\text{ mm}^2$$

A_{net} for section 1-2

$$A_{net} = 140 \times 10 - (2 \times 17.5 \times 10) + \frac{(25)^2}{4\times 70}\times 10 = 1072.32\text{ mm}^2$$

Hence minimum of the two areas will be the **net area.**

$$\therefore \quad A_{net} = 1072.32\text{ mm}^2$$

(ii) The **maximum tensile force** carried by the tension member may be given by

$$P_{max} = A_{net} \times \text{permissible stress in tension}$$

$$= A_{net} \times \sigma_{st} = \frac{1072.32 \times 150}{10^3} = 160.85\text{ kN}$$

2.2 An I-section with two cover plates is used as tension member and is subjected to a load of 120 t over a length of 16 m. If the allowable tensile stress is 15000 kg/cm^2, is the member design safe? If so, find the factor of safety. Use ISWB 300@ 48.1 kg/m, A = 61.33 cm^2, I_{xx} = 9821.6 cm^4, I_{yy} = 990.1 cm^4. There are two rows of fasteners in each flange, thus making atleast two holes in one section. The fastener diameter is 22 mm, cover plate size 260 mm × 8 mm, flange thickness of I-section = 10 mm. The center of fastener is located at 60 mm from center line of the I-section.

[10 marks : 2006]

Solution:

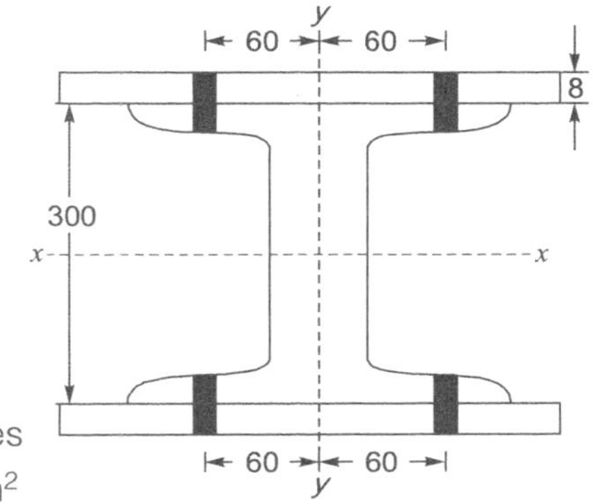

Given

$$P = 120\ t = 1200 \text{ kN}$$

Permissible tensile stress, σ_{at} = 1500 kg/cm^2 = 150 MPa

$\therefore$ Net area required $= \dfrac{P}{\sigma_{at}} = \dfrac{1200 \times 10^3 \text{N}}{150 \text{ MPa}}$

$= 8000 \text{ mm}^2$

Net area of I-section provided = Gross area – area of fastener holes

$= 6133 - 4 \times (23.5) \times 10 = 5193 \text{ mm}^2$

Net area provided by plates $= [260 \times 8 - 2 \times 23.5 \times 8] \times 2 = 3408 \text{ mm}^2$

Total net area provided $= 5193 + 3408 = 8601 \text{ mm}^2 > 8000 \text{ mm}^2$ (safe)

Actual load carrying capacity $= (A_{net})_{provided} \times \sigma_{at} = \dfrac{8601 \times 150}{10^3} = 1290.15 \text{ kN}$

$\therefore$ Factor of safety $= \dfrac{\text{Actual load carrying capacity}}{\text{Load applied}} = \dfrac{1290.15}{1200} = 1.075$

2.3 **Design a single equal-angle tension member. It is 4 m long and subjected to a tensile load of 25 t. It is connected to a gusseted plate through one leg only. Assume rivet diameter of 18 mm.**

σ_t = 1500 kg/cm^2

Check for slenderness ratio. Given, r_{min} = 1.94 cm

[15 marks : 2010]

Solution:

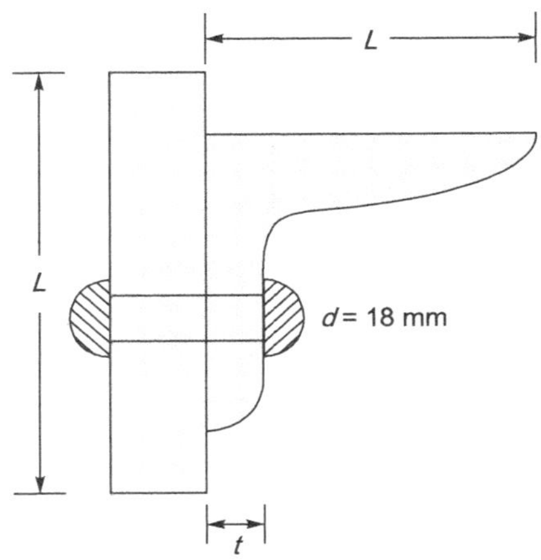

Rivet diameter, d = 18 mm

Hole diameter, d' = 18 + 1.5 = 19.5 mm

Permissible tensile stress, σ_t = 1500 kg/cm^2

Minimum radius of gyration, r_{min} = 1.94 cm

Slenderness ratio $= \dfrac{L_{eff}}{r_{min}} = \dfrac{400}{1.94}$

$= 206.185 < 350$

Hence for slenderness ratio, the tension member is safe.

Area required $= \dfrac{P}{\sigma_{at}} = \dfrac{25 \times 10^3}{1500}$

$= 16.67 \text{ cm}^2 = 1667 \text{ mm}^2$

Area of connected leg, $A_1 = \left(L - \dfrac{t}{2} - d\right) \times t$

Assuming thickness of angle section as 8 mm

$$A_1 = (L - 4 - 19.5) \times 8 = (8L - 188)$$

and

$$A_2 = (L - 4) \times 8 = (8L - 32)$$

$$\therefore \quad k = \frac{3A_1}{3A_1 + A_2} = \frac{3(8L - 188)}{3(8L - 188) + (8L - 32)} = \frac{24L - 564}{24L - 564 + 8L - 32}$$

$$\therefore \quad k = \frac{24L - 564}{32L - 596}$$

$$\therefore \quad A_{\text{net}} = (A_1 + kA_2) = (8L - 188) + \left(\frac{24L - 564}{32L - 596}\right) \times (8L - 32)$$

$$1667 = \frac{(8L - 188)(32L - 596) + (24L - 564)(8L - 32)}{(32L - 596)}$$

$$448L^2 - 16064L + 130096 = 1667(32L - 596)$$

$$\Rightarrow \quad 448L^2 - 69408L + 1123628 = 0$$

$$L = 136.56 \text{ mm and } 18.36 \text{ mm}$$

Hence take $L = 140$ mm

$\therefore$ The angle is 140 × 140 × 8 mm.

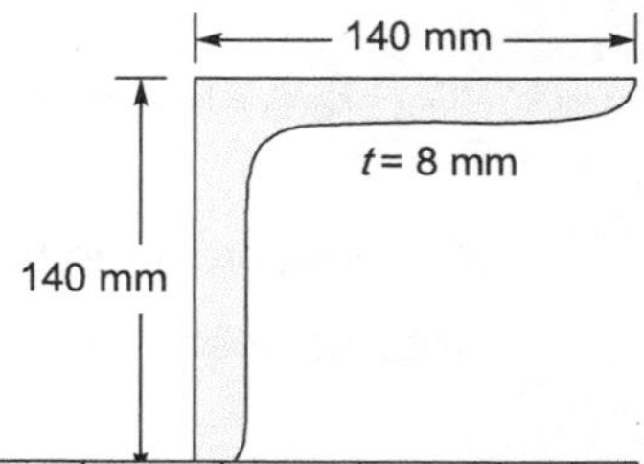

2.4 Design an unequal angle section to serve as a tie member of 1.6 m length in a roof truss. It has to carry an axial load of 118 kN. The gusset plate is connected to the longer leg of the angle. Also design the fillet weld.

Permissible stress in weld = 108 N/mm²

Permissible stress in axial tension = 150 N/mm²

Thickness (t), sectional area (A), weight/m run (w), distance of centres of gravity along x-axis and y-axis. (c_x and c_y), maximum and minimum radius of gyration r_{max} and r_{min} are given for some angles for design.

[14 marks : 2011]

Angle	t	A	w	c_x	c_y	r_{max}	r_{min}
80 × 50	6	746	5.9	26.4	11.6	26.9	10.7
	8	978	7.7	27.3	12.4	26.6	10.6
	10	1202	9.4	28.1	13.2	26.3	10.6
90 × 60	6	865	6.8	28.7	13.9	30.7	12.8
	8	1137	8.9	29.6	14.8	30.4	12.7
	10	1401	11.0	30.4	15.5	30.1	12.7

Solution:

$$A_{net\,\text{required}} = \frac{118 \times 1000}{150} = 786.67 \text{ mm}^2$$

Choosing an angle section having a sectional area about 10% more than the A_{net} required i.e.

$$786.67 \times 1.1 = 865.34 \text{ mm}^2.$$

Using ISA 90 × 60 × 6 mm thick having a sectional area of 865 mm² with the longer leg welded to gusset plate.

$$A_{net} = A_1 + A_2 \times k_1$$

where

$$A_1 = \left(90 - \frac{6}{2}\right) \times 6 = 522 \text{ mm}^2$$

$$A_2 = \left(60 - \frac{6}{2}\right) \times 6 = 342 \text{ mm}^2$$

$$k_1 = \frac{3 \times 522}{3 \times 522 + 342} = 0.82$$

$$\therefore \quad A_{net} = 522 + 0.82 \times 342$$
$$= 802.44 \text{ mm}^2 > 786.67 \text{ mm}^2 (= A_{net\ reqd.})$$

Check for slenderness ratio

Effective length, $l = 160$ cm

Least radius of gyration, $r_{min} = 1.28$ cm

Slenderness ratio, $\frac{l}{r_{min}} = \frac{160}{1.28} = 125 < 350$

Hence, the slenderness ratio is within the permissible limit.

Therefore, use ISA 90 × 60 × 6 mm thick.

$$\text{Size of weld} = \left(\frac{3}{4}\right) \times 6 = 4.5 \text{ mm}$$

$$\text{Permissible stress in weld} = 108 \text{ MPa}$$

$$\text{Force in weld per millimetre length} = \frac{108}{1000} \times 0.7 \times 4.5 = 0.34 \text{ kN}$$

$$\text{Effective length of weld required} = \frac{118}{0.34} = 347.06 \text{ mm (say 350 mm)}$$

Distributing weld in such a way that the c.g. of the weld coincides with that of the angle section from figure.

$$x + y + 90 = 350 \text{ mm} \quad \text{...(i)}$$

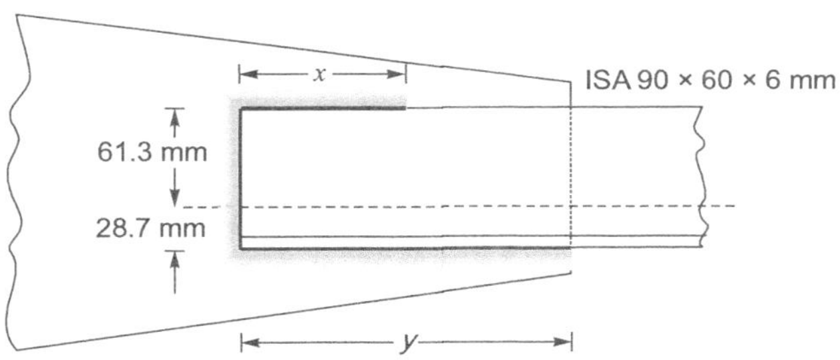

Taking moment of forces about the top face

$$0.34 \times y \times 90 + 0.34 \times 90 \times \frac{90}{2} = 118 \times 61.3$$

or $y = 191.39$ mm = 192 mm (say)

Substituting in equation (i) $x = 68$ mm

2.5 **What is the principle of design of a splice in a steel member subjected to an axial tensile force? Explain with the help of a neat sketch?**

[5 marks : 2014]

Solution:

A tension member is spliced when the length of section available is less than that of tension member required. Tension splices are also used when size of section change over different segments. The force is then carried through these covers across the joint and is transferred to the other portion of the member through the connection.

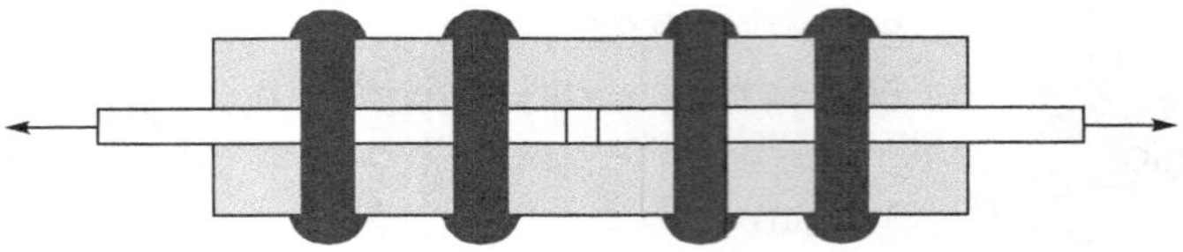

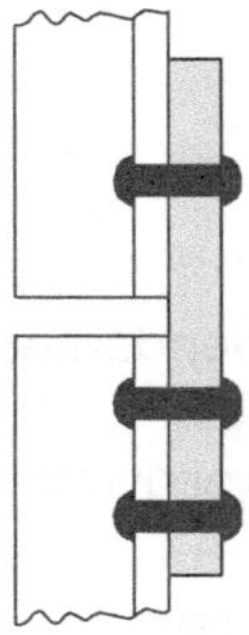

Design: Splices connection for axial tensile force is usually the riveted or bolted connections. As per IS specification, the splice cover and its connection should be designed to develop the net tensile strength of the main member. When member of different thickness are spliced, packing is required to fill the gap.

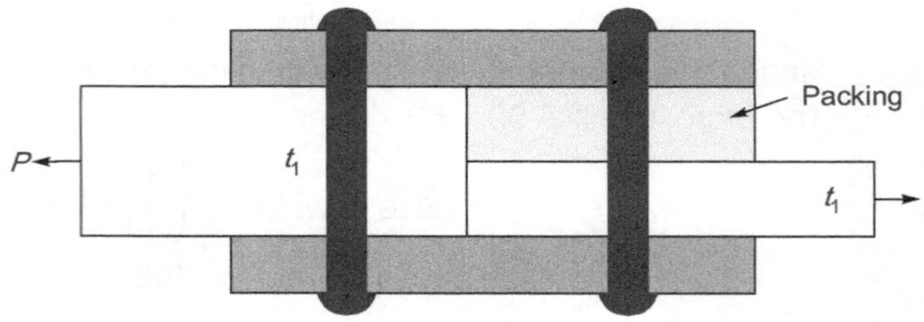

If the difference of thickness between two plates being joined ($t_1 - t_2$) is greater than 6 mm, then additional rivets shall be provided on packing extension. The no. of additional rivets shall be 2.5% of actual no. of rivets obtained from normal calculation per 2 mm thickness of packing.

2.6 **What is the principle of design of splice in a steel member subjected to an axial force? Explain with the help of neat sketches.**

[5 marks : 2015]

Solution:

Splice design: Splice is provided where a joint is coming. Whenever a splice occurs, the force is transferred from member to splice through connection. This force is again transferred to other connecting member through connection.

Depending on the nature of axial force, splice can either be tension splice or the compression splice.

Generally the exact load carried by a splice is uncertain. Thus tension splice is designed for a design load or 0.3 times the member design capacity, whichever is more. It is always beneficent to design the splice for the strength of the member.

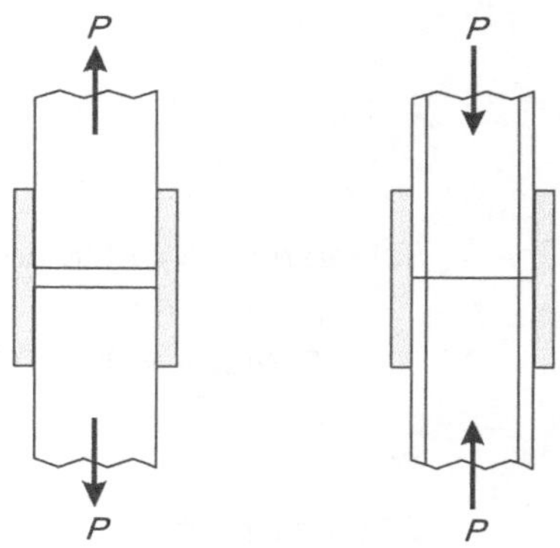

2.7 **If a roof truss, a diagonal consists of an ISA 60 mm × 60 mm × 8 mm (ISA 6060 @ 0.07 kN/m) and it is connected to gusset plate by one leg only by 18 mm diameter rivets in one chain line along the length of the member. Determine tensile strength of the member, if yield stress for steel is 250 MPa.**

[12 marks : 2019]

Solution:

Diameter of rivet hole = 18 + 1.5 = 19.5 mm

Effective area, $A_{eff} = A_1 + kA_2$

A_1 = Net area of connected leg

A_2 = Area of outstanding leg

If angle connected by one leg only,

$$k = \frac{3A_1}{3A_1 + A_2} \text{ or } \beta = \text{shear lag factor} = 0.7$$

$$A_1 = \left(60 - \frac{8}{2} - 19.5\right) \times 8 = 292 \text{ mm}^2$$

$$A_2 = \left(60 - \frac{8}{2}\right) \times 8 = 448 \text{ mm}^2$$

$$k = \frac{3 \times 292}{3 \times 292 + 448} = 0.662 \text{ or } \beta = 0.7$$

$$A_{eff} = 292 + 0.662 \times 448 = 588.576 \text{ mm}^2$$

or $A_{eff} = 292 + 0.7 \times 448 = 605.6 \text{ mm}^2$

Allowable tensile stress, $f_a = 0.6 \times f_y \times 150$ MPa

Tensile strength = $150 \times 588.578 \times 10^{-3} = 88.29$ kN

or Tensile strength = $150 \times 605.6 \times 10^{-3} = 90.84$ kN

2.8 Determine the tensile strength of an ISMC 175 when it is connected to gusset plates through the two flanges by two rows of 16 mm bolts with a connection length 200 mm as shown in the figure. Use grade Fe410 steel (fy = 250 MPa). Also given $\gamma_{m0} = 1.1$ and $\gamma_{m1} = 1.25$.

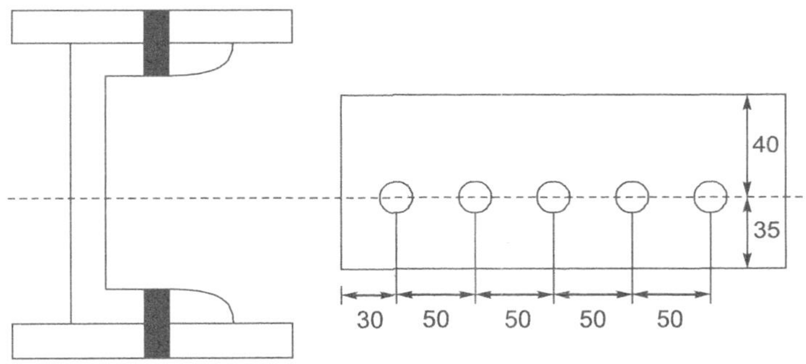

Properties of ISMC 175:
$A = 2490$ mm^2, h = 175 mm, b_f = 75 mm, t_f = 10.2 mm and t_w = 6 mm.

[12 marks : 2022]

Solution:

Gross strength of channel section,

$$T_{dg} = \frac{A_g f_y}{\gamma_{m0}} = \frac{2490 \times 250}{1.1} \text{ N} = 565.91 \text{ kN}$$

Net strength of channel section,
Here the unconnected part is web

$$A_{go} = (175 - 2 \times 10.2) \times 6 = 927.6 \text{ mm}^2$$

Connected part is flange $A_{nc} = \left(75 - \frac{6}{2} - 18\right) \times 10.2 \times 2 = 1101.6 \text{ mm}^2$

$$T_{dn} = \frac{0.9 A_{nc} f_u}{\gamma_{m1}} + \beta \frac{A_{go} f_y}{\gamma_{m0}}$$

$$\beta = 1.4 - 0.076\left(\frac{w}{t_w}\right)\left(\frac{b_s}{L_c}\right)\left(\frac{f_y}{f_u}\right)$$

$$0.7 \le \beta \le 0.9\frac{f_u}{f_y}\frac{\gamma_{m0}}{\gamma_{m1}}$$

For the given section, $w = \frac{175}{2}$ mm = 87.5 mm (shear lag is considered in web)

Shear lag distance, $b_s = w + g - t$

$$= \frac{175}{2} + 40 - \left(\frac{10.2+6}{2}\right) = 119.4 \text{ mm}$$

$$\beta = 1.4 - 0.076 \times \left(\frac{87.5}{6}\right)\left(\frac{119.4}{200}\right)\left(\frac{250}{410}\right)$$

$\Rightarrow$ $\beta = 0.9965$

$$0.7 < \beta = (0.9965) < 0.9 \times \frac{410}{250} \times \frac{1.1}{1.25}$$

$$0.7 < 0.9965 < 1.299$$

$$T_{dn} = \left(\frac{0.9 \times 410 \times 1101.6}{1.25} + \frac{0.996 \times 250 \times 927.6}{1.1}\right) \times 10^{-3}$$

$$= 325.192 \text{ kN} + 210.08 \text{ kN}$$

$$= 535.27 \text{ kN}$$

Block shear strength of connection,

T_{db} = Minimum of T_{db1} and T_{db2}

$$A_{vg} = 230 \times 10.2 = 2346 \text{ mm}^2$$

$$A_{tn} = \left(35 - \frac{18}{2}\right) \times 10.2 = 265.2 \text{ mm}^2$$

$$A_{vn} = \left(230 - 4 \times 18 - \frac{18}{2}\right) \times 10.2 = 1519.8 \text{ mm}^2$$

$$A_{tg} = 35 \times 10.2 = 357 \text{ mm}^2$$

$$T_{db1} = \left[\frac{A_{vg} f_y}{\sqrt{3}\gamma_{m0}} + \frac{0.9 A_{tn} f_u}{\gamma_{m1}}\right] \times 2$$

$\Rightarrow$
$$T_{db1} = \left(\frac{2346 \times 250}{\sqrt{3} \times 1.1} + \frac{0.9 \times 265.2 \times 410}{1.25}\right) \times 2 \text{ N}$$

$$= 772.24 \text{ kN}$$

$$T_{db2} = \frac{0.9 A_{vn} f_u}{\sqrt{3}\gamma_{m1}} + \frac{A_{tg} f_y}{\gamma_{m0}}$$

$$= \left(\frac{0.9 \times 1519.8 \times 410}{\sqrt{3} \times 1.25} + \frac{357 \times 250}{1.1}\right) \times 2 \text{ N}$$

$$= 680.32 \text{ kN}$$

$$T_{db} = \text{minimum } [772.24, 680.32]$$

$$= 680.32 \text{ kN}$$

$\therefore$ Tensile strength of given channel section is 535.27 kN.

3. Compression Member

3.1 Design a suitable built up section using two channels for a steel column to carry an axial load of 1200 kN. The effective length of the column is 5.8 m. Provide suitable single lacing with 45° angles of inclination. Adopt 22 mm diameter rivets. For channel ISMC 350

Area of cross section = 5366 mm^2

$I_{xx} = 100.08 \times 10^6$ mm^4

$I_{yy} = 4.306 \times 10^6$ mm^4

$C_x = 24.4$ mm

Permissible stress in axial compression for steel with $\sigma_y = 260$ N/mm^2

l/r	40	50	60	100	110	120	130	140	150
Permissible stress N/mm^2	145	136	126	82	73	64	57	51	46

[28 marks : 2001]

Solution :

Given that, $P = 1200$ kN, Area of section = 2 × 5366 = 10732 mm^2

$$\therefore \quad \sigma_{ac} \text{ provided} = \frac{1200 \times 10^3}{10732} = 111.82 \text{ N/mm}^2$$

$$\lambda \text{ (for } \sigma_{ac} = 111.82 \text{ N/mm}^2) = 60 + \left(\frac{100-60}{126-82}\right) \times (126 - 111.82) = 72.9$$

But
$$\lambda \le \frac{l_{eff}}{r_{min}}$$

$$\Rightarrow \quad r_{min} \ge \frac{l_{eff}}{\lambda} \ge \frac{5.8 \times 10^3}{72.9} \ge 79.56 \text{ mm}$$

$$\therefore \quad I_{min} \ge \text{Area of section} \times (r_{min})^2 = 2 \times 5366 \times (79.56)^2 = 67.93 \times 10^6 \text{ mm}^4$$

Let the spacing between the two channel section be S.

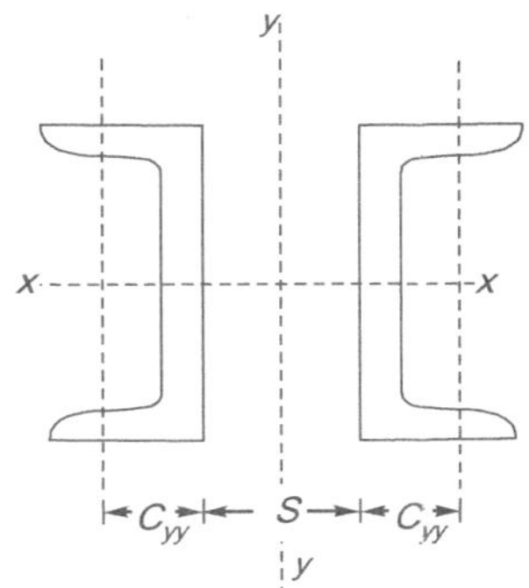

For most efficient use of material I_{xx} of combination = I_{yy} of combination

In this case the compression member will carry maximum force

$$I_{xx} \text{ of combination} = 2 \times I_{xx} \text{ of single section}$$
$$= 2 \times 100.08 \times 10^6 = 200.16 \times 10^6 \text{ mm}^4$$

$$I_{yy} \text{ of combination} = 2 \times I_{yy} \text{ of one section} + 2 \times A \times \left(24.4 + \frac{S}{2}\right)^2$$

But for the loading of 1200 kN, the design will be economical when

$$I_{min} = I_{yy} \text{ of combination}$$

$$\therefore \quad 2 \times 4.306 \times 10^6 + 2 \times 5366 \times \left(24.4 + \frac{S}{2}\right)^2 = 67.93 \times 10^6$$

$$\Rightarrow \quad 24.4 + \frac{S}{2} = 74.35$$

$$\Rightarrow \quad S = 99.9 \text{ mm}$$

$$\Rightarrow \quad S = 100 \text{ mm}$$

DESIGN OF LACING

Assuming the rivet are connected at the CG of the channel section Given that $\theta = 45°$, then

$$\frac{C}{2} = \frac{100+24.4+24.4}{\tan 45°}$$

$$\Rightarrow \qquad C = 297.6 \text{ mm}$$

We have to adopt 22 mm diameter rivets, which will give the width of lacing bar as 65 mm.

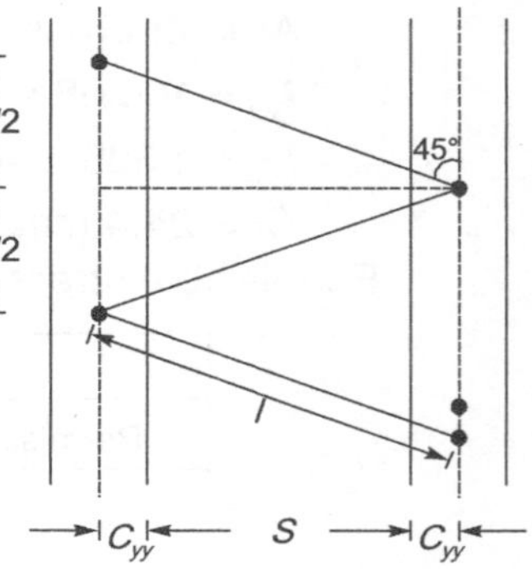

$$\therefore \qquad \sin 45° = \frac{100+24.4+24.4}{l}$$

$$\Rightarrow \qquad l = 210.43 \text{ mm}$$

Thus thickness of lacing bar, $t = \frac{l}{40} = \frac{210.43}{40} = 5.26 \text{ mm}$

Adopting thickness of lacing bar, $t = 8$ mm

$$\text{Slenderness ratio of lacing bar} = \frac{l_{eff} \text{ of lacing bar}}{t/\sqrt{12}} = \frac{210.43}{8/\sqrt{12}}$$

$$= 91.12 < 145 \text{ (Hence ok)}$$

CHECK FOR TENSION

$$F = \frac{V}{2\sin\theta}$$

where $V = \frac{2.5}{100} \times \text{applied load}$

$$= \frac{2.5}{100} \times 1200 = 30 \text{ kN}$$

$$\therefore \qquad F = \frac{30}{2\sin 45°} = 21.21 \text{ kN}$$

CHECK FOR COMPRESSION

$$F = 21.21 \text{ kN}$$

$$\sigma_{ac} \text{ required} = \frac{F}{\text{Area of lacing bar}} = \frac{21.21\times10^3}{65\times8}$$

$$\Rightarrow \qquad \sigma_{ac} \text{ required} = 40.79 \text{ N/mm}^2$$

Now, λ for lacing bar $= 91.12$

$$\therefore \quad \sigma_{ac} \text{ provided (for } \lambda = 91.12) = 126 - \left(\frac{126-82}{100-60}\right)\times(91.12-60)$$

$$= 91.77 \text{ N/mm}^2 > 40.79 \text{ N/mm}^2$$

This σ_{ac} provided is more than σ_{ac} required (Ok)

If bar is safe in compression, it is safe in tension also.

CHECK FOR RIVETS

Resultant force on rivet $= V\cot\theta = 30 \cot 45° = 30$ kN

Assuming power driven shop rivets

$$\text{Shearing strength of rivet} = \frac{\pi}{4}\times(23.5)^2\times\frac{100}{1000} = 43.37 \text{ kN}$$

$$\text{Bearing strength of rivet} = 23.5\times8\times\frac{300}{1000} = 56.4 \text{ kN}$$

Hence rivet value, $R_v = 43.37$ kN > Resultant force on rivet (Hence ok)

$$\text{No. of rivets} = \frac{30}{R_v} = \frac{30}{43.37} = 0.69$$

Hence provide one rivet each at the ends of lacing bar.

3.2 Check the top chord member of a bridge truss for its safety. It is 6 m long and is subjected to an axial compressive force of 600 kN (60*t*). The cross sectional details of the top chord members are as follows :

2 ISMC 175, cover plate of 24 cm × 0.8 cm at the top of the channel sections. The channels are facing back to back and spacing of the channels is 9 cm.

Area of the entire cross section = 67.96 cm²

Neutral Axis depth from the top = 6.96 cm

I_{xx} = 1223 cm⁴, I_{yy} = 121 cm⁴- with reference moment of inertia about their own axis (ISMC 175), C_{yy} = 2.20 cm.

Check the safety of the section and design single lacing system and sketch the elevation.

λ	60	60	70	80	90	10	200	210	220
σ_{ac}	117.2	113	107.5	100.7	92.8	84	27	24.3	21.9

[15 marks : 2005]

Solution:

The brief schematic of the top chord member is shown below:

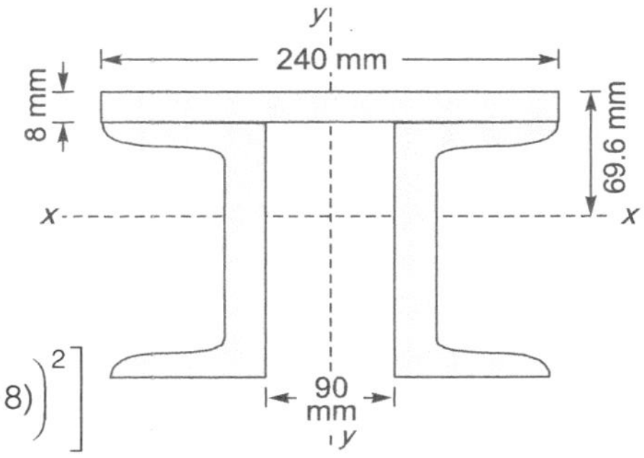

Since *y-y* axis is the axis of symmetry of the section, hence I_{min} will be either about yy axis or *x-x* axis.

$$I_{xx} = \frac{240\times 8^3}{12} + 240\times 8(69.6-4)^2$$

$$+2\left[1223\times 10^4 + \frac{(6796-240\times 8)}{2}\times\left(\frac{175}{2}-(69.6-8)\right)^2\right]$$

$$= \frac{240\times 8^3}{12} + 240\times 8\times(65.6)^2 + 2[1223\times 10^4 + 2438\times(25.9)^2]$$

$$= 36\times 10^6 \text{ mm}^4$$

The moment of inertia of the section about *y-y* axis is

$$I_{yy} = \frac{8\times 240^3}{12} + 2\left[121\times 10^4 + 2438\times\left(\frac{90}{2}+22\right)^2\right] = 33.52\times 10^6 \text{ mm}^4$$

Hence $\quad I_{yy} = I_{min}$

$$\therefore \quad r_{min} = \sqrt{\frac{I_{min}}{A}} = \sqrt{\frac{33.52\times 10^6}{6796}} = 70.23 \text{ mm}$$

$$\text{Slenderness ratio,} \quad \lambda = \frac{l_{eff}}{r_{min}} = \frac{6\times 10^3}{70.23} = 85.43$$

The value of σ_{ac} can be interpolated from the table given

$$\therefore \quad \sigma_{ac}\,(\text{for } \lambda = 85.43) = 100.7 - \frac{(100.7-92.8)}{(90-80)}\times(85.43-80)$$

$$\Rightarrow \quad \sigma_{ac} = 96.41 \text{ N/mm}^2$$

The maximum load that can be taken by the top chord member is

$$P = \sigma_{ac} \times A = \frac{96.41 \times 6796}{10^3} = 655.20 \text{ kN} > 600 \text{ kN (Hence ok)}$$

DESIGN OF SINGLE LACING

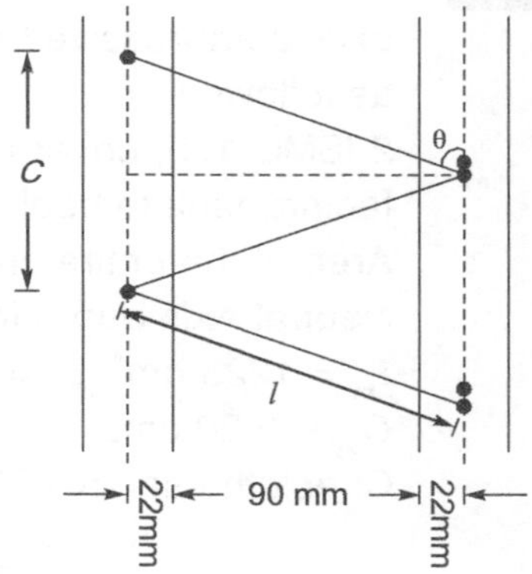

Assuming the rivets are connected at the CG of channel section.

Assuming $\theta = 45°$, we get

$$\frac{C}{2} = \frac{90+22+22}{\tan 45°}$$

$$\Rightarrow \quad C = 268 \text{ mm}$$

Let us adopt rivets of 20 mm diameter. This will give the width of lacing as 60 mm

$$\therefore \quad \sin 45° = \frac{90+22+22}{l}$$

$$\Rightarrow \quad l = 134\sqrt{2}$$

$$\Rightarrow \quad l = 189.5 \text{ mm}$$

Thickness of lacing,

$$t = \frac{l}{40} = \frac{189.5}{40} = 4.74 \text{ mm}$$

Hence adopting $t = 8$ mm

Slenderness ratio of lacing bar

$$= \frac{l_{eff}}{r_{min}} = \frac{l_{eff}}{t/\sqrt{12}} = \frac{189.5 \times \sqrt{12}}{8}$$

$$= 82.06 < 145 \text{ (Hence ok)}$$

CHECK FOR TENSILE FORCE

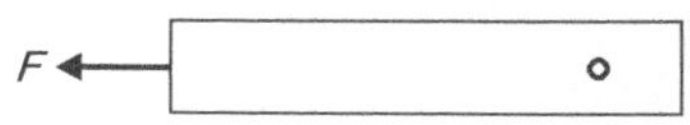

$$F = \frac{V}{2\sin\theta}$$

and $$V = \frac{2.5}{100} \times \text{applied load} = \frac{2.5}{100} \times 600 = 15 \text{ kN}$$

$$\therefore \quad F = \frac{15}{2\sin 45°} = 10.61 \text{ kN}$$

CHECK FOR COMPRESSIVE FORCE

$$F = 10.61 \text{ kN}$$

$$\sigma_{ac} \text{ required} = \frac{F}{\text{Area of lacing}} = \frac{10.61 \times 10^3}{60 \times 8} = 22.10 \text{ N/mm}^2$$

$$\lambda \text{ for lacing bar} = 82.06$$

$$\therefore \quad \sigma_{ac} \text{ provided (for } \lambda = 82.05) = 100.7 - \left(\frac{100.7 - 92.8}{90 - 80}\right) \times (82.06 - 80) = 99.07 \text{ N/mm}^2$$

Hence σ_{ac} provided is more than σ_{ac} required (ok)

If it is safe in compression, it is safe in tension also.

CHECK FOR RIVETS

Resultant force on rivet= $V \cot 45° = 15 \times 1 = 15$ kN

Assuming power driven shop rivets

$$\text{Shearing strength of rivet} = \frac{\pi}{4}\times(21.5)^2\times\frac{100}{1000} = 36.3 \text{ kN}$$

$$\therefore \quad \text{Bearing strength of rivet} = 21.5\times 8\times\frac{300}{1000} = 51.6 \text{ kN}$$

$$\therefore \quad \text{Rivet value, } R_v = 36.3 \text{ kN}$$

Resultant force on rivet < Rivet value, R_v (Hence OK)

$$\text{No. of rivets} = \frac{15}{36.3} = 0.41\text{, hence providing one rivet}$$

3.3 **An unequal angle section 200 mm × 150 mm × 15 mm is to be used in a truss as a strut of length 4.5 m. The cross-sectional properties of the section are as follows :**
Area of cross-section = 5025 mm²; I_{xx} = 2 × 10⁷ mm⁴; I_{yy} = 9.7 × 10⁶ mm⁴ I_{xy} = – 8.3 × 10⁶ mm⁴
Using the table of permissible compressive stresses given below, determine the safe load on the member.

Slenderness ratio	100	110	120	130	140	150	160	170	180
Permissible Compressive Stress MPa	80	72	64	57	51	45	41	37	33

[20 marks : 2009]

Solution :

Since the given angle section is unequal, the moment of inertia about x-x and y-y axis are not principal moment of inertia.

The **principal moment of inertia** may be calculated as given below:

$$I_{uu} = \frac{I_{xx}+I_{yy}}{2}\pm\sqrt{\left(\frac{I_{xx}-I_{yy}}{2}\right)^2+\left(I_{xy}\right)^2}$$

$$= \frac{2\times10^7+0.97\times10^7}{2}\pm\sqrt{\left(\frac{2\times10^7-0.97\times10^7}{2}\right)^2+\left(-0.83\times10^7\right)^2}$$

$$= 1.485\times10^7\pm\sqrt{0.265225\times10^{14}+0.6889\times10^{14}}$$

$$= 1.485\times10^7\pm0.977\times10^7$$

$$\therefore \quad I_{uu} = (1.485+0.977)\times10^7 = 2.462\times10^7 \text{ mm}^4$$

$$I_{vv} = (1.485-0.977)\times10^7 = 0.508\times10^7 \text{ mm}^4$$

Thus the value of least moment of inertia i.e. $I_{min} = I_{vv} = 0.508\times10^7$ mm⁴

∴ Minimum radius of gyration,

$$r_{min} = \sqrt{\frac{I_{min}}{A}} = \sqrt{\frac{0.508\times10^7}{5025}} = 31.8 \text{ mm}$$

Assuming the length of the strut as its effective length

$$\therefore \quad l_{eff} = 4.5 \text{ m}$$

$$\text{But, slenderness ratio,} \quad \lambda = \frac{l_{eff}}{r_{min}} = \frac{4.5\times10^3}{31.8} = 141.51$$

The value of permissible compressive stress (σ_{ac}) may be calculated with the help of given table by interpolation.

$$\therefore \qquad \sigma_{ac}(\text{for } \lambda = 141.51) = 51 - \left(\frac{51-45}{150-140}\right) \times (141.51-140) = 50.094 \text{ MPa}$$

$$\text{Safe load on the member} = \sigma_{ac} \times A = 50.094 \times 5025 \times 10^{-3} = 251.72 \text{ kN}$$

3.4 Explain why the design of compression member is done by indirect method.

[5 marks : 2010]

Solution:

In case of tension member, stress is constant which is independent of the cross sectional area. It means for a given load we can determine the required area directly as given below,

Required area, $$A_{reqd.} = \frac{\text{Applied load}}{\text{Permissible tensile stress}}$$

while in compression member, permissible stress σ_{ac} depends upon slenderness ration (λ) and so on the area itself. Hence, we have to assume area first and then we check whether the applied force is less than the permissible force or not.

$$P < \sigma_{ac} \times \text{Area}_{assumed}$$

Hence, in case of compression member, we use indirect way to design.

3.5 Rolled steel section ISWB 300 is used as a column of height 6 m, fixed at base and pinned at top. Find the permissible compressive load on the column using the table of permissible compressive stresses as given in the table below:

Cross-section properties of ISWB 300 section are as follows:

Area of cross-section = 6133 mm^2
Flange width = 200 mm
Flange thickness = 10 mm
Web thickness = 7.4 mm
I_{xx} = 98.216×10^6 mm^4
I_{yy} = 9.9×10^6 mm^4

λ	50	60	70	80	90	100	110	120
σ_{ac}	132	122	112	101	90	80	72	64

λ	13	140	150	160	170	180	190	200
σ_{ac}	57	51	45	41	37	33	30	28

λ = slenderness ratio
σ_{ac} = allowable compressive stress in MPa
Use linear interpolation for intermediate values of λ.

[15 marks : 2012]

Solution:

The load carrying capacity of the column is given by

$$P = \sigma_{ac} \times A$$

where,

P = Permissible load on the column
σ_{ac} = Permissible stress in column section
A = Area of column section

But σ_{ac} depends on slenderness ratio which in turn depends on r_{min}

Now, as we know $$r_{min} = \sqrt{\frac{I_{min}}{A}}$$

$$I_{min} = I_{yy} = 9.9 \times 10^6 \text{ mm}^4$$

$$\therefore \quad r_{min} = \sqrt{\frac{9.9 \times 10^6}{6133}} = 40.177 \text{ mm}$$

Now, slenderness ratio, $\lambda = \dfrac{l_{eff}}{r_{min}}$

For a column fixed at base and pinned at top

$$l_{eff} = 0.8 \times l = 0.8 \times 6 = 4800 \text{ mm}$$

$$\therefore \quad \lambda = \frac{4800}{40.177} = 119.47$$

The value of σ_{ac} can be interpolated from the table given

$$\therefore \quad \sigma_{ac}(\text{for } \lambda = 119.47) = 72 - \left(\frac{72-64}{120-110}\right) \times (119.47 - 110)$$

$$\Rightarrow \quad \sigma_{ac} = 64.424 \text{ N/mm}^2$$

The permissible compressible load on the column

$$P = \sigma_{ac} \times A$$
$$= 64.424 \times 6133 = 395112.392 \text{ N} = 395.112 \text{ kN}$$

3.6 A steel column consisting of ISMB 400 has one end restrained against translation and rotation, while the other end is restrained against translation only. Its unsupported length is 5 m. Determine its axial load carrying capacity at service loads using the limit state design of IS : 800-2007.

Design compressive strength $f_{cd} = \dfrac{f_y / \gamma_{mo}}{\phi + [\phi^2 - \lambda^2]^{0.5}}$

$$\phi = 0.5(1 + \alpha(\lambda - 0.2) + \lambda^2)$$

$$\lambda = \sqrt{\frac{f_y}{f_{cc}}}$$

Euler buckling stress f_{cc}

$$f_{cc} = \frac{\pi^2 E}{\left(\frac{kL}{r}\right)^2}$$

and $\alpha = 0.34$ for buckling class b

$f_y = 300$ MPa, $\gamma_{mo} = 1.1$, $\gamma_f = 1.5$

$E = 200$ GPa

Minimum radius of gyration for ISMB 400 = 28.2 mm, area of cross-section = 7846 mm^2.

[10 marks : 2014]

Solution:

$$L_{effective}(kL) = 0.8 \text{ L}$$

$$\text{kL} = 0.8 \times 5 = 4 \text{ m or } 4000 \text{ mm}$$

Euler buckling stress, $f_{cc} = \dfrac{\pi^2 E}{\left(\frac{kL}{r}\right)^2}$

where, r = minimum radius of gyration = 28.2 mm (given)

$$\therefore \quad f_{cc} = \frac{\pi^2 \times 2 \times 10^5}{\left(\frac{4000}{28.2}\right)^2} = 98.11\text{ N/mm}^2$$

Hence,
$$\lambda = \sqrt{\frac{f_y}{f_{cc}}} = \sqrt{\frac{300}{98.11}}$$
$$\lambda = 1.749$$

Hence, $\phi = 0.5(1 + \alpha(\lambda - 0.2) + \lambda^2)$

Here given, $\alpha = 0.34$ for buckling class b.

$$\therefore \quad \phi = 0.5[1 + 0.34(1.749 - 0.2) + (1.749)^2]$$
$$\phi = 2.29$$

Hence, design compressive strength,

$$f_{cd} = \frac{f_y/\gamma_{m0}}{\phi + \left[\phi^2 - \lambda^2\right]^{0.5}} = \frac{300/1.1}{2.29 + \left[2.29^2 - 1.749^2\right]^{0.5}}$$

$$= \frac{272.72}{2.29 + 1.478} = 72.4\text{ N/mm}^2$$

$\therefore$ Load carrying capacity of column, $P = A \times f_{cd}$

$$P = 7846 \times 72.4\text{ N} = 568.05\text{ kN}$$

Service load, $$P = \frac{568.05}{1.5} = 378.7\text{ kN}$$

3.7 A column is made of 4 Nos. ISA 100 × 100 × 10 as shown. The effective length of column is 6 m. The lacing of column consists of 60 mm × 10 mm flat bars arranged in a single laced system by bolts and inclined at an angle of 45°. Determine the factored load on the column and check the local buckling of column angles. [ISA 100 × 100 × 10 : A = 1903 mm², I_{XX} = = I_{YY} = 177 × 10⁴ mm⁴, C_{XX} = C_{YY} = 28.4 mm]. Assume gauge length = 55 mm and r_{vv}= 19.4 mm.

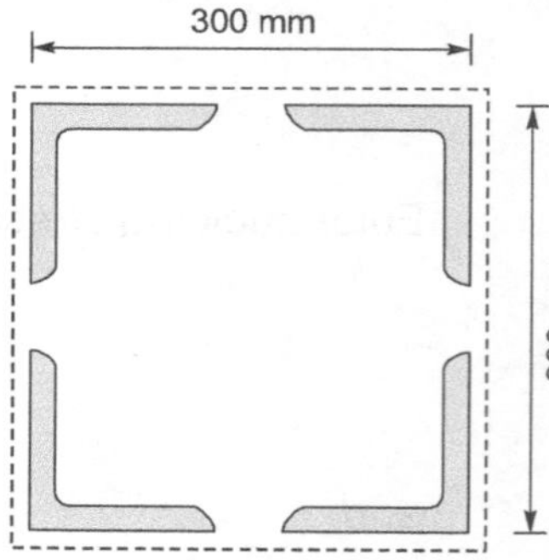

KL/r	30	40	50	60	70	80	90
f_{cd} (MPa)	204	185	167	150	133	118	105

[20 marks : 2015]

Solution:

Given effective length of column (l) = 6 m

$\therefore$ This problem has to be solved using **working stress method of steel design.**

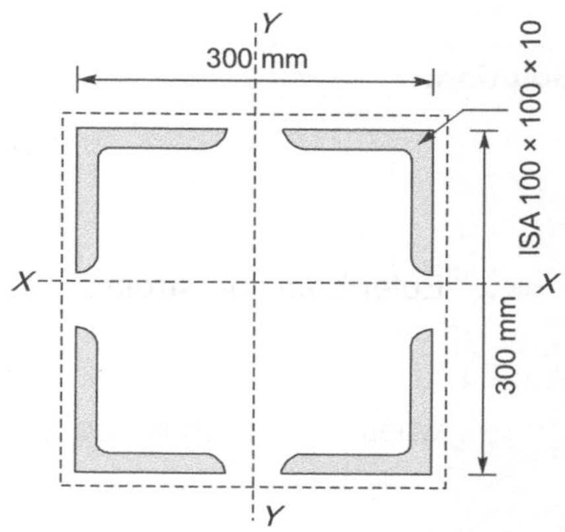

For the given column section and arrangement of angles,

$$I_{xx} = I_{yy}$$

But
$$I_{xx} = 4\left(177 \times 10^4 + 1903\left(\frac{300}{2} - 28.4\right)^2\right)$$
$$= 119.635 \times 10^6\text{ mm}^4$$

Also, $$I_{xx} = I_{yy} = Ar_{xx}^2 = Ar_{yy}^2 = Ar^2 \text{ (say)}$$ where r = radius of gyration

$$\therefore \quad r = r_{xx} = r_{yy} = \sqrt{\frac{119.635 \times 10^6}{4 \times 1903}} = 125.366\text{mm}$$

$$\therefore \text{ Slenderness ratio, } \lambda = \frac{l}{r} = \frac{6000}{125.366} = 47.8598$$

∴ Allowable compressive stress

$$f_{cd} = 185 + \frac{167 - 185}{50 - 40}(47.8598 - 40) = 170.852 \text{ N/mm}^2$$

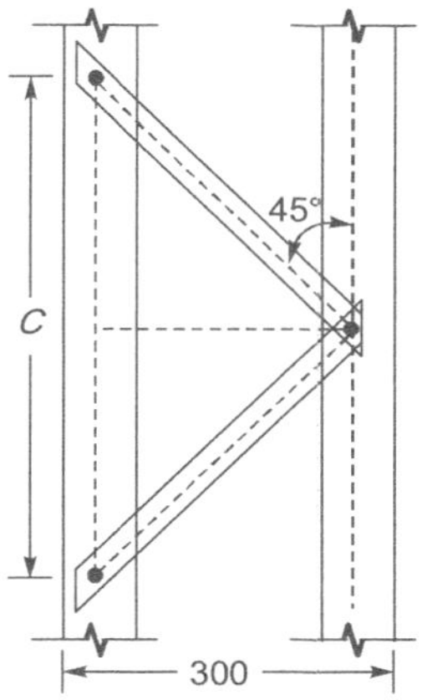

∴ Allowable factored compressive load,

$$P_d = f_{cd} \cdot A = 170.852 \,(4 \times 1903)\text{N}$$
$$= 1300.525 \text{ kN} \simeq 1300 \text{ kN (say)}$$

Distance between two lacing bolts on one angle

$$C = 2[300 - 2 \times 55]\tan 45° = 380\text{mm}$$

Section of lacing flat = 60 × 10 mm

$r_{xx} = r_{yy}$ of component member i.e. angle section 100 × 100 × 10

$$= \sqrt{\frac{177 \times 10^4}{1903}} = 30.498\text{mm}$$

$$0.7\lambda = 0.7\,(47.8598) = 33.502$$

$$\frac{C}{r_{xx}} = \frac{C}{r_{yy}} = \frac{380}{30.498} = 12.46 < 50 < 0.7\lambda$$

Thus $$\frac{C}{r_{xx}} = \frac{C}{r_{yy}} < 50 \text{ (O.K.)} < 0.7\lambda \text{ (O.K.)}$$

Thus, **column is safe against local buckling.**

3.8 **A 6 m long built-up column carries an axial compressive force and consists of 4 nos. ISMC 350 sections as shown in figure. A lacing is provided in each of the four planes. Explain the principle of design of lacing with the help of a neat sketch.**

[12 marks : 2017]

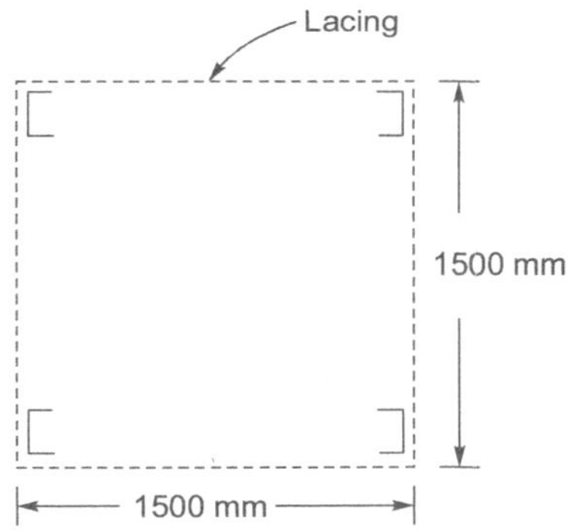

Solution :

Principle of design of lacing:

1. The angle of inclination θ of the lacing bar with the longitudinal axis of the component member should be kept between 40°-70°. As an initial trial value θ may be assumed to be 45° and spacing of the lacing bars is calculated.

2. The maximum spacing of the lacing bars 'a_1' should be such that minimum slenderness ratio of the component member a_1/r_y is not greater than 50 or 0.70 times the slenderness ratio of the member as a whole where a_1 is the length of the component member and r_y is the radius of gyration about *yy*-axis of the component member.

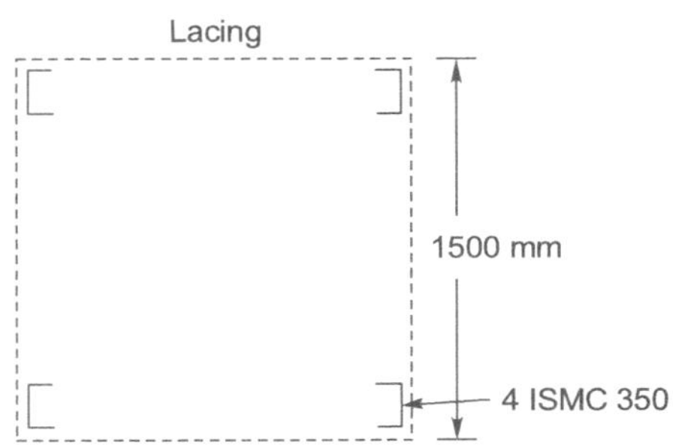

3. The lacing for compression member should be proportioned

to resist a total transverse shear 'V_t' equal to 2.5% of the axial force in the column. The shear V_t is divided equally in all parallel planes N in which there is shear resisting elements such as lacing or continuous flats. Hence, V_t/N is transverse force to which lacing is subjected.

4. The section of the lacing flat is initially assumed and then checked for safety. The minimum flat width $\geq 3 \times$ nominal diameter of end connector. For double lacing system, the thickness $t \nless 1/60 \times$ Effective length.
5. The minimum radius of gyration of lacing flat i.e., $r_{min} \geq t/\sqrt{12}$.
6. The slenderness ratio of the lacing bar is computed which should be < 145.
7. The design compressive strength of the lacing flat is computed for the calculated value of slenderness ratio of lacing flat.
8. The tensile strength of the flat is calculated and it should be more than the force in the lacing flat.

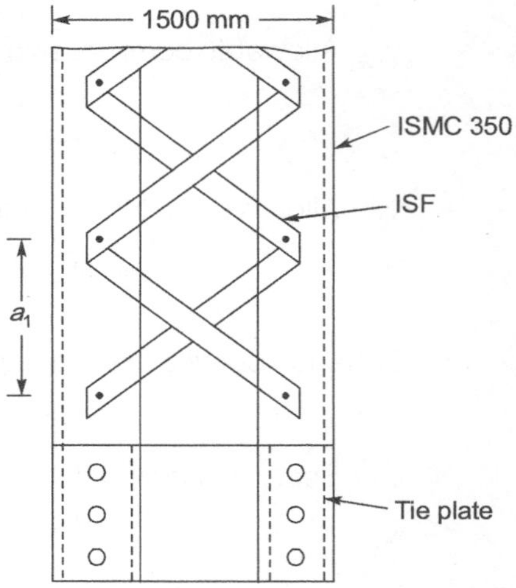

9. To check the distortion of the column section at the column ends tie plates (battens) are provided at the ends of the lacing system and these should be able to resist the forces to which the lacing flat is subjected. The plate is designed as battened plate.

3.9 **An ISMB 400 beam is spliced at a section carrying factored bending moment of 120 kNm and factored shear force of 80 kN. The splice is to be designed so that the flange splice will carry the bending moment and the web splice will carry the shear force. Field welding with 8 mm fillet will be used. Determine the size of 100 mm wide flange plate using the following data: t_f = 16 mm; t_w = 8.9 mm and b_f = 140 mm.**

[12 marks : 2018]

Solution:

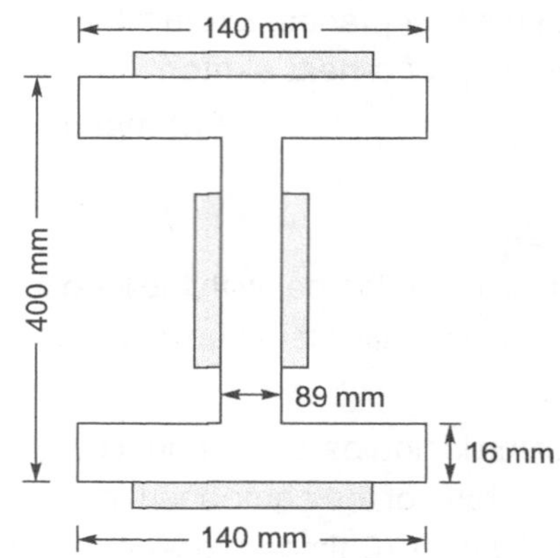

$$I_{xx} = \left[\frac{140 \times 400^3}{12} - \frac{131.1 \times 368^3}{12}\right] = 202.21 \times 10^6 \text{ mm}^4$$

Given,
$$M_u = 120 \text{ kNm}$$

$\therefore$
$$\sigma_{top} = \frac{M}{I} y_{top} = \frac{120 \times 10^6}{202.2 \times 10^6} \times 200 = 118.69 \text{ N/mm}^2$$

$$\text{Force borne by flange} = \frac{\sigma_{max}}{y_{max}} \times A\bar{y} = \frac{118.69}{200} \times (140 \times 16 \times 192) = 255.2 \text{ kN}$$

Let thickness of flange splice place = t.

$$\therefore \quad \frac{A_g f_y}{1.1} = 255.2 \text{ kN}$$

$$(100 \times t)\frac{250}{1.1} = 255.2 \times 10^3$$

$$t = 11.23 \text{ mm} \approx 12 \text{ mm}$$

Let Length = L over the flange, with weld size 8 mm

$$\therefore \quad 255.2 \times 10^3 = \frac{f_u}{\sqrt{3} \times 1.50} \times L \times 2 \times 0.7 \times 8$$

$$\therefore \quad L = 144.38 \text{ mm} \simeq 150 \text{ mm}$$

$$\text{Flange splice plate} = 100 \times 150 \times 12$$

Web spice will carry shear only

$$\therefore \quad \frac{V_u}{ht} < \frac{f_u}{\sqrt{3}\gamma_{max}}$$

$$\frac{80 \times 1000}{100t} < \frac{410}{\sqrt{3} \times 1.5}$$

$$t > 5.1 \text{ mm}$$

$$\gamma_{m0} = 1.50$$

$\because$ Field welding

$$\gamma_{max} = 1.25 \text{ is shop welding}$$

$$\therefore \quad \text{provide, } t = 6 \text{ mm}$$

Let Length = L_m, in web splice

$$2 \times L \times \frac{f_u}{\sqrt{3}\gamma_{max}} 0.7s = V_u$$

$$\therefore \quad L = \frac{\sqrt{3} \times 1.50 \times 80000}{2 \times 410 \times 0.7 \times 8} = 45 \text{ mm}$$

Provide web splice [100 × 50 × 6].

3.10 A laced column consisting of two ISMC 300 channels placed back-to-back is subjected to factored axial load of 1100 kN. The 10 m long column is restrained in translation but not in rotation in ends. Single lacing at 45° is provided, and connected to flanges by bolts. Verify the capacity of the selected section and determine their spacing. Also determine the size of 50 mm wide lacing rods considering only compressive force in them. Take f_y = 250 MPa and gauge length of lacing rods = 50 mm. The properties of ISMC 300 are as follows : A = 4630 mm^2, B = 90 mm, t = 7.8 mm, T = 13.6 mm, α = 96°, C_y = 23.5 mm, I_{xx} = 6.42 × 10^7 mm^4, I_{yy} = 3.13 × 10^6 mm^4, r_x = 118.0 mm, r_y = 26.0 mm. Do not design the connection of lacing rod to the channel member. Table 9(c) of IS : 800 is enclosed for reference.

[20 marks : 2018]

Solution:

Given: Factored axial load = 1100 kN

Unsupported length = 10 m

Effective length = $1.0 \times L = 10$ m because column is hinged at both the ends.

Spacing of channels:

While providing spacing of the channels, our aim is to avoid the buckling in place of lacing. For that I_{yy} should be slightly greater than I_{xx}.

$I_{xx} = 2 \times I_{xx} = 12.84 \times 10^7$ mm^4.

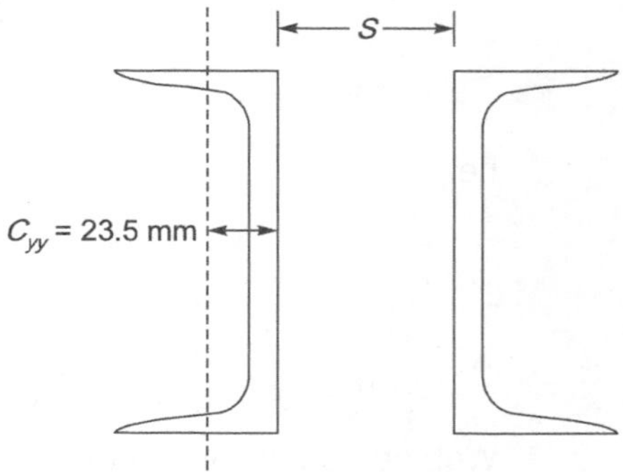

$$I_{yy} = 2 \times \left[I_{yy} + A_g \times \left(\frac{S}{2} + 23.5 \right)^2 \right]$$

Equating, $I_{xx} = I_{yy}$

We get, $S = 183$ mm

∴ Providing, $S = 185$ mm

Since, $I_{yy} > I_{xx}$,

Least radius of gyration = $r_x = 118$ mm

$$\therefore \quad \lambda = 1.05 \frac{kL}{r_{min}} = 88.98$$

Effective length is increased by 5% for laced column.

As per table 9c of IS 800 : 2000

For $\lambda = 88.98$,

$f_{CD} = 122.5$ N/mm^2

∴ Design compressive strength $P_{CD} = f_{CD} A_g = 1134.3$ kN > 1100 kN

Hence the column is safe.

Spacing and capacity of section:

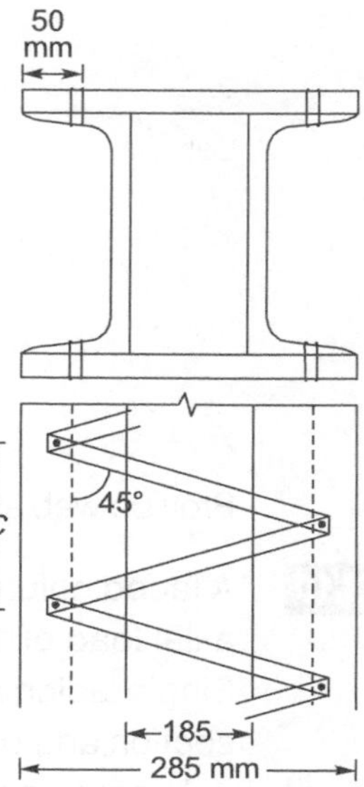

Horizontal spacing between connection

= 185 + 50 + 50 = 285 mm

⇒ Length of lacing = 285 cosec 45°

$l_1 = 403.05$ mm

$l_{eff} = 1.\, l_1$ [∵ Single lacing]

$$\Rightarrow \quad \text{Min thickness} = \frac{l}{40} = 10.07 \simeq 12 \text{ mm}$$

∴ Choose (50 × 12) lacing flat,

Since, Inclination = 45°

$C = 2\, l_1 \cos 45° = 569.99 \simeq 570$ mm

$$\therefore \quad \frac{C}{r_{min}} = \frac{570}{26} = 21.92 < 50 < 0.7\, \lambda_{whole} = 62.29$$

Hence spacing and section are decided.

Maximum shear, $V_t = \dfrac{2.5}{100} \times 1100 = 27.5$ kN

Transverse shear in each panel $= \dfrac{V_t}{N} = \dfrac{27.5}{2} = 13.75$ kN

Compressive force in lacing bars $= \frac{V_t}{N} \text{cosec}\theta = 19.445 \text{ kN}$

Minimum radius of gyration of lacing flat section,

$$r = \frac{t}{\sqrt{12}} = 3.46 \text{ mm}$$

$$\therefore \quad \frac{l}{r} = \frac{375}{3.46} = 108.38 \leq 145 \qquad \text{[OK]}$$

For, $\frac{l}{r} = 108.38$

$f_y = 250$ MPa (design compressive stress)

$f_{CD} = 96.6 \text{ N/mm}^2$

Design compressive strength $= A_g \times f_{CD} = 96.6 \times 50 \times 12$

$P_{CD} = 57.96$ kN

$\therefore \quad P_{CD} > 19.44$ kN (safe in compression)

3.11 Check the adequacy of a HB 450 @ 0.872 kN/m rolled steel beam section for a column to carry an axial load of 1100 kN. The column is 4 m long and restrained in position but not in direction at both ends. Allowable axial stress in compression is 105 MPa. The sectional properties of the given section are as follows:

$$A = 11114 \text{ mm}^2,\ r_{xx} = 187.8 \text{ mm},\ r_{yy} = 51.8 \text{ mm}$$

[12 marks : 2019]

Solution:

Applied load = 1100 kN

1. $$\lambda = \frac{kL}{r} = \frac{1 \times 4000}{51.8} = 77.22 < 180$$

2. $$\sigma_{ac} \propto \frac{1}{\lambda} = \frac{0.6 f_{cc} f_y}{\left(f_y^n + f_{cc}^n\right)^{1/n}} \qquad \text{(where } n = 1.4)$$

$$f_{cc} = \frac{\pi^2 E}{\left(\frac{kL}{r}\right)^2}$$

$E = 2 \times 10^5 \text{ N/mm}^2$

$f_{cc} = 331.03 \text{ N/mm}^2$

$$\sigma_{ac} = \frac{0.6 \times 331.03 \times 250}{\left(250^{1.4} + 331.03^{1.4}\right)^{\frac{1}{1.4}}} = 103.77 \text{ N/mm}^2 < 105 \text{ MPa}$$

3. $P_{safe} = A_g \times \sigma_{ac} = 11114 \times 103.77 = 1153.3$ kN

4. $P_{safe} > P_{applied}$

Hence design is safe.

3.12 Calculate the safe load of a compression member made up of a welded box section as shown in the figure. The effective lengths along major axis and minor axis are 5 m and 2 m respectively. The yield stress of the material is 340 MPa.

$\frac{KL}{r}$	Design Compressive Stress f_{cd} (MPa)
20	299
30	278
40	256

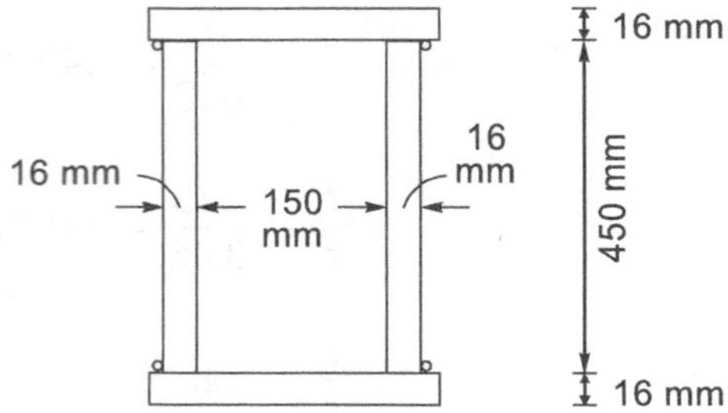

[12 Marks : 2021]

Solution:

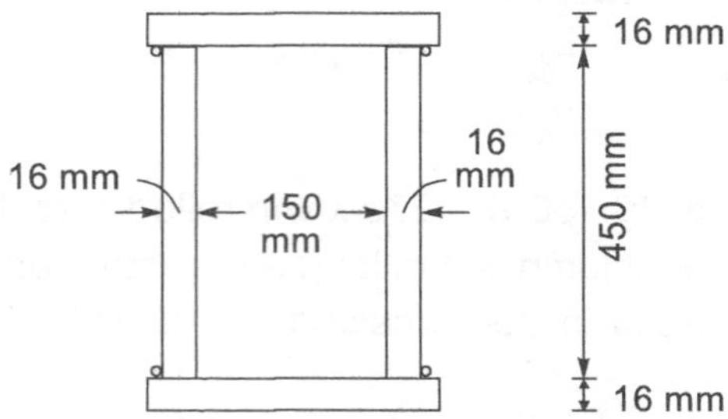

(i) Moment of Inertia about principle axis.

$$I_{xx} = \left[\frac{182\times16^3}{12} + (182\times16)(225+8)^2\right]\times2 + \left[\frac{16\times450^3}{12}\right]\times2$$

$$= 559.304\times10^6\ \text{mm}^4$$

$$I_{yy} = \left[\frac{16\times182^3}{12}\right]\times2 + \left[\frac{450\times16^3}{12} + (450\times16)(75+8)^2\right]\times2$$

$$= 115.585\times10^6\ \text{mm}^4$$

As $I_{xx} > I_{yy}$

Major principle axis $\rightarrow$ x-x axis.

Minor principle axis $\rightarrow$ y-y axis.

(ii) Maximum slenderness ratio

about x-x axis

$$\lambda_{xx} = \frac{L_{ex}}{r_{xx}}$$

Where, $$r_{xx} = \sqrt{\frac{I_{xx}}{A}} = \sqrt{\frac{559.304\times10^6}{(450\times16)\times2+(182\times16)\times2}} = 166.3\ \text{mm}$$

$$\lambda_{xx} = \frac{5000}{166.3} = 30.066$$

About y-y axis $$\lambda_{yy} = \frac{L_{ey}}{r_{yy}}$$

Where, $r_{yy} = \sqrt{\frac{I_{yy}}{A}} = \sqrt{\frac{115.585 \times 10^6}{20224}} = 75.6 \text{ mm}$

Hence, $\lambda_{yy} = \frac{2000}{75.6} = 26.455$

As, $\lambda_{xx} > \lambda_{yy}$,

$\lambda_{max} = \lambda_{xx}$

Also, Buckling strength will be based on *x-x* axis.

(iii) Compressive strength

$\frac{KL}{r}$	f_{cd} (MPa)
20	299
30	278
40	256

(a) Using linear interpolation, to determine f_{cd} corresponding to $\lambda = 30.066$

$$\frac{30.066 - 30}{40 - 30} = \frac{f_{cd} - 278}{256 - 278}$$

$$f_{cd} = 277.855 \text{ MPa}$$

(b) Design compressive strength,

$$P_d = A_e f_{cd} = 20244 \times 277.855 = 5619.34 \text{ kN}$$

(c) Safe compressive load $= \frac{(P_d)}{\gamma_F} = \frac{5619.34}{1.5} = 3746.23 \text{ kN}$

3.13 **An upper storey column ISHB 300@577 N/m (b_f = 250 mm, t_f = 10.6 mm) carries a factored load of 1200 kN and a factored moment of 12 kNm. It is to be spliced with lower storey column ISHB 400@806 N/m (b_f = 250 mm, t_f = 12.7 mm). Design a suitable splice. Given that f_y = 250 MPa, γ_{m0} = 1.1, γ_{m1} = 1.25 and $\beta_{pk} = (1 - 0.0125t_{pk})$. Also sketch connection details. Use bolts of Grade 5.8.**

[20 marks : 2022]

Solution:

Given data:

$$f_y = 250 \text{ MPa}$$
$$\gamma_{m0} = 1.10$$
$$\gamma_{m1} = 1.25$$

Bolt grade 5.8

$$f_{ub} = 520 \text{ MPa}$$
$$f_{yb} = 400 \text{ MPa}$$

Properties of ISHB 400 @ 806 N/m

$$b_F = 250 \text{ mm}$$
$$t_F = 12.7 \text{ mm}$$

Properties of ISHB 350 @ 577 N/m

$$b_F = 250 \text{ mm}$$
$$t_F = 10.6 \text{ mm}$$

1. Packing plate thickness,

   As the two sections to be spliced are of different depths, packing plate will be provided.

Thickness of packing plate $= \dfrac{400-300}{2} = 50$ mm

2. Bearing plate design,

A bearing plate is provided between two column sections.

$$\text{Length of plate} = 400 \text{ mm}$$

$$\text{Width of plate} = 250 \text{ mm}$$

$$\text{Direct load on each flange} = \frac{1200}{2} = 600 \text{ kN}$$

Distance between the CG of column flanges of ISHB 400

$$= 400 - 12.7 = 387.3 \text{ mm}$$

Distance between the CG of column flanges of ISHB 300

$$= 300 - 10.6 = 289.4 \text{ mm}$$

Distance between the lire of action of forces on the flanges of the two column sections

$$= \frac{387.3-289.4}{2} = 48.95 \text{ mm}$$

$$\text{Moment due to the couple} = 600 \times 48.95 \times 10^{-3} = 29.37 \text{ kNm}$$

$$\text{Total moment} = 29.37 + 12 = 41.37 \text{ kNm}$$

Equating the bending moment to the moment of resistance of the plate section,

$$41.37 \times 10^6 = \frac{1}{6} \times 250 \times t^2 \times \frac{250}{1.1}$$

$$\Rightarrow \quad t = 66.09 \text{ mm} \simeq 70 \text{ mm (say)}$$

Provide a bearing plate of size 400 mm × 250 mm × 70 mm

3. Splice plate design

Assuming the column ends are made flush. Splices will be designed for 50% of the load on one flange.

$$P = \frac{1}{2} \times 1200 \times 0.5 = 300 \text{ kN}$$

Assuming 6 mm thickness of splice plate.

$$\text{Load on splice due to moment} = \frac{12 \times 10^3}{(400+6)} = 29.55 \text{ kN}$$

$$\text{Total load} = 300 + 29.55 = 329.55 \text{ kN}$$

Cross-sectional area of the splice plate required

$$= \frac{329.55 \times 10^3}{250} = 1318.2 \text{ mm}^2$$

$$\text{Width of splice plate} = \text{Width of column} = 250 \text{ mm}$$

$$\therefore \quad \text{Thickness of splice plate} = \frac{1318.2}{250} = 5.27 \simeq 6 \text{ mm (say)}$$

∴ Provide splice plate of size 250 mm × 6 mm

Length of splice plate.

Let us provide 20 mm diameter bolts of grade 5.6.

$$\text{Strength of bolt in single shear} = \frac{520}{\sqrt{3}\times 1.25}\times 0.78\times\frac{\pi}{4}\times 20^2 \text{ N} = 58.85 \text{ kN}$$

Strength of bolt in bearing (Assume $k_b = 0.5$)

$$= \frac{2.5k_b dt f_y}{\gamma_{mb}} = \frac{2.5\times 0.5\times 20\times 6\times 410}{1.25} \text{ N} = 49.2 \text{ kN}$$

$\therefore$ Strength of bolt, $V_{db} = 49.2$ kN

Number of bolts required for lower column,

$$n = \frac{329.55}{49.2} = 6.698 \simeq 8$$

For bolts in upper column shear strength of bolt will be reduced.

$$\text{Shear strength of bolt} = \beta_{pkg}\times V_{dsb}$$
$$= (1 - 0.0125 \times 50) \times 58.85 = 22.07 \text{ kN}$$

Bearing strength of bolt = 49.2 kN

No. of bolts required for upper column

$$= \frac{329.55}{22.07} = 14.93 \simeq 16$$

$\therefore$ Provide 8, 20 mm diameter bolts to connect splice plate with the flange of lower storey column and 16, 20 mm diameter bolts to connect the splice plates to the flanges of upper storey column.

Provide the bolts in 2 vertical rows at a pitch of 50 mm and edge distance of 35 mm as shown in figure.

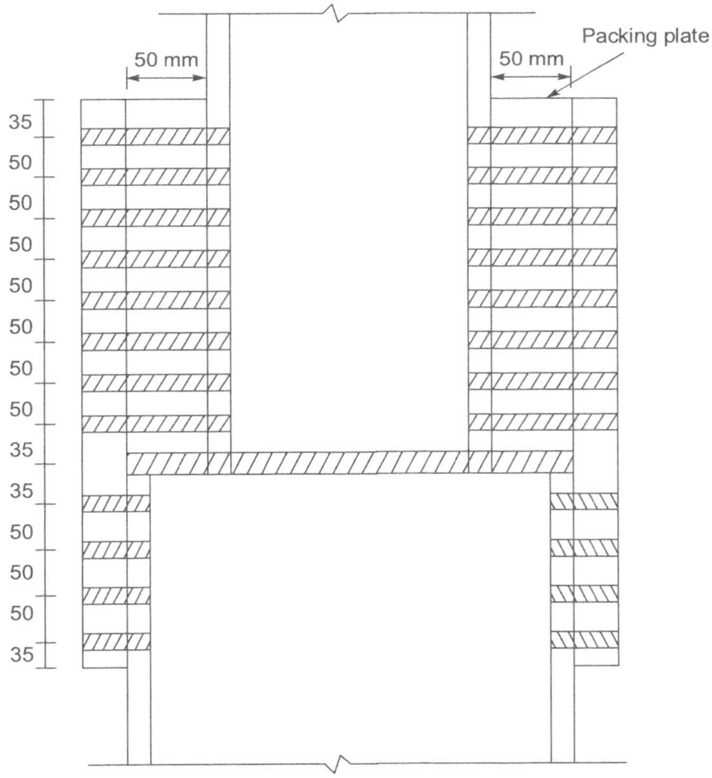

Length of splice plate = $50 \times 10 + 35 \times 4 = 640$ mm

$\therefore$ Provide splice plate of size = 640 mm × 250 mm × 6 mm

3.14 A member of a transmission tower is composed of two angles ISA 75 × 75 × 8 in star configuration as shown in the figure below. The angles are tack welded by a gusset plate of thickness 10 mm suitably. Find the axial compressive load carrying capacity of the member under dead and live load condition. The nodal length of the member is 3 m. Assume K = 0.85 and E250 grade of steel. The properties of ISA 75 × 75 × 8 are as follows :

$A = 1140\ mm^2$, $I_{yy} = I_{zz} = 59\ cm^4$, $C_y = C_z = 21.4\ mm$,
$r_{uu} = 28.8\ mm$, $r_{vv} = 14.5\ mm$

Given:

$\frac{KL}{r}$	60	70	80	90	100	110	120	130
f_{cr}(MPa)	168	152	136	121	107	94.6	83.7	74.3

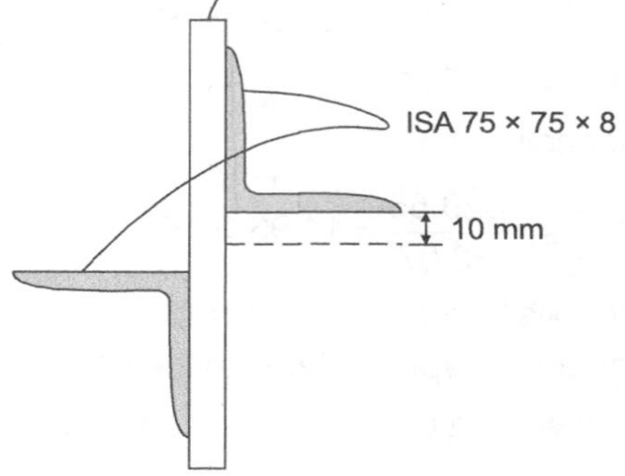

Use limit state method.

[20 marks : 2023]

Solution:

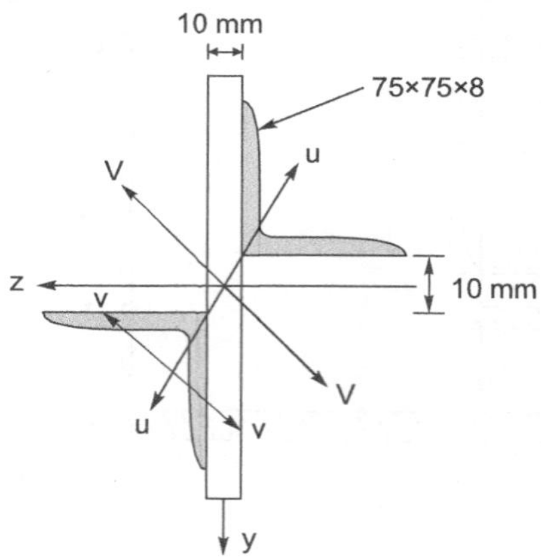

$$A = 1140\ mm^2$$
$$I_{yy} = I_{zz} = 59\ cm^4$$
$$c_y = c_z = 21.4\ mm$$
$$r_{uu} = 28.8\ mm$$
$$r_{vv} = 14.5\ mm$$

$$I_{zz} = I_{yy} = \left[I_{yy} + A\left(c_y + \frac{10}{z}\right)^2\right] \times 2$$

$$= \left[59 + 11.4\left(2.14 + \frac{1}{2}\right)^2\right] \times 2 = 276.907 \text{ cm}^4$$

$$r_{zz} = r_{yy} = \sqrt{\frac{I_{yy}}{A}} = \sqrt{\frac{276.907}{2 \times 11.4}}$$

$$= 3.485 \text{ cm or } 34.85 \text{ mm}$$

$$r_{vv} = \sqrt{r_{vv}^2 + 2\left(c_y + \frac{10}{2}\right)^2}$$

$$= \sqrt{14.5^2 + 2\left(21.4 + \frac{10}{2}\right)^2} = 40.05 \text{ mm}$$

$r_{uu} = r_{UU} = 28.8$ mm (As there is not shift in *uu* axis)

∴ Least radius of gyration : $r_{uu} = 28.8$ mm

Slenderness ratio, $\lambda = \dfrac{KL}{r_{min}} = \dfrac{0.85 \times 3000}{28.8} = 88.541$

From table

λ	f_{cr}
80	136
88.541	x
90	121

$$\frac{88.541 - 80}{90 - 80} = \frac{x - 136}{121 - 136}$$

∴ $x = 123.188 \text{ N/mm}^2$

So, Compressive strength $= f_{cr} \times A$

$= 123.19 \times 140 \times 2 = 280.87$ kN

4. Beams

4.1 A stanchion of effective length of 6m consists of a built-up twin box section using an ISMB250 with two plates of 260 mm × 10 mm size welded each to the tips of the two flanges of ISMB250 on both sides of 250 mm depth with 4 mm size fillet weld continuous throughout the height. The geometrical properties of ISMB 250 are

$h = 250$ mm $b = 125$ mm,

$A = 47.55 \text{ cm}^2$ $I_{xx} = 5131.6 \text{ cm}^4$

$I_{yy} = 334.5 \text{ cm}^4$

Determine the maximum permissible compressive load including self weight for the stanchion, if the permissible compressive stress σ_{ac} given by the following formula :

$$\sigma_{ac} = \frac{0.60 \times f_{cc} \times f_y}{[(f_{cc})^{1.4} + (f_y)^{1.4}]^{0.7}}$$

where yield stress $f_y = 250$ MPa,

$f_{cc} = \dfrac{\pi^2 E}{\lambda^2}, \lambda = \dfrac{l}{r}$ = radius of gyration,

$E = 2 \times 10^5 \text{ N/mm}^2$.

[10 marks : 2004]

Solution:

The brief schematic of the stanchion is given below:

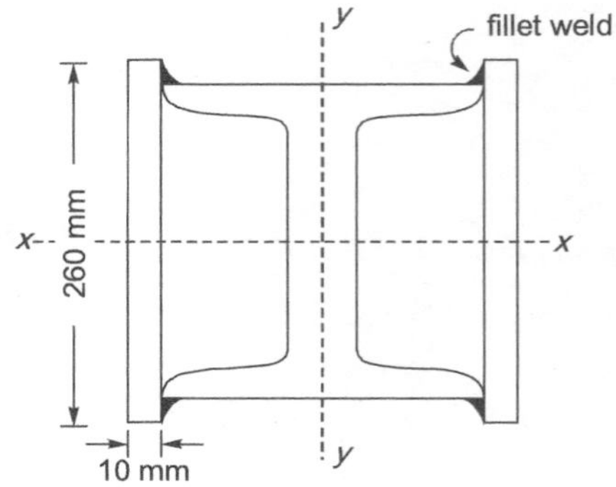

The moment of inertia of the stanchion about x-x axis is

$$I_{xx} = 5131.6 \times 10^4 + 2\left[\frac{10\times(260)^3}{12}\right]$$

$$= 80.61 \times 10^6 \text{ mm}^4$$

The moment of inertia of the stanchion y-y axis is

$$I_{yy} = 334.5 \times 10^4 + 2\left[\frac{260\times10^3}{12} + 260\times10\times\left(\frac{125}{2}+\frac{10}{2}\right)^2\right]$$

$$= 334.5 \times 10^4 + 2\left[\frac{260\times10^3}{12} + 2600\times(67.5)^2\right] = 27.08 \times 10^6 \text{ mm}^4$$

Since $I_{yy} < I_{xx}$,

therefore $I_{min} = I_{yy} = 27.08 \times 10^6 \text{ mm}^4$

$$\therefore \quad r_{min} = \sqrt{\frac{I_{min}}{A}} = \sqrt{\frac{27.08\times10^6}{[4755+(2\times260\times10)]}} = 52.16 \text{ mm}$$

Hence slenderness ratio,

$$\lambda = \frac{l_{eff}}{r_{min}} = \frac{6\times1000}{52.16} = 115.03$$

$$\therefore \quad f_{cc} = \frac{\pi^2 E}{\lambda^2} = \frac{\pi^2\times2\times10^5}{(115.03)^2} = 149.18 \text{ N/mm}^2$$

$$\therefore \quad \sigma_{ac} = \frac{0.60\times f_{cc}\times f_y}{[(f_{cc})^{1.4}+(f_y)^{1.4}]^{0.7}}$$

$$= \frac{0.60\times149.18\times250}{[(149.18)^{1.4}+(250)^{1.4}]^{0.7}} = 75.77 \text{ N/mm}^2$$

The maximum permissible load for the stanchion is

$$P = \sigma_{ac}\times A = \frac{75.77\times[4755+(2\times260\times10)]}{10^3} = 754.29 \text{ kN}$$

4.2 A beam MB 600 @ 123 kgf is supported over an effective span of 9 m. Two floor joists transmit floor loads at a distance of 1.5 m on either side of the midspan. Determine the safe load which the two floor joists can transmit on the beam if the beam is effectively restrained laterally by the floor joists. Properties of MB 600 @123 kgf are :

Overall depth D = 600 mm

Width of flange b = 210 mm

Depth of web d_1 = 558.4 mm

Mean thickness of flange t_f = 20.8 mm

Thickness of web t_w = 12 mm
Radius of gyration about *yy*-axis r_y = 41.2 mm
Modulus of section about *xx*-axis Z_x = 3060×10^3 mm^3
Modulus of section about *yy*-axis Z_y = 252×10^3 mm^3

Maximum permissible bending stress σ_{bc}(N/mm^2) for $t_f / t_w < 2$ and $d_1 / t_w < 85$ in given in the table below as a function of slenderness ratio (l/r) and D/t_f :

l/r \ D/t_f	20	25	30	35	40
60	151	150	149	149	149
65	148	147	146	146	145
70	146	144	143	142	142
75	143	141	140	139	138
80	140	138	136	135	134

[12 marks : 2011]

Solution:

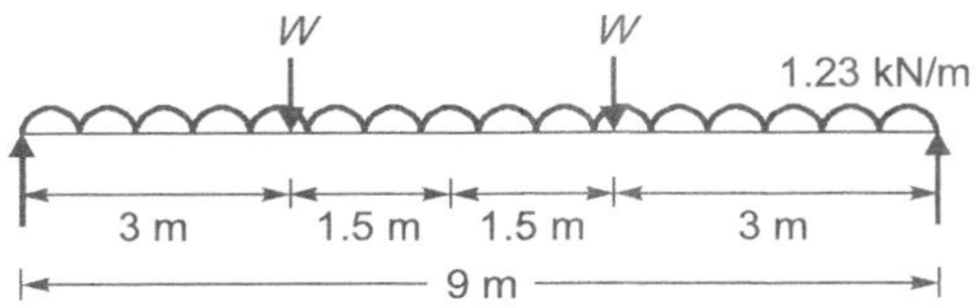

Check:

$$\frac{t_f}{t_w} = \frac{20.8}{12} = 1.73 < 2$$

$$\frac{d_1}{t_w} = \frac{558.4}{12} = 46.5 < 85$$

So using the table provided in the question to calculate maximum permissible bending stress in beam (σ_{bc}).

$$\frac{D}{t_f} = \frac{600}{20.8} = 28.85$$

Effective length of flange = 3 m = 3000 mm

$$\therefore \quad \text{Slenderness ratio} = \frac{l}{r_y} = \frac{3000}{41.2} = 72.82$$

From table for $\frac{l}{r_y} = 72.82$ and $\frac{D}{t_f} = 25$

$$\sigma_{bc} = 144 - \left(\frac{144-141}{75-70}\right) \times (72.82 - 70) = 142.31 \text{ MPa}$$

$\therefore$ For $\frac{l}{r_y} = 72.82$ and $\frac{D}{t_f} = 30$

$$\sigma_{bc} = 143 - (143 - 140)\frac{2.82}{5} = 141.3 \text{ MPa}$$

$\therefore$ For $\qquad \dfrac{l}{r_y} = 72.82 \quad$ and $\quad \dfrac{D}{t_f} = 28.8$

$$\sigma_{bc} = 142.3 - (142.3 - 141.3)\frac{(28.8 - 25)}{(30 - 25)} = 141.54 \text{ MPa}$$

Moment of resistance of the beam,

$$\text{M.R.} = \sigma_{bc} \times Z_x = 141.54 \times 3060 \times 10^3$$
$$= 4.33 \times 10^8 \text{ N-mm} = 4.33 \times 10^2 \text{ kN-m}$$

Equating the bending moment to the moment of resistance,

$$\text{BM at centre} = \frac{1.23 \times 9^2}{8} + W \times 4.5 - W \times 1.5 = 12.45 + 3\,W$$

$$3\,W + 12.45 = 433 \text{ kN-m}$$

or $\qquad W = 140.18$ kN

4.3 What are the various modes of failure for a steel beam?

[6 marks : 2019]

Solution:

Failure mode	Description	Illustration	Comments
Excessive bending	Beam is adequately braced in the lateral plane. Plate components are not too Components are not too thin (compact section) Beam fails due to excessive deformation in the plane of loading.	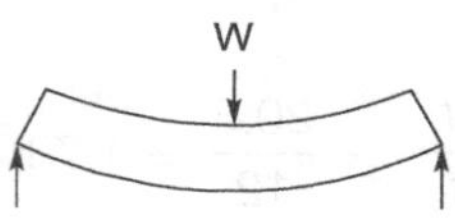	This is the mode of failure if all other modes of failure are prevented.
Lateral torsional buckling	In this case failure occurs due to lateral deflection as well as twist. The load at which this failure occurs depends upon the proportion of the beam, manner of loading and support conditions.	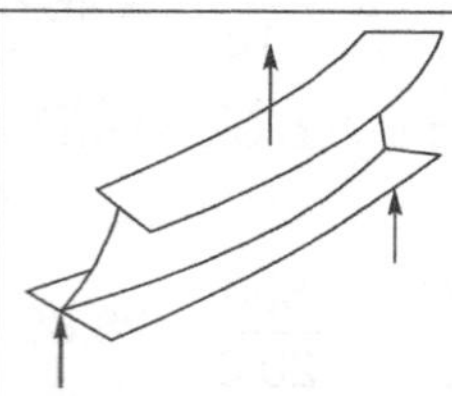	By providing suitable lateral bracing this can be prevented.
Local buckling	Flange may buckle due to compression. Web may buckle due to shear or due to combined effect of shear and bending or due to direct vertical compression under a concentrated load.	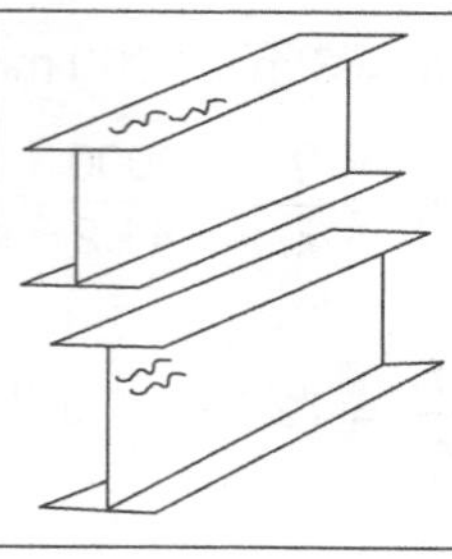	This failure is unlikely for hot rolled sections since their proportioning are made suitably. However in plate girders web stiffening may be necessary. Bearing stiffeners are provided at supports and under point loads.
Local Failure	This may occur due to the following: (i) Web may yield due to shear (ii) Web may suffer local crushing (iii) Local failure around any openings if present in the web	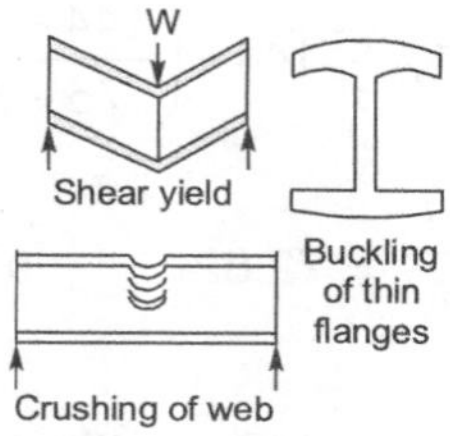	These are likely in short span or deep beam. Suitable web stiffening should be done. Regions surrounding web holes may be strengthened by local reinforcement.

4.4 Find the web buckling and web crippling strength of a beam (ISLB 350) simply supported at both ends. Assume the stiff bearing length 100 mm and grade of steel E 250.
Section properties of ISLB 350 : t_w = 7.4 mm, t_f = 11.4 mm, R = 16 mm
R = radius of root.
Given: Design compressive stress f_{cd} (N/m²)

KL/r	f_{cd}
90	121
100	107
110	94.6

[12 marks : 2020]

Solution:

ISLB 350 t_f = 11.4 mm, R = 16 mm

(i) **Web buckling strength:**

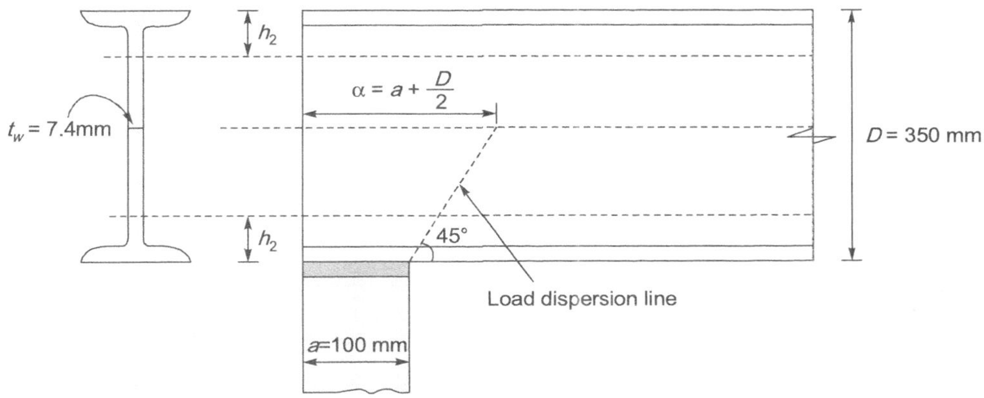

$$h_2 = \text{Depth of fillet}$$
$$= t_f + R = 11.4 + 16 = 27.4 \text{ mm}$$
$$a = \text{Stiff bearing length} = 100 \text{ mm}$$

$$\text{Width of web at N.A.} = \alpha = a + \frac{D}{2} = 100 + \frac{350}{2} = 275 \text{ mm}$$

Section area of imaginary column $= A_g = \alpha \times t_w = 275 \times 7.4 = 2035 \text{ mm}^2$

Slenderness ratio of imaginary column

$$= \frac{kL}{r_{min}} = \frac{2.425d}{t_w}$$

d = depth of web between toes of fillets = $D - 2h_2 = 350 - 2 \times 27.4$

$$d = D - 2h_2 = 350 - 2 \times 27.4 = 295.2 \text{ mm}$$

So, for
$$\frac{kL}{r_{min}} = \frac{2.425d}{t_w} = \frac{2.425 \times 295.2}{7.4} = 96.73$$

$$f_{cd} = 121 - \frac{6.73}{10} \times (121 - 107) \quad \text{(from the given table)}$$
$$= 111.58 \text{ N/mm}^2$$

So, imaginary column strength = web buckling strength

$$= f_{cd} \times A_g = 111.58 \times 2035$$
$$= 227.06 \text{ kN}$$

(ii) **Web crippling strength:**

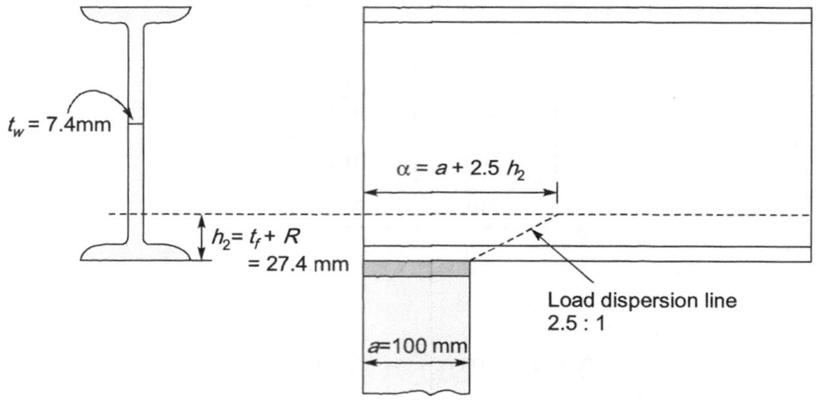

$$\alpha = a + 2.5\,h_2 = 100 + 2.5 \times 27.4 = 168.5 \text{ mm}$$

Web bearing strength $= P_b = (\alpha \times t_w) \times \sigma_P$

at toe of fillet $\quad \sigma_P$ = Design bearing strength $= \dfrac{f_y}{1.1}$

$$P_b = (168.5 \times 7.4) \times \frac{250}{1.1} = 283.386 \text{ kN}$$

4.5 A rolled steel joist ISMB 450 is used as beam for the roof of a hall 7.5 m x 12 m. Thickness of RC slab is 125 mm. The rolled steel joists are spaced at 3 m centre to centre. The floor finishing load is 1.5 kN/m^2 and the roof slab has to support a live load of 4 kN/m^2. Assume the self-weight of the beam as 1 kN/m. Take the width of bearing for the beam as 300 mm. The limiting deflection for the beam is span/240. γ_{mo} = 1.1, f_y = 250 MPa. Check the adequacy of the section against any two modes of failure.

Properties of ISMB 450:

Elastic section modulus = 30390.8 × 10^3 mm^3

Plastic section modulus, Z_p = 1533.36 × 10^3 m^3

Depth of section, h = 450 mm

Width of flange, b_f =150 mm

Thickness of flange, t_f = 17.4 mm

Thickness of web, t_w = 9.4 mm

Radius at root = 15 mm

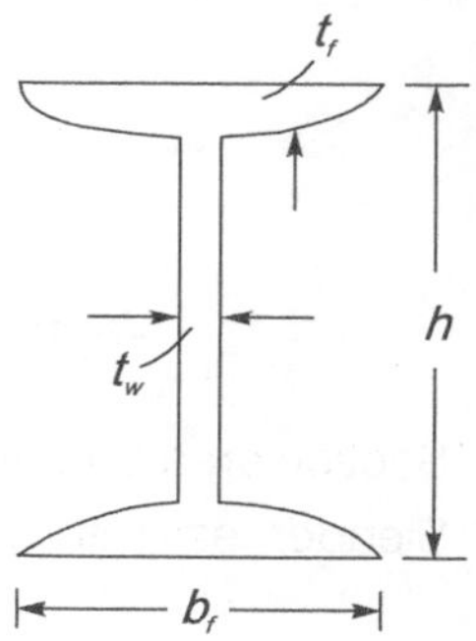

Shear capacity, $V_d = \dfrac{A_v \cdot f_y}{\sqrt{3}\,\gamma_{mo}}$

Design bending strength, $M_d = \dfrac{\beta_b \cdot Z_p \cdot f_y}{\gamma_{mo}}$

Slenderness ratio	Design Compressive Stress (f_{cd})
100	107
110	94.6
120	83.7

Buckling strength $F_b = (b_1 + n_1)\, t_w \cdot f_{cd}$

Capacity of web at its connection to the flange, $F_w = \dfrac{(b_1 + n_2) t_w \cdot f_{yw}}{\gamma_{mo}}$

The section ISMB 450 may be assumed as plastic.

[20 Marks : 2021]

Solution:

In this question, Z_e value given in question is greater than Z_p value. Which is not correct. Still solution is done using same values given in question.

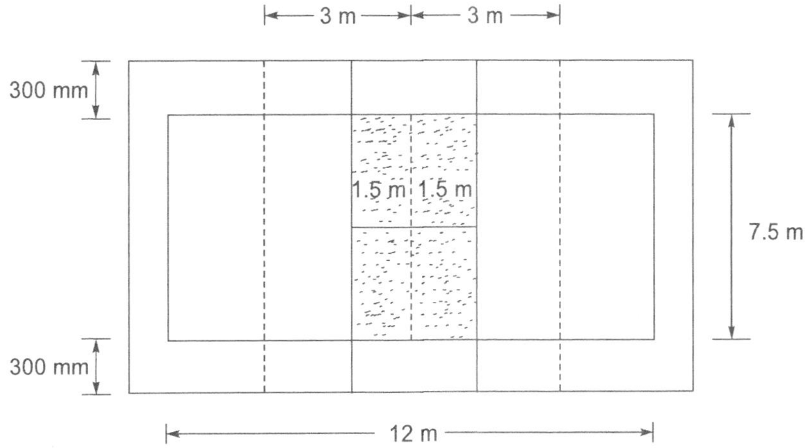

Shaded region is tributory area for which joist will be designed is:

(1) **Load's on Beam**

$w_s \rightarrow$ Self - *wt* of slab $= 0.125 \times 3 \times 24$ kN/m^3

(Assuming unit-*wt* of *RCC* slab $= 24$ kN/m^3)

$= 9$ kN/m

$w_f \rightarrow$ Weight of finishing $= 1.5$ kN/m$^2 \times 3$ m $= 4.5$ kN/m

$w_b \rightarrow$ Self - *wt* of Beam $= 1$ kN/m

$w_L \rightarrow$ Live load $= 4$ kN/m$^2 \times 3$ m $= 12$ kN/m

Total service load, $w = w_s + w_t + w_b + w_L = 9 + 4.5 + 1 + 12 = 26.5$ kN/m.

Factored load or design load

$$w_u = \gamma_F \times w = 1.5 \times 26.5 = 39.75 \text{ kN/m}$$

(2) Maximum bending moment and maximum shear force

Effective span of beam = c/c distance between support

$$= 7.5 + \frac{0.3}{2} + \frac{0.3}{2} = 7.8 \text{ m}$$

$$\text{Maximum BM} = (\text{B.M})_{max} = M_u = \frac{w_u \cdot L^2}{8} = \frac{39.75 \times 7.8^2}{8}$$

$$= 302.3 \text{ kNm}$$

Maximum applied shear force,

$$V_u = \frac{w_u \cdot L}{2} = \frac{39.75 \times 7.8}{2} = 155.025 \text{ kN}$$

(3) Adequacy check for ISMB 450.

(i) Shear strength of the beam

Shear capacity $$V_d = \frac{A_v \cdot f_y}{\sqrt{3} \cdot \gamma_{mo}}$$

$$= \frac{(ht_w) \cdot f_y}{\sqrt{3}\,\gamma_{mo}} = \frac{(450 \times 9.4) \times 250}{\sqrt{3} \times 1.1} = 555.044 \text{ kN}$$

As $V_d > V_u$, safe in shear

Also, $$0.6V_d = 0.6 \times 555.044 = 333.03 \text{ kN}$$

$$V_u < 0.6V_d$$

⇒ Low shear case

(ii) Bending strength of beam

$$M_d = \frac{\beta_b Z_p f_y}{\gamma_{mo}} \leq 1.2 Z_e \frac{f_y}{\gamma_{mo}}$$

$$= \frac{1 \times 1533.36 \times 10^3 \times 250}{1.1} \leq \frac{1.2 \times 30390.8 \times 10^3 \times 250}{1.1}$$

$$= 348.49 \text{ kNm} \leq 8288.4 \text{ kNm}.$$

Also, $V_u < 0.6\ V_d$. So no reduction in bending strength required as it is a low shear case.

Hence, $M_d > M_u$ (Beam is safe in Bending)

(iii) Web buckling strength

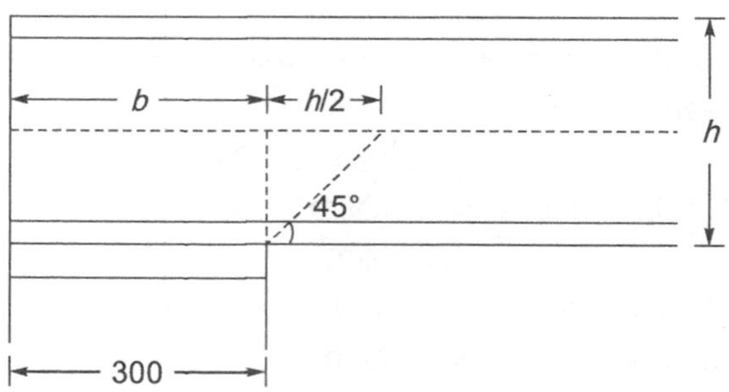

$$P_{WB} = A_e \cdot f_{cd} = (b + n_1) \cdot t_w \cdot f_{cd}$$

Where, f_{cd} is design compressive stress which depends upon slenderness ratio of web:

$$\lambda = 2.45\frac{d}{t_w} = 2.45 \times \frac{h - 2(t_F + R_1)}{t_w}$$

$$= 2.45 \times \frac{450 - 2(17.4 + 15)}{9.4} = 100.4$$

Using table given

Slenderness Ratio	f_{cd} (N/mm^2)
100	107
110	94.6
120	83.7

Using linear interpolation to determine f_{cd} value corresponding to $\lambda = 100.4$.

$$\frac{100.4-100}{110-100} = \frac{f_{cd}-107}{94.6-107}$$

$$f_{cd} = 106.54 \text{ N/mm}^2$$

$$\therefore \quad P_{WB} = \left(300+\frac{450}{2}\right)\times 9.4\times 106.54 = 525.6 \text{ kN}$$

As $P_{WB} > V_u$, safe in web buckling.

(iv) Web crippling strength

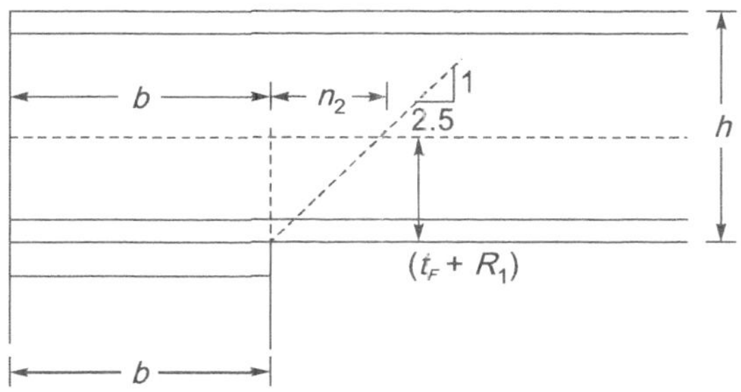

$$P_{wc} = A_e\cdot\frac{f_y}{\gamma_{mo}} = \frac{(b+n_2)\,t_w\,f_y}{\gamma_{mo}}$$

$$= \frac{[300+2.5(17.4+15)]\times 9.4\times 250}{1.1} = 813.96 \text{ kN}$$

As $P_{wc} > V_u$, safe in web crippling.

(v) Check for deflection

$$\delta_{cal} = \frac{5}{384}\frac{wL^4}{EI} \quad (\text{Where } I = Z_e\times\frac{h}{2})$$

$$= \frac{5}{384}\times\frac{26.5\times(7.8\times 10^3)^4}{2\times 10^5\times 30390.8\times 10^3\times\frac{450}{2}} = 0.934 \text{ mm}$$

$$\text{Limiting deflection} = \frac{\text{Span}}{240} = \frac{7800}{240} = 32.5 \text{ mm}$$

As δ_{cal} < limiting deflection.

Hence beam is safe in deflection.

Conclusion: ISMB 450 satisfies all the criteria's hence, it is Adequate.

4.6 A simply supported beam of 6 m span carries a udl of 70 kN/m including its self weight. The available section is ISMB 450. Cover plates of 8 mm thick may be provided, if required, only on the top flange. Design the steel beam and check for shear and deflection.

Take $\gamma_f = 1.5$; $\gamma_{m0} = 1.1$. Assume that the section is plastic,
$f_y = 250$ MPa.
Sectional area of ISMB 450; $A = 9227$ mm^2.
Width of flange = 150 mm.
Thickness of flange = 17.4 mm.
Thickness of web = 9.4 mm.
Moment of inertia $I_{zz} = 30390.8$ cm^4
Radius at root = 15 mm
Plastic modulus $Z_{pz} = 1533.4$ cm^3

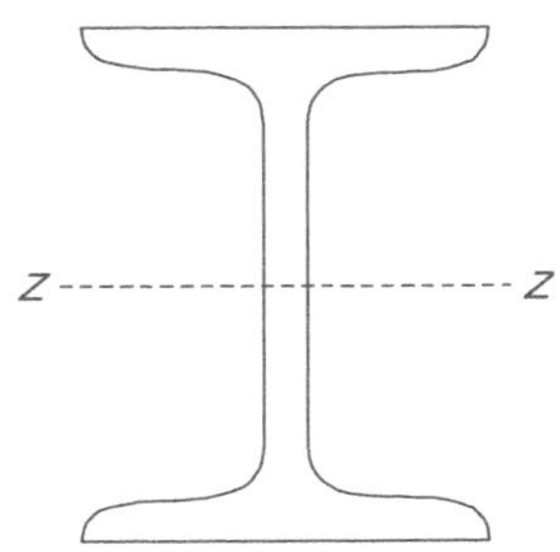

[20 marks : 2022]

Solution:

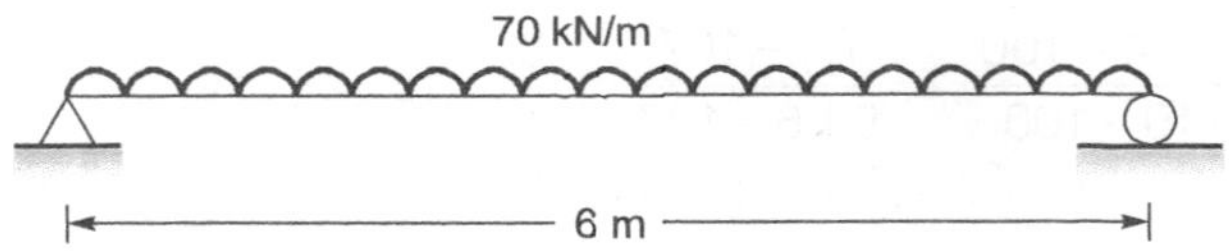

Given: Beam cross-section ISMB 450

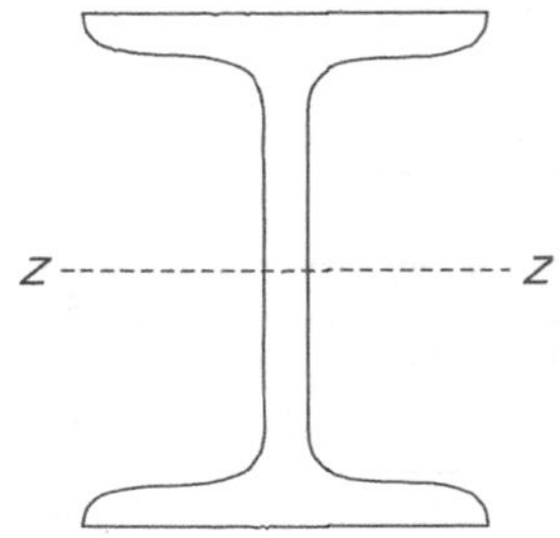

$A = 9227\text{ mm}^2$, $h = 450$ mm, $b = 150$ mm, $t_f = 17.4$ mm, $t_w = 9.4$ mm
$I_{zz} = 30390.8\text{ cm}^4$, $R_1 = 15$ mm, $Z_{pz} = 1533.4\text{ cm}^3$, Section is plastic
Factored bending moment

$$M_{uz} = \gamma_f \frac{w.L^2}{8} = \frac{1.5\times70\times6^2}{8} = 472.5 \text{ kNm}$$

Maximum factored shear force,

$$V_u = \gamma_f \frac{wL}{2} = \frac{1.5\times70\times6}{2} = 315 \text{ kN}$$

Section modulus required

$$(Z_{pz})_{\text{reqd.}} = \frac{M_{uz}}{\left(\frac{f_y}{\gamma_{m0}}\right)} = \frac{472.5\times10^6}{\left(\frac{250}{1.1}\right)} = 2079000 \text{ mm}^3 = 2079\times10^3 \text{ mm}^3$$

$$(Z_p)_{\text{rolled}} = 1533.4\times10^3 \text{ mm}^3$$

$$\therefore \quad (Z_p)_{\text{rolled}} < (Z_p)_{\text{reqd.}}$$

Let us provide two cover plates of size 310 mm × 8 mm at top flange.

Properties of built-up section:

Thickness of cover plate provided = 8 mm

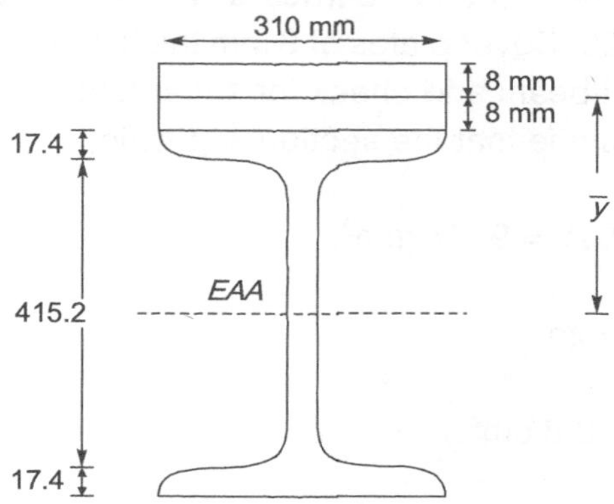

Equal area axis

$$A_1 = A_2$$

$$2\times(310\times 8) + 150\times\bar{y} = 150(17.4-\bar{y})+415.2\times 9.4+150\times 17.4$$

$$\Rightarrow \quad \bar{y} = 13.876 \text{ mm}$$

$$(Z_{pz})_{\text{built-up}} = \frac{A}{2}(\bar{y}_1+\bar{y}_2)$$

$$\bar{y}_1 = \frac{(310\times 8\times 2)(8+13.876)+(150\times 13.876)(13.875/2)}{7041.4}$$

$$= \frac{108504.96+14440.75}{7041.4} = 17.46 \text{ mm}$$

$$\bar{y}_2 = \frac{(150\times 3.524)\times\frac{3.524}{2}+(415.2\times 9.4)\left(\frac{415.2}{2}+3.524\right)+(150\times 17.4)\left(\frac{17.4}{2}+415.2+3.524\right)}{7039.6}$$

$$= \frac{931.4+823991.63+1115576.64}{7039.6} = 275.65 \text{ mm}$$

$$Z_p = \frac{14187}{2}(17.46+275.65)$$

$$= 2079.18\times 10^3 \text{ mm}^3$$

It is greater than $(Z_{pz})_{reqd}$. Hence, safe

Centrodial axis from top

$$\text{C.G distance} = \frac{(310\times 8\times 2)\times 8+9227\times 241}{(310\times 8\times 2)+9227}$$

$$= \frac{2263387}{14187} = 159.54 \text{ mm}$$

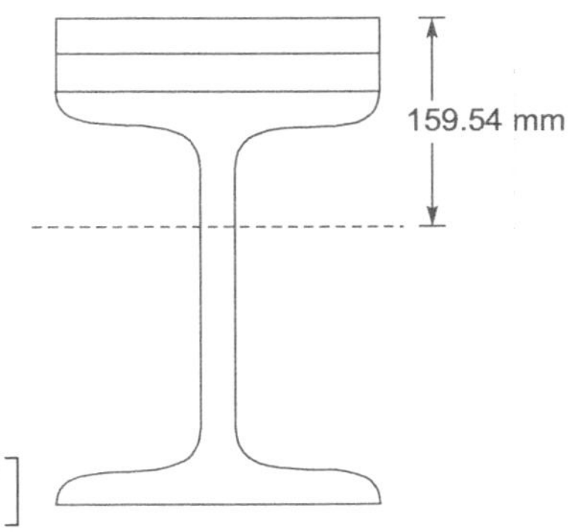

$$I_{CG} = \left[\frac{310\times 16^3}{12}+310\times 16(159.54-8)^2\right] + \left[30390.8\times 10^4+9227(241-159.54)^2\right]$$

$$= (11400.91+36513.58)\times 10^4 \text{ mm}^4$$

$$= 47914.5 \text{ cm}^4$$

Shear strength check for shear buckling

$$\frac{d}{t_w} = \frac{450-2(t_f+R)}{9.4} = \frac{450-2(17.4+15)}{9.4}$$

$$= 40.98 < 67\epsilon \qquad \text{where } \epsilon = \sqrt{\frac{250}{f_y}} = 1$$

Hence, web will not fail in shear buckling

$\therefore$ Shear strength, $$V_d = \frac{f_y}{\sqrt{3}.\gamma_{m0}} \times A_v = \frac{250}{\sqrt{3} \times 1.1} \times 450 \times 9.4 \text{ N} = 555.04 \text{ kN} > V_u$$

Hence safe in shear.

Check for deflection

Deflection are calculated at service load

$$\Delta_{cal} = \frac{5}{384}\frac{WL^4}{EI} = \frac{5 \times 70 \times (6000)^4}{384 \times 47914.5 \times 10^4 \times 2 \times 10^5} = 12.327 \text{ mm}$$

$$\Delta_{max} = \frac{\text{Span}}{300} = \frac{6000}{300} = 20 \text{ mm}$$

As $\Delta_{cal} < \Delta_{max}$

Hence safe in deflection.

5. Plate Girders and Industrial Roofs

5.1 The joint *A* of a roof truss is shown below, where the main rafter and the main tie intersect. It is a hinged joint. It is subjected to end reactions as shown below: (1) Dead load + Live load and (2) Dead load + Wind load. Effective length of the main rafter is 2.83 m. Check the assumed section of the main rafter of its safety against compression. Properties of the section assumed—double-angle back-to-back, angle being 80 mm × 50 mm × 10 mm having an area of 12.02 cm² each, r_{xx} of the double-angle section = 2.49 cm.

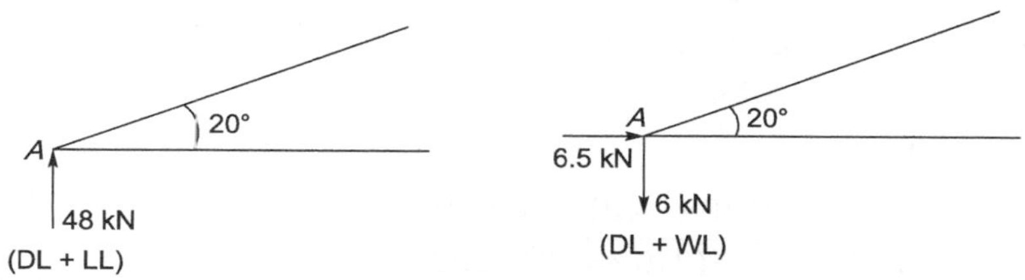

l/r	0	10	20	30	40	50	60	70	80	90	100	110	120	130
f_c (kg/cm²)	1250	1246	1239	1224	1203	1172	1130	1075	1007	928	840	753	671	597
l/r	140	150	160	170	180	190	200	210	220	230	240	250	300	350
f_c (kg/cm²)	531	474	423	377	336	300	270	243	219	199	181	166	109	76

[12 marks : 2010]

Solution:

Case-1: For Deal load + Live load

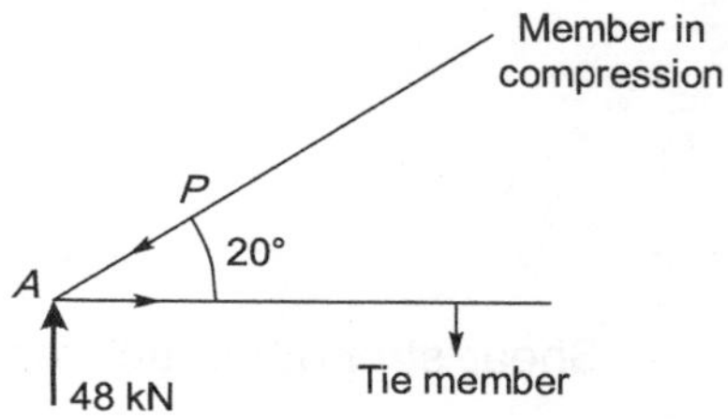

$$P \sin 20° = 48$$

$$\Rightarrow \quad P = \frac{48}{\sin 20°} = 140.34 \text{ kN}$$

Case-2: For Dead load + Wind load

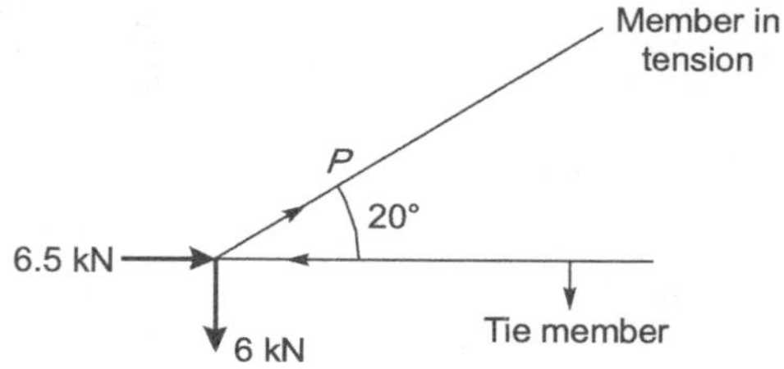

$$P \sin 20° = 6$$

$$P = 17.54 \text{ kN}$$

In the load combination *DL* + *LL*, only the rafter is in compression hence, for this case only, we will check the safety of main rafter in compression.

r_{xx} of combination = 2.49 cm

Let us assume that r_{xx} as the r_{min} value

$$L_{eff} = 2.83 \text{ m} = 283 \text{ cm}$$

$$\text{Slenderness ratio, } \lambda = \frac{L_{eff}}{r_{min}} = \frac{283}{2.49} = 113.65$$

$\therefore$ σ_{ac} can be determined by interpolation

$$\sigma_{ac} = 671 + \frac{753-671}{120-110} \times (120-113.65) = 723.07 \text{ kg/cm}^2$$

$$\text{Force that can be resisted} = 723.07 \times 12.02 \times 2 = 17382.60 \text{ kg}$$

$$= 173.82 \text{ kN} > 140.34 \text{ kN}$$

Hence the section is safe.

5.2 **Design a suitable section for steel plate beam girder carrying a uniformly distributed load of 60 kN/m (including self weight) over an effective span of 15 m. Check for shear and deflection also. Use limit state method of design. Use ISWB 600 @ 145.1 kg f/m with I_{xx} = 106198.5 cm^4 and I_{yy} = 4702.5 cm^4.**
[20 marks : 2013]

Solution:

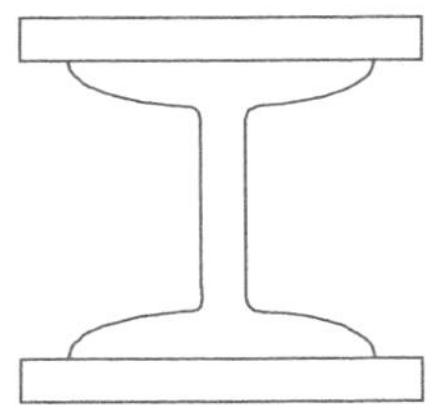

$$W = 60 \text{ kN/m}$$

$\Rightarrow$ Factored load $(W_u) = 1.5 \times 60 = 90$ kN/m, $l = 15$ m

Factored BM $$(M_u) = \frac{W_u l^2}{8} = \frac{90 \times 15^2}{8} = 2531.25 \text{ kNm}$$

Assuming section qualifies for atleast '*c*' class (semi-compact section) and

$$t_w = 11.6 \text{ mm}$$

$$\frac{600}{11.6} = \frac{d}{t_w} = 51.72 < 67\,\varepsilon$$

Where $\varepsilon = \sqrt{\frac{250}{f_y}} = 1$

(No interaction of shear and bending stresses in web)

$$M_d = Z_e\left(\frac{f_y}{1.1}\right) \quad \text{(for semi-compact section)}$$

As $$M_d \geq M_u$$

$$Z_{e\,reqd.}\left(\frac{f_y}{1.1}\right) \geq 2531.25 \text{ kNm}$$

$$Z_{e\,reqd.} \geq 11.1375 \times 10^6 \text{ mm}^3 = 11.1375 \times 10^3 \text{ cm}^3$$

$$Z_{d\,(ISWB\,600)} = \frac{I_{xx}}{y_{max}} = \frac{106198.5 \text{ cm}^4}{30 \text{ cm}} = 3539.95 \text{ cm}^3 = 3.54 \times 10^3 \text{ cm}^3$$

As $Z_{e\,reqd.} > Z_{d\,member}$ of ISWB 600

Plates are required.

Area of each plates, $$A = \frac{Z_{e\,reqd.} - Z_d}{D} = \frac{(11.14-3.54) \times 10^3 \text{cm}^3}{60 \text{cm}} = 126.66 \text{ cm}^2$$

$$\text{Choosing plate thick} = 2.0 \text{ cm} = 20 \text{ mm}$$
$$\text{Width} = 80 \text{ cm} = 800 \text{ mm}$$
Area of Each plate, $A = 16000 \text{ mm}^2 = 160 \text{ cm}^2$

$$Z = \frac{106198.5 \text{ cm}^4 + 2\times160(30+1)^2}{32 \text{ cm}} = 12.93 \times 10^3 \text{ cm}^3 > Z_{reqd.}$$

Check for Shear

$$V_u = \frac{w_u L}{2} = \frac{90\times15}{2} = 675 \text{ kN}$$

$$V_d = A_v\left(\frac{f_y}{\gamma_{mo}}\right)\frac{1}{\sqrt{3}}$$

A_v = area under shear, $\gamma_{mo} = 1.1$

$$\frac{f_y}{\gamma_{mo}\times\sqrt{3}} = 0.5248 f_y$$

$$A_v = D\times t_w = (600 + 2\times20)\times11.6 = 7424 \text{ mm}^2$$
$$V_d = 7424\times0.5248\times250 \text{ N} = 974.03 \text{ kN}$$

As $V_u < V_d$ (OK safe in shear)

Check for Deflection

Deflection is serviceability limit state, hence calculated on the basis of unfactored loads.

$$\delta_{act} = \frac{5}{384}\frac{wl^4}{EI} = \frac{5}{384}\times\frac{(60 \text{ kN/m})(15 \text{ m})^4}{(2\times10^{11} \text{ Pa})(413718.5 \text{ cm}^4)}$$

$$I = 413718.5 \text{ cm}^4 \text{ (of combined section)}$$

$$\delta_{act} = \frac{5}{384}\times\frac{(60\times10^3 \text{ N/m})(15 \text{ m})^4}{(2\times10^{11} \text{ Pa})(413718.5\times10^{-8} \text{ m}^4)} = 0.047799 \text{ m}$$

$$\delta_{act} = 47.8 \text{ mm}$$

$$\delta_{allowable} = \frac{l}{300} = \frac{15\times10^3}{300} = 50 \text{ mm}$$

$$\delta_{act} < \delta_{allowable} \quad \text{OK.}$$

5.3 **Two wheels, placed at a distance of 2.5 m part, with a load of 200 kN on each of them, are moving on a simply supported girder (I-section) of span 6.0 m from left to right. The top and bottom flanges of the I-section are 200 × 20 mm and the size of web plate 800 × 6 mm.**
If the allowable bending compressive, bending tensile and average shear stresses are 110 MPa, 165 MPa and 100 MPa respectively, check the adequacy of the section against bending and shear stress. Self weight of the girder may be neglected.

[20 marks : 2015]

Solution:

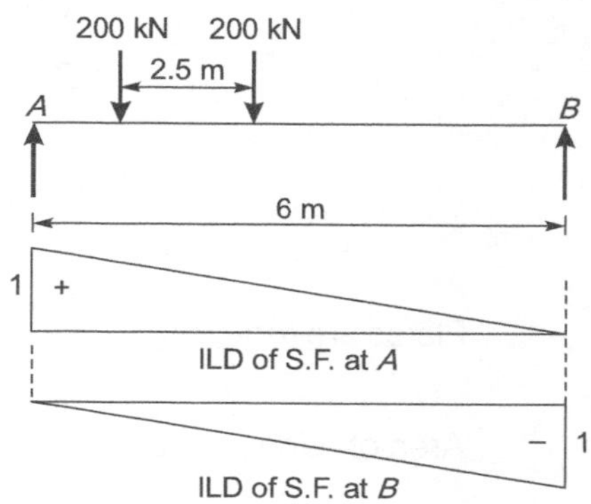

Maximum S.F. will occur when left 200 kN load is at A or right 200 kN load is at B.

$$\Rightarrow \quad R_A\times6 = 200\times6 + 200(6-2.5)$$
$$\therefore \quad R_A = 316.67 \text{ kN}$$
$$\Rightarrow \quad \text{Maximum S.F.} = 316.67 \text{ kN}$$

Maximum B.M. will occur when one of 200 kN load and resultant of both 200 kN loads are equidistant from center of span as shown.

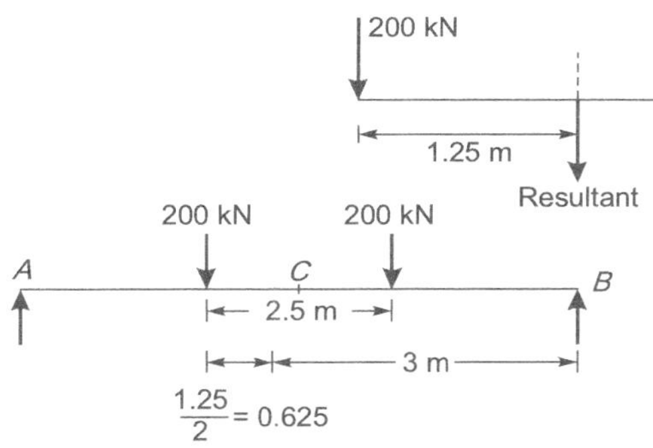

$$\Rightarrow \quad \Sigma M_B = 0$$

$$\Rightarrow R_A \times 6 - 200 \times (3 + 0.625) - 200\,(3 + 0.625 - 2.5) = 0$$

$$\Rightarrow \quad R_A = 158.33 \text{ kN}$$

$\therefore$ Max. BM under left 200 kN load $= R_A(3 - 0.625) = 158.33\,(2.375) = 376.034$ kNm

$$I_{xx} = 200 \times \frac{840^3}{12} - 194 \times \frac{800^3}{12} = 1601.067 \times 10^6 \text{ mm}^4$$

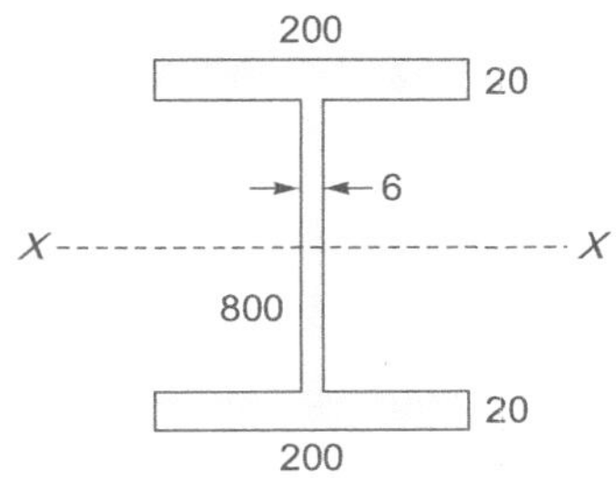

Maximum bending comp. stress $= \dfrac{376.034 \times 10^6}{1601.067 \times 10^6} \times 420$

$= 98.64 \text{ N/mm}^2 < 110 \text{ N/mm}^2$ (OK)

Maximum bending tensile stress $= \dfrac{376.034 \times 10^6}{1601.067 \times 10^6} \times 420$

$= 98.64 \text{ N/mm}^2 < 165 \text{ N/mm}^2$ (OK)

Maximum shear stress $= \dfrac{F}{8I}\left[\dfrac{B}{t_w}\left(D^2 - d^2\right) + d^2\right]$

$$= \frac{316.67 \times 10^3}{8 \times 1601.067 \times 10^6}\left[\frac{200}{6}\left(840^2 - 800^2\right) + 800^2\right]$$

$= 69.88 \text{ N/mm}^2 < 100 \text{ N/mm}^2$ (OK)

$\therefore$ Section is safe against bending and shear.

5.4 The bottom chord of the truss is composed of 2 Nos. ISA 60 × 60 × 6 with gusset thickness of 10 mm (back to back and both sides of the gusset). Calculate the factored load carrying capacity of the member under compression only.
(ISA 60 × 60 × 6 : $A = 684 \text{ mm}^2$, $I_{XX} = I_{YY} = 226000 \text{ mm}^4$, $C_{XX} = C_{YY} = 16.9$ mm)

KL/r	80	90	100	110	120	130	140	150
f_{cd} (MPa)	118	105	92.6	82.1	73	65.2	58.4	52.6

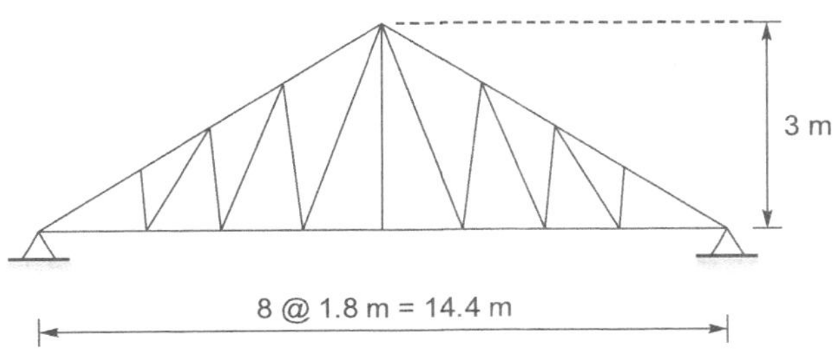

Longitudinal ties are provided at alternate node of the bottom chord.
What are high strength friction grip bolts? State their advantages?

[15+5 marks : 2015]

Solution:

Compressive load carrying capacity,

$$P_c = f_{cd} \times A_g$$

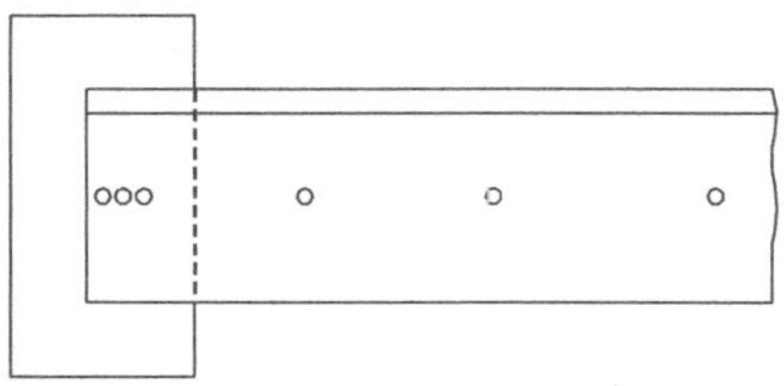

Buckling about x-x axis.

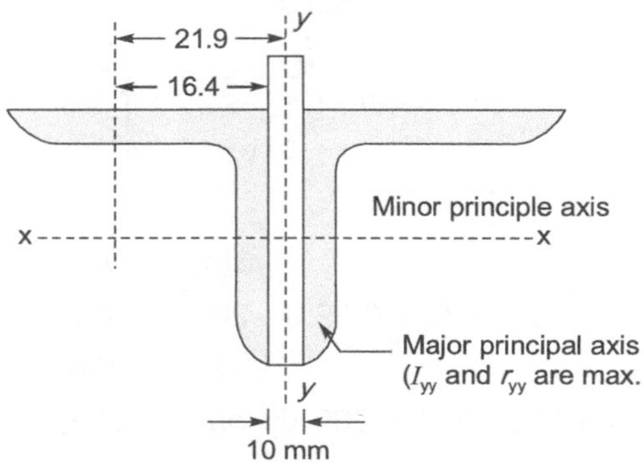

Buckling about x-x axis

$$I_{xx} = 2\,(226000) = 452000 \text{ mm}^4$$

$$I_{yy} = 2\,(226000 + 684 \times 214^2) = 1108106.48 \text{ mm}^4$$

$\therefore$ $$I_{xx} < I_{yy}$$

But $$r_{xx} = \sqrt{\frac{I_{xx}}{A}} = \sqrt{\frac{0.452 \times 10^6}{2 \times 684}} = 18.177 \text{mm}$$

$$r_{yy} = \sqrt{\frac{1108106.48}{2 \times 684}} = 28.46 \text{ mm}$$

Effective length of strut = 1.8 m

$\therefore$ $$\lambda_{xx} = \frac{l_{xx}}{r_{xx}} = \frac{1.8 \times 1000}{18.177} = 99.026$$

$$\lambda_{yy} = \frac{l_{yy}}{r_{yy}} = \frac{2 \times l_x}{r_{yy}} = \frac{2 \times 1800}{28.46} = 126.44$$

$\because$ Calculate f_{cd} by interpolation

$$f_{cd} = 73 - \frac{6.4}{10}(73 - 65.2) = 68 \text{ N/mm}^2$$

$$P_c = f_{cd} \times A_g = 68 \times (2 \times 684) = 93 \text{ kN}$$

High Strength Friction Grip bolts: These bolts (abbreviated as HSFG bolts) are provided where member length is small or the members are required to carry large forces. The resistance against the applied load is provided by the friction developed between the members due to tightened bolt. The tension developed in the bolt is about 90% of proof load. Under service loads, load transfer occurs due to friction but at ultimate loads, the slipping between the members occurs and HSFG bolts behave just like ordinary bolts.

Advantages of HSFG bolts:

1. These bolts provide rigid joint with no slipping between the members.
2. The frictional resistance is effective outside the bolt hole only and thus less amount of load gets transferred through net section and failure possibility at net section gets reduced.
3. HSFG bolts provide high static strength to the joint.
4. Bolts do not get subjected to shear bearing.

5.5 Show various components of a steel plate girder through neat sketches. Explain design criteria of each and the corresponding limit state.

[10 marks : 2016]

Solution:

Components of a Steel Plate Girder:

(a) Web plate

(b) Flange angles with or without cover plates for riveted/bolted plate girders and only flange plates for welded plate girders.

(c) Stiffeners-Bearing, transverse and longitudinal

(d) Splices for web and flange

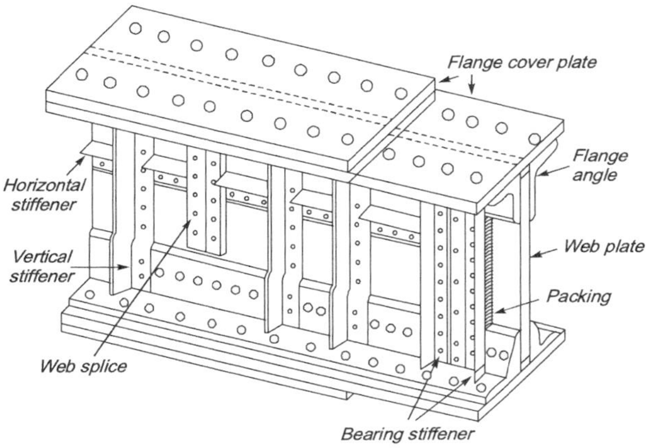

Typical elements of a riveted/bolted plate girder

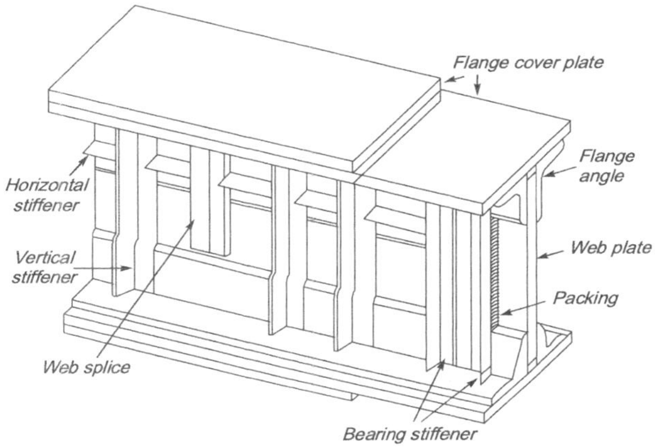

Typical elements of welded plate girder

Functions of Various Components of a Steel Plate Girder:

Flanges: Flanges of required width and thickness are required to resist the flexural moment due to dead and superimposed loads coming on the plate girder by developing compressive force in one flange and tensile force in the other flange i.e. by developing an opposite resisting couple.

Web: Web of required depth and thickness is required to keep the flange plates at the requisite distance and to resist shear in the beam.

Stiffeners: Stiffeners are provided to safeguard the web against local buckling failure. These stiffeners are provided either in the longitudinal direction called as **horizontal stiffeners** or in the transverse direction called as **vertical stiffeners** or in both the directions.

Transverse (vertical) stiffeners are of two types:

(a) **Bearing stiffener**: At the support regions, a certain portion of web acts as column i.e. compression member and there is a possibility of web buckling and hence bearing stiffeners are required.

(b) **Intermediate stiffener:** If concentrated loads are also acting on the girder (may be due to cross beams etc.) then intermediate stiffeners are also required to prevent the web from buckling.

Economical Depth of web:

$$d = \sqrt[3]{\frac{6M}{t_w f_y}}$$

For a trial girder section, d/t_w for the web may be taken as any value ranging from 135 to 240. However, the following provision of **IS 800 : 2007** are also useful:

If $d/t_w < 67\varepsilon$ then the plate girder may be designed as ordinary beam , where $\varepsilon = \sqrt{250/f_y}$

Web Thickness: Initially the web thickness can be assumed as 6 mm (if painted) or 8 mm (if unpainted)

(a) Minimum web thickness based on serviceability requirement

As per **Cl. 8.6.1.1** of **IS 800 : 2007,**

(i) When transverse stiffeners are not provided and web connected to flanges along both longitudinal edges,

$$\frac{d}{t_w} \leq 200\,\varepsilon_w$$

(ii) When transverse stiffness are not provided and web connected to flanges along one longitudinal edge only

$$\frac{d}{t_w} \leq 90\,\varepsilon_w$$

(iii) When only transverse stiffeners are provided,

Case-1: $3d \geq c \geq d$ $\quad \frac{d}{t_w} \leq 200\,\varepsilon_w$

Case-2: $0.74d \leq c \leq d$ $\quad \frac{d}{t_w} \leq 200\,\varepsilon_w$

Case-3: $c < 0.74\,d$ $\quad \frac{d}{t_w} \leq 270\,\varepsilon_w$

Case-4 $:c > 3d$

Here the web shall be considered as unstiffened.

(iv) When transverse stiffeners are provided along with longitudinal stiffener at one level only i.e. at 0.2d from the compression flange.

Case-1: $2.4d \geq c \geq d$ $\quad \frac{d}{t_w} \leq 250\,\varepsilon_w$

Case-2: $0.74d \leq c \leq d$ $\quad \frac{c}{t_w} \leq 250\,\varepsilon_w$

Case-3: $c < 0.74d$ $\quad \frac{d}{t_w} \leq 340\,\varepsilon_w$

(v) When a second longitudinal stiffener is also provided (at neutral axis of the section)

$$\frac{d}{t_w} \leq 400\,\varepsilon_w \qquad \text{where, } \varepsilon_w = \sqrt{250/f_{yw}}$$

(b) Minimum web thickness based on compression flange buckling requirement

As per **Cl. 8.6.1.2** of **IS 800 : 2007,** in order to avoid buckling of compression flange, the web thickness shall comply with the following requirements:

(i) When transverse stiffeners are not provided

$$\frac{d}{t_w} \leq 345\,\varepsilon_f^2, \qquad \text{where, } \varepsilon_f = \sqrt{250/f_{yf}}$$

(ii) When transverse stiffeners are provided

Case-1 $c \geq 1.5d$ $\quad \frac{d}{t_w} \leq 345\,\varepsilon_f^2$

Case-2 $c < 1.5d$ $\quad \frac{d}{t_w} \leq 345\,\varepsilon_f$

Size of Flanges:

Flanges of riveted/bolted plate girders consist of pair of angles with or without cover plates.

The flange angles should form as large a part of area of flange as possible and preferably not less than 1/3rd of the calculated flange area in order to keep the center of gravity of flange within the back of the angles and not in the flange cover plates otherwise the stability gets affected.

Angle section as flanges:

It is assumed that design moment is resisted entirely by the flanges and using partial factor of safety for material for plastic section,

$$M = \frac{A_f f_y d}{\gamma_{m0}} = \frac{A_f f_y d}{1.1}$$

Thus, area of flange A_f can be determined. Select **$9.4\varepsilon < t_f < 13.6\ b_f\varepsilon$** so that flexural strength can be found by the formula for semi-compact section as per **Cl. 8.2.1.2** of **IS 800:2007**as,

$$A_f = b_f t_f$$

5.6 **A factory shed is to be constructed using truss having a span of 12 m. The spacing of the truss is 4 m. Galvanished iron sheets are to be placed over purlins located at the nodes as shown in figure. Determine the design load at the nodes due to live load and wind load for an intermediate truss. Assume design wind pressure = 1500 N/m². C_{pe} = –0.7 (for wind angle 90°), C_{pi} = ±0.2. Loads are to be shown in the diagram separately.**

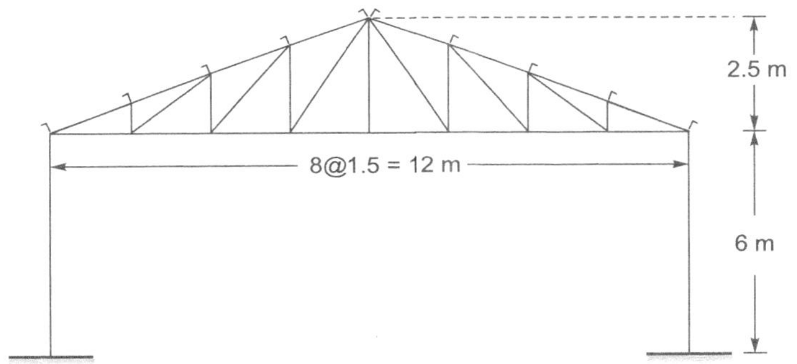

[20 marks : 2017]

Solution :

Design loads at nodes due to live load:

Let θ be inclination of roof with horizontal

$$\tan\theta = \frac{2.50}{6.0} = 0.4167$$

$$\Rightarrow \quad \theta = 22°37' = 22.62°$$

Let us assume that no access is provided to the roof. The live load is reduced by 20 kN/m² for each degree above 10° slope.

$$\text{Live load} = 750 - 20\,(22.62° - 10°) = 497.60 \text{ N/m}^2$$

$$\text{The load on each intermediate panel} = 497.60 \text{ N/m}^2$$

$$\text{Length of rafter} = \sqrt{\left(\frac{12}{2}\right)^2 + (2.5)^2} = 6.50 \text{ m}$$

$$\text{Length of each panel} = \frac{6.50}{4} = 1.625 \text{ m}$$

Panel length in plan = 1.625 cos 22.62° = 1.5 m

∴ Load on each intermediate panel = 497.60 × 4 × 1.5 N = 2.98 kN [∵ spacing of truss = 4.0 m]

The load on each end panel point = $\frac{2.98}{2}$ kN = 1.49 kN

The resultant panel loads due to live loads are shown below.

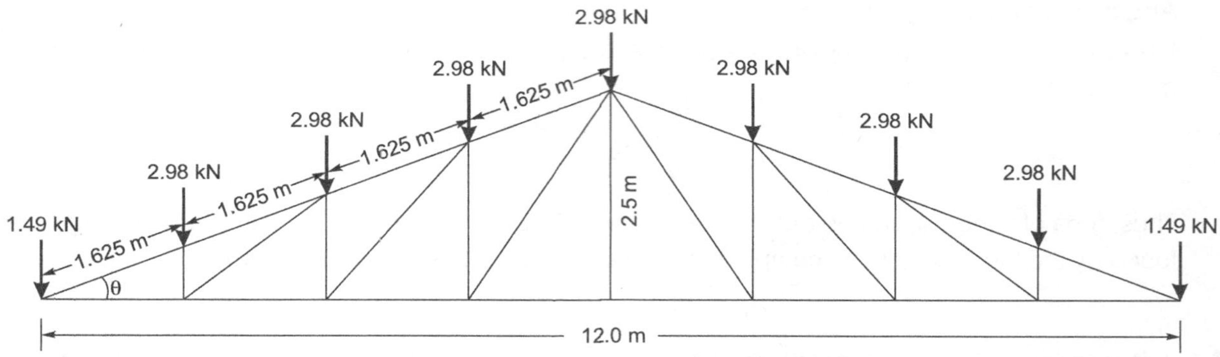

Wind loads:

Given, Design wind pressure = 1500 N/m² = 1.5 kN/m² = p_d

$$C_{pe} = -0.70 \text{ (for wind angle 90°)}$$

$$C_{pi} = \pm 0.20$$

$$\text{Wind load} = (C_{pe} - C_{pi}) \times p_d A$$

$$= (-0.70 - 0.20) \times 1.5 \times 1.625 \times 4$$

$$= -8.775 \text{ kN (up lift)}$$

$$\text{Wind load per unit length of purlin} = \frac{8.775}{4} = 2.19375 \text{ kN/m} = 2193.75 \text{ N/m}$$

Wind load on panel points

Windward side/Leeward side (Assuming C_{pe} same)

$$F = (C_{pe} - C_{pi}) P_d A$$

$$= -8.775 \text{ kN}$$

Wind load on each intermediate panel point = –8.775 kN

$$\text{Wind load on each end panel point} = \frac{-8.775}{2} = -4.3875 \text{ kN}$$

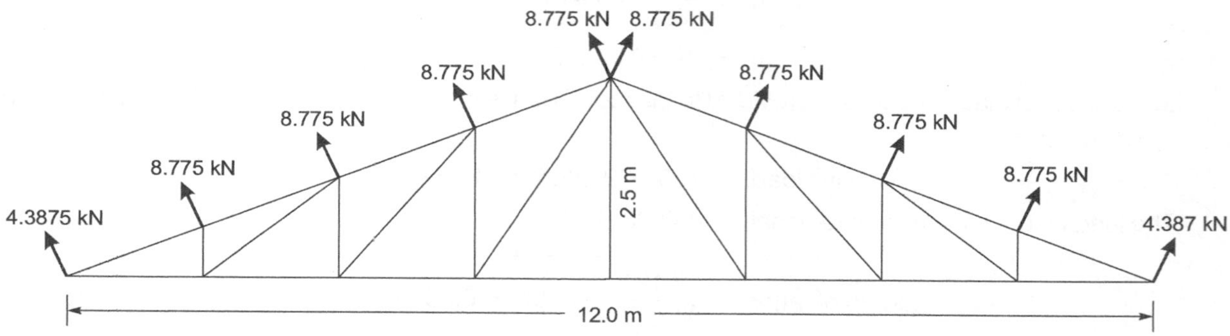

5.7 A welded plate girder of span (simply supported) 20 m is subjected to a uniformly distributed load of 100 kN/m including its self-weight. The cross-section of the girder consists of top and bottom flanges (500 mm × 50 mm) and web plate (1000 mm × 15 mm). The section is safe against flexure and shear. Design the connection between the flange and web only/ Use γ_{mw} = 1.25, and E 250 grade of steel. Use the limit state method of design.

[20 marks : 2017]

Solution :

Given, E250 steel, $f_y = f_{yp} = f_{yw} = 250$ MPa

$$E = 2 \times 10^5 \text{ MPa}$$

Size of web plate, $d \times t_w = 1000 \text{ mm} \times 15 \text{ mm}$

Size of flange plates, $b_f \times t_f = 500 \text{ mm} \times 50 \text{ mm}$

Total uniform load including self weight = 100 kN/m

Factored load, $W_u = 1.5 \times 100 = 150$ kN/m

Span of plate girder, $L = 20$ m

∴ Maximum shear force,
$$V = \frac{W_u L}{2} = \frac{150 \times 20}{2} = 1500 \text{ kN}$$

Flange to web connection:

There will be two weld lengths along the span for each flange to web connection.

∴
$$q_w = \frac{V A_f \bar{y}}{2 I_z} = \text{Horizontal shear force/unit length of weld}$$

$$I_z = \frac{b_f D^3}{12} - \frac{(b_f - t_w) d^3}{12}$$

where, $D = 1000 + 2t_f = 1100$ mm [∴ $t_f = 50$ mm]

⇒
$$I_z = \frac{500 \times (1100)^3}{12} - \frac{(500-15) \times (1000)^3}{12}$$

⇒
$$= 5.545833 \times 10^{10} - 4.041667 \times 10^{10} = 1.504166 \times 10^{10} \text{ mm}^4$$

∴
$$q_w = \frac{1500 \times 500 \times 50 \times (500 + 50/2)}{2 \times 1504166 \times 10^4} = 0.6545 \text{ kN/mm}$$

Let us provide weld of size, $S = 10$ mm

Strength of shop weld per unit length

$$f_{wd} = \frac{f_y}{\sqrt{3}\gamma_{mw}}(kS) = \frac{0.70 \times 10 \times 250 \times 10^{-3}}{1.25 \times \sqrt{3}}$$

$$= 0.808 \text{ kN/mm} > 0.6544 \text{ kN/mm} \quad \text{(OK)}$$

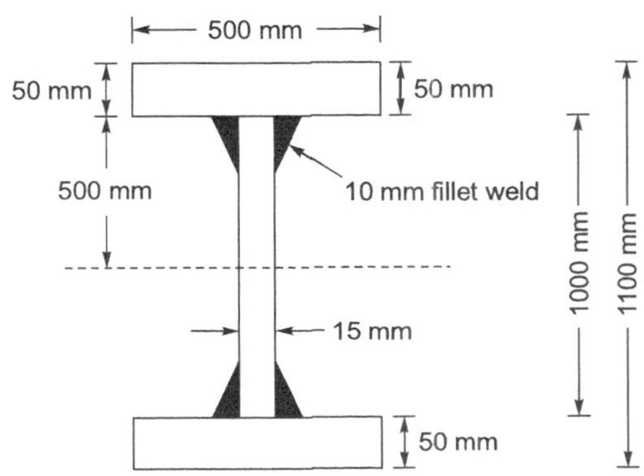

5.8 Determine the ultimate bending moments and forces due to vertical and horizontal loads that act on a simply-supported gantry girder. Use the following data:

1. Simply supported span = 6 m
2. Distance between crane wheels = 3.6 m
3. Self-weight of girder = 1.5 kN/m
4. Maximum crane wheel load (static) = 220 kN
5. Weight of crab/trolley = 60 kN
6. Maximum hool load = 200 kN

Take impact factor of 25% and assume double flanged wheels e = 0.15 m while the girder depth, D = 0.60 m.

[20 marks : 2018]

Solution:

Load on Gantry Girder are expressed here in the figure.

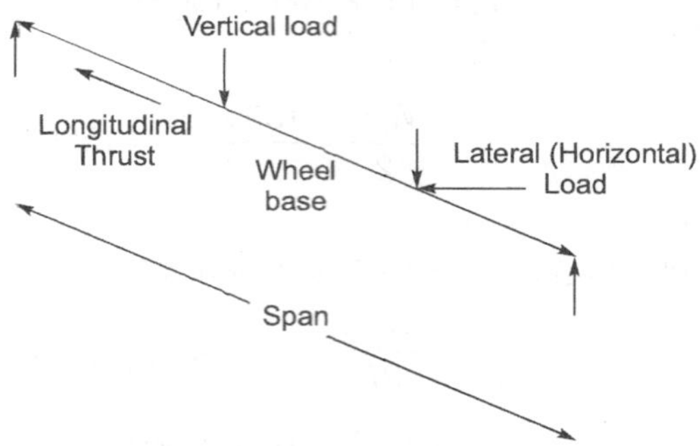

Since the impact factor is 25% crane is electric overhead.

	For electrically operated OT crane		Additional load
1.	Vertical forces transferred to rails	-	25% of maximum static wheel load.
2.	Lateral loads (transverse to the rails)	-	10% of weight of the crab and weight lifted on the crane.
3.	Longitudinal loads (along the rails)	-	5% of the static wheel load.

$$\text{Maximum permissible deflection} = \frac{L}{750}$$

For Fe 410 grade of steel, f_u = 410 MPa, $f_y = f_{yw} = f_{yf}$ = 250 MPa

Given:

- Simply supported span = 6 m
- Distance between crane wheels = 3.6 m
- Self weight of girder = 1.5 kN/m.
- Maximum crane wheel load = 220 kN
- Trolly weight = 60 kN
- Maximum hook load = 200 kN
- Diameter of crane wheel = 150 mm
- Depth of girder = 600 mm

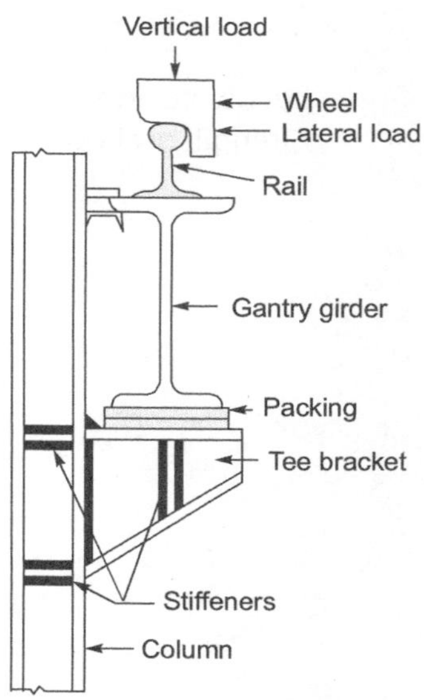

Loads depicted on a gantry girder supporting light cranes

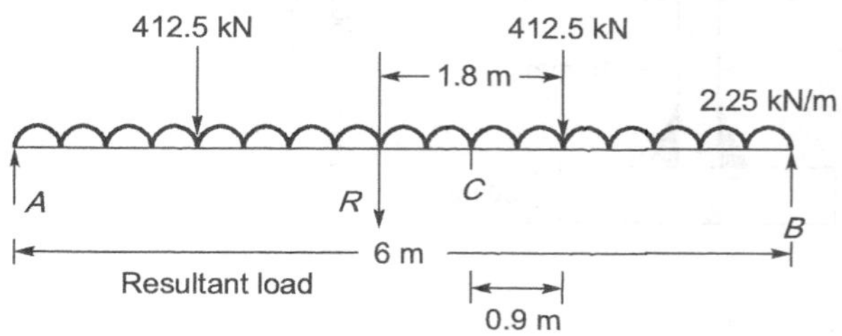

1. **Maximum bending moment:**

Self weight of girder = 1.5 kN/m

Neglecting self weight of rail section,

$\therefore$ Factored DL = 1.5 × 1.5 = 2.25 kN/m

Factored maximum crane wheel load (static) = 1.5 × 220 = 330 kN

25% addition load due to impact, w_u = 1.25 × 330 = 412.5 kN

Note: *For maximum bending moment, keep the loads such that resultant of loads and chosen load are equidistant from the centre. Maximum BM will occur under the chosen load.*

Calculating reactions:

$$R_A \times 6 - 2.25 \times \frac{6^2}{2} - 825 \times 3.9 = 0$$

$$R_A = 543 \text{ kN}$$

$$R_B = 295.5 \text{ kN}$$

$$\text{BM}_{max} = \text{Under right wheel}$$

$$= R_B \times 2.1 - 2.25 \times \frac{2.1^2}{2}$$

$$= 615.59 \text{ kNm (Including impact effect)}$$

2. **Maximum shear force:**

It consists of maximum shear due to movement of wheel loads and self weight of girder.

For maximum shear force the wheel loads shall be placed as shown.

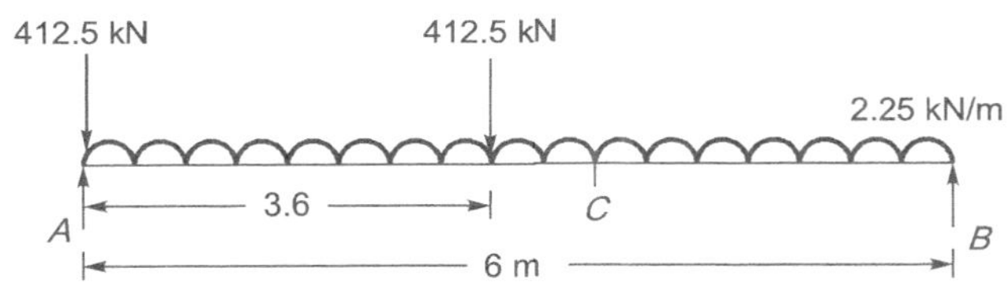

Note: *One wheel should be at the support.*

$$R_A \times 6 = 412.5 \times 6 + 2.25 \times \frac{6^2}{2} + 412.5 \times 2.4$$

$$R_A = 584.25 \text{ kN}$$

Hence maximum shear force due to wheel loads = 584.25 kN

3. **Lateral forces:**

1. Lateral force transverse to rail = 10% of weight of crab and weight lifted.

$$\text{Lateral force transverse to rails} = \frac{10}{100} \times (200 + 60) = 26 \text{ kN}$$

Factored lateral force transverse to rails = 1.5 × 26 = 39 kN

$$\therefore \quad \text{Lateral force on each wheel} = \frac{39}{2} = 19.5 \text{ kN}$$

$$\text{Ultimate horizontal bending moment} = \frac{WL}{4} = \frac{19.5 \times 6}{4} = 29.25 \text{ kNm}$$

3. Longitudinal loads

$$\text{Longitudinal loads along the rail} = \frac{5}{100} \times \text{Static wheel load} \times 1.25 \times 1.5 = 20.625 \text{ kN}$$

$$\text{Reaction force due to drag force} = \frac{20.625 \times (0.3 + 0.15)}{6} = 1.547 \text{ kN}$$

$$\text{Maximum BM due to drag force} = R\left(\frac{L}{2} - \frac{C}{4}\right) = 1.547\left(\frac{6}{2} - \frac{3.6}{4}\right) = 3.25 \text{ kNm}$$

Maximum ultimate design vertical moment = 615.59 + 29.25 + 3.25 = 648.09 kNm

Maximum vertical reaction = 584.25 + 1.547 = 585.8 kN

5.9 **A pitched roof is to be provided for a workshop of effective span 18 m. The trusses are spaced at 4 m centre to centre and purlins at 1.6 m centre to centre. The pitch of the roof is 28°, weight of the roofing material is 0.162 kN/m, normal wind pressure is 1.2 kN/m² and permissible bending stress is 165 MPa. Check the suitability of ISLB 12575 @ 0.119 kN/m section for purlins, if I_{XX} = 406.8 cm⁴ and I_{YY} = 43.4 cm⁴ for given section.**

[14 marks : 2019]

Solution:

Part-1: Load analysis

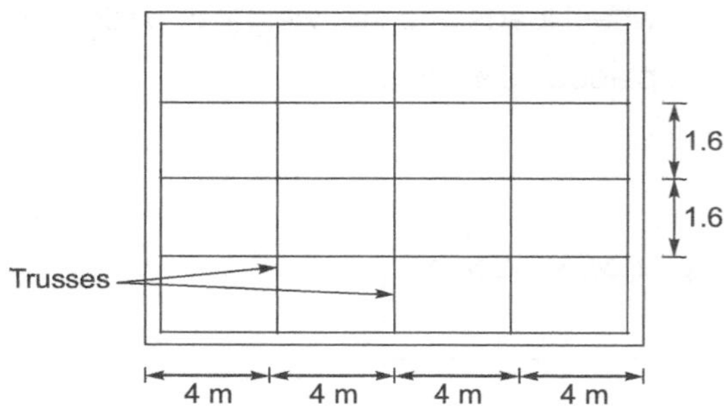

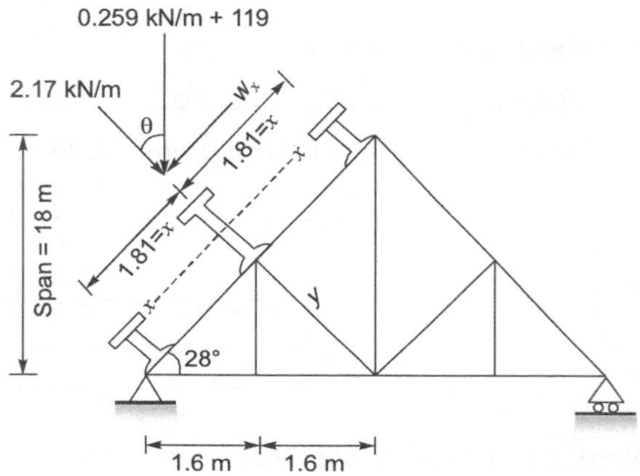

Wind load

Wind process acts normal to surface of roof.

Wind load = 120 kg/m² = 1200 N/m² = 1.2 kN/m²

$$\cos 28° = \frac{1.6}{x} = x = 1.81 \text{ m}$$

Wind load/m width = 1.2 × 1.81 m = 2.17 kN/m [If acts along y-axis]

Weight of roofing material = 162 N/m² of plan area

= 0.162 kN/m² of plan area

Weight of roofing material/m run=0.162 × 1.6 m = 0.259 kN/m

[If acts vertically downwards]

Self weight of purlin = 11.9 kg/m = 0.119 kN/m

[If acts vertically downward]

Total vertically downward load = 0.119 + 0.259 = 0.378 kN/m

Total UDL along y-axis = 2.17 + (0.378 × cos 28°)

w_y = 2.51 kN/m

Total UDL along x-axis = w_x = 0.378 × sin 28° = 0.178 kN/m

$$M_{xx} = \text{BM about } x\text{-axis} = \frac{w_y l^2}{10} = \frac{2.51 \times 4^2}{10} = 4 \text{ kNm}$$

$$M_{yy} = \text{BM about y-axis} = \frac{w_x l^2}{10} = \frac{0.178 \times 4^2}{10} = 0.28 \text{ kNm}$$

$$Z_{xx} = \frac{I_{xx}}{y} = \frac{406.8 \times 10^4}{(125/2)} \text{ mm}^4 = 65.1 \times 10^3 \text{ mm}^3$$

$$Z_{yy} = \frac{I_{yy}}{y} = \frac{43.4 \times 10^4}{(75/2)} = 11.57 \times 10^3 \text{ mm}^3$$

$$\sigma_{bc,\ cal} \text{ at } A = \frac{M_{xx}}{Z_{xx}} + \frac{M_{yy}}{Z_{yy}} \leq \sigma_{bc} = 165 \text{ MPa}$$

$$= \frac{4 \times 10^6}{65.1 \times 10^3} + \frac{0.28 \times 10^6}{11.57 \times 10^3} = 85.6 \text{ N/mm}^2$$

$$< \sigma_{bc} = 165 \text{ MPa}$$

Hence, purlin is safe in bending.

5.10 In an industrial shed, it is proposed to provide a hot rolled section ISMB 500 to carry a two-wheeled system crab on it. The crab can move over the flange of the beam from one end to another end and each wheel of the crab is capable to carry a maximum vertical load of 60 kN (including self weight of wheel).

The centre to centre distance between two supporting ends of the beam is 6 m and the end of the beams are restrained against torsion.

The space between two wheels = 2.4 m. Take impact factor for vertical load as 25%.

Verify the capability of the beam to carry the bending moment developed due to vertical load only. Assume the section is plastic. Grade of steel E 250.

Properties of ISMB 500:

b_f = 180 mm, t_f = 17.2 mm, t_w = 10.2 mm

r_z = 202.1 mm, r_y = 35.2 mm

Z_{ex} = 1808.7 cm³, Z_{pz} = 2074.67 cm³

Critical stress,

$f_{cr,b}$ (MPa)

$\frac{KL}{r}$	$\frac{h}{t_f}$	
	25	30
170	136.7	121.3
180	127.1	112.2

Design bending compressive stress to lateral buckling f_{bd}, for f_y = 250 MPa.

$f_{cr,b}$	f_{bd} (MPa)
150	106.8
100	77.3

[20 marks : 2020]

Solution:

Part - 1 : Factored applied BM ($M_{u,\ cal}$)

Each wheel load reaction = 60 kN

Impact factor = 1.25

Partial safety factor for loads = 1.5

Factored wheel load = W = 1.5 × 1.25 × 60 = 112.5 kN

Since ends of beam are restrained against torsion, effective length of compression flange of beam

l = length of beam = 6 m

To get maximum B.M, keep 2 loads as shown in figure.

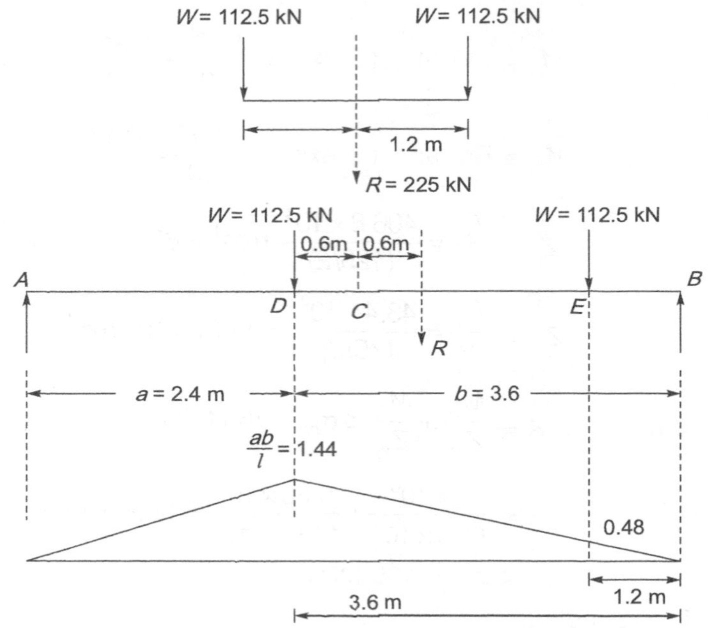

ILD for BM at D

$$\frac{ab}{l} = \frac{2.4 \times 3.6}{6} = 1.44 \text{ m}$$

So, maximum applied factored B.M.,

$$M_{u,\text{cal}} = 112.5\,(1.44 + 0.48)$$

$$M_{u,\text{cal}} = 216 \text{ kN-m}$$

Part - 2 : Factored moment carrying capacity of beam (M_u)

For $$\frac{kL}{r_{\min}} = \frac{6000 \text{ mm}}{35.2 \text{ mm}} = 170.45$$

$$\frac{h}{t_f} = \frac{465.6 \text{ mm}}{17.2 \text{ mm}} = 27.07$$

$\frac{kL}{r_{\min}}$	h/t_f		
	25	27.07	30
170	136.7	↓	121.3
170.45→	136.27	129.9	120.89
180	127.1		112.2

$f_{cr,\,b} = 129.9 \text{ N/mm}^2$ $\qquad F_{bd} = ?$

$f_{cr,b}$	f_{bd}
150	106.8
129.9→	?
100	77.3

$h = D - 2t_f$
$= 550 - 2 \times 17.2$
$= 465.6$ mm

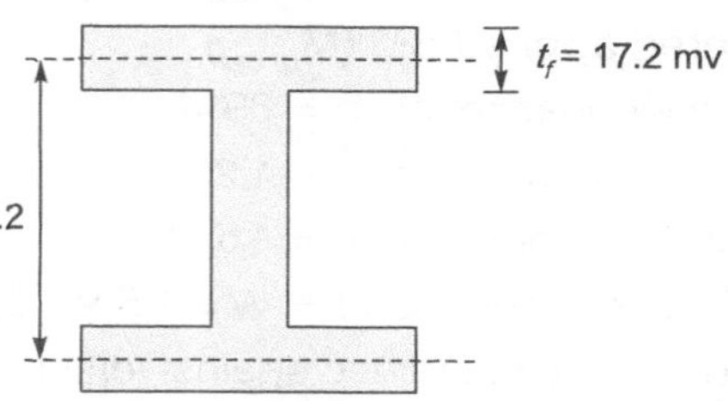

h = Centre of centre distance between flanges

$$f_{bd} = 106.8 - \frac{20.1}{50}(106.8 - 77.3) = 94.94 \text{ N/mm}^2$$

M_u = Factored moment carrying capacity

$$= f_{bd} \times Z_P \quad \text{(for plastic section)}$$
$$= 94.94 \times 2074.67 \times 10^3 = 196.9 \text{ kNm}$$

Since $M_{u,cal} > M_u$, the beam is unsafe against bending.

5.11 A rafter member of a roof truss caries 40 kN compressive load (DL + LL) and 67 kN tensile load (DL + WL). The effective nodal length of the member is 2.1 m. A circular tube section of nominal bore diameter of 50 mm is used. Check the adequacy of the section. Grade of steel = E 250, Young's modulus E = 200 GPa.

Given:

Sectional properties of the section A = 523 mm^2, r = 20.3 mm

Outside diameter = 60.3 mm

Stress reduction factor $\chi = \dfrac{1}{\phi + \sqrt{\phi^2 - \lambda^2}}$

$\phi = 0.5[1 + \alpha(\lambda - 0.2) + \lambda^2]$

$\lambda = \sqrt{\dfrac{f_y}{f_{cc}}}, \; f_{cc} = \dfrac{\pi^2 E}{\left(\dfrac{KL}{r}\right)^2}$

Buckling class	a	b	c	d
α	0.21	0.34	0.49	0.76

[20 marks : 2020]

Solution:

Compressive load = 40 kN (DL + LL)

Tensile load = 67 kN (DL + WL)

kL = 2.1 m

Cross-section of column ⇒ Hollow circular section

Most efficient cross-section so, buckling class is 'a'

A_g = 523 mm^2

r = 20.3 mm

Outside diameter = 60.3 mm

So, the thickness of hollow circular section is $\dfrac{10.3}{2} = 5.15$ mm

P_c = Factored compressive of load carrying capacity of member

$= f_{cd} \times A_g$

$$f_{cc} = \frac{\pi^2 E}{\left(\frac{kL}{r}\right)^2} = \frac{\pi^2 \times 2 \times 10^5}{\left(\frac{2100}{20.3}\right)^2} = 184.45 \text{ N/mm}^2$$

$$\lambda = \sqrt{\frac{f_y}{f_{cc}}} = \sqrt{\frac{250}{184.45}} = 1.164$$

Note: *There is a printing mistake in question in formula for stress reduction factor.*

$$\phi = 0.5[1+0.21(1.164-0.2)+1.164^2] = 1.279$$

(For hollows circular section buckling class is a so, $\alpha = 0.21$)

$$\chi = \frac{1}{\phi+\sqrt{\phi^2-\lambda^2}} = \frac{1}{1.279+\sqrt{(1.279)^2-(1.164)^2}} = \frac{1}{1.81}$$

$$= 0.552$$

$$f_{cd} = \chi\left(\frac{f_y}{1.1}\right) = 0.552\times\left(\frac{250}{1.1}\right) = 125.56 \text{ N/mm}^2$$

$$P_c = f_{cd}\times A_g = 125.56\times 523 = 65.67 \text{ kN}$$

$$P_u = \text{Factored comp. load} = 1.5\times 40$$

$$= 60 \text{ kN}$$

Since $P_u < P_c$, member is safe in compressive

Tensile load carrying capacity = $P_T = A_g \times \dfrac{f_y}{1.1}$

$$P_T = 523\times\frac{250}{1.1} = 118.86 \text{ kN}$$

Factored tensile force, $P_{T,\text{cal}} = 1.5\times 67 = 100.5$ kN

Partial safety factor for DL + WL = 1.5

Since $P_{T,\text{cal}} < P_T$, member is safe in tension also.

5.12 **How are the various types of loads estimated for the design of roof truss?**

[12 marks : 2022]

Solution:

Various type of loads on Roof Truss:

1. **Dead load:** The dead loads on roof truss consists of (i) weight of roof covering; (ii) weight of purlins; (iii) weight of bracings; and (iv) self-weight of the trusses.

   *(i) **Weight of roof covering***

	Type of covering	Weight per m² of plan area
1.	Trafford asbestos sheets	159 N/m²
2.	20 gauge CGI sheets	112.7 N/m²

   *(ii) **Weight of purlins:*** The load due to weight of purlins per square meter of plan area, may be assumed as 70 to 120 N for glazed roofing, 60 to 90 N for G.I. sheeting and 90 to 150 N for A.C. sheeting

   *(iii) **Weight of bracings:*** The load due to the weight of bracings may be assumed as 12 to 15 N/m² of plan area.

   Generally, the load due to self-weight of the truss is estimated from the following empirical expression applicable for pitch equal to 1 in 4 and spacing of 4 m, with corrugated G.I. sheets.

$$w = 10\left(\frac{L}{3}+5\right)$$

where, w = load per square meter of plan area, due to weight of the truss, in N/m².

L = span of the truss, in meters.

2. **Imposed load**: IS 875 recommends that for the roofs with slope upto and including 10°, live load measured on plan should be taken as 1500 N/m^2 if access to roof in provided, and as 750 N/m^2 if access to roof is not provided except for the maintenance purpose. For sloping roofs with slope greater than 10°, the live load may be taken as 750 N/m^2 less 20 N/m^2 for every degree increase in slope over 10°, subject to a minimum of 400 N/m^2 of the plan area. For members supporting the roof members and roof purlins, such as trusses, beam, girders etc, live load may be taken equal to 2/3rd of the above load.
3. **Snow load:** IS 875 recommends snow load of 2.5 N/m^2 per mm depth of snow. No snow load may be considered if the slopes are greater than 50°.
4. **Wind load:** The load due to wind is one of the most important loads to be considered in the design of roof trusses and other types of pitched roofs. The design wind pressure p_z as per IS code 875 Part-3 2015 is given by

$$p_z = 0.6\ V_z^2 = 0.6\ (k_1\ k_2\ k_3\ k_4 V_b)^2$$

where, p_z = wind pressure at any height z above MSL

V_b = basic wind speed in m/s at 10 m height

k_1 = Probability factor (or risk coefficient)

k_2 = Terrain, height and structure size factor

k_3 = Topography factor

k_4 = Important factor

Design wind pressure $p_d = k_d\ k_a\ k_c \cdot p_z$

where, k_d = wind directionality factor

k_a = area averanging factor

k_c = combination factor

The wind force F acting in a direction normal to the individual structural element or cladding unit is

$$F = (C_{pe} - C_{pi})\ Ap_d$$

where C_{pe} = external pressure coefficient

C_{pi} = internal pressure coefficient

5.13 **Design a welded plate girder 24 m in span and laterally restrained throughout. It has to support a uniform load of 100 kN/m throughout the span, exclusive of self weight. Design the girder cross section without intermediate transverse stiffeners. The steel for the flange and web plates is of grade Fe410. (μ = 0.30 and $E = 2 \times 10^5$ MPa). Given that partial safety factor, $\gamma_{m\omega}$ = 1.5 (for site welding) and $\gamma_{m\omega}$ = 1.25 (for shop welding) and λ_ω = 1.2.**

[20 marks : 2022]

Solution:

For Fe410 grade of steel:

$$f_u = 410 \text{ MPa},\ f_y = f_{yp} = f_{yw} = 250 \text{ MPa}$$

$$\mu = 0.3$$

$$E = 2 \times 10^5 \text{ MPa}$$

Partial safety factor, γ_{mw} = 1.50 (for site welding)

= 1.25 (for shop welding)

$$\varepsilon = \varepsilon_w = \varepsilon_f = \sqrt{\frac{250}{f_y}} = \sqrt{\frac{250}{250}} = 1$$

Design forces

Total superimposed load = 100 kN/m

Factored superimposed load = 1.5 × 100 = 150 kN/m

$$\text{Let, self-weight of plate girder} = \frac{WL}{400} = \frac{(100\times24)\times24}{400} = 144 \text{ kN}$$

$$\text{Self-weight of plate girder per meter length} = \frac{144}{24} = 6 \text{ kN/m}$$

Factored self-weight = 1.5 × 6 = 9 kN/m

Total uniform factored load = 150 + 9 = 159 kN/m

$$\text{Maximum bending moment} = \frac{159\times24^2}{8} = 11448 \text{ kNm}$$

$$\text{Maximum shear force} = \frac{159\times24}{2} = 1908 \text{ kN}$$

Design of web

Optimum depth of plate girder,

$$d = \left(\frac{M_z k}{f_y}\right)^{(0.33)}$$

When intermediate transverse stiffeners are not to be provided:

$$\frac{d}{t_w} \leq 200\varepsilon \text{ i.e., } 200 \text{ (from serviceability criteria)}$$

and $\leq 345\,\varepsilon_f^2$ i.e., 345 (from flange buckling criteria)

$$\text{Let us assume } k = \frac{d}{t_w} = 180$$

$$\therefore \quad d = \left(\frac{11448\times10^6\times180}{250}\right)^{0.33} = 1871.9 \text{ mm} \simeq 1800 \text{ mm}$$

$$\text{Optimum web thickness,} \quad t_w = \left(\frac{M_z}{f_y k^2}\right)^{0.33}\left(\frac{11448\times10^6}{250\times180^2}\right)^{0.33} = 10.95 \text{ mm} \simeq 12 \text{ mm (say)}$$

(Thickness provided is more since intermediate transverse stiffeners are not to be provided)

Let us try web plate 1800 × 12 mm in size.

Design of flanges

Let us assume that bending moment will be resisted by the flanges and shear by the web.

$$\text{Required area of flange,} \quad A_f = \frac{M_z\gamma_{m0}}{f_y d} = \frac{11448\times10^6\times1.10}{250\times1800} = 27984 \text{ mm}^2$$

Assuming width of flange equal to 0.3 times depth of girder.

$$b_f = 0.3 \times 1800 = 540 \text{ mm} \simeq 560 \text{ mm}$$

Thickness of flange, $t_f = \dfrac{27984}{560} = 49.97 \simeq 50 \text{ mm}$

Let us try 560 × 50 mm flange plate.

Classification of flanges

For the flanges to be classified as plastic $\dfrac{b}{t_f} \le 8.4\varepsilon$

The outstand of flange, $b = \dfrac{b_f - t_w}{2} = \dfrac{560-12}{2} = 274 \text{ mm}$

$$\frac{b}{t_f} = \frac{274}{50} = 5.48 < 8.4 \; (84\varepsilon = 8.4 \times 1 = 8.4)$$

Hence, the flanges are plastic. ($\beta_b = 1.0$)

Check for bending strength

The trial section of the plate girder is shown in figure. The plastic section modulus of the section.

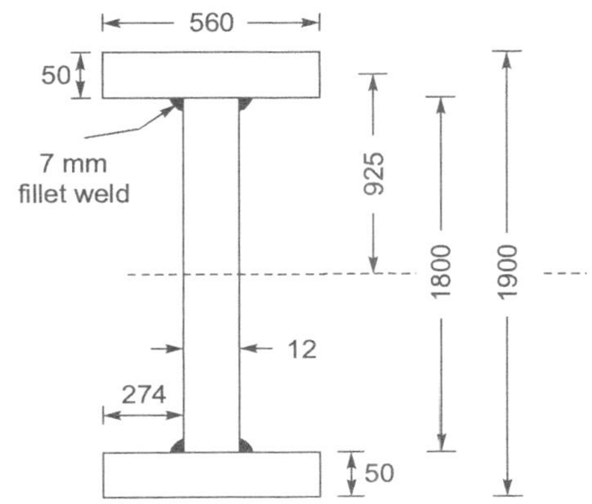

(All dimensions in mm)

$$Z_{pz} = 2b_f t_f \frac{(D-t_f)}{2} = 2 \times 560 \times 50 \times \frac{1900-50}{2}$$

$$= 51.80 \times 10^6 \text{ mm}^3$$

Moment capacity:

$$M_d = \beta_b Z_{pz} \frac{f_y}{\gamma_{m0}} = 1.0 \times 51.80 \times 10^6 \times \frac{250}{1.10} \times 10^{-6} \text{ kNm}$$

$$= 11772.7 \text{ kNm} > 11448 \text{ kNm}$$

which is safe.

Shear capacity of web

Let us use simple post-critical method.

$$\frac{d}{t_w} = \frac{1800}{12} = 150 < 200 \qquad (200\varepsilon = 200 \times 1 = 200)$$

and also, < 345 $\qquad (345\varepsilon_f^2 = 345 \times 1 = 345)$

which is all right.

Elastic critical shear stress,

$$\tau_{cr,\,e} = \frac{k_v \pi^2 E}{12(1-\mu^2)\left(\frac{d}{t_w}\right)^2}$$

Transverse stiffeners will be provided at supports only.

Hence, $k_v = 5.35$

$$\tau_{cr,\,e} = \frac{5.35\times\pi^2\times2\times10^5}{12\times(1-0.3^2)\times150^2} = 42.98 \text{ N/mm}^2$$

The non-dimensional web slenderness ratio for shear buckling stress.

$$\lambda_w = \sqrt{\frac{f_{yw}}{\sqrt{3}\,\tau_{cr,e}}} = \sqrt{\frac{250}{\sqrt{3}\times42.98}} = 1.83 \simeq 1.80 > 1.20$$

Shear stress corresponding to buckling (for $\lambda_w > 1.20$)

$$\tau_b = \frac{f_{yw}}{\sqrt{3}\lambda_w^2} = \frac{250}{\sqrt{3}\times1.80^2} = 44.55 \text{ N/mm}^2$$

Shear force corresponding to web buckling,

$$V_{cr} = dt_w\tau_b = 1800 \times 12 \times 44.55 \times 10^{-3} = 962.28 \text{ kN} < 1908 \text{ kN}$$

which is unsafe.

Let us revise the web thickness from 12 mm to 16 mm.

New value of $\tau_{cr,\,e}$, λ_w, τ_b, and V_{cr} will be as follows

$$\frac{d}{t_w} = \frac{1800}{16} = 112.5$$

$$\tau_{cr,\,e} = \frac{5.35\times\pi^2\times2\times10^5}{12\times(1-0.3^2)\times(112.5)^2} = 76.41 \text{ N/mm}^2$$

$$\lambda_w = \sqrt{\frac{f_{yw}}{\sqrt{3}\,\tau_{cr,e}}} = \sqrt{\frac{250}{\sqrt{3}\times76.41}} = 1.374 > 1.2$$

$$\tau_b = \frac{f_{yw}}{\sqrt{3}\lambda_w^2} = \frac{250}{\sqrt{3}\times1.37^2} = 76.90 \text{ N/mm}^2$$

$$V_{cr} = dt_w\tau_b = 1800 \times 16 \times 76.90 \times 10^{-3} = 2214.7 \text{ N} > 1908 \text{ kN}$$

which is safe.

Check for lateral-torsional buckling

Since the compression flange of the girder is laterally restrained throughout, the possibility of lateral-torsional buckling is not there and this check is not required.

Flange to web connection

There will be two weld length along the span for each flange to web connection.

$$q_w = \frac{VA_f\bar{y}}{2I_z}$$

$$I_z = \frac{b_f D^3}{12} - \frac{(b_f - t_w)d^3}{12}$$

$$= \frac{560 \times 1900^3}{12} - \frac{(560-16)1800^3}{12}$$

$$= 55702.7 \times 10^6 \text{ mm}^4$$

$$q_w = \frac{1908 \times 560 \times 50 \times \left(900 + \frac{50}{2}\right)}{2 \times 55702.7 \times 10^6} = 0.4436 \text{ kN/mm}$$

Let us provide weld of size, $S = 10$ mm

$\therefore \quad kS = 0.7 \times 10 = 7.0$ mm

Strength of shop weld per unit length,

$$f_{wd} = \frac{4.2 \times 250 \times 10^{-3}}{\sqrt{3} \times 1.25} = 0.4849 \text{ kN/mm} > 0.4436 \text{ kN/mm}$$

which is all right.

6. Plastic Analysis

6.1 **Determine the value of fully plastic moment M_p for the frame shown in figure below. Plastic moment capacity of beam = 1/2 × plastic moment capacity of column.**

[12 marks : 2001]

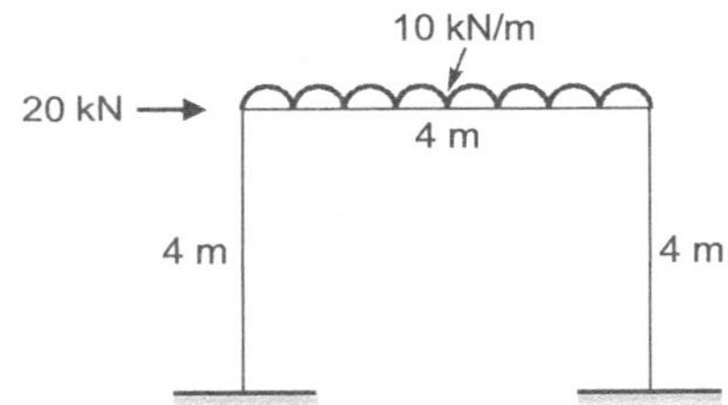

Solution :

Number of possible plastic hinges, $n = 5$

Degree of static indeterminacy, $r = 6 - 3 = 3$

Number of independent mechanism, $i = n - r = 5 - 3 = 2$

Beam Mechanism :

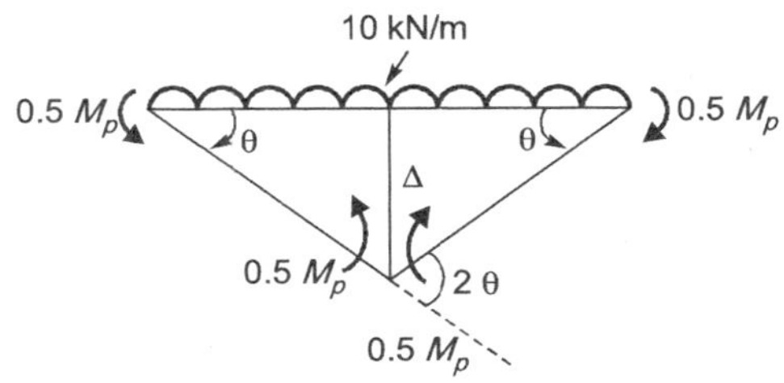

Applying principle of virtual work, we get

$$10 \times \frac{1}{2} \times 4 \times \Delta - 0.5\, M_p\theta - 0.5\, M_p(\theta + \theta) - 0.5\, M_p\theta = 0$$

$\Rightarrow \quad 20 \times \Delta = 2\, M_p\theta \quad [2\theta = \Delta]$

$\Rightarrow \quad 2M_p = 20 \times 2$

$\Rightarrow \quad M_p = 20$ kN-m

Panel Mechanism :

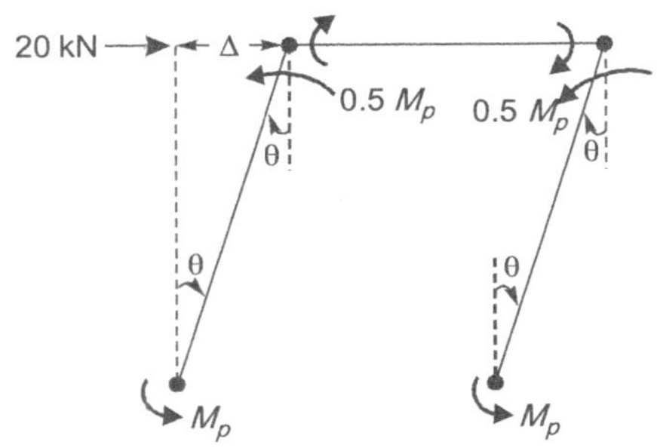

Applying principle of virtual work, we get

$20 \times \Delta - M_p\theta - 0.5M_p\theta - 0.5M_p\theta - M_p\theta = 0 \quad [\Delta = 4\theta]$

$\Rightarrow \quad 20 \times 4\theta = 3M_p\theta$

$\Rightarrow \quad M_p = \frac{80}{3}$

$\Rightarrow \quad M_p = 26.67$ kN-m

Composite Mechanism:

Applying principle of virtual work, we get

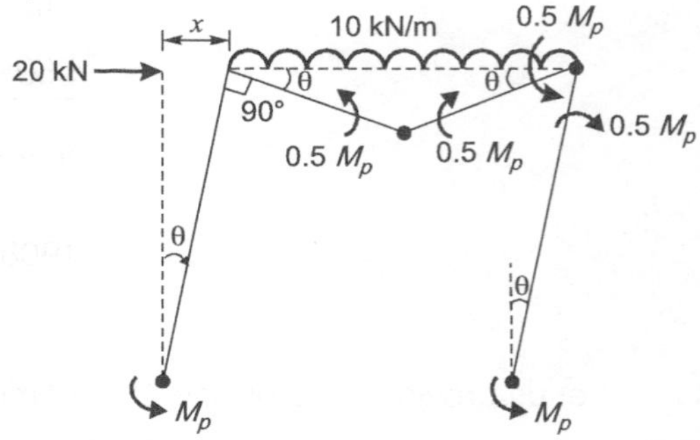

$$20 \times 4\theta + 10 \times \frac{1}{2} \times x\theta \times 4 - M_p\theta - M_p\theta - 0.5\,M_p(\theta + \phi) - 0.5\,M_p(\theta + \phi) = 0$$

$$\Rightarrow \quad 80\theta + 20x\theta = 3\,M_p\theta + M_p\phi$$

$$\Rightarrow \quad (80 + 20x)\theta = 3M_p\theta + M_p \times \frac{\theta x}{4-x}$$

$$\Rightarrow \quad 80 + 20x = 3M_p + \frac{M_p x}{4-x}$$

$$\Rightarrow \quad M_p\left(3 + \frac{x}{4-x}\right) = 80 + 20x$$

$$\Rightarrow \quad M_p\left(\frac{12 - 3x + x}{4-x}\right) = 80 + 20x$$

$$\Rightarrow \quad M_p = \frac{(4-x)(80+20x)}{12-2x}$$

$$\Rightarrow \quad M_p = \frac{(4-x)(40+10x)}{(6-x)}$$

$$\Rightarrow \quad M_p = \frac{160 - 10x^2}{6-x}$$

For maximum value of M_p, $\quad \frac{\partial M_p}{dx} = 0$

$$\Rightarrow \quad [(6-x) \times (-20x)] - (160 - 10x^2) \times (-1) = 0$$

$$\Rightarrow \quad 10x^2 - 160 + 120x - 20x^2 = 0$$

$$\Rightarrow \quad -10x^2 + 120x - 160 = 0$$

$$\Rightarrow \quad x^2 - 12x + 16 = 0$$

$$\Rightarrow \quad x = 1.53 \text{ m}$$

$$\therefore \quad M_p = \frac{160 - 10 \times (1.53)^2}{6 - 1.53}$$

$$\Rightarrow \quad M_p = 30.56 \text{ kN-m}$$

Thus the maximum value of M_p is chosen as the plastic moment capacity of the portal frame i.e. $M_p = 30.56$ kN-m.

6.2 **Discuss the concept of plastic hinge. Obtain the shape factor for a circular section.**

[15 marks : 2002]

Solution:

A **plastic hinge** is a zone of yielding due to flexure in a structural member. Although hinges do not actually form, it can be seen that large changes of slope occur over small lengths of the member at positions of maximum moments. A strain hardening action usually occurs at these hinges so that large deflections are accompanied by a slight increase in loads.

A structure can support the computed ultimate load due to the formation of plastic hinges at certain critical sections. The members remain elastic until the moment reaches a value M_p, the maximum moment of resistance of a fully yielded cross-section. Any additional moment will cause the beam to rotate with little increase in stress. The rotation occurs at a constant moment, M_p. The zone acts as if it was hinged except with a constant restraining moment M_p. The plastic hinge, therefore can be defined as a yielded zone due to flexure in a structure in which infinite rotation can take place at a constant restraining moment M_p of the section. The value of the moment at the adjacent sections of the yield zone for a certain length, is more than the yield moment. This length depends upon the loading and geometry of the section.

Sometimes a plastic hinge is called a "rusty hinge" or "friction hinge". In a framed structure with prismatic members plastic hinge may occur at points of concentrated loads, at the ends of each member meeting at a connection involving change in geometry and at the point of zero shear in a span under distributed load.

Where two members meet a hinge will form in the member whose capacity is less. Where three members meet, plastic hinge will form in all members irrespective of their capacities. Same is the case in a joint where four members meet.

Plastic hinges are reached first at sections subjected to greatest curvature.

Formation of a sufficient number of plastic hinges converts the structure into a mechanism.

A plastic hinge is formed completely when the stresses throughout the section in the plastic zone are equal to the yield stress.

We know that plastic section modulus is given by

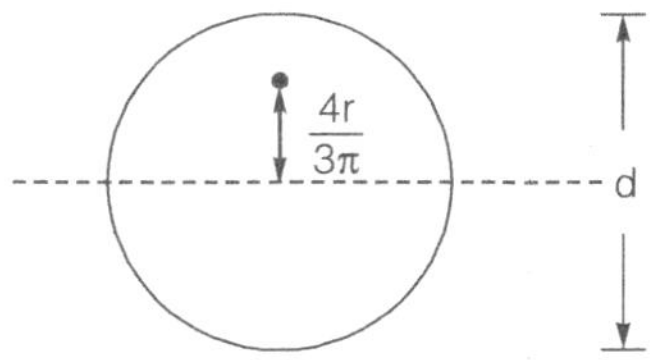

$$Z_p = A_c y_c + A_t y_t$$

In case of a circular section, $A_c = A_t = \frac{A}{2} = \frac{1}{2} \times \frac{\pi d^2}{4} = \frac{\pi d^2}{8}$

$$y_c = y_t = \frac{4r}{3\pi} = \frac{4 \times d}{2 \times 3\pi} = \frac{2d}{3\pi}$$

$$\therefore \quad Z_p = \frac{\pi d^2}{8}\left[\frac{2d}{3\pi} + \frac{2d}{3\pi}\right]$$

$$\Rightarrow \quad Z_p = \frac{\pi d^2}{8} \times \frac{4d}{3\pi} = \frac{d^3}{6}$$

The elastic section modulus is given by

$$Z_e = \frac{I}{y_{max}} = \frac{\pi d^4}{64} \times \frac{2}{d} = \frac{\pi d^3}{32}$$

Shape factor, $S = \frac{Z_p}{Z_e} = \frac{d^3/6}{\pi d^3/32} = \frac{32}{6\pi} = 1.698$

6.3 **Determine the shape factor for the cross-section given in the figure below:**

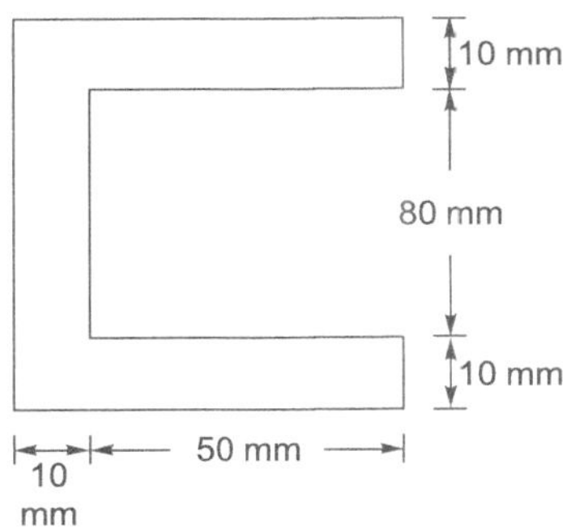

[8 marks : 2003]

Solution :

We know that in plastic analysis, neutral axis divides the total area into two equal parts.

Total area, $A = 2 \times 50 \times 10 + 10 \times 100 = 2000 \text{ mm}^2$

$$\therefore \quad \frac{A}{2} = \frac{2000}{2} = 1000 \text{ mm}^2$$

Half of the total area (1000 mm^2) is more than the area of top flange (60 × 10 = 600 mm). Hence, the neutral axis will lie in the web.

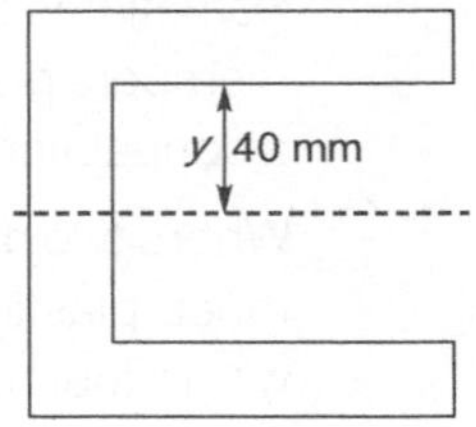

Let the neutral axis lies below the bottom of top flange at a distance y.

$$\therefore \quad 10 \times y = 1000 - 600$$

$$\Rightarrow \quad y = \frac{400}{10} = 40 \text{ mm}$$

But, plastic section modulus is given by

$$Z_p = A_t y_t + A_c y_c \qquad [\because A_t = A_c = \frac{A}{2} = 1000 \text{ mm}^2]$$

$$= 2 A_t y_t \qquad [\because y_t = y_c]$$

$$= 2 \times [(60 \times 10 \times 45) + (40 \times 10 \times 20)] = 70000 \text{ mm}^3$$

To find elastic section modulus, taking moment of areas about the top of top flange

$$2000\, \bar{y} = 60 \times 10 \times 5 + 80 \times 10 \times 50 + 60 \times 10 \times 95$$

$$\Rightarrow \quad \bar{y} = \frac{100000}{2000} = 50 \text{ mm}$$

The moment of inertia of the section about its centroidal axis is given by

$$I_{CG} = 2 \times [\frac{60 \times 10^3}{12} + 60 \times 10 \times (50-5)^2 + \frac{10 \times 40^3}{12} + 10 \times 40 \times (50-30)^2]$$

$$\Rightarrow \quad I_{CG} = 2 \times 1433333.33 = 2866666.67 \text{ mm}^4$$

But, $$Z_e = \frac{I_{CG}}{y_{max}} = \frac{2866666.67}{50} = 57333.33 \text{ mm}^3$$

Shape factor, $$S = \frac{Z_p}{Z_e} = \frac{70000}{57333.33} = 1.22$$

6.4 A portal frame *ABCD* fixed at the base carries an ultimate concentrated load of 2.4 P_u at the centre of the beam *BC* and ultimate horizontal load P_u acting at B from left to right. The length of the columns *AB* and *CD* are 2a and 3a respectively and beam is 2a long. Each column has a fully plastic moment of M_{p1} and fully plastic moment of beam is M_{p2}. Assuming $M_{p1} < M_{p2}$ determine M_{p1} and M_{p2} for the conditions that collapse may just occur whatever may be the direction of horizontal force.

[20 marks : 2003]

Solution :

The portal frame with loadings is shown below:

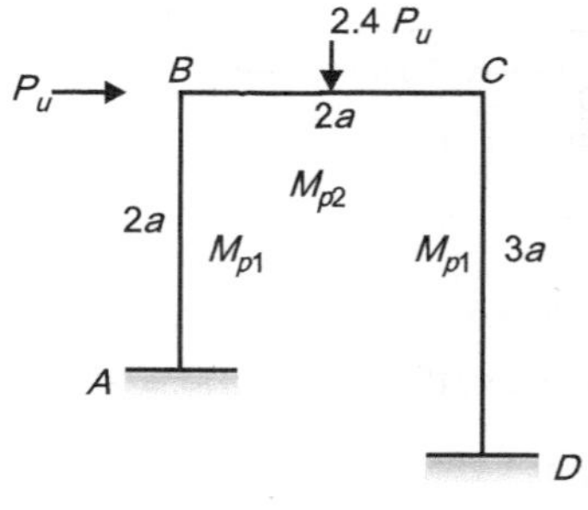

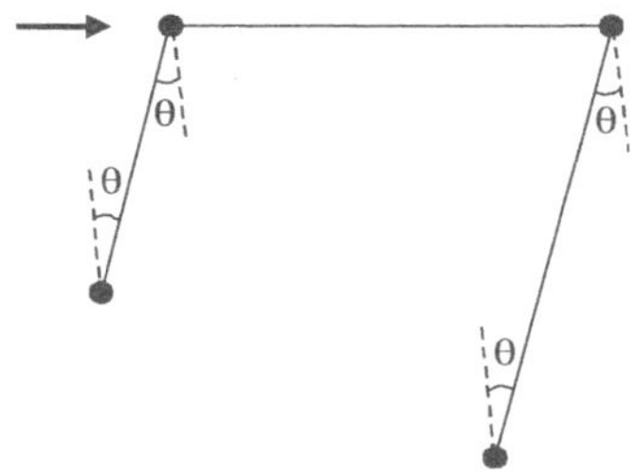

If we consider only sway mechanism, only M_{p1} can be found. Hence we have to consider composite mechanism for both the directions of sway.

(i) **When the sway force acts from left to right**

Applying principle of virtual work, we get

$$P_u \times 2a\theta + 2.4\, P_u \times a\theta - M_{p1}\theta - M_{p2} \times 2\theta - M_{p1}\theta - M_{p1}\phi - M_{p1}\phi = 0$$

$$\Rightarrow \quad 4.4\, P_u a\theta = 2M_{p1}\theta + 2M_{p2}\theta + 2M_{p1}\,\phi \qquad [\because 2a\theta = 3a\phi]$$

$$\Rightarrow \quad 4.4\, P_u a\theta = 2M_{p1}\theta + 2M_{p2}\theta + 2M_{p1} \times \frac{2\theta}{3}$$

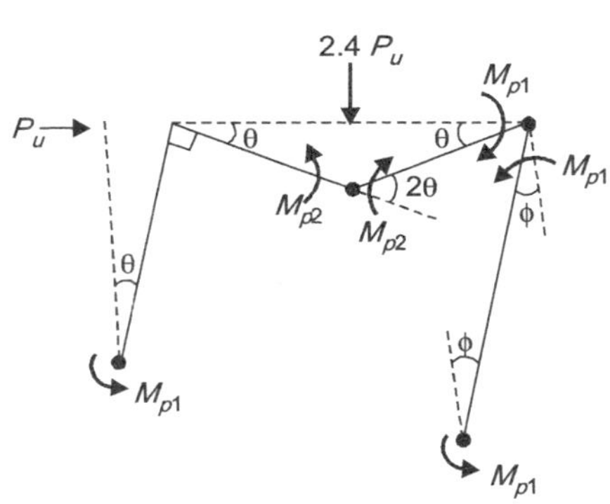

$$\Rightarrow \quad 4.4\, P_u a = \left(2 + \frac{4}{3}\right) M_{p1} + 2M_{p2}$$

$$\Rightarrow \quad \frac{10}{3} M_{p_1} + 2M_{p_2} = 4.4\, P_u a$$

$$\Rightarrow \quad 10\, M_{p1} + 6\, M_{p2} = 13.2\, P_u a$$

$$\Rightarrow \quad 5\, M_{p1} + 3\, M_{p2} = 6.6\, P_u a \qquad \text{...(i)}$$

(ii) **When the sway force acts from right to left**

Applying principle of virtual work, we get

$$P_u \times 3a\theta + 2.4\, P_u \times a\theta - M_{p1}\,\theta - M_{p2} \times 2\theta - M_{p1}\,\theta - M_{p1}\,\phi - M_{p1}\,\phi = 0$$

$$\Rightarrow \quad 5.4\, P_u\, a\theta = 2M_{p1}\,\theta + 2M_{p2}\,\theta + 2M_{p1}\,\phi \quad [3a\theta = 2a\phi]$$

$$\Rightarrow \quad 5.4\, P_u\, a\theta = 2M_{p1}\,\theta + 2M_{p2}\,\theta + 2M_{p1} \times \frac{3\theta}{2}$$

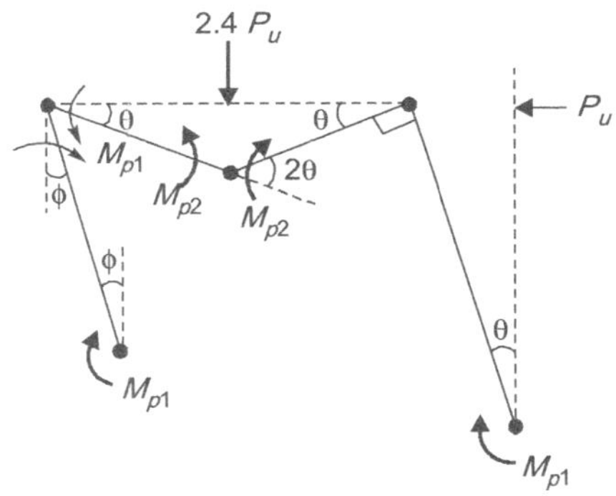

$$\Rightarrow \quad 5\, M_{p1} + 2\, M_{p2} = 5.4\, P_u a \qquad \text{...(ii)}$$

Subtracting (ii) and (i), we get

$$M_{p2} = 6.6\, P_u a - 5.4\, P_u a$$

$$\Rightarrow \quad M_{p2} = 1.2\, P_u a$$

Substituting value of M_{p2} in (i), we get

$$M_{p1} = \frac{6.6 P_u a - 3 \times 1.2 P_u a}{5}$$

$$\Rightarrow \quad M_{p1} = 0.6 P_u a$$

6.5 For a mild steel beam of rectangular cross-section with b = 200 mm and d = 100 mm, yield stress f_y = 250 MPa, modulus of elasticity, $E = 2.0 \times 10^5$ MPa, determine the value of bending moment M_y at initial yield of the extreme fibre and also the value of fully plastic moment M_p.
Determine also the curvature K_y at initial yield and also the curvature K_z. When half the cross-section has yielded along with the corresponding value of the bending moment M_z.

[10 marks : 2004]

Solution :

The cross-section of the rectangular beam is shown below :

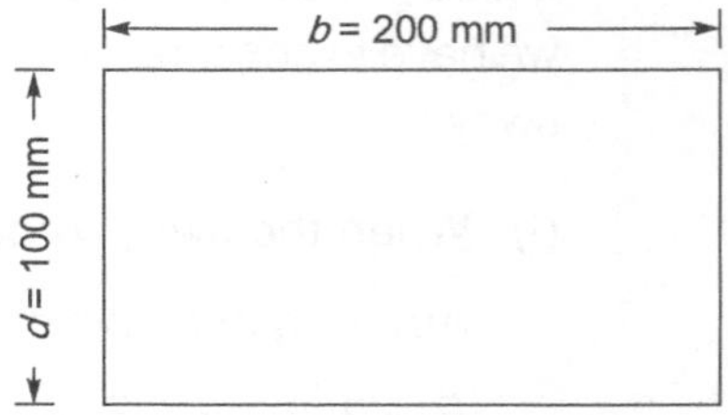

Elastic section modulus, $Z_e = \dfrac{I}{y_{max}} = \dfrac{200 \times 100^3}{12 \times 50}$

$= \dfrac{1}{3} \times 10^6 \text{ mm}^3$

Bending moment at initial yield, $M_y = \sigma_y Z_e = 250 \times \dfrac{1}{3} \times 10^6 \times 10^{-6}$

$= 83.33$ kN-m

Shape factor for rectangular section, $K_s = 1.5$

But $K_s = \dfrac{Z_p}{Z_e}$

$\Rightarrow$ $Z_p = K_s \times Z_e = 1.5 \times \dfrac{1}{3} \times 10^6 = 5 \times 10^5 \text{ mm}^3$

Plastic moment capacity, $M_p = \sigma_y Z_p = 250 \times 5 \times 10^5 \times 10^{-6} = 125$ kN-m

From theory of pure bending we have

$$\frac{M}{I} = \frac{\sigma}{y} = \frac{E}{R}$$

$\therefore$ $\dfrac{\sigma}{y} = \dfrac{E}{R}$

$\Rightarrow$ $\dfrac{1}{R} = \dfrac{\sigma}{yE}$ where $\dfrac{1}{R}$ is known as curvature

If curvature at initial yield is K_y, then

$K_y = \dfrac{\sigma_y}{y_{max} E}$ $\left[\because y_{max} = \dfrac{d}{2} = \dfrac{100}{2} = 50 \text{ mm}\right]$

$= \dfrac{250}{50 \times 2 \times 10^5}$

$= 2.5 \times 10^{-5}$ per mm

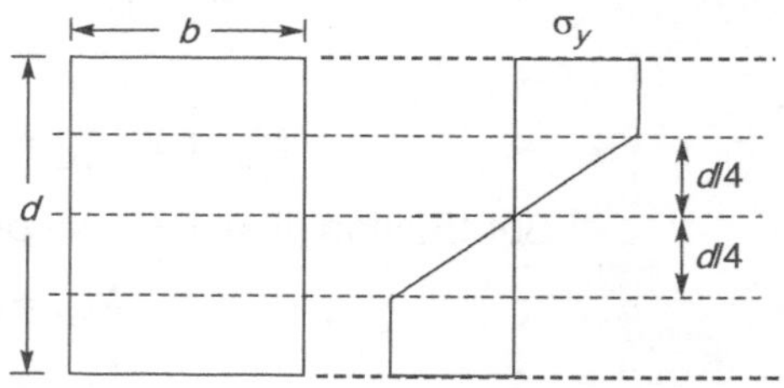

Similarly, $K_z = \dfrac{\sigma_y}{yE}$ $\left[\because y = \dfrac{d}{4} = \dfrac{100}{4} = 25 \text{ mm}\right]$

$= \dfrac{250}{25 \times 2 \times 10^5} = 5 \times 10^{-5}$ per mm

Bending moment when half of the cross-section has yielded will be given by

$M_z = M_y \left[K_s - \dfrac{1}{2} \times \left(\dfrac{K_y}{K_z}\right)^2\right]$ where K_s is shape factor

$= \dfrac{250}{3} \times \left[1.5 - \dfrac{1}{2} \times \left(\dfrac{2.5 \times 10^{-5}}{5 \times 10^{-5}}\right)^2\right]$

$= \dfrac{250}{3} \times [1.5 - 0.125] = 114.58$ kN-m

Alternative $$M_z = \left[\sigma_y \times \frac{d}{4} \times b\right] \times LA_1 + \left[\frac{1}{2} \times \sigma_y \times \frac{d}{4} \times b\right] \times LA_2$$

Where, $$LA_1 = \frac{3d}{4} \text{ and } LA_2 = \frac{d}{3}$$

$\Rightarrow$ $$M_z = 114.58 \text{ kNm}$$

6.6 A mild steel T-section has the following cross-sectional dimensions :
Total depth = 200 mm Width of flange = 120 mm
Thickness of flange = 20 mm Thickness of web = 20 mm
If the yield stress, σ_y = 250 MPa, determine the plastic moment capacity of the section. Also calculate the shape factor of the section.

[20 marks : 2009]

Solution:

We know that in plastic analysis, the neutral axis divides the total area into two equal parts.

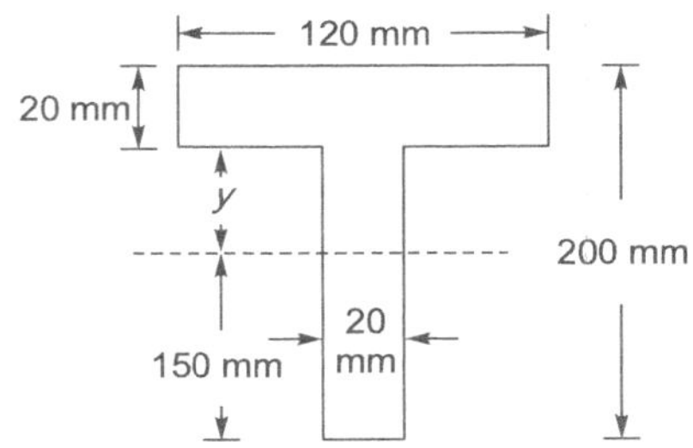

Total area, $$A = 120 \times 20 + (200 - 20) \times 20 = 6000 \text{ mm}^2$$

$\therefore$ $$\frac{A}{2} = \frac{6000}{2} = 3000 \text{ mm}^2$$

Thus, we find that half of the total area (3000 mm²) is greater than the area of flange (120 × 20 = 2400 mm²). It means that the neutral axis will lie in the web.

Let neutral axis lies at a distance y from the bottom of the flange, then

$$(120 \times 20) + 20\,y = 3000\ (= A/2)$$

$$20 \times y = 3000 - 2400$$

$\Rightarrow$ $$y = \frac{600}{20} = 30 \text{ mm}$$

But, plastic section modulus is given by

$$Z_p = A_t y_t + A_c y_c$$

$$= 150 \times 20 \times \frac{150}{2} + [(120 \times 20 \times 40) + (30 \times 20 \times 15)]$$

$$= 225000 + 105000 = 330000 \text{ mm}^3.$$

To find elastic section modulus, Z_e

Taking moment of areas about the top of flange

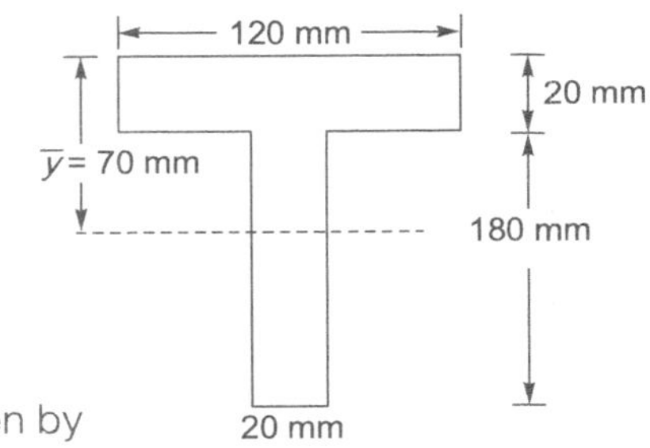

$$6000 \times \bar{y} = (120 \times 20 \times 10) + (20 \times 180 \times 110)$$

$\Rightarrow$ $$\bar{y} = 70 \text{ mm from top}$$

The moment of inertia of the T-section about its centroidal axis is given by

$$I_{CG} = \frac{120 \times (20)^3}{12} + 120 \times 20 \times (70 - 10)^2 + \frac{20 \times 180^3}{12} + 20 \times 180 \times (110 - 70)^2$$

$\Rightarrow$ $$I_{CG} = 2420 \times 10^4 \text{ mm}^4$$

$$\therefore \quad Z_e = \frac{I_{CG}}{y_{max}} = \frac{2420\times10^4}{200-70} = 18.615 \times 10^4 \text{ mm}^3$$

But shape factor,
$$S = \frac{Z_p}{Z_e} = \frac{330000}{18.615\times10^4} = 1.773$$

∴ Plastic moment capacity, $M_p = \sigma_y Z_p$

$$= 250 \times 330000 \times 10^{-6}$$
$$= 82.5 \text{ kN-m.}$$

6.7 **(i) Distinguish between yield moment and plastic moment capacity of a section.**
(ii) Determine the shape factor from first principles for a triangular section of base width *b* and height *h*.

[2 + 12 = 14 marks : 2011]

Solution:

(i) **Yield Moment capacity** of any section is the maximum moment of resistance when the extreme fibre of the cross-section is just about to yield while all inside fibre remain being unyielded. It is denoted by M_y.

Plastic Moment Capacity of any section is the maximum moment of resistance when all the fibres of the cross-section are fully yielded. It is the maximum moment capacity of any section. It is denoted by M_p.

(ii) **To find shape factor of a triangular section of base width '*b*' and height '*h*'**

Moment of Inertia about centroid of triangle

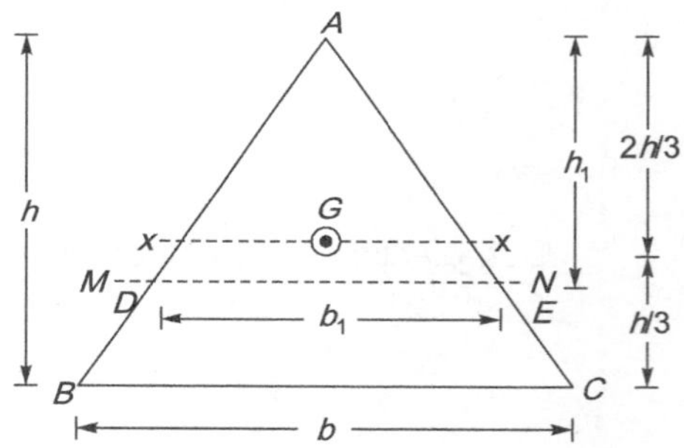

$$I_{xx} = \frac{bh^3}{36}$$

Section modulus,
$$Z = \frac{bh^3}{36} / \frac{2}{3}h = \frac{bh^2}{24}$$

Plastic section modulus,
$$Z_P = \frac{A}{2}(\bar{Y}_1 + \bar{Y}_2)$$

Let MN be the equal area axis, then area of (ΔADE) = $\frac{1}{2}$ area of (ΔABC)

$$\Rightarrow \quad \frac{1}{2}b_1h_1 = \frac{1}{2}\left(\frac{bh}{2}\right)$$

$$\Rightarrow \quad b_1h_1 = \frac{bh}{2} \quad \text{...(i)}$$

From similar ΔADE and ΔABC
$$\frac{h_1}{b_1} = \frac{h}{b} \quad \text{...(ii)}$$

From equation (i) and (ii)

$$h_1 = \frac{h}{\sqrt{2}} \quad \text{and} \quad b_1 = \frac{b}{\sqrt{2}}$$

and
$$A = \frac{bh}{2}$$

Distance of centroid of ΔADE from base DE

$$\bar{Y}_1 = \frac{h_1}{3} = \frac{h}{3\sqrt{2}}$$

Distance of centroid of trapezium $ECBD$ from DE

$$\bar{Y}_2 = \left(\frac{h-h_1}{3}\right)\times\left(\frac{b_1+2b}{b_1+b}\right)$$

$$= \left(\frac{h-\frac{h}{\sqrt{2}}}{3}\right)\times\left(\frac{\frac{b}{\sqrt{2}}+2b}{\frac{b}{\sqrt{2}}+b}\right) = \frac{h\left(\sqrt{2}-1\right)}{3\sqrt{2}}\times\frac{(2\sqrt{2}+1)}{(\sqrt{2}+1)}$$

$$= \frac{0.4645h}{3} = 0.15483\,h$$

$$Z_p = \frac{A}{2}\left(\bar{Y}_1+\bar{Y}_2\right) = \frac{1}{2}\times\frac{bh}{2}\left[\frac{h}{3}\times\frac{1}{\sqrt{2}}+\frac{h}{3}\times 0.4645\right]$$

$$= \frac{bh^2}{12}[0.7071+0.4645] = 0.0976\,bh^2$$

$$\text{Shape factor} = \frac{Z_p}{Z} = \frac{0.0976bh^2}{bh^2/24} = 2.343$$

6.8 ISMB 450 is used as a propped cantilever beam of span 12 m. Assuming σ_y = 250 MPa determine the factored uniformly distributed load q_u the beam can carry including self weight, if the load is to be applied over the entire span.

The properties of ISMB 450 are as follows:

Weight/metre	: 72.4 kg
area of cross-section	: 9227 mm^2
width of flange	: 150 mm
thickness of flange	: 17.4 mm

$$I_{xx} = 3.039\times10^8\ \text{mm}^4$$
$$I_{yy} = 8.34\times10^6\ \text{mm}^4$$

[20 marks : 2012]

Solution:

As there is no factored load term defined in working stress method, so only limit state method has to be adppted.

In LSM, the section needs to be defined first.

Calculation for thickness of the web,

$$A = 2b_f\,t_f + (D-2t_f)\times t_w$$

$$\Rightarrow \quad 9227 = 2\times150\times17.4 + (450-2\times17.4)\times t_w$$

$$\Rightarrow \quad t_w = 9.65\ \text{mm}$$

$$\Rightarrow \quad \epsilon = \sqrt{\frac{250}{f_y}} = \sqrt{\frac{250}{250}} = 1$$

As per IS 800 : 2007

$$\frac{b}{t_f} = \frac{150}{17.4} = 8.62 < 9.4\epsilon$$

$$\frac{d}{t_w} = \frac{(450-2\times17.4)}{9.65} = 43.02 < 84\epsilon$$

So, the given section is plastic in nature

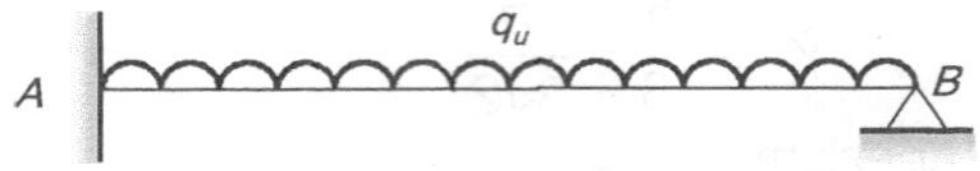

From the plastic analysis of the given propped cantilever beam

$$q_u = \frac{11.656\, M_p}{L^2}$$

$$M_p = \frac{\beta z_p f_y}{1.1}$$

Since, for plastic section $\beta = 1$

$\therefore$ $z_p = \text{Shape factor} \times z_e$

$$z_p = 1.12 \times \frac{3.039 \times 10^8}{225}$$

$$z_p = 1.51 \times 10^6 \text{ mm}^3$$

Therefore, $M_p = 1 \times 1.51 \times 10^6 \times \frac{250}{1.1}$

$$M_p = 343.81 \text{ kN-m}$$

$\therefore$ $q_u = \frac{11.656 \times 343.81}{12^2}$

$$q_u = 27.829 \text{ kN-m}$$

6.9 **Determine the collapse load of the beam as shown below. EI = Constant.**

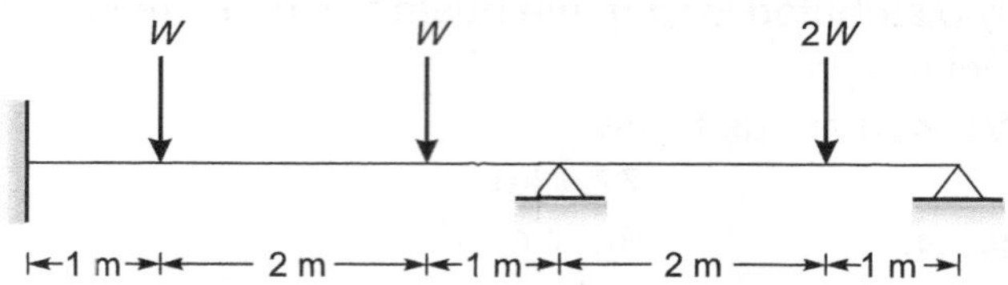

[10 marks : 2013]

Solution:

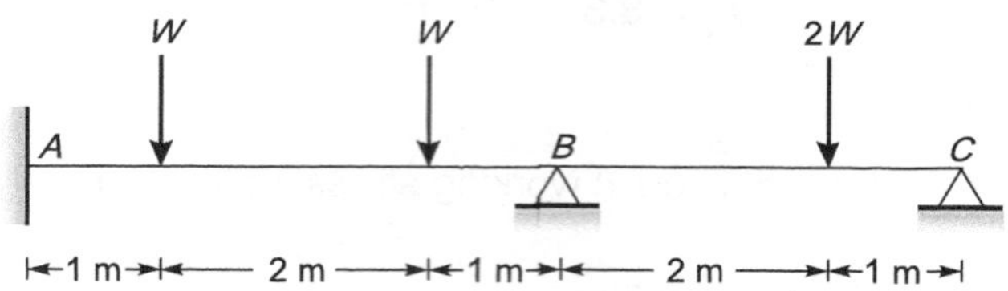

Possible member of plastic hinges = 5

Degree of redundancy = 2

$\therefore$ Number of possible independent mechanism = 5 – 2 = 3

Mechanism-1

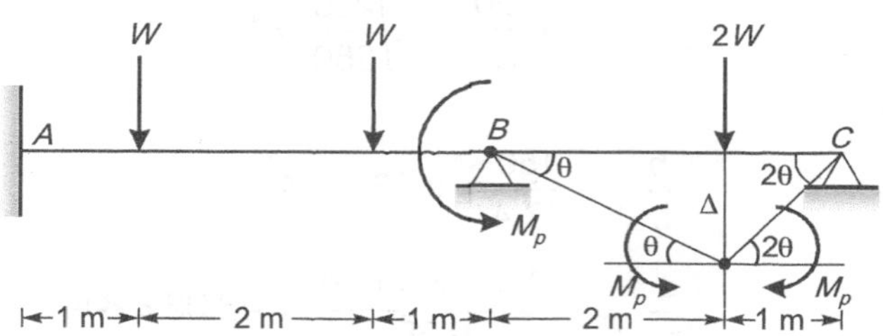

$$\therefore \quad M_p\theta + M_p(\theta + 2\theta) = 2W(\Delta)$$
$$= 2W(2\theta) = 4W\theta$$
$$\Rightarrow \quad 4M_p\theta = 4W\theta$$
$$\Rightarrow \quad M_p = W$$

Mechanism-2

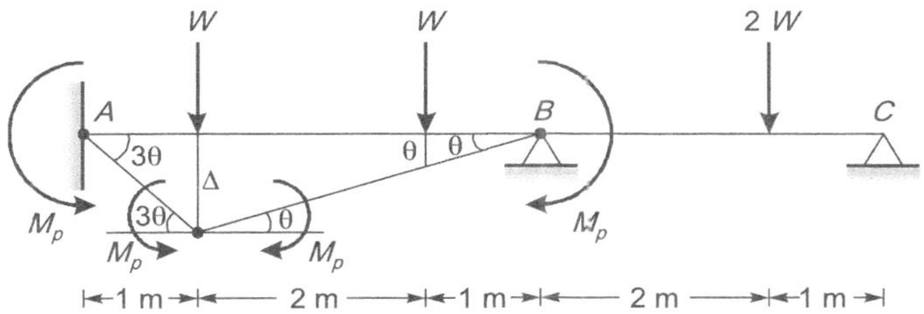

$$\therefore \quad M_p(3\theta) + M_p(4\theta) + M_p\theta = W.\Delta + W(\theta.1)$$
$$= W(3\theta) + W\theta$$
$$= 4W\theta$$
$$\Rightarrow \quad 8M_p\theta = 4W\theta$$
$$\Rightarrow \quad M_p = W/2 \text{ or } W = 2M_p$$

Mechanism-3

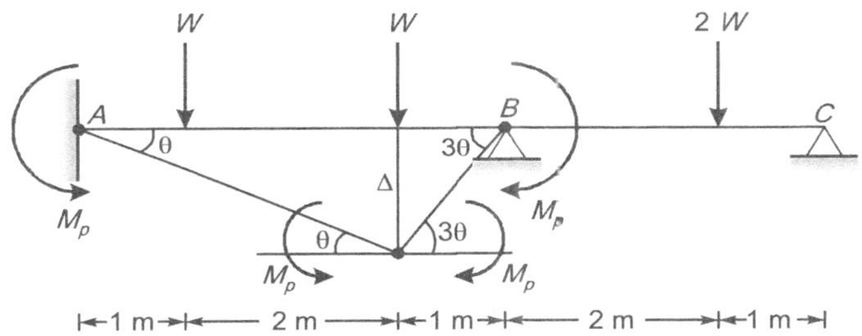

$$\therefore \quad M_p\theta + M_p4\theta + M_p3\theta = W\theta + W3\theta$$
$$= 4W\theta$$
$$\Rightarrow \quad 8M_p\theta = 4W\theta$$
$$\Rightarrow \quad M_p = W/2 \text{ or } W = 2M_F$$
$$\therefore \quad \text{Collapse load} = W = M_p$$

6.10 Draw the different mechanisms for the frame as shown in the figure for plastic analysis.

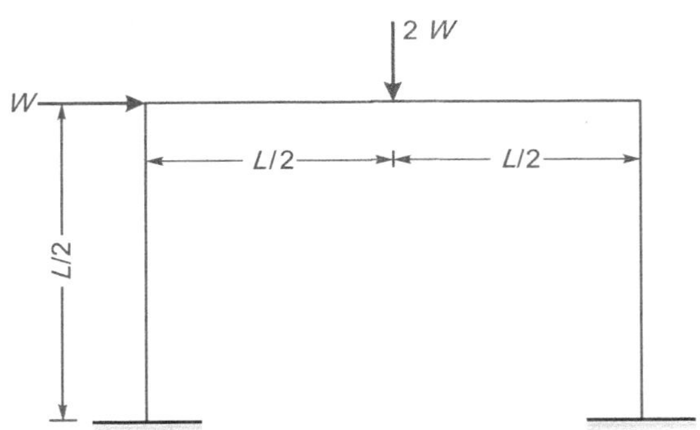

[4 marks : 2013]

Solution:

The frame is labeled as shown:

Possible locations for the formation of plastic hinges are at *A*, *B*, *C*, *D* and *E*.

Thus no. of plastic hinges

$$= 5(A, B, C, D, E)$$

Degree of static indeterminacy for the frame is

$$D_s = (r_e - 3) + 3C$$
$$= 6 - 3 + 3 \times 0 = 3$$

∴ No. of plastic hinges required for complete collpase

$$= 3 + 1 = 4$$

Thus possible no. of independent mechanism

$$= 5 - 3 = 2$$

These two independent mechanism will be "beam mechanism" and "sway mechanism".

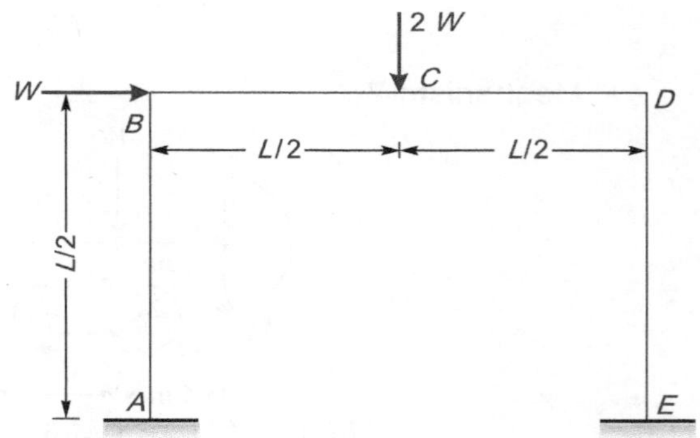

Beam mechanism:

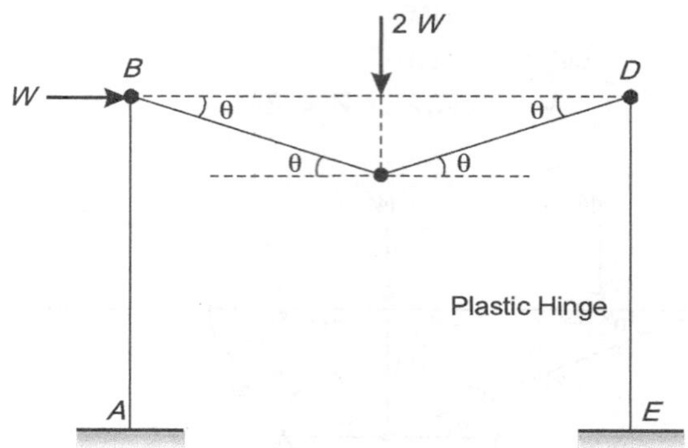

Sway Mechanism

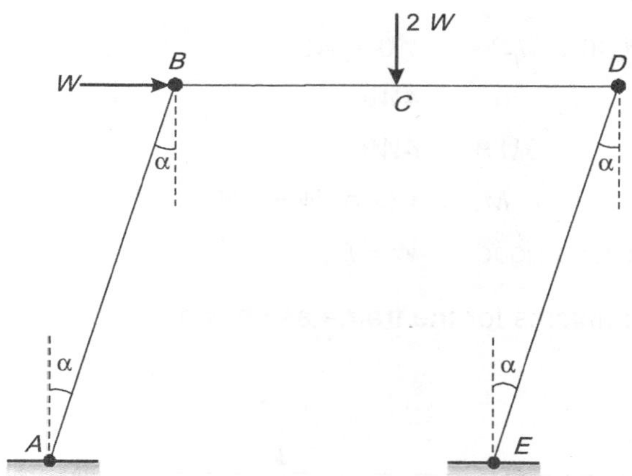

Combined Mechanism

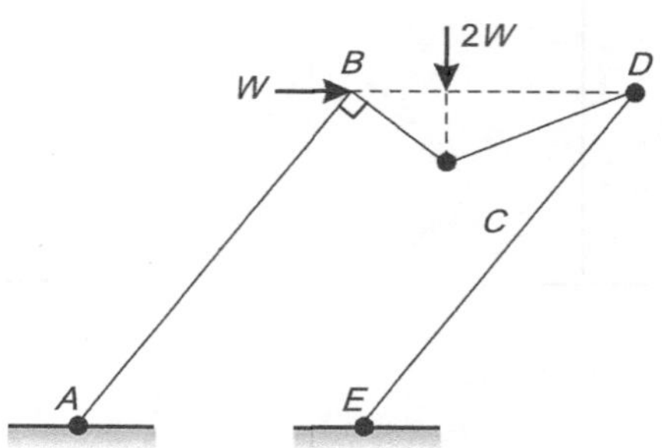

Out of '*I*' and '*II*' only two is possible because while combining beam and sway mechanism, hinge won't form at *B* (as at *B* hinge is hogging in case of beam mechanism and sagging in case of sway mechanism, hence both effects will cancel out in case of combined mechanism). Hence only *II* is possible with '4' hinges for full failure/collapse.

6.11 **Discuss the stress-strain curve of steel of grade Fe 500 and its application under Plastic Analysis.** **[5 + 5 marks : 2015]**

Solution:

As shown in figure, for Fe 500 grade of steel, there is no definite yield point and the stress is directly proportional to strain upto a stress of $0.87f_y$ only. There after the curve becomes inelastic. In this case, the yield stress is taken as 0.2% proof stress.

Application of Fe 500 in Plastic Analysis: In general, as the strength of steel increases, its ductility decreases. Thus mild steel (Fe 250) is more ductile than cold deformed bars (Fe 415 and Fe 500). Plastic design requires a large yield plateau which is not available considerably with cold deformed bars but is available with mild steel. Large yield plateu beyond yield point ensures sufficient elongation of the member beyond yield stress i.e. large ductility. Thus Fe 500 is not suitable for plastic analysis and design.

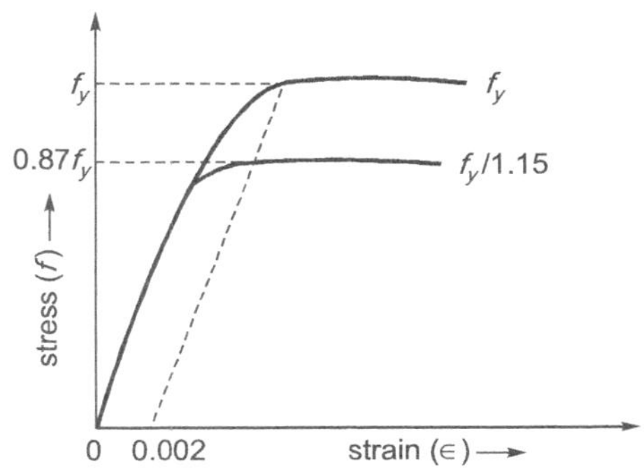

6.12 **Find the shape factor for a triangular section as shown in the figure.**

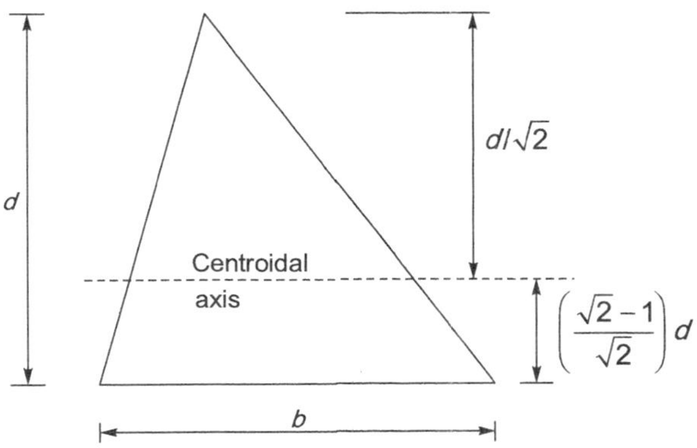

[10 marks : 2015]

Solution:

Here centroidal axis is not at $\left(\frac{d}{3}\right)$ from base *AB*.

$$\therefore \qquad I_{xx} \neq \frac{bd^3}{36}$$

Consider an elementary strip of width 'b_y' and thickness 'd_y' at a distance '*y*' from base *AB*.

$$\therefore \qquad \frac{b_y}{d-y} = \frac{b}{d}$$

$$\Rightarrow \qquad b_y = \frac{b(d-y)}{d}$$

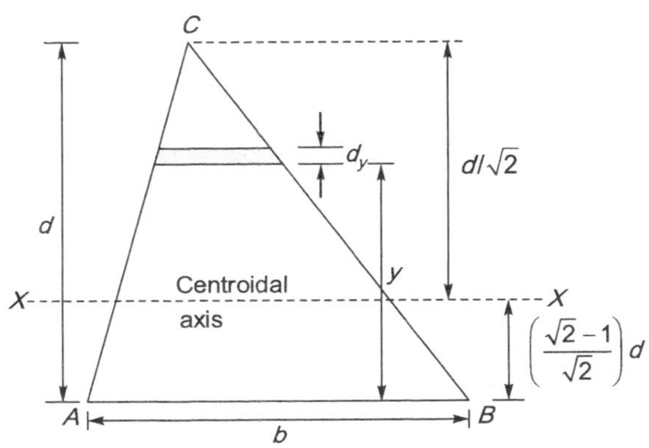

$\therefore$ Area of elementary strip, $dA = b_y.dy = \frac{b(d-y)}{d}\cdot dy$

$$\therefore \quad I_{AB} = \int y^2 dA = \int_0^d \frac{b(d-y)}{d} y^2 \cdot dy$$

$$= \frac{b}{d}\int_0^d \left(dy^2 - y^3\right)dy = \frac{b}{d}\left[\frac{dy^3}{3} - \frac{y^4}{4}\right]_0^d$$

$$= \frac{b}{d}\left[\frac{d^4}{3} - \frac{d^4}{4}\right] = \frac{bd^3}{12}$$

From parallel axis theorem, $I_{AB} = I_{xx} + A\left(\frac{\sqrt{2}-1}{\sqrt{2}}\right)^2 d^2$

$$\Rightarrow \quad I_{xx} = I_{AB} - A\left(\frac{\sqrt{2}-1}{\sqrt{2}}\right)^2 d^2$$

$$= \frac{bd^3}{12} - \frac{bd}{2}\left(\frac{3-2\sqrt{2}}{2}\right)d^2$$

$$= \frac{bd^3}{12} - bd^3\left(\frac{3-2\sqrt{2}}{4}\right)$$

$$= \frac{bd^3}{12} \times \left(1 - 3\left(3-2\sqrt{2}\right)\right)$$

$$= \frac{bd^3}{12} \times \left(1 - 9 + 6\sqrt{2}\right) = \frac{bd^3}{12}\left(6\sqrt{2} - 8\right)$$

$$= \left(\frac{3\sqrt{2}-4}{6}\right)bd^3$$

$\therefore$ Elastic section modulus $(z) = \frac{I_{xx}}{d/\sqrt{2}} = \left(\frac{3\sqrt{2}-4}{6}\right)\frac{bd^3}{d}\sqrt{2}$

$$= \left(\frac{3-2\sqrt{2}}{3}\right)bd^2$$

Plastic section modulus

Let $X_1 X_1$ be the equal area axis distant 'y' from apex C.

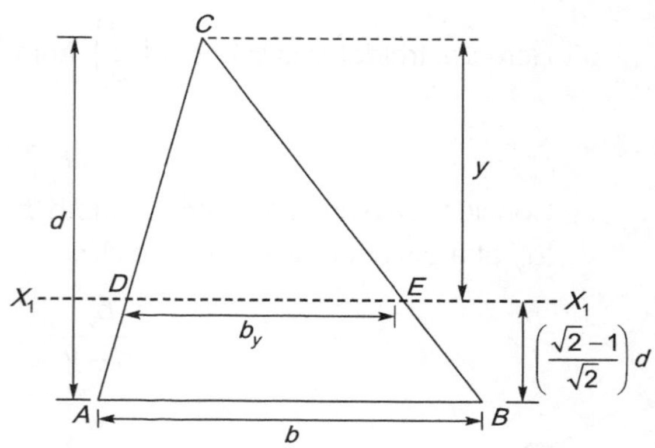

$$\therefore \quad \frac{b_y}{b} = \frac{y}{d}$$

$$\Rightarrow \quad b_y = \frac{b}{d}y$$

$$\therefore \quad \text{ar.}(\Delta CDE) = \frac{1}{2}ar.(\Delta CAB)$$

$$\Rightarrow \quad \frac{1}{2}\cdot b_y \cdot y = \frac{1}{2}\left(\frac{1}{2}bd\right)$$

$$\Rightarrow \qquad \frac{b}{d}y^2 = \frac{1}{2}bd$$

$$\Rightarrow \qquad y = \frac{d}{\sqrt{2}} \text{ from apex } c$$

$$\therefore \qquad b_y = \frac{b}{d}\cdot y = \frac{b}{\sqrt{2}}$$

$$\therefore \qquad \text{Plastic section modulus } (z_p) = \frac{A}{2}(\bar{y}_1 + \bar{y}_2)$$

where

$$\bar{y}_1 = \frac{y}{3} = \frac{d}{3\sqrt{2}}$$

$$\bar{y}_2 = \left(\frac{d-y}{3}\right)\left(\frac{b_y + 2b}{b_y + b}\right) = \left(\frac{d - d/\sqrt{2}}{3}\right)\left(\frac{b/\sqrt{2} + 2b}{b/\sqrt{2} + b}\right)$$

$$= \left(\frac{\sqrt{2}-1}{3\sqrt{2}}\right)d\left(\frac{1+2\sqrt{2}}{1+\sqrt{2}}\right) = \frac{(\sqrt{2}-1)(2\sqrt{2}+1)}{3\sqrt{2}(1+\sqrt{2})}d$$

$$\therefore \qquad z_p = \frac{A}{2}(\bar{y}_1 + \bar{y}_2) = \frac{1}{2}\cdot\frac{bd}{2}\left[\frac{d}{3\sqrt{2}} + \frac{(\sqrt{2}-1)(2\sqrt{2}+1)}{3\sqrt{2}(\sqrt{2}+1)}\right]$$

$$= \frac{bd^2}{4}\left(\frac{1}{3\sqrt{2}}\right)\left[1 + \frac{(\sqrt{2}-1)(2\sqrt{2}+1)}{\sqrt{2}+1}\right]$$

$$= \frac{bd^2}{12\sqrt{2}}\left[\frac{4}{\sqrt{2}+1}\right] = \frac{bd^2}{3\sqrt{2}(1+\sqrt{2})}$$

$$\therefore \qquad \text{Shape factor} = \frac{z_p}{z} = \frac{bd^2 \times 3}{3\sqrt{2}(1+\sqrt{2})(3-2\sqrt{2})bd^2}$$

$$= \frac{1}{\sqrt{2}(\sqrt{2}-1)} = 1.707$$

6.13 What are the assumptions for plastic analysis of steel sections?

[5 marks : 2016]

Solution:

Following assumptions are made in the plastic analysis of steel sections:

(i) In simple plastic theory for steel sections, it is assumed that deflections of the structure are not the critical design criteria.

Thus structure as a whole or its parts can be analysed/designed on the basis of collapse load.

(ii) Plastic analysis is based on idealization of stress-strain curve of mild steel as elastic-perfectly plastic curve.

(iii) Width-to-thickness ratio of plate elements is small so that local buckling does not occur i.e. sections are plastic sections.

6.14 Show that in a rectangular section subjected to axial force N and bending moment M.

$$\frac{M}{M_p}+\left(\frac{N}{N_p}\right)^2=1$$

where N_p = axial yield capacity
M_p = plastic moment capacity

[10 marks : 2016]

Solution:

1. **At the beginning of yielding.**

At the beginning to yielding,

$$f_y=\frac{N}{A}+\frac{M}{I}\times y \text{ at the extreme-fibre}$$

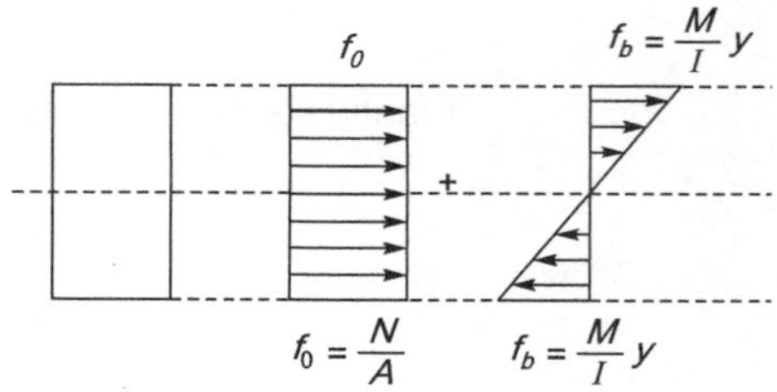

Fig. A

2. **After yielding of top fibre,**

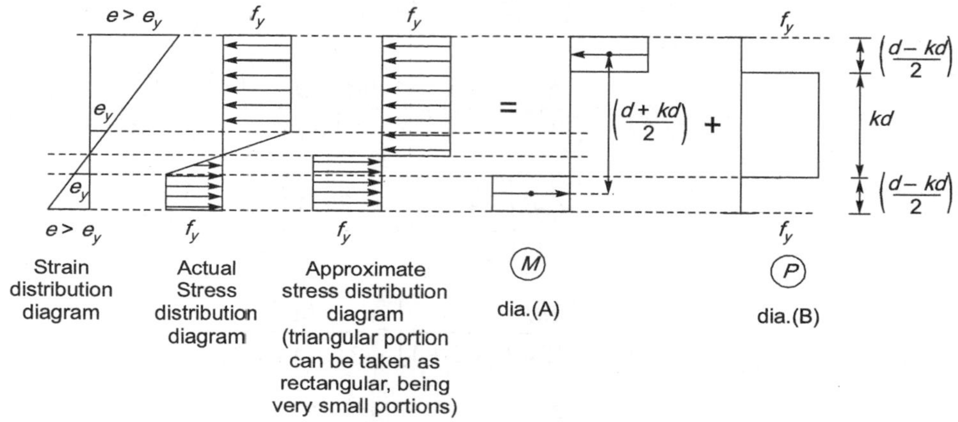

Fig. B

From the stress distribution diagram (A), k are get

$$M = C\times \text{lever arm}$$

$$M=\left[f_y\cdot\frac{b(d-kd)}{2}\right]\left(\frac{d+kd}{2}\right) = f_y\cdot\frac{bd^2}{4}(1-k^2) \quad \text{...(i)}$$

From stress distribution diagram (B), we can get

$$N = f_y(b\times kd) \quad \text{...(ii)}$$

If $k=0$, $N=0$ and $M=f_y\cdot\frac{bd^2}{4}$ which is M_P of cross-section.

If $k=1$, $M=0$ and $N=f_y\times bd$, which is (N_p) axial yield capacity of c/s.
So, equations (i) and (ii) can be written as

$$\frac{M}{M_P}=(1-k^2) \quad \text{and} \quad \frac{N}{N_P}=k\text{, eliminating } k \text{ gives, } \frac{M}{M_P}+\left(\frac{N}{N_P}\right)^2=1$$

6.15 Determine the collapse load in a three-span continuous beam shown in figure. It has a moment capacity of M_P over the exterior spans AB and CD, and 1.5 M_P over the interior span BC.

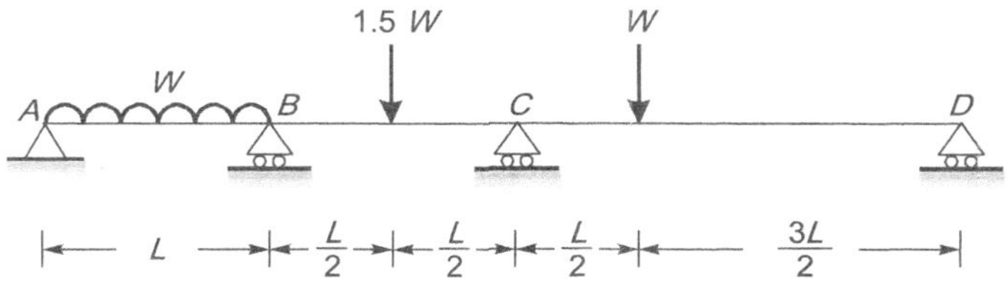

[20 marks : 2017]

Solution :

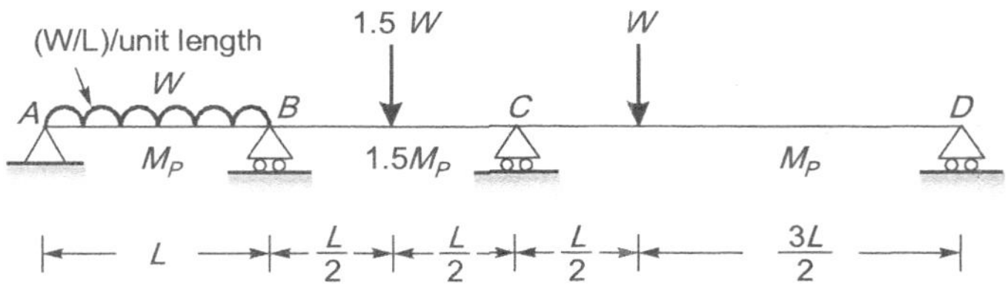

Using kinematic theorem, the collapse load for the beam will be the minimum of the collapse loads for the various spans.

Span AB

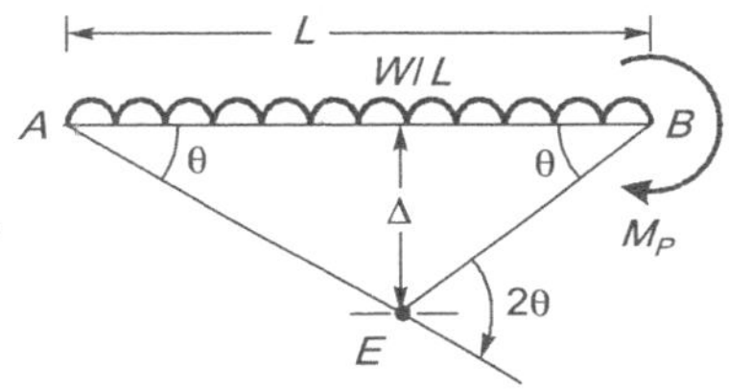

The end is simple support and end B is intermediate support.

Therefore, the beam AB acts as propped cantilever with B as fixed end.

Plastic hinges will form at centre of AB and at B.

By principal of virtual work,

External work done = Internal work done

$\Rightarrow$ Load × deflection = Moment × rotation

$$\Rightarrow \frac{W}{L} \times \frac{1}{2} \times L \times \frac{L}{2}\theta = M_P(\theta + \theta) + M_P\theta$$

$$\Rightarrow \frac{WL}{4}\theta = 3M_P\theta$$

$$\Rightarrow W = \frac{3 \times 4M_P}{L} = \frac{12M_P}{L}$$

Span BC

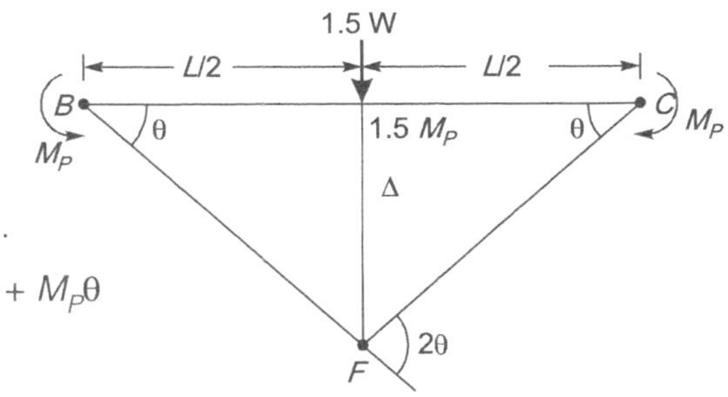

Plastic hinge will form at B, C and under the load 1.5 W

External work done = Internal work done.

$$\Rightarrow 1.5\, W \times \Delta = M_P\theta + 1.5\, M_P(2\theta) + M_P\theta$$

$$\Rightarrow 1.5W \times \frac{L}{2}\theta = 5\, M_P\theta$$

$$\Rightarrow W_u = \frac{6.67M_P}{L}$$

Span CD

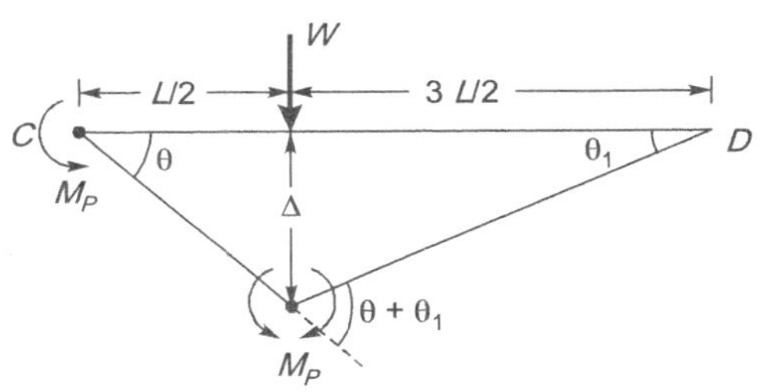

$$\Delta = \frac{L}{2}\theta = \frac{3L}{2}\theta_1$$

$$\Rightarrow \theta = 3\theta_1$$

Plastic hinge will form at C and under the load W.

$\therefore$ External work done = Internal work done

$$\Rightarrow \quad W \times \Delta = M_p\theta + M_p(\theta + \theta_1)$$

$$\Rightarrow \quad W \times \frac{L}{2}\theta = 2M_p\theta + M_p\theta_1$$

$$\Rightarrow \quad \frac{3WL\theta_1}{2} = 2M_p \times 3\theta_1 + M_p\theta_1$$

$$\Rightarrow \quad W = \frac{2}{3L} \times (6M_P + M_P) = \frac{14M_P}{3L} = \frac{4.67M_P}{L}$$

The collapse load is least of three ultimate loads.

Hence, collapse load, $W = \dfrac{4.67M_P}{L}$

6.16 **Find the shape factor of a triangular section of base b and height h for bending about an axis parallel to the base.**

[12 marks : 2018]

Solution:

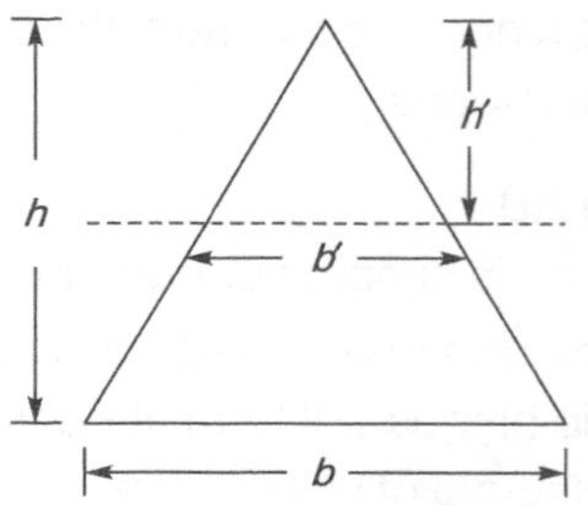

$$\text{Shape factor} = \frac{z_p}{z_e};$$

z_p = Plastic section modulus, z_e = Elastic section modulus

As, $$\frac{b'h'}{2} = \frac{1}{2}\left(\frac{bh}{2}\right)$$

...($\because$ Equal area axis is present at b' level)

$$\Rightarrow \quad b' = \frac{bh}{2h'} \quad \text{...(i)}$$

Also, by similar Δ's,

$$\frac{b'}{b} = \frac{h'}{h} \quad \text{...(ii)}$$

From (i) and (ii),

$$\frac{bh}{2h'} \times \frac{1}{b} = \frac{h'}{h}$$

$$\Rightarrow \quad h' = \left(\frac{h}{\sqrt{2}}\right)$$

$$\Rightarrow \quad b' = \frac{bh}{2h'} = \frac{bh\sqrt{2}}{2h}$$

$$\therefore \quad b' = \left(\frac{b}{\sqrt{2}}\right)$$

Now, $$z_p = \frac{A}{z}(y_t + y_b) = A_1\bar{y}_1 + A_2\bar{y}_2$$

$$\Rightarrow \quad z_p = \frac{b'h'}{2}\left(\frac{h'}{3}\right) + \frac{1}{2}\left(\frac{bh}{2}\right)\left(\frac{2b+b'}{b+b'}\right)\left(\frac{h-h'}{3}\right)$$

$$\Rightarrow \quad z_p = \frac{bh}{4}\left[\frac{h}{3\sqrt{2}} + \left(\frac{2b+\frac{b}{\sqrt{2}}}{b+\frac{b}{\sqrt{2}}}\right)\left(\frac{h-\frac{h}{\sqrt{2}}}{3}\right)\right]$$

$$\Rightarrow \qquad Z_p = \frac{bh}{4}\left[\frac{h}{3\sqrt{2}} + \left(\frac{2\sqrt{2}+1}{\sqrt{2}+1}\right)\left(\frac{\sqrt{2}-1}{3\sqrt{2}}\right)h\right]$$

$$\Rightarrow \qquad Z_p = 0.09763\, bh^2 \qquad \text{...(iii)}$$

Now, for elastic section modulus,

$$Z_e = \frac{I}{y} = \frac{bh^3/36}{2h/3} = \frac{bh^2}{24} \qquad \text{...(iv)}$$

Now, Shape factor $= \dfrac{Z_p}{Z_e}$

From (iii) and (iv),

$$\text{Shape factor} = \frac{0.9763}{1/24} = 2.343$$

6.17 **Find the collapse load for a frame of uniform cross-section shown in figure under the applied forces. Also determine the minimum section in steel required to resist the applied forces.**

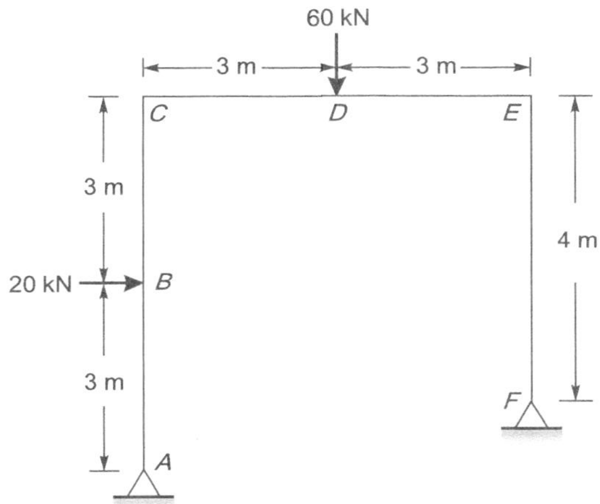

[20 marks : 2018]

Solution:

As the loads are given in the question minimum M_P required has to be calculated to get the minimum section required to resist applied loads.

Static indeterminacy D_S.

$$\Rightarrow \qquad D_S = \text{Number of reaction} - \text{Number of equilibrium equation}$$
$$= 4 - 3 = 1$$

$\Rightarrow$ Number of possible plastic hinges = 4 at B, C, D, E

$\therefore$ Number of independent mechanism= $N - D_S = 4 - 1 = 3$

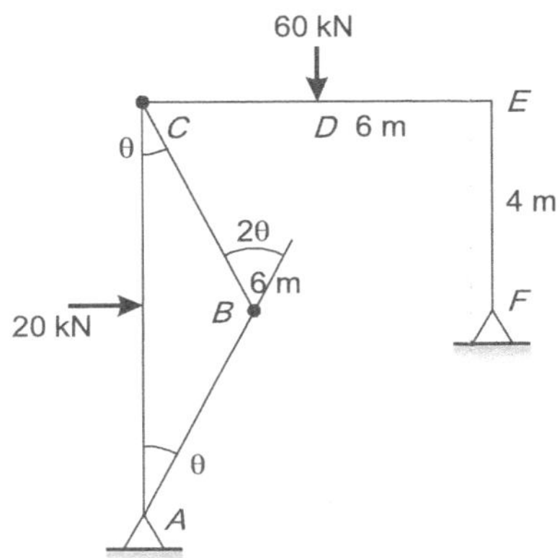

Case 1:

Hinges at B and C.

Let M_P be the plastic moment capacity of the steel section.

Equating external and internal work done

$$20 \times \Delta = M_P 2\theta + M_P\theta$$
$$20 \times 3\theta = 3M_P\theta$$
$$M_P = 20 \text{ kNm}$$

Case 2:

Hinges at C and E

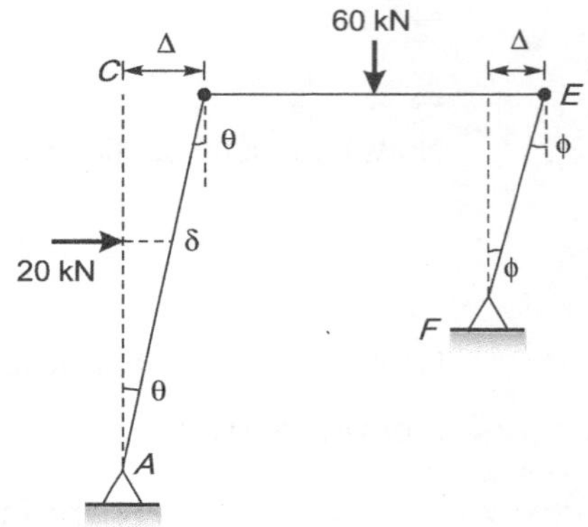

Since, $\Delta_C = \Delta_E$

$\therefore$ $6\theta = 4\phi$

$\therefore$ $\phi = 1.5\theta$

Equating external and internal work done

$$20 \times \delta = M_p\theta + M_p\phi$$

$$20 \times 3\theta = 2.5\, M_p\theta$$

$\Rightarrow$ $M_p = 24$ kNm

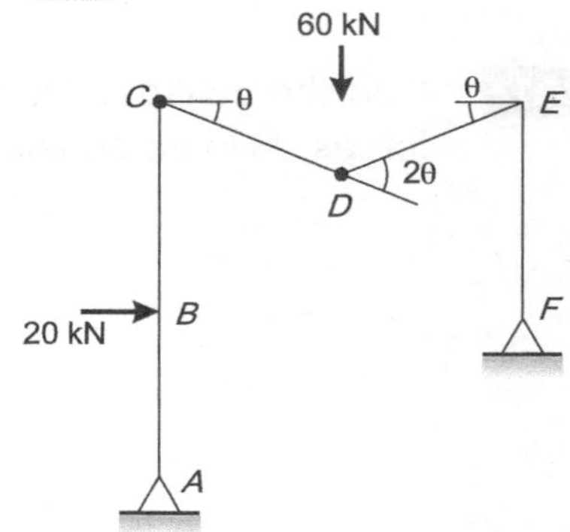

Another possibility with hinges at C, D and E.

Equating external and internal workdone.

$$60 \times 3\theta = M_p\theta + 2M_p\theta + M_p\theta$$

$$M_p = 45 \text{ kNm}$$

Case 3:

Hinges at B and E

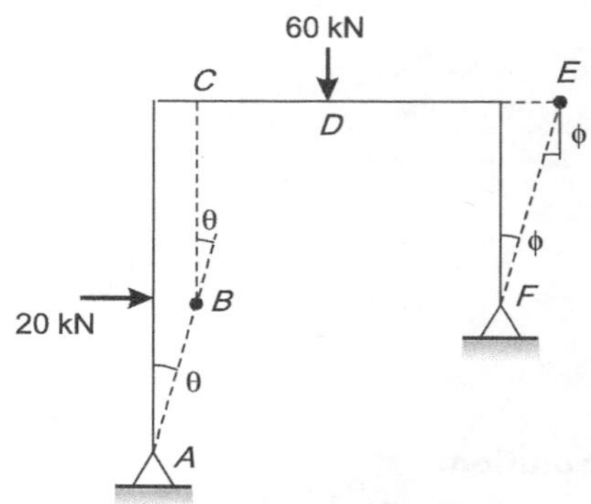

Since, $\Delta_C = \Delta_E$

$\therefore$ $3\theta = 4\phi$

$\therefore$ $\phi = 0.75\theta$

Equating internal and external work done

$$20 \times 3\theta = M_p\theta + M_p\phi$$

$$60\theta = 1.75\, M_p\theta$$

Case 4:

Combined mechanicsm

Hinges at D and E

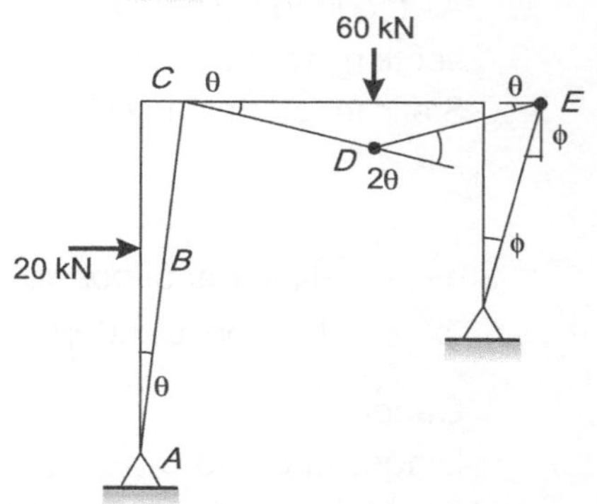

Equating external and internal work

Also $\Delta_C = \Delta_E$

$\therefore$ $6\theta = 4\phi$

$\therefore$ $\phi = 1.5\theta$

$$60 \times 3\theta + 20 \times 3\theta = M_p 2\theta + M_p\theta + M_p\phi$$

$$M_p = 53.33 \text{ kNm}$$

$\therefore$ Maximum $M_p = 53.33$ kNm

$\therefore$ Atleast this much strength should be provided to prevent collapse.

$\therefore$ Plastic modulus Z_P required $= \left(\dfrac{M_P}{f_y}\right) = 213320 \text{ mm}^3$.

$$Z_e = \frac{Z_P}{SF} = \frac{213320}{1.12} = 190464 \text{ mm}^3$$

6.18 Find the designed plastic moment for the portal frame as shown in the figure below under collapse condition for the factored (applied) loads. Assume that the frame has uniform cross-section. Also, find the minimum section required for the frame for E250 grade of steel:
Given:

Section	ISMB 125	ISMB 150	ISMB 175	ISMB 200	ISMB 225	ISMB 250	ISMB 300	ISMB 350
Plastic section modulus (cm^3)	81	110	184	255	348	466	651	889

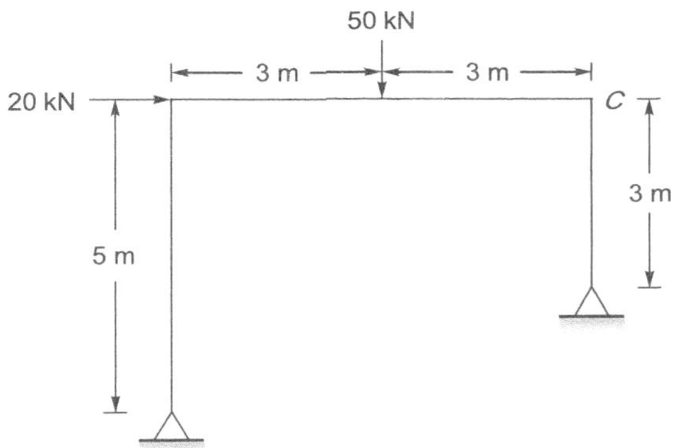

[20 marks : 2023]

Solution:

The possible location of plastic hinges are B, C and D.
Number of possible plastic hinge, $N = 3$
Degree of redundancy, $r = 4 - 3 = 1$
$\therefore$ No. of possible independent mechanism,

$$n = N - r = 3 - 1 = 2$$

The two independent mechanisms are
1. Beam mechanism
2. Sway mechanism

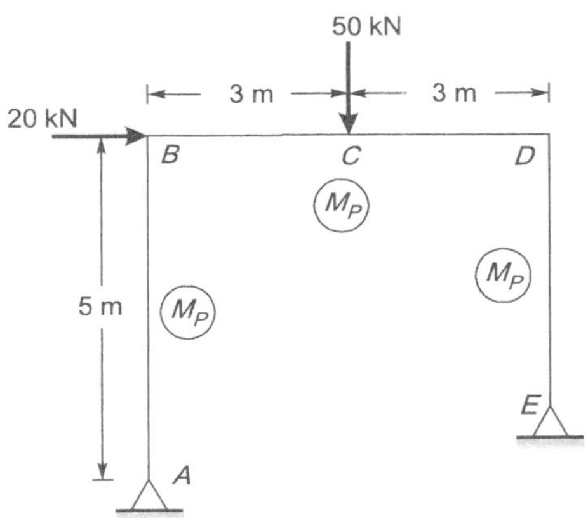

Mechanism-1:

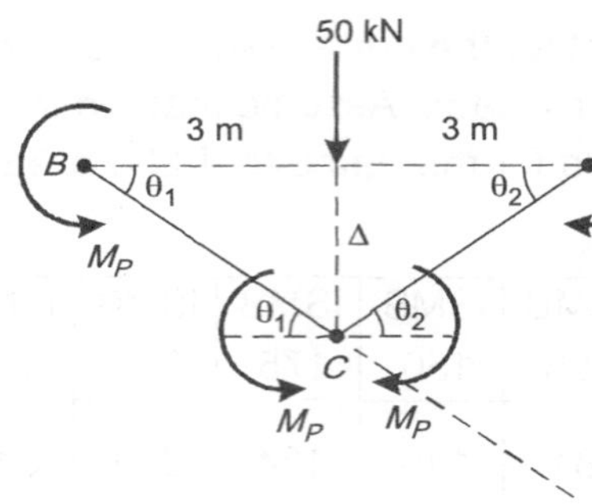

$\therefore$ $$\Delta = 3 \times \theta_1 = 3\theta_2$$

$\Rightarrow$ $$\theta_1 = \theta_2$$

By principle of virtual work

External work done = Internal work done

$\Rightarrow$ $$M_P\theta_1 + M_P\theta_1 + M_P\theta_2 + M_P\theta_2 = 50 \times \Delta$$

$\therefore$ $$4M\theta_1 = 50 \times 3\theta_1 \qquad [\Delta = 3\theta_1 \text{ and } \theta_1 = \theta_2]$$

$\therefore$ $$M_P = \frac{150}{4} = 37.5 \text{ kNm}$$

Mechanism-2:

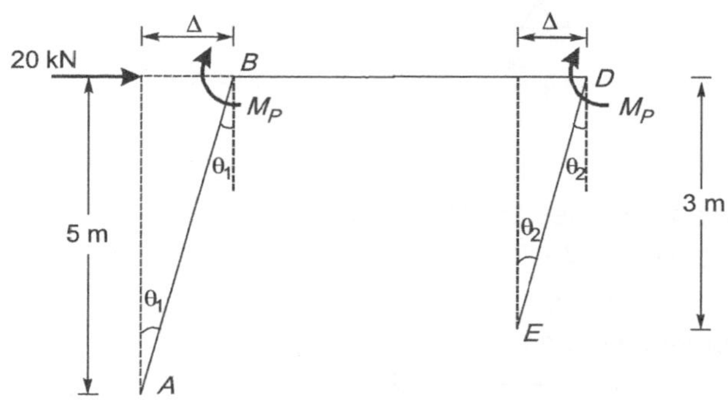

$$\Delta = 5 \times \theta_1 = 3\theta_2$$

$\therefore$ $$\theta_1 = \frac{3\theta_2}{5}$$

By principal of virtual work,

External work done = Internal work done

$\Rightarrow$ $$M_P\theta_1 + M_P\theta_2 = 20 \times \Delta$$

$\Rightarrow$ $$M_P\left(\frac{3\theta_2}{5}\right) + M_P\theta_2 = 20 \times 3\theta_2 \qquad \left[\because \theta_1 = \frac{3\theta_2}{5}\right]$$

$\therefore$ $$M_P = \frac{60 \times 5}{8} = 37.5 \text{ kNm}$$

Mechanism-3:

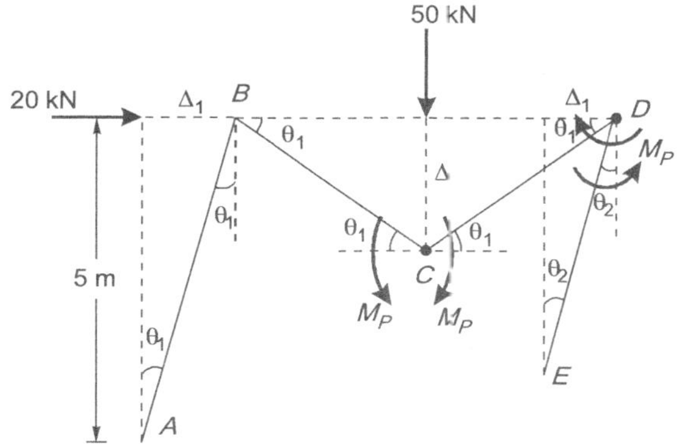

$$\Delta_1 = 5\theta_1 = 3\theta_2$$

$$\Rightarrow \qquad \theta_2 = \frac{5\theta_1}{3}$$

$$\Delta = 3\theta_1$$

By principal of virtual work,

External work done = internal work done

$$\Rightarrow \qquad M_P\theta_1 + M_P\theta_1 + M_P\theta_1 + M_P\theta_2 = 20 \times \Delta_1 + 50 \times \Delta$$

$$\Rightarrow \qquad 3M_P\theta_1 + M_P\left(\frac{5\theta_1}{3}\right) = 20 \times 5\theta_1 + 50 \times 3\theta_1$$

$$\therefore \qquad M_P = \frac{(100+150)}{(14/3)} = 53.571 \text{ kNm}$$

Therefore, the design plastic moment for the frame = 53.571 kN

Let us assume, the plastic section modulus of section is z_p.

$$\therefore \qquad M_P = z_p \times f_y$$

$$\Rightarrow \qquad 53.571 \times 10^6 = z_p \times 250$$

$$\therefore \qquad z_p = 214284 \text{ mm}^3$$

$$z_p = 214.284 \text{ cm}^3$$

So, from the given table, minimum section required for the frame is ISMB 200 (z_p = 255 cm^3)

5 Design of Concrete and Masonry Structures

Revised Syllabus of ESE : *Limit state design for bending, shear, axial compression and combined forces; Design of beams, Slabs, Lintels, Foundations, Retaining walls, Tanks, Staircases; Principles of pre-stressed concrete design including materials and methods; Earthquake resistant design of structures; Design of Masonry Structure.*

---

1. Fundamentals of RCC

1.1 Explain 'under-reinforced', 'balanced' and 'over-reinforced' sections in the ultimate load theory.

[10 marks : 1999]

Solution:

The stress block parameters are shown below:

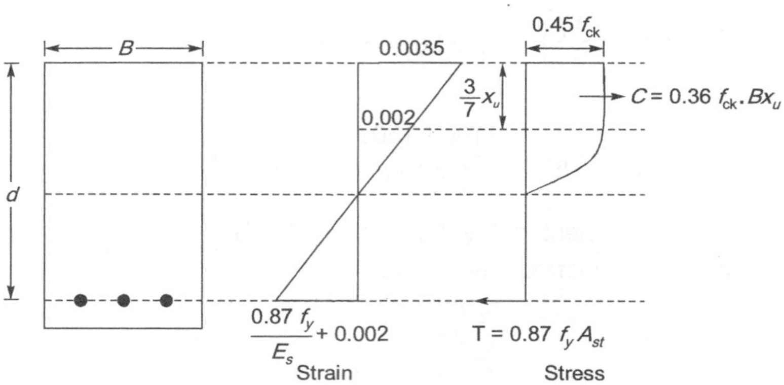

Actual depth of neutral axis is given by

$$C = T$$

$$\Rightarrow \quad 0.36\ f_{ck}\ Bx_u = 0.87\ f_y\ A_{st}$$

$$\frac{x_u}{d} = \frac{0.87}{0.36} \cdot \frac{f_y}{f_{ck}} \cdot \frac{A_{st}}{Bd}$$

Also limiting value of neutral axis is given by.

$$\frac{x_{u,max}}{d} = \frac{700}{1100 + 0.87\ f_y}$$

Now

(i) when $\frac{x_u}{d}$ is less than the limiting value, it means that the reinforcement provided is less than its limiting value. Thus steel reinforcement reaches its yield stress before ultimate strain is reached in concrete. Such a section is called a **under reinforced section.**

(ii) When $\frac{x_u}{d}$ is equal to the limiting value, the case corresponds to balanced section design in which the steel reinforcement reaches its yield stress at the same instant when ultimate strain is reached in concrete. Such a section is called **balanced section.**

(iii) When $\frac{x_u}{d}$ is greater than the limiting value, the case corresponds to the beam in which percentage of steel is sufficient to ensure that steel yield does not take place. Failure occurs when the strain in extreme fibres in concrete reaches its ultimate value. Thus failure takes place due to crushing of concrete while strain in steel remains below the yield strain. Such a section is called **over reinforced section.** It should be noted that compression failure is sudden and therefore not desirable.

1.2 Derive the expression for limiting percentage of tensile reinforcement in a flexural *RC* member.

[5 marks : 2002]

Solution:

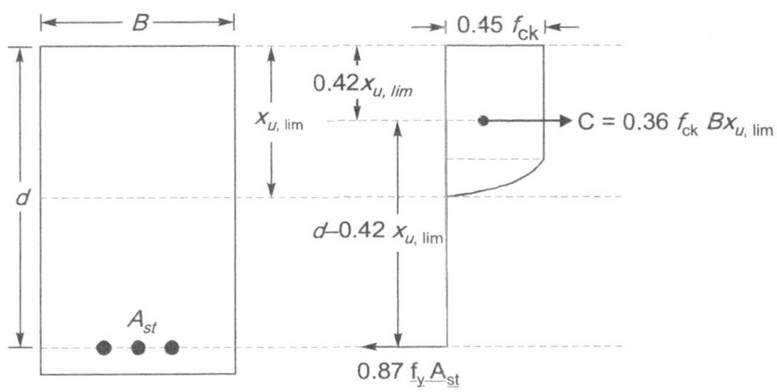

Limiting percentage of tensile steel is calculated for a **balanced section**. Since, in flexure total compressive force is equal to total tensile force, we get

$$C = T$$

$$0.36 \times f_{ck} \times B\,x_{u,\,lim} = 0.87\, f_y\, A_{st}$$

$$\Rightarrow \quad x_{u,\,lim} = \frac{0.87\, f_y\, A_{st}}{0.36\, f_{ck}\, B}$$

$$\Rightarrow \quad \frac{x_{u,\,lim}}{d} = \frac{0.87\, f_y}{0.36\, f_{ck}} \times \frac{A_{st}}{Bd}$$

$$\Rightarrow \quad \frac{A_{st}}{Bd} = \frac{0.36 f_{ck}}{0.87 f_y} \times \frac{x_{u,\,lim}}{d}$$

$$\Rightarrow \quad \frac{A_{st}}{Bd} \times 100 = \frac{0.36}{0.87} \times 100 \times \frac{f_{ck}}{f_y} \times \frac{x_{u,\,lim}}{d}$$

$$\Rightarrow \quad p_t = 41.38 \times \frac{f_{ck}}{f_y} \times \frac{x_{u,\,lim}}{d}\,\%$$

If M 20 concrete and Fe 415 steel is used, then

$$p_t = 41.38 \times \frac{20}{415} \times \frac{0.48\,d}{d}$$

$$p_t = 0.957\%$$

1.3 IS 456-2000 suggests use of a certain stress-strain curve of concrete in the absence of actual experimental data. The code also allows use of an expression $\left(5000\sqrt{f_{ck}}\right)$ to estimate the modulus of elasticity of concrete (E_c). Draw a neat representation of that curve, briefly explain its salient features. The suggested value represents the value of E at which, if any, of the point(s) on the stress-strain curve.

[5 marks : 2013]

Solution:

Stress-strain curve of concrete.

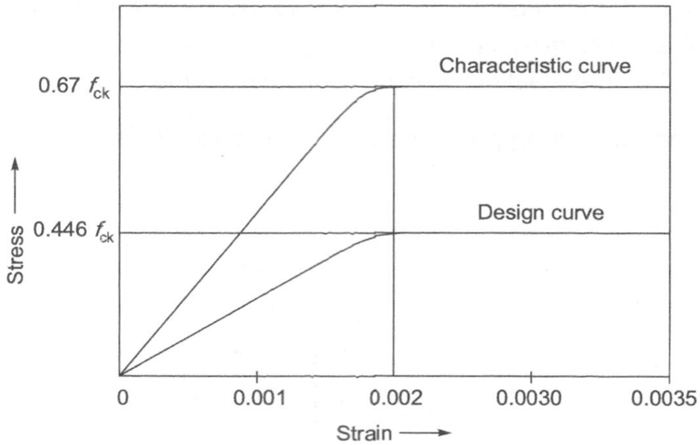

1.4 State the assumption in the limit state of collapse in compression in flexure regarding strain at the highly compressed extreme fibre in concrete. Show with the help of a neat sketch.

[5 marks : 2014]

Solution:

As per IS 456: 2000 Clause (39.1) design of the limit state of collapse in compression shall be based on following assumptions:

(a) Plane section normal to the axis remain plane after bending.

(b) The maximum compressive strain in concrete in axial compression is taken as 0.002.

(c) The maximum compressive strain at the highly compressed extreme fibre in concrete subjected to axial compression and when there is no tension on the section shall be 0.0035 minus 0.75 times the strain at the least compressed extreme fibre.

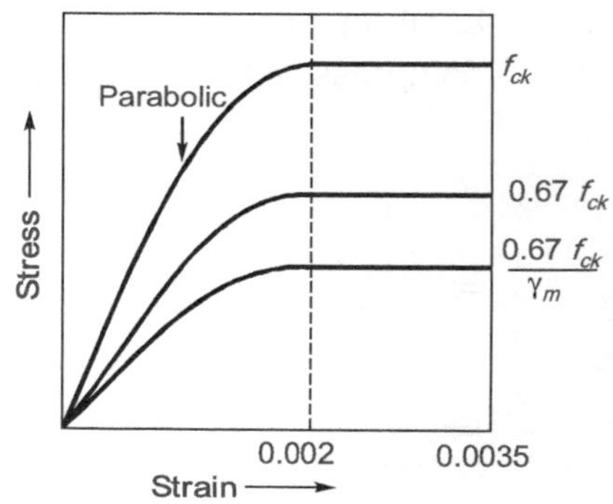

Strain curve for concrete

(d) The relationship between the compressive stress distribution in concrete and the strain in concrete may be assumed to be rectangle, trapezoid, parabola or any other shape which results in prediction of strength in substantial agreement with the result of test. An acceptable stress-strain curve is given in figure. For design purpose, the compressive strength of concrete in structure shall be assumed to be 0.67 times the characteristic strength. The partial safety factor γ_m = 1.5 shall be applied in addition to this.

1.5 Explain under reinforced and over reinforced failure of a reinforced concrete beam.

[5 marks : 2014]

Solution:

Under reinforced failure: An 'under reinforced section' is one in which the area of tension steel (A_{st}) such that the permissible stress is reached in the steel before the permissible stress is reached in the extreme fibre of the concrete.

In this case,

(i) $x_a < x_c$

(ii) $C_a < \sigma_{cbc}$

(iii) $t_a = \sigma_{st}$

(iv) $A_{st\ provided} < A_{st\ balanced}$

A slight increase in the load, at this stage causes the beam to yield and elongate. In this case failure is due to failure of steel. It is known as tension failure or ductile failure. This give sufficient warning before failure. Hence an under reinforced section is always preferred.

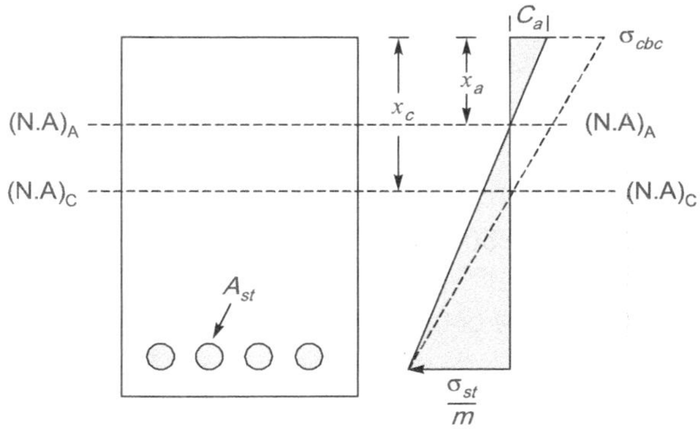

Over reinforced failure: An 'over reinforced' section is one in which the area of tension steel (A_{st}) is such that permissible stress in extreme fibre of concrete is reached before the permissible stress in the steel.

In this case

(i) $x_a > x_c$

(ii) $C_a = \sigma_{cbc}$

(iii) $t_a < \sigma_{st}$

(iv) $A_{st\ provided} > A_{st\ balanced}$

The concrete fails in compression before the steel reaches its permissible stress. Hence this type of failure is known as compression failure or Brittle failure. The failure occurs without warning, hence an over reinforced section is always undesirable.

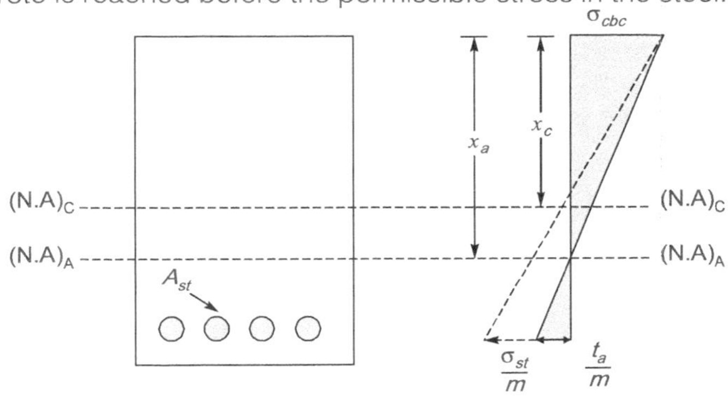

1.6 **"In limit state method of design or working stress method of design, balanced neutral axis depth is not dependent on compressive strength of concrete." Justify the statement with the help of suitable formula.**

[5 marks : 2015]

Solution:

LSM: For balanced section, maximum compressive strain in concrete (= 0.0035) and yield tensile strain in steel $\left(= \frac{0.87f_y}{E_s} + 0.002\right)$ will reach simultaneously.

$$\therefore \quad \frac{\epsilon_{cu}}{x_{u\lim}} = \frac{\epsilon_{st}}{d - x_{u\lim}}$$

$$\Rightarrow \quad \left(\frac{d}{x_{u\lim}} - 1\right) = \frac{\epsilon_{st}}{\epsilon_{cu}}$$

$$\Rightarrow \quad \frac{d}{x_{u\lim}} = \frac{\epsilon_{cu} + \epsilon_{st}}{\epsilon_{cu}}$$

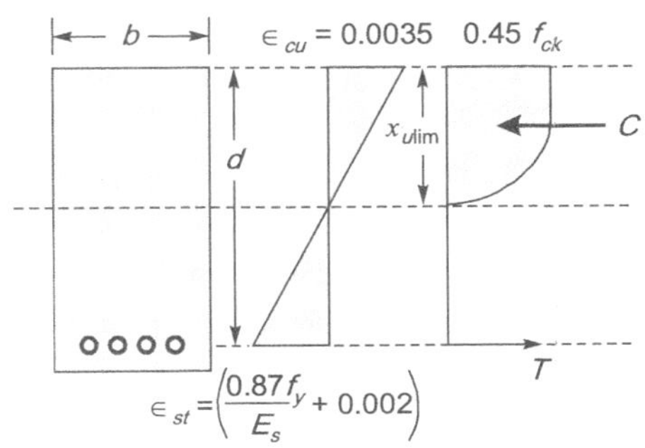

$$\Rightarrow \quad \frac{x_{u\lim}}{d} = \frac{\epsilon_{cu}}{\epsilon_{cu} + \epsilon_{st}} = \frac{0.0035}{\frac{0.87 f_y}{E_s} + 0.0055} \text{ which is independent of } f_{ck}$$

For Fe 250, $\frac{x_{u\lim}}{d} = 0.531$

For Fe 415, $\frac{x_{u\lim}}{d} = 0.479$

For Fe 500, $\frac{x_{u\lim}}{d} = 0.456$

WSM: For balanced section, both concrete and steel will reach their permissible stress values simultaneously.

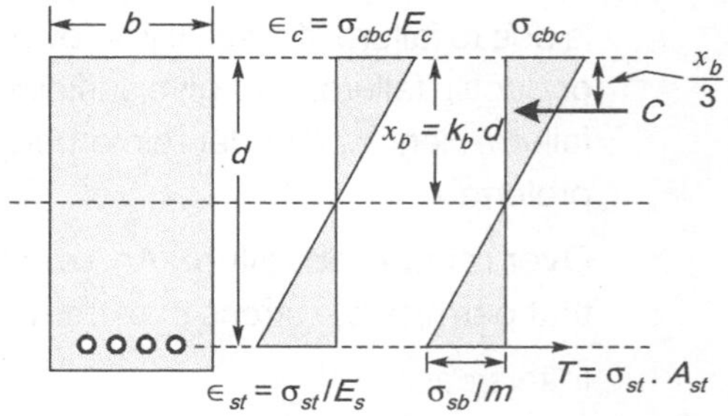

$$\therefore \quad \frac{\sigma_{cbc}}{x_b} = \frac{\sigma_{st}/m}{d - x_b}$$

$$\Rightarrow \quad \left(\frac{d}{x_b} - 1\right) = \frac{\sigma_{st}/m}{\sigma_{cbc}}$$

$$\Rightarrow \quad \frac{d}{x_b} = \frac{\sigma_{st}/m + \sigma_{cbc}}{c_{cbc}}$$

$$\Rightarrow \quad \frac{x_b}{d} = k_b = \frac{\sigma_{cbc}}{\sigma_{st}/m + \sigma_{cbc}}$$

But $$m = \frac{280}{3\sigma_{cbc}}$$

$$\therefore \quad \frac{x_b}{d} = k_b = \frac{\sigma_{cbc}}{\frac{3\sigma_{st} \cdot \sigma_{cbc}}{280} + \sigma_{cbc}} = \frac{93.33}{\sigma_{st} + 93.33}$$

which is also independent of f_{ck}.

1.7 What do you understand by the term characteristic strength of a material? Please elaborate.

[8 marks : 2016]

Solution:

Characteristic strength of material: Characteristic strength of a material is the probabilistic approach of defining the strength of a material. It is defined as that strength below which not more than a certain percentage of test results conducted on the material are expected to fall. This limiting percentage is generally taken as 5%.

Example: For M20 concrete, it implies its characteristic strength is 20 N/mm^2, that means, if 100 samples of this concrete are tested for compressive strength than not more than 5 samples (5% of 100) should have a compressive strength less than 20 N/mm^2.

To calculate characteristic strength, tests are conducted on several specimen of material and they are arranged in increasing order of the strength. A graph is then plotted between number of samples and calculated stress. The point on the graph which gives value of stress for which not more than 5% of the total samples have failed, is called the characteristic strength of the material.

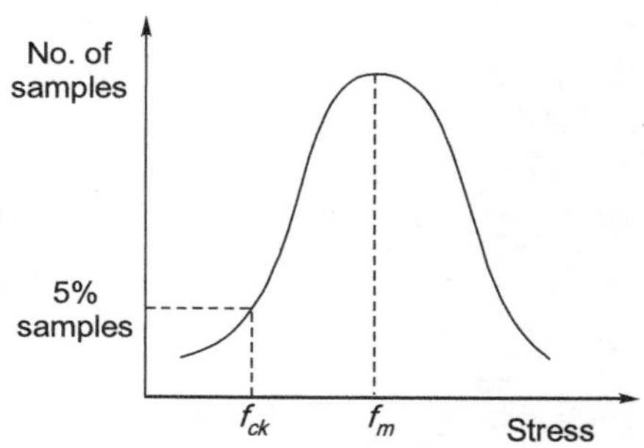

f_{ck} = Characteristic strength

f_m = Target mean strength

1.8 What do you understand by the term limit state?

[5 marks : 2016]

Solution:

Limit State: For a structure, limit state is a state of impending failure, beyond which the structure cannot perform its intended function satisfactorily.

Beyond limit state, the structure cannot be used either due to collapse of structure or its unserviceability.

There are two types of limit state:

1. Ultimate limit state (limit state of collapse):

   It is due to failure of structure in any of the following modes

   (a) flexure (b) compression

   (c) torsion (d) shear

2. Serviceability Limit State:

   In this structure becomes unserviceable due to attainment of the limit state in any of the following:

   (a) excessive deflection (b) excessive cracking

   (c) vibration (d) corrosion

1.9 Explain briefly with an example the Acceptance Criteria for Concrete as per IS 456-2000.

[8 marks : 2019]

Solution:

Acceptance Criteria for Concrete

As per 16, 1(a) / IS 456 / 2000 Page 30

1. **For all concrete: M15 grade and above**

   The average strength of four (4) non overlapping consecutive test result shall be not less than

   For M15 or higher:

   $$f_{average} \leq (f_{ck} + 0.825\ \sigma)\ \text{N/mm}^2$$

   (Rounded off to 0.5 N/mm²)

   or $(f_{ck} + 3)$ N/mm²

   (whichever is more)

2. **As per 16, 1(b) / IS 456 / 2000 Page 30**

   Individual test results of any sample

   $$\text{ITR} \geq (f_{ck} - 3)\ \text{N/mm}^2$$

3. **As per 15, 4 of IS 456 / 2000 Page 29**

   The test result of the sample shall be average of the strength of three specimen and the individual variation shall be not more than ±15% of average strength. If more the test result of the sample are invalid.

2. Beams and Slabs

2.1 (i) Calculate the ultimate moment of resistance of an RC rectangular beam with the following data:

Breadth of beam = 230 mm

Overall depth of beam = 550 mm

Tension steel consist of 4 numbers of 20 mm diameter bars of grade Fe 415

Clear cover = 30 mm, M20 concrete grade.

(ii) Hence determine the intensity of safe superimposed load (excluding self weight) this above beam can carry on a simply supported span of 5 m.

[25 + 5 = 30 marks : 1999]

Solution:

(i) $A_{st} = 4 \times \frac{\pi}{4} \times 20^2 = 1256.64\ \text{mm}^2$

(a) Limiting depth of NA

$$x_{u,\text{lim}} = 0.48\,d = 0.48 \times (550 - 30 - 20/2) = 0.48 \times 510 = 244.8\ \text{mm}$$

(b) Actual depth of NA

$$C = T$$

$$0.36\,f_{ck}\,B\,x_u = 0.87\,f_y\,A_{st}$$

$$\Rightarrow \quad x_u = \frac{0.87 \times 415 \times 1256.64}{0.36 \times 20 \times 230} = 273.98\ \text{mm};\ x_u > x_{u,\,\text{lim}}$$

Hence the section is over reinforced and since over reinforced section is not allowed in LSM, the depth of actual NA is limited up to $x_{u,\,\text{lim}} = 244.8$ mm

(c) Moment of Resistance,

$$MR = 0.36\,f_{ck}\,Bx_{u,\,\text{lim}}\,(d - 0.42\,x_{u,\,\text{lim}})$$

$$= 0.36 \times 20 \times 230 \times 244.8\,(510 - 0.42 \times 244.8)$$

$$MR = 165.06\ \text{kN-m}.$$

(ii) If 'w' is the net super imposed load including self weight of the beam, then the bending moment for the beam will be

$$M_{\max} = \frac{w_u l^2}{8} = MR$$

$$\Rightarrow \quad \frac{w_u \times 5^2}{8} = 165.06$$

$$\Rightarrow \quad w_u = \frac{165.06 \times 8}{25} = 52.82\ \text{kN/m}$$

But self weight of beam $= 0.230 \times 0.550 \times 1 \times 25 = 3.1625$ kN/m

Hence **safe super imposed load** $= \left(\frac{52.82}{1.5}\right) - 3.1625 = 32.05$ kN/m

2.2 Use limit state method to design a *RC* rectangular beam having an effective simply supported span of 6 m. The beam is required to support live service and super imposed (dead) loads of 14 kN/m and 9.5 kN/m, respectively. The materials to be used are M20 grade concrete and HYSD bars of grade Fe 415. The unit weight of concrete is 25 kN/m³. Adopt d/b ratio as 2. For the given materials $p_{t,\,\text{lim}} = 0.955\%$.

[15 marks : 2000]

Solution:

Given data: $f_{ck} = 20$ N/mm², (Live load) = 14 kN/m, $f_y = 415$ N/mm², (Super imposed dead load) = 9.5 kN/m

$\frac{d}{b} = 2$ (effective span), $l = 6$ m

(i) Assuming, **overall depth of the beam** as $\frac{1}{10}$ of effective span

i.e. $$D = \frac{1}{10} \times 6000 = 600\ \text{mm}$$

Assume effective cover of 50 mm

$$d = 600 - 50 = 550\ \text{mm}$$

$$\therefore \quad b = \frac{550}{2} = 275\ \text{mm}$$

(ii) Load calculations

$$\text{Live load} = 14 \text{ kN/m}$$

$$\text{Super imposed dead load} = 9.5 \text{ kN/m}$$

$$\text{Self weight of the beam} = 0.275 \times 0.6 \times 1 \times 25 = 4.125 \text{ kNm}$$

$$\text{Total load} = 14 + 9.5 + 4.125 \text{ kN/m} = 27.625 \text{ kN/m}$$

Factored load, $w_u = 1.5 \times w = 1.5 \times 27.625 = 41.4375$ kN/m

(iii) Maximum bending moment,

$$M_{max} = \frac{w_u l^2}{8} = \frac{41.4375 \times 6^2}{8} = 186.47 \text{ kN-m.}$$

(iv) Equating M_{max} with MR of the balanced section.

$$\frac{d}{b} = 2 \Rightarrow b = \frac{d}{2} \text{ and } x_{u,\,lim} = 0.48\,d$$

$$M_{max} = MR = 0.36 f_{ck} \times b \times x_{u,\,lim}\,(d - 0.42\,x_{u,\,lim})$$

$$\Rightarrow \quad 0.36 \times 20 \times \frac{d}{2} \times 0.48d\,(d - 0.42 \times 0.48 \times d) = 186.47 \times 10^6$$

$$\Rightarrow \quad d^2\,(d - 0.2016\,d) = 107.911 \times 10^6$$

$$\Rightarrow \quad d^3 = 135.16 \times 10^6$$

$$\Rightarrow \quad d = 5.132 \times 10^2 = 513.2 \text{ mm}$$

Adopting $d = 520$ mm

$D = 520 + 50 = 570$ mm < 600 mm (assumed) Hence OK.

Now, $b = \dfrac{d}{2} = \dfrac{520}{2} = 260$ mm

(v) Area of steel,

$$A_{st} = \frac{MR}{0.87 f_y \left(d - 0.42\,x_{u,\,lim}\right)}$$

$$= \frac{186.47 \times 10^6}{0.87 \times 415 \times (520 - 0.42 \times 0.48 \times 520)}$$

$$A_{st} = 1244 \text{ mm}^2$$

$$p_t = \frac{A_{st}}{bd} \times 100 = \frac{1244}{260 \times 520} \times 100 = 0.92\% < 0.955\% \text{ Hence OK.}$$

2.3 The size of a *RC* beam is restricted to 250 mm × 500 mm. It carries a super imposed load of 25 kN/m over a span of 6 m. Determine the reinforcements for the beam by LSM method. M20 concrete and Fe 415 steel are used. Effective cover to steel = 40 mm.
Salient points on design stress-strain curve for steel
[20 marks : 2001]

Stress	Strain
$0.800\,f_{yd}$	0.00144
$0.850\,f_{yd}$	0.00163
$0.900\,f_{yd}$	0.00192
$0.950\,f_{yd}$	0.00241
$0.975\,f_{yd}$	0.00276
$1.000\,f_{yd}$	0.00380

Solution:

(i) Load calculations

$$\text{Self weight of the beam} = 0.25 \times 0.5 \times 1 \times 25 = 3.125 \text{ kN/m}$$

$$\text{Super imposed load} = 25 \text{ kN/m}$$

$$\text{Total load, } w = 28.125 \text{ kN/m}$$

$$d = 500 - 40 = 460 \text{ mm}$$

(ii) Maximum bending moment

$$M_{max} = \frac{wl^2}{8} = \frac{28.125 \times 6^2}{8} = 126.56 \text{ kN-m}$$

$$\text{Factored } M_{max} = 1.5 \times 126.56 = 189.84 \text{ kN-m}$$

(iii) Calculating moment of resistance for balanced section:

$$M_{u,\,lim} = 0.36\, f_{ck}\, B\, x_{u,\,lim}\,(d - 0.42\, x_{u,\,lim})$$

$$= 0.36 \times 20 \times 250 \times 0.48 \times 460\,(460 - 0.42 \times 0.48 \times 460)$$

$$M_{u,\,lim} = 145.96 \text{ kN-m.}$$

$\because M_{max} > M_{u,\,lim}$, hence section is over reinforced. But over reinforced section is not allowed in LSD. Hence either the maximum bending moment is restricted to $M_{u,\,lim}$ or a doubly reinforced section is provided.

(iv) Now taking $M_{u,\,lim} = M_1$ and providing **area of steel** A_{st_1} as per the value of M_1

$$\therefore \quad A_{st_1} = \frac{M_1}{0.87\, f_y\left(d - 0.42\, x_{u,\,lim}\right)} = \frac{145.96 \times 10^6}{0.87 \times 415 \times (460 - 0.42 \times 0.48 \times 460)}$$

$$= 1100.75 \text{ mm}^2$$

(v) Now, **remaining moment which is to be resisted** is

$$M_2 = M_{max} - M_1 = 189.84 - 145.96 = 43.88 \text{ kN-m}$$

(vi) Providing **tension steel** as per the value of M_2

$$A_{st_2} = \frac{M_2}{0.87\, f_y\,(d - d_c)} = \frac{43.88 \times 10^6}{0.87 \times 415 \times (460 - 40)} = 289.37 \text{ mm}^2$$

$$A_{st} = A_{st_1} + A_{st_2} = 1100.75 + 289.37 = 1390.12 \text{ mm}^2$$

(vii) But as we are designing a **doubly reinforced section**. Hence the compression steel is also to be provided as per the value of M_2

$$A_{sc} = \frac{M_2}{(f_{sc} - 0.45 f_{ck})(d - d_c)}$$

f_{sc} can be found with the help of strain in compression steel.

$$\varepsilon_{sc} = \left[\frac{x_{u,\,lim} - d_c}{x_{u,\,lim}}\right] \times 0.0035 = \left[\frac{0.48 \times 460 - 40}{0.48 \times 460}\right] \times 0.0035 = 0.00287$$

$$f_{sc} - 0.975\, f_{yd} = \frac{1.000 f_{yd} - 0.975 f_{yd}}{0.0038 - 0.00276}(0.00287 - 0.00276)$$

$$f_{yd} = 0.87\, f_y = 0.87 \times 415 = 361.05$$

$$f_{sc} \text{ for } 0.00287 = 352.02 + \frac{(361.05 - 352.02)}{(0.00380 - 0.00276)} \times (0.00287 - 0.00276)$$

$$= 352.02 + 0.9551 = 352.975 \text{ N/mm}^2$$

$$\therefore \quad A_{sc} = \frac{43.88 \times 10^6}{(352.975 - 0.45 \times 20)(460 - 40)} = \frac{43.88 \times 10^6}{144469.54} = 303.73 \text{ mm}^2$$

2.4 Design the RC floor slab for a room of internal dimensions of 4.0 m × 9.5 m. Assume the slab to be simply supported on 230 mm thick masonry walls. The slab is to support live load of 4.0 kN/m^2 and surface finish of 1 kN/m^2. Use M20 grade concrete, HYSD steel of Fe 415 grade. Draw reinforcement details.

[20 marks : 2002]

Solution:

(i) Calculation of design constants

For M 20 concrete, $m = 13.33,\ c = 7\ \text{N/mm}^2 = \sigma_{cbc}$

For Fe 415 steel, $t = 230\ \text{N/mm}^2 = \sigma_{st}$

$$\therefore \quad k = \frac{mc}{t + mc} = \frac{13.33 \times 7}{230 + (13.33 \times 7)} = 0.2886$$

$$j = 1 - \frac{k}{3} = 1 - \frac{0.2886}{3} = 0.9038$$

$$Q = \frac{1}{2} cjk = \frac{1}{2} \times 7 \times 0.9038 \times 0.2886 = 0.9129$$

(ii) As per the vertical deflection criterion, the span to effective depth ratio for spans upto 10 m for a simply supported slab is given by

$$\frac{l}{d} = 20$$

$$\Rightarrow \quad d = \frac{l}{20} = \frac{4000}{20} = 200\ \text{mm}$$

(iii) Effective span

(a) clear span + effective depth = 4.00 + 0.2 = 4.2 m

(b) centre to centre distance between supports = 4.0 + 0.23 = 4.23 m

Hence lesser of the above two will be adopted i.e. $l_e = 4.2$ m

(iv) Bending moment and shear force

Assuming a nominal cover of 20 mm, the total depth of the slab will become 200 + 20 = 220 mm.

(a) Load calculations:

Load due to self weight of slab =	0.22 × 1 × 1 × 25	= 5.5 kN/m^2
Superimposed live load		= 4.0 kN/m^2
Load due to surface finishing		= 1.0 kN/m^2
	Total	= 10.5 kN/m^2

(b) Bending moment per meter run of slab

$$M = \frac{wl_e^2}{8} = \frac{10.5 \times (4.2)^2}{8} = 23.15\ \text{kN-m}$$

(c) Shear force

$$V = \frac{wl_e}{2} = \frac{10.5 \times 4.2}{2} = 22.05\ \text{kN}$$

(v) Design of section

$$d = \sqrt{\frac{M}{QB}} = \sqrt{\frac{23.15 \times 10^6}{0.9129 \times 1000}} \qquad [B = 1000\ \text{mm}]$$

$$\Rightarrow \quad d = 159.24\ \text{mm} < 200\ \text{mm. Hence OK}$$

Now taking $d = 160$ mm and nominal cover = 20 mm

we get $D = d + 20 = 160 + 20 = 180$ mm.

(vi) Main reinforcement

$$A_{st} = \frac{M}{\sigma_{st}\, jd} = \frac{23.15 \times 10^6}{230 \times 0.9038 \times 160} = 696 \text{ mm}^2$$

Minimum reinforcement should be 0.15% of total cross sectional area

$$= \frac{0.15}{100} \times 180 \times 1000 = 270 \text{ mm}^2$$

$\therefore$ A_{st} is more than minimum reinforcement.

Spacing of bars, $s = \dfrac{1000 \times A_\phi}{A_{st}}$

Adopting 10 mm ϕ bars, we get

$$A_\phi = \frac{\pi}{4} \times 10^2 = 78.54 \text{ mm}^2$$

$$\therefore \quad s = \frac{1000 \times 78.54}{696} = 112.84 \text{ mm}$$

Adopting s = 110 mm c/c, we get actual

$$A_{st} = \frac{1000 \times 78.54}{110} = 714 \text{ mm}^2$$

Spacing is less than 3d and 300 mm both.

Bend every third bar at the support, at a distance of $\dfrac{l_e}{5} = \dfrac{4.2}{5} = 0.84$ m from the edge of support

A_{st} at support $= \dfrac{2}{3} \times 714 = 476$ mm^2 which is more than minimum reinforcement (0.15%).

(vii) Check for development length at the support

The code stipulates that at simple supports, the diameter of the reinforcement be such that

$$L_d \le 1.3\frac{M_1}{V} + L_0$$

$$M_1 = \sigma_{st} \,.\, j \,.\, d \times A_{st} \text{ at support}$$
$$= 230 \times 0.9038 \times 160 \times 476$$
$$= 15.832 \times 10^6 \text{ N-mm}$$
$$V = 22.05 \text{ kN}$$

Assuming a clear cover of 25 mm is provided at the side (end) and providing a U-hook

Given width of support = 230 mm = l_s

$$\therefore \quad L_0 = \frac{l_s}{2} - x' + 13\,\phi = \frac{230}{2} - 25 + 13 \times 10 = 220 \text{ mm}$$

$$L_d = \frac{\phi\, \sigma_{st}}{4\, \tau_{bd}} = \frac{10 \times 230}{4 \times 1.6 \times 1.2} = 299.48 \text{ mm (60\% increase for HYSD bars)}$$

Now $$1.3\frac{M_1}{V} + L_0 = \frac{1.3 \times 15.832 \times 10^6}{22.05 \times 10^3} + 220$$

$= 1153.4 \text{ mm} > L_d$ Hence ok.

The reinforcement should extend by a length equal to $\frac{L_d}{3} = \frac{299.48}{3} = 99.83 \text{ mm} \approx 100 \text{ mm}$ beyond the face of support.

(viii) Check for shear

p = % reinforcement at support $= \frac{476}{180 \times 1000} \times 100 = 0.264\%$

For $p = 0.15\%$, $\tau_c = 0.28$

For $p = 0.25\%$, $\tau_c = 0.36$

For $p = 0.5\%$, $\tau_c = 0.48$

$\therefore$ For $p = 0.264\%$, $\tau_c = 0.36 + \left(\frac{0.48 - 0.36}{0.5 - 0.25}\right)(0.264 - 0.25) = 0.3667 \text{ N/mm}^2$

Now $\tau_v = \frac{V}{Bd} = \frac{22.05 \times 10^3}{1000 \times 160} = 0.138 \text{ N/mm}^2$

$\because$ $\tau_v < \tau_c$ Hence safe.

(ix) Distribution reinforcement

$$A_{st} = \frac{0.15\,BD}{100} = \frac{0.15 \times 1000 \times 180}{100} = 270 \text{ mm}^2$$

Using 6 mm ϕ bars, each having

$$A_\phi = \frac{\pi}{4} \times 6^2 = 28.27 \text{ mm}^2$$

$$\text{Spacing of bars} = \frac{1000 \times A_\phi}{A_{sd}} = \frac{1000 \times 28.27}{270} = 104.7 \approx 105 \text{ mm}$$

This is less than 5d and 450 mm both. Hence provide 6 mmϕ bars @ 105 mm C/C. Near the edge of the support, the distribution reinforcement may be provided both at top as well as the bottom. The details of the reinforcement are shown below.

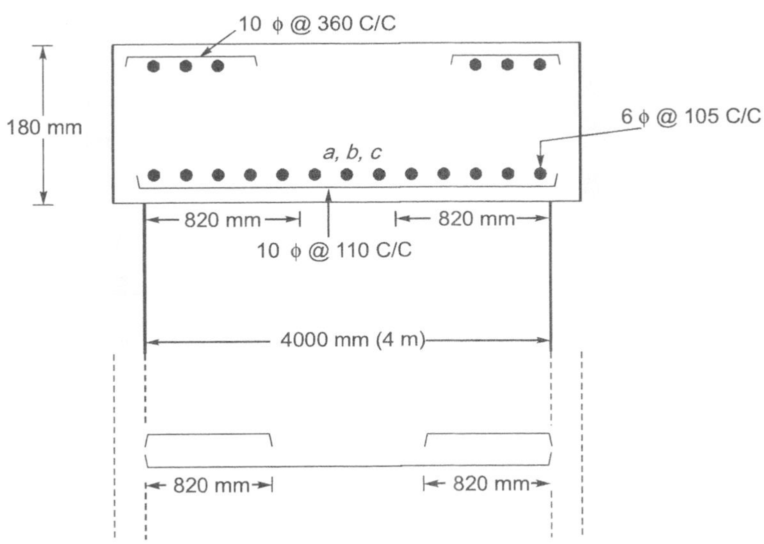

2.5 Determine ultimate moment capacity of a doubly reinforced beam with B = 300 mm, D = 600 mm, A_{st} = 2060 mm², A_{sc} = 804 mm² and effective cover of 50 mm for both tension and compression steels. The materials used are M20 concrete and HYSD steel of grade Fe 415.
The salient points on stress-strain curve are:

[15 marks : 2003]

Stress, MPa	Strain
288	0.00144
306	0.00163
324	0.00192
342	0.00241
351	0.00276
360	0.00380

Solution:

Given Data: $B = 300$ mm, $D = 600$ mm, $A_{st} = 2060$ mm², $A_{sc} = 804$ mm², $d_c = 50$ mm,
$d = D - d_c = 600 - 50 = 550$ mm, $M\,20\,/\,Fe\,415$

(i) Limiting depth of neutral axis

$$x_{u,\,\text{lim}} = 0.48\text{ d} = 0.48 \times 550 = 264 \text{ mm}$$

(ii) Actual depth of neutral axis:

$$\text{Total compressive force} = \text{Total tensile force}$$

$$0.36\, f_{ck}\, B\, x_u + A_{sc}(f_{sc} - 0.45\, f_{ck}) = 0.87\, f_y\, A_{st}$$

$$\Rightarrow 0.36 \times 20 \times 300 \times x_u + 804(f_{sc} - 0.45 \times 20) = 0.87 \times 415 \times 2060$$

$$\Rightarrow 2160\, x_u + 804\,(f_{sc} - 9) = 743763$$

$$\Rightarrow 2160\, x_u + 804\, f_{sc} - 7236 = 743763$$

$$\Rightarrow 2160\, x_u + 804\, f_{sc} = 750999$$

$$\Rightarrow x_u = \frac{750999 - 804\, f_{sc}}{2160}$$

Trial 1:

$$\text{Assuming } f_{sc} = 350 \text{ MPa.}$$

We get,

$$x_u = \frac{750999 - 804 \times 350}{2160} = 217.407 \text{ mm}$$

$$\text{Value of } \varepsilon_{sc} = \frac{0.0035}{x_u}(x_u - d_c) = \frac{0.0035}{217.407}(217.407 - 50)$$

$$= 0.002695$$

$$f_{sc} \text{ for } 0.002695 = 342 + \frac{(351 - 342)}{(0.00276 - 0.00241)}(0.002695 - 0.00241)$$

$$= 342 + 7.328 = 349.328 \text{ MPa}$$

Trial 2:

$$x_u = \frac{750999 - 804 \times 349.328}{2160} = 217.657 \text{ mm}$$

$$\text{Value of } \varepsilon_{sc} = \frac{0.0035}{217.657}(217.657 - 50) = 0.002695$$

Hence $f_{sc} = 349.328$ MPa and $x_u = 217.657$ mm (adopted)

$\because x_u < x_{u,\,\text{lim}}$, hence section is under reinforced.

(iii) Moment of Resistance

$$MR = 0.36 f_{ck} B x_u \times (d - 0.42 x_u) + (f_{sc} - 0.45 f_{ck}) A_{sc} (d - d_c)$$
$$= [0.36 \times 20 \times 300 \times 217.657 \times (550 - 0.42 \times 217.657)]$$
$$+ [(349.328 - 0.45 \times 20) \times 804 \times (550 - 50)]$$
$$= 352.409 \text{ kN-m}$$

2.6 A simply supported 18 m effective span RCC rectangular beam of 500 mm × 1500 mm (overall depth) section is reinforced throughout with 21 numbers of 25 mm diameter bars in three layers of 7 bars each at a clear cover of 37.5 mm on tensile face. The reinforcement in the compression face is 4 numbers 25 mm diameter and 1 number 20 mm diameter bars in one layer at an effective cover of 50 mm. The clear cover between the different layers on tension face is 25 mm. M25 grade concrete and Fe 415 grade steel bars are used in the beam throughout. The beam is laterally restrained throughout the span.

(a) Find out the maximum bending moment M_u at limit state of collapse for the beam section assuming the stress in compression reinforcement $\sigma_{sc} = 0.9566 f_y$ corresponding to tensile stress f_y in tensile reinforcement at the limit state.

(b) What should be the superimposed uniformly distributed load '*w*', the beam can carry at working conditions?

(c) Design the shear reinforcement at support if design shear strength of concrete τ_c is given as follows for different values of $p = \dfrac{100 A_s}{bd}$

p	1.25	1.5	1.75
τ_c (MPa)	0.70	0.74	0.78

[10 × 3 = 30 marks : 2004]

Solution:

(a)

$$f_y = 415 \text{ N/mm}^2$$
$$x_{u,\,lim} = 0.48 d = 0.48 \times 1400 = 672 \text{ mm}$$
$$B = 500 \text{ mm}$$
$$f_{ck} = 25 \text{ N/mm}^2$$
$$d = 1400 \text{ mm}$$
$$d_c = 50 \text{ mm}$$
$$f_{sc} = \sigma_{sc} = 0.9566 \times 0.87 \times 415 = 345.38 \text{ N/mm}^2$$
$$A_{st} = 21 \times \frac{\pi}{4} \times 25^2 = 10308.35 \text{ mm}^2$$

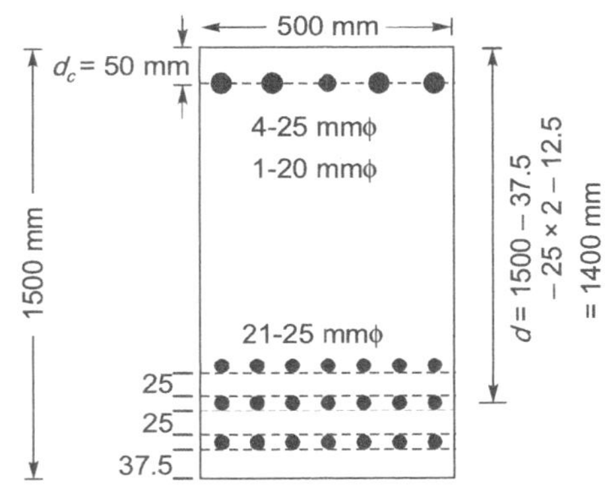

$$A_{sc} = 4 \times \frac{\pi}{4} \times 25^2 + 1 \times \frac{\pi}{4} \times 20^2 = 625\pi + 100\pi = 2277.65 \text{ mm}^2$$

Firstly we need to determine limiting NA depth

$$x_{u,\,lim} \text{ for Fe415} = 0.48 d = 0.48 \times 1400 = 672 \text{ mm}$$

So, $C_1 + C_2 = T$

Also stress in tensile reinforcement $= 0.87 f_y$ assuming $x < x_{u,\,lim}$

$$0.36 f_{ck} B x + (f_{sc} - 0.45 f_{ck}) A_{sc} = 0.87 f_y A_{st}$$

$$4500x + 761031.2 = 3721829.768$$

$$x = 657.955 \text{ mm}$$

$$MR = 0.36 f_{ck} B x (d - 0.42x) + (f_{sc} - 0.45 f_{ck}) A_{sc} (d - d_c)$$

$$= 0.36 \times 25 \times 500 \times 657.955(1400 - 0.42 \times 657.955)$$

$$+ (345.38 - 0.45 \times 25) \times 2277.65 \times (1400 - 50)$$

$$= 4353.15 \text{ kN-m.}$$

(b)
$$M_u = \frac{w_u l^2}{8}$$

$$\Rightarrow \quad \frac{4353.15 \times 8}{18 \times 18} = w_u$$

$$\Rightarrow \quad w_u = 107.48 \text{ kN/m}$$

$$\therefore \quad w_{total} = \frac{w_u}{1.5} = \frac{107.48}{1.5} = 71.65 \text{ kN/m}$$

w_{total} is super imposed UDL along with the self weight of the beam. Hence, the self weight of beam should be deducted.

$$\text{Self weight of beam} = 0.5 \times 1.5 \times 1 \times 25 = 18.75 \text{ kN/m}$$

Hence,
$$w = w_{total} - 18.75 = 71.65 - 18.75 = 52.9 \text{ kN/m}$$

(c) $\because$
$$w_u = 107.48 \text{ kN/m}$$

$$V_u = \frac{w_u l}{2} = \frac{107.48 \times 18}{2} = 967.32 \text{ kN}$$

$\therefore$ **Nominal shear stress**
$$\tau_v = \frac{V_u}{Bd} = \frac{967.32 \times 10^3}{500 \times 1400} = 1.382 \text{ N/mm}^2$$

Now
$$p = 100 \times \frac{A_{st}}{Bd} = \frac{100 \times 10308.35}{500 \times 1400} = 1.47\%$$

For $p = 1.25\%$, we have $\tau_c = 0.70$ N/mm^2

For $p = 1.50\%$, we have $\tau_c = 0.74$ N/mm^2

$$\therefore \quad \tau_c \text{ for } 1.47\% = 0.70 + \frac{(0.74 - 0.70)}{(1.50 - 1.25)}(1.47 - 1.25) = 0.70 + 0.0352$$

$$\tau_c = 0.7352 \text{ N/mm}^2$$

$$\therefore \quad V_c = \tau_c Bd = 0.7352 \times 500 \times 1400 = 514.64 \text{ kN}$$

$\because$ $\tau_v > \tau_c$, hence shear reinforcement is required.

$$V_{us} = V_u - V_c = 967.32 - 514.64 = 452.68 \text{ kN}$$

Now shear reinforcement at supports should be provided according to **IS:456 – 2000** i.e.

(d) **Spacing,**
$$S_v = \frac{0.87 f_y A_{sv} d}{V_{us}}$$

Adopting 2-legged 10 mm ϕ bars for vertical stirrups.

$$A_{sv} = 2 \times \frac{\pi}{4} \times 10^2 = 157.08 \text{ mm}^2$$

$$\therefore \quad S_v = \frac{0.87 \times 415 \times 157.08 \times 1400}{452.68 \times 1000} = 175.4 \text{ mm}$$

Hence, providing 170 mm spacing at supports.

Minimum shear reinforcement,

$$\frac{A_{sv}}{bS_v} \geq \frac{0.4}{0.87 f_y}$$

$$\frac{157.08}{500 \times S_v} \geq \frac{0.4}{0.87 \times 415}$$

$$S_v \leq 283.57$$

Provide $S_v = 280$ mm

Hence, at mid span spacing between stirrups can be 280 mm c/c.

Maximum spacing $= 0.75d = 0.75 \times 1400 = 1050$ mm

Hence provided reinforcement is OK.

2.7 Find the moment of resistance of the hollow box section RC beam using WSM. M20 concrete and Fe 415 steel are used. The walls of the box are 50 mm thick. It is reinforced with 4 numbers of 16 mm diameter bars as shown in figure. The effective cover to main steel is 40 mm. The hollow portion of the beam is 300 mm x 300 mm and outer dimensions of beam are 400 mm x 400 mm. Allowable stress in concrete = 7 MPa, allowable stress in stress in steel = 230 MPa. Modular ratio is 13.33 and 20 in tension and compression zones.

[25 marks : 2005]

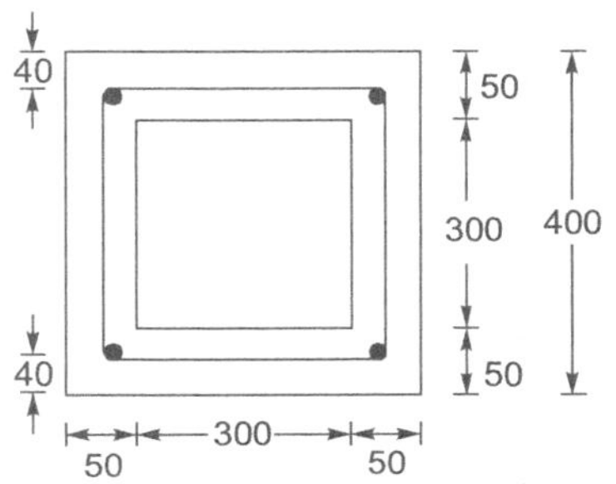

All dimensions are in mm.

Solution:

Given: $A_{st} = 2 \times \frac{\pi}{4} \times 16^2 = 402.12$ mm^2, $A_{sc} = 2 \times \frac{\pi}{4} \times 16^2 = 402.12$ mm^2,

$\sigma_{cbc} = 7$ N/mm^2, $\sigma_{st} = 230$ N/mm^2, $m = 13.33$ (in tension), $1.5\,m = 20$ (in compression)

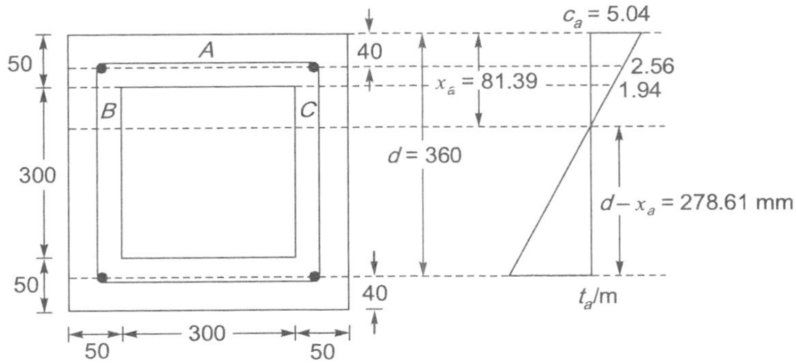

All dimensions are in mm.

(i) Critical neutral axis

$$x_c = \left(\frac{mc}{t + mc}\right) d = \left(\frac{13.33 \times 7}{230 + (13.33 \times 7)}\right) \times 360 = 103.9 \text{ mm.}$$

(ii) Actual depth of neutral axis

Equating the moment of area about NA for both sides

$$\Rightarrow \quad 400 \times 50 \times (x_a - 25) + 2\left[50 \times (x_a - 50)\frac{(x_a - 50)}{2}\right] + (1.5\,m - 1) \times 402.12 \times (x_a - 40)$$

$= 13.33 \times 402.12 \times (360 - x_a)$

$\Rightarrow \quad 400 \times 50 \times (x_a - 25) + [50 \times (x_a - 50)^2] + 19 \times 402.12 \times (x_a - 40)$

$= 13.33 \times 402.12 \times (360 - x_a)$

$\Rightarrow \quad 20000\,(x_a - 25) + 50(x_a^2 + 2500 - 100\,x_a) + 7640.28\,(x_a - 40) = 5360.26\,(360 - x_a)$

$\Rightarrow \quad 20000\,x_a - 5 \times 10^5 + 50\,x_a^2 + 125000 - 5000\,x_a + 7640.28\,x_a - 305611.2$

$+ 5360.26\,x_a - 1929693.6 = 0$

$\Rightarrow \quad 50\,x_a^2 + 28000.54\,x_a - 2610304.8 = 0$

$\Rightarrow \quad x_a^2 + 560.01\,x_a - 52206.10 = 0$

$\Rightarrow \quad x_a = 81.39$ mm

$\because \quad x_a < x_c$ section is under reinforced.

$\because \quad t_a = \sigma_{st} = 230$ MPa

but $c_a < \sigma_{cbc}$

From similar triangles. $\dfrac{c_a}{x_a} = \dfrac{t_a/m}{d - x_a}$

$$c_a = \left(\frac{x_a}{d - x_a}\right) \times \frac{t_a}{m}$$

$$= \frac{81.39}{278.61} \times \frac{230}{13.33} = 5.04 \text{ N/mm}^2$$

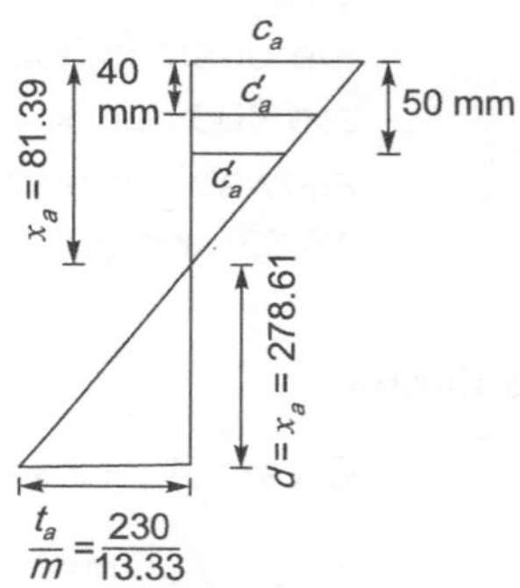

Compressive stress $c_{a'}$ at the level of steel

$$\frac{c_{a'}}{81.39 - 40} = \frac{c_a}{81.39}$$

$$c_{a'} = \frac{5.04 \times 41.39}{81.39}$$

$$c_{a'} = 2.56 \text{ N/mm}^2$$

Compressive stress c_{a_1} at 50 mm from the top

$$\frac{c_{a_1}}{81.39 - 50} = \frac{c_a}{81.39}$$

$$c_{a_1} = \frac{5.04 \times 31.39}{81.39} = 1.94 \text{ N/mm}^2$$

(iii) (a) Compressive force for part *A*,

$$C_1 = 400 \times 50 \times \frac{1}{2} \times (5.04 + 1.94) = 69800 \text{ N}$$

$$y_1 \text{ from top} = \left[\frac{5.04 + 2 \times 1.94}{5.04 + 1.94}\right] \times \frac{50}{3} = 21.3 \text{ mm}$$

$$\text{Lever Arm} = d - y_1 = 360 - 21.3 = 338.70 \text{ mm}$$

$$M_1 = 69800 \times 338.70 = 23.64 \text{ kN-m}$$

(b) Compressive force for part *B* and *C* is equal.

$$C_2 = 50 \times (81.39 - 50) \times \frac{1.94}{2} = 1522.41 \text{ N}$$

$$y_2 \text{ from top} = 50 + \frac{31.39}{3} = 60.46 \text{ mm}$$

$$\text{Lever Arm} = 360 - 60.46 = 299.54 \text{ mm}$$

$$M_2 = 1522.41 \times 299.54 = 0.456 \text{ kN-m}$$

$\therefore \qquad M_3 = 0.456$ kN-m.

Hence MR of the section for concrete = $M_1 + M_2 + M_3 = 23.64 + 0.456 + 0.456 = 24.55$ kN-m

Compressive force in steel

$$C_4 = (1.5\,m - 1)\,A_{sc} \times c_{a'} = (20 - 1) \times 402.12 \times 2.56 = 19558.144 \text{ N}$$

$$\text{Lever Arm} = 360 - 40 = 320 \text{ mm}$$

$$MR_{st} = 19558.144 \times 320 = 6.259 \text{ kN}$$

Hence total MR of the section = 24.55 + 6.259 = 30.81 kN-m

2.8 **A reinforced concrete beam 500 mm wide and 600 mm deep has a 100 mm wide and100 mm deep groove in the compression side, the reinforcement being on the tension side only consisting of 4 numbers of 20 mm diameter bar. Determine position of neutral axis and calculate the stresses developed for a bending of 85000 N-m. Assume m = 15.**

[15 marks : 2006]

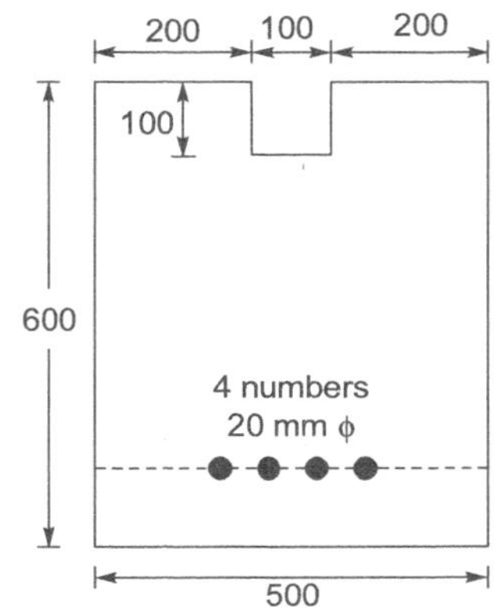

All dimensions are in mm

Solution:

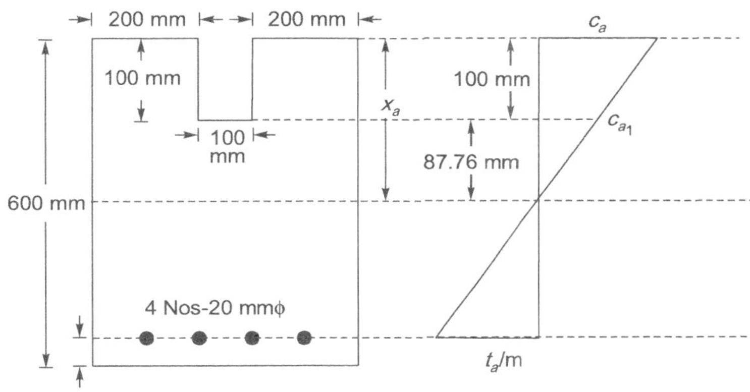

Assuming effective cover = 50 mm

Hence, $\qquad d = 600 - 50 = 550$ mm

$$A_{st} = 4 \times \frac{\pi}{4} \times 20^2 = 1256.63 \text{ mm}^2$$

(i) Finding **actual depth of NA** (Assuming x_a below the groove):

Taking moment of area about NA on both sides

$$\Rightarrow \qquad 500 \times x_a \times \frac{x_a}{2} - 100 \times 100 \times (x_a - 50) = 15 \times 1256.63\,(550 - x_a)$$

$$\Rightarrow \qquad 250\,x_a^2 - 10^4 x_a + 5 \times 10^5 + 18849.45\,x_a - 10367197.5 = 0$$

$$\Rightarrow \qquad 250\,x_a^2 + 8849.45\,x_a - 9867197.5 = 0$$

$$\Rightarrow \qquad x_a^2 + 35.39 x_a - 39468.79 = 0$$

$$\Rightarrow \qquad x_a = 181.76 \text{ mm}$$

(ii) Now if the **compressive stress in top most fibre is c_a** then from similar triangles, we have

$$\frac{c_a}{x_a} = \frac{c_{a_1}}{(x_a - 100)}$$

$$\Rightarrow \qquad c_{a_1} = \left(\frac{181.76 - 100}{181.76}\right) c_a$$

$$\Rightarrow \qquad c_{a_1} = 0.45\, c_a$$

Compressive force without considering any groove.

$$C_1 = 500 \times \frac{1}{2} \times c_a \times x_a = 250 \times 181.76 \times c_a = 45440\, c_a \text{N}$$

$$\text{Lever Arm} = d - \frac{x_a}{3} = 550 - \frac{181.76}{3} = 489.41 \text{ mm}$$

$$M_1 = 45440\, c_a \times 489.41 = 22.239\, c_a \text{ kN-m}$$

Compressive force in groove,

$$C_2 = 100 \times \frac{1}{2} \times 100\,(c_a + c_{a_1})$$

$$= 50 \times 100 \times (c_a + 0.45\, c_a) = 5000 \times 1.45\, c_a = 7250\, c_a$$

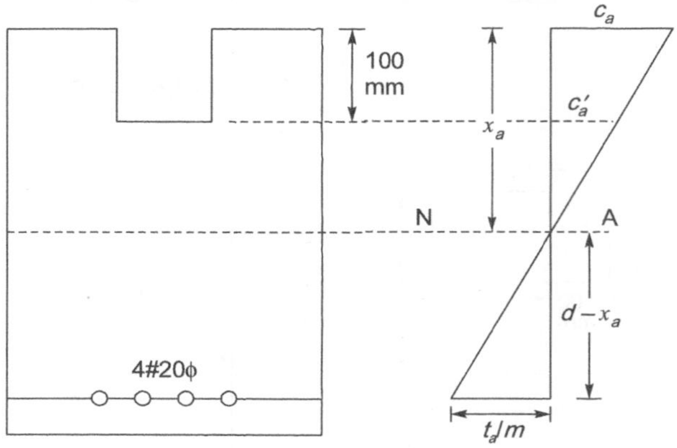

Taking y from top

$$y = \left(\frac{c_a + 2c_{a_1}}{c_a + c_{a_1}}\right) \times \frac{100}{3} = 43.7 \text{ mm}$$

$$\text{Lever Arm} = d - y = 550 - 43.7 = 506.3 \text{ mm}$$

$$M_2 = 7250\, c_a \times 506.3 \text{ mm} = 3.67\, c_a \text{ kN-m}$$

$$\text{Total MR of the section} = 22.239\, c_a - 3.7\, c_a$$

$$\text{MR} = 18.569\, c_a \text{ kN-m}$$

But $\qquad \text{BM} = 85000 \text{ N-m} = 85 \text{ kN-m}$

$$\therefore \qquad 18.569\, c_a = 85$$

$$\Rightarrow \qquad c_a = 4.577 \text{ N/mm}^2$$

From similar triangles, we have

$$\frac{c_a}{x_a} = \frac{t_a/m}{d - x_a}$$

$$\Rightarrow \qquad t_a = \frac{mc_a\,(d - x_a)}{x_a} = \frac{15 \times 4.577 \times (550 - 181.76)}{181.76} = 139.1 \text{ N/mm}^2$$

2.9 A simply supported beam of 5 m effective span is subjected to 24 kN/m live load. $f_{ck} = 20$ N/m^2; $f_y = 415$ N/mm^2. The overall depth of the beam is 400 mm and width is 250 mm. Design the reinforcement of the beam if $k = 0.138$, $j = 0.80$.

[10 marks : 2006]

Solution:

b = 250 mm, D = 400 mm, L = 5 m

Assuming unit weight of concrete as 25 kN/m^3 and effective cover as 50 mm.

1. **Load and BM calculation:**

$$\text{Dead load on the beam} = \frac{250}{1000}\times\frac{400}{1000}\times 1\times 25 = 2.5 \text{ kN/m}$$

$$\text{Live load on the beam} = 24 \text{ kN/m}$$

Factored load as per **IS : 456-2000**

$$w_u = 1.5(2.5 + 24) = 39.75 \text{ kN/m}$$

$$\text{Factored moment at mid-span, } M_u = \frac{w_u L^2}{8} = \frac{39.75\times 5^2}{8} = 124.22 \text{ kN-m}$$

For balanced section,

$$M_{u,\,\text{lim}} = 0.36 f_{ck}\times\left(\frac{x_{u,\,\text{lim}}}{d}\right)\times\left(1-0.42\times\frac{x_{u,\text{lim}}}{d}\right)bd^2$$

But for Fe415,

$$\frac{x_{u,\text{lim}}}{d} = 0.48$$

$$\therefore \quad M_{u,\,\text{lim}} = 0.36\times 20\times 0.48\times(1-0.42\times 0.48)\times 250\times(400-50)^2$$

$$= 84.5 \text{ kN-m}$$

Since $M_{u,\,\text{lim}} < M_u$, hence the section is a over-reinforced section

2. **A_{st} calculation**

$$m_{u,\,\text{lim}} = 0.87 f_y A_{st1}\,(d - 0.42\,x_{u,\,\text{lim}})$$

$$84.5\times 10^6 = 0.87\times 415\times A_{st1}\times(350 - 0.42\times 0.48\times 350)$$

$$A_{st1} = 837.53 \text{ mm}^2$$

Assuming 20 mm ϕ bars for compression steel

$$d' = 30 \text{ mm clear cover} + 8 \text{ mm stirrup} + \phi/2$$

$$= 30 + 8 + \frac{20}{2} = 48 \text{ mm}$$

$$A_{st2} = \frac{M_u - M_{u,\text{lim}}}{0.87 f_y (d-d')} = \frac{(124.22-84.5)\times 10^6}{0.87\times 415\times(350-48)} = 364.28 \text{ mm}^2$$

$$\therefore \quad A_{st} = A_{st\,1,\,\text{lim}} + A_{st2} = 837.53 + 364.28 = 1201.81 \text{ mm}^2$$

Using 4 bars,

$$\phi_{\text{required}} = \sqrt{\frac{1201.81/4}{\pi/4}} = 19.55 \text{ mm}$$

Providing 4 nos. 20 mm ϕ bars

$$A_{st} = 4\times\frac{\pi}{4}\times(20)^2 = 1256.64 \text{ mm}^2$$

3. **A_{sc} calculation**

$$\text{Actual } d = 400 - \left(30 + 8 + \frac{20}{2}\right) = 352 \text{ mm}$$

Strain in compression steel is given by

$$\epsilon_{sc} = 0.0035\times\left(1-\frac{d'}{x_{u,\,\text{max}}}\right)$$

$$= 0.0035\times\left(1-\frac{48}{0.48\times352}\right) = 0.00251$$

$$\in_y = \frac{0.87f_y}{E_s} = \frac{0.87\times415}{2\times10^5} = 0.00181 < 0.00251$$

$\therefore$ f_{sc} can be assumed as $0.87\ f_y = 0.87\times415 = 361.05\ \text{N/mm}^2$

$$\therefore \quad 0.87\ f_y\ A_{st2} = A_{sc}\,(f_{sc} - 0.45\ f_{ck})$$

$$0.87\times415\times364.28 = A_{sc}\,(361.05 - 0.45\times\ 20)$$

$$A_{sc} = 373.6\ \text{mm}^2$$

2.10 An RCC beam is of 230 mm width, 500 mm overall depth. Effective cover to compression steel and tension steel is 40 mm. Compression reinforcement consists of 2 nos. 16 mm diameter bars and tension steel consists of 2 nos. 25 mm diameter bars. This doubly reinforced beam is made of M 20 concrete and Fe 415 steel. Find the moment of resistance of this beam by LIMIT STATE METHOD.

Table: Stress in Compression Reinforcement in MPa

Grade of Steel MPa	*d'/d* Values			
Fe415	0.05	0.10	0.15	0.20
	355	353	342	329

[20 marks : 2007]

Solution:

Given: $b = 230$ mm, $D = 500$ mm, $d' = 40$ mm, $f_{ck} = 20$ MPa, $f_y = 415$ MPa

Effective depth, $d = D - d' = 500 - 40 = 460$ mm

$$A_{st} = 2\times\frac{\pi}{4}\times(25)^2 = 981.75\text{mm}^2$$

$$A_{sc} = 2\times\frac{\pi}{4}\times(16)^2 = 402.12\ \text{mm}^2$$

For, Fe 415, $\quad x_{u,\,max} = 0.48\ d = 0.48\times460 = 220.8$ mm

Here $\quad \dfrac{d'}{d} = \dfrac{40}{460} = 0.087$

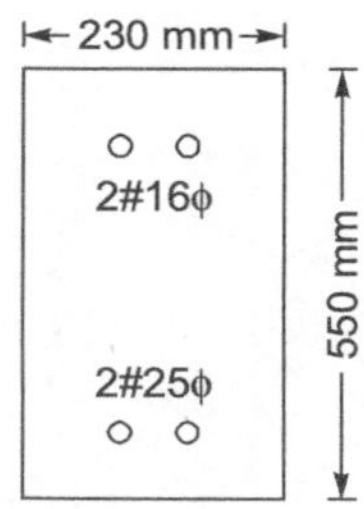

From the above table, the stress in compression reinforcement can be interpolated.

$$\therefore \quad f_{sc} = 355 - \left(\frac{355-353}{0.10-0.05}\right)\times(0.087-0.05)$$

$$= 355 - 1.48 = 353.52\ \text{MPa.}$$

$$F_T = F_C$$

$$0.87\times415\times981.75 = 0.36\times230\times20\times x_u + (353.52 - 0.45\times20)\times402.12$$

$$x_u = 130.388\ \text{mm}$$

The **moment of resistance** is given by

$$M_u = 0.36\ f_{ck}\ b\ x_u\,(d - 0.42\ x_u) + (f_{sc} - 0.45\ f_{ck})\ A_{sc}\,(d - d')$$

$$= 0.36\times20\times230\times130.388\times(460 - 0.42\times130.388)$$
$$+ (353.52 - 0.45\times20)\times402.12\times(460 - 40)$$

$$= 145.7\ \text{kN-m}$$

2.11 A simply supported beam of 4.5 m effective span is carrying a live load of 25 kN/m. The size of the beam has to be restricted to 250 mm × 380 mm depth. Design the beam for bending using limit state method. The design coefficients are $K = 0.138$, $J = 0.80$, $K_u = 0.479$. Use M20 grade concrete and Fe415 steel.

[15 marks : 2008]

Solution:

Given: $b = 250$ mm, $D = 380$ mm, $L = 4.5$ m

Assuming unit weight of concrete as 25 kN/m^3 and effective cover as 50 mm

1. Load calculation

$$\text{Dead load on the beam} = \frac{250}{1000}\times\frac{380}{1000}\times 25 = 2.375 \text{ kN/m}$$

$$\text{Live load on the beam} = 25 \text{ kN/m}$$

Factored load as per **IS : 456 – 2000,**

$$w_u = 1.5(2.375 + 25) = 41.0625 \text{ kN/m}$$

2. BM calculation

$$\text{Factored moment at mid-span,} \quad M_u = \frac{w_u L^2}{8} = \frac{41.0625\times(4.5)^2}{8} = 104 \text{ kN-m}$$

$$\text{For balanced section,} \quad M_{u,\,lim} = 0.36 f_{ck}\times\left(\frac{x_{u,lim}}{d}\right)\times\left(1-0.42\times\frac{x_{u,lim}}{d}\right)bd^2$$

$$\text{But for Fe 415,} \quad \frac{x_{u,lim}}{d} = K_u = 0.479 \text{ (given)}$$

$$\therefore \quad M_{u,\,lim} = 0.36\times 20\times 0.479\times(1-0.42\times 0.479)\times 250\times(380-50)^2$$

$$= 75 \text{ kN-m}$$

Since $M_{u,\,lim} < M_u$, hence the section is a doubly reinforced section.

3. A_{st} calculation

$$m_{u,\,lim} = 0.87 f_y A_{st1}\,(d - 0.42\,x_{u,lim})$$

$$75\times 10^6 = 0.87\times 415\times A_{st1}\,(330 - 0.42\times 0.48\times 330)$$

$$A_{st1} = 788.42 \text{ mm}^2$$

Assuming 20 mm ϕ bars for compression steel,

$$d' = (30 \text{ mm clear cover} + 8 \text{ mm stirrup} + \phi/2)$$

$$= 30 + 8 + \frac{20}{2} = 48 \text{ mm}$$

$$A_{st_2} = \frac{M_u - M_{u,lim}}{0.87 f_y (d - d')} = \frac{(104-75)\times 10^6}{0.87\times 415\times(330-48)} = 284.83$$

$$A_{st} = 788.42 + 284.83 = 1073.25 \text{ mm}^2$$

$$\text{Using 3 bars,} \quad \phi_{required} = \sqrt{\frac{1073.25/3}{\pi/4}} = 21.3 \text{ mm}$$

Providing 3 nos. 22 mm ϕ

$$A_{st} = 3\times\frac{\pi}{4}\times(22)^2 = 1141 \text{ mm}^2$$

$$\text{Actual } d = 380 - \left(30 + 8 + \frac{22}{2}\right) = 331 \text{ mm}$$

(b) In compression zone

Strain in the compression steel is given by

$$\in_{sc} = 0.0035 \times \left(1 - \frac{d'}{x_{u,\max}}\right)$$

$$= 0.0035 \times \left(1 - \frac{48}{0.48 \times 331}\right) = 0.00244$$

$$\in_y = \frac{0.87 f_y}{E_s} = \frac{0.87 \times 415}{2 \times 10^5} = 0.00181 < 0.00244$$

$\therefore$ f_{sc} can be assumed as $0.87 f_y = 0.87 \times 415 = 361.05 \text{ N/mm}^2$

$$A_{sc} = \frac{0.87 f_y \times A_{st_2}}{f_{sc} - 0.45 f_{ck}}$$

$$= \frac{0.87 \times 415 \times 284.83}{0.87 \times 415 - 0.45 \times 20} = 292.11 \approx 293 \text{ mm}^2$$

2.12 **A reinforced concrete beam of rectangular cross-section (600 mm × 300 m) is shown in the figure. Assume M20 grade concrete and Fe 415 grade steel. Permissible compressive stress of concrete in bending σ_{cbc} = 7.0 MPa for M20 concrete. Compute the maximum stresses in concrete and steel when a moment of 50 kN-m is applied to the cross-section.**

[25 marks : 2009]

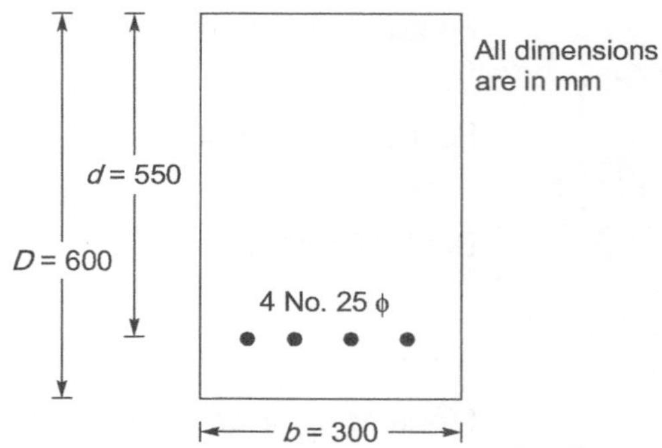

Solution:

Area of tension steel, $A_{st} = 4 \times \frac{\pi}{4} \times (25)^2 = 1963.5 \text{ mm}^2$

Modular ratio, $m = \frac{280}{3\sigma_{cbc}} = \frac{280}{3 \times 7} = 13.33$

Area of transformed section $A = A_g - A_{st} + mA_{st}$

$= bD + (m-1)A_{st}$

Equating moment of areas of the transformed section about the top edge.

$$A\bar{y} = bD \times \frac{D}{2} + (m-1)A_{st} \times d$$

$$\Rightarrow \quad A\bar{y} = 300 \times 600 \times \frac{600}{2} + (13.33 - 1) \times 1963.5 \times 550$$

$$\Rightarrow \quad \bar{y} = \frac{67315475.25}{(300 \times 600) + (13.33 - 1) \times 1963.5}$$

$$\Rightarrow \quad \bar{y} = 329.6 \text{ mm}$$

Distance of neutral axis from extreme compression fibre, $y_t = 329.6$ mm

Distance of neutral axis from extreme tension fibre, $y_b = 600 - 329.6 = 270.4$ mm

Distance of neutral axis from centroid of reinforcing steel, $y_s = 550 - 329.6 = 220.4$ mm

Transformed moment of inertia,

$$I = \frac{b(y_t)^3}{3} + \frac{b(y_b)^3}{3} + (m-1)A_{st} \times (y_s)^2$$

$$= \frac{300 \times (329.6)^3}{3} + \frac{300 \times (270.4)^3}{3} + (13.33 - 1) \times (1963.5) \times (220.4)^2$$

$$= 6.733 \times 10^9 \text{ mm}^4$$

Modulus of rupture of M 20 concrete, $f_{cr} = 0.7\sqrt{20} = 3.13$ MPa

Cracking moment,

$$M_{cr} = f_{cr} \times \frac{I}{y_b} = 3.13 \times \frac{6.733 \times 10^9}{270.4} \times 10^{-6} = 77.937 \text{ kN-m}$$

But given value of bending moment, $M = 50$ kN-m

Since $M < M_{cr}$, hence the given section is an uncracked section.

Maximum compressive stress in concrete,

$$f_c = \frac{M}{I} \times y_t = \frac{50 \times 10^6}{6.733 \times 10^9} \times 329.6 = 2.45 \text{ MPa}$$

Maximum tensile stress in concrete,

$$f_{ct} = \frac{M}{I} \times y_b = \frac{50 \times 10^6}{6.733 \times 10^9} \times 270.4 = 2.01 \text{ MPa}$$

Tensile stress in concrete at the level of steel,

$$f_{cs} = \frac{M}{I} \times y_s = \frac{50 \times 10^6}{6.733 \times 10^9} \times 220.4 = 1.64 \text{ MPa}$$

∴ **Tensile stress in steel,** $f_{st} = mf_{cs} = 13.33 \times 1.64 = 21.86$ MPa

2.13 **Design a T-beam for commercial complex with reference to the data as stated below by working stress method.**

Clear span of T-beam = 10 m

Spacing of T-beam = 2.5 m

Live load = 4 kN/m²

Thickness of the slab = 15 cm

Materials = M20 grade concrete and Fe415 HYSD bars

Effective span of the beam = 10.5 m

Assume width of rib = 300 mm

[15 marks : 2010]

Solution:

As it is a T-beam, according to IS : 456

$$b_f = \frac{l_o}{6} + b_w + 6D_f$$

this b_f should be used in evaluation of neutral axis, moment of resistance etc.

$$b_f \leq \text{Spacing of T-beam (Used in load calculation)}$$

$$b_f = \frac{10000}{6} + 300 + 6 \times 150 = 2866.67 \text{ mm}$$

We take $b_f = 2500$ mm

1. **Load calculation**

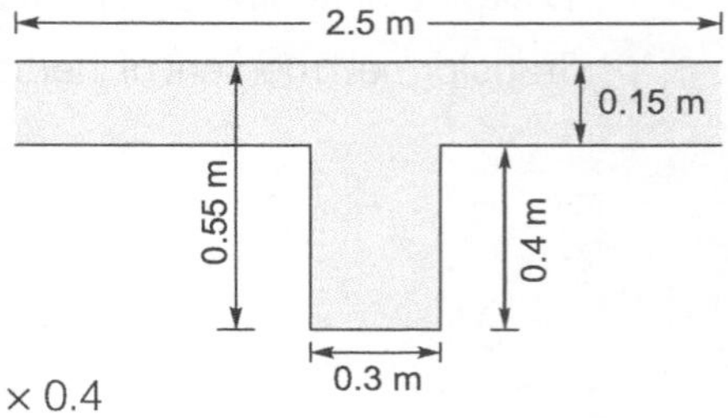

$$\text{Take overall depth} = \frac{1}{20}\text{effective span}$$

$$= \frac{1}{20}\times 10500 = 525 \text{ mm}$$

Take $D = 550$ mm, d = 500 mm

$$DL = 25\times 2.5\times 0.15 + 25\times 0.3\times 0.4$$

$$= 12.375 \text{ kN/m}$$

$$LL = 4\times 2.5 = 10 \text{ kN/m}$$

$$\text{Surface finishing} = 1\times 2.5 = 2.5 \text{ kN/m}$$

$$\text{Total load } (w) = 24.875 \text{ kN/m}$$

$$M = \frac{wl_{eff}^{\ 2}}{8} = \frac{24.875\times 10.5^2}{8} = 342.81 \text{ kN-m}$$

Assumptions to be kept in mind while designing

(a) Section should be balanced

(b) Lever arm = 0.9 d

(c) BM is effectively resisted by flange only.

2. **A_{st} calculation**

For M20,

$$\sigma_{cbc} = C = 7 \text{ MPa}$$

$$m = 13$$

$$x_{bal} = \frac{mc}{t+mc}d = \frac{13\times 7}{13\times 7+230}d = 0.2835d$$

Stress at level of junction of flange and web

$$= \sigma_{cbc}\times\left(\frac{x-150}{x}\right)$$

$$\text{Force} = \frac{1}{2}\left[\sigma_{cbc}+\sigma_{cbc}\left(\frac{x-150}{x}\right)\right]\times 2500\times 150$$

$$\text{Moment of resistance} = \text{Force}\times\text{L.A.}$$

$$M = \frac{1}{2}\left[7+7\times\left(\frac{0.2835d-150}{0.2835d}\right)\right]\times 2500\times 150\times 0.9d$$

$$M = \frac{1}{2}[3.969d-1050]\times\frac{2500\times 150\times 0.9}{0.2835}$$

$$M = 595238.0952\,(3.969\,d-1050)$$

We get, $d = 409.65$ mm

and we have taken $d = 500$ mm

So, OK.

Now provide $d = 450$ mm, $D = 500$ mm [effective cover = 50 mm]

$$BM = \sigma_{st}\times A_{st}\times 0.9\text{ d}$$

$$342.81\times 10^6 = 230\times A_{st}\times 0.9\times 450$$

$$A_{st} = 3680.2 \text{ mm}^2$$

Provide 8 Nos. 25 mm diameter bars.

2.14 A simply supported T-beam of span 9 m of reinforced concrete has the following dimensions:

Flange width = 2000 mm Flange thickness = 150 mm

Overall depth = 750 mm Rib width = 300 mm

The beam is provided with 6 No. 32 mm diameter HYSD bars of grade Fe 500.

Concrete used is of grade M 25.

Find the moment of resistance of the beam using limit state method.

Also find the magnitude of two point loads at 3 m distance from the ends.

[12 marks : 2012]

Solution:

Given: Flange width, b = 2000 mm; Span, l_0 = 9m = 9000 mm; Flange thickness, D_f = 150 mm; Overall depth, D = 750 mm; Rib width, b_w = 300 mm

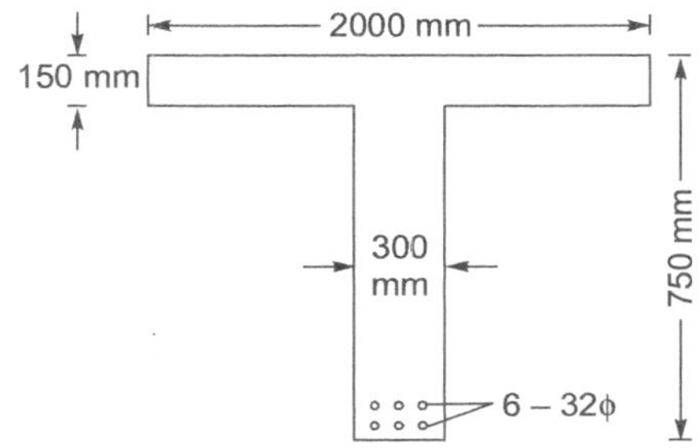

$$A_{st} = 6\times\frac{\pi}{4}\times 32^2 = 4825.486 \text{ mm}^2$$

Effective width of flange for an isolated T-beam is given as

$$b_f = \frac{l_0}{\left(\frac{l_0}{b}\right)+4} + b_w = \frac{9000}{\frac{9000}{2000}+4} + 300$$

$\Rightarrow$ b_f = 1358.82 mm < b

Now to calculate the moment of resistance of the beam, we need to calculate the actual depth of neutral axis.

Assuming that the NA lies in the flange.

Force of compression, $C = 0.36\, f_{ck} b_f x_u$

Force of tension, $T = 0.87\, f_y A_{st}$

But $C = T$

$\therefore$ $0.36 f_{ck}\, b_f\, x_u = 0.87 f_y\, A_{st}$

$$\Rightarrow \quad x_u = \frac{0.87 f_y A_{st}}{0.36 f_{ck} b_f} = \frac{0.87\times 500\times 4825.486}{0.36\times 25\times 1358.82}$$

$\Rightarrow$ x_u = 171.64 mm

$\therefore$ $x_u > D_f$

Hence our assumption is wrong.

$\therefore$ NA lies in the web.

As we can see from the stress diagram, depth of rectangular portion of stress diagram

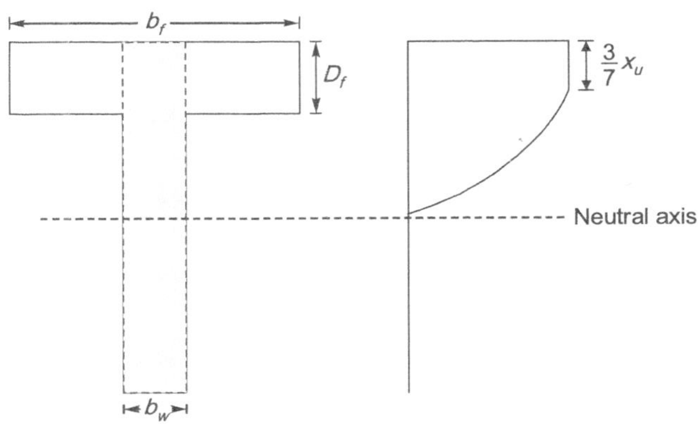

$$= \frac{3}{7}x_u = \frac{3}{7}\times 171.64 = 73.56 \text{ mm}$$

$$\therefore \; D_f > \frac{3}{7}x_u$$

It means that depth of flange is greater than the rectangular portion of stress diagram.

In this case it is assumed that depth of rectangular portion of the stress block is equal to y_f

where, $y_f = 0.15x + 0.65D_f \not> D_f$

$\therefore$ The given T-beam can be shown as

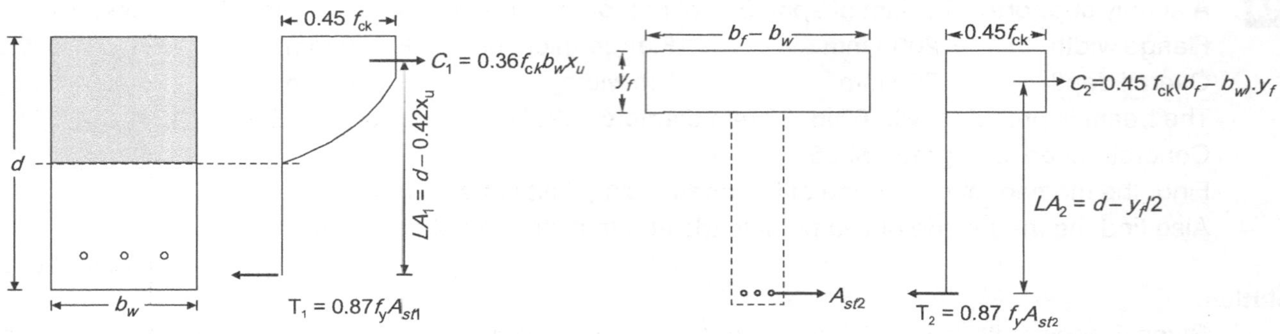

$\therefore \quad C = C_1 + C_2$

$= 0.36 f_{ck} b_w x_u + 0.45 f_{ck}(b_f - b_w) y_f$

$T = T_1 + T_2$

$= 0.87 f_y A_{st1} + 0.87 f_y A_{st2} = 0.87 f_y A_{st}$

But $\quad C = T$

$\therefore \quad 0.36 f_{ck} b_w x_u + 0.45 f_{ck}(b_f - b_w) y_f = 0.87 f_y A_{st}$

$\Rightarrow \quad 0.36 \times 25 \times 300 \times x_u + 0.45 \times 25 \times (1358.82 - 300) y_f$

$= 0.87 \times 500 \times 4825.486$

$\Rightarrow \quad 2700 x_u + 11911.725 \times (0.15 x_u + 0.65 \times 150)$

$= 2099,086.41$

Solving $\quad x_u = 209 \text{ mm}$

$\therefore \quad x_u > D_f$ (OK)

and $\quad \frac{3}{7} x_u = \frac{3}{7} \times 209 = 89.57 \text{ mm}$

$\therefore \quad D_f > \frac{3}{7} x_u$ (OK)

$\therefore$ Moment of the resistance of the beam with respect to concrete is given by

$M_R = C_1 \times L_1 + C_2 \times L_2$

$= 0.36 f_{ck} b_w x_u \times (d - 0.42 x_u) + 0.45 f_{ck}(b_f - b_w) y_f (d - y_f/2)$

Assuming effective cover = 50 mm

$\therefore$ Effective depth, $\quad d = D - 50$

$= 750 - 50 = 700 \text{ mm}$

$y_f = 0.15 x_u + 0.65 D_f$

$= 0.15 \times 209 + 0.65 \times 150 = 128.85 \text{ mm}$

$\therefore \quad M_R = 0.36 \times 25 \times 300 \times 209 \times (700 - 0.42 \times 209)$

$+ 0.45 \times 25 \times (1358.82 - 300) \times (0.15 \times 209$

$+ 0.65 \times 150) \times (700 - 64.425)$

$= 345,475,746 + 975,496,886.4$

$= 132,097,2632 \text{ N-mm}$

$= 1320.973 \text{ kN-m}$

Now, to calculate the magnitude of point load, equating the bending moment to moment of resistance

$$W_d = (2 \times 0.15 + 0.3 \times 0.75) \times 25 = 13.125 \text{ kN/m}$$

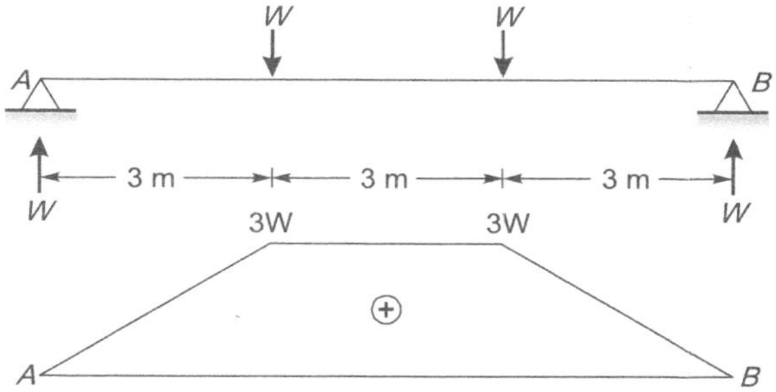

$$\frac{M_u}{1.5} = M_w = 880.65 \text{ kN-m}$$

$$880.65 = \frac{W_d l^2}{8} + \frac{Wl}{3} = \frac{13.125 \times 9^2}{8} + \frac{Wl}{3}$$

$$W = 249.253 \text{ kN}$$

2.15 Determine the moment of resistance of a Tee-section having the following properties:
Flange width = 2000 mm
Flange depth = 100 mm
Web width = 250 mm
Effective depth = 750 mm
Area of steel = 8 bars of 20 mm diameter
Material used = M 25 grade of conc. and Fe 415 HYSD bars

[4 marks : 2013]

Solution:

Given:

$$b = 2000 \text{ mm} = b_f$$
$$D_f = 100 \text{ mm}$$
$$b_w = 250 \text{ mm}$$
$$d = 750 \text{ mm} = d$$

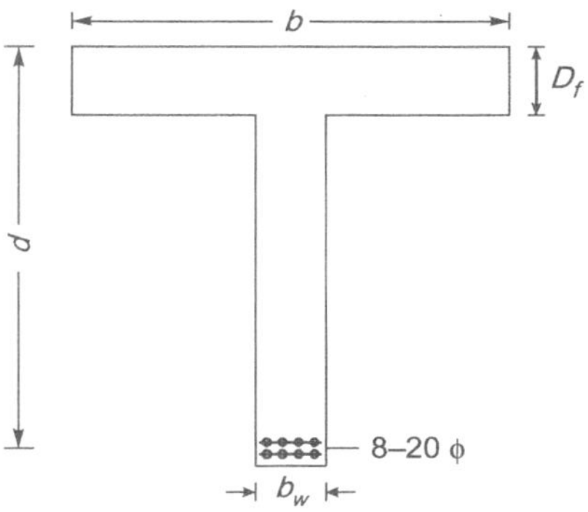

Effective width of flange for T-beam section is

$$b_f = \frac{l_0}{6} + b_w + 6D_f$$

Here the length of span is not given so we can consider width of flange as effective width of flange.

$$A_{st} = 8 \times \frac{\pi}{4}(20)^2 = 2513.27 \text{ mm}^2$$

For M 25, Fe 415

$$x_{u\max} = 0.48d = 0.48\,(750) = 360 \text{ mm}$$

Let $x_u \le x_{max}$ and $x_u \le D_f$

$$C_u = T_u$$

$\Rightarrow$ $$0.36\, f_{ck}\, b_f x_u = 0.87\, f_y A_{st}$$

$\Rightarrow$ $$0.36(25)(2000)x_u = 0.87(415)\,2513.27$$

$\Rightarrow$ $$x_u = 50.412 \text{ mm} < x_{u\max}$$
$$< D_f$$

$\therefore$ Neutral axis lies in the flange of T-beam

$\therefore$ $$\text{MOR} = 0.87\, f_y A_{st}(d - 0.42\, x_u)$$
$$= 0.87\,(415)\,(2513.27)\,(750 - 0.42 \times 50.412) = 661.35 \text{ kNm}$$

Alternatively $$\text{MOR} = 0.36\, f_{ck}\, b x_u\,(d - 0.42\, x_u)$$
$$= 0.36\,(25)(2000)\,(50.412)\,(750 - 0.42 \times 50.412)$$
$$= 661.35 \text{ kNm Which is same as above}$$

$\therefore$ Moment of resistance of Tee-section is 661.35 kNm.

2.16 A reinforced concrete beam, having a simply supported span of 6 m, carries a dead load of 15 kN/m (incl. its dead load) and an imposed load of 20 kN/m at service. Design the cross-section of the beam at the mid-span only for flexure and shear at the limit state of collapse. Assume moderate exposure condition and grade of steel as Fe415. Draw a neat sketch showing the reinforcement details.

[15 marks : 2014]

Solution:

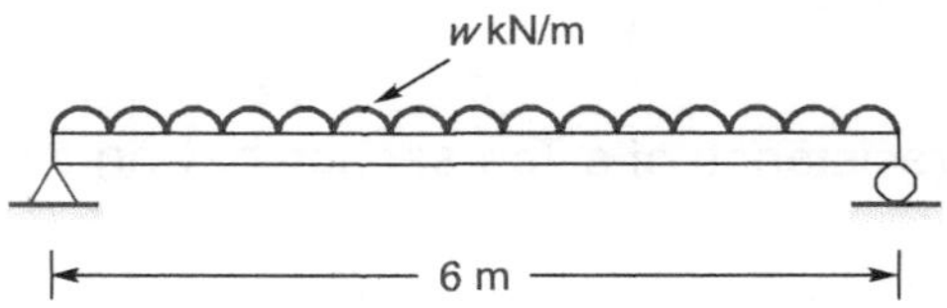

Total load on beam, $w = 15\text{ kN/m} + 20\text{ kN/m} = 35\text{ kN/m}$

Factored load, $w_u = 1.5 \times 35 = 52.5\text{ kN/m}$

Maximum factor BM at mid span is given by,

$$M_u = \frac{w_u l^2}{8} = \frac{52.5 \times 6^2}{8} = 236.25\text{ kN-m}$$

and

$$V_u = \frac{w_u l}{2} = \frac{52.5 \times 6}{2} = 157.5\text{ kN}$$

1. **Design for flexure:**

For design, $BM_u = MR$

$$236.25 \times 10^6 = 0.36 f_{ck} B x_{u\,lim} (d - 0.42\, x_{u\,lim}) \qquad \text{...(i)}$$

- For moderate exposure minimum grade of concrete should not be less than M25
- $x_{u\,lim} = 0.48\text{ d for Fe-415}$
- Generally, $\frac{d}{B} = 2$

From (i), we have

$$236.25 \times 10^6 = 0.36 \times 25 \times \frac{d}{2} \times 0.48d(d - 0.42 \times 0.48d)$$

$$236.25 \times 10^6 = 0.36 \times 25 \times \frac{d}{2} \times 0.48d \times 0.7984d$$

$$236.25 \times 10^6 = 1.724\, d^3$$

or $d = 515.56\text{ mm}$

keep $d = 520\text{ mm}$

Hence, $B = \frac{d}{2} = \frac{520}{2} = 260\text{ mm}$

Overall depth of beam, $D = d + \text{Effective cover}$

Providing effective cover of 30 mm

$$= 520 + 30 = 550\text{ mm}$$

Thus the dimension of beam section are 260 mm × 550 mm

Area of steel:

$$A_{st} = \frac{BM_u}{0.87 f_y \left(d - 0.42 x_{u\,lim}\right)}$$

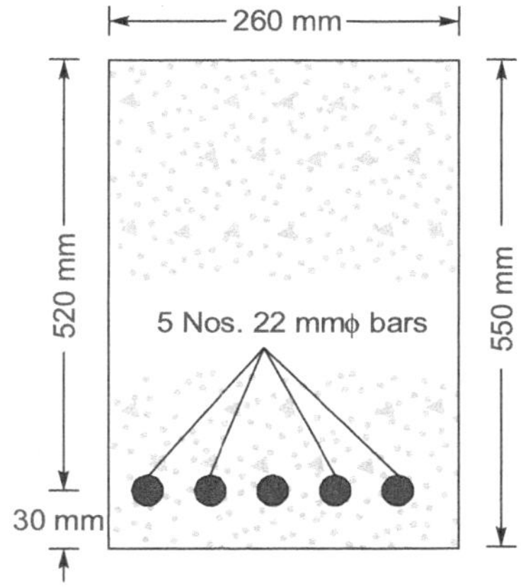

$$A_{st} = \frac{236.25 \times 10^6}{0.87 \times 415(520 - 0.42 \times 0.48 \times 520)}$$

$$= 1576 \text{ mm}^2$$

Using 22 mmϕ steel bars,

$$\text{Area of single bar} = \frac{\pi\phi^2}{4} = \frac{\pi \times 22^2}{4} = 380.13 \text{ mm}^2$$

$$\text{No. of bars} = \frac{A_{st}}{380.13} = \frac{1576}{380.13} = 4.14 = 5 \text{ No.}$$

2. **Design for shear**

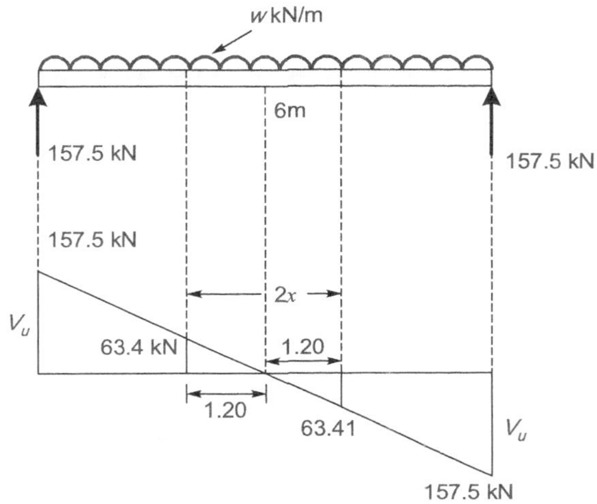

Nominal shear stress (τ_{vu}),

$$\tau_{vu} = \frac{V_u}{Bd} = \frac{157.5 \times 10^3}{260 \times 520} = 1.16 \text{ N/mm}^2$$

For M25, $\tau_{c\max} = 3.1 \text{ N/mm}^2$

Here $\tau_{vu} < \tau_{c\max}$ OK

At mid span shear force is zero. It mean section at mid span remain safe in shear but for safety in central zone, minimum shear reinforcement may be provided.

$$\% P_t = \frac{A_{st}}{Bd} \times 100 = \frac{1576}{260 \times 520} \times 100 = 1.16\%$$

$$\tau_c = 0.64 + \frac{0.67 - 0.64}{1.25 - 1.00} \times (1.16 - 1.00)$$

$$= 0.659 \text{ N/mm}^2$$

$\%P_t$	τ_c (N/mm²) for M_{25}
≤ 0.75	0.57
1.00	0.64
1.25	0.67
1.50	0.68

Net shear force for which shear reinforcement is required,

$$V_{us} = V_u - V_{Cu} = 157.5 - \tau_c Bd$$

$$= 157.5 - 0.659 \times 260 \times 520 \times 10^{-3} = 68.4032 \text{ kN}$$

Using 2-legged 8 mmϕ stirrups,

$$S_v = \frac{0.87 f_y A_{sv} d}{V_{us}} = \frac{0.87 \times 415 \times 100.53 \times 520}{68403.2} = 275.924 \text{ mm}$$

Check: (1) 0.75 d = 390 mm, (2) 300 mm

$$A_{sv} = \frac{2 \times \pi \times 8^2}{4} = 100.53 \text{ mm}^2$$

Maximum spacing of minimum shear reinforcement in central zone of $2x = 1.20 \times 2 = 2.4$ m

$$\frac{A_{sv}}{Bs_v} = \frac{0.4}{0.87 f_y}$$

$$S_v = \frac{0.87 f_y A_{sv}}{0.4 \times B} = \frac{0.87 \times 415 \times 100.53}{0.4 \times 260} = 349 \text{ mm} \not> 300 \text{ mm}$$

Thus maximum spacing of shear reinforcement in central zone should be 270 mm.

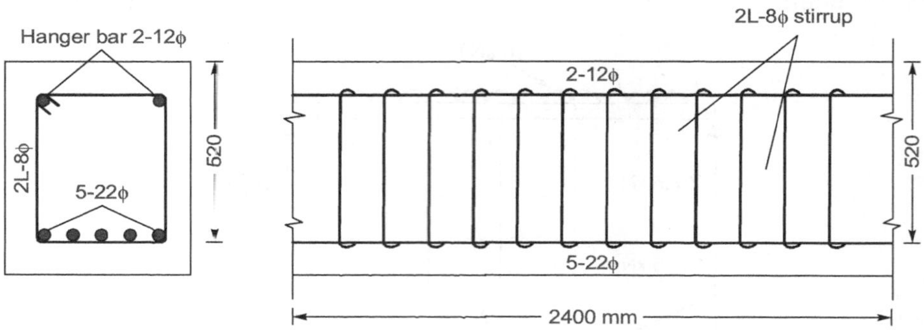

2.17 Determine the reinforcement required to resist a factored bending moment of 40 kN-m acting on a beam of hollow cross section as shown, using M 25 grade of concrete and Fe 415 steel. Effective depth of the beam is 360 mm. Adopt limit state method of design.

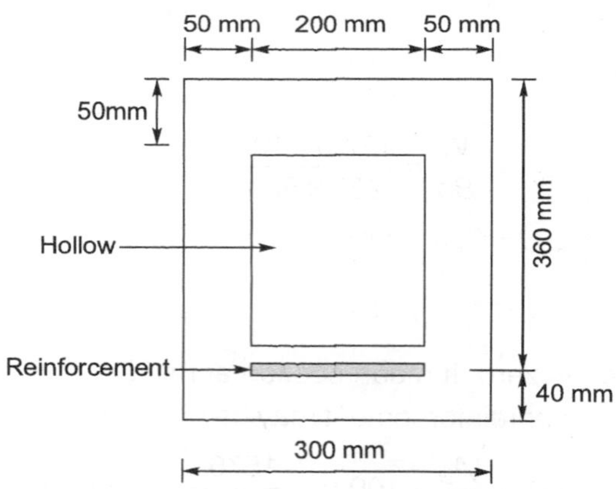

[10 marks : 2015]

Solution:

Given: M25 concrete, $f_{ck} = 25$ N/mm^2

Fe415 steel, $f_y = 415$ N/mm^2

Width of beam, $b = 300$ mm

Overall depth of beam, $D = 400$ mm

Effective depth of beam, $d = 360$ mm

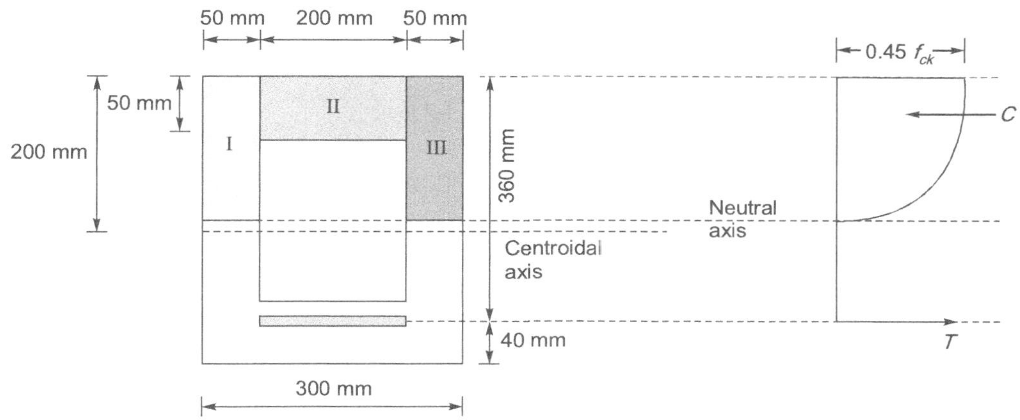

Limiting depth of neutral axis (x_{ulim}) = $0.48d = 0.48\,(360) = 172.8$ mm

Depth of rectangular portion of compressive stress block = $\frac{3}{7}x_{u\lim} = \frac{3}{7}(172.8) = 74.06\,\text{mm}$

Divide the area above the balance neutral axis in portions I, II and III as shown.

Parts I and III are each of width 50 mm and are under the influence of complete compressive stress block for a depth of x_{ulim} (= 74.06 mm).

Part II is of width 200 mm which is under the influence of compressive stress block for a depth of 50 mm

$$\left(< \frac{3}{7}x_{u\lim} = 74.06\,\text{mm}\right)$$

Let C_1 = Compressive force in part I

$\therefore$
$$C_1 = C_3 = 0.36\, f_{ck}\, b\, x_{ulim} = 0.36\,(25)\,(50)\,(172.8)\,\text{N}$$
$$= 77.76\ \text{kN}$$

C_2 = Compressive force in part II
$$= 0.45\, f_{ck}\, b\,(50) = 0.45\,(25)\,(200)\,(50)\ \text{N} = 112.5\ \text{kN}$$

For statical equilibrium, $C = T$

$\Rightarrow$ $C_1 + C_2 + C_3 = T$

$\Rightarrow$ $77.76 + 112.5 + 77.76 = T$

$\Rightarrow$ $T = 268.02$ kN

Limiting moment of resistance corresponding to balanced section is given by,

$$M_{u\lim} = (C_1 + C_3)(d - 0.42x_{u\lim}) + C_2\left(d - \frac{50}{2}\right)$$

$$= 2 \times 77.76 \times 10^3\,(360 - 0.42 \times 172.8) + 112.5 \times 10^3\left(360 - \frac{50}{2}\right)$$

$$= 44.7 \times 10^6 + 37.69 \times 10^6\ \text{N.mm} = 82.39\ \text{kNm}$$

Given factored bending moment, $M = 40\ \text{kNm} < M_{ulim}$ (= 82.39 kNm)

$\therefore$ Section is under-reinforced and $x_u < x_{ulim}$

$\Rightarrow$ Reinforcement steel will yield first i.e. $f_{st} = 0.87\, f_y$.

Determination of actual depth of neutral axis (x_u)

Let neutral axis lies wholly in the top concrete portion without intercepting the hollow portion of beam i.e., $x_u \leq 50$ mm.

$\therefore \quad M = C(d - 0.42x_u)$

$\Rightarrow \quad 40 \times 10^6 = 0.36 f_{ck} b x_u (d - 0.42 x_u)$

$\Rightarrow \quad 40 \times 10^6 = 0.36\,(25)\,(300) x_u (360 - 0.42 x_u)$

$\Rightarrow \quad 0.42x_u^2 - 360x_u + 14814.815 = 0$

$\therefore \quad x_u = 43.344$ mm < 50 mm (Neglecting $x_u = 813.8$ mm)

$\therefore$ Our assumption is correct.

$\therefore \quad M = T(d - 0.42x_u)$

$\Rightarrow \quad 40 \times 10^6 = 0.87 f_y A_{st} (d - 0.42 x_u)$

$\Rightarrow \quad 40 \times 10^6 = 0.87\,(415)\,A_{st}(360 - 0.42 \times 43.344)$

$\Rightarrow \quad A_{st} = 324.14 \text{ mm}^2$

$\therefore$ Reinforcement required to resist the factored moment of 40 kNm is 324.14 mm^2

2.18 **Design a simply supported roof slab (bending and shear only) for a room 3.5 m x 8 m clear in size if the superimposed dead and live load is 5 kN/m^2. Use M25 mix of Fe415 grade steel. Show reinforcement details. Shear strength of concrete = 0.35 MPa.**

[12 marks : 2017]

Solution :

Assume slab to be supported on 230 mm thick masonary wall.

(i) Calculation of design constant

For M25 concrete, $m = \dfrac{280}{3\sigma_{cbc}} = \dfrac{280}{3 \times 8} = 11.67 = \text{m}, \quad c = 8 \text{ N/mm}^2 = \sigma_{cbc}.$

For Fe415, $\sigma_{st} = 230 \text{ N/mm}^2$

$$\therefore \quad k = \frac{m\sigma_{cbc}}{\sigma_{st} + m\sigma_{cbc}} = \frac{11.67 \times 8}{230 + 11.67 \times 8} = 0.2887$$

$$j = 1 - \frac{k}{3} = 1 - \frac{0.2887}{3} = 0.9038$$

$$Q = \frac{1}{2} cjk = \frac{1}{2} \times 8 \times 0.9038 \times 0.2887 = 1.044$$

(ii) As per vertical deflection criteria, the span to effective depth ratio for spans upto 10 m for a simply supported slab is given by;

$$\frac{l}{d} = 20$$

$$\Rightarrow \quad d = \frac{l}{20} = \frac{3500}{20} = 175 \text{ mm}$$

(iii) Effective span

(a) Clear span + effective depth = 3.5 + 0.175 = 3.675 m

(b) Centre to centre distance b/w supports = 3.5 + 0.23 = 3.73 m

Hence, lesser of the above two will be adopted i.e. $l_e = 3.675$ m

(iv) Bending moment and shear force

Assuming nominal cover of 20 mm, the total depth of slab will become 175 + 20 = 195 mm

(a) Load calculations

Load due to self weight of slab = 0.195 × 1 × 1 × 25 = 4.875 kN/m²

Superimposed dead and live load = 5 kN/m²

Total load = 9.875 kN/m²

(b) Bending moment per metre run of slab

$$M = \frac{wl_e^2}{8} = \frac{9.875 \times (3.675)^2}{8} = 16.67 \text{ kNm}$$

(c) Shear force, $$V = \frac{wl_e}{8} = \frac{9.875 \times 3.675}{2} = 18.145 \text{ kN}$$

(v) Design of section

$$d = \sqrt{\frac{M}{QB}} = \sqrt{\frac{16.67 \times 10^6}{1000 \times 1.044}} = 126.36 < 175 \text{ mm Hence, OK.}$$

Now, taking d = 130 mm and nominal over = 20 mm

We get $D = d + 20 = 150$ mm

(vi) Main reinforcement

$$A_{st} = \frac{M}{\sigma_{st} jd} = \frac{16.67 \times 10^6}{230 \times 0.9038 \times 150} = 534.62 \text{ mm}^2$$

Minimum r/f should be 0.15% of total cross-sectional area.

$$= \frac{0.15}{100} \times 150 \times 1000 = 225 \text{ mm}^2$$

∴ A_{st} is more than minimum reinforcement.

Spacing of bars, $$S = \frac{1000 A_\phi}{A_{st}}$$

Adopting 10 mm ϕ bars, we get $A_\phi = \frac{\pi}{4} \times (10)^2 = 78.54 \text{ mm}^2$

∴ $$S_v = \frac{1000 \times 78.54}{534.62} = 146.91 \text{ mm}$$

Adopt spacing 145 mm

We get, $$\text{actual } A_{st} = \frac{1000 \times 78.54}{145} = 541.66 \text{ mm}^2$$

Spacing is less than $3d$ and 300 mm both.

(vii) Check for shear,

$$\tau_c = 0.35 \text{ N/mm}^2$$

$$\tau_v = \frac{V_u}{bd} = \frac{18.145 \times 10^3}{1000 \times 150} = 0.12 \text{ N/mm}^2$$

∵ $\tau_v < \tau_c$. Hence, safe.

(viii) Distribution reinforcement

$$A_{st} = \frac{0.15BD}{100} = \frac{0.15 \times 1000 \times 150}{100} = 225 \text{ mm}^2$$

Using 6 mm ϕ bars, each having,

$$A_\phi = \frac{\pi}{4} \times (6)^2 = 28.27 \text{ mm}^2$$

Spacing of bars,
$$S_v = \frac{1000 A_\phi}{A_{st}} = \frac{1000 \times 28.27}{225} = 125.64 \text{ mm}$$

Hence, provide 6 mm ϕ bars @ 125 mm c/c.

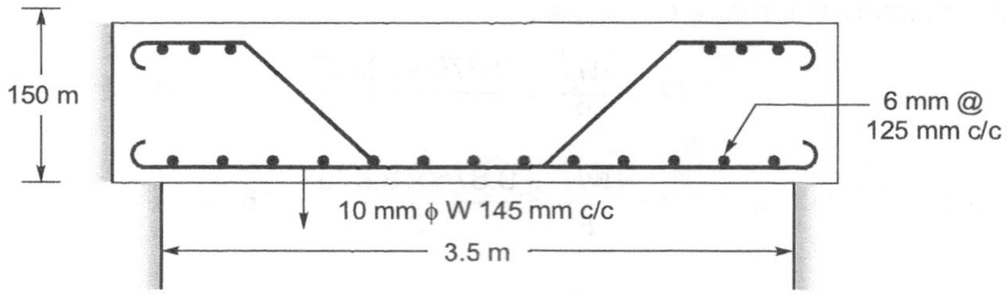

Reinforcement details

2.19 Design the reinforcement at critical section only for a beam for which Bending moment and shear force diagram has been shown in figure. Following parameter may be used for design:

1. M-20 grade concrete
2. Fe-415 grade steel
3. Nominal concrete cover : 30 mm
4. (Depth/width) ratio of beam : (02)
5. Diameter of reinforcing bar : 20 mm for flexure reinforcement and 8 mm for shear reinforcement
6. Shear strength of concrete = 0.6 N/mm²
7. $(M_u, \lim/f_{ck} bd^2) = 0.138$
8. $(p_t, \lim f_y/f_{ck}) = 19.82$

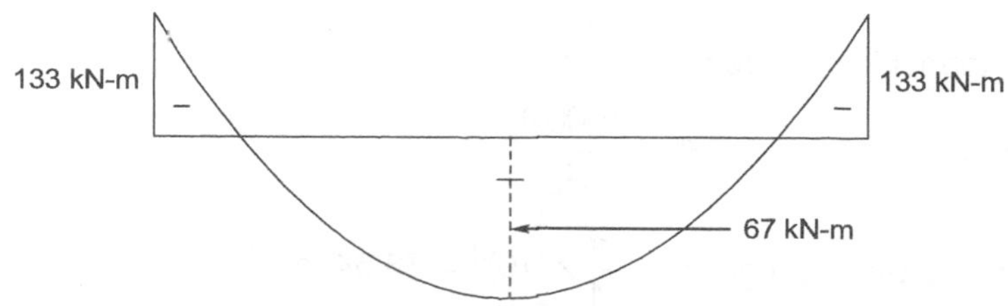

Service State Bending Moment Diagram

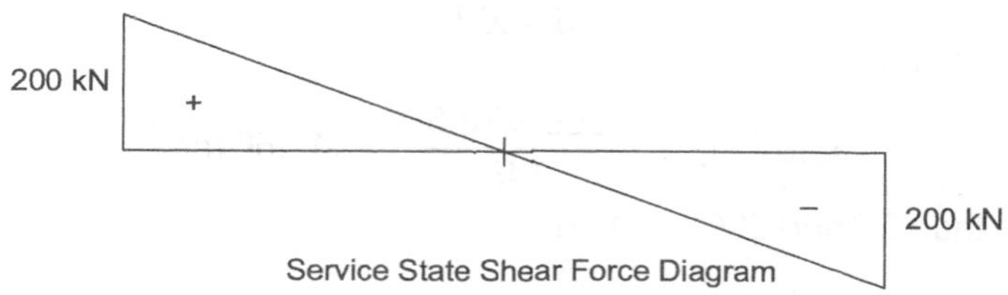

Service State Shear Force Diagram

[20 marks : 2018]

Solution:

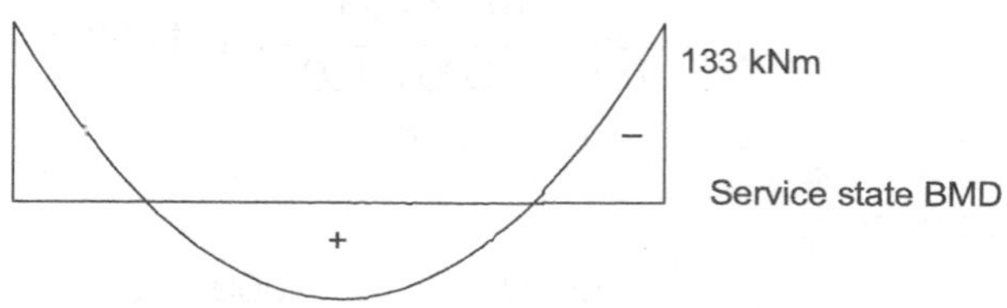

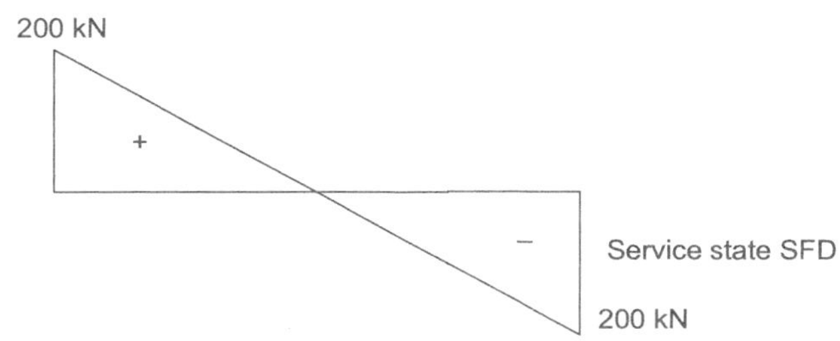

$$\text{Design } BM = 1.5 \times 133 = 199.5 \text{ kNm}$$
$$\text{Design, } SF = 1.5 \times 200 = 300 \text{ kN}$$
$$\text{Design, } D = 2 \times \text{Width } B$$
$$D = 2B$$

Given: Nominal cover = 30 mm

Equating design BM to m_{ulim}

$$0.138\, f_{ck}\, Bd^2 = 199.5 \times 10^6$$

Let $d \approx 2B$

$$2.76 \times B \times (2B)^2 = 199.5 \times 10^6$$
$$B = 262.4 \text{ mm}$$
$$d = 525 \text{ mm}$$
$$D\text{, overall depth} = \left(d = \frac{20}{2} + 8 + 30 + 525\right) = 573 \text{ mm}$$

Let $D = 580$ mm, $B = 290$ mm

$$\therefore \quad d = 580 - 30 - 8 - \frac{20}{2} = 532 \text{ mm}$$

Reinforcement

At support

$$M_{ulim} = 0.138\, f_{ck} Bd^2$$
$$= 0.138 \times 20 \times 290 \times 532^2$$
$$= 226.5 \text{ kNm} > M_u$$

Hence beam will be under reinforced.

$$A_{st,\text{ support}} = 0.5 \frac{f_{ck}}{f_y}\left[1 - \sqrt{1 - \frac{4.6 \times 199.5 \times 10^6}{20 \times 290 \times 532^2}}\right] \times 290 \times 532$$
$$= 1249 \text{ mm}^2$$

Using 20 mm bar,

$$\text{Number of bars} = \frac{1249}{\frac{\pi}{4} \times 20^2} = 4$$

At midspan

$$M_u = 1.5 \times 67 = 100.5 \text{ kNm}$$

$$\therefore \quad A_{st} = 0.5 \frac{f_{ck}}{f_y}\left[1 - \sqrt{1 - \frac{4.6 \times M_u}{f_{ck} \times Bd^2}}\right] \times Bd$$

$$= 0.5 \frac{20}{415}\left[1 - \sqrt{1 - \frac{4.6 \times 100.5 \times 10^6}{20 \times 290 \times 532^2}}\right] \times 290 \times 532$$

$$= 566.67 \text{ mm}^2$$

$$\text{Number of bars required} = \frac{566.67}{\frac{\pi}{4} \times 20^2} = 1.80 \approx 2.0 \text{ bars}$$

Design of shear reinforcement

$$\text{Design, SF} = 300 \text{ kN}$$

$$\tau_v = \frac{V_u}{Bd} = \frac{300 \times 10^3}{290 \times 532} = 1.94 \text{ MPa} < \tau_{c\,max} \qquad (\therefore \text{ safe})$$

$$\tau_v > \tau_c$$

$\therefore$ Shear reinforcement is necessary.

$$V_{us} = [V_u - \tau_c Bd]$$
$$= 300 \times 10^3 - 0.6 \times 290 \times 532$$
$$= 207.432 \text{ kN}$$

$\therefore$
$$V_{us} = 0.87 f_y A_{sv} \times \frac{d}{s_v}$$

Using 2-legged 8 mm stirrups.

$$s_v = \frac{0.87 \times 415 \times 2 \times \frac{\pi}{4} \times 8^2 \times 532}{207.432 \times 1000}$$
$$= 93.04 \text{ mm}$$

$\therefore$ Provide 2-legged 8 mm ϕ @ 90 mm c/c.

Also
$$\frac{0.87 \times f_y \times A_{sv}}{Bs_v} \geq 0.4$$

$\therefore$
$$s_v \leq 312.74 \text{ mm} \qquad (\therefore \text{ 90 mm spacing is satisfactory})$$

And also
$$s_v \leq 300 \text{ mm}$$

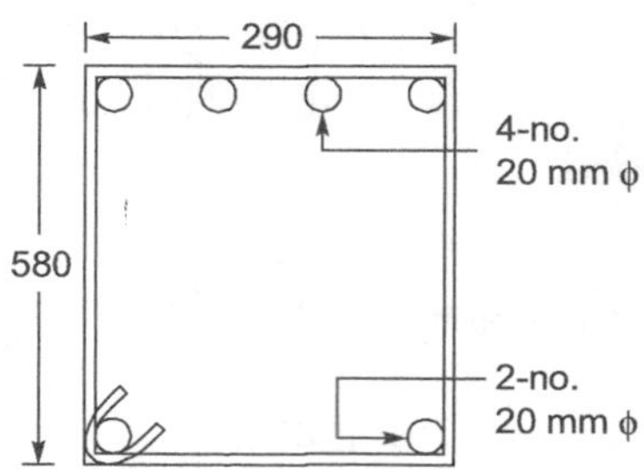

2.20 **Design a cantilever slab shown in figure for flexure only. Sketch the reinforcement also. Following parameters may be used for design, applying different checks and detailing the reinforcement:**

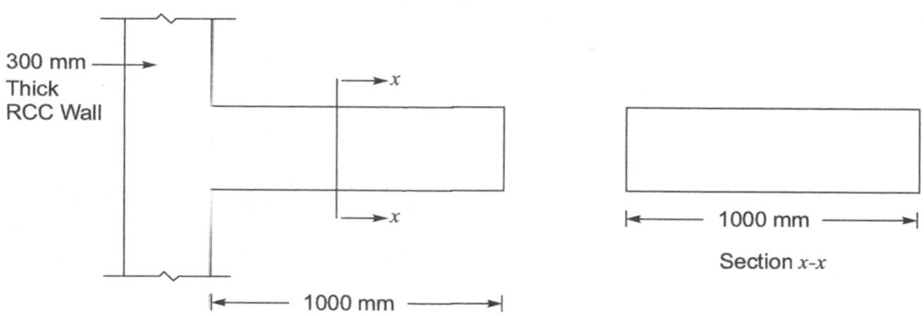

1. Span to effective depth : 10 (maximum)
2. Mild exposure condition : Nominal concrete cover 20.0 mm.
3. 2.0 Hours of fire resistance : Nominal concrete cover 25.00 mm.
4. Maximum live load : 30 kN/m^2
5. Load combination : 1.5 × Dead load + 1.5 × Live load
6. Effective length : length to the face of support plus half the effective depth.
7. Grade of concrete M-20.
8. Grade of reinforcing steel Fe415.
9. Unit weight of RCC : 25 kN/m^3
10. Development length in Tension : 48 × diameter of reinforcing bar
11. Development length in compression : 37 × diameter of bar
12. Minimum reinforcement : 0.12% of total cross-sectional area
13. Maximum spacing of main reinforcement : 3 × effective depth
14. Maximum spacing of distribution reinforcement : 5 × effective depth
15. Diameter of main reinforcement bar : 10 mm
16. $(M_u, \lim/f_{ck}bd^2) = 0.138$
17. $(pt, \lim f_y/f_{ck}) = 19.82$

[20 marks : 2018]

Solution:

Given: Span = 1 m

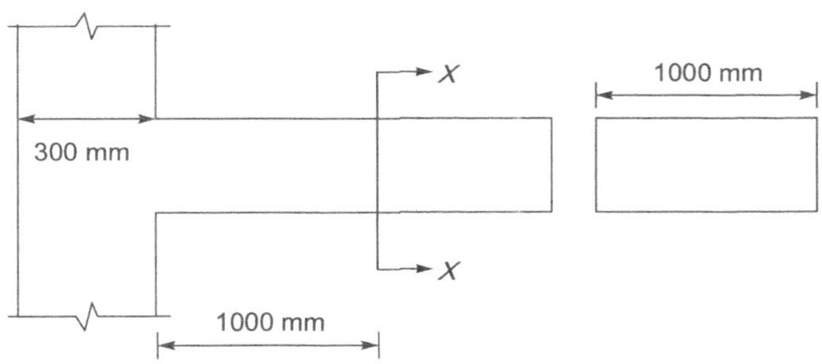

Step 1: Depth calculation

$$\text{Span/effective depth} \leq 10$$

$$\therefore \quad \text{Effective depth} \geq 100 \text{ mm}$$

Providing 150 mm overall depth, with 2 mm clear cover/nominal cover.

$$\text{Hence,} \quad \text{Effective cover} = 25 + \frac{10}{2} = 30 \text{mm}$$

$$\therefore \quad d = D - \text{Effective cover} = 120 \text{ mm}$$

Step 2: Depth calculation

$$\text{Maximum live load} = 30 \text{ kN/m}^2 \times 1 = 30 \text{ kN/m}$$

$$\text{Factored live load} = 1.5 \times 30 = 45 \text{ kN/m}$$

$$\text{Maximum dead load} = 25 \times 1 \times 0.15 = 3.75 \text{ kN/m}$$

$$\text{Maximum dead load} = 1.5 \times 3.75 = 5.625 \text{ kN/m}$$

$$\therefore \quad \text{Load combination} = 1.5 \times DL + 1.5 \times LL = 50.625 \text{ kN/m}$$

Step 3: Effective length

$$L_{\text{effective}} = L_0 + \frac{d}{2} = 1 + \frac{0.12}{2} = 1.06 \text{ m}$$

Step 4: Depth check

$$\text{Ultimate load } w_u = 50.625 \text{ kN/m}$$

$$\therefore \quad BM_u = \frac{w_u l^2}{2} = \frac{50.625 \times 1.06^2}{2} = 28.44 \text{ kNm}$$

$\therefore$ Equating BM_u to $M_{u,\,lim}$

$$0.138\, f_{ck}\, bd^2 = BM_u$$

$$0.138 \times 20 \times 1000 \times d^2 = 28.44 \times 10^6$$

$$\Rightarrow \quad d \geq 101.5 \text{ mm}$$

Hence $\quad d_{provided} > d_{required}$

$\therefore$ Safe

Step 5: Reinforcement calculation:

$$A_{st\,main} = \frac{0.5 f_{ck}}{f_y}\left[1 - \sqrt{1 - \frac{4.6 M_u}{f_{ck} bd^2}}\right] \times bd$$

$$= 0.5 \times \frac{20}{415}\left[1 - \sqrt{1 - \frac{4.6 \times 28.44 \times 10^6}{20 \times 1000 \times 120^2}}\right] \times 1000 \times 120$$

$$= 755.4 \text{ mm}^2$$

Also, $\quad A_{st\,min} = \frac{0.12}{100} \times bD = 180 \text{ mm}^2 < A_{st\,main}$ $\quad$ [$\therefore$ OK]

Using 10 mm bars, $\quad \text{Spacing} = \dfrac{1000}{\dfrac{A_{st}}{\frac{\pi}{4} \times 10^2}} = 103.9 \text{mm}$

$\therefore$ Provide 10 mm bars @ 100 mm c/c.

Distribution steel 8 mm bars, spacing $= \dfrac{1000}{180/\frac{\pi}{4} \times 8^2} = 279$ mm

$\therefore$ Provide 8 mm bars @ 250 mm c/c

Maximum spacing of main reinforcement = $3d$ or 300 mm = 360 mm [Least]

Maximum spacing of distribution steel = $5d$ or 450 mm = 600 mm [Least] [$\therefore$ Safe]

Step 6: Development length check

Required development length in tension = 48ϕ = 480 mm

As per the figure given, additional anchorage will be required to provide 480 mm development length beyond the junction/face.

Assuming clear cover of 30 mm.

$$\text{Additional anchorage} = 480 - (300 - 30) = 210 \text{ mm}$$

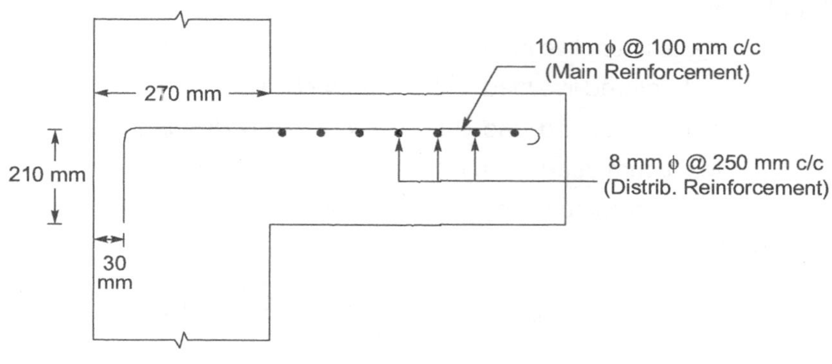

2.21 Find the moment of resistance of a beam 300 x 600 mm deep if it is reinforced with 3 Nos. of 20 mm dia. bars in compression and tension, each at an effective cover of 40 mm. Use M 20 grade concrete and steel grade Fe415.

Points on stress-strain curve for Fe415 steel.

Stress level	Fe415 grade	
	Strain	Stress (N/mm^2)
0.80 f_y	0.00144	288.7
0.85 f_y	0.00163	306.7
0.90 f_y	0.00192	324.8
0.85 f_y	0.00241	342.8
0.975 f_y	0.00276	351.8
1.00 f_y	0.00380	360.9

[12 marks : 2019]

Solution:

$$B = 300 \text{ mm}, \quad D = 600 \text{ mm}$$

$$A_{sc} = 3\times\frac{\pi}{4}\times 20^2 = 942.48 \text{ mm}^2$$

$$A_{st} = 3\times\frac{\pi}{4}\times 20^2 = 942.48 \text{ mm}^2$$

Effective cover = 40 mm

$$f_{ck} = 20 \text{ N/mm}^2$$

$$f_y = 415 \text{ N/mm}^2$$

Let us assume trial depth of NA = 100 mm

Strain at the level of compression steel,

$$\varepsilon_{sc} = \frac{0.0035}{100}(100-40) = 0.0021$$

$$f_{sc} = 324.8 + \frac{342.8-324.8}{0.00241-0.00192}(0.0021-0.00192)$$

$$= 331.4 \text{ N/mm}^2$$

Depth of NA

$$0.36\, f_{ck}Bx_u + (f_{sc} - 0.45f_{ck}) \times A_{sc} = 0.87\, f_yA_{st}$$

$$0.36 \times 20 \times 300 \times x_u + (331.4 - 0.45 \times 20) \times 942.48$$

$$= 0.87 \times 415 \times 942.48$$

$$x_u = 16.86 \text{ mm} \neq (x_u) \text{ assumed.}$$

Assuming new depth of NA,

$$x_u = 50 \text{ mm}$$

Now,
$$\varepsilon_{sc} = \frac{0.0035}{50}\times(50-40) = 0.0007$$

Assuming, linear variation upto strain 0.00144

So,
$$f_{sc} = 0.007 \times E_s$$

$$= 0.007 \times 2.1 \times 10^5$$

$$= 147 \text{ N/mm}^2$$

Now, new depth of NA

$$0.36\, f_{ck}\, Bx_u + (f_{sc} - 0.45\, f_{ck})\, A_{sc} = 0.87\, f_y\, A_{st}$$

$0.36 \times 20 \times 300 \times x_u + (147 - 0.45 \times 20) \times 942.48$
$= 0.87 \times 415 \times 942.48$

$x_u = 97.32 \text{ mm} \neq (x_u) \text{ assumed}$

Now, assuming, $x_u = 60 \text{ mm}$

$$\varepsilon_{sc} = \frac{0.0035}{60}(60 - 40) = 0.001167$$

So, $f_{sc} = 0.001167 \times E_s$
$= 0.001167 \times 2.1 \times 10^5$
$= 245 \text{ N/mm}^2$

$0.36 \times 20 \times 300 \times x_u + (245 - 0.45 \times 20) \times 942.48$
$= 0.87 \times 415 \times 942.478$

$x_u = 54.56 \text{ mm} \simeq 60 \text{ mm}$

So, taking average, $$x_u = \frac{54.56 + 60}{2} = 57.28 \text{ mm}$$

$M_u = 0.36 f_{ck} B x_u (d - 0.42 x_u) + (f_{sc} - f_{ck}) A_{sc} (d - d')$
$= 0.36 \times 20 \times 300 \times 57.28 + (560 - 0.42 \times 57.28)$
$+ (245 - 0.45 \times 20) \times 942.48 \times (560 - 40)$

$M_u = 181.971 \text{ kNm}$

2.22 Design a two way slab for an office room 5.8 m x 4.2 m clear in size if the superimposed load is 4 kN/m². Use M25 grade of concrete and steel grade Fe415. The bending moment coefficients for two-way slabs simply supported on four sides is given below:

l_y/l_x	1.0	1.1	1.2	1.3	1.4	1.5	1.75	2.0
a_x	0.062	0.074	0.084	0.093	0.099	0.104	0.113	0.118
a_y	0.062	0.061	0.059	0.055	0.051	0.046	0.037	0.029

Assume the edges simply supported and the corners not held down. Assume the shape factor for shear $k = 1.3$.

Design shear strength of concrete of M25 grade.

$100A_{st}/bd$	τ_c N/mm²
0.25	0.36
0.50	0.49
0.75	0.57
1.00	0.64

[20 marks : 2019]

Solution:

Short span = 4.2 m

So total depth = $\frac{4200}{20} = 210 \text{ mm}$

Effective depth = 210 − 30 = 180 mm

$L_{ex} = 4.2 + 0.18 = 4.33 \text{ m}$

$L_{ey} = 5.8 + 0.18 = 5.98 \text{ m}$

$$\gamma = \frac{L_y}{L_{ex}} = \frac{5.98}{4.38} = 1.37 \not> 2$$

$$\alpha_x = 0.093 + \frac{0.099 - 0.093}{1.4 - 1.3}(1.37 - 1.3) = 0.0972$$

$$\alpha_y = -\frac{0.055 - 0.051}{1.4 - 1.3}(1.4 - 1.37) = 0.055 = 0.0539$$

Load calculation:

Self weight of slab = $0.18 \times 25 = 4.5$ kN/m^2

Live load = 4 kN/m^2

Total factored load, w_u = 1.5 [4.5+ 4] = 12.75 kN/m^2

$$M_x = \alpha_x w_u L_x^2$$

$$= 0.0972 \times 12.75 \times 4.38^2 = 23.78 \text{ kNm}$$

$$M_y = \alpha_y w_u L_x^2$$

$$= 0.0539 \times 12.75 \times 4.38^2 = 13.18 \text{ kNm}$$

Check for depth:

$$d = \sqrt{\frac{23.78 \times 10^6}{0.138 \times 25 \times 1000}} = 83.02 \not> 180 \text{ mm} \qquad \text{(OK)}$$

Steel calculation:

Along shorter span,

$$A_{st_x} = \frac{0.5 f_{ck}}{f_y}\left[1 - \sqrt{1 - \frac{4.6\, M_{ux}}{f_{ck}\, Bd^2}}\right] Bd$$

$$= \frac{0.5 \times 5}{415}\left[1 - \sqrt{1 - \frac{4.6 \times 23.78 \times 10^6}{25 \times 1000 \times 180^2}}\right] \times 1000 \times 180$$

$$= 379.36 \text{ mm}^2$$

$$\text{Spacing of 8 mm bar} = \frac{1000 \times \frac{\pi}{4} \times 8^2}{379.36} = 132.5 \text{ mm c/c}$$

(So provide 130 mm c/c)

For longer span:

$$A_{st_y} = \frac{0.5 \times 25}{415}\left[1 - \sqrt{1 - \frac{4.6 \times 13.18 \times 10^6}{25 \times 1000 \times 180^2}}\right] 1000 \times 180$$

$$= 206.85 \text{ mm}^2$$

$$A_{st\,min} = \frac{0.12 \times 210 \times 100}{100} = 252 \text{ mm}^2$$

$$\text{Spacing of 8 mm bar} = \frac{1000 \times \frac{\pi}{4} \times 8^2}{252} = 199.36 \text{ mm}$$

(So provide 190 mm c/c)

Check of shear:

$$\text{Maximum shear force} = \frac{wL_x}{2} = \frac{12.75 \times 4.2}{2} = 26.78 \text{ kN}$$

$$\text{Nominal shear stress} = \frac{26.78 \times 10^3}{1000 \times 180} = 0.15 \text{ MPa}$$

$$P_t\% = \frac{100 \times \frac{\pi}{4} \times 8^2 \times \frac{1000}{130}}{1000 \times 180} = 0.22\%$$

$$\tau_c = 0.3 \times 1.3 = 0.39 \text{ MPa} > 0.15 \text{ MPa}$$

(So safe in shear)

2.23 The opening of a masonry building is 3 m and 3.5 m high. The ceiling of the roof is 4.5 m above the floor. The space between top of lintel and bottom of roof is filled with brick masonry. The roof transmits a total load of 25 kN/m run to the lintel. Design the lintel supported on brick walls of width 300 mm. Use M20 grade concrete and steel grade of Fe415. Assume the unit weight of the brick masonry is 20 kN/m³ and that of concrete is 25 kN/m³. The design shear strength of concrete is given in Table.

$\frac{100 A_s}{bd}$	τ_c N/mm² M 20
≤0.15	0.28
0.25	0.36
0.50	0.48
0.75	0.56
1.0	0.62
1.25	0.67

The design bond stress for MS bars is given by τ_{bd} = 1.2 N/mm² for M20 grade of concrete

[20 marks : 2019]

Solution:

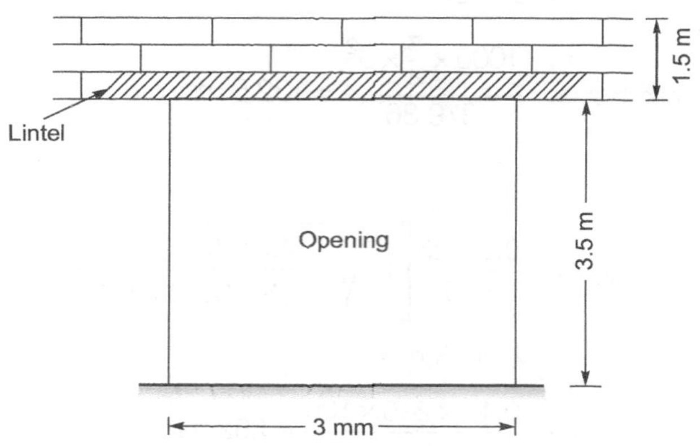

$$\text{Dead weight of brick} = 1 \times 0.3 \times 20 = 6 \text{ kN/m}$$

$$\text{Effective depth of lintel} = \frac{3000}{10} = 300 \text{ mm}$$

$$\text{Total depth} = 300 + 30 = 330 \text{ mm}$$

$$\text{Dead weight of lintel} = 0.33 \times 0.3 \times 25 = 2.475 \text{ kN/m}$$

$$\text{Total load} = 25 + 6 + 2.475 = 33.475 \text{ kN/m}$$

$$w_u = 33.475 \times 1.5 = 50.21 \text{ kN/m}$$

Effective length = 3 + 0.3 = 3.3 m

$$M_u = \frac{w_u L^2}{8} = \frac{50.21 \times 3.3^2}{8} = 68.35 \text{ kN/m}$$

Check depth:

$$d = \sqrt{\frac{68.35 \times 10^6}{0.138 \times 20 \times 300}} = 284.32 < 300 \text{ mm} \quad \text{(OK)}$$

$$A_{st} = \frac{0.5 \times 20}{415}\left[1 - \sqrt{1 - \frac{4.6 \times 68.35 \times 10^6}{20 \times 300 \times 300^2}}\right] \times 300 \times 300$$

$$= 766.97 \text{ mm}^2$$

Provide 4 bars of 16 mm ϕ.

Check for shear

$$SF = \frac{wL}{2} = \frac{50.21 \times 3}{2} = 76.82 \text{ kN}$$

$$\tau_v = \frac{76.82 \times 10^3}{300 \times 300} = 0.854 \text{ N/mm}^2$$

$$P_t = \frac{100 \times \frac{\pi}{4} \times 16^2 \times 4}{300 \times 300} = 0.9\%$$

$$\tau_c = 0.56 + \frac{0.62 - 0.56}{1 - 0.75}[0.92 - 0.75] = 0.60 \text{ MPa}$$

$$\tau_v - \tau_c = 0.854 - 0.6 = 0.254 < 0.4 \text{ MPa}$$

So minimum shear reinforcement must be provided.

Spacing of 8 mm 2-legged bar

$$S_v = \frac{0.87 f_y A_{sv} d}{0.4 bd} = 300.875 \text{ mm}$$

$$= 0.75 \times 300 = 225 \text{ mm}$$

So provide 2L–8 mm ϕ @ 220 mm c/c

Check for bond:

$$\frac{1.3 M_u}{V_1} \geq L_d$$

$$\frac{1.3 \times 68.35 \times 1000}{76.82} \geq \frac{0.87 \times 415 \times 16}{4 \times 1.6 \times 1.2}$$

$$11.56.66 \text{ mm} \geq 752.19 \text{ mm} \quad \text{(OK)}$$

Safe in bond.

2.24 A floor of an old building consists of 150 mm thick RC slab monolithic with the beam of width 300 mm and total depth 400 mm. The beams are spaced 3.5 m c/c and their effective span (simply supported) is 7 m. The beams are reinforced with 5 Nos. 28 ϕ bars as tension reinforcement. Determine the moment carrying capacity of the beams. Use M25 and Fe500. Adopt limit state method of design. Nominal cover = 30 mm. Diameter of the stirrups = 8 mm.

[12 marks : 2020]

Solution:

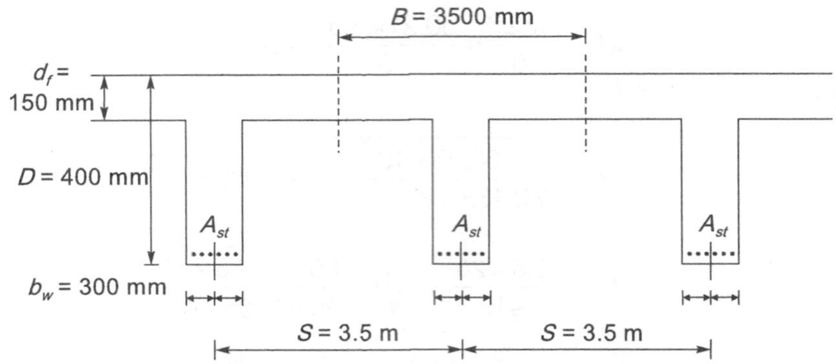

Given: Grade of concrete = M25

Grade of steel = Fe500

$\phi_s = 8$ mm

So, effective depth, $d = 400 - 30 - 8 - \frac{28}{2} = 348$ mm

Effective flange width, $b_f = b_w + 6D_f + \frac{l_o}{6} \not> \frac{l_1 + l_2}{2}$

$$= 300 + 6 \times 150 + \frac{7000}{6} \not> \frac{3500 + 3500}{2}$$

$= 2366.67 \not> 3500$ mm

So, $b_f = 2366.67$ mm

Assuming neutral axis depth $x < D_f$

From $C = T$

$$0.36\, f_{ck}\, b_f x = 0.87\, f_y A_{st}$$

$$\Rightarrow \quad 0.36 \times 25 \times 2366.67 \times x = 0.87 \times 500 \times 5 \times \frac{\pi}{4} \times 28^2$$

$x = 62.88$ mm < 150 mm (OK)

Thus moment capacity, $\text{MOR} = 0.36\, f_{ck}\, b_f x\,(d - 0.42x)$

$= 0.36 \times 25 \times 2366.67 \times 62.88 \times (348 - 0.42 \times 62.88)$

$M_u = 430.72$ kNm

$$\text{Working MOR} = \frac{M_u}{1.5} = \frac{430.72}{1.5} = 287.15 \text{ kNm}$$

2.25 **A simply supported reinforced concrete beam of size 300 mm × 500 mm is reinforced with 5 Nos. 16 φ bars as tension reinforcement. Two bars are curtailed at quarter span from both ends. Find out the load carrying capacity (UDL) of the beam having effective span of 6 m. Also design the beam against shear force. Use M25 and Fe415. Nominal cover = 30 mm. Use limit state method of design. Show the reinforcement detail (cross-section) also. Use 2 Nos. 12 φ bars as hanger bars.**

$\frac{M_u}{bd^2}$	2	2.5	2.75	3	3.25	3.5
P_t	0.51	0.61	0.74	0.83	0.91	–

P_t	0.25	0.5	0.75	1.0	1.25
τ_c MPa	0.36	0.49	0.57	0.64	0.7

[20 marks : 2020]

Solution:

A.

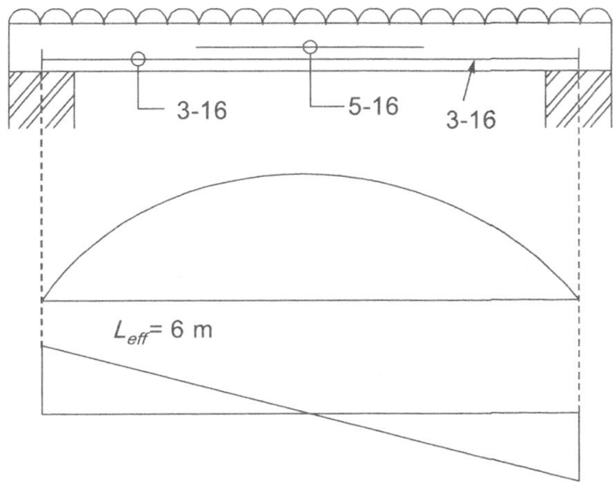

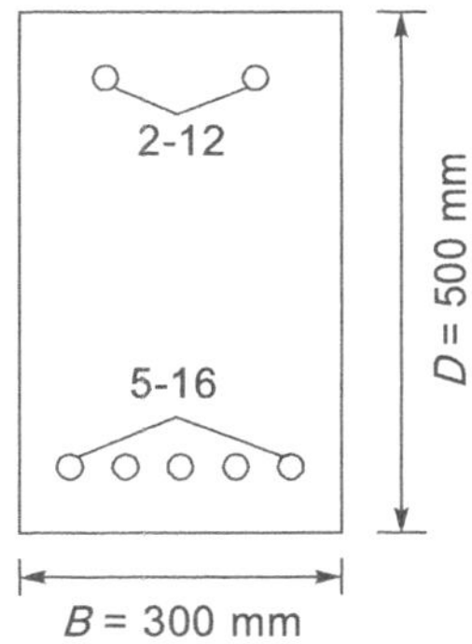

$\begin{bmatrix} \frac{M25}{Fe415} \end{bmatrix}$ $$A_{st} = 5 \times \frac{\pi}{4}(16)^2 = 1005.31 \text{ mm}^2$$

1. **Load carrying capacity of beam.**

(neglecting hanger bars)

(a) Assuming 8 mm ϕ stirrups

$$\text{Effective cover} = \text{Nominal cover} + \text{Dia. of stress} + \frac{1}{2} \text{ dia. of main}$$

$$= 30 + 8 + \frac{16}{2} = 46 \text{ mm}$$

$$d = 500 - 46 = 454 \text{ mm}$$

(b) $$x_{u,\,lim} = 0.48 \times d = 0.48 \times 454$$

$$= 217.92 \text{ mm}$$

(c) $$x_u = \frac{0.87 f_y A_{st}}{0.36 f_{ck} B} = \frac{0.87 \times 415 \times 5 \times \frac{\pi}{4}(16)^2}{0.36 \times 25 \times 300}$$

$$= 134.43 \text{ mm}$$

$x_u < x_{u,\,lim}$ (Under reinforcement section)

(d) Moment of resistance

$$M_u = 0.36 f_{ck} B x_u (d - 0.42 x_u)$$

$$= 0.36 \times 25 \times 300 \times 134.43 \times \frac{(454 - 0.42 \times 134.43)}{10^6}$$

$$= 144.30 \text{ kN-m}$$

(e) Working MR, $$MR = \frac{M_u}{1.50} = \frac{144.30}{1.50} = 96.20 \text{ kNm}$$

(g) From table given in equation

$$P_t\% = \frac{A_{st}}{Bd} \times 100 = \frac{1005.31}{300 \times 454} \times 100 = 0.738$$

$$\frac{M_u}{Bd^2} = 250 + \frac{2.75 - 2.50}{(0.74 - 0.61)} \times (0.738 - 0.61) = 2.746$$

$$M_u = \frac{2.746 \times 300 \times 454^2}{10^6} = 169.80 \text{ kNm}$$

$$MR_w = \frac{169.80}{1.50} = 113.20 \text{ kNm}$$

(h) Considering minimum of above two (as per limit state method)

MR = 96.20 kNm

(i) So total load on beam

$$\text{MR} = \text{BM} = \frac{wL^2}{8} = 96.20$$

$$w = \frac{96.20 \times 8}{L_{eff}^2} = \frac{96.20 \times 8}{6^2} = 21.38 \text{ kN/m}$$

Deducting self weight = 0.30 × 0.50 × 1 × 25 = 3.75 kN/m

Net self imposed load capacity of the beam

= 21.38 – 3.75 = 17.63 kN/m

B. Design of shear reinforcement

Total load on beam = 21.38 kN/m

(width of support is not given)

So considering, $L_d = L_{eff} = 6.0$ m

(i) Maximum shear force

$$V_u = 1.50 \times \frac{21.38 \times 6}{2} = 96.21 \text{ kN}$$

(ii)
$$\tau_v = \frac{V_u}{Bd} = \frac{96.21 \times 1000}{300 \times 454} = 0.706 \text{ N/mm}^2$$

(iii)
$$P_t\% = \frac{A_{st}}{Bd} = \frac{3 \times \frac{\pi}{4} \times (16)^2}{300 \times 454} \times 100 = 0.44\%$$

(2 bars are curtailed)

(iv)
$$\tau_c = 0.36 + \frac{(0.49 - 0.36)}{(0.50 - 0.25)} \times (0.44 - 0.25) = 0.46 \text{ N/mm}^2$$

(v)
$$V_c = \tau_c Bd$$

$$= 0.46 \times 300 \times \frac{454}{1000} = 62.65 \text{ kN}$$

(vi)
$$V_s = V_u - V_c = 96.21 - 62.65 = 33.56 \text{ kN}$$

(vii) Spacing of 2 Leg – 8 mm ϕ

$$S_v = \frac{A_{sv} \times 0.87 f_y \times d}{V_s} = \frac{2 \times \frac{\pi}{4}(8)^2 \times 0.87 \times 415 \times 454}{33.56 \times 1000} = 491 \text{ mm}$$

(viii) Minimum shear reinforcement

$$\frac{A_{SV}}{BS_V} \geq \frac{0.40}{0.87 f_y}$$

$$S_v \leq \frac{0.87 \times f_y \times A_{sv}}{0.40 \times B} \leq \frac{0.87 \times 415 \times 2 \times \frac{\pi}{4}(8)^2}{0.40 \times 300}$$

$$\leq 302 \text{ mm}$$

(ix) Maximum spacing

(i) $0.75\,d = 0.75 \times 454 = 340$

(ii) 300 mm

So provide 8ϕ - 2 leg - @ 300 mm c/c

Throughout the span of the beam

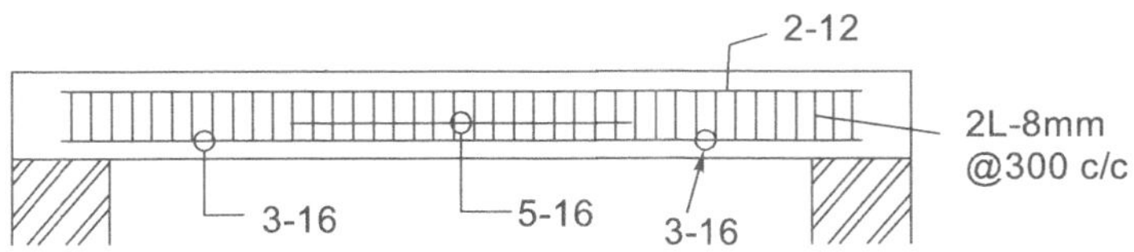

2.26 A simply supported rectangular beam (cross-section 300 mm × 400 mm) with effective span of 6 meters is carrying the following characteristic load:

(i) Characteristic Dead load (including self-weight) = 15 kN/m

(ii) Characteristic Imposed load (not fixed) = 10 kN/m

(iii) Characteristic equivalent wind load (acting downward) = 5 kN/m

Calculate the Design Bending Moment and Design shear force for most critical load combination for limit state of collapse and limit state of serviceability.

Partial Safety Factor (γ_f) for load

Load Combination	Limit State of Collapse			Limit State of Serviceability		
	Dead load	Imposed load	Wind load	Dead load	Imposed load	Wind load
Dead load + Imposed Load	1.5	1.5	—	1.0	1.0	—
Dead Load + Wind load	0.9	—	1.5	1.0	—	1.0
Dead Load + Imposed Load + Wind load	1.2	1.2	1.2	1.0	0.8	0.8

[12 Marks : 2021]

Solution:

Most critical load combination for limit state of collapse:

	Load combination	Factored load (in kN/m)
Case-1	DL + IL =	1.5 DL + 1.5 IL
		= 1.5 × 15 + 1.5 × 10 = 37.5
Case-2	DL + WL =	0.9 × DL + 1.5 WL
		= 0.9 × 15 + 1.5 × 5 = 21
Case-3	DL + IL + WL =	1.2 DL + 1.2 IL + 1.2 WL
		= 1.2 × (15 + 10 + 5) = 36

Most critical is case -1

(a) Maximum bending moment for limit state of collapse

$$\text{For simply supported beam} = \frac{w_u \times L_e^2}{8} = \frac{37.5 \times 6^2}{8}$$

$$= 168.75 \text{ kN-m}$$

(b) Maximum SF. $V_u = \dfrac{w_u \times L_{cl}}{2}$

(Note: Maximum SF shall be calculated from clear span = $L_{cl} = (L_{eff} - d)$)

$$D = 400 \text{ mm}$$

Assuming, Effective cover = 50 mm

$\therefore$ $d = 350$ mm

$$L_{clear} = 6.0 - 0.35 = 5.65 \text{ m}$$

Maximum SF at face of support,

$$V_u = \frac{w_u \times L_{cl}}{2} = \frac{37.5 \times 5.65}{2} = 105.94 \text{ kN}$$

Load combinations for limit state of serviceability:

	Load combination	Service load (kN/m)
Case-1	DL + IL	= 1.0 DL + 1.0 IL
		= 1.0 × (15 + 10) = 25
Case-2	DL + WL	= 1.0 × DL + 1.0 × WL
		= 1.0 × (15 + 5) = 20
Case-3	DL + IL + WL	= 1.0 × DL + 0.8 IL + 0.8 WL
		= 1.0 × 15 + 0.8 × (10 + 5) = 27

$\therefore$ Most critical case is load combination DL + IL + WL

For simply supported beam:

(a) Maximum bending moment $= \dfrac{wL_e^2}{8} = \dfrac{27 \times 6^2}{8} = 121.5$ kNm

(b) Maximum SF. $= \dfrac{wL_{cl}}{2} = \dfrac{27 \times 5.65}{2} = 76.275$ kN

2.27 **A solid RCC slab is having effective span of L_x = 3.0 m and L_y = 4.5 m. Position of slab S - 1 is shown in the figure below.**

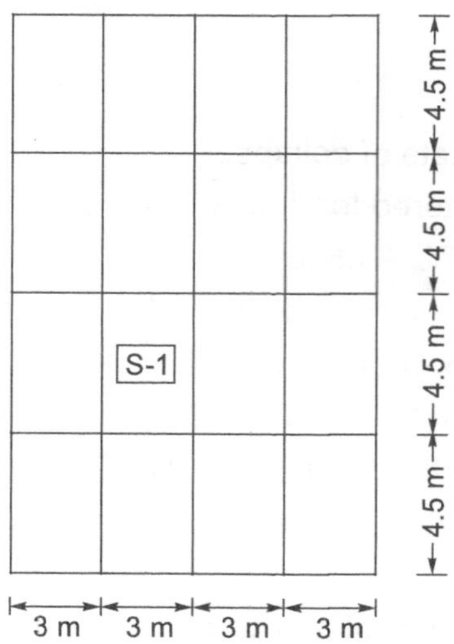

Centre line Top plan of slab panels

Slab carrying total design load of 20 kN/m². In slab S-1, calculate and provide the spacing of all main reinforcement only, by limit state method of design, for 8 mm diameter high strength deformed bars. Check these spacings must not exceed the standard guidelines of IS 456 : 2000.

Effective thickness of slab = 120 mm

Grade of concrete M 20

Grade of reinforcement Fe 415

$\frac{M}{bd^2}$	0.30	0.40	0.512	0.60	0.65	0.662
p_t	0.085	0.114	0.143	0.172	0.187	0.203

p_t is the percentage of reinforcement.

Table: Bending moment coefficients for rectangular pannels supoorted on four sides with provision of torsion at corners.

Case No.	Type of Panel and Moments Considered	Short Span Coefficient α_x (Values of l_y/l_x)								Long Span Coefficients α_y for All Values of l_y/l_x
		1.0	1.1	1.2	1.3	1.4	1.5	1.75	2.0	
(1)	(2)	(3)	(4)	(5)	(6)	(7)	(8)	(9)	(10)	(11)
1.	*Interior Panels:*									
	Negative moment at continuous edge	0.032	0.037	0.043	0.047	0.051	0.053	0.060	0.065	0.032
	Positive moment at mid-span	0.024	0.028	0.032	0.036	0.039	0.041	0.045	0.049	0.024
9.	*Four Edges: Discontinuous:*									
	Positive moment at mid-span	0.056	0.064	0.072	0.079	0.085	0.089	0.100	0.107	0.056

[20 Marks : 2021]

Solution:

Given: Materials used

M20, Fe415

$$\Rightarrow \quad f_{ck} = 20 \text{ N/mm}^2;$$

$$f_y = 415 \text{ N/mm}^2;$$

$$L_{xeff} = 3.0 \text{ m}$$

$$L_{yeff} = 4.5 \text{ m}$$

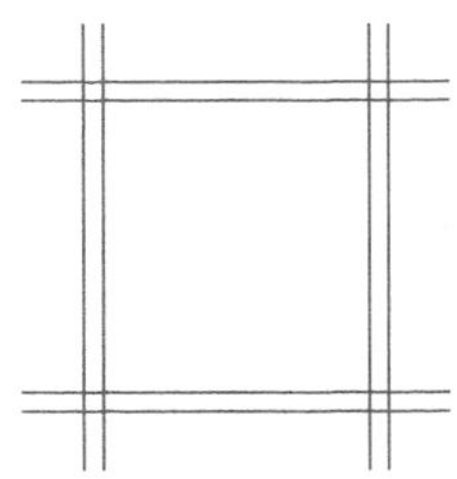

Step 1

$$\frac{L_{yeff}}{L_{xeff}} = \frac{4.5}{3.0} = 1.50 < 2.0$$

It is a two way slab.

It is a Interior panel.

Step 2

$$\text{Total Design load} = 20 \text{ kN/m}^2$$

$$\text{factored load} = W_u = 1.5 \times 20$$

$$= 30 \text{ kN/m}^2$$

Step 3 Effective depth of slab (given) = 120 mm.

Step 4

Bending Moment Calculations

	α value	BM_u (kN-m)	$\frac{M_u}{B.d^2}$	P_t%
$M_{x(-)}$	0.053	0.053 × 270 = 14.31	0.99	Not given
$M_{x(+)}$	0.041	0.041 × 270 = 11.07	0.77	Not given
$M_{y(-)}$	0.032	0.032 × 270 = 8.64	0.60	0.172
$M_{y(+)}$	0.024	0.024 × 270 = 6.48	0.45	Can be calculated from given values

$$\text{Moments} = \alpha \cdot W_u \cdot L_{xe}^2$$

$$= \alpha \times 30 \times 3.0^2 = 270\,\alpha$$

$$\frac{BM_u}{B \cdot d^2} = \frac{BM_u \times 10^6}{1000 \times 120^2}$$

Values for 0.99 and 0.77 not given in table

Use

$$\text{Ast} = \frac{0.5 f_{ck}}{f_y}\left[1 - \sqrt{1 - \frac{4.6 M_u}{f_{ck} B d^2}}\right] \times Bd$$

$$\text{Ast}_{x(-)} = \frac{0.5 \times 20}{415} \times \left[1 - \sqrt{1 - \frac{4.6 \times 14.31 \times 10^6}{20 \times 1000 \times 120^2}}\right] \times 1000 \times 120 = 352 \text{ mm}^2$$

$$\text{Ast}_{x(+)} = \frac{0.5 \times 20}{415} \times \left[1 - \sqrt{1 - \frac{4.6 \times 11.07 \times 10^6}{20 \times 1000 \times 120^2}}\right] \times 1000 \times 120 = 268 \text{ mm}^2$$

by same formula.

$$Ast_{y(-)} = \text{for } BM_u = 8.64 \text{ kN-m}$$

$$= 206.92 \text{ say } 207 \text{ mm}^2$$

Also, from table given for $\frac{M_u}{bd^2} = 0.60$,

$$P_t\,\% = 0.172\%$$

$$\therefore \quad A_{st,\,y(-)} = \frac{0.172}{100} \times 1000 \times 120 \simeq 206.4 \text{ mm}^2$$

$$= 206.4 \text{ say } 207 \text{ mm}^2$$

$A_{st,y(+)}$ for $BM_u = 6.48$ kN-m

For $\frac{M_u}{bd^2} = 0.45$

$$P_t(\%) = 0.114 + \frac{0.143 - 0.114}{0.512 - 0.4} \times (0.45 - 0.4) = 0.127\%$$

$$\therefore \quad A_{st,y(+)} = \frac{0.127}{100} \times 1000 \times 120 = 152.4 \text{ mm}^2$$

Step 5 Spacing of reinforcement using 8 mm ϕ.

$$A_{st\,x(-)} = \frac{1000}{A_{st}} \times \frac{\pi}{4}(8)^2$$

$$= \frac{1000}{352} \times \frac{\pi}{4}(8)^2 = 142 \text{ mm} = \text{say } 140 \text{ mm}$$

$$A_{stx(+)} = \frac{1000}{268} \times \frac{\pi}{4}(8)^2 = 187 \text{ mm} = \text{say } 180 \text{ mm}$$

$$A_{st\,y(-)} = \frac{1000}{207} \times \frac{\pi}{4}(8)^2 = 242.83 \text{ mm} = \text{say } 240 \text{ mm}$$

$$A_{st\,y(+)} = \frac{1000}{152.4} \times \frac{\pi}{4}(8)^2 = 329.83 \text{ mm} = \text{say } 270 \text{ mm}$$

Step 6 Minimum Reinforcement

Considering 30 mm eff cover

$$D = 120 + 30 = 150 \text{ mm}$$

$$A_{st\,min} = \frac{0.12}{100} \times 1000 \times 150 = 180 \text{ mm}^2$$

Spacing of 8 mm ϕ shall not exceed

(a) Minimum steel $= \frac{1000}{180} \times \frac{\pi}{4}(8)^2 = 279$ mm

(b) 3d = 3 × 120 = 360 mm

(c) 300 mm

only in $Ast_{y(+)}$ – steel – spacing is more than 279 mm

Provide $Asy_{y(+)}$ = 8 mm ϕ @ 270 mm c/c

Step 7 Provided steel:

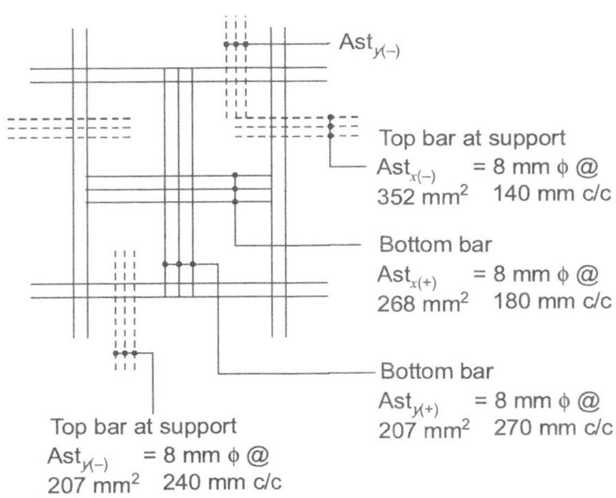

2.28 Design bending moment and shear force diagram have been given below for a two span continuous beam. Effective span of beam is 6.0 m each. Design a rectangular, singly reinforced RCC beam section at support 'B' only by LSM of design.

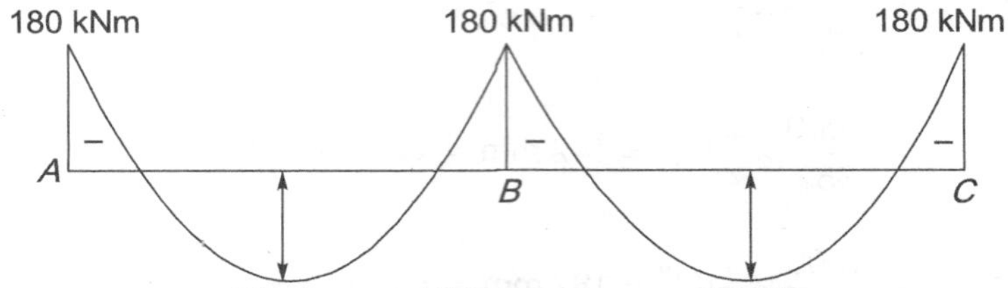

BM Diagram

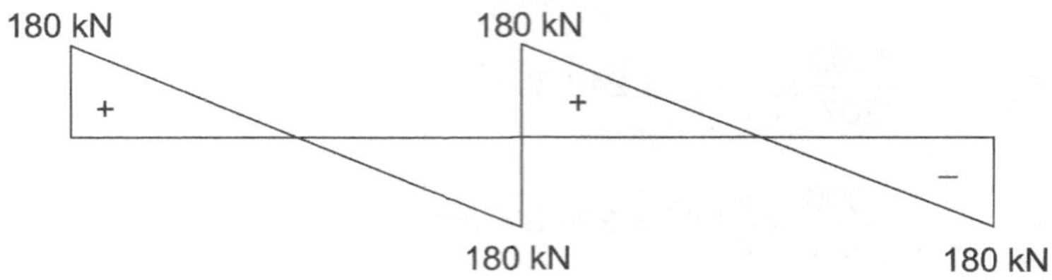

SF Diagram

The following parameters may be used for design. All notations are as per IS 456 : 2000.

(1) Grade of concrete M 20

(2) Grade of reinforcement Fe 415

(3) Nominal cover to reinforcement – 25 mm

(4) (Effective depth/width) ratio – 2

(5) Diameter of flexural bar – 20 mm

(6) Diameter of shear reinforcement – 8 mm (CS area 50 mm^2)

(7) Design shear strength of concrete $\tau_c = 0.88$ N/mm^2

(8) Maximum shear stress $\tau_{c,\max} = 2.8$ N/mm^2

(9) Round off effective (calculated) to next higher multiple of 100

(10) $M_{u,lim}/f_{ck}\, bd^2 = 0.138$

(11) $\left(\frac{x}{d}\right) = 1.2 - \left[(1.2)^2 - \frac{6.68\, M_u}{f_{ck}\, bd^2}\right]^{1/2}$

(12) Lever arm, $z = d\left(1 - 0.42\frac{x}{d}\right)$

(13) $A_{st} = \frac{M_u}{0.87\, f_y \cdot z}$

(14) $V_{us} = \frac{0.87\, f_y \cdot A_{sv} \cdot d}{S_v}$

(15) Minimum shear reinforcement $= \frac{A_{sv}}{b \cdot S_v} = \frac{0.40}{087\, f_y}$

Check all spacing of reinforcement as per standard guidelines.

[20 Marks : 2021]

Solution:

Beam is to be designed at support 'B' only by LSM where

$$\left.\begin{aligned} BM &= 180 \text{ kN-m} \\ SF &= 180 \text{ kN} \end{aligned}\right\}$$

Factored values :

$$BM_u = 1.5 \times 180 = 270 \text{ kN-w}$$

$$SF = V_u = (1.5 \times 180) = 270 \text{ kN}$$

(a) Design for Bending Moment

Using the value given in question,

$$\frac{d}{B} = 2.0$$

Materials used M20, Fe415,

$$f_{ck} = 20 \text{ N/mm}^2$$

$$f_y = 415 \text{ N/mm}^2$$

1. Equating for a limiting section.

$$BM_U = 0.138 \times f_{ck} \cdot B \cdot d^2$$

$$270 \times 10^6 = 0.138 \times 20 \times B \times (2B)^2$$

$$B = 290.30 \text{ mm}$$

$$d = 2 \times 290.30 \text{ mm} = 580.5 \text{ mm}$$

Let us provide,

$$b = 300 \text{ mm}$$

$$d = 600 \text{ mm}$$

Keeping effective cover

$$= \text{Clear cover} + \phi_{st} + \frac{1}{2}\phi_m = 25 + 8 + \frac{1}{2} \times 20 = 43 \text{ mm}$$

(say 50 mm)

$$\text{Overall depth} = d + \text{effective cover}$$

$$D = 600 + 50 = 650 \text{ mm}$$

2. Area of steel

$$\frac{x}{d} = 1.2 - \left[(1.2)^2 - \frac{6.68 \times M_u}{f_{ck} \cdot bd^2}\right]^{1/2}$$

$$= 1.20 - \sqrt{(1.2)^2 - \frac{6.68 \times 270 \times 10^6}{20 \times 300 \times 600^2}} = 0.42218$$

$$x = 0.42218 \times 600 = 253.31 \text{ mm}$$

$$\text{LA (Lever arm)} = Z = d\left(1 - 0.42 \times \frac{x}{d}\right)$$

$$= 600 \times (1 - 0.42 \times 0.42218) = 493.60 \text{ mm}$$

$$A_{st} = \frac{BM_u}{0.87 \times f_y \times z} = \frac{270 \times 10^6}{0.87 \times 415 \times 493.60} = 1515 \text{ mm}^2$$

Number of 20 mmϕ bars

$$= \frac{1515}{\frac{\pi}{4}(20)^2} = 4.82 \quad \text{Provide 5 - 20 mm } \phi$$

Provide 5 - 20 mm ϕ bars at top face of beam for negative BM at 'B'.

3. **Check for Shear**

Maximum shear force = $V_u = 270$ kN

Nominal shear stress

$$\tau_v = \frac{V_u}{Bd} = \frac{270 \times 10^3}{300 \times 600} = 1.50 = 0.88 \text{ N/mm}^2$$

Since, $\tau_v > \tau_c$

Shear reinforcement needs to be provided.

Shear force for design of shear reinforcement

$$V_{us} = (\tau_v - \tau_c) \text{ B.d}$$

$$= (1.50 - 0.88) \times \frac{300 \times 600}{1000}$$

$$= 111.60 \text{ kN}$$

Using 2-legged 8 mm ϕ stirrups

Spacing: $$S_V = \frac{A_{sv} \times 0.87 f_y \times d}{V_{us}}$$

$$= \frac{2 \times 50 \times 0.87 \times 415 \times 600}{111.60 \times 1000}$$

$$= 194.11 \text{ mm}$$

Spacing as per minimum shear steel

$$S_v = \frac{0.87 f_y \times A_{sv}}{0.4 \times B} = \frac{0.87 \times 415 \times 2 \times 50}{0.4 \times 300} = 300.8 \text{ mm}$$

So provide 2 legged - 8 mm ϕ @ 190 mm c/c near support

Provide 2-10 mm bars at bottom also as hanger bars.

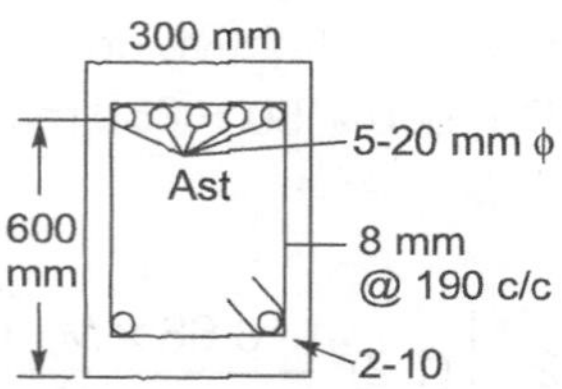

4. **Check for spacing of reinforcement**

(a) Maximum spacing of shear reinforcement

(i) $0.75d = 0.75 \times 600$

$= 450$ mm

(ii) 300 mm

Provided spacing of 190 mm is ok

(b) Spacing of Main bars

$B = 300$ mm

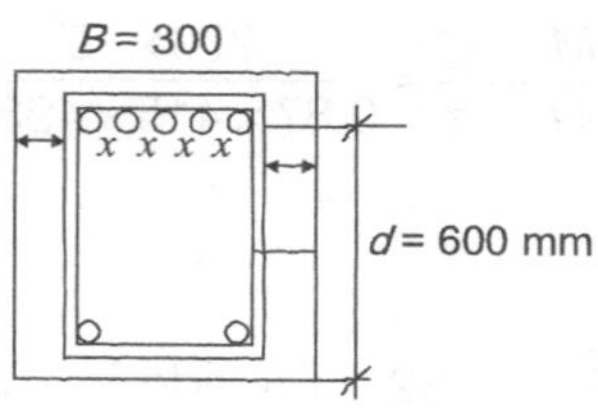

$$x = \frac{(300 - 2 \times 25 - 2 \times 8 - 5 \times 20)}{4} = 33.5 \text{ mm}$$

Maximum spacing of horizontal reinforcement

(i) Maximum dia of main bar = 20 mm

(ii) 5 mm + maximum size of aggregates

= 5 + 20 = 25 mm

So, provided spacing of 33.5 mm is ok.

2.29 The floor of a hall is 9 m × 12 m to the centre of supports. The beams are spaced at 3 m centre to centre. Thickness of the floor slab is 150 mm. Live load on the floor is 3 kN/m^2. Load due to floor finish and plaster being 1.5 kN/m^2, design one intermediate T-beam for flexure and shear as per Limit State Design. Sketch the reinforcement details. Density of concrete is 25 kN/m^3. Assume an effective cover to the main reinforcement is 50 mm. Width of rib = 300 mm.

Effective width of Flange $b_f = \frac{l_0}{6} + b_w + 6D_f$

For $Fe500, \left(\frac{x_{u,max}}{d}\right) = 0.46$

Strength of shear reinforcement $V_{us} = \frac{0.87 f_y A_{sv} d}{S_v}$

$\frac{100A_s}{bd}$	Design shear strength of concrete τ_c N/mm^2
0.75	0.57
1.00	0.64
1.25	0.70
1.50	0.74

[20 marks : 2022]

Solution:

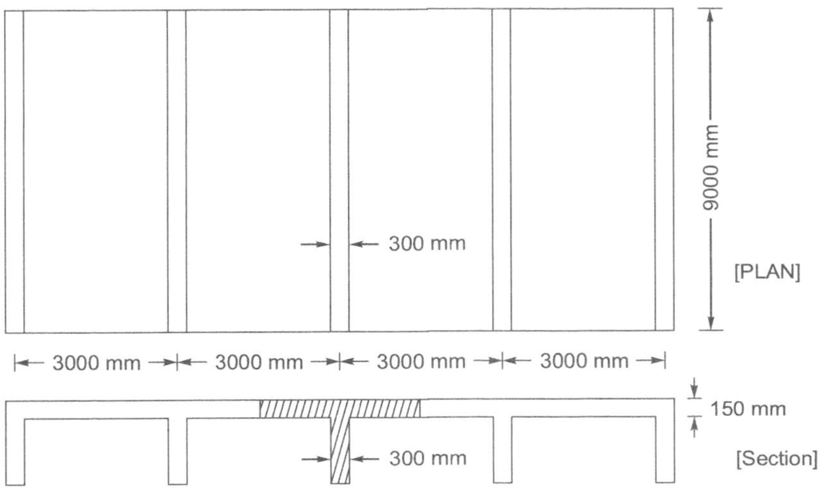

Dead load of slab = 0.15 × 25 = 3.75 kN/m^2

Live load on floor = 3 kN/m^2

Load due to floor finish and plaster = 1.5 kN/m^2

Total load = 8.25 kN/m^2

Load per m run of beam = Load on slab per unit area × c/c distance between beam
= 8.25 × 3 = 24.75 kN/m

Effective width of flange, $b_f = \frac{l_0}{6} + b_w + 6D_f = \frac{9000}{6} + 300 + 6 \times 150 = 2700$ mm

Let us adopt overall depth D of beam equal to 550 mm

Effective cover = 50 mm

∴ Effective depth, $d = 550 - 50 = 500$ mm

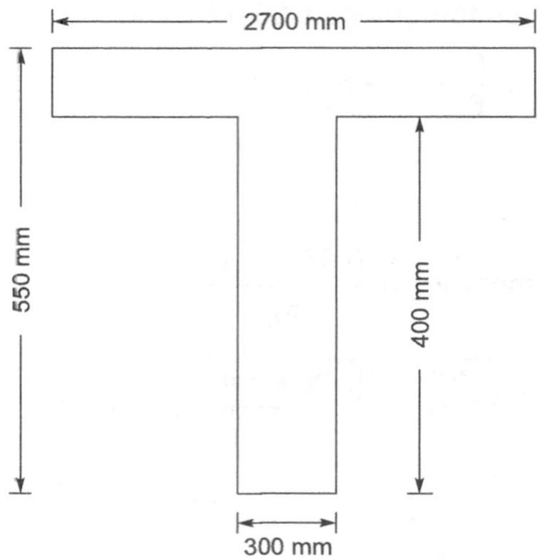

Dead load of web of beam = Width of web × Depth of web × Concrete density
= 0.3 × 0.40 × 25 = 3 kN/m

Total load on beam per meter run = 24.75 + 3 = 27.75 kN/m

$$\text{Factored (BM)}_u = \frac{w_u \times l^2}{8} = \frac{1.5 \times 27.75 \times 9^2}{8} = 421.45 \text{ kNm}$$

Assume grade of concrete to be M25.

Let us assume that N.A. lies in flange, that is

$$x = \frac{0.87 \times f_y \times A_{st}}{0.36 \times f_{ck} \times b_f} = \frac{0.87 \times 500 \times A_{st}}{0.36 \times 25 \times 2700}$$

⇒ $x = 0.0179\, A_{st}$ ...(i)

Now, $(BM)_u = 0.87 \times f_y \times A_{st}\,[d - 0.472\, x_u]$

⇒ $421.45 \times 10^6 = 0.87 \times 500 \times A_{st}\,[500 - 0.42 \times 0.0179\, A_{st}]$

⇒ $421.45 \times 10^6 = 217500\, A_{st} - 3.27\, A^2_{st}$

⇒ $A_{st} = 1997.7 \text{ mm}^2$

Now, from eq. (i), $x_u = 35.75$ mm

∴ N.A. lies in the flange.

Let's provide 2 bars of 28 mmϕ and 2 bars of 25 mmϕ.

$$\text{Minimum area of steel} = \frac{0.85 \times b_w d}{100} = \frac{0.85}{100} \times 300 \times 500 = 1275 \text{ mm}^2 \quad \text{(OK)}$$

Check for shear:

Factored shear force, $V_u = \dfrac{w_u \times L}{2} = \dfrac{1.5 \times 27.75 \times 9}{2} = 187.31 \text{ kN}$

Nominal shear stress, $\tau_v = \dfrac{V_u}{b_w d} = \dfrac{187.31 \times 10^3}{300 \times 500} = 1.248 \text{ N/mm}^2$

Maximum shear stress for M25:

$$\tau_{max} = 0.625\sqrt{f_{ck}} = 0.625\sqrt{25} = 3.125 \text{ N/mm}^2$$

$\therefore$ $\tau_v < \tau_{max}$ (OK)

Percentage tensile reinforcement

$$p_t = \frac{A_{st}}{b_w d} \times 100 = \frac{1997.7}{300 \times 500} \times 100 = 1.3318\%$$

From the table given in the question,

For, $\dfrac{100 A_s}{bd} = 1.3318\%$

$$\tau_c = 0.7 + \left(\frac{0.74 - 0.7}{1.5 - 1.25}\right)(1.3318 - 1.25) = 0.713 \text{ N/mm}^2$$

$\because$ $\tau_v > \tau_c$

Now, shear reinforcement is designed for a shear force of

$$\begin{aligned} V_{us} &= V_u - \tau_c\, bd \\ &= 187.31 - 0.713 \times 300 \times 500 \times 10^{-3} \\ &= 80.36 \text{ kN} \end{aligned}$$

Using 2-legged 8 mm diameter stirrups,

$$A_{sv} = 2 \times \frac{\pi}{4} \times 8^2 = 100.53 \text{ mm}^2$$

Spacing of shear reinforcement,

$$S_v = \frac{0.87 \times f_y \times A_{sv} \times d}{V_{us}}$$

$$S_v = \frac{0.87 \times 500 \times 100.53 \times 500}{80.36 \times 10^3} = 272.09 \text{ mm}$$

Thus, provide 2-legged 8 mm diameter stirrups @ 270 mm c/c.

Check for spacing:

$$S_v \le \begin{cases} 0.75d = 0.75 \times 500 = 375 \text{ mm} \\ 300 \text{ mm} \end{cases}$$ (OK)

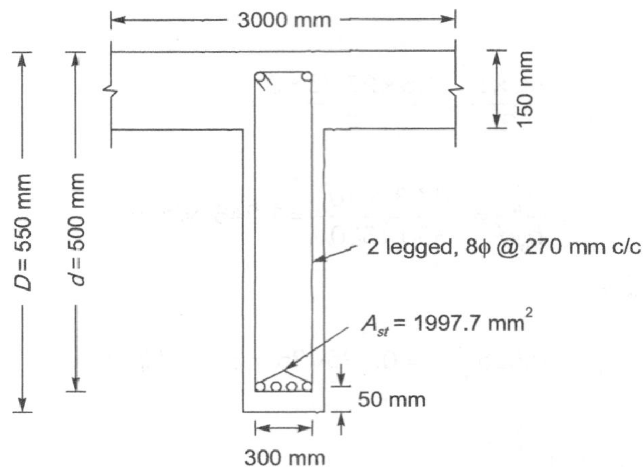

2.30 A simply supported one way slab 178 mm thick having an effective span of 4.88 m is reinforced with 12 mm diameter rebars at 125 mm centre to centre. The nominal concrete cover to the main reinforcement is 20 mm. The slab is subjected to a live load of 4 kN/m^2 and surface finish of 1.2 kN/m^2. Use M25 concrete and Fe500 grade steel. Compute only the short-term deflection and deflection due to shrinkage. Shrinkage strain is 0.0003. Density of concrete is 25 kN/m^3.

$$E_c = 5000\sqrt{f_{ck}}\ ,\ E_s = 2\times 10^5 \text{ MPa}$$

$$P_t = \frac{100A_{st}}{bd}\ ;\ P_c = \frac{100A_{sc}}{bd}$$

[20 marks : 2022]

Solution:

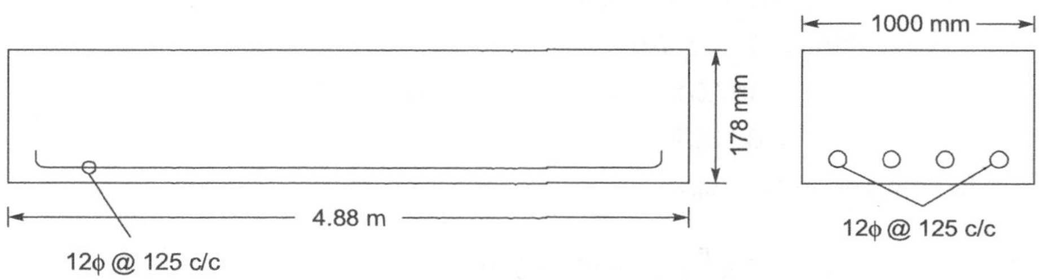

Nominal cover = 20 mm

$$\therefore \quad \text{Effective cover} = 20 + \frac{12}{2} = 26 \text{ mm}$$

Live load = 4 kN/m^2

Surface finish = 1.2 kN/m^2

Dead load of slab = 25 × 0.178 = 4.45 kN/m^2

∴ Total load, w = 9.65 kN/m^2

Short term deflection is given by,

$$\delta = \frac{5}{384}\frac{wl^4}{EI}$$

where $E = E_c = 5000\sqrt{f_{ck}} = 5000\sqrt{25} = 25000 \text{ N/mm}^2$

$$I = I_{eff} = \frac{I_r}{1.2 - \frac{M_r}{M}\times\frac{z}{d}\left(1-\frac{x}{d}\right)\frac{b_w}{b}}$$

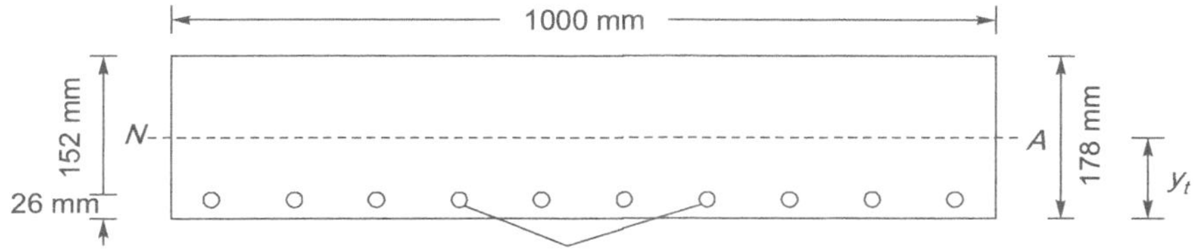

Uncracked section

$$y_t = \frac{178}{2} = 89 \text{ mm}$$

$$I_{gr} = \frac{100 \times 178^3}{12} = 4.7 \times 10^8 \text{ mm}^4$$

For cracked section, $f_{cr} = 0.7\sqrt{f_{ck}} = 0.7\sqrt{25} = 3.5 \text{ N/mm}^2$

$$M_r = \frac{f_{cr} I_{gr}}{y_t} = \frac{3.5 \times 4.7 \times 10^8}{89} \text{ Nm} = 18.48 \text{ kNm}$$

$$E_s = 2 \times 10^5 \text{ N/mm}^2$$

$$m = \frac{E_s}{E_c} = \frac{2 \times 10^5}{25000} = 8$$

Taking moments about NA,

$$\frac{bx^2}{2} = mA_{st}(d - x)$$

$$\Rightarrow \quad 1000\frac{x^2}{2} = 8 \times \left(\frac{1000 \times \frac{\pi}{4} \times 12^2}{125}\right)(152 - x)$$

$$\Rightarrow \quad 500x^2 = 7238.23\,(152 - x)$$

$$\Rightarrow \quad x^2 + 14.476x - 2200.42 = 0$$

$$\therefore \quad x = 40.225 \text{ mm}$$

$\therefore$ Lever arm, $z = d - \frac{x}{3} = 152 - \frac{40.225}{3} = 138.59 \text{ mm}$

$$\therefore \quad I_{cr} = \frac{1000 \times 40.25^3}{3} + 8 \times \left(\frac{1000 \times \frac{\pi}{4} \times 12^2}{125}\right)(152 - 40.225)^2$$

$$= 112167748.7 \text{ mm}^4$$

$$M = \frac{wl^2}{8} = \frac{9.65 \times 4.88^2}{8} = 28.726 \text{ kNm}$$

$$\therefore \quad I_{\text{eff}} = \frac{I_{cr}}{1.2 - \frac{M_r}{M} \times \frac{z}{d}\left(1 - \frac{x}{d}\right)\frac{b_w}{b}}$$

$$= \frac{112167748.7}{1.2 - \frac{18.48}{28.726} \times \left(\frac{138.59}{152}\right)\left(1 - \frac{40.225}{152}\right)(1)} = 145925685.9 \text{ mm}^4$$

$\therefore$ Short term deflection, $$\delta = \frac{5}{384} \times \frac{9.65 \times (4880)^4}{25000 \times 145925685.9} \text{ mm} = 19.533 \text{ mm}$$

Deflection due to shrinkage

$$k_4 = \frac{0.72(p_t - p_c)}{\sqrt{p_t}}$$

$$p_t = \frac{1000 \times \frac{\pi}{4} \times \frac{12^2}{125}}{1000 \times 152} \times 100 = 0.595\%$$

$$p_c = 0$$

$$\therefore \quad k_4 = \frac{0.72(0.595)}{\sqrt{0.595}} = 0.5554$$

$$\psi_{cs} = \frac{k_4 E_{cs}}{D} = \frac{0.5554 \times 0.0003}{178} = 9.3606 \times 10^{-7}$$

$$k_3 = 0.125 \text{ for simply supported members}$$

$$\therefore \quad \alpha_{cs} = \psi_{cs} k_3 l^2 = (9.3606 \times 10^{-7})\, 0.125 \times (4880)^2$$

$$= 2.786 \text{ mm}$$

2.31 A room of effective span 16.50 m × 11.00 m is surrounded by brick walls. In order to lay the RCC slab over it, the room is divided in four equal panels by providing two central beams. The slab is simply supported on all the four walls as shown in the figure below. Using limit state design, determine and provide main reinforcement in a single panel, using 12 mm diameter steel bars of Fe-415 grade. Consider the grade of concrete as M-20. Draw the reinforcement detail of a panel. Use the following additional data:

Total factored load on slab (dead load + live load) = 16 kN/m².

Thickness of slab - 175 mm

Effective depth of slab = 150 mm

Note: Use table given below for finding the moments at different locations.

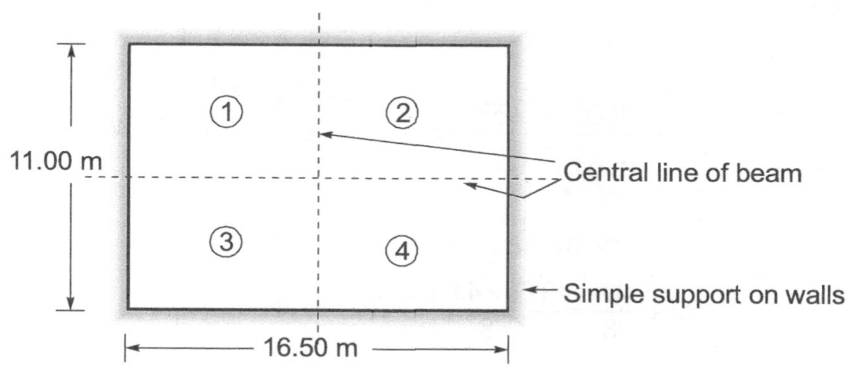

(1)	Type of panel and moments considered (2)	Short span coefficients α_x (Values of l_y/l_x) 1.0 (3)	1.1 (4)	1.2 (5)	1.3 (6)	1.4 (7)	1.5 (8)	1.75 (9)	2.0 (10)	Long span coefficients α_y for all values of l_y/l_x (11)
1.	Interior panels:									
	Negative moment at continuous edge	0.032	0.037	0.043	0.047	0.051	0.053	0.060	0.065	0.032
	Positive moment at mid-span	0.024	0.028	0.032	0.036	0.041	0.041	0.045	0.049	0.024
2.	One short edge continuous:									
	Negative moment at continuous edge	0.037	0.043	0.048	0.051	0.055	0.057	0.064	0.068	0.037
	Positive moment at mid-span	0.028	0.032	0.036	0.039	0.041	0.044	0.048	0.052	0.028
3.	One long edge discontinuous:									
	Negative moment at continuous edge	0.037	0.044	0.052	0.057	0.063	0.067	0.077	0.085	0.037
	Positive moment at mid-span	0.028	0.033	0.039	0.044	0.047	0.051	0.059	0.065	0.028
4.	Two adjacent edges discontinuous:									
	Negative moment at continuous edge	0.047	0.053	0.060	0.065	0.071	0.075	0.084	0.091	0.047
	Positive moment at mid-span	0.035	0.040	0.045	0.049	0.053	0.056	0.063	0.069	0.035
5.	Two short edges discontinuous:									
	Negative moment at continuous edge	0.045	0.049	0.052	0.056	0.059	0.060	0.065	0.069	–
	Positive moment at mid-span	0.035	0.037	0.040	0.043	0.044	0.045	0.049	0.052	0.035
6.	Two long edges discontinuous:									
	Negative moment at continuous edge	–	–	–	–	–	–	–	–	0.045
	Positive moment at mid-span	0.035	0.043	0.051	0.057	0.063	0.068	0.080	0.088	0.035
7.	Three edges discontinuous:									
	(One long edge continuous)									
	Negative moment at continuous edge	0.057	0.064	0.071	0.076	0.080	0.084	0.091	0.097	–
	Positive moment at mid-span	0.043	0.048	0.053	0.057	0.060	0.064	0.069	0.073	0.043
8.	Three edges discontinuous:									
	(One short edge continuous)	–	–	–	–	–	–	–	–	0.057
	Negative moment at continuous edge	0.043	0.051	0.059	0.065	0.071	0.076	0.087	0.096	0.043
	Positive moment at mid-span									
9.	Four edges discontinuous:									
	Positive moment at mid-span	0.056	0.064	0.072	0.079	0.085	0.089	0.100	0.107	0.056

[20 marks : 2023]

Solution:

Considering one slab of the structure

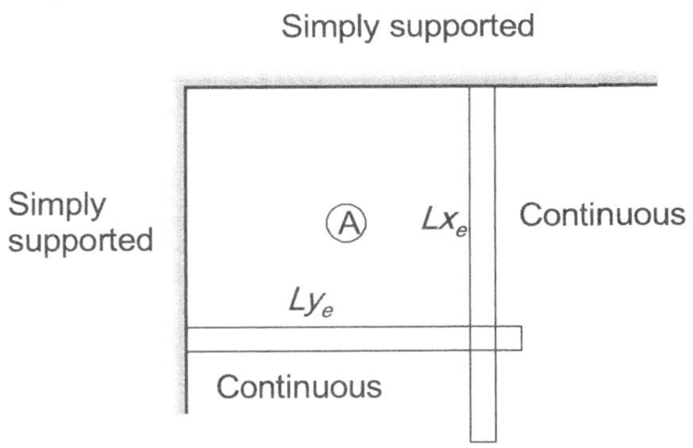

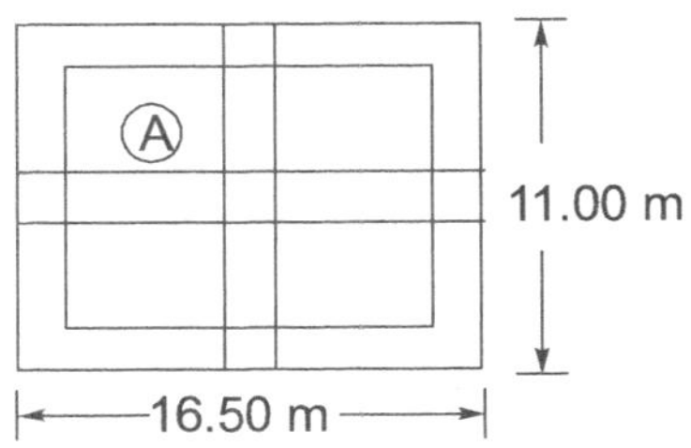

1. Considering effective span

$$L_{xe} = \frac{11.0}{2} = 5.50\text{m}$$

$$L_{ye} = \frac{16.50}{2} = 8.25\text{m}$$

$$r = \frac{L_{ye}}{L_{xe}} = \frac{8.25}{5.50} = 1.50 < 2.0$$

$\Rightarrow$ So, it is a two way slab.

2. Depth of slab, $D = 175$ mm

 Effective depth of slab, $d = 150$ mm

3. Loads

 Total factored load, $w_u = 16$ kN/m^2

4. Moment coefficients for r = 1.50

 The slab is a type of panel number of 4 of table 26 of IS 456 : 2000. So moment coefficient are as following:

$$\alpha_{x(-)} = 0.075 \qquad \alpha_{y(-)} = 0.047$$
$$\alpha_{x(+)} = 0.056 \qquad \alpha_{y(+)} = 0.035$$

5. Moments are given as, $M = \alpha w_u L_e$

 Now,
$$M_{x(-)} = 0.075 \times 16 \times 5.50^2 = 36.3 \text{ kNm}$$
$$M_{x(+)} = 0.056 \times 16 \times 5.50^2 = 27.10 \text{ kNm}$$
$$M_{y(-)} = 0.047 \times 16 \times 5.50^2 = 22.75 \text{ kNm}$$
$$M_{y(+)} = 0.035 \times 16 \times 5.50^2 = 16.94 \text{ kNm}$$

6. Area of steel required.

 For x-direction,

$$Ast_{x(-)} = \frac{0.5\times 20}{415}\times\left[1-\sqrt{1-\frac{4.6\times 36.3\times 10^6}{20\times 1000\times 150^2}}\right]\times 1000\times 150$$
$$= 748 \text{ mm}^2$$

$$Ast_{x(+)} = \frac{0.5\times 20}{415}\times\left[1-\sqrt{1-\frac{4.6\times 27.10\times 10^6}{20\times 1000\times 150^2}}\right]\times 1000\times 150$$
$$= 541.15 \text{ mm}^2$$

 Use 12 mm diameter bars.

 For y-direction, $d_y = d_x - 12 = 150 - 12 = 138$ m

 where d_y and d_x are effective depth of slab in y and x direction respectively,

Now,
$$Ast_{y(-)} = \frac{0.5\times 20}{415}\times\left[1-\sqrt{1-\frac{4.6\times 22.75\times 10^6}{20\times 1000\times 138^2}}\right]\times 1000\times 138 = 493.4 \text{ mm}^2$$

$$Ast_{y(+)} = \frac{0.5\times 20}{415}\times\left[1-\sqrt{1-\frac{4.6\times 16.94\times 10^6}{20\times 1000\times 138^2}}\right]\times 1000\times 138 = 359.6 \text{ mm}^2$$

 Use 12 mm diameter bars.

7. Spacing(S) of 12 mm diameter bars.

 (a) For $Ast_{x(-)}, S = \frac{1000}{748}\times\frac{\pi}{4}(12)^2 = 151.2$mm, provide 12mm@ 150mm c/c

 (b) For $Ast_{x(+)}, S = \frac{1000}{541.15}\times\frac{\pi}{4}(12)^2 = 209$ mm, provide 12mm@ 200mm c/c

 (c) For $Ast_{y(-)}, S = \frac{1000}{493.4}\times\frac{\pi}{4}(12)^2 = 229.22$ mm, provide 12 mm@ 220 mm c/c

 (c) For $Ast_{y(+)}, S = \frac{1000}{360}\times\frac{\pi}{4}(12)^2 = 314.16$ mm, provide 12mm@ 300mm c/c

 Note: Spacing should be less than 300 mm.

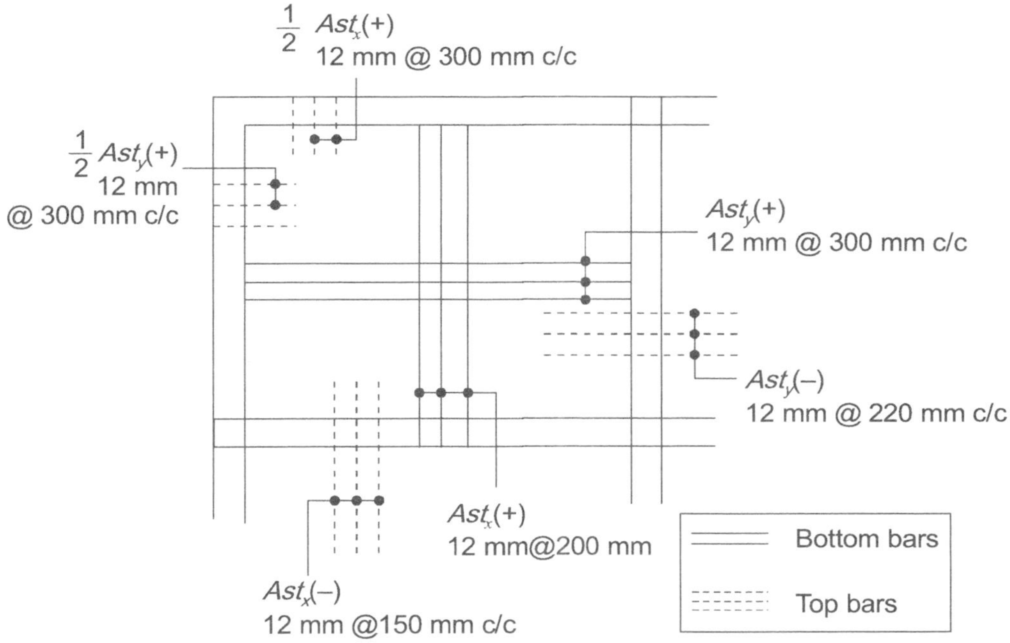

2.32 A T beam is continuous over a span of 10 m. The sectional parameters of the beam are as below:
Width of web = 250 mm
Width of flange = 1100 mm
Effective depth of beam = 460 mm
Area of steel in tension = 1800 mm^2
Area of steel in compression = 1000 mm^2
Use M 20 grade of concrete and Fe-415 grade of steel. Estimate the safety of the beam for deflection control using the empirical method given in the Code IS 456 : 2000. The corresponding graphs are reproduced below.

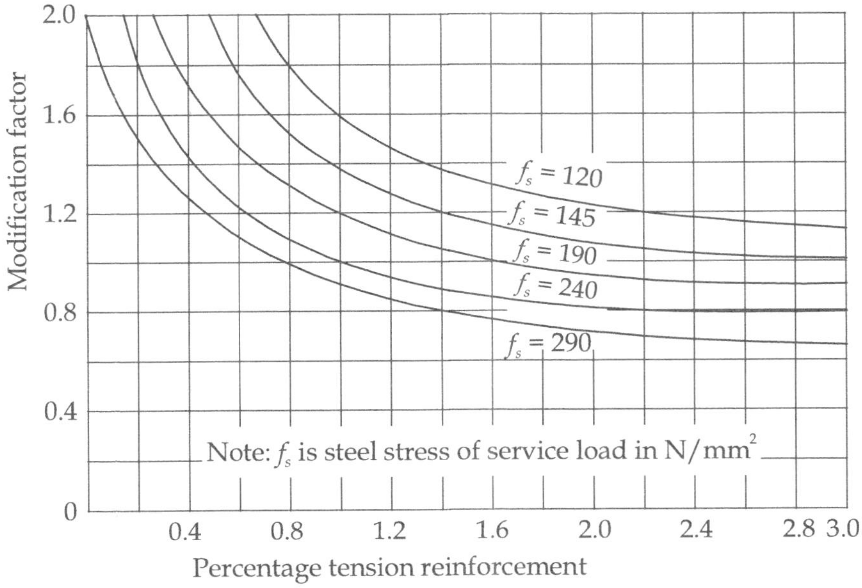

$$f_a = 0.58 f_y \frac{\text{Area of cross-section of steel required}}{\text{Area of cross-section of steel provided}}$$

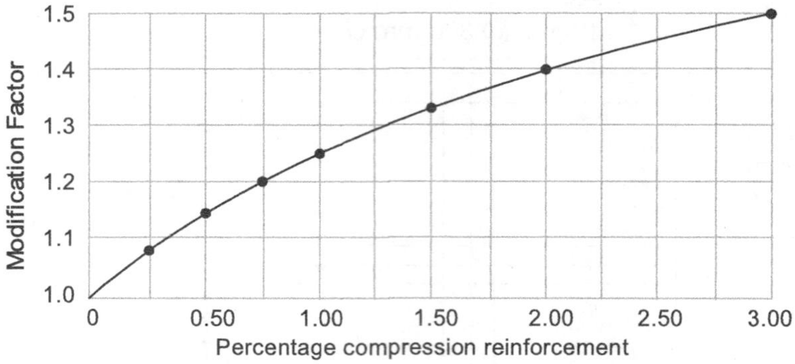

Modification factor for compression reinforcement

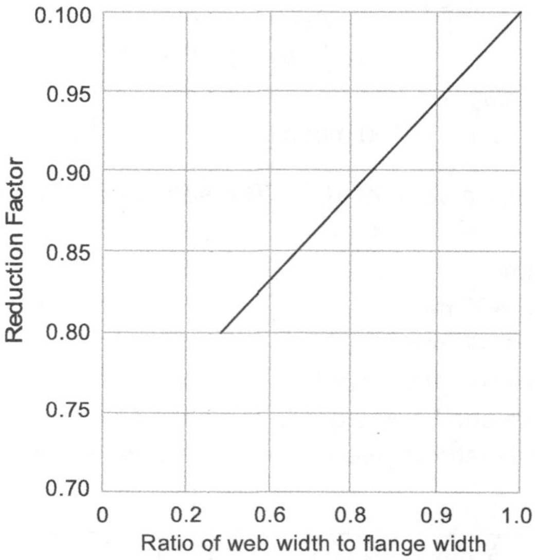

Reduction factor for ratios of span to effective depth for flanged beams

[12 marks : 2023]

Solution:

Effective depth of T-beam required :

$$d = \frac{\text{Effective span}}{A\text{-Value} \times MF_t \times MF_C \times RF_T}$$

(i) Beam is continuous so, A – value = 26

(ii)
$$P_t\,\% = \frac{A_{st}}{b_w d} \times 100 = \frac{1800}{250 \times 460} \times 100 = 1.565\%$$

$$f_s = 0.58 f_y \times \frac{A_{st\ (\text{required})}}{A_{st\ (\text{provided})}}$$

$$= 0.58 \times 415 \times \frac{1800}{1800} = 240.7$$

Modification factor based on tension steel
(Using figure 4)

$$MF_t = 0.90 \text{ (approx.)}$$

(iii) $$P_c\,\% = \frac{A_{sc}}{b_w \cdot d} \times 100 = \frac{1000}{250 \times 460} \times 100 = 0.87\%$$

Modification factor based on compression steel (using figure 5)

$$MF_C = 1.22$$

(iv) Reduction factor for flange section

$$\text{Ratio of } \frac{b_w}{B_f} = \frac{250}{1100} = 0.227$$

$$RF_{\text{flanged}} = 0.80 \qquad \text{(Using figure 6)}$$

(iv) Effective depth of T-beam required

$$d = \frac{\text{Effective span}}{A \times MF_t \times MF_C \times RF_{\text{flanged}}}$$

$$= \frac{10000}{26 \times 0.9 \times 1.22 \times 0.80} = 438 \text{ mm}$$

Since, Effective depth provided = 460 mm > 438 mm
So, Beam is safe for deflection.

3. Shear, Bond, Torsion and Development Length

3.1 A 5 m effective span simply supported beam is subjected to a load of 40 kN/m including its self weight. The size of the beam is 250 mm × 500 mm. The beam is reinforced with 4-20 ϕ at bottom (out of which two bars are curtailed) and 2-12 ϕ at top. Design the beam against shear force and show the reinforcement details. Use M 20 and Fe 415. Use limit state method of design.

% tension steel	0.15	0.25	0.5	0.75	1.0
τ_c (MPa)	0.28	0.36	0.48	0.56	0.62
% tension steel	1.25	1.5	2.0	2.5	3 & above
τ_c (MPa)	0.67	0.72	0.79	0.82	0.82

[20 marks : 2013]

Solution:

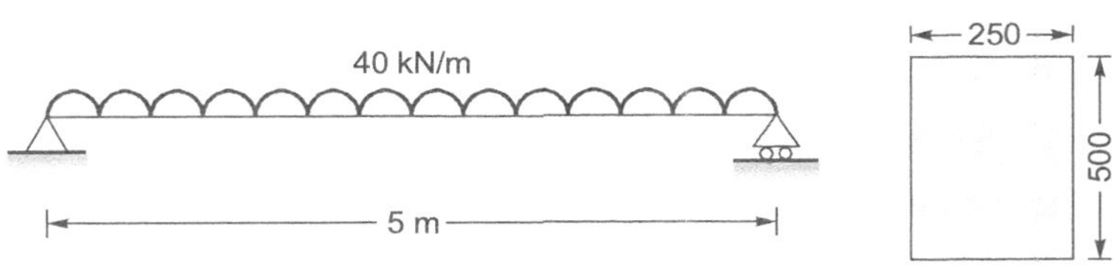

$$W = 40 \text{ kN/m}$$

Factored load $$W_u = 1.5 \times 40 = 60 \text{ kN/m}$$

Max. SF, $$V_u = \frac{W_u l}{2} = \frac{60 \times 5}{2} = 150 \text{ kN}$$

Effective depth, $d = 500 - 25 - \frac{20}{2} = 465 \text{ mm}$

Nominal shear stress, $\tau_v = \frac{V_u}{bd} = \frac{150 \times 10^3}{250 \times 465}$

$= 1.29 \text{ N/mm}^2 < \tau_{c\,max} = 2.8 \text{ N/mm}^2$ (OK)

Design shear strength of concrete:

$$\%P_t = \frac{A_{st} \text{ at critical section}}{Bd} \times 100$$

At supports only 2-20ϕ bars are present on tension side.

$$\therefore \quad \%P_t = \frac{2 \times \frac{\pi}{4} \times (20)^2}{250 \times 465} \times 100 = 0.54$$

$$\tau_c = 0.48 + \frac{0.56 - 0.48}{0.75 - 0.5} \times (0.54 - 0.5) = 0.4928 \text{ N/mm}^2$$

SF to be resisted by shear stirrups,

$$V_{us} = V_u - \tau_c bd = 150 - 0.4928 \times 250 \times 465 \times 10^{-3} = 92.712 \text{ kN}$$

Using 2-legged 8 mm bars as shear stirrups

$$A_{Sv} = 2 \times \frac{\pi}{4} \times 8^2 = 100.53 \text{ mm}^2$$

Spacing of 2-legged 8 mm stirrups

$$S_v = \frac{A_{Sv} \times (0.87 f_y) d}{V_{us}} = \frac{100.53 \times 0.87 \times 415 \times 465}{92.712 \times 10^3} = 182.04 \text{ mm}$$

Hence adopt spacing near support = 180 mm

Max. spacing corresponding to min. shear reinforcement

$$\frac{A_{Sv}}{B.S_v} = \frac{0.4}{0.87 f_y}$$

$$S_v = \frac{0.87 \times 415 \times 100.53}{0.4 \times 250} = 362.96$$

Max. spacing should not exceed.

(i) $0.75\,d = 0.75 \times 465 = 348.75$ mm

(ii) 300 mm

Hence max. spacing = 300 mm

Location of min. shear reinforcement (Max. spacing)

$$\frac{x}{92.712} = \frac{2.5}{150}$$

$$x = \frac{2.5 \times 92.712}{150} = 1.545 \text{ m from mid}$$

$\because$ Location of max. spacing, $2x = 3.09$ m

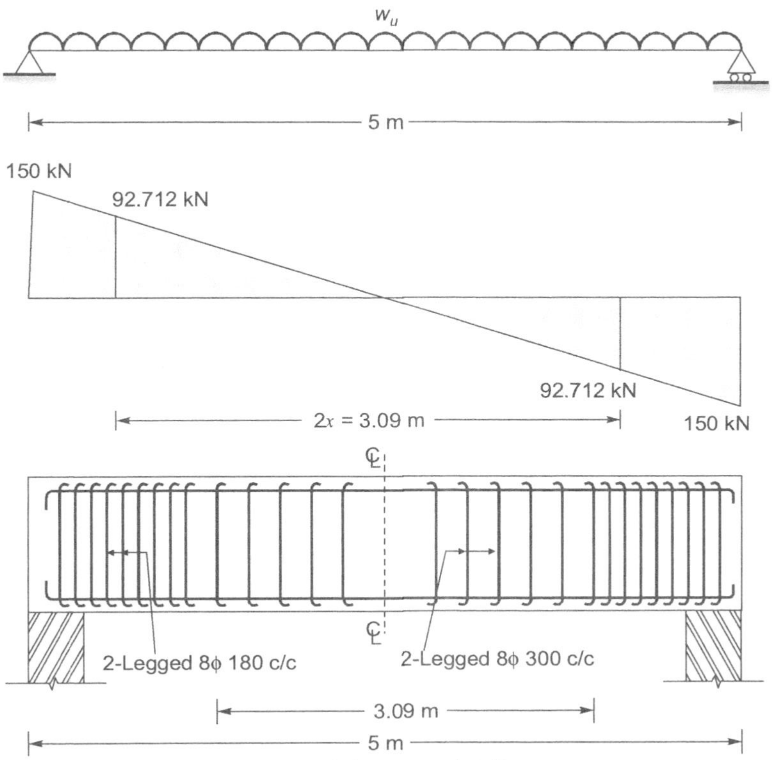

Shear reinforcement detailing

3.2 Show that development length of a steel bar of dia ϕ embedded in concrete is given by

$$L_d = \frac{0.87\,\sigma_y\,\phi}{4\,\tau_{bd}}$$

where, τ_{bd} = bond strength of concrete, σ_y = yield strength of steel, ϕ = bar dia

[4 marks : 2014]

Solution:

Development length is minimum length of embedment of steel reinforcement in concrete so that the strength of bond is equals to strength of reinforcement.

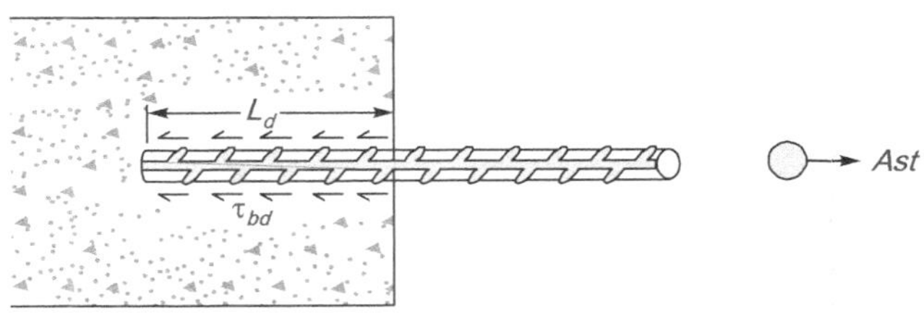

$$L_d \times (\pi\phi) n \times \tau_{bd} = n \times \frac{\pi}{4}\phi^2 \times 0.87\sigma_y$$

$$\therefore \qquad L_d = \frac{0.87\sigma_y\phi}{4\tau_{bd}}$$

3.3 A simply supported beam is 25 × 50 cm deep and has 2-20 mm Fe 415 grade steel bars going into the support shown in figure. If the shear force at the centre of support is 110 kN at service loads, determine the anchorage length. Assume M20 mix. Bond stress for mild steel for M20 concrete is 1.2 MPa in limit state. Take clear cover to steel = 25 mm.

[20 marks : 2016]

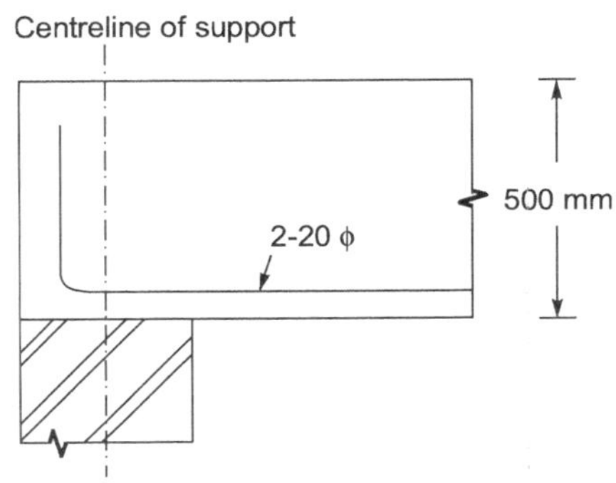

Solution:

For calculation of anchorage length,

$$L_d \le \frac{1.3M_1}{V} + L_0$$

Now,
$$L_d = \frac{0.87 f_y \phi}{4\,\tau_{bd}} = \frac{0.87 \times 415 \times 20}{4 \times (1.6 \times 1.2)} = 940.23 \text{ mm}$$

$$V = 1.5 \times 110 = 165 \text{ kN}$$

$$M_1 = MR \text{ of section at centreline of support}$$

Position of NA:

$$C = T$$

$$\Rightarrow \quad 0.36\, f_{ck}\, x_u\, b = 0.87\, f_y\, A_{st}$$

$$\Rightarrow \quad 0.36 \times 20 \times x_u \times 250 = 0.87 \times 415 \times 2 \times \frac{\pi}{4} \times 20^2$$

$$\Rightarrow \quad x_u = 126.03 \text{ mm} \qquad \left[x_{u,\max} = 0.48\, d = 0.48 \times \left[(500-25) - \frac{20}{2}\right]\right.$$

Now,
$$M_1 = 0.87\, f_y\, A_{st}\,(d - 0.42\, x_u)$$

$$= 0.87 \times 415 \times 2 \times \frac{\pi}{4} \times 20^2 \left[\left(500 - 25 - \frac{20}{2}\right) - 0.42 \times 126.03\right]$$

$$M_1 = 93.48 \text{ kNm}$$

For anchorage length,
$$L_d \le \frac{1.3M_1}{V} + L_0$$

$$\Rightarrow \quad 940.23 \le \frac{1.3 \times 93.48}{165} \times 10^3 + L_0$$

$$\Rightarrow \quad L_0 \ge 203.721 \text{ mm}$$

4. Staircase, Wall, Columns and Footings

4.1 A reinforced concrete column of effective length of 3.0 m is to be designed to support a factored load of 2400 kN. Using limit state design method determine the cross-sectional dimensions of the column and reinforcement when one side of column is restricted to 350 mm. The concrete used is of grade M20 and reinforcement is of HYSD steel of Fe415.

[25 marks : 2000]

Solution:

Given: One side of column is 350 mm

Effective length of column is $l_{eff} = 3000$ mm

$$\text{Slenderness ratio in first direction} = \frac{l_{eff}}{D} = \frac{3000}{350} = 8.57 < 12$$

Eccentricity in first direction,
$$e_1 = \frac{l}{500} + \frac{D}{30} = \frac{3000}{500} + \frac{350}{30} = 17.67 \text{ or } 20 \text{ mm}$$

$$e_{1\,\min} = 20 \text{ mm}$$

Assume another side is B,
$$e_2 = \frac{l}{500} + \frac{B}{30}$$

Take $e_2 = 20$ mm as minimum eccentricity

$$e_2 = \frac{l}{500} + \frac{B}{30} = 20 = \frac{3000}{500} + \frac{B}{30} = 20$$

$$B = 420 \text{ mm}$$

Hence adopt another side as 420 mm

$$\text{Slenderness ratio in another direction} = \frac{l}{420} = \frac{3000}{420} = 7.14 < 12$$

Hence design the column as short column.

$$P_u = 0.4 f_{ck} A_c + 0.67 f_y A_{sc}$$

$$2400 \times 1000 = 0.4 \times 20 \times (350 \times 420 - A_{sc}) + 0.67 \times 415 \times A_{sc}$$

$$A_{sc} = 4532.494 \text{ mm}^2$$

Provide 28 mm diameter of reinforcement

$$\text{No. of Bar required} = \frac{A_{sc}}{\frac{\pi}{4} \times 28^2} = \frac{4532.494}{\frac{\pi}{4} \times 28^2} = 7.361 \simeq 8$$

Hence provide 8 Bars of 28 mm ϕ of reinforcement

Check for maximum reinforcement

$$= \frac{\frac{\pi}{4} \times 28^2 \times 8}{420 \times 350} \times 100\% = 3.35\% < 6\% \qquad \text{(Ok, as per IS 456 2000)}$$

Design of lateral ties:

$$\text{Diameter of lateral ties} \geq \begin{cases} (i) \frac{\phi_{min}}{4} = \frac{28}{4} = 7 \text{ mm} \\ (ii) \; 6 \text{mm} \end{cases}$$

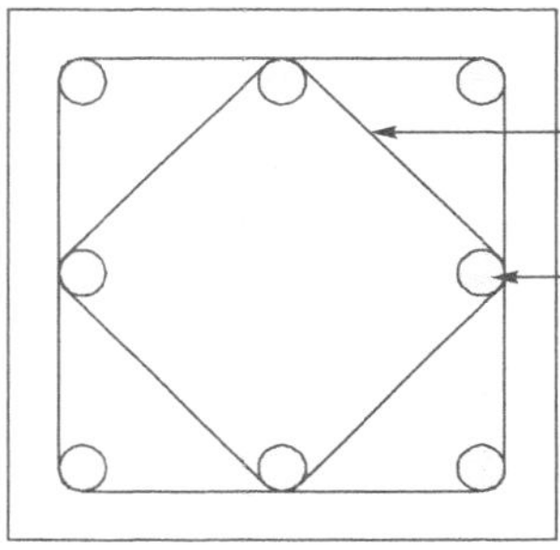

Hence provide bar of 8 mm as lateral ties.

Spacing (pitch) of lateral ties (minimum of)

(i) least lateral dimension = 350 mm

(ii) $16 \times \phi_{max} = 16 \times 28 = 448$ mm

(iii) 300 mm

Hence provide 8 mm dia of lateral ties at 300 mm c/c spacing.

4.2 **Design a sloped square footing of a *RC* column 300 mm × 300 mm size carrying an axial load of 320 kN. The safe bearing capacity of soil is 150 kN/m². M20 concrete and Fe 415 steel are used. Use working stress method of design.**

p_t	0.25	0.50	0.75	1.00	1.25	1.50
τ_c, N/mm²	0.22	0.30	0.35	0.39	0.42	0.45

Allowable shear stress data

[20 marks : 2001]

Solution:

(i) Design constants

$$\therefore \qquad k = \frac{mc}{t + mc} = \frac{13 \times 7}{230 + (13 \times 7)} = 0.2835$$

$$j = 1 - \frac{k}{3} = 1 - \frac{0.2835}{3} = 0.9055$$

$$Q = \frac{1}{2} \times c \times j \times k = \frac{1}{2} \times 7 \times 0.9055 \times 0.2835 = 0.90$$

(ii) **Design of size**

$$\text{Load from column} = 320 \text{ kN}$$

$$\text{Foundation weight} = \frac{10}{100} \times 320 = 32 \text{ kN}$$

(Assuming 10% column load),

$$P_T = 352 \text{ kN}$$

$$\therefore \quad \text{Area} = \frac{P_T}{q_0} = \frac{352}{150} = 2.35 \text{ m}^2$$

Since the footing is a square footing, therefore $B = \sqrt{A}$

$$\Rightarrow \quad B = \sqrt{2.35} = 1.53 \approx 1.55 \text{ m}$$

$$\therefore \quad A_p = 1.55 \times 1.55 = 2.4045 \text{ m}^2$$

$$\therefore \text{ Net soil pressure } w = \frac{P}{A} = \frac{320}{2.4045} = 133.20 \text{ kN/m}^2$$

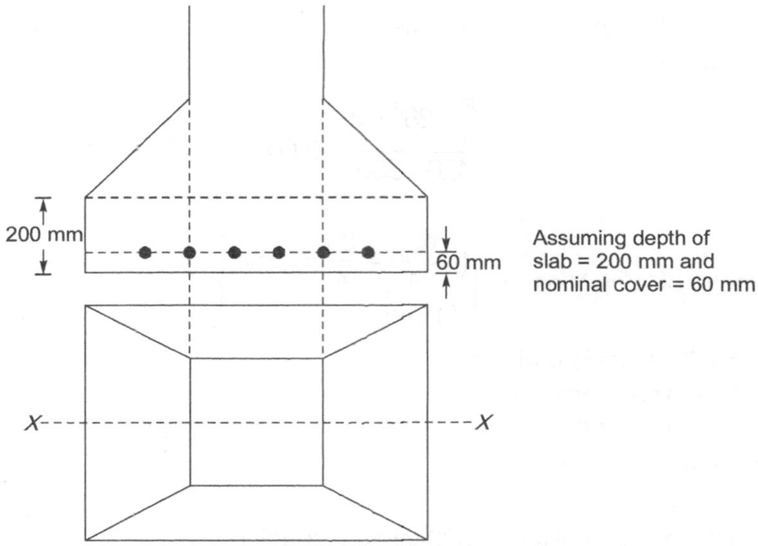

(iii) **Depth of foundation** for bending moment critical section is at the face of column section

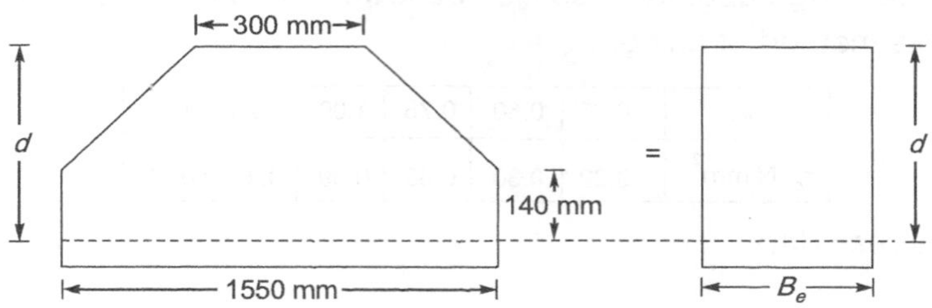

Here $b = 300$ mm, $B = 1550$ mm

Width of equivalent section,

$$B_e = b + \frac{1}{8}(B - b) = 300 + \frac{1}{8}(1550 - 300)$$

$$= 456.25 \text{ mm (It is used only for depth calculation)}$$

(iv) Bending Moment

Overhang, $O_x = \dfrac{1.55 - 0.3}{2} = 0.625$ m

$$\text{Bending moment} = \text{W.B.}\frac{O_x^2}{2} = 133.2 \times 1.55 \times \frac{(0.625)^2}{2} = 40.32 \text{ kN-m}$$

$$\therefore \quad d = \sqrt{\frac{BM}{QB_e}} = \sqrt{\frac{40.32 \times 10^6}{0.9 \times 456.25}} = 313.40 \text{ mm}$$

Effective cover = 60 mm

∴ Total depth, $D = 313.40 + 60 = 373.40$ mm

Adopting $D = 400$ mm

∴ Effective depth, $d = 400 - 60 = 340$ mm

(v) One way shear

Overhang, $O_x = \dfrac{1.55 - 0.3}{2} - 0.34 = 0.285$ m

$V = w \times B \times O_x = 133.20 \times 1.55 \times 0.285 = 58.84$ kN

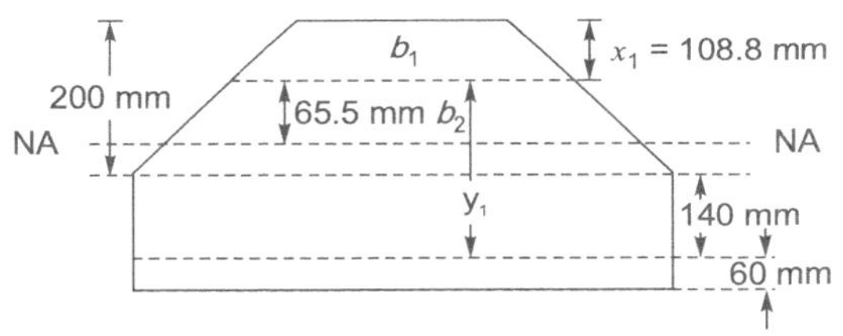

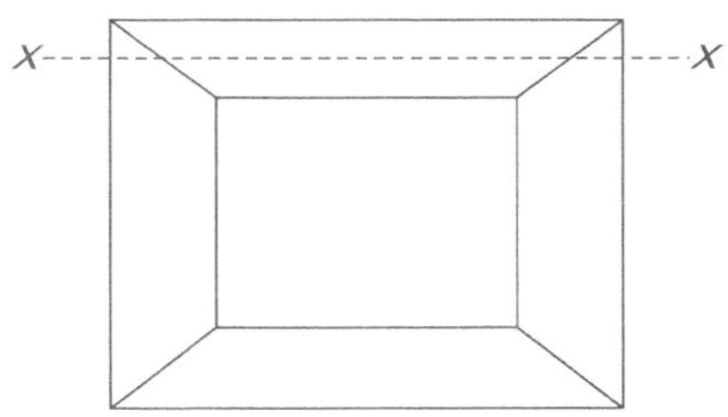

Shear stress, $\tau_v = \dfrac{V}{B \times d} = \dfrac{V}{b_2 \times d}$

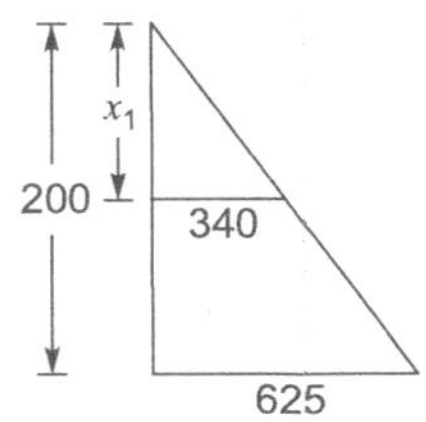

$b_1 = b + 2d = 300 + (2 \times 340) = 980$ mm

$$x_1 = \frac{200}{625} \times 340 = 108.8 \text{ mm}$$

$$y_1 = 340 - 108.8 = 231.2 \text{ mm}$$

Depth of neutral axis = $kd = ky_1 = 0.2835 \times 231.2 = 65.5$ mm

Now, width of the section of NA = b_2

$$\therefore \quad b_2 = 980 + \left[\frac{1550 - 980}{91.2}\right] \times 65.5$$

$$b_2 = 1389.4 \text{ mm}$$

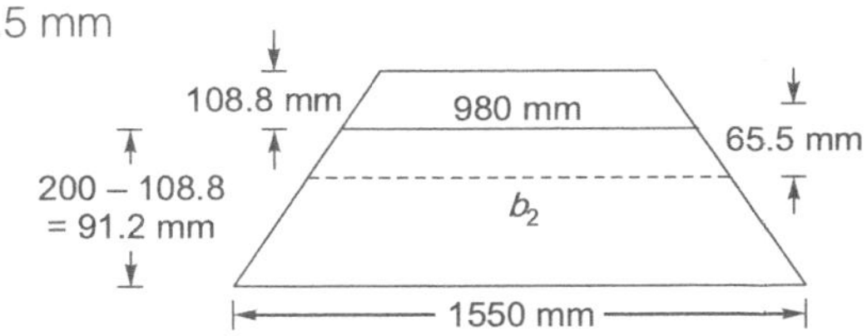

$$\text{Shear stress} = \frac{V}{b_2\, y_1} = \frac{58.84 \times 10^3}{1389.4 \times 231.2} = 0.18 \text{ N/mm}^2 < 0.22 \ (\min \tau_{c,\,min})$$

Hence the section is safe in single shear.

(vi) Check for **punching shear** (double shear)

Net punching force $= P - w(a + d)(b + d)$

$= 320 - 133.2\,(0.3 + 0.34)\,(0.3 + 0.34)$

$P_{net} = 265.44$ kN

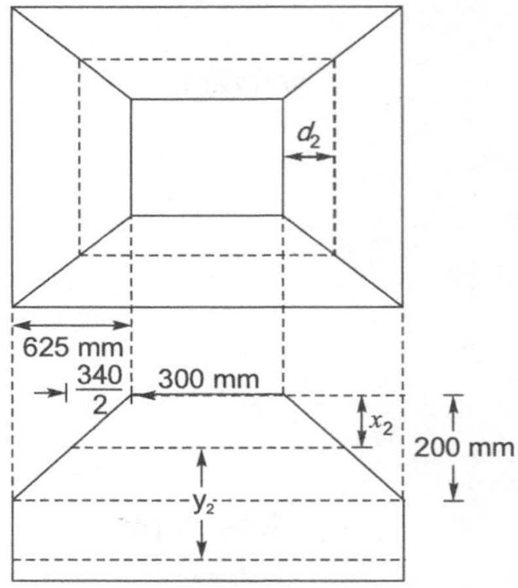

$$x_2 = \frac{200}{625} \times 170 = 54.4 \text{ mm}$$

$$y_2 = 340 - x_2 = 340 - 54.4$$

$$= 285.6 \text{ mm}$$

$$\text{Punching shear stress} = \frac{\text{Net Punching Force}}{\text{Perimeter} \times \text{Depth}}$$

$$= \frac{265.44 \times 10^3}{2[(a+d)+(b+d)] \times y_2}$$

$$= \frac{265.44 \times 10^3}{2[(0.3+0.34)+(0.3+0.34)] \times 285.6 \times 10^3}$$

$$= 0.36 \text{ N/mm}^2$$

Permissible punching stress $= k_c \times \tau_c$

$$= 1 \times 0.16\sqrt{f_{ck}} = 1 \times 0.16\sqrt{20}$$

$$= 0.7155 \text{ N/mm}^2 > 0.36 \text{ N/mm}^2$$

$$\begin{bmatrix} \tau_c = 0.16\sqrt{f_{ck}} \text{ in WSM} \\ \tau_c = 0.25\sqrt{f_{ck}} \text{ in LSM} \end{bmatrix}$$

Hence section is safe in punching shear.

(vii) Area of steel

$$M = 40.32 \text{ kN-m}$$

$$A_{st} = \frac{M}{t\,.\,jd} = \frac{40.32 \times 10^6}{230 \times 0.9050 \times 340} = 569.725 \text{ mm}^2$$

$$\text{Number of 10 mm } \phi \text{ bars} = \frac{569.725}{\frac{\pi}{4} \times (10)^2} = 7.25 \approx 8 \text{ bars}$$

Minimum percentage of steel $= 0.12\%$

$$= \frac{0.12}{100} \times B_e \times D = \frac{0.12}{100} \times 456 \times 400$$

$$= 218.88 \text{ mm}^2 < 569 \text{ mm}^2$$

∴ Provide 8 No's 10 mm ϕ bars in both directions

(viii) Check for **development length** after providing 60 mm cover length available

$$= \frac{1}{2}(1550 - 300) - 60 = 565 \text{ mm}$$

τ_{bd} for M20 $= 1.2$ N/mm^2,

$\sigma_{st} = 230$ N/mm^2

$$L_d = \frac{\sigma_{st}\phi}{4\,\tau_{bd}} = \frac{230}{4 \times 1.6 \times 1.2}\phi = 29.95\,\phi = 29.95 \times 10$$

$$= 299.5 \text{ mm} < 565 \text{ mm}$$

Hence ok.

4.3 A RC column of size 460 mm × 600 mm having effective length of 3.6 m is to be designed using LSM to support an axial service load of 2500 kN. Use M20 grade concrete and HYSD steel of Fe 415 grade.

[15 marks : 2002]

Solution:

(i) $$\frac{\text{Effective length}}{\text{Least lateral dimension}} = \frac{3600}{460} = 7.83 < 12$$

Hence, the column will be designed as **short column**.

(ii) Adopting $e_{min} = 20$ mm

But e_{min} shall not exceed 0.05 B and 0.05 D

$\therefore$ $20 \le 0.05 \times 460$ and $20 \le 0.05 \times 600$

$20 \le 23$ and $20 \le 30$ Hence OK

Hence, the equation $P_u = 0.4\, f_{ck} A_c + 0.67\, f_y A_{sc}$ can be used

(iii) Service load = 2500 kN

Factored load, $P_u = 1.5 \times 2500 = 3750$ kN

$f_{ck} = 20$ N/mm², $A_c = A - A_{sc}$

where $A = 460 \times 600 = 276000$ mm²

$f_y = 415$ N/mm²

$\therefore$ $P_u = 0.4\, f_{ck} A_c + 0.67\, f_y A_{sc}$

$\Rightarrow$ $3750 \times 10^3 = 0.4 \times 20 \times (276000 - A_{sc}) + 0.67 \times 415 \times A_{sc}$

$$\Rightarrow \quad A_{sc} = \frac{3750 \times 10^3 - 2208000}{270.05}$$

$\Rightarrow$ $A_{sc} = 5710$ mm²

Adopting 36 mm bars, we get number of bars $= \dfrac{5710}{\frac{\pi}{4} \times (36)^2} = 5.61 \approx 6$ bars

(iv) Diameter of lateral ties

(a) $\frac{1}{4} \times 36 = 9$ mm

(b) 6 mm

$\therefore$ Provide 10 mm bars as lateral ties

(v) Spacing of lateral ties

(a) Least lateral dimension = B = 460 mm

(b) $16\,\phi = 16 \times 36 = 576$ mm

(c) 300 mm

lesser of the above three will be provided as the spacing. Hence 10 mm ϕ bars will be provided at 300 mm c/c

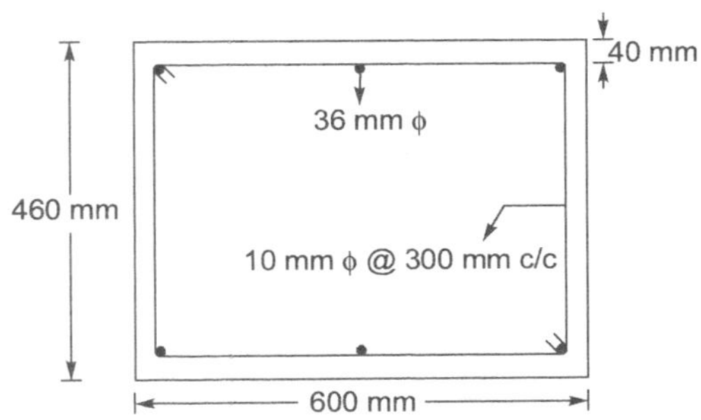

4.4 Determine the thickness of the footing slab of uniform thickness for a 400 mm square column transmitting an axial load of 1100 kN. The safe bearing capacity of soil is 150 kN/m². The materials to be used are M20 concrete and HYSD steel of grade Fe 415. Use LSM.

[15 marks : 2003]

Solution:

(i) Size of foundation

$$\text{Load from column} = 1100 \text{ kN}$$

$$\text{Foundation load} = \frac{10}{100} \times 1100 = 110 \text{ kN}$$

(Assume 10% of column load), $P_T = 1210$ kN

$$\therefore \quad \text{Area of foundation} = \frac{P_T}{q_0} = \frac{1210}{150} = 8.07 \text{ m}^2$$

Let us design a square footing

$$\therefore \quad A = B^2$$

$$\Rightarrow \quad B^2 = 8.07$$

$$\Rightarrow \quad B = \sqrt{8.07}$$

$$\Rightarrow \quad B = 2.84 \text{ m} \approx 2.9 \text{ m}$$

Now, final area selected $= B \times B = 2.9 \times 2.9 = 8.41$ m²

$$\therefore \text{ Net soil pressure, } w = \frac{P}{A} = \frac{1100}{8.41} = 130.8 \text{ kN/m}^2$$

$$\text{Factored soil pressure} = 1.5 \times 130.8 = 196.2 \text{ kN/m}^2$$

(ii) Design of depth as per bending moment

Overhang $$O_x = \frac{B-a}{2} = \frac{2.9-0.4}{2} = \frac{2.5}{2} \text{ m}$$

$$M_{yy} = 196.2 \times 2.9 \times \frac{(2.5)^2}{8} = 444.51 \text{ kN-m}$$

Overhang, $$O_y = \frac{B-a}{2} = \frac{2.5}{2} \text{ m}$$

$$\therefore \quad M_{xx} = 444.51 \text{ kNm}$$

$$Q = 0.36\, f_{ck} \frac{x_{u,\max}}{d}\left(1 - 0.42 \frac{x_{u,\max}}{d}\right)$$

$$= 0.36 \times 20 \times 0.48\,(1 - 0.42 \times 0.48) = 2.76 \text{ N/mm}^2$$

Now, depth, $$d = \sqrt{\frac{M_{xx} \text{ or } M_{yy}}{Q \times B}} = \sqrt{\frac{444.51 \times 10^3}{2.76 \times 2.9}} = 235.66 \text{ mm}$$

Adopt effective depth = 240 mm

Assuming effective cover = 60 mm

$$\therefore \quad \text{Total depth} = 240 + 60 = 300 \text{ mm}$$

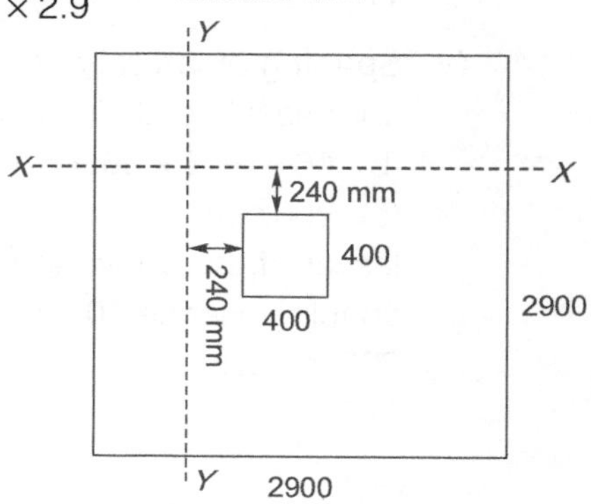

(iii) Check for shear

Overhang $$O_{x_1} = O_{y_1} = \frac{2900 - 400}{2} - 240$$

$$= 1010 \text{ mm} = 1.01 \text{ m}$$

Shear force $V_{ux} = V_{uy} = 196.2 \times 2.9 \times 1.01 \simeq 574.67$ kN

$$\tau_{vx} = \tau_{vy} = \frac{V_u}{B \times d} = \frac{574.67 \times 10^3}{2900 \times 240} = 0.83 \text{ N/mm}^2$$

But τ_c for minimum percentage of steel (0.15%) = 0.28 N/mm²

$$\therefore \qquad \text{Depth required, } d = \frac{V_u}{B\,\tau_c} = \frac{574.67 \times 10^3}{2900 \times 0.28} = 707.72 \text{ mm}$$

There should be a balance between overhang and shear force.

Hence, using $d = 600$ mm

$$\text{Overhang} = \frac{2900 - 400}{2} - 600 = 0.650 \text{ m}$$

$$V_u = 196.2 \times 2.9 \times 0.65 = 369.84 \text{ kN}$$

$$\therefore \qquad \tau_v = \frac{369.84 \times 10^3}{2900 \times 600} = 0.21 < 0.28 \text{ Hence OK.}$$

(iv) Check for punching shear

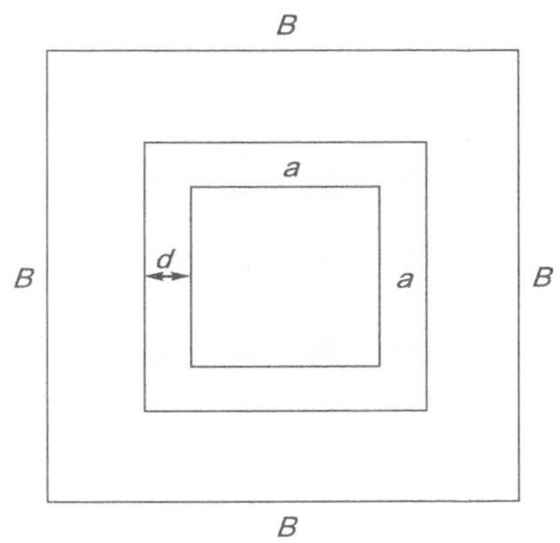

Net punching shear, $P_u = P_c\,(\text{design}) - w_u\,(a + d)\,(a + d)$

$$\Rightarrow \qquad P_u = 1.5 \times 1100 - 196.2\,(0.4 + 0.6)^2$$

$$\Rightarrow \qquad P_u = 1453.8 \text{ kN}$$

$$\text{Punching shear stress} = \frac{\text{Net Punching Force}}{2\left[(a+d)+(a+d)\right] \times d}$$

$$= \frac{1453.8 \times 10^3}{2\left[(0.4+0.6)+(0.4+0.6)\right] \times 10^3 \times 600}$$

$$= 0.61 \text{ N/mm}^2$$

Maximum permissible punching shear stress = $k_s\,\tau_c$

$$\text{where} \qquad k_s = \left(0.5 + \frac{b}{a}\right) = 0.5 + \frac{a}{a} = 1.5$$

but k_s shall not exceed 1.0

$$\tau_c = 0.25\sqrt{f_{ck}} = 0.25\sqrt{20} = 1.118 \text{ N/mm}^2$$

$\therefore$ Permissible punching shear stress

$$= 1.0 \times 1.118 = 1.118 > 0.61, \text{ hence OK.}$$

$\therefore$ Effective depth provided as per all the three criteria = 600 mm

Effective cover = 60 mm

Total depth = 600 + 60 = 660 mm

(v) Area of steel

Since the final effective depth is far larger than that required from bending moment criteria, hence the foundation becomes under reinforced and thus A_{st} will be calculated as

$$A_{st} = 0.5\,\frac{f_{ck}}{f_y}\left[1 - \sqrt{1 - \frac{4.6\,M_u}{f_{ck}\,Bd^2}}\right] Bd \qquad [m_u = m_{yy} = m_{xx}]$$

$$= 0.5 \times \frac{20}{415}\left[1-\sqrt{1-\frac{4.6 \times 444.51 \times 10^6}{20 \times 2900 \times 600^2}}\right] \times 2900 \times 600$$

A_{st} = 2105.84 mm²

Adopting bar diameter = 12 mm

$$\therefore \text{ Number of bars required} = \frac{2105.84}{\frac{\pi}{4}\times(12)^2} = 18.62 \approx 19 \text{ bars}$$

$$\text{Spacing of bars} = \frac{2900}{19} = 152.63 \text{ mm}$$

16 bars will be provided in each direction at 150 mm c/c

$$\text{Minimum area of steel} = \frac{0.12}{100} \times 2900 \times 660 = 2296.8 \text{ mm}^2 > 2105.84 \text{ mm}^2$$

Hence providing 12 mm ϕ bars

$$\text{Number of bars} = \frac{2296.8}{\frac{\pi}{4}\times(12)^2} = 20.31 \approx 21 \text{ bars}$$

$$\text{Spacing} = \frac{2900}{21} = 138.09 \text{ mm}$$

Provide 21 No's 12 mm ϕ bars @ 130 mm c/c.

(vi) Check for development length

$$L_d = \frac{\phi\,\sigma_{st}}{4\,\tau_{bd}} = \frac{(0.87\times 415)\times 12}{4\times 1.6\times 1.2} = 564.14 \text{ mm}$$

(60% increase for HYSD bars)

$$\text{Providing 60 mm side cover, length of bars available} = \frac{1}{2}(B-a)-60$$

$$= \frac{1}{2}(2900-400)-60 = 1190 \text{ mm} > 564.14 \text{ mm (Hence OK)}$$

4.5 **Design a circular column with helical reinforcement subjected to a working load of 1500 kN. Diameter of the column is 450 mm. The column has unsupported length of 3.5 m and is effectively held in position at both ends but not restrained against rotation. Use limit state design method. Use M25 concrete and HYSD Fe415 steel.**

[20 marks : 2011]

Solution:

Given: P = 1500 kN, P_u = 1.5 × 1500 = 2250 kN, Circular dia. = 450 mm, L_{eff} = 3.5 m

$$\text{Check for slenderness ratio} \quad \frac{L_{eff}}{D} = \frac{3500}{450} = 7.78 < 12$$

Therefore, column can be designed as short column

Check for minimum eccentricity $e_{min} = \frac{L_{eff}}{500} + \frac{D}{30}$ or 20 mm whichever is greater

$$e_{min} = \frac{3500}{500} + \frac{450}{30} = 22 \text{ mm}$$

As, $0.05\,D = 22.5$ mm

$$\approx e_{min}$$

$\Rightarrow$ Codal formula for axially compressed short column may be used

$$P = 1.05\,(0.4\,f_{ck}\,A_c + 0.67\,f_y A_{sc})$$

Area of concrete $A_c = \frac{\pi}{4} \times 450^2 - A_{sc}$

A_{sc} = Area of steel

$$2250 \times 10^3 = 1.05\left(0.4 \times 25 \times \left(\frac{\pi}{4} \times 450^2 - A_{sc}\right) + 0.67 \times 415 A_{sc}\right)$$

$$A_{sc} = 2060.9 \text{ mm}^2$$

$$A_{sc,min} \text{ at } 0.8\% \text{ of } A_g = \frac{0.8}{100} \times \frac{\pi}{4} \times 450^2 = 1272.35 \text{ mm}^2$$

Providing 8 bars of 20 mm ϕ

$\Rightarrow$ A_{sc} provided $= 8 \times \frac{\pi}{4} \times 20^2 = 2513.27 \text{ mm}^2 > 2060.9 \text{ mm}^2$ (OK)

Design of spiral reinforcement

Assuming a clear cover of 40 mm over spirals,

Core diameter = 450 – (40 × 2) = 370 mm

Assuming a spiral bar diameter of 6 mm and pitch S_t

$$p_s = \frac{\text{Volume of spiral reinforcement}}{\text{Volume of core}} \text{ per unit length of column}$$

$$= \frac{\left(\frac{\pi}{4} \times 6^2\right) \times \pi \times \frac{(370-6)}{S_t}}{\left(\frac{\pi}{4} \times 370^2\right)} = \frac{0.3007}{S_t}$$

As per code: $p_s \geq 0.36\left[\frac{A_g}{A_{core}} - 1\right]\left[\frac{f_{ck}}{f_y}\right]$

$$\Rightarrow \quad \frac{0.3007}{S_t} \geq 0.36\left[\frac{\frac{\pi}{4} \times 450^2}{\frac{\pi}{4} \times 370^2} - 1\right]\left[\frac{25}{415}\right]$$

$$\Rightarrow \qquad \frac{0.3007}{S_t} \geq 0.36 \times (0.479) \times (0.0602)$$

$$\Rightarrow \qquad S_t \leq 28.94 \text{ mm}$$

As per code
$$S_t < \begin{cases} 75 \text{ mm} \\ \text{Core } \dfrac{\text{dia.}}{6} = 61.67 \text{ mm} \end{cases}$$

and
$$S_t > \begin{cases} 25 \text{ mm} \\ 3\phi_h = 18 \text{ mm} \end{cases}$$

Providing 6ϕ spiral at 28 mm c/c pitch.

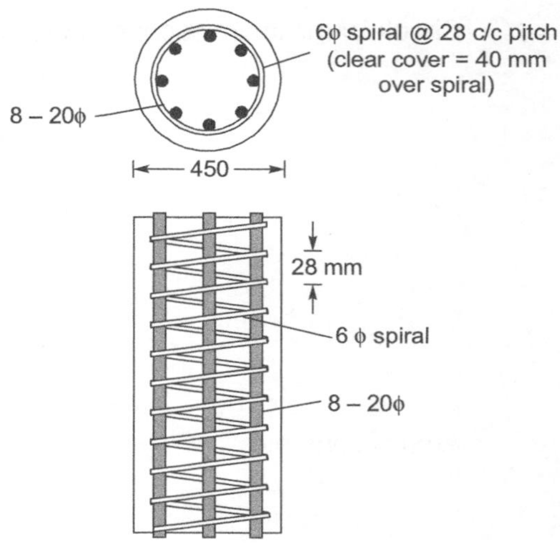

4.6 **Design a constant thickness footing for a reinforced concrete column of 300 mm × 300 mm. The column is carrying an axial working load of 600 kN. The BC of soil is 200 kN/m². Use M-25 concrete and HYSD Fe-415 bars. Use limit state design method.**

$\frac{100 A_{st}}{bd}$	0.15	0.25	0.50	0.75	1.0
τ_c (N/mm²)	0.19	0.36	0.49	0.57	0.64

[10 marks : 2011]

Solution:

Given: $P = 600$ kN, $q_a = 200$ kN/m²

Assuming the weight of the footing and backfill to be 10% of the column load of 600 kN.

$$\therefore \qquad \text{Base area required} = \frac{600 \times 1.1}{200} = 3.3 \text{ m}^2$$

$$\text{Designing a square footing of size} = \sqrt{3.3} = 1.82 \text{ m}$$

Assuming size of the footing to be 2 m × 2 m

Thickness of footing slab based on shear:

Net soil pressure at ultimate load with a load factor of 1.5.

$$q_u = \frac{600 \times 1.5}{2 \times 2} = 225 \text{ kN/m}^2 = 0.225 \text{ N/mm}^2$$

(a) One way shear

Critical section is at a cistance of 'd' from the column face. 'd' is the effective depth of the footing

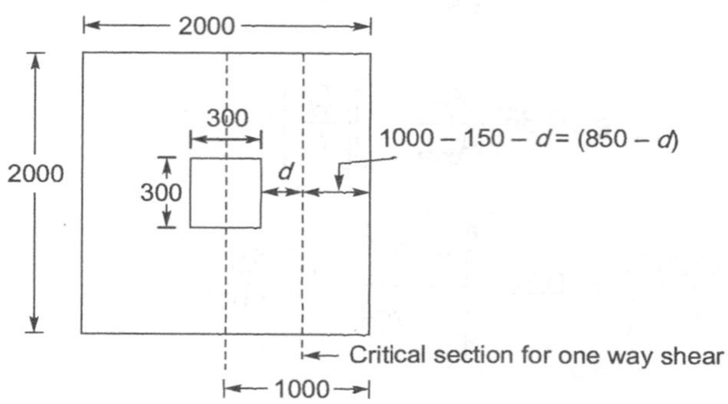

Factored shear force:

$$V_{u_1} = 0.225 \times (2000 \times (850 - d))\text{N}$$

$$V_{u_1} = (382500 - 450\,d)\text{ N}$$

Let percentage of steel reinforcement to be provided in tension be = 0.25

From table provided : $\tau_c = 0.36\text{ N/mm}^2$

One way shear resistance, $V_{c_1} = 0.36 \times (2000 \times d) = 720\text{ d N}$

For footing to be safe in one way shear : $V_{u_1} \le V_{c_1}$

or, $382500 - 450\,d \le 720\,d$

or, $d \ge 326.92\text{ mm}$ ...(1)

(b) Two-way shear

Critical section is at $d/2$ from the periphery of the column.

$\Rightarrow$ Factored shear force $= V_{u_2} = 0.225 \times \left[2000^2 - (300 + d)^2\right]$

Taking depth of footing as 330 mm

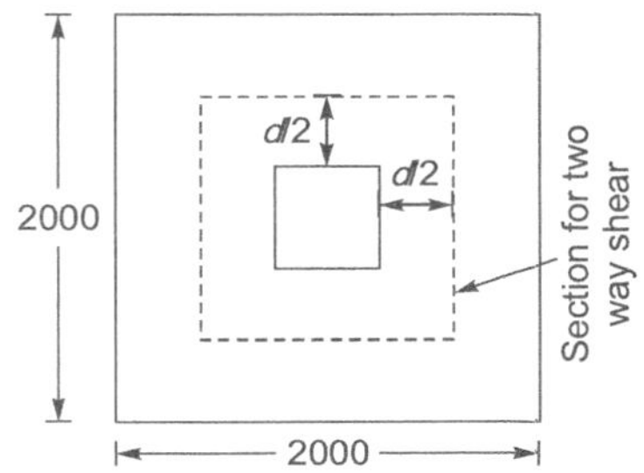

$\Rightarrow$ $V_{u_2} = 810.7 \times 10^3\text{ N}$

Two way shear resistance : $V_{c_2} = k_s\tau_c[4 \times (300 + d) \times d]$

where , $k_s = 1$ for a square column

and $\tau_c = 0.25\sqrt{f_{ck}} = 1.25\text{ N/mm}^2$

$\Rightarrow$ $V_{c_2} = 1 \times 1.25 \times [4(300 + d) \times d] = 5d\,(300 + d)$

For footing to be safe in two-way shear : $V_{u_2} \le V_{c_2}$

or $810.7 \times 10^3 \le 5d^2 + 1500\,d$

or $5d^2 + 1500\,d - 810.7 \times 10^3 \ge 0$

on solving, $d \ge 279.7\text{ mm}$

So, providing a clear cover of 50 mm and 12 ϕ bars in both directions with an effective depth of 330 mm as per one way shear criteria

Overall thicknes $= D = 330 + 50 + 12/2 = 386\text{ mm} \approx 400\text{ mm}$

and Effective thickness $= d = 400 - 50 - 12/2 = 344\text{ mm}$

Design of flexural reinforcement

Critical section for moment is at column face

$$\Rightarrow \quad M_u = (0.225 \times 2000) \times \frac{(1000 - 150)^2}{2} = 162.56\text{ kN-m}$$

$$A_{st} \doteq \frac{0.5 f_{ck}}{f_y}\left[1 - \sqrt{1 - \frac{4.6 M_u}{f_{ck} B d^2}}\right] Bd$$

$$A_{st} = \frac{0.5 \times 25}{415} \times \left[1 - \sqrt{\frac{4.6 \times 162.56 \times 10^6}{25 \times 2000 \times 344^2}}\right] 2000 \times 344$$

$$A_{st} = 1353.71\text{ mm}^2$$

$$p_t\% = \frac{1353.71}{2000 \times 344} \times 100$$

$$p_{t,\,reqd}\% = 0.1967\%$$

or $p_{t,reqd} = 0.1967\%$ which is less than p_t for one way thickness.

$$p_t = 0.25\% > \text{minimum } p_t = 0.12\%$$

$\therefore$ $$A_{st,reqd} = \frac{0.25}{100} \times (2000 \times 344) = 1720 \text{ mm}^2$$

Number of bars required using 12 mm ϕ bars

$$= \frac{1720}{\frac{\pi}{4} \times 12^2} = 15.21 \approx 16$$

$$\text{Spacing of bars} = S = \frac{(2000 - (50 \times 2) - 12 \times 16)}{16 - 1} = 113.87 \text{ mm}$$

Provide 12 mm ϕ bar @ 110 mm c/c both ways.

4.7 **Check the safety and stability of the retaining wall shown in figure. Safe bearing capacity of soil = 80 kN/m², coefficient of friction between soil and concrete = 0.55, angle of repose of earth = 35°, density of earth = 19 kN/m³.**

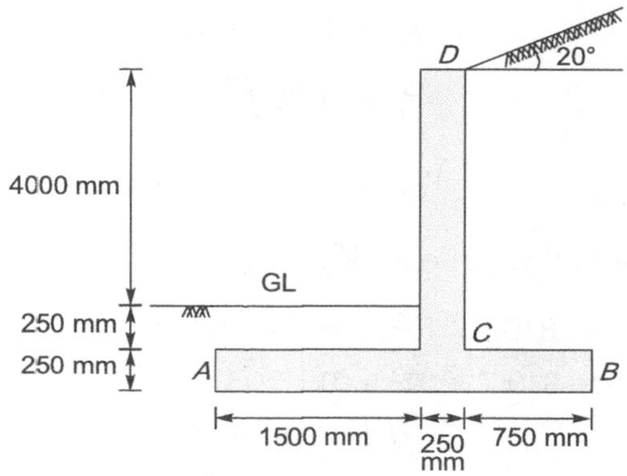

[20 marks : 2017]

Solution :

Given, safe bearing capacity of soil = 80 kN/m²

Coefficient of friction between soil and concrete = 0.55

Angle of repose = 35°

Density of earth = 19 kN/m³

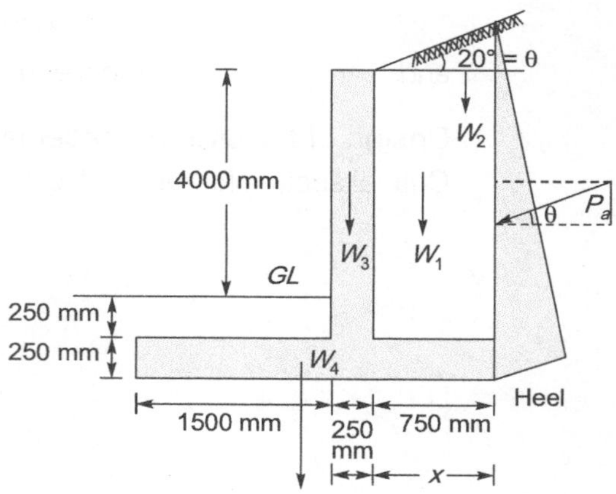

1. **Stability against overturning:**

Force due to active pressure,

$$P_a = \frac{1}{2} k_a \gamma h'^2$$

where $$h' = h + x \tan\theta$$

$$= (4000 + 250 + 250) + 750 \tan 20° = 4773 \text{ mm}$$

$$k_a = \left\{ \frac{\cos\theta - \sqrt{\cos^2\theta - \cos^2\phi}}{\cos\theta + \sqrt{\cos^2\theta - \cos^2\phi}} \right\} \cos\theta$$ [where, $\theta = 20°$ and $\phi = 35°$]

$$\Rightarrow \qquad k_a = 0.322$$

$$\therefore \qquad P_a = 0.322 \times 19 \times (4.773)^2/2 = 69.69 \text{ kN/metre length of wall}$$

$$\therefore \qquad P_a \cos\theta = 69.69 \cos 20° = 65.49 \text{ kN}$$

$$P_a \sin\theta = 69.69 \sin 20° = 23.84 \text{ kN}$$

Overturning moment, $M_a = (P_a \cos\theta)\dfrac{h'}{3} = 65.49 \times \dfrac{4.773}{3} = 104.19$ kNm/m length of wall

Force (kN)	Distance from heel (m)	Moment (kNm)
$W_1 = 19 \times 4.25 \times 0.750 = 60.5625$	0.375	22.71
$W_2 = \frac{1}{2} \times (4.773 - 4.5) \times 0.750 \times 19 = 1.945$	0.25	0.4863
$W_3 = (4.5 - 0.250) \times 25 \times 0.25 = 26.56$	0.875	23.24
$W_4 = 25 \times 2.5 \times 0.25 = 15.625$	1.25	19.53
$P_a \sin\theta = 23.84$ kN	0	0
$W = 128.53$ kN		65.97 kNm

∴ Distance of resultant vertical force from heel

$$x_w = \frac{M_w}{W} = \frac{65.97}{128.53} = 0.5133 \text{ m}$$

Stabilising moment about toe

$$M_r = W(L - x_w)$$

$$= 128.53\,(2.50 - 0.5133) = 255.35 \text{ kNm (per unit length of wall)}$$

$$\therefore \qquad (FS)_{overturning} = \frac{0.9M_r}{M_a} = \frac{0.9 \times 255.35}{104.19} = 2.21 > 1.40 \qquad \text{(OK)}$$

2. **Soil pressure at footing base**

Resultant vertical reaction,

$$R = W = 128.53 \text{ kN/m length of wall.}$$

Distance of R from heel, $L_R = \left(\dfrac{M_w + M_a}{R}\right) = \left(\dfrac{65.97 + 104.19}{128.53}\right) = 1.324$ m

Eccentricity, $e = L_R - \dfrac{L}{2} = 1.324 - \dfrac{2.5}{2} = 0.074 \text{ m} < \left(\dfrac{2.5}{6} = 0.417 \text{ m}\right)$

Hence, resultant lies within the middle third of base,

$$\therefore \qquad \frac{6e}{L} = \frac{6 \times 0.074}{2.50} = 0.1776$$

$$\therefore \qquad q_{max,\,min} = \frac{R}{L}\left(1 \pm \frac{6e}{L}\right) = \frac{128.53}{2.5}\left\{1 \pm \frac{6 \times 0.074}{2.5}\right\} = 51.41\,(1 \pm 0.1776)$$

$$\therefore \qquad q_{max} = 60.54 \text{ kN/m}^2 < (q_a = 80 \text{ kN/m}^2) \qquad \text{(OK)}$$

and $\qquad q_{min} = 42.28 \text{ kN/m}^2 > 0$

∴ There is no tension in the footing.

3. **Stability against sliding:**

$$\text{Sliding force} = P_a \cos\theta = 65.49 \text{ kN}$$

Resisting force (ignoring passive earth pressure)

$$F = \mu R$$

$$\Rightarrow \quad F = 0.55 \times 128.53 = 70.69 \text{ kN}$$

$$\therefore \quad (FS)_{\text{sliding}} = \frac{0.9F}{P_a \cos\theta} = \frac{0.9 \times 70.69}{65.49} = 0.971 < 1.4$$

$\Rightarrow$ Not safe against sliding

Hence, a shear key may be provided to counteract the balance force through passive resistance.

4.8 **A separate two-flight staircase is to be constructed for an old residential house as shown in figure. The floor-to-floor height of the building is 3 m. There are 10 nos. of risers of each 150 mm and treads of 250 mm. Four columns are located at *A*, *B*, *C*, *D*. The beams *AB*, *CD* and *AC* are located at floor level and the beam *BD* is located at half landing level. The thickness of waist slab and landing is 150 mm. Live load = 3 kN/m². Half landing slab is cantilevered from the beam *BD*. Neglect the line load along the edge of the half landing. Draw the bending moment diagram for the flight for design purpose.**

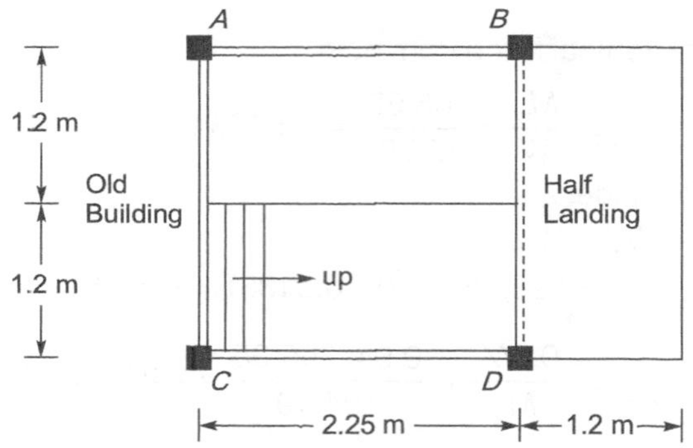

[20 marks : 2017]

Solution :

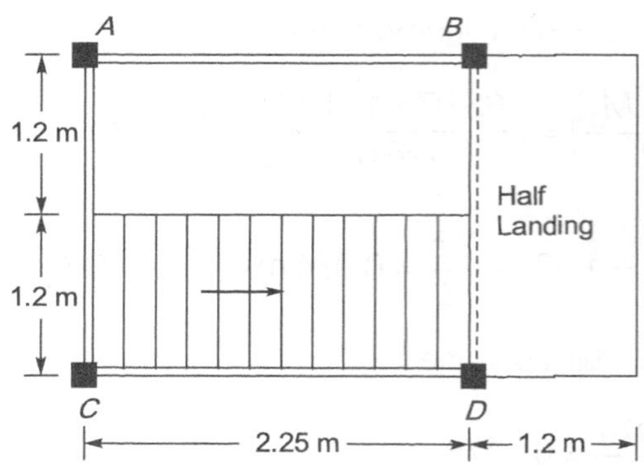

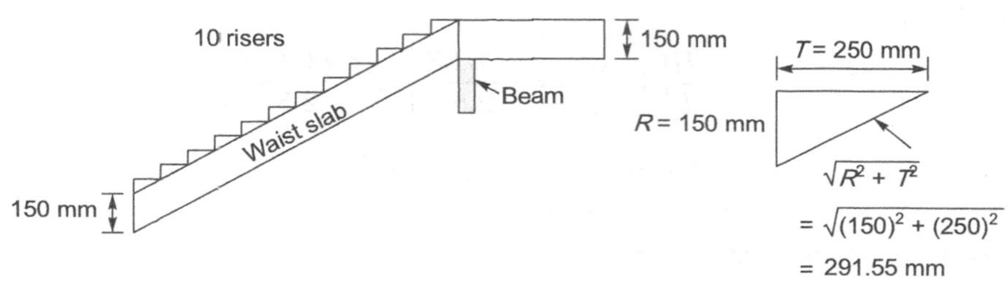

Given R = 150 mm and T = 250 mm
Slab thickness in landing region = 150 mm
Waist slab thickness = 150 mm

Loads on going flight on projected plan area

1. Self weight of waise slab $= 25 \times 0.150 \times \frac{291.55}{250} = 4.37$ kN/m^2
2. Self weight of steps $= 25 \times \frac{1}{2} \times 0.15 = 1.875$ kN/m^2
3. Live loads = 3.0 kN/m^2

Total load = 9.245 kN/m^2

Factored load = 1.5 × 9.245 kN/m^2 = 13.87 kN/m^2

As per IS 456 : 2000 loads on half landing can be taken as one half in each direction asshown in figure

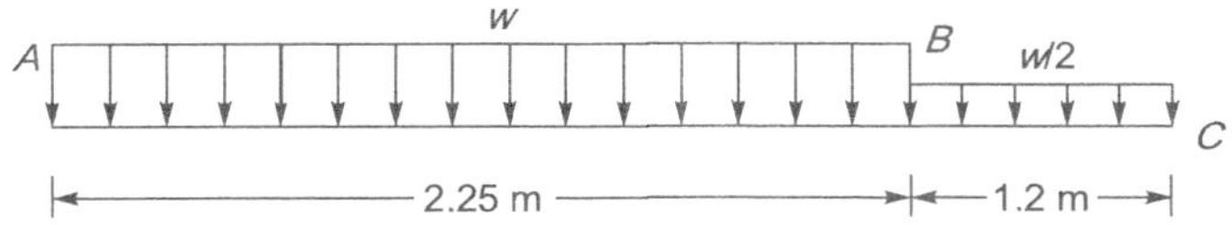

Design moment:

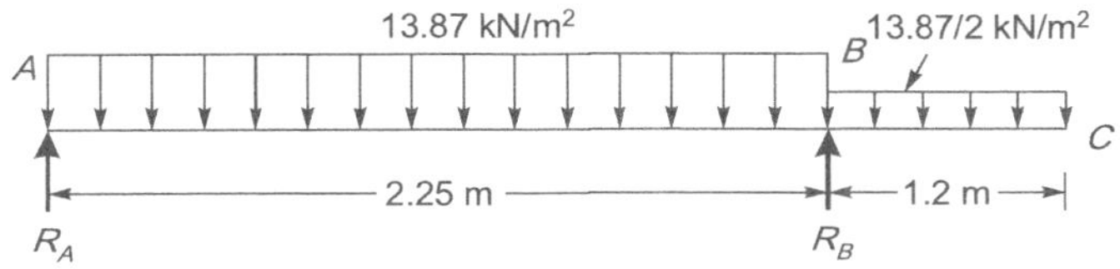

For portion BC;
x measured from C $(0 < x < 1.2)$

$$M_x = \frac{-6.935 \times x^2}{2} = -3.4675x^2$$

At x = 1.2 m $\quad M_B = 5$ kNm/m

$\Sigma M_A = 0 \quad R_B \times 2.25 = 13.87 \times \frac{(2.25)^2}{2} + \left(\frac{13.87}{2}\right) \times 1.2 \times (2.25 + 0.6)$

$\Rightarrow \quad R_B = 26.145$ kN/m

$$R_A = 13.87 \times 2.25 + \frac{13.87}{2} \times 1.2 - 26.145 = 13.38 \text{ kN/m}$$

Design bending moment at centre;

$$M_x = 13.38x - \frac{13.87x^2}{2}$$

$\therefore \quad \frac{dMx}{dx} = 0$

$\Rightarrow \quad 13.38 - 13.87x = 0 \Rightarrow x = 0.965$ m

$$M_{max} = 13.38 \times 0.965 - \frac{13.87 \times (0.965)^2}{2} = 6.45 \text{ kNm/m}$$

At x = 1.125 m, $\quad M_{centre} = 6.275$ kN/m

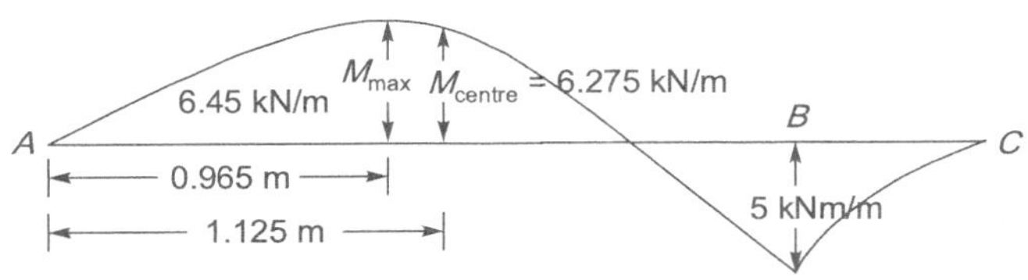

Bending moment diagram

4.9 Design a square footing (bending and shear only) for a column load of 1400 kN at service from a 400 mm square column containing 8 nos. 20 mm bars. The bearing capacity of soil is 100 kN/m^2 at 1 m depth below ground level. The unit weight of the Earth is 20 kN/m^3. Use M25 grade concrete and Fe 415 grade steel, load factor = 1.5. Show reinforcement details. Shear strength of concrete = 0.35 MPa.

[20 marks : 2017]

Solution:

Size of footing:

Given, $P = 1400$ kN, $q_0 = 100$ kN/m^2 at 1 m depth

Assuming the weight of footing + backfill to be 10% of load P.

$$\therefore \quad \text{Footing area required} = \frac{1400 \times 1.1}{100} = 15.40 \text{ m}^2$$

$$\text{Minimum size of footing} = \sqrt{15.40} = 3.92 \text{ m}$$

Assume a 4 m × 4 m footing size.

Thickness of footing slab based on shear:

Net soil pressure at ultimate loads (Load factor = 1.5)

$$\therefore \quad q_u = \frac{1400 \times 1.50}{4.0 \times 4.0} = 131.25 \text{ kN/m}^2 = 0.131 \text{ N/mm}^2$$

(a) One-way shear:

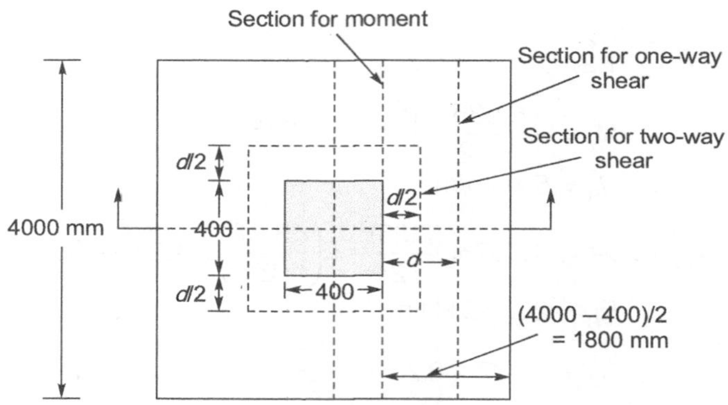

Plan

The critical section for one way shear is at distance d from column face.

$\therefore$ Factored shear force, $V_{u1} = 0.131 \times 4000\,(1800 - d) = 943200 - 524d$

Given, $\tau_c = 0.35$ MPa

$\therefore$ One way shear resistance, $V_{c1} = 0.35 \times 4000 \times d = 1400\,d$

$$\therefore \quad V_{u1} \le V_{c1}$$

$$\Rightarrow \quad 943200 - 524d \le 1400\,d$$

$$\Rightarrow \quad d \ge 490.23 \text{ mm} \simeq 491 \text{ mm}$$

(b) Two way shear:

The critical section for two way shear is act $d/2$ from column periphery

$$\therefore \quad \text{Factored shear force, } V_{u2} = 0.131[(4000)^2 - (400 + d)^2]$$

$$\Rightarrow \quad V_{u2} = 0.131\,[(4000)^2 - (400 + 491)^2]$$

$$\Rightarrow \quad = 1992 \times 10^3 \text{ N}$$

Two way shear resistance, $V_{c2} = k\tau_c[4 \times (400 + d)]d \quad k = 1$ (square column)

$$\tau_c = 0.25\sqrt{f_{ck}} = 0.25\sqrt{25} = 1.25 \text{ N/mm}^2$$

$\therefore$ $$V_{c2} = 1.25 \times 1 \times 4d(400 + d)$$

$\Rightarrow$ $$V_{c2} = (2000d + 5d^2)\text{N}$$

$\therefore$ $$V_{u2} \le V_{c2}$$

$\Rightarrow$ $$1992 \times 10^3 \le 2000d + 5d^2$$

$\Rightarrow$ $$d \ge 462.12 \text{ mm}$$

Evidently, one way shear governs the thickness. Assuming clear cover of 75 mm and 16 ϕ bars in both directions with an average $d = 491$ mm.

$\therefore$ Total thickness, $D = (491 + 75 + 16/2) = 574 \text{ mm} \simeq 590 \text{ mm}$

Now, $d_{eff} = 590 - 75 - 16/2 = 507$ mm

Assuming unit weight of concrete = 24 kN/m^3

Given: Unit weight of soil = 20 kN/m^3

$\therefore$ $$q = \frac{1400}{4 \times 4} + 24 \times 0.59 + 20 \times (1 - 0.59) = 109.86 > 100 \text{ kN/m}^2 \text{ (unsafe)}$$

$\therefore$ Increase footing size to say 4.25 m × 4.25 m

$\therefore$ $$q = \frac{1400}{4.25 \times 4.25} + 24 \times 0.59 + 20(1 - 0.59)$$
$$= 99.87 \text{ kN/m}^2 < 100 \text{ kN/m}^2 \quad \text{(OK)}$$

Design for flexural reinforcement:

$$M_u = \frac{0.131 \times 4250 \times (1925)^2}{2} = 1031.55 \times 10^6 \text{Nmm}$$

$\Rightarrow$ $$R_u = \frac{M_u}{Bd^2} = \frac{1031.55 \times 10^6}{4250 \times (507)^2} = 0.944 \text{ MPa}$$

$\therefore$ $$\frac{A_{st}}{Bd} = \frac{f_{ck}}{2f_y}\left\{1 - \sqrt{1 - \frac{4.6R_u}{f_{ck}}}\right\}Bd$$

$\Rightarrow$ $$(A_{st})_{required} = \frac{25}{2 \times 415}\left[\left\{1 - \sqrt{1 - \frac{4.6 \times 0.944}{25}}\right\}\right] \times 4250 \times 507$$

$\Rightarrow$ $$(A_{st})_{req} = 5905.27 \text{ mm}^2$$
$$(A_{st})_{min} = 0.0012\,BD = 0.0012 \times 4250 \times 590 = 3009 \text{ mm}^2$$

$\therefore$ Provide, $A_{st} = 5905.27 \text{mm}^2$

Using 16 ϕ bars, number of bars required $= \dfrac{5905.27}{\frac{\pi}{4} \times 16^2} = 29.37 \simeq 30$

Corresponding spacing, $s = \dfrac{\{4250 - 75 \times 2 - 30 \times 16\}}{(30 - 1)} = 124.83 \text{ mm} \simeq 120 \text{ mm}$

Provide 16 mm 30 bars both ways.

Check for development length.

$$L_d = \frac{\phi f_{st}}{4\tau_{bd}} = \frac{0.87 \times 415}{4 \times 1.2 \times 1.6} = 47\phi$$

For 16 ϕ bars in footing, $L_d = 47 \times 16 = 752$ mm

Length available = (1925 – 75) = 1850 mm > 752 mm. Hence OK.

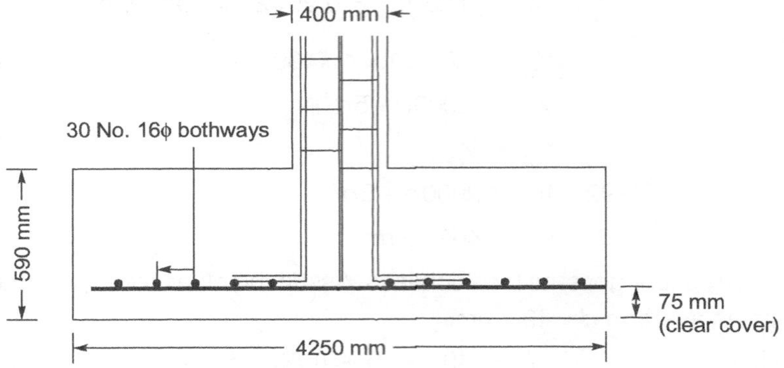

Section *X-X*

4.10 An isolated footing it transferring load from a column (300 mm × 300 mm) as shown in figure. Arrange the plan dimensions of footing so that there will be uniform soil pressure intensity. Following parameters may be used.

1. Column size : 300 mm × 300 mm
2. Safe Bearing capacity : 100 kN/m^2
3. Dead weight of footing and soil weight over it may be taken as 10% of vertical load of column.

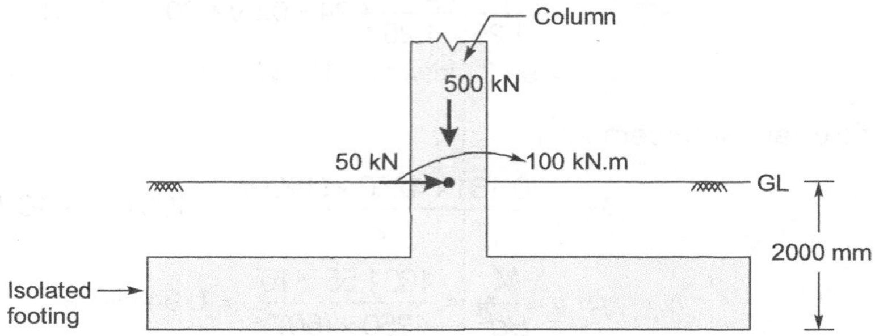

[12 marks : 2018]

Solution:

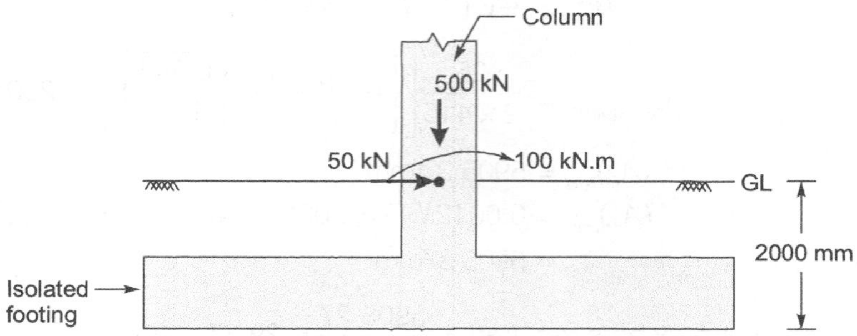

Given:

1. Column size : 300 mm × 300 mm
2. Safe bearing capacity, SBC = 100 kN/m^2
3. Dead weight of footing = 10% of column load

$$\text{Weight of footing} = 0.1 \times 500 \text{ kN}$$

$$\therefore \quad \text{Total load} = 550 \text{ kN}$$

$$\text{Factored moment} = [100 + (50 \times 2)] = 200 \text{ kNm}$$

$$\text{Area of footing} = \frac{\text{Total load}}{\text{Safe bearing capacity}} = \frac{550}{100} = 5.5 \text{ m}^2$$

$$\therefore \quad \text{Size of foundation} = \sqrt{5.5} = 2.34 \text{ provide } 2.5 \text{ m}$$

Provide 2.5 m × 2.5 m footing.

The footing will be so provided that the centre of gravity of the column load will coincide with the centre of gravity of the footing area. The footing is placed symmetrical with respect to the XX axis of the column.

$$\text{Eccentricity} = \frac{\text{Moment}}{\text{Load}} = \frac{200 \times 1000}{550} = 363.6 \text{ mm}$$

Projection of the footing beyond the column faces will be

$$1250 - 363.6 - 150 = 736.4 \text{ mm}$$

and

$$1250 + 363.6 - 150 = 1463.6 \text{ mm}$$

$$\text{Net upward uniform soil pressure intensity} = \frac{550}{2.5 \times 2.5} = 88 \text{ kN/m}^2$$

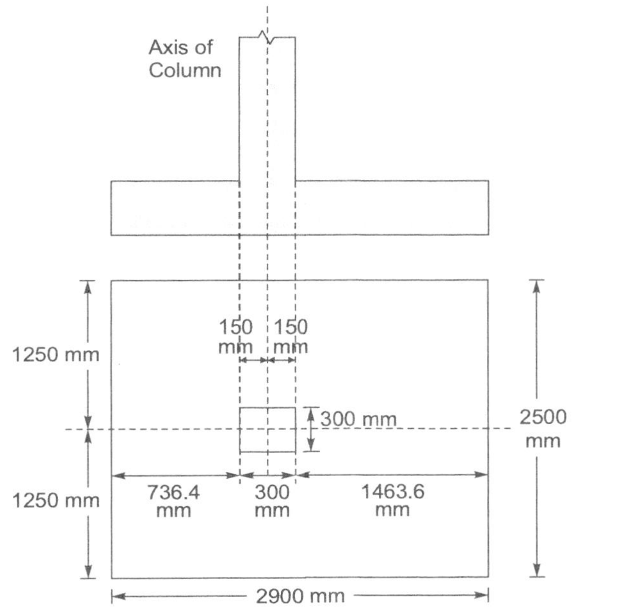

4.11 **Design and sketch the reinforcement in vertical wall, Toe slab and Heel slab (for maximum Bending moment and maximum shear force only) for the cantilever retaining structure shown in figure.**

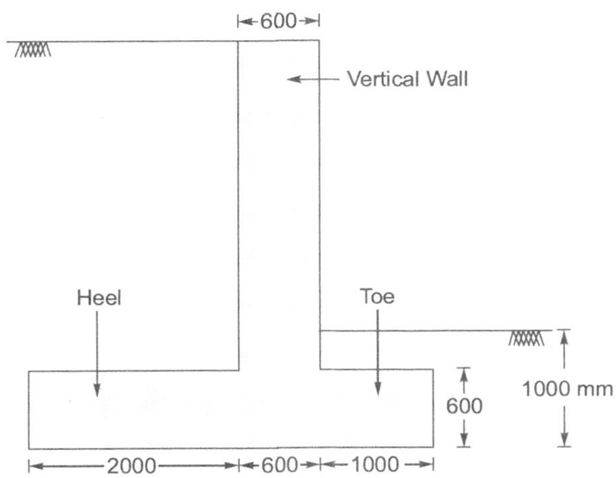

(All dimensions are in mm)

Following parameters may be used for design and sketch:

1. Maximum bending moment in vertical wall : 400 kN-m, Toe : 160 kN-m and in Heel : 200 kN-m.
2. Maximum shear force in vertical wall : 200 kN, Toe : 120 kN and in Heel : 160 kN.
3. Grade of concrete M-20.
4. Grade of reinforcing bar Fe415.
5. Nominal concrete cover 20 mm.
6. Development length : 47 × diameter of bar.
7. Diameter of main bar in vertical wall : 25 mm, in Heel : 25 mm and in Toe : 25 mm.
8. (M_u, lim/$f_{ck} bd^2$) = 0.138.
9. (p_t, lim f_y/f_{ck}) = 19.82.
10. Shear strength of concrete = 0.60 N/mm^2.

[20 marks : 2018]

Solution:

Part A: Vertical wall is designed as cantilever fixed at base.

$$M_u = 1.5 \times \text{Max. BM in vertical wall} = 1.5 \times 400 = 600 \text{ kNm}$$

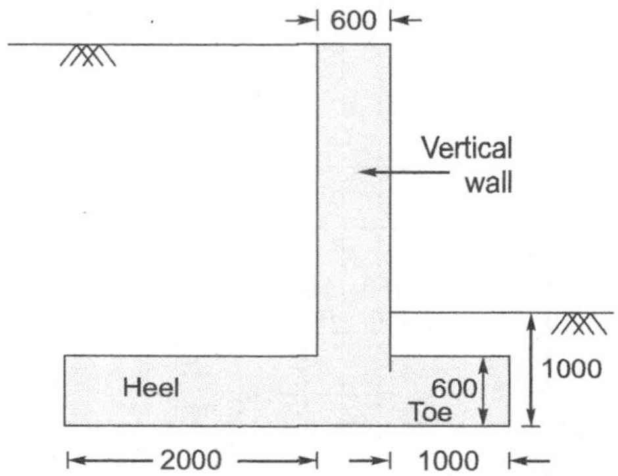

$$\text{Effective thickness} = \left[600 - \text{Nominal cover} - \frac{\phi}{2}\right] = 567.5 \text{ mm}$$

Check whether under reinforcement or over reinforced.

$$M_{ulim} = 0.138 \times f_{ck} bd^2 = 0.138 \times 20 \times 1000 \times 567.5^2 = 888.875 \text{ kNm}$$

∴ $M_u < M_{ulim}$

∴ Under - reinforced

$$A_{st} = 0.5\frac{f_{ck}}{f_y}\left[1-\sqrt{1-\frac{4.6\,M_u}{f_{ck}bd^2}}\right]bd$$

$$= 0.5\frac{20}{415}\left[1-\sqrt{1-\frac{4.6 \times 600 \times 10^6}{20 \times 1000 \times 567.5^2}}\right] \times 1000 \times 567.5$$

$$= 3336.92 \text{ mm}^2$$

Using 25 mm bars, $\quad \text{Spacing} = \dfrac{1000}{A_{st} / \frac{\pi}{4} \times 25^2} = 147.10 \text{ mm}$

provide 25 mm bars @ 140 mm c/c.

$$\text{Development length} = 47\phi = 1175 \text{ mm}$$

$$\text{Distribution bars} = \frac{0.12}{100} \times bd = \frac{0.12}{100} \times 1000 \times 600 = 720 \text{ mm}^2$$

Using 12 mm bars

$$\text{Spacing} = \frac{1000}{A_{st} / \frac{\pi}{4} \times 12^2} = 157 \text{ mm say } 150 \text{ mm}$$

Provide 12 mm bars @ 300 mm c/c, near each face.OR
or We may give 10 mm bars @ 200 mm c/c near each face.

Shear check

$$\text{Factored shear force} = 1.5 \times 200 = 300 \text{ kN}$$

$$\tau_v = \frac{300 \times 10^3}{bd} = \frac{300000}{1000 \times 567.5} = 0.529 \text{ N/mm}^2$$

$\tau_{v\,given} > \tau_v$. Safe in shear.

Part B:

Toe slab reinforcement design:

$$\text{Factored bending moment, } M_u = 1.5 \times 160 = 240 \text{ kNm.}$$

$$\text{Effective depth} = 567.5 \text{ mm}$$

As already shown $M_u < M_{ulim}$

$\therefore$ Under reinforced.

$$A_{st} = 0.5\frac{20}{415}\left[1 - \sqrt{1 - \frac{4.6 \times 240 \times 10^6}{20 \times 1000 \times 567.5^2}}\right] \times 1000 \times 567.5$$

$$= 1227 \text{ mm}^2$$

Using 25 mm bars,

$$\text{Spacing} = \frac{1000}{1227 / \frac{\pi}{4} \times 25^2} = 399.8 \text{ mm} \not> 300 \text{mm}$$

$\therefore$ Provide 25 mm bars @ 300 mm c/c

Provide 12 mm bars @ 300 mm c/c as distribution bars on each face.

Shear check

$$\text{Factored shear force} = 1.5 \times 120 = 180 \text{ kN}$$

$$\tau_v = \frac{180 \times 10^3}{bd} = \frac{180000}{1000 \times 567.5} = 0.317 \text{ N/mm}^2$$

$\tau_v < \tau_c$. Safe in shear.

Part C:

Heel slab reinforcement design:

$$\text{Factored, BM} = 1.5 \times 200 = 300 \text{ kNm.}$$

$$\text{Effective depth} = 567.5 \text{ mm}$$

$$A_{st} = 0.5\frac{20}{415}\left[1 - \sqrt{1 - \frac{4.6 \times 300 \times 10^6}{20 \times 1000 \times 567.5^2}}\right] \times 1000 \times 567.5$$

$$= 1553 \text{ mm}^2$$

Using 25 mm bars,

$$\text{Spacing} = \frac{1000}{1553 / \frac{\pi}{4} \times 25^2} 315.9 \text{ mm} \not> 300 \text{mm}$$

$\therefore$ Provide 25 mm bars @ 300 mm c/c

Provide 12 mm bars @ 300 mm c/c as distribution bars on each face.

Shear check

$$\text{Factored shear force} = 1.5 \times 160 = 240 \text{ kN}$$

$$\tau_v = \frac{240 \times 10^3}{bd} = \frac{240000}{1000 \times 567.5} = 0.423 \text{ N/mm}^2$$

$\tau_v < \tau_v$. Safe in shear.

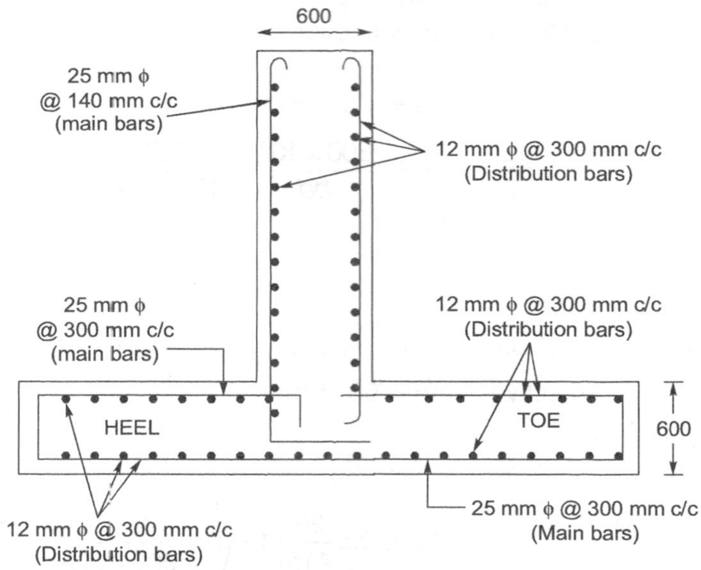

Reinforcement Detailing

4.12 A square column (500 mm x 500 mm) carries load from two beams, which are mutually perpendicular as shown in figure. Overhang portion of beams carry a total load of 615.0 kN/m (include self-weight). Design the column at top of footing level. Footing is fully embedded in very hard rock. Beams are restrained against rotation at Beam-column junction. The minimum eccentricity is less than 0.05 times the lateral dimension of column. Sketch all details required at column cross-section.

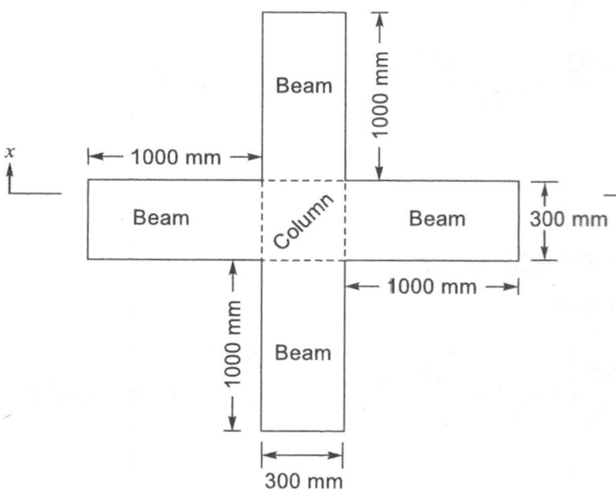

Top Plan

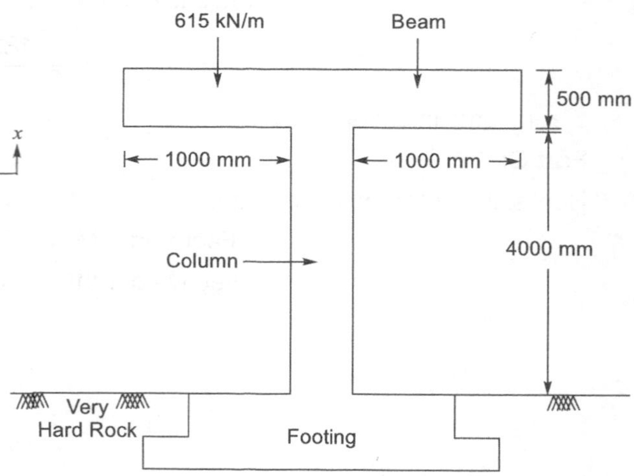

Sectional Elevation x-x

Use: M-20 grade concrete, Fe-415 grade reinforcing bars, Appropriate coefficient from 1.0/1.20/1.50/2.0, Main reinforcing bar : 32 mm diameter.

[20 marks : 2018]

Solution:

$$\text{Total load on column} = 4 \times 615 = 2460 \text{ kN}$$

Self weight is included

$$\therefore \quad W_u = 1.5 \times 2460 \text{ kN} = 3690 \text{ kN}$$

Min eccentricity conditions are satisfied.

Since, beam are restrained against rotation at beam column junction and the column is embedded into very hard rock

$$\therefore \quad L_{eff} = 0.65L = 2.6 \text{ m}$$

$$\therefore \quad \frac{L_{eff}}{D} = \frac{2600}{500} = 45.2 < 12 \qquad \text{Hence short column.}$$

$$A_g = 500 \times 500$$

$$P_u = 0.4 f_{ck} A_c + 0.67 f_y A_{sc}$$

$$3690 \times 1000 = 0.4 \times 20 \times (A_g - A_{sc}) + 0.67 \times 415 \times A_{sc}$$

$$\therefore \quad A_{sc} = 6258.1 \text{ mm}^2$$

$$\text{Minimum steel} = 0.8\% A_g = 2000 \text{ mm}^2 \qquad \text{OK}$$

Using 32 mm bars as specified

$$\text{Number of bars} = \frac{A_{sc}}{\frac{\pi}{4}32^2} = 7.28 \approx 8 \text{ bars}$$

∴ Providing 8 bars of 32 mm diameter.

Design of lateral ties:

According to IS456 : 2000

$$\left.\begin{aligned} \text{Diameter of lateral ties} &\nless \left(\frac{\phi_{main}}{4}\right) = 8 \text{ mm} \\ &\nless 6 \text{ mm} \end{aligned}\right\} \text{max.}$$

∴ Using 8 mm lateral ties.

$$\left.\begin{aligned} \text{Spacing of lateral ties} &\ngtr b \\ &\ngtr 16\phi \\ &\ngtr 300 \text{ mm} \end{aligned}\right\} \text{min.}$$

∴ Give 8 mm ties @ 300 mm c/c.

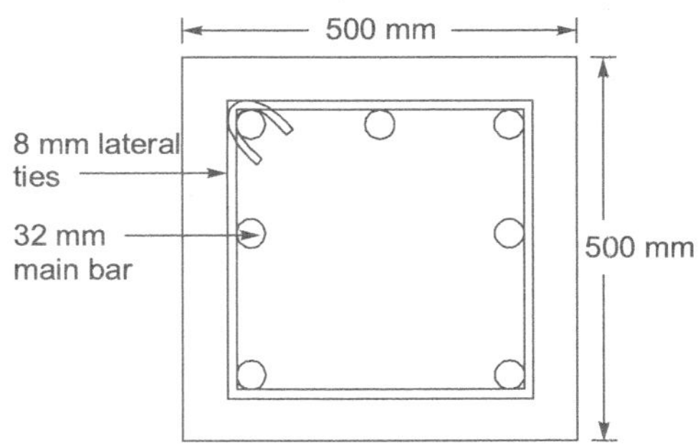

4.13 **Design a square column of height 3 m subjected to an axial load of 1500 kN under dead and live load condition. Use limit state method of design. Assume effective length factor = 1.2. Size of the column is fixed at 400 mm × 400 mm. Show the reinforcement detail (cross-section). Use M25 and Fe500.**

[20 marks : 2020]

Solution:

Step-1: Checking the column as short or long

Unsupported length of column

$$L_0 = 3.0 \text{ m} = 3000 \text{ mm}$$

$$(l_e)_x = (l_e)_y = 3000 \times 1.2 = 3600 \text{ mm}$$

$$D_x = D_y = 400 \text{ mm}$$

Slenderness ratio, $\lambda = \frac{l_e}{D} = \frac{3600}{400} = 9 < 12$

Slenderness ratio is less than 12 so the column is short.

Step-2: Calculation of minimum eccentricity

$$(e_x)_{min} = (e_y)_{min} = \max\begin{cases} \frac{l_x}{500} + \frac{D}{30} = \frac{3000}{500} + \frac{400}{30} = 19.33 \text{ mm} \\ 20 \text{ mm} \end{cases}$$

$$0.05B = 0.05 \times 400 = 20 \text{ mm}$$

So, $e_{min} \not> 0.05B$

Since, the minimum eccentricities are less than 0.05 times the least lateral dimensions, the following formula given by IS 456 : 2000 can be used of the design of axially loaded short columns.

Step-3: Use of column design formula

$$P_u = 0.4f_{ck}A_g + (0.67f_y - 0.4f_{ck})A_{sc}$$

Factored axial load, $P_u = 1.5 \times 1500 = 2250$ kN

$$2250 \times 10^3 = (0.4 \times 25 \times 400^2) + (0.67 \times 500 - 0.4 \times 25)\, A_{sc}$$

$$A_{sc} = 2000 \text{ mm}^2$$

Step-4: Reinforcement design

Provide (4-16 mm) + (4 + 20 mm) bars with 20 mm diameter bars at the corner and 16 mm diameter bars at the face.

$$(A_{sc})_{provided} = 2060.9 \text{ mm}^2 > 2000 \text{ mm}^2$$

$$\text{Percentage of R/F provided} = 2060.9 \times \frac{100}{400^2} = 1.288\%$$

> Minimum R/F required (= 0.8%)

< Maximum reinforced (= 6%)

Step-5: Lateral ties

Diameter of tie, $\phi_t \geq \begin{cases} \frac{\phi_{max}}{4} = \frac{20}{4} = 5 \text{ mm} \\ 6 \text{ mm} \end{cases}$

Adopt tie members as 6 mm

Spacing of ties, $S_t \leq \begin{cases} \text{Least lateral dimension of column} = 400 \text{ mm} \\ 16 \times \phi_m = 16 \times 16 = 256 \text{ mm} \\ 300 \text{ mm} \end{cases}$

Provide 6 mm ϕ @ 250 mm c/c

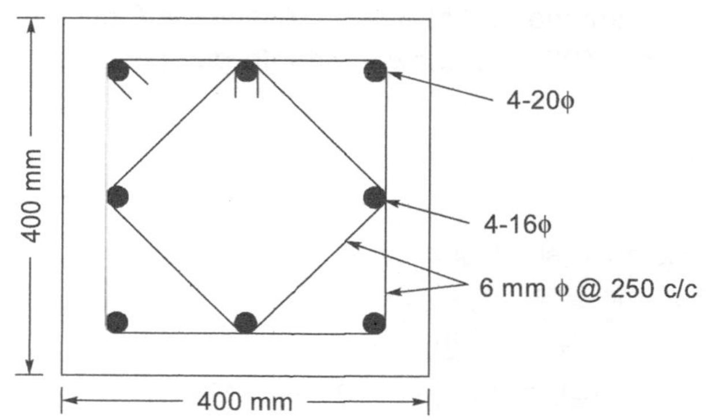

4.14 A combined footing is to be provided for two columns (size 300 × 300) spaced at 3 m c/c. Axial load on each of the columns is 350 kN. The width of the footing is fixed at 1.4 m. A foundation beam of 400 mm × 800 mm is provided along the length. Design the foundation slab using M25 and Fe500. Assume the thickness of the slab varies from 250 mm to 150 mm. Also show the reinforcement detail (in cross-section) of the footing slab. Use limit state method of design. Bearing capacity of the soil is 100 kN/m^2.

$\frac{M_u}{bd^2}$	0.3	0.35	0.4	0.45	0.5
P_t	0.070	0.082	0.094	0.106	0.118

P_t	< 0.15	0.25	0.5	0.75	1.0
τ_c, MPa	0.29	0.36	0.49	0.57	0.64

[20 marks : 2020]

Solution:

Load on each column, $P = 350$ kN

Size of column = 300 × 300 mm

Width of footing, $B = 1.40$ m

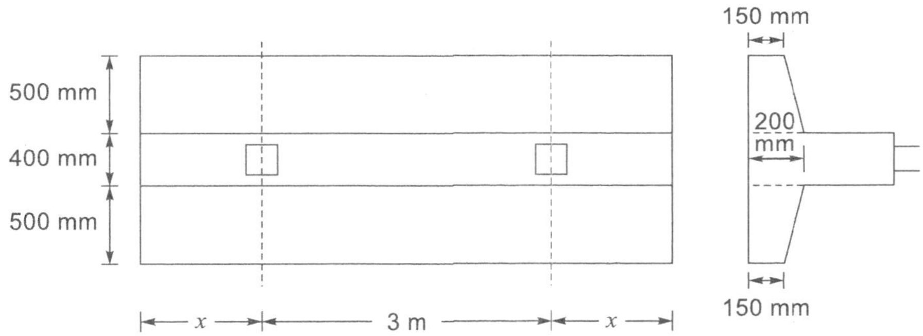

A. **Size of footing**

Total load from column = (2 × 350) = 700 kN

Load of beam and foundation slab (20% P)

= 140 kN

Total $P_T = 840$ kN

Area of footing required,

$$\frac{P_T}{q_0} = \frac{840}{100} = 8.40 \text{ m}^2$$

Length of footing, $\frac{A}{B} = \frac{8.40}{1.40} = 6$ m

Load on both column is same. So projection on two side of column will be same

$L = 3 + 2x = 6$ m

So, $x = 1.50$ m

Design soil pressure

$$w_u = \frac{P_T}{A} = \frac{700}{6 \times 1.40} = 83.33 \text{ kN/m}^2$$

B. Design of slab

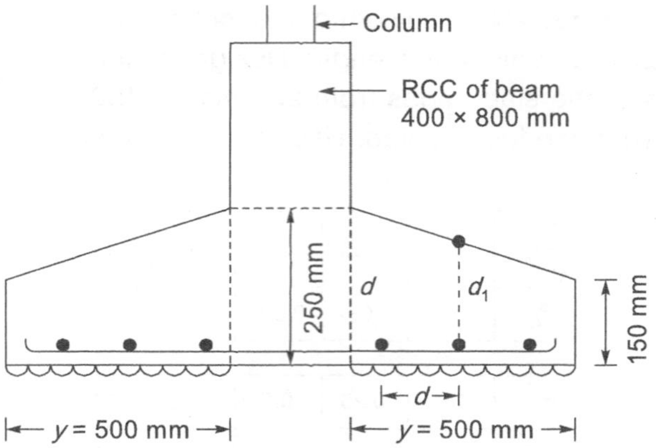

- Thickness of slab is already given in question, the slab on two sides of beam shall be designed as cantilever projection.
- Depth at face of beam = 250 mm
- Consider 1 m width of cantilever slab for design.
- Effective cover = $50+\frac{16}{2} = 58$ mm $\simeq$ 60 mm (say)
- Effective depth, $d = 250 - 60 = 190$ mm at face of beam

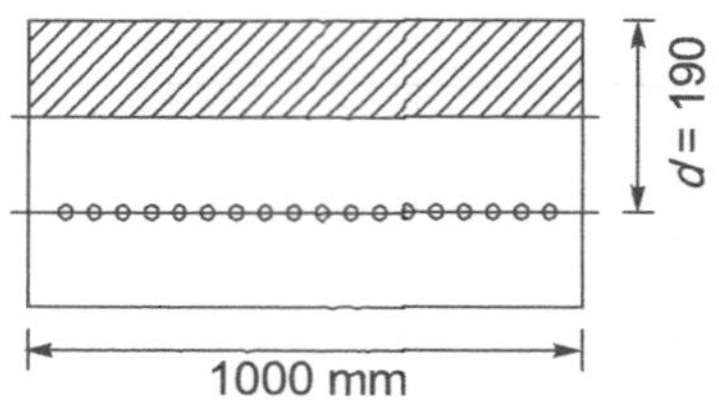

(i) Maximum bending moment (factored)

$$M_u = 1.50\times w_0 \times \frac{y^2}{2} = 1.50\times 83.33\times \frac{0.50^2}{2} = 15.62 \text{ kNm}$$

(ii) $$\frac{M_u}{Bd^2} = \frac{15.62\times 10^6}{1000\times 190^2} = 0.433$$

From table given in question

$$P_t\% = 0.094+\frac{(0.106-0.094)}{(0.45-0.40)}\times(0.433-0.40) = 0.102\%$$

$$A_{st} = \frac{0.102}{100}\times 1000\times 190 = 194 \text{ mm}^2$$

(iii) As per IS code

Minimum area of steel

$$A_{st\,min} = \frac{0.12}{100}\times BD = \frac{0.12}{100}\times 1000\times 250 = 300 \text{ mm}^2 \text{(Provide)}$$

Spacing of 10 mm ϕ bars

$$= \frac{1000}{300}\times\frac{\pi}{4}(10)^2 = 261 \text{ mm}$$

Provide 10 mm ϕ @ 260 mm c/c

(iv) Same minimum reinforcement may be provided as distribution bars. 10 mm ϕ @ 260 mm c/c

(v) Check for shear

Maximum shear force at 'd' distance from face of beam c/s of shear

Projection $y_2 = y - d = 0.50 - 0.19 = 0.31$ m

Maximum shear force,

$$V_u = 1.50 \times w_0 \times y_2 \times 1 = 1.50 \times 83.33 \times 0.31 \times 1 = 38.75 \text{ kN}$$

Total depth of slab at 'd' distance

$$= 250 - \frac{(250-150)}{500} \times 190 = 212 \text{ mm}$$

Effective depth, $d_1 = 212 - 60 = 152$ mm

Shear stress, $\frac{V_u}{BD_1} = \frac{38.75 \times 1000}{1000 \times 152} = 0.255 \text{ N/mm}^2$

$P_t\% = 0.12\%$ provided.

$\tau_c = 0.29$ $\tau_V < \tau_C$ = safe in shear.

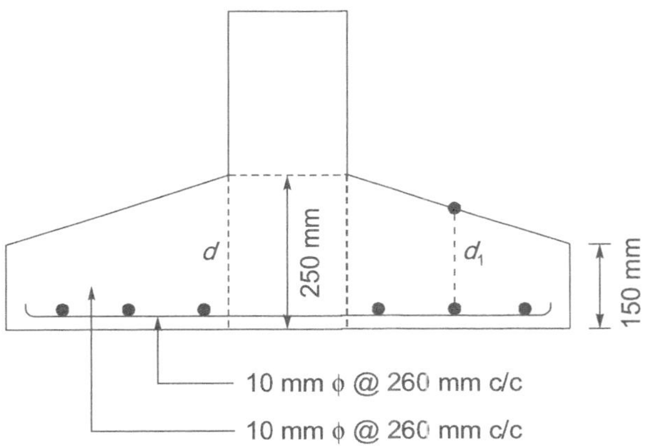

(Beam reinforcement not asked to design in question)

4.15 An RCC cantilever retaining wall is to be designed to support the soil as shown below. Design and sketch the reinforcement for vertical wall only. Also sketch the position of main reinforcement in Toe and Heel slab. (Do not design Toe and Hell slab). Neglect the effect of passive earth pressure and self-weight of vertical wall. Water table is not affecting the moisture condition of retained soil.

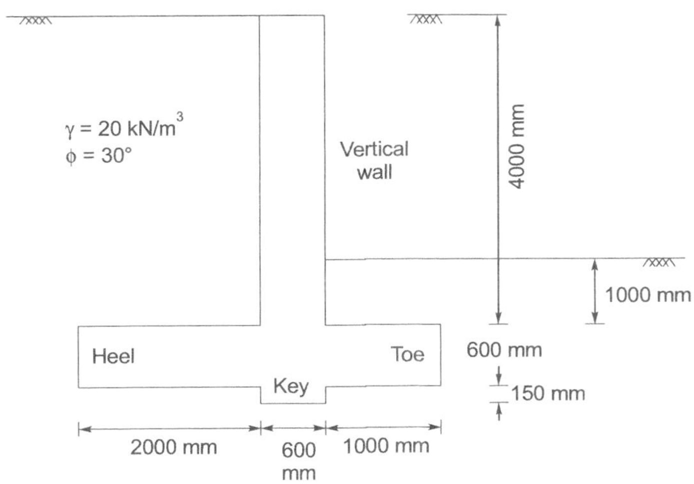

(1) M 20 grade of concrete
(2) Fe 415 grade of reinforcement
(3) Diameter of main and distribution reinforcement : 8 mm
(4) Minimum effective thickness required : 400 mm
(5) $K_a = \frac{1-\sin\phi}{1+\sin\phi}$

For M 20 and Fe 415

Percentage reinforcement p_t in %

$\frac{M_u}{bd^2}$	0.4	0.5	0.6	0.667
p_t	0.114	0.142	0.172	0.204

Design shear strength of concrete τ_c in N/mm^2

p_t	0.15	0.25	0.5	0.75
τ_c	0.28	0.36	0.48	0.56

[20 Marks : 2021]

Solution:

Design of vertical wall of retaining wall----stem.

$$\text{Height of wall} = 4.0 \text{ m}$$

Coefficient of active earth pressure

$$k_a = \frac{1-\sin\phi}{1+\sin\phi} = \frac{1-\sin 30}{1+\sin 30} = \frac{1}{3}$$

Active earth pressure at bottom

$$= k_a \cdot \gamma \cdot H = \frac{1}{3} \times 20 \times 4 = 26.67$$

Total active earth pressure thrust on the wall:

$$P_a = \frac{1}{2} k_a \cdot \gamma \cdot H^2 = \frac{1}{2} \times \frac{1}{3} \times 20 \times 4^2 = 53.33 \text{ kN/m}$$

Maximum BM (factored)

$$BM_u = 1.5 \times P_a \times \frac{H}{3} = 1.5 \times 53.33 \times \frac{4}{3} = 106.67 \text{ kN/m} \quad \text{Say } 107 \text{ kN-m}$$

Let us provide thickness of

$$d = 400 \text{ mm as given in question}$$

$$\text{Total depth} = d + 60 \text{ mm effective cover}$$

$$= 400 + 60 = 460 \text{ mm}$$

Eff. depth required

$$d = \sqrt{\frac{BM_u}{Q.B}} = \sqrt{\frac{107 \times 10^6}{0.138 \times 20 \times 1000}}$$

$$= 197 \text{ mm} < 400 \text{ mm available} \quad \text{safe}$$

$$A_{st\,min} = \frac{0.12}{100} \times 1000 \times 460 = 552 \text{ mm}^2$$

$$\frac{BM_u}{Bd^2} = \frac{107 \times 10^6}{1000 \times 400^2} = 0.668$$

From table given:

$$P_t\% = 0.204\%$$

$$A_{st} = \frac{0.204}{100} \times 1000 \times 400 = 816\,\text{mm}^2$$

$$\text{Spacing of 8 mm } \phi = \frac{1000}{816} \times \frac{\pi}{4}(8)^2 = 61.6\,\text{mm}$$

Provide 8 mm ϕ @ 60 mm c/c.

Vertical steel = 8 mmϕ @ 60 mm c/c

Horizontal steel = Minimum = 552 mm²

Spacing of 8 mmϕ steel.

$$= \frac{1000}{552} \times \frac{\pi}{4}(8)^2 = 91\,\text{mm}$$

Provide 8 mmϕ @ 90 mm c/c

Check for shear:

Maximum shear force at base= $V_u = 1.5\,P_a = 1.5 \times 53.33 = 80$ kN

$$\tau_v = \frac{V_u}{Bd} = \frac{80000}{1000 \times 400} = 0.20\ \text{N/mm}^2$$

$$P_t\% = 0.204\%$$

$$\tau_c = 0.28 + \frac{(0.36 - 0.28)}{(0.25 - 0.15)} \times (0.204 - 0.15) = 0.32\ \text{N/mm}^2.$$

$$\tau_v < \tau_c \quad \text{Safe}$$

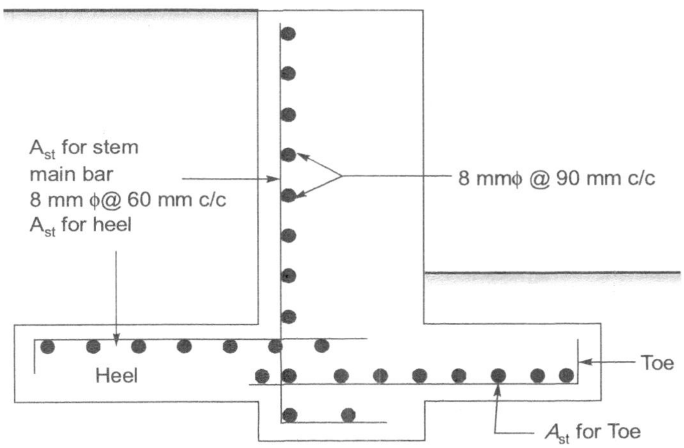

4.16 Design an RCC square column with the following data. Provide the main reinforcement and stirrups. Check the standard guidelines of IS 456 : 2000.

(i) Column size : 500 mm × 500 mm

(ii) Concrete grade M 20

(iii) Steel reinforcement Fe 415

(iv) Design load: 1600 kN

(v) Design moment $M_{ux} = M_{uy} = 100$ kNm

(vi) Eccentricities $e_x = e_y = 20$ mm

(vii) Minimum reinforcement : 1.2%

(viii) Check $\left[\frac{M_{ux}}{M_{ux_1}}\right]^{\alpha_n} + \left[\frac{M_{uy}}{M_{uy_1}}\right]^{\alpha_n} < 1.0$, where $\alpha_n = \frac{P_u}{P_{uz}}$

(ix) Diameter of longitudinal reinforcement = 20 mm

(x) Diameter of lateral ties = 8 mm

(xi) Nominal cover to reinforcement = 40 mm.

Compression with bending-Rectangular
Section-Reinforcement Distributed equally on four sides

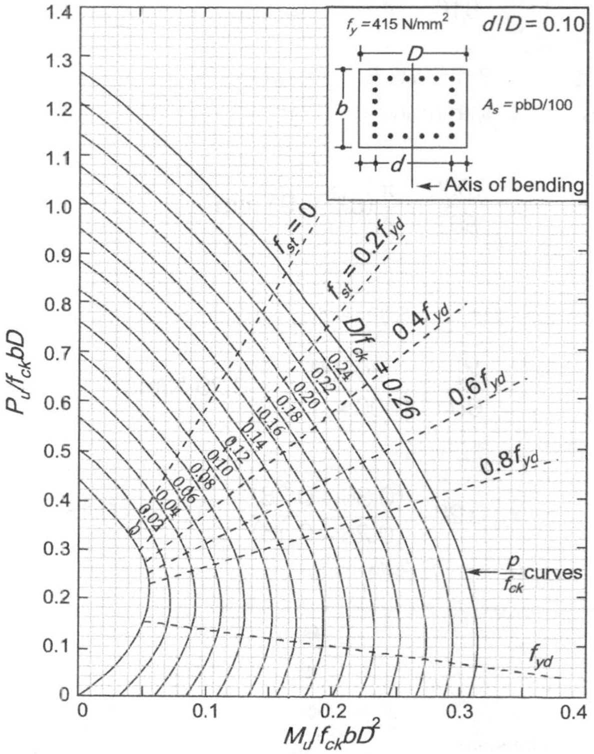

f_y
250
415
500
f_{ck}
15
20
25
30

Values of P_{uz} for compression members

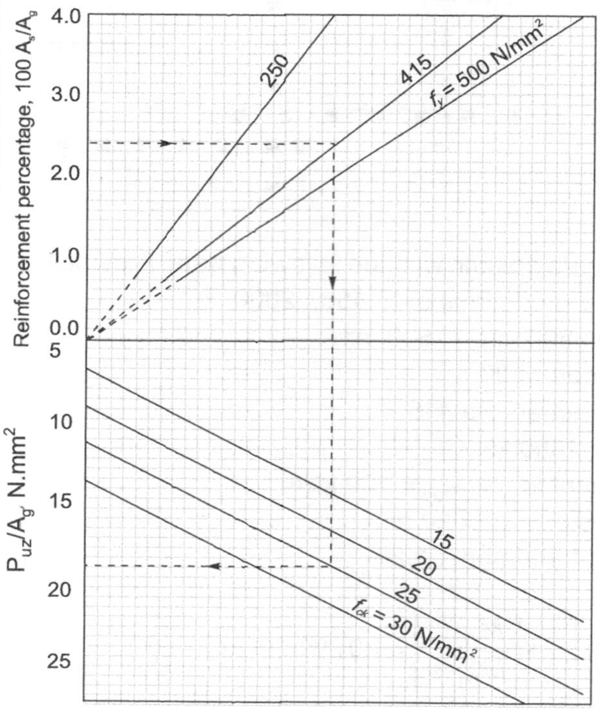

[20 Marks : 2022]

Solution:

Design of column

$$B = 500 \text{ mm} - \text{square}$$
$$P_u = 1600 \text{ kN} = \text{Consider factored}$$
$$M_{ux} = M_{uy} = 100 \text{ kNm}$$

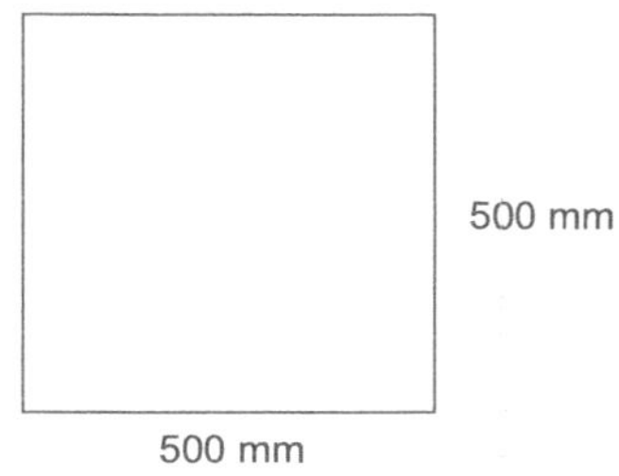

Minimum eccentricity

$$e_{x,\min} = e_{y,\min} = \left(\frac{L}{500} + \frac{B}{30}\right) \text{ or } 20 \text{ mm}$$
$$= 20 \text{ mm} \quad \text{(given in question)}$$

Minimum moments, $M_{ux} = M_{uy} = P_u \cdot e_{\min}$

$$= 1600 \times \frac{20}{1000} = 32 \text{ kN-m } (< 100 \text{ kNm}) \quad \text{OK}$$

Trial 1 Consider A_{st} = Minimum = 1.20% (given in question)

$$A_{sc} = \frac{1.2}{100} \times 500 \times 500 = 3000 \text{ mm}^2$$
$$P_{uz} = 0.45 \times f_{ck} \cdot A_c + 0.75 f_y \cdot A_{sc}$$
$$= [0.45 \times 20 \times (500 \times 500 - 3000) + 0.75 \times 415 \times 3000]/1000$$
$$= 3156.75 \text{ kN}$$

P_{uz} can be found from figure also.

For $P_t\% = 1.20\%$ (M20 | Fe 415)

$$\frac{P_{uz}}{Ag} = 12.50$$
$$P_{uz} = 12.50 \times A_g$$
$$= 12.50 \times \frac{500^2}{1000} = 3125 \text{ kN}$$
$$\frac{P_u}{p_{uz}} = \frac{1600}{3156.75} = 0.507$$
$$\alpha_n = \text{Calculated based on } \frac{P_u}{P_{uz}}$$
$$= 1.0 + \frac{(2.0 - 1.0)}{(0.8 - 0.2)} \times (0.507 - 0.20) = 1.5116$$
$$\frac{P_t}{f_{ck}} = \frac{1.20}{20} = 0.06$$
$$\frac{P_u}{f_{ck}Bd} = \frac{1600 \times 1000}{20 \times 500 \times 500} = 0.32$$

From interaction curve - value of

$$\frac{M_{ux1}}{f_{ck}BD^2} = 0.10$$
$$M_{ux1} = 0.10 \times 20 \times 500 \times \frac{500^2}{10^6} = 250 \text{ kN-m}$$

Since column is square,

$$M_{ux1} = M_{uy1}$$

$$\text{Check}\left(\frac{M_{ux}}{M_{ux1}}\right)^{\alpha_n} + \left(\frac{M_{uy}}{M_{uy1}}\right)^{\alpha_n} = \left(\frac{100}{250}\right)^{1.5116} + \left(\frac{100}{250}\right)^{1.5116}$$

$$= 0.500 < 1.0 - \text{Safe}$$

So we can provide minimum steel 1.2% as suggested in question.

$$\text{Area of steel} = A_{sc} = 3000 \text{ mm}^2$$

Number of 20 mmϕ bars

$$n = \frac{3000}{\frac{\pi}{4}(20)^2} = 9.50$$ Provide 10 bars of 20 mmϕ.

Lateral ties:

$$\text{Diameter} = 8 \text{ mm} \quad \text{(given in question)}$$

$$\left[\frac{20}{4} = 5 \text{ mm}\right]$$

Spacing:

$$\text{min.}\begin{cases} \text{Least lateral dimension} = 500 \text{ mm} \\ 16\phi_{\text{longi.min}} = 16 \times 20 = 320 \text{ mm} \\ = 300 \text{ mm} \end{cases}$$

So, provide 8 mmϕ @ 300 mm c/c.

Reinforcement details:

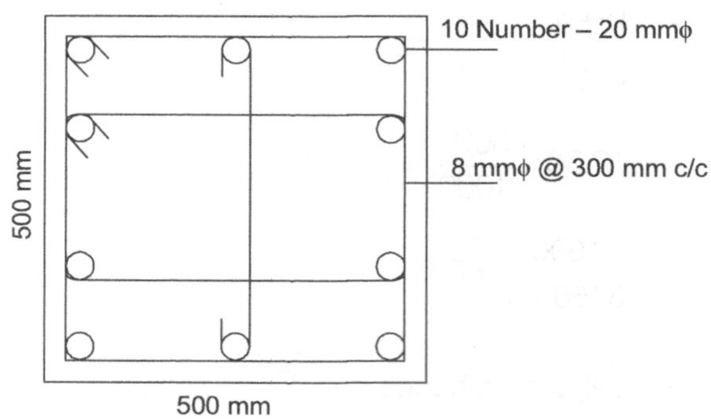

4.17 A short column of 3 m effective height is subjected to an ultimate load of 1400 kN. Cross section of the column is 300 × 600 mm. The reinforced concrete column consists of 8 longitudinal reinforcing bars of 20 mm diameter as shown. M25 grade concrete and Fe500 grade are used. Obtain its ultimate moment carrying capacity only about its major axis. Effective cover to the longitudinal reinforcing bars is 60 mm.

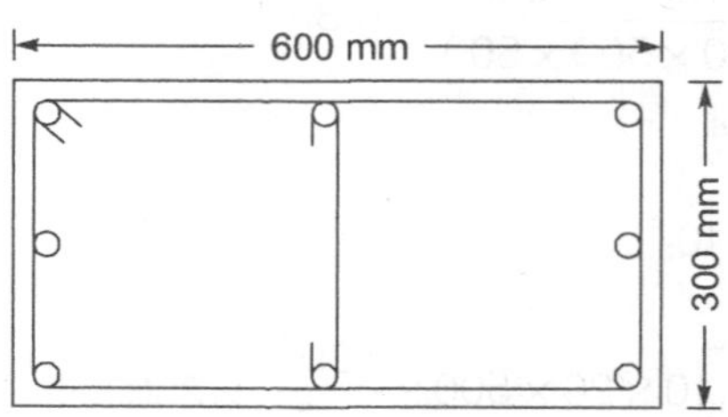

[12 marks : 2022]

Solution:

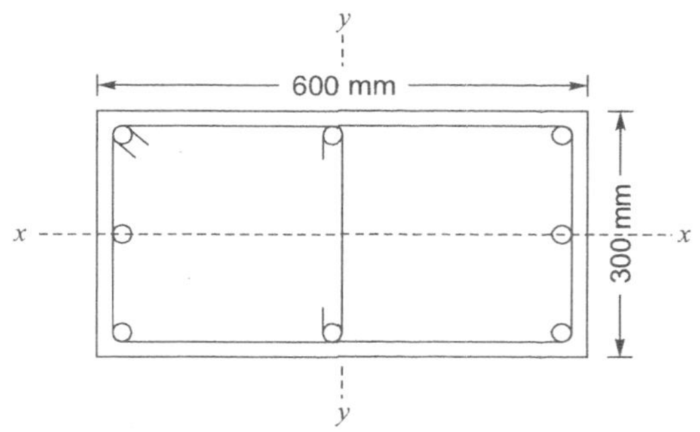

$$A_{sc} = 8 \times \frac{\pi}{4} \times 20^2 = 2513.27 \text{ mm}^2$$

$\because$ Moment of inertia of column section is maximum about *y-y* axis and thus y-y axis is the major axis of column section.

$$\therefore \quad I_{yy} = \frac{300 \times 600^3}{12} = 54 \times 10^8 \text{ mm}^4$$

Effective cover, $d' = 60$ mm

Ultimate load, $P_u = 1400$ kN

$$\therefore \quad \frac{P_u}{f_{ck}bD} = \frac{1400 \times 10^3}{25 \times 300 \times 600} = 0.311$$

$$\frac{d'}{D} = \frac{60}{600} = 0.1$$

Percentage reinforcement, $p = \frac{A_{sc}}{bD} \times 100 = \frac{2513.27}{300 \times 600} \times 100 = 1.396\%$

$$\therefore \quad \frac{p}{f_{ck}} = \frac{1.396}{25} = 0.05584 \simeq 0.056 \text{ (say)}$$

$\therefore$ Using chart 48, for $\frac{P_u}{f_{ck}bD} = 0.311$ and $\frac{p}{f_{ck}} = 0.056$

$$\frac{M_u}{f_{ck}bD^2} = 0.10$$

$$\Rightarrow \quad M_u = 0.10 \times 25 \times 300 \times 600^2 \text{ Nmm} = 270 \text{ kNm}$$

$\therefore$ Ultimate moment carrying capacity of column about major axis is 270 kNm.

4.18 Explain the specific reasons for the following:
Helically reinforced circular columns have better compressive strength than that of similar columns with lateral ties.

[6 marks : 2023]

Solution:

Load carrying capacity of helically reinforced column are generally found more than column with normal lateral ties.

IS 456 : 2000 recommends to consider compressive strength of helically reinforced column 5% more than normal column with lateral ties.

Reasons:

1. Helical reinforcement is provided like a spring at very close pitch (value ≯ 75 mm). The reinforcement provided is continuous providing better lateral strength to concrete.
2. In case of helically reinforced column concrete, inside core is subjected to triaxial compression so chance of failure of such column is reduced.
3. Column with helical reinforcement are generally used in circular column. These columns are found more ductile than column with normal lateral ties, due to less pitch and continuity of steel around main bars.
4. Columns reinforced with helical reinforcement are considered more effective for earthquake resistant structure.

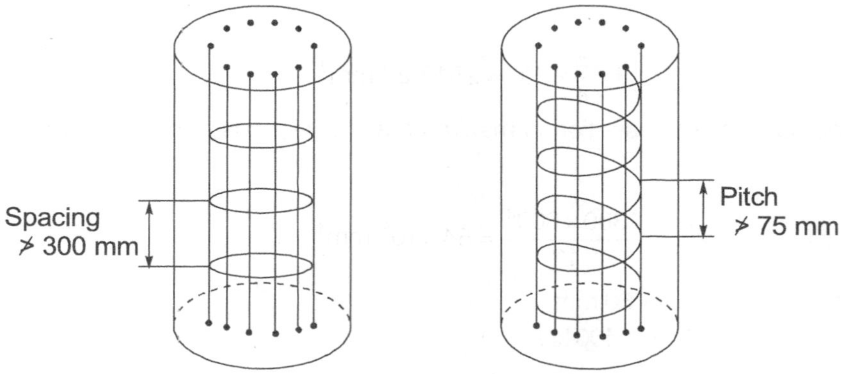

Column with seperate lateral ties (Less strength) (Less ductile)

Column with helical reinforcement (More strength) (More ductile)

4.19 A short braced reinforced concrete column has unsupported length of 3.5 m and size of 300 mm × 360 mm. Verify the applicability of simplified formula of P_u (i.e., ultimate load-carrying capacity of a short axially loaded column) as given in the Code. Also, determine the design moments due to minimum eccentricity to be considered for this case if the column is subjected to an ultimate axial load of 1600 kN."

[10 marks : 2023]

Solution:

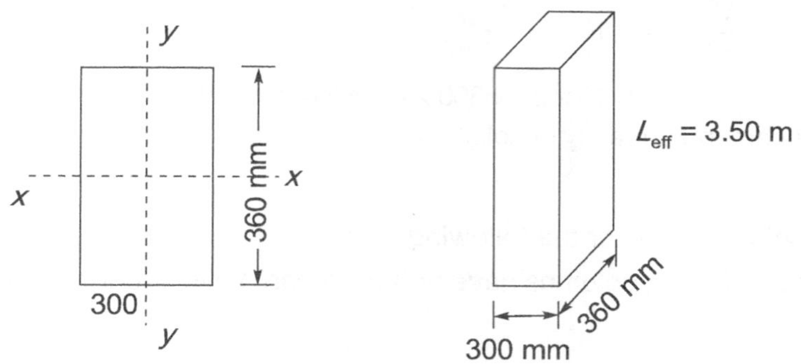

Before using I.S. code formula, as per provision number 39.3.

$$P_u = 0.40\, f_{ck} \cdot A_C + 0.67\, f_y \cdot A_{sc}$$

Following conditions need to be checked:

(a) The column is axially loaded.

$$P_u = 1600 \text{ kN} \qquad \text{(OK)}$$

(b) The column is short column.

$$\frac{L_e}{D} = \frac{3500}{360} = 9.72 < 12$$

$$\frac{L_e}{B} = \frac{3500}{300} = 11.67 < 12$$

$\therefore$ It is a short column (OK)

(c) As per *Cl.* 25.4 minimum eccentricity should be less than or equal to 0.05 times lateral dimension.

1. About x-x

$$e_{min;\,x} = \frac{L_{x0}}{500} + \frac{D}{30} = \frac{3500}{500} + \frac{360}{30}$$

$$= 19 \text{ mm or } 20 \text{ mm (Maximum of two)}$$

$$\therefore \quad e_{min;\,x} = 20 \text{ mm}$$

$$0.05\,D = 0.05 \times 360 = 18 \text{ mm}$$

$$e_{min,\,x} > 0.05D - \text{Condition not satisfied.}$$

2. About y-y axis

$$e_{min;\,y} = \frac{L_{y0}}{500} + \frac{B}{30} = \left(\frac{3500}{500} + \frac{300}{30}\right)$$

$$= 17 \text{ mm} \quad \text{or } 20 \text{ mm}$$

$$\therefore \quad e_{min,\,y} = 20 \text{ mm} \quad \text{(Maximum of two)}$$

$$0.05B = 0.05 \times 300 = 15 \text{ mm}$$

$$e_{min,\,y} > 0.05B - \text{Condition not satisfied.}$$

So, the formula $P_u = 0.40\, f_{ck} \cdot A_c + 0.67 f_y \cdot A_{SC}$ should not be used.

(d) Due to minimum eccentricity, minimum design moments:

$$e_{min,\,xx} = 20 \text{ mm} \quad \text{(as calculated above)}$$

$$e_{min,\,yy} = 20 \text{ mm}$$

$$M_{u,\,min,\,xx} = P_u \cdot e_{min,\,x}$$

$$= 1600 \times \frac{20}{1000} = 32 \text{ kN-m}$$

$$M_{u,\,min,\,yy} = P_u \cdot e_{min,\,y}$$

$$= 1600 \times \frac{20}{1000} = 32 \text{ kN-m}$$

4.20 A trapezoidal combined footing supports two columns of sizes 450 mm × 450 mm and 550 mm × 550 mm carrying service load of 750 kN and 1250 kN respectively. The CG of smaller column lies at 0.45 m from the property line. The centre-to-centre distance of two columns is 4.4 m. The total length of footing is to be restricted to 5.6 m. Determine and show the layout plan of the above footing. Consider safe bearing capacity of soil = 155 kN/m^2.

[10 marks : 2023]

Solution:

$$\text{Column 1} = 450 \times 450 \text{ mm}$$

$$P_1 = 750 \text{ kN}$$

$$\text{Column 2} = 550 \times 550 \text{ mm}$$

$$P_2 = 1250 \text{ kN}$$

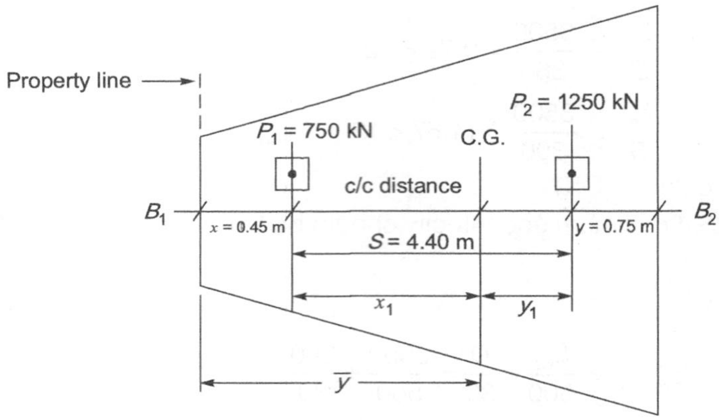

(1) Distance of short column centre to property line

$$x = 0.45 \text{ m}$$

c/c distance, $S = 4.40$ m

Total length, $L = 5.60$ m

$$y = L - x - S$$
$$= 5.60 - 0.45 - 4.40 = 0.75 \text{ m}$$

(2) Size of footing required

$$\text{Total load} = P_1 + P_2$$
$$= 750 + 1250 = 2000 \text{ kN}$$

weight of footing [assuming 20% of $(P_1 + P_2)$] = 400 kN

Assuming weight of soil = [10% of $(P_1 + P_2)$] = 200 kN

$$\text{Total load} = 2600 \text{ kN}$$

$$\text{Area of footing required, } A = \frac{P_{total}}{q_0} = \frac{2600}{155} = 16.775 \text{ say } 16.80 \text{ m}^2$$

$$\therefore \quad \left(\frac{B_1 + B_2}{2}\right) \times L = 16.80$$

$$B_1 + B_2 = \frac{16.80 \times 2}{5.60} = 6.0 \text{ m}$$

(3) C.G of footing:

$$\bar{y}_{CG} = \left(\frac{B_1 + 2B_2}{B_1 + B_2}\right) \times \frac{L}{3}$$

$$= \frac{B_1 + 2B_2}{6.0} \times \frac{5.60}{3.0} = (B_1 + 2B_2) \times \left(\frac{5.60}{18.0}\right)$$

C.G of load from P_1:

$$x_1 = \frac{P_2 \cdot S}{P_1 + P_2} = \frac{1250 \times 4.4}{2000} = 2.8125 \text{ m}$$

$$\bar{y}_{CG} = x_1 + x$$
$$= 2.8125 + 0.45 = 3.2625 \text{ m}$$

So, $$(B_1 + 2B_2) \times \frac{5.60}{18.0} = 3.2625$$

$$\Rightarrow \quad (B_1 + B_2 + B_2) = \frac{3.2625 \times 18}{5.60}$$

$$\Rightarrow \quad 6.0 + B_2 = 10.4866 \text{ m} = 10.487 \text{ m say}$$

$$\Rightarrow \quad B_2 = 10.487 - 6.0 = 4.487 \text{ m}$$

$$\therefore \quad B_1 = (B_1 + B_2) - B_2$$

$$= 6.0 - 4.487 = 1.513 \text{ m}$$

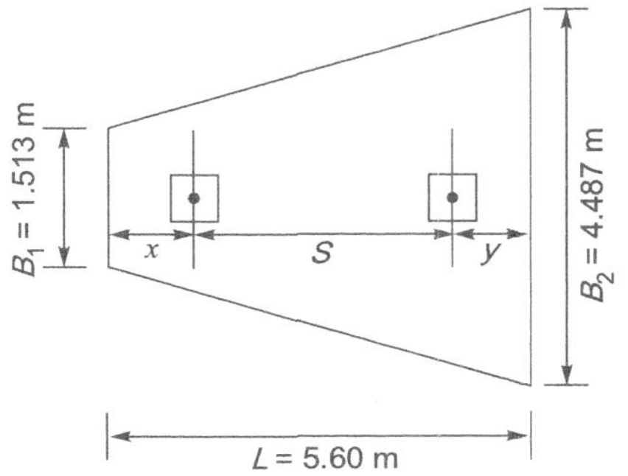

5. Prestress Concrete

5.1 **List various methods used for post tensioning of concrete structures. Describe salient features of Magnel-Blaton system of post-tensioning.**

[10 marks : 2000]

Solution:

Various methods used for post tensioning of concrete structures are:

(i) Freyssinet System
(ii) Magnel Blaton System
(iii) Gifford Udall System
(iv) PSC Monowire System
(v) CCL Standard System
(vi) Lee-Mccall System
(vii) Electrical Prestressing

Magnel Blaton System: In this system a cable of rectangular section is provided, which contains layers of wires 5 mm to 8 mm diameter. The wires are arranged with four wires per layer. The wires in the same layer and adjacent layers are separated with a clearance of 4 mm. The geometrical pattern of the wires is maintained in the same form throughout the length of the cable by providing grills or spacers at regular intervals. The grills do not offer any appreciable frictional resistance to the wires which can be moved relative to each other during the tensioning process.

The wires are anchored by wedging, two at a time into sandwich plates. The sandwich plates are about 25 mm thick and are provided with two wedge shaped grooves on its two faces. The wires are taken two in each groove and tightened. Then a steel wedge is driven between the tightened wires to anchor them against the plate. A complete anchorage unit may consist of one to eight sandwich plates, the number of wires depending on the number of wires in the cable. Each plate can anchor eight wires. The various sandwich plates forming a unit are arranged one above the other against a distribution plate. The wires are tightened by jacking two wires at a time.

5.2 **A simply supported post tensioned concrete beam of span 15 m has a rectangular cross section 300 mm × 600 mm. The prestress at ends is 1150 kN with zero eccentricity at the supports. The eccentricity at the mid span is 200 mm. The cable profile is parabolic. Assuming friction coefficient k_f = 0.15 per 100 m and μ = 0.35, determine the loss due to friction at the centre of the beam.**

[10 marks : 2001]

Solution:

Given data: Span = 15 m, B = 300 mm, D = 600 mm Initial prestressing force, P_0 = 1150 kN, Eccentricity at supports = 0, Eccentricity at mid span = 200 mm, k_f = 0.15 per 100 m = $\frac{0.15}{100}$ per m = 0.0015 per meter

μ = 0.35

Since the cable profile is parabolic, the equation of profile if

$$y = \frac{4h}{l^2} x (l - x)$$

$$\Rightarrow \quad \frac{dy}{dx} = \frac{4h}{l^2}(l - 2x)$$

$$\Rightarrow \quad \theta\,(x = 0) = \frac{4h}{l}$$

$$\Rightarrow \quad \theta = \frac{4 \times 200}{15000} = 0.0533 \text{ radians}$$

Since, prestressing is being done on the both ends, so $\alpha = \theta$

$$\therefore \quad \alpha = 0.0533 \text{ radians}$$

$$\therefore \quad P_x = P_0 e^{-(k_f x + \mu\alpha)} \quad \left[\text{Distance of centre from one end, } x = \frac{15}{2} = 7.5 \text{ m}\right]$$

$$\Rightarrow \quad P_x = 1150\, e^{-(0.0015 \times 7.5 + 0.35 \times 0.0533)}$$

$$\Rightarrow \quad P_x = 1116.12 \text{ kN}$$

$$\therefore \quad \text{Loss of prestress} = P_0 - P_x = 1150 - 1116.12 = 33.88 \text{ kN}$$

$$\% \text{ loss of prestress} = \frac{33.88}{1150} \times 100 = 2.946\%$$

5.3 **A pretensioned beam of size 225 mm x 300 mm deep is prestressed by 12 wires 5 mm diameter initially stressed at 1100 MPa. The centroid of the prestressing wires is located at 100 mm from bottom. Estimate the loss of prestress due to elastic deformation, creep, shrinkage and relaxation for the following stipulations:**

Grade of concrete = M40

Relaxation of steel = 5%

E_s = 2 x 10^5 MPa

Creep coefficient = 1.6

Residual shrinkage strain = 3 x 10^{-4}

[10 marks : 2003]

Solution:

Area of beam section, $A = 225 \times 300 = 67500 \text{ mm}^2$

Moment of Inertia, $I = \frac{BD^3}{12} = \frac{225 \times 300^3}{12} = 5.06 \times 10^8 \text{ mm}^4$

Eccentricity, $e = 150 - 100 = 50$ mm

Modulus of Elasticity of concrete, $E_c = 5000\sqrt{f_{ck}} = 5000\sqrt{40} = 31622.78$ MPa

$\therefore$ Modular ratio, $m = \frac{E_s}{E_c} = \frac{2 \times 10^5}{31622.78} = 6.32$

Area of wires, $A_{st} = 12 \times \frac{\pi}{4} \times (5)^2 = 235.62 \text{ mm}^2$

$\therefore$ Initial prestressing force, $P_0 = 1100 \times 235.62 = 259.18$ kN

Since span is not given, therefore dead load moment is ignored. Now, stress in concrete at the level of steel,

$$f_c = \frac{P_0}{A} + \frac{P_0 e^2}{I}$$

$$\Rightarrow \quad f_c = \frac{259.18 \times 10^3}{67500} + \frac{259.18 \times 10^3 \times (50)^2}{5.06 \times 10^8} = 5.12 \text{ MPa}$$

(i) Loss of prestress due to **elastic shortening** = $mf_c = 6.32 \times 5.12 = 32.36$ MPa

(ii) Loss due to **creep** in concrete = $\phi m f_c = 1.6 \times 6.32 \times 5.12 = 51.78$ MPa

(iii) Loss due to **shrinkage** in concrete = shrinkage strain $\times E_s = 3 \times 10^{-4} \times 2 \times 10^5 = 60$ MPa

(iv) Loss due to **relaxation** in steel = 5% of initial stress = $\frac{5}{100} \times 1100 = 55$ MPa

Total loss = 32.36 + 51.78 + 60 + 55 = 199.14 MPa

$$\% \text{ loss} = \frac{199.14}{1100} \times 100 = 18.10\%.$$

5.4 **Enumerate the losses in prestress in prestressed concrete beams. Why are there no similar losses in reinforcement concrete beams?**

[10 marks : 2004]

Or

Explain various types of "losses of prestress" briefly.

[5 marks : 2005]

Solution:

Loses of prestress take place in a prestressed concrete member due to many causes. In general these losses may be classified into the following:

(i) Loss of prestress during tensioning process

(ii) Loss of prestress at anchoring stage

(iii) Losses occurring subsequently

Loss of prestress during tensioning process: There always exist a certain amount of friction in the jacking and anchorage system and on the walls of the duct where the wires fan out at the anchorages with the result, the actual stress in the tendon is less than what is indicated by the pressure gauge. The losses due to friction in the jack and at the anchorage are different for different systems of prestressing. Considerable friction loss takes place due to friction between tendon and material surrounding it, namely the concrete or the sheathing. This loss due to friction may be classified into (a) loss due to length effect and (b) loss due to curvature effect. Thus the combined loss of prestress due to above two effects is given as

$$\text{Loss of prestress} = P_0 (Kx + \mu\alpha)$$

where P_0 = Prestressing force in the prestressed steel at the tensioning end.

K = friction coefficient for wave effect whose value ranges between 15×10^{-4} per meter

μ = Coefficient of friction in curve

α = the cumulative angle in radians through which the tangents to the cable profile has turned between any two points under consideration.

x = distance of any point from jack

It may be noted that this loss occurs only in post tensioned members.

Loss of prestress at anchoring stage: This loss is due to the fact that the anchorage fixtures themselves are subjected to a stretch. It is also possible that the friction wedges holding the wires may slip a little. The **loss of prestress due to anchorage slip** is given by

$$\Delta f_s = \frac{\Delta a}{L} E_s$$

where Δa = anchorage slip

E_s = Young's modulus of elasticity

L = Length of the tendon

This loss is applicable to post tensioned members only.

Losses occurring subsequently: The losses which occur subsequent to prestressing are the following:

(a) Loss of prestress due to shrinkage of concrete

∴ Loss of prestress = Shrinkage strain × E_s

Value of shrinkage strain

for pretensioned members = 3×10^{-4}

for post tensioned members = $\dfrac{2 \times 10^{-4}}{\log_{10}(T+2)}$

where T is the age of concrete at transfer in days.

(b) Loss of stress due to creep of concrete

loss of prestress due to creep = $\phi m f_c$

where ϕ = creep coefficient = $\dfrac{\text{creep strain}}{\text{elastic strain}}$

m = modular ratio = $\dfrac{E_s}{E_c}$

f_c = original prestress in concrete

ϕ	age of loading
2.2	7
1.6	28
1.1	1 year

(c) Loss of stress due to creep of steel (stress relaxation)

Steel undergoes loss of prestress due to creep. Generally it is given as a percentage of initial stress. It varies from 1% – 5%. It may be normally adopted as 3%.

(d) Loss of prestress due to elastic shortening.

(i) For **pretensioned member**, loss of prestress is given by

$$\Delta f_c = m f_c$$

where m = modular ratio

f_c = stress in concrete at the level of steel

(ii) **Post tensioned members:**

The members get shortened due to compressive force applied over the surface at the time of tensioning. So if only one reinforcement is provided in post tensioned members, there will be no loss of prestress due to elastic shortening of the member because elastic shortening of concrete takes place before anchoring the steel. When there a number of bars tensioned simultaneously, then also loss of prestress due to elastic shortening is zero. But if they are tensioned one by one, the loss of prestress will occur in every bar except one which under goes tension in last.

5.5 **A prestressed concrete beam 400 mm wide and 600 mm deep has a span of 6 m. The beam is prestressed with a tendon bent as shown. There is a central concentrated load of 180 kN acting on the beam. Effective prestressing force = 1200 kN.**
Calculate the extreme fibre stresses at mid span taking into account the self weight of beam also.

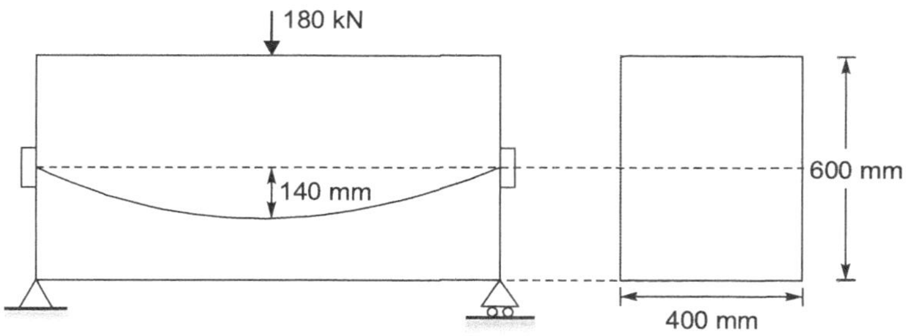

[10 marks : 2005]

Solution:

Bending moment due to concentrated load at mid span

$$M_1 = \frac{PL}{4} = \frac{180 \times 6}{4} = 270 \text{ kN-m}$$

Self weight of beam = $0.4 \times 0.6 \times 1 \times 25 = 6$ kN/m

∴ Bending moment due to self weight at mid span,

$$M_2 = \frac{6 \times (6)^2}{8} = 27 \text{ kN-m}$$

Area of cross section, $A = 400 \times 600 = 240000 \text{ mm}^2$

Section modulus, $Z = \frac{BD^2}{6} = \frac{400 \times (600)^2}{6} = 2.4 \times 10^7 \text{ mm}^3$

Effective prestressing force,

$$P_0 = 1200 \text{ kN}$$

Assuming compressive stresses as positive.

Stress at top, $f_{top} = \frac{P_0}{A} - \frac{P_0\, e}{Z} + \frac{M_2}{Z} + \frac{M_1}{Z}$

Stress at bottom, $f_{bot} = \frac{P_0}{A} + \frac{P_0\, e}{Z} - \frac{M_2}{Z} - \frac{M_1}{Z}$

$$\frac{P_0}{A} = \frac{1200 \times 10^3}{240000} = 5 \text{ MPa}$$

$$\frac{P_0\, e}{Z} = \frac{1200 \times 10^3 \times 140}{2.4 \times 10^7} = 7 \text{ MPa}$$

$$\frac{M_2}{Z} = \frac{27 \times 10^6}{2.4 \times 10^7} = 1.125 \text{ MPa}$$

$$\frac{M_1}{Z} = \frac{270 \times 10^6}{2.4 \times 10^7} = 11.25 \text{ MPa}$$

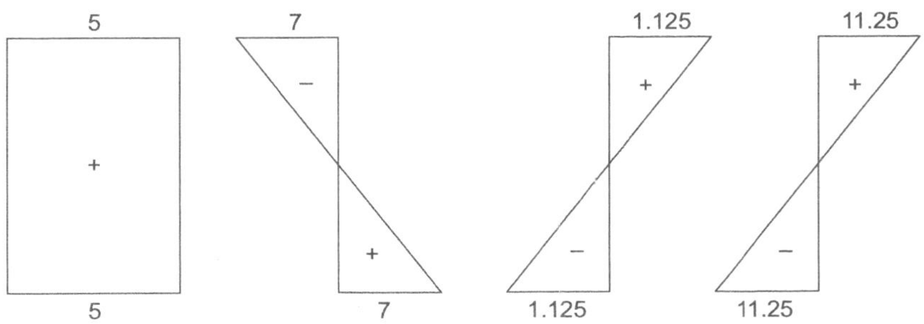

$$f_{top} = 5 - 7 + 1.125 + 11.25 = 10.375 \text{ MPa}$$
$$f_{bot} = 5 + 7 - 1.125 - 11.25 = -0.375 \text{ MPa}$$

5.6 A prestressed concrete beam with a cross section 300 mm width and 600 mm depth, is 12 m long. It carries a live load of 12 kN/m in addition to its self weight. It is prestressed with 2000 mm^2 high tensile steel located at 175 mm from soffit. The cable profile is straight for full length of the beam and it is stressed to a level of 800 N/mm^2. It is bonded to concrete. Determine the location of thrust line in the beam and plot its position at end, and at mid span section.

Assume $\frac{E_s}{E_c} = 6.0$

[15 marks : 2006]

Solution:

Bending Moment m at mid span, $M = \frac{wl^2}{8} = \frac{12 \times (12)^2}{8} = 216$ kN-m

Prestressing force, $P = A_{st} \times f_s = 2000 \times 800 = 16 \times 10^5 \text{ N} = 1600$ kN

Shift of thrust line or pressure line from cable line at mid span is given by

$$a = \frac{M}{P} = \frac{216 \times 10^6}{1600 \times 10^3} = 135 \text{ mm}$$

Eccentricity $= 300 - 175 = 125$ mm

Eccentricity of pressure line $= 135 - 125 = 10$ mm.

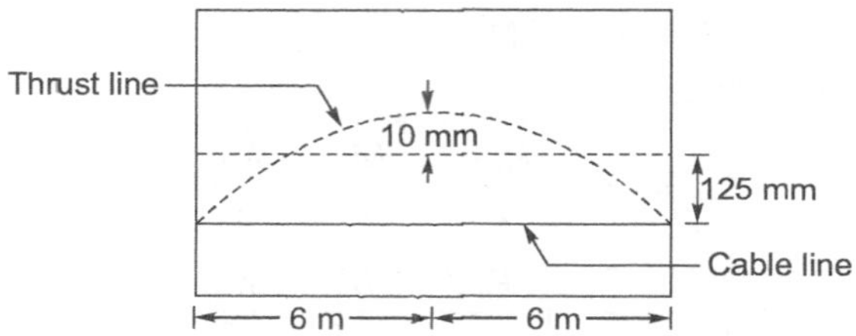

5.7 A post tensioned prestressed beam of rectangular section 240 mm wide is to be designed for a live load of 25 kN/m, uniformly distributed on an effective span of 12 m. The stress is concrete must not exceed 17 MPa in compression or 1.4 MPa in tension at any time and the loss of prestress shall be assumed as 15 per cent.

(i) Calculate the minimum possible depth of beam.

(ii) For the section designed above, calculate the minimum prestressing force and the corresponding eccentricity.

[20 marks : 2007]

Solution:

Imposed load, $q = 25$ kN/m, $h = 1 - \frac{15}{100} = 0.85$

Width of section, $b = 240$ mm, $f_{ct} = f_{cw} = 17$ MPa

$f_{tt} = f_{tw} = -1.4$ MPa, Span = 12 m

Let overall depth of the section be h.

Live load moment, $M_q = \frac{qL^2}{8} = \frac{25 \times 12^2}{8} = 450 \; kN\text{-}m$

Dead load moment, $M_g = \frac{bh \times 25 \times 12^2}{8} = 450bh$ N-mm

Range of stress at bottom fibre, $f_{br} = \eta f_{ct} - f_{tw}$

$$= 0.85 \times 17 - (-1.4) = 15.85 \text{ MPa}$$

(i) **Minimum section modulus is given by**

$$Z_b = \frac{M_q + (1-\eta)M_g}{f_{br}}$$

$$\Rightarrow \quad \frac{bh^2}{6} = \frac{450\times10^6 + (1-0.85)\times 450bh}{15.85}$$

$$\Rightarrow \quad \frac{240\times h^2}{6} = \frac{450\times10^6 + 0.15\times450\times240\times h}{15.85}$$

$$\Rightarrow \quad 634\,h^2 = 450\times10^6 + 16200\,h$$

$$\Rightarrow \quad 634h^2 - 16200h - 450\times10^6 = 0$$

$$\Rightarrow \quad h = 855.35 \text{ mm}$$

Take $h = 860$ mm

(ii) **For the section provided b = 240 mm and h = 860 mm**

Area of section, $A = 240 \times 860 = 206.4 \times 10^3 \text{ mm}^2$

$$Z_b = Z_t = \frac{bh^2}{6} = \frac{240\times860^2}{6} = 29.6\times10^6 \text{ mm}^3$$

Self weight moment, $M_g = 450bh = 450 \times 240 \times 860 = 92.8 \times 10^6$ N-mm

$$f_{sup} = f_{tt} - \frac{M_g}{Z_t} = -1.4 - \frac{92.8\times10^6}{29.6\times10^6} = -4.53 \text{ MPa}$$

$$f_{inf} = \frac{f_{tw}}{\eta} + \frac{M_q+M_g}{\eta Z_b} = -\frac{1.4}{0.85} + \frac{(450+92.8)\times10^6}{0.85\times29.6\times10^6} = 19.93 \text{ MPa}$$

Minimum prestressing force is given by

$$P = \frac{A(f_{inf}Z_b + f_{sup}Z_t)}{Z_b+Z_t} = \frac{206.4\times10^3(19.93-4.53)\times29.6\times10^6}{2\times29.6\times10^6}$$

$$= 1589280 \text{ N} = 1589.28 \text{ kN}$$

Corresponding eccentricity is given by

$$e = \frac{Z_tZ_b(f_{inf}-f_{sup})}{A(f_{inf}Z_b+f_{sup}Z_t)} = \frac{(29.6\times10^6)^2\times(19.93+4.53)}{206.4\times10^3\times(19.93-4.53)\times29.6\times10^6}$$

$$= 227.78 \text{ mm}$$

5.8 The prestressed concrete beam shown in the figure, provided with 9 nos. of 4 mm diameter wires as shown is subjected to a prestress of 800 N/mm^2. Determine the sagging moment that can be applied to the section so that the maximum compressive stress in concrete shall not exceed 14 N/mm^2 and the maximum tensile stress in the concrete shall not exceed 1.4 N/mm^2. Neglect the losses in prestress.

[15 marks : 2008]

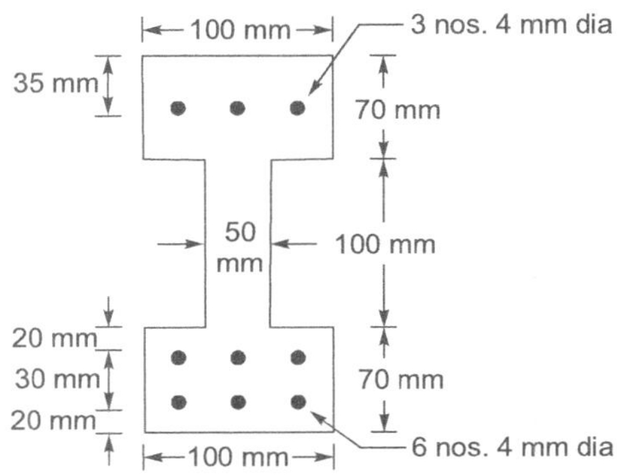

Solution:

Area of steel wires in the top flange $= 3\times\frac{\pi}{4}\times4^2 = 37.7 \text{ mm}^2$

Area of steel wires in the bottom flange $= 6 \times \frac{\pi}{4} \times 4^2 = 75.4 \text{ mm}^2$

Total area of steel $= 37.7 + 75.4 = 113.1 \text{ mm}^2$

Prestressing force applied, $P = \frac{113.1 \times 800}{10^3} = 90.48 \text{ kN}$

Area of the beam section, $A = 2 \times 100 \times 70 + 50 \times 100 = 19000 \text{ mm}^2$

Moment of inertia of the beam section,

$$I = \frac{100 \times 240^3}{12} - \frac{50 \times 100^3}{12} = 1.1103 \times 10^8 \text{ mm}^4$$

Section modulus, $Z = \frac{1.1103 \times 10^8}{120} = 925250 \text{ mm}^3$

Direct stress, $\frac{P}{A} = \frac{90.48 \times 10^3}{19000} = 4.76 \text{ N/mm}^2$

Extreme stress due to eccentricity of the prestressing force

$$= \mp\left[\frac{75.4 \times 800 \times 85}{925250} - \frac{37.7 \times 800 \times 85}{925250}\right]$$

$$= \mp 2.77 \text{ N/mm}^2$$

Let the maximum sagging bending moment for the section be M.

(i) Limiting the extreme compressive stress at top at 14 N/mm²

$$\therefore \quad 4.76 - 2.77 + \frac{M}{925250} = 14$$

$$\Rightarrow \quad M = 11.11 \text{ kN-m}$$

(ii) Limiting the extreme tensile stress at bottom to 1.4 N/mm²

$$\therefore \quad 4.76 + 2.77 - \frac{M}{925250} = -1.4$$

$$\Rightarrow \quad M = 8.26 \text{ kN-m}$$

$\therefore$ Maximum sagging bending moment that can be applied will be 8.26 kN-m.

5.9 **Explain the losses in prestress due to shrinkage and creep.**

[10 marks : 2008]

Solution:

Loss of prestress due to shrinkage: Shrinkage means contraction of concrete due to chemical changes and drying. The shrinkage of concrete in prestressed members result in a shortening of tensioned wires and hence contributes to the loss of stress. The shrinkage of concrete is influenced by the type of cement and aggregates and the method of curing used. Use of high strength concrete with low water cement ratios result in reduction in shrinkage and consequent loss of prestress. The primary cause of drying shrinkage is the progressive loss of water from concrete. The differential shrinkage between the interior and surface of large members may result in strain gradients leading to surface cracking. Hence, proper curing is essential to prevent shrinkage cracks in prestressed members.

In the case of pretensioned members, generally moist curing is resorted to in order to prevent shrinkage until the time of transfer. Consequently, the total residual shrinkage strain will be larger in pretensioned members after transfer of prestress in comparison with post-tensioned members, where a portion of shrinkage will have already taken place by the time of transfer of stress.

The loss of stress in steel due to the shrinkage of concrete is estimated as

$$\text{Loss of stress} = \epsilon_{cs} \times E_s$$

where ϵ_{cs} is total residual shrinkage strain having values of 300×10^{-6} for pretensioning and $\frac{200 \times 10^{-6}}{\log_{10}(t+2)}$ for post-tensioning where t is age of concrete at transfer in days.

This value may be increased by 50 per cent in dry atmospheric conditions, subject to a maximum value of 300×10^{-6} units. E_s is modulus of elasticity.

Loss of prestress due to creep: Creep of concrete means the deformation of concrete which depends upon the interval of time to which the member is loaded.

The ratio, $\phi = \frac{\text{creep strain}}{\text{elastic strain}}$ is called creep coefficient.

The magnitude of the creep coefficient ϕ, varies depending upon the humidity, concrete quality, duration of applied loading and the age of concrete when loaded. It is taken from 1.5 to 2. The lower value may be taken in the case of post-tensioned members due to late application of prestress. But in the case of pretensioned members which receive the prestress early, the coefficient of creep is found to be equal to 2 and in a few cases, slightly greater than 2.

The **loss of prestress due to creep** = $\phi m f_c$

where ϕ is creep coefficient, f_c is original prestress in concrete and m is modular ratio.

5.10 A simply supported high strength concrete beam of rectangular cross-section ($b \times D$), shown in the figure, supports uniformly distributed dead and live loads of intensities 'p' and 'q' per unit length respectively. The beam is prestressed by a straight tendon carrying a prestressing force P at an eccentricity 'e'. Show clearly and neatly the stress distributions, through the beam depth, due to eccentric prestressing, dead and live loads at a cross-section where maximum stresses occur.

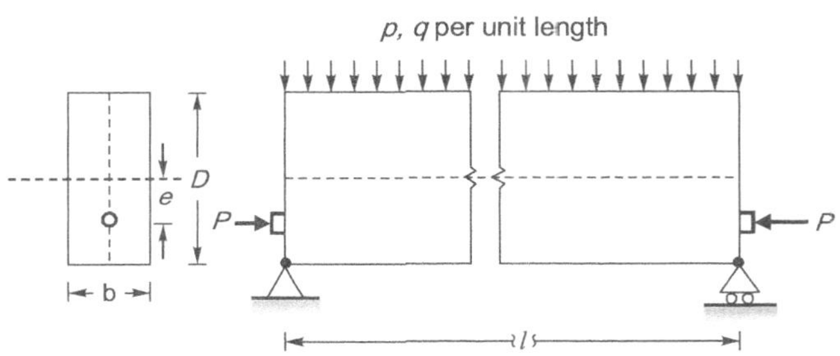

[15 marks : 2009]

Solution:

The stresses on the section consist of the following:

Assuming compressive stress as positive and tensile stress as negative.

(i) **Direct stress** due to prestressing force.

$$= +\frac{P}{A} = +\frac{P}{bD}$$

(ii) **Extreme stresses** due to the **eccentricity** of the prestressing force

$$= \mp\frac{Pe}{Z} = \mp\frac{6Pe}{bD^2} \quad \text{[for rectangular section } Z = bD^2/6]$$

(iii) **Extreme stresses due to dead and live load** will occur at a section where bending moment is maximum. The bending moment due to dead and live loads will be maximum at the centre of span.

$\therefore$ Bending moment, $M = \dfrac{(p+q)L^2}{8}$

Extreme stresses due to dead and live loads

$$= \pm\frac{M}{Z} = \pm\frac{6(p+q)L^2}{8\times bD^2} = \pm\frac{3}{4}\frac{(p+q)L^2}{bD^2}$$

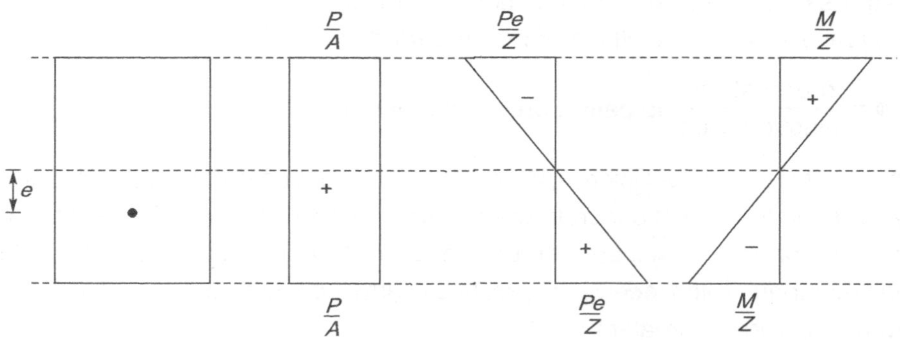

Direct stress | Stress due to eccentricity | Stress due to dead and live loads

$$\text{Final stress at the top fibre} = \frac{P}{bD} - \frac{6Pe}{bD^2} + \frac{3}{4}\frac{(p+q)L^2}{bD^2}$$

$$\text{Final stress at the bottom fibre} = \frac{P}{bD} + \frac{6Pe}{bD^2} - \frac{3}{4}\frac{(p+q)L^2}{bD^2}$$

5.11 **A Pretensioned beam of size 250 mm × 350 mm prestressed by 12 wires of 7 mm dia is initially stressed to 1200 N/mm^2 with the eccentricity located 100 mm from soffit. Estimate thye final loss of prestress due to elastic deformations, creep, shrinkage and relaxation.**

Given: Relaxation of steel stress = 90 N/mm^2

E_S = 210 kN/mm^2

E_C = 35 kN/mm^2

Creep coefficient (ϕ) = 1.6

Residual shrinkage strain = 3 × 10^{-4}

[20 marks : 2010]

Solution:

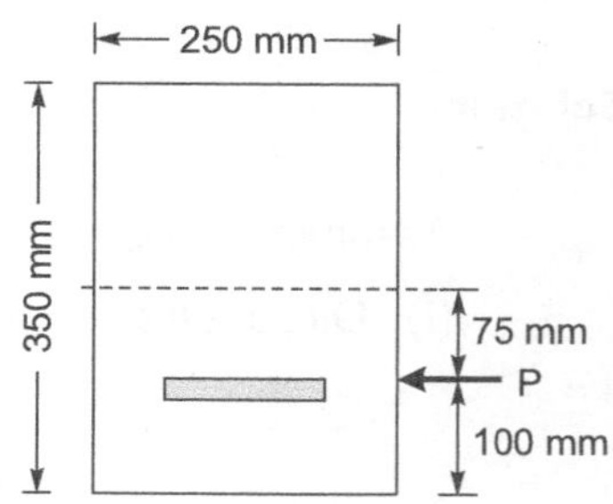

Initial prestressing force. $P = f \times A_{st}$

$$P = 1200 \times 12 \times \frac{\pi}{4} \times 7^2 \times 10^{-3}$$

$$P = 554.18 \text{ kN}$$

$$e = \frac{350}{2} - 100 = 75 \text{ mm}$$

Stress in concrete at level of steel

$$f_c = \frac{P}{A} + \frac{Pe.y}{I}$$

$$f_c = \frac{P}{A} + \frac{Pe^2}{I}$$

$$f_c = \frac{554.18 \times 10^3}{250 \times 350} + \frac{554.18 \times 1000 \times 75^2}{250 \times \frac{350^3}{12}}$$

$$f_c = 6.33 + 3.489$$

$$f_c = 9.819 \simeq 9.82 \text{ MPa}$$

1. Loss due to elastic shortening = mf_c

$\because$ $$m = \frac{E_S}{E_C} = \frac{210 \times 10^3}{35 \times 10^3} = 6$$

$$f_L = 6 \times 9.82 = 58.92 \text{ MPa}$$

2. Loss due to creep = $\phi m f_c$

$$= 1.6 \times 6 \times 9.82 = 94.272 \text{ MPa}$$

3. Loss due to shrinkage = εE_s

$$= 3 \times 10^{-4} \times 210 \times 10^3 = 63 \text{ MPa}$$

4. Relaxation of steel stress = 90 MPa

$$\text{Total loss} = \Sigma\text{All losses}$$

$$= 58.92 + 94.272 + 63 + 90 = 306.192 \text{ MPa}$$

$$\% \text{ loss} = \frac{306.192}{1200} \times 100 = 25.516\%$$

5.12 What are the various methods of pre-stressing? Also discuss the most widely used method of pre-tensioning and post-tensioning of concrete elements.

[10 marks : 2011]

Solution:

The various methods of pre-stressing are:

Pre-tensioning System

(i) Hoyer System

Post-tensioning System

(i) The Freyssinet System
(ii) The Magnel Blaton System
(iii) Gifford Udall System
(iv) P.S.C. Monowire System
(v) C.C.L. Standard System
(vi) Lee-Mccall system

For mass production of pretensioned elements, the long-line process developed by Hoyer is generally used in a factory. In this method the tendons are stretched between two bulk heads several hundred metres apart so that a number of similar units may be cast along the same group of tensioned wires as shown in figure. The concrete is now poured so that a number of beams can be produced in one line, by providing suitable shuttering between them. After the concrete has hardened, the wires are released from the bulkheads. The wires between the different units in one line of beams will also be cut off. The prestress is transferred to the concrete through bond between the tendons and the concrete.

The transfer of prestress to concrete is usually achieved by large hydraulic or screw jacks by which all the wires are simultaneously released after the concrete attains the requisite compressive strength. The bond of prestressing wires may be considerably improved by forming surface indentations and by helical crimping of the wires. Strands have considerably better bond characteristics than plain wires of equal cross-sectional area. Supplementary anchoring devices are required when single wires of larger diameter (exceeding 7 mm) are used in the pretensioned units.

The most commonly used devices are the 'Weinberg clip' and the 'Dorland clip'. These clips are clamped on to the tensioned wires close to the end diaphragms of the units before concreting operations.

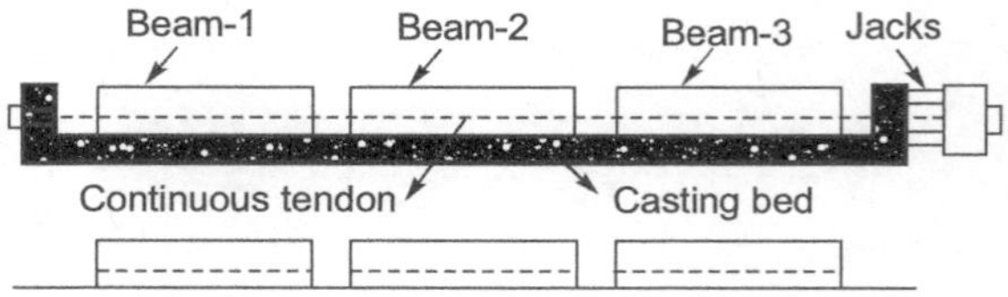

In post-tensioning, the concrete units are first cast by incorporating ducts or grooves to house the tendons. When the concrete attains sufficient strength, the high-tensile wires are tensioned by means of jack bearing on the end face of the member and anchored by wedges or nuts.

The post-tensioning systems based on wedge-action include the Freyssinet, Gifford-Udall, Anderson and Magnel-Blaton anchorages. The Freyssinet anchorage system, is widely used in Europe and India.

High tension steel wires 5 mm to 8 mm diameter about 12 in number are arranged to form a group into a cable with a spiral spring inside. The spiral spring provides the means for a proper clearance between the wires and thus provides a channel which can be cement grouted. It further assists to transfer the reaction to concrete. The whole thing is enclosed in a thin metal steel.

The anchorage consists of a cylinder of ordinary good quality concrete and is provided with corrugations on the outside. It has a central conical hole and is provided with heavy hoop reinforcement. These cylinder are kept in the proper position and the conical plugs are pushed into the conical holes after the cables are tightened. The central hole passing axially through the plug permits cement grout to be injected through it. In this way the space between the wires will be filled with the grout. This provides additional restraint against the slipping of the tendons.

5.13 **Distinguish clearly between pre-tensioned and post-tensioned prestressed concrete bringing out all the operations involved.**

[8 marks : 2012]

Solution:

Pre-tensioning : Pre-tensioning is accomplished by stressing wires or strands, called tendons to predetermined amount by stretching them between the two anchorages prior to placing the concrete as shown in the figure. The concrete is then placed and tendons become bonded to concrete throughout their length. After the concrete has hardened, the tendons are released by cutting them at the anchorages. The tendons tend to regain their original length by shortening and in this process transfer through bond a compressive stress to the concrete. The tendons are usually stressed by the use of hydraulic jacks. The stress in tendons is maintained during the placing and curing of concrete by anchoring ends of the tendons to abutments that may be as much as 200 m apart.

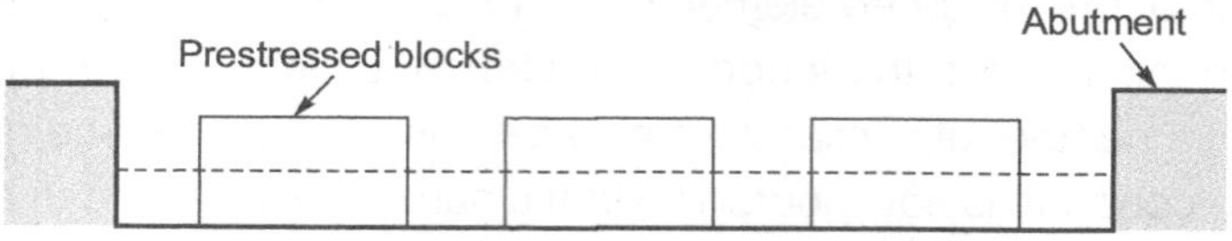

Sections for Pre-tensioning

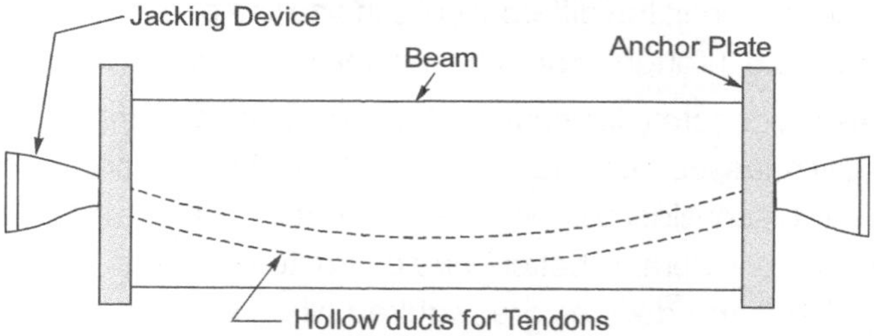

Sections for Post-tensioning

Post-tensioning: The alternative to pre-tensioning is post-tensioning. In a post-tensioned beam, tendons are stressed and each end is anchored to the concrete section after the concrete has been cast and has attained sufficient strength to safely withstand the prestressing force. In post-tensioning method, tendons are coated with grease or a bituminous material to prevent them from becoming bonded to concrete. Another method used in preventing the tendons from bonding to concrete during placing and cutting of concrete is to encase the tendon in a flexible metal hose before placing it in the forms. The metal hose is referred to as sheath or duct and remains in the structure . After the tendons has been stressed, the void between the tendons and the sheath is filled with grout. Thus, tendons becomes bonded to concrete and corrosion of steel is prevented.

5.14 A rectangular simply supported prestressed concrete beam of cross-section 200 mm wide and 300 mm deep is prestressed by 15 No. 5 mm diameter wires located at 65 mm from soffit and 3 No. 5 mm diameter wires at 25 mm from the top. Assuming effective stress in steel wires as 840 N/mm^2,

(i) calculate the stresses in concrete at the extreme fibres at mid-span section due to prestress and its own self weight over a span of 6 m.

(ii) If a uniformly distributed working load of 6 kN/m is imposed on the beam, obtain the maximum compressive stress in concrete.

(iii) if the modulus of rupture of concrete is 6.5 N/mm^2, estimate the load factor against cracking. Assume density of concrete = 24 kN/m^3.

[20 marks : 2012]

Solution:

Given,

$$\text{Area of cross-section of beam} = 200 \times 300 \text{ mm}^2 = 60{,}000 \text{ mm}^2$$

$$\text{Area of steel wires at top} = 3 \times \frac{\pi}{4} \times 5^2 = 58.905 \text{ mm}^2$$

$$\text{Area of steel wires at bottom} = 15 \times \frac{\pi}{4} \times 5^2 = 294.524 \text{ mm}^2$$

$$\text{Total area of steel} = 58.905 + 294.524 = 353.429 \text{ mm}^2$$

(i) **Prestressing force applied** $= \dfrac{840 \times 353.429}{10^3} = 296.88$ kN

$$\text{Direct stress, } \frac{P}{A} = \frac{296.88 \times 10^3}{200 \times 300} = 4.948 \text{ N/mm}^2$$

$$\text{Eccentricity of the bottom steel} = 150 - 65 = 85 \text{ mm}$$

$$\text{Prestressing force at bottom} = 294.524 \times 840 = 247.4 \text{ kN}$$

$$Z = \frac{bd^2}{6} = \frac{200 \times 300^2}{6} = 3 \times 10^6 \text{ mm}^3$$

$$\therefore \quad \text{Extreme stress} = \frac{Pey}{I} = \frac{Pe}{z} = \mp \frac{247.4 \times 1000 \times 85}{3 \times 10^6} \text{N/mm}^2 = \mp 7 \text{ N/mm}^2$$

$$\text{Prestressing force at top} = 58.905 \times 840 = 49.480 \text{ kN}$$

$$\text{Eccentricity} = 150 - 25 = 125 \text{ mm}$$

$$\therefore \text{ Extreme stress, } \frac{Pe}{z} = \pm \frac{49.48 \times 1000 \times 125}{3 \times 10^6} = \pm 2.062 \text{ N/mm}^2$$

Now, self weight of the beam $= 0.2 \times 0.3 \times 24 = 1.44$ kN/m

$$\therefore \quad \text{Moment due to self weight} = \frac{1.44 \times 6^2}{8} = 6.48 \text{ kN-m}$$

Extreme stress due to self weight

$$\Rightarrow \qquad \pm\frac{M}{I}y = \pm\frac{6.48\times10^6\times6}{200\times300^2} = 2.16 \text{ N/mm}^2$$

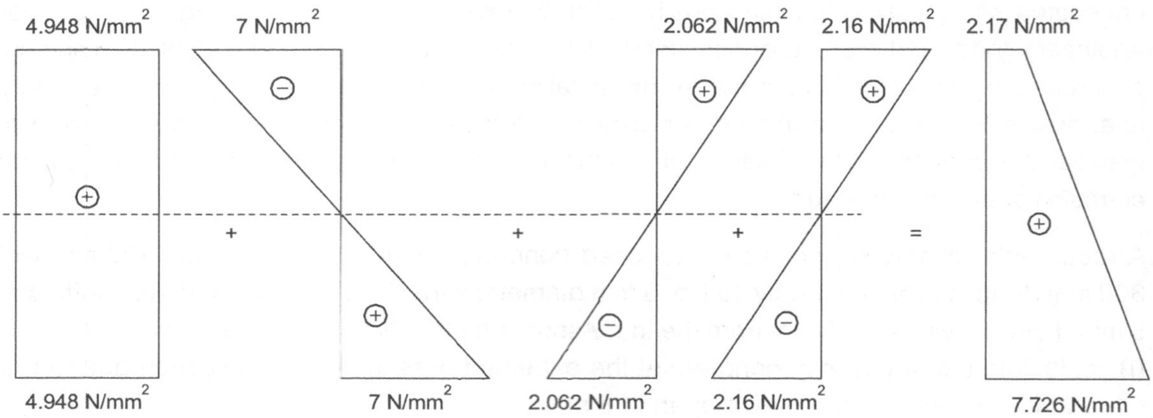

Hence,

Extreme stress in concrete at mid-span at top, $f_t = 4.948 - 7 + 2.062 + 2.16 = 2.17 \text{ N/mm}^2$

Stress at bottom, $f_b = 4.948 + 7 - 2.062 - 2.16 = 7.726 \text{ N/mm}^2$

(ii) When a beam is subjected to a working load of 6 kN/m

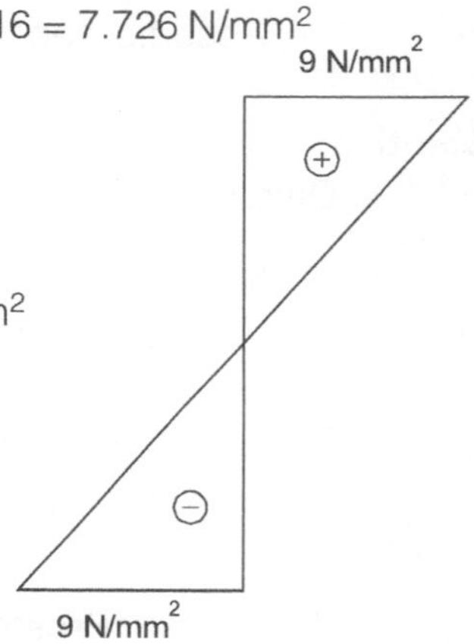

$$\text{Moment due to working load} = \frac{wl^2}{8} = \frac{6\times6^2}{8} = 27 \text{ kN/m}$$

$$\therefore \quad \text{Extreme stress due to live load} = \pm\frac{M}{I}y = \pm\frac{27\times10^6\times6}{200\times300^2} = 9 \text{ N/mm}^2$$

$\therefore$ Final stress at top $= 2.172 + 9 = 11.172 \text{ N/mm}^2$

Final stress at bottom $= 7.728 - 9 = -1.272 \text{ N/mm}^2$

$\therefore$ Maximum compressive stress in concrete $= 11.172 \text{ N/mm}^2$

(iii)

$$A = 200\times300 = 60000 \text{ mm}^2$$

$$Z = \frac{200\times300^2}{6} = 3\times10^6 \text{ mm}^3$$

$$\text{D.L. of the beam} = 0.2\times0.3\times24 = 1.44 \text{ kN/m}$$

$$\text{Total load on the beam} = 1.44 + 6 = 7.44 \text{ kN/m}$$

$$\therefore \quad \text{Maximum working moment} = \frac{7.44\times6^2}{8} = 33.48 \text{ kNm}$$

Let M' = additional moment so that the tensile stress at soffit reaches 6.5 N/mm²

$$\frac{P}{A}+\frac{P_{bottom}e_{bottom}}{Z}-\frac{P_{top}e_{top}}{Z}-\frac{\text{working moment}}{Z}-\frac{M'}{Z} = -6.5$$

$$\therefore 4.948+7-2.062-\frac{33.48\times10^6}{3000000}-\frac{M'}{3000000} = -6.5$$

$$\Rightarrow \quad \frac{M'}{3000000} = 9.888 - 11.16 + 6.5$$

$$\Rightarrow \quad M' = 15.678\times10^6 \text{ N-mm} = 15.678 \text{ kN-m}$$

$$\text{Cracking moment} = \text{Working Moment} + M'$$

$$= 33.48 + 15.678 = 49.158 \text{ kNm}$$

$$\text{Load factor against cracking} = \frac{\text{Cracking moment}}{\text{Working moment}} = \frac{49.158}{33.48} = 1.468$$

5.15 A concrete beam simply supported at both ends with a rectangular section 300 mm × 600 mm is prestressed by 2 post tensioned cables of area 500 mm^2 each. The cables are located at a constant eccentricity of 100 mm throughout the beam of span 8 m. The cables are stressed to 1600 MPa initially. Calculate the deflection of the beam (maximum) when it carries an imposed load of 20 kN/m allowing 20% loss in prestress. Assume the modulus of elasticity of concrete and steel are 30000 MPa and 2 × 10^5 MPa respectively. Neglect the effect of shrinkage and creep.

[10 marks : 2013]

Solution:

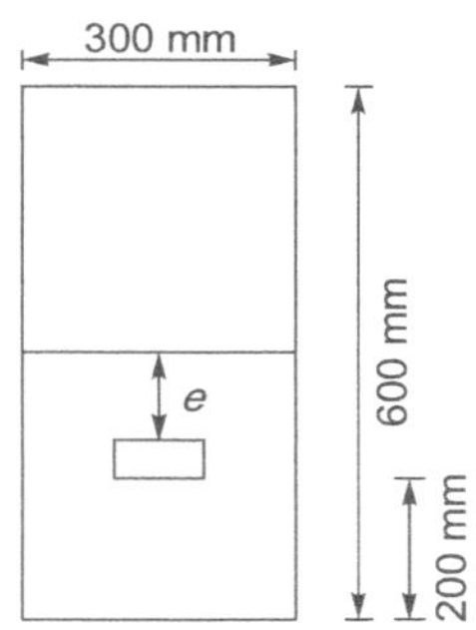

Magnitude of prestressing force, $P = 2 \times 500 \times 1600 = 1600 \times 10^3$ N

Eccentricity, $e = 100$ mm

Self weight of the beam $= 0.3 \times 0.6 \times 1 \times 24 = 4.32$ kN/m

Live load of the beam $= 20$ kN/m

Total load $W = 20 + 4.32 = 24.32$ kN/m

Deflection due to 'W' ⇒ $\delta_1 = \dfrac{5}{384}\dfrac{Wl^4}{\text{EI(both of concrete)}}$

$$E_{conc} = 3 \times 10^{10} \text{ Pa}$$

$$I = \frac{300 \times 600^3}{12} = 5.4 \times 10^9 \text{ mm}^4$$

(approx. as steel area is not considered)

$$\delta_1 = \frac{5}{384}\frac{(24.32 \times 10^3)(8^4)(10^3)}{(3 \times 10^{10})(5.4 \times 10^9 \times 10^{-12})}\text{mm}$$

$$\delta_1 = \frac{5}{384} \times \frac{24.32 \times 8^4}{3 \times 5.4} \times \frac{1}{10}\text{mm} = 8 \text{ mm}(\downarrow)$$

'δ_2' due to prestressing force/moment

$$\delta_2 = \frac{Ml^2}{8EI}(\uparrow)$$

$$M = kPe$$

$$M = 0.8 \times 1600 \times 10^3 \times 0.1 = 128 \times 10^3 \text{ Nm} = 128 \times 10^6 \text{ Nmm}$$

$$\delta_2 = \frac{(128 \times 10^6)8000^2}{8(30000) \times (5.4 \times 10^9)}\text{mm} = 6.32 \text{ mm}(\uparrow)$$

$$\text{Net downward} = \delta_1 - \delta_2 = 8 - 6.32 = 1.68 \text{ mm } (\downarrow)$$

5.16 List types of losses of pre-stress in pre-tensioning and post-tensioning systems.

[5 marks : 2015]

Solution:

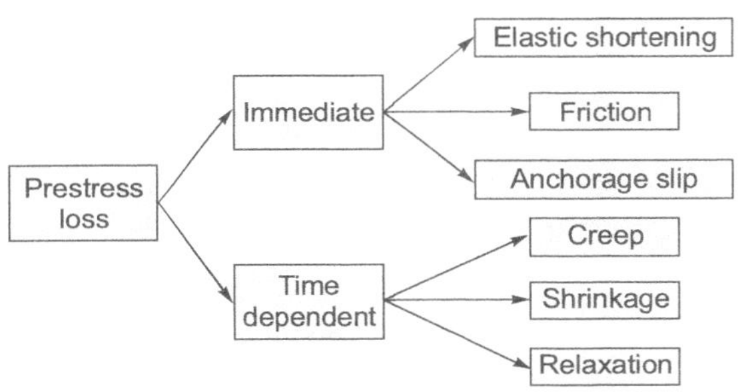

Types of losses in pre-tensioning and post tensioning system

S. No.	Pre-tensioning system	Post tensioning system
1.	Elastic shortening of concrete	No loss due to elastic shortening of concrete if all the tendons are stretched simultaneously
2.	Shrinkage of concrete	Shrinkage of concrete
3.	Stress relaxation in steel	Stress relaxation in steel
4.	Creep of concrete	Creep of concrete
5.		Anchorage slip
6.		Friction

5.17 A PSC simply supported beam of length 4 m, width 200 mm and depth 200 mm is prestressed with a prestressing force of 200 kN. The beam is subjected to two concentrated loads of 10 kN each located 1 m from each support. Neglecting the dead weight of the beam, sketch the cable profile of the tendons for load balancing condition. Determine the maximum stress produced in the concrete.

[10 marks : 2015]

Solution:

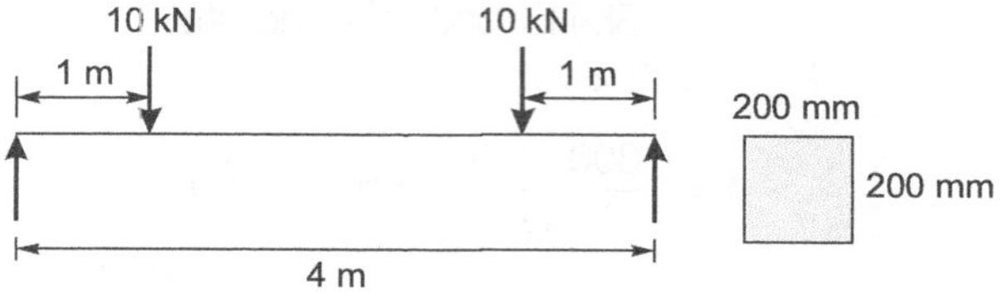

Prestressing force, $P = 200$ kN

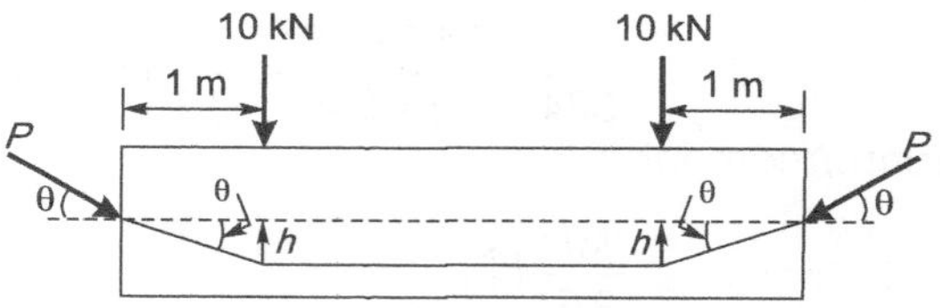

Let θ = Inclination of cable with horizontal at the ends.

$\therefore$ $\tan\theta = \dfrac{h}{l} \simeq \sin\theta$

$\therefore$ Total upward force due to prestress $= 2P\sin\theta$

Total downward force $= 20$ kN

(Neglecting self weight of beam as given in question)

For load balancing, $2P\sin\theta = 20$

$\Rightarrow$ $2 \times 200 \sin\theta = 20$

$\Rightarrow$ $\sin\theta = 0.05 = \dfrac{h}{l} = \dfrac{h}{1\text{m}}$

$\Rightarrow$ $h = 0.05$ m

$= 50$ mm

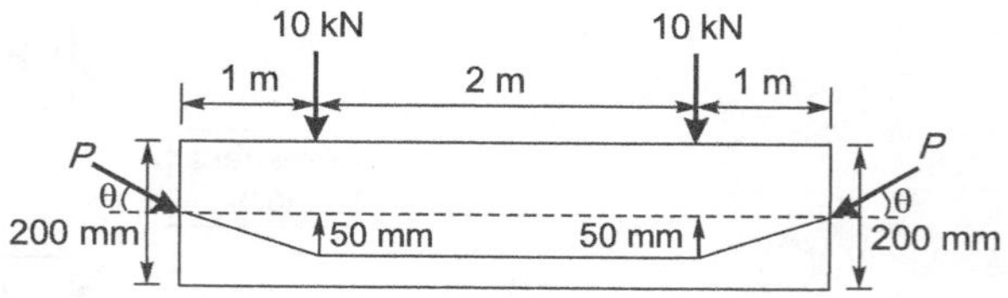

Cable Profile

Direct stress due to prestress $= +\dfrac{P}{A}$ (Compressive)

$$= +\frac{200 \times 1000}{200 \times 200} = +5 \text{ N/mm}^2$$

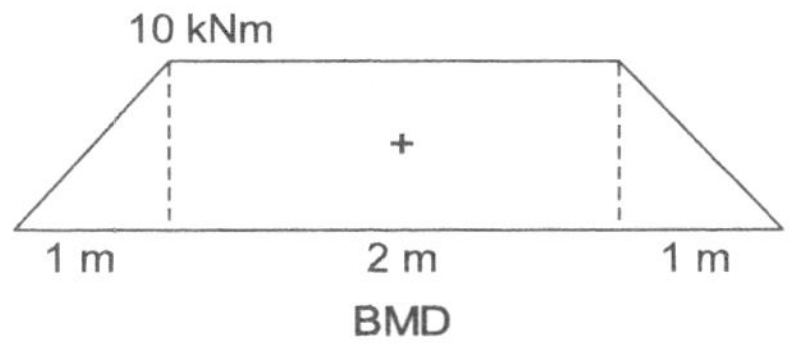

$$\text{Maximum bending stress} = \pm\frac{M}{Z}$$

$$= \pm\frac{10 \times 10^6}{200 \times 200^2/6} = \pm 7.5 \text{ N/mm}^2$$

$\therefore$ Maximum stress at top $= +5 + 7.5 = +12.5$ N/mm^2

Maximum stress at bottom $= +5 - 7.5 = -2.5$ N/mm^2

5.18 A prestressed concrete beam of 150 mm × 300 mm cross-section supports a live load of 5 kN/m over a simple span of 8 m. It has a parabolic cable having an eccentricity of 75 mm at the mid-span and zero at the ends. Determine the force of prestress if the net resultant stress at the bottom fibre at mid-span is zero under the action of self-weight, live load and prestress force.

[12 marks : 2017]

Solution :

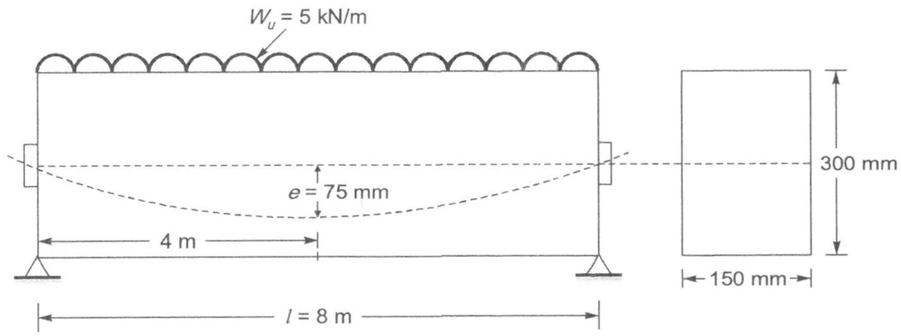

Let bending moment due to uniformally distributed live lcad $= M_l$

$$\therefore \quad M_l = \frac{w_l l^2}{8} = \frac{5 \times (8)^2}{8} = 40 \text{ kNm}$$

Self weight of beam, $w_{DL} = 0.150 \times 0.30 \times 1 \times 25 = 1.125$ kN/m

BM due to dead weight at mid span, $M_d = \frac{w_{DL} l^2}{8} = \frac{1.125 \times 8^2}{8} = 9$ kNm

Area of cross-section, $A = 150 \times 300 = 45000$ mm^2

Section modulus, $Z = \frac{bd^2}{6} = \frac{150 \times (300)^2}{6} = 2.25 \times 10^6 \text{mm}^3$

Let P_0 be the effective prestressing force.

Given, at mid span,

Stress at bottom fibre, $\sigma_{bottom} = 0$ ...(i)

Assuming compressive stress as positive.

Stress at bottom fibre in middle of span,

$$\sigma_{bottom} = \frac{P_0}{A} + \frac{P_0 e}{Z} - \frac{(M_d + M_l)}{Z} = 0$$

Substituting respective values in above equation, we get

$$\frac{P_0 \times 10^3}{45000} + \frac{P_0 \times 75 \times 10^3}{2.25 \times 10^6} = \frac{(40 + 9) \times 10^6}{2.25 \times 10^6}$$

$$\Rightarrow \quad \frac{P_0}{45} + \frac{75P_0}{2250} = 21.778$$

$$\Rightarrow \quad 0.0222P_0 + 0.0333P_0 = 21.778$$

$$\Rightarrow \quad 0.05555P_0 = 21.778$$

$$\Rightarrow \quad P_0 = 392 \text{ kN}$$

$\therefore$ Required prestressing force, $P_0 = 392$ kN

5.19 A prestressed concrete beam supports an imposed load of 6.5 kN/m over an effective span of 12 m. The beam has a rectangular section of width 250 mm and depth of 700 mm. Find the effective prestressing force in the cable if it is parabolic with an eccentricity of 110 mm at the centre and zero at the ends, for the following conditions:

(i) if the bending effect of the prestressing force is nullified by the imposed load for the mid-span section (neglecting self weight of the beam).

(ii) if the resultant stress due to self-weight, imposed load and prestressing force is zero at the soffit of the beam for the mid-span section. Assume the density of concrete is 24 kN/m^3.

[12 marks : 2019]

Solution:

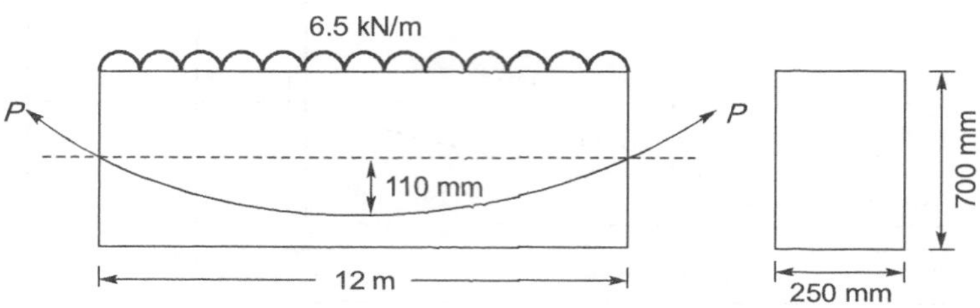

(i) If the bending effect of the prestressing force in nullified by the imposed load for the mid-span section (neglecting self weight)

Moment due to prestressing force = Moment due to imposed load

$$P \times e = \frac{wl^2}{8}$$

$$P = \frac{wl^2}{8e} = \frac{6.5 \times 12^2}{8 \times 0.11}$$

$$P = 1063.64 \text{ kN}$$

(ii) Stress due to imposed load, self-weight and prestressing force is zero at the soffit of the beam for the mid-span section.

Self weight of beam = 24 × 0.25 × 0.7 = 4.2 kN/m

Total load = 4.2 + 6.5 = 10.7 kN/m

$$\Rightarrow \quad \text{Moment} = \frac{wl^2}{8} = \frac{10.7 \times 12^2}{8} = 192.6 \text{ kNm}$$

Stress of soffit = 0

$$\frac{P}{A} + \frac{Pe}{Z} - \frac{M}{Z} = 0$$

$$P\left(\frac{1}{A}+\frac{e}{Z}\right) = \frac{M}{Z}$$

$$P\left[\frac{1}{250\times700}+\frac{110\times6}{250\times700^2}\right] = \frac{192.6\times10^6\times6}{250\times700^2}$$

$$P = 849.7 \text{ kN}$$

5.20 A simply supported prestressed concrete beam of width 100 mm, depth 200 mm and span 10 m, carries a UDL of intensity 'w'. If the member is prestressed with a parabolic cable having zero eccentricity at the ends and 60 mm eccentricity at mid, determine the value 'w' for the following conditions, for effective prestressing force of 125 kN.

(i) Load balancing case.
(ii) For no tensile stress condition at mid span.
(iii) For cracking condition taking the tensile strength of concrete as 1.5 N/mm^2.

For all cases neglect the weight of concrete.

[20 marks : 2020]

Solution:

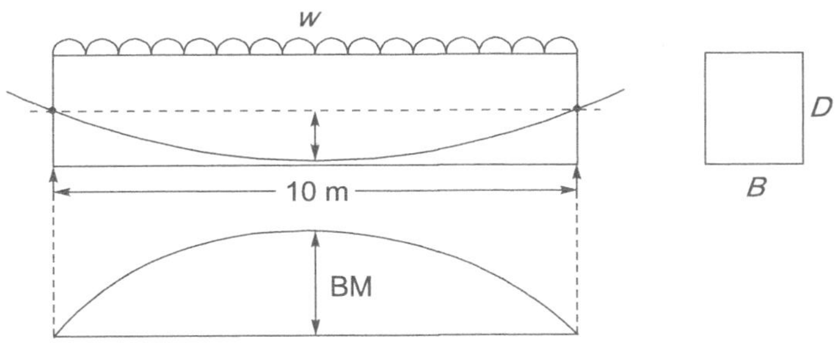

$$B = 100 \text{ mm}, \quad D = 200 \text{ mm}$$

$$\text{Span} = 10 \text{ m}, \quad \text{UDL} = w$$

Parabolic cable, e_1 = 0 at ends

e_2 = 60 mm at midspan

P-force = 125 kN (Neglect weight of concrete)

For all cases we have to neglect weight of concrete.

1. For load balancing, $Pe = \dfrac{wL^2}{8}$

So, $$w = \frac{8Pe}{L^2} = \frac{8\times125\times0.06}{10^2} = 0.6 \text{ kN/m}$$

2. For no tensile stress condition at mid span (at bottom)

$$\frac{P}{A}+\frac{Pe}{Z}-\frac{M}{Z} = 0$$

$$\frac{125000}{100\times200}+\frac{125000\times60}{100\times\frac{200^2}{6}} = \frac{M}{100\times\frac{200^2}{6}}$$

$$(6.25+11.25)\times\frac{100\times200^2}{6\times10^6} = \text{BM}$$

So $$\text{BM} = 11.67 \text{ kNm} = \frac{wL^2}{8}$$

$$w = \frac{8 \times 11.67}{10^2} = 0.933 \text{ kN/m}$$

3. For cracking condition

   Stress at bottom fibre = –1.5 N/mm^2

$$\frac{P}{A} + \frac{Pe}{Z} - \frac{M}{Z} = -1.50$$

$$6.25 + 11.25 + 1.50 = \frac{M}{Z}$$

$$M = 19 \times Z = \frac{19 \times 100 \times 200^2}{6 \times 10^6}$$

$$\text{BM} = 12.67 \text{ kNm} = \frac{wL^2}{8}$$

$$w = \frac{8 \times 12.67}{10^2} = 1.0133 \text{ kN/m}$$

5.21 **A rectangular prestressed concrete beam has a cross-section of 200 mm × 300 mm. Its effective span is 8 meters. This beam is prestressed by a straight cable, 50 mm below the central longitudinal axis. This beam supports an imposed load of 20 kN/m.**

Find the magnitude of prestressing force which can balance the stresses due to dead load and imposed load at bottom fibre of mid-span. Unit weight of concrete : 25 kN/m^3.

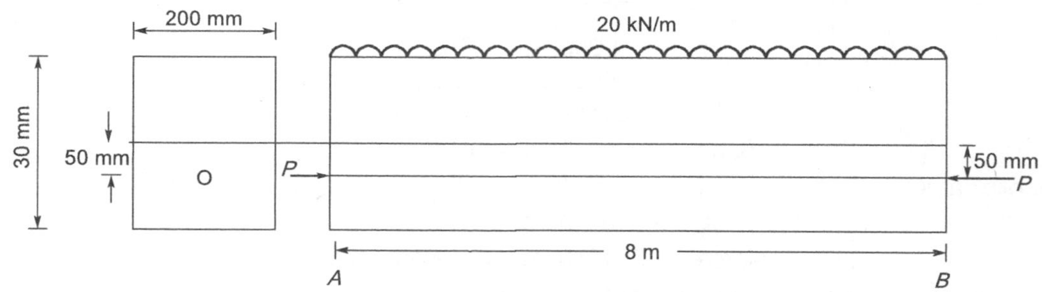

[12 Marks : 2021]

Solution:

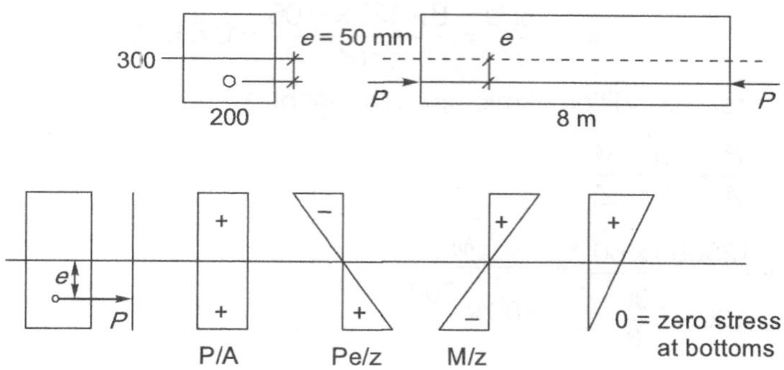

1. Self wt. of beam, DL = 0.20 × 0.30 × 1.0 × 25 = 1.5 kN/m
2. Imposed load w_L = 20 kN/m

At mid span:

$$\text{Maximum BM} = \frac{wL^2}{8} = (1.5+20)\times\frac{8^2}{8} = 172 \text{ kN-m}$$

Let P be the prestressing force which can balance the stresses due to dead load and imposed load at bottom fibre of mid span:

i.e. $\sigma_b = 0$

$$\Rightarrow \quad \frac{P}{A}+\frac{Pe}{z}-\frac{M}{z} = 0$$

$$\Rightarrow \quad P\left(\frac{1}{A}+\frac{e}{z}\right) = \frac{M}{z}$$

$$\Rightarrow \quad P = \frac{M}{\left(\frac{z}{A}+e\right)} = \frac{M}{\left(\frac{D}{6}+e\right)} \qquad \left[\because \frac{Z}{A} = \frac{bD^2}{6\times bD} = \frac{D}{6}\right]$$

$$= \frac{172\times10^6}{\left(\frac{300}{6}+50\right)\times1000} = \frac{172\times10^6}{100\times10^3} = 1720 \text{ kN}$$

5.22 A prestressed concrete beam, 300 × 800 mm, is simply supported over an effective span of 12 m. It is prestressed with a cable whose eccentricity varies linearly from 50 mm at the supports to 250 mm at $\left(\frac{1}{3}\right)^{rd}$ span points from either support. Effective prestress in the cable is 900 kN. Determine,

(i) Net deflection due to prestress and self weight.

(ii) Central concentrated load required to caused a maximum downward deflection of $\left(\frac{\text{Span}}{400}\right)$.

Use M40 grade concrete. Derive only the expressions for deflections due to prestress. $E_c = 5000\sqrt{f_{ck}}$. Density of concrete = 25 kN/m³.

[20 marks : 2022]

Solution:

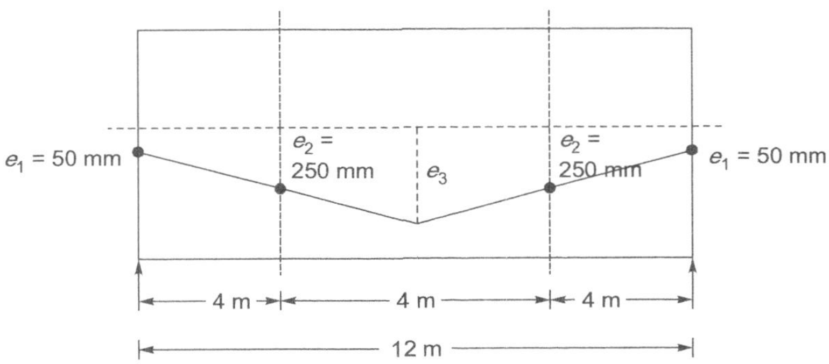

Given: b = 300 mm, d = 800 mm, l_{eff} = 12 m

e_1 = 50 mm (at support), e_2 = 250 mm (at middle third, varying linearly, so cable is straight cable)

$$E_c = 5000\times\sqrt{f_{ck}} = 5000\times\sqrt{40} \text{ N/mm}^2$$

$$E_c I_c = \in = \sqrt{\frac{250}{f_y}} \approx 1 = 4.0477 \times 10^{14} \text{ N.mm}^2$$

1. Net deflection due to prestress and self weight.

(a) Deflection due to self weight

$$DL = 0.3 \times 0.8 \times 1.0 \times 25 = 6.0 \text{ kN/m} = 6.0 \text{ N/mm}$$

$$\delta_1 = \frac{5}{384} \times \frac{wL^4}{E_c I_c} = \frac{5}{384} \times \frac{6.0 \times 12000^4}{4.0477 \times 10^{14}}$$

$$= 4.002 \text{ mm} \downarrow \text{downward}$$

(b) Deflection due to prestressing force

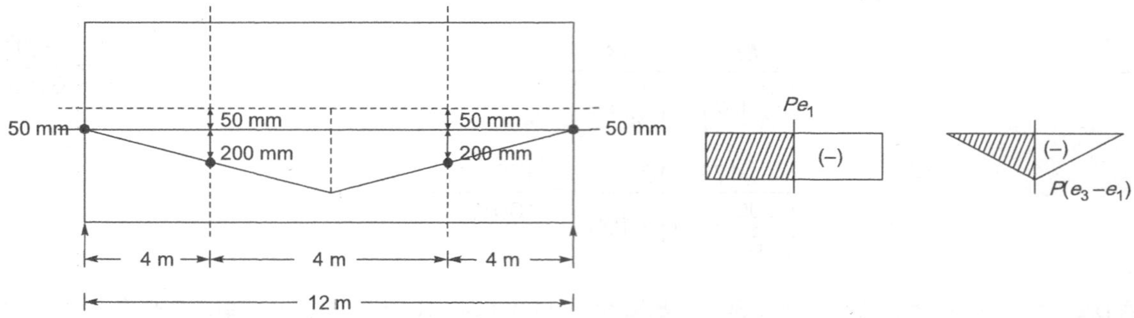

Eccentricity at 4 m = 250 mm

Eccentricity at end = 50 mm

$$\text{Eccentricity at mid span} = 50 + \frac{250-50}{4} \times 6 = 350 \text{ mm}$$

Deflection at centre due to prestressing cable (upward)

$$= (-)\left(\frac{P.e_1 L^2}{8E_c I_c} + \frac{P(e_3 - e_1)L^2}{12 E_c I_c}\right)$$

$$= (-)\left(\frac{900 \times 10^3 \times 50 \times 12000^2}{8 \times 4.0477 \times 10^{14}} + \frac{900 \times 10^3 \times (350-50) \times 12000^2}{12 \times 4.0477 \times 10^{14}}\right)$$

$$= (-)(2.001 + 8.005) = (-)\ 10.006 \text{ mm}$$

Net deflection at mid-span

$$= 4.002 - 10.006 = -6.004 \text{ mm}$$

$$\text{say} \simeq -6.0 \text{ mm (say)}$$

(ii) Central concentrated load required for maximum deflection of $\left(\frac{\text{span}}{400}\right)$ is given by:

$$\text{Deflection required} = \frac{\text{span}}{400} = \frac{12000}{400} = 30 \text{ mm}$$

$$\text{Deflection due to point load} = \frac{WL^3}{48EI}$$

$$\therefore \quad \frac{WL^3}{48EI} - 6.003 = 30 \text{ mm}$$

$$\Rightarrow \quad \frac{W \times 12000^3}{48 \times 4.0477 \times 10^{14}} = 36.003$$

$$\Rightarrow \quad W = 404803 \text{ N} = 404.803 \text{ kN}$$

5.23 Explain the specific reasons for the following:

For prestressed concrete, the code recommends to use high tensile steel and high strength of concrete.

[6 marks : 2023]

Solution:

(i)

1. **Reasons for using high strength steel:** In steel, large amount of loss of stresses are observed after prestressing of cable. Loss of stress due to various reasons are around 200 – 300 N/mm^2. Therefore, high strength steel is used of initial strength of about 1000 – 2000 N/mm^2, so that even after losses, about 80 to 90% of initial prestress are left in steel for effective working of prestressed concrete member.

$$\text{Loss of stress} \not> 10 \text{ to } 20\% \text{ of } p_0 (200 \text{ N/mm}^2)$$

where p_0 is initial stress in steel.

So, $$p_0 \text{ required} \not< 1000 \text{ to } 2000 \text{ N/mm}^2$$

2. **Reasons for using high strength concrete:**
   (a) To bear large stresses developed due to high value of force.
   (b) To bear high bond stresses developed in pre tensioned members.
   (c) To bear high bearing stresses developed in post tensioned members.
   (d) To get high value of E_C.
   (e) To keep the loss of stress less in steel due to various reasons.

5.24 A rectangular beam of size 300 mm × 600 mm is used over a simply supported effective span of 7 m. The beam supports a live load of 12 kN/m. A straight tendon is provided at an eccentricity of 100 mm below the centroid of the beam section. Find the minimum prestressing force required for no tension condition at mid-span under live load. Also, show the stress distribution under self-weight only at mid-span and at the ends of the member.

[20 marks : 2023]

Solution:

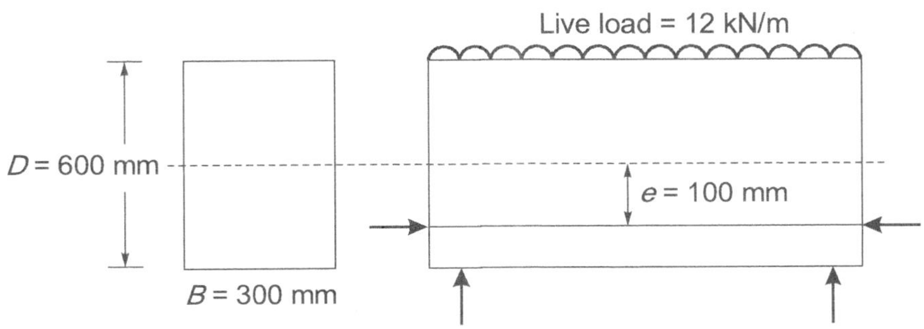

1. Calculation of load:

$$\text{Dead load} = 0.30 \times 0.60 \times 1.0 \times 25$$
$$= 4.50 \text{ kN/m}$$
$$\text{Live load} = 12.00 \text{ kN/m}$$
$$\text{Total load} = 16.50 \text{ kN/m}$$

Bending moment at centre,

$$M_x = \frac{wL^2}{8} = \frac{16.50 \times 7^2}{8} = 101.0625 \text{ kN-m}$$

2. For no tension condition at mid span under live load.
   Note: Live load will always act with dead load.

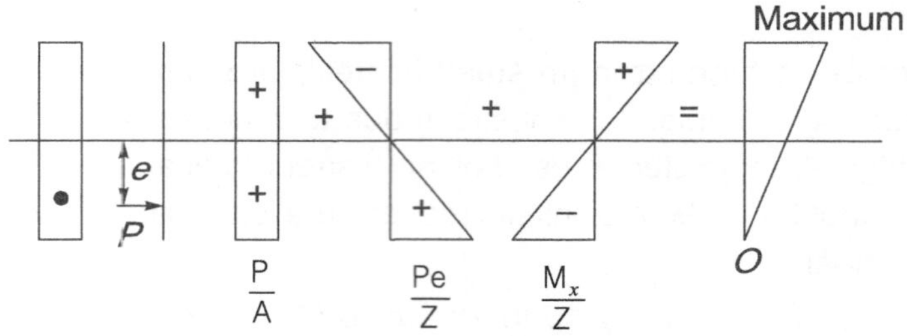

Stress at bottom fibre, $(f_c)_{\text{bottom}} = \frac{P}{A} + \frac{Pe}{Z} - \frac{M_x}{Z}$

where e is eccentricity and Z is section modulus.

For no tension at bottom, $(f_c)_{\text{bottom}} = 0$

So, $$\frac{P}{A} + \frac{Pe}{Z} = \frac{M_x}{Z}$$

$$\Rightarrow \quad P\left(\frac{Z}{A} + e\right) = M_x$$

Now, $$\frac{Z}{A} = \frac{BD^2}{6 \times BD}$$

$$\Rightarrow \quad P = \frac{M_x}{\left(\frac{D}{6} + e\right)}$$

$$= \frac{101.0625 \times 10^6}{\left(\frac{600}{6} + 100\right) \times 1000} \text{kN} = 505.3125 \text{ kN}$$

3. Stresses developed only due to DL.
   (a) At end, moment = 0:

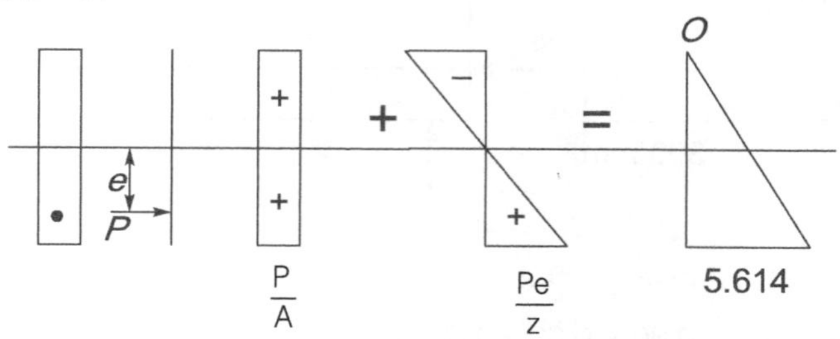

$$\frac{P}{A} = \frac{505.3125 \times 1000}{300 \times 600} = 2.807 \text{ N/mm}^2$$

$$\frac{Pe}{Z} = \frac{505.3125 \times 1000 \times 100}{\left(\frac{300 \times 600^2}{6}\right)} = 2.807 \text{ N/mm}^2$$

$$\text{At top} = \frac{P}{A} - \frac{Pe}{Z} = 2.807 - 2.807 = 0 \text{ N/mm}^2$$

$$\text{At bottom} = \frac{P}{A} + \frac{Pe}{Z} = 2.807 + 2.807 = 5.614 \text{ N/mm}^2$$

(b) At mid span:

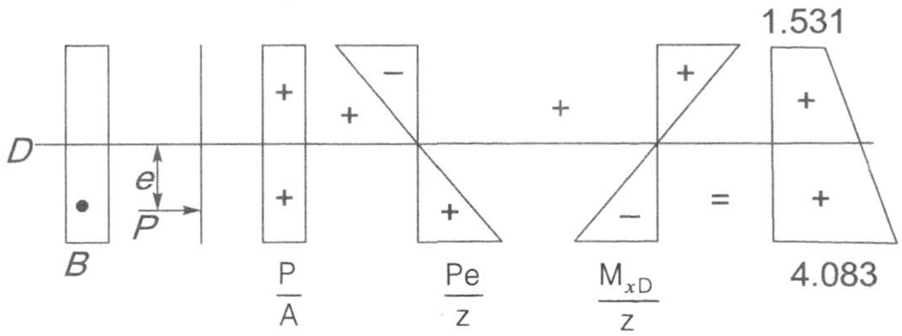

Due to DL $$M_{xD} = \frac{w_d \cdot L^2}{8} = \frac{4.50 \times 7^2}{8} = 27.5625 \text{ kN-m}$$

$$\frac{M_{xD}}{Z} = \frac{27.5625 \times 10^6}{\left(300 \times \frac{600^2}{6}\right)} = 1.531$$

$$\text{Stress at Top} = \frac{P}{A} - \frac{Pe}{Z} + \frac{M_{xD}}{Z}$$

$$= 2.807 - 2.807 + 1.531 = 1.531 \text{ N/mm}^2$$

$$\text{Stress at bottom} = \frac{P}{A} + \frac{Pe}{z} - \frac{M_{xD}}{z}$$

$$= 2.807 + 2.807 - 1.531 = 4.083 \text{ N/mm}^2$$

6. Miscellaneous

6.1 Explain with appropriate sketches how a 'tremie' is used to place concrete underwater.

[5 marks : 2013]

Solution:

Tremie:

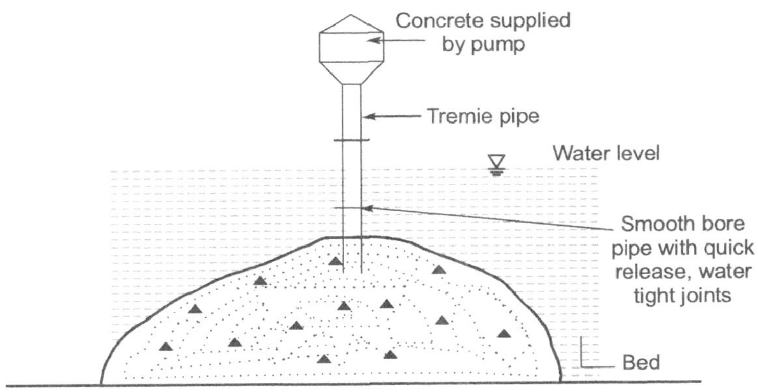

A tremie is a watertight pipe with funnel shaped hopper at upper end and loose plug at bottom/discharge end.

Tremie is supported on a working platform above water level.

During concreting there should not be any air and water in the tremie and this is done by keeping the pipe full of concrete at all the time.

As the hopper is filled with fresh concrete, it forces the plug at tremie bottom.

As concreting progress, tremie pipe is raised slightly and concrete flows outwards. Tremie should never be used laterally.

6.2 **Corrosion of reinforcement in reinforced concrete construction is manner of serious concern to civil engineers. List and briefly discuss some of the provisions made in codes (such as IS 456-2000) to address this problem during the design and construction of structures likely to be subjected to such deterioration.**

[10 marks : 2013]

Solution:

Corrosion of reinforcement in RCC: In general, concrete provides a high degree of protection to steel against corrosion due to alkaline environment provided by concrete. Also well proportioned concrete has very low permeability for oxygen, chloride ions, CO_2 and water etc.

In highly aggressive environments, other corrosion resisting measures are employed like cathodic protection of steel, coatings on steel, use of corrosion inhibitors etc.

Some of the codal provisions for corrosion protection of reinforcement are:

(i) For reinforced concrete members, totally immersed in sea water, the cover should be increased by 40 mm beyond what is specified for normal conditions. However total cover should not exceed 75 mm.

(ii) Chloride ions in concrete should be limited to its critical value. As per IS : 456, total chloride in concrete $\leq$ 0.15% by mass of cement in concrete.
For prestressed concrete, this limit is 0.06%.

(iii) In excessively aggressive environment where cover and concrete quality cannot be assumed, special protection systems are used. Corrosion inhibitors can also be added in concrete.

6.3 **As part of routine quality control, a sample of three cubes is taken from a site using M 30 concrete. The cubes are appropriately cured and tested for their compressive strength. The values obtained for the three cubes are: 31 MPa, 28 MPa and 28 MPa.**
Based on the provision of IS 456-2000, answer the following:

(i) Does the above constitute a valid set of results from the point of view of acceptance? Justify you answer.

(ii) Is the concrete represented by the results acceptable? Clearly indicate if any additional information may be needed before a final decision on acceptance is made.

[10 marks : 2013]

Solution:

(i) For M 30, 95% of sample should have strength more or equal to 30. Hence only 5% samples can fail. Here 2 out of 3 samples (i.e., 66%) fail, hence concrete can't be accepted. This may be either due to less concrete strength or due to lack of adequate sample representation.

(ii) Further more tests should be done, and if same results appear again, then concrete is rejected.

6.4 **The floor of a water tank is subjected to a direct pull of 50 kN/m and bending moment of 0.5 kN/m in the vertical plane. A section of the floor is shown in figure 10. Find the maximum stresses in concrete and steel. Assume M 30 concrete. Permissible bending stress in concrete = 10 MPa. Use Fe 415.**

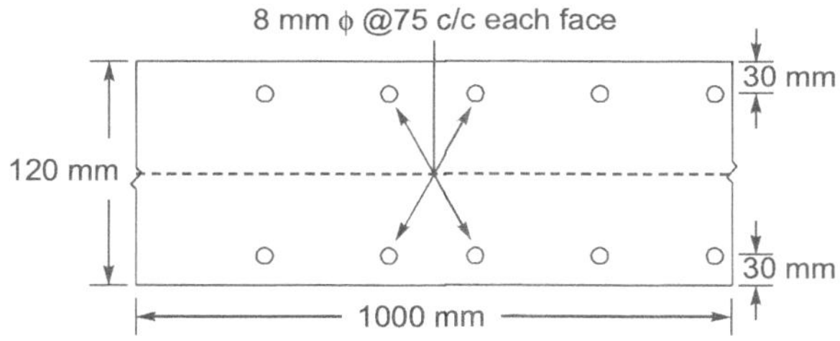

Figure 10

[20 marks : 2017]

Solution :

$$M = \frac{280}{3\sigma_{cbc}} = \frac{280}{3\times 10} = 9.33$$

For balanced section, $Q = 0.35$

$$M = QbD^2$$

$\Rightarrow$ $0.5\times 10^6 = 0.35\,(1000)\,D^2$

$\Rightarrow$ $D = 37.8 \text{ mm} < 120 \text{ mm}$ (OK)

Assume uncracked condition.

Effective depth, $a = 120 - 30 = 90$ mm

Area of steel provided, $A_s = \dfrac{1000A_\phi}{s} = \dfrac{1000\times\frac{\pi}{4}\times 8^2}{75} = 670.21 \text{ mm}^2/\text{m}$

Let x = Depth of neutral axis from top face.

$\therefore$ Taking moments about the neutral axis.

$$1000(x)\left(\frac{x}{2}\right) + A_s(M-1)(x) = 1000(D-x)\frac{(D-x)}{2} + A_s(M-1)(d-x)$$

$$\Rightarrow \quad 500x^2 + 670.21(9.33-1)x = 1000(120-x)\frac{(120-x)}{2} + 670.21(9.33-1)(90-x)$$

$$\Rightarrow \quad 500x^2 + 5582.85x = 500(120-x)^2 + 502456.437 - 5582.85x$$

$$\Rightarrow \quad 500x^2 = 500x^2 + 7200000 - 120000x + 502456.37 - 11165.7\,x$$

$$\Rightarrow \quad x = 58.723 \text{ mm}$$

$$\therefore \quad D - x = 120 - 58.723 = 61.277 \text{ mm}$$

$$\therefore \quad I_{xx} = \frac{1000}{3}\left[(58.723)^3 + (61.277)^3\right] + 9.33\times 670.21(90-58.723)^2$$

$$+9.33\times 670.21\,(58.723-30)^2$$

$$= 155.47\times 10^6 \text{ mm}^4$$

Now, $M = \dfrac{f_{cbt}I_{xx}}{y}$

where, σ_{cbt} = Stress in concrete in bending tension

$$\Rightarrow \quad 0.5\times 10^6 = \frac{f_{cbt}\times 155.47\times 10^6}{61.277}$$

$\Rightarrow$ $f_{cbt} = 0.197 \text{ N/mm}^2 < 2 \text{ N/mm}^2$ (OK)

($\because$ for M30, $\sigma_{cbt} = 2 \text{ N/mm}^2$)

Equivalent total area of section, $A = bD + (M-1)A_{sc} + (M-1)A_{st}$

$= 120 \times 1000 + (9.33 - 1)\,670.21 + (9.33 - 1)\,670.21$

$= 131.17 \times 10^3 \text{ mm}^2$

Direct tensile stress in concrete, $f_{ct} = \frac{T}{A} = \frac{50 \times 10^3}{131.17 \times 10^3} = 0.38 \text{ N/mm}^2 < 1.5 \text{ N/mm}^2$ (OK)

($\because$ for M30, $\sigma_{ct} = 1.5 \text{ N/mm}^2$)

Check: $\frac{f_{ct}}{\sigma_{st}} + \frac{f_{cbt}}{\sigma_{cbt}} = \frac{0.38}{1.5} + \frac{0.177}{2} = 0.342 < 1$ (OK)

Now, $T = f_{st} A_s$

$\Rightarrow$ $50 \times 10^3 = f_{st}(2 \times 670.21)$

($\because$ Both tension and compression steel will resist direction pull)

$\Rightarrow$ $f_{st} = 37.3 \text{ N/mm}^2 < 130 \text{ N/mm}^2$ (OK)

6.5 **A cylidrical water tank of capacity 500 m^3 is resting on ground and have a free-flexible joint at base (vertical wall-base slab connection). Overall height of tank is restricted to 4.3 m (it includes a free board of 0.3 m). Design the vertical cylindrical wall of tank only. Following parameters may be used for design, if required:**

1. $\sigma_{cbc} = 10 \text{ N/mm}^2$
2. $\sigma_{cbt} = 2.0 \text{ N/mm}^2$
3. $\sigma_{ct} = 1.5 \text{ N/mm}^2$
4. $\sigma_{st} = 130 \text{ N/mm}^2$
5. $\gamma_w = 10 \text{ kN/m}^3$
6. Main reinforcing bar diameter : 16 mm
7. $m = 10$

[12 marks : 2018]

Solution:

Given data:

Tank capacity = 500 m^3

Total height (including free board) = 4.3 m;

Free board = 0.3 m

$\sigma_{cbc} = 10 \text{ N/mm}^2$, $\sigma_{cbt} = 2.0 \text{ N/mm}^2$, $\sigma_{ct} = 1.5 \text{ N/mm}^2$, $\sigma_{st} = 130 \text{ N/mm}^2$, $m = 10$

Main reinforcing bar diameter = 16 mm

Dimensions:

Depth of water inside the tank = 4.3 – 0.3 = 4 m

D = Diameter of tank

$\therefore$ $\frac{\pi}{4} D^2 \times 4 = 500$

$\Rightarrow$ $D = 12.616$ m

$\therefore$ Provide diameter, $D = 13$ m

Design:

Consider the bottom 1 m wall of tank.

Pressure intensity corresponding to its centre.

$p = \gamma_{wh} = 10 \times 3.5 = 35 \text{ kN/m}^2$

Hoop tension, $T = \frac{pD}{2} = \frac{35 \times 13}{2} = 227.5$ kN/m height at base.

$$\text{Steel required for 1 m height} = \frac{227.5 \times 1000}{\sigma_{st}} = \frac{227.5 \times 1000}{130} = 1750 \text{ mm}^2\text{/m height.}$$

For 16 mm diameter bar,

$$\text{Spacing} = \frac{1000 \times \frac{\pi}{4} \times 16^2}{1750} = 114.9 \text{ mm}$$

Hence, provide 16 mmϕ bars@ 110 mm c/c.

$$\text{Actual area of steel provided} = \frac{1000 \times \frac{\pi}{4} \times 16^2}{110} = 1827.84 \text{ mm}^2$$

Thickness of wall should not be less than

1. 150 mm
2. 30 mm per metre depth +50 mm

$\therefore$ $t = (30 \times 4) + 50 = 170$ mm

3. Thickness required to limit the tensile stress in concrete to 1.2 N/mm^2.

$$\therefore \quad \sigma_{ct} = \frac{T}{1000t + (m-1)A_{st}}$$

$$\Rightarrow \quad 1.5 = \frac{227.5 \times 1000}{1000t + (10-1)(1827.84)}$$

$$t = 135.216 \text{ mm}$$

Choosing maximum of the above 3 values,

Provide thickness, $t = 170$ mm

Spacing of hoops can be increased near the top.

Minimum reinforcement at top = 0.3%

$$\therefore \quad A_{sh} = \frac{0.3}{100} \times (1000 \times 170) = 510 \text{ mm}^2$$

$$\text{Spacing of hoops} = \frac{1000 \times \frac{\pi}{4} \times 16^2}{510} = 394.24 \text{ mm}$$

$$\text{Maximum spacing} = 3t = 3 \times 170 = 510 \text{ mm}$$

$\therefore$ Keep spacing of 390 mm c/c at the top.

Now, as thickness of wall is less than 225 mm, reinforcement is provided at the centre of the wall. If thickness > 225 mm, half hoops will be placed near the inner and half at the outer face, at double spacing.

Vertical reinforcement:

0.3% for 100 mm thickness

0.2% for 450 mm thickness

$\therefore$ For 170 mm thick wall,

$$\text{Vertical reinforcement,} \quad A_{st} = 0.3 + \left(\frac{0.2 - 0.3}{450 - 100}\right)(170 - 100) = 0.28\%$$

$$\therefore \quad A_{st} = \frac{0.28}{100} \times 170 \times 1000 = 476 \text{ mm}^2$$

$$\text{Using 10 mm}\phi \text{ bar, spacing} = \frac{1000 \times \frac{\pi}{4} \times 10^2}{476} = 165 \text{ mm}$$

∴ Provide 10 mmϕ bars @ 160 mm c/c in vertical direction.

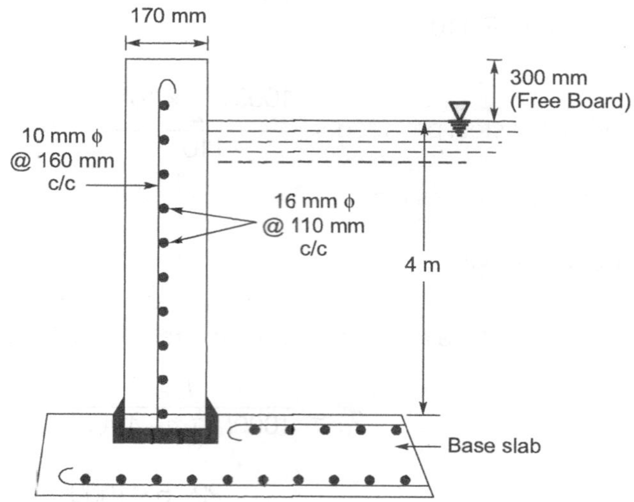

6.6 Design the counterforts of a retaining wall to retain earth for a height of 6.5 m above the ground level. The unit weight of soil is 16 kN/m³ and the angle of repose of soil is 30°. The safe bearing capacity of soil is 180 kN/m². Use M20 grade concrete and steel of grade Fe415. The cross-section of the retaining wall is given below. The spacing of counterfort is taken as 3.5 m. Assume a cover of 40 mm for counterforts.

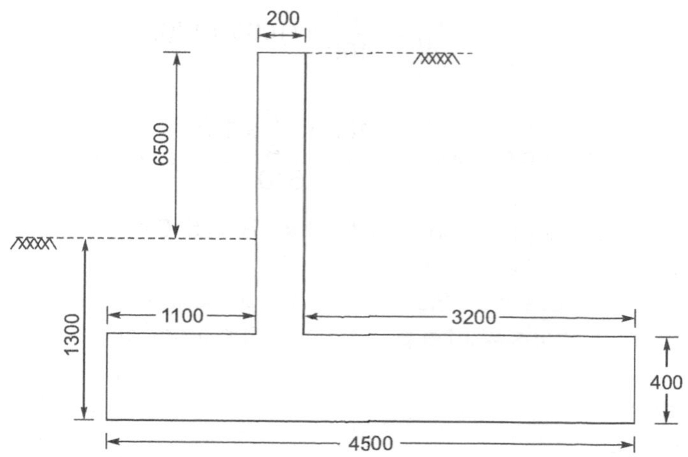

All dimension are in mm.

Assume the maximum pressure at toe end is 166.05 kN/m² and the minimum pressure at the heel end is 38.92 kN/m². Sketch the reinforcement details.

[20 marks : 2019]

Solution:

Design of counterfort

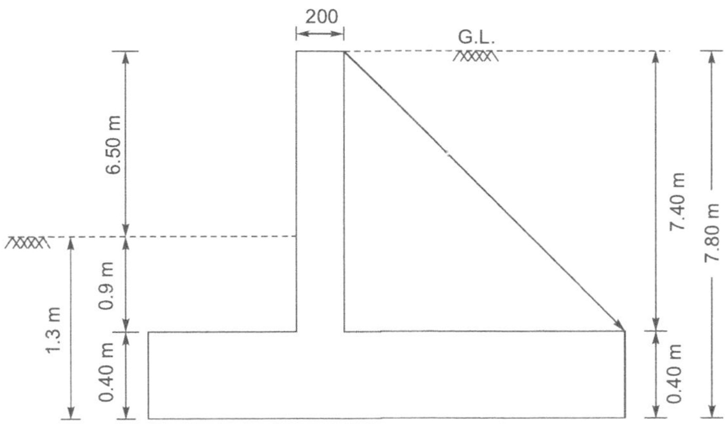

(a) Unit weight of soil, $\gamma = 16$ kN/m^3

Angle of repose, $\phi = 30°$

Safe bearing capacity of soil = 180 kN/m^2

M20, Fe415

Spacing of counterfort = 3.50 m

Cover = 40 mm

Each counterfort will receive a total pressure from 3.50 m width of soil.

Earth pressure, $$K_a = \frac{1-\sin\phi}{1+\sin\phi} = \frac{1-\sin 30°}{1+\sin 30°} = \frac{1}{3}$$

(b) Earth pressure at base of stem

$$= \frac{1}{3} \times 16 \times 7.40 = 39.50 \text{ kN/m}^2$$

(c) Maximum bending moment at base of stem to be considered for each counterfort

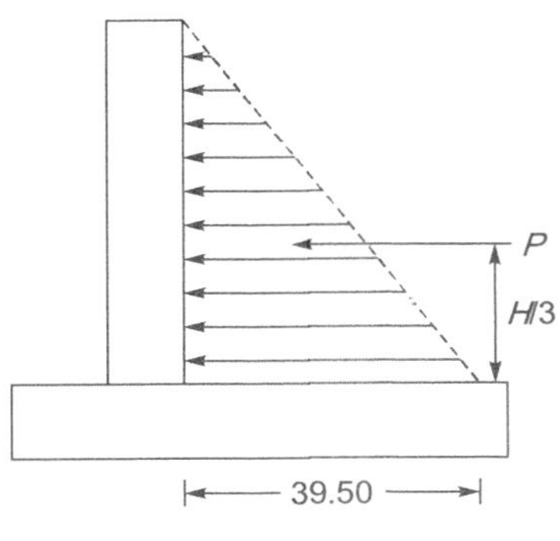

$$\text{BM} = \frac{1}{2} \times 39.50 \times \frac{7.40^2}{3} = 360.503 \text{ kNm}$$

$$M_u = 1.50 \times 360.503 \times 3.50 = 1892.80 \text{ kN/m}$$

The counterforts are designed as T-beams with stem portion acting as slab and counterforts.

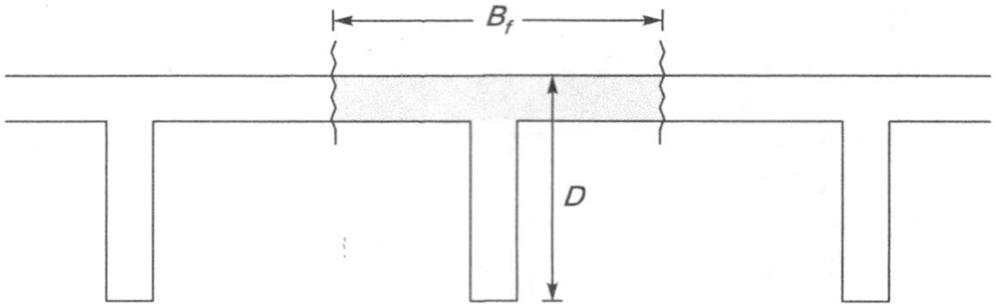

(d) Assume width of counterforts = 500 mm

Effective width of flange,

$$B_f = \frac{L_0}{6} + b_w + 6d_f = \frac{7400}{6} + 500 + 6 \times 200 = 2933 \text{ mm}$$

$B_{f\,max}$ = 3500 mm (Consider B_f = 2933 mm)

(c/c distance between counterforts)

(e)

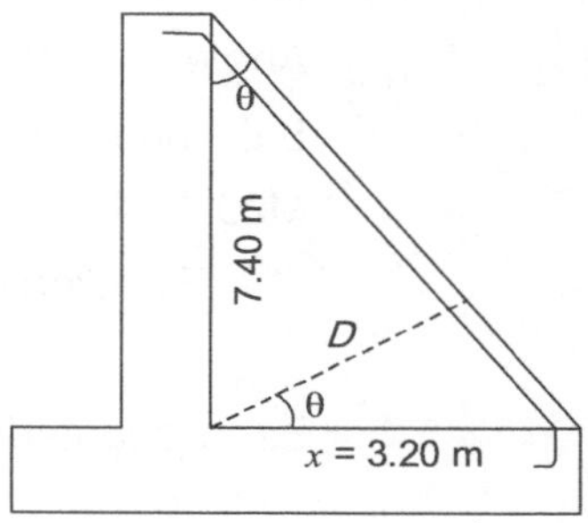

$$\tan\theta = \frac{3200}{7400} = 0.4324$$

$$\theta = 23.4°$$

Overall depth of section perpendicular to reinforced.

$$D = x\cos\theta$$

$$= 3200 \times \cos 23.4° = 2937 \text{ mm}$$

$$\text{Effective cover} = \frac{20}{2} + 40 = 50 \text{ mm}$$

$$d = 2937 - 50 = 2887 \text{ mm}$$

(g) Depth of NA

Assuming NA lies in flange portion of T-beam (Stem)

Equating

$$BM_u = 0.36 \times f_{ck} \times B_f \times x_u\,(d - 0.42\,x_u)$$

$$1892.8 \times 10^6 = 0.36 \times 20 \times 2933 \times x_u \times (2887 - 0.42\,x_u)$$

$$0.42x_u^2 - 2887x_u + 89631.4 = 0$$

$$x_u = 31.2 \text{ mm}, < 200 \text{ mm}$$

Assumption is correct.

(h) Area of steel

$$A_{st} = \frac{BM_u}{0.87f_y(d - 0.42x_u)}$$

$$= \frac{1892.8 \times 10^6}{0.87 \times 415(2887 - 0.42 \times 31.2)} = 1824.2 \text{ mm}^2$$

$$\text{Nos. of 28 mm } \phi \text{ bars} = \frac{2957}{\frac{\pi}{4} \times (28)^2} = 4.80$$

$$\left[\text{Minimum steel} = \frac{0.85}{415} \times 2887 \times 500 = 2957 \text{ mm}^2\right]$$

Say 5-28 mm ϕ bars.

(i) **Check for shear**

Maximum shear force at base

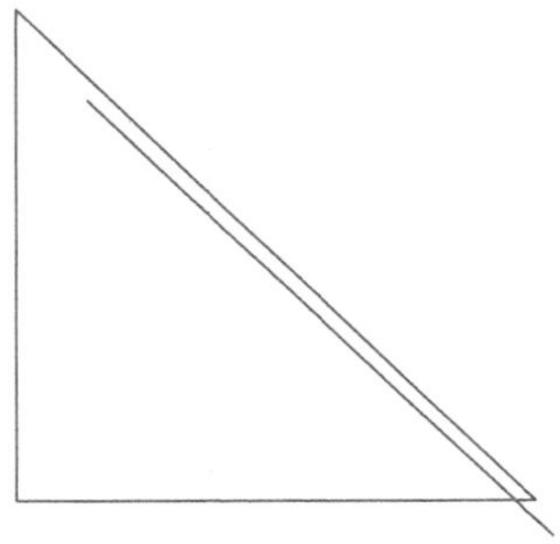

$$V_u = 1.5\,P$$

$$= 1.5 \times \frac{1}{2} \times 39.50 \times 7.40 \times 3.50$$

$$= 767.30 \text{ kN}$$

The section is varying, (depth and moment are measuring in same direction)

$$V_{u_e} = V_u - \frac{M_u}{d}\tan\theta$$

$$= 767.30 - \frac{1892.80}{2.887} \times \tan 23.4°$$

$$= 483.6 \text{ kN}$$

Nominal shear stress, $\tau_v = \dfrac{V_{ue}}{Bd} = \dfrac{483.6 \times 1000}{2887 \times 500} = 0.34 \text{ N/mm}^2$

$$\text{Percentage of steel} = \frac{5 \times \frac{\pi}{4} \times (28)^2}{2880 \times 500} \times 100 = 0.205\%$$

$$\tau_c = 0.28 + \frac{(0.36 - 0.28)}{(0.25 - 0.15)} \times (0.205 - 0.15)$$

$$= 0.324 \text{ N/mm}^2$$

$(\tau_v - \tau_c) = 0.34 - 0.32 = 0.02$ (Very less)

So provide minimum shear reinforced as per

$$\frac{A_{sv}}{BS_v} \geq \frac{0.40}{0.87 f_y}$$

Spacing 2-legged 10 mm ϕ stirrups.

$$\frac{2 \times \frac{\pi}{4} \times (10)^2}{500 \times S_v} \geq \frac{0.40}{0.87 \times 415}$$

$$S_v \leq 283 \text{ mm}$$

Provide 2-legged 10 mm ϕ @ 280 mm c/c

(j) **Tie connection**

Tension force on bottom of counterfort

$$T = 39.50 \times 3.50 = 138.25 \text{ kN/m}$$

$$T_u = 1.5 \times 138.25 = 207.40 \text{ kN}$$

Area of steel required

$$A_{st_v} = \frac{207.40 \times 100}{0.87 \times 415} = 574.4 \text{ mm}^2$$

Spacing of 10 mm ϕ bars in two layer.

$$= \frac{1000}{(A_{st}/2)} \times \frac{\pi}{4} \times (10)^2 = \frac{1000}{\left(\frac{574.4}{2}\right)} \times \frac{\pi}{4} \times (10)^2 = 273 \text{ mm}$$

Provide 10 mm ϕ @ 270 mm c/c.

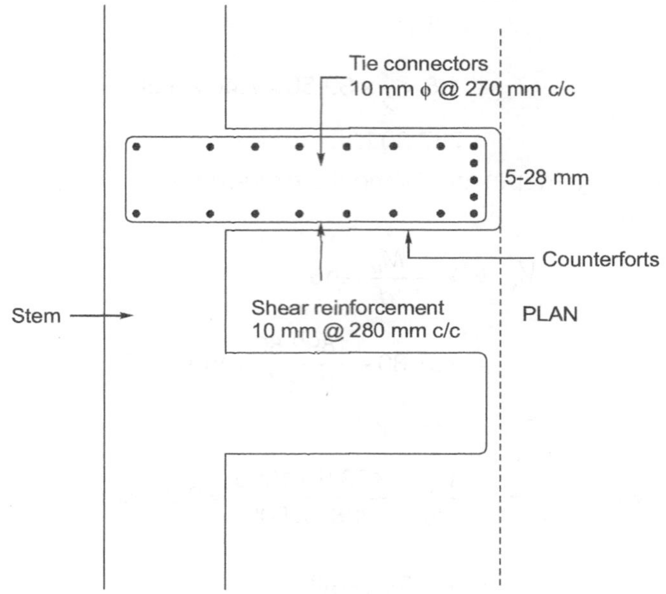

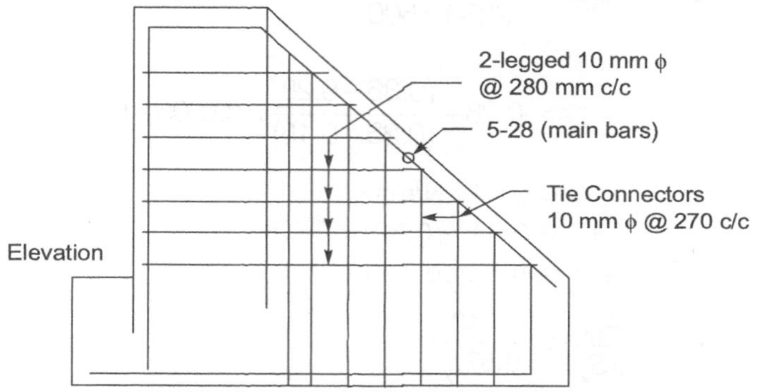

6.7 Design the side walls of an underground tank of size 12 m × 3 m × 3 m deep. The of repose of soil is 30°. The density of soil is taken as 17 kN/m³. Assume the soil is saturated. Use M25 grade of concrete and Fe415 grade of steel. Take Q = 1.156 N/mm² and J = 0.87.

[20 marks : 2019]

Solution:

Design of long wall

The bottom 1 m (> H/4) of short walls will be designed as cantilever while the top portion will designed as slab supported by long walls.

(a) Tank empty with pressure of saturated soil from outside

(i) Top Portion

At $h = 1$ m above the base of short wall

$$p_a = k_a \gamma'(H - h) + \gamma_w(H - h)$$

$$= \frac{1}{3} \times 7190 \times 2 + 9810 \times 2 = 24413 \text{ N/mm}^2$$

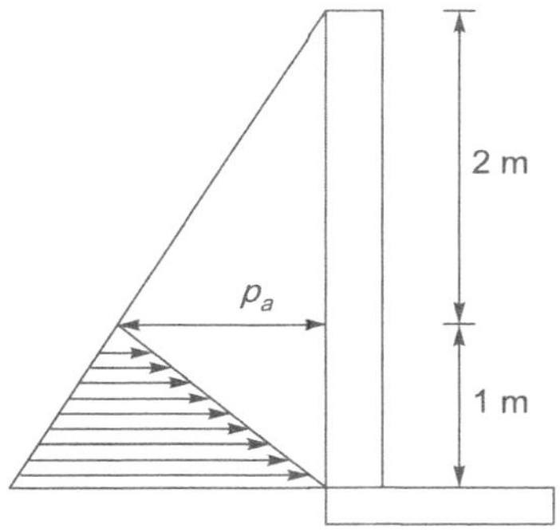

$$M_f \text{ at support} = \frac{p_a L^2}{12} = \frac{24413 \times 3^2}{12} = 18309.75 \text{ Nm} \quad \text{(Causing tension outside)}$$

$$M \text{ at centre} = \frac{p_a L^2}{8} - m_f = 9154.375 \text{ Nm} \quad \text{(Casing tension inside)}$$

$$d = 260 - (25 + 16 + 8) = 211 \text{ mm}$$

At support, $$A_{st} = \frac{18309.75 \times 1000}{130 \times 0.8 \times 211} = 767.25 \text{ mm}^2$$

Using 12 mm ϕ bars, $$S = \frac{1000 \times 113.1}{767.25} = 147.41 \text{ mm c/c}$$

So provide 12 mm ϕ @ 140 mm c/c at outer face at 2 m below top.

At mid span, $$A_{st} = \frac{767.25}{2} \simeq 384 \text{ mm}^2$$

Minimum = 660.4 mm^2 (as found earlier)

$$\text{Spacing of 12 mm } \phi = \frac{1000 \times 113.1}{660.4} = 171.3 \text{ mm}$$

Hence provide 12 mm ϕ @ 170 mm c/c at inner face.

(ii) **Bottom Portion**

The bottom 1 m will bend as cantilever intensity of earth pressure at bottom.

$$P_a = 36620 \text{ N/mm}^2$$

$$M = \frac{1}{2} \times 36620 \times 1 \times \frac{1}{3} = 6103 \text{ Nm} \quad \text{(with tension at outer face)}$$

$$A_{st} = \frac{6103 \times 1000}{0.87 \times 130 \times 225} = 207 \text{ mm}^2$$

Minimum steel = 660.4 mm^2

Spacing of 12 mm ϕ = 170 mm c/c at outer face.

(b) **Tank full with water and no earthfill outside**

(i) **Top Portion**

Top 2 m acts as slab supporting on long walls and bottom 1 m acts as cantilever.

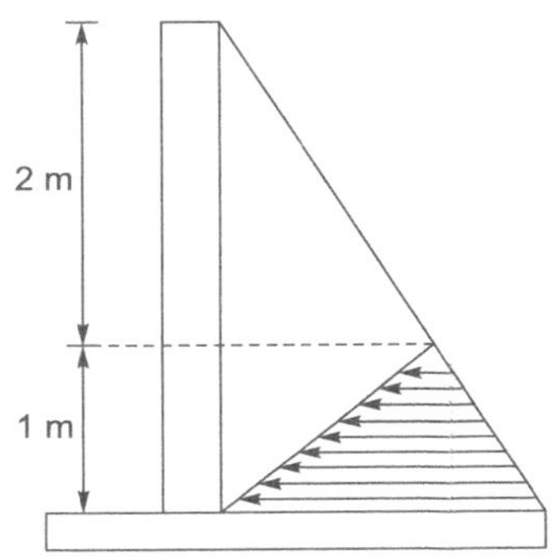

At $$h = 1 \text{ m}$$

$$p = \gamma_w (H - h)$$

$$= 9810 \times 2 = 19620 \text{ N/m}^2$$

$$M_f \text{ at support} = \frac{pB^2}{12} = \frac{19620 \times 3^2}{12} = 14715 \text{ Nm}$$

(Causing tension at the inside)

$$M_c \text{ at center } = \frac{pB^2}{211} = 7357.5 \text{ Nm} \quad \text{(Causing tension at outside)}$$

Direct tension in short wall, due to water pressure on the end one meter width of long walls is

$$P_B = \gamma_w (H - h) \times 1 = 19620 \text{ N}$$

Effective depth, d for horizontal steel = 211 mm

$$\therefore \text{ Distance} \quad x = d - \frac{T}{2} = 211 - 130 = 81 \text{ mm}$$

Hence net BM = $m - P_B x$

$$A_{st1} = \frac{m - P_B x}{\sigma_{st} \tau d} \quad \text{and} \quad A_{st2} = \frac{P_B}{\sigma_s}$$

At the inside face (end of short wall)

$$A_{st1} = \frac{m_f - P_B x}{\sigma_{st} d \tau} = \frac{1000 \times 14715 - 19620 \times 81}{130 \times 0.87 \times 211} = 550.02 \text{ mm}^2$$

$$A_{st2} = \frac{P_B}{\sigma_{st}} = \frac{19620}{130} = 150.92 \text{ mm}^2$$

$$A_{st} = A_{st1} + A_{st2} = 700.94 \text{ mm}^2$$

$$\text{Using 12 mm } \phi, \quad \text{spacing} = \frac{1000 \times 113.1}{700.94} = 161.35 \text{ mm}$$

Hence provide 12 mm ϕ @ 160 mm c/c at inner face.

At the outside face (Middle of short wall)

$$A_{st1} = \frac{m_c - P_B x}{\sigma_{st} \tau d} = \frac{7357.5 \times 1000 - 19620 \times 81}{130 \times 0.87 \times 211} = 241.7 \text{ mm}^2$$

$$A_{st2} = \frac{19620}{130} = 150.92 \text{ mm}^2$$

$$A_{st} = 241.7 + 150.92 = 392.62 \text{ mm}^2$$

But $\quad A_{st\ min} = 660.5 \text{ mm}^2$

So provide 12 mm ϕ @ 170 mm c/c at outside face.

(ii) Bottom Portion

$$p \text{ at bottom} = 9810 \times 3 = 29430 \text{ N/m}^2$$

$$m = \frac{1}{2} \times 29430 \times 1 \times \frac{1}{3} = 4905 \text{ Nm (with tension at inside face)}$$

$$A_{st} = \frac{4905 \times 1000}{130 \times 0.87 \times 225} = 192.75 \text{ mm}^2 \not> 660.5 \text{ mm}^2$$

So provide 12 mm ϕ @ 170 mm c/c at inside face.

6.8 **A symmetrical reinforced concrete frame building 25 m × 25 m in plan is located in seismic zone IV on hard soil. The height of the building is 30 m. Determine the base shear due to earthquake.**
Given:
$Z = 0.24$, $I = 1.5$
Total dead load = 150000 kN
Total live load (effective) = 50000 kN

$$T = 0.09 \frac{h}{\sqrt{d}}$$

$$\frac{S_a}{g} = \begin{cases} 1+15T & 0 \le T \le 0.10 \\ 2.5 & 0.10 \le T \le 0.40 \\ \frac{1}{T} & 0.40 \le T \le 4.0 \end{cases}$$

[12 marks : 2020]

Solution:

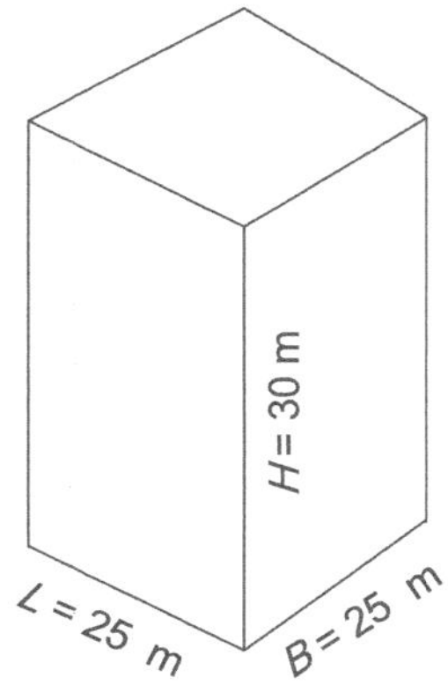

I = 1.5m, Z = 0.24, R = 5 (Assuming special RC frame)

Height of building, $h = 30$ m

Base dimension, $D = 25$ m

So, time period,
$$T = 0.09\frac{h}{\sqrt{D}}$$
$$= 0.09 \times \frac{30}{\sqrt{25}} = 0.54 \text{ s}$$

For $T = 0.54$ s
$$\frac{S_a}{g} = \frac{1}{T} = \frac{1}{0.54} = 1.852$$

Thus horizontal seismic coefficient,
$$A_h = \left(\frac{Z}{2}\right)\left(\frac{I}{R}\right)\left(\frac{S_a}{g}\right) = \left(\frac{0.24}{2}\right)\left(\frac{1.5}{5}\right)(1.852) = 0.06667$$

Response reduction factor

$R = 5.0$ (For framed RCC building with ductile detailing)

Note: *In zone IV ductile detailing is must for all buildings as per IS 13920.*

Total seismic weight of building = DL + 50% LL = 150000 + 50% of 50000 = 175000 kN

Total base shear, $V_B = W \times A_h = 0.0667 \times 175000 = 11667.6$ kN

6.9 A four-storey reinforced concrete (RC) office building located in seismic zone IV is shown in the figure. The RC frames are infilled with brick masonry. The lumped weights due to dead loads is 15 kN/m^2 on the floors and 12 kN/m^2 on the roof. The floors have to cater to a live load of 4 kN/m^2 on the floors and 1.5 kN/m^2 on The roof. Calculate the design seismic load on the structure at different floors using Linear static (Equivalent static) analysis, along x-axis.

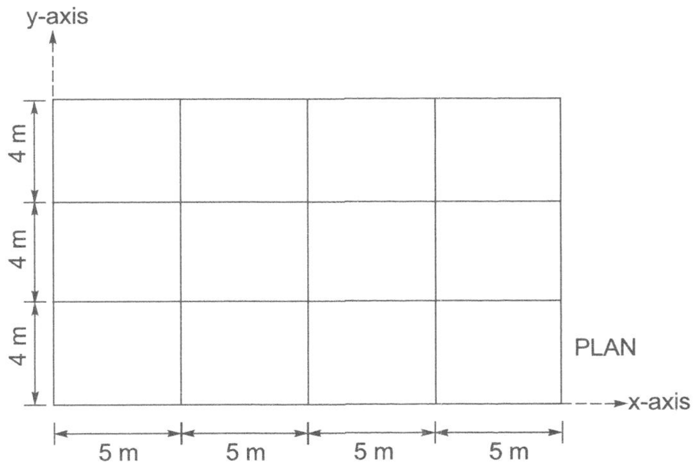

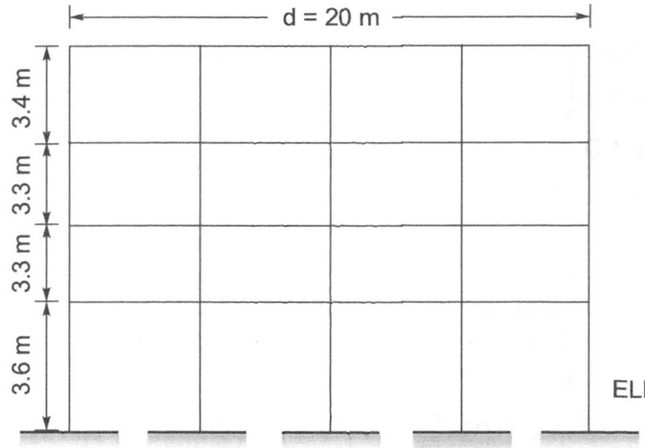

Zone factor Z = 0.24

Importance factor I = 1.2

Response Reduction factor R = 5

Fundamental translational natural period $T_a = \dfrac{0.09h}{\sqrt{d}}$

$$\frac{S_a}{g} = \begin{cases} 2.5 & 0 < T \le 0.55\ s \\ \dfrac{1.36}{T} & 0.55 \le T \le 4.0s \\ 0.34 & T > 4.0\ s \end{cases}$$

[20 Marks : 2021]

Solution:

Given: Since building is located in Zone-IV,

Zone factor (Z) = 0.24

Importance factor, I = 1.20

Response reduction factor, R = 5.00

$h = 3.6 + 3.3 + 3.3 + 3.4 = 13.60$ m

Along x-axis, $d = L = 4 \times 5$ m = 20 m

$$\text{Fundamental time period, } T_a = \frac{0.09 \times h}{\sqrt{d}} = \frac{0.09 \times 13.6}{\sqrt{20}} = 0.274\, sec$$

From table given for, T_a = 0.274 sec

$$\frac{S_a}{g} = 2.5$$

Calculation of base shear on floor 0.50% of live load is to be considered and on roof, no live load needs to be considered for the computation of seismic weight.

Seismic weights on floors

$$W_1 = (12\text{ m} \times 20\text{ m}) \times (\text{DL} + \text{Part of LL})$$
$$= 12 \times 20 \times (15 + 4 \times 0.5)$$
$$= 4080\text{ kN} = W_1 = W_2 = W_3$$

Seismic weight of roof $W_4 = 12 \times 20 \times (12 + 0) = 2880$ kN

Total seismic weight of building $= W_1 + W_2 + W_3 + W_4$

$= 4080 + 4080 + 4080 + 2880 = 15120$ kN

Total lateral force due to earthquake

$$V_B = A_H \times \Sigma W$$

$$= \left(\frac{Z}{2}\right) \times \left(\frac{I}{R}\right) \times \left(\frac{S_a}{g}\right) \times \Sigma W$$

$$= \frac{0.24}{2} \times \frac{1.20}{5.0} \times 2.50 \times 15120$$

$$V_B = 1088.64 \text{ kN}$$

Say 1089 kN

Seismic load on each floor:

$$\Sigma Wh^2 = 4080 \times 3.60^2 + 4080 \times 6.9^2 + 4080 \times 10.2^2 + 2880 \times 13.6^2$$

$$= 1204293.60$$

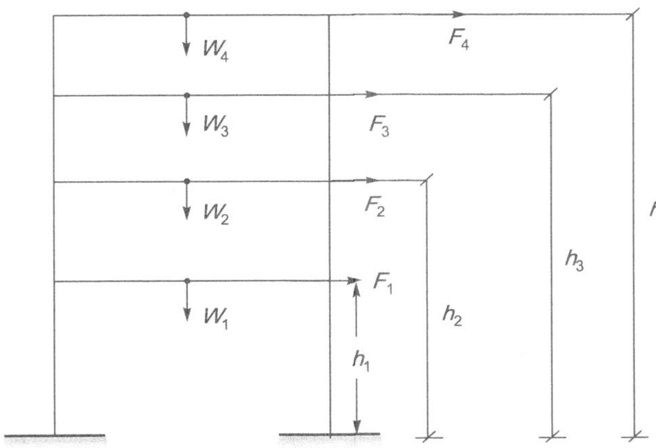

Distribution of lateral force $= \dfrac{W_i h_i^2}{\Sigma W_i h_i^2} \times V_B$

Floor	W_i	h_i	$W_i h_i^2$	Distribution of lateral forces
1	4080	3.6	52876.8	47.80 kN = F_1
2	4080	6.9	194248.8	175.6 kN = F_2
3	4080	10.2	424483.5	383.7 kN = F_3
4	2880	13.6	532684.8	481.8 kN = F_4
		ΣWh^2	1204293.6	Total = 1089 kN

6.10 Design an open cylindrical water tank of 350 m^3 capacity. This tank will rest on ground and have a free-flexible joint at base. Overall height of tank is 4.0 m, including the free board of 200 mm. Design the vertical cylindrical wall of tank and sketch the details. Consider only maximum hoop tension for entire height.

The following parameters may be used for design, if required.

(i) Permissible direct tensile stress of concrete $\sigma_{ct} = 1.2$ MPa

(ii) Permissible tensile stress of steel σ_{st} = 150 MPa up to 225 mm from water fall = 190 MPa beyond 225 mm from water fall.
(iii) Molecular ratio, $m = 13$
(iv) Minimum thickness required = 180 mm
(v) Minimum % of reinforcement up to 100 mm thickness 0.24% linearly reduces to 0.16% for more than 450 mm thickness
(vi) Use 16 mm diameter, high strength deformed bar for hoop reinforcement
(vii) Use 8 mm diameter, high strength deforced bar for vertical distribution reinforcement
(viii) Minimum diameter of tank required = 11.0 m
(ix) Unit weight of water = 10 kN/m^3

(x) Tensile stress in concrete = $\dfrac{T}{A_c + (m-1)A_s}$

[20 Marks : 2021]

Solution:

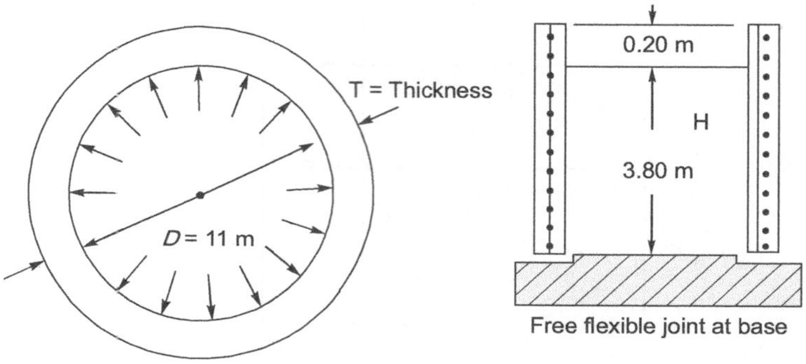

1. Minimum dia. of tank = 11 m

$$D = 11 \text{ m}$$

Volume, $$V = 350 \text{ m}^3 = \frac{\pi}{4}(D)^2 \times H$$

$$H = \frac{350}{\frac{\pi}{4}(11)^2} = 3.68 \text{ m} \quad \text{say } 3.70 \text{ m}$$

Tank height given = 4.0 m
free board = 0.2 m
H available = 4.0 − 0.2 = 3.80 m
> 3.70 m required (OK)

Let us design for $D = 11$ m
and, $H = 3.80$ m

2. Maximum hoop tension in tank at bottom.

$$= T_H = \frac{p.D}{2} = \frac{\gamma_w H.D}{2}$$

$$T_H = \frac{10 \times 3.80 \times 11}{2} = 209 \text{ kN}$$

3. Area of steel required. (Considered cracked)

$$A_{st} = \frac{T_H}{f_{st}} = \frac{209000}{150} = 1393.33 \simeq 1394 \text{ mm}^2$$

Using 16 mm ϕ bars for hoop height,

$$\text{Spacing} = \frac{1000}{1394} \times \frac{\pi}{4}(16)^2 = 144 \text{ mm}$$

Provide 16 mm ϕ @ 140 mm c/c in single layer.

4. Minimum thickness of tank = 180 mm

Check, stress developed in concrete if

T = 180 mm is provided

$$f_{ct} = \frac{T_H}{1000 \times T + (m-1)A_{st}} \quad \text{[Considering uncracked section]}$$

$$= \frac{209000}{\left(1000 \times 180 + (13-1) \times 1394\right)}$$

$$= 1.06 \text{ N/mm}^2 < (f_{ct}\text{(pev)}) = 1.20 \text{ N/mm}^2$$

So safe in direct tension.

5. Minimum reinforcement, for T = 180 mm

$$P_t(\%) = 0.24 - \frac{(0.24 - 0.16)}{(450 - 100)} \times (180 - 100) = 0.222\%$$

$$A_{st} = \frac{0.222}{100} \times 1000 \times 180$$

$$= 399.6 \text{ say } 400 \text{ mm}^2$$

Spacing of 8 mm ϕ for vertical reinforcement.

$$= \frac{1000}{399.6} \times \frac{\pi}{4} 8^2 = 125.7 \text{ mm}$$

Provide 8 mmϕ @ 125 mm c/c

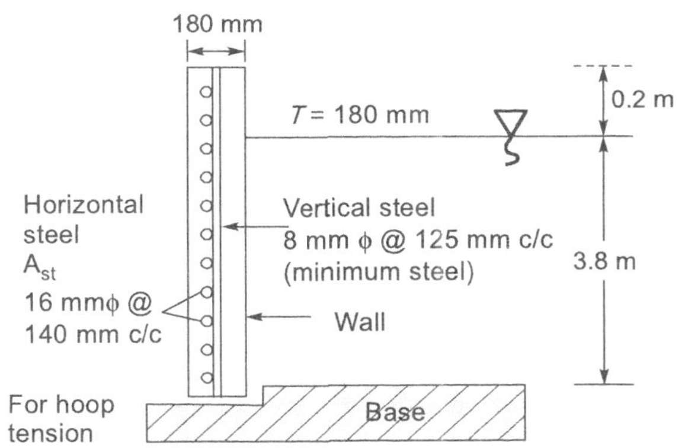

6.11 A simply supported post-tensioned concrete beam of 10 m span, 230 mm wide and 400 mm deep is prestressed with a straight cable, having cross sectional area 385 mm², located at 60 mm from the soffit of the beam subjected to an initial stress of 1200 N/mm². Using M50 grade concrete and the following data, estimate the percentage loss of prestress.

$$E_c = 5000\sqrt{f_{ck}}$$

Modulus of elasticity of steel = 2.1×10^5 N/mm²
Relaxation of stress in steel = 4.5%
Shrinkage strain of concrete = 0.0003
Creep coefficient of concrete = 1.6
Friction coefficient for wave effect = 0.0025 per metre
Slip at anchorage = 2 mm
Loss of prestress due to friction = $p_0(\mu\alpha + kx)$

[12 marks : 2022]

Solution:

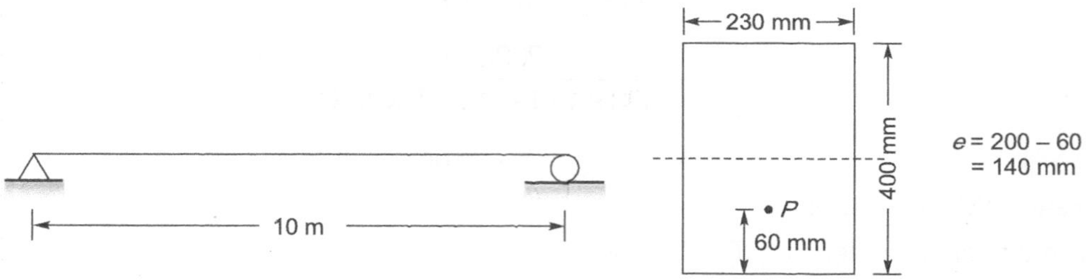

Cross-sectional area of cable = 385 mm²
Initial prestress = 1200 N/mm²
Tension in cable [P] = $1200 \times 385 \times 10^{-3}$ kN = 462 kN

$$E_c = 5000\sqrt{50} = 35355.34 \text{ MPa}$$

Stress in concrete at the level of steel

$$f_c = \frac{P}{A} + \frac{P.e}{I} \times e = \frac{462\times10^3}{230\times400} + \frac{462\times10^3\times140}{\frac{230\times400^3}{12}}\times140$$

$$= 5.02 + 7.38 = 12.4 \text{ N/mm}^2$$

Loss of prestress in post-tensioned beam:

(a) Loss of prestress due to elastic shortening of concrete will be zero because there is only one cable.

In case of postensioned member, if there is only one tendon, there is no loss of prestress because the applied prestress is recorded after the elastic shortening of member.

(b) Loss of prestress due to creep of concrete:

$$= \theta.m.f_c = 1.6\times\frac{E_s}{E_c}\times f_c$$

$$= 1.6\times\frac{2.1\times10^5}{35355.34}\times12.4 = 117.844 \text{ N/mm}^2$$

(c) Loss of prestress due to shrinkage of concrete = $\epsilon_c \times E_s$

$$= 0.0003 \times 2.1 \times 10^5 = 63 \text{ N/mm}^2$$

(d) Loss of prestress due to anchorage slip:

$$= \frac{\Delta L}{L} E_s = \frac{2}{10{,}000} \times 2.1 \times 10^5 = 42 \text{ N/mm}^2$$

(e) Loss of prestress due to relaxation in steel:

$$= \frac{4.5}{100} \times 1200 = 54 \text{ N/mm}^2$$

(f) Loss of prestress due to friction:

$$= p_0[\mu\alpha + k.x]$$

For straight cable, $\alpha = 0$

Let us assume that the cable is anchored from one end only,

$\therefore$ $\Delta f = 1200[\mu \times 0 + 0.0025 \times 10] = 30 \text{ N/mm}^2$

Let us assume that the cable is anchored on both ends,

$\therefore$ $\Delta f = 1200[\mu \times 0 + 0.0025 \times 5] = 15 \text{ N/mm}^2$

$\therefore$ Total loss of prestress when cable is anchored on one side:

$117.844 + 63 + 42 + 54 + 30 = 306.844 \text{ N/mm}^2$

$\therefore$ Percentage loss $= \frac{306.844}{1200} \times 100 = 25.6\%$

Total loss of prestress when cable is anchored one both side

$$= 306.844 - 30 + 15 = 291.844 \text{ N/mm}^2$$

Percentage loss $= \frac{291.844}{1200} \times 100 = 24.3\%$

■■■■

6 Construction Practice, Planning and Management

Revised Syllabus of ESE : *Construction-Planning, Equipment, Site investigation and Management including Estimation with latest project management tools and network analysis for different Types of works; Analysis of Rates of various types of works; Tendering Process and Contract Management, Quality Control, Productivity, Operation Cost; Land acquisition; Labour safety and welfare.*

---

1. Project Management and Network Techniques

1.1 Explain various steps involved in the development of networks.

[10 marks : 2006]

Solution:

Various steps involved in the development of networks are as follows:

(i) **Objective:** During the planning of a project, the first and foremost step is to define the project and to decide the way in which it is to be carried out. The task to be undertaken requires to be set down as specific, definite, complete and well defined verbal statement. Specific verbal statement means the specific description of particular dimensions, type of materials, plants, etc. necessary for the project. Objective specifies the task to be undertaken and policy of its execution. This specification defines the project and determines the way in which it is to be carried out.

(ii) **Plan breakdown:** After establishing objective of the task, the planner has to adopt either forward planning or backward planning (or mixed planning) to achieve the goal. This backward to forward thinking will give a list of activities or jobs to be performed to achieve the task and also stages in the project execution.

(iii) **Sequencing:** In the second step we have obtained a general list of various activities and events necessary for the completion of the project. This general list is to be reviewed so that in each of the main group, those with definite similarities can be put in suitable subgroups.

(iv) **Location of nodes:** Now the events listed above are required to be located on paper so that a visual effect of movement along a time scale is obtained. Events should be located in such a way that they represent initial picture of the relation amongst them. This relationship results from the proposed use of manpower, money, material and other resources during a particular period of time.

(v) **Drawing Arrows:** Events having close and direct relationship are joined to each other by arrows representing activity to be performed for passing from one stage of the project to the other. These activities should fall in logical sequence.

(vi) **Checking:** At this stage, the diagram is checked with respect to content, sequence and sense and degree of detail.

It is essential to check the diagram for events and activities in respect of logic and accuracy. Particular attention should be paid to multiple events, i.e., those events at which more than one arrows enters and/or more than one arrows leave, since it is at this point that errors are most likely to occur. The checking ensures that the network correctly represents the sequence.

It should be ensured that network does not contain loops or cycles. If located these should be removed. Also, it should be checked whether there is any event (other than first) which has only outgoing arrows, or whether there is any event (other than last one) which has only incoming arrows. Such situation, if found, should be rectified. There should be no dead ends left.

An arrow should always represent singular situation but an event may represent commencement of more than one operations. In respect of sufficient detail, a ratio, known, as *E/A* ratio given as

$$\frac{E}{A} = \frac{\text{Total number of events}}{\text{Total number of activities}}$$

In a good network, its value should lie between 1 to 1.6.

(vii) **Redraw:** The errors found in the previous steps are removed and the diagram is redrawn by introducing unique dummies, if necessary.

(viii) **Number:** After drawn the final network, the events are numbered using Fulkerson's rule.

1.2 Draw a network for the following situations;
(i) A is prerequisite on D
(ii) B is prerequisite on D and E
(iii) C is prerequisite on E
(iv) D is prerequisite on F
(v) E is prerequisite on F

[10 marks : 2016]

Solution:

Network diagram will be as shown:

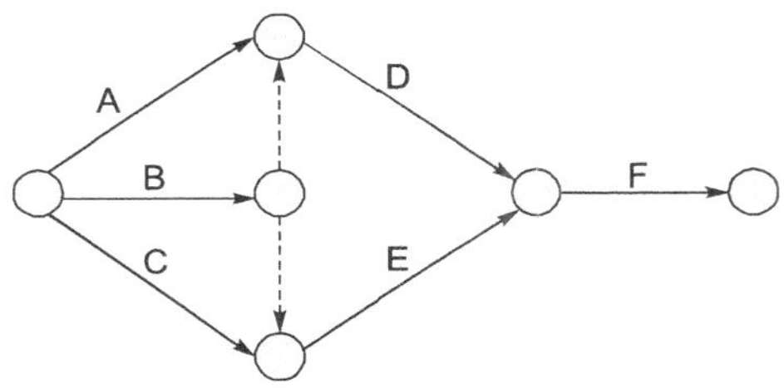

1.3 Briefly answer the following:
(i) How can an existing bar chart be modified to depict the project progress made?
(ii) Differentiate between 'Forward Planning' and 'Backward Planning' for network construction.

[10 marks : 2017]

Solution:

(i) A bar chart doesn't show the progress of work and hence it cannot be used as a control device. Controlling is essential for rescheduling the remaining activities. However, an existing bar chart can be modified to depict the progress made. This can be done by showing the progress of each of each activity, by hatched lines along the corresponding bar of the activity. Generally, hatching is done in half the width of the bar.

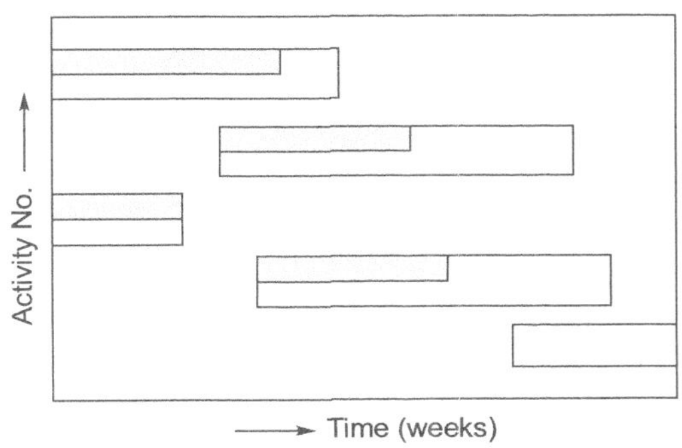

(ii) **'Forward Planning':** In this method, the planner starts from the initial event and builds up the events and activities logically and sequentially until the end event is reached. In this method, while considering activity, a planner asks himself the following questions:

What event comes next?

What are dependent events?

What events can take place concurrently?

Backword planning: In this method, the planner starts with the end event and arranges the events and activities until the initial event is reached. Keeping the goal in view, the planner asks himself if we want to achieve this, what events or activities should have taken place.

1.4 **What is a work breakdown structure in Construction Project Management? Define and explain in brief. Further, how Work Breakdown Structure is classified into different levels for making the job convenient? Explain with an example.**

[12 marks : 2018]

Solution:

(i) **Work breakdown structure in construction project management:**

A work breakdown structure is a key project deliverable that organizes the team's work into manageable sections. The project management body of knowledge (PMBOK) defines the work breakdown structure as a 'deliverable oriented hierarchical decomposition of work to be executed by the project team'. Work breakdown structure visually defines the scope into manageable chunks that a project team can understand, as each level of work breakdown structure provides further definition and detail.

Figure below explains how work breakdown structure is classified into different levels for making the job convenient.

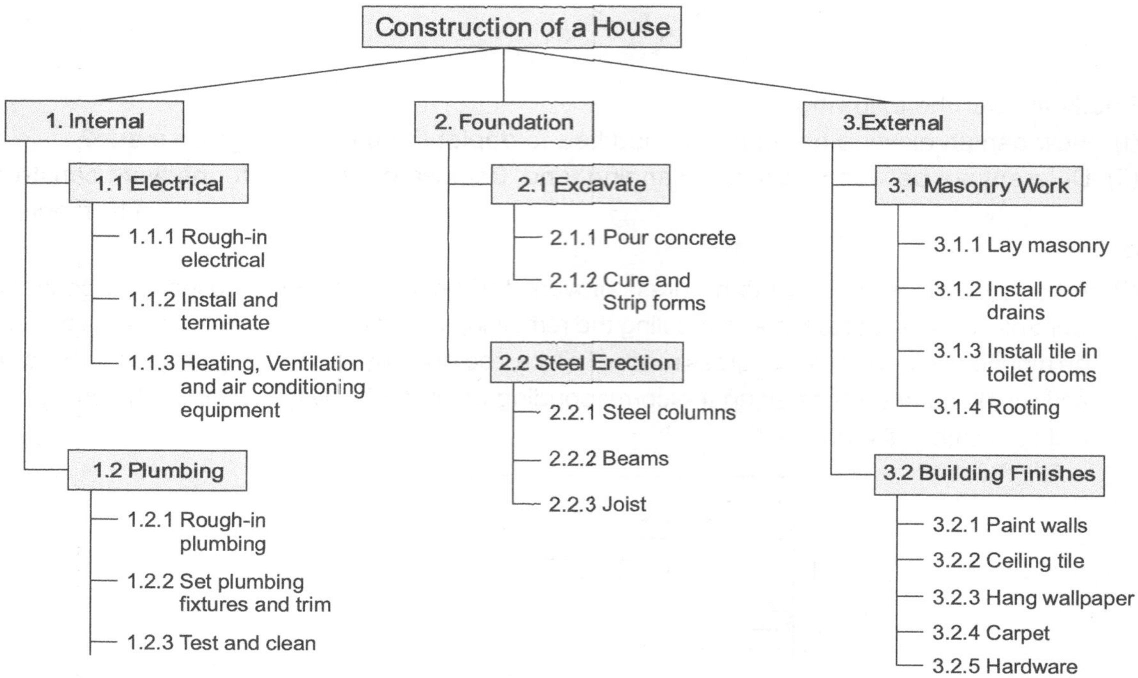

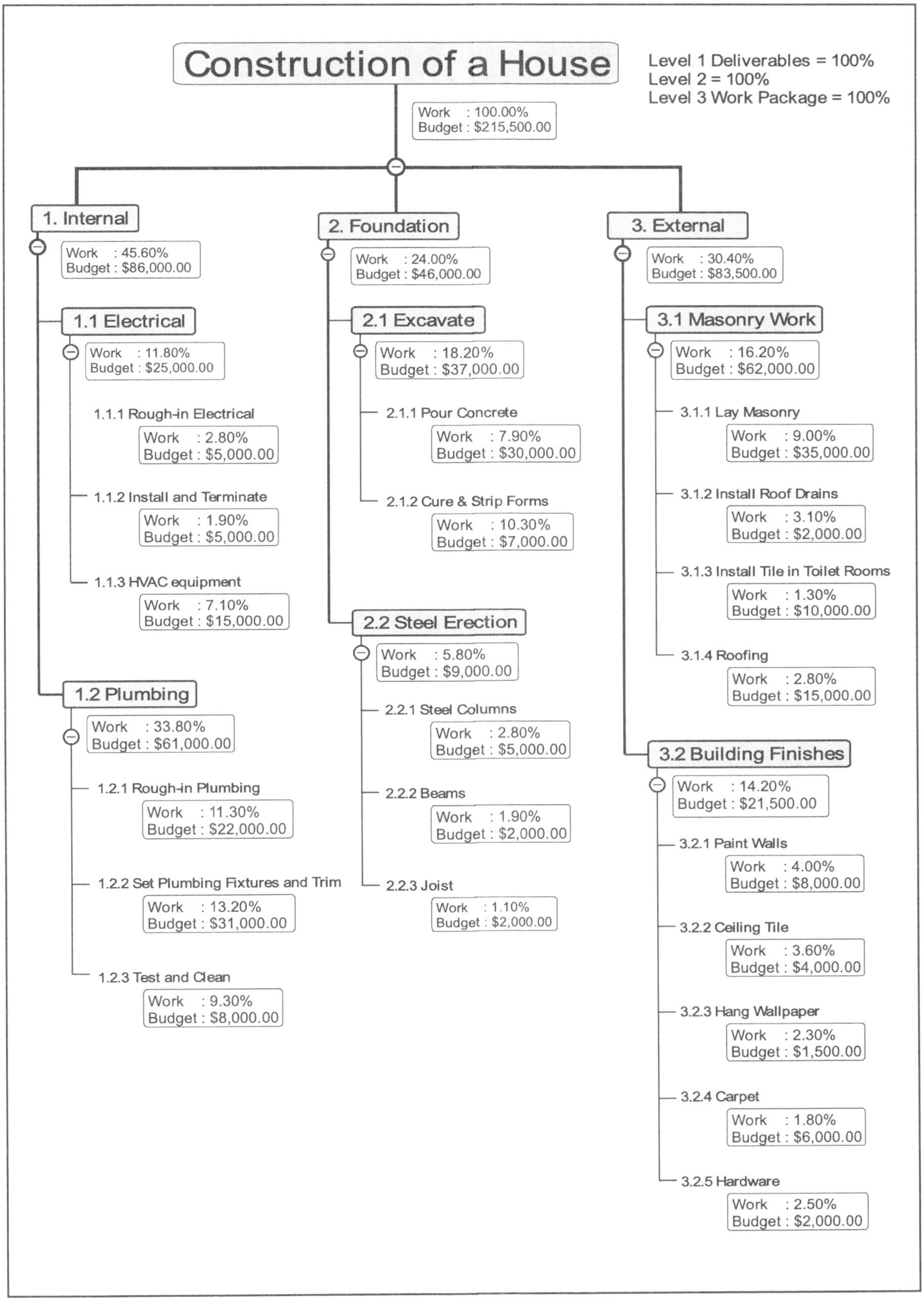
Construction of a House
Level 1 Deliverables = 100%
Level 2 = 100%
Level 3 Work Package = 100%
Work : 100.00%
Budget : $215,500.00
1. Internal
Work : 45.60%
Budget : $86,000.00
1.1 Electrical
Work : 11.80%
Budget : $25,000.00
1.1.1 Rough-in Electrical
Work : 2.80%
Budget : $5,000.00
1.1.2 Install and Terminate
Work : 1.90%
Budget : $5,000.00
1.1.3 HVAC equipment
Work : 7.10%
Budget : $15,000.00
1.2 Plumbing
Work : 33.80%
Budget : $61,000.00
1.2.1 Rough-in Plumbing
Work : 11.30%
Budget : $22,000.00
1.2.2 Set Plumbing Fixtures and Trim
Work : 13.20%
Budget : $31,000.00
1.2.3 Test and Clean
Work : 9.30%
Budget : $8,000.00
2. Foundation
Work : 24.00%
Budget : $46,000.00
2.1 Excavate
Work : 18.20%
Budget : $37,000.00
2.1.1 Pour Concrete
Work : 7.90%
Budget : $30,000.00
2.1.2 Cure & Strip Forms
Work : 10.30%
Budget : $7,000.00
2.2 Steel Erection
Work : 5.80%
Budget : $9,000.00
2.2.1 Steel Columns
Work : 2.80%
Budget : $5,000.00
2.2.2 Beams
Work : 1.90%
Budget : $2,000.00
2.2.3 Joist
Work : 1.10%
Budget : $2,000.00
3. External
Work : 30.40%
Budget : $83,500.00
3.1 Masonry Work
Work : 16.20%
Budget : $62,000.00
3.1.1 Lay Masonry
Work : 9.00%
Budget : $35,000.00
3.1.2 Install Roof Drains
Work : 3.10%
Budget : $2,000.00
3.1.3 Install Tile in Toilet Rooms
Work : 1.30%
Budget : $10,000.00
3.1.4 Roofing
Work : 2.80%
Budget : $15,000.00
3.2 Building Finishes
Work : 14.20%
Budget : $21,500.00
3.2.1 Paint Walls
Work : 4.00%
Budget : $8,000.00
3.2.2 Ceiling Tile
Work : 3.60%
Budget : $4,000.00
3.2.3 Hang Wallpaper
Work : 2.30%
Budget : $1,500.00
3.2.4 Carpet
Work : 1.80%
Budget : $6,000.00
3.2.5 Hardware
Work : 2.50%
Budget : $2,000.00

Hence, various levels of WBS can show the various details at a glance:

1. What the various elements of project are,
2. How the necessary work is distributed between the elements of the project.
3. How the cost or budget is distributed between the elements of the project.
4. How the larger elements of the project are sub-divided into smaller ones.

WBS is a systematic approach to scooping the project work in which logical hierarchical pattern is devised in a Top-Down Approach.

1.5 Define the terms activity, event and Network.

[12 marks : 2019]

Solution:

Event

- An event is either start or completion of an activity.
- Events are significant points in a project which act as control points of the project.
- An event is an instant of time and it does not require time or resources.

Following are examples of an event:

1. All parts assembled
2. A budget prepared
3. Construction completed

Following can not be events:

1. Prepare budget
2. Assemble parts
3. Excavate trench

- Events are represented by nodes in a network. It may have any of the following shapes.

(i) Circular *(ii) Square* *(iii) Rectangular* *(iv) Oval*

Different Shapes for Events

Most commonly adopted shape for events is circular shape.

- **Tail event or the start event:**

  It makes the beginning of an activity.

  If it is the first event of project then known as "initial as start event".

  It has only outgoing arrow.

  e.g: Event 10 is a tail event. Arrows represent job or activity of the project.

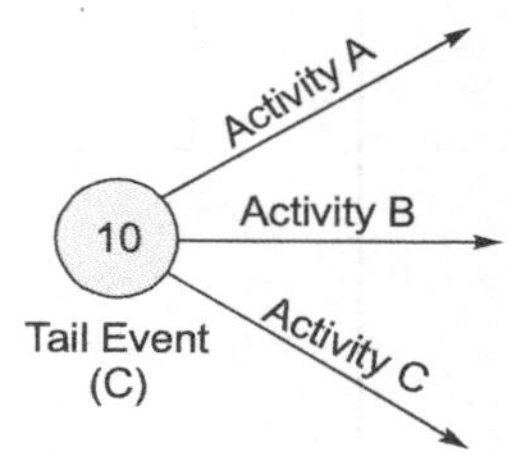

Tail Event

- **Head event or the final event :**

  The event which marks the completion of an activity is known as "head event".

  If this event represents completion of entire project then it is called "Finish event".

  It has only incoming arrows.

  e.g.: Event 20 is a head event.

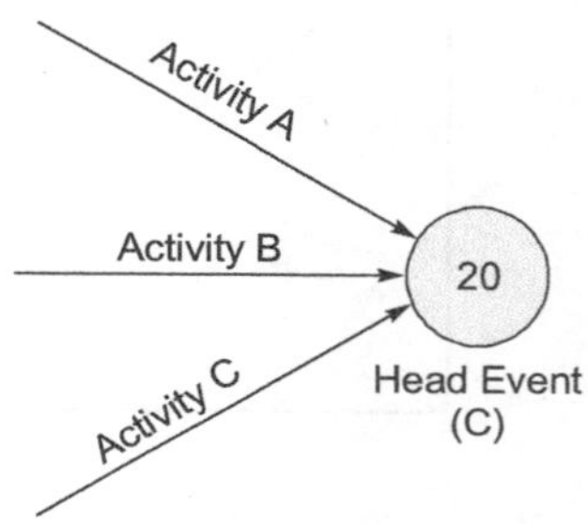

Head Event

- **Dual role events** : All events except the first and the last event of a project are dual role events. They have both incoming and outgoing arrows.

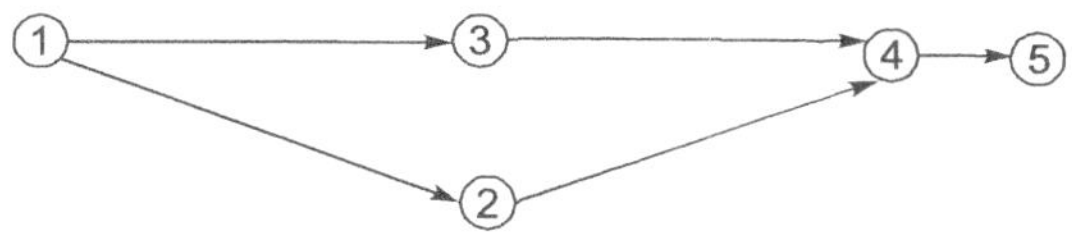

Dual Role Events

e.g.: Events 2, 3 and 4, are dual role events.

- **Successor events :** The event or events that follow another event are called successor events to that event.

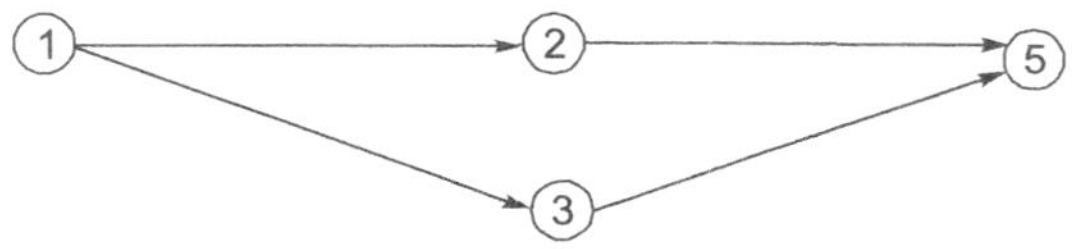

Successor Events

e.g.: Event 2 and 3 are successor events of event 1.

- **Predecessor events :** The event or events that occur before another event are called predecessor event to that event.
  In above figure, events 2, 3 are predecessor to event 5.

Network

- It is the flow of diagram consisting of activities and events connected logically and sequentially.
- Network diagram are of two types:
  (i) Activity-on-Arrow Network (A-O-A)
  (ii) Activity-on-Node Network (A-O-N)
- It is an outcome of the improvements in the previous method (discussed further).
- They are called by various names such as PERT, CPM, UNETICS, LESS, TOPS and SCANS.
- However all these have emerged from the two major network systems viz.:
  1. PERT
  2. CPM

Advantages of network method over bar chart and milestone chart (discussed further):

1. Inter-relationships between activities and events of a project are clearly shown.
2. The project can be treated as an integrated whole with all its sub-activities clearly related with each other. It helps in controlling the project.
3. Network method is useful for very complicated projects having large number of activities.
4. It indicates the time required in between two activities in which rescheduling of a project is possible.
5. Time uncertainty is accounted for and so it is also useful for research and development projects.

1.6 **Explain major activities involved in different stages of planning for a construction project.**

[20 marks : 2019]

Solution:

Planning is the most important technique of management . In a construction project plan includes the estimates, the budget and time schedule and sequences of completion of each part of the project, manpower planning and the plant and equipments.

Following steps are involved in effective planning:

(i) **Crystalizing the opportunity or problem:** The first step in planning would be to find out the problem or identify the opportunity to be seized, this is necessary to be able to formulate practical and realistic objectives.

(ii) **Securing and analyzing necessary information:** Adequate information is required on course of action possible. It is necessary to determine the nature of the information required and where this information will be available. This information must be analysed to establish the relationship and tabulate them for adequate interpretation.

(iii) **Establishing planning premises and constraints:** An analysis of the data so collected will result in the formulation of certain assumptions on the basis of which the plan will be made through a process of forecasting. Constraints such as government control will also exist. Planning will be in the backdrop of such premises and constraints which must be watched to detect changes and their effect on the plans.

(iv) **Ascertaining alternative course of action or plan:** Based on the above analysis, possible alternative course of action will be identified and examined. Generally, every situation will have more than one course of action. Exploitation of the right course will depend to a large content on experience, ingenuity and imagination of the planner.

(v) **Selecting optimum plan:** An evaluation of the above alternate course of action can be carried out either by judgement alone or with the help of quantitative techniques and staff assistant, to best suit the interest of the organisation.

(vi) **Determining derivative plan:** The above selected plan will form the basic plan from which other plans will develop to support it. For example, basic marketing plan may have been evolved which may result in other derivative plans such as the advertising plan.

(vii) **Fixing the timing of introduction:** The question of timing-who will do, what will have to be decided and an appropriate time schedule drawn up with the details of construction work for communication.

(viii) **Arranging future evaluation of effectiveness of the plan:** Since the ultimate aim of the plan to achieve the objective, result or goal, an evaluation at the earliest possible opportunity necessary to evaluate the adequacy of cost and time and determining whether the planned objectives are reached as desired.

1.7 **What is Work Breakdown Structure (WBS) with respect to construction planning and management? How is WBS classified into different levels?**

[12 marks : 2020]

Solution:

The first major step in the planning process after project requirements definition is the development of the Work Breakdown Structure (WBS). A WBS is a product-oriented family tree subdivisions of the hard ware, services and data required to produce the end product. The WBS is structure in accordance with the way the work will be performed and reflects the way in which project costs and data will be summarized and eventually reported. Preparation of the WSB also considers other that required structured data, such as scheduling, configuration management, contract funding, the technical performance parameters.

The work breakdown structure acts as a vehicle for breaking the work down into smaller elements thus providing a greater probability that every major and minor activity will be accounted for Although a variety of work breakdown structure exist, the most common is the six-level indented structure shown below :

	Level	Description		Level	
Managerial levels	1	Total program	Technical levels	1	Subtask
	2	Project		2	Work package
	3	Task		3	Level of effort

Level 1 is the total program and is composed of a set of projects. The summation of the activities and costs associated with each project must equal the total program. Each project, however can be broken down into task where the summation of all tasks equal the summation of all projects, which in turn, comprises the total program. The reason for this subdivision of effort is simply case of control. Program management therefore becomes synonymous with the integration of activities, and project manager acts as the integrator, using the work breakdown structure as the common framework.

The upper three levels of the WBS are normally specified by the customer (if part of an RPF/RFQ) as the summary level for reporting purposes. The lower levels are generated by the contractor for in house control. Each level serves a vital purpose. Level 1 is generally used for the authorization and release of all work budgets are prepared at level 2 and schedules are prepared at level 3.

2. PERT and CPM

2.1 **What are the main advantages of Activity on the Node networks over Activity on the Arrow networks?**

[15 marks : 1999]

Solution:

In AON system sometimes called precedence diagrams also, the nodes represent the activities and the arrow, their interdependence or precedence relationships.

Nodes are usually represented by squares or rectangles but circles and other conventional geometrical shapes may also be used. One of the most common types of node representation is shown in figure:

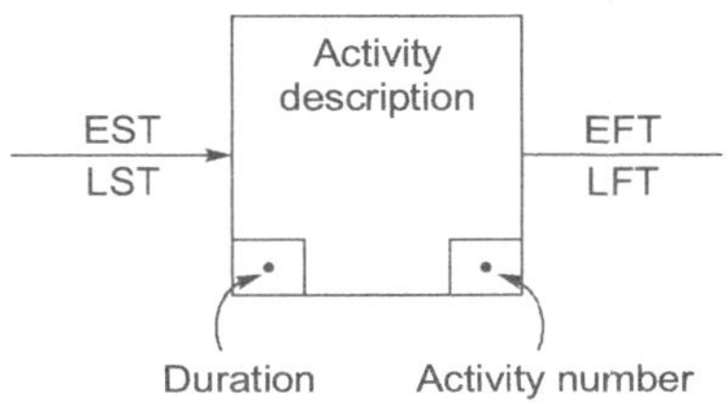

Main advantages of *AON* system over *AOA* system are:

(i) *AON* system of the network completely eliminates the use of dummy activities.

(ii) This system can show activity which should be allowed to overlap each other or must be separated by a time delay.

(iii) This system is self sufficient as it contains all activity times (*EST*, *LST*, *EFT*, *LFT*) on the diagram itself. This facilitates efficient scheduling and control.

(iv) Revisions and modifications can be carried out easily without affecting most of the activities i.e. few activities only have to be altered to incorporate changes in duration, logic and activity sequence.

(v) Pre-operations and post-operations of the activity under consideration are distinctly visible.

(vi) This system adopts simple notations similar to the engineering flow charts and hence can be easily understood by non-specialists also.

(vii) It is more advantageous in projects involving highly repetitive nature of series of operations.

2.2 **With the help of suitable diagrams compare and contrast Activity on the Arrow networks with Activity on the Node networks (Precedence Diagrams).**

[10 marks : 1999]

Solution:

CPM has two systems, namely *AOA* system i.e. Activity on Arrow Network System and *AON* system i.e. Activity on Node Network System.

1. **AOA System:** In this system an activity is graphically represented by an arrow bounded by circle at each end. Activity name or symbol is printed on one side of the arrow and time duration of the activity on the other side of the arrow. A circle situated at the tail end of the arrow is called tail event and represents the instant of time when the activity will start and the circle situated at the head end or arrow end is called head event and represents the instant of time when the activity will finish as shown in figure.

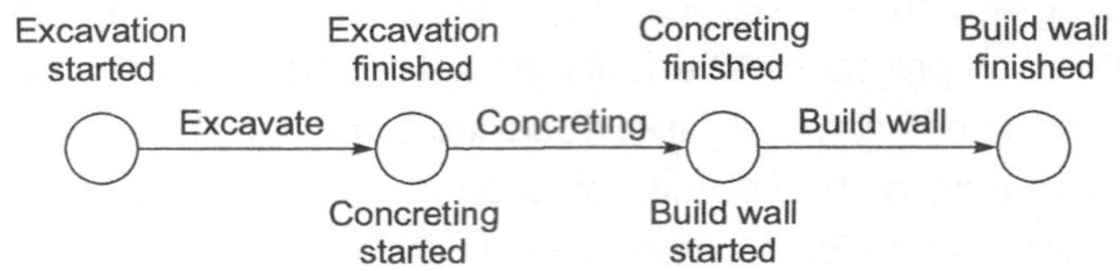

The AOA system uses dummies for the following purposes:

(i) To maintain a unique numbering system for different activity.

(ii) To keep the logical sequence of activity showing correct inter-relationship.

Consider the network shown below. Both activities namely fixing electrical circuit and plumbing start at one event and both ends at another event. So one dummy was introduced just to identify each activity by its own $i-j$ number.

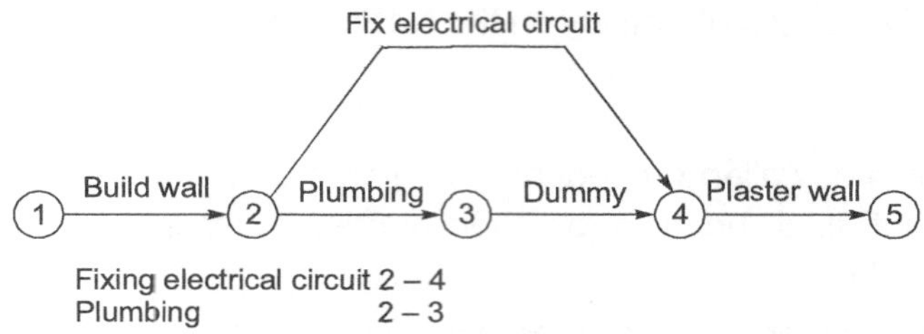

Consider the network shown in figure. Let *E*, *F*, *G* represents the shuttering of three bays of a retaining wall and *L*, *M* and *N*, the corresponding concreting operation. In view of the nature of the activities, shuttering has to precede concreting.

As regarding the labour, only one forward crew for shuttering and one forward crew for concreting are available. Here dummies have been used to represent the constraint and consequent interdependence. But number of dummies must be minimum for the efficient network.

2. **AON System:** In this system, the nodes represent the activities and the arrow, their interdependence or precedence relationships. Nodes are usually represented by squares or rectangles but circles and other conventional shapes may also be used. One of the most common types of node representation is shown in figure.

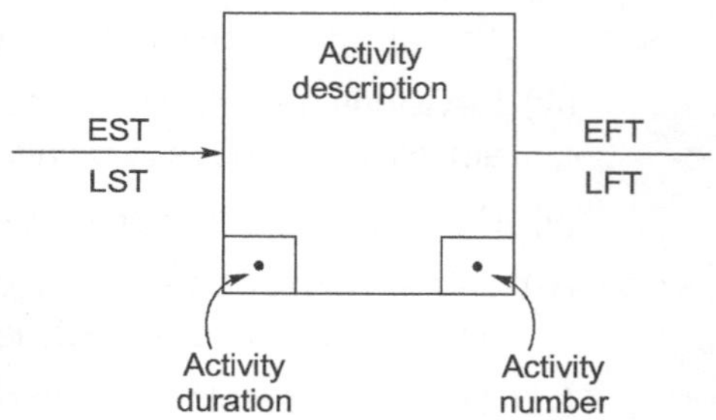

The AON system also called precedence diagram eliminates the use of dummies. Events have no place while activities have their places with durations. Start of an activity must be linked to the end of all activities which must be completed before that start may take place. The connecting lines runs in the left right direction.

Some examples are shown below:

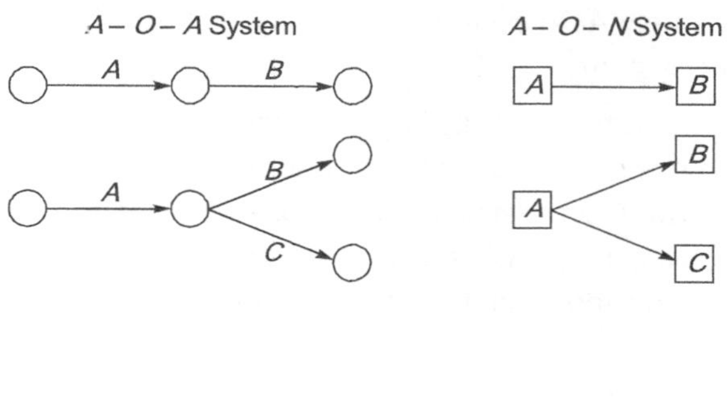

2.3 What is the difference between 'Free Float' and 'Total Float'? Explain the significance of each term.

[15 marks : 1999]

Solution:

Total Float: It is the time span by which the starting (or finishing) of an activity can be delayed without delaying the completion of the project. In certain activities, it will be found that there is a difference between maximum time available and the actual time required to perform the activity. This difference is known as the total float.

Effectively, it is an inbuilt reserve of a resource (time) which is available for use in certain tail event and head event. Now, maximum time available to perform events can therefore be considered as setting the limits.

Consider an activity *i-j*. The time duration available for this activity is equal to the difference between its earliest start time $\left(T_E^i\right)$ and the latest finish time $\left(T_L^j\right)$.

$$\therefore \quad \text{Maximum time available} = T_L^j - T_E^i$$

$$\text{activity time required} = t^{ij}$$

$$\therefore \quad \textbf{Total float } (F_T) = \textbf{time available} - \textbf{time required}$$

$$F_T = \left(T_L^j - T_E^i\right) - t^{ij}$$

or

$$F_T = T_L^j - \left(T_E^i + t^{ij}\right) = LFT - EFT$$

or

$$F_T = \left(T_L^j - t^{ij}\right) - T_E^i = LST - EST$$

It should be clearly noted that the total float for each activity is a measure of its particular relationship to all other activities in the project, since the earliest start time ties in all preceding activities and the latest finish time ties in all succeeding activities.

Free Float: It is that portion of positive total float that can be used by an activity without delaying any succeeding activity (or without affecting the total float of the succeeding activity). The concept of free float is based on the possibility that all the events occur at their earliest times (i.e. all activities start as early as possible).

To get a clear concept of the free float, consider activity *i-j* and its successor activity *j-k*. Events *i* and *j* has earliest occurrence times as T_E^i and T_E^j. Earliest start time for activity *i - j* will be T_E^i, while EST for *j-k* will be T_E^j. However if t^{ij} is the activity time, activity *i-j* will be complete by $\left(T_E^i + t^{ij}\right)$ time, while activity *j-k* cannot start because its EST $\left(T_E^j\right)$ is greater than $\left(T_E^i + t^{ij}\right)$. The difference between the two is the free float for *i-j*.

$$F_F \text{ for } i\text{-}j = T_E^j - \left(T_E^i + t^{ij}\right)$$

But $\left(T_E^i + t^{ij}\right)$ is the earliest finish time (*EFT*) of the activity *i-j*, while T_E^j is the early start time for activity *j-k*.

$$F_F \text{ for } i\text{-}j = EST \text{ of successor activity} - EFT \text{ of present activity}$$

or

$$\mathbf{F_F \text{ for } i\text{-}j = T_E^j - EFT}$$

Hence free float for an activity i-j is the difference between its earliest finish time and the earliest start time of its successor activity.

2.4 A typical serious candidate for services examination take no less than 25 minutes to answer a question and sometimes as much as 48 minutes, 35 minutes times are most frequent than any other time. If this performance were an activity in a PERT project, estimate

(i) would be expected time to answer a question,

(ii) its variance,

(iii) the time to be allocated for answering a question in scheduling the project.

[10 marks : 2000]

Solution:

Given data:

Optimistic time, t_o = 25 minutes, Most likely time, t_m = 35 minutes, Pessimistic time, t_p = 48 minutes

(i) **Expected time** to answer the question can be given as

$$t_e = \frac{t_o + 4t_m + t_p}{6} = \frac{25 + (4 \times 35) + 48}{6} = 35.5 \text{ minutes}$$

(ii) **Variance** can be given as

$$\sigma^2 = \left(\frac{t_p - t_o}{6}\right)^2 = \left(\frac{48 - 25}{6}\right)^2 = 14.69 \text{ minutes}^2$$

$\therefore$ Standard deviation $\sigma = \sqrt{14.69} = 3.83$ minutes

(iii) **Time to be allocated** for answering the question

$$= t_e \pm \sigma = 35.5 \pm 3.83$$

$$= 39.33 \text{ (max) or } 31.63 \text{ (min)}$$

2.5 Determine the total float, free float and critical path for the installation project represented by the network shown in the figure below. The estimated duration in days for the activities are shown in the figure.

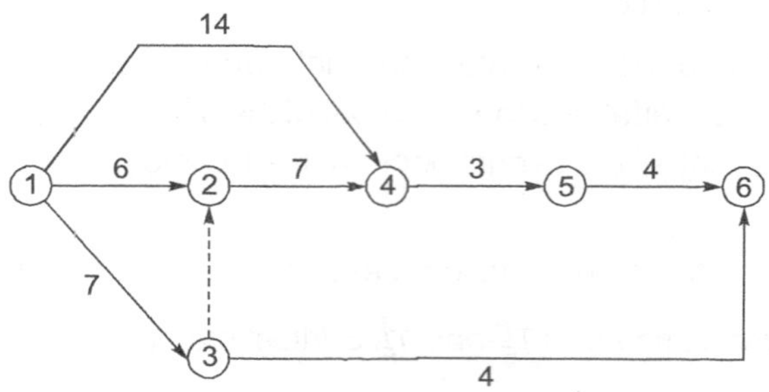

[20 marks : 2000]

Solution:

The earliest expected time of each event can be calculated as shown below.

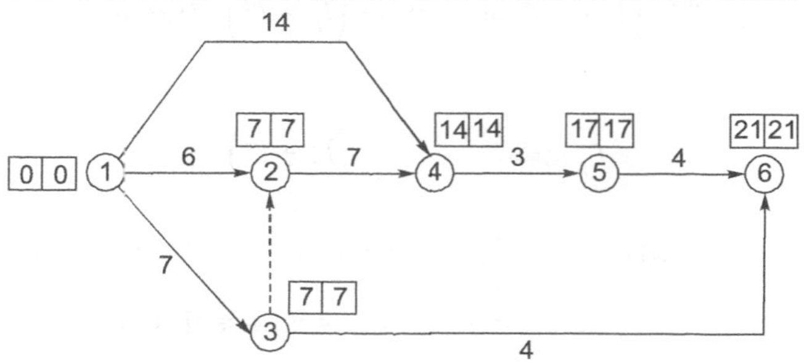

The project completion time will be 21 days.

Now, we know that: $EST = T_E^i$

$EFT = T_E^i + t^{ij}$

$$LST = T_L^j - t^{ij}$$
$$LFT = T_L^j$$
$$F_T = LST - EST = LFT - EFT$$
$$F_F = EST \text{ for successor activity} - EFT \text{ of present activity}$$
$$= F_T - S_j \text{ (head event slack)}$$

The calculations are tabulated below:

Activity	Duration	*EST*	*EFT*	*LFT*	*LST*	F_T	F_F	Remarks
1–2	6	0	6	7	1	1	1	
1–3	7	0	7	7	0	0	0	Critical
1–4	14	0	14	14	0	0	0	Critical
2–4	7	7	14	14	7	0	0	Critical
3–2	0	7	7	7	7	0	0	Critical
3–6	4	7	11	21	17	10	10	
4–5	3	14	17	17	14	0	0	Critical
5–6	4	17	21	21	17	0	0	Critical

There will be **two critical paths** in the above network.
(i) 1 – 3 – 2 – 4 – 5 – 6
(ii) 1 – 4 – 5 – 6

2.6 A network is formed by the following activities. The duration of the activities are given below.

Activity	Preceded by	Duration (days)
A	Starting	4
B	*A*	2
C (terminal)	*D*	5
D	*E*	2
E	*A*	6
F	*B*	1
G	*B*	2
H (terminal)	*E, G*	3
I (terminal)	*F*	2

Draw the network, calculate the project duration and identify the critical path.

[20 marks : 2001]

Solution:

The network may be developed using the given data, as shown below.

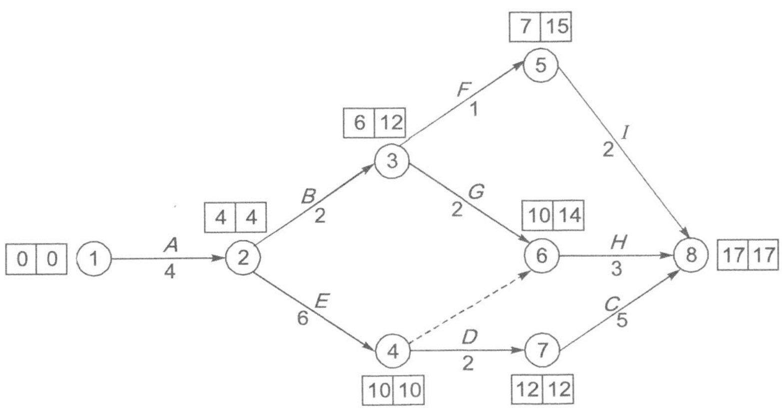

Since H is preceded by both *E* and *G*, therefore to maintain the logical relationship between these activities, a dummy is inserted between *E* and *H*. Also the earliest event times and latest allowable occurrence time can also be calculated as

$$T_E^j = T_E^i + t^{ij}$$

$$T_L^i = T_L^j - t^{ij}$$

Also, the events can also be numbered by using Fulkerson's rule. Thus the earliest occurrence time of last event (8) is 17 days which means that it is the project completion time.

Now we know that

$$EST = T_E^i$$

$$EFT = T_E^i + t^{ij}$$

$$LST = T_L^j - t^{ij}$$

$$LFT = T_L^j$$

$$\text{Total float, } F_T = LST - EST = LFT - EFT$$

These calculations are tabulated below:

Activity	Duration	*EST*	*EFT*	*LFT*	*LST*	F_T	Remarks
A	4	0	4	4	0	0	Critical
B	2	4	6	12	10	6	
C	5	12	17	17	12	0	Critical
D	2	10	12	12	10	0	Critical
E	6	4	10	10	4	0	Critical
F	1	6	7	15	14	8	
G	2	6	8	14	12	6	
H	3	10	13	17	14	4	
I	2	7	9	17	15	8	

Hence, the critical activities will be the one with total float $F_T = 0$. Thus, the critical path in terms of activities will be *A – E – D – C*. In other words, the critical events are 1, 2, 4, 7 and 8. So, the critical path in terms of events is 1 – 2 – 4 – 7 – 8.

2.7 **The time estimates in days for optimistic, most likely and pessimistic times of various activities of a project are shown below in the network diagram. Obtain the expected time and the latest allowable time for each event. Draw the critical path and estimate the approximate value of standard deviation for the network.**

[25 marks : 2002]

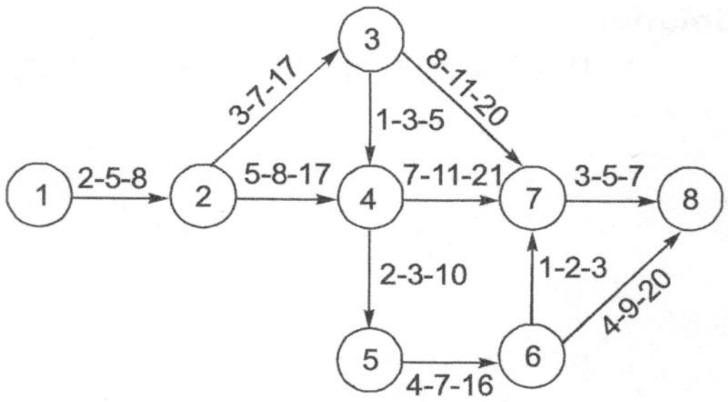

Solution:

We know that

$$\text{Expected time, } t_e = \frac{t_o + 4t_m + t_p}{6}$$

$$\text{Standard deviation, } \sigma = \frac{t_p - t_o}{6}$$

Earliest expected time, $T_E = T_E$(Predecessor event) + t_e(activity)

Latest allowable time, $T_L = T_L$(Successor event) – t_e(activity)

Activity	t_o	t_m	t_p	Expected time, t_e	Standard deviation, σ
1 – 2	2	5	8	5	1
2 – 3	3	7	17	8	2.33
2 – 4	5	8	17	9	2
3 – 4	1	3	5	3	0.67
3 – 7	8	11	20	12	2
4 – 7	7	11	21	12	2.33
4 – 5	2	3	10	4	1.33
5 – 6	4	7	16	8	2
6 – 7	1	2	3	2	0.33
6 – 8	4	9	20	10	2.67
7 – 8	3	5	7	5	0.67

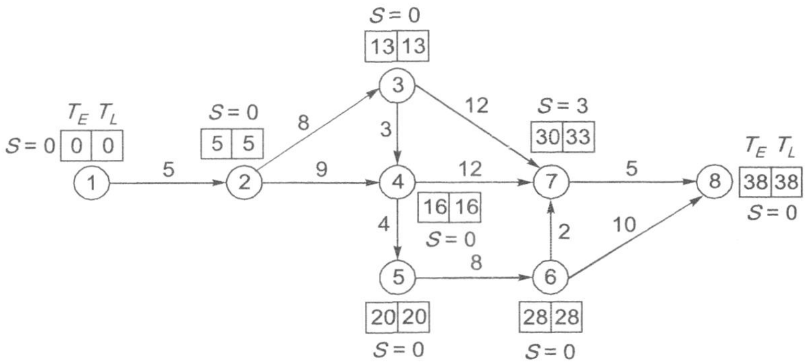

It is evident that the critical path will be through events which have same values of T_E and T_L i.e. Zero slack. Therefore the critical path will be 1 – 2 – 3 – 4 – 5 – 6 – 8.

Standard deviation for the network may be given as

$$\sigma = \sqrt{\sigma_{1-2}^2 + \sigma_{2-3}^2 + \sigma_{3-4}^2 + \sigma_{4-5}^2 + \sigma_{5-6}^2 + \sigma_{6-8}^2}$$

$$\Rightarrow \quad \sigma = \sqrt{1^2 + 2.33^2 + 0.67^2 + 1.33^2 + 2^2 + 2.67^2}$$

$$\Rightarrow \quad \sigma = 4.45 \text{ days.}$$

2.8 Define and explain the terms:

(i) Most likely time

(ii) Mean time and

(iii) Expected time as related to various activities of a project.

[15 marks : 2002]

Solution:

(i) **Most likely time:** This is the time required to complete the activity if normal conditions prevail. This time estimate lies between the optimistic and the pessimistic time estimates. Thus this time estimate reflects a situation where normal conditions are prevailing, things are as usual and there is nothing exciting. This time is denoted by t_m.

(ii) **Mean time:** Mean time t_{avg} (or mean of the distribution) is given by the sum of the time durations t_1, t_2, t_3, t_4 ..... t_n, taken by activities of a particular type for their completion divided by the number of activities n. Thus,

$$t_{avg} = \frac{t_1 + t_2 + t_3 + t_4 + \ldots.. + t_n}{n} = \frac{\Sigma t}{n}$$

(iii) **Expected time:** The three time estimates viz., the optimistic time t_o, the pessimistic time t_p and the most likely time t_m are identified on the beta probability distribution curve and by using t_p and t_o, the variance and the standard deviation can be calculated. The next step is to obtain the average or mean time taken for completion of an activity. The average of or the mean time taken for the completion of an activity is commonly called expected time and is denoted by t_e. If the exact shape of the probability distribution curve is known, the expected time t_e could be accurately calculated. However since the precise curves as generally not available, the use of an approximation is made. It has been indicated by the statisticians that in beta distribution the average or the expected time is obtained by adding together one sixth of the optimistic time, two thirds of most likely time and one sixth of the pessimistic time. Thus a large weightage given to most likely time t_m is justified because the chance of completion of the activity in the most likely time t_m is much more than in the optimistic time t_o or in the pessimistic time t_p.

$$t_e = \frac{t_o + 4t_m + t_p}{6}$$

2.9 **A small project consists of seven activities with the following details:**

Activity	DURATION, WEEKS			Immediate predecessor
	Most likely	Optimistic	Pessimistic	
A	4	2	8	–
B	7	3	15	A
C	4	4	4	A
D	11	5	23	B, C
E	8	4	16	B
F	6	3	15	D, E
G	5	5	5	D

(i) Draw the network, find critical path, the expected completion time and the next most critical path.

(ii) What project duration will have 95% confidence of completion. Take Z = 1.64.

[20 marks : 2003]

Solution:

The expected time and standard deviation of each activity are tabulated below:

Activity	t_o	t_m	t_p	$t_e = \frac{t_o + 4t_m + t_p}{6}$	$\sigma = \frac{t_p - t_o}{6}$
A	2	4	8	4.33	1
B	3	7	15	7.67	2
C	4	4	4	4	0
D	5	11	23	12	3
E	4	8	16	8.67	2
F	3	6	15	7	2
G	5	5	5	5	0

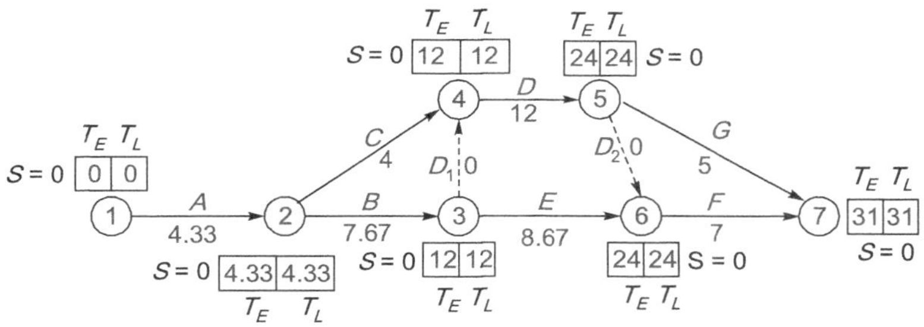

Various critical paths can be possible for the network above.

(i) 1 – 2 – 4 – 5 – 7 = 4.33 + 4 + 12 + 5 = 25.33 weeks

(ii) 1 – 2 – 4 – 5 – 6 – 7 = 4.33 + 4 + 12 + 0 + 7 = 27.33 weeks

(iii) 1 – 2 – 3 – 4 – 5 – 7 = 4.33 + 7.67 + 0 + 12 + 5 = **29** weeks

(iv) 1 – 2 – 3 – 4 – 5 – 6 – 7 = 4.33 + 7.67 + 0 + 12 + 0 + 7 = **31** weeks

(v) 1 – 2 – 3 – 6 – 7 = 4.33 + 7.67 + 8.67 + 7 = 27.67 weeks

Hence the most critical path is 1 – 2 – 3 – 4 – 5 – 6 – 7 which gives the expected completion time as 31 weeks. The next most critical path is the one with the expected completion time as 29 weeks i.e. 1 – 2 – 3 – 4 – 5 – 7.

Now the **standard deviation for the critical path**

$$\sigma = \sqrt{(\sigma_{1-2})^2 + (\sigma_{2-3})^2 + (\sigma_{3-4})^2 + (\sigma_{4-5})^2 + (\sigma_{5-6})^2 + (\sigma_{6-7})^2}$$

$$= \sqrt{1^2 + 2^2 + 0^2 + 3^2 + 0^2 + 2^2} = 4.24 \text{ weeks}$$

Given $Z = 1.64$

$$\therefore \quad \frac{T_S - T_E}{\sigma} = Z$$

$$\Rightarrow \quad \frac{T_S - 31}{4.24} = 1.64$$

$$\Rightarrow \quad T_S = 31 + (4.24 \times 1.64) = 37.95 \text{ weeks.}$$

2.10 **Draw the network for the following project and indicate the event times and critical path. Also find the project duration and total float for all activities.**

Activity	*A*	*B*	*C*	*D*	*E*	*F*	*G*	*H*	*K*	*L*
Duration (days)	4	2	2	6	6	1	1	2	5	3
Depends Upon	–	*A*	*A, B, F*	*C, L*	*D, G, H*	*A*	*F*	*G, L*	*A*	*F, K*

[20 marks : 2004]

Solution:

The earliest occurrence times of events and their latest allowable occurrence times can be calculated as

$$T_E^j = T_E^i + t^{ij}$$

$$T_E^i = T_L^j - t^{ij}$$

Also, the events can also be numbered using **Fulkerson's rule. The network for the given data is shown below:**

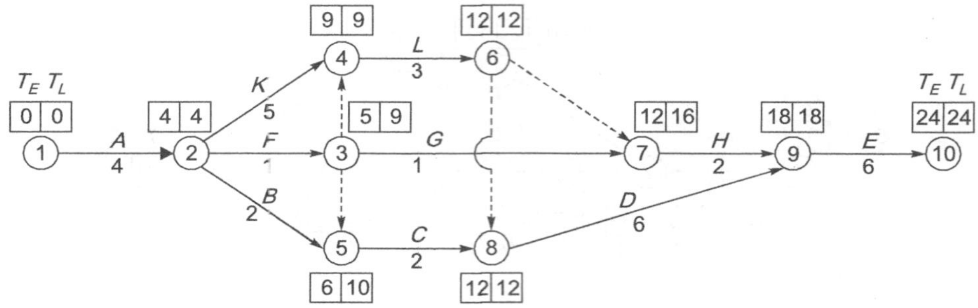

The earliest occurrence time of the last event is the project duration i.e. 24 days.

Now we know that

$$EST = T_E^i$$

$$EFT = T_E^i + t^{ij}$$

$$LST = T_L^j - t^{ij}$$

$$LFT = T_L^j$$

also Total float $F_T = LST - EST = LFT - EFT$

The calculations are tabulated below:

Activity	Duration	*EST*	*EFT*	*LFT*	*LST*	F_T	Remarks
A	4	0	4	4	0	0	Critical
B	2	4	6	10	8	4	
C	2	6	8	12	10	4	
D	6	12	18	18	12	0	Critical
E	6	18	24	24	18	0	Critical
F	1	4	5	9	8	4	
G	1	5	6	16	15	10	
H	2	12	14	18	16	4	
K	5	4	9	9	4	0	Critical
L	3	9	12	12	9	0	Critical

Activities *A*, *D*, *E*, *K* and *L* are critical, since the total float of these activities is zero. Thus the critical path will be *A* – *K* – *L* – *D* – *E*. In other words in terms of events 1 – 2 – 4 – 6 – 8 – 9 – 10 is the critical path.

2.11 **Draw a comparison between PERT and CPM.**

[10 marks : 2005]

Solution:

Some points of comparison between PERT and CPM are as follows:

(i) In CPM emphasis is laid on activities and in PERT on events. Thus, CPM network is activity based and PERT network is event based.

(ii) In CPM, activity times are deterministic i.e. time estimates for completion of activities are fairly accurate, whereas in the case of PERT, activity times are stochastic or probabilistic and are assumed to follow a Beta distribution. Thus, CPM is suitable when time estimates can be determined with fair degree of accuracy and costs can be calculated in advance. On the other hand PERT is suitable when there is extreme degree of uncertainty in the various aspects of the project and control over time outweighs control over costs.

(iii) CPM puts an emphasis on optimizing resource allocation and minimizing overall cost for a given project execution time and therefore it is cost biased. On the other hand in PERT, the emphasis is on shortening and monitoring project execution time without to much concern for cost and therefore it is time biased.

(iv) In CPM, a critical path is the timewise longest path in a network. However, in PERT, a critical path is the path that joins the critical events.

(v) In CPM only one time estimate is required for each activity whereas in PERT three time estimates are required for each activity.

(vi) In CPM critical path is obtained by using the concept of float and in PERT critical path is obtaind by using the concept of slack.

2.12 **(i) With the help of data given below, draw the network diagram and find the project completion time based on expected time.**

Preceeding event node number	Succeeding event node number	Optimistic time in weeks	Most likely time in weeks	Pessimistic time in weeks
10	20	10	12	20
10	40	5	15	19
20	30	10	15	26
20	70	15	20	25
30	60	5	10	15
40	50	4	8	12
50	60	5	10	15
60	80	2	4	6
70	80	2	4	6

(ii) If the project in (i) is to be completed within 40 weeks, draw a table showing the original early event time, late event time, original slack and revised early event time, revised late event time and revised slack time. Draw the revised network diagram.

[20 + 10 = 30 marks : 2005]

Solution:

(i) The network diagram is shown below:

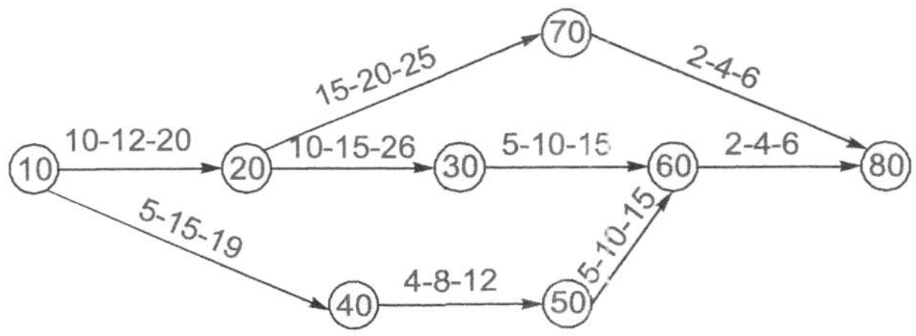

Activity	t_o	t_m	t_p	Expected time $t_e = \frac{t_o + 4t_m + t_p}{6}$
10 – 20	10	12	20	13
10 – 40	5	15	19	14
20 – 30	10	15	26	16
20 – 70	15	20	25	20
30 – 60	5	10	15	10
40 – 50	4	8	12	8
50 – 60	5	10	15	10
60 – 80	2	4	6	4
70 – 80	2	4	6	4

Now, the earliest expected time for any event is given by $T_E^j = T_E^i + t^{ij}$ and latest allowable time is given by $T_L^i = T_L^j - t^{ij}$

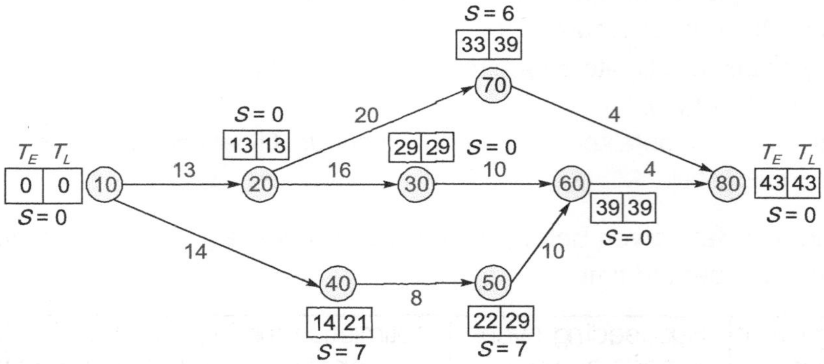

Hence, the earliest expected time of the last event is 43 weeks. So it will be the project completion time also. Also the critical path is 10 – 20 – 30 – 60 – 80.

(ii) The boxes at each event shows early event times and late event times respectively.

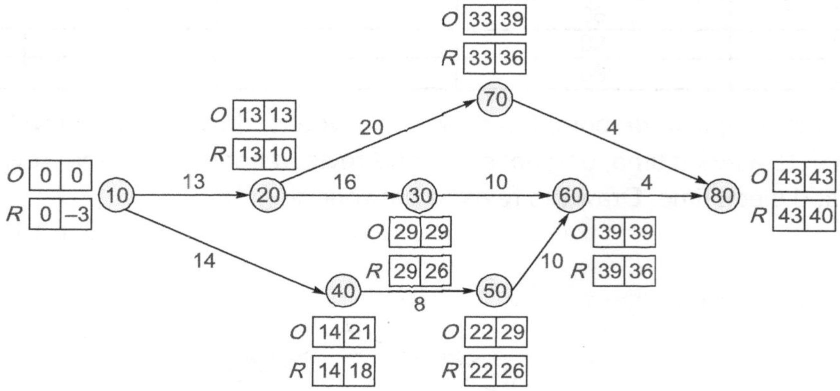

Event	Original T_E	Original T_L	Original Slack $T_L - T_E$	Revised T_E	Revised T_L	Revised Slack $T_L - T_E$
10	0	0	0	0	–3	–3
20	13	13	0	13	10	–3
30	29	29	0	29	26	–3
40	14	21	7	14	18	4
50	22	29	7	22	26	4
60	39	39	0	39	36	–3
70	33	39	6	33	36	3
80	43	43	0	43	40	–3

2.13 For the network shown, the estimated time in days for each activity is shown in the network diagram. Determine the critical path. Also determine the probability of completing the work in 35 days. The values of Z and corresponding probability are given in the table.

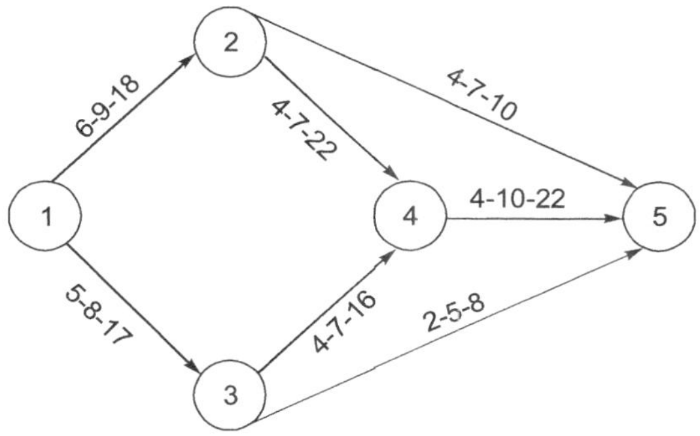

Z	1.0	1.1	1.2	1.5	2.0	3.0
Probability (%)	84.13	86.43	88.9	93.92	97.92	99.87

[20 marks : 2006]

Solution:

The expected time and standard deviation of each activity are calculated and tabulated below:

Activity	t_o	t_m	t_p	$t_e = \dfrac{t_o + 4t_m + t_p}{6}$	$\sigma = \dfrac{t_p - t_o}{6}$
1 – 2	6	9	18	10	2
1 – 3	5	8	17	9	2
2 – 4	4	7	22	9	3
2 – 5	4	7	10	7	1
3 – 4	4	7	16	8	2
3 – 5	2	5	8	5	1
4 – 5	4	10	22	11	3

The earliest expected time (T_E) and latest occurrence time T_L can be given as $T_E^j = T_E^i + t_e^{ij}$ and $T_L^i = T_L^j - t_E^{ij}$ respectively.

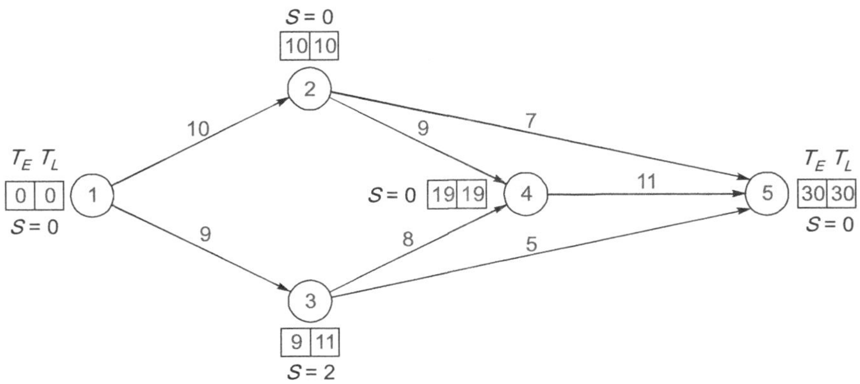

The earliest expected time of the last event is 30 days. Thus, this will be the project duration. Now the critical path will be the one which joins the events with same value of T_E and T_L. In other words, events with zero slack will be the most critical events. Hence the critical path will be 1 – 2 – 4 – 5.

Now the **standard deviation for the critical path will be**

$$\sigma^2 = (\sigma_{1-2})^2 + (\sigma_{2-4})^2 + (\sigma_{4-5})^2$$

$$\Rightarrow \qquad \sigma = \sqrt{2^2 + 3^2 + 3^2}$$

$$\Rightarrow \qquad \sigma = 4.69 \text{ days}$$

We know that $Z = \dfrac{T_S - T_E}{\sigma}$

Given $T_S = 35$ days

$T_E = 30$ days

$\sigma = 4.69$ days

$$\therefore \qquad Z = \frac{35-30}{4.69} = 1.07$$

For $Z = 1$; probability = 84.13%

For $Z = 1.1$; probability = 86.43%

$\therefore$ Probability at $Z = 1.07$ will be

$$= 84.13 + \left(\frac{86.43-84.13}{1.1-1.0}\right)(1.07-1.0) = 85.74\%.$$

2.18 Information on the activities required for a medium-size Civil Engineering Project is as follows:

Activity	A	B	C	D	E	F	G	H	I	J	K
Node No.	1-2	1-3	1-4	2-5	3-5	3-6	3-7	4-6	5-7	6-8	7-8
Duration (in months)	02	07	08	03	06	10	04	06	02	05	06

Draw the network and calculate the following :

(i) Earliest and latest event time

(ii) Earliest and latest start time

(iii) Earliest and latest finish time

(iv) Total, free and independent float

(v) Critical path

[20 marks : 2008]

Solution:

The network of the given activities is shown in figure:

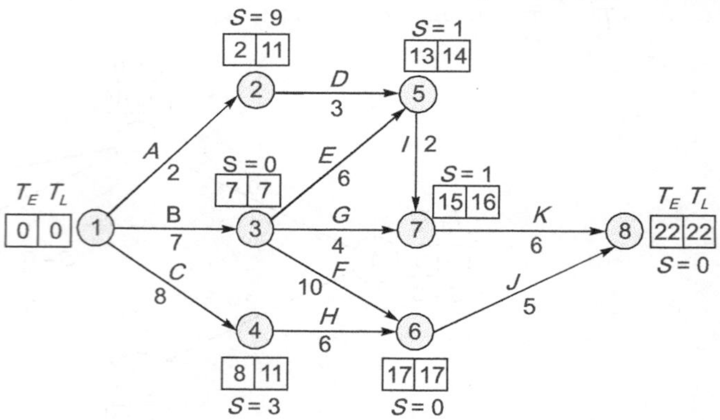

The earliest event time and latest event time have been calculated as:

$$T_E^j = T_E^i + t^{ij}$$

$$T_L^i = T_L^j - t^{ij}$$

$$S = T_L - T_E$$

We also know that

$$EST = T_E^i$$

$$EFT = T_E^i + t^{ij}$$

$$LST = T_L^j - t^{ij}$$

$$LFT = T_L^j$$

Total Float, $F_T = LFT - EFT = LST - EST$

Free Float, F_F = *EST* of successor activity – *EFT* of present activity

$= F_T - S_j$

Independent Float, $F_{ID} = F_F - S_i$

The values of activity times, total float, free float and independent float are tabulated below:

Activity	Duration	*EST*	*EFT*	*LST*	*LFT*	F_T	F_F	F_{ID}	Remarks
A	2	0	2	9	11	9	0	0	
B	7	0	7	0	7	0	0	0	Critical
C	8	0	8	3	11	3	0	0	
D	3	2	5	11	14	9	8	–1	
E	6	7	13	8	14	1	0	0	
F	10	7	17	7	17	0	0	0	Critical
G	4	7	11	12	16	5	4	4	
H	6	8	14	11	17	3	3	0	
I	2	13	15	14	16	1	0	–1	
J	5	17	22	17	22	0	0	0	Critical
K	6	15	21	16	22	1	1	0	

The activities on the critical path are those activities that have total float (F_T) equal to zero. Thus *B – F – J* will be the critical path.

2.14 Define Optimistic, Pessimistic and Most Likely Time Estimate.

[5 marks : 2008]

Or

(i) Enlist and explain time estimates in PERT.

(ii) Explain Resources Allocation.

[6 + 4 marks : 2009]

Solution:

(i) There are three kinds of time estimates generally made for each activity in PERT which are as indicated below:

(a) **Optimistic Time Estimate:** This is the shortest possible time in which an activity can be completed under ideal conditions. Thus, this time estimate represents the time in which the activity would be completed if everything went perfectly well without any problems or adverse conditions being developed during the execution of the activity. In arriving at this time estimate no provisions are made for delays or setbacks and better than normal conditions are assumed to prevail during the execution of the activity. This time estimate is denoted by t_o.

(b) **Pessimistic Time Estimate:** This is the maximum possible time that would be required to complete the activity. Thus, this time estimate represents the time required to complete the activity if everything went wrong and abnormal situations prevailed. However, this estimate does not include the possible effects of major catastrophes such as floods, fires, earthquakes, etc. or that of labour strikes or unrest etc. This time estimate is denoted by t_p.

(c) **Most Likely Time Estimate:** This is the time required to complete the activity if normal conditions prevail. This time estimate lies between the optimistic and the pessimistic time estimates. Thus, this time estimate reflects a situation where normal conditions are prevailing, things are as usual and there is nothing exciting. This time estimate is denoted by t_m.

(ii) **Resources Allocation**: When the resources (e.g. men, materials, machines, space, money) are limited and the demands are made for the same type of resource by several activities at the same time, then it becomes imperative to adopt a systematic method for the allocation of resources. For proper management of resources it is essential to asses the requirement of various resources and also to know the times at which they are needed on various activities. Further the various activities of a project should be scheduled in such a way that the demand for various resources is more or less uniform for the entire duration of the project. This can be achieved by the following two approaches.

(a) **Resource Smootheing:** In resource smoothing the total project duration is not changed but the activities having flcats are rescheduled so that a uniform demand for the resources is achieved. Thus in this case there is constraint on project duration. There is however, no constraint on the resources.

(b) **Resource Levelling:** In resource levelling, the activities are so rescheduled that the maximum or peak resources requirement does not cross the limit of available resources. In rescheduling the available floats are used first. If by doing so, the resources demand is more than the available resources, the duration of some of the activities is increased so that the resources requirements for these activities is decreased. Thus in this process, the project duration initially planned might be increased.

2.15 A construction work consists of activities with PERT durations in days as given below:

Activity	P	Q	R	S	T	U	W	Y	Z
Predecessors	–	P, T	Q	–	S	–	S	S	U, W
t_o(in days)	3	4	4	3	8	1	2	4	6
t_m(in days)	6	8	5	3	14	4	5	7	15
t_p(in days)	9	9	9	3	17	7	14	13	30

Determine: (i) The probability of completing the job in 32 days and
(ii) The completion time with 50% probability.

Z	−1.5	−1.3	−1.0
Probability	0.07	0.10	0.16

[10 marks : 2009]

Solution:

The expected duration and standard deviation for each activity are calculated below:

Activity	t_o	t_m	t_p	$t_e = \frac{t_o + 4t_m + t_p}{6}$	$\sigma = \frac{t_p - t_o}{6}$	σ^2
P	3	6	9	6	1	1
Q	4	8	9	7.5	0.833	0.694
R	4	5	9	5.5	0.833	0.694
S	3	3	3	3	0	0
T	8	14	17	13.5	1.5	2.25
U	1	4	7	4	1	1
W	2	5	14	6	2	4
Y	4	7	13	7.5	1.5	2.25
Z	6	15	30	16	4	16

The network diagram of the PERT activities is drawn below:

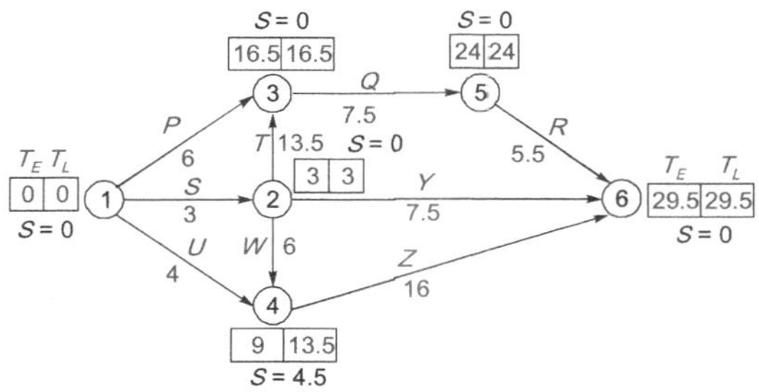

The events at which slack ($T_L - T_E$) is zero will be critical. Hence, the critical path will be $S-T-Q-R$.

The standard deviation of the critical path will be

$$\sigma = \sqrt{(\sigma_S)^2 + (\sigma_T)^2 + (\sigma_Q)^2 + (\sigma_R)^2} = \sqrt{0 + 2.25 + 0.694 + 0.694}$$

$$\Rightarrow \qquad \sigma = 1.91 \text{ days}$$

(i) We know that $\quad Z = \dfrac{T_S - T_E}{\sigma}$

Given that $\quad T_S = 32$ days, $T_E = 29.5$ days, $\sigma = 1.91$ days

$$\therefore \qquad Z = \frac{32 - 29.5}{1.91} = 1.31$$

From given table, we have

for $Z = -1.0$, $p = 0.16$ i.e. for $Z = 1.0$, $p = 1 - 0.16 = 0.84$

for $Z = -1.3$, $p = 0.10$ i.e. for $Z = 1.3$, $p = 1 - 0.10 = 0.90$

for $Z = -1.5$, $p = 0.07$ i.e. for $Z = 1.5$, $p = 1 - 0.07 = 0.93$

$\therefore$ Probability at $Z = 1.31$ will be

$$= 0.90 + \left(\frac{0.93 - 0.90}{1.5 - 1.3}\right) \times (1.31 - 1.30) = 0.9015 = 90.15\%$$

(ii) **For 50% probability,** $Z = 0$, i.e.

$$\therefore \qquad T_S = T_E = 29.5 \text{ days}$$

2.16 A contractor intends to bid for a small-size civil engineering project consisting of the following activities and corresponding different time limits. Work out (i) critical path and standard deviation for the whole network, (ii) Z-factor for completing the project in 42 weeks, (iii) completion time duration for which the contractor should bid considering 93% probability assuming Z-factor as 1.5 and (iv) *TF*, *FF* and *IF*:

Activity	Optimistic time (in weeks)	Most likely time (in weeks)	Pessimistic time (in weeks)
1 – 2	1	3	5
1 – 3	3	6	15
2 – 3	2	5	14
2 – 4	5	7	9
3 – 4	2	4	12
2 – 5	6	9	18
4 – 5	1	2	3
3 – 8	3	12	15
4 – 6	4	6	8
5 – 7	1	2.5	7
6 – 8	3	4	5
6 – 7	0	0	0
7 – 9	1	8	9
9 – 10	1	3	5
8 – 10	0	0	0
8 – 11	1	9	11
10 – 11	3	4.5	9
9 – 11	4	9	20

[15 marks : 2010]

Solution:

We know that the expected time and standard deviation of each activity are given as

$$t_e = \frac{t_o + 4t_m + t_p}{6} \text{ and } \sigma = \frac{t_p - t_o}{6}$$

Activity	t_o	t_m	t_p	t_e	σ
1 – 2	1	3	5	3	0.67
1 – 3	3	6	15	7	2
2 – 3	2	5	14	6	2
2 – 4	5	7	9	7	0.67
3 – 4	2	4	12	5	1.67
2 – 5	6	9	18	10	2
4 – 5	1	2	3	2	0.33
3 – 8	3	12	15	11	2
4 – 6	4	6	8	6	0.67
5 – 7	1	2.5	7	3	1
6 – 8	3	4	5	4	0.33
6 – 7	0	0	0	0	0
7 – 9	1	8	9	7	1.33
9 – 10	1	3	5	3	0.67
8 – 10	0	0	0	0	0
8 – 11	1	9	11	8	1.67
10 – 11	3	4.5	9	5	1
9 – 11	4	9	20	10	2.67

(i) The network for the given activities is shown below:

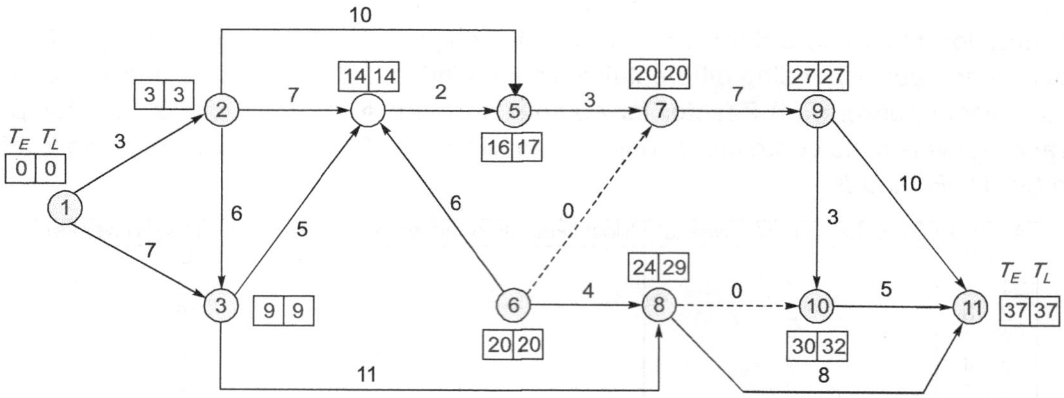

The critical path is the path with zero slack events i.e. $T_L - T_E = 0$

Thus the critical path is 1 – 2 – 3 – 4 – 6 – 7 – 9 – 11

The standard deviation for the whole project is given by

$$\sigma = \sqrt{(\sigma_{1-2})^2 + (\sigma_{2-3})^2 + (\sigma_{3-4})^2 + (\sigma_{4-6})^2 + (\sigma_{6-7})^2 + (\sigma_{7-9})^2 + (\sigma_{9-11})^2}$$

$$= \sqrt{(0.67)^2 + (2)^2 + (1.67)^2 + (0.67)^2 + 0 + (1.33)^2 + (2.67)^2}$$

$$= 4.07 \text{ weeks}$$

(ii) We know that **Z-factor** is given by

$$Z = \frac{T_S - T_E}{\sigma} \quad \text{[But } T_S = 42 \text{ weeks]}$$

$$\Rightarrow \quad Z = \frac{42 - 37}{4.07}$$

$$\Rightarrow \quad Z = 1.23$$

(iii) Given that $Z = 1.5$

$$\therefore \quad Z = \frac{T_S - T_E}{\sigma}$$

$$\Rightarrow \quad 1.5 = \frac{T_S - 37}{4.07}$$

$$\Rightarrow \quad T_S = 37 + (1.5 \times 4.07)$$

$$\Rightarrow \quad T_S = 43.1 \text{ weeks}$$

(iv) The **earliest event time and latest event time** have been calculated as:

$$T_E^j = T_E^i + t^{ij};\ T_L^i = T_L^j - t^{ij};\ S = T_L - T_E$$

We also know that $EST = T_E^i;\ EFT = T_E^i + t^{ij};\ LST = T_L^j - t^{ij};\ LFT = T_L^j$

Total Float; $F_T = LFT - EFT = LST - EST$

Free Float, $F_F = EST$ of successor activity $- EFT$ of present activity

$= F_T - S_j$

Independent Float, $F_{ID} = F_F - S_i$

The value of activity times, total float, free float and independent float are tabulated below:

Activity	t_e	EST	EFT	LST	LFT	F_T	F_F	F_{ID}
1 – 2	3	0	3	0	3	0	0	0
1 – 3	7	0	7	2	9	2	2	2
2 – 3	6	3	9	3	9	0	0	0
2 – 4	7	3	10	7	14	4	4	4
3 – 4	5	9	14	9	14	0	0	0
2 – 5	10	3	13	7	17	4	3	3
4 – 5	2	14	16	15	17	1	0	0
3 – 8	11	9	20	18	29	9	4	4
4 – 6	6	14	20	14	20	0	0	0
5 – 7	3	16	19	17	20	1	1	0
6 – 8	4	20	24	25	29	5	0	0
6 – 7	0	20	20	20	20	0	0	0
7 – 9	7	20	27	20	27	0	0	0
9 – 10	3	27	30	29	32	2	0	0
8 – 10	0	24	24	32	32	8	6	1
8 – 11	8	24	32	29	37	5	5	0
10 – 11	5	30	35	32	37	2	2	0
9 – 11	10	27	37	27	37	0	0	0

2.17 The data for planning a project by network technique is given below

Activity	Duration (in weeks)	Activities immediately Preceding	Activities immediately Following
A	4	—	*B*, *C*
B	4	*A*	*D*, *F*
C	6	*A*	*E*, *G*
D	3	*B*	*E*, *G*
E	5	*C*, *D*	*H*, *I*
F	6	*B*	*I*
G	5	*C*, *D*	*I*
H	7	*E*	—
I	4	*E*, *F*, *G*	—

For the above data, answer the following :

(i) Draw the CPM diagram.

(ii) Prepare a CPM schedule and calculate F_T, F_F and F_{IN}.

(iii) State the critical path

(iv) Indicate the project duration

[10 marks : 2011]

Solution:

(i) CPM diagram

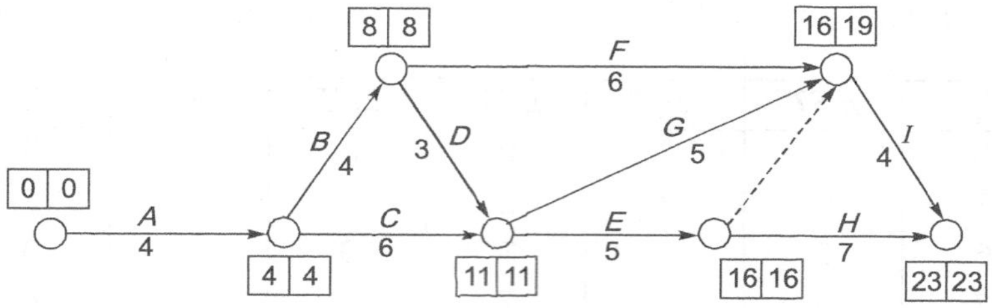

Total project duration = 23 weeks

(ii)

Activity (*A*)	Duration (t_{ij})	*EST*	*EFT*	*LST*	*LFT*	*TF* (F_T)	*FF* (F_F)	IN_F (F_{IN})
A	4	0	4	0	4	0	0	0
B	4	4	8	4	8	0	0	0
C	6	4	10	5	11	1	1	0
D	3	8	11	8	11	0	0	0
E	5	11	16	11	16	0	0	0
F	6	8	14	13	19	5	2	3
G	5	11	16	14	19	3	0	3
H	7	16	23	16	23	0	0	0
I	4	16	20	19	23	3	3	0

$$EFT = EST + t_{ij}$$
$$LST = LFT - t_{ij}$$
$$TF = (LFT - EFT) = (LST - EST) = (T_L^j - T_E^i) - t_{ij}$$
$$FF = TF - \text{head event slack}$$
$$IF = TF - FF = F_T - F_F = S_j = \text{head event slack}$$

(iii) Along critical path $TF = 0$

Therefore, critical path = $A - B - D - E - H$

(iv) Total project duration = 23 days

2.18 The data for planning a certain Civil Engineering project by CPM-Network analysis is given below. Draw the network and establish the critical path.

Also determine the following:

(i) Prepare a CPM schedule and calculate total float, free float and independent float.

(ii) Compute the project duration:

Activity	Duration in weeks	Activity immediately Preceding	Activity immediately Following
A	03	–	*E*
B	04	–	*D, F, G*
C	14	–	*H*
D	03	*B*	*H*
E	05	*A*	–
F	06	*B*	–
G	04	*B*	I
H	01	*C, D*	I
I	01	*G, H*	–

[10 marks : 2012]

Solution:

(i) The network may be developed using the given data, as shown below:

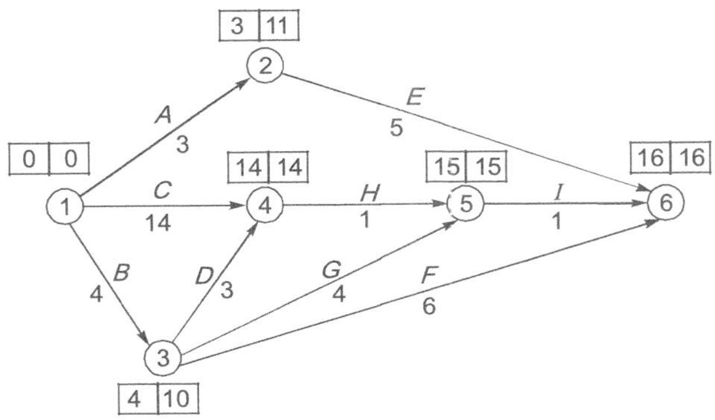

The earliest event times and latest allowable occurrence time can be calculated as

$$T_E^j = T_E^i + t^{ij}$$
$$T_L^i = T_L^j - t^{ij}$$

Also, the events can be numbered by fulkerson's rule

$$\therefore \quad EST = T_E^i$$
$$EFT = \text{EST} + t^{ij} = T_E^i + t^{ij}$$
$$LST = T_L^j - t^{ij}$$
$$LFT = T_L^j$$

Total float, $F_T = LST - EST$ or $LFT - EST$

Also, we know that

Free float, F_F = *EST* of successor activity – *EFT* of present activity

$= F_T - S_j$

Activity	t_E	*EST*	*EFT*	*LST*	*LFT*	F_T	F_F	F_{ID}
A	03	0	3	8	11	8	0	0
B	04	0	4	6	10	6	0	0
C	14	0	14	0	14	0	0	0
D	03	4	7	11	14	7	7	1
E	05	3	8	11	16	8	8	0
F	06	4	10	10	16	6	6	0
G	04	4	8	11	15	7	7	1
H	01	14	15	14	15	0	0	0
I	01	15	16	15	16	0	0	0

Independent float, $F_{ID} = (T_E^j - T_L^i) - t^{ij}$

$= F_F - S_i$

= Free float – Tail event slack

The value of activity times, total float, free float and independent float are tabulated below:

Hence, the critical activities are *C*, *H* and *I* and events joining these activities will be the critical events. Thus the critical path will be 1 – 4 – 5 – 6.

(ii) Therefore project duration = 14 + 1 + 1 = 16 weeks.

2.19 **PERT calculations indicate that duration of a given project is 72 weeks. With the variance of 15, work out number of weeks within which the project is expected to be completed with probability of 50%, 80% and 98%. Take *Z*-values of 0.89 and 2.1 for probability of 80% and 98% respectively**

[5 marks : 2012]

Solution:

Given, Variance, $\sigma^2 = 15$

∴ Standard deviation, $\sigma = \sqrt{15}$

As we know $Z = \dfrac{T_S - T_E}{\sigma}$

T_E = 72 weeks (given)

For **50% probability,** $Z = 0$

∴ $T_S = \sigma.Z + T_E = T_E = 72$ weeks

For **80% probability,** $Z = 0.89$

∴ $T_S = \sigma.Z + T_E = \sqrt{15} \times 0.89 + 72 = 75.447$ weeks

For **98% probability,** $Z = 2.1$

∴ $T_S = \sigma.Z + T_E = \sqrt{15} \times 2.1 + 72 = 80.133$ weeks

2.20 For an activity of casting a raft foundation of a High rise building, three engineers *A*, *B* and *C* have given the time estimates as follows. State who is more certain about the time of completion of job. Also calculate expected time of completion of each engineer.

[5 marks : 2012]

Engineer	Times in week		
	Optimistic	Most likely	Pessimistic
A	05	07	09
B	04	06	07
C	03	05	08

Solution:

The degree of uncertainty is indicated by the variance of time estimates.

The variance of time estimate given by engineer A is

$$\sigma_A^2 = \left(\frac{t_P - t_0}{6}\right)^2 = \left(\frac{9-5}{6}\right)^2 = 0.444$$

The variance of time estimate given by engineer B is

$$\sigma_B^2 = \left(\frac{t_P - t_0}{6}\right)^2 = \left(\frac{7-4}{6}\right)^2 = 0.25$$

The variance of time estimates given by engineer C is

$$\sigma_C^2 = \left(\frac{t_P - t_0}{6}\right)^2 = \left(\frac{8-3}{6}\right)^2 = 0.694$$

Thus, the variance of time estimates given by engineer B is least. Since less the variance, greater will be certainty. Hence engineer B's time estimates have more certainty.

Also expected time of completion is given by,

$$T_E = \frac{t_0 + 4t_m + t_p}{6}$$

$\therefore$ Expected time of engineer A, B and C are

$$T_{E_A} = \frac{5+4\times 7+9}{6} = 7 \text{ weeks}$$

$$T_{E_B} = \frac{4+4\times 6+7}{6} = 5.833 \text{ weeks}$$

$$T_{E_C} = \frac{3+4\times 5+8}{6} = 5.167 \text{ weeks}$$

2.21 In a small project, three activities, A, B and C, are on the critical path. There optimistic, most likely and pessimistic time durations (weeks) is given below:

Activity	Optimistic duration	Most likely duration	Pessimistic duration
A	6	8	11
B	4	6	8
C	11	15	18

Assuming that the noncritical activities are completed within that time:

(i) What is the probability that the project can be completed in 29 weeks?

(ii) What is the probability that the project can be completed in 27 weeks?

For a probability of 10%, 20% and 30% the Z values may be taken to be 1.26, 0.84 and 0.53, respectively.

[10 marks : 2013]

Solution:

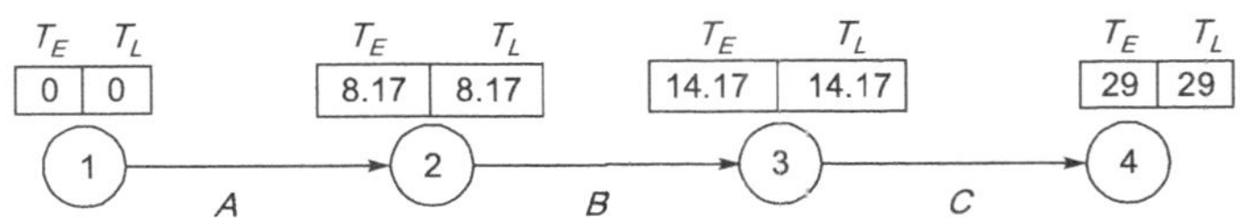

Activity	t_o	t_m	t_P	$t_E = \dfrac{t_0 + 4t_m + t_p}{6}$	$\sigma^2 = \left(\dfrac{t_p - t_0}{6}\right)^2$
A	6	8	11	8.17	0.6938
B	4	6	8	6	0.445
C	11	15	18	14.83	1.362

Mean time of project completion,

$$T_E = 8.17 + 6 + 14.83 = 29 \text{ weeks}$$

Variance of project = Σ(variance of activities)

Variance of project, $\sigma^2 = \left(\frac{11-6}{6}\right)^2 + \left(\frac{8-4}{6}\right)^2 + \left(\frac{18-11}{6}\right)^2 = 2.5008$ weeks

Standard deviation of project= $\sqrt{2.5008} = 1.5814$ weeks

Given:

Probability	Z
10%	1.26
20%	0.84
30%	0.53

(i) Z value = $\frac{29-29}{1.58} = 0$, Hence prob. = 50%

Therefore probability that the project can be completed in 29 weeks is 50%.

(ii) Z value = $\frac{27-29}{1.58} = -1.26$

⇒ $P = 50 - 10 = 40\%$

∴ Probability that the project can be completed in 27 weeks in 40%.

2.22 **In PERT network, how is the "expected time" of completion of a activity related to most likely, optimistic and the pessimistic times of completion of that activity?**

[4 marks : 2014]

Solution:

The average time taken for the completion of an activity is known as **'expected time'**.

If the exact shape of the probability distribution curve is known, the average time or expected time can be accurately calculated. However for non-repetitive activities, the precise curve are never available. For approximate curves, the expected time is calculated by weighted average with weight 1 to optimistic time (t_0), weight 4 to most likely time (t_m) and weight 1 to most pessimistic time (t_P)

Thus,

$$t_E = \frac{t_0 + 4t_m + t_P}{6}$$

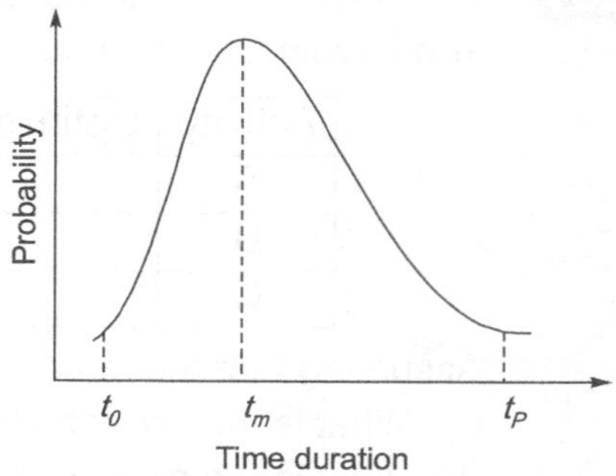

Beta distribution

2.23 **For the network of a construction project with various activities shown below, determine the project completion time. Determine the total float of each activity. Mention the critical activities.**

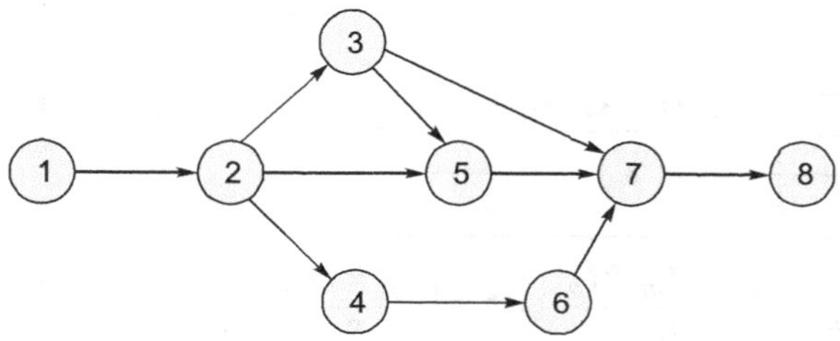

Activity	Duration (weeks)
1-2	16
2-3	24
2-4	36
2-5	8
3-5	12
4-6	40
3-7	32
5-7	8
6-7	16
7-8	12

[20 marks : 2015]

Solution:

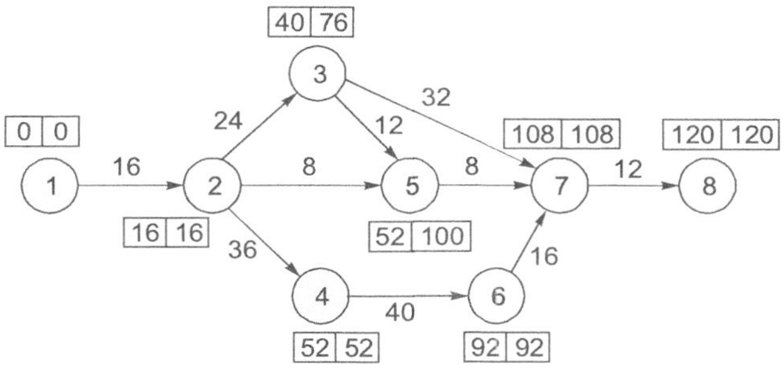

Activity	Duration (weeks)	EST	EFT	LST	LFT	Total Float	Remarks
1 - 2	16	0	16	0	16	0	Critical
2 - 3	24	16	40	52	76	36	
2 - 4	36	16	52	16	52	0	Critical
2 - 5	8	16	24	92	100	76	
3 - 5	12	40	52	88	100	48	
4 - 6	40	52	92	52	92	0	Critical
5 - 7	8	52	60	100	108	48	
6 - 7	16	92	108	92	108	0	Critical
3 - 7	32	40	72	76	108	36	
7 - 8	12	108	120	108	120	0	Critical

$$\text{Total float } (TF) = LFT - EST - t_{ij}$$
$$= LFT - (EST + t_{ij})$$
$$= LFT - EFT$$

For critical path activities, $TF = 0$

∴ Critical path of the network is 1 - 2 - 4 - 6 - 7 - 8

The project completion time is 120 weeks.

2.24 (i) Define critical path.
(ii) What is direct cost slope?
(iii) Explain FLOAT and slack.

[10 marks : 2015]

Solution:

(i) **Critical Path:** In a project involving a sequence of activities, critical path is the longest path in the network for the project to get completed on due date.

(ii) **Direct Cost Slope:** Direct cost in a project is the cost of materials, equipment, labour etc. that are directly involved to complete an activity. When time duration of a project decreases, the direct cost increases. The slope of this cost curve is called as direct cost slope which is given by,

$$\text{Direct cost slope} = \frac{\text{Crash cost} - \text{Normal cost}}{\text{Normal time} - \text{Crash time}}$$

(iii) **Slack:** Slack is defined as the difference of latest event time and the earliest event time.

For head event,

$$\text{Slack} = \text{Latest allowable occurrence time of event } j - \text{Earliest occurrence time of event } j = L_j - E_j$$

For tail event, $\text{Slack} = L_i - E_i$

Float: Float is the difference of latest start and the earliest start or the latest finish and earliest finish time of an activity. Float gives an idea about the time by which the start or finish of an activity can be delayed without affecting the total project duration. Total float for all the activities lying on critical path is always zero. There are four types of floats viz. total float, independent float, free float and interfering float.

2.25 **What are the main advantages of Activity on the Node (A-O-N) network over Activity on the Arrow (A-O-A) networks?**

[8 marks : 2016]

Solution:

In AON system sometimes called precedence diagrams also, the nodes represent the activities and the arrow, their interdependence or precedence relationships.

Nodes are usually represented by squares or rectangles but circles and other conventional geometrical shapes may also be used. One of the most common types of node representation is shown in figure:

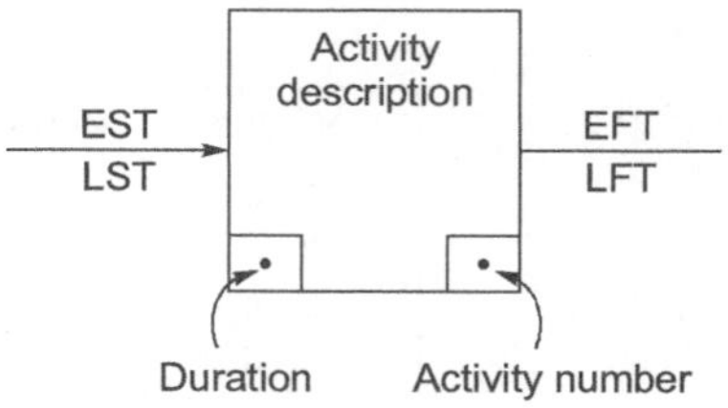

Main advantages of *AON* system over *AOA* system are:

(i) *AON* system of the network completely eliminates the use of dummy activities.

(ii) This system can show activity which should be allowed to overlap each other or must be separated by a time delay.

(iii) This system is self sufficient as it contains all activity times (*EST*, *LST*, *EFT*, *LFT*) on the diagram itself. This facilitates efficient scheduling and control.

(iv) Revisions and modifications can be carried out easily without affecting most of the activities i.e. few activities only have to be altered to incorporate changes in duration, logic and activity sequence.

(v) Pre-operations and post-operations of the activity under consideration are distinctly visible.

(vi) This system adopts simple notations similar to the engineering flow charts and hence can be easily understood by non-specialists also.

(vii) It is more advantageous in projects involving highly repetitive nature of series of operations.

2.26 **In a CPM network, the critical path includes five activities. Their durations are tabulated as follows:**

	Duration (Days)		
Activity	Optimistic Time (T_o)	Most Likely Time (T_m)	Pessimistic Time (T_m)
A	2	4	7
B	5	8	14
C	4	6	8
D	2	2	2
E	7	10	21

Compute the following values:

(i) The probability that the project will finish by the end of day 32.

(ii) The probability that the project will finish by the end of day 34.

(ii) The probability that the project will finish before day 30.

(iv) The probability that the project will finish on the 32nd day.

(v) The probability that the project will finish no later than the 35th day.

Z =	0	0.34	0.67	1.01
Probability	50%	63.33%	74.9%	84.4%

[20 marks : 2016]

Solution:

Activity	Optimistic Time (T_o)	Most likely Time (T_m)	Pessimistic Time (T_P)	Expected Time (T_E)	$\sigma^2 = \left(\frac{t_p - t_0}{6}\right)^2$
A	2	4	7	4.167	0.6944
B	5	8	14	8.5	2.25
C	4	6	8	6	0.444
D	2	2	2	2	0
E	7	10	21	11.33	5.444
				ΣT_E=32 days	$\Sigma\sigma^2$=8.8324

$$\text{Expected time } (T_E) = \frac{T_0 + 4T_m + T_P}{6}$$

$$\therefore \quad \sigma = \sqrt{\Sigma\sigma^2} = \sqrt{8.8324} = 2.9719$$

Earliest expected time of completion, T_E = 32 days

(i) **Probability of completion in 32 days,** $Z = \frac{T_S - T_E}{\sigma} = \frac{32-32}{2.9719} = 0$

$\therefore$ Probability = 50%

(ii) **Probability of completion in 34 days,** $Z = \frac{34-32}{2.9719} = 0.673$

$\therefore$ Probability = 74.984% $\simeq$ 75% (By interpolation from the table given)

(iii) **Probability of completion before day 30** = Probability of completion in 29 days

$$Z = \frac{29-32}{2.9719} = -1.01$$

Probability = 15.6%

(iv) **Probability of completion on 32nd day** = Probability of completion in 32 days – Probability of completion in 31 days

$\because$ Probability of completion in 32 days = 50%

and for probability of completion in 31 days,

$$Z = \frac{31-32}{2.9719} = -0.3365$$

$\because$ Probability of completion in 31 days = 36.81%

Probability of completion on 32nd day = 50% – 36.81% = 13.19%

(v) **Probability that the project will finish no later than the 35 days** = Probability of completion in 35 days

$$Z = \frac{35-32}{2.9719} = 1.01$$

Probability = 84.4%

2.27 **Briefly answer the following:**
(i) Differentiate between the terms 'Activity' and Dummy'.
(ii) Differentiate between 'Optimistic time estimate' and 'Pessimistic time estimate'.

[10 marks : 2017]

Solution:

(i) **Activity:** An activity is the actual performance of a task. It is the work required to complete a specific event. An activity is a recognizable part of a project that requires time and resources (manpower, material, space, facilities) for its completion.

A significant activity must be

(a) a positive, specific and tangible and meaningful effort.
(b) such that the primary responsibility of effort can be determined
(c) having a time span.

Dummy: A dummy is a type of operation in the network which neither requires any time nor any resources but is merely a device to identify a dependences among activities. A dummy is thus a connecting link for control purposes for maintaining uniqueness of activity.

A dummy is also represented by arrow but since it is not an activity, it is represented by dashed arrow. A dummy is identified by the numbers of the terminal nodes.

(ii) **'Optimistic time estimate' & Pessimistic time estimate:**

- **Optimistic time estimate:** This is the shortest possible time in which an activity can be completed under ideal conditions. This particular time estimate represents the time in which we could complete the activity or job if everything went along perfectly, with no problems or adverse conditions. Better than normal conditions are assumed to prevail this time estimate is denoted by t_o.
- **Pessimistic time estimate:** It is the best guess of the maximum time that would be required to complete the activity. The particular time estimate represents the time it might take us to complete a particular activity if everytime went wrong and abnormal situations prevailed. However, this estimate does not include possible effects of highly unusual catastrophies such as earthquakes, floods, fires etc. This estimate is denoted by t_p.

2.28 **Calculate the expected time t_E and critical path for the following network shown in figure. The optimisitc, most likely and pessimistic time for each activity is as shown. 1 is the initial event and 10 is the end event.**

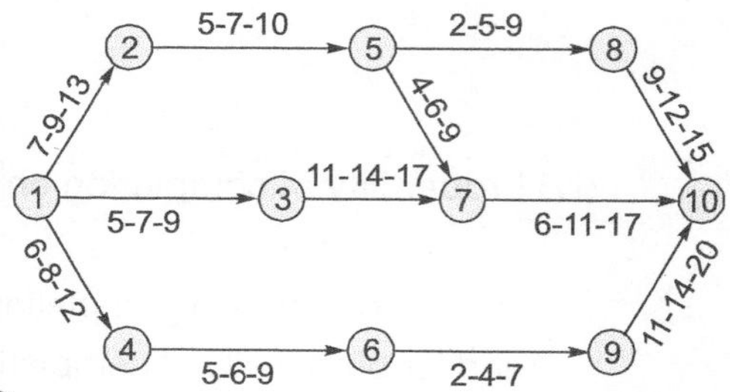

[12 marks : 2017]

Solution:

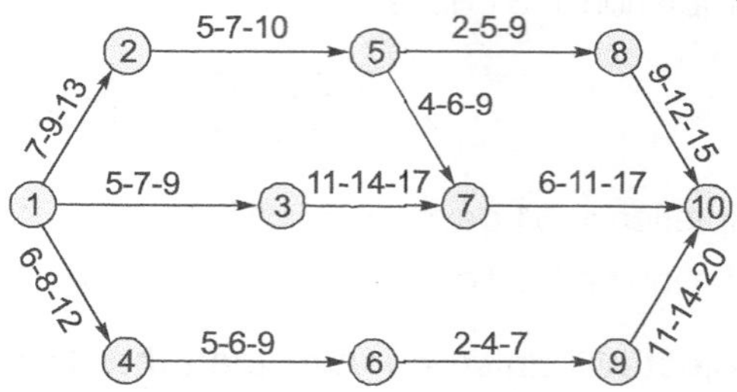

There are following four paths from the starting event to the end event.

Path A : 1 – 2 – 5 – 8 – 10
Path B : 1 – 2 – 5 – 7 – 10
Path C : 1 – 3 – 7 – 10
Path D : 1 – 4 – 6 – 9 – 10

Path	Activity	t_0	t_L	t_P	$t_E = \frac{t_0 + 4t_L + t_P}{6}$	Σt_E
A	1 – 2	7	9	13	9.33	33.664
	2 – 5	5	7	10	7.167	
	5 – 8	2	5	9	5.167	
	8 – 10	9	12	15	12	
B	1 – 2	7	9	13	9.33	33.831
	2 – 5	5	7	10	7.167	
	5 – 7	4	6	9	6.167	
	7 – 10	6	11	17	11.167	
C	1 – 3	5	7	9	7	32.167
	3 – 7	11	14	17	14	
	7 – 10	6	11	17	11.167	
D	1 – 4	6	8	12	8.33	33.327
	4 – 6	5	6	9	6.33	
	6 – 9	2	4	7	4.167	
	9 – 10	11	14	20	14.5	

$\therefore$ Path B (1 – 2 – 5 – 7 – 10) is the critical path as Σt_E for this path is the maximum, and expected time of project is 33.831 unit.

2.29 A construction project has the following characteristics:

Activity	Preceding Activity	Duration (weeks)
A	None	5
B	A	2
C	A	6
D	B	12
E	D	10
F	D	5
G	D	5
H	B	9
I	C, E	1
J	G	2
K	F, I, J	3
L	K	9
M	H, G	7
N	M	9

(i) Draw a network for this project.
(ii) Find various paths and the critical path as well as the project completion time.
(iii) Prepare an activity schedule showing Earliest Start Time, Earliest Finish Time, Latest Start Time, Latest Finish Time and float for each activity.
(iv) Will the critical path change if activity G takes 10 weeks instead of 5 weeks? If so, what will be the new critical path?

[12 marks : 2018]

Solution:

(i)

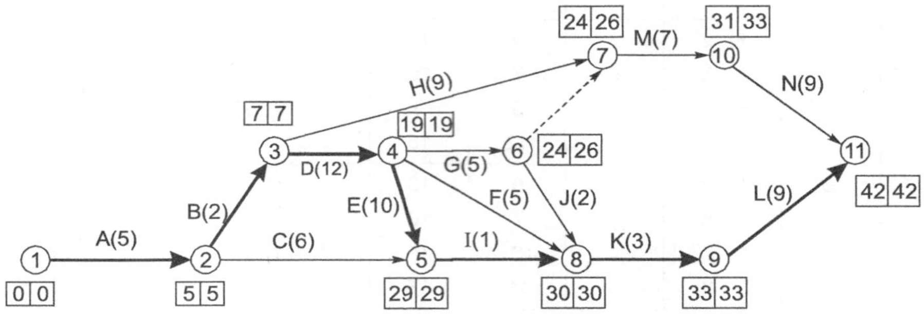

(ii)

	Paths	Completion time (weeks)
(i)	A - B - H - M - N	5 + 2 + 9 + 7 + 9 = 32
(ii)	A - B - D - G - M - N	5 + 2 + 12 + 5 + 7 + 9 = 40
(iii)	A - B - D - E - I - K - L	5 + 2 + 12 + 10 + 1 + 3 + 9 = 42
(iv)	A - C - I - K - L	5 + 6 + 1 + 3 + 9 = 24
(v)	A - B - D - G - J - K - L	5 + 2 + 12 + 5 + 2 + 3 + 9 = 38
(vi)	A - B - D - F - K - L	5 + 2 + 12 + 5 + 3 + 9 = 36

Critical path is : A – B – D – E – I – K – L

Project completion time = 42 weeks.

Critical path is shown by bold lines on the network diagram.

(iii) Earliest start time, EST = T_E^i

Earliest finish time, EFT = EST + t_{ij}

Latest finish time, LFT = T_L^j

Latest start time, LST = LFT – t_{ij}

$$\text{Total float, } F_T = \left(T_{Lj} - T_{Ei} - t_{ij}\right)$$
$$= \text{LFT} - \text{EFT}$$

Activity	Duration (Weeks)	EST	EFT	LFT	LST	F_T
A	5	0	5	5	0	0
B	2	5	7	7	5	0
C	6	5	11	29	23	18
D	12	7	19	19	7	0
E	10	19	29	29	19	0
F	5	19	24	30	25	6
G	5	19	24	28	23	4
H	9	7	16	26	17	10
I	1	29	30	30	29	0
J	2	24	26	30	28	4
K	3	30	33	33	30	0
L	9	33	42	42	33	0
M	7	24	31	33	26	2
N	9	31	40	42	33	2

(iv) If activity *G* takes 10 weeks instead of 5 weeks:

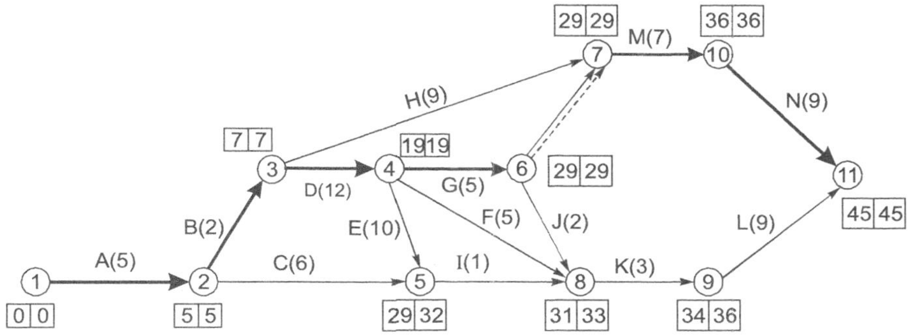

Critical path has changed:

New critical path is A – B – D – G – Dummy - M – N

It is shown by bold lines in the network diagram.

2.30

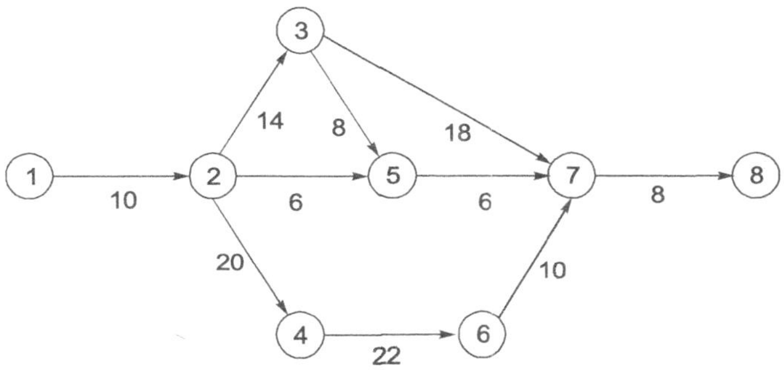

Identify the critical path in the network as shown in figure and determine the project completion time. The duration are in weeks.

[20 marks : 2019]

Solution:

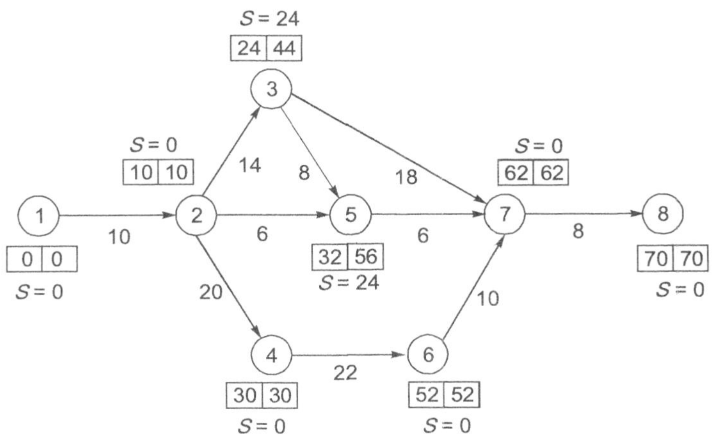

Notation T_E | T_L

$$\text{Slack} = S = T_L - T_E$$

$$\text{Project duration} = 70 \text{ weeks}$$

$$\text{Critical path} = 1\text{–}2\text{–}4\text{–}6\text{–}7\text{–}8$$

2.31 The following table gives the activities in a construction project and other relevant information.

Activity	Duration (Days)
1 – 2	20
1 – 3	25
2 – 3	10
2 – 4	12
3 – 4	6
4 – 5	10

(i) Draw the network for the project.
(ii) Find the critical path.
(iii) Find free, total and independent floats for each activity.

[20 marks : 2020]

Solution:

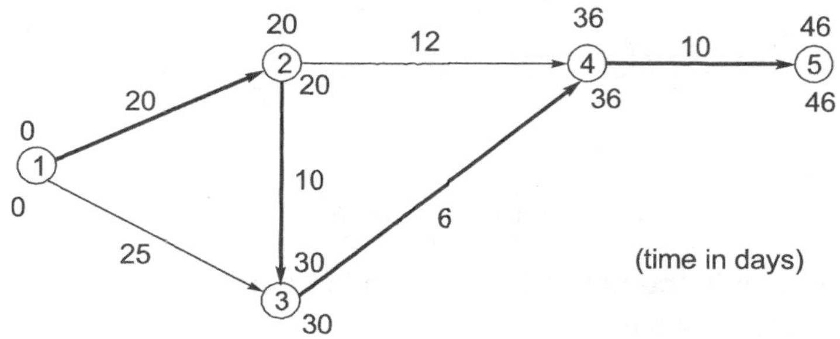

Critical path (1-2-3-4-5) (Duration = 46 days)

Activity	Time	EST	EFT	LST	LFT	TF	FF	IF	Remarks
1-2	20	0	20	0	20	0	0	0	Critical
1-3	25	0	25	5	30	5	5	5	
2-3	10	20	30	20	30	0	0	0	Critical
2-4	12	20	32	24	36	4	4	4	
3-4	6	30	36	30	36	0	0	0	Critical
4-5	10	36	46	36	46	0	0	0	Critical

Note:

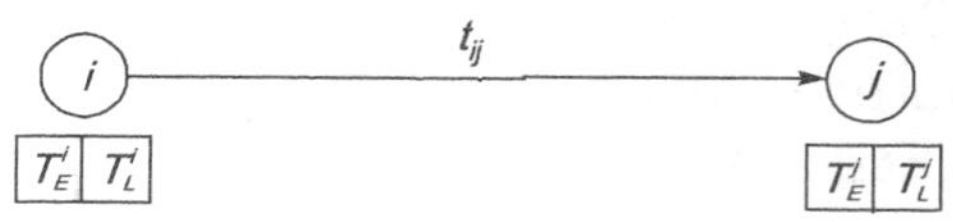

$EST = T_E^i$

$$LST = T_L^j - t_{ij}$$
$$EFT = T_E^i + t_{ij}$$
$$LFT = T_L^j$$

Total float, $TF = LST - EST = LFT - EFT$

Free float $FF = TF - S_j$

Independent float, $IF = FF - S_i$

2.32 **Differentiate between optimistic time estimate and pessimistic time estimate in a PERT network. A construction company has an opportunity to submit a bid for the construction of a residential building and a commercial building. The 3 time estimates (in months) for completion of each building are as follows:**

	Optimistic time (in months)	Most likely time (in months)	Pessimistic time (in months)
Residential Building	3	4	6
Commercial Building	4	6	8

Determine the expected time for completion of each building. Also analyse which building has more reliable time estimate.

[20 Marks : 2021]

Solution:

Optimistic Time (t_o) :

This is the shortest possible time in which an activity can be completed, under ideal conditions, This particular time estimates represents the time in which we could complete the activity or job if everything went along perfectly with no problems or adverse conditions. Better than normal conditions are assumed to prevail.

This time estimate is demanded by to.

Pessimistic time (t_p):

It is the best given of maximum time that would be required to complete the activity. This particular time estimate represents the time, it might take us to complete a particular activity if everything went wrong and abnormal situations prevailed. However, this estimates does not include possible effects of highly unusual catastrophic conditions such as earthquakes, floods, drawings, fire etc.

This time is denoted by t_p.

Expected completion time for residential building is given by

$$t_{e1} = \frac{t_0 + 4t_L + t_p}{6} = \frac{3 + 4 \times 4 + 6}{6} = 4.167 \text{ months}$$

Variance for residential building is given by

$$\sigma_1^2 = \left(\frac{t_p - t_o}{6}\right)^2 = \left(\frac{6-3}{6}\right)^2 = 0.25 \text{ months}^2$$

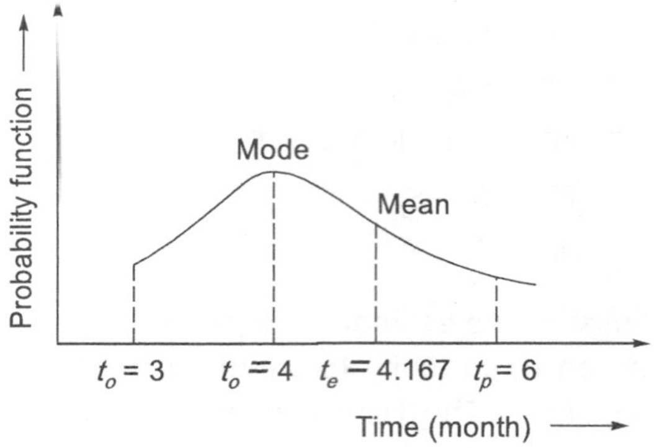

Expected completion time for commercial building is given by

$$t_{e2} = \frac{t_0 + 4t_L + t_p}{6} = \frac{4 + 4\times 6 + 8}{6} = 6 \text{ months}$$

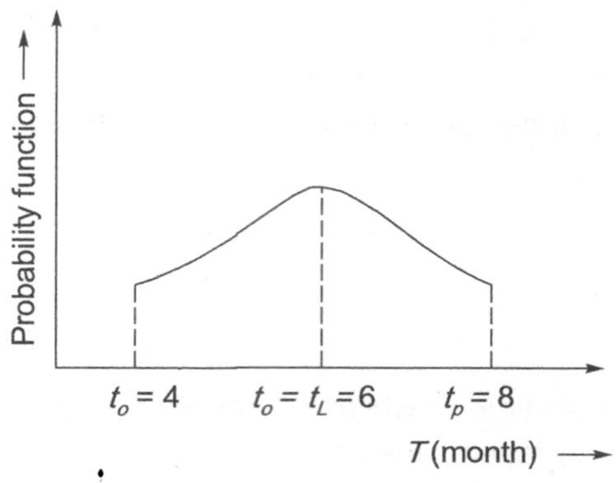

Variance for commercial building is given by

$$\sigma_2^2 = \left(\frac{t_p - t_0}{6}\right)^2 = \left(\frac{8-4}{6}\right)^2 = \left(\frac{4}{6}\right)^2 = 0.444 \text{ month}^2$$

As variance for residential builcing is less than variance for commercial building hence uncertainty is less for completion of project.

Hence, time estimates for residential building is more reliable.

2.33 The following table shows the normal, shortest and longest durations for each activity to be completed for the completion of the project. The table also shows the slope (increase/decrease) in cost per unit duration of each activity. The contract includes penalty clause of ₹10000/per unit time and bonus of ₹5000/- per unit time for the completion of project with respect to the normal duration of the project. The overhead (indirect) cost per unit time is ₹16000/-. Calculate the optimum cost of project completion. The cost (direct) of completing all the eight activities in normal duration is ₹ 650000/-.

Activity	Normal duration	Minimum duration	Maximum duration	Slope (±)
1–2	6	4	6	8000/-
1–3	8	4	11	Rs. 4000/- for first 2 unit time and Rs. 9000/- subsequently
1–4	5	3	6	3000/-
2–4	3	3	3	—
2–5	5	3	9	8000/-
3–6	12	8	14	Rs. 8000/- for first 3 unit time and Rs. 20000/- subsequently
4–6	8	5	10	5000/-
5–6	6	6	6	—

[20 marks : 2022]

Solution:

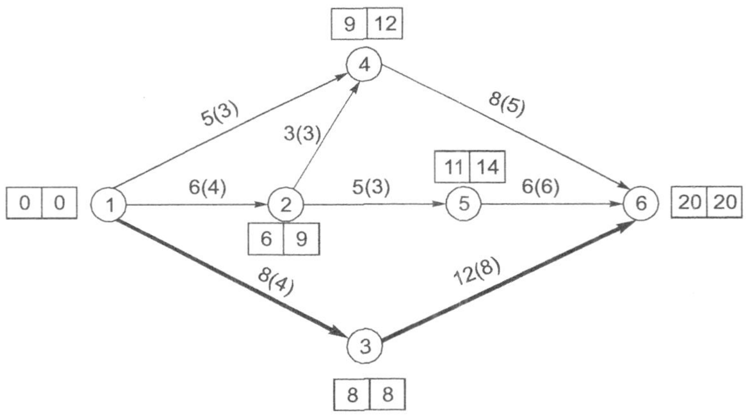

Total project duration is 20 days and critical path is 1–3–6.

Total project cost for 20 days = ₹(650000 + 16000 × 20) = ₹ 970000

Crashing activity 1–3 by 2 days,

Total project cost for 18 days = ₹ (650000 + 16000 × 18 + 4000 × 2 – 5000 × 2)

= ₹ 936000

Crashing activity 3–6 by 1 day,

Total project for 17 days = ₹ (650000 + 16000 × 17 + 8000 – 5000 × 3)

= ₹ 915000

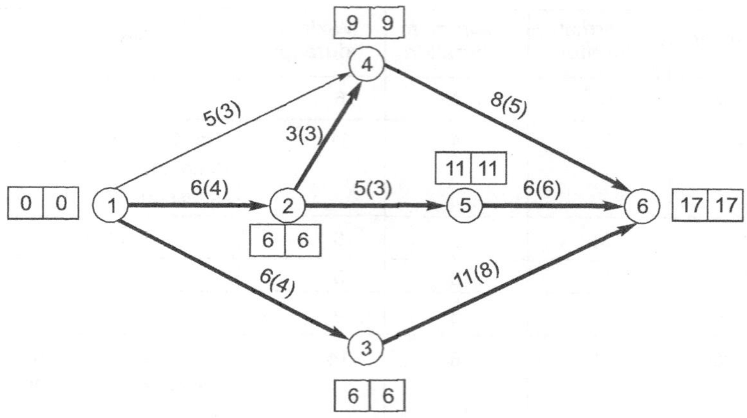

Crashing activity 1–2 by 2 days and activity 1–3 by 2 days

Total project cost for 15 days = ₹ 650000 + 16000 × 15 + 2 × (8000 + 9000) – 5000 × 5

= ₹ 897000

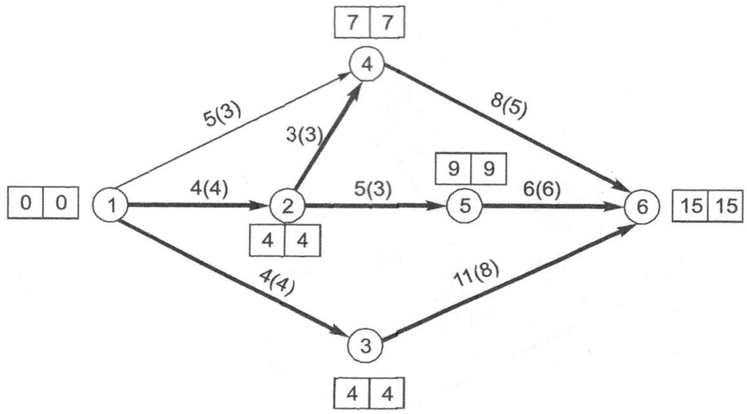

Maximum cost of project,

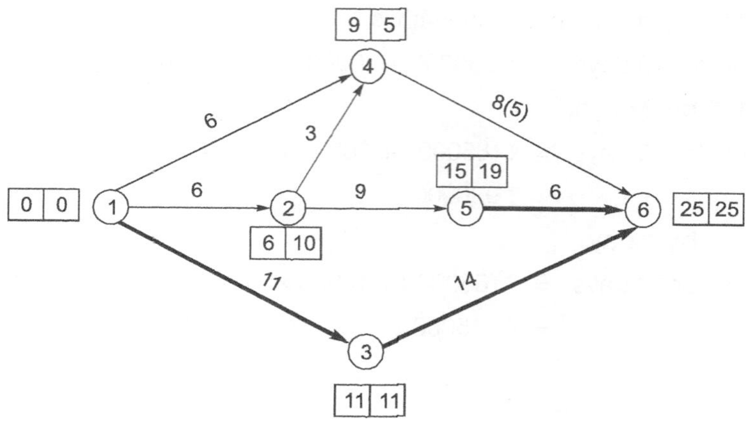

Maximum cost of project is corresponding to 25 days,

= ₹ 650000 + 16000 × 25 – (2 × 4000 + 9000 + 2 × 8000) + 5 × 10000

= ₹ 1067000

2.34 The following table gives the details of various activities of a construction project:

Activity	Optimistic time (months)	Most likely time (months)	Pessimistic time (months)
1 - 2	2	2	8
1 - 3	2	5	8
1 - 4	3	3	9
2 - 5	2	2	4
3 - 5	3	6	15
4 - 6	3	6	9
5 - 6	4	7	16
6 - 7	2	2	2

(i) Draw the network for the project.

(ii) Find the expected duration and variance of each activity.

(iii) What is the expected project length?

(iv) What is the probability that the project will be completed at least 3 months earlier than expected?

(v) What will be the time required for 95% probability of its completion?

[20 marks : 2023]

Solution:

(i) Network diagram for the project is as shown below:

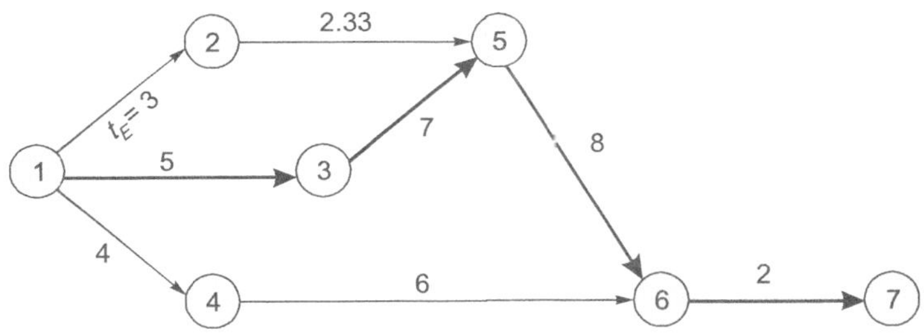

(ii)

Activity	Optimistic time (t_o) months	Most likely time (t_m) months	Pessimistic time (t_p) months	Expected time (t_E) months $t_E = \left(\frac{t_o + 4t_m + t_p}{6}\right)$	Variance $\sigma^2 = \left(\frac{t_p - t_o}{6}\right)^2$ (months)^2
1 - 2	2	2	8	3	1
1 - 3	2	5	8	5	1
1 - 4	3	3	9	4	1
2 - 5	2	2	4	2.33	0.11
3 - 5	3	6	15	7	4
4 - 6	3	6	9	6	1
5 - 6	4	7	16	8	4
6 - 7	2	2	2	2	0

(iii) In the figure shown below, S_j denotes the slack of event and critical path is the path joining events of zero slack. So, critical path is along 1 - 3 - 5 - 6 - 7 and expected projected length is 22 months.

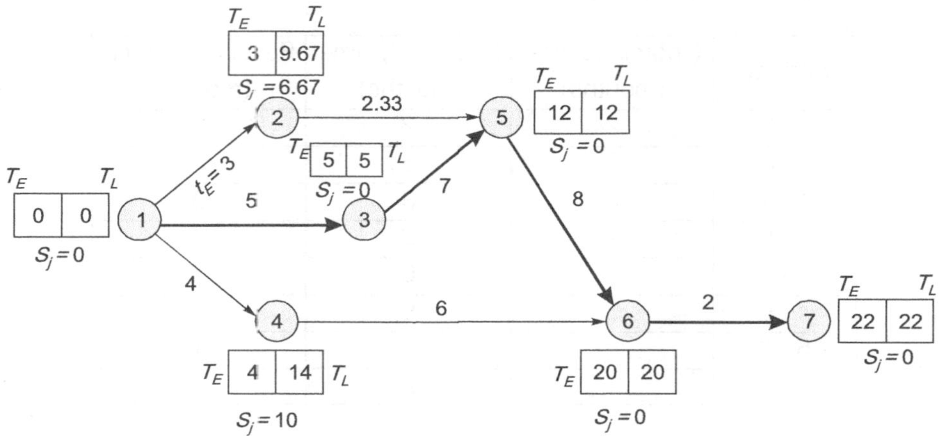

(iv) Probability that the project will be completed at least 3 months earlier than expected i.e. probability that the project will be completed in 19 months.

$$Z = \frac{T_S - T_E}{\sigma}$$

where, Z = Probability factor

T_S = 19 months, T_E = 22 months

σ = Standard deviation of project

$$\sigma = \sqrt{1+4+4+0} = 3$$

Now, $$Z = \frac{19-22}{3} = -1$$

Now, probability of completion is given as

Z	P
0	50%
−1	15.87%

P = 15.87%

(v) Time required for 95% probability completion of its completion is given as below:

$$Z = \frac{T_S - T_E}{\sigma}$$

Probability factor (Z) for 95% probability is given as

Z	P
1	84.13%
?	95%
2	99.87%

Now from interpolation,

$$\frac{99.87-84.13}{2-1} = \frac{95-84.13}{Z-1}$$

$$Z = 1.691$$

Now, $$Z = \frac{T_S - 22}{3}$$

$$1.691 = \frac{T_S - 22}{3}$$

$$T_S = 27.073 \text{ months}$$

Hence, time required for 95% probability of its completion is 27.073 months.

3. Crashing, Resource Allocation, Updating and Engineering Economy

3.1 A, B, C and D are the activities. Their normal and crash durations and associated costs are given in the table below:

Activity	Duration in days		Direct cost in Rs.	
	Normal	Crash	Normal	Crash
A	8	4	6000	12000
B	4	2	2000	14000
C	10	4	4000	8000
D	6	4	4000	8000

For the entire project the indirect cost is Rs. 1000 per day. *A* and *B* are starting activities; *C* follows *B*; *D* follows *A* and *C*; *D* is the finishing activity. Draw *CPM* Network. Calculate points for *PTC* graph and plot the same. Determine the optimum cost and optimum duration for the project. *PTC* is Project-Time-Cost-Trade-Off graph.

[25 marks : 2007]

Solution:

The CPM network for the given activities is shown below:

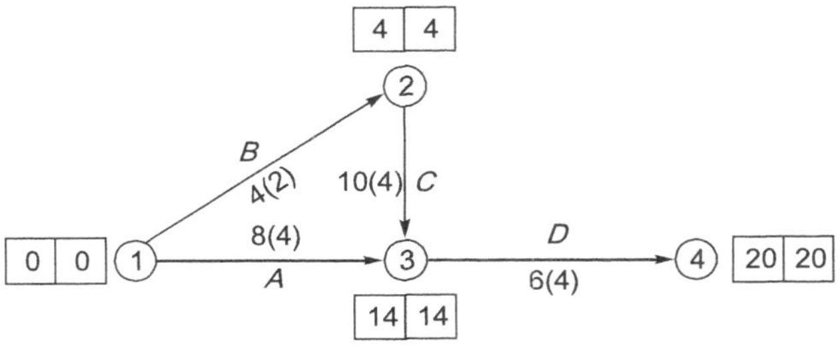

Normal duration of the project = 20 days

Total direct cost of the project = 6000 + 2000 + 4000 + 4000 = 16000

Total indirect cost of the project = $1000 \times 20 = 20000$

∴ Total project cost = Total direct cost + Total indirect cost

= 16000 + 20000 = 36000

Activity	Duration in days		Direct cost in Rs.		Cost slope = $(C_c - C_n)/(t_n - t_c)$
	Normal	Crash	Normal	Crash	
A	8	4	6000	12000	1500
B	4	2	2000	14000	6000
C	10	4	4000	8000	666.67
D	6	4	4000	8000	2000

First stage of crashing: As it is evident from the network diagram above that the critical path is *B*-*C*-*D*. Among the critical activities minimum cost slope is of activity *C*. Hence, it should be crashed first.

The activity *C* can be crashed by 6 days.

Increase in direct cost = 666.67 × 6 = 4000

Decrease in indirect cost = 1000 × 6 = 6000

Net change = 4000 – 6000 = –2000 (decrease)

∴ Total cost of project after first stage of crashing = 36000 – 2000 = 34000

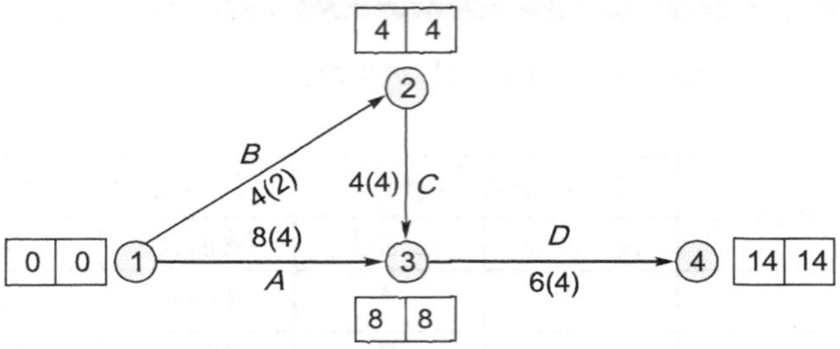

Total project duration after first stage of crashing = 14 days

Second Stage of crashing: After first stage of crashing the number of critical paths are increased i.e. either *B* – *C* – *D* or *A* – *D* may be the critical path. In the second stage possible alternatives for crashing are activity *D* which can be crashed by 2 days and activity *A* and *B* combined which can be crashed by 2 days each. But the cost slope *A* and *B* combined is (1500 + 6000 = 7500), which is more than cost slope of activity *D*. Hence activity *D* will be crashed by 2 days.

∴ Increase in direct cost = 2000 × 2 = 4000

Decrease in indirect cost = 1000 × 2 = 2000

Net change = 4000 – 2000 (increase)

∴ Total cost of project after second stage of crashing = 34000 + 2000 = 36000

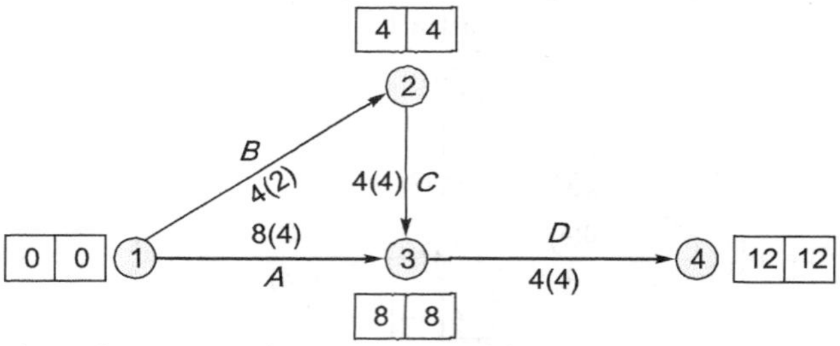

∴ Total project duration after second stage of crashing = 12 days

Third stage of crashing: After second stage of crashing, the number of critical path will remain same i.e. *B–C–D* and *A–D*.

In third stage only possible alternative is to crash activity *B* and *A* simultaneously by 2 days.

∴ Increase in direct cost = (1500 + 6000) × 2 = 15000

Decrease in indirect cost = 1000 × 2 = 2000

Net change = 15000 – 2000 = 13000 (increase)

∴ Total cost of the project after third stage of crashing = 36000 + 13000 = 49000

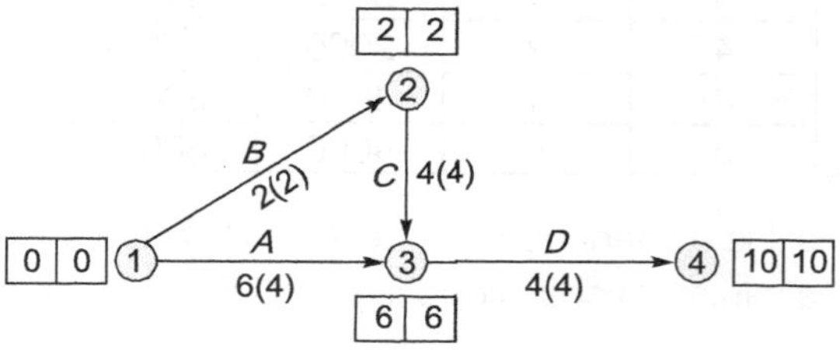

$\therefore$ Total project duration after third stage of crashing = 10 days

Now, *A* can further be crashed by 2 days but it cannot happen because activity *B* and *C* are parallel to *A* and they have been already crashed to their maximum limits.

Optimum cost of the project = Rs. 34000

Optimum duration of the project completion = 14 days

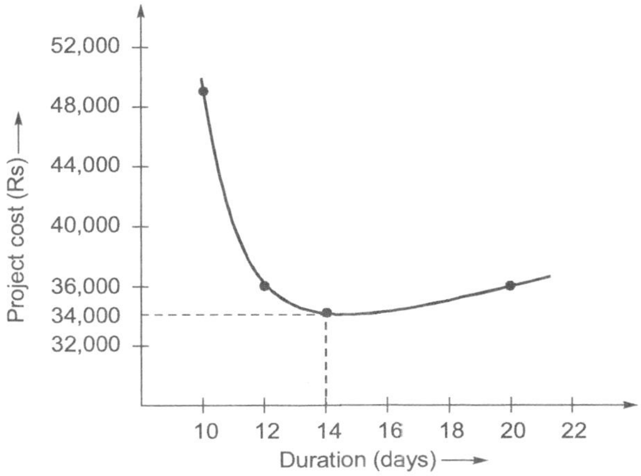

3.2 Calculate optimum cost and optimum duration for jobs of network given in the table below:

Activity	Normal duration in days	Normal cost Rs.	Crash duration in days	Crash cost Rs.
1-2	5	4000	4	5000
1-3	7	8000	3	10000
2-3	6	6000	2	8400

Indirect cost = Rs. 1000 per day. Sketch project time - cost diagram.

[10 marks : 2009]

Solution:

The network for the given activities is shown below:

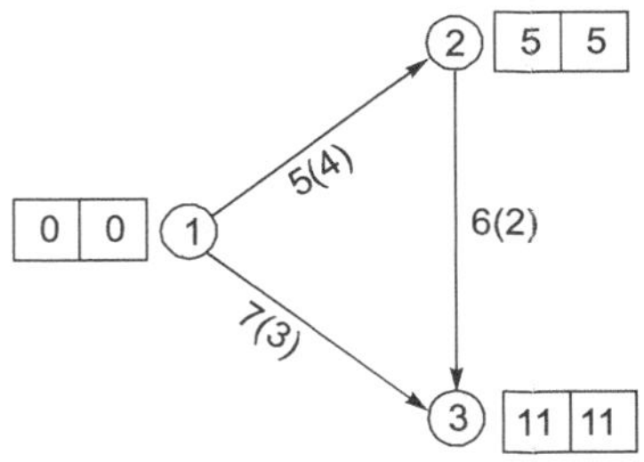

Normal duration of the project = 11 days

Total direct cost of the project = 4000 + 8000 + 6000 = 18000

Total indirect cost of the project = 1000 × 11 = 11000

Total cost of the project = 18000 + 11000 = Rs. 29000

We know that

$$\text{Cost slope} = \frac{\text{Crash cost} - \text{Normal cost}}{\text{Normal time} - \text{Crash time}}$$

$$\therefore \quad \text{Cost slope for } 1-2 = \frac{5000-4000}{5-4} = 1000$$

$$\text{Cost slope for } 1-3 = \frac{10000-8000}{7-3} = 500$$

$$\text{Cost slope for } 2-3 = \frac{8400-6000}{6-2} = 600$$

First stage of crashing: It is evident from the network diagram above that the critical path is 1 – 2 – 3. Among the critical activities, minimum cost slope is of activity 2 – 3. Hence it should be crashed first. The activity 2 – 3 can be crashed by 4 days.

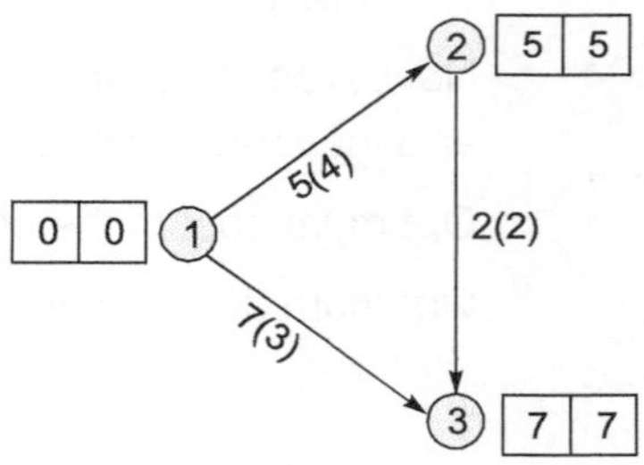

∴ Increase in direct cost = [cost slope of 2 – 3] × 4 = 600 × 4 = 2400

Decrease in indirect cost = 1000 × 4 = 4000

Net change = 2400 – 4000 = – 1600 (Decrease)

∴ Total cost of project after first stage of crashing = 29000 – 1600 = 27400

Total project duration of project after first stage of crashing = 7 days

Second stage of crashing: After first stage of crashing, the number of critical paths are increased i.e. either 1 – 2 – 3 or 1 – 3 may be the critical path. In second stage of crashing, only possible alternative is to crash activity

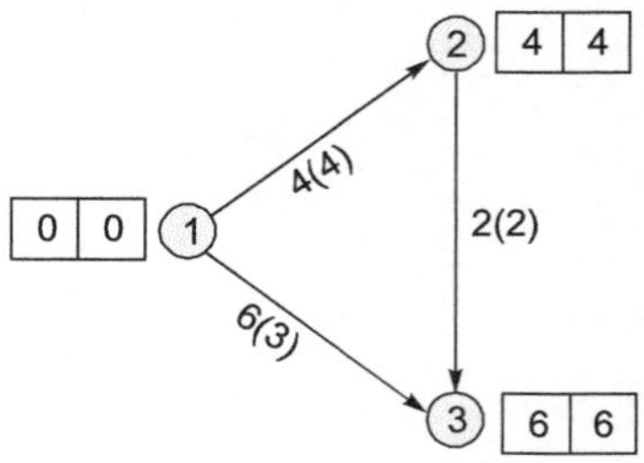

1 – 2 and 1 – 3 simultaneously by 1 day.

Combined cost slope of activity 1 – 2 and 1 – 3 = 1000 + 500 = 1500

Increase in direct cost = 1500 × 1 = 1500

Decrease in indirect cost = 1000 × 1 = 1000

Net change = 1500 – 1000 = 500 (increase)

∴ Total cost of project after second stage of crashing = 27400 + 500 = 27900

Total project duration after second stage of crashing = 6 days

Now, activity 1 – 3 can be further crashed by 3 days but it cannot happen because activity 1 – 2 is parallel to activity 1 – 3 and it has been already crashed to its maximum limit.

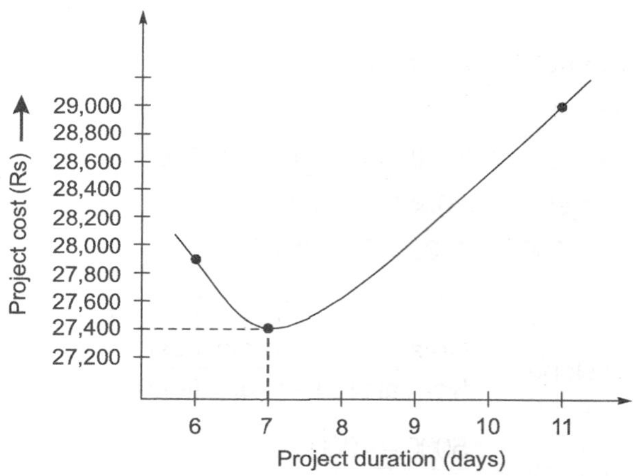

Thus the optimum cost and duration of the project is Rs. 27400 and 7 days respectively.

3.3 **Determine the minimum cost and optimum duration for a project network. The duration and cost of each activity of the project is given in the following table. The indirect cost of the project is Rs. 800 per day. Draw the time-scaled version of the network.**

Activity	Normal duration (in days)	Normal cost (in Rs.)	Crash duration (in days)	Crash cost (in Rs.)
A – B	2	10,000	2	10,000
A – C	7	5,000	3	9,000
B – C	6	3,000	3	4,200
B – D	5	2,000	4	2,500
C – D	0	0	0	0
C – E	9	6,000	4	9,000
D – F	11	6,000	6	10,000
E – F	6	7,000	3	9,100

[15 marks : 2010]

Solution:

The CPM network for the given activities is shown below:

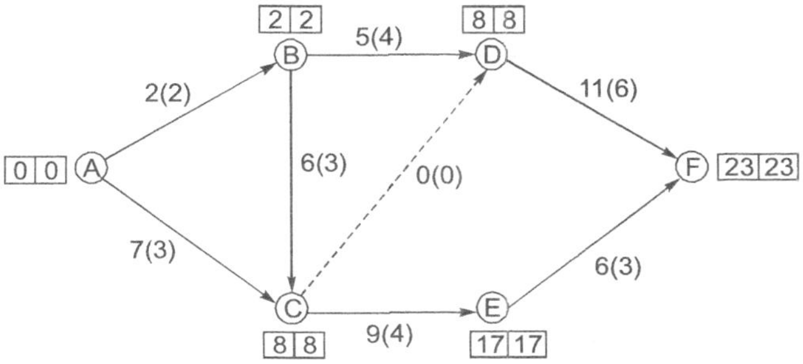

Normal duration of the project = 23 days

Total direct cost of the project = 10000 + 5000 + 3000 + 2000 + 0 + 6000 + 6000 + 7000 = 39000

Total indirect cost of the project = $23 \times 800 = 18400$

Total project cost = Total direct cost + Total indirect cost

= 39000 + 18400

= 57400

Activity	Duration in days		Direct cost (in ₹)		Cost slope $\frac{C_c - C_n}{t_n - t_c}$
	Normal	Crash	Normal	Crash	
A-B	2	2	10,000	10,000	0
A-C	7	3	5,000	9000	1000
B-C	6	3	3,000	4200	400
B-D	5	4	2,000	2500	500
C-D	0	0	0	0	0
C-E	9	4	6,000	9000	600
D-F	11	6	6,000	10,000	800
E-F	6	3	7,000	9100	700

1st stage of crashing: As it is evident from the network diagram above that the critical path is A-B-C-E-F. Among the critical activities {A-B, B-C, C-E, E-F} minimum cost slope is of activity B-C. {A-B activity has zero potential to crash}.

Here activity B-C can be crashed by 1 day because B-D is parallel activity and that activity BD will also become critical activity.

Hence activity B-C can be crashed by 1 day then

Increase in direct cost = 400 × 1 = 400

Decrease in indirect cost = 800 × 1 = 800

Total cost after first stage crashing = 57,400 + 400 – 8000

= 57,000

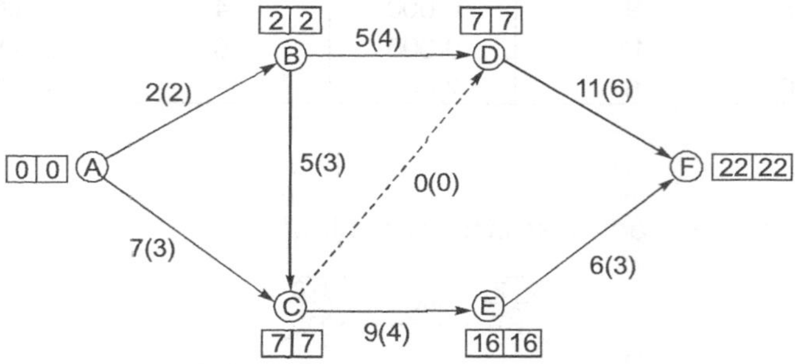

2nd stage crashing: After first stage of crashing the number of critical paths are increased i.e., either A-B-C-E-F or A-C-E-F are the critical paths.

In the second stage possible alternatives for crashing are activity C-E and E-F. But activity C-E has least cost slope. Activity C-E can be crashed by 5 days but D-F is parallel activity that will also become critical.

Hence activity C-E can be crashed by 4 days.

Hence Increase in direct cost = 4 × 600 = 2400

Decrease in indirect cost = 4 × 800 = 3200

Total cost after 2nd stage crashing = 57,000 + 2,400 – 3,200 = 56,200

Now project duration is 18 days.

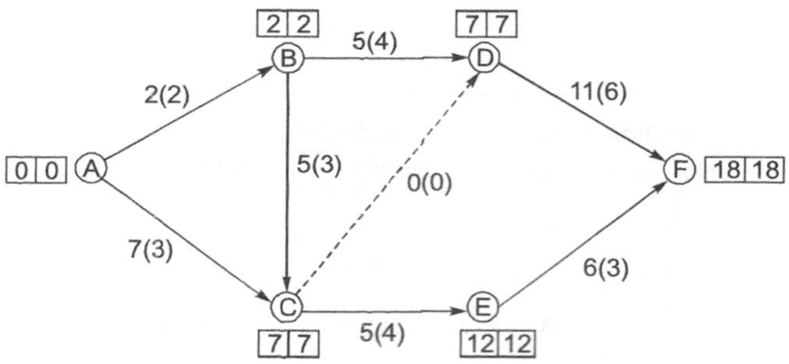

3rd stage crashing: After second stage crashing, the number of critical path will be 5

(i) A-B-C-E-F (ii) A-B-C-D-F (iii) A-B-D-F

(iv) A-C-D-F (v) A-C-E-F

Now combined slope of A-C, B-D and B-C activity = 1000 + 400+500 = 1900

Combined slope of D-F and E-F activity = 800 + 700 = 1500

Combined slope of C-E and D-F activity = 600 + 800 = 1400

Thus crashing of C-E and D-F simultaneously by 1 day.

Increase in direct cost = 1400 × 1 = 1400

Decrease in indirect cost = 800 × 1 = 800

Total cost after 3rd stage crashing = 56200 + 1400 – 800 = 56800

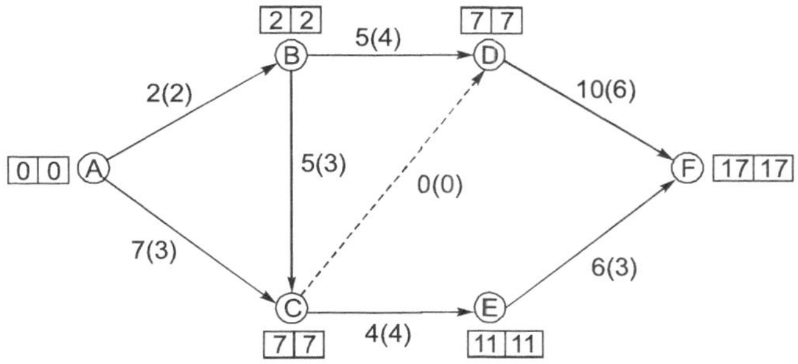

4th stage crashing: After 3rd stage crashing no. of critical path will be same but project duration has been decreased to 17 days.

Now we can crash D-F and E-F activity by '3' day.

Increase in direct cost = 1500 × 3 = 4500

Decrease in indirect cost = 800 × 3 = 2400

Total cost of project after 4th stage of crashing = 56800 + 4500 – 2400 = 58900

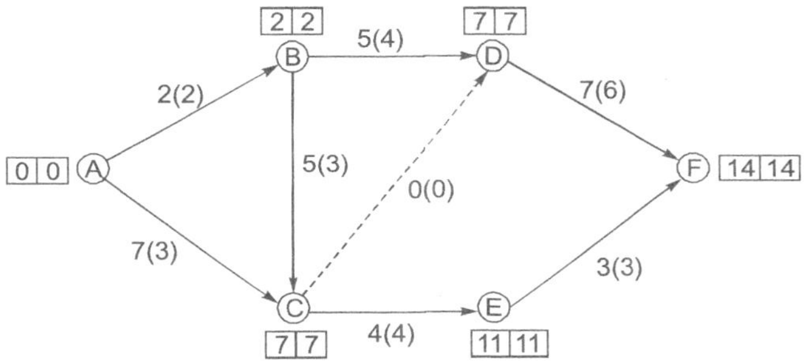

5th stage crashing: Now we have options for crashing only activities A-C, B-D, B-C which have combine cost slope = 1900

We can crash there activities by 1 day.

Increase in direct cost = 1900

Decrease in indirect cost = 800

Total cost of project = 58900 + 1900 – 800 = 60000

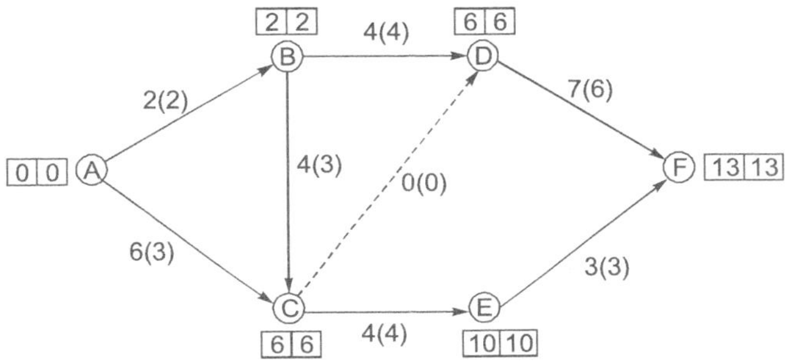

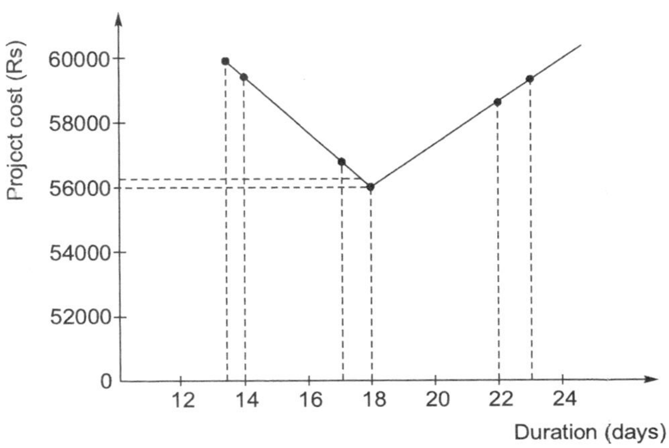

Thus optimum cost and duration of project is Rs. 56200 and 18 days respectively.

3.4 Explain the various methods of comparison of civil engineering project alternatives.

[10 marks : 2011]

Solution:

Methods of comparison of civil engineering project alternatives are as follows:

A. The Present Worth Amount

B. The Annual Equivalent Amount

C. The Capitalized Amount

D. The Rate of Return Method

A. **The Present Worth Amount**

(i) Indicate all expenses and benefits at appropriate point of time in the form of cash flow separately for each alternative. Represent downward arrows for expenses and upward arrow for benefits or savings.

(ii) Shift the amount at zero time by using proper interest formula and find total present worth amount of each alternative and compare.

B. **The Annual Equivalent Amount:** The annual equivalent amount is another basis for comparison that has characteristic similar to the present worth amount. The cash flow is converted into a series of equal annual amount by first calculating the present worth amount for the original series and then multiplying it with capital recovery factor $\left(\frac{A}{P}, i, n\right)$. The comparison is made by seeing the annual equivalent amount of each alternative.

C. **The Capitalized Amount**: The annual equivalent amount obtained is assumed to extend for infinite period and the capitalized amount is obtained by using A/i. The equivalent annual amount is calculated for replacement and maintenance cost and added to the initial cost of the alternative.

D. **Rate of Return Method:** It is defined as the interest rate that reduces the present worth amount of a series of receipts and disbursements to zero for each alternative. In economic terms, the fundamental concept of rate of return is that it is the rate of interest earned on the unrecovered balance of an investment so that the unrecovered balance is zero at end of investment's life.

3.5 Differentiate resource levelling and resource smoothing.

[5 marks : 2011]

Solution:

Resource Smoothing:

- In resource smoothing the total project duration is not changed but some of the activity start times are shifted by their available float so that more or less uniform demand is generated.
- Resource in this case is considered unlimited and critical activities remain unchanged.
- The total project completion time also remain unchanged.

Resource Levelling:

- In resource levelling, the activity start times are so rescheduled that the peak demand does not exceed the available limit of resource. If it does not give the desirable result by consideration of floats, the total project duration to minimum extent may be extended.
- Resource in this case is considered limited
- Critical path as well as the project completion time may be altered.

3.6 Details of construction project comprising of three activities are given in the following table:

S.N.	Activity	Unit	Estimated Quantity	Estimated rate per unit	Rate of award
1.	*A*	M^3	5000	1000	850
2.	*B*	*MT*	4500	4000	4200
3.	*C*	M^2	7000	5000	4800

Based on the information provided in the table, answer the following questions:

(i) What should be the cost of the project for which an "approval" is obtained from the competent authority before proceeding with the advertisement for the job etc.?

(ii) If at a certain point in time, the work, done for the activities *A*, *B* and *C* is 2700, 3000 and 4000 in the corresponding units, what is the percentage of the financial completion of the project?

(iii) Clearly state the assumptions in calculating (ii) above.

[2 + 4 + 2 = 8 marks : 2014]

Solution:

(i) Cost of project of approval,

$$= (5000 \times 850) + (4500 \times 4200) + (7000 \times 4800)$$
$$= \text{Rs. } 56{,}750{,}000$$

(ii)

S.N.	Activity	Unit	Estimated Quantity	Finished work	Rate	Revenue (lakh)
1.	*A*	M^3	5000	2700	850	22.95
2.	*B*	*MT*	4500	3000	4200	126.00
3.	*C*	M^2	7000	4000	4800	192.00

At a certain point of time total revenue earn

$$R = 22.95 + 126.00 + 192.00 = 340.95 \text{ Lakhs}$$

$\therefore$ Financial completion of the project

$$= \frac{\text{Revenue}}{\text{Total cost}} \times 100 = \frac{340.95}{567.50} \times 100 = 60.07\%$$

(iii) Assumptions:

1. Rate of award remains constant with respect to time and also the quantity of finished work.
2. Percentage of final completion is based as the amount of work completed rather than the actual completion of work i.e.

$$\text{Percentage of final completion at any point and time} = \frac{\text{Finished work} \times \text{Rate}}{\text{Estimated Quantity} \times \text{Rate}} \times 100$$

3.7 (i) Briefly explain five important factors which should be considered for selection of equipment for a construction project.

(ii) The operating time per hour for a bulldozer working in a construction project is 55 minutes to push sandy loam top soil, having swell factor as 1.24, for a hauling distance of 45 m. What will be the output of the bulldozer per hour? Assume the forward speed and the reverse speed of the bulldozer as 2.7 kmph and 5.4 kmph respectively. Rated mold board capacity in loose volume may be assumed as 3.6 cum. The gear shifting time of 0.2 minute may be assumed.

[10 marks : 2017]

Solution:

(i) **Factors which affects selection of equipment for construction project:**

1. **Operating cost :** The most efficient and therefore the most economical equipment is one whose operating cost is the minimum.
2. **Availability of equipment :** As far as practicable the equipment which is easily available in the market should be selected for the purpose because any delay in delivery may increase the cost of construction or cost of production substantially.
3. **Cost benefit analysis :** For various alternatives, cost benefit analysis must be made and selection is based on economics only.
4. **Study of site condition :** Topographical condition, type of soil, existing approach roads and other working conditions must be studied before making any final decision.
5. **Size of equipment :** Size of equipment should meet the demand of work. It is better to use more than one equipment of small size than using of one of large size.
6. **Suitability of equipment for future :** The equipment should be of versatile nature so that it can be used for other purposes which will mean higher demand and will bring high resale value.

(ii) Given,

Actual operating time per hour = 55 minutes

Swell factor = 1.24

Haul distance, $D = 45$ m

Forward speed, $F = 2.7 \text{ kmph} = \dfrac{2.7 \times 1000}{60} = 45 \text{ m/minute}$

Reverse speed, $R = 5.4 \text{ kmph} = 90 \text{ metre/minute}$

Gear shifting time, $G = 0.2$ minutes

Rated mold beard capacity in loose volume = 3.6 m^3

∴ Time required per trip in minutes (cycle time)

$$= \left(\frac{D}{F} + \frac{D}{R} + G\right) = \left(\frac{45}{45} + \frac{45}{90} + 0.20\right)$$

$$= 1 + 0.5 + 0.20 = 1.7 \text{ minutes}$$

We know,

Output of bulldozer in bank measure volume (cum/hr)

$$= \left\{\frac{\text{Rated mold board capacity in loose volume}}{\text{Swell factor}}\right\} \times \left\{\frac{\text{Actual operating time in minutes/hour}}{\text{Time required per trip in minutes}}\right\}$$

$$= \frac{3.60}{1.24} \times \frac{55}{1.7} = 93.93 \text{ cum/hr (bank measure)}$$

3.8 **What is Resource Levelling in Construction Project Management and how it is different than Resource Loading?**

[8 marks : 2018]

Solution:

Resource Levelling: In project management, resource levelling is defined by A guide to the project management body of knowledge (PMBOK guide) as "A technique in which start and finish dates are adjusted based on resource constraints with the goal of balancing demand for resources with the available supply.

In Resource levelling, the main constraint would be the resources and if the maximum demand on any resource is not to exceed a certain limit, the activities will then have to be rescheduled so that the total demand on the resource at any time will be within limit. The project duration is consequently exceeded.

Resource Loading: Resource loading also known as resource allocation simply means deciding what resources each activity of the project requires. For this resource usage profile i.e. histogram is drawn either on earliest start of an activity or latest start of each activity basis.

The process of resource loading consists of fitting activities into pattern of resource availability within stipulated time period and it is achieved by

1. Resource smoothening
2. Resource levelling
3. Main difference between resource loading and resource levelling is that resource levelling relates project requirements with available resource while resource loading is the process of allocation of resources to various activities.
4. In allocation/resource loading, we have to deal with manpower only while both time and resources are involved in resource levelling.

3.9 **Calculate the hourly tyre cost that should be part of machine operating cost if a set of tyres can be expected to last 5000 hr. Tyres cost ₹40000/- (per set of four). Tyre repair cost is estimated to average 16% of the straight-line tyre depreciation. The machine has a service life of 4 yr and operates 2500 hr per year. The discount rate is 8%. Compare both the cases, i.e., without time value and with time value of money.**

[20 marks : 2022]

Solution:

Hourly tyre cost (by not considering time value of money).

The hourly tyre cost is equal to the sum of hourly tyre use (replacement) cost and hourly tyre repair cost. The hourly tyre use cost is obtained by dividing the cost of a set of tyres by the life of tyres in hours. The hourly tyre repair cost is equal to 15% of the hourly depreciation (straight-line) cost of tyres. The total depreciation cost of a set of tyres over its estimated life is equal to its initial cost as its salvage value is assumed as zero.

$$\text{Hourly tyre cost} = (\text{hourly tyre use cost}) + (\text{hourly tyre repair cost})$$

$$= \frac{40000}{5000} + \frac{0.16 \times 40000}{5000} = ₹9.28/\text{hr}$$

Hourly tyre cost (considering time value of money)

$$\text{Machine service life} = 4 \times 2500 \text{ hr} = 10000 \text{ hr}$$

$$\text{Tyre life} = 5000 \text{ hr}$$

A second set of tyres will be purchased at the end of $\frac{10000}{5000} = 2$ years

$$\text{Hourly tyre repair cost} = \frac{0.16 \times 40000}{5000} = ₹1.28/\text{hr}$$

$$A_{1st\ set} = P\left[\frac{i(1+i)^n}{(1+i)^n - 1}\right] = 40000\left[\frac{0.08(1+0.08)^4}{(1+0.08)^4 - 1}\right]$$

$$= ₹12076.83/\text{years}$$

$$\text{First set cost per hour} = \frac{A_{1^{st}\ set}}{2500} = \frac{12076.83}{2500} = ₹\ 4.83/\text{hr}$$

$$A_{2nd\ set} = P'\left[\frac{i(1+i)^n}{(1+i)^n - 1}\right] = \frac{P}{(1+i)^2}\left[\frac{i(1+i)^n}{(1+i)^n - 1}\right]$$

$$= \frac{40000}{(1+0.08)^2}\left[\frac{0.08(1+0.08)^4}{(1+0.08)^4 - 1}\right]$$

$$= ₹\ 10353.94/\text{years}$$

$$\text{Second set cost per hour} = \frac{10353.94}{2500} = ₹ 4.14/\text{hr}$$

$$\text{Hourly tyre cost} = \text{Repair cost} + \text{1st set} + \text{2nd set}$$

$$= ₹\ (1.28 + 4.83 + 4.14)\ /\ \text{hr}$$

$$= ₹\ 10.25/\text{hr}$$

4. Construction Equipments

4.1 **Discuss in brief the working of concrete pumps.**

[10 marks : 2000]

Or

Enumerate and explain briefly the various types of concrete pumps widely used in the construction industry. Explain briefly the important guidelines to be followed for successful pumping of concrete.

[8 marks : 2008]

Solution:

There are three types of pump which are widely used in concrete industry i.e. piston pumps, pneumatic pumps and squeeze pressure pumps.

(i) **Piston pump:** It is a heavy duty, simple two-stroke mechanical reciprocating pump consisting of a receiving hopper, inlet and outlet valves, a piston and a cylinder. The pumping action starts with the suction stroke drawing concrete into cylinder as the piston moves backwards. During this operation the outlet valve is closed. On the forward or pressure stroke, the inlet valve closes and outlet valve opens to allow concrete to be pushed into delivery pipe. The pipeline is completely filled and concrete moves uniformly. The pump capacity can range from 15 m^3/h to 150 m^3/h. The normal distance to which concrete can be pumped is about 400 m, horizontally and 80 m vertically.

(ii) **Pneumatic pump:** The pneumatic pump normally uses a reblending discharge box at the discharge end to bleed off the air and tc prevent segregation and spraying.

(iii) **Squeeze Pressure pump:** It basically consists of three parts: a concrete receiving hopper, a controlling valve system and concrete transmission system. The concrete placed in the receiving hopper is fed by rotating blades into the flexible pipe connected to the pumping chamber, which is under vacuum of about 600 mm of mercury. The rotating rollers progressively squeeze the flexible pipes and force the concrete to move through the delivery pipe in a continuous flow.

The important guidelines to be followed for successful pumping of concrete are:

(i) A combined gradation of coarse and fine aggregates should be used that ensures no gaps in sizes that will allow paste to be squeezed through the coarser particles under the pressures induced in the line. This is the most often overlooked aspect of good pumping.

(ii) A minimum pipe diameter of 5 inch should be used.

(iii) The line should always be lubricated with cement paste or mortar before beginning the pumping of concrete.
(iv) A steady uniform supply of concrete should be ensured with a slump of between 50 mm and 125 mm, as it enters the pump.
(v) The aggregates should be presoaked always before mixing them in the concrete to prevent their soaking up mix water under the imposed pressure.
(vi) The use of reducers in the conduit line should be avoided.
(vii) Aluminium lines should never be used in concrete pumping as the aluminium particles will be scraped from the inside of the pipe as the concrete moves through and will become part of the concrete. Aluminium and cement react, liberating hydrogen gas that can rupture the concrete with disastrous results.

4.2 **List out the basic parts and operations of a Hoe and state its applications.**

[10 marks : 2001]

Solution:

Hoe is an excavating equipment of the power shovel group and is used primarily to excavate below the natural surface of the ground on which the machine rests A hoe is sometimes referred to by other names such as back hoe or back shovel.

Basic parts and operations of a hoe:

The basic parts of a hoe are illustrated in the figure shown.

The machine is placed in operation by setting the boom at the desired angle. The dipper is moved to the desired position. The free end of the boom is lowered down by releasing the tension in the hoist cable until the dipper teeth engages the material to be dug. As the cable is pulled in, the dipper is filled up. Then the boom is raised and swing to the dumping position.

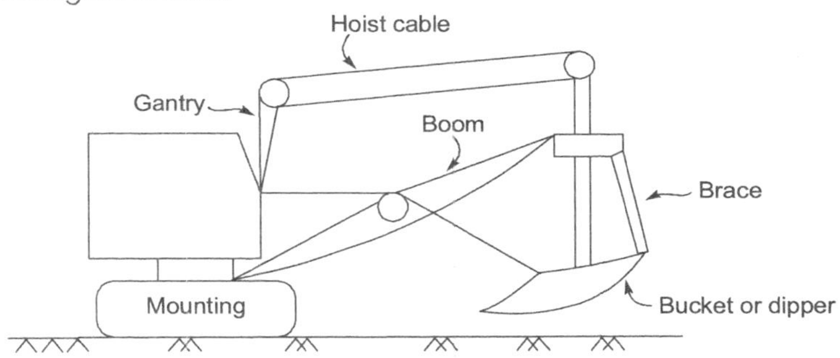

Applications of hoe:

1. It can be used to excavate below the natural surface of the ground on which the machine rests.
2. It can dig trenches, footings or basements and general grading work which requires precise control of depths.
3. It can operate on close range work and dump into trucks.
4. It can penetrate easily into toughest materials to be dug. However, these are not production excavation machines. They are designed for mobility and general purpose work.

4.3 **List out the basic parts and explain operation of power shovel with a neat sketch.**

[10 marks : 2003]

Or

Describe a basic shovel by means of a simple diagram and name the main parts. Also indicate its suitability in field application.

[10 marks : 2004]

Solution:

Power shovel is a construction equipment whose purpose is to excavate the earth and load it into trucks or other hauling equipments waiting nearby. The basic parts of a power shovel consist the mounting (crawler truck or rubber tyred wheel), cab, boom, dipper stick, dipper and hoist line. These parts are illustrated in the figure.

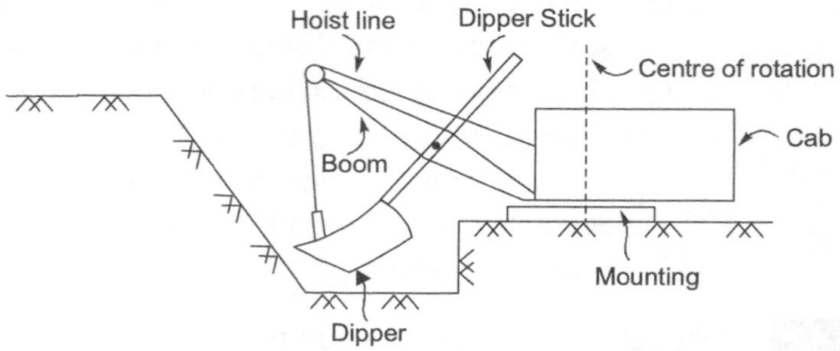

For operation, first step is to release the hoist to bring the dipper down and move the stick in a vertical plane. Now the dipper is moved forward and downward with the cutting teeth pointing towards the face to be excavated. The downward force is applied to the tip of the dipper through the dipper stick.

The depth of cut and angle of digging is controlled by the job conditions. When the dipper is filled with the load, the next step is to pullout the dipper by applying the required tension through the hoisting line and also to hoist it till the machine can be swing around for dumping. During swinging the position of the dipper is so adjusted that it comes in most suitable dumping position. To empty the dipper, the door is opened by unlatching it through the trip arrangement. The cycle is thus repeated again and again.

Power shovels are used predominantly for hard digging above track level and for load haul units. They are capable of excavating all classes of earth, except the solid rock without prior loosening.

4.4 **List four different types of vibrators used for compaction of concrete and briefly explain each of them.**

[10 marks : 2003]

Or

Briefly describe five different types of vibrators used in cement concrete making industry.

[15 marks : 2007]

Solution:

Vibrations are invariably used in compaction of cement concrete. Some of the vibrators are described below:

(i) **Internal vibrator:** It is the most commonly used vibrator in cement concrete making industry. It is also called 'needle vibrator', immersion vibrator' or 'poker vibrator. It is consisted for a power unit, a flexible shaft and a needle. The power unit may be electrically driven or operated by patrol engine or air compressor. The vibrations are caused by eccentric weights attached to the shaft or the motor or to the rotor of a vibrating element. The frequency of vibration varies upto 12,000 cycles of vibration per minute. The needle diameter varies from 20 mm to 75 mm and its length varies from 25 cm to 90 cm. The bigger needle is used in the construction of mass concrete dam. Internal vibrators are portable and can be shifted from place to place very easily during concreting operation. They can also be used in difficult positions and situations.

(ii) **External Vibrator:** It is also known as form work vibrator. These are used for concreting columns, thin walls or in the casting of precast units. The machine is clamped on to the external wall surface of the formwork. The vibration is then given to the formwork so that the concrete in the vicinity of the shutter gets vibrated. This method of vibrating concrete is particularly useful and adopted where reinforcement, lateral ties and spacers interfere to much with the internal vibrator. These give a good finish to the concrete surface. Since the vibration is given to the concrete in directly through the formwork, these consume more power and the efficiency of external vibrator is lower than the efficiency of internal vibrator.

(iii) **Table Vibrator:** This is the special case of formwork vibrator, where the vibrator is clamped to the table or table is mounted on springs which are vibrated transferring the vibration to the table. They are commonly used for vibrating concrete cubes. This is adopted mostly in the laboratories and in making small but precise prefabricated RCC members.

(iv) **Platform Vibrator:** It is nothing but a table vibrator, but is larger in size. This is used in the manufacture of large prefabricated concrete elements such as electric poles, railway sleepers, prefabricated roofing elements, etc. sometimes, the platform vibrator is also coupled with jerking or shock giving arrangements such that a thorough compaction is given to the concrete.

(v) **Surface Vibrator:** Surface vibrators are sometimes known as 'screed board vibrators'. A small vibrator placed on the levelling of thin concrete members such as floor slabs, roof slabs and road surface. Mostly, floor slabs and roof slabs are so thin that internal vibrator or any other type of vibrator cannot be easily employed. In such cases, the surface vibrator can be effectively used. In general surface vibrators are not effective beyond about 15 cm. In the modern construction practices like vacuum dewatering technique or slip form paving technique, the use of screed board vibrator are common feature.

4.5 Discuss in brief the various operations involved in a concrete construction project.

[10 marks : 2004]

Solution:

The various operations in production of concrete include:

(i) **Storing of materials:** In general, cement should be stored in such a way that it does not come in contact with moisture. All aggregates must be clean and free from dust, dirt, mud, silt, clay etc. There should not be presence of vegetable matter and animal refuse. These impurities do not allow the proper bond between cement paste and aggregates, consequently the strength of concrete is reduced. The aggregate should be stored on hard and dry base or on platforms of planks, GI sheets etc. Different size of aggregates should be piled separately by constructing different compartments. Water for concrete mixing at site should be stored in masonry built tanks or other clean containers. The tanks should be always thoroughly and regularly cleaned and water is not stored over longer period but stored a day in advance only.

(ii) **Batching or measurement of materials:** A proper and accurate measurement of all the materials used in the production of concrete is essential to ensure uniformity of proportions and aggregate grading in successive batches. For most of the large and important jobs the batching of materials is usually done by weighing. In weigh batching, the weight of surface water carried by the wet aggregate must be taken into account. The factors affecting the choice of proper batching system are: (i) size of job, (ii) required production rate, and (iii) required standards of batching performance. The batching equipment falls into three general categories, namely, manual, semiautomatic and fully automatic systems.

(iii) **Mixing of concrete materials:** The object of mixing is to coat the surface of all aggregate particles with cement paste and to blend all the ingredients of concrete into a uniform mass. The mixing action of concrete thus involves two operations : (i) a general blending of different particle sizes of the ingredients to be uniformly distributed throughout the concrete mass, and (ii) a vigorous rubbing action of cement paste on to the surface of the inert aggregate particles. Concrete mixing is normally done by mechanical means called mixer, but sometimes the mixing of concrete is done by hand. Machine mixing is more efficient and economical compared to hand mixing.

(iv) **Transportation of concrete:** Concrete from the mixer should be transported to the point where it has to be placed as rapidly as possible by a method which prevents segregation or loss of ingredients. The concrete has to be placed before setting has commenced. Attempts have been made to limit the time lapse between mixing and compaction within the forms. The specifications, however, permit a maximum of two hours between the introduction of mixing water to the cement and aggregates, and the discharge, if the concrete is transported in a truck mixer or agitator. In the absence of an agitator, this figure is reduced to one hour only. The principal methods of transporting concrete are pans, wheel barrows, power buggies, tower bucket, crane bucket, dump truck, chutes, concrete pumps, transit mixer, agitating truck and belt conveyor.

(v) **Placing of concrete:** The methods, used in placing concrete in its final position have an important effect on its homogeneity, density and behaviour in service. The same care which has been used to secure homogeneity in mixing and the avoidance of segregation in transporting must be exercised to preserve homogeneity in placing. The concrete should be placed in its final position rapidly so that it is not too stiff to work.

(vi) **Compaction of concrete:** The process of inducing a close packing of the solids particles in freshly mixed concrete during placement by the reduction of volume of voids usually by tamping, vibration, centrifugation, subjecting concrete to high pressure or shock is termed as compaction. Compaction removes the entrapped air which is responsible for segregation.

(vii) **Finishing of concrete:** Finishing is defined as the process of leveling and smoothing the top surface of freshly placed concrete to achieve the desired appearance. The various methods, used for the purpose of finishing the top surface are screeding, floating and trowelling.

(viii) **Curing of concrete:** Curing is defined as the process of maintaining humidity and moisture at favourable temperature of freshly placed concrete during some definite period following placing, casting and finishing to attain full hydration of the cement.

Some methods of curing are:

(a) Ponding of water over the concrete surface after it has set.
(b) Covering the concrete with wet straw or damp earth.
(c) Sprinkling of water
(d) Covering the surface with water proof paper
(e) Membrane curing of concrete
(f) Chemical curing
(g) Steam curing

4.6 **Explain about tilting type of mixers and pan type of mixers used in concrete mixing.**

[10 marks : 2006]

Solution:

Tilting Mixers: The size of a tilting type mixer is designated by a number representing its nominal mix batch capacity in litres per batch followed by letter '*T*' i.e. 85*T*, 100*T*, 140*T*, 200*T*, etc. In the tilting type mixer, the chamber (drum) which is generally bowl-shaped or double-conical frustum type is tilted for discharging. The efficiency of the mixing operation depends upon the shape and design of the vanes (blades) fixed inside the drum. These vanes direct the concrete into tracing a circulatory path. In addition, there is vertical free falling action due to gravity. The mixed concrete is discharged from the open top of the drum by tilting it downwards. The discharge action is always good as all concrete can be tipped out rapidly under gravity in an unsegregated mass as soon as the drum is tilted. For this reason tilting drum mixers are preferable for the mixes of low workability and for those containing large-size aggregates. The only disadvantage seems to be that a certain amount of mortar adheres to the drum and is left out in the drum itself during discharging operation. Therefore, before the beginning of mixing the first batch of concrete, a certain amount of mortar is mixed in the mixer. This process is called **buttering** the mixer.

Pan Mixers: The pan mixer consists essentially of a circular pan rotating about a vertical axis. One or two stars of paddles also rotate in the pan about a vertical axis not coincident with the axis of the pan.

In some types, the pan is static and the axis of star trave s along a circular path about the axis of the pan. In other types, the paddles are stationary and the pan rotates about the vertical axis. In either case, the relative movement between the paddles and the concrete is same, and concrete in every part of the pan is thoroughly mixed. There is another set of blades called scraper blades, which prevents the sticking of the mortar to the pan sides by continuous scrapping. The paddle height can be adjusted to prevent a permanent coating of mortar forming on the bottom of pan. The mix is discharged through a central hole at the bottom of the pan. The pan mixer is generally not mobile and is therefore used either as a central mixing plant on a large concrete project or at a precast concrete factory. This mixer is particularly efficient with stiff and cohesive mixes. Pan mixers are extensively used in the laboratories for mixing small quantities of concrete of consistent quality, because of the efficient scrapping arrangement. A pan mixer consisting of a bowl-and-stirrer working principle of the cake mixer is sometimes used for mixing the mortar.

4.7 **Suggest any one or more of the following compacting equipments, which are most suitable for effective compaction in respect of each one of the gravel, sand, silt and clay:**

(i) Sheepsfoot Roller **(ii) Tamping Roller**

(iii) Vibratory Roller **(iv) Pneumatic Tired Roller**

Also indicate which method of compaction is adopted in these four compacting equipments.

[7 marks : 2008]

Solution:

(i) **Sheeps foot roller:** It is suitable for compacting all fine grained materials like silt and clay, but is generally not suitable for use on cohesionless granular materials like gravel and sand. Sheepsfoot rollers produce kneading action and a pressure to mix and compact the soil from the bottom to the top layer.

(ii) **Tamping roller:** It is suitable for compacting all fine-grained soils like silt and clay. As a tamping roller moves over the surface, the feet penetrate the soil to produce a kneading action and a pressure to mix and compact the soil from the bottom to the top of the layer. With repeated passages of the roller over the surface, the penetration of the feet decreases until the roller is said to walk out of the fill. Because the pads are tapered, a tamping foot roller can walk out of the lift without fluffing the soil.

(iii) **Vibratory roller:** Certain types of soil such as sand, gravel and relatively large shot rock respond quite well to compaction produced by a combination of pressure and vibration. When these materials are vibrated, the particles shift their positions and nestle more closely with adjacent particles to increase the density of the mass. Vibratory rollers create impact forces and these forces result in greater compacting power than an equivalent static load. The impact forces are higher than the static forces because the vibratory drum converts potential energy into kinetic energy.

(iv) **Pneumatic roller:** Because of the heavy loads and high tyre pressures, they are capable of compacting all types of soils to greater depths. They apply the principle of kneading action to effect compaction below the surface.

4.8 **(i) Briefly explain the different types of bulldozers according to their uses.**

(ii) Explain the Derrick crane with a neat sketch.

[6 + 4 marks : 2009]

Solution:

(i) **Classification of bull dozers:**

(a) **According to the direction of blades:** When the cutting blade is set perpendicular to the direction of travel, it is called a bull dozer. Bull dozer pushes the earth forward and dump it to some place. When the cutting blade is set at an angle with the direction of travel it is called an angle dozer. Angle dozer is used for pushing the earth to the side, right or left.

(b) **According to control of blade:** Based on the method of controlling (lowering or raising) of blade, a bull dozer may be classified as cable controlled bull dozer or hydraulically controlled bull dozer. Cable controlled bull dozer is simple to install and operate. It is also easy to repair the controls of cable controlled bull dozer. Hydraulically controlled bulldozer is able to produce a high down pressure on the blades to force them into the ground. It is also able to maintain a precise setting of the position of the blade.

(c) **According to Mountings:** Based on mountings, a bull dozer may be classified as crawler mounted bull dozer and wheel mounted bull dozer. Crawler mounted bull dozer is able to deliver greater tractive force especially when operating on loose or muddy soil. It is able to travel over very soft soil. The speed of crawler mounted bull dozer does not exceed 12 kmph normally. Wheel mounted bull dozer can travel at high speeds sometimes exceeding 50 kmph. It is able to travel on paved highways without damaging the surfaces.

(ii) **Derrick cranes** are also known as **stationary cranes**. Derrick cranes consist of a mast, a boom and a bull wheel on which the boom rotates about a vertical axis and guys or supporting members. These cranes are either electrically operated, diesel operated or diesel-electrically operated. A typical arrangement of a derrick crane is shown in figure:

The **power driven cranes are mainly of two types i.e. (a) Guy derrick (b) stiff-leg derrick.** The guy-derrick has a small mast supported by a number of guys and a boom pivoted at the lower end of the mast. The boom can revolve through 360°. This crane is used for heavy loads upto 200 tonnes. When the loads become less than 50 tonnes, guy ropes are replaced by trussed structure which becomes stiff-leg derrick. It consists of a mast, which can rotate and a boom. A stiff leg derrick is used for loads from 7 to 50 tonnes.

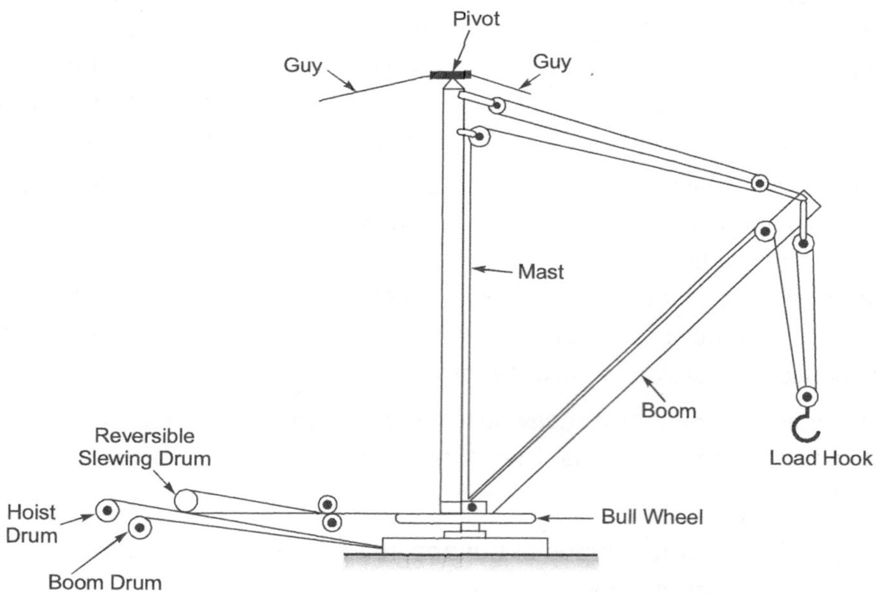

4.9 A 1/2 cubic metre short boom dragline having ideal output of 150 m^3/hr is to be used to excavate hard tough clay. The effect of the depth of cut of 5.0 m and angle of swing of 120° shall be 0.89. The operating factor shall be 50 min/hour. Determine the probable output of the dragline.

[4 marks : 2010]

Solution:

$$\text{Ideal output of dragline} = 150 \text{ m}^3/\text{hr}$$
$$\text{Depth-Swing factor of dragline} = 0.89$$

Operating factor of dragline = 50 min/hour

∴ Probable output of dragline = Ideal output × Depth & swing factor × Operating factor

$$= 150 \times 0.89 \times \frac{50}{60} = 111.25 \text{ m}^3/\text{hr}$$

4.10 **Indicate the performance of power shovel, back hoe, dragline and clam-shell for the following conditions in terms of very good, good, fair or poor:**

(i) Excavation in hard soil or rock

(ii) Excavation in wet soil or mud

(iii) Loading efficiency

[6 marks : 2010]

Solution:

Comparison between the different types of excavating equipment is tabulated below:

S.No.	Condition	Power shovel	Back hoe	Dragline	Clamshell
(i)	Excavation in hard soil	Good	Good	Poor	Poor
(ii)	Excavation in wet soil or mud	Poor	Poor	Fair	Fair
(iii)	Loading efficiency	Very Good	Good	Fair	Poor

4.11 **Describe the method of conveying materials with belt conveyors. List the components. State the advantages.**

[5 marks : 2011]

Solution:

Conveying is the continuous transportation of materials as distinct from intermittent transportation which implies transport in a successive stages. Conveyor is an equipment which is capable of carrying material in a continuous stream usually having as its distinguishing feature same kind of an endless chain or belt which by its motion constitutes the continuous carrying arrangement of the conveyor. Aerial transportation through ropeways or aerial tramways is also sometimes included in conveying equipment. Conveying may involve a horizontal or inclined or vertical movement. Generally when the movement is along either horizontal or inclined, the equipment is called conveyor and if it's motion is predominantly vertical it is called elevator.

Belt conveyor system provides the most satisfactory and economical method for handling and transporting materials like earth, sand, gravel, crushed stones, ores, cement, concrete etc. This system provides continuity of supply from the source to the work-site there by eliminating the use of trucks and the haulage roads.

A conveyor system for transporting material for a short distance may be portable or a fixed limit depending upon the requirements. If a belt conveyor is required to transport materials over a considerable distance upto several kilometers, the system should consist of a number of different flights as there is a limit to the maximum length of the belt (approx 335 m). Each flight is a complete conveyor unit which discharges its load onto the trail end of the succeeding unit.

Components of a belt conveyor are:

(a) Continuous belt
   - (i) Standard
   - (ii) Shock pad
   - (iii) Stepped pad
   - (iv) Stepped ply

(b) Idlers

(c) Driving and tail pulleys

(d) feeders

(e) Trippers

Advantages of belt conveyor:

(i) Continuous and uniform haulage
(ii) Low maintenance cost
(iii) Low labour requirement
(iv) High reliability
(v) Ability to cross adverse terrain
(vi) Excellent safety records.

4.12 List the different types of equipments used for excavation work.

[5 marks : 2011]

Solution:

Types of equipment used for excavation work

(a) Tractors
   (i) Crawler tractor (ii) Wheel tractor

Two wheel tractor or four wheel tractor

(b) Bulldozer
(c) Power shovel
(d) Dragline
   (i) Crawler mounted dragline (ii) Wheel mounted dragline
   (iii) Truck mounted dragline
(e) Clamshells
(f) Hoe
(g) Scraper
(h) Trenching machines
   (i) Wheel type trenching machine (ii) Ladder type reaching machine

4.13 Calculate the time required to grade and finish 30 km of road formation of 9.0 m width for two-lane road with motor grader having width of 3.0 m, using six passes with speed for each of the successive two passes as 5 kmph, 7 kmph and 9 kmph respectively. Assume machine efficiency based on operator skill, machine characteristics and working conditions as 80%.

[6 marks : 2012]

Solution:

Machine efficiency, $\eta = 80\% = 0.8$

Distance = 30 km

Width = 9 m

Motor grader width = 3 m

Speed for successive two passes as 5 kmph, 7 kmph and 9 kmph

$\therefore$ Number of passes = 6

Now, time required to complete a grade and finish 30 km of road formation of width 9 m is

$$= \frac{3}{\eta}\times\left[\frac{30}{5}\times 2+\frac{30}{7}\times 2+\frac{30}{9}\times 2\right] = \frac{3}{0.8}\times[27.238] = 102.143 \text{ hrs}$$

4.14 Enlist major concreting equipments required to carry out following operations:
Mixing, transportation delivery and compacting equipment.

[4 marks : 2012]

Solution:

The **mixing operation** consists essentially of rotation or stirring, the objective being to coat the surface of all aggregates particles with cement paste, and to blend all the ingredients of concrete into a uniform mass. The usual type of mixer is batch mixer.

There are four types of batch mixers:

(a) Tilting drum mixer
(b) Non-tilting drum mixer
(c) Pan mixer
(d) Dual drum mixer

Major equipments to carry out transportation are:

(a) By buckets
(b) By dump trucks
(c) By belt conveyors
(d) By monorail
(e) Chutes
(f) Truck-mixers
(g) Concrete dumpers

Effective compaction of material is an essential part of construction engineering operations. The equipments required for these works range from smallest vibrating rammers to the largest rollers of vibrating type.

Types of compactors:

(a) Static smooth wheeled rollers
(b) Sheep foot or Pad foot roller
(c) Pneumatic tyred rollers
(d) Vibratory rollers
(e) Tampers and vibrating plates

4.15 Calculate number of transit mixers (TM) required for transporting concrete from central batching plant to site. The cycle time data of a 6 m^3 typical transit mixers is given below:
Loading time of TM = 6.0 minutes.
Travel time of loaded TM to site = 30 minutes
Average waiting time at site = 5.0 minutes
Discharge time of concrete at site through concrete pump = 15 minutes.
Travel time for return trip = 24 minutes.
If the central batching plant having average output of 60 m^3/hr is to run continuously, work out the requirement of no. of concrete pumps and TM.

[6 marks : 2012]

Solution:

$$\text{Output of central batching plant} = 60 \text{ m}^3/\text{hr}$$

$$\text{Total cycle time of a 6 m}^3 \text{ transit mixer} = 6 + 30 + 5 + 15 + 24 = 80 \text{ minutes}$$

$$\therefore \quad \text{Output of transit mixer per hour} = \frac{6}{80} \times 60 \text{ m}^3/\text{hr} = 4.5 \text{ m}^3/\text{hr}$$

$$\text{Hence, number of transit mixers required} = \frac{60}{4.5} = 13.33 \simeq 14 \text{ TM}$$

4.16 Name various types of Earth Excavating Equipments and give their corresponding digging depth.

[4 marks : 2012]

Solution:

The various type of earth excavating equipments are:

1. **Bulldozer:** Bulldozers are very efficient excavating tools for short haul applications upto 100 m and as auxiliary machine to other construction equipments. In addition to excavating and hauling, many other functions are also performed by bulldozers. Therefore they are called versatile equipments.
2. **Power Shovel:** It is a construction equipment whose purpose is to excavate the earth and load it into trucks or other hauling equipment nearby they are capable of excavating all classes of earth except the solid rock without prior loosening.

3. **Dragline:** Since the basic character of the machine is dragline the bucket against the material to be dug, it is called as dragline. A dragline is used to excavate the earth and load it into hauling units such as trucks or tractor pulled wagons or to deposit it into dams/embankments or spoil banks near the pit from which it is excavated. Primarily, the functions of a dragline are same as that of a power shovel but in some projects, a dragline has distinct advantages compared with a power shovel because of its long light boom.

   The dragline can operate in a depth approximately upto 1/3 of its boom length. Its boom length varies from 21 m to 36 m.
4. **Clamshells:** Clamshell is a machine having most of the characteristics of dragline and crane in common. Digging is done like a dragline and once the bucket is filled, it works like a crane.
5. **Hoe:** Hoe is an excavating equipment of the power shovel group. Since the digging mechanism resembles to an ordinary garden hoe, it is named as Hoe. They are fitted with buckets and their corresponding digging depth capability is from 5 m to a maximum of 9.5 m.
6. **Scraper:** Scraper is a machine which can scrape the ground and load it simultaneously, transport it over the required distance, dump at the desired place and then spread the dumped material over the required area in required lvel and return to the pit for the next cycle.
7. **Trenchers:** Trenchers are the equipments used for excavating trenches or ditches of variable width and depth.

4.17 **Describe briefly five different types of earthwork equipment (rollers) used for compacting soils.**

[5 marks : 2014]

Solution:

Soil compaction is achieved by rolling, ramming or by vibration.

The following equipments are used for compacting soils:

1. **Smooth wheeled rollers:** The smooth wheeled rollers are suitable to roll a wide range of soils, preferably granular soils and pavement materials for the various layers. These are particularly found to be useful in compacting soils and other materials where a crushing action is advantageous.
2. **Pneumatic tyre roller:** In Pneumatic tyre rollers, in addition to the direct pressure due to rolling, also a slight kneading action on the soil. Hence these are considered to be most suitable to compacting nonplastic silts and fine sands.
3. **Sheepfoot rollers:** Sheepfoot rollers are considered more suitable to compact clayey soils. During rolling operation the soil under the projecting feet is compacted and also there is a considerable kneading action on the soil.
4. **Rammers:** Rammers are useful to compact relatively small areas and where the rollers cannot operate such as compaction of trenches, foundation and slopes.
5. **Vibrators:** Vibrators are most suited for compacting dry cohesionless granular materials. There are also vibrator mounted rollers to give the combined effects of rolling and vibration. Vibratory rollers are advantageously used in compacting a wide range of materials.

4.18 **Write in brief the principles of Dragline and Clamshell used as excavation equipments, the detail of their components and neat sketches showing their parts. How both equipments can compared?**

[20 marks : 2018]

Solution:

Dragline: The basic character of the machine is dragging the bucket against the material to be dug; it is called as dragline.

- Dragline in used to excavate the earth and load it into hauling units such as trucks or tractor pulled wagons or to deposit it into banks, embankments near the pit from which it is excavated.

- Dragline has a distinct advantage compared to power shovel because of its light boom.

Advantages of Dragline:

- A dragline usually doesn't have to go into a pit or hole for excavating the earth. It is more useful when earth is it to be removed from a ditch, canal or pit containing water.
- When the excavated earth is to be deposited on near by banks or dams, it is better to use dragline with a long boom.
- Dragline is excellent for excavating trenches without shoring.

  The size of dragline is expressed by the size of its bucket, measured in cubic meter.
- Clamshells is a machine having most of the characteristics of dragline and crane in common.
- Digging is done like a dragline and once the bucket is filled it works like a crane.
- It consists of two halves which are hinged at the top.
- It is classified into light and heavy bucket.
- It is primarily used for handling loose materials such as sand, gravel, crushed, stone, coal, etc.
- It is suited for vertical lifting materials from one location to another.
- The limits of vertical movement depends on the length of the crane boom.

Comparison between Clamshell and Dragline

S.No.	Parameter	Clamshell	Dragline
1.	Cycle time	Less	Relatively more
2.	Capacity	$20T$ to $40T$	$5T$ to $30T$
3.	Depth of excavation	Low	High
4.	Operation	Left	Pull
5.	Efficiency	Good	Relatively low
6.	Suitability in hard soils	Poor	Poor

4.19 Briefly explain at least five different types of vibrators used in cement concrete making industry.
[20 marks : 2019]

Solution:

Vibrators

These are the mechanical devices which are used to compact concrete in the formwork. The advantages of vibrators over hand methods are as follows:

(i) It is possible by means of vibrators to make a harsh and stiff concrete mix, with a slump of about 40 m or less, workable.

(ii) The quality of concrete can be improved by use of vibrators as less water will be required or in other way, economy can be achieved by adopting a leaner mix when vibrators are used.

(iii) The use of vibrators results in the reduction of consolidation time. Hence the vibrators are used where the rapid progress of work is of great importance.

(iv) With the help of vibrators, it is possible to deposit concrete in small openings or places where it will be difficult to deposit concrete by hand methods.

Following are the four types of vibrators:

1. Internal or immersion vibrators
2. Surface vibrators
3. Form or shutter vibrators
4. Vibrating tables

1. **Internal or immersion vibrators:** These vibrators consist of a steel tube which is inserted in fresh concrete. This steel tube is called the poker and it is connected to an electric motor or a petrol engine through a flexible tube. They are available in sizes varying from 40 mm to 100 mm diameters and the size is decided by keeping in mind the spacing between reinforcing bars in concrete. The frequency of vibration is about 300 to 600 rpm.

   The poker vibrates while it is being inserted. The internal vibrators should be inserted and withdrawn slowly and they should be operated continuously while they are being withdrawn. Otherwise holes will be formed inside the concrete. The vibrator can be placed vertically or at a slight inclination not exceeding 10° to the vertical with a view to avoid flow of concrete due to vibration into the mould and consequent scope of segregation. Hence skilled and experienced men should handle internal vibrators. Theses vibrators are more efficient than other types of vibrators and hence they are most commonly used.

2. **Surface vibrators:** These vibrators are mounted on platform or screeds. They are used to finish concrete surfaces such as bridge floors, road slabs, station platform, etc. These vibrators are found to be more effective for compacting very dry concrete mixes because the vibration acts in the same direction of gravity and the concrete is compacted in a confined zone. These vibrators also cause movement of fine material to the top and it aids in finishing operations. However the movement of excess fine material at top will not be desirable for plastic mixes as the wearing resistance of such fine material is very low.

3. **Form or shutter vibrators:** These vibrators are attached to the formwork and external centering of walls, columns, etc. The vibrating action is conveyed to the concrete through the formwork during transmission of vibrations. Hence, they are not generally used. But they are very much helpful for concrete sections which are too thin for the use of internal vibrators.

   These vibrators require more power because of loss of some power in vibrating the rigid shutters. They are also heavy and hence they cannot be clamped at as many points as possible for uniform compaction of concrete. The compaction by these vibrators is found to be effective only upto a distance of about 450 mm form the face of the formwork.

4. **Vibrating tables:** These are in the form of a rigidly built steel platform mounted on flexible springs and they are operated by electromagnetic action or electric motors. They are found to be very effective in compacting stiff and harsh concrete mixes and hence they are invariably used in the preparation of pre-cast structural products in factories and test specimens in laboratories.

   The tables are vibrated either mechanically or by placing the springs under the supports of tables. The frequency of vibrations varies form 3000 to 7200 vibrations per minute. The two parameters of vibrations are frequency and time and they are related as follows:

   $$\text{Frequency} \propto \frac{1}{\text{Time}}$$

   It means that the frequency is inversely proportional to the time of vibration. In other words, if frequency is more, the consolidation of concrete will be achieved in less time and vice-versa.

4.20 What is a crane? How is it used in the construction industry? Briefly explain three different types of cranes that are being used in construction works.

[20 marks : 2020]

Solution:

A crane is a tower or derrick that is equipped with cables and pulleys that are used to lift and lower material. They are commonly used in the construction industry and in the manufacturing of heavy equipment. Cranes for construction are normally temporary structures, either fixed to the ground or mounted on a built vehicle.

The crane is an essential piece of equipment for heavy construction work and lifting tasks of all kinds. Equipped with cables and pulleys, and based upon the application of fundamental mechanical principles, a crane can lift and lower loads well beyond the capabilities of human construction workers.

Crane design has evolved to meet the demands of a variety of industrial needs, and modern cranes often coordinate simple systems to achieve complex lifting tasks – sometimes in environments which would be dangerous for human workers.

Types of Cranes:

1. **Tower Crane:**

   The tower crane is a form of balance crane which is commonly used on urban construction sites. This machine is anchored to the ground and provides an optimum blend of height and lifting capability which is often deployed to erect multi-storey city buildings.

   Two horizontal arms jut from a central tower, with one used to suspend the heavy loads to be lifted, and the other fitted with heavy concrete blocks as a counter-weight. A tower crane is controlled by a driver who either sits high above in a small cabin located at the top of the tower, or else uses a remote-control system to operate his machine from the ground.

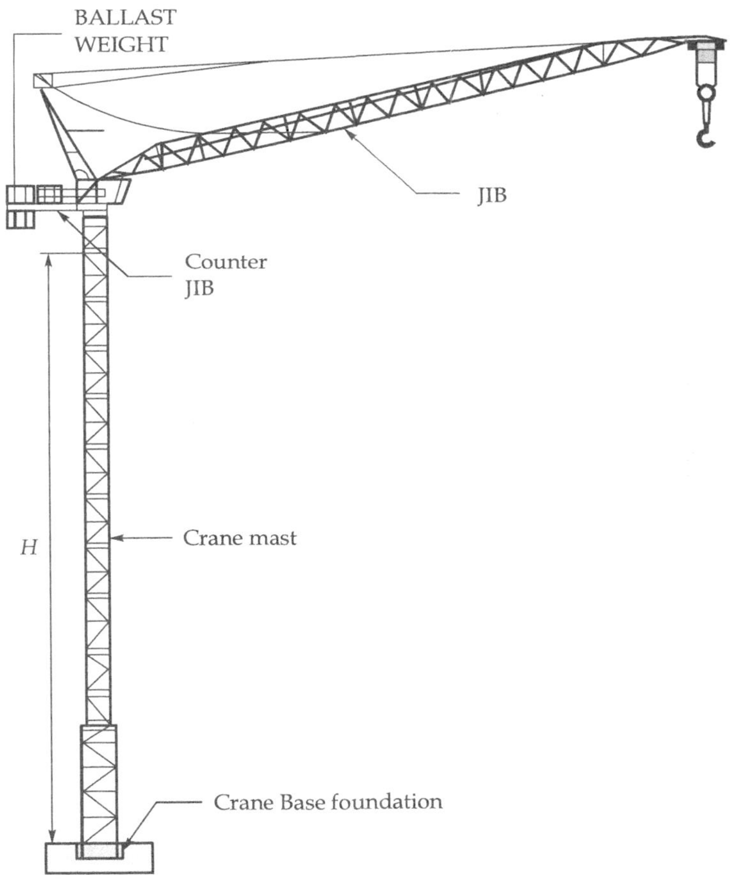

2. **Mobile Cranes:**

   Mobile cranes are commonly mounted on wheeled vehicles, but cranes used for railway work are adapted to travel on rail tracks, and various floating cranes can be attached to barges when used for construction work on bridges and waterways.

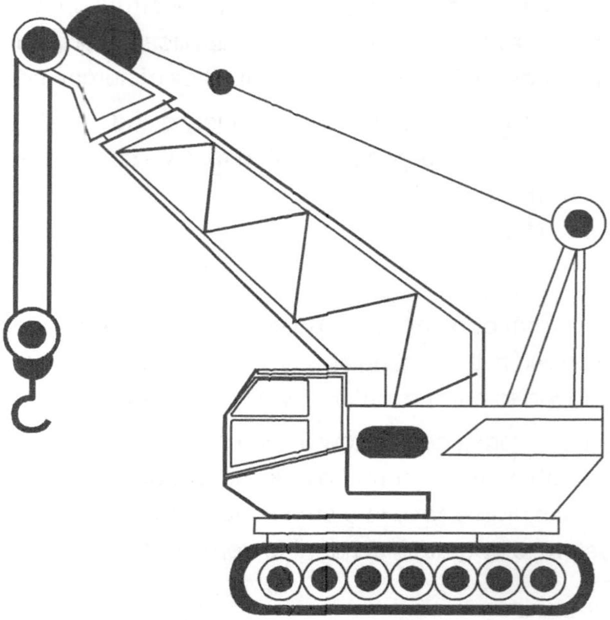

Many types of crane are mobile, including Mobile Tower Cranes. Different types of crane serve a temporary purpose, and a mobile crane may be little more than a robust steel boom fitted to a transportable platform.

The lifting arm is typically hinged to allow it to be hoisted and lowered as required. This is usually achieved by cable systems or hydraulic mechanisms, and the whole mobile structure can be fitted with outriggers to provide further stability during on-site operations.

3. **Gantry Cranes:**

Using a hoist installed in a fixed machinery housing, or otherwise able to slide along a rail framework, this crane employs a strong overhead gantry to lift and maneuver extremely heavy industrial loads.

Gantry cranes and other so-called 'overhead' cranes – which also carry suspended loads in a similar fashion – are widely used in factories and shipyards and similar commercial locations where their robust qualities make them essential equipment.

4.21 **(i) Briefly discuss various factors affecting the output of power shovel to excavate earth.**
(ii) What are the precautions to be taken for labour safety during excavation works?

[10 +10 = 20 Marks : 2021]

Solution:

(i) **The various factors affecting the output of power shovel:**

1. **Class of Material:** Production decreases with increase in hardness of material.
2. **Height of Cut:** If the depth of fall from which a shovel is excavating material is too shallow, cycle time increases and it will be difficult or impossible to fill the bucket in one pass up the face, consequently the production is reduced.
3. **Angle of Swing :** The angle of swing of a shovel is the horizontal angle, expressed in degrees, between the position of the bucket when it is excavating and the position where it discharges the load. The total time in a cycle includes digging, swinging to the dumping position, dumping, and returning to the digging position. If the angle of swing is increased, the time for a cycle will be increased, whereas if the angle of swing is decreased, the time for a cycle will be decreased.

4. Operator's Skill
5. Condition of the Shovel
6. **Haul Unit Exchange:** Haul-unit exchange refers to the total time required for a loaded truck to clear its loading position under the excavator and for the next empty truck to be positioned for loading.
7. **Size of Hauling Units:** Size of hauling units like trucks affect the output of a shovel, for optimum output, size of hauling units should be governed by size of a power shovel. If the shovel used is of smaller size, the size of hauling units also must be small while for a large shovel, hauling units must be of larger size.
8. **Handling of Oversize Material:** When handling shot rock, carefully evaluate the amount of oversize material to be moved. A machine with a bucket whose bite width and pocket are satisfactory for the average-size pieces may spend too much time handling individual oversize pieces. A large bucket, or a larger machine, or changing the blasting pattern should be considered when there is a large percentage of oversize material.

(ii) Safety measures for excavation

(a) Prior to excavation work, a complete knowledge of underground structures such as sewers, waterpipes, gas mains etc. is essential so as to proper precaution to prevent any accidents.
(b) The workers must be provided with all protective devices.
(c) When depth of excavation exceeds 2 m, the trenches should be securely shored and timbered.
(d) Sheathing should be placed against the side of the trench. In case of loose or soft soil, sheathing should be driven into the bottom of the trench.
(e) Excavated materials should be kept away from the edge of the trench.
(f) Heavy equipments such as excavating machineries, trucks etc. should be kept away at a suitable distance from the excavated sides.
(g) A fence or barricade should be erected and at night the area must be properly lighted.

4.22 **(i) To put large quantity of concrete in short duration with accuracy, which pump should be used and why? Compare it with other available pumps which are used for concreting.**
(ii) What safety measures should be taken in case of placing the concrete using pumps?
[6 + 6 = 12 marks : 2022]

Solution:

(i) To put large quantity of concrete in short duration with accuracy, **boom concrete pump** is used. The reasons for using boom concrete pump are as follows:

- Boom concrete pumps can be used for inaccessible areas where *line concrete pumps* cannot be used.
- Very minimal amount of skilled personnel are required.
- Boom concrete pumps have robotic arms which can be moved precisely.
- Boom concrete pumps can be used for greater heights as well.

The other available concrete pump is **line concrete pump.**

Comparison between boom and line concrete pumps:

Boom concrete pump	Line concrete pump
Can be used in inaccessible areas.	Cannot be used in inaccessible areas.
Very less supervision required due to robotic arms of the pump.	Constant supervision is required.
Can be used for large projects.	Cannot be used for large projects.

(ii) Although the method of transporting and placing concrete by pumps is fast and efficient, a small part of unpumpable mix in hopper can block the pump, leading to delay while the pump is stripped down. The blockage is indicated by an increase in the pressure shown on the pressure gauge. Most blockage occur at the tapered sections at the pump end. The reasons include unsuitability of concrete mix, pipeline and joint deficiencies, careless use of hose end, and operator's errors. High temperatures, may also cause blockage. Chances of blockage are least in continuous pumping. A pipeline not well cleaned after previous operation, uncleared and worn-out hoses, too many or too sharp bend and use of worn-out joints add to the problem of blockage.

Great attention is required in the design of mix, for a minor variation in the concrete mix is sufficient to make an otherwise pumpable mix completely unpumpable. At the end of the run, the pipeline must be cleared of concrete by inserting a plunger at the pipe end and forcing it through under pressure. After the concrete is cleared, the pipeline is washed out to leave a smooth clean surface ready for next day's work. The minor blockage may be cleared by forward and reverse pumping.

4.23 Discuss the parameters influencing the degree of compaction achieved by a vibratory roller.

[10 marks : 2023]

Solution:

The various parameters influencing the degree of compaction achieved by a vibratory roller are as follows:

(a) **Operating frequency:** Frequency controls the number of impacts per meter between the drum and mat. Each material has natural frequency under which it vibrates or show maximum displacement of particles. Maximum compaction occurs when it vibrates near resonant frequency. Typically, vibratory rollers operate in the range of 1500 to 3000 vibrations per minute.

(b) **Amplitude of vibration:** It is the vertical displacement of the roller drum. Amplitude of vibration is directly linked to the impact force. Higher amplitude results in achieving higher thickness of compacted material at minimum number of passes, thereby improving productivity.

(c) **Speed of travel:** Slow speed results in closeness of impacts, whereas high speed results in increase in the spacing between blows. Working speed from 2.0 to 6.0 kmph will provide the best results. The machine should be able to move forward and reverse with vibrations to achieve higher productivity.

(d) **Roller weight:** Heavier rollers exert greater force on the surface, which helps in compressing the material more efficiently. Higher rollers may be suitable for finishing or surface compaction, while heavier rollers are often used for base or sub base compaction.

(e) **Number of roller passes:** Density of soil increases with the number of passes of roller but after optimum number of passes, further increase in density is insignificant.

4.24 On a homogenous embankment, compacting rollers are used to compact silty clay soil. Determine the quantity of earth compacted if the sheeps-foot roller travels at 4 km/hr, time of rolling is 50 min, length of drum is 2.4 m, number of drums is one, fraction of overlap is $\frac{1}{8}$, layer thickness is 0.45 m and number of passes given is 5.

[10 marks : 2023]

Solution:

Given:

$$\text{Speed} = 4 \text{ km/hr} = \frac{4 \times 1000}{60} \text{ m/min}$$

$$\text{Time of rolling, } t = 50 \text{ min}$$

$$\text{Length of drum, } l = 2.4 \text{ m}$$

$$\text{Number of drums, } n = 1$$

Fraction of overlap, $p = \frac{1}{8}$

Thickness of layer, $h = 0.45$

Number of passes, $N = 5$

Now, Output of roller $(Q) = \frac{S \cdot h \cdot l \cdot n(1-p) \cdot t}{N}$

$$= \frac{\left(\frac{4000}{60}\right) \times 0.45 \times 2.4\left(1-\frac{1}{8}\right) \times 50}{5} = 630 \text{ m}^3$$

So, Quantity of earth compacted = 630 m^3

5. Miscellaneous

5.1 **Discuss and differentiate line organisation and staff organisation.**

[5 marks : 2011]

Solution:

Line organisation is the most oldest and simplest method of administrative organisation. According to this type of organisation, the authority flows from top to bottom. The line of command is carried out from top to bottom. This is the reason for calling this organisation as scalar organisation which means scalar chain of command is a part and parcel of this type of administrative organisation. In this type of organisation the line of command flows on an even basis without any gaps in communication and coordination taking place.

Features of line organisation are as follows:

(i) It is the most simplest form of organisation.

(ii) Line of authority flows from top to bottom.

(iii) Specialised and supportive secessions do not take place in these organisation.

(iv) Unified control by the line officers can be maintained since they can independently take decisions in their areas and spheres.

(v) This kind of organisation always helps in bringing efficiencey in communication and bringing stability to a concern.

In **staff organization** the nature of relationship is advisory. Staff people assist line people at lower levels over whom they have no direct authority. Similarly they may also assist people at senior level who have no direct authority over them.

Features of staff organization are:

(i) Services of specialists are available to more than one departments. This becomes valuable when the workload within individual department is not sufficient for one person.

(ii) Staff specialists can work more effectively for long term performance improvement unhampered by the pressures of day to day problems.

5.2 **Safety precautions to be taken at a construction site are related to the nature of the construction activity. Discuss this statement with appropriate examples.**

[10 marks : 2013]

Solution:

Safety precautions to be taken at site are invariably dependent on the type of construction activity. However contain general precautions are followed in all construction activities irrespective of their nature. For instance, *PPE* **(Personal Protective Equipments)** will definitely differ for underground and over-ground construction activities. Similarly in repair works, certain repairs are done with the help of resins. 'Epoxy', a commonly used resin is highly toxic and irritant to skin. Thus this will require a different set of PPE ranging from full face mask, goggles and adequate fine protection.

5.3 **Explain briefly the highlighted terms in the following sentence. (not more than 20 words)**
The contractor and the client agreed to send the matter for ARBITRATION.

Or

The contractor was informed that the plea of FORCE MAJEURE for the delay in the completion was not acceptable.

[5 marks : 2013]

Solution:

Arbitration: In matters relating generally to technical disputes, it is expected that the judge is well conversant with the technicalities of the dispute. To achieve this goal, both the parties appoint a person of common choice as ARBITRATOR and this process of dispute resolution is called as arbitration.

Force Majeure: These are those conditions which are beyond the control of human beings such as earthquake, flood, draught etc. It is also generally referred to as **"By an act of GoD"**.

5.4 **All parts carry equal marks:**

(a) Cement is made of solid complexes of oxides of calcium etc. Write the names of the four well known complexes along with the notation used to represent these.

(b) Write the full form of the following in context of civil engineering construction:
(i) RMC (ii) GIS (iii) BOLT (iv) BOQ

[4 + 4 = 8 marks : 2013]

Solution:

(a) The four common oxides of calcium in variably found in cement are:

		Chemical formula	Notation
(i)	Tri-calcium silicate:	$3CaO.SiO_2$	C_3S
(ii)	Dicalcium silicate:	$2CaO.SiO_2$	C_2S
(iii)	Tri-calcium Aluminate:	$3CaO.Al_2O_3$	C_3A
(iv)	Tetra-calcium alumino Ferrite:	$4CaO.Al_2O_3.Fe_2O_3$	C_4AF

(b) (i) **RMC** – Ready Mixed Concrete
(ii) **GIS** – Geographical Information System
(iii) **BOLT** – Build, Own, Lease and Transfer
(iv) **BOQ** – Bill of Quantities

5.5 **What is the difference between "security deposit" and "mobilization advance" in a construction contract?**

[4 marks : 2014]

Solution:

Security deposit: On acceptance of the tender, the contractor has to deposit 10% of the tendered amount as security money with the department which is inclusive of the earnest money already deposited. This amount is kept as a check so that the contractor fulfills all the terms and condition of contract and carries out the work satisfactorily according to the specifications and complete the work in time.

If the contractor fails to fulfil the terms of contract his whole or part of the security money is forfeited by the department.

Mobilization advance: In some countries where the government is actively encouraging small contractors, it is possible to get a "mobilization advance", which is really a short-term loan from the client, if he gets a government tender (contract), to help to start up.

The client will require some collateral or other evidence of security and will not actually give you the money out right, but will pay for materials or equipment bought for the contract. The client recovers the loan in monthly installments from the money he earns on the contract. All the money will have to be repaid before the contract is completed.

5.6 Define the following terms briefly in the context of construction contracts:

(i) Beta distribution in PERT (ii) EPC contract

(iii) PPP (iv) Escalation

[4 + 2 + 2 + 2 = 10 marks : 2014]

Solution:

(i) **Beta distribution in PERT:** The Beta distribution is a typical type of probability distribution curve, which is not symmetrical about its apex. Figure shows two beta distributions, one having skew to the left and other having skew to the right.

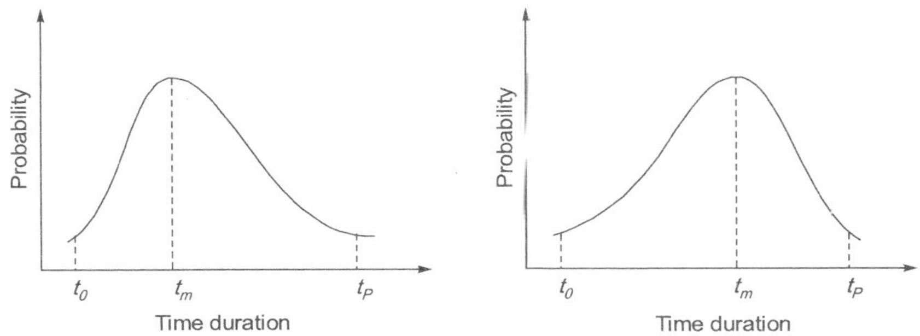

Where, t_0 = Optimistic time, t_m = Most likely time and t_p = Pessimistic time

Properties of β-distribution curve:

1. It have a small probability of reaching the most optimistic time (shortest)
2. It has a small probability of reaching the most pessimistic time (longest time)
3. It has one and only one most likely time.
4. In β-distribution the amount of uncertainty in estimation can be measured easily

(ii) **EPC contracts:** EPC stands for engineering, procurement and construction. It is common form of contracting arrangement within construction industry.

There are following functions of EPC contracts:

(a) **Engineering function:** Initiation, implementation, planning, estimating, valuation and design.

(b) **Procurement functions:** Purchasing, expending, receiving, invoicing (billing) and reconciliation.

(c) **Construction functions:** Construction schedule, material handling, on site client communications, valuation cash flow.

(iii) **PPP:** PPP stands for public private partnership. It is a joint venture which is funded and operated through a partnership of government and one or more private sector companies.

(iv) **Escalation:** In construction, this is an allowance to change the contracted price for labour, material etc. based on a change in market prices or a Consumer Price Index (CPI) as a bench mark provision in a contract.

5.7 What do you understand by multiskilling of labour? What are the benefits of multiskilling?

[8 marks : 2016]

Solution:

Multi-skilled workers have training and skills in more than one area of a business. For example, in a manufacturing environment, multi-skilled workers may receive training in all aspects of building product, as well as the ability to perform quality inspections. Such accommodations allow a firm to move workers where they're needed from one moment to another. Companies cross-train employees to increase productivity and reduce labour costs.

Advantages of Multi-skilling:

Flexibility: A company with multi-skilled employees has a flexible workforce, which provides the employer with the ability to schedule and arrange workers to best suit the needs of the business. Workers are able to fill in for absent employees and work in any area of the business that requires increased manpower at any time, and for any duration.

Decreased Labour Costs: A business with a multi-skilled labour force can operate with a reduced number of employees necessary to conduct business. Workers who are skilled in only one area of the business may sit idle while waiting for work to become available. A multi-skilled workforce moves with the workload instead of waiting for the work to come to them.

Efficiency in Planning: Planning and scheduling workers can make changes to the production schedule to meet customer demand without a loss of productivity. For example, when a customer requests a faster delivery, planners can adjust the production schedule to meet the new date without disruption to production because all workers can focus on the new customer demand.

Employee Satisfaction: Multi-skilled workers are not threatened by obsolescence when new technology changes the method of production, as workers used to learn new skills consistently and can adapt to changes in production. Employee satisfaction improves morale in a business, which leads to increases in productivity and employee retention rates.

5.8 Explain the different types of contracts adopted in construction.

[20 marks : 2019]

Solution:

Project Contract

- A contract is a legal agreement. It is an exchange of promises by two or more persons/organisations.
- Section 2 *(b)* of the Indian Contract Act, 1872 defines contract as "an agreement enforceable by law". The definition given by the Indian Contract Act has two distinct parts, *viz.,*
  (a) a contract is an agreement
  (b) the agreement must be enforceable by law.
- Though contract is necessarily an agreement, but all agreements are not contracts. Only those agreements that are enforceable by law are contracts. For an agreement to become enforceable by law, it must satisfy certain essential elements called "essentials" of valid contract".
- The contractor is a person, a firm or a company who undertakes contract.

Type of Contracts

Broadly contracts can be divided into two types. They are:

(a) Turnkey contracts

(b) Non-turnkey contracts

(a) Turnkey Contracts

- In a turnkey contract, the entire responsibility of project execution is entrusted to the contractor. It is as if the owner comes into the picture only when the project is completed and he turns the key of the plant to start production. Till such time the project reaches a 'ready-to-start' stage, the contractor takes care of all aspects of project execution.
- Since the contractor of a turnkey project is expected to do everything right from scratch, the scope of contract covers all areas of the project, *viz.,* design, engineering, construction, structural work, supply and erection of plant and machinery, supply of spares, testing and commissioning.

Since the entire work is done by a single contractor, turnkey contract agreements invariably have a clause on performance guarantee.

- Turnkey contracts may be for the whole project or even for a sub-unit of a main project. Turnkey contract for a sub-unit is undertaken when such sub-unit is a separately identifiable component of the total project and whose performance can be assessed independent of other units of the project.

(b) Non-turnkey Contracts

- Non-turnkey contracts are preferred when the projects are small sized, the know-how for the project is available with the promoters and when there is a strong, competent and capable project team available with the organisation.
- In non-turnkey arrangement, the total project is divided into suitable work packages. The packages that need to be given to contractors are identified. A contractor may be entrusted with one or more work packages. However, the responsibility for overall supervision and performance of the project will rest with the in-house project department.

The following are some of the non-turnkey contracts:

1. **Lumpsum contract (Fixed price):** This is a single fixed price contact. In this contact, contractor agrees to perform specified job for fixed sum. The owner provides the contractor exact specification of the work. In this contract following are the advantages of the fixed price contract.
   (a) Owner in aware of the cost of the project before the project construction starts.
   (b) It avoids a lot of details and accounting by both owner and contractor
   (c) Contractor gets free hand to execute the work
   (d) In this contract is used with design contract method of delivery, contractor gets opportunity to use value engineering.

   In this contract following are the disadvantages of the fixed price contract.
   (a) It is very difficult to accommodate any change and specification.
   (b) This contract is as good as the accuracy of the contract document. If errors exist in the contract document, the contract need to be renegotiated and hence more risk in involved from the owner side.
   (c) In the case of unforeseen hazard during the construction, contractor may be put in adverse situation.

   Payment by the owner can be carried out in lump-sum contract as total amount at time or percentage of total cost after finishing certain amount of work. For example suppose a water tank was awarded as lump-sum contract. It can be said that 15% of amount will be paid after construction of foundation 50% of total amount will be paid after construction of staging and 80% of the total amount will be after the construction of the tank. Rest will be paid only when the tank becomes operational. Lump-sum contract in often used in sub-contracting for labour contract. In our country many labourers for excavation, plastering work with this method.

2. **Unit price contract:** In this type of contract, the price is paid per unit of the work carried out. There are different variations of this type of contract. Some of them are mentioned below.

   Bill of quantities contract: In this type of contract owner provide the drawing, quantities of work to be done and specification. The contractor bid based on the unit cost of the items of construction. The contractor overhead, profit and other expanses can be included in the unit cost of the item of work. Sometimes contractor quotes the unit price of the work and lump-sum amount separately as profit overhead. The estimated quantities of the work to be done called Bill of the quantities is fixed.

   This type of construction is usually followed in government sector for large infrastructure construction. This type of contract provides owner a competitive bid. Disadvantages of the methods are:
   (a) Owner needs to measure the quantity of work done in the field, hence requires owner presence at the site.

(b) Final price of the construction is not known precisely until last price of work is completed. If there is significant difference between the estimated quantities and the reality of the situation, owner is put in adverse situation. Mistaken quantities is called unbalanced bid. Significant unbalanced bid now considered as unethical.

Schedule of rate contract: Many a time, the quantity of work to be executed is not known before. Contract is signed based on the unit cost of the item of work. Generally more items are inserted in the contract than to be executed because it becomes sometimes difficult to exactly specify all the items. There is no guarantee that all the items mentioned will be used in the construction. This type of contact are widely used in underground work, flood control and road constructions. Advantages and disadvantages of this type of contract in the same as the bill qualities contract. There are other variations of schedule of rate contract where unit price plus profit is charged as the cost.

(c) **Cost plus contract:** In this contract, the payment is made based on the work carried out plus the free which includes overhead, profit etc. Sometimes a cap is put on the type of contract by provided maximum and minimum cost limit such as guaranteed maximum cost contract. If project cost exceed this limit, contractor is responsible for that. Sometimes incentive clause is also included if the contractor bring the project before certain specified limit.

The advantages of this type of contract is that considerable overlap is provide between design and construction. Hence the project can be executed in the fast-tract basis. This contract is suitable for the work where it is difficult to define the task to be done before the awarding the contract.

3. **Item rate contract:** Item rate contract is also known as unit price contract or schedule contract. A contractor undertakes the execution of work on an item rate basis. He is required to quote rate for individual item of work on the basis of schedule of quantities (i.e., bill of quantities) furnished by the department. The amount to be received by the contractor, depends upon the quantities of work actually performed. The payment to the contractor is made on the basis of the detailed measurements of different items of work actually executed by him.

Suitability: The item rate contract is most commonly used for all types of engineering works of the government undertakings including railway department. It is suitable for works which can be distinctly split into various items and quantities under each item can be estimated accurately.

5.9 **(i) What information is generally needed to be provided in tender documents?**
(ii) Differentiate between 'Lump sum contract' and 'Unit price contract'.

[6 + 6 = 12 Marks : 2021]

Solution:

(i) Tender documents are meant to keep the tenders informed about the general and specific conditions applicable for tenderers. Tender documents usually consist of the following:

(a) A letter of invitation to the tenderers
(b) Specimen tender form
(c) General instruction to the tenders
(d) Details of civil/structural work along with complete set of civil/structural drawings.
(e) Details and specifications of machinery/equipment to be supplied, if any
(f) Draft contract agreement
(g) Arbitration authority who will decide in case of dispute
(h) Time schedule for completion of work
(i) Amount of earnest money to be deposited and the form in which it is to be deposited.

(ii) **Lumpsum contract (Fixed price):** This is a single fixed price contact. In this contact, contractor agrees to perform specified job for fixed sum. The owner provides the contractor exact specification of the work. In this contract following are the advantages of the fixed price contract.

(a) Owner in aware of the cost of the project before the project construction starts.

(b) It avoids a lot of details and accounting by both owner and contractor

(c) Contractor gets free hand to execute the work

(d) In this contract is used with design contract method of delivery, contractor gets opportunity to use value engineering.

Unit price contract: In this type of contract, the price is paid per unit of the work carried out. There are different variations of this type of contract. Some of them are mentioned below.

Bill of quantities contract: In this type of contract owner provide the drawing, quantities of work to be done and specification. The contractor bid based on the unit cost of the items of construction. The contractor overhead, profit and other expanses can be included in the unit cost of the item of work. Sometimes contractor quotes the unit price of the work and lump-sum amount separately as profit overhead. The estimated quantities of the work to be done called Bill of the quantities is fixed.

This type of construction is usually followed in government sector for large infrastructure construction. This type of contract provides owner a competitive bid. Disadvantages of the methods are:

(a) Owner needs to measure the quantity of work done in the field, hence requires owner presence at the site.

(b) Final price of the construction is not known precisely until last price of work is completed. If there is significant difference between the estimated quantities and the reality of the situation, owner is put in adverse situation. Mistaken quantities is called unbalanced bid. Significant unbalanced bid now considered as unethical.

5.10 **Estimate the following quantities for the given structure (plan and section):**

(i) Earthwork in excavation in foundation.

(ii) Concrete in foundation.

(ii) Brick work in super structure

All windows, doors and shelves have the dimensions (1.2 m × 1.2 m), (1.2 m × 2.1 m) and (1 m × 1.5 m × 0.15 m) respectively. All dimensions are in m, if not mentioned.

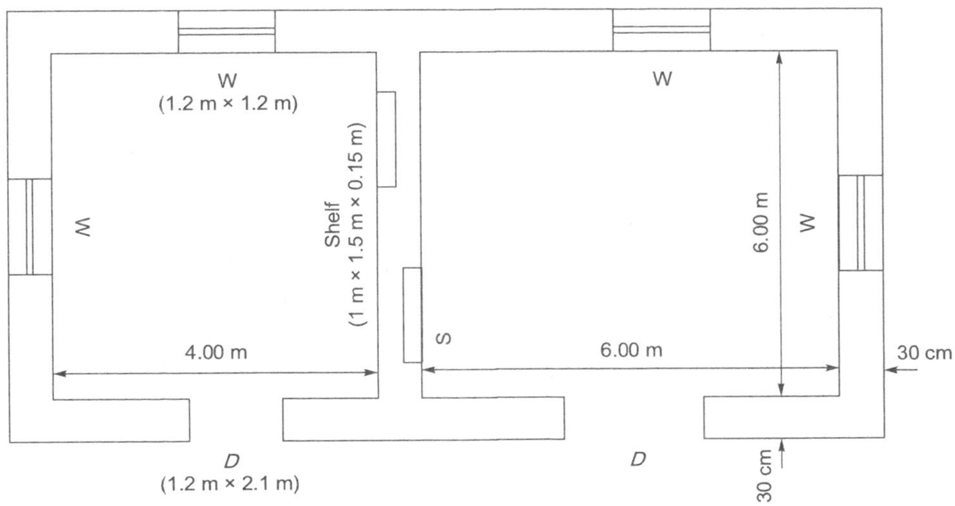

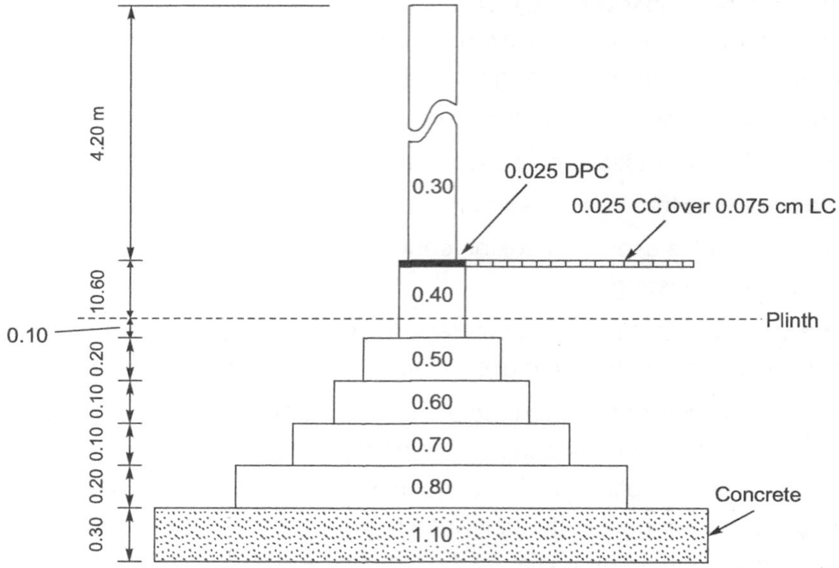

[20 marks : 2022]

Solution:

Item No.	Particulars of items	No.	Length	Breadth	Height of Depth	Quantity	Explanatory note
							Long wall, c/c. length = 4 + 6 + 0.30 + 2 × (0.30/2) = 10.60 m Short and inner wall, c/c. length = 6 + 2 × (0.30/2) = 6.30 m
1.	Earthwork in excavation in foundation						
	Long wall	2	11.70 m	1.10 m	1.00 m	25.74	L = 10.60 + 1.10 = 11.70 m
	Short wall	3	5.20 m	1.10 m	1.00 m	17.16	L = 6.30– 1.10 = 5.20 m
					Total	42.90 cu m	
2.	Concrete in foundation						
	Long wall	2	11.70 m	1.10 m	0.30 m	7.72	Length same for excavation
	Short wall	3	5.20 m	1.10 m	0.30 m	5.15	
					Total	12.87 cu m	
3.	1st class brick-work in lime mortar in superstructure						
	Long wall	2	10.90 m	0.30 m	4.20 m	27.47	L = 10.60 + 0.30 = 10.90 m
	Short wall	3	6.00 m	0.30 m	4.20 m	22.68	L = 6.30 – 0.30 = 6.00 m
					Total	50.15 cu m	

5.11 What are the precautions to be taken for labour safety during formwork construction?

[10 marks : 2023]

Solution:

(i) Safety measurers for scaffolding, ladders formwork and other equipments are as follows:

(a) All scaffolds and working platforms should be securely fastened to the building or structure. If independent of a building, they should be braced or guyed properly.

(b) In case, scaffolds are to be kept for a long period, a regular plank stair-way, wide enough to allow two persons to pass, should be erected with handrails on both sides.

(c) When work is being carried out over a scaffold platform, a protective overhead covering should be provided for the men working on the scaffold.

(d) All wooden ladders or bamboo ladders, must be strong enough.

(e) Ladders in heavy duty work should not exceed 6 m in length, for light work it should not exceed 8 m in length.

(f) Dismantling of scaffold should be in a proper sequence.

(g) No un-insulated electric wires should exist within 3 m of working platform.

(h) The supporting ballas should be individually strong and properly braced and fastened.

(i) All persons handling construction equipment should be fully acquainted with the safety aspects of the machines and their operation.

5.12 Explain different types of contract. Discuss the importance of each type of contract.

[10 marks : 2023]

Solution:

Item rate contract is also known as Unit Price contract or Schedule contract. A contractor undertakes the execution of work on an item rate basis. It is required to quote rate for individual item of work on the basis of schedule of quantities (i.e., Bill of Quantities) furnished by the department by contractor. The amount to be received by the contractor, depends upon the quantities of work actually performed. The payment to the contractor is made on the basis of the detailed measurements of different items of work actually executed by him.

Merits:

(a) This method ensures a very detailed analysis of cost and payment to the contractor and also is based upon detailed measurements of each item actually done, so this method is more scientific.

(b) Changes in drawings and quantities of individual item can be made as per requirements within agreed limits.

(c) There is no urgency of providing detailed drawings at the time of awarding the contract. It can be prepared later on.

(d) A contractor is asked to write down the rate of individual in figures and words both so it is not easy to form a cartel during item submission of tender.

(e) An engineer can compare the rates quoted by the contractor with the schedule of rates prepared by the department to find out whether tender is unbalanced or not.

PERCENTAGE RATE CONTRACT:

In this form of contract, the department draws up 'item rate tender' i.e., Bill of quantities with-rate, amount and total amount. The contract are required to offer to carry out the work as per with the rates shown inspecific price schedule or some percentage above or some percentage be the rates indicated in the schedule of work attached with the tender. The percentage above or below or at par is applicable on the over all cost of work also.

Merits:

(i) The ranking amongst the contractors is easily known just on the opening of the tenders.

(ii) As there is no provision to quote contractor's own rate for an individual item, benefit due to increased quantity with a beneficial rate can not be availed by the contractor. The chance of unbalanced tender gets eliminated.

LUMP-SUM CONTRACT:

In this type of contract, a contractor is required to quote a fixed sum for execution of complete work in all respect in the stipulated time according to the drawing, design and specifications supplied to him with the tender.

The departmental schedule of rates for various items of work are also provided which regulates the payment of the contractor in respect of any additions and alterations which are made over the original work. On the completion of the work, no detailed measurement of different items of work is needed but the whole work done must be compared and checked with the drawings and specifications.

Merits:

(i) As the total cost of the work is known before hand, the owner can arrange the fund in time.

(ii) Detailed measurements of the work done are not required except in respect of additions and alterations.

(iii) The contractor's profit mainly lies in the completion time. Hence for getting more profit the contractor tries to complete the work as early as possible. This concept on the part of contractor coincides with the objective of the owner also.

(iv) As it is difficult to get intermediate payment, the contractor tries to finish as early as possible to get the full payment at earliest and go to another venture. The owner also benefits from the project being completed early or in time.

(v) Due to competition among the contractors, the contractor tries to take the work up even at the less profit, resulting in low cost of the work.

COST PLUS PERCENTAGE RATE CONTRACT:

In this type of contract, a contractor agrees to take the work of construction on the actual cost of work plus on agreed percentage in addition, for his services. It is generally adopted when the labour and material cost are liable to fluctuate heavily in the market. The contractor arranges materials and labour at his cost and keeps proper account which is paid by the department or owner with certain percentage (say 10%) of the cost of construction as his profit. An agreement is made accordingly in advance.

Merits:

(i) The contract can be quickly drawn up and agreed and work can be completed in the shortest possible time. It is well suited during war period or calamity-period when the shortest possible project time is the main criteria in place of the cost involved.

(ii) It is particularly suitable when work can not be executed by other types of contracts due to uncertainty and fluctuations in the market rates of labour and materials.

COST PLUS FIXED FEE CONTRACT:

In this type of contract, the contractor is paid by the owner an agreed fixed lump-sum amount over and above actual cost of the work. This fixed fee will include overhead and profit to the contractor. The fee does not vary with the actual cost of the work as in the case of cost plus percentage rate contract.

Merits:

(i) Since the fixed fee covers the contractor's overhead charges and the contractor will try to finish the work as early as possible, so the owner gets the advantage of early completion.

(ii) For any intermediate payment, the value of work done should be less than the payment being made.

(iii) It is suitable form of contract where considerable amount of addition and alterations are expected.

LABOUR CONTRACT:

In labour contract, the contractor undertakes contract for the labour portion only excluding the materials which are arranged and supplied at the work site by the department/owner. The contractor engages the required labour and gets the work done as per drawings and specifications. It is a item rate basis for labour portion only and the contractor is paid for the quantities of work done on measurement of different items of work at the stipulated rate in the contract agreement.

Merits:

(i) The materials stored by the department are thus utilized.

(ii) The work done through labour contract is of superior quality as better quality materials are arranged by the owner.

(iii) The overall cost of construction may be less, as no profit is paid on the cost of materials.

(iv) This system is very convenient for private building construction.

■■■■